D1574538

COMPREHENSIVE ANALYTICAL CHEMISTRY

ELSEVIER SCIENCE B.V.
Sara Burgerhartstraat 25
P.O. Box 211, 1000 AE Amsterdam, The Netherlands

First edition 2003

Library of Congress Cataloging in Publication Data
A catalog record from the Library of Congress has been applied for.

British Library Cataloguing in Publication Data
Analysis and fate of surfactants in the aquatic environment. –
  (Wilson & Wilson's comprehensive analytical chemistry; v. 40).
  1. Surface active agents – Analysis   2. Water – Pollution –
  Measurement   3. Surface active agents - Environmental aspects
  I. Knepper, T. P.   II. Barcelo, D. (Damia)   III. Voogt, Pim de
628.1'61

ISBN: 0-444-50935-6
ISSN: 0166-526X

♾ The paper used in this publication meets the requirements of ANSI/NISO Z39.48-1992 (Permanence of Paper).
Printed in The Netherlands.

COMPREHENSIVE ANALYTICAL CHEMISTRY

ADVISORY BOARD

Wilson & Wilson's

COMPREHENSIVE ANALYTICAL CHEMISTRY

Edited by

D. BARCELÓ

Research Professor
Department of Environmental Chemistry
IIUQAB-CSIC
Jordi Girona 18-26
08034 Barcelona
Spain

Wilson & Wilson's

COMPREHENSIVE ANALYTICAL CHEMISTRY

VOLUME XL

ANALYSIS AND FATE OF SURFACTANTS IN THE AQUATIC ENVIRONMENT

Edited by

T.P. KNEPPER
D. BARCELÓ
P. DE VOOGT

2003

ELSEVIER
Amsterdam – Heidelberg – Boston – London – New York – Oxford – Paris
San Diego – San Francisco – Singapore – Sydney – Tokyo

CONTRIBUTORS TO VOLUME XL

Damia Barceló
Institute of Chemical and Environmental Research, IIQAB-CSIC, Department of Environmental Chemistry, 18-26 c / Jordi Girona, 08034 Barcelona, Spain

José Luis Berna
PETRESA (Petroquimica Espapñola, S.A.), Apartado 40, San Roque, E-11360 Cádiz, Spain

Julian Blasco
Consejo Superior de Investigaciones Cientificas, Campus Univ. Rio San Pedro, 11510 Puerto Real (Cádiz), Spain

Lea S. Bonnington
Institute of Chemical and Environmental Research, IIQAB-CSIC, Department of Environmental Chemistry, 18-26 c / Jordi Girona, 08034 Barcelona, Spain

Jürgen Büsing
DG Research, Env. & Sustainable Development Programme, EU, Square de Meeus, 8, 1050 Brussels, Belgium

Peter Eichhorn
Institute of Chemical and Environmental Research, IIQAB-CSIC, Department of Environmental Chemistry, 18-26 c / Jordi Girona, 08034 Barcelona, Spain

M.F. Ferández
Laboratorio de Investigaciones Médicas, Dept. de Radiologia, Universidad de Granada, 18071 Granada, Spain

Imma Ferrer
National Water Quality Laboratory, U.S. Geological Survey, P.O. Box 25046, MS 407, Denver Federal Center, Denver, CO 80225-0046, USA

Edward T. Furlong
National Water Quality Laboratory, U.S. Geological Survey, P.O. Box 25046, MS 407, Denver Federal Center, Denver, CO 80225-0046, USA

Abelardo Gómez-Parra
Departamento de Química Física, Facultad de Ciencias del Mar, University of Cádiz, Polígono Rio San Pedro s / n Puerto Real 11510 (Cádiz), Spain

Eduardo González-Mazo
Depart. de Quimica Fisica, Facultad de Ciencias del Mar, Universidad de Cádiz, Polígono Rio San Pedro S / N E-11510 Puerto Real, (Cádiz), Spain

Miriam Hampel
Consejo Superior de Investigaciones Cientificas, Campus Univ. Rio San Pedro. 11510 Puerto Real (Cádiz), Spain

William Henderson
Chemistry Department, University of Waikato, Private Bag 3105, Hamilton, New Zealand

Niels Jonkers
Department of Environmental and Toxicological Chemistry (ETC), University of Amsterdam, Nieuwe Achtergracht 166, 1018 WV Amsterdam, The Netherlands

Thomas P. Knepper
ESWE-Institute for Water Research and Water Technology, Kurfürstenstr. 6, 65203 Wiesbaden, Germany

Victor M. León
Depart. de Quimica Fisica, Facultad de Ciencias del Mar, Universidad de Cádiz, Polígono Rio San Pedro S/N E-11510 Puerto Real, Cádiz, Spain

Ignacio Moreno-Garrido
Consejo Superior de Investigaciones Cientificas, Campus Univ. Rio San Pedro, 11510 Puerto Real (Cádiz), Spain

Nicolas Olea
Laboratorio de Investigaciones Médicas, Dept. de Radiologia, Universidad de Granada, 18071 Granada, Spain

F. Olea-Serrano
Laboratorio de Investigaciones Médicas, Dept. de Radiologia, Universidad de Granada, 18071 Granada, Spain

Mira Petrovic
Institute of Chemical and Environmental Research, IIQAB-CSIC, Department of Environmental Chemistry, 18-26 c/Jordi Girona, 08034 Barcelona, Spain

Monica Saez Ribas
Depart. de Quimica Fisica, Facultad de Ciencias del Mar, Universidad de Cádiz, Polígono Rio San Pedro S/N E-11510 Puerto Real, Cádiz, Spain

A. Rivas
Laboratorio de Investigaciones Médicas, Dept. de Radiologia, Universidad de Granada, 18071 Granada, Spain

Horst F. Schroeder
Institut für Siedlungswasserwirtschaft, Templergraben 55, 52056 Aachen, Germany

Francesc Ventura
AGBAR (Sociedad General de Aguas de Barcelona), Passeig de Sant Joan 39, 080009 Barcelona, Spain

Pim de Voogt
Department of Environmental and Toxicological Chemistry (ETC), University of Amsterdam, Nieuwe Achtergracht 166, 1018 WV Amsterdam, The Netherlands

Jerzy A. Zabkiewicz
New Zealand Forest Research Institute Ltd, Private Bag 3020, Rotorua 3200, New Zealand

WILSON AND WILSON'S

COMPREHENSIVE ANALYTICAL CHEMISTRY

VOLUMES IN THE SERIES

Contents

Series Editor's Preface

Obviously this is not just another book in the series: my involvement as co-editor of the book, together with my two old friends, Thomas Knepper and Pim de Voogt, makes this book rather special to me. Everything started in 1996 when I first met separately with Thomas and afterwards with Pim and we decided to establish a consortium covering the analysis, behaviour and toxicity of surfactants in the aquatic environment.

Our basic idea was to prepare a project proposal to be submitted to the European Union for funding. Surfactants were selected as the main pollutants as there was no project at that time that comprehensively covered the topic. There were many projects dealing with pesticides, phenols, persistent organic pollutants but none targeted surfactants. The project acronym, PRISTINE, is explained in the Foreword to this volume. Clearly the project was a big success and will undoubtedly be known for many years as the European Union surfactant project. I should add as well that the scientific officer involved in the development of the project, Dr Juergen Buesing, played an extremely active role. He was always constructive and communicative, so this book is also partly due to his efforts and I would like to acknowledge here his full support of the initiative. He was responsible for our accomplishing the objectives of the submitted proposal. With this book we even exceeded the objectives. We believe that this is an ideal situation for European Union project management: to have not only the work done at the laboratory level, with the various primary literature publications that are a result of the project, the associated conferences and the technology transfer between the different research institutions, and the bringing together of policy makers, industry representatives and end users, but also to have a final summation of the project in the form of a book from which the larger community can benefit.

In addition to a general introduction to surfactants, the book comprises a comprehensive variety of analytical techniques, including sample handling, for the analysis of surfactants in the aquatic environment. Sample preparation includes automated solid phase extraction and pressurised liquid extraction for water and solid samples, respectively. Analytical determinations of surfactants are based either on gas chromatography-mass spectrometry or liquid chromatography-mass spectrometry. Readers will find all the necessary information for analysing the different groups of surfactants, with special emphasis on transformation products. Quality assurance is also reported on in detail; it is obviously a necessary practice in order to ensure good quality data. One point that needs to be mentioned is the lack of reference materials on surfactants, which makes it even more difficult to ensure the reliability of inter-laboratory data in surfactant analysis. Chapters on toxicity and risk assessment are also included and give a complete perspective on the surfactants problem in the aquatic environment. Overall this book gives an extensive coverage of surfactant analysis in the aquatic environment and I am convinced that it will become the "surfactants book" for the analytical and environmental chemistry community.

Finally, I would like to thank not only my co-editors but also the various authors, most of whom are good friends and co-workers for several years, for their contributions to this comprehensive work on surfactants.

D. Barceló

Editor's Preface

Commercial mixtures of surfactants comprise several tens to hundreds of homologues, oligomers and isomers of anionic, nonionic, cationic and amphoteric compounds. Therefore, their identification and quantification in the environment is complicated and cumbersome. The requirement of more specific analytical methods has prompted a replacement of many of the separate steps in traditional methods of analysis, usually non-chromatographic, by chromatographic tools.

Detection, identification and quantification of these compounds in aqueous solutions, even in the form of matrix-free standards, present the analyst with considerable challenges. Even today, the standardised analysis of surfactants is not performed by substance-specific methods, but by sum parameter analysis on spectrophotometric and titrimetric bases. These substance-class-specific determination methods are not only very insensitive, but also very unspecific and therefore can be influenced by interference from other compounds of similar structure. Additionally, these determination methods also often fail to provide information regarding primary degradation products, including those with only marginal modifications in the molecule, and strongly modified metabolites.

High polarity is one of the reasons why both the ionic and amphoteric surfactants, and especially their metabolites, are difficult to detect. This property, however, is important for the application tasks of surface-active compounds, but is also the reason for their high water solubility. Due to this fact, their extraction and concentration from the water phase, which can be carried out in a number of very different ways, is not always straightforward. Furthermore, they are often not volatile without decomposition, which thus prevents application of gas chromatographic (GC) separation techniques combined with appropriate detection. This very effective separation method in environmental analysis is thus applicable only for short-chain surfactants and their metabolites following derivatisation of the various polar groups in order to improve their volatility.

Derivatisation is often substance-specific, and thus although volatile derivatives of the expected compounds will be produced according to the procedure adopted, discrimination of other surfactants that are simultaneously present but differ in structure – and thereby do not react with the derivatisation reagent – can occur, preventing simultaneous detection.

Many separation and detection methods applied in combination with liquid chromatography (LC) that are described in the literature for the determination of surfactants are not specific to the detection of these compounds at trace levels. Even ultraviolet (UV) spectra obtained from diode array detectors often give only limited information. Furthermore, non-reproducible retention behaviour as well as coelution interference effects are frequently observed during the separation of surfactant-containing extracts. This is recognised, however, only in those cases where specific detection methods such as mass spectrometry (MS) are applied.

In order to study simultaneously the behaviour of parent priority surfactants and their degradation products, it is essential to have accurate and sensitive analytical methods that enable the determination of the low concentrations generally occurring in the aquatic environment. As a result of an exhaustive review of the analytical methods used for the quantification within the framework of the three-year research project *Priority surfactants and their toxic metabolites in wastewater effluents: An integrated study* (PRISTINE), it is concluded that the most appropriate procedure for this purpose is high-performance (HP) LC in reversed phase (RP), associated with preliminary techniques of concentration and purification by solid phase extraction (SPE). However, the complex mixtures of metabolites and a lack of reference standards currently limit the applicability of HPLC with UV- or fluorescence (FL-) detection methods.

In off-line coupling of LC and MS for the analysis of surfactants in water samples, the suitability of desorption techniques such as Fast Atom Bombardment (FAB) and Desorption Chemical Ionisation was well established early on. In rapid succession, new interfaces like Atmospheric Pressure Chemical Ionisation (APCI) and Electrospray Ionisation (ESI) were applied successfully to solve a large number of analytical problems with these substance classes. In order to perform structure analysis on the metabolites and to improve sensitivity for the detection of the various surfactants and their metabolites in the environment, the use of various MS-MS techniques has also proven very useful, if not necessary, and in some cases even high-resolution MS is required.

Following their use in aqueous systems as surface-active compounds, surfactants often reach wastewater treatment plants (WWTP). Although the legislator prescribes their biodegradability, they often cannot be completely mineralised because of short retention times in the WWTP. Therefore, together with their biochemical degradation products (metabolites), they arrive in surface waters serving as receiving water. In addition, relevant compounds adsorb to the biological sewage sludge in the WWTP, and then, in original or partly degraded form, are disposed of along with excess sludge when this is removed.

The particularly persistent compounds and their metabolites reach the raw water serving for drinking water treatment. Some of them are even found in drinking water. Today these effects are not observed as often as in the past, as the non-degradable compounds of the branched alkylbenzene sulfonate type (ABS), formerly widely applied, have been replaced by compounds that are more easily degradable. Nevertheless some surfactants can still be found in relatively remote, estuarine sediments across Western Europe and in relatively high concentrations in effluents of wastewater treatment plants. Therefore it is important that the fate of the corresponding metabolites should also be studied. The most widely used anionic and nonionic surfactants should have the highest priority.

The environmental behaviour of LAS, as one of the most widely-used xenobiotic organic compounds, has aroused considerable interest and study. As a result, it has been determined that, under certain conditions, LAS compounds are completely biodegradable; however, in the marine environment their degradation is known to be slower. The presence of metabolites of the anionic LAS surfactants, the long and short chain SPC derivatives, in the aqueous environment is well known, and as such these degradation intermediates needed to be monitored (and tested for their toxic effects).

Not only the extreme amounts of surfactants emitted into sewage plants and thereby into surface waters, but also the broad variety of the chemical structures combined with their excellent water solubility, their surface-active nature and the persistence of some of the known metabolites make them a group of environmental pollutants that need to be addressed with high priority. For some of the compounds, an enormous impact on the ecosystem, e.g. on fish, can be shown.

In this book, methods and data covering the state-of-the-art of modern analysis and environmental fate of the entire synthetic surfactant spectrum is provided. The first part deals with the analysis of surfactants and consists of three chapters. The chapter dealing with the unequivocal detection of surfactants is mainly devoted to highly sophisticated and established hyphenated mass spectrometric methods, such as LC–MS and LC–MS–MS. In addition, examples are given for the use of solid phase micro extraction (SPME) coupled to (GC)–MS and capillary electrophoresis (CE)–MS. Sample preparation methods have been thoroughly evaluated for all four groups of surfactants (anionic, nonionic, cationic and amphoteric) including their major metabolites. One chapter also addresses quality assurance and interlaboratory studies.

The second part, comprising four chapters, centers on an extensive set of data regarding the different environmental levels of all analyzed surfactants and their major degradation products in various countries. The aerobic biodegradation of surfactants is treated extensively, with emphasis on metabolic routes and novel and persistent metabolites formed. In addition, anaerobic degradation and sorption is also covered.

The presence of both surfactants and their degradation products in different aquatic matrices, such as wastewater, surface water and marine water besides biota is discussed and its importance for the environment evaluated.

By the nature of its content, with contributions from experienced practitioners, the book aims to serve as a practical reference for researchers, post docs, PhD-students and postgraduates as well as risk assessors working on surfactants in environmental laboratories, environmental agencies, the surfactant industry, the water industry and sewage treatment facilities. Each chapter includes extensive references to the literature and also contains detailed investigations. The broad spectrum of the book and its application to environmental priority compounds makes it unique in many ways.

We thank the authors for their time and effort in preparing their chapters. Without their co-operation and engagement this volume would certainly not have been possible.

T.P. Knepper
Wiesbaden, Germany

D. Barceló
Barcelona, Spain

P. de Voogt
Amsterdam, NL
December 20, 2002

Foreword

An understanding of the fate and behaviour of organic chemicals, such as surfactants, in the environment is a prerequisite for the sustainable development of human health and ecosystems. Surfactants are important in daily life in households as well as in industrial cleansing processes. As they are being produced in huge amounts, it is important to have a detailed knowledge of their lifetime in the environment, their biodegradability in wastewater treatment plants and in natural waters, and their ecotoxicity. Parameters relevant for the assessment of the long-term behaviour, such as interactions with hormonal systems, also need to be understood in order to avoid unexpected adverse effects on future generations of people and the environment.

The EU-funded PRISTINE project (*Priority surfactants and their toxic metabolites in wastewater effluents: An integrated study; ENV4-CT97-0494*) brought together a critical mass of scientists and stakeholders from the field of analytical environmental chemistry, toxicology and industry in order to treat wastewater heavily contaminated with surfactants, and created a momentum which triggered new RTD activities in the following priority areas:

- Development of tools in order to trace and measure, in particular, extremely water-soluble and toxic chemicals as well as their degradation products
- Improvement to the functioning of Wastewater Treatment Plants
- Improved risk assessment methodologies regarding surfactants and their degradation products.

The project was one of five European research projects within the WASTEWATER CLUSTER of the EU Environment & Climate Programme, the result of a focused approach within the area of *Environmental Technologies*. The objective was to improve the understanding of the transformation, fate and toxicity of selected groups of industrial pollutants discharged into the water resources by using complementary sampling and advanced measuring techniques.

With the knowledge gained, strategies can be developed to protect sensitive aquatic freshwater and coastal ecosystems and to ensure a sustainable supply of quality drinking water. The lessons learnt show that analytical methods have to become more and more sophisticated to achieve the detection of "close to zero pollution". There is a need to develop operational goal-driven monitoring strategies supporting the final aim of pollution prevention and protection of drinking water resources. This requires the development of sensitive detectors and monitoring techniques capable of measuring relevant changes at all different levels and to give warning signals at the earliest possible stage. In general, sustainable development must strike a balance between economic, social and environmental objectives of society in order to maximise current standards of living, without compromising that of future generations.

With knowledge regarding concentrations and the fate of surfactants it should be possible to identify areas at risk and to design the most appropriate countermeasures. By applying the precautionary principle, the protection of sensitive aquatic freshwater and coastal ecosystems and of a sustainable supply of drinking water in areas with high population densities should be ensured.

The PRISTINE project, and thus the content of the present book, provides policy makers and industry with detailed information on analysis and concentrations of surfactants and their degradation products in the environment. Furthermore, the book provides relevant information to all groups working in the field of surfactants in environmental laboratories, environmental agencies, the surfactant industry, water industry and sewage treatment facilities.

Jürgen Büsing
Scientific Officer
European Commission
DG RTD 1.03
Brussels, Belgium

Thomas P. Knepper
Co-ordinator PRISTINE
ESWE-Institute for Water
Research and Water Technology
Wiesbaden, Germany

Chapter 1

Surfactants: properties, production, and environmental aspects

Thomas P. Knepper and José Luis Berna

1.1 INTRODUCTION: FROM SOAP TO MODERN SURFACTANTS

Surfactants are a broad group of chemicals that play important and vital roles in a great variety of fields. Applications as diverse as cleaning, food, metallurgy, pharmacy, medicine, paints and varnishes, mining, and many others utilise the characteristic properties provided by the surfactants. The human body with its continuous and complex chemical and biological reactions is also an extraordinary receptor of the benefits and characteristics of surfactants. Soap was the first man-made surface-active agent (surfactant) known. Soap made from animal fat and wood ashes seems to have been used only for medical purposes in antiquity and it was not until the 2nd century AD that it was recognised as a cleaning agent. Although soap was produced domestically for laundry purposes in the European middle ages and early modern times, cake soap was a luxury product that came into common use only in the 19th century. Early soapmakers dispersed wood or plant ashes in water, added fat to the solution, then boiled it, whilst adding further ashes as the water evaporated. During the process, the neutral fat underwent a chemical reaction with the ash in a process called saponification.

Comprehensive Analytical Chemistry XL
T.P. Knepper, D. Barceló and P. de Voogt (Eds)

1

The first synthetic detergents were developed in the 19th century when sulfonation of olive and almond oils (1831) and castor oil (1875) yielded appropriate auxiliaries for textile dyeing. The castor oil derivative was effectively used in promoting dyeing with Turkey Red and hence became commonly known as Turkey Red Oil [1]. In 1898 an American patent claimed the manufacture of sodium salts of petroleum sulfonates [2]. Pure sodium cetyl sulfonate was found to have similar properties to soap (1913), and it remained the only important surface-active agent for a long time to follow. Fat shortages during World Wars I and II stimulated the search for substitutes, yielding for example, the short-chain alkyl naphthalene sulfonates, produced by sulfonation of naphthalene coupled with alcohols [3]. The rapid development of the chemical industry and the increasing importance of surface-active substances in technical and domestic applications led to the commercial introduction of the first synthetic detergent produced entirely on the basis of petrochemical feedstocks in the early 1930s. These anionic branched alkylbenzene sulfonates (ABSs) rapidly became the major surfactants in the US, and later on in other countries around the world where advanced chemical technology allowed their production [4]. The replacement of soap by ABS in most laundry products was essentially driven by two factors: the undesired formation of insoluble calcium and magnesium salts of soap, which precipitated in the washing liquor and deposited on the clean surface of the laundry, and the lower production costs of ABS.

The widespread use of ABS, mainly in laundry detergents, and their subsequent discharge into the sewer, however, led to the unexpected effect of strong foam formation in sewage water, treated sewage and even in river water [5,6]. This observation was directly related to the physical properties of the surfactant that had originally been responsible for its great success.

Shortly after the appearance of the described environmental problems it was recognised that ABS derived from tetrapropylene was quite resistant to biodegradation due to the presence of a branched alkyl chain in the hydrophobic moiety. Legal restrictions introduced in the mid-1960s, in Germany, e.g. mandating a certain degree of primary degradation for anionic surfactants [7], or voluntary industrial bans as in the US, led to the development and introduction of alternative anionic surfactants into the detergent market, the so-called linear alkylbenzene sulfonates (LASs). A modification in the chemical structure consisting of substitution of the branched alkyl chain by a linear one profoundly enhanced the biodegradability. Quickly after the changeover from ABS to LAS in the mid-1960s foam-related environmental problems had almost disappeared. Along with this, the levels of surfactants in aquatic environments dropped significantly [8].

Since that time an enormous number of surfactants covering a wide range of chemical and physicochemical properties have been developed for quite universal as well as specific tasks in domestic and industrial applications. The criteria for selection of a surfactant for industrial production is directly connected with the feasibility of large-scale production. This is determined by several factors including availability and costs of raw materials, cost of manufacture, and performance of the finished products. In addition to these aspects, environmental considerations likewise play an increasingly important role.

1.2 GENERAL PROPERTIES OF SURFACTANTS

Surfactants are surface-active agents that reduce surface tension and exhibit a tendency to form micelles in solvents. They concentrate at interfaces between bodies or droplets of water and those of oil, or lipids, to act as emulsifying or foaming agents. This mode of action is directly related to their chemical structure. All surfactants have in common an asymmetric skeleton with a hydrophobic and a hydrophilic moiety (Fig. 1.1). The hydrophobic part generally consists of a linear or a branched alkyl chain, which is then linked to a hydrophilic group. Due to this bifunctionality both parts of the surfactant molecule interact differently with water, the most commonly used solvent. The hydrophilic group is surrounded by water molecules which results in good solubility. On the other hand, the hydrophobic moieties are repulsed by strong interactions between the water molecules. Hence the molecules are repelled out of the inner phase and accumulate at the interfaces. These are the fundamentals for one of the most important uses of surfactants, at least the most popular one, as key ingredients in detergent and cleaning preparations that benefit from these properties in order to formulate all types of household, industrial and institutional products as well as cosmetic and body-care applications.

The two distinctive affinities in the surfactant molecule mentioned above serve as the basis for the commonly accepted definition of surfactant groups. According to the charge of their hydrophilic moiety, surfactants can be classified into four categories: anionic, non-ionic, cationic and amphoteric.

3

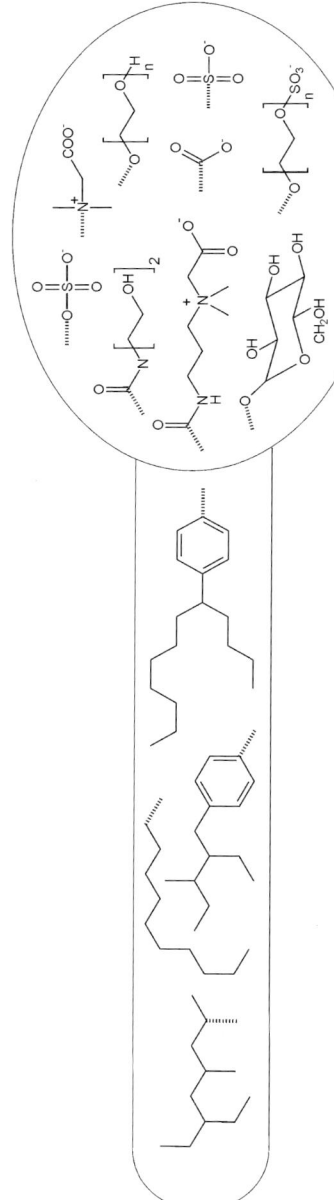

Fig. 1.1. Diagrammatic tail–head model of surfactant molecule showing examples of some of the important hydrophobic and hydrophilic groups used.

Anionic surfactants are known as the products having one or more functional groups that ionise in water to produce negatively charged surface-active organic ions. The anionic surfactants constitute the most important group in terms of consumption on a worldwide basis and among the products belonging to this family are the following (see also Table 1.1):

- Sulfonated surfactants (i.e. those for which the hydrophilic group is a sulfonic ion)
    - Linear alkylbenzene sulfonates (LASs)
    - α-Olefin sulfonates (AOSs)
    - Secondary alkane sulfonates (SASs)
    - Methylester sulfonates (MESs)
- Sulfated surfactants (i.e. those for which the hydrophilic group is a sulfate ion)
    - Alkyl sulfates (FASs or ASs)
    - Alkylether sulfates (AESs)

TABLE 1.1

Chemical structures of common anionic surfactants

Surfactant type	Structure
Soap	$R-COO^-Na^+$; R: $C_{11}-C_{17}$
Linear alkylbenzene sulfonates (LASs)	$CH_3-(CH_2)_x-CH-(CH_2)_y-CH_3$ with benzene ring bearing SO_3^-; $x+y$: 7-10
Alkyl sulfate (AS)	$C_nH_{2n+1}-O-SO_3^-Na^+$; n: 12–16
Alkylether sulfates (AESs)	$C_nH_{2n+1}-O-(CH_2CH_2O)_m-SO_3^-Na^+$; n: 12–16, m: 0–12
α-Olefin sulfonates (AOSs)	Commercial mixture of $R-CH=CH-CH_2CH_2-SO_3^-Na^+$ and $R-CHOH-CH_2CH_2-SO_3^-Na^+$
Secondary alkane sulfonates (SASs)	$CH_3-CH_2CH-CH_2-CH_2-[CH_2]_nCH_2-CH_3$ with $SO_3^-Na^+$

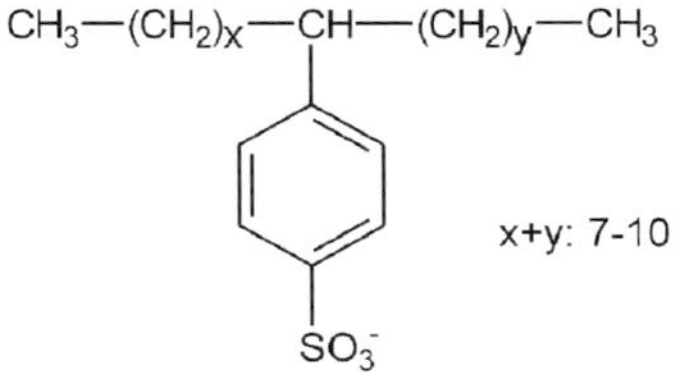

5

- Carboxylated surfactants (i.e. those for which the hydrophilic group is a carboxylic ion)
    Soap (fatty acid derivatives)
- Sulfosuccinates

The *non-ionic surfactants* do not produce ions in aqueous solution. The solubility of non-ionic surfactants in water is due to the presence of functional groups in the molecules that have a strong affinity for water. Similarly to the anionic surfactants, and any other group of surfactants, they also show the same general property of these products, which is the reduction of the surface tension of water.

This is a very important group of products and most of them are adducts of long-chain alcohols or alkylphenols with a number of ethylene oxide (EO) units. Adducts with propylene oxide (PO) and copolymers of ethylene and PO are also used although they are less important in terms of usage volume than the pure ethoxylate derivatives.

The most representative non-ionic surfactants are the alkyl (alcohol) ethoxylates. These are adducts of a long-chain alcohol (12–18) with a variable number of EO units (3–11). Other non-ionic surfactants are derived from carbohydrates (glucoside and glucamide derivatives), organosilicones, fatty alcohols, and amides. Products in this category are as follows (compare also Table 1.2):

- Alkylphenol ethoxylates (APEOs)
- Alcohol ethoxylates (AEs)
- Fatty acid diethanolamides (FADAs)
- Fatty acid ethoxylates
- Amine ethoxylates
- Alkanol amides
- Alkyl polyglucosides (APGs)
- Alkyl glucamides (AGs)
- Organosilicones

Cationic surfactants are surface-active agents that have one or more functional groups in their molecule that ionise in aqueous solution to produce positively charged organic ions. The most representative cationic surfactants are quaternary ammonium derivatives in which the N atom is bonded to four alkyl groups. For many years, ditallow dimethylammonium chloride (DTDMAC) has been the most widely used product of this family. Its recalcitrance to biodegradation, however, has

TABLE 1.2

Chemical structure of common non-ionic surfactants

Surfactant type	Structure

Alkylphenol ethoxylates (APEOs)

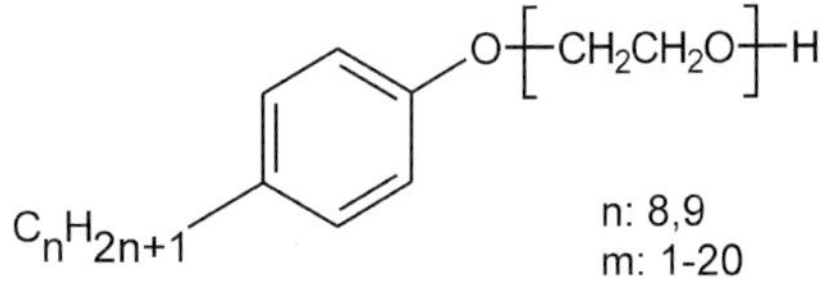

C_nH_{2n+1} —⟨benzene ring⟩— O—[CH_2CH_2O]—H

n: 8,9
m: 1-20

Alcohol ethoxylates (AEs)

C_nH_{2n+1}—O—$(CH_2CH_2O)_m$—H, n: 10–16, m: 0–9

Fatty acid diethanolamides (FADAs)

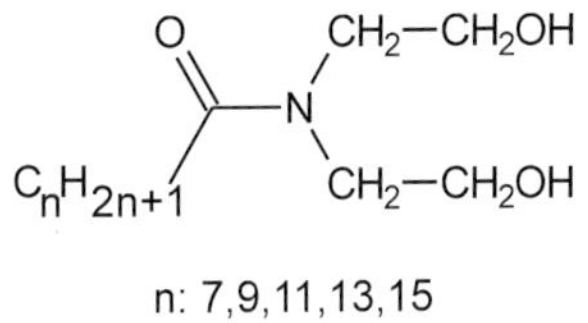

CH_2—CH_2OH
N
CH_2—CH_2OH
C=O
C_nH_{2n+1}

n: 7,9,11,13,15

Alkyl polyglucosides (APGs)

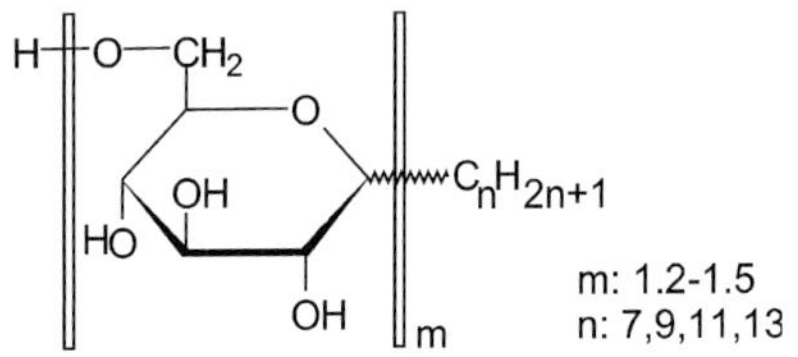

C_nH_{2n+1}

m: 1.2-1.5
n: 7,9,11,13

(continued on next page)

7

TABLE 1.2 (*continued*)

Surfactant type	Structure
Alkyl glucamides (AGs)	
Organosilicones	

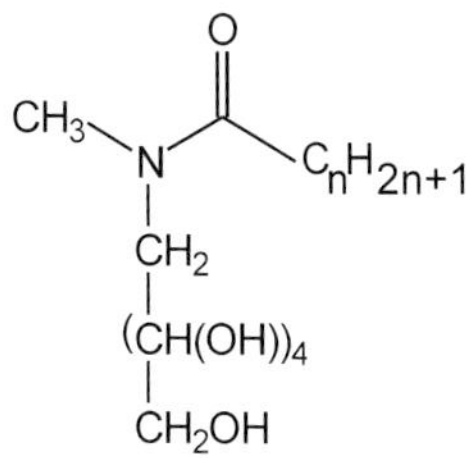

Alkyl glucamides (AGs):

$$\underset{CH_3}{\overset{O}{\underset{\displaystyle N}{\parallel}}}\!\!-\!\!C\!\!-\!\!C_nH_{2n+1}$$

with N also bearing:
CH$_2$—(CH(OH))$_4$—CH$_2$OH

n: 11,13

Organosilicones:

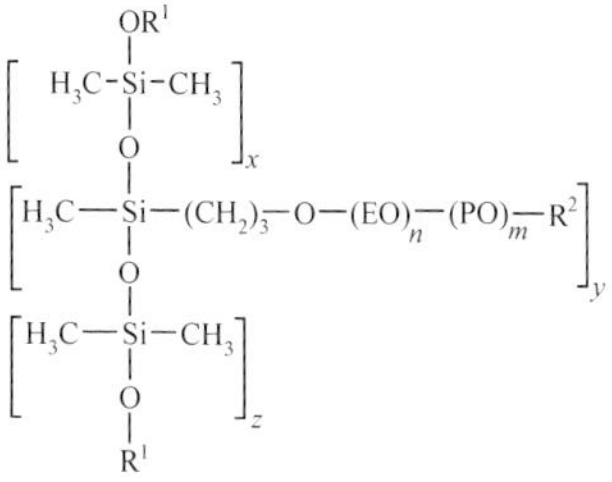

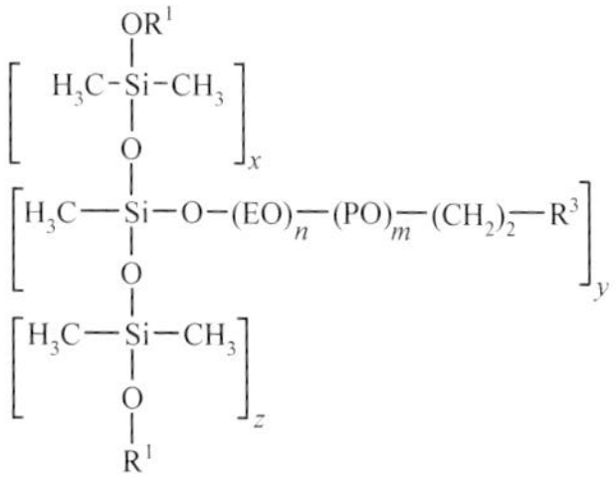

$$\left[H_3C\!-\!\underset{\underset{O}{|}}{\overset{OR^1}{\underset{|}{Si}}}\!-\!CH_3\right]_x$$

$$\left[H_3C\!-\!\underset{\underset{O}{|}}{Si}\!-\!(CH_2)_3\!-\!O\!-\!(EO)_n\!-\!(PO)_m\!-\!R^2\right]_y$$

$$\left[H_3C\!-\!\underset{\underset{R^1}{|}}{\overset{CH_3}{\underset{|}{Si}}}\!-\!O\right]_z$$

$$\left[H_3C\!-\!\underset{\underset{O}{|}}{\overset{OR^1}{\underset{|}{Si}}}\!-\!CH_3\right]_x$$

$$\left[H_3C\!-\!\underset{\underset{O}{|}}{Si}\!-\!O\!-\!(EO)_n\!-\!(PO)_m\!-\!(CH_2)_2\!-\!R^3\right]_y$$

$$\left[H_3C\!-\!\underset{\underset{R^1}{|}}{\overset{CH_3}{\underset{|}{Si}}}\!-\!O\right]_z$$

$x = 0 - 50$; $y = 1 - 100$; $n = 3 - 30$; $m = 0, 3 - 30$;

$R^1 = Si(CH_3)_3$, alkylpolyether; $R^2 = H, CH_3, COCH_3$; $R^3 = OH, OCH_3$

promoted the development of alternative materials that are more easily biodegradable; the so-called esterquats. In the new family of esterquats the ester linkage can be easily hydrolysed to release hydroxyalkyl ammonium compounds which are more amenable to further microbial breakdown by commonly occurring microorganisms in the environment. Products in this group are:

- Ditallow ammonium compounds (see also Table 1.3)
- Imidazolinium salts

Amphoteric surfactants have two or more functional groups which, depending on the conditions of the medium, can be ionised in aqueous solutions to give the compound the characteristics of either an anionic

TABLE 1.3

Chemical structure of common cationic surfactants

Surfactant type	Structure
Ditallow dimethylammonium chloride (DTDMAC)	
Ditallow ester of di(hydroxyethyl)dimethyl ammonium chloride (DEEDMAC)	
Ditallow ester of 2,3-dihydroxy-propanetrimethyl ammonium chloride (LDEQ)	
Ditallow ester of tri(hydroxyethyl)methyl ammonium chloride (EQ)	

TABLE 1.4

Example of the chemical structure of an amphoteric surfactant

Surfactant	Structure
Cocamidopropyl betaine	$CH_3-(CH_2)n-C(=O)-NH-CH_2CH_2CH_2-N^{\oplus}(CH_3)_2-CH_2-C(=O)-O^{\ominus}$ n: 6,8,10,12

or a cationic surfactant. By choosing the right conditions they can be compatible with either anionic or cationic surfactants in detergent preparations. The most significant amphoteric surfactants are derivatives of betaine. Examples of products in this family are:

- Betaines (see also Table 1.4)
- Imidazolines

There are many more surfactants, not mentioned above, used in the various fields and applications but for the purpose of this book the products listed represent the most important chemicals of this group.

1.3 PRODUCTION RATES AND USE OF RELEVANT SURFACTANTS

Surfactants—excluding soap—are considered as high-volume chemicals, with a global production rate of about 11×10^6 t [9]. It is estimated that, in addition, approximately 9×10^6 t of soap in various forms is consumed, and thus this is the single most important surfactant by far. However, because of the widespread use of these products it is difficult to make an accurate estimation of the various markets. Soap is frequently not accounted for in major statistical studies covering surfactants and thus accurate figures and consumption patterns in the various geographical areas are not available. As such the following sections deal with non-soap surfactants.

There are two major uses; household products and industrial processing aids, which together account for more than 85% of consumption. Other major categories include personal care products and industrial and institutional products (I&I). An average annual growth of 3% per year is expected for the next decade, which will thus result in an estimated total market of nearly 14 million tons in the year 2010.

Because of the variation in compositions observed for commercially available surfactants, statistical figures must be interpreted with caution unless the figures are specifically expressed with respect to the active content. We have tried to provide the most accurate data available and the data in the tables below refer to products as 100% active matter, unless otherwise specified.

Council of European Surfactant Producers (CESIO) statistics [10] indicate a total surfactant production of 2.4 million tons in Europe for 1999, which were distributed in the categories shown in Table 1.5. Other information sources [11] indicate a surfactant consumption of 2.1 million tons in Europe for 1998, which compares very well with the CESIO figure. Europe is a net exporter of surfactants and a precise figure of actual European consumption is thus difficult to estimate, although information from CESIO provides data on total sales and captive use.

The major surfactant families in use in the European market in volume terms are the non-ionics followed by anionic derivatives. A detailed breakdown of the total surfactant consumption for Western Europe can be found in Tables 1.6 and 1.7, which show the estimated proportions for the various uses of surfactants in the European market. Probably the most important final use of surfactants in Europe is in detergent manufacture, as demonstrated by the figures summarised in Table 1.8.

The worldwide consumption of major surfactants classified according to structural type for 1998 is summarised in Table 1.9, with the breakdown by final use and geographical areas shown in Table 1.10.

In Table 1.11, a summary of the estimated world surfactant consumption by regions is presented. The two major consumption areas are Asia and North America followed by Europe.

The value of the global surfactant market for 1998 was estimated at US$ 14 bn. While the market in industrialised countries, with a very limited number of private surfactant companies, only shows slight growth, the demand is rapidly increasing in other nations, particularly in the Asian-Pacific region [12]. The sales revenue in China, one of the strongest expanding markets, for example, is predicted to reach US$ 1.44 bn by 2005. The expected growth in the developing countries

TABLE 1.5

Surfactant consumption in Western Europe for the period 1998–1999, in 10^3 t [10]

Surfactants	Production[a] (10^3 t)		Sales and captive use[a] (10^3 t)	
	1998	1999	1998	1999
Anionics				
LAS	409	409	324	310
SAS	69	77	57	62
AS	107	114	99	102
AES	246	279	228	237
Other anionics	79	91	66	72
Total anionics	910	970	774	783
Non-ionics				
AE	638	772	572	640
Other ethoxylates	259	273	206	202
Amine oxides	12	16	10	16
Other non-ionics	217	184	186	147
Total non-ionics	1126	1245	974	1005
Cationics				
Esterquats and imidazolinium salts	115	121	80	80
Other cationics	52	58	39	39
Total cationics	167	179	119	119
Amphoterics				
Betaines	35	30	32	26
Imidazolines	3	3	3	3
Other amphoterics	17	24	15	21
Total amphoterics	55	57	50	50
Total surfactants	2258	2451	1917	1957

[a]These data represent volume expressed as 100% active substances.

is mainly due to the expansion in the household cleaning and laundry product sectors, driven by the improvements in the standard of living for consumers.

In general, the change in surfactant consumption is directly coupled to the usage in detergent products, which makes up about 55% of the entire production. Cleaning applications cover a multitude of fields comprising household detergents, consumer products, personal care

TABLE 1.6

Total surfactant consumption for Western Europe in 1998 by type [11]

Surfactant type	Consumption (10^3 t)
AE	353.3
AES	269.3
AS	118.6
Amine oxides	9.9
Betaines	50.1
LAS	393.7
Quaternary ammonium compounds	161.4
APEO	91.6
AG	13.7
Other	687.5
Total[a]	2149.1

[a]Does not include soap.

products and I&I uses. I&I products are used in the professional cleaning of objects and surfaces by trained personnel or machines, particularly outside of the private sector. They may contain the same ingredients as used in household products, where a relatively small number of 'commodity' surfactants are dominant. For that, large-scale production at a competitive cost is a prerequisite. The remaining fraction (approximately 45%) are used as process aids to utilise aspects of the surface-active properties, e.g. foaming, wetting and emulsification. In areas such as textile processing, pesticide emulsifying, oilfield demulsifying, paint and ink production, surfactants with more specialised functions can be used. They are likewise also consumed in other fields, such as coatings, polymers and food and feed industries.

TABLE 1.7

Surfactant consumption for Western Europe in 1998 by end use area [11]

End use area	Consumption (10^3 t)
Household	1118.7
Personal care	162.6
I&I cleaners	191.9
Industrial applications	675.9
Total[a]	2149.1

[a]Does not include soap.

TABLE 1.8

Classification of major household detergents for Western Europe in 1998 according to end use area [11]

Major household detergents	Consumption (10^3 t)
Heavy duty powders	2810
Heavy duty liquids	385
Light duty liquids	1020
Fabric softeners	845
Others	
Hard surface cleaners	1180
Fine fabric (includes soap)	353
Scourers	208
Auto dish products[a]	325
Miscellaneous	92
Subtotal others	2158
Total	7218

[a]Includes automatic dishwashing powders and liquids, rinse aids and hand dishwashing powders.

TABLE 1.9

Major worldwide surfactant consumption for 1998 classified according to type [11]

Surfactant type	Consumption (10^3 t)
LAS	3027.1
b-ABS	197.8
AES	910.6
AS	478.6
Alkylglyceryl ether sulfonates	18.3
Monoalkyl phosphates	15.2
AEO	849.1
APEO	700.7
Amine oxides	50.4
Quaternary ammonium compounds	434.3
Betaines	85.4
Others[a]	3093.0
Total	9860.5

[a]Does not include soap.

TABLE 1.10

Total world surfactant consumption in 1998 by end use [11]

Surfactant use	Consumption (10^3 t)
Household detergents	
North America	1114.5
Western Europe	1118.7
Asia	1598.5
Other regions	1234.9
Total	5066.6
Personal care	
North America	209.5
Western Europe	162.6
Asia	194.4
Other regions	155.3
Total	721.8
I&I Cleaners	
North America	311.4
Western Europe	191.9
Asia	75.3
Other regions	52.8
Total	631.4
Industrial process aids	
North America	1316.4
Western Europe	675.9
Asia	1084.7
Other regions	363.7
Total	3440.7
Grand total[a]	9860.5

[a]Does not include soap.

Anionic surfactants have the largest market share and are used especially in (household) detergent applications, where they exhibit excellent cleaning properties and are usually good foaming agents. Non-ionics are the other high volume surfactant class since they offer a broad spectrum of properties including wetting, dispersion, emulsification and detergency. Cationic and amphoteric surfactants are produced in much smaller quantities. Cationics are applied as bactericides and disinfectants and above all in fabric conditioners due to their ability to adhere to

TABLE 1.11

Total world surfactant consumption in 1998 by region [11]

Region	Consumption (10^3 t)
North America	2951.8
Western Europe	2149.1
Asia	2952.9
Other regions	1806.7
Total	9860.5

surfaces. Depending on the pH, amphoterics may behave as cationic, non-ionic or anionic surfactants. Because of their mildness towards skin they are generally found in toiletries and cosmetic formulations.

The predominance of anionic and non-ionic surfactants is reflected by their sales figures for the Western Europe market, shown in Table 1.5. The major representatives of each surfactant group are discussed in terms of application, manufacturing and composition in the following sections.

1.3.1 Anionic surfactants

Soap can be considered as a natural anionic surfactant, which is used in detergents, coatings and cosmetics. The addition of 1–5% soap to modern detergents inhibits excessive foaming caused by other surface-active ingredients. Unlike synthetic surfactants, soap usage has remained almost constant since the early 1980s, mainly because of its relatively poor performance. Since the invention of industrial soda production by Leblanc, soap has been manufactured by saponification of fats with alkali [13]. Soap derived from palm kernel or tallow contains an even number of carbon atoms ranging from C_{12} to C_{18}. The alkyl chain is linear but may contain one or more unsaturated bonds.

From the anionic surfactants (Table 1.1) the most relevant is LAS with an annual global production volume of more than 3×10^6 t in 2001. LAS has a wide application because of its excellent detersive properties and cost–performance ratio. Commercial LAS is applied mainly in the formulation of powder and liquid laundry detergents. The calcium salts are used as an emulsifier in pesticide formulations; their amine salts are used in dry cleaning and as degreasing agents in the metal industry [14].

Fatty alcohol sulfates (or ASs) are applied in powder and liquid detergents. In these formulations they are used in addition to, or as substitutes for, LAS. Furthermore, they are also used in cosmetics and body-care preparations [15].

Another important group of anionic surfactants are AESs, which are used in laundry and cleaning detergents as well as in cosmetic products. Characteristic properties of AESs are the ability to function in hard water, high foam capacity and low irritation to skin [16].

1.3.2 Non-ionic surfactants

Due to their distinctive physico-chemical properties, non-ionic surfactants are applied in the fields of industry, processing technology and science, wherever their interfacial effects of detergency, (de)foaming, (de)emulsification, dispersion or solubilisation can enhance product or process performance. The characteristics of non-ionic surfactants that make them beneficial for detergents include their relatively low ionic sensitivity and their sorptive behaviour [17].

APEOs are used in domestic and industrial applications. They are applied as detergents, emulsifiers, wetting agents, dispersants or solubilisers. APEO derived from nonylphenol (NP), i.e. nonylphenol ethoxylates (NPEOs) comprises about 80% of the total market volume, while octylphenol-derived surfactants (OPEOs) account for 15–20%. Because of the persistence and toxicity of some degradation intermediates, their use has been reduced in several countries either through voluntary bans by the chemical industry or by legal regulations. However, excellent properties in combination with comparably low production costs hampers their complete replacement with other more environmentally acceptable alternatives.

AEs have found application in all kinds of domestic and institutional detergents, and cleaning agents. Their low foaming characteristics, better electrolyte compatibility and degreasing capacity relative to anionic surfactants make them especially attractive for use in I&I products [18]. Furthermore, they are applied in cosmetics, agriculture, and in the textile, paper and oil industries. They have become increasingly important in more recent years, due to efforts to replace APEO.

FADAs are important ingredients in formulated detergents and cosmetics, since they enhance the emulsifying properties of anionic and non-ionic surfactants. They also contribute to foam stability in the presence of soil, aid in the control of viscosity of aqueous detergent solutions and act as anti-static and anti-corrosive agents [19].

APGs have been known since 1893 but it is only in the last decade that they have been manufactured at the industrial scale. Due to favourable properties in terms of foaming performance, synergism with other surfactants [20], skin compatibility, and very low toxicity [21], they are being used increasingly as cleaning agents and detergents. This group of non-ionic surfactants consists of a multiplicity of homologues and isomers, including stereoisomers, binding isomers and ring isomers within the glucose moiety (Table 1.2). The global production rate in 1997 amounted to 80 000 t. Even if they are unlikely to become a main surfactant such as soap, LAS, AES or AEO, they play an important role as co-surfactants.

AG are next in importance in the list of non-petrochemically based surfactants, with an annual production rate of 50 000 t. They find application in laundry detergents and dishwashing agents where they exhibit very good foaming power and excellent degreasing capacity [22].

1.3.3 Cationic surfactants

Cationic surfactants show a high affinity for negatively charged surfaces making them suitable for industrial applications and as components for consumer products where they are used as disinfectants, foam depressants, and first and foremost as textile softeners [23]. Due to the possible formation of ion-pair associates they are usually not formulated together with anionic surfactants.

DTDMAC was the first widely used cationic surfactant produced at the industrial scale since the 1960s. Its main application was as a fabric softening active agent. Due to its physico-chemical properties, largely determined by the positively charged head group and the long alkyl chains, it adsorbs onto fabrics and makes textiles feel soft.

However, since DTDMAC also shows a pronounced resistance to biodegradation (see Chapter 1.7), it has been almost completely replaced by the more environmentally acceptable esterquats.

From the esterquats introduced onto the surfactant market in the early 1990s, DEEDMAC, LEQ and EQ (Table 1.3) have gained widespread use [24−26].

1.3.4 Amphoteric surfactants

Even if this class covers the smallest market segment, amphoteric surfactants still remain useful because of their unique properties, which justifies their comparably high manufacturing costs. Since they have partial anionic and cationic character, they can be compatible, under specific conditions, with both anionic and cationic surfactants. They can function in acid or basic pH systems and, at their isoelectric point, they exhibit special behaviour. Many amphoteric surfactants demonstrate exceptional foaming and detergency properties combined with anti-static effects.

Cocamidopropyl betaine (Table 1.4) is the most prominent representative of the class of amphoteric surfactants. Due to the synergism with other surfactants and its gentleness to the skin and mucous membranes, cocamidopropyl betaine performs well in shampoo and cosmetics where its dosage lies in the order of 1–5% [27].

1.4 RAW MATERIALS

Surfactants can be produced from both petrochemical resources and/or renewable, mostly oleochemical, feedstocks. Crude oil and natural gas make up the first class while palm oil (+kernel oil), tallow and coconut oil are the most relevant representatives of the group of renewable resources. Though the worldwide supplies of crude oil and natural gas are limited—estimated in 1996 at 131×10^9 t and 77×10^9 m^3, respectively [28]—it is not expected that this will cause concern in the coming decades or even until the next century. In this respect it should be stressed that surfactant products only represent 1.5% of all petrochemical uses. Regarding the petrochemically derived raw materials, the main starting products comprise ethylene, n-paraffins and benzene obtained from crude oil by industrial processes such as distillation, cracking and adsorption/desorption. The primary products are subsequently converted to a series of intermediates like α-olefins, oxo-alcohols, primary alcohols, ethylene oxide and alkyl benzenes, which are then further modified to yield the desired surfactants.

The total world production of natural oils and fats in 1997 amounted to 100×10^6 t, of which 80×10^6 t were of vegetable and 20×10^6 t of animal origin [28]. From these oils and fats, 80% are suitable for human nutrition and 14% end up in so-called oleochemical uses, among them the synthesis of surfactants. The worldwide capacity of the basic oleochemicals, fatty acids and fatty alcohols, was estimated in 1998 at 5.2×10^6 t [29]. The most important C_{12}/C_{14} alkyl derivatives are found in coconut oil and palm kernel oil. The C_{16} and C_{18} homologues are obtained from palm oil and tallow.

A certain shift from petrochemically-based to oleochemically-based surfactants, namely on the alcohol derivatives field, has occurred in recent years in industrialised countries. Figure 1.2 represents the key derivatives and major intermediates used in the production of surfactants.

The most relevant raw materials utilised and chemical synthetic pathways adopted are covered in the following sections.

1.4.1 Linear alkylbenzene

Linear alkylbenzene (LAB) is produced by alkylating benzene with either alkylhalides or mono-olefins, the second option being the most widely used in commercial processes. The characteristics of final LAB, namely the isomer distribution, depend on the alkylation catalyst used; HF catalyst produces a LAB known a 'Low 2-phenyl' with a 2-phenyl isomer content around 17%, while fixed bed and AlCl$_3$ processes produce a 'High 2-phenyl' LAB, which consists of approximately 30% of this isomer. Commercial LAB is a mixture of C_{10}, C_{11}, C_{12} and C_{13} homologues with all positional isomers except the terminal one

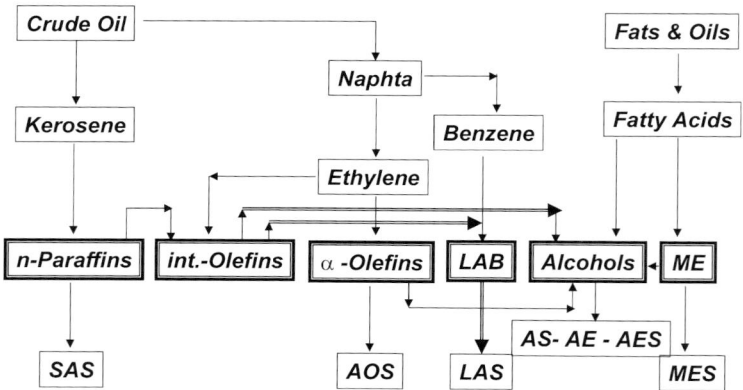

Fig. 1.2. Key derivatives and major intermediates used in surfactant production.

(1-phenyl). The various available technologies are summarised in Fig. 1.3 and their relative importance in the overall market are represented in Fig. 1.4. Thereby the predominance of the HF process is clearly demonstrated. It starts, in most cases, with the production of linear paraffins extracted from kerosene using selective molecular sieves. The linear paraffins ($C_{10}-C_{14}$ range) are used to produce linear olefins by dehydrogenation or chloroparaffins by chlorination. In the first case, the mono-olefins obtained are subsequently used to alkylate the benzene with HF or fixed bed catalysts as shown in Figs. 1.5 and 1.6. Chlorinated paraffins are used in combination with $AlCl_3$ as the catalyst for the alkylation of benzene to produce LAB. This was the first technology used for the commercial production of LAB (Fig. 1.7) although today it represents only a small portion of the total LAB produced worldwide.

The most modern technology used today is the fixed bed alkylation process based on a non-corrosive solid catalyst, which leads to considerable simplification of the overall process as compared with the existing HF or $AlCl_3$ alternatives. The process, known as *Petresa-UOP DETAL*, was first used in 1995 and since then the new LAB plants build in different countries have all been based on this technology.

1.4.2 Linear paraffins

Linear paraffins are key raw materials for the production of LAB as well as for long chain alcohols, which in turn are transformed into a group of

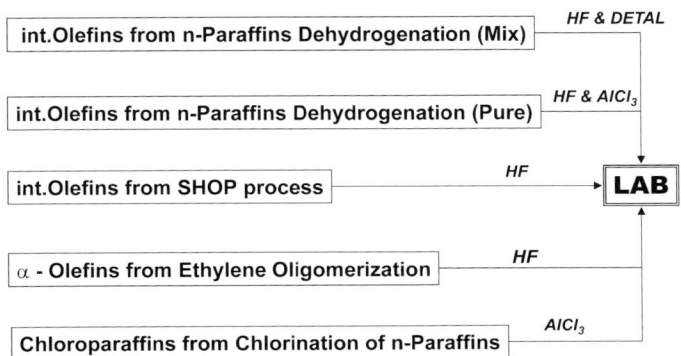

Fig. 1.3. Currently used technologies for the synthesis of LAB.

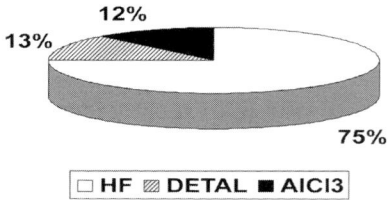

Fig. 1.4. World LAB production according to preparation methodology.

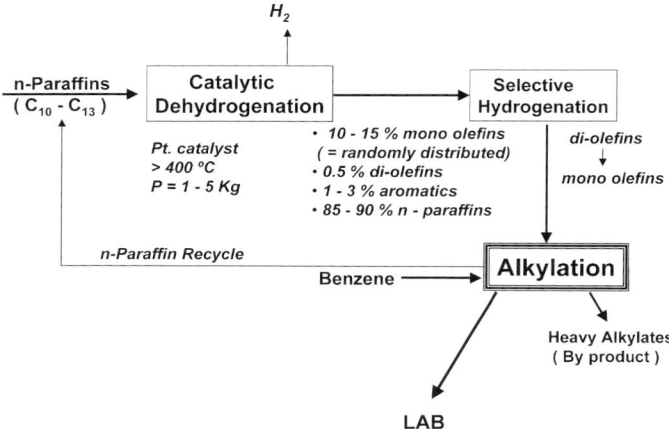

Fig. 1.5. Scheme for the production of LAB starting with n-paraffins.

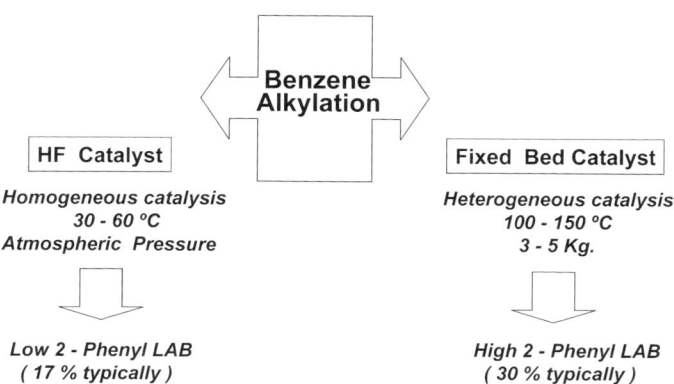

Fig. 1.6. Formation of different LAB types depending on the choice of the alkylation process.

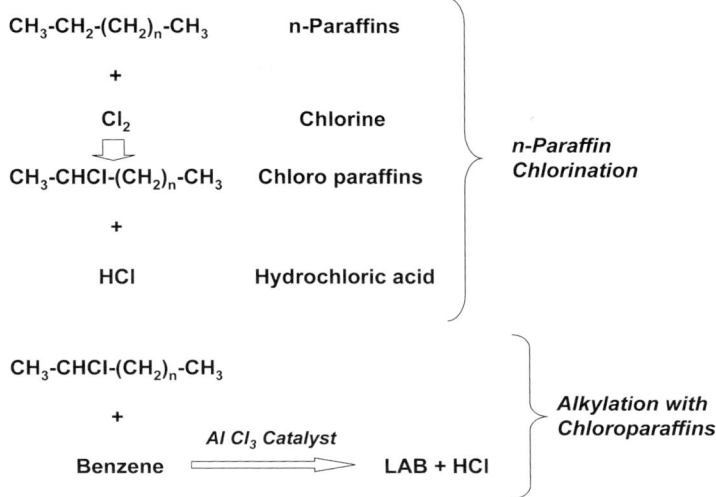

Fig. 1.7. Formation of chloroparaffins and subsequent alkylation as main steps in the LAB production.

very important surfactants. This is the most important final use of linear paraffins on a worldwide basis. Linear paraffins, however, can also be directly transformed into a particular surfactant; SAS. For this particular use, a $C_{16}-C_{18}$ cut is used and transformed into SAS using a specific photosulfonation reaction (UV catalysed) or a sulfo-oxidation reaction. Linear paraffins are in all cases extracted from a selected crude oil fraction; i.e. kerosene or gas oil, using specific molecular sieves (zeolites) in a continuous process. The final product is a highly linear (>98%) mix of the various homologues contained in the feedstock used, generally in the range of $C_{10}-C_{14}$ for kerosenes and from C_{14} up to C_{18+} when a heavier feed is used (gas oil).

1.4.3 Ethylene

Ethylene is obtained by catalytic cracking of naphtha. It is one of the key petrochemical commodities worldwide used mostly in the production of polyethylene, ethyl benzene, ethylene oxide and others. The consumption of ethylene for the production of alcohols and other surfactant raw materials represents less than 10% of the total end uses of ethylene on a worldwide basis.

1.4.4 α-Olefins

α-Olefins are important intermediates used in the production of the sulfonated surfactants (AOSs: C_{16}–C_{18} range) or as raw material for the production of long chain alcohols (C_{12}–C_{18} range) used for subsequent ethoxylation or sulfation. α-Olefins are also used in some cases for the production of LAB although this is the most expensive route and it is only adopted when internal olefins, obtained by dehydrogenation of linear paraffin, are not available.

α-Olefins are produced by ethylene oligomerisation using different technologies:

(a) Ziegler process (Fig. 1.8)
(b) Modified Ziegler process, using two steps in order to minimise the production of light olefins. In each case the final olefin product is a polydisperse mixture, the first exhibiting a 'Poisson' distribution and the second a 'peak' distribution more centred around the desired carbon chain length (Fig. 1.9).
(c) Shell higher olefin process (SHOP). It uses a non-Ziegler catalyst system and it produces both alpha as well as internal olefins (Fig. 1.10).

1.4.5 Linear alcohols

Alcohols in the range C_{12}–C_{18} are important raw materials for the production of a key group of surfactants; ethoxylates, sulfates and ethoxysulfates among others. Alcohols used in the surfactant industry are primary, linear, or with different degrees of branching, and they can be produced from either petrochemical sources (ethylene or linear paraffins) or from oleochemical products (animal fats and vegetable oils).

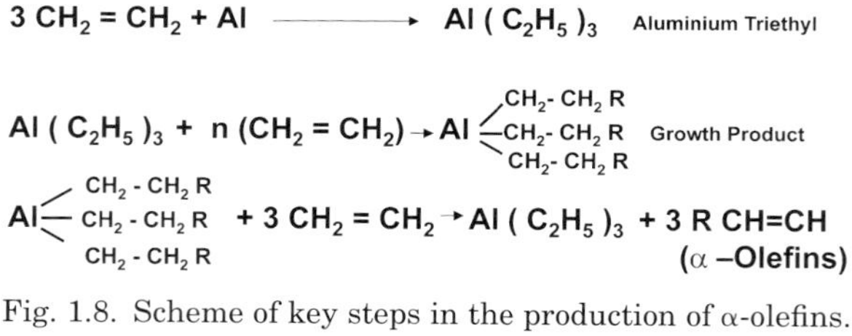

$$3\ CH_2 = CH_2 + Al \longrightarrow Al\ (\ C_2H_5\ )_3 \quad \text{Aluminium Triethyl}$$

$$Al\ (\ C_2H_5\ )_3 + n\ (CH_2 = CH_2) \rightarrow Al \underset{CH_2\text{-}CH_2\ R}{\overset{CH_2\text{-}CH_2\ R}{\underset{}{\diagdown}}} \quad \text{Growth Product}$$

$$Al \underset{CH_2\text{-}CH_2\ R}{\overset{CH_2\text{-}CH_2\ R}{\diagdown}} + 3\ CH_2 = CH_2 \rightarrow Al\ (\ C_2H_5\ )_3 + 3\ R\ CH{=}CH$$
$$(\alpha\text{ –Olefins})$$

Fig. 1.8. Scheme of key steps in the production of α-olefins.

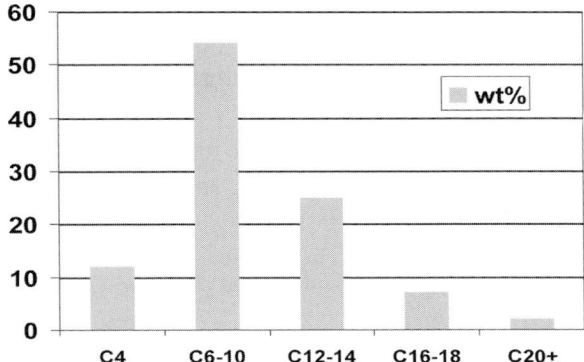

Fig. 1.9. POISSON product distribution during synthesis of the modified AMOCO-Ziegler α-olefins.

Oleochemical alcohols are primary, even carbon-numbered structures with a high linearity (>95%), while the petrochemical derivatives can be even or odd numbered, and depending on the process, their linearity can be as high as the oleochemicals (Ziegler alcohols) or can exhibit variable branching (30% modified OXO–60% standard OXO).

The various steps introduced in the production of alcohols from ethylene are represented in Fig. 1.11. Similar to the production of α-olefins, in the production of alcohols using the Ziegler process, the final product is a mix showing a typical Poisson distribution with alcohols from C_4 up to C_{28}. Alcohols are also obtained from n-paraffin

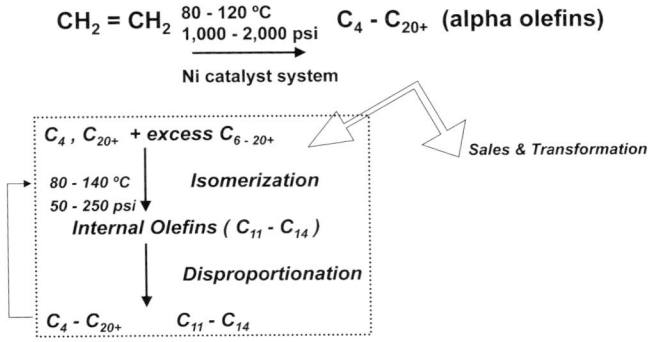

Fig. 1.10. Scheme of the (SHOP) utilizing a non-Ziegler catalyst system.

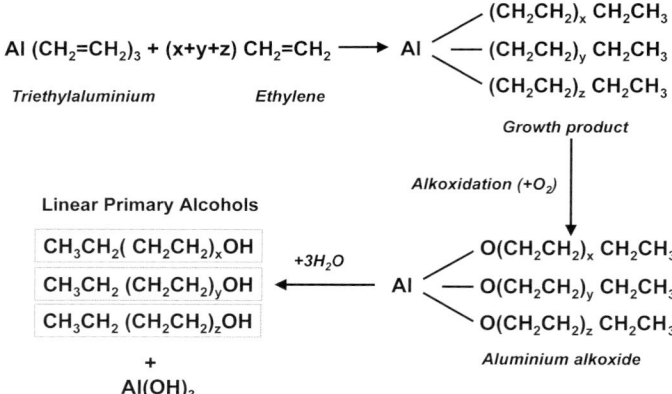

Fig.1.11. Scheme for the formation of linear primary alcohols out of ethylene according to the Ziegler process.

dehydrogenation (internal olefins) or olefins from ethylene using hydroformylation reaction (OXO process) as shown in Figs. 1.12 and 1.13.

Alcohols are also widely produced from fats and oils, and in some cases one of the oldest techniques utilised in the manufacture of soap, fat splitting, is also used as part of the process. Vegetable oils as well as animal fats are all triglycerides and consequently a key step of the various technologies used in the alcohol manufacture involves the separation of glycerine. Oleochemical alcohols are produced using two

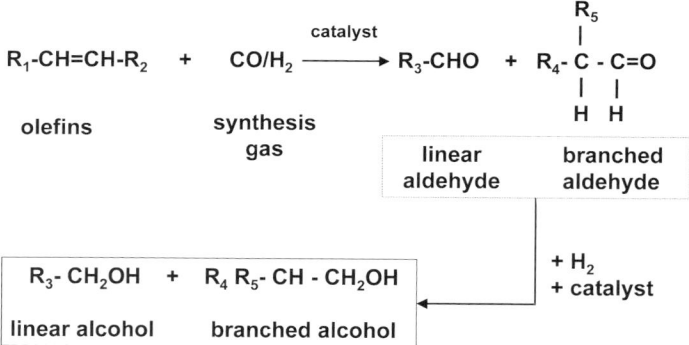

Fig. 1.12. Scheme for the formation of linear and branched primary alcohols starting with olefins (hydroformylation) according to the OXO standard process reactions.

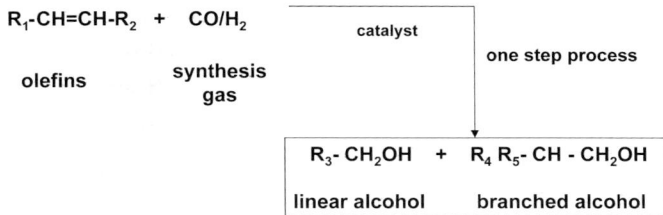

Fig. 1.13. Scheme for the formation of linear and branched primary alcohols starting with olefins according to the modified OXO (Shell) standard process reactions.

basic routes, fatty acid or methylester hydrogenation. The different steps involved for both cases are presented in Fig. 1.14.

Oleochemical alcohols, sometimes known as natural alcohols, are also identified by the carbon range; $C_{12}–C_{14}$ lauric, $C_{16}–C_{18}$ tallow, regardless of the origin of the raw material. $C_{16}–C_{18}$ alcohols were predominantly produced in the past from tallow, hence their name, although today they are also widely produced from palm oil. Lauric range alcohols are produced from either coconut oil or from palm kernel oil.

Secondary alcohols, produced previously in small quantities from linear paraffin oxidation, have today almost disappeared from the market. The difficulties in producing the corresponding derivatives (ethoxylates, etc.) were a major drawback for their potential development.

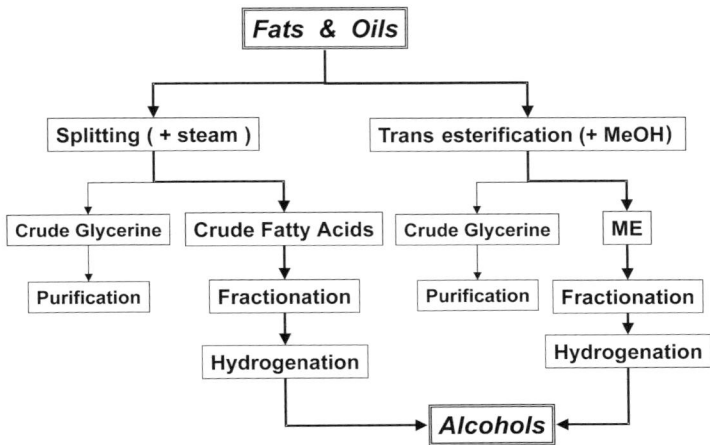

Fig. 1.14. Main processing steps in the formation of alcohols starting from fats and oils.

Other alcohols used in the surfactant industry, although of minor importance, are the 'Guerbet' alcohols obtained by self-condensation of lower alcohols to yield products in the range of 16–26 carbons with a high degree of branching.

1.4.6 Alkylphenol

Octyl- and nonylphenol are well known raw materials used in the surfactant industry since the early 1960s, mainly for the production of their corresponding ethoxylated derivatives (APE). Today, these products have lost considerable importance in this industry as a consequence of substantial environmental threats, resulting from their relatively slow biodegradation, toxicity of their biodegradation metabolites and positive endocrine-disrupting reactions.

Alkylphenols (AP) are produced by alkylation of phenol using either octene or nonene mixtures obtained from pyrolysis of gasoline and other petrochemical sources. These are highly branched olefins and consequently the corresponding final alkylated phenol is also a branched derivative.

1.4.7 Methylesters

In addition to their transformation into linear alcohols by hydrogenation, methyl esters obtained in the transesterification of fats and oils can also be used as raw materials for sulfonation, yielding a family of derivatives known as methyl ester sulfonates (MES). These products are only significant in some southeast Asian markets, Japan and to a much lesser extent in other markets such as the US.

1.4.8 Other surfactant raw materials

There are many other products used either as main raw materials or involved to different degrees in the production of surfactants. Some of the most significant ones are:

- benzene; used in the production of LAB,
- phenol; used for AP production,
- ethylene oxide; used for ethoxylation of alcohols and AP,
- propylene oxide; used in the propoxylation of alcohols and AP,

- sulphur; transformed into SO_3 for sulfonation and sulfation reactions,
- amines, amides, caustic soda, ammonia, glucose, hydrogen, methanol, and others.

1.5 SYNTHESIS OF SURFACTANTS

Various chemical reactions are utilised in the preparation of surfactants and an extensive description of them is not feasible within the limited scope of this book. Only the most important ones in this industry, sulfonation–sulfation and ethoxylation, used to generate a broad variety of sulfonated, sulfated and ethoxysulfated surfactants, will be described here. These key processes provide the fundamental change in the chemical structure of the molecule to convert it into a surface-active product.

1.5.1 Sulfonation and sulfation

LAS is produced by sulfonating LAB with either sulphuric acid, oleum or SO_3 gas (Fig. 1.15). The first two methods were predominant in the market until the 1970s and, although still utilised in some places, they have been largely replaced by the most modern film sulfonation techniques using SO_3 gas under various design possibilities [30].

LAB is sulfonated in the 'para' position and the resulting sulfonic acid is then neutralised with caustic soda to produce the sodium salt, which is the most widely used form of LAS worldwide. Other salts such as ammonia and triethanol amine are also used, although in substantially lower amounts compared with NaLAS.

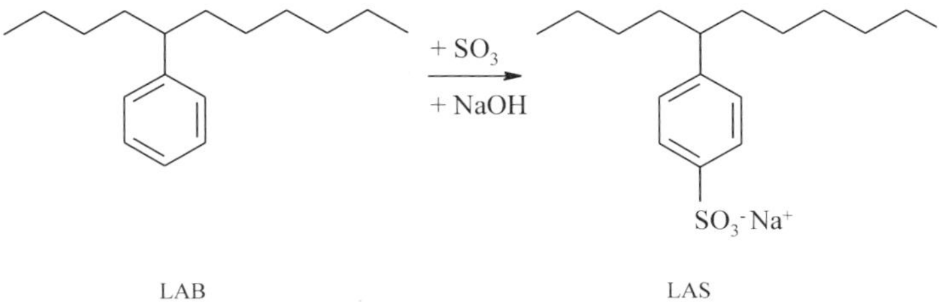

$$+ SO_3$$
$$+ NaOH$$

LAB   LAS

Fig. 1.15. Schematic pathway for the synthesis of LAS out of LAB.

LAS generally consists of a mixture of homologues and isomers. Individual components are identified by the length of the alkyl substituent and by the position of attachment of the benzene ring. In formulations, the alkyl chain length varies from C_{10} to C_{13} with an average approximate distribution of C_{12}. Isomer configurations range from the 2-phenyl position to the 7-phenyl position depending on the alkyl chain length. Commercial blends usually contain some sulfonated impurities ($<1\%$), namely dialkyl tetralinsulfonates (DATSs) and slightly branched alkylbenzene sulfonates (ABSs). The relative amounts of these impurities depend on the synthetic route used. Alkylation of benzene in the presence of $AlCl_3$ tends to yield higher DATSs than either HF or fixed bed alkylation processes [31,32]. The amount of ABS formed is similar in all alkylation technologies.

LAB sulfonic acid, also known as HLAS, is a stable material and it can be stored and transported for further neutralisation. Final neutralised LAS, usually as the sodium salt, is easily obtained as a white paste, which does not require any additional post-treatment.

This is not the case when sulfating alcohols (AS), ethoxylated alcohols (AES), AOS or MES, as the corresponding sulfonic acids are unstable and they can be easily decomposed. In these cases the neutralisation is done simultaneously with the sulfonation–sulfation, in order to produce the final sulfated, sulfonated and/or ethoxysulfated derivative.

ASs are made up of alkyl chains numbering 12 to 18 carbon atoms which are bound to a sulfate group. ASs are obtained by sulfation of fatty alcohols derived from palm oil, kernel oil and coconut oil, as well as from petrochemical raw materials.

Commercial AESs are produced via sulfation of AEs, which are also directly used as non-ionic surfactants (see below). AES preparations typically also contain some level of alkyl sulfates. The majority of technical AES blends are obtained from AE feed stocks that have alkyl chains in the range of 12–15 and an average degree of ethoxylation of three.

AOS and MES are also produced using sulfonation methodologies similar to that for LAS (falling film sulfonation) although they require some additional post-treatment steps. When sulfonating α-olefins to produce AOS, a final hydrolysis step is required in order to eliminate sulfone by-products formed during the sulfonation.

In the case of methyl ester sulfonation to produce MES, the colour of the final sulfonated product is usually very dark, and the process thus includes a bleaching step with hydrogen peroxide in order to obtain an acceptable quality.

In all the cases described above, the sulfonating agent, SO_3 gas, is usually prepared on site by burning sulphur to SO_2, followed by subsequent conversion to SO_3 using a standard V_2O_5 catalytic converter.

Sulfonation of *n*-paraffins to produce SAS is a totally different process involving ultraviolet (UV) catalysed reactors and much more complex processing steps than the standard sulfonation processes documented above. This has limited the availability and accessibility of this technology and consequently its broader use. Sulfonated paraffins are only produced in Europe in limited amounts.

A special sulfonation technique is also used for the production of sulfosuccinates (Fig. 1.16). These are derived from maleate esters or half esters using a suitable sulfonation agent, most commonly sodium metabisulfite.

1.5.2 Ethoxylation

Ethoxylation of alcohols, alkylphenols, fatty acids, and many other organic raw materials is also a very important reaction for the surfactant industry, used to produce a broad variety of surfactants, most of which belong to the non-ionics group. The reaction with propoxylation (PO) is also practised, although to a lesser degree. The product used for ethoxylation is EO, or PO in the case of propoxylation, and is conducted using alkaline catalysts such as NaOH or $NaOCH_3$.

Because of the difficulty and dangers of using EO in a continuous process, ethoxylation is usually conducted in batch operations. The required amount of the EO and alcohols are measured and pumped into the reactor in order to complete the reaction.

There is a great variety of ethylene oxide derivatives with differing numbers of EO oligomers produced, depending on the desired properties of the final AEs used in the detergent or cleaning preparation. When the final product is transformed into an ethoxy sulfate, the AES usually

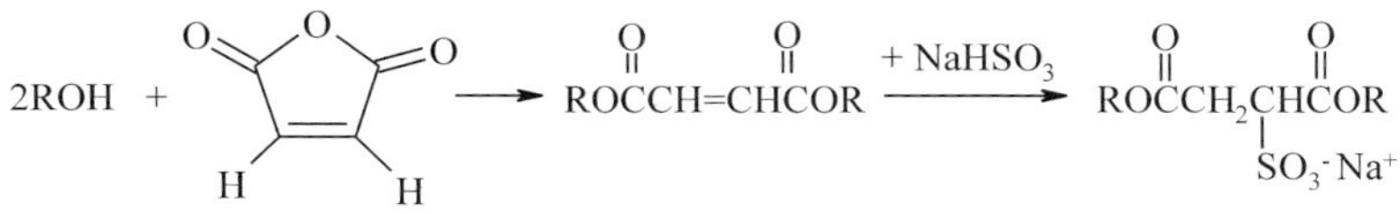

Fig. 1.16. Schematic pathway for the synthesis of sulfosuccinates.

contains a low number of EO units (1–3); however, products used as non-ionic surfactants can contain on average from 5 up to more than 20 EO units. The proportions of alcohol and EO are calculated to produce a given adduct, but the final ethoxylated product obtained comprises a typical Poisson distribution of various oligomers, the average being the desired molecular weight. The ethoxylation reaction, in general terms, is as follows:

$$ROH + [CH_2OCH_2]_n \xrightarrow{cat.} RO[CH_2CH_2O]_n H \qquad (1.1)$$

There are technologies available that produce the so-called 'peaked ethoxylates', which increase the amount of the desired adduct and minimise the formation of lighter and heavier adducts.

1.5.3 Synthesis of other surfactants

FADAs are nitrogen derivatives of coconut oil synthesised from fatty acid and diethanolamine. Equimolar amounts of the two starting compounds yield water-insoluble monoethanolamides, whereas the reaction of two moles of diethanolamine with one of the acids results in water-soluble FADA possessing the typical alkyl chain distribution with the C_{12}/C_{14} homologues prevailing [33].

AGs and APGs are produced completely with renewable feedstocks such as glucose and fatty alcohols derived from starch and palm kernel oil [34]. AGs, which are mainly the two homologues C_{12}- and C_{14}–N-methyl glucamide, are manufactured by reductive amination of glucose followed by acylation with fatty acid derivatives [35].

The synthesis of organosilicones and organosilicone surfactants has been well described elsewhere [36–39] and hence only a brief review is given here. Industrially the manufacture of silicones is performed stepwise via the alkylchlorosilanes, produced through the reaction of elemental silicon with methyl chloride (the Müller–Rochow Process) [40,41]. Inclusion of HCl and/or $H_{2(g)}$ into the reaction mixture, as in Eq. (1.2), yields CH_3HSiCl_2, the precursor to the organofunctional silanes, and therefore the silicone surfactants:

$$CH_3Cl_{(g)} + Si_{(s)} + HCl \rightarrow CH_3HSiCl_{2(g)}... \qquad (1.2)$$

The product mixture also contains $(CH_3)_3SiCl$, $(CH_3)_2SiCl_2$, CH_3SiCl_3, CH_3H_2SiCl, CH_3H_3Si, $(CH_3)_2HSiCl$, CH_3H_2SiCl, CH_3HSiCl_2, etc., which are purified by fractional distillation, and is the main reason for the relatively high cost of the silicone surfactants. Hydrolysis of the chlorosilanes yields the corresponding silanols, which are subsequently condensed to give the silicone backbone products. The organofunctional silicones are then made by addition of the appropriate alkene across the silane bond via hydrosilylation, for which hexachloroplatinic acid (Speier's catalyst) [42] is the most commonly used catalyst. Chloroalkyl functionalised silanes, i.e. $R_3Si(CH_2)_xCl$, are commonly adopted precursors for further organofunctional silanes. The propyl radical ($x = 3$) is one of the most common linking groups used, due to its solvolytic and thermal stability, and relative ease of preparation [36]. Other methods for the preparation of silicone surfactants include *trans*-etherification, which yields surfactants possessing an Si–O–C linkage, as used in polyurethane foam production, and two-step syntheses using reactive intermediates such as epoxy- and amino-functionalised silicones [43].

Cocamidopropyl betaine is manufactured by a two-step reaction: coconut oil-derived fatty acid is reacted with dimethylaminopropyl-amine to yield the cocamide that is subsequently converted to the betaine by the addition of monochloroacetate. The acyl group in the amide linkage ranges in length from 8 to 16 carbon atoms with C_{12} and C_{14} as the predominant components [44].

DTDMAC (Table 1.3) is produced at the industrial scale by combining a suitable amine with an alkylating agent.

1.6 ANALYSIS OF SURFACTANTS AND THEIR DEGRADATION PRODUCTS

Despite the large amounts of surfactants discharged into wastewater, the qualitative and quantitative determination in water or sludge samples according to the compound groups causes problems to the analysts even today. The property of surface activity, which imparts to surfactant molecules non-volatility, high polarity and therefore an excellent water-solubility, poses the analyst considerable problems.

The application of sum or group-specific parameters to quantify low amounts of surfactants in environmental samples and to pursue their fate therein does not require the employment of advanced analytical techniques; however, it also does not provide detailed information on the individual metabolisms. In addition to this shortcoming, the sum parameter analysis on a spectrophotometric and titrimetric basis, e.g. methylene blue active substances (MBAS) for the determination of anionic surfactants, is very insensitive and quite unspecific, since interference with other anionic substances can occur yielding a positive bias to the quantification. Underestimation is also possible when cationic compounds compete with the methylene blue reagent by forming stable complexes with the anionic surfactant. Furthermore, group-specific parameters do not give any additional information about the individual components occurring in commercial blends. The applicability of group-specific parameters for the determination of surfactants is even more questionable in the analysis of complex mixtures such as environmental samples like sewage influent, effluent or surface water samples.

The determination of bismuth activity as an indicator of non-ionic surfactants also suffers from interference in environmental samples. Substance group specific methods also failed to detect different types of fluorine-containing anionic, cationic and non-ionic surfactants. Already marginal modifications in the precursor surfactant due to primary degradation or advanced metabolisation implicated their lack of detection [45].

Since surfactants are commercially produced by means of large-scale chemical processes, complex mixtures of homologues and isomeric compounds, e.g. non-ionics of the alkylethoxylate type that may differ in length of alkyl as well as polyether chains, can result. The determination and differentiation of the products in quality control during production and trade is a somewhat easier task. However, more difficulties arise in the analysis of the compounds of these mixtures and formulations in environmental samples.

Thus, more sophisticated methods including preconcentration, chromatographic separation and sensitive and accurate detection are required for the compound-specific analysis of the broad range of surfactants. The request for more specific methods is further increased when the investigations not only centre on the parent compound, but also aim at the qualitative and quantitative determination of degradation intermediates, often formed at low concentrations, during the wastewater treatment process.

The sample pretreatment steps for surfactant analysis such as extraction and concentration can be carried out in a variety of ways. Most of the common and well tested procedures have been described by Schmitt [46]. Solid phase extraction (SPE) with C_2, C_8 or C_{18} and divinyl benzene resins as well as special phases, e.g. graphitised carbon black (GCB), can be adequate methods, especially for SPE of metabolites from water-samples [47,48]. Hyphenated procedures such as solid-phase micro-extraction (SPME), accelerated solvent extraction (ASE) and supercritical fluid extraction (SFE) have also been used for the extraction and concentration of surfactants from water, wastewater, sediment or sludge samples.

Liquid chromatography (LC) either using thin-layer (TLC) or high performance column chromatography (HPLC) under normal phase (NP) and reversed phase (RP) conditions, ion chromatography (IC) or capillary zone electrophoresis (CZE) can be used for this purpose according to the compounds to be separated [46]. The most appropriate procedure for the analysis of surfactants and their metabolites is, by far, HPLC. A major advantage of this technique is its ability to separate single components of complex surfactant mixtures. The use of optical detection systems like UV or fluorescence (FL) detection is only feasible when the molecule contains a chromophore. The lack of such a group can be compensated for if a mass spectrometer is coupled to the liquid chromatographic separation unit. Furthermore, this detector can be a very valuable tool in assigning a definitive identity to a chromatographic peak with relatively high sensitivity. The fragmentation patterns can also provide structural information of unknown compounds such as metabolites resulting from surfactant degradation.

The high water-solubility of surfactants and their, often more polar, metabolites prevents direct application of gas chromatographic separation (GC) with appropriate detection. The necessary volatilisation without thermal decomposition can be achieved by derivatisation of the analytes, but these manipulations are time- and manpower-consuming and can be susceptible to discrimination. Additionally, each derivatisation step in environmental analysis is normally target-directed to produce volatile derivatives of the compounds to be determined. Unknown surfactants that are simultaneously present, but differ in structure and therefore cannot react with the derivatisation reagent, are discriminated under these conditions.

The greatest disadvantage of all detector systems such as, e.g. FID, UV, diode array detection (UV-DAD), FL, refractory index (RI), light scattering detector (LSD) or conductivity, applied in combination with GC, LC or CZE, is that they only provide an electric signal at the detector. The retention time alone of standard compounds, if available, is not sufficient for a reliable identification. LC separation of surfactant-containing extracts may often result in non-reproducible retention behaviour and only in those cases where substance-specific detection methods, like mass spectrometry (MS), are applied will this be evident.

As a consequence, for unequivocal identification of the constituents of complex mixtures found in surfactant blends and also in the analyses of surfactants and their metabolites in environmental samples, MS and tandem mass spectrometry (MS–MS) have proved to be more advantageous and are discussed thoroughly in Chapter 2. To optimise the output of reliable results and to save manpower and time certain procedures in sample preconcentration, clean-up and separation prior to MS examinations are inevitable. These are discussed in the present book in more detail in Chapter 3.

1.7 FATE OF SURFACTANTS AND ENVIRONMENTAL PROBLEMS

1.7.1 Introduction

Environmental problems related to detergents can be subdivided into four generations according to Giger et al. [49]. The first was characterised by foaming in natural and wastewaters produced by poorly degradable surfactants based on highly ABS. Avoiding foaming and a strict legislation related to biodegradation of surfactants led ABS to be replaced by LAS, which is at present the main anionic surfactant. The second phase is related to the phosphate component of detergents, which is responsible for eutrophication of natural waters. A ban of these compounds and substitution with other, more environmentally acceptable, products such as nitrilotriacetic acid (NTA), were adopted. A third phase is characterised by the exhaustive detection of toxic metabolites of surfactants in wastewater effluents and in digested sludges [50]. The accumulation of LAS in digested sludges, the worldwide presence of NP, a recalcitrant metabolite from NPEO with recent evidences of its estrogenic character [51], the replacement of the cationic surfactant DTDMAC because of its toxicity to aquatic organisms and its accumulation in sewage sludges by highly biodegradable-related compounds can illustrate the recent concerns about detergent-related environmental problems. The fourth phase according to Giger[1] is characterised by the attempt to avoid these problems by means of preventive action.

Awareness of these environmental problems caused by surfactants, associated with their fate and toxicity, has led to a series of changes and developments in the surfactant market since the introduction of the first generation of synthetic surfactants in the 1950s. Whereas in the beginning, economical and application aspects governed the design of industrially manufactured surfactants, later on their fate after usage also had to be taken into account. Massive amounts of domestically and industrially applied surfactants are discharged after use into the water and hence, in most instances, to treatment plants. As such they make up one of the most relevant organic pollutants of anthropogenic origin with the potential to enter the environment.

The fate of surfactants (Fig. 1.17) in wastewater treatment plants (WWTP) is determined by, among other factors, the composition of the wastewater, the operational type of the WWTP and the physico-chemical properties of the surfactant molecule itself. In general, WWTPs provide a primary treatment, which is simply the removal of solid materials by mechanical means.

Secondary treatments are accomplished either by an activated sludge process or a trickling filter, in which organic components of the wastewater are subjected to aerobic degradation by microbial biocoe-nosis [52]. Surfactants can be metabolised by adapted micro-organisms

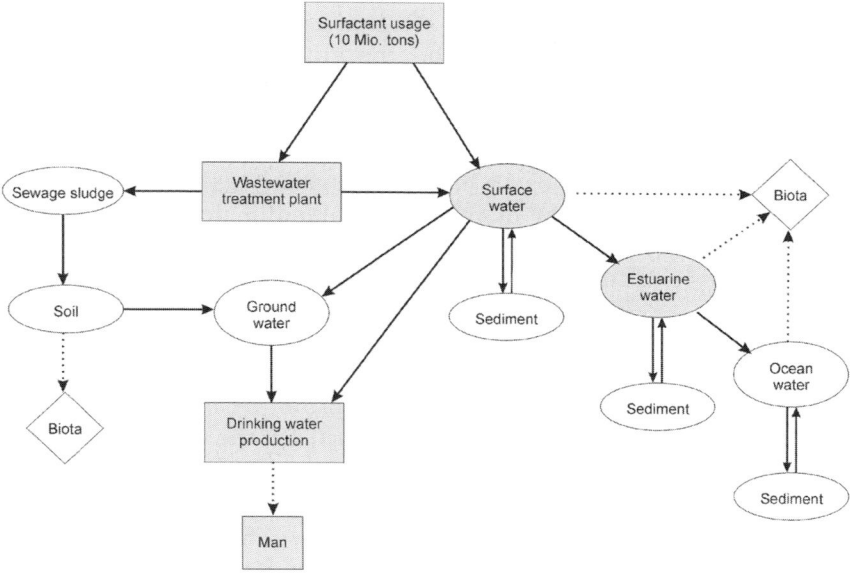

Fig. 1.17. Fate of surfactants in the environment after discharge into sewers.

using them as carbon and/or energy sources. The initial step yielding alterations to the molecular structure is referred to as primary degradation (see Chapter 5). For a series of surfactants this is associated with the loss of their surface-active properties. The metabolism ideally results in the complete mineralisation of a compound (ultimate degradation) yielding carbon dioxide and water and inorganic substances such as sulfate, or complete conversion into bacterial biomass. However, this end point is not completely attained in some instances because of insufficient retention time in the WWTP, a shock load emitted into the WWTP or slow degradation kinetics of the parent compound or its intermediates (see Chapter 6.1). Hence, surfactants together with their water-soluble biochemical degradation products (metabolites) often reach surface water bodies serving as receiving water from treated WWTP effluents (see Chapters 6.2–6.5).

Depending upon the physical properties of a surfactant (component), removal from the mixed liquor is further possible through precipitation of insoluble salts and adsorption onto solids or bacterial flocs, which, in turn, are subsequently withdrawn with the excess sludge [53]. In particular, intact or partly degraded low water-soluble surfactants are eliminated by this route.

The removed excess sludge is often subsequently decomposed by anaerobic digestion to reduce the content of organic compounds, resulting primarily in the formation of carbon dioxide and methane. Among other alternative routes for sludge treatment the most prominent are composting and aerobic digestion. The recovered digested sludge is usually further stabilised by drying, centrifugation or thermal treatment and finally incinerated or disposed of to landfill deposits. The high content of organic material and essential nutrients means that it is also utilised for agricultural purposes as a fertiliser and soil conditioner. Surfactants not degraded during the wastewater treatment operations are further degraded in soils as a consequence of the sludge-amended practices.

The potential for leaching or adsorption by plants has been thoroughly studied for LAS but not so much for the other surfactants (see Chapter 6.5). Thus, it is still a topic of recent studies, since this path represents a potential source for surfactants and their metabolites, which have not been entirely destroyed during sludge processing, to find a way into the terrestrial environment and from there into ground and surface water by leaching or run-off [53,54].

As regards monitoring data of surfactants in European ground waters, nothing is known, even for LAS, but according to a Danish study, LAS does not leach to underground water [55,56]. The effect on plants growing on soils amended with sludge containing LAS was found to be negligible. Mortensen et al. [57] reported accurate data for degradation of LAS in sludge-amended soil under realistic field conditions. LAS was not taken up by plants and its degradation in soil was increased by the presence of crop plants.

Residues of incompletely degraded surfactants can also enter the aquatic environment via WWTP effluents, and these can follow various fates. More hydrophobic species with low water solubilities are prone to bind to suspended particles or to sediments (see Chapter 6.2.1) [58–60] and in very rare cases may cross the water–gas phase boundary to enter the atmosphere [61].

In addition, bioaccumulation in aquatic organisms has been studied, again mainly for LAS [62], the non-ionic surfactants and their degradation products (see Chapter 7.2). A close relation with hydrophobicity can be drawn from studies performed with pure homologues of LAS.

During the wastewater treatment, an oxidative conversion of the surfactant molecules leads predominantly to the formation of polar compounds. They display a particularly high solubility and mobility in the aqueous medium and, therefore, transportation over relatively long distances can occur if they are not further degraded, resulting in the wide dissemination of these pollutants in riverine systems and thus also to estuaries, coastal regions and ultimately the marine environment (see Chapters 6.2 and 6.3). In the latter, the final levels will mainly be influenced by dilution effects and physical removal by precipitation or adsorption [63] because of relatively low microbial activity in this ecosystem compared with fresh water environments [64].

The mobility of very slowly degradable compounds or persistent metabolites present in surface water or bank filtration-enriched ground water is of particular interest for the production of potable water. In common with many other compounds, certain surfactants, and especially their polar metabolites, have the potential to bypass the technical purification units used, which may include flocculation (active charcoal) filtration, ozonation or chlorination, and thus can be found ultimately in drinking water destined for human consumption (see Chapter 6.4).

Due to their unique surface-active properties, surfactants possess the capacity to remobilize non-polar contaminants, such as polycyclic aromatic hydrocarbons or heavy metals, which have accumulated in sediments or digested sludges. This ability can be positively exploited, e.g. in the remediation of contaminated soils [76], but it is undesirable in cases where the surfactants are present in digested sewage sludge used to amend soil.

1.7.2 Toxicity of surfactants and risk assessment

The fate of surfactants under environmental conditions merits special attention not only because of their widespread use, but also the possible exposure of water organisms to these relatively highly toxic compounds. Although the toxicity of intact surfactant molecules is substance-specific, with cationics even performing in some cases as disinfectants, the inherent surface-active properties of surfactants in general contributes to noxious interactions such as disruption of biomembranes and denaturation of proteins [52]. The aquatic toxicity varies strongly among living organisms such as fish, daphnia, algae or bacteria (see Chapter 7.1) [66,67]. Among these species bacteria appear to be the least sensitive species in terms of acute and chronic toxicity. This feature is an indispensable requirement for their degradation capacity in biological sewage treatment processes that receive high loads of surfactant-contaminated wastewater [68]. In instances where primary degradation leads to the loss of surface-activity and the formation of more polar compounds, the toxicity is notably reduced. Thus the toxicities of the biodegradation intermediates of LAS, the SPC, is several orders of magnitude lower than that of the parent molecule [69].

In contrast, an enhanced toxicity of some metabolites relative to the parent compounds has likewise been detected when the intermediates show a higher lipophilicity than the intact substances, such as e.g. NP [70,71].

The above described toxicity issues related to the usage of surfactants will be discussed in the following sections using three different surfactant types, which have been used abundantly in the past. These examples clearly demonstrate the way in which the behaviour and (ultimate) fate can be decisive elements for the success or failure of commercially marketed surfactants.

Branched alkylbenzene sulfonates
The wide application of the first industrial anionic surfactant, (ABS) led to unexpected environmental problems. Strong foam formation in

WWTPs receiving domestic wastewater contaminated with ABS arose, because of their poor biodegradability. By 1960 they were present in sewage plant effluents in the range from 5 to 10 mg L^{-1} (data based upon solvent extraction and colorimetric detection) [52]. Consequently high amounts of ABS reached surface waters, e.g. in rivers from the US values of up to 1 mg L^{-1} were reported [72]. The levels of ABS were strongly related to the flow rates of the river and the amounts of discharged wastewater. No further marked elimination was observed to occur. In Germany and other European countries metre-high foam was observed during the dry summer in 1959 in WWTPs and rivers, reaching such an extent that even inland water transportation was disabled. In some instances, the levels in surface waters exceeded 1 mg L^{-1} [6].

The presence of ABS, however, was not limited to surface waters and they could also be detected in ground water supplies receiving effluents from cesspools or septic tanks. A monitoring study carried out in 1959 in 32 major cities throughout the US demonstrated that ABS in surface water-borne potable water averaged between 15 and 34 $\mu g\,L^{-1}$ [73]. Residues of ABS in rivers likewise posed a threat to municipal drinking water suppliers using surface water as a resource for drinking water production.

Investigation of the pronounced resistance of ABS to microbial degradation demonstrated that the branched alkyl chain derived from tetrapropylene was responsible for the longevity of ABS in the aquatic environment [74]. Shortly after this discovery, the surfactant industry reacted and made attempts—in some countries additionally forced by legal restrictions—to provide an alternative surfactant with comparable functional properties, but with inherent biodegradability. The outcome was the introduction of LAS on the detergent market. After the switch from ABS to LAS in almost all nations in the mid 1960s, a substantial drop in the levels of ABS was observed [8].

The continued use of ABS in some small markets, for example the South East Asia region, despite the advantages offered by LAS in terms of performance characteristics such as foam generation and detergency [75,76], seems to be linked to political situations (oleochemical producer pressure) rather than to any other economical justification.

Ditallow dimethyl ammonium chloride
Another surfactant that gave rise to environmental concern was the cationic DTDMAC. This compound has been used since the 1960s

mainly as a fabric softener. As for ABS, the environmental relevance of DTDMAC is directly associated with its slow biodegradation kinetics, but unlike ABS, which generated the above described foam phenomenon, the drawbacks in DTDMAC usage were less apparent.

After discharge into the sewage water, the extremely low solubility surfactant can be removed during the treatment process by forming complexes with dissolved and suspended organic materials, particularly with anionic surfactants and humic substances. Depending on the operational type of the treatment plant, removal rates of approximately 30% for mechanical treatment only, 70% for trickling filter treatment and 95% for activated sludge treatment have been reported [77,78]. However, only a minor portion of this removal is due to biodegradation, and instead DTDMAC is adsorbed onto the sewage sludge and withdrawn together with the excess sludge from the mixed liquor. In subsequent treatment, including anaerobic digestion, no marked reduction via biotic processes occurs and DTDMAC hence accumulated in treated sludge. In two studies reporting on the enrichment of DTDMAC in digested sludge, the levels in dried sludge ranged from 3.8 to $5.1 \, \text{g kg}^{-1}$ and 8.9 to $9.2 \, \text{g kg}^{-1}$, respectively [78,79]. Residual concentrations of the cationic surfactant likewise ended up in river waters receiving effluents from WWTPs. Depending on the dilution factor at the specific site, mean levels of DTDMAC in surface waters in the proximity of WWTP effluents ranged from 17 to $51 \, \text{mg L}^{-1}$ [80]. The subsequent fate of the compound in the aquatic environment was determined by its tendency to adsorb onto surfaces such as suspended particles and sediments. Analysis of the cationic surfactant in river sediments yielded values of $<3-67 \, \text{mg kg}^{-1}$ [81] and $6-69 \, \text{mg kg}^{-1}$ [23]. The occurrence of DTDMAC was also described in coastal sediments sampled near to the outlet of a WWTP [82].

Taking into consideration its physico-chemical properties, removal efficiencies, low biodegradability, predicted environmental levels, toxicity, and the need to provide sufficient safety margins for aquatic organisms, the demand for alternative cationic surfactants arose. Since 1991, DTDMAC has been replaced in some European countries due to producer's voluntary initiatives with new quaternary ammonium compounds, the esterquats. These contain an ester function in the hydrophobic chain (Table 1.3) that can be easily cleaved, releasing intermediates that are susceptible to ultimate degradation [24–26]. The effects of the phasing-out and replacement of DTDMAC can be demonstrated by the results of a Swiss study, where the surfactant industry voluntarily replaced DTDMAC in 1991. The levels in anaerobically stabilised sewage sludge dropped approximately 94% over the period 1991–1994 (from 3.67 to 0.21 g kg^{-1}) concomitant with a decrease in consumption of 1140 t in 1989 to 100 t in 1994 [83].

Nonylphenol ethoxylates
A third surfactant class meriting special attention with respect to environmental issues is the non-ionic alkylphenol ethoxylates, especially the most prevalently used NPEO. Contrary to ABS and DTDMAC where the poor biodegradability of the surfactant molecule itself was the mitigating factor, in this instance the formation of undesirable metabolites was the reason for concern. It was discovered in 1984 that NPEO yielded breakdown products that were more toxic to aquatic organisms than the intact parent compounds [70,71]. The biodegradation of NPEO in WWTPs led to the build-up of NP, which turned out to be very resistant to further destruction. Due to the low water solubility and high lipophilicity of NP [84,85], it has a strong tendency to adsorb onto sewage sludge. During anaerobic digestion of the removed excess sludge no substantial breakdown of NP is observed, instead the levels can even be increased during this process due to further formation of the phenolic compound from short-chain NPEO, which have also been withdrawn with the sludge. It is estimated that 92–96% of the NP formed is discharged from the WWTP via digested sludge and only 4–8% via secondary effluents [86]. Residual concentrations of NP in Dutch treatment facility effluents were reported to reach values of up to 1.2 μg L^{-1} [87], while in a German survey the levels ranged from 0.3 to 1.0 μg L^{-1} [88].

After release into natural waters, NP and its precursors might be fixed in riverine sediments or exert adverse effects on aquatic organisms. Apart from the fact that these derivatives are more toxic than the initial NPEO, it is also claimed that some cause estrogenic effects (see Chapter 7.3) [89] and demonstrate a considerable potential for bioaccumulation [90].

In response to the problems associated with the use of NPEO, measures were taken in a series of countries to reduce the application volume of these non-ionic surfactants. In 1986, Germany initiated voluntary restrictions and in Switzerland the use of NPEO in laundry detergents was completely banned. Throughout northern Europe, a voluntary application ban in household cleaning agents began in 1995.

However, NPEO is still being used in substantial amounts in industrial applications and in institutional cleaners, where their replacement by other surfactants, such as e.g. AEs, is hampered due to the excellent properties and competitive production costs offered by NPEO. In contrast, no regulatory action has been taken in the US, despite the range of environmental problems presented by these compounds.

The assessment of whether a substance presents a risk to the receiving environmental compartment is based on a comparison of the measured or predicted environmental concentration (PEC) of the chemical of concern with the predicted no effect concentration (PNEC) to organisms in the ecosystem. This is briefly discussed in Chapter 7.4. Studies carried out so far, e.g. by Vandepitte and Feijtel [91], show that the risk of anionic surfactants such as LAS, AE and AES for the aquatic environment is low, since the PECs are always lower than the maximum permissible concentrations.

1.8 Conclusion

Surfactants are surface-active compounds, which are used in industrial processes as well as in trade and household products. They have one of the highest production rates of all organic chemicals. Commercial mixtures of surfactants consist of several tens to hundreds of homologues, oligomers and isomers of anionic, non-ionic, cationic and amphoteric compounds. Therefore, their identification and quantification in the environment is complicated and cumbersome. Detection, identification and quantification of these compounds in aqueous solutions, even in the form of matrix-free standards, still poses the analyst considerable problems.

The present book contains chapters written by researchers in the field of analytical environmental chemistry, toxicology and industry who are interested in improving knowledge of the fate and toxicity data of surfactants and metabolites formed.

Analytical methodologies, which allow the determination of various surfactants and identification of new metabolites are discussed, especially in Chapter 2. The use of multistep sample preparation methods together with advanced techniques like LC–MS (including various types of interfacing systems) for the final identification of surfactants is described in Chapters 3 and 4. Chapter 5 covers the degradation pathway of the major surfactants, which has been studied in different aquatic matrices and in model studies in order to determine their possible impact on the health of the ecosystem. The levels of relevant surfactants as well as their persistent degradation products have been monitored worldwide in wastewater, freshwater, marine water, drinking water, sediments, soil and biota. These data are presented in Chapter 6. A correlation of the analytical data of the main surfactants has been performed in Chapter 7 with their toxicity, e.g. endocrine disruptor effects. In this respect there was especially a lack of metabolite information, their toxicity and their levels in the aquatic environment.

The information summarised in this book will provide monitoring and risk/hazard evaluation on surfactants for the aquatic environment.

REFERENCES

1 L. Chalmers, *Domestic and Industrial Chemical Specialities*, Leonard Hill, London, 1966.
2 U.S. Patent 601,603.
3 A. Davidsohn and B.M. Milwidsky, *Synthetic Detergents*, 6th edn., George Godwin Ltd, London, 1978.
4 E. Jungermann, *J. Am. Oil Chem. Soc.*, 56 (1979) 827A.
5 H.L. Webster and J. Halliday, *Analyst*, 84 (1959) 552.
6 W.K. Fischer and K. Winkler, *Vom Wasser*, 47 (1976) 81.
7 Anonymous. German Detergent Act, Verordnung über die Abbaubarkeit von Detergentien in Wasch- und Reinigungsmitteln. BGBl I, 12.12.1962, pp. 698–701.
8 W.T. Sullivan and R.D. Swisher, *Environ. Sci. Technol.*, 2 (1968) 194.
9 H.G. Hauthal, *Tenside Surf. Det.*, 37 (2000) 320.
10 CESIO (Council of European Surfactant Producers) Statistics 1999; internet page: http://www.cefic.be/sector/cesio/cesio.
11 Petresa, (Petroquímica Española, S.A.), Cadiz, Spain internal report, 2000.
12 P. Feng and G. Hua, *Chem. Ind.*, 4 (1999) 302.
13 H. Brüschweiler, F. Schwager and H. Gämperle, *Seifen, Öle, Fette, Wachse*, 114 (1988) 301.
14 Environmental fate and behavior of LAS, literature review, Bongerts, Kuyer and Huiswaard, Consulting Engineers, Delft, The Netherlands, 1993.
15 H.A. Painter, In: N.T. de Oude (Ed.), *The Handbook of Environmental Chemistry*, vol. 3, Springer Verlag, Heidelberg, Germany, 1992, part F.

16 Nederlandse Vereniging van Zeepfabrikanten NVZ, Environmental data review of alkylether sulphates (AES), Bongerts, Kuyer and Huiswaard, consulting engineers, Delft, The Netherlands, 1994.

17 J. Falbe, *Surfactants in Consumer Products*, Springer Verlag, Berlin, 1987.

18 C.-D. Hager, *Tenside Surf. Det.*, 36 (1999) 334.

19 T.H. Kritchevsky, *Soap Chem. Spec.*, 34 (1958) 48.

20 D. Balzer, *Tenside Surf. Det.*, 28 (1991) 419.

21 F.A. Hughes and B.W. Lew, *Am. Oil Chem. Soc.*, 47 (1970) 162.

22 K. Hill, DECHEMA Kolloquium, March, 18th 1999. Frankfurt, Germany.

23 R.S. Boethling, In: J. Cross and E.J. Singer (Eds.), *Cationic Surfactants*, vol. 53, Marcel Dekker, Inc, New York, NY, 1994, p. 95.

24 S.T. Giolando, R.A. Rapaport, R.J. Larson, T.W. Federle, M. Stalmans and P. Masscheleyn, *Chemosphere*, 30 (1995) 1067.

25 R. Puchta and P. Krings, *Tenside Surf. Det.*, 30 (1996) 186.

26 J. Waters, H.H. Kleiser, M.J. How, M.D. Barrett, R.R. Birch, R.J. Fletcher, S. Haigh, S.G. Hales, S.J. Marshall and T.C. Pestell, *Tenside Surf. Det.*, 25 (1991) 460.

27 F. Balaguer, F. Comelles, J.M. Puig, M.M. Recasens, J. Coll, J. Sánchez and C. Pelejero, *NCP documenta*, 232 (1998) 5.

28 European Oleochemicals Allied Products Group, http://www.apag.org/oleo.

29 B.D. Dobson, *Presented at the 5th World Surfactant Congress—CESIO 2000, Firenze, Italy*, 2000.

30 K. Kosswig, In: E. Bartholomé, E. Biekert, H. Hellmann, H. Ley, W.M. Weigert and E. Weise (Eds.), *Ullmanns Enzyklopädie der technischen Chemie*, Verlag Chemie, Weinheim, Germany, 1982, pp. 455–515.

31 A. Moreno, J. Bravo and J.L. Berna, *J. Am. Oil Chem. Soc.*, 65 (1988) 1000.

32 J.C. Drozd and W. Gorman, *J. Am. Oil Chem. Soc.*, 65 (1988) 398.

33 R.A. Reck, *J. Am. Oil Chem. Soc.*, 62 (1985) 355.

34 R. Tsushima, *Proceedings of the 4th World Surfactants Congress* 43, Barcelona, vol. I., 1995.

35 H. Kelkenberg, *Tenside Surf. Det.*, 1 (1988) 8.

36 T.C. Kendrick, B. Parbhoo and J.W. White, In: S. Patai and Z. Rappoport (Eds.), *The Silicon–Heteroatom Bond*, John Wiley and Sons Ltd, 1991, Ch. 3 pp. 67–150, .

37 L.H. Sommer, *Stereochemistry, Mechanism and Silicon*, McGraw-Hill Inc, 1965.

38 M.G. Voronkov, V.P. Milesshkevich and Y.A. Yuzhelevskii, *The Siloxane Bond: Physical Properties and Chemical Transformations*, Consultants Bureau, New York, 1978.

39 G.E. LeGrow and L.J. Petroff, In: R.M. Hill (Ed.), *Silicone Surfactants*, Surfactant Science Series, vol. 86., Marcel Dekker Inc, New York, 1999, Ch. 2, pp. 59–64, .

40 R. Müller, D.R. (Germany) Patent, 1942, 5348.

41 E.G. Rochow, U.S. Patent, 1941, 2380995.

42 J.L. Speier, J.A. Webster and G.H. Barnes, *J. Am. Chem. Soc.*, 79 (1957) 974.

43 R.M. Hill, In: R.M. Hill (Ed.), *Silicone Surfactants*, Surfactant Science Series, vol. 86, Marcel Dekker Inc, New York, 1999, Ch. 1, pp. 1–47, .

44 J.E. Hunter and J.F. Fowler Jr., *J. Surf. Det.*, 1 (1998) 235.

45 H.-Q. Li and H.Fr. Schroeder, *Water Sci. Technol.*, 42 (2000) 391.

46 T.M. Schmitt, *Analysis of Surfactants*, Marcel Dekker, New York, 1992.

47 A. DiCorcia, R. Samperi and A. Marcomini, *Environ. Sci. Technol.*, 28 (1994) 850.

48 H.Fr. Schroeder, In: D. Barceló (Ed.), *Techniques and Instrumentation in Analytical Chemistry; Sample Handling and Trace Analysis of Pollutants–Techniques, Applications and Quality Assurance*, vol. 21, Elsevier, Amsterdam, 2000, p. 828.

49 W. Giger, A.C. Alder, P. Fernandez and E. Molnar, *EAWAG News*, 10 (1994) 24.

50 W. Giger, *EAWAG News*, 28F (1989) 8.

51 S. Jobling, D. Sheahan, J.A. Osborne, P. Matthiiesen and P. Sumpter, *Environ. Toxicol. Chem.*, 15 (1996) 194.

52 R.D. Swisher, *Surfactant Biodegradation*, Marcel Dekker Inc, New York, 1987.

53 C.J. Krueger, L.B. Barber, D.W. Metge and J.A. Field, *Environ. Sci. Technol.*, 32 (1998) 1134.

54 E.M. Thurman, L.B. Barber and D. LeBlanc, *J. Contam. Hydrol.*, 1 (1986) 143.

55 VKI report 10566, 1999-06-25/TKH. s Hörsholm, Denmark.

56 K. Figge and P. Schöberl, *Tenside Surf. Det.*, 26 (1989) 122.

57 G.K. Mortensen, H. Elsgaard, P. Ambus, E.S. Jensen and C. Groen, *J. Environ. Quality*, 30 (2001) 1266.

58 D.M. John, W.A. House and G.F. White, *Environ. Toxicol. Chem.*, 19 (2000) 293.

59 J.C. Westall, H. Chen, W. Zhang and B.J. Brownawell, *Environ. Sci. Technol.*, 33 (1999) 3310.

60 B.J. Brownawell, H. Chen, W. Zhang and J.C. Westall, *Environ. Sci. Technol.*, 31 (1997) 1735.

61 J. Dachs, D.A. Van Ry and S.J. Eisenreich, *Environ. Sci. Technol.*, 33 (1999) 2676.

62 J. Tolls, M. Haller, W. Seinen and D.T.H.M. Sijm, *Environ. Sci. Technol.*, 34 (2000) 304.

63 E. González-Mazo, J.M. Forja and A. Gómez-Parra, *Environ. Sci. Technol.*, 32 (1998) 1636.

64 R.J. Shimp, *Tenside Surf. Det.*, 26 (1989) 390340.

65 B.N. Aronstein and M. Alexander, *Environ. Toxicol. Chem.*, 11 (1992) 1227.

66 W. Guhl and P. Gode, *Tenside Surf. Det.*, 26 (1989) 282.

67 N. Scholz, *Tenside Surf. Det.*, 34 (1997) 229.

68 C. Verge and A. Moreno, *Tenside Surf. Det.*, 33 (1996) 323.

69 R.A. Kimerle and R.D. Swisher, *Water Res.*, 11 (1977) 31.

70 W. Giger, P.H. Brunner and C. Schaffner, *Science*, 225 (1984) 623.

71 M. Ahel, W. Giger and C. Schaffner, *Water Res.*, 28 (1994) 1143.

72 J.D. Fairing and F.R. Short, *Anal. Chem.*, 28 (1956) 1826.

73 R.C. Jente, *J. Am. Water Works Assoc.*, 53 (1961) 297.

74 H.J. Heinz and W.K. Fischer, *Fette Seifen und Anstrichmittel*, 64 (1962) 270.

75 K.L. Matheson, *J. Am. Oil Chem. Soc.*, 62 (1985) 1269.

76 D. Wharry, K.L. Matheson and J.L. Berna, *Chem. Age India*, 38 (1987) 87.

77 B.W. Topping and J. Waters, *Tenside Surf. Det.*, 19 (1982) 164.

78 E. Matthijs, P. Gerike, H. Klotz, J.G.A. Kooijman, H.G. Karber and J. Waters, Removal and Mass Balance of the Cationic Fabric Softener Di(Hydrogenated)Tallow Dimethyl Ammonium Chloride in Activated Sludge Sewage Treatment Plants; European Association of Surfactant Manufacturers (AIS/CESIO), Brussels, Belgium, 1992.

79 H. Hellmann, *Z. Wasser-Abwasser-Forsch.*, 22 (1989) 131.

80 D.J. Versteeg, T.C.J. Feijtel, C.E. Cowan, T.E. Ward and R.A. Rapaport, *Chemosphere*, 24 (1992) 641.

81 M.A. Lewis and V.T. Wee, *Environ. Toxicol. Chem.*, 2 (1983) 105.

82 P. Fernández, M. Valls, J.M. Bayona and J. Albaigés, *Environ. Sci. Technol.*, 25 (1991) 547.

83 P. Fernández, A.C. Alder, M.J.-F. Suter and W. Giger, *Anal. Chem.*, 68 (1996) 921.

84 M. Ahel and W. Giger, *Chemosphere*, 26 (1993) 1461.

85 M. Ahel and W. Giger, *Chemosphere*, 26 (1993) 1471.

86 M. Ahel, W. Giger and M. Koch, *Water Res.*, 28 (1994) 1131.

87 A.C. Belfroid, T. Murk, P. de Voogt, A. Schäfer, G.B.J. Rijs and D. Vethaak, Hormoonontregelaars in water. Oriënterende studie naar de aanwezigheid van oestrogeen-actieve stoffen in watersystemen en afvalwater in Nederland. RIZA/RIKZ-rapport 99.007/99.024, Lelystad/Den Haag, The Netherlands, 1999.

88 P. Spengler, W. Körner and J.W. Metzger, *Vom Wasser*, 93 (1999) 141.
89 A.M. Soto, H. Justicia and J.W. Wray, *Environ. Health Persp.*, 92 (1991) 167.
90 R. Ekelund, A. Bergman, A. Granmo and M. Berggren, *Environ. Pollut.*, 64 (1990) 107.
91 V. Vandepitte and T.C.J. Feijtel, *Tenside Surf. Det.*, 37 (2000) 35.

Chapter 2

Separation and detection

2.1   GC AND GC–MS DETERMINATION OF SURFACTANTS

Francesc Ventura and Pim de Voogt

2.1.1   Introduction

Gas chromatography (GC) has developed into the most powerful and versatile analytical separation method for organic compounds nowadays. A large number of applications for the analysis of surfactants have emerged since the early 1960s when the first GC papers on separation of non-ionics were published. The only major drawback for application of GC to surfactants is their lack of volatility. This can be easily overcome by chemical modification (derivatisation), examples of which will be discussed extensively in the following paragraphs. This chapter focuses on surfactant types, and in addition discusses some structural aspects of alkylphenol ethoxylates (APEOs) that are important for, as well as illustrative of, aspects of separation and identification that are linked to the complexity of the mixtures of surfactants that are involved.

2.1.2   Anionic surfactants

Linear alkylbenzene sulfonates (LASs)
GC and GC–mass spectrometry (GC–MS) have been used for the determination of LASs in the environment, and for those studies involving information about the distribution of alkyl homologues and

T.P. Knepper, D. Barceló and P. de Voogt (Eds)

phenyl positional isomers, which are almost completely resolved. Thus, capillary GC of derivatised (or desulfonated) LAS allows separation according to alkyl chain length, with the highest chain lengths giving longer retention times and with internal isomers of each homologue eluting first. The higher separation efficiency of GC over high pressure liquid chromatography (HPLC) has been particularly useful for studying the mechanism of biodegradation of homologues and isomers in laboratory-scale experiments. However, the major drawback of the analysis of anionic surfactants by GC or GC–MS is that, as non-volatile compounds, they cannot be analysed directly and pretreatment is required to form volatile derivatives prior to GC or GC–MS analysis.

Figure 2.1.1 shows the most common derivatisation methods for anionic surfactants reported in the literature [1]. The first method of LAS determination by GC consisted of a microdesulfonation procedure in which LASs were desulfonated in boiling phosphoric acid at high temperature [2–4] and the corresponding alkylbenzenes analysed. The microdesulfonation method was further improved by introducing additional concentration and clean-up steps [5–11], which allowed the determination of LAS in influent, effluent and river water samples at low $\mu g\ L^{-1}$ levels [7,8] and sediment and sludge samples [8] at $\mu g\ g^{-1}$. In addition to the desulfonation procedure, several derivatisation techniques have been used to make LAS analysis amenable to GC.

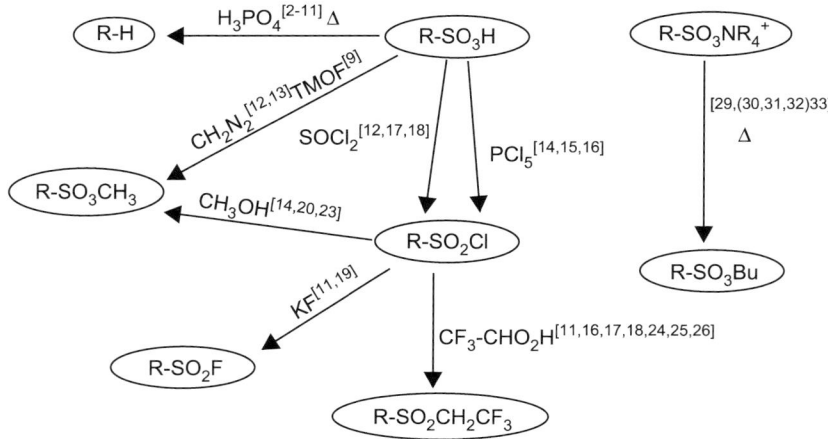

Fig. 2.1.1. Derivatisation scheme for GC and GC–MS analysis of LAS. Adapted from Reemtsma [1]. Reproduced with permission. © 1996 by Elsevier.

Thus, esterification with diazomethane [12] or trimethylorthoformate [9] to render methylsulfonate esters of LAS and further GC–flame ionisation detection (FID) analysis has been employed. The procedure utilising diazomethane has been applied for the determination of LAS in industrial wastewater [13]. Sulfonyl chloride derivatives of LASs have been extensively used in the determination of LAS. They are formed by reaction from free sulfonic acid with phosphorous pentachloride [14–16] or thionyl chloride [12,17,18]. Sulfonyl chlorides can be further converted into their corresponding methylsulfonate esters by reaction with methanol [14], although no advantage over the GC analysis of sulfonyl chlorides was observed [15], or into the most volatile sulfonyl fluoride compounds by reacting them with an aqueous potassium fluoride solution [11,19]. The mass spectra of LAS methyl esters under electron impact (EI) conditions show the m/z 185 as base peak and fragmentation patterns comparable with those of alkylbenzenes [14]. The procedure developed by Hon-Nami and Tanya [14] with a detection limit of about 3 μg L^{-1} has been applied for the determination of LAS in river water and sediment [20], estuarine water [21], lagoon water [22] and industrial wastewater [23]. Sulfonyl chlorides can also be transformed into trifluoroethyl derivatives by esterification with trifluoroethanol and further analysis by GC–MS under negative [16, 17,24] or positive [18] chemical ionisation (CI) or EI [11,25,26] modes. Figure 2.1.2 shows an example of the application of GC–MS in the positive CI mode for the determination of LAS in sediments [18].

This method is also suitable for simultaneously studying the presence of related compounds of LAS, such as impurities from the manufacturing process, for example dialkyltetralin sulfonates (DATSs). The two ions usually monitored in the GC–MS negative (N)CI mode are the parent ion and the [M-100]$^-$ ion, which corresponds to the loss of trifluoroethanol, whereas the m/z 163 ion is characteristic of all the derivatised sulfonates [16]. LAS and DATS can also be monitored in the EI mode. Major fragments for LASs are m/z 267 and 253; the former is the major fragment ion for 2-(sulfophenyl)alkanes, whereas for internal isomers the major fragment ion is m/z 253. The ions monitored for DATS are m/z 307, a significant fragment when one of the alkyl side chains is a methyl group, and 279.

Other biodegradation intermediates, such as sulfophenyl carboxylates (SPCs) and alkyltetralin carboxylates, can also be analysed by this method. In the latter case, the carboxylates are converted into their acid chlorides and, in a second step, into their trifluoroethyl derivatives as

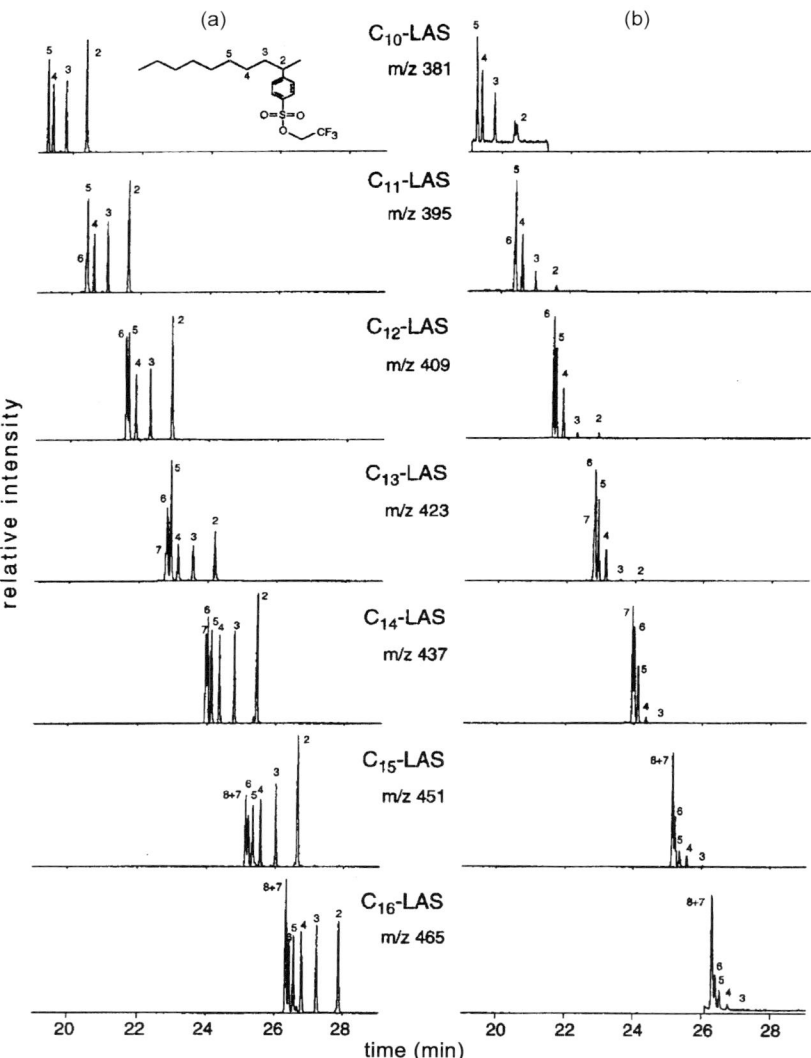

Fig. 2.1.2. Mass chromatograms in the SIM mode using protonated molecular ions of LAS trifluoroethyl esters. (a) LAS standard mixture and (b) LAS found in a sediment. Peak labels indicate the position of the benzene sulfonate moiety on the linear alkyl chain. Reproduced with permission from Ref. [18]. © 1997 by American Chemical Society.

described above. The method has been successfully applied to study all these compounds in river water and sediment achieving a detection limit of 1 $\mu g \, L^{-1}$ (for a 50 mL sample) [16] and LAS and SPC in sewage effluents and river water with an estimated detection limit of 0.01 $\mu g \, L^{-1}$ (for a 100 mL sample) [17,27]. A gas chromatographic tandem mass spectrometric method to differentiate LAS and branched alkylbenzene sulfonates (ABSs), which can be applied to study their presence in recent sediments [28], has been reported. Another derivatisation process [29] that is less time-consuming and avoids the use of hazardous reagents (i.e. diazomethane) consists of the extraction of LAS as its tetraalkyl ammonium ion pair (i.e. tetrabutyl ammonium $RSO_3^- N(Bu)_4^+$) and their conversion into butyl esters [RSO_3Bu] in the hot temperature GC injection port. The method has been successfully applied for the determination of LAS from sewage sludges and effluents [30,31], river water [32] with an estimated limit of detection of 0.1 $\mu g \, L^{-1}$ for 200 mL samples, and it has also been suitable for their determination in sediments [33].

Secondary alkane sulfonates (SASs), alkyl sulfates (ASs) and alkylether sulfates (AESs)

SASs, which are manufactured especially in Europe, usually contain homologues in the $C_{13}-C_{17}$ range and each homologue comprises several isomers defined by the carbon atom to which the sulfonate group is attached. The analysis of SAS by GC as described above for LAS requires derivatisation prior to GC analysis. Commercial blends of SAS have been analysed as their methyl esters by derivatisations with diazomethane [34] or trimethyl orthoformate [9]. Very few papers deal with the determination of SAS in environmental samples. Field et al. determined SAS by GC–MS of their butylesters by ion pair derivatisation as described above, in sewage sludges [30] and wastewater [35], and studied their fate during municipal wastewater treatment [36] with an estimated detection limit of 1 $\mu g \, L^{-1}$. Mass spectra of butylesters of all SAS components under EI conditions give an intense homologue specific ion of [M-138]$^+$, corresponding to the loss of sulfonate butyl ester [30]. Figure 2.1.3 shows, as an example, the selected ion chromatograms of m/z 196, 210, 224 and 238 that correspond to the [M-138]$^+$ ions for the mixture $C_{14}-C_{17}$ SAS in the primary and secondary effluents collected from a Swiss WWTP [35]. Commercial ASs consist of a long alkyl chain group ($C_{12}-C_{18}$) and a terminal sulfate

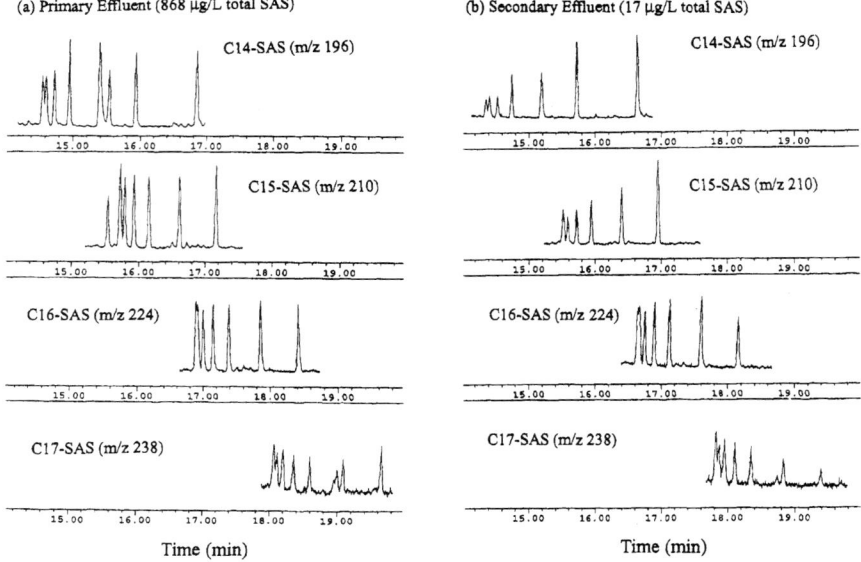

(a) Primary Effluent (868 µg/L total SAS)

C14-SAS (m/z 196)

C15-SAS (m/z 210)

C16-SAS (m/z 224)

C17-SAS (m/z 238)

Time (min)

(b) Secondary Effluent (17 µg/L total SAS)

C14-SAS (m/z 196)

C15-SAS (m/z 210)

C16-SAS (m/z 224)

C17-SAS (m/z 238)

Time (min)

Fig. 2.1.3. Selected ion chromatograms for C_{14}–C_{17}-SAS homologues in (a) primary effluent and (b) secondary effluent collected from a sewage treatment plant. Reproduced with permission from Ref. [35]. © 1994 by American Chemical Society.

group, whereas the AESs with the same structure contain a variable ethoxylate (EO) chain (E_1–E_{12}). Both types of products have been determined in product formulations [37] and water samples [38] by cleavage of the sulfate group followed by GC of the alkyl halides formed by reaction with HI [37] or HBr [38]. The method was suitable for determining AS in water samples at the low µg L^{-1} level [38] but was unsuitable for resolving AS from AES. To overcome this problem, Fendinger et al. [39] described a specific GC–FID method to determine AS in natural waters that involves conversion of the AS into their trimethylsilyl ethers by reaction with dimethyl formamide and BSTFA [N,O-bis(trimethylsilyl) trifluoroacetimide] achieving a detection limit of 5 µg L^{-1}. The determination of alkyl chain distribution and ethoxylation degree of AESs in commercial products and dishwashing liquids [40] has proved to be suitable by hydrolysis, further derivatisation with BSTFA/pyridine to get the trimethylsilyl derivatives and followed by high temperature GC (HTGC) for separation and effective carbon number theory for quantification. AESs have also been analysed by GC after chemical decomposition by using the mixed anhydrides of acetic and *p*-toluenesulfonic acids to give alkyl acetates and the ethoxylated units are converted into ethylene glycol diacetate [41]. Reviews covering chromatographic methods for the analysis of anionic surfactants in commercial products and environmental samples have appeared recently [42,43].

2.1.3 Non-ionic surfactants

Early attempts to apply GC for the detection of non-ionic surfactants date back to the 1960s with reports describing the separation of alkylethoxylates with up to two EO units [44], and APEO with up to seven EO units [45]. The retention behaviour of the non-ionics on apolar columns is primarily dependent on the number of EO units. Further, within a peak cluster corresponding to a certain number of EO units, separation of alkyl isomers may be achieved if the column's plate number is sufficiently high (typically for capillary columns > 25 m). Due to the decreasing volatility of the oligomers with increasing number of EO units, response factors also decrease, making quantitative analysis impossible for the higher molecular weights (e.g. above $6-7$ EO units in APEO [46], see Fig. 2.1.4). In his extensive review of methods of analysis, Cross has reviewed the early literature up to 1987 [47].

Alcohol ethoxylates (AEs)
The analysis of underivatised AEs by GC is restricted due to the low volatility of the high molecular weight oligomers of AE. The complexity of determining the oligomer distribution must be also taken into account because of the different FID response between the lower and higher oligomers. Analysis of underivatised AEs using capillary GC columns has been performed by several authors [48–51] although the short lifetime of GC columns when high temperature is used has limited its application. It has been shown that split injection gives higher results for low ethoxylated oligomers while the on-column injection mode is more suitable for the higher homologues [49]. Different derivatisation reagents have been employed to increase the volatility and additionally to improve the separation of AE oligomers. The most common ones, trimethylsilyl ethers [34,40,52–56] (see Fig. 2.1.5), allow the analysis of up to 20 EO units. An estimation of the FID response for the different oligomers of derivatised AE has also been calculated [56]

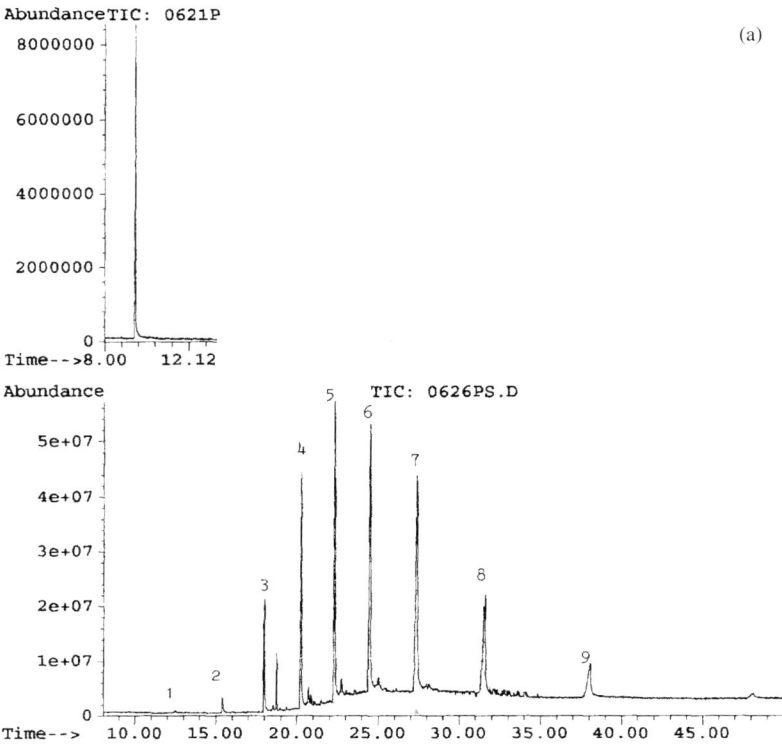

Fig. 2.1.4. (a) Total ion chromatograms obtained with HRGC–LRMS of *t*-octylphenol and a mixture of *t*-octylphenol ethoxylates: top, *t*-OP; bottom, OPEO mixture with an average of 8–9 ethoxylate units. Number of ethoxylate units indicated above each peak. For conditions, see (c). Note that oligomers with number of ethoxylate units >6 in bottom chromatogram tend to 'disappear' (see text). Compare with (b) and (c) to see absence of peak clusters. (b) Total ion chromatograms obtained with HRGC–LRMS of mixtures of nonylphenol and nonylphenol ethoxylates: top, NP mixture; bottom, NPEO mixture with an average of two ethoxylate units. For conditions see (c). Note the clusters of peaks (due to differently branched isomers) present at each oligomer (number of ethoxylate units indicated above peak cluster), and compare with (a). (c) Total ion chromatograms obtained with HRGC–LRMS of mixtures of nonylphenol ethoxylates: top, NPEO mixture with an average of four ethoxylate units; bottom, NPEO mixture with an average of 10 ethoxylate units. Column DB-5, 30 m, temperature program: 100°C, 10°C/min to 300°C. Injection 2 μL. Detection: EI-quadrupole MS, 70 eV. Note that oligomers with number of ethoxylate units >6 in bottom chromatogram tend to 'disappear' (see text). Modified from Ref. [46].

Reproduced with permission. © 1997 by Elsevier.

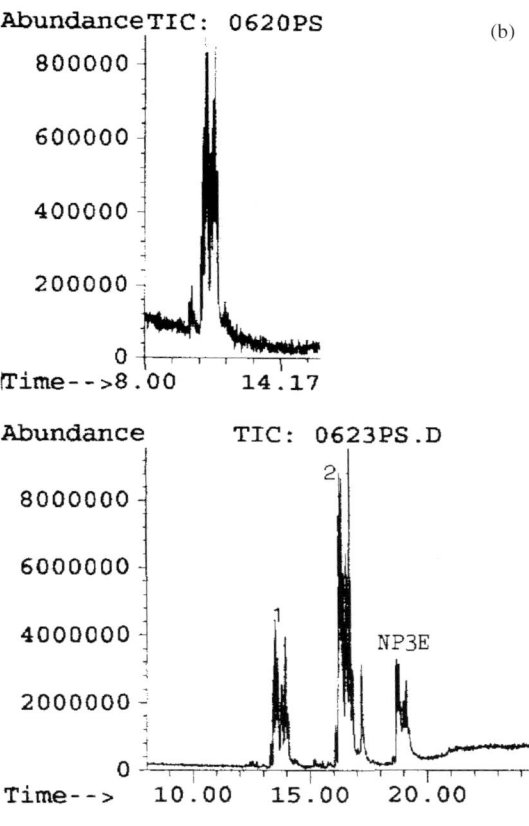

Fig. 2.1.4. (*continued*)

using the effective carbon number concept. Acetate [57] and trifluoroacetate [55] derivatives of AEs formed by reaction with the corresponding anhydride have also been described, but care should be taken since cleavage between the fatty alcohol and the ethoxylated chain can arise [58]. Another approach, which is more suitable for environmental samples because it simplifies the chromatographic profiles, consists of the cleavage of the AE molecule by HBr [59–61] leading to the formation of alkyl bromides derived from the fatty alcohols and dibromoethane. The alkyl bromide derivatives are more suitable for GC and do not discriminate against high molecular weight components, but a previous cleanup is necessary in environmental samples [62] to

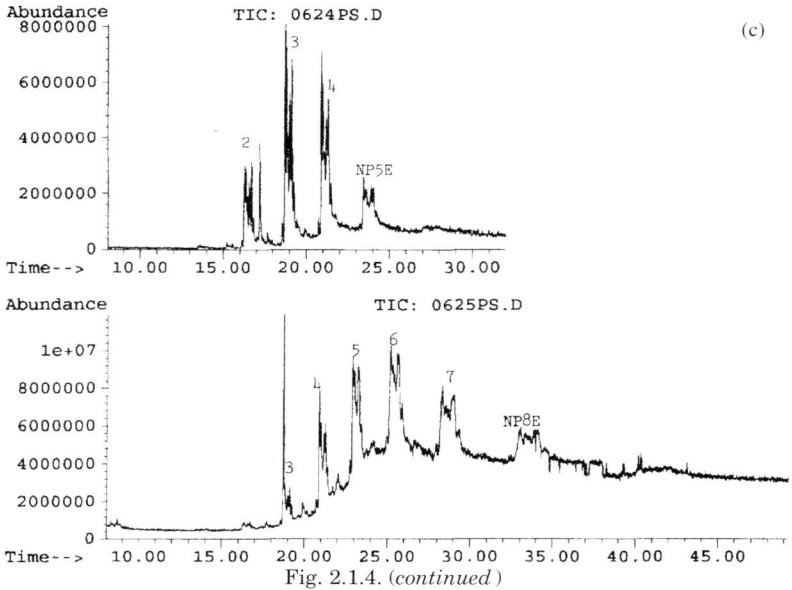

(c)

remove interferences that can be generated from the reaction of HBr with glycols, alcohols, etc. The method has been successfully applied to wastewater [59,62] and river and wastewater samples [63]. In the latter case, the authors analysed the derivatives by GC–MS in the single ion

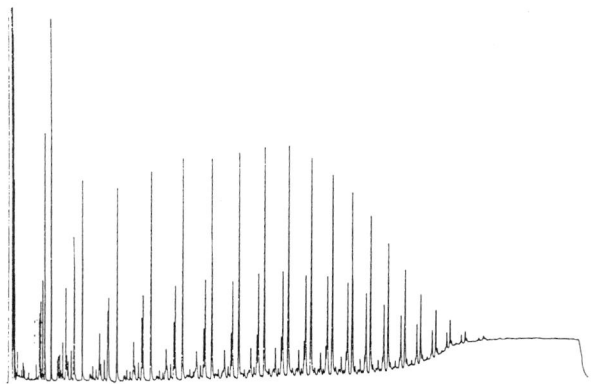

Fig. 2.1.5. High temperature GC analysis of silylated tridecanol/pentadecanol polyethoxylates. Reproduced with permission from Ref. [34]. © 1990 by Dr Alfred Huethig Publishers.

monitoring (SIM) mode. The ions, monitored for all chains were m/z 135 and 137 because the EI spectra for alkyl bromides are similar. The former was used for quantification purposes and the 137 ion was used as a qualifier ion and should be present at approximately the same abundance as the 135 ion ($^{79}Br/^{81}Br$ 1:1).

The chromatographic determination of non-ionic surfactants of the AE type in the aquatic environment has been described in recent reviews [43,64,65].

Alkylphenol ethoxylates (APEOs)

The same general considerations about the low volatility and the need for derivatisation of APEOs as described above for AEs are valid. Their low volatility is caused by the high molecular weight, the intrinsic polarity due to large numbers of EO units, and the hydroxyl moiety that has hydrogen-bonding capacity with stationary phase polar groups. The analysis by packed and capillary GC of both underivatised and trimethylsilyl and other derivatives of APEO commercial products has been reviewed [43]. The widespread use of APEOs, particularly nonylphenol ethoxylates (NPEO), in many different applications has resulted in their presence in different environmental compartments. Recent concerns regarding the toxicity to aquatic organisms [66] and estrogenic properties [67] of their degradation products, especially nonylphenol (NP) remaining after deethoxylation and the lower NP ethoxylates ($NPEO_1$ and $NPEO_2$) formed under anaerobic conditions, and to nonylphenol carboxylates (NPEC and NPE_2C) produced under aerobic conditions [68,69] have received much attention. The same degradation pathway is followed by octylphenol ethoxylates (OPEOs) [70] and the estrogenic potency of octylphenol (OP) has been estimated to be even better than NP [71].

Thus, reviews on trace analysis and environmental behaviour of APEO [43,46,72] and toxicity [66], analytical methods [73], occurrence [74] and persistence [75] of NP metabolites in the aquatic environment have been published. GC and GC–MS are the methods of choice for the analysis of environmental matrices containing APEO oligomers with six or less EO units.

High resolution (HR)-GC chromatograms of NP show a complicated isomer profile [46] with several tens of peaks (see Fig. 2.1.5(b), top). In contrast, commercial standards of OP are usually made up from *t*-octylphenol [*p*-(1,1,3,3-tetramethylbutyl)phenol], which is essentially a single compound (see Fig. 2.1.5(a), top). Hence, gas chromatograms of NPEOs show distinct groups of peaks for each oligomer (see Fig. 2.1.5(b), bottom and C), whereas those of OPEOs show one major peak for each oligomer only (cf. Fig. 2.1.5(a), bottom). The isomeric composition of alkylphenols is further discussed below.

The influence of the low volatility of the higher molecular weight ethoxymers on GC analysis of APEO is demonstrated in Fig. 2.1.5(c) (bottom). Thus GC analysis without derivatisation is limited to oligomers with up to 4–5 EO units only. Higher ethoxymers can be detected only by GC after derivatisation (see also below) and often high temperatures, requiring dedicated columns and instrumentation, are necessary.

One drawback of high-temperature GC analysis is that sample degradation for the high molecular weight AEs and APEOs might occur. High-temperature capillary columns are coated with a stabilised bonded polysiloxane film, which allows a column oven temperature of up to 400°C.

An orthogonal array design and electronic pressure programming have been described for the optimisation of the GC determination of NP [76]. The characterisation of isomers of technical NP previously reported by Giger et al. [77] was further improved by GC–MS and GC–FTIR [78] identifying 14 isomers. On the other hand, Wheeler et al. [79] indicated the presence of five different groups of isomers of the 22 separated isomers. The isomeric composition of APEO mixtures and their structure elucidation are discussed in more detail in the next section.

Mass spectra of OPEOs, their acidic metabolites, OPEC and halogenated derivatives in EI and CI modes have been reported [80–82]. The most prominent ions formed under CI resulted from alkyl ion displacement and olefin displacement, and those formed under EI resulted from benzylic cleavage [82].

Analysis of underivatised APEOs by GC–EI–MS, has been used for the determination of the most volatile biodegradation metabolites in environmental samples. Thus, the presence of APEO$_n$ ($n < 4$) in wastewater and sewage effluents [77,82–85], river water [27,86–93], drinking water [88,94,95], sediments [96] and air [97,98] has been widely reported.

Analyses of derivatives of NPEOs have been carried out with a large number of derivatisation reagents. Among them, polyfluorinated reagents, such as pentafluorobenzoyl chloride and heptafluorobutyric anhydride, have been used to determine NPEOs in water, biota and sludge [99,100], whereas pentafluorobenzyl bromide was employed in the analysis of sludges and sediments [101]. By using these reagents, NPEO derivatives can be analysed by GC–ECD and GC–MS in EI or NCI modes. The latter showed the lowest detection limit (0.3 pg g^{-1} d.w.) when employed in the SIM mode [101]. The mass spectra of pentafluorobenzyl (PFB) derivatives of NP showed, under EI conditions, a base peak at m/z 181, which corresponds to the cleavage of the PFB group but is unspecific for NP because it is common to all PFB derivatives. On the other hand, under NCI conditions, the base peak at m/z 219 corresponding to the alkylphenolate anion, which is the complementary fragment at m/z 181 found in the EI mode, was employed for SIM mode [101].

Another approach consists of an in-situ acetylation and extraction of NPEOs and further analysis of the acetyl derivatives. The method has been applied to analyse effluent water and sewage sludges [102,103], sediments [104] and river waters [105]. Silylated derivatives [106] using BSA or BSTFA have also been used to determine NPEO ($n \leq 6$) in seawater [107] and wastewater [107,108], sediments [109] and sludges from wool scour effluents [110]. Halogenated derivatives of alkylphenols (AP) can also be formed as a result of chlorination practices in water treatment or wastewater if bromide is present. Brominated OPs and NPs (BrAPEOs) have been identified by GC–MS in sewage [111] and tap water [89], respectively.

Acidic biodegradation products of APEOs, namely NPE$_1$C, and NPE$_2$C have been commonly identified in environmental samples by GC–MS after derivatisation. Methylation with diazomethane has been employed for the determination of brominated APEC in wastewater [112] and sewage [111,113], whereas a solution of BF$_3$ in methanol [86, 114,115] or 1N HCl/MeOH has also been used for the determination of APECs in sewage [86], river and tap water [89] and sludge [115]. A procedure involving elution from the solid phase extraction disks and derivatisation using methyl iodide in a single step has been carried out to analyse NPECs ($n \leq 4$) by GC–CI–MS in sewage, river water, paper mill effluents [116] and industrial and municipal sludges [117]. Silylation was used for the determination of NPEC in sludges from wool scour effluents [106,110]. A method involving propylation by a propanol/acetyl chloride derivatisation procedure and further analysis by GC–MS in EI and CI modes has been employed for the determination of NPECs ($n \leq 3$) and carboxynonylphenol ethoxy carboxylates (CNPECs) ($n \leq 2$) in wastewater effluents [118] and river waters [27]. An alternative method employs direct derivatisation in the GC injection port using ion-pair reagents by reacting the carboxylic acid groups with tetraalkyl ammonium salts to form carboxylate ion pairs [RCOO⁻N(Bu)₄⁺], which are converted upon introduction in the GC injection port into their corresponding butyl esters [RCOOBu] [119]. The method has been successfully used for the determination of NPEC metabolites in river water and industrial effluents allowing their quantification at the 0.1 µg L⁻¹ level but failed for trace determination of CNPEC residues [119]. Figure 2.1.6 shows the on-line derivatisation

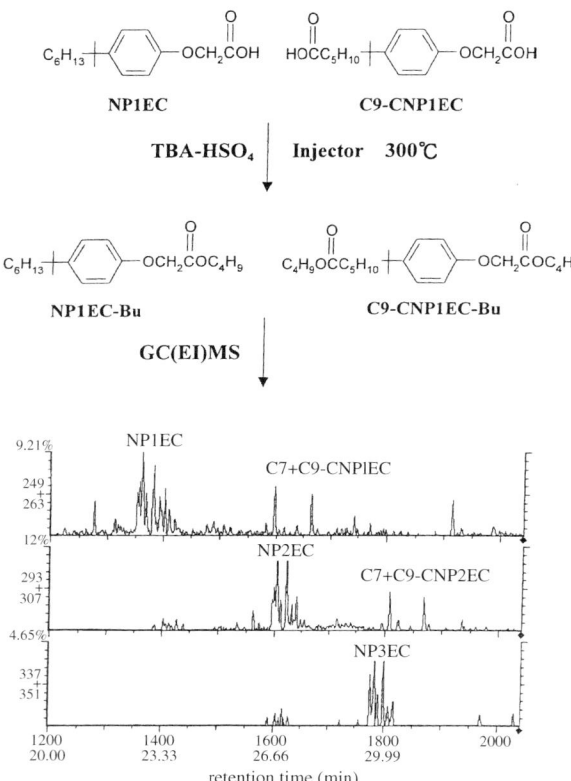

Fig. 2.1.6. On-line derivatisation and the selected characteristic mass chromatograms of butylated residues isolated from river water polluted by an industrial effluent. Reproduced with permission from Ref. [119]. © 1999 by Elsevier.

procedure and the mass chromatographic profiles of butylated NPEC and CNPEC metabolites from river water polluted by an industrial effluent. Mass spectra of butylated NPECs and CNPECs in the EI mode showed the characteristic ions produced by benzylic cleavage of [M-71]$^+$ and [M-85]$^+$. The molecular ions of CNPEC have higher intensity than NPEC and can be used to distinguish between both types of residues [119].

Supercritical fluid chromatography (SFC) can, in principle, be a very suitable means of analysis due to the high solubility of non-ionics in CO_2 [120]. SFC is even capable of analysing higher molecular weight AE [121] but has the disadvantage of not fully resolving the higher alkyl homologues [122]. SFE/SFC has proved to be successful in selectively extracting and determining AEs and APEs from aqueous samples [121]. Oligomer separation of nonylphenols by SFC was shown to be more complete than by HPLC, with shorter elution times [123]. However, to our knowledge these techniques have not yet been applied to environmental samples.

Isomeric composition of alkylphenols

APs and APEO oligomers are mixtures of isomers of a single compound (e.g. NP, $OPEO_2$). The isomers result from the various branching possibilities of the alkyl chain, the various possible positions of the phenyl ring in the alkyl skeleton, as well as from different (*ortho, meta, para*) substitutions relative to the phenol moiety on the phenyl ring. In addition, many chiral centres can occur in such molecules. Theoretically, therefore, several hundreds of isomers and stereoisomers exist just for a single compound like 4-nonylphenol. Figure 2.1.7 represents a set of possible structures for one of the 4-nonylphenol isomers, viz. 2(*p*-phenol)-methyloctane, illustrating that six positional and 13 additional stereoisomers exist for this molecule.

The synthesis of commercial nonylphenols generally proceeds through an alkylation reaction of phenols with propylene trimer [124], which itself has a spread of isomers [47]. The alkylation occurs preferentially at the *para* position of the phenol, but 3–6% of *ortho*-nonylphenol can be found in commercial products [125]. Also *meta* and disubstituted species have been mentioned [47]. A common feature of the isomers appears to be a dimethyl substitution of the α-carbon atom [83].

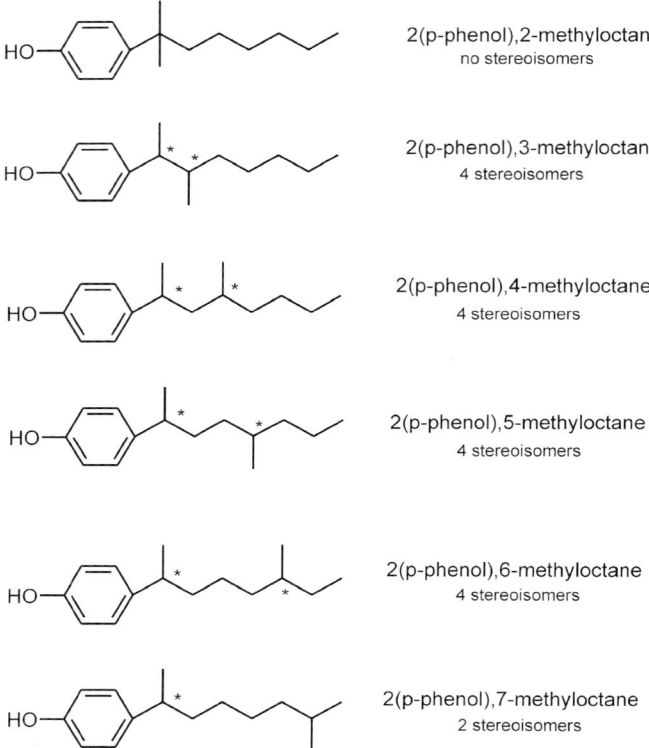

Fig. 2.1.7. Possible structures for the nonylphenol isomers 2(p-phenol)-methyloctane.
* indicates asymmetric carbon atom.

For commercial octylphenol products the synthesis is usually performed with the starting material 1,1,3,3-tetramethylbutane, thus resulting in t-OPs with only a single major compound for each possible (*ortho*, *meta* (minor), or *para*-substituted) phenyl positional isomer. High-resolution chromatograms of OPs are therefore much less complicated than those of NPs (compare Fig. 2.1.4(a)–(c)), although the presence of 11 minor constituents in *para*(t-octyl)-phenol has been reported [126].

It is quite obvious that the structural variety implies differences in chromatographic as well as environmental behaviour of many isomers. Highly branched isomers are more poorly degradable [124], have higher octanol/water partition coefficients [127], and shorter retention times than more linear ones [127].

Structure elucidation of APs has been the subject of only a few papers until now. Campbell et al. [128] used ^{1}H NMR to distinguish *ortho*, *meta* and *para*-substituted alkylphenols. GC was used some 30 years ago for separation and characterisation of APs, albeit without identification of isomers [129]. Szymanowski et al. [126] detected 29 components in commercial *p*-NP, without fully resolving their identities. Bhatt et al. [124] and Wheeler et al. [125] have used GC–MS for high-resolution separation of the isomers of NP and FTIR for subsequent structure confirmation, thus identifying 18 and 22 isomers, respectively. Resolution in liquid chromatography is usually not sufficient to detect single isomers. Recently 22 isomers have been resolved in an HPLC study using a graphitic carbon column [127], although without further identification.

The work of Wheeler et al. [125] has resulted in the differentiation of NP isomers into five distinct groups based on their fragmentation patterns in EI–MS, viz. α-dimethyl, α-methyl/α-ethyl/β-primary, α-methyl/β-methyl, α-methyl, and α-methyl/α-propyl.

Miscellaneous non-ionic surfactants

The analysis of underivatised alkyl polyglycosides (APG) by HTGC allows separation according to chain length and, in addition, into glucofuranoside and glucopyranoside isomers and into α- and β-anomers [130,131]. Characterisation of APGs in pure substances and industrial products has also been carried out by GC–MS of their trimethylsilylether derivatives [132] that can be identified by characteristic mass fragments (i.e. 204 and 217), which are the base peaks of pyranosides and furanosides. However, there are few papers dealing with the presence of APG in environmental matrices and their determination using GC [131]. Figure 2.1.8 shows the total ion chromatogram of alkyl polyglycosides in a basic washing agent after silylation [132].

The GC analysis of silylated derivatives of ethoxylated amines and fatty acid diethanolamides, among other non-ionic surfactants, has been described for many years and reviewed recently [43]. However, most of the determinations deal with the use of old packed columns for the characterisation of commercial blends and there are no applications of GC to their study in the environment.

68

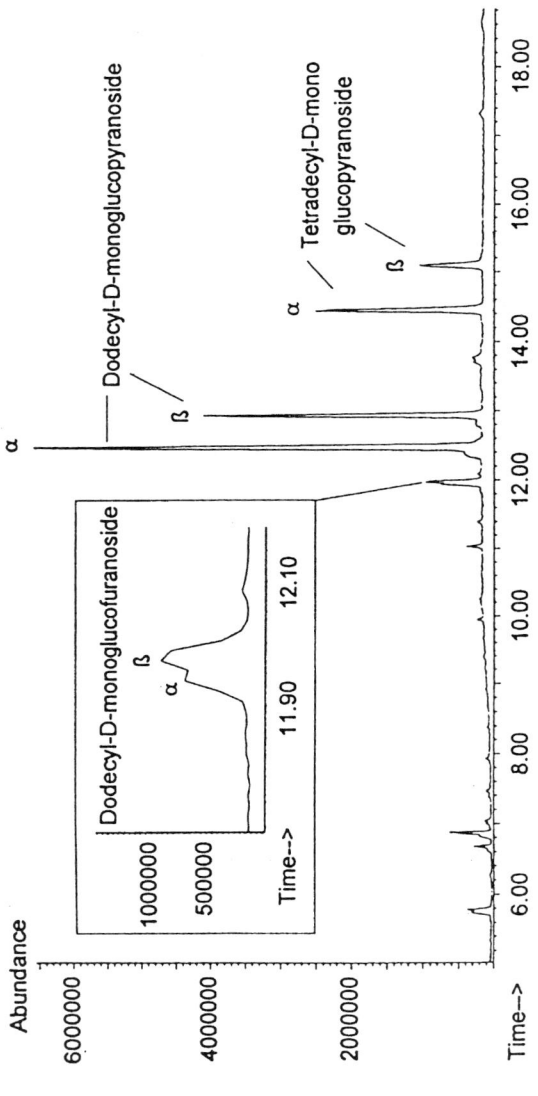

Fig. 2.1.8. Mass chromatogram of silylated alkyl polyglycosides. Reprinted with permission from Ref. [132]. © 1998 by Carl Hanser Verlag.

2.1.4 Cationic surfactants

Due to the lack of volatility of quaternary ammonium salts, GC analysis requires derivatisation or degradation to convert them into volatile compounds [133]. Thus, alkylbenzyl dimethyl ammonium halides (ABDACs) have been derivatised as trichloroethyl carbamates and cyanamides for GC–ECD and GC–NPD [134] analysis and the method was applied to their determination in natural waters [135]. The chain length of the raw material fatty acids of quaternized triethanolamines (esterquats) was determined by GC by reacting with methanol/sulfuric acid to yield the fatty acid methyl esters [136]. Several authors have used the Hofmann degradation method, which converts a quaternary ammonium hydroxide into a tertiary amine and an olefin for the analysis of quaternary amines by GC. Thus, alkyltrimethyl- and dialkyldimethyl-ammonium halides refluxed with sodium methoxide in N,N-dimethylformamide gave the expected dimethyldialkylamines [137], but the method was further improved by diluting the sample in methanolic potassium hydroxide with the degradation occurring in the GC injection port [138]. The alkyl chain distributions of commercial benzalkonium chlorides have also been determined [139].

Determination of ABDAC in wet wipes by debenzylation after alkaline degradation with potassium *tert*-butoxide has been reported and the conditions of the degradation reaction have been studied [140]. Recently, ABDACs have been determined in river water and sewage effluents by SPE followed by GC–MS of alkyldimethyl amines formed from the SPE eluate by the Hofmann degradation with potassium *tert*-butoxide [141]. The method allows their quantitation at ≤ 0.1 $\mu g\,L^{-1}$ in 500 mL of water. Figure 2.1.9 shows as an example the extracted ion chromatogram of dialkylamines present in river water, (a) prior to debenzylation and (b) after Hofmann degradation, as well as their EI mass spectra whose base peak at $m/z = 58$ corresponds to $[CH_2=N(CH_3)_2]^+$. The simultaneous analysis of alkyltrimethyl-, dialkyldimethyl- and trialkylmethylammonium salts by applying pyrolysis GC and selective detection using a thermionic detector has been reported [142].

The determination of cationic surfactants in the environment by GC and GC–MS is very scarce, as described above, but this is not the case for several compounds related to cationic surfactants. Thus, long chain tertiary amines that are amenable to direct analysis by GC have been

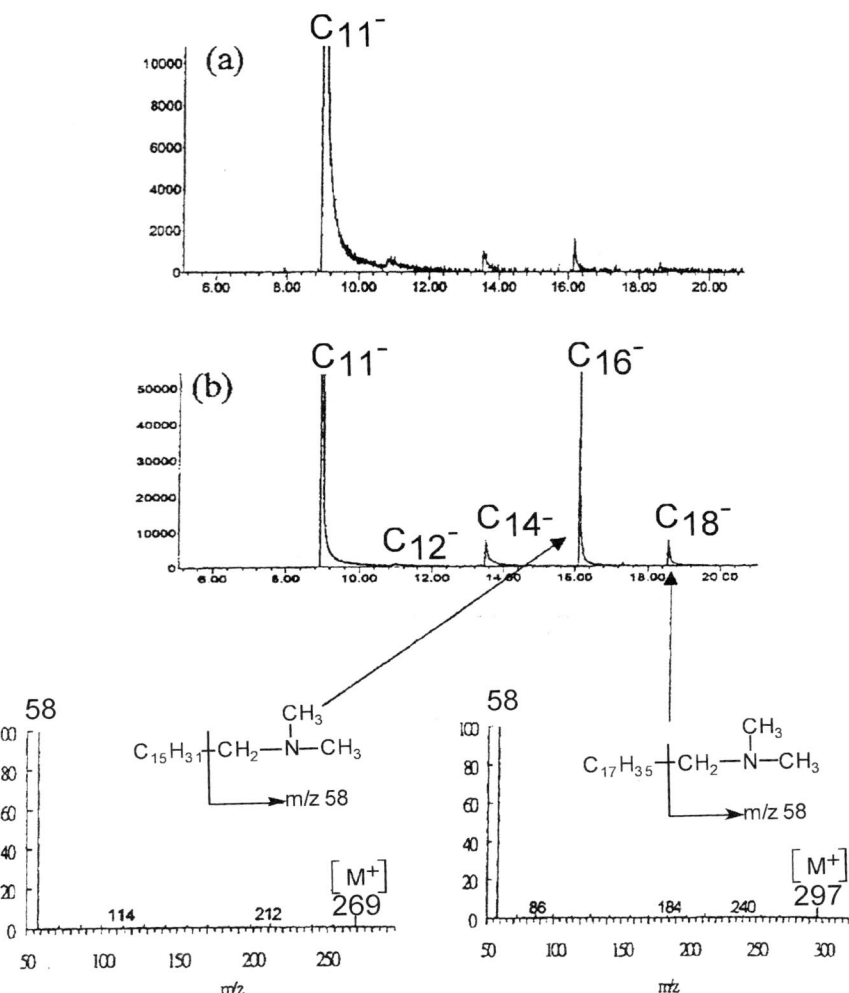

Fig. 2.1.9. Mass chromatograms (m/z 58) of alkylbenzyldimethylammonium chlorides as their corresponding alkyldimethylamines and the internal standard (undecyldimethyl-amine) (a) prior to debenzylation and (b) after Hofmann degradation with potassium *tert*-butoxide; and EI mass spectra detected in river water. Reproduced with permission from Ref. [141]. © 2001 by American Chemical Society.

identified in different environmental samples including sludges, coastal waters and sediments [107,109,143–145], seawater [146] and biota [147]. These trialkylamines come from trace impurities in quaternary ammonium salts used as fabric softeners in household laundry detergents and present a homologous distribution of odd–even carbon number alkyl derivatives having 14–18 carbons in the alkyl chain. Long chain alkylnitriles, which are intermediates in the production of cationic surfactants, have also been identified as impurities in formulations and were found in urban coastal environments [107,143,147].

2.1.5 Conclusion

Numerous applications have been shown to exist that overcome the general problems of lack of volatility and instability at higher temperatures that principally hamper direct analysis of surfactants by GC methods. Thus, a whole suite of derivatisation techniques are available for the gas chromatographist to successfully determine anionic, non-ionic and cationic surfactants in the environment. This enables the analyst to combine the high-resolution chromatography that capillary GC offers with sophisticated detection methods such as mass spectrometry. In particular, for the further elucidation of the complex mixtures, which is typical for the composition of many of the commercial surfactant formulations, the high resolving power of GC will be necessary.

REFERENCES

1 T. Reemtsma, *J. Chromatogr. A*, 733 (1996) 473.
2 E.A. Setzkorn and A.B. Carel, *J. Am. Oil Chem. Soc.*, 40 (1963) 57.
3 R.D. Swisher, *J. Am. Oil Chem. Soc.*, 43 (1966) 137.
4 W.T. Sullivan and R.D. Swisher, *Environ. Sci. Technol.*, 20 (1986) 376.
5 International Organization for Standardization. ISO 6841. Geneva (Switzerland), 1972.
6 H. Leidner, R. Gloor and K. Wuhrmann, *Tenside Surf. Det.*, 13 (1976) 122.
7 J. Waters and J.T. Garrigan, *Water Res.*, 171 (1983) 1549.
8 Q.W. Osburn, *J. Am. Oil Chem. Soc.*, 63 (1986) 257.
9 D. Louis and J.M. Talbot, *Tenside Surf. Det.*, 32 (1995) 347.

10 T. Uchiyama and A. Kawauchi, *J. Surf. Det.*, 2 (1999) 331.

11 E. Buse, A. Gollock and M. Muzic, *Tenside Surf. Det.*, 33 (1996) 404.

12 J.J. Kirkland, *Anal. Chem.*, 32 (1960) 1388.

13 D.K. Nguyen, A. Bruchet and P. Arpino, *J. High Resolut. Chromatogr.*, 17 (1994) 153.

14 H. Hon-Nami and T. Hanya, *J. Chromatogr.*, 161 (1978) 205.

15 J. McEvoy and W. Giger, *Environ. Sci. Technol.*, 20 (1986) 376.

16 M.L. Trehy, W.E. Gledhill and R.G. Orth, *Anal. Chem.*, 62 (1990) 2581.

17 W.H. Ding, J.H. Lo and S.H. Tzing, *J. Chromatogr. A*, 818 (1998) 270.

18 R. Reiser, H.O. Toljander and W. Giger, *Anal. Chem.*, 69 (1997) 4923.

19 J.S. Parsons, *J. Gas Chromatogr.*, 5 (1967) 254.

20 H. Hon-Nami and T. Hanya, *Jpn. J. Limnol.*, 41 (1980) 1.

21 H. Hon-Nami and T. Hanya, *Water Res.*, 14 (1980) 1251.

22 Y. Yamin and M. Ashraf-Khorassani, *J. High Resolut. Chromatogr.*, 17 (1994) 634.

23 B. Bastian, T.P. Knepper, P. Hoffmann and H.M. Ortner, *Fresenius J. Anal. Chem.*, 348 (1994) 674–679.

24 J.A. Field, J.A. Leenheer, K.A. Thorn, L.B. Barber, C. Rostad, D.L. Macalady and S.R. Daniel, *J. Contam. Hydrol.*, 9 (1992) 55.

25 C.F. Tabor and L.B. Barber, *Environ. Sci. Technol.*, 30 (1996) 161.

26 P. Kolbener, A. Ritter, F. Corradini, U. Baumann and A.M. Cook, *Tenside Surf. Det.*, 33 (1996) 149.

27 W.H. Ding, S.H. Tzing and J.H. Lo, *Chemosphere*, 38 (1999) 2597.

28 M.J. Suter, R. Reiser and W. Giger, *J. Mass Spectrom.*, 31 (1996) 357.

29 A. Heywood, A. Mathias and A.E. Williams, *Anal. Chem.*, 42 (1970) 1272.

30 J.A. Field, D.J. Miller, T.M. Field, S.B. Hawthorne and W. Giger, *Anal. Chem.*, 64 (1992) 3161.

31 C.J. Krueger and J. Field, *Anal. Chem.*, 67 (1995) 3363–3366.

32 W.H. Ding and C.T. Chen, *J. Chromatogr. A*, 857 (1999) 359.

33 W.H. Ding and J.C.H. Fann, *Anal. Chim. Acta*, 408 (2000) 291.

34 P. Sandra and F. David, *J. High Resolut. Commun.*, 13 (1990) 414.

35 J.A. Field, T.M. Field, T. Poigner and W. Giger, *Environ. Sci. Technol.*, 28 (1994) 497.

36 J.A. Field, T.M. Field, T. Poigner, H. Siegrist and W. Giger, *Water Res.*, 29 (1995) 1301.

37 E.L. Sones, J.L. Hoyt and A.J. Sooter, *J. Am. Oil Chem. Soc.*, 56 (1979) 689.

38 T.A. Neubecker, *Environ. Sci. Technol.*, 26 (1992) 61.

39 N.J. Fendinger, W.M. Begley, D.C. McAvoy and W.S. Eckhoff, *Environ. Sci. Technol.*, 26 (1992) 2493.

40 H.T. Rasmussen, A.M. Pinto, M.W. DeMouth, P. Touretzky and B.P. McPherson, *J. High Resolut. Chromatogr.*, 17 (1994) 593.

41 K. Tsuji and K. Konishi, *J. Am. Oil Chem. Soc.*, 52 (1975) 106.

42 E. Matthijs, In: J. Cross (Ed.), *Anionic Surfactants. Analytical Chemistry. 2nd Edition, Revised and Expanded*, Surfactant Science Series, Vol. 73, Ch. 3, Marcel Dekker, 1998.

43 T.M. Schmitt, *Analysis of Surfactants, 2nd Edition, Revised and Expanded*, Surfactant Science Series, Vol. 96., Marcel Dekker, 2001.

44 M.F. Puthoff and H.G. Nadeau, *Anal. Chem.*, 36 (1964) 1914. J.H. Bendedict, *Anal. Chem.*, 33 (1961) 1884.

45 H.G. Nadeau, *Anal. Chem.*, 36 (1964) 1914.

46 P. de Voogt, K. de Beer and F. van der Wielen, *Trends Anal. Chem.*, 16 (1997) 584.

47 In: J. Cross (Ed.), *Nonionic surfactants, chemical analysis*, Surfactant Science Series, Vol. 19, Marcel Dekker, New York, 1987, pp. 4–169.

48 U. Vettori, S. Issa, R. Maffei Facino and M. Carini, *Biomed. Environ. Mass Spectrom.*, 17 (1988) 193.

49 M. Motz, B. Fell, W. Meltzow and Z. Xia, *Tenside Surf. Det.*, 32 (1995) 17.

50 H.P.M. Van Lieshout, H.G. Janssen and C.A. Cramers, *Am. Lab.*, 27 (1995) 40.

51 K.B. Sherrard, P.J. Marriott, R.G. Amiet, R. Colton, M.J. McCormick and G.C. Smith, *Environ. Sci. Technol.*, 29 (1995) 2235.

52 A.H. Silver and H.T. Kalinoski, *J. Am. Oil Chem. Soc.*, 69 (1992) 599.

53 C.H. Asmussen and H.J. Stan, *J. High Resolut. Chromatogr.*, 21 (1998) 597.

54 J. Finke, U. Kobold, T. Dülffer, S. Pongratz, A. Puhlmann and A. Rudolphi, *Tenside Surf. Det.*, 35 (1998) 478.

55 K. Komarek, J. Minar and S. Skvarenina, *J. Chromatogr. A*, 727(8) (1996) 131.

56 F.W. Jones, *J. Chromatogr. Sci.*, 36 (1998) 223.

57 L. Gildenberg and J.R. Trowbridge, *J. Am. Oil Chem. Soc.*, 42 (1965) 69.

58 K. Tsuji and K. Konishi, *J. Am. Oil Chem. Soc.*, 51 (1974) 55.

59 V.T. Wee, In: L.H. Keith (Ed.), *Advances in the Identification and Analysis of Organic Pollutants in Water*, Ann Arbor Science Publishers, Ann Arbor, MI, 1981, p. 467.

60 B.G. Luke, *J. Chromatogr.*, 84 (1973) 43.

61 R.S. Tobin, F.I. Onuska, B.G. Brownlee, D.H.J. Anthony and M.E. Comba, *Water Res.*, 10 (1976) 529.

62 AIS-CESIO, *Proceedings of the Workshop Development of Analytical Methodologies for Alcohol Ethoxylates in Environmental Samples*, Brussels, 1992.

63 N.J. Fendinger, W.M. Begley, D.C. McAvoy and W.S. Eckhoff, *Environ. Sci. Technol.*, 29 (1995) 856.

64 A.T. Kiewiet and P. de Voogt, *J. Chromatogr. A*, 733 (1996) 185.

65 A. Marcomini and M. Zanette, *J. Chromatogr. A*, 733 (1996) 193.

66 M.R. Servos, *Water Qual. Res. J. Can.*, 34 (1999) 123.

67 S. Jobling and J.P. Sumpter, *Aquat. Toxicol.*, 27 (1993) 361.

68 M. Ahel, W. Giger and M. Koch, *Water Res.*, 28 (1994) 1131.
69 N. Jonkers, T. Knepper and P. de Voogt, *Environ. Sci. Technol.*, 35 (2001) 335.
70 H.A. Ball, M. Reinhard and P.L. Mc Carthy, *Environ. Sci. Technol.*, 23 (1989) 951.
71 A.C. Nimrod and W.H. Benson, *Crit. Rev. Toxicol.*, 26 (1996) 335.
72 B. Thiele, K. Günther and M.J. Schwuger, *Chem. Rev.*, 97 (1997) 3247.
73 H.B. Lee, *Water Qual. Res. J. Can.*, 34 (1999) 3.
74 D.T. Bennie, *Water Qual. Res. J. Can.*, 34 (1999) 79.
75 R.J. Maguire, *Water Qual. Res. J. Can.*, 34 (1999) 37.
76 K.K. Chee, M.K. Wong and H.K. Lee, *J. Microcol.*, Sep. 8 (1996) 29.
77 W. Giger, E. Stephanou and C. Schaffner, *Chemosphere*, 10 (1981) 1253.
78 B.D. Bhatt, J.V. Prasad, G. Kalpana and S. Ali, *J. Chromatogr. Sci.*, 30 (1992) 203.
79 T.F. Wheeler, J.R. Heim, M.R. LaTorre and A.B. Janes, *J. Chromatogr. Sci.*, 35 (1997) 19.
80 E. Stephanou, *Chemosphere*, 13 (1984) 43.
81 E. Stephanou, *Org. Mass Spectrom.*, 19 (1984) 510.
82 E. Stephanou, M. Reinhard and H.A. Ball, *Biomed. Environ. Mass Spectrom.*, 15 (1988) 275.
83 E. Stephanou and W. Giger, *Environ. Sci. Technol.*, 16 (1982) 800.
84 N. Paxeus, *Water Res.*, 30 (1996) 1115.
85 H. Shirashi, A. Otsuki and K. Fuwa, *Biomed. Mass Spectrom.*, 12 (1985) 86.
86 M. Ahel, T. Conrad and W. Giger, *Environ. Sci. Technol.*, 21 (1987) 697.
87 L.S. Sheldon and R. Hites, *Environ. Sci. Technol.*, 12 (1978) 1188.
88 L.S. Sheldon and R. Hites, *Environ. Sci. Technol.*, 13 (1979) 574.
89 F. Ventura, A. Figueras, J. Caixach, I. Espadaler, J. Romero and J. Guardiola, J. Rivera, *Water Res.*, 22 (1988) 1211.
90 J. Rivera, F. Ventura, J. Caixach, M. De Torres, A. Figueras and J. Guardiola, *Int. J. Environ. Anal. Chem.*, 29 (1987) 15.
91 M.A. Blackburn and J. Waldock, *Water Res.*, 29 (1995) 1623.
92 W.H. Ding and S.H. Tzing, *J. Chromatogr. A*, 824 (1998) 79.
93 C.M. Lye, C.L.J. Frid, M.E. Gill, D.W. Cooper and D.M. Jones, *Environ. Sci. Technol.*, 33 (1999) 1009.
94 A. Guardiola, F. Ventura, L. Matia, J. Caixach and J. Rivera, *J. Chromatogr.*, 562 (1991) 481.
95 L.B. Clark, R.T. Rosen, T.G. Hartman, J.B. Louis, I.H. Suffet, R.L. Lippincott and J.D. Rosen, *Int. J. Environ. Anal. Chem.*, 47 (1992) 167.
96 R.C. Hale, C.L. Smith, P.O. de Dur, E. Harvey, E.O. Bush, M.J. LaGuardia and G.G. Vadas, *Environ. Toxicol. Chem.*, 19 (2000) 946.
97 J. Dachs, D.A. Van Ry and S.J. Eisenreich, *Environ. Sci. Technol.*, 33 (1999) 2676.

98   D.A. Van Ry, J. Dachs, C.L. Gigliotti, P.A. Brunciak, E.D. Nelson and S.J. Eisenreich, *Environ. Sci. Technol.*, 34 (2000) 2410.

99   C. Wahlberg, L. Renberg and U. Wideqvist, *Chemosphere*, 20 (1990) 179.

100 J. Ejlertsson, M.L. Nilsson, H. Kylin, Å. Bergman, L. Karlson, M. Öquist and B.H. Svensson, *Environ. Sci. Technol.*, 33 (1999) 301.

101 N. Chalaux, J.M. Bayona and J. Albaigés, *J. Chromatogr. A*, 686 (1994) 275.

102 H.B. Lee and T.E. Peart, *Anal. Chem.*, 67 (1995) 1976.

103 M. Hawrelak, E. Bennet and C. Metcalfe, *Chemosphere*, 39 (1999) 745.

104 E.R. Bennet and C.D. Metcalfe, *Environ. Toxicol. Chem.*, 17 (1998) 1230.

105 D.T. Bennie, C.A. Sullivan, H.B. Lee, T.E. Peart and R.J. Maguire, *Sci. Total Environ.*, 193 (1997) 263.

106 E. Stephanou, In: A. Bjørseth, and G. Angeletti (Eds.), *Organic micro-pollutants in the aquatic environment*, Reidel Publishers, Dordrecht, 1986, p. 155.

107 M. Valls, J.M. Bayona and J. Albaigés, *Int. J. Environ. Anal. Chem.*, 39 (1990) 329.

108 R.A. Rudel, S.J. Melly, P.W. Geno, G. Sun and J.G. Brody, *Environ. Sci. Technol.*, 32 (1998) 861.

109 N. Chalaux, J.M. Bayona, I. Venkatesan and J. Albaigés, *Mar. Pollut. Bull.*, 24 (1992) 403.

110 F.W. Jones and D.J. Westmoreland, *Environ. Sci. Technol.*, 32 (1998) 2623.

111 H.A. Ball, M. Reinhard and P.L. McCarty, *Environ. Sci. Technol.*, 23 (1989) 951.

112 M. Reinhard, N. Goodman and K.E. Mortelmans, *Environ. Sci. Technol.*, 16 (1982) 351.

113 H.A. Ball, M. Reinhard, In: R.L. Jolley, R.J. Bull, W.P. Davis, S. Katz, M.H. Roberts, and V.A. Jacobs (Eds.), *Water chlorination chemistry, environmental impact and health effects*, Lewis Publishers Inc., Chelsea, MI, 1985, p. 1505.

114 H. Brüschweiler, H. Gamperle and F. Schwager, *Tenside Surf. Det.*, 20 (1983) 317.

115 H.B. Lee, T.E. Peart, D.T. Bennie and R.J. Maguire, *J. Chromatogr. A*, 785 (1997) 385.

116 J.A. Field and R.L. Reed, *Environ. Sci. Technol.*, 30 (1996) 3544.

117 J.A. Field and R.L. Reed, *Environ. Sci. Technol.*, 33 (1999) 2782.

118 W.H. Ding, Y. Fujita, R. Aeschiman and M. Reinhard, *Fresenius J. Anal. Chem.*, 354 (1996) 48.

119 W.H. Ding and C.T. Chen, *J. Chromatogr. A*, 862 (1999) 113.

120 K.A. Consani and R.D. Smith, *J. Supercrit. Fluids*, 3 (1990) 51.

121 M. Kane, J.R. Dean, S.M. Hitchen, C.J. Dowle and R.L. Tranter, *Anal. Proc.*, 30 (1993) 399.

122 A.H. Silver and H.T. Kalinoski, *J. Am. Oil Chem. Soc.*, 69 (1992) 599.

123 Z. Wang and M. Fingas, *J. Chromatogr. Sci.*, 31 (1993) 509.

124 B.D. Bhatt, J.V. Prasad, G. Kalpana and S. Ali, *J. Chromatogr. Sci.*, 30 (1992) 203.

125 T.F. Wheeler, J.R. Heim, M.R. LaTorre and E.B. Janes, *J. Chromatogr. Sci.*, 35 (1997) 19.

126 J. Szymanowski, H. Scewyk and J. Hepter, *Tenside Surf. Det.*, 18 (1981) 333.

127 J.L. Gundersen, *J. Chromatogr. A*, 914 (2001) 161–168.

128 C.B. Campbell, R.W. Lozier and A. Onopchenko, *Anal. Chem.*, 64 (1992) 1502.

129 R. Vittorio and S. Piergenido, *J. Chromatogr.*, 153 (1978) 181.

130 R. Spilker, B. Menzenbach, U. Schneider and I. Venn, *Tenside Surf. Det.*, 33 (1996) 21.

131 H. Waldhoff, J. Scherler, M. Schmitt, *Proceedings of 4th World Surfactant Congress*, 1996, Barcelona, p. 507.

132 P. Billian and H.J. Stan, *Tenside Surf. Det.*, 35 (1998) 181.

133 H.T. Rasmussen and B.P. McPherson, In: J. Cross, and E.J. Singer (Eds.), *Cationic surfactants. Analytical and biological evaluation*, Surfactant Science Series, Vol. 53., Marcel Dekker, New York, 1986.

134 S.L. Abidi, *J. Chromatogr.*, 200 (1980) 216.

135 S.L. Abidi, *J. Chromatogr.*, 213 (1981) 463.

136 A.J. Wilkes, C. Jacobs, G. Walraven and J.M. Talbot, *Proceedings of 4th World Surfactant Congress*, 1996, Barcelona, p. 389.

137 S. Takano, C. Takasaki, K. Kunihiro and M. Yamanaka, *J. Am. Oil Chem. Soc.*, 54 (1977) 139.

138 S. Takano, M. Kuzukawa and M. Yamanaka, *J. Am. Oil Chem. Soc.*, 54 (1977) 484.

139 L.K. Ng, M. Hupe and A.G. Harris, *J. Chromatogr.*, 351 (1986) 554.

140 S. Suzuki, Y. Nakamura, M. Kaneko, K. Mori and Y. Watanabe, *J. Chromatogr.*, 463 (1989) 188.

141 W.H. Ding and Y.H. Liao, *Anal. Chem.*, 73 (2001) 36.

142 F. David and P. Sandra, *J. High Resolut. Chromatogr. & Chromatogr. Commun.*, 11 (1988) 897.

143 P. Fernandez, M. Valls, J.M. Bayona and J. Albaigés, *Environ. Sci. Technol.*, 25 (1991) 547.

144 M. Valls, J.M. Bayona and J. Albaigés, *Nature*, 337 (1989) 722–724.

145 M. Valls and J.M. Bayona, *Fresenius J. Anal. Chem.*, 339 (1991) 212.

146 C. Maldonado, J. Dachs and J.M. Bayona, *Environ. Sci. Technol.*, 33 (1999) 3290.

147 M. Valls, P. Fernandez and J.M. Bayona, *Chemosphere*, 19 (1989) 1819.

2.2 CAPILLARY ELECTROPHORESIS IN SURFACTANT ANALYSIS

Mira Petrovic and Damia Barceló

2.2.1 Introduction

In recent years, capillary electrophoresis (CE) has been increasingly used for the analysis of surfactants, which are widely used in industrial and household products, and frequently present in environmental and wastewater matrices [1]. The main advantages of CE are high separation efficiency, good selectivity, low solvent consumption, short analysis time, low cost and rapid method development. However, the main drawback is the lower sensitivity compared with high-pressure liquid chromatography (HPLC), due to the small volumes injected (several nL). Because of the limited sensitivity, CE is mainly used for the determination of surfactants in industrial blends, wastewaters and sewage sludge while its application to trace analysis of environmental samples is rarely reported.

CE analysis of surfactants is problematical for several reasons. Surfactant monomers tend to form micelles (pseudo-stationary phase) in which the hydrophobic groups are oriented to the centre and the charged groups toward the outer surface. Micelle formation causes severe tailing of chromatographic peaks, insufficient separation efficiency and irreproducible migration times during electrophoretic separation, since the micelles have different electrophoretic mobility to the individual surfactant molecules. This property complicates the analytical application of CE to the analysis of surfactants, but is used extensively to control electro-osmotic flow and undesirable sorption of analytes in micellar electrokinetic capillary chromatography (MEKC). Another problem is the strong adsorption of the surfactants, mainly cationic, to the inner surface of the fused silica capillary. Both phenomena, micelle formation and adsorption onto the capillary wall, can be inhibited by the addition of an organic modifier, such as acetonitrile, acetone, tetrahydrofuran (THF) or methanol, in concentrations greater than 20%, to electrophoretic buffer.

Since CE is generally coupled to an ultraviolet (UV)/diode array detector, molecules to be analysed should have a chromophore. CE has

T.P. Knepper, D. Barceló and P. de Voogt (Eds)

been mainly applied for the separation of ionic surfactants (mainly anionic), whereas non-ionics are rarely analysed using this technique, since they do not posses electrophoretic mobility. The classes of anionic surfactants analysed using CE include alkyl sulfates (ASs), alkyl sulfonates, alkyl ethoxy sulfates, linear alkylbenzene sulfonates (LASs) and their degradation products sulfophenyl carboxylates (SPCs). Non-ionic surfactants are neutral molecules and therefore not mobile in an electrostatic field. In order to analyse them, additional interactions with an ionic buffer constituent are required. Mobility of non-ionic surfactants and their separation can be induced by adding sodium dodecylsulfate (SDS) to the electrophoretic buffer at concentrations above the critical micelle concentration (CMC). However, applying the pure mechanism of MEKC, non-ionic surfactants are difficult to separate because of their low water solubility and high partition coefficients into the micellar phase [2]. Therefore, the separations are performed using a high percentage of organic solvent that prevents micelle formation, but not the interaction between the surfactants (SDS) and non-ionic analyte. This interactive, solvophobic association, results in different electrophoretic mobility due to the differences in the strength of analyte–surfactant association complexes.

2.2.2 Ionic surfactants

Linear alkylbenzene sulfonates and their polar degradation products
The applicability of CE to the determination of LAS has been investigated very thoroughly in numerous works [3–10]. Using aqueous buffers (usually phosphate, acetate or borate), without adding organic solvents, all LAS homologues and isomers elute in a single peak, which allows rapid analysis of total LAS. Using organic modifiers, LAS homologues can be separated according to the alkyl chain length. Acetonitrile, added at concentrations of 15–40% [3,10], and isopropanol (30%) [9] were found to resolve LAS homologues in short migration times, with minor peak-splitting due to partial resolution of isomers. It was found that using acetonitrile, analytes could be analysed in less than 10 min but resolution was not as good as with isopropanol and, in some real sample analyses, there were some impurities co-eluting with the target compounds [8]. Figure 2.2.1 shows an electropherogram of a LAS commercial mixture obtained by

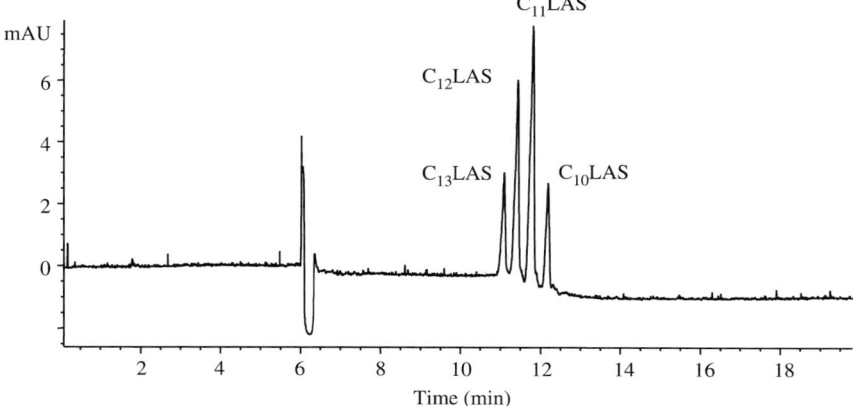

Fig. 2.2.1. CE–UV (220 nm) electropherogram of LAS commercial mixture. Fused-silica capillary 75 μm i.d.; buffer 30 mM ammonium acetate + 40% acetonitrile, pH 5.3; voltage 25 kV.

CE using fused-silica capillary and the buffer solution containing 30 mM ammonium acetate in water at pH 5.3 with 40% acetonitrile. Under cathodic detection conditions, LAS homologues elute in order of decreasing alkyl chain length.

Several authors [7,11] have reported the application of non-aqueous systems that offer advantages of adjusting relative migration rates via changes in solvation and ion pairing. LASs were separated in methanol and methanol/acetonitrile mixtures using tetramethylammonium as a counter ion. Using pure organic solvents, electro-osmotic flow is significantly lowered and polarity is often switched [12]. In this case, LAS homologues elute in reverse order and detection occurs at the anode. Non-aqueous systems were also found to be efficient in the separation of low-mobility surfactants, such as long-chain ASs.

Commercial LASs are complex mixtures of four individual LASs (C_{10}–C_{13}) with 20 possible positional isomers. Isomeric separation can be achieved by solvophobic association with SDS or host–guest interaction with cyclodextrins. Complete resolution of 19 isomers was achieved using 10 mM phosphate buffer (pH 6.8) containing 40 mM SDS and 30% acetonitrile [4]. LAS isomers in technical products were separated using α-cyclodextrin, but complete resolution of all isomers was not achieved [5].

Main biodegradation products of LAS, sulfophenyl carboxylates (SPCs), were separated by CE using α-cyclodextrin as the chiral selector [6]. The best separation of enantiomers was achieved with 60 mM α-cyclodextrin in a 20 mM citrate buffer at pH 4.0 with an uncoated fused-silica capillary. The method was applied for the qualitative and quantitative analysis of SPC in primary sewage effluents with a detection limit of 1 μg L^{-1}.

LASs are UV absorbing compounds and detection is usually performed using an UV detector operating at a wavelength of 214 nm. Limits of detection (LOD) of about 1 μg L^{-1} [8] in spiked ground water after 200-fold enrichment and 10 μg L^{-1} [4] for wastewater (100-fold enrichment) were achieved.

Although CE is generally coupled to UV detection, the coupling with a mass spectrometric (MS) detector allows unequivocal determination of target compounds in complex matrices. Recently, CE–MS, using electrospray ionisation (ESI), was applied for the determination of LAS and SPC in sewage treatment plants and coastal waters [8,10]. Figure 2.2.2 shows total ion current electropherogram of the simultaneous separation of LAS and SPC and Fig. 2.2.3 shows the selected

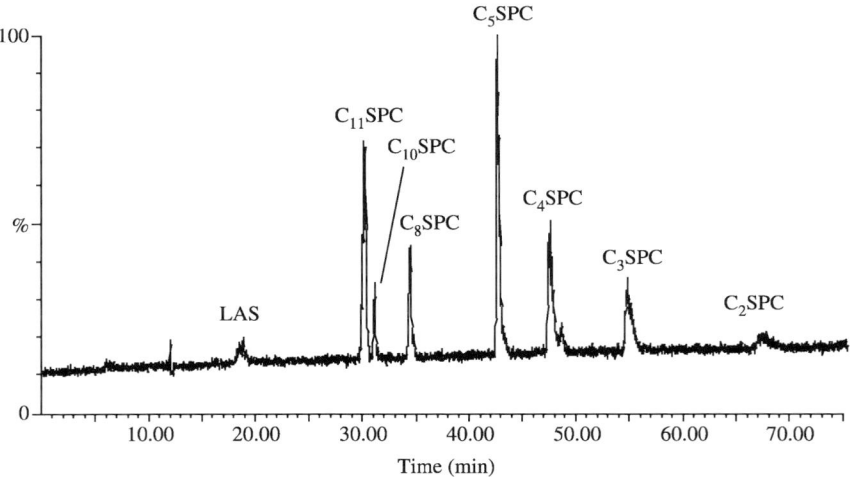

Fig. 2.2.2. Total ion current (TIC) CE–ESI–MS electropherogram of the simultaneous separation of LAS and SPC in an extract from the influent of a STP. For experimental conditions see Ref. [10]. Reprinted with permission from Ref. [10] © 2001 The Royal Society of Chemistry.

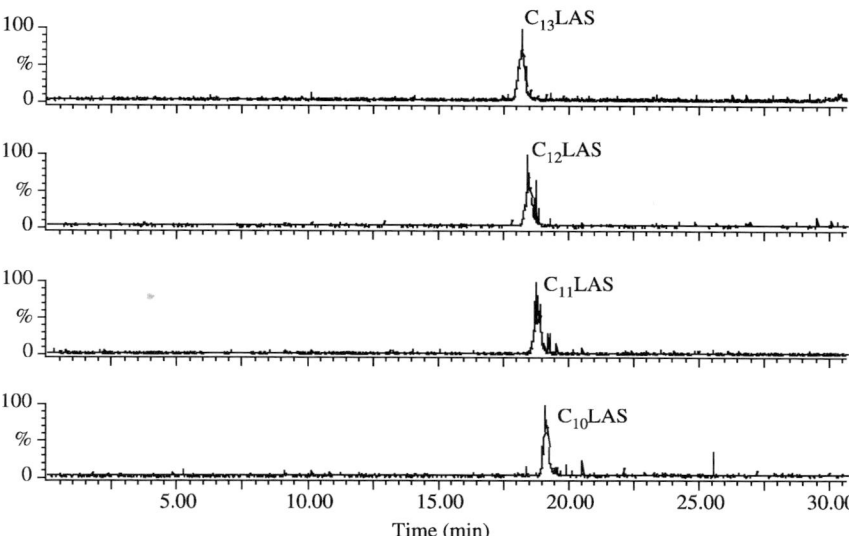

Fig. 2.2.3. Selected ion electropherograms of individual LASs detected in the STP influent by CE–ESI–MS. Ions monitored corresponded to m/z 297 for $C_{10}LAS$, m/z 311 for $C_{11}LAS$, m/z 325 for $C_{12}LAS$ and m/z 339 for $C_{13}LAS$. Reprinted with permission from Ref. [10] © 2001 The Royal Society of Chemistry.

ion electropherograms of individual LASs in influent wastewater of a sewage treatment plant. Positional isomers were not detected in the CE–MS chromatographic trace, probably due to the lower sensitivity of CE–MS versus CE–UV. The elution of SPC at higher retention times than the LAS is due to the fact that the SPCs possess two moieties with anionic properties and they are double charged at the working pH. The drawback of this method is the long analysis time needed for the simultaneous determination of LAS and SPC. The disadvantage of CE as a fast technique is due to the design of the interface that requires a very long capillary, which increases analysis time. However, new designs of CE–MS interfaces already use shorter capillaries, thus avoiding some problems, but further work is needed to improve the stability of the CE–MS system. The lack of sensitivity can also be partially overcome by the combination of effective techniques for the pre-concentration of samples (e.g. automatic solid phase extraction (SPE)), but is still the major problem of CE–MS. The LODs of CE–ESI–MS ranged from 4.4 to 23 $\mu g\, L^{-1}$ for individual LAS homologues, which is significantly higher than values obtained by CE–UV ($1 \ \mu g \ L^{-1}$) [8] or by LC–ESI–MS ($0.05–0.15 \ \mu g \ L^{-1}$) [13].

Other anionic surfactants
ASs and alkyl ether sulfates (AESs) can be readily separated according to alkyl chain length [7]. Because of the lack of absorbing constituents in the analytes, indirect photometric detection is usually performed using sodium *p*-toluene sulfonate [11], *p*-toluene sulfonic acid [11], naphthalene monosulfonate [14] or adenosine monophosphate [15] as an electrolyte additive. Under cathodic detection, ASs and alkyl sulfonates elute according to decreasing alkyl chain length and sulfonates elute prior to sulfates having the same alkyl chain length. Using aqueous systems, band broadening, because of low solubility of long-chain homologues, was observed. High contents of organic solvent or use of completely non-aqueous solvents can be used to improve the separation efficiency. However, the method still lacks sensitivity and LODs were in the $0.5–3 \ mg \ L^{-1}$ range [15].

Separation of alkyl ethoxy sulfates [7], alkylphenol ethoxy sulfates and phosphates [16] using aqueous and non-aqueous CE are also reported. Commercial blends of alkyl ethoxy sulfates are complex mixtures of homologues, with different alkyl chain lengths and oligomers, with different number of ethoxy (EO) units. With an increasing number of EO units, the electrophoretic mobility is decreased and migration to the detector increased. In aqueous CE, electrophoretic selectivity between different chain lengths within the different surfactant groups (AS, AES, LAS, SAS) exceeds the poor selectivity between different functionalities at a given chain length, which makes peak identification and quantitative analysis of complex samples almost impossible. To overcome this problem, several authors proposed non-aqueous CE, where significant differences in the mobilities of the different types of anionic surfactants can be obtained [11,17]. Figure 2.2.4 depicts the separation of a complex mixture of AS, AES, LAS (assigned as branched alkylbenzene sulfonates (ABSs)) and SAS by non-aqueous CE.

Capillary gel electrophoresis (CGE) is a specialised form of CE that employs a fused silica capillary filled with an immobilised and cross-linked polyacrylamide gel. This gel contains a network of pores, which allows separation of ions based on differences in molecular size. CGE was applied to the separation of ionic alkylphenol ethoxy sulfates and

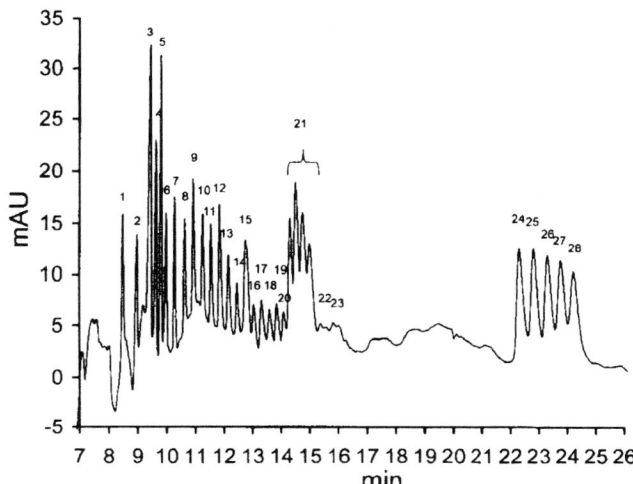

Fig. 2.2.4. Separation of AS, AES, ABS, and SAS with 15 mM naphthalene sulfonic acid, 15 mM triethylamine, and 0.001% Polybrene in ACN/MeOH/THF 80/18/2 v/v/v as the BGE. Peak identification: 1, C_8 AS; 2, C_{10} AS; 3, C_{12} AS; 4, C_{12} AES × 1 EO; 5, C_{14} AS; 6, C_{14} AES × 1 EO; 7, C_{12} AES × 2 EO; 8, C_{14} AES × 2 EO; 9, C_{12} AES × 3 EO; 10, C_{14} AES × 3 EO; 11, C_{12} AES × 4 EO; 12, C_{14} AES × 4 EO; 13, C_{12} AES × 5 EO; 14, C_{14} AES × 5 EO; 15, C_{12} AES × 6 EO; 16, C_{14} AES × 6 EO; 17, C_{12} AES × 7 EO; 18, C_{14} AES × 7 EO; 19, C_{12} AES × 8 EO; 20, C_{14} AES × 8 EO; 21, ABS + $C_{12/14}$ AES × 9 EO; 22, C_{12} AES × 10 EO; 23, C_{12} AES × 11 EO; 24, C_8 SAS; 25, C_{10} SAS; 26, C_{12} SAS; 27, C_{14} SAS; 28, C_{16} SAS. Reprinted with permission from Ref. [17] © 2002 Wiley–VCH Verlag GmbH.

phosphates [16]. This method offers very efficient resolution of oligomers. For example, a mixture of phosphated alkylphenol ethoxylate containing an average 40 mol ethylene oxide was successfully separated and more than 54 oligomers distinguished in a CGE chromatogram. However, the price for the high-resolution of CGE is time, with about 75 min required for this separation.

Cationic surfactants
The main difficulties in CE analysis of cationic surfactants arise from their strong adsorption to the capillary wall and their ability to form micelles at low concentrations. The addition of organic modifiers in high amounts or separation in absolutely non-aqueous media disrupt micelle formation within the sample and also effectiveness of the organic modifier to disrupt micelles of alkylbenzyl dimethyl ammonium

compounds decreases in the order acetonitrile > THF > acetone > methanol. CE was mainly applied to the analysis of quaternary alkylammonium compounds and alkyl pyridinium salts, used in a variety of commercial formulations, e.g. textile softeners, cosmetics and pharmaceuticals. The separation of different homologues is easily performed. These compounds elute from the capillary column in order of decreasing charge-to-mass ratio, i.e. in order of increasing alkyl chain length.

Heinig et al. [18] reported the separation of quaternary alkyl and dialkyl ammonium compounds using alkylbenzyl dimethylammonium compounds as background chromophores for indirect UV detection. High separation efficiencies could be obtained in a completely non-aqueous media (e.g. mixture of acetonitrile and ethylene glycol) [19]. However, a comparison with HPLC determination showed that CE is characterised by significantly higher LODs (1.9 versus 0.17 mg L^{-1} for C_{12}-benzyldimethylammonium chloride) and poorer reproducibility. Shamsi and Danielson [15] obtained LODs in the $0.25-1$ mg L^{-1} range for $C_{12}-C_{18}$ dialkyldimethyl quaternary ammonium compounds using tetrazolium violet as the electrolyte additive for indirect photometric detection. Piera et al. [20] applied CE to the analysis of an industrial mixture of C_n-benzyl surfactants using a tetrahydrofuran-containing buffer and direct UV detection (LOD of 4 mg L^{-1}).

2.2.3 Non-ionic surfactants

Neutral non-ionic surfactants do not possess an electrophoretic mobility. They can be analysed by CE after transformation to ionicderivatives or after inducing salvophobic association with SDS. Derivatisation of alcohol ethoxylates (AEOs), polyethylene glycols (PEGs) or alkylphenol ethoxylates (APEOs) is usually performed with phthalic anhydride [2,16,21]. To force the reaction and to obtain a single series of products 10-fold molar excess of phthalic anhydride is needed. As a consequence, the large peak that corresponds to phthalic acid will elute at the beginning of the electropherogram.

Another possibility is to perform CE separation in electrolytes containing high amounts of organic solvents and anionic surfactants (SDS), based on the formation of association complexes between analytes and SDS without micelle formation. As shown in Fig. 2.2.5, SDS concentrations of $40-100$ mmol L^{-1} and acetonitrile contents of

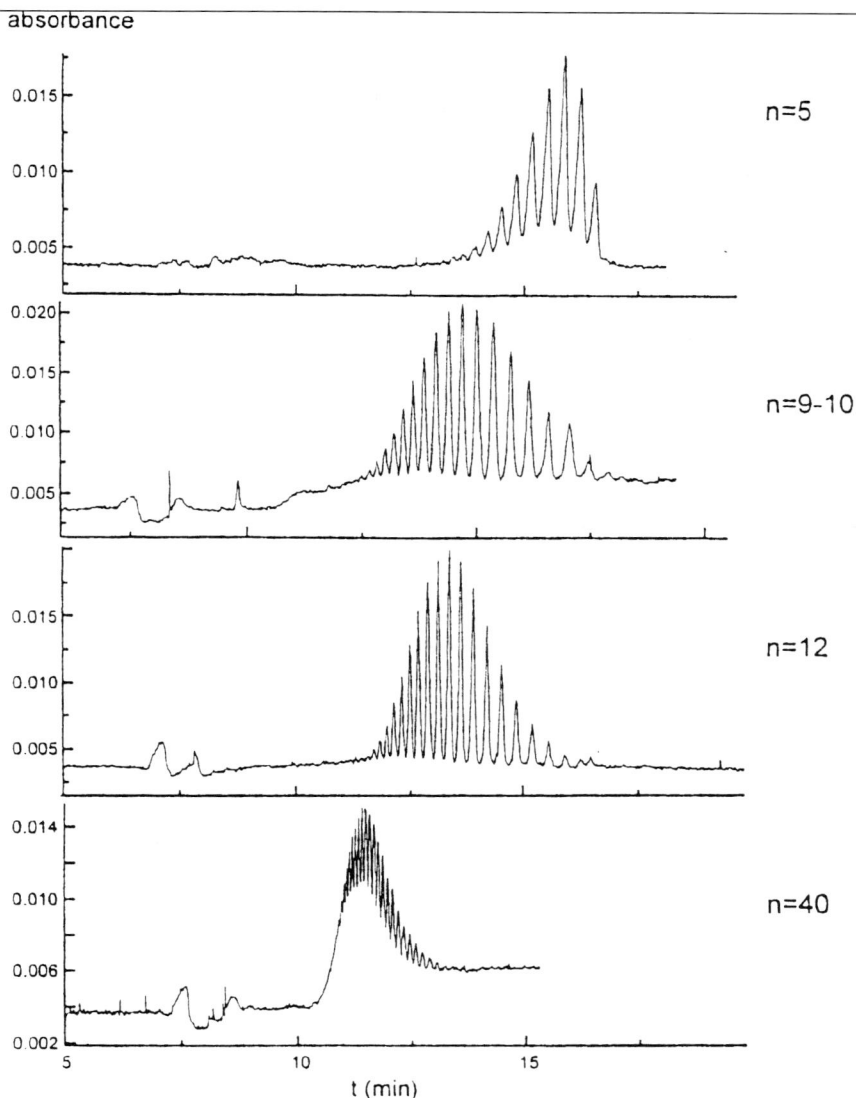

Fig. 2.2.5. Migration behaviour of octylphenol ethoxylates with dependence on degree of ethoxylation. Buffer; 10 mmol L^{-1} phosphate, pH 6.8, 70 mmol L^{-1} SDS, 35% acetonitrile. Samples: Igepal CA-520 ($n = 5$); Triton X-100 ($n = 9$–10); Igepal CA-720 ($n = 12$); Triton X-405 ($n = 40$). Reprinted with permission from Ref. [22] © 1997 Springer Verlag.

20–40% in phosphate or borate electrolytes were found to separate octylphenol ethoxylates containing 10–12 EO units [22]. However, the lower and higher ethoxylated products could not be separated with base line resolution, and the method is proposed as an alternative to HPLC for rapid 'fingerprinting' of ethoxylates in technical products and household formulations.

CE based on the formation of association complexes with SDS, gave a good separation efficiency for APEO, but for AEO insufficient peak resolution and irreproducible peak pattern were obtained [2]. This is explained by the higher hydrophobicity of AEO (longer alkyl chain) and smaller differences in association strength for individual oligomers. AEO derivatisation to ionic species was found to be a more useful technique. Additionally, derivatisation with phthalic anhydride gives UV detectable products. Using 50% acetonitrile and 150 mm borate buffer at pH 8.5, AEOs, LASs and alkyl polyglucosides were simultaneously detected in laundry detergent and shower gel samples [2]. LODs of CE were one order of magnitude higher than those obtained with HPLC, while the reproducibility of peak area was comparable with that of HPLC (0.5–2% RSD).

PEG and APEO were also separated using CGE [16]. CGE provided baseline resolution of PEG oligomers consisting of at least 120 ethylene oxide units. The main drawbacks are long analysis time (e.g. 110 min for the separation of non-ionic APEO containing 30 EO units) and fragile and expensive columns. Additionally, sensitivity is very limited because the polyacrylamide gels absorb UV light appreciably below 240 nm.

2.2.4 Conclusion

CE is a fast and efficient method that offers good selectivity and low operating costs. Due to the high separation power of CE, it can be used as an alternative to HPLC for samples with a less complex matrix and high surfactant concentration. The main drawback is lower sensitivity (at least one order of magnitude lower than of HPLC) that limits its application to trace analysis of surfactants in environmental matrices. However, it can be used for the rapid analysis of ionic and non-ionic surfactants in industrial and household formulations and also for environmental samples with high surfactant concentrations, e.g. sewage sludge.

Buffer parameters such as type of buffer used, its pH and ionic strength, and organic solvent content can be varied to influence the electrophoretic mobility of target compounds leading to an improved peak resolution.

The use of MS detection has been demonstrated to be an indispensable tool in the analysis of complicated matrices because it can overcome the large number of interfering substances.

REFERENCES

1   J. Riu and D. Barceló, In: D. Barcelo (Ed.), *Sample Handling and Trace Analysis of Pollutants: Techniques, Applications and Quality Assurance*, Elsevier, New York, 2000, Ch. 17.
2   K. Heining, C. Vogt and G. Werner, *Anal. Chem.*, 70 (1998) 1885.
3   P.L. Desbene, C. Rony, B. Desmazieres and J.C. Jacquier, *J. Chromatogr.*, 608 (1992) 375.
4   K. Heinig, C. Vogt and G. Werner, *Analyst*, 123 (1998) 349.
5   C. Vogt, K. Heinig, B. Langer, J. Mattusch and G. Werner, *Fresenius J. Anal. Chem.*, 352 (1995) 508.
6   C. Kanz, M. Nolke, T. Fleischmann, H.P.E. Kohler and W. Giger, *Anal. Chem.*, 70 (1998) 913.
7   K. Heinig, C. Vogt and G. Werner, *J. Cap. Elec.*, 5 (1996) 261.
8   J. Riu, P. Eichhorn, J.A. Guerrero, Th.P. Knepper and D. Barceló, *J. Chromatogr. A*, 889 (2000) 221.
9   D. Barceló and J. Riu, *Biomed. Chromatogr.*, 14 (2000) 14.
10   J. Riu and D. Barceló, *Analyst*, 126 (2001) 825.
11   H. Salimi-Moosavi and R.M. Cassidy, *Anal. Chem.*, 68 (1996) 293.
12   T.M. Schmitt, *Analysis of Surfactants*, Marcel Dekker, Inc., New York, 2001, pp. 402–409.
13   P. Eichhorn, M. Petrovic, D. Barceló and Th.P. Knepper, *Vom Wasser*, 95 (2000) 245.
14   S.A. Shamsi and N.D. Danielson, *Anal. Chem.*, 67 (1995) 4210.
15   S.A. Shamsi and N.D. Danielson, *J. Chromatogr. A*, 739 (1996) 405.
16   R.A. Wallingford, *Anal. Chem.*, 68 (1996) 2541.
17   M. Grob and F. Stainer, *Electrophoresis*, 23 (2002) 1921.
18   K. Heinig, C. Vogt and G. Werner, *J. Chromatogr. A*, 781 (1997) 17.
19   K. Heinig, C. Vogt and G. Werner, *Fresenius J. Anal. Chem.*, 358 (1997) 500.
20   E. Piera, P. Era and M.R. Infante, *J. Chromatogr.*, 757 (1997) 275.
21   J. Bullock, *J. Chromatogr. A*, 645 (1993) 169.
22   K. Heinig, C. Vogt and G. Werner, *Fresenius J. Anal. Chem.*, 357 (1997) 695.

2.3 LC DETERMINATION USING CONVENTIONAL DETECTORS

Mira Petrovic, Eduardo Gonzalez-Mazo, Pim de Voogt and Damia Barceló

2.3.1 Introduction

Commercial mixtures of surfactants consist of several tens to hundreds of homologues oligomers and isomers. Their separation and quantification is complicated and a cumbersome task. Detection, identification and quantification of these compounds in aqueous solutions, even in the form of matrix-free standards, present the analyst with considerable problems. The low volatility and high polarity of some surfactants and their metabolites hamper the application of gas-chromatographic (GC) methods. GC is directly applicable only for surfactants with a low number of ethylene oxide groups and to some relatively volatile metabolic products, while the analysis of higher-molecular-mass oligomers is severely limited and requires adequate derivatisation.

High-performance liquid chromatography (HPLC) permits the separation of complex mixtures with a wide distribution of alkyl and ethoxy homologues without discrimination of long-chain homologues. Today, HPLC, combined with various detection systems, such as ultraviolet (UV), fluorescence (FL), mass spectrometric (MS), refractive index (RI), evaporative light scattering detection (ELSD) or conductivity is used as a routine method for the analysis of ionic and non-ionic surfactants and their degradation products in industrial blends, wastewaters and various environmental matrices.

A number of studies have reported the application of different HPLC methods for the analysis of surfactants in wastewaters, surface waters, sediments, sludges and biological samples and several comprehensive reviews have been published on this issue [1–3].

This section covers only applications of HPLC with detection systems other than MS, giving special attention to specific issues related to the separation and detection of different groups of ionic and non-ionic surfactants.

Comprehensive Analytical Chemistry XL
T.P. Knepper, D. Barceló and P. de Voogt (Eds)

89

2.3.2 Ionic surfactants

Anionic surfactants

Linear alkylbenzene sulfonates (LASs) and their by-products and metabolites. The first application of HPLC in the analysis of LAS started in the mid-1970s. Its principal advantage over other techniques (e.g. GC) is that it allows direct analysis without any necessity to derivatise the surfactant. The technique was first applied by Kunihiro et al. [4] using silica gel as the stationary phase and acidified hexane/ethanol (80/20) as the mobile phase. The chromatogram thus obtained presented a single peak, which provided no information on the different homologues and positional isomers present. This first experiment did, nevertheless, enable the LAS to be determined without the prior need to modify its structure. Subsequent research was directed towards finding improved chromatographic conditions for the separation and quantification, separately, of the various homologues and isomers present in the LAS in a similar way to GC analysis. In 1977, Nakae et al. [5] achieved the complete separation of the homologues of alkylbenzyl dimethylammonium chlorides and of alkylpyridinium halides without sample pre-treatment, employing a methanolic solution of inorganic acids, or of their salts, as the mobile-phase. For the stationary phase, they employed a column filled with spherical microparticles of a co-polymer of styrene and divinylbenzene. Then Nakae and Kunihiro [6] separated a series of homologues of LAS synthesised in the laboratory, using the same chromatographic conditions described in the previous study. However, this method could not be applied for the determination of trace quantities of LAS in environmental samples.

This problem was resolved by Nakae et al. [7] using non-polar octadecylsilica as the stationary phase and a solution of 0.1 M of sodium perchlorate in methanol/water (80:20) as the mobile phase. The ternary system (water−alcohol−salt), previously used by Fudano and Konishi [8] as an eluent for the separation of ionic surfactants at higher concentrations, induced the so-called 'salting out effect'. The addition of the organic solvent to the water modified the polarity of the eluent and produced a good separation within a short period of time [9]. It also has the function of dissociating the surfactant micelles in individual molecules that are dissolved in the eluent [8]. The presence of the salt ($NaClO_4$) in the mobile phase has a considerable influence on the elution of the homologues, increasing the retention capacity of the homologues and thereby improving their separation. The authors [7], also introduced other interesting innovations: for example, they observed that, by adding dodecyl sodium sulphate to the samples, the loss of LAS by adsorption in the injection system was largely avoided. By following this procedure, the separation of all the homologues constituting the alkylic chain was obtained, and some of the positional isomers were also separated.

A more efficient separation of the homologues and isomers was achieved using 0.1 M $NaClO_4$ in acetonitrile/water (45/55) as the mobile phase [10].

The quantification of the total concentration of LAS and of its individual homologues (eluting all the positional isomers of every homologue in a single peak) can be achieved rapidly and specifically using a C_8 column and working under isocratic conditions [11–13]. By using a C_{18} analytical column, the retention time is increased by a factor of 2.5 over that obtained with a C_8 column and as a result of this, two peaks appear for each homologue of the LAS [9]. Other authors continued to use the C_{18} chromatographic column but under an isocratic regime with less polar solvents. Holt et al. [14] and Matthijs and Stalmans [15] employed a solution of 0.0875 M of $NaClO_4$ in methanol/ water (84/16) as the mobile phase. A complete separation of all the isomers and homologues of the LAS is obtained by using a long column (e.g. 250 mm × 4 mm i.d.) of octadecylsilica (C_{18}) with a particle size of 3 μm and applying a consistent elution gradient in two mixtures of acetonitrile and water at different proportions (A = 33:67 and B = 80:20) to which 10 g L^{-1} sodium perchlorate is added [16,17].

Tong and Tan [18] have performed a simultaneous determination of branched alkylbenzene sulfonates (ABSs) and LASs by HPLC using a C_{18} column. The elution of ABS produced one large wide peak representing the elution of the compounds with high molecular weight, with the same retention time of the C_{11}LAS.

The recent use of HPLC for the analysis of sulfophenyl carboxylates (SPCs) has been one of the most interesting applications of this technique for the study of the environmental behaviour of anionic surfactants. SPCs are separated by reversed-phase ion-paired chromatography, in which a hydrophobic stationary phase is used and the mobile phase is eluted with aqueous buffers containing a low concentration of the counter-ion [19].

The first applications of this technique were carried out by Eksborg et al. [20] and Wahlund [21], although Swisher et al. [22], were the first to develop it for the direct determination of some of the SPCs that are formed during the biodegradation of LAS. Subsequently Eggert et al. [23] modified the initial methodology by performing a derivatization of the SPC, followed by the separation and GC–MS identification of the SPC. A mixture of internal standards of SPC, synthesised in the laboratory, was separated by Taylor and Nickless [24], and compared with that produced by treating a natural environmental sample of seawater containing LAS that had been subjected to the same conditions under which it biodegrades. Chromatography was achieved by using a mobile phase containing cetyltrimethyl ammonium ions. The analysis of the compounds of partially degraded LAS revealed the existence of at least 13 meta-stable intermediates that proved difficult to identify.

Marcomini et al. [25] performed the protonation of the carboxylic groups of the SPC. This technique termed 'ion suppression effect' chromatography enabled the individual elution of the SPC contained in a primary sludge from a wastewater treatment plant. The quantification of the SPC was performed under the assumption that the signals detected by fluorescence before the elution of the LAS (with the exception of NPEC) correspond solely to SPC.

The simultaneous determination of all the monocarboxylic SPC and LAS homologues has been performed using a gradient regime [26], based on previous methods [27,28]. The mobile phase was $MeOH/H_2O$ (80/20) with 1.25 mM tetraethylammonium hydrogenosulfate and water. The different homologues were separated using a Lichrosorb RP-8 column (250 mm × 4.6 mm i.d.) with a particle size of 10 μm. Figure 2.3.1 shows chromatograms obtained using this procedure showing efficient separation of each LAS and SPC homologue as a unique peak, under which all the positional isomers are eluted [26].

The simultaneous determination of LAS and its potential biodegradation intermediates of low molecular weight (C < 8), as sulfonated and non-sulfonated, has been performed by RP–HPLC–UV using a gradient regime with the mobile phase at pH 2.2 [29]. Compound elution with the same number of carbon atoms in the carboxylic chain was as follows: SPCs, hydrophenyl carboxylic acids and finally phenylcarboxylic acids.

The simultaneous determination of LAS and their co-products (dialkyltetralin sulfonates (DATSs) and alkylbenzene sulfonates with

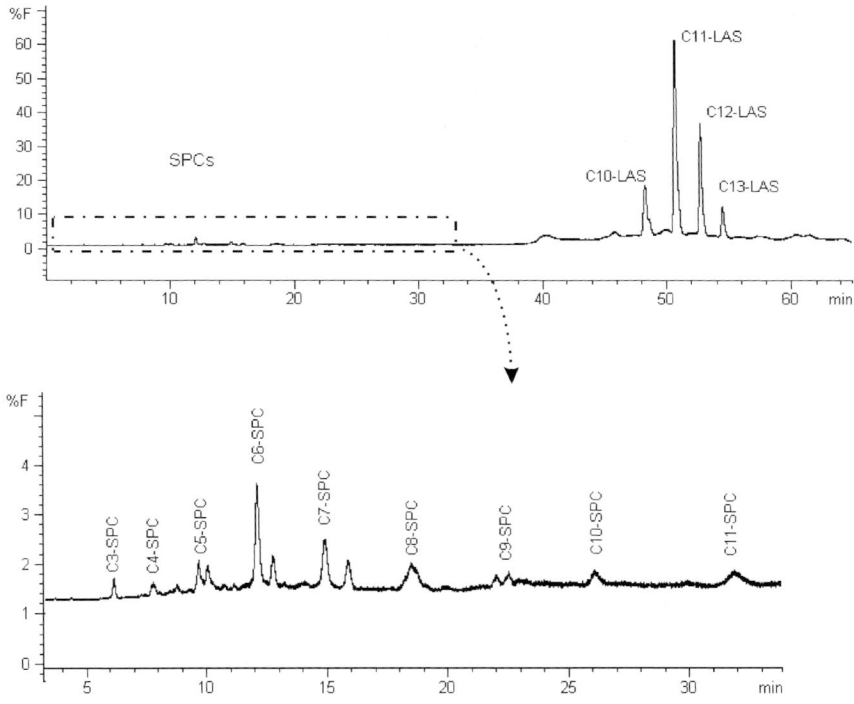

Fig. 2.3.1. HPLC–FL chromatogram obtained from a standard of LAS and SPC homologues using a Lichrosorb RP-8 column of 250×4.6 mm^2 (10 μm).

a branched methyl (iso-LAS)) have received much attention. It was claimed that the HPLC technique is unsuitable for determining simultaneously LAS and DATS due to problems of co-elution [30]. However, Crescenzi et al. [31] developed a method based on HPLC with fluorometric detection for determining LAS and DATS in aqueous environmental samples. Figure 2.3.2 shows the comparison of chromatograms obtained using C_8 and C_{18} columns, respectively. It appears that the C_8 column is more suitable than the C_{18} not only for lumping isomers of any single LAS or DATS homologue into a single peak but also for separating LAS from DATS.

Detection systems most often used to analyse LAS and their metabolites or by-products included direct or indirect UV and FL. Kikuchi et al. [32] found that in LAS determination, the chromatograms

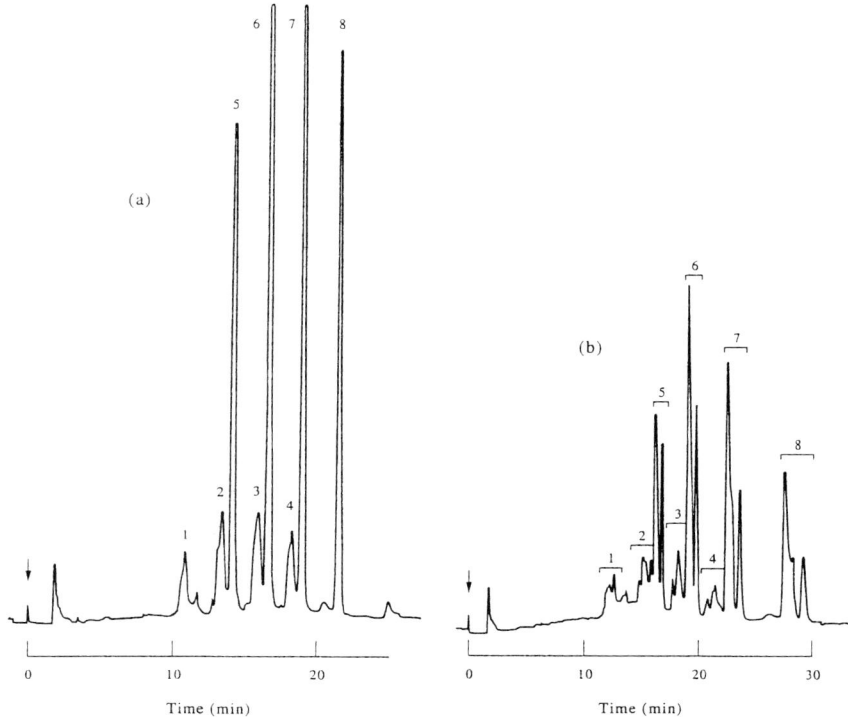

Fig. 2.3.2. LC chromatograms of a standard solution containing LAS and DATS obtained by eluting them on (a) C_8 column and (b) C_{18} column. In both cases, peaks 1–4 are, respectively, for C_{10}–C_{13} DATS, while peaks 5–8 are, respectively, for C_{10}–C_{13} LAS. Reprinted with permission from Ref. [31] © 2001 Elsevier.

obtained by FL detection are far more specific than UV-absorption detection. This conclusion is shared by other authors [33,34], who found that the use of octadecylsilica minicolumns as the purification stage was not sufficient to eliminate interference with UV-detection. RP–HPLC with fluorescence detector was first used by Nakae et al. [7] for the analysis of LAS from river water and subsequently has been extensively used for direct analysis in a variety of environmental matrices of LAS [12,13,32,35,36]. The method shows a high reproducibility, expressed as the relative standard deviation of results: 6% for Marcomini and Giger [37]; 1.4–4% for Castles et al. [36]; and 2–3% for González-Mazo et al. [13]. Therefore for LAS determination, fluorescence detection (excitation at 225 nm, emission at 290 nm) must be preferred for selectivity and sensitivity [7,11,13,15,32,36,38]; nevertheless, UV detection can sometimes be used satisfactorily [35,38]. The simultaneous determination of LAS and all the monocarboxylic acids present in samples has also been performed by RP–HPLC with fluorescence detection [26]. However, the complexity of SPC mixtures and the lack of reference standards currently limit the applicability of HPLC with UV-fluorescence [39].

Other detection systems, such as conductivity detector or refractive index detection are generally applicable for the determination of common anionic surfactants [1]. However, they are less sensitive than other techniques and are used more often for the characterisation of pure surfactants, than for their determination at low concentrations.

Alkylether sulfates (AESs) and alkyl sulfates (ASs). Typically, ion-interaction chromatography utilising a RP column with an aromatic ion-pairing agent has been a popular approach for the separation and detection of aliphatic anionic surfactants, although anion exchange columns are also used for lower molecular weight compounds. Shamsi and Danielson [40] used mixed-mode chromatography (RP phenyl/anion exchange) with a naphthalene disulfonate mobile-phase to separate C_6–C_{18} sulfates and C_6–C_{18} sulfonates.

Zhou and Pietrzyk [41] found that increasing the mobile-phase ionic strength not only increases the retention of AS and AES on a reversed stationary phase, but also improves the resolution since the peak widths are significantly reduced. The authors achieved baseline separation of a multicomponent alkane sulfonate and alkyl sulfate mixture from C_2 to C_{18} using a mobile-phase gradient whereby acetonitrile concentration increases and LiOH concentration decreases.

Although AS and AES can be detected at a low UV wavelength, sensitivity is lacking and a more suitable detection was achieved using indirect photometric detection, post-column colour formation reactions, or a pre-column derivatisation, suppressed conductivity detection and refractive index detection [1,42,43]. A comparison of detection limits for the determination of these anionic surfactants shows that photometric and conductivity detectors are better (picomole or nanogram range) than refractive index or fluorometry detectors by about a factor of 1000 [40].

There are a number of HPLC methods available for determining AES. Generally, normal-phase systems are used for the determination of EO homologue distribution, while reversed-phase systems are used to separate AS and AES on the basis of alkyl chain length [1]. Some authors reported on dual separation according to both hydrophobe and hydrophile distribution, which proved to be quite easy to achieve since the degree of ethoxylation in AES is much smaller than with non-ionic ethoxylates.

For AES, a HPLC method using ELSD was developed and the response was slightly lower than for the non-ionic ethoxylates [44].

Cationic surfactants

During the last 30 years, several techniques have been developed for the analysis of cationic surfactants, most of them based on HPLC employing either normal-phase or reversed-phase separation. Normal-phase HPLC seems to be the more promising method because the analysis is performed directly without prior derivatisation or modification of the quaternary structure, simultaneously providing information on the chain length distribution of the fatty moieties and the nature of the quaternary substitution [45]. However, separation based on reversed-phase chromatography is widely used. The most frequently used columns are cyanoamino-, aminopropyl-, or polyvinylalcohol-bonded silica operated in normal-phase mode or cyano-, octyl-, octadecyl- and phenyl-bonded phases used in reversed-phase mode. Efficient separation could be obtained by varying the ionic strength of the mobile phase, adjusting pH by using appropriate buffers (i.e. phosphate buffers) or by changing the counter-ion. Using normal-phase separation elution of compounds creates a reverse order of alkyl character and under the proper conditions, partial resolution by chain length can be obtained [45], while using reversed-phase separation elution creates an increasing alkyl character.

One of the main problems to be solved in the analysis of cationic surfactants is the strong adsorption of the surfactant to glassware, tubing and apparatus. To avoid losses, the solvent system used should contain a substantial percentage of organic solvent. Additionally, mobile phases containing more than 20–25% methanol will help to inhibit micelle formation [46].

Ion-exchange chromatography techniques that provide cationic surfactant residues suitable for a conductivity analysis are also developed [47,48]. They usually employ strong cation exchange columns and a mobile-phase based on an organic solvent (acetonitrile or methanol) and water containing salt, such as ammonium formate with the pH adjusted to a suitable value.

For the cationic surfactants, the available HPLC detection methods involve direct UV (for cationics with chromophores, such as benzylalkyldimethyl ammonium salts) or for compounds that lack UV absorbance, indirect photometry in conjunction with a post-column addition of bromophenol blue or other anionic dye [49], refractive index [50,51], conductivity detection [47,52] and fluorescence combined with post-column addition of the ion-pair [53] were used. These modes of detection, limited to isocratic elution, are not totally satisfactory for the separation of quaternary compounds with a wide range of molecular weights. Thus, to overcome the limitation of other detection systems, the ELS detector has been introduced as a universal detector compatible with gradient elution [45].

For the analysis of ditallowdimethylammonium chloride (DTDMAC), the most widely used active ingredient in fabric softeners, conventional detection techniques using UV or direct FL are not applicable, because of the lack of chromophores. Therefore, methods utilising conductivity detection [47,52] or the post-column ion-pair formation with phase separation and UV/vis or fluorescence detection [53–55] are used to monitor DTDMAC in surface water, wastewater, sludge and marine sediments. Shibukawa et al. [56] reported the separation and determination of long alkyl chain quaternary ammonium compounds: dodecyltrimethyl ammonium chloride (DTMA), tetradecyldimethylbenzyl ammonium chloride (TDDBA), cetryltrimethyl ammonium bromide (CTMA) and stearyltrimethyl ammonium chloride (STMA). They were separated with a column packed with a hydrophilic polymer packing using a water–acetonitrile mixture containing 4,4'-bipyridyl and hydrochloric acid to depress hydrophobic adsorption of the quaternary compounds and to increase the sensitivity of the conductometric detection with a micromembrane suppressor. Figure 2.3.3 shows a HPLC chromatogram of the quaternary ammonium ions separated.

2.3.3   Non-ionic surfactants

Commercially, the most important non-ionic surfactants (APEOs and AEOs) are amphipathic molecules consisting of a hydrophilic (ethylene oxide chains of various length) and a hydrophobic (alkyl phenols, fatty acids, long chain linear alcohols, etc.) part. The polyethoxylated

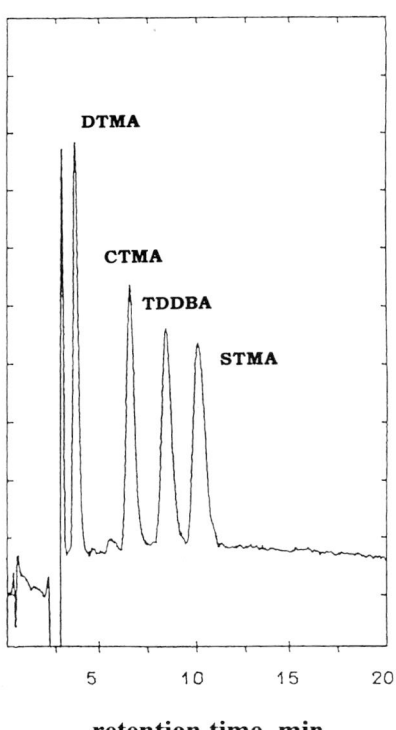

retention time, min

Fig. 2.3.3. Chromatogram of quaternary alkylammonium compounds. Columna, Shodex Asahipak GF-310 HQ (100 × 4.6 mm^2); mobile-phase, 0.4 mM 4,4'bipyridyl, 0.8 mM HCl, 27% (m/v) acetonitrile; flow rate 0.6 ml min^{-1}; analyte concentration 1.0 μM; injection loop, 50 μL. Reprinted with permission from Ref. [56] © 1999 Elsevier.

surfactants produced at the industrial scale have large numbers of impurities, and represent a polydisperse mixture of compounds with different lengths of ethylene oxide chains approaching Poisson distribution of different oligomers. Using a normal-phase separation mechanism on polar or moderately polar supports (silica, CN modified silica, NH$_2$ modified silica, diol, and methylated silica), polyethoxylated surfactants can be separated according to the length of the ethylene oxide chain.

On the other side, reversed-phase chromatography on non-polar supports (C$_1$, C$_2$, C$_8$, C$_{18}$, polyethylene-coated silica or alumina) is apparently unsuitable for the separation of different oligomers, because of its large selectivity towards alkyl groups. However, this technique is particularly suitable for the separation of surfactants containing various hydrophobic moieties (homologue-by-homologue separation), and according to both the alkyl and polyethoxy chain lengths and side products (polyethylene glycols).

Alkylphenol ethoxylates (APEOs)
Separation of APEO has been attempted using both normal-phase and reversed-phase systems combined with absorption or fluorescence detection.

In the normal-phase system, the APEOs are separated according to increasing number of ethylene oxide units, while corresponding oligomers with the same number of ethoxy units but different alkyl substituents co-elute, as shown in Fig. 2.3.4 for NPEO and OPEO and in Fig. 2.3.5 for NP and OP [57]. This approach usually employs polar or moderately polar supports, such as bare silica gel, amino- or cyano-bonded silica, diol columns, methylated silica support (C_1) or alumina. As industrial APEO blends usually contain a wide distribution of ethoxy units, they can be successfully separated only by using binary or ternary gradient elution, while isocratic separation is successful over a limited oligomer range. By variation of the composition of the mobile-phase, it is possible to extend the range of oligomers that can be determined (up to 25 mol adducts). The information about the exact distribution of individual oligomers in an environmental extract is rather important, since it permits identification of the major source of APEOs, as well as assessment of the degree of degradation of the parent surfactant.

Ahel et al. [58−60] developed NP−HPLC and UV detection (277 nm) for the determination of APEOs and their lipophilic metabolites in sewage effluents. To determine the concentrations of individual oligomers in a wide molecular weight range ($n_{EO} = 0-20$), an aminosilica column was used as the stationary phase, and binary mixtures of n-hexane-2-propanol and 2-propanol-water as the mobile-phase with linear and hyperbolic gradient profiles. The application of short column and 3 or 5 µm aminosilica material allowed satisfactory separation of NP from NP_1EO and NP_2EO under isocratic conditions, while columns packed with irregularly shaped 10 µm particles of aminosilica material allowed separation of oligomers in a wide molecular weight range ($n_{EO} = 3-20$). Aminosilica is the most frequently used material [61−64], although several methods using cyano-modified

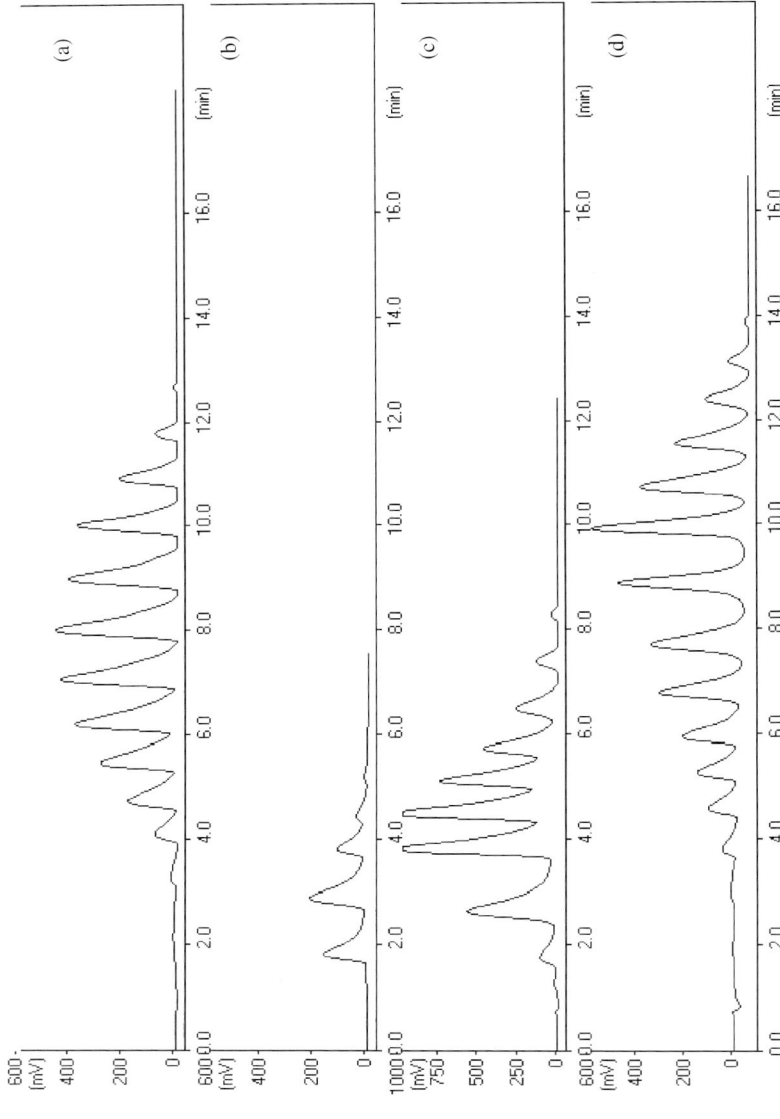

Fig. 2.3.4. Normal-phase chromatograms of commercial mixtures of alkylphenol ethoxylates: (a) OP[EO]$_{8/9}$; (b) NP[EO]$_2$; (c) NP[EO]$_4$; (d) NP[EO]$_{10}$. Column: 100×4.6 mm^2 Hypersil 3 NH$_2$ (3 μm), gradient elution with n-hexane-2-propanol-H$_2$O, detection: fluorescence, excitation 230 nm, emission 290 nm.

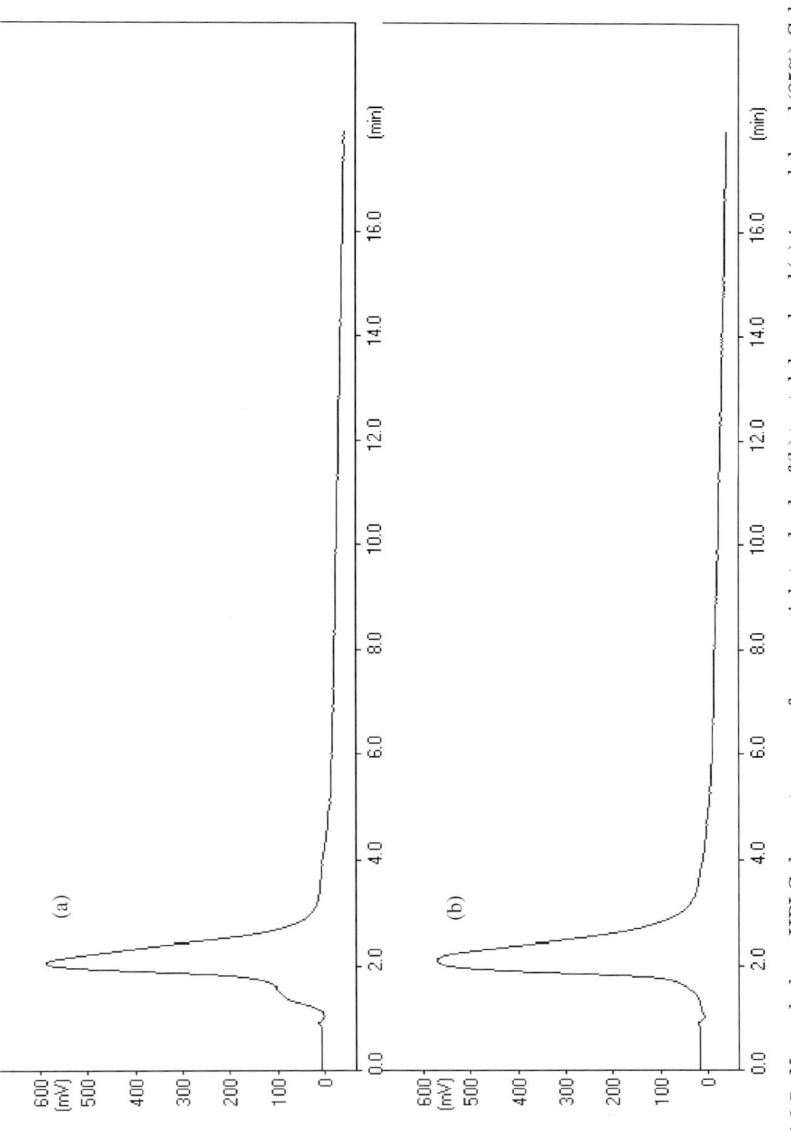

Fig. 2.3.5. Normal-phase HPLC chromatograms of commercial standards of (b) t-octylphenol and (a) 4-nonylphenol (85%). Column: 100×4.6 mm^2 Hypersil 3 NH$_2$ (3 μm), gradient elution with n-hexane-2-propanol-H$_2$O, detection: fluorescence, excitation 230 nm, emission 290 nm.

silica [65,66], silica [67] or diol columns [68] were also employed for the analysis of NPEOs in wastewaters or in technical blends.

Zhou et al. [68] used a diol column and UV detection at 275 nm for the determination of NPEOs in industrial blends achieving detection limits of 50–130 ng (absolute value) of individual NPEOs. Reported detection limits using NP–HPLC and UV are generally in the range of 0.5–5 $\mu g\,L^{-1}$ for an individual NPEO [59,69]. Although the UV absorbance detector is widely used for the detection of NPEOs, much higher sensitivity and better selectivity could be achieved with a fluorescence detector. The application of fluorescence detection also provides a significant improvement with respect to the detection of NP_1EO and NP_2EO in the presence of early eluting compounds of lower polarity, such as PAHs and phthalates [64]. Fluorescence detection at wavelengths of 225–230 nm (excitation) and 290–310 nm (emission) gave detection limits of individual oligomers below the $\mu g\,L^{-1}$ level. For example, Bennie et al. [70] reported a detection limit of 0.02 $\mu g\,L^{-1}$ for NP_1EO and NP_2EO in river water, while Ahel et al. [60] reported a determination limit (S/N = 10) for an individual oligomer in the range of 0.2–0.6 $\mu g\,L^{-1}$, depending on its response factor (R_F). The authors showed that relative R_F (relative to R_F of $NP_1EO = 1$) of individual oligomers increase linearly with increasing number of ethoxy groups. They were determined by quantifying the exact composition of commercial mixtures of NPEOs using 2,4,6-trimethylphenol or 4-t-butylphenol, respectively, as internal standards and then calculating individual response factors [60].

Reversed-phase HPLC on C_8, C_{18} silica gel columns, alumina-based C_{18} or polyethylene-coated alumina allows separation according to the character of the hydrophobic moiety and therefore provides complementary qualitative data, particularly on homologous alkylphenolic compounds [71]. In this case, the length of the ethylene oxide chain does not influence the separation and the various oligomers containing the same hydrophobic moiety elute in one peak (Fig. 2.3.6) [72]. Eluting all the oligomers into one peak has the advantage of increasing the peak intensity and, therefore, increasing the sensitivity of determination. This approach is particularly suitable for routine determination of OPEOs and NPEOs since the quantification is rather simplified. However, in this case, certain overlapping of homologues, e.g. linear octylphenol (OP) and NPEOs can occur (Fig. 2.3.7), because of the shift in retention time of the OP compound due to the change in alkyl chain shape [57]. RP–HPLC was applied for the simultaneous determination

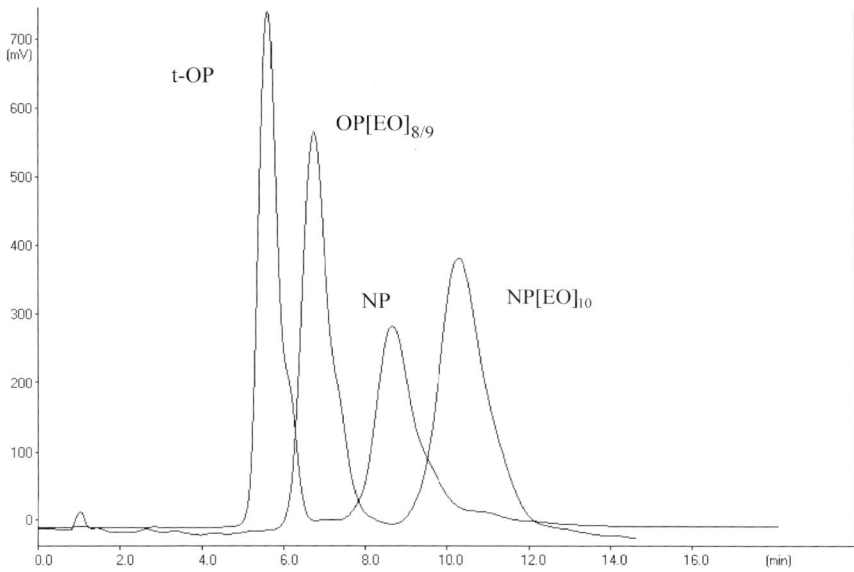

Fig. 2.3.6. Overlay of reversed-phase chromatograms of four commercial mixtures of alkylphenols and alkyl phenolethoxylates: *t*-OP, OP[EO]$_{8/9}$, 4-NP, and NP[EO]$_{10}$. Column: 125×4 mm^2 Lichrospher RP-18 (5 μm), isocratic elution 1.0 mL min^{-1}, mobile-phase 8:2 (v/v) methanol/water.

of APEOs and LAS using an elution gradient with acetonitrile and water-containing NaClO$_4$ [37], and for simultaneous determination of their corresponding degradation products (NPEC and SPC) using fluorometric detection (225 nm excitation and 295 nm emission) [73].

Alcohol ethoxylates (AEOs) and poly(ethylene glycol)s (PEGs)
Aliphatic AEOs, considered as environmentally safe surfactants, are the most extensively used non-ionic surfactants. The commercial mixtures consist of homologues with an even number of carbon atoms ranging typically from 12 to 18 or of a mixture of even−odd linear and α-substituted alkyl chains with 11−15 carbons. Furthermore, each homologue shows an ethoxymer distribution accounting typically for 1−30 ethoxy units with an average ethoxylation number in the range 5−15. The separation of the AEO complex mixtures was achieved by reversed-phase and normal-phase chromatographic systems [74−76].

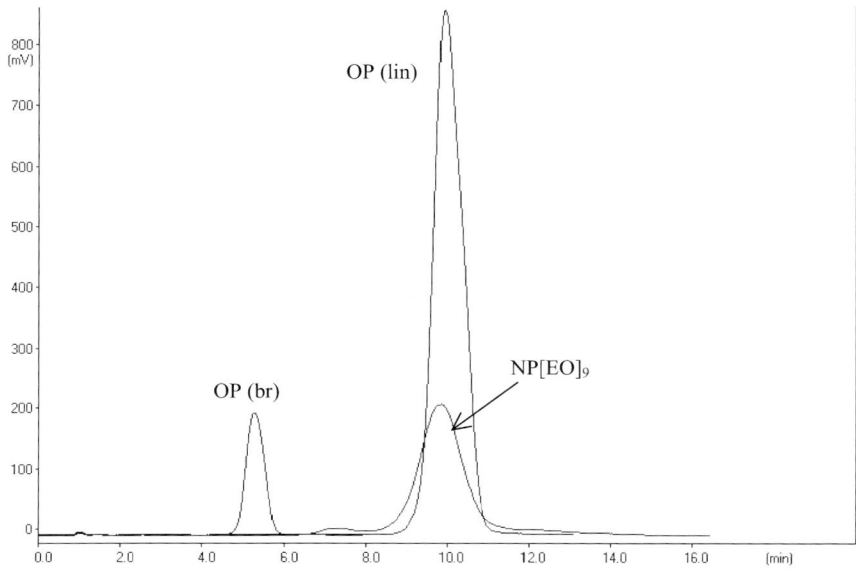

Fig. 2.3.7. Overlay of reversed-phase HPLC chromatograms of branched and linear octylphenol, and NP[EO]$_9$. Column: 125×4 mm^2 Lichrospher RP-18 (5 μm), isocratic elution 1.0 mL min^{-1}, mobile-phase 8:2 (v/v) methanol/water.

While the alkyl chain distribution is determined on a non-polar RP8 and RP18, EO homologue distribution is determined using a polar phase. AEOs are not UV-absorbing species, so they cannot be directly determined by HPLC followed by standard optical detection systems (UV and FL), unless suitable derivatives are prepared [2]. Because of this, methods based on liquid chromatography–mass spectrometry [77–79] are currently considered as the benchmark procedure that gives sufficiently high selectivity and sensitivity.

AEOs have been analysed by HPLC and UV or fluorescence detection after suitable derivatisation. The derivatising agents proposed so far are phenyl isocyanate [80,81], 1-anthroylnitrile [82], 3,5-dinitrobenzoyl chloride [83], naphthyl isocyanate [84] and naphthoyl chloride [84]. However, the lack of fluorescence activity and the need for synthesis through a multistep reaction for some derivatising agents limits their application in a real-world analysis. In fact, only a few of them were applied in the determination of AEOs in environmental samples. Zanette et al. [84] developed derivatisation and separation procedures for the routine determination of AEOs and PEGs in aqueous environmental samples. The AEO derivatives were separated into homologue-by-homologue and ethoxymer-by-ethoxymer by reversed-phase HPLC−fluorescence after derivatisation with naphthyl isocyanate and naphthoyl chloride, respectively. The ethoxymer-by-ethoxymer separation of PEGs was obtained by RP−HPLC−FL after derivatisation with naphthoyl chloride. The chromatograms of the naphthyl isocyanate derivatives of AEOs and NPEOs, and the naphthoyl chloride derivatives of PEGs in the influent, and final effluent of a sewage treatment plant are shown in Fig. 2.3.8.

Kiewiet et al. [81] used phenyl isocyanate for the derivatisation of $C_{12}EO−C_{18}EO$ in raw influents, settled influents and effluents of seven WWTP in the Netherlands, while Heining et al. [85] used the same method for 'fingerprinting' of technical products and household formulations. These derivatisation procedures have been shown to be suitable for the analysis of real-world samples, providing information on alkyl and/or ethoxylate distribution, but still lack the appropriate sensitivity. The sensitivity of these techniques is rather low, 5−10 ng of injected AEOs and PEGs ($5 \mu g L^{-1}$ for the pre-concentration of 1 L samples) using naphthyl isocyanate and naphthoyl chloride, and $3 \mu g L^{-1}$ using phenyl isocyanate (pre-concentration of 4.7 L of WWTP samples).

Refractometric [86] and ELSD [85,87] are alternative methods, which do not require derivatisation of AEOs.

A differential refractometric detector was applied to simultaneously analyse mixtures of AEOs and an anionic surfactant (sodium dodecyl-sulphate (SDS)) in water using ion-pair reversed-phase HPLC [86]. Eluents containing 55% acetonitrile in water with tetraethylammo-nium as the ion-pairing reagent and NaCl allowed reproducible separation of SDS and polydisperse $C_{12}EO$ surfactant mixtures on a C_8 column (chromatogram shown in Fig. 2.3.9). The disadvantages of this detection method are baseline drift when solvent gradients are used and their quite high detection limit.

The ELSD is not plagued with the problem of baseline shift and is significantly more sensitive however; it presents other disadvantages, such as the strong non-linearity of the detector response and the possibility of interference from non-volatile compounds in the sample matrix [87]. The HPLC method using traditionally reversed-phase solvents and a hybrid column/pre-column has demonstrated the separation of ethoxylated homologues of broadly distributed linear AEOs

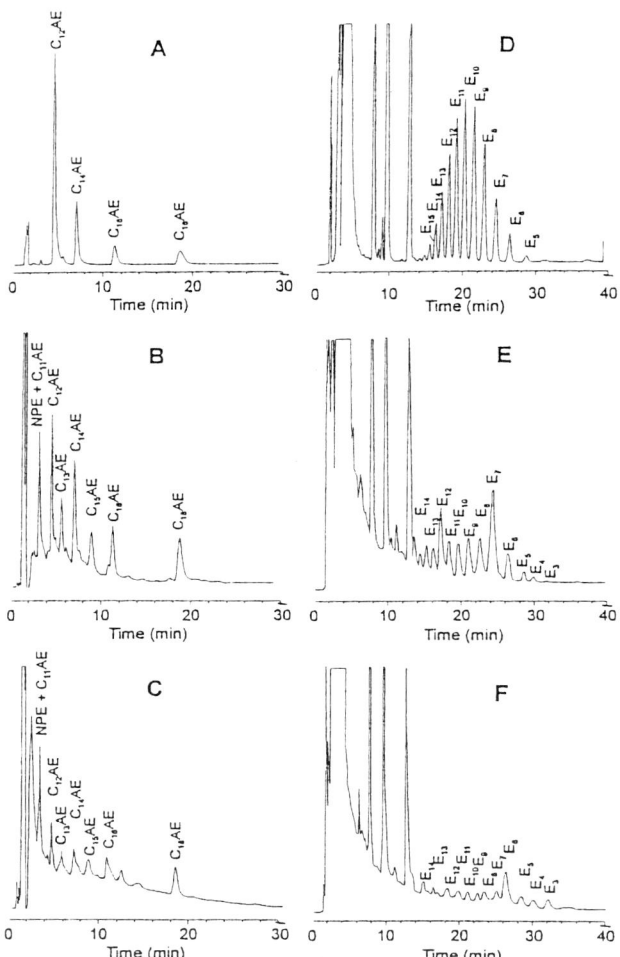

Fig. 2.3.8. Chromatogram of the NIC derivatives of oleochemical AEOs in a standard solution (a); AEOs and NPEOs in extracts of the influent (b) and final effluent (c) of a sewage treatment plant. Chromatogram of the NC derivatives of PEGs in a standard solution (d); and extracts of the influent (e) and final effluent (f). Stationary phase: C_{18} column; mobile-phase: methanol–acetonitrile (a)–(c) and acetonitrile–water (d)–(f).

Reprinted with permission from Ref. [84] © 1996 Elsevier.

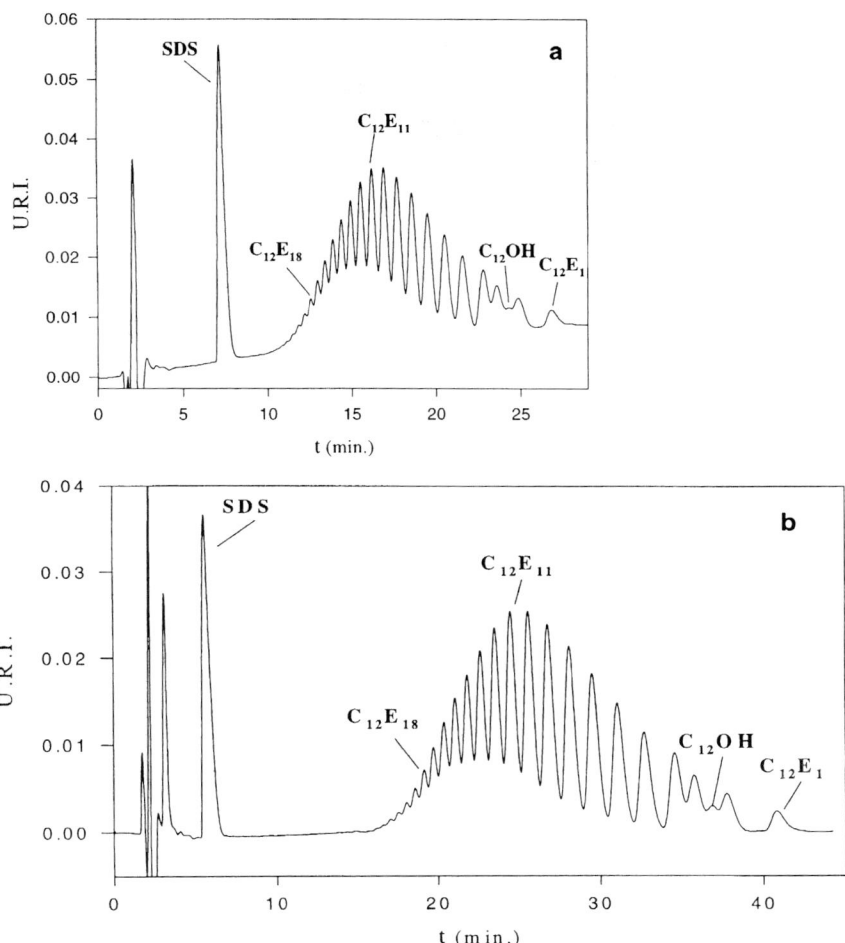

Fig. 2.3.9. Analysis of a mixture of SDS (0.2 g L^{-1}) and C$_{12}$E$_{11}$ (2 g L^{-1}) on C Ultraspher column (250 × 4.6 mm^2, d 5 μm) under the following operating conditions: (a) mobile-phase CH$_3$CN–water (58:42, v/v) + (C$_3$H$_7$)NBr 2 × 10^{-2} mol L^{-1} NaCl 4 × 10 mol L^{-1}, flow-rate 1.12 mL min^{-1} and temperature 29°C; (b) mobile-phase CH$_3$CN–water (55:45, v/v) + (C$_3$H$_7$)NBr 5 × 10 mol L^{-1} + NaCl 10^{-2} mol L^{-1}, flow-rate 1 mL min^{-1} and temperature 26°C. Reprinted with permission from Ref. [86] © 2000 Elsevier.

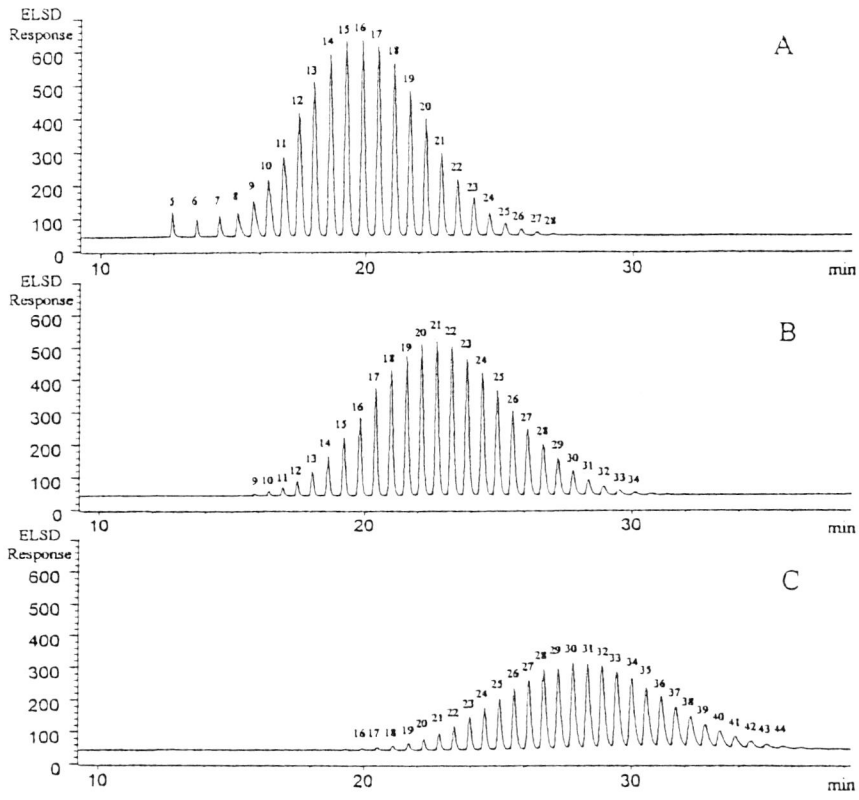

Fig. 2.3.10. Separation of pure-hydrophobe ethoxylated dodecyl alcohol surfactants. (a) $C_{12}E_{15.3}$ (867 mg L^{-1}); (b) $C_{12}E_{20.6}$ (859 mg L^{-1}); and (c) $C_{12}E_{30.0}$ (861 mg L^{-1}). ELSD detection for all chromatograms. Reprinted with permission from Ref. [87] © 1996 Elsevier.

with numbers of EO groups ranging from 3 to 50. Figure 2.3.10(a)–(c) shows separation of several pure-hydrophobe ethoxylated dodecyl alcohol surfactants, demonstrating the wide range of applicability of this method.

Native PEGs represent a rather difficult class of compounds for either RP–HPLC or NP–HPLC due to their high polarity [3]. On polar stationary phases (silica, diol, CN, aminopropyl) they are very strongly retained and strong solvents like methanol or ethanol are needed to elute them. However, such an elution leads to very poor resolution of individual oligomers, which usually elute in one single peak. In contrast, for hydrophobic sorbents (C_8 and C_{18}) the retention of these polar compounds is very weak. Despite this apparent drawback, several HPLC methods were proposed and excellent separation of PEG oligomers were reported [3].

Other non-ionic surfactants
Octadecyl silica and poly(styrene-divinylbenzene) were found to be well suited as a stationary phase in reversed-phase HPLC separation of fatty acid mono- and diethanol amides (FAMA and FADA). Using methanol–water mobile-phase homologues containing C_{10}–C_{18} alkyl groups were eluted, according to the increasing alkyl chain length. These compounds, widely used as foam boosters and foam stabilisers in liquid detergents and cosmetic products, can be detected using UV (210 nm), refractive index or nitrogen-specific chemiluminescence detection [1].

Ethoxylated alkyl amines in pesticide formulations were separated using two different columns: a cyano-modified silica column to determine the alkyl distribution and an amino-modified column to determine the ethylene oxide distribution [88]. The detection, specific for ethoxylated amine, was performed with a post-column ion-pair extraction system and fluorescence detection.

Commercial alkyl polyglycosides (APGs) are complex mixtures of alkyl homologues, oligomers of different hydrophilic sugar moiety, anomers (α- and β-anomeric forms with respect to the bonding between the sugar and the alkyl chain) and isomers (furanosides and pyranosides). Isocratic RP–HPLC, combined with an ELSD, has proven to be a reliable method that allows monitoring of the kinetics of alkyl glycoside synthesis, which includes simultaneous monitoring of the initial carbohydrate and the surfactant formed [89]. An HPLC method using C_8 column and an eluent mixture of water/acetonitrile (60:40, v/v) was able to detect 30 ppm APG using an RI detector, while in simultaneous UV detection a detection threshold of 100 ppm was achieved [90].

Gradient elution RP–HPLC allows the separation of APGs according to their alkyl chain length and simultaneous separation of alkyl ethoxy glucosides with regard to the length of the ethoxy chain. NP–HPLC with isooctane-ethyl acetate-2-propanol in the gradient mode allows the separation of alkyl glycosides into α- and β-anomers [91].

2.3.4 Conclusion

LC with conventional detectors (e.g. UV and fluorescence detection systems) has been extensively used for the separation and quantitative determination of ionic and non-ionic surfactants in both industrial products and various environmental samples. This method is a readily available technique in many laboratories and has proved to be a versatile technique directly applicable to different surfactants with a chromophore group and to their degradation products. However, the application of UV or FL detectors in the analysis of fatty acids, fatty alcohol derivatives, as well as some cationic surfactants, requires a time-consuming derivatisation step prior to LC analysis. Furthermore, the complexity of surfactant mixtures consisting of various isomers, homologues and oligomers and the lack of reference standards limit the applicability of conventional detectors. For the analysis of complex environmental samples, the LC technique using conventional detectors is often unsuitable for trace level determination due to low sensitivity and lack of selectivity. Co-elution of interfering matrix components may additionally hamper the accurate quantification of target compounds.

At present, within modern analytical techniques, only GC and LC combined with MS and tandem MS, respectively, provide sufficient selectivity and inherent sensitivity in analysing surfactants in complex samples.

REFERENCES

1   T.M. Schmitt, *Analysis of Surfactants*. Marcel Dekker Inc., New York, 2001, p. 197.
2   A. Marcomini and M. Zanette, *J. Chromatogr. A*, 733 (1996) 193.
3   K. Rissler, *J. Chromatogr. A*, 742 (1996) 1.
4   K. Kunihiro, A. Nakae and G. Muto, *Bunseki Kagaku*, 24 (1975) 188.
5   A. Nakae, K. Kunihiro and G. Muto, *J. Chromatogr.*, 134 (1977) 459.
6   A. Nakae and K. Kunihiro, *J. Chromatogr.*, 152 (1978) 137.
7   A. Nakae, K. Tsuji and M. Yamanaka, *Anal. Chem.*, 52 (1980) 2275.
8   S. Fudano and K. Konishi, *J. Chromatogr.*, 93 (1974) 467.
9   L. Cavalli and M. Lazzarin, *La Rivista delle Sostanze Grasse*, 64 (1987) 263.
10 A. Nakae, K. Tsuji and M. Yamanaka, *Anal. Chem.*, 53(12) (1981) 1818.
11 A. Di Corcia, M. Marchetti, R. Samperi and A. Marcomini, *Anal. Chem.*, 63 (1991) 1179.

12 E. González-Mazo, J.M. Quiroga, D. Sales and A. Gómez-Parra, *Toxicol. Environ. Chem.*, 59 (1997) 77.

13 E. González-Mazo, J.M. Forja and A. Gómez-Parra, *Environ. Sci. Technol.*, 32 (1998) 1636.

14 M.S. Holt, E. Matthijs and J. Waters, *Water Res.*, 6 (1989) 749.

15 E. Matthijs and M. Stalmans, *Tenside Surf. Det.*, 30 (1993) 29.

16 S. Terzic, D. Hrsak and M. Ahel, *Mar. Pollut. Bull.*, 24 (1992) 199.

17 S. Terzic and M. Ahel, *Bull. Environ. Contam. Toxicol.*, 50 (1993) 241.

18 S.L. Tong and C.B. Tan, *Int. J. Environ. Anal. Chem.*, 50 (1993) 73.

19 R. Gloor and E.L. Johnson, *J. Chromatogr. Sci.*, 13 (1977) 413.

20 S. Eksborg, P. Lagerström, R. Modin and G. Schill, *J. Chromatogr.*, 83 (1973) 99.

21 K. Wahlund, *J. Chromatogr.*, 115 (1975) 411.

22 R.D. Swisher, W.E. Gledhill, R.A. Kimerle and T.A. Taulli, *Proceedings of VII International Congress on Surface-Active Substances*, Moscow, 1978, p. 218.

23 C.R. Eggert, R.G. Kaley and W.E. Gledhill, In: A.W. Bourquiny and P.H. Pritchard (Eds.), *Proceedings of the Workshop—Microbial Degradation of Pollutants in Marine Environments*, USA, 1979, p. 451.

24 P.W. Taylor and G. Nickless, *J. Chromatogr.*, 178 (1979) 259.

25 A. Marcomini, S. Busetti, A. Sfriso, S. Capri, T. La Noce and A. Liberatori, In: G. Angeletti and A. Bioset (Eds.), *Organic Micropollutants in the Aquatic Environment*, Kluwer, Dorchrecht, 1991, p. 294.

26 V.M. León, E. González-Mazo and A. Gómez-Parra, *J. Chromatogr. A*, 889 (2000) 211.

27 E. González-Mazo, M. Honing, D. Barceló and A. Gómez-Parra, *Environ. Sci. Technol.*, 31 (1997) 504.

28 L. Cavalli, G. Cassani and M. Lazzarin, *Tenside Surf. Det.*, 33 (1996) 158.

29 L. Sarrazin, A. Arnoux and P. Rebouillon, *J. Chromatogr. A*, 760 (1997) 285.

30 J. Field, J.A. Leenheer, K.A. Thorn, L.B. Barber II, C. Rostad, D.L. Macalady and S.R. Daniel, *J. Contam. Hydrol.*, 9 (1992) 55.

31 C. Crescenzi, A. Di Corcia, E. Marchiori, R. Samperi and A. Marcomini, *Water Res.*, 30 (1996) 722.

32 M. Kikuchi, A. Tokai and T. Yoshida, *Water Res.*, 20(5) (1986) 643.

33 K. Inaba and K. Amano, *Int. J. Environ. Anal. Chem.*, 34 (1988) 203.

34 M.S. Holt, E. Matthijs and J. Waters, *Water Res.*, 6 (1989) 749.

35 E. Matthijs and H. De Henau, *Tenside Det.*, 24 (1987) 193.

36 M.A. Castles, B.L. Moore and S.R. Ward, *Anal. Chem.*, 61 (1989) 2534.

37 A. Marcomini and W. Giger, *Anal. Chem.*, 59 (1987) 1709.

38 I. Fujita, Y. Ozasa, T. Tobino and T. Sugimura, *Chem. Pharm. Bull.*, 38 (1990) 1425.

39 J.A. Field, J.A. Leenheer, K.A. Thorn, L.B. Barber II, C. Rostad, D.L. Macalady and S.R. Daniel, *J. Contam. Hydrol.*, 9 (1992) 55.

40 S.A. Shamsi and N.D. Danielson, *J. Chromatogr. Sci.*, 33 (1995) 505.

111

41 D. Zhou and D.J. Pietrzyk, *Anal. Chem.*, 64 (1992) 1003.

42 D.J. Pietrzyk, P.G. Rigas and D. Yuan, *J. Chromatogr. Sci.*, 27 (1989) 485.

43 S.A. Maki, J. Wangsa and N.D. Danielson, *Anal. Chem.*, 64 (1992) 583.

44 G.R. Bear, *J. Chromatogr. A*, (1988) 91.

45 J.A. Wilkes, G. Walraven and J. Talbot, *J. Am. Oil Chem. Soc.*, 69 (1992) 609.

46 J. Fisher and P. Jandera, *J. Chromatogr. B*, 681 (1996) 3.

47 P. Gerike, H. Klotz, J.G.A. Kooijman, E. Matthijs and J. Waters, *Water Res.*, 28(1) (1994) 147.

48 D.G.P.A. Breen, J.M. Horner, K.D. Bartle, A.A. Clifford, J. Waters and J.G. Lawrence, *Water Res.*, 30 (1996) 476.

49 P. Helboe, *J. Chromatogr.*, 261 (1983) 117.

50 K. Nakamura and Y. Morikawa, *J. Am. Oil Chem. Soc.*, 59 (1982) 64.

51 F. Spagnolo, M.T. Hatcher and B.K. Faulseit, *J. Chromatogr. Sci.*, 25 (1987) 399.

52 V.T. Wee and J.M. Kennedy, *Anal. Chem.*, 54 (1982) 1631.

53 S.M. Gort, E.A. Hogendoorn, R.A. Baumann and P. Van Zoonen, *Int. J. Environ. Anal. Chem.*, 53 (1993) 289.

54 M. Schoester and G. Kloster, *Vom Wasser*, 77 (1991) 13.

55 P. Fernandez, A.C. Alder, M.J.F. Suter and W. Giger, *Anal. Chem.*, 68 (1996) 921.

56 M. Shibukawa, R. Eto, A. Kira, F. Miura, K. Oguma, H. Tatsumoto, H. Ogura and A. Uchiumi, *J. Chromatogr. A*, 830 (1999) 321.

57 P. De Voogt, personal communication.

58 M. Ahel and W. Giger, *Anal. Chem.*, 57 (1985) 1577.

59 M. Ahel and W. Giger, *Anal. Chem.*, 57 (1985) 2584.

60 M. Ahel, W. Giger, E. Molnar and S. Ibric, *Croat. Chem. Acta*, 73 (2000) 209.

61 M. Ahel, T. Conrad and W. Giger, *Environ. Sci. Technol.*, 21 (1987) 697.

62 A.M. Rothman, *J. Chromatogr.*, 253 (1982) 283.

63 A.A. Boyd-Boland and J.B. Pawliszyn, *Anal. Chem.*, 68 (1996) 1521.

64 M.S. Holt, E.H. McKerrel, J. Perry and R.J. Watkinson, *J. Chromatogr.*, 362 (1986) 419.

65 E. Kubeck and C.G. Naylor, *J. Am. Oil Chem. Soc.*, 67 (1990) 400.

66 M.J. Scarlett, J.A. Fisher, H. Zhang and M. Ronan, *Water Res.*, 28 (1994) 2109.

67 M. Kudoh, H. Ozawa, S. Fudaro and K. Tsuji, *J. Chromatogr.*, 287 (1984) 337.

68 C. Zhou, A. Bahr and G. Schwedt, *Anal. Chim. Acta*, 236 (1990) 273.

69 D. Brown, H. De Henau, J.T. Garrigan, P. Gerike, M. Holt, E. Kunkel, E. Matthijs, E. Keck, J. Waters and R.J. Watkinson, *Tenside Surf. Det.*, 24 (1987) 14.

70 D.T. Bennie, C.A. Sullivan, J.B. Lee, T.E. Peart and R.J. Maguire, *J. Sci. Total Environ.*, 193 (1997) 263.

71 W. Giger, P.H. Brunner and C. Schaffner, *Science*, 225 (1984) 623.

72 P. de Voogt, K. de Beer and F. van der Wielen, *Trends Anal. Chem.*, 16 (1997) 584.

73 A. Di Corcia, A. Marcomini and R. Samperi, *Environ. Sci. Technol.*, 28 (1994) 850.

74 R. Spilker and H.B. Mekelburger, *Tenside Surf. Det.*, 35 (1998) 415.

75 K. Lemr, *J. Chromatogr. A*, 732 (1996) 299.

76 T. Okada, *J. Chromatogr.*, 609 (1992) 213.

77 K.A. Evans, S.T. Dubey, L. Kravetz, J. Gumulka, R. Mueller and J.R. Stork, *Anal. Chem.*, 55 (1994) 699.

78 M. Petrovic and D. Barcelo, *Anal. Chem.*, 72 (2000) 4560.

79 C. Crescenzi, A. Di Corcia, R. Samperi and A. Marcomini, *Anal. Chem.*, 67 (1995) 1797.

80 A.M. Rothman, *J. Chromatogr.*, 253 (1982) 283.

81 A. Kiewiet, J.M.D. van der Steen and J.R. Parsons, *Anal. Chem.*, 67 (1995) 4409.

82 P.L. Desbene, B. Desmazeires, J.J. Basselier and A. Desbene Monvernay, *J. Chromatogr.*, 461 (1989) 305.

83 A. Nozava and T. Ohnuma, *J. Chromatogr.*, 187 (1980) 261.

84 M. Zanette, A. Marcomini, E. Marchiori and R. Samperi, *J. Chromatogr. A*, 756 (1996) 159.

85 K. Heining, C. Vogt and G. Werner, *Anal. Chem.*, 70 (1998) 1885.

86 F.I. Portet, C. Treiner and P.L. Desebène, *J. Chromatogr. A*, 878 (2000) 99.

87 T.C.G. Kibbey, T.P. Yavaraski and K.F. Hayes, *J. Chromatogr. A*, 752 (1996) 155.

88 R.H. Schreuder, A. Martijn, H. Poppe and J.C. Kraak, *J. Chromatogr.*, 368 (1986) 339.

89 P. Chaimbault, C. Elfakir, M. Lafosse, F. Blanchard and I. Marc, *J. High Resol. Chromatogr.*, 22 (1999) 188.

90 H.S. Klaffke, T. Neubert and L.W. Kroh, *Tenside Surf. Det.*, 35 (1998) 2.

91 G. Czichocki, H. Fiedler, K. Haage, H. Much and S. Weidner, *J. Chromatogr. A*, 943 (2002) 241.

2.4 ATMOSPHERIC PRESSURE IONISATION MASS SPECTROMETRY—I. GENERAL ASPECTS

Thomas P. Knepper and Peter Eichhorn

2.4.1 Introduction

In comparison with the detection systems described in Chapters 2.1–2.3, a mass spectrometric (MS) detector, selecting ions in a sample according to their mass-to-charge ratio, offers unequalled selectivity and sensitivity and provides valuable information regarding molecular weight and species analysis. The generated mass spectra may also provide useful information on the structure of the individual analytes and may therefore assist in structural identification of unknown compounds. These features are highly valuable in the trace analysis of surfactants and their degradation products, especially in the challenging and complex matrices common to environmental samples.

2.4.2 Ionisation methods in LC–MS coupling

In combining liquid chromatography (LC) separation with MS detection, three major instrumental difficulties can be encountered, namely:

- flow-rate incompatibility, which results from the need to introduce the mobile phase, flowing up to 1 mL min^{-1}, into the high vacuum of the mass analyser;
- LC eluent incompatibility, due to frequently used non-volatile additives such as buffers or ion pair (IP) reagents;
- low amenability of non-volatile and/or thermally labile compounds to produce gas phase ions.

These problems have largely been solved by the development of a wide variety of powerful LC–MS interfaces (reviewed in Refs. [1–3]). In the following paragraphs, the two most widely used atmospheric pressure ionisation (API) systems, namely atmospheric pressure chemical ionisation (APCI) and electrospray ionisation (ESI), are briefly described, along with the older technique of thermospray ionisation

Comprehensive Analytical Chemistry XL
T.P. Knepper, D. Barceló and P. de Voogt (Eds)

(TSI), whose applications have decreased in recent years. All the three API techniques have in common that the liquid inlet system is combined with the ionisation source. However, the main differences are as follows:

- *APCI*. The column effluent is nebulised into an atmospheric-pressure ion source. Through a corona discharge, electrons initiate the reactant gas-mediated ionisation of the analytes. Proton transfers are typical reactions generating $[M + H]^+$ or $[M - H]^-$ ions, although radical ion formation is possible as in high vacuum chemical ionisation (CI). The ions formed are injected into the high vacuum of the mass spectrometer. APCI typically accepts flow rates of up to 2 mL min^{-1}.
- *ESI*. The LC eluent enters the ionisation chamber at atmospheric pressure through a narrow capillary tube, the end of which is maintained at a high positive or negative potential (Fig. 2.4.1). The strong electric field creates a mist of small highly charged droplets. Through vaporisation of the solvents, the droplets become rapidly smaller until they reach a critical surface charge density. At this point, ions begin to desorb from the surface of the droplets into the gas phase and are subsequently extracted into the MS analyser.
- *TSI*. The liquid is converted into a vapour jet and small droplets are generated with the help of a heated vapouriser tube. A buffer dissolved in the eluent assists the ionisation process through the formation of adduct ions, which are produced via statistical charging of individual droplets. Due to the 'softness' of the procedure, no structurally characteristic fragments, which could aid identification of unknown compounds, are formed.

TSI, APCI and ESI are generally considered soft ionisation techniques, as they usually generate pseudo-molecular ions, adduct ions and clusters, which thereby yield molecular weight information. The ion source parameters of the two latter modes can, however, be adjusted in such a manner that some in-source fragmentation, also termed as in-source collision-induced dissociation (CID), is induced (for a discussion of CID relevant to MS–MS see Chapter 2.4.3). By increasing the voltage applied to the sample cone, which extracts the ions from the atmospheric pressure region and transmits them into the first vacuum region, the ions are accelerated more quickly through the low vacuum region resulting in collisions with other solvent or gas molecules. By these means, structurally informative fragmentation can

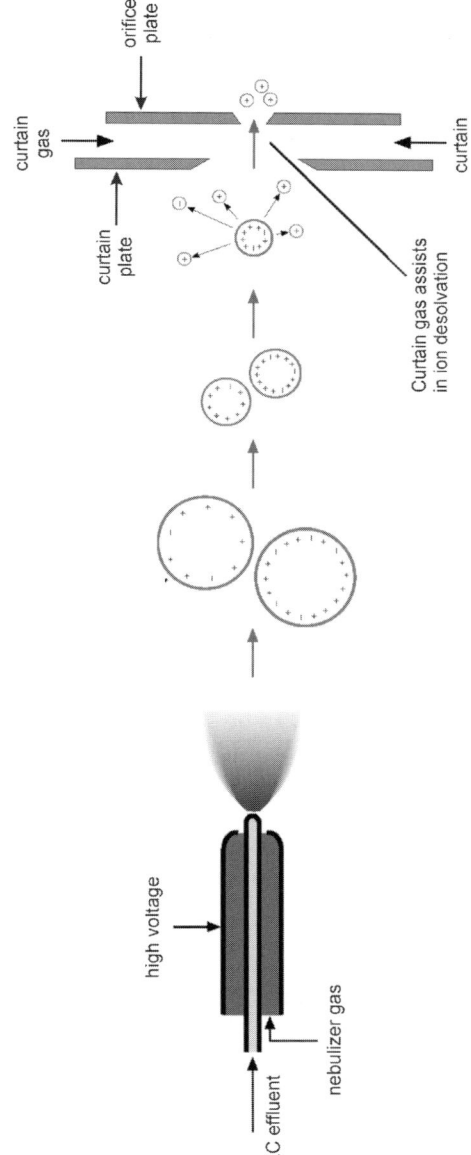

Fig. 2.4.1. Process of ion generation in ESI interface. Adapted from Perkin–Elmer Sciex instruction manual of LC–MS–API 150.

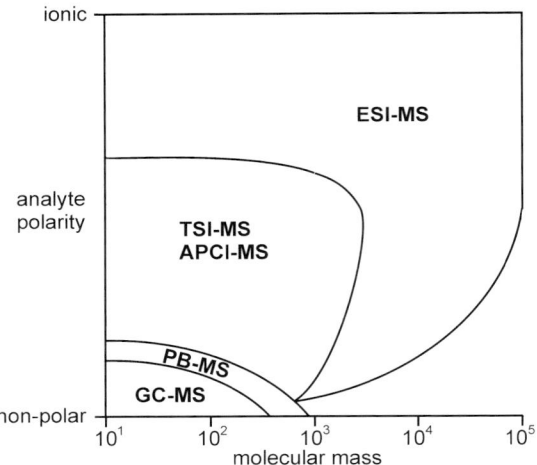

Fig. 2.4.2. Application ranges of different LC−MS interfaces (adapted from Ref. [3]).

be obtained. Even if the extent and type of fragmentation are different from and less extensive than conventional electron impact techniques, as is usually the case, in-source CID can be a very valuable tool in identifying unknown compounds.

The suitability of each interface for a particular analysis is determined by the polarity of the analyte and its molecular mass. From the scheme presented in Fig. 2.4.2, it can be seen that APCI is advantageous for less to moderately polar compounds, while the most suitable choice for charged analytes, or those for which solution-phase charging can be readily obtained is ESI. A unique feature of ESI is that molecules with high molecular masses appear as multiple charged ions allowing the detection of large biomolecules even on instruments with a rather limited m/z range.

2.4.3 Mass analysers and detection modes

In addition to the diversity of ionisation techniques available, mass spectrometers offer a selection of mass analyser configurations. Of note are single (MS) and triple quadrupole (MS−MS) instruments, ion trap analysers (MS)^n, time-of-flight (ToF) analysers, sector field analysers, and Fourier transform-ion cyclotron resonance (FTICR) instruments.

Mass resolution generally increases for instruments in the listed order. Hybrid variants utilising various composites of mass analysers are likewise available and have become commonplace (e.g. quadrupole-ToF).

The instruments may be operated in different detection modes depending on the objectives of the work. Table 2.4.1 gives an overview of the distinct modes of quadrupole mass spectrometers in terms of application, content of information and sensitivity.

The potential for generating CID spectra by triple quadrupole instruments has played an important role in the identification of surfactants in complex mixtures analysed by TSI instruments. Since the softness of this ionisation process does not induce in-source fragmentation—this feature is offered solely by ESI and APCI interfaces (see above)—structural information can only be gathered by fragmentation of selected precursor ions in the collision cell, thus yielding daughter ion spectra [4−7]. The high selectivity of this procedure has been exploited for screening purposes of surfactants occurring in wastewaters by performing flow-injection analysis (FIA) of sample extracts, i.e. their direct infusion into the ion source. Time-consuming chromatographic separations could therefore be circumvented. However, the application of FIA−TSI−MS−MS, and also FIA−MS−MS, methods in general have their limitations since ambiguous results may be obtained when CID spectra of two or more isobaric precursor ions are present in the extract and are thus simultaneously recorded (see Chapter 2.5). This may lead to misinterpretations in assigning chemical structures and precludes any quantitative assessment.

2.4.4 Analysis of surfactants in environmental samples by LC−MS

LC−API−MS as an analytical method has been recognised as a robust analytical technique in environmental chemistry [8] and dramatically changed the potential for analysing organic micropollutants and their degradation products in water analysis [9]. Of the various techniques existing for surfactant analysis in the aquatic environment, more recently the method of choice is quite often API−MS, generally coupled with LC. Until the mid-1990s, when robust instruments fitted with soft API sources (APCI and ESI) became available at reasonable costs, GC−MS methods with time-consuming derivatisation steps, HPLC with

TABLE 2.4.1

Detection modes of single and triple quadrupole mass spectrometers

Mode	Detection and application	Information	Sensitivity
Scan	Full scan spectra; identification; structure elucidation	High	Low
Selected ion monitoring (SIM)	Detection of selected molecular ions or fragments; quantitation	Low	High
Multiple reaction monitoring (MRM)[a]	Detection of selected mass transfers; quantitation	Low	High
Product ion scan (PIS)[a]	CID spectra of selected ions	High	Medium
Precursor ion scan[a]	Detection of ions with identical fragments	High	Medium
Neutral loss scan[a]	Scan with constant loss of neutral group; identification of functional groups	High	Medium

[a]Modes only available on triple-quad instruments.

relatively insensitive ultraviolet and fluorescence detection and TSI-MS approaches were predominantly used. The main reasons for applying the LC–MS coupling as a cornerstone in the analytical methodologies are due to the following requirements:

- A flexible and efficient chromatographic separation, offering the possibility for separating complex mixtures of compounds with a range of polarities, was needed. This preferably should be accomplished without tedious derivatisation steps, and should allow direct analysis of aqueous samples.
- High sensitivity, essential for the detection of trace amounts of analytes, in combination with high selectivity, required for the examination of challenging matrices, is also necessary.
- The detection system should assist in the structural elucidation of unknown organic compounds. Moreover, accurate quantitative determinations should be feasible.

The trend of discovering the analytical field of environmental analysis of surfactants by LC–MS is described in detail in Chapters 2.6–2.13 and also reflected by the method collection in Chapter 3.1 (Table 3.1.1), which gives an overview on analytical determinations of surfactants in aqueous matrices. Most methods have focused on high volume surfactants and their metabolites, such as the alkylphenol ethoxylates (APEO, Chapter 2.6), linear alkylbenzene sulfonates (LAS, Chapter 2.10) and alcohol ethoxylates (AE, Chapter 2.9). Surfactants with lower consumption rates such as the cationics (Chapter 2.12) and esterquats (Chapter 2.13) or the fluorinated surfactants perfluoro alkane sulfonates (PFAS) and perfluoro alkane carboxylates (PFAC) used in fire fighting foams (Chapter 2.11) are also covered in this book, but have received less attention.

2.4.5 Conclusion

Many chapters in this book, especially those dealing exclusively with LC–MS determination of surfactants and their metabolites, illustrate the power of MS state-of-the-art techniques for accurate determinations of polar organic pollutants in environmental samples. LC–ESI–MS(MS) is of unparalleled value in identifying and characterising degradative products and is a highly valuable tool in elucidating metabolic pathways of surfactants as well as for investigating the occurrence of surfactants and their metabolites in the environment. This is of particular importance in view of strongly polar and recalcitrant degradation products, which are reported to find their ways from wastewaters into various environmental compartments, including drinking waters.

REFERENCES

1   J. Abian, *J. Mass. Spectrom.*, 34 (1999) 157.
2   W.M.A. Niessen, *J. Chromatogr. A*, 794 (1998) 407.
3   W.M.A. Niessen and A.P. Tinke, *J. Chromatogr. A*, 703 (1995) 37.
4   H.F. Schröder, *Vom Wasser*, 79 (1992) 193.
5   H.F. Schröder, *J. Chromatogr.*, 643 (1993) 145.
6   H.F. Schröder, *Vom Wasser*, 81 (1993) 299.
7   H.F. Schröder, *Fresenius J. Anal. Chem.*, 353 (1995) 93.
8   D. Barceló (Ed.), Applications of LC-MS in Environmental Chemistry, *J. Chromatogr. Library*, 59 (1996).
9   T. Reemtsma, *Trends Anal. Chem.*, 20 (2001) 500.

2.5 ATMOSPHERIC PRESSURE IONISATION MASS SPECTROMETRY—II. FLOW INJECTION ANALYSIS—MASS AND TANDEM MASS SPECTROMETRY IN THE ANALYSIS OF SURFACTANTS—ADVANTAGES AND DISADVANTAGES

Horst Fr. Schröder

2.5.1 Introduction

The space of time from the first MS–MS (tandem mass spectrometer) instrument built by Thomson [1] to the coupling of liquid chromatography with mass (MS) and MS–MS spectrometers lasted about 60 years. The initial step in liquid chromatography coupled with MS (LC–MS) in the early 1970s, the introduction of liquids and liquid mixtures through a narrow capillary into a high-vacuum system, was first undertaken by Talroze et al. [2]. This step developed into an approach that combined a high resolution liquid chromatographic technique 'high-performance liquid chromatography' (HPLC) and the universal detector 'mass spectrometer' (MS). With this development, the first step was taken in the direction of flow injection analysis (FIA) being coupled on-line to MS and MS–MS instruments.

Several years later, the next step in the application of MS–MS for mixture analysis was developed by Hunt et al. [3–5] who described a master scheme for the direct analysis of organic compounds in environmental samples using soft chemical ionisation (CI) to perform product, parent and neutral loss MS–MS experiments for identification [6,7]. The breakthrough in LC–MS was the development of soft ionisation techniques, e.g. desorption ionisation (continuous flow-fast atom bombardment (CF-FAB), secondary ion mass spectrometry (SIMS) or laser desorption (LD)), and nebulisation ionisation techniques such as thermospray ionisation (TSI), and atmospheric pressure ionisation (API) techniques such as atmospheric pressure chemical ionisation (APCI), and electrospray ionisation (ESI).

One of the most serious drawbacks that has been observed in the ionisation processes with soft ionisation techniques is the very soft generation of ions. This process, which predominately leads to molecular ions or adduct ions but no fragments for identification, however, was also used to improve and speed-up MS analysis.

Comprehensive Analytical Chemistry XL
T.P. Knepper, D. Barceló and P. de Voogt (Eds)

The technique FIA–MS–MS is based on the mixture analysis technique [6,7]. Applying this technique, the generation of fragments for the unequivocal identification of compounds is also essential, but the follow-up in the ionisation process is quite different from ionisation observed with fragment generating interfaces. While ionisation techniques such as electron impact (EI) in gas chromatography–mass spectrometric analysis (GC–MS) and particle beam (PB) in LC–MS result in fragments just from the beginning, in the mixture analysis approach, soft ionising interfaces are pre-requisite. The fragmentation then is achieved by collision induced dissociation (CID, also termed collision activated dissociation (CAD)), which can be briefly described as follows. After ionisation of molecular or adduct ions a selection process by mass filter is performed and ions selected out of the mixture because of their defined mass per charge ratio (m/z), are allowed to fly into the collision chamber where they collide with noble gas atoms. Through collision processes, the parent ions (molecular or adduct ions) are destroyed, resulting in fragment ions (product ions), which are recorded as product ion spectra. Using soft ionisation interfaces (APCI, ESI, TSI and FAB) prior to fragmentation, the entire spectrum of ionisable parent ions available as an overview spectrum will be generated from the mixture of compounds. Selected ions defined by their m/z ratios are allowed to enter the collision cell where they are fragmented by CID. If fragment-generating PB ionisation is applied, a separation prior to ionisation becomes essential because multiple fragment ions from all mixture components together will be observed from the very beginning. Therefore, identification of defined mixture constituents will be impossible. Thus, fragment generating ionisation can only be used in the analysis of mixtures after a successful GC or LC separation. For identification in MS–MS mode, the soft ionising API methods are successfully amenable for both, FIA and LC.

2.5.2 Information provided by FIA–MS and MS–MS

MS–MS is a general analytical method for the analysis of unknown compounds. This analytical technique must be applied if soft ionisation techniques are performed and fragment information for identification is needed. With the different types of data obtained by MS–MS application, e.g. product ion scan, parent ion scan, neutral loss scan, and single-reaction monitoring, a large amount of information concerning the compounds analysed will be provided.

The first step on the way to accumulating information from unknown mixtures, which contain dissolved analytes of concern, is the insertion of this liquid mixture into the MS instrument. By applying FIA–MS in combination with soft ionisation techniques, and by by-passing analytical columns as described in the literature [8,9], this procedure presents a survey of the compounds ionisable under these conditions. The series of signals obtained under soft API conditions in the overview spectra are molecular or cluster ions, which represent the compound ions and provide the molar mass information of ionisable compounds; however, they are not suitable for identification. The application of FIA–MS–MS, which generates fragments of selected parent ions by CID—the so-called product ions—is therefore essential for identification. The diagnostic parent ion scan and neutral loss experiments may support characterisation and identification. Here, the unique analytical aspects of MS–MS can be exploited to screen unknown mixtures quite rapidly for compound classes in the parent or neutral loss mode [10,11] while confirmation of functional group identity can be obtained by the generation of product ion spectra. Therefore, it is not necessary to identify every compound in a mixture if only selected compounds, e.g. surfactants containing polyethylene glycol (PEG), sulfonated surfactants, etc. are of interest. Selectivity and speed can be achieved in this mode.

The present chapter contains advantages of FIA–MS and MS–MS which, on the basis of theoretical considerations and real applications, have resulted from this method, and also the disadvantages related to the application of FIA–MS and MS–MS, especially for the examination of unknown matrices.

2.5.3 Application of FIA–MS and MS–MS

The soft API techniques such as APCI and ESI, which are predominantly applied nowadays fulfil the most desirable criteria for an ionisation method in MS–MS, especially for mixture analysis, where each compound contained in the mixture produce as few—ideally only one—ions of different mass-to-charge ratio (m/z) as possible upon ionisation. The result would be a quite simple FIA–MS overview spectrum with few interferences in the parent ion to be selected and fragmented by CID. Under these conditions, an increased sensitivity with lower detection limits is achieved, since the ion current for a given compound is 'concentrated' in a single signal with a defined m/z ratio, not broadened by LC retention. Therefore, molecular or adduct ions of compounds—solvents or reagents for ionisation support—present in the solution during the ionisation process will be generated. Adduct ions of compounds consisting of the molecules together with sodium or other alkaline ions and ammonium ions in the positive mode can be observed, respectively. Acetonitrile and methanol, acetic or formic acid as neutrals or ion-pairing reagents such as methane sulfonic acid or acetate [12] in the positive or negative mode, respectively, build-up adduct ions which then can be observed. So first molecular weight information is provided if molecular ions are generated or if the clustering ions in the adduct ions generated are known.

With the commercially available MS–MS instruments in the 1990s, MS–MS helped to overcome these identification obstacles via CID in MS–MS mode or via ion trap in MS^n mode. In parallel, the applicability of MS–MS and MS^n by CID on tandem or ion trap MS become much easier and more informative with new MS spectrometric hardware, which is supported by the implemented improved computer systems. This combination of high-end data systems with more automated MS–MS and MS^n instruments facilitates the application of MS–MS and leads to a tremendous increase in the information output.

One disadvantage is that product ion libraries amenable to MS–MS results obtained either by FIA or LC are not commercially available. Despite the efforts made in the past [13], existing laboratory-made libraries generated by triple quad instruments [14] have not been adaptable to mass spectrometers other than of the same type. New perspectives in identification, using library data bases, can be expected from libraries elaborated by ion trap instruments that provide the MS^n option. The differences in fragment ion patterns observable in the results obtained from fragment generating interfaces and techniques or by application of soft ionisation techniques in combination with MS–MS, e.g. EI, PB, in-source CID, CI or CID spectra generated by triple quad MS or by ion trap, respectively, might be quite impressive.

Depending on the compound mixtures to be analysed, time-saving with FIA–MS and MS–MS can be impressive. So, in industrial applications, FIA–MS alone often meets the need in product control because of the information about the sample available prior to analysis. In the identification of products of competitors, the speed and specificity of FIA–MS and MS–MS are of paramount importance. These applications of MS in the solution of problems here, however, quite often remain unpublished to avoid releases of successful strategies for examinations to competitors.

In environmental analytical applications where analyte concentrations, e.g. surfactants or their metabolites, are quite low, extraction and concentration steps become essential. Solid phase extraction (SPE) with cartridges, disks or SPME fibres (solid phase micro extraction) because of its good variety of SP materials available has become the method of choice for the analysis of surfactants in water samples in combination with FIA as well as LC–MS analysis. SPE followed by sequential selective elution provides far-reaching pre-separations if eluents with different polarities and their mixtures are applied. The compounds under these conditions are separated in the MS spectrometer by their m/z ratios providing an overview of the ionisable compounds contained in a sample. Identification in the sense it has been mentioned before, however, requires the generation of fragments.

From the FIA–MS overview spectrum, speculation that there can be more than just one structurally defined molecule type behind an observable signal i.e. the presence of isobaric compounds, cannot be excluded whenever one signal defined by the m/z-ratio is examined in FIA–MS spectra. Consequently, the information obtained by FIA–MS is quite limited whenever we deal with complex mixtures of environmental pollutants rather than the analysis of pure products or formulations with a known range of ingredients. LC separation is inevitable when mixtures of isomeric compounds should be identified with MS–MS. Therefore, in FIA–MS–MS special attention has to be paid to avoid the generation of mixed product ion spectra from isomeric parent compounds. This would block identification by library search and may lead to misinterpretations of product ion spectra because of the fragmentation behaviour observed.

2.5.4 The determination of surfactants in blends, formulations and environmental samples using FIA–API combined with MS and MS–MS

MS and MS–MS techniques, meanwhile, have developed has the method of choice in the analysis of unknown organic compounds present as polar or non-polar constituents in all types of mixtures. This development in the analyses of surfactants contained in industrial blends, household formulations and environmental samples, started in the late 1980s with the successful combination of LC and MS. In parallel, spectrophotometric and titrimetric determinations according to the 'Standard Methods' [15,16] were used in the 1990s and are still performed today. Qualitative and quantitative information obtained using these unspecific methods often did not reflect the truth, but, nevertheless, if the results were of judicial relevance these methods had to be applied. With the recognition of the environmental relevance of surfactants that are synthesised as polar compounds at the million-ton scale and mainly discharged with wastewaters the analytical methods applied for the determination changed from unspecific group parameter analysis to substance-specific mass spectrometric determination methods. In parallel to the matrix, the surfactants contained in, e.g. industrial blends, formulations and environmental samples, the difficulties in qualitative and quantitative determination increase. So, according to the degree of complexity of the sample matrix, the analytical methods applied for their determination should be selected. While the analysis of surfactants in blends and formulations normally is quite easy and therefore the objective of identifying surfactants can be reached by the application of the FIA–MS technique, the analysis of surface active compounds of anthropogenic or biogenic origin in environmental samples needs more comprehensive examinations. Their substance-specific determination can be performed by several fundamentally different methods coupling LC to MS and MS–MS spectrometry. Two of them, the most informative substance-specific ones and the most often applied ones in qualitative and quantitative analysis of blends, formulations and environmental samples, are presented, compared and discussed here following selected examples:

1. Mixture analysis (FIA–MS–MS) with a preceding screening step using flow FIA–MS [5,6,8,9,17] in combination with soft ionisation techniques

compared with

2. HPLC in combination with MS and MS–MS.

Advantages as well as disadvantages can be observed working with both techniques. The greatest advantage of FIA–MS, using all possibilities available today with actual MS, MS–MS and MS^n hardware and software compared with (HP)LC, is the fast analysis. The continuous pressure to shorten analysis time has the scientists in its grasp. Within a few minutes, information about the molecular weight and even the fragmentation behaviour with the follow-up of product ions in the MS^n mode of selected compounds that are contained in the mixture can be received. For this purpose, working in the FIA mode, the application of MS–MS or MS^n after selection of ions of interest to obtain their product ion spectra is essential for identification.

The FIA–MS screening approach using soft ionisation interfaces prior to any CID procedure provides an overview of the MS separation procedure, which is based on the different m/z ratios of the molecular or cluster ions generated. With the help of this very fast screening method—positive or negative FIA–MS by-passing the analytical column—the surfactant chemist is able to characterise complex blends and formulations without difficulty (Fig. 2.5.1) while the experienced analyst is able to make initial statements about the presence of frequently used and therefore most important surfactants in environmental samples (Fig. 2.5.2) despite the presence of complex matrices. The information provided by ESI or APCI–FIA–MS overview spectra for a first characterisation [8,17–19], which were also available with non-API soft ionising interfaces such as FAB [20] or TSI [9] in industrial blends as well as environmental samples, were obtained from:

1. Equidistant or clustered signals, characteristic of some anionic, non-ionic or cationic surfactants (cf. Fig. 2.5.1(a) and (b). So the presence of non-ionic surfactants of alkylpolyglycolether (alcohol ethoxylate) type (AE) (structural formula: C_nH_{2n+1}–O–(CH$_2$–CH$_2$–O)$_x$–H) could be confirmed in the formulation (Fig. 2.5.1(a)) applying APCI–FIA–MS in positive mode. AE compounds with high probability could also be assumed in the heavily loaded environmental sample because of the patterns of $\Delta m/z$ 44 equally spaced ammonium adduct ions ([M + NH$_4$]$^+$) shown in its FIA–MS spectrum in Fig. 2.5.1(b).

2. The selectivity of negative or positive ionisation for most of the anionic or non-ionic surfactants, respectively. So, positive ionisation of this environmental extract made compounds recognisable, which were assumed to be AE surfactants because of their equally spaced signals of ions (cf. Fig. 2.5.1(a)). Negative ionisation of the same extract, however, proved the presence of the anionic linear alkylben-

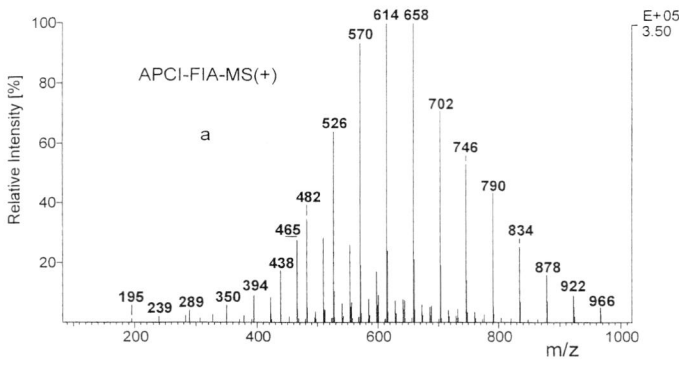

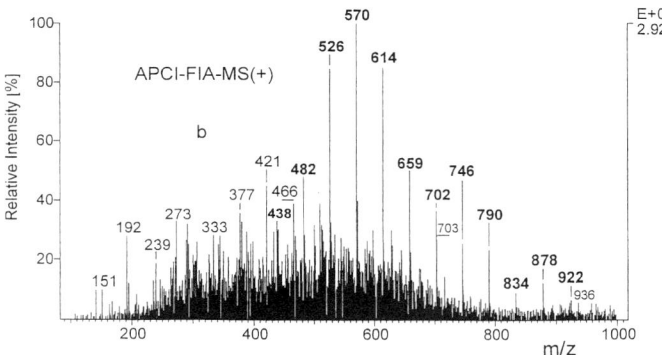

Fig. 2.5.1. APCI–FIA–MS(+) screening of (a) surfactant blend containing characteristic $\Delta\,m/z$ 44 equally spaced ion patterns of non-ionic alkylpolyethyleneglycolether (AE) homologues (C_nH_{2n+1}–O–(CH_2–CH_2–O)$_n$–H) and (b) APCI–FIA–MS(+) screening of C_{18}–SPE of environmental sample containing AE compounds with the same ion patterns as in (a).

zene sulfonate surfactants (LAS; C_nH_{2n+1}–C_6H_4–SO_3^-) (Fig. 2.5.2), too, recognisable from their [M − H]$^-$ ion cluster at m/z 297, 311, 325 and 339 corresponding to the $n = 10$–13 LAS homologues. Additional information from ESI or APCI FIA–MS overview spectra sometimes become available if spectra were recorded.

3. Either in ESI or APCI ionisation mode. Differences in the ionisation yields of both API techniques as observed with some surfactants later on can be taken into account to get additional information to the substance-specific information by fragmentation. So the negative ionisation of an alkylether sulfate blend (AES) using either ESI or

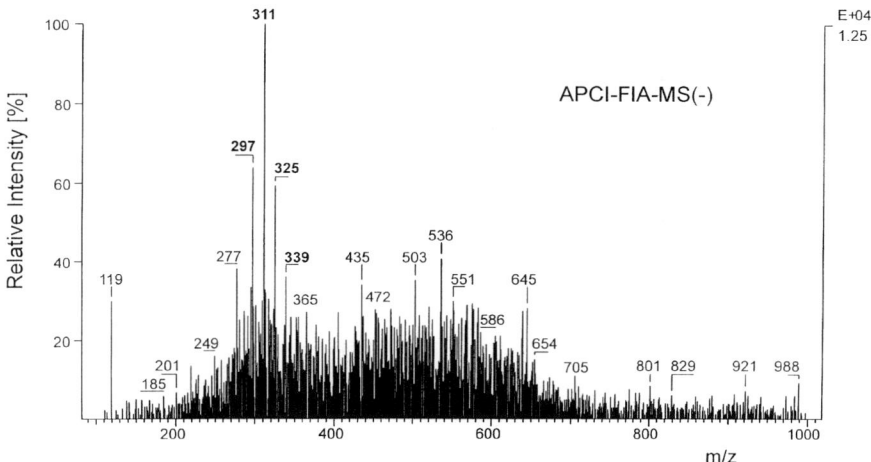

Fig. 2.5.2. APCI–FIA–MS(−) screening of environmental sample as in Fig. 2.5.1(b) containing the negatively charged anionic LAS compounds (C_nH_{2n+1}–C_6H_4–SO_3^-) with their characteristic ions at m/z 297, 311, 325 and 339.

APCI shows considerable differences in the ionisation yields of AES homologues if FIA–MS spectra were compared (cf. Fig. 2.5.3(a) and (b)).

4. The destructive positive ionisation of some anionic polyether compounds such as AES, alkylaryl ethersulfates, -carboxylates, and -phosphates resulting in alkylpolyglycolether or alkylarylpolyglycolether ions (cf. Fig. 2.5.4(a) and (b)), respectively, may also support the identification of the constituents of complex mixtures [8]. So nonylphenol polyglycolether sulfates (C_9H_{19}–C_6H_4–O–$(CH_2$–CH_2–$O)_x$–SO_3^-; NPEO–SO$_4$) will be negatively ionised in the form of $[M - H]^-$ ions while positive ionisation resulted in positively charged nonylphenol ethoxylate (NPEO) ions ($[M - SO_3 + NH_4]^+$). These differences in the behaviour of NPEO–SO$_4$ homologues under positive or negative ionisation can be used for their confirmation. With this procedure the absence of NPEO homologues in a mixture can only be excluded when LC was performed.

These methods provide information about the distribution of the surfactant homologues as observed from, e.g. AE or alkylarylpolyglycolethers (alkylphenol ethoxylates, APEO), LAS, secondary alkane

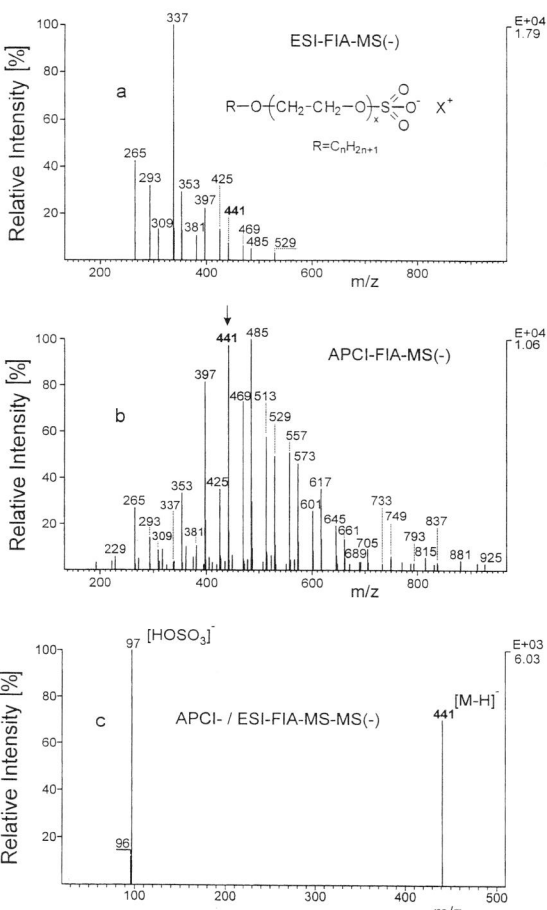

Fig. 2.5.3. (a) ESI–FIA–MS(−) with the general structural formula in the inset and (b) APCI–FIA–MS(−) overview spectrum of an AES surfactant blend (C_nH_{2n+1}–O–$(CH_2$–CH_2–$O)_x$–SO_3^-), and (c) ESI–FIA–MS–MS(−) (CID) product ion mass spectrum of $[M − H]^-$ parent ion at m/z 441 of the anionic AES blend as in (a) ($n = 12$, $x = 4$) collision voltage: $−30$ eV. Collision gas: Argon. Collision gas pressure: 2.0 mTorr [8].

sulfonates (SAS) or AES. Additional structural information, however, such as the degree of branching, the distribution of homologues and isomers, the location of insaturation and side chains requires the application of MS–MS.

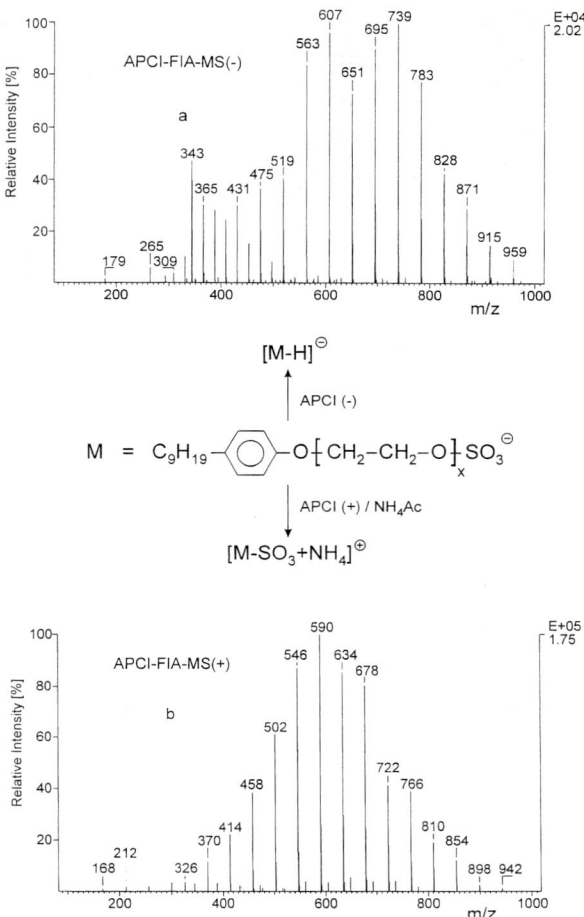

Fig. 2.5.4. (a) APCI–FIA–MS(−) and (b) APCI–FIA–MS(+) overview spectrum combined with general structural formula demonstrating differences in the ionisation behaviour of NPEO–SO$_4$ (C$_9$H$_{19}$–C$_6$H$_4$–O–(CH$_2$–CH$_2$–O)$_x$–SO$_3^-$).

2.5.5 The application of FIA–MS and MS–MS to surfactant blends and their mixtures

To demonstrate the problems and advantages observable with the application of FIA–MS and MS–MS in the analysis of surfactants contained in household detergents, FIA–MS spectra of selected non-ionic, anionic and cationic surfactants blends were recorded. The results obtained will be explained presenting these FIA–MS and MS–MS spectra. Therefore, the procedure in the follow-up of the qualitative and quantitative determination of surfactants applying FIA in both API modes will become obvious. The large variability in the ionisation potential of these two interfaces that use the same basic principle of API ionisation for the same surfactant compounds on the one hand can be very helpful in the characterisation and identification process. On the other hand, the results proved that neither of the API interfaces could be used as a universal interface ionising the whole spectrum of compounds contained in the samples. This will become obvious when the advantages offered by the selection between ESI applied in the conventional way or the pneumatically assisted version ('ion spray') can be elaborated. An improvement in the ionisation efficiency especially observed with the ionisation of highly polar compounds, e.g. preferentially ionic compounds, is worth mentioning. The application of APCI, however, shows an improved ability to ionise the more medium polar compounds.

These observations obtained from the application of different API techniques are determinative for qualitative and quantitative FIA results in the analysis of non-ionic and ionic surfactants. Therefore, both ionic surfactant types, anionic and cationic surfactant blends, besides a non-ionic AE surfactant blend were examined, recording their FIA–MS and MS–MS spectra from the blends before the spectra were generated from the mixture of all blends. The results, which show considerable variation, will be presented and discussed as follows.

The complex APCI–FIA–MS(+) overview spectrum of a non-ionic surfactant blend of AE type which contains C_{10}, C_{12}, C_{14}, C_{16}, and C_{18} homologues [21] is presented in Fig. 2.5.5(a) together with its general structural formula. The patterns of ions in the APCI spectrum was equivalent to the patterns observable in the ESI–FIA–MS(+) spectrum. An improved sensitivity in APCI mode compared with ESI mode ionisation, however, could be observed. The addition of ammonium acetate for ionisation support resulted in a different series of equally spaced $[M + NH_4]^+$ cluster ions. The homologues that differ in the alkyl chain length can be observed with $\Delta m/z$ 28 ($-CH_2-CH_2-$) equally spaced signals while the polyether chain homologues differ by $\Delta m/z$ 44 ($-CH_2-CH_2-O-$), both determining the pattern of ions. The application of CID recording the FIA–MS–MS(+) product ion spectra of

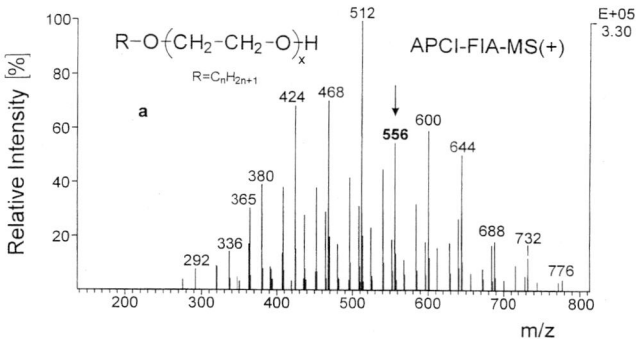

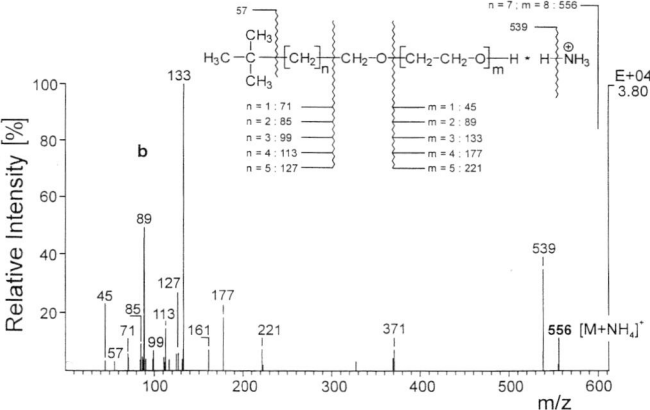

Fig. 2.5.5. (a) APCI–FIA–MS(+) overview spectrum and general structural formula of non-ionic AE blend (C_nH_{2n+1}–O–$(CH_2$–CH_2–O$)_n$–H) containing the alkyl chain homologues C_{10}, C_{12}, C_{14}, C_{16}, and C_{18} (R = C_nH_{2n+1}; n = 10, 12, 14, 16 and 18) [21]. Compounds ionised as $[M + NH_4]^+$ ions. (b) Product ion spectrum of selected $[M + NH_4]^+$ parent ion at m/z 556 from AE blend as in (a) generated by APCI- or ESI–FIA–MS–MS(−) applying CID; inset: fragmentation scheme. Collision voltage: −25 eV. Collision gas: Argon. Collision gas pressure: 2.0 mTorr (1 Torr = 133.322 Pa).

selected signals in the overview spectrum resulted in the spectrum shown in Fig. 2.5.5(b). This product ion spectrum contains the AE characteristic alkyl-chain and polyether-chain fragments as explained by the fragmentation scheme in the inset of Fig. 2.5.5(b). Variations in collision energy and/or collision gas pressure led to a variation of the product ion intensities.

As an example of an anionic surfactant mixture frequently contained in detergent formulations, an AES blend with the general formula $C_nH_{2n+1}-O-(CH_2-CH_2-O)_x-SO_3^-$ was examined in the negative FIA–MS mode. Because of the considerable differences observed between both API ionisation mode overview spectra, the ESI–FIA–MS(−) and the APCI–FIA–MS(−) spectra are reproduced in Fig. 2.5.3(a) and (b), respectively. Ionisation of this blend in the positive APCI–FIA–MS mode, not presented here, leads to the destruction of the AES molecules by scission of the $O-SO_3$ bond. Instead of the ions of the anionic surfactant mixture of AES, ions of AE can then be observed imaging the presence of non-ionic surfactants of AE type.

In this AES blend, C_{12} and C_{14} alkyl chains of alcohols were coupled with polyglycolether chains, which vary in the number of polyether units resulting in homologue $[M - H]^-$ ions. The ions observed started at m/z 309 ($C_{12}-EO_1$) and 337 ($C_{14}-EO_1$) differing by $\Delta m/z$ 44 ($-CH_2-CH_2-O-$) as listed in Table 2.5.1. Here, the ions observable were characterised indicating the number of alkyl and polyether chain links. In parallel, the precursor compounds from synthesis, the alkylsulfates ($x = 0$), were still present and can be observed with ions at m/z 265 and 293.

Identification in CID mode performed by FIA–MS–MS(−) using APCI or ESI interface and applied to all AES parent ions ($[M - H]^-$) resulted in the $[HOSO_3]^-$ ion at m/z 97 as presented in Fig. 2.5.3(c). Compared with product ion spectra of AE compounds positively generated (cf. Fig. 2.5.5(b)) this negative product ion spectrum of an AES homologue is less informative because only one negative ion, the $[HOSO_3]^-$ ion at m/z 97, will be generated. In addition not only AES but also some more organic sulfate compounds in CID(−) mode result in this type of ion. For unequivocal identification, the product ion spectra of AES generated in negative FIA–MS mode are not quite characteristic and therefore need further confirmation such as LC separation or parent ion examination of m/z 97 to recognise $\Delta m/z$ 44 equally spaced ions. To characterise branching of the alkyl chain moiety in AES molecules, the generation of positively charged product ion spectra obtained by positive ionisation in the form of $[M - SO_3 + NH_4]^+$ ions may be quite helpful resulting in a more informative product ion pattern indicating tertiary, secondary, and primary carbon atoms of the alkyl chains.

The third industrial blend presented here as an example belongs to the cationic surfactants of fatty acid polyglycol amine type with

TABLE 2.5.1

List of AES (C_nH_{2n+1}–O–$(CH_2$–CH_2–O$)_x$–SO$_3$H; $n = 12$ or 14) homologues contained in an industrial blend ionised by ESI or APCI interface in negative mode (cf. Fig. 2.5.3(a) and (b))

n (R = C_nH_{2n+1})	x –$(CH_2CH_2O)_x$–	(m/z)	
		ESI($-$) [M $-$ H]$^-$	APCI($-$) [M $-$ H]$^-$
12	0	265	265
12	1	309	309
12	2	353	353
12	3	397	397
12	4	441	441
12	5	485	485
12	6	529	529
12	7	n.d.	573
12	8	n.d.	617
12	9	n.d.	661
12	10	n.d.	705
12	11	n.d.	749
12	12	n.d.	793
12	13	n.d.	837
12	14	n.d.	881
12	15	n.d.	925
14	0	293	293
14	1	337	337
14	2	381	381
14	3	425	425
14	4	469	469
14	5	n.d.	513
14	6	n.d.	557
14	7	n.d.	601
14	8	n.d.	645
14	9	n.d.	689
14	10	n.d.	733

n.d. = not detected.

the general formula (R–N$^{\oplus}$H$((CH_2$–CH_2–O$)_{x,y}$H$)_2$X$^-$), which nowadays are more often observed in the environment. Applying APCI- or ESI–FIA–MS(+) the polyglycol amines are ionised as [M + H]$^+$ ions resulting in equal-spaced signals (Δ m/z 44) of homologue compounds because of a varying number of polyether chain links. In parallel to the [M + H]$^+$ ions observed in the FIA–MS spectrum starting with m/z 406 and ending at 1198, the [M + H–(H$_2$O)]$^+$ ions with Δ m/z 18 are generated from this blend when APCI was applied (Fig. 2.5.6(a)).

137

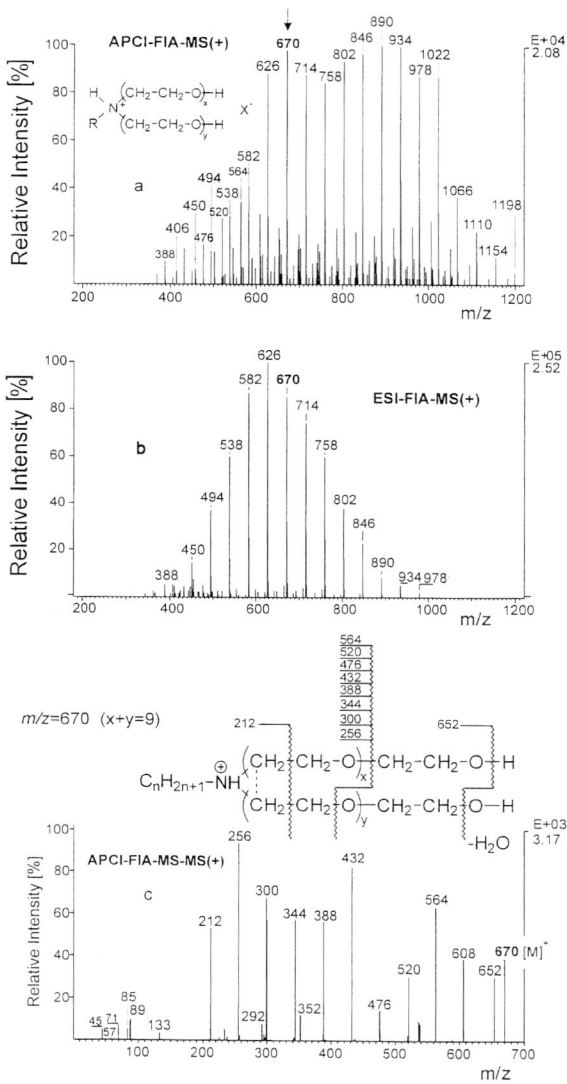

Fig. 2.5.6. (a) APCI–FIA–MS(+) overview spectrum and general structural formula of fatty acid polyglycol amine blend $(R-N^{\oplus}H((CH_2-CH_2-O)_{x,y}H)_2X^-))$ and (b) ESI–FIA–MS(+) overview spectrum of mixture as in (a). Compounds ionised as $[M]^+$ [8]. (c) APCI–FIA–MS–MS(+) product ion spectrum and fragmentation scheme of selected polyglycol amine homologue m/z 670.

Under ESI–FIA–MS(+) conditions, predominantly [M + H]$^+$ ions besides small amounts of impurities could be observed (Fig. 2.5.6(b)). Compared with APCI ionisation, the ESI interface predominantly ionised the shorter chain homologues whereas the longer chain compounds contained in the blend were discriminated.

The application of CID(+) to the polyglycol amines in the APCI–FIA–MS–MS(+) mode showed a product ion pattern with ions equally spaced with $\Delta m/z$ 44 starting at m/z 212 as shown in Fig. 2.5.6(c). In addition, the inset also contains the fragmentation scheme of the fatty acid polyglycol amine parent ion at m/z 670.

These results obtained from the analyses of industrial blends proved that the identification of the constituents of the different surfactant blends in the FIA–MS and MS–MS mode can be performed successfully in a time-saving manner only using the product ion scan carried out in mixture analysis mode. The applicability of positive ionisation either using FIA–MS for screening and MS–MS for the identification of these surfactants was evaluated after the blends examined before were mixed resulting in a complex surfactant mixture (cf. Fig. 2.5.7(a)). Identification of selected mixture constituents known to belong to the different blends used for mixture composition was performed by applying the whole spectrum of analytical techniques provided by MS–MS such as product ion, parent ion and/or neutral loss scans.

First a screening in the APCI–FIA–MS(+) and APCI–FIA–MS(−) mode was carried out. From the positively generated overview spectrum as presented in Fig. 2.5.7(a) for identification, characteristic parent ions already known from the FIA–MS spectra of the pure blends and examined by MS–MS now were selected for MS–MS examination of the mixture. From the composed surfactant mixture, the ions at m/z 380, 556 and 670 were submitted to CID in positive APCI–FIA–MS–MS mode. Product ion spectra of these ions are presented in Fig. 2.5.7(b)–(d).

The negative APCI–FIA–MS screening of the composed surfactant mixture resulted in the FIA–MS(−) spectrum shown in Fig. 2.5.8(a). All CID(−) experiments with selected ions, e.g. m/z 265, 441 or 513, contained in the overview spectrum always led to a single product ion at m/z 97 as already presented in Fig. 2.5.3(c).

Provided that the compounds analysed in positive mode are not known, the selected and fragmented compound ions (cf. Fig. 2.5.7(b)–(d)) after computer aided product ion library search were described as AE (Fig. 2.5.7(b) and (c)) and polyglycol amines

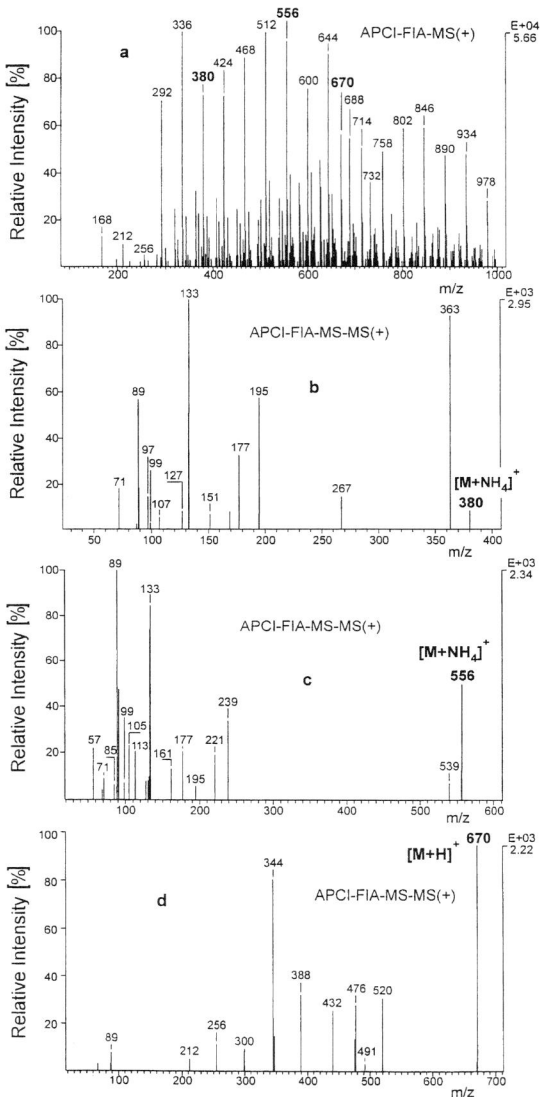

Fig. 2.5.7. (a) APCI–FIA–MS(+) screening recording an artificial formulation mixed from surfactant blends as presented in Figs. 2.5.3 (AES), 2.5.5 (AE) and 2.5.6 (polyglycol amine blend). Product ion spectra of selected parent ions m/z 380, 556 and 670 of surfactant formulation as in (a) obtained by APCI–FIA–MS–MS(+).

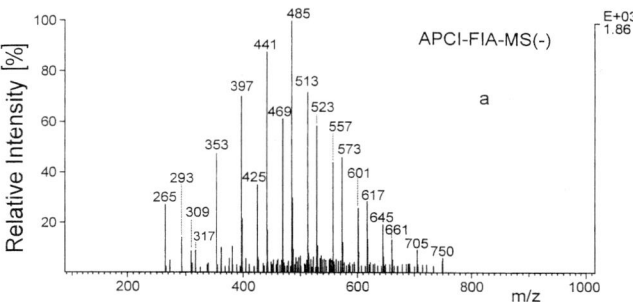

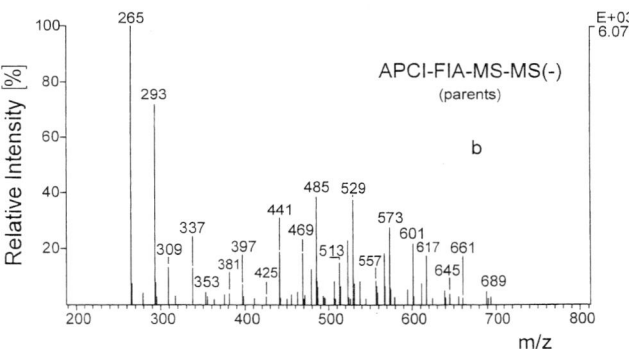

Fig. 2.5.8. (a) APCI–FIA–MS screening of mixture as in Fig. 2.5.7(a) recorded in negative mode presenting AS and AES homologues. (b) APCI–FIA–MS–MS(−) parent ion mass spectrum generated in CID mode from product ion (m/z 97) as observed in APCI–FIA–MS–MS(−) product ion spectrum in Fig. 2.5.3(d) confirming the presence of AS and AES homologues in the mixture in (a).

(Fig. 2.5.7(d)). From the series of ions equidistant by $\pm\Delta\,m/z$ 44 observable in negative ionisation mode under CID in FIA–MS–MS(−) mode, only a single product ion at m/z 97 was generated, known as [HOSO$_3$]$^−$ product ion. This product ion is not quite characteristic because it will also be obtained by CID(−) from anionic sulfate compounds such as AES or APEO–SO$_4$. For reconfirmation, the negative parent ions of m/z 97 were generated vice versa resulting in the parent ion spectrum as shown in Fig. 2.5.8(b). With this the presence of anionic sulfate compounds could be confirmed. While in the parent ion spectrum, the $m/z \geq 309$ ions were observed with decreased intensities, the intensities of the ions at m/z 265 and 293 belonging to the C_{12} and C_{14} alkylsulfates (AS), respectively, increased compared with the FIA–MS overview spectrum (cf. Fig. 2.5.8(a)). With this the probability that the series of equally spaced ions starting at m/z 309 or 337 ($\Delta m/z$ 44) were AES homologues also increased.

The result of polyglycol amine identification could be confirmed by the same procedure recording the CID parent ion scan of m/z 344 ($\pm\Delta m/z$ 44) by APCI–FIA–MS–MS(+). The information obtained from the surfactant mixture in Fig. 2.5.7(a) by product ion scan of m/z 670 (cf. Fig. 2.5.7(d)) and the spectrum from parent ion scan of m/z 344 as presented in Fig. 2.5.9 allows the identification of this ion as a polyglycol amine surfactant.

Applying product and parent ion scans in the FIA–MS–MS(+) mode, the unequivocal identification of the AE blend constituents is not possible. The reason for this failure is the simultaneous presence of C_{12} and C_{14} AE and AES compounds in the mixture that, under positive APCI ionisation conditions, were ionised with the same patterns of $[C_nH_{2n+1}O(CH_2–CH_2–O)_x + NH_4]^+$ ions ($\Delta m/z$ 44) and the same ion masses according to n, the alkyl chain lengths, and x, the polyether chain lengths. The destructive ionisation process of AES in APCI(+) mode therefore imagines a differentiation of the $\Delta m/z$ 44 compounds but results are not reliable.

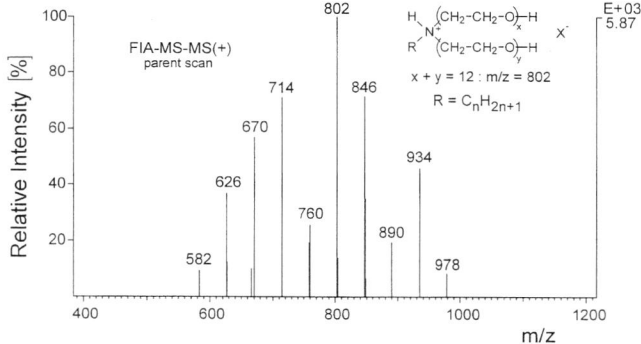

Fig. 2.5.9. APCI–FIA–MS–MS(+) parent ion mass spectrum from product ion (m/z 344) observed in APCI–FIA–MS–MS(+) product ion spectrum as in Fig. 2.5.6(c) (inset: general structure of polyglycol amine).

With the simultaneous presence of AES and AE homologues in the composed surfactant mixture, even the use of FIA–MS–MS for differentiation is not particularly easy despite the various possibilities, e.g. positive or negative parent ion mode scanning or recording parents in $+/-$ mode, provided by the classical mixture analysis approach of FIA–MS–MS. Here, the application of a preceding RP-C_{18} LC separation combined with APCI–MS–MS($+/-$) and taking different RTs of AES and AE into account can help to overcome these difficulties in the analysis of AES and AE simultaneously present. It has to be taken into account, however, that the application of LC performed by RP-C_{18} separation cannot be regarded as panacea. Because RP-C_{18} will not grasp polyglycol amine surfactants, the separation and determination of the other constituents can easily be performed.

2.5.6 The application of FIA–MS and MS–MS to unknown surfactant formulations

While the surfactant mixture composed by mixing the different blends could be cleared up by FIA–MS and MS–MS to a great extent, these methods failed in the identification of most constituents contained in a commercially available household detergent formulation. The limitations of mixture analysis became obvious with the application of the API methods such as ESI and APCI in FIA–MS–MS mode and are described here by means of examples.

FIA–MS spectra recorded by APCI or ESI in the positive (Fig. 2.5.10(a) and (b)) or negative (Fig. 2.5.10(c) and (d)) mode spectra are located altogether in Fig. 2.5.10. The survey on the compounds that could be ionised in the different modes and using APCI or ESI, respectively, gave an impression of the complexity of the detergent formulation and the content of surfactant homologues, which were separated by their m/z ratios. Estimations about the surfactant classes can be made, although without MS–MS reliable characterisations, and furthermore, identifications were impossible. So at first glance, identification seemed to be possible only after the application of FIA–MS–MS and/or LC separations by RT comparison for confirmation.

By the application of API–FIA–MS–MS($+/-$) to the detergent formulation the main ingredients could be identified in product ion mode. MS–MS spectra available in a product ion library were used for comparison. So AE compounds with their $[M + NH_4]^+$ ions were

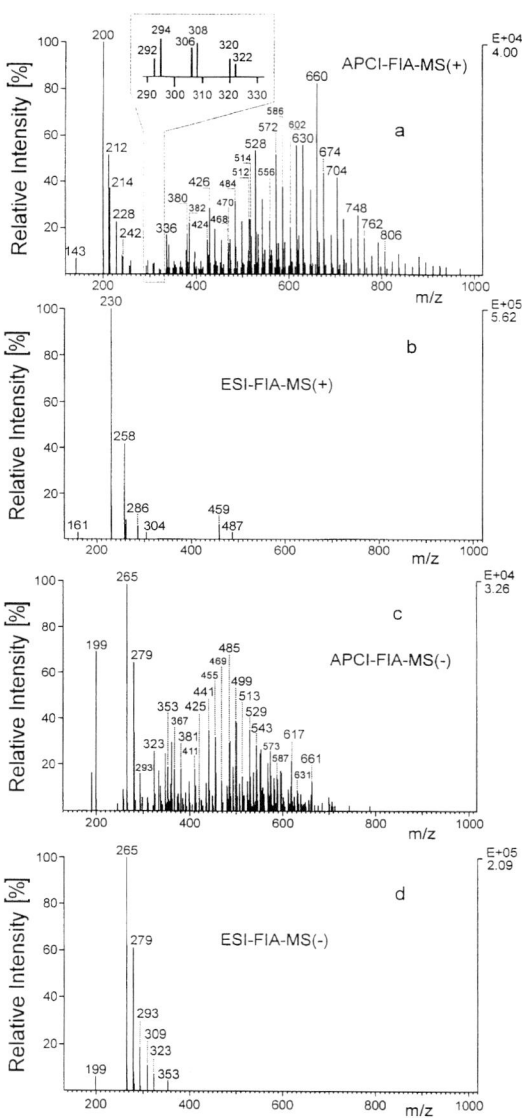

Fig. 2.5.10. (a) APCI–FIA–MS(+), (b) ESI–FIA–MS(+), (c) APCI–FIA–MS(−) and (d) ESI–FIA–MS(−) overview spectra all recorded from the methanolic solution of the same commercially available household detergent mixture containing different types of unknown surfactants.

confirmed in Fig. 2.5.10(a). AE homologues with C_9–C_{11} carbon chains could be observed starting at m/z 294, 308 and 322. In parallel, ions of C_{12}–C_{14} AES homologues ionised in a destructive process as $[M - SO_3 + NH_4]^+$ ions were generated starting at m/z 292, 306 or 320, all equally spaced by $\Delta\, m/z$ 44. These AE ions generated by the abstraction of SO_3 from C_{12}–C_{14} AES homologue molecules are equivalent to $[M - H]^-$ ions of AES ($C_nH_{2n+1}O$–$(CH_2$–CH_2–$O)_x$–SO_3^-) observable after APCI($-$) with ions at m/z 353 ($n = 12$; $x = 1$), 367 ($n = 13$; $x = 1$) or 381 ($n = 14$; $x = 1$) (cf. Fig. 2.5.10(c)).

From the ESI–FIA–MS(+) spectrum in Fig. 2.5.10(b), the amphoteric amine oxide surfactants ($[M]^+$ ions at m/z 230, 258 and 286) and their dimeric ions ($[2M - H]^+$ at m/z 459 and 487 (230 combined with 258)) could be recognised. The identity of the amine oxides was confirmed by recording product ions of the $[M]^+$ ion at m/z 230 before the parent ion scan of fragment m/z 58 and vice versa was recorded for confirmation. This spectrum contained the $\Delta\, m/z$ 28 equally spaced characteristic amine oxide homologue ions at m/z 230, 258 and 286.

The APCI–FIA–MS($-$) in Fig. 2.5.10(c) also contained impurities of alkyl sulfates ($[M - H]^-$ ions at m/z 265, 279 or 293), AES compounds ($C_nH_{2n+1}O$–$(CH_2$–CH_2–$O)_x$–SO_3–) with ions at m/z 353, 367 or 381 and their homologue ions ($x > 1$) equally spaced by $+\Delta\, m/z$ 44. A short chain secondary alkylaryl sulfonate with its $[M - H]^-$ ion at m/z 199 was also observed. This alkylaryl sulfonate impurity could be identified and confirmed by its fragmentation behaviour in negative product ion mode resulting in the 'LAS characteristic product ion' at m/z 183 ($[CH_2{=}CH$–C_6H_4–$SO_3]^-$) besides a $[SO_3]^-$ ion at m/z 80.

In Fig. 2.5.10(d), under ESI–FIA–MS($-$) conditions a discrimination of polyethoxylated AES molecules became obvious. So only the precursor alkylsulfates with ions at m/z 265, 279 or 293 together with the mono-ethoxylated C_{12} and C_{13} and the bis-ethoxylated C_{12} AES presenting the $[M - H]^-$ ions at m/z 309 and 323 besides 353, respectively, were observable.

Comparison of the results in the ionisation of AES obtained from the different API techniques in positive or negative mode, which were presented in Fig. 2.5.10 proved that neither APCI nor ESI can be termed as a universal interface.

To obtain further insights in this complex detergent formulation and for validation and confirmation of results obtained by FIA–MS and MS–MS, a comprehensive LC–MS examination was also performed. So recognition and interpretation of failure of the FIA–MS and MS–MS approach will become obvious. Therefore, the results of different API ionisation techniques—APCI and ESI—applied in positive or negative mode and using different stationary or mobile phases, are reproduced as reconstructed total ion current traces (RIC) in Fig. 2.5.11. Separations were performed under RP conditions with and without ion-pairing and, in addition, ion chromatography combined with APCI and ESI ionisation was applied for separation. Under a variety of conditions, we obtained from this mixture of surfactants ionised in the positive mode, RICs that were quite different from APCI (Fig. 2.5.11(a)) or ESI

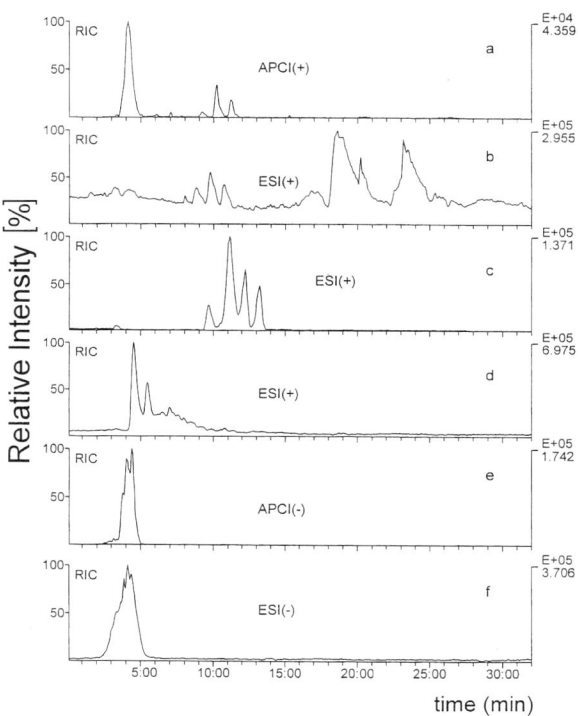

Fig. 2.5.11. (a) APCI–LC–MS(+), (b) ESI–LC–MS(+), (c) ESI–LC–MS(+), (d) ESI–LC–MS(+), (e) APCI–LC–MS(−) and (f) ESI–LC–MS(−) reconstructed ion chromatograms (RIC) of methanolic solution of the household detergent mixture as in Fig. 2.5.2. Chromatographic conditions: (a), (b), (e), and (f) RP-C_{18}, methanol/water gradient elution; (c) ion-pairing RP-C_{18} using trifluoro acetic acid (TFA) (5 mmol), methanol/water gradient elution; (d) isocratic elution performed on PLRP-column, eluent: methanol/water methane sulfonic acid (5 mmol).

(Fig. 2.5.11(b)–(d)) ionisation. Negative ionisation resulted in the RICs in Fig. 2.5.11(e) and (f). The same LC separation conditions were successfully applied when LC–MS–MS for identification was performed.

Results obtained by LC should be discussed in comparison with the identification that could be achieved by the time-saving FIA–MS and MS–MS technique. With the application of time-consuming LC separations, many problems observed in the determination and identification of the constituents of complex mixtures can be solved. The problems that were still observable under these improved possibilities of column separation applied to this commercially available household detergent formulation become obvious with the discussion of results presented in Fig. 2.5.11.

The MS detection after a separation under RP-C_{18} gradient conditions and ionised either positively by APCI (Fig. 2.5.11(a)) or ESI (Fig. 2.5.11(b)) led to quite different TICs. So the signal observed in the TIC(+) in Fig. 2.5.11(a) with a RT of 3.8–5.0 min after APCI(+) ionisation imagined the presence of AE compounds (C_nH_{2n+1}–O–(CH_2–CH_2–O)$_x$–H) ionised in the form of [M + NH_4]$^+$ ions. All ions were equally spaced with $\Delta m/z$ 44, starting at m/z 292, 306 or 320 and ending at 644, 438 or 496 according to alkyl chains with $n = 12$, 13 and 14, respectively. The compounds with m/z 292, 306 or 320 could be assigned to AE homologues with two EO-units while the longest PEG chain contained 10 units. Later ESI and APCI(−) results, however, reported that these compounds had been anionic surfactants of the AES type prior to ionisation. These AESs in LC separation were separated like anionics but in the successive APCI(+) ionisation process they were ionised via an SO_3 abstraction as non-ionics of AE type as also described in the literature [8].

Furthermore, the positive APCI trace (Fig. 2.5.11(a)) contained signals originating from AE compounds with the general formula C_nH_{2n+1}–O–(CH_2–CH_2–O)$_x$–H. These AE ionised as [M + NH_4]$^+$ ions, too, were already contained in the mixture but were not generated by SO_3 abstraction from AES. The compounds had been separated and were recorded at a retention time of 9–11 min. These AE ions belonged to a series of C_9–C_{11} homologues of non-ionics starting at m/z 294, 308 and 322 and ending at m/z 910, 924 and 938, respectively. The AE alkyl chains contained $n = 9$, 10 and 11 units while x, the number of EO units in all cases, varied between 3 and 17.

Applying the same chromatographic conditions but for ESI(+) ionisation (Fig. 2.5.11(b)), the AES compounds with the general structure

$C_nH_{2n+1}-O-(CH_2-CH_2-O)_x-SO_3^-$ now could be observed as $[M + NH_4]^+$ ions at a RT of 3.8–5.0 min. Signal intensity was quite low but a wide range of alkyl ethoxylated sulfate compounds, starting at m/z 372 and ending at 680 and AES ammonium adduct ions with 3–10 EO units were recorded. With an elevated intensity as observed under APCI(+) ionisation, the $[M + NH_4]^+$ ions of the non-ionics AE homologue mixture were observed in Fig. 2.5.11(b) at a RT of 7–10 min.

This ESI(+) TIC, however, is dominated by strong and broad signals that eluted between 17 and 31 min, neither observable under APCI(+/−) nor ESI(−) conditions. Even under gradient RP-C_{18} conditions a strong tailing effect was observed while isocratic RP-C_{18} failed. The information obtained by ESI–LC–MS(+) was that the compounds could be ionised in the form of $[M]^+$ ions at m/z 230, 258 and 286. ESI–LC–MS–MS(+) resulted in product ion spectra which, by means of a MS–MS library, were found to be characteristic for the amphoteric amine oxide surfactants. These compounds not yet observed in household formulations will be presented later on with the RIC of LC separation (cf. Fig. 2.5.11(d)). After identification as amine oxides, the separation and detection of this compound mixture now could be achieved by an isocratic elution using a PLRP-column and methane sulfonic acid and ESI(+) ionisation with the result of sharp signals (RT = 4–6 min) as presented in Fig. 2.5.11(d).

The application of trifluoroacetic acid (TFA) for ion pairing purposes in LC separation led to a retention shift of all constituents contained in the formulation. The AS and AES that eluted first, but were delayed compared with RP-C_{18} gradient elution, could be observed with a tremendous improvement in peak shape (RT = 9.0–12.5 min) in the total ion mass trace after ESI(+) ionisation (Fig. 2.5.11(c)). AE and amine oxides were not observed during the recording time of 30 min.

The TICs recorded in the negative APCI (Fig. 2.5.11(e)) or ESI (Fig. 2.5.11(f)) ionisation mode in both cases contained only one signal. The relatively sharp signal with an RT of 3–5 min in negative APCI TIC after RP-C_{18} (Fig. 2.5.11(e)) contained the AES which now could be recorded in their original form of AES ions ($[C_nH_{2n+1}-O-(CH_2-CH_2-O)_x-SO_3]^-$). The TIC contained the signals of the $[M]^-$ ions of C_{12}, C_{13} and C_{14} AES homologues and the precursor compounds, the analogous C_{12}, C_{13} and C_{14} AS observed as $[M]^-$ ions at m/z 265, 279 and 293.

The TIC recorded by negative ESI after RP-C_{18} (Fig. 2.5.11(f)) also contained just one broad signal with an RT of 2.0–5.5 min. Here, the anionic AES formed the $[M]^-$ ions as also observed under APCI(−)

conditions besides $[M]^-$ ions of non-ethoxylated C_{12}, C_{13} and C_{14} AS as precursors from chemical synthesis of AES. Under these ESI conditions only the short EO chain AES (EO_3 and EO_4) were ionised and shorter EO-chain lengths were observed under APCI($-$) and ESI($+$) ionisations [8]. The reason for this signal broadening observed in the ESI($-$) RIC (Fig. 2.5.11(f)) was the low sensitivity of the APCI ionisation for AS as observable in Fig. 2.5.11(e). On the contrary, sensitivity of APCI for AES ionisation was comparable with ESI. So a signal suppression of AS in the APCI($-$) RIC plot was observed while AS eluted between 2.0 and 3.5 min in ESI($-$) with a quite good response.

The results of this stock-taking of the surfactants present in a household detergent formulation demonstrated that the peak shapes of negatively recorded TICs (e) and (f) look quite similar, while the APCI($+$) and ESI($+$) TICs recorded under the same chromatographic conditions (Fig. 2.5.11(a) and (b)) are quite different. As expected, the use of different chromatographic conditions resulted in considerable variations, but APCI or ESI applied under the same LC conditions proved the selectivity of both interface types for specific compounds. This effect sometimes will be supported by the selection and application of highly specific and selective LC conditions.

Consequently, the application of both API interfaces, APCI and ESI, are absolutely necessary if reliable results are to be obtained. From these examinations it became obvious that the most important problem is to ionise the constituents of mixtures in a quite complete manner without influencing each other.

By summing up the results of qualitative analysis of surfactant blends and formulations achieved by FIA–MS and MS–MS in comparison with LC–MS and MS–MS applied for identification, considerable differences could be observed and should be taken into consideration prior to analysis:

1. The differences in ionisation efficiencies, however, not only result from the use of FIA or LC but, as mentioned before, depend also on the application of the APCI or ESI interface for ionisation. Therefore the application of both API methods, APCI and ESI, is the only way to overcome discrimination problems because of interface type selection. Even the use of the 'ion spray' technique instead of conventional ESI may influence the ionisation efficiency considerably.
2. The selectivity and specificity of FIA–MS–MS can be improved by the use of specific analytical MS–MS techniques in the multiple

reaction monitoring mode (MRM), if product, parent or neutral loss scans are performed.

3. The time investment to perform LC instead of FIA analyses for identifying surfactants in blends and formulations increased with the use of LC in MS and MS–MS mode at least by a factor of five.

4. With the tremendous increase in matrix complexity as observed in the analysis of surfactants in environmental samples, the application of LC–MS and MS–MS, will become the only reliable technique while even highly specific and selective FIA–MS and MS–MS techniques fail.

2.5.7 Influence of matrix compounds with the application of FIA–MS and MS–MS

The phenomenon of discrimination induced by matrix effects is quite disadvantageous in qualitative analysis, but in the quantification procedure its influence is devastating. Several techniques applying MS or MS–MS, however, can help to overcome this crucial influence. The most common way out of this dilemma is the application of column separation to minimise or even exclude matrix effects entailed with the disadvantage of long-lasting separations. In parallel, the difference between the time needed for the performance of LC examinations and the time needed for FIA becomes obvious and will be pointed out later on.

From quantification performed in the FIA mode, applying mixture analysis by FIA–MS–MS, reliable results can be achieved if product-, parent- or neutral loss scans (MRM mode), each of them of quite outstanding specificity, are recorded. Compared with FIA–MS, the time for recording has to be expanded marginally compared with column separation techniques.

The standard addition procedure is another method for recognising and overcoming potential matrix effects in quantification. Both alternatives, FIA–MS or FIA–MS–MS, can be performed using this procedure. Despite the increased expenditure because of a multiplication in analyses, the FIA approach combined with standard addition remains the faster technique even with the application of specific analytical MS–MS techniques such as product-, parent- or neutral loss scans applying selected reaction monitoring (SRM). The greatest drawback of this technique is that the compounds to be quantified must be available for standard addition prior to analysis. However, quantification in LC−MS mode makes standard compounds essential, too.

Regarding the phenomenon of discrimination of analytes because of ion suppression, reactions induced by complex organic and/or inorganic matrices either in FIA−MS [22–26] or LC−MS [27–32] or MS−MS mode were reported.

Several types of potential matrix effects are known in MS−MS and can be described as follows [33]:

1. Changes in the composition of a mixture of solvent and sample may result in different degrees of protonation during ionisation, altering the quantitative results that might be obtained.
2. Influence of matrix compounds to the analytes of interest by changing their structures or reactivities.
3. Increase or decrease in ionisation efficiency for the analytes of interest because of a change in mixture composition introduced into the source.
4. Changes in the absolute amounts vapourised into the ion source may change instantaneous pressure resulting in a decrease of fragmentation reactivity through collision stabilisation.

With the application of FIA in the mixture analytical mode for the analysis of environmental samples and after a marginal sample pre-treatment by SPE, matrix effects are a high probability. But, as cited previously [27–31], matrix effects were not only observed with FIA but also in LC−MS and MS−MS modes. Advice to overcome these problems by, e.g. an improved sample preparation, dilution of the analyte solution, application of stable isotopic modification of LC conditions [29] or even application of two-dimensional LC separations [27], post-column standard addition [29], addition of additives into the mobile phase (e.g. propionic acid, ammonium formate) [34,35] or even matrix compounds [32] were proposed and discussed.

2.5.8 Interferences induced by organic matrix compounds

For examination of these interferences during the ionisation of surfactants in the FIA−MS mode we tried to verify and quantify potential effects in the ionisation process. Ionisation efficiencies of selected pure surfactant blends were compared with those of their mixtures by the application of pattern recognition comparing selected ion mass traces.

Therefore, a C_{13}-AE, a cationic (quaternary ammonium) surfactant (quat), an amphoteric C_{12}-alkylamido betaine, and the non-ionic fatty acid diethanol amide (FADA) as presented with their FIA$-$MS spectra in Fig. 2.5.12(a)$-$(d) were analysed as pure blends and as mixtures always obtained from two blends in FIA$-$MS multiple ion detection mode (MID). Mixtures as well as pure blends contained identical concentrations of surfactant homologues. For AE quantitation the mass traces of all $\Delta\, m/z$ 44 equally spaced homologues (m/z 306$-$966) of the C_{13}-AE were recorded. The cationic (quaternary ammonium) surfactant, the amphoteric C_{12}-alkylamido betaine, and the non-ionic FADA were quantified recording the mass traces at m/z 214 and 228, or 184, 212, 240, 268, 285, 296, 313, 324 and 341, or 232, 260, 288, 316 and 344, respectively.

To recognise ion suppression reactions, the AE blend was mixed together either (Fig. 2.5.13(a) and (b)) with the cationic quaternary ammonium surfactant, (c, d) the alkylamido betaine compound, or (e, f) the non-ionic FADA, respectively. Then the homologues of the pure blends and the constituents of the mixtures were quantified as presented in Fig. 2.5.13. Ionisation of their methanolic solutions was performed by APCI($+$) in FIA$-$MS mode. The concentrations of the surfactants in the mixtures were identical with the surfactant concentrations of the blends in the methanolic solutions. Repeated injections of the pure AE blend (A; 0$-$4.0 min), the selected compounds in the form of pure blends (B; 4.0$-$8.8 min) and their mixtures (C; 8.8$-$14.0 min) were ionised and compounds were recorded in MID mode. For recognition and documentation of interferences, the results obtained were plotted as selected mass traces of AE blend (A; b, d, f) and as selected mass traces of surfactant blends (B; a, c, e). The comparison of signal heights (B vs. C and A vs. C) provides the information if a suppression or promotion has taken place and the areas under the signals allow semi-quantitative estimations of these effects. In this way the ionisation efficiencies for the pure blends and for the mixture of blends that had been determined by selected ion mass trace analysis as reproduced in Fig. 2.5.13, could be compared and estimated quite easily.

If N-containing surfactants, e.g. quats, alkylamido betaine or FADA, were ionised together with AE a depression was expected for AE compounds because of the proton affinity of the amino nitrogen

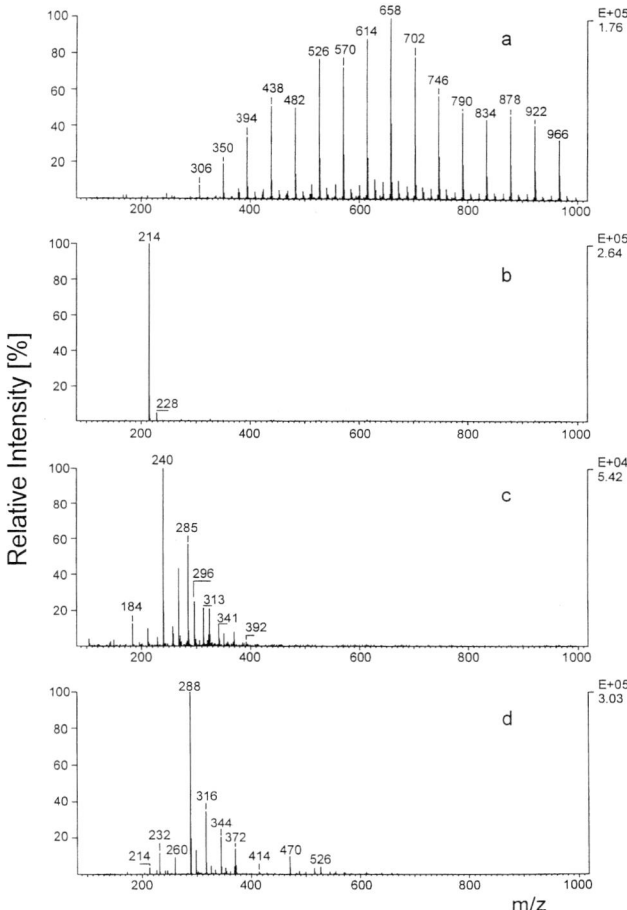

Fig. 2.5.12. APCI–FIA–MS(+) overview spectra of industrial surfactant blends used as pure blends or mixtures in the examination of ionisation interferences. (a) C_{13}-AE, (b) cationic (alkyl benzyl dimethyl ammonium quat) surfactant, (c) amphoteric C_{12}-alkylamido betaine, and (d) non-ionic FADA all recorded from methanolic solutions.

present in all types of selected surfactant molecules other than AE compounds. However, what was observed was the exact opposite of what was expected: a reduction of 18–30% in ionisation efficiencies of N-containing surfactants was found while the ionisation of AE was either not or only marginally influenced in an unexpected

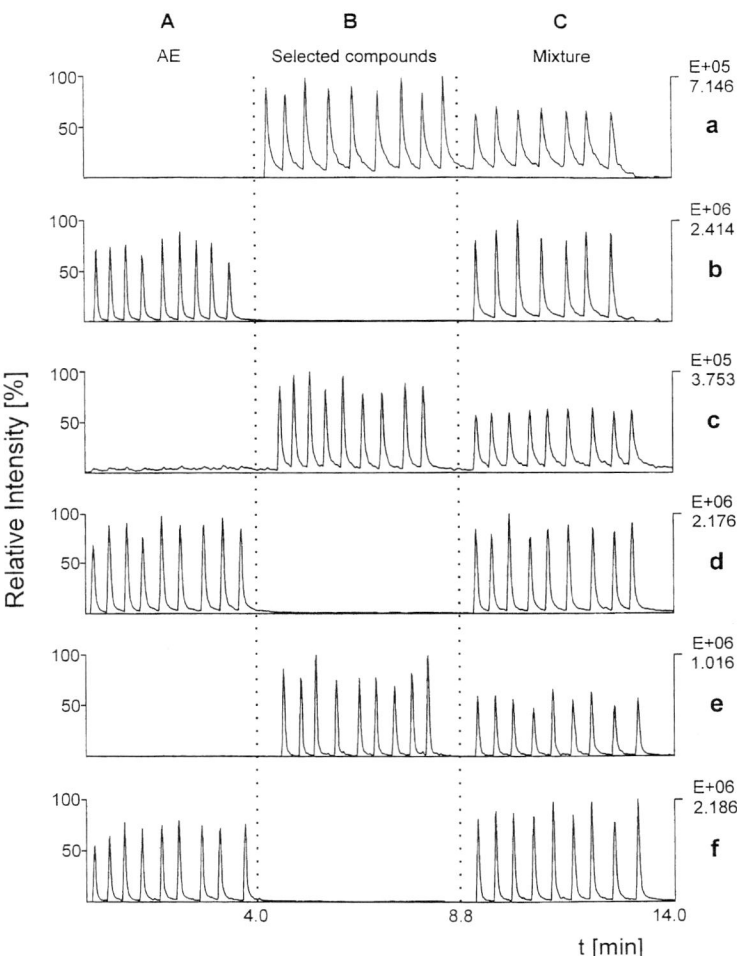

Fig. 2.5.13. Ion current traces of APCI–FIA–MS(+) examinations of compound interferences performed in multiple ion detection mode applied for quantification of pure AE blend (A: AE b,d,f; recorded between 0 and 4.0 min), pure quat (B a), betaine (B c) and FADA (B e) blend (cf. B: Selected compounds a,c,d; recorded between 4.0 and 8.5 min). In mixtures of AE with quat, betaine or FADA, respectively, all constituents were determined by MID (cf. C: Mixture: AE/quat (a,b), AE/betaine (c,d), and AE/FADA (e,f) (recorded between 8.5 and 14.0 min, respectively). Ions recorded in MID mode for quantification: AE (all ions starting at m/z 306 + Δ 44 and ending at 966), quat m/z 214 and 220, betaine m/z 184, 212, 240, 268, 285, 296, 313, 324 and 341, FADA m/z 232, 260, 288, 316 and 344.

manner—a promotion in ionisation efficiency of 7% for AE was found (C; Fig. 2.5.13(f)) if the mixture of AE and FADA was analysed while in parallel the FADA ionisation efficiency was reduced by 28% (C; Fig. 2.5.13(e)). Because of these results obtained by FIA–MS in a minimum of time by using pure blends and their mixtures, FIA–MS was not found to provide very reliable quantification results. Since time investment, however, is a decisive argument in the decision of the analytical approach to be applied for comparison, the different quantification techniques provided by FIA were performed using the same materials. Then quantification of these surfactants was performed in FIA mode and the time needed for the application of the different techniques—MS with standard addition or MS–MS with and without standard addition, respectively—was calculated and compared with quantification performed in FIA–MS mode.

Obstacles of ionisation interferences in the quantitative determination of the N-containing surfactants from industrial blends dissolved together with AE compounds in methanol applying FIA–MS can be minimised or even eliminated if quantification was performed in the standard addition mode. So the standard deviations (SD) observed now reached a maximum of $\pm 7\%$ for N-containing compounds whereas AE could be quantified with a SD of $\pm 4\%$. In parallel, the time investment for FIA–MS quantification in the standard addition mode, however, increased considerably and reached a factor of 3–4.

An improved specificity was observed when FIA–MS–MS in product or parent ion mode was used to perform quantification of the surfactants in the methanolic mixtures. The ions selected for quantitation in product or parent ion mode were: C13-AE: m/z 71, 85, 99, 113, and 127 from alkyl chain together with 89, 133, and 177 from PEG chain generated from parent ions m/z 394, 526, 658, 790 and 922; alkylbenzyl dimethyl ammonium quat: m/z 91 and 58 generated from parent ion m/z 214; FADA: m/z 88, 106 and 227 generated from parent ions m/z 232, 260, 288, 316, 344 and 372 while the alkylamido betaine was quantified generating the parent ion m/z 343 obtained from product ion at m/z 240.

The quantification of N-containing surfactants under MS–MS conditions but without standard addition resulted in a SD of $\pm 14\%$. In parallel, an increase in time by a factor of 1.5 was found compared with FIA–MS analysis without standard addition. The results obtained with MS–MS using product, parent or neutral loss scan are more specific but as ionisation takes place in the ionisation chamber and therefore prior to MS–MS ionisation cannot be influenced [31].

The alternative to FIA–MS performed in the standard addition mode is the quantification in FIA–MS–MS mode combined with a standard addition procedure. Here, the analytical specificity and the approach of a rapid analysis is combined, however, analysis time is expanded by a factor of 4.5 compared with FIA–MS.

For comparison of time necessary to quantify the compounds examined by FIA–MS and MS–MS, LC was also examined. API–LC–MS studies with RP-C_{18} (AE, betaine, FADA) or PLRP (quat) column applied for separation were performed. Quantitative results could be obtained with SDs $< 5\%$, while the invested time in parallel increased by a factor > 12 compared with FIA–MS.

2.5.9 Interferences induced by inorganic matrix compounds

Besides the interferences observed between organic analytes of relevance and organic matrix, inorganic matrix compounds often are also present (e.g. insufficient elimination of inorganic salt matrix during SPE, sodium ions from glassware) and may influence ionisation efficiencies considerably. An examination using a surfactant blend under controlled conditions was performed in APCI–FIA–MS(+) mode. So inferences could be observed while a continuous flow of a methanol/water (1:1; v/v) solution of AE (10 mg L^{-1}), which in addition contained ammonium acetate (5 mM), was pumped by a syringe pump into the APCI source while in parallel 10 μL of the same solutions, although containing either potassium or sodium acetate (5 mM), were injected into this continuous flow of the AE/NH_4-acetate solution. Ionisation was performed in the FIA–MS(+) mode and the ion current recorded is reproduced in Fig. 2.5.14. With the infusion (↓) (cf. Fig. 2.5.14) of both types of alkaline ions, potassium and sodium, an ion current considerably reduced in intensity was observed for the $[M + NH_4]^+$ of the AE blend. This result supported the demand for a minimum of sample pretreatment, e.g. SPE to get rid of inorganic matrix compounds.

2.5.10 Conclusion

In the analyses of polar anthropogenic surfactant compounds, the application of substance-specific MS-detection in combination with

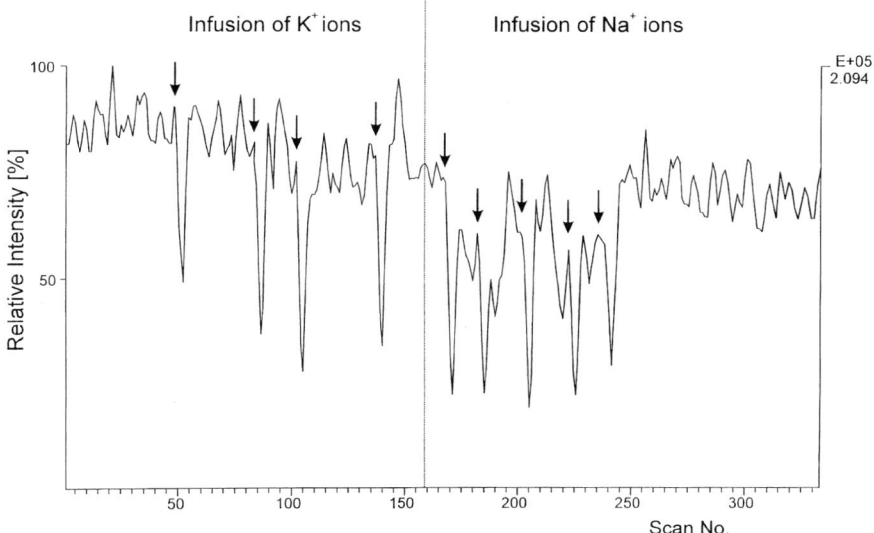

Fig. 2.5.14. Total ion current trace of APCI–FIA–MS(+) examinations for interferences induced by inorganic compounds (K^+ and Na^+) injected during the ionisation of an continuously infused AE/NH_4Ac solution.

liquid insertion had become the method of choice. The lack of commercially available standard materials had gravely limited the application of unspecific detector systems, e.g. UV, fluorescence or light scattering detectors. In addition, chromophores and fluorophores of most surfactant compounds are either weak or absent. So the fact that the availability of standard materials of anthropogenic surfactants is limited and that their biochemical degradation products are not available at all, had promoted this development. In order to obtain reliable results in the identification of unknown compounds without standard compounds, the application of substance-specific detector systems, e.g. MS, is essential. The quantification of surfactants by MS without any standard either to evaluate their MS response or for standard addition procedure to eliminate influence of matrix, however, is hopeless.

In the qualitative analyses of surfactants, the FIA–MS screening method applying both soft ionising API interface types, APCI and ESI, provides the overview spectra that contain the molecular ions or adduct ions of a complete pattern of isomers and homologue. It gives an impression about the molecular weight distribution of the compounds contained in the samples, e.g. in industrial blends, in household formulations or in environmental samples. The alternatives to ionise in positive or negative mode help to recognise non-ionic or ionic compounds and the application of APCI or ESI supported differentiation between more non-polar or polar analytes. For identifying compounds, the fragmentation of the molecular or adduct ions by generation of product ions, however, is essential and therefore FIA–MS–MS in CID mode or MS^n by ion-trap had to be performed in product ion mode.

FIA–MS–MS in parent or neutral loss mode on triple quad instruments can also be applied to screen mixtures of unknown compounds quite rapidly, so that compound classes can be recognised. Yet despite the information about molecular weight and the structural information by product ions, MS-data systems in their commercial form up to the mid-1990s provided no structural information for identification purposes in the form of libraries comparable with the NIST-library of EI spectra in GC–MS analysis. It can be hoped that troubles arising out of the lack of computer-searchable library data for identification will be overcome with the gradual increase in trap applications in MS–MS mode. The situation in identification 'is set to change' [36].

If the computerised surfactant analysis can be applied to samples with a low complexity, the generation of product ions sometimes will become obsolete [37]. So FIA–MS–MS in product ion, parent ion and neutral loss mode can easily be used for functional-group-specific analysis [10,11]. Of some disadvantage is that isobaric compounds with the same molar mass examined by medium or even high resolution can neither be recognised nor identified in FIA–MS and FIA–MS–MS mode. Moreover, a confirmation of these results by column chromatographic methods, e.g. LC, is essential, whenever there arises any doubt about their interpretation. The insight that LC separation after the application of the FIA–MS and MS–MS approach must precede any attempt to characterise compounds is essential for preventing mis-interpretations. So LC separations in combination with MS have developed into the most important procedure in the analysis of complex samples, e.g. environmental samples, despite all advantages and the new developments of FIA–MS, FIA–MS–MS, multiple-stage MS (MS^n) and progress in library data processing for identification during the last years.

Interferences in the quantitative determination of surfactant compounds in FIA–MS mode could be induced by organic and inorganic compounds applied as a surrogate matrix and were confirmed in the examinations of mixtures with known surfactant content. SD in quantification observed under these conditions was not tolerable. To overcome these obstacles, the application of the standard addition technique combined with MS and MS–MS, however, led to more reliable results in a fraction of the time that is necessary to perform LC–MS—provided the standards that are essential in both techniques, FIA and LC are available. Despite the fact that FIA–MS and FIA–MS–MS in standard addition technique minimised matrix interference, the suppression of interaction among compounds in the mixtures during the FIA–MS ionisation process can be the decisive argument for applying the long-winded analytical procedure of LC–MS or even LC–MS–MS. These techniques enable us to determine the quality and quantity of surfactants in most complex matrices. In both cases, ionisation takes place in the ion source and therefore selectivity can be improved by MS–MS; interference by matrix, however, cannot be compensated for by MS–MS. The ultimate alternative to get rid of matrix compounds, the application of two-dimensional LC, was not necessary with environmental samples containing surfactants.

Since the lack of commercially available standards will make quantification impossible, the potential of LC–MS, which is often regarded as panacea in the analysis of polar compounds like surfactants, nevertheless remains limited, also.

REFERENCES

1   J.J. Thomson, *Phil. Mag.*, VI(18) (1910) 824.
2   V.L. Talroze, V.E. Skurat, I.G. Gorodetskii and N.B. Zolotai, *Russ. J. Phys. Chem.*, 46 (1972) 456.
3   D.F. Hunt, J. Shabanowitz, T.M. Harvey and M.L. Coates, *J. Chromatogr. Rev.*, 271 (1983) 93.
4   D.F. Hunt, J. Shabanowitz and A.B. Giordani, *Anal. Chem.*, 52 (1980) 386.
5   D.F. Hunt, J. Shabanowitz, T.M. Harvey and M. Coates, *Anal. Chem.*, 57 (1985) 525.
6   J. Johnson and R. Yost, *Anal. Chem.*, 57 (1985) 758A.
7   H. Schwarz, *Nachr. Chem. Tech. Lab.*, 29 (1981) 687.

8   H.Fr. Schröder and F. Ventura, In: D. Barceló (Ed.), *Techniques and Instrumentation in Analytical Chemistry*, Sample Handling and Trace Analysis of Pollutants—Techniques, Applications and Quality Assurance, Vol. 21, Elsevier, Amsterdam, The Netherlands, 2000, p. 828.

9   H.Fr. Schröder, In: D. Barceló (Ed.), *Applications of LC/MS in Environmental Chemistry*, Vol. 59, Elsevier, Amsterdam, The Netherlands, 1996, p. 263.

10 H.Fr. Schröder, *J. Chromatogr. A*, 777 (1997) 127.

11 H.Fr. Schröder, *Water Sci. Technol.*, 34 (1996) 21.

12 H. Fr. Schröder, In: A. Kornmüller and T. Reemtsma (Eds.), Geminitenside Schriftenreihe "Biologische Abwasserreinigung" 2. Kolloquium im Sonderforschungsbereich 193 der DFG an der Technischen Universität Berlin "Anwendung der LC–MS in der Wasseranalytik", Berlin, Germany, 2001, Vol. 16, pp. 223–247, ISBN 3 7983 1852 2.

13 P. Kienhuis, "How to built and use an universal CI-CID spectrum-library for spectra generated by chemical ionization and CID-fragmentation". Proceedings 15th LC–MS Symposium, Montreux, Switzerland, 1998.

14 H.Fr. Schröder, *Vom Wasser*, 81 (1993) 299.

15 L.S. Clesceri, A.E. Greenberg and A.D. Eaton (Eds), Standard Methods for the Examination of Water and Wastewater. 20th Edition, American Public Health Association, Washington, DC, 1998, pp. 5–47 (5540 C, Anionic Surfactants as MBAS) and pp. 5–49 (5540 D, Non-ionic Surfactants as CTAS).

16 German Standard Methods for the Examination of Water, Waste Water and Sludge; General Measures of Effects and Substances (Group H). (a) Determination of Anionic Surfactants by Measurement of the Methylene Blue Index MBAS (H 24), (b) Determination of Bismut Active Substances (H 23-2), (c) Determination of the Disulfine Blue Active Substances (H 20), WILEY/VCH, Weinheim, Germany and Beuth, Berlin, Germany, 1998.

17 I. Ogura, D.L. DuVal, S. Kawakami and K. Miyajima, *J. Am. Oil Chem. Soc.*, 73 (1996) 137.

18 J. Efer, U. Ceglarek, T. Anspach, W. Engewald and W. Pihan, *gwf Wasser, Abwasser*, 138 (1997) 577.

19 H.Fr. Schröder and K. Fytianos, *Chromatographia*, 50 (1999) 583.

20 F. Ventura, In: D. Barceló (Ed.), *Environmental Analysis: Techniques, Applications and Quality Assurance*, Vol. 13, Elsevier, Amsterdam, The Netherlands, 1993, p. 481.

21 H.Fr. Schröder, *Vom Wasser*, 73 (1989) 111.

22 R. King, R. Bonfiglio, C. Fernandez-Metzler, C. Miller-Stein and T. Olah, *J. Am. Soc. Mass Spectrom.*, 11 (2000) 942.

23 X. Tong, I.E. Ita, J. Wang and J.V. Pivnichny, *J. Pharm. Biomed. Anal.*, 20 (1999) 773.

24 J.S. Salau, M. Honing, R. Tauler and D. Barceló, *J. Chromatogr. A*, 795 (1998) 3.

25 D.L. Buhrman, P.I. Price and P.J. Rudewicz, *J. Am. Soc. Mass Spectrom.*, 7 (1996) 1099.

26 M.G. Ikonomou, A.T. Blades and P. Kebarle, *Anal. Chem.*, 63 (1991) 1989.

27 B.K. Choi, D.M. Hercules and A.I. Gusev, *J. Chromatogr. A*, 907 (2001) 337.

28 S.D. Clarke, H.M. Hill, T.A.G. Noctor and D. Thomas, *Pharm. Sci.*, 2 (1996) 203.

29 H. Allmendinger, In: Th. Reemtsma and A. Kornmüller, (Eds), Wie beeinflusst die Matrix die Quantifizierung mit LC/MS?, Schriftenreihe "Biologische Abwasserreinigung" 1. Kolloquium im Sonderforschungsbereich 193 der DFG an der Technischen Universität Berlin "Anwendung der LC–MS in der Wasseranalytik", Berlin, Germany, 1999, Vol. 11, pp. 69, ISBN 3-7983-1796-8.

30 M. Stüber, Th. Reemtsma and M. Jekel, In: A. Kornmüller and T. Reemtsma (Eds.), Die Crux mit der Matrix: Matrixeffekte in der quantitativen Analyse mit LC–MS, Schriftenreihe "Biologische Abwasserreinigung" 2. Kolloquium im Sonderforschungsbereich 193 der DFG an der Technischen Universität Berlin "Anwendung der LC–MS in der Wasseranalytik", Berlin, Germany, 2001, Vol. 16, pp. 157, ISBN 3 7983 1852 2.

31 Th.P. Knepper, In: A. Kornmüller and T. Reemtsma (Eds.), Richtig gemessen oder falsch geschätzt: Ergebnisse und Erkenntnisse von quantitativen Untersuchungen mit LC–MS, Schriftenreihe "Biologische Abwasserreinigung" 2. Kolloquium im Sonderforschungsbereich 193 der DFG an der Technischen Universität Berlin "Anwendung der LC–MS in der Wasseranalytik", Berlin, Germany, 2001, Vol. 16, pp. 146, ISBN 3 7983 1852 2.

32 B.K. Choi, D.M. Hercules and A.I. Gusev, *Fresenius J. Anal. Chem.*, 369 (2001) 370.

33 K.L. Bush, G.L. Glish and S.A. McLuckey, *Mass Spectrometry / Mass Spectrometry, Techniques and Applications of Tandem Mass Spectrometry*, VCH Verlagsgesellschaft mbH, Weinheim, Germany, 1988.

34 D. Temesi and B. Law, *LC–GC*, 17 (1999) 626.

35 A. Apffel, S. Fischer, G. Goldberg, P.C. Goodley and F.E. Kuhlmann, *J. Chromatogr. A*, 712 (1995) 177.

36 J.M. Halket, In: A. Kornmüller and T. Reemtsma, (Eds.), Mass spectral libraries for LC–MS, Schriftenreihe "Biologische Abwasserreinigung" 2. Kolloquium im Sonderforschungsbereich 193 der DFG an der Technischen Universität Berlin "Anwendung der LC–MS in der Wasseranalytik", Berlin, Germany, 2001, Vol. 16, pp. 173, ISBN 3 7983 1852 2.

37 M. Hazos, C. Piemonti, J. Bruzual and A. Mendez, Proc. 44th ASMS Conf. on Mass Spectrom and Allied Topics, Portland, 1996, p. 1271.

2.6   ATMOSPHERIC PRESSURE IONISATION MASS
      SPECTROMETRY—III. NON-IONIC SURFACTANTS: LC–MS
      AND LC–MS–MS OF ALKYLPHENOL ETHOXYLATES AND
      THEIR DEGRADATION PRODUCTS

Mira Petrovic, Horst Fr. Schröder and Damia Barceló

2.6.1   Alkylphenol ethoxylates (APEOs)

Liquid chromatography–mass spectrometry (LC–MS) analysis of
APEOs has been attempted using both normal-phase (NP) and
reversed-phase (RP) systems. Using NP-LC, the APEOs are separated
according to the increasing number of ethylene oxide units, while
corresponding oligomers with the same number of ethoxy units, but
different alkyl substituents; e.g. nonylphenol ethoxylates (NPEOs)
and octylphenol ethoxylates (OPEOs), co-elute. RP-LC on C_8, C_{18} silica
gel columns, alumina-based C_{18} or polyethylene-coated alumina
allows separation according to the character of the hydrophobic
moiety and it is particularly well-suited to separate surfactants
containing various hydrophobic moieties (homologue-by-homologue
separation). In this case, the length of the ethylene oxide chain does
not influence the separation and the various oligomers containing the
same hydrophobic moiety elute in one peak. However, concentration of
individual oligomers (e.g. $NPEO_1$, $NPEO_2$, $NPEO_3$, etc.) can be readily
obtained by extracting total ion chromatograms for characteristic m/z
values (Fig. 2.6.1). This approach is suitable for routine determination
of OPEOs and NPEOs since the quantification is simplified and
eluting all the oligomers into one peak has the advantage of
increasing the peak intensity and therefore, increasing the sensitivity
of determination.

   APEOs can be detected using both electrospray (ESI) and atmospheric pressure chemical ionization (APCI), under positive ionization
(PI) conditions. Generally, it was found that ESI offers better sensitivity
and specificity for a higher range of oligomers than does APCI [1].
However, it is important to mention that the overall sensitivity of MS
detection—as well as the formation of adducts and fragmentation—
depends strongly on the source design and can vary significantly
depending on the particular instrument used. Specific issues related to

Comprehensive Analytical Chemistry XL
T.P. Knepper, D. Barceló and P. de Voogt (Eds)

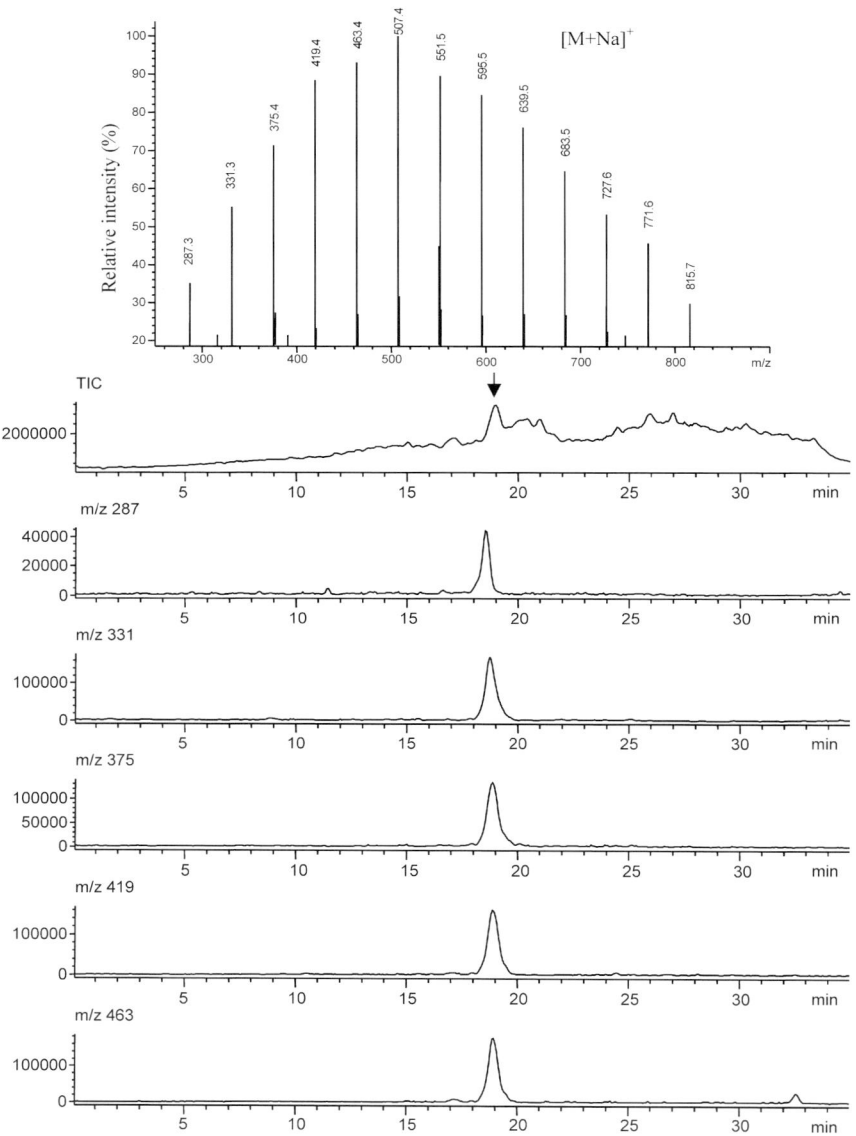

Fig. 2.6.1. RP–LC–ESI–MS analysis of flocculation sludge from a Barcelona drinking water treatment plant. Column C_{18} LiChrolute 250×4.6 mm, 5 μm, gradient elution with ACN–water. Upper trace total ion current (TIC), lower traces extracted ion chromatograms for NPEOs, $n_{EO} = 1-5$. Inset: ESI mass spectrum of NPEO oligomeric mixture.

the formation of distinct adducts with ions originating from the buffer, the sample and/or the introduction system (e.g. H^+, Na^+, K^+, Cl^-, CH_3COO^-), the formation of cluster and multiple charged ions, ionisation suppression caused by co-elution of other analytes or natural matrix components or due to the presence of ion-pairing agents or high salt concentrations in the mobile phase buffer used, as well as issues related with the response factors of different oligomers are discussed in detail in Chapter 4.3. (Quantitation in surfactant analysis.)

Several authors have reported the identification and determination of APEOs in industrial blends, wastewaters and environmental samples by APCI–MS using flow injection analysis (FIA) [2,3] or preceded by LC, using both RP [4–9] and NP [10,11] separation.

APCI–MS analysis of OPEO in river water and effluents of wastewater treatment plants (WWTP) applying FIA and ionisation promoted by applying ammonium acetate, resulted in equal spaced (Δ m/z 44) $[M + NH_4]^+$ ions starting from m/z 312 and ending at m/z 972 [12]. The oligomeric distribution followed a Gaussian curve with a maximum at $n_{EO} = 9$. LC–APCI–MS was also applied for the determination of NPEOs in tannery wastewaters [8]. Besides NPEO compounds, biochemical oxidation products of PEGs (carboxylated PEGs), resulting from degradation of non-ionic surfactants were determined.

The alicyclic homologue of Triton X-114, an octylcyclohexylethoxylate ('reduced' octylphenolethoxylate) mixture (C_8H_{17}–C_6H_{10}–O–(CH_2–CH_2–O)$_m$–H) and the aromatic OPEOs, were examined. Their APCI–FIA–MS(+) spectra are presented in Fig. 2.6.2(a,b), respectively. APCI–LC–MS(+) on a RP-C_{18} column was applied to separate and differentiate the constituents of this mixture; UV-DAD spectra were recorded in parallel to distinguish the aromatic and the alicyclic compounds. In Fig. 2.6.2 (6) the reconstituted ion current (RIC) and mass traces of aromatic (2 and 3) and alicyclic (4 and 5) homologues both ionised as $[M + NH_4]^+$ ions equally spaced with $\Delta m/z$ 44 u are presented. The UV trace 220 nm is plotted in Fig. 2.6.2 (1). Here the aromatic ring system serving as chromophore can be recognised, whereas the alicyclic compounds without chromophoric groups cannot be observed [13].

ESI–MS methods, combined with NP-LC [14–16] or RP-LC [17–21] have been frequently used for the quantitative analysis of ethoxylates in environmental and wastewater samples.

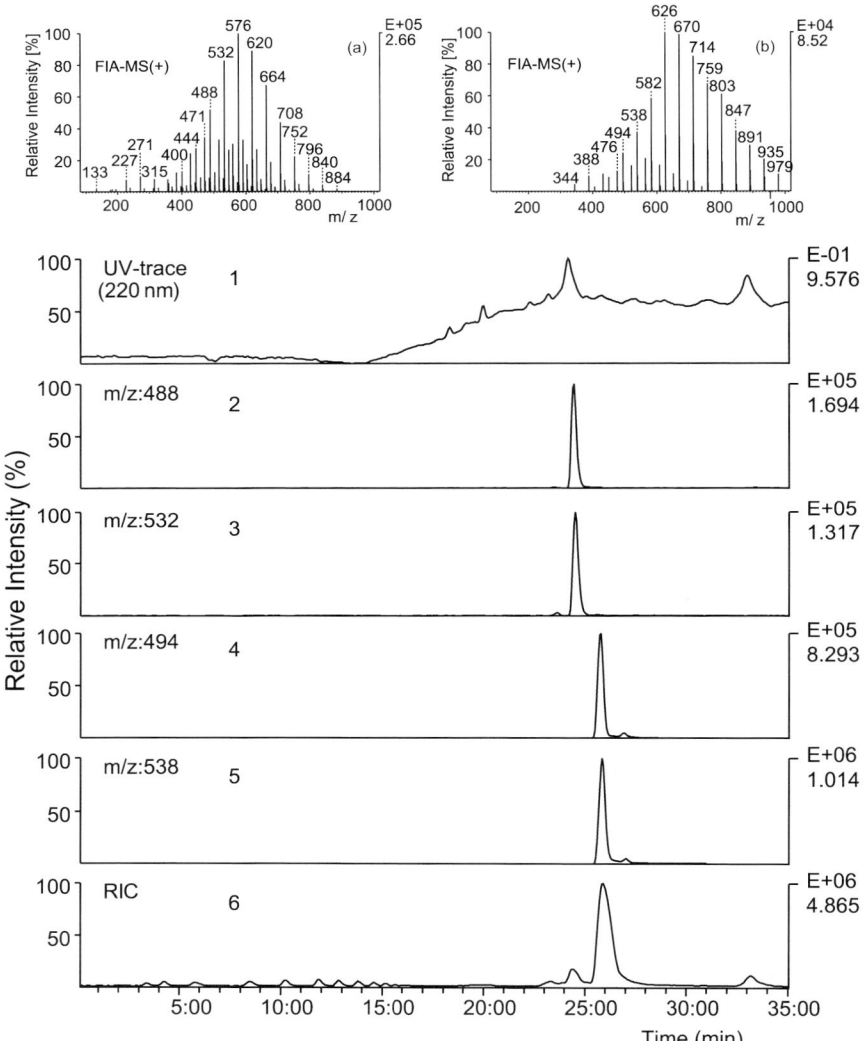

Fig. 2.6.2. APCI−LC−MS(+) TIC chromatogram (6) and selected mass traces of (2) and (3) of octylphenolethoxylates (OPEOs) together with (4) and (5) their alicyclic homologues, the octylcyclohexanolpolyglycolether; (1) UV trace 220 nm presenting OPEO signals [13].

Normal-phase LC–ESI–MS was used for the quantitative determination of individual NPEOs in complex environmental matrices (e.g. marine sediment) [14]. The best chromatographic separation of NPEO oligomers was achieved under NP-LC conditions using non-polar solvents as mobile phases (e.g. gradient elution with toluene as solvent A and 0.5 mM NaOAc in toluene–methanol–water (10:88:2, v/v) as solvent B). However, such mobile phases are not compatible with ESI, and post-column addition of a polar solvent and a modifier is required to facilitate ionisation of the target analytes and evaporation of the pre-formed ions into the gas phase via the electrospray process and thus to enhance the signal and system stability. High oligomer NPEOs benefit most from the post-column addition (mobile phase B at a flow-rate ratio of 0.75:1) with over 10-times signal intensity enhancement, whereas low oligomers ($NPEO_1$, $NPEO_2$) show approximately a three times increase in signal intensity compared with results from no post-column addition.

Using reversed-phase LC–ESI–MS, NPEOs ($n_{EO} = 1$–20) were detected in samples of raw and treated wastewater, river and drinking (tap) water after their pre-concentration by solid phase extraction (SPE) using GBC material and differential elution [22]. RP-LC separation of the NPEOs and electrospray application for MS coupling allowed sensitive detection of higher ethoxylates ($n_{EO} > 3$). However, the method gave very low sensitivity for $NPEO_1$ and $NPEO_2$, and they were quantified using fluorescence (FL) detection.

NPEOs and OPEOs ($n_{EO} = 3$–10) as industrial blends or standard compound (Triton X-100), respectively, were separated together with linear alkylbenzene sulfonates (LASs) on a C_1-RP column [10]. The intensive ions that could be observed in the spectra were mono-, di- and tri-sodium adduct ions $[M + Na]^+$ (m/z 581), $[M + 2Na]^+$ (m/z 604) and $[M + 3Na]^+$ (m/z 626) of the EO_7 homologue. The intensity of the molecular $[M + H]^+$-ion, however, was small compared with the sodium adduct ions. The compounds had been concentrated prior to separation on C_{18} and SAX SPE cartridges. Samples from river water were handled in the same way.

Ferguson et al. [18] reported on the application of a mixed-mode HPLC separation, coupled with ESI–MS for the comprehensive analysis of NPEOs and nonylphenol (NP) concentrations and distributions in sediment and sewage samples. The mixed-mode separation, which operates with both size-exclusion and reversed-phase mechanisms, allows the resolution of NPEO ethoxymers prior to introduction to the MS using a solvent system that is readily compatible with ESI. In this method, elution of NPEOs is reversed relative to normal-phase chromatography, with smaller, less ethoxylated compounds, including NP, eluting last (Fig. 2.6.3). The separation allows all NPEOs and NP to be quantified in a single chromatographic run, while removing the effects of isobaric interferences and co-analyte electrospray competition.

Recently, Shao et al. [23] reported on the complete separation of individual NPEO ($n_{EO} = 2$–20), which was achieved combining a C_{18} pre-column with a silica analytical column, using acetonitrile–water as eluent followed by ESI–MS detection. The method used ramped cone voltage from 25 to 70 V, which suppressed the appearance of doubly charged adducts effectively and increased the response of oligomers of relatively high molecular mass.

Several methods, applying MS–MS techniques, for the identification and characterization of APEOs and their metabolites, alkylphenoxy carboxylates (APECs) and alkylphenols (APs) in environmental liquid and solid samples and industrial blends are reported. APCI–MS–MS

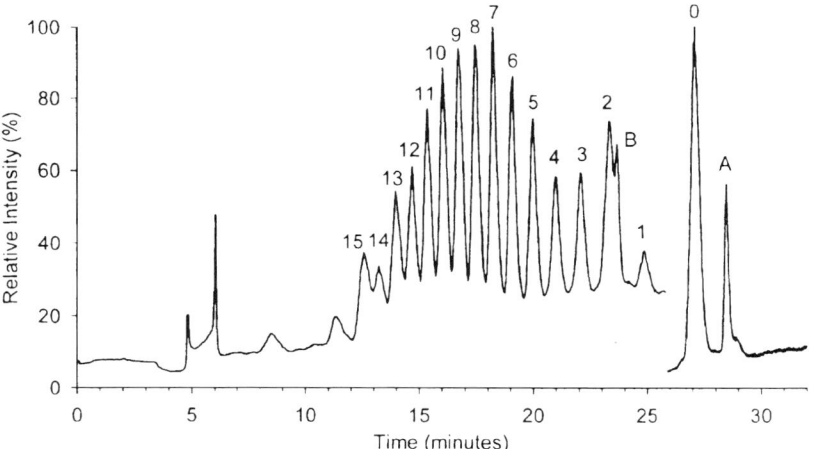

Fig. 2.6.3. Mixed-mode HPLC–ESI–MS summed ion chromatogram of a sediment extract (32–36 cm depth core slice) showing resolution of NPEOs and NPs. Numbered peaks correspond to NPEOs and $[^{13}C_6]$NPEOs with the indicated number of ethoxy groups (0 = NP, $[^{13}C_6]$NP; 1 = NPEO$_1$, $[^{13}C_6]$NPEO$_1$, etc.). Peaks 'A' and 'B' are the internal standards, n-NP and n-NPEO$_3$, respectively. (Note the discontinuity at retention time 25.8 min, corresponding to the shift in MS polarity from positive to negative ion mode.) Reprinted with permission from Ref. [18] © 2001 Elsevier.

showed ethoxy chain fragments at m/z 89, 133, 177 and the diagnostic fragment of OPEOs at m/z 277 and 291 for NPEOs, which can be explained as shown in Fig. 2.6.4 [12,24–26]. Mixture analysis by FIA–MS–MS(+) was applied for the confirmation using the diagnostic precursor scans of m/z 277 and 291 for the detection and identification of OPEOs and NPEOs, respectively [25].

The precursor ion scanning of m/z 121 and 133 and multiple reaction monitoring (MRM) applying APCI–FIA–MS–MS(+) were used for a rapid screening of NPEOs as contained in the industrial blend Igepal CP-720. The precursor scan (PS) of 121, characteristic for ethoxylates with 1–4 chain units (EO_1–EO_4), and the PS of 133, characteristic for EO_5–EO_{16}, demonstrated a preferential elimination of the EO_5–EO_{16} NPEOs from wastewater samples. The PS alone was not characteristic for these compounds because linear alcohol ethoxylates resulted in the same precursors; monitoring 16 MRM transitions allowed the confirmation of results [27].

For confirmation of low concentrations of NPEO homologues in complex samples from the Elbe river, APCI–LC–MS–MS(+) was applied to record the substance-characteristic ion mass trace of m/z 291. The SPE isolates contained complex mixtures of different surfactants. The presence of NPEOs in these complex samples was confirmed by generating the precursor ion mass spectrum of m/z 291 applying MS–MS in the FIA–APCI(+) mode. This spectrum, showed in Fig. 2.6.5, presents the characteristic series of ions of NPEOs at m/z 458, 502,...,678, all equally spaced with $\Delta m/z$ 44 u. Besides the NPEOs, small amounts of impurities could be observed because of the very low concentrations of NPEOs in the water sample [25]. In the foam sample, the identity of NPEOs could be easily confirmed by APCI–FIA–MS–MS(+) because of their high concentrations in this matrix. The LC–

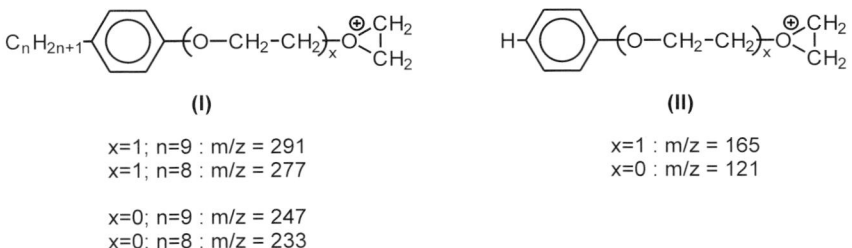

(I)

x=1; n=9 : m/z = 291
x=1; n=8 : m/z = 277

x=0; n=9 : m/z = 247
x=0; n=8 : m/z = 233

(II)

x=1 : m/z = 165
x=0 : m/z = 121

Fig. 2.6.4. Substance specific fragments for characterisation of APEOs [13,26].

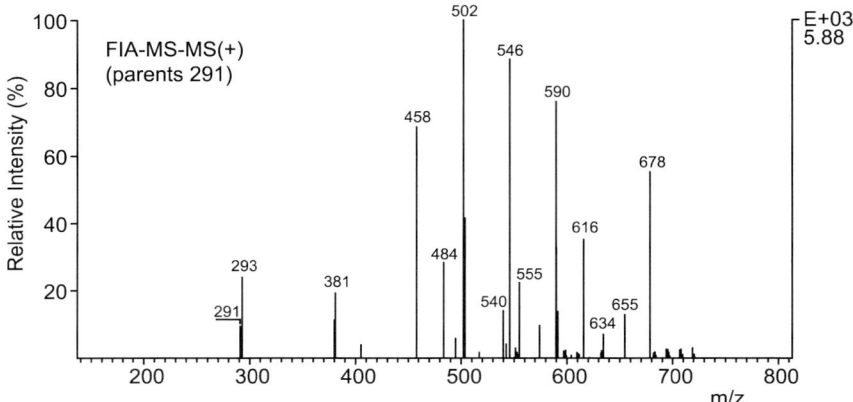

Fig. 2.6.5. APCI–FIA–MS–MS(+) (CID) parent ion mass spectrum of product ion at m/z 291 of C_{18}-SPE surface water extract from the river Elbe presenting NPEO characteristic ions at m/z 458, 502, 546, 590, 634 and 678, all equally spaced with $\Delta m/z$ 44 [25].

MS–MS(+) product ion spectrum of the NPEO homologue $[M + NH_4]^+$ ion at m/z 678 with $n_{EO} = 10$ together with its fragmentation scheme is presented in Fig. 2.6.6.

The alicyclic isomer of Triton X-114, an octylcyclohexylethoxylate (reduced octylphenolethoxylate) mixture $(C_8H_{17}-C_6H_{10}-O-(CH_2-CH_2-O)_n-H)$ was examined by APCI–FIA–MS–MS(+). The equally spaced ions ($\Delta m/z$ 44 u) could be fragmented as presented in the fragmentation scheme together with the daughter ion spectrum in Fig. 2.6.7. No fragmentation could be observed in the alkyl chain; however, unlike aromatic homologues, a bond scission took place between the alcoholic oxygen and the alicyclic ring system. PEG ions as base peak ions with the general structure $HO-(CH_2-CH_2-O)_x-H$ were generated according to the PEG chain length of the precursor parent ion while a neutral loss of 193 could be observed in parallel [28].

Methods applying LC–MS–MS for the quantitative determination of APEO were seldom reported. Houde et al. [29] recently described a LC(ESI)–MS–MS method for the determination of NPEO and nonylphenoxy carboxylates (NPEC) in surface and drinking water using a reversed-phase column (C_8) with isocratic elution. Transitions from the $[M + NH_4]^+$ ions to different product ions (m/z 127, 183, 227, 271, 315, 359 for NPEO $n_{EO} = 1-6$ and m/z 291 for $n_{EO} = 7-17$) were monitored, yielding detection limits from 10 to 50 ng L^{-1}.

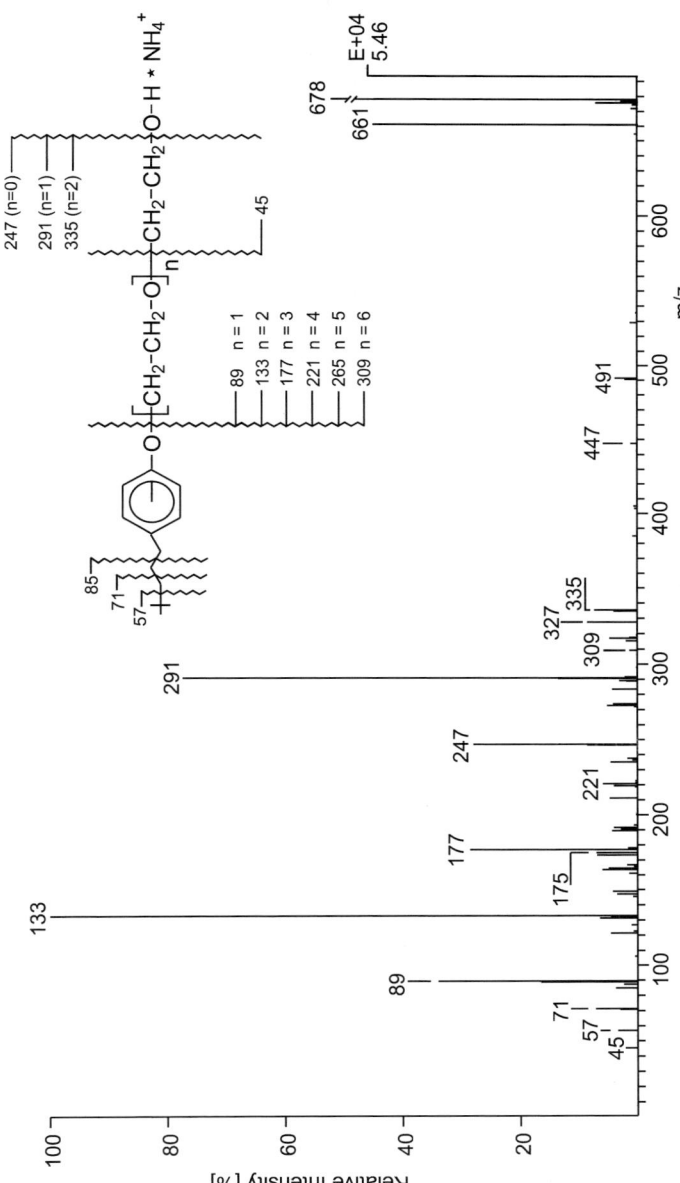

Fig. 2.6.6. APCI–LC–MS–MS(+) (CID) daughter ion mass spectrum of $[M + NH_4]^+$ ion at m/z 678 generated from C_{18}-SPE of foam sample. Compound could be identified as non-ionic surfactant NPEO $(C_9H_{19}-C_6H_4-O-(CH_2-CH_2-O)_m-H)$; (inset) fragmentation scheme under CID conditions [28].

171

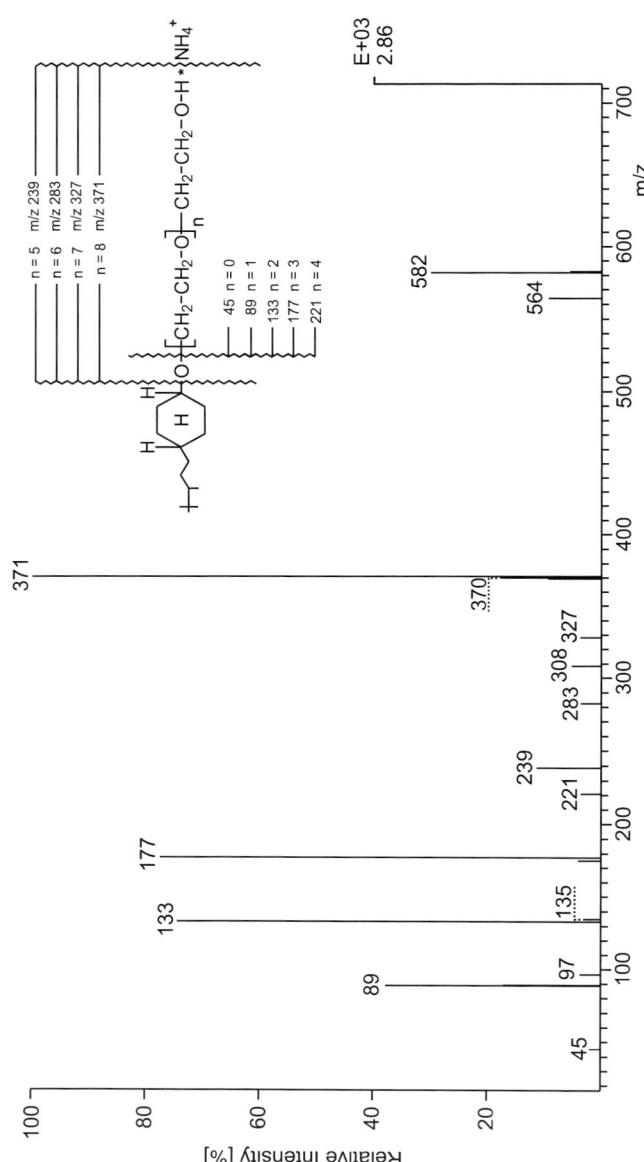

Fig. 2.6.7. APCI–FIA–MS–MS(+) daughter ion mass spectrum of $[M + NH_4]^+$ ion of octylcyclohexylethoxylate homologue at m/z 582 from 'reduced' octylphenolethoxylate mixture $(C_8H_{17}-C_6H_{10}-O-(CH_2-CH_2-O)_m-H)$; (inset) fragmentation scheme under CID conditions [28].

2.6.2 Alkylphenols (APs)

Octylphenol (OP) and NP, were detected under negative ionisation (NI) conditions, using both APCI and ESI interfaces. The sensitivity of detection, using an ESI source was approximately 40–50 times higher than that obtained with an APCI source [30]. In contrast to the GC–MS analysis that reveals the presence of 22 isomers of the alkyl chain in the technical mixture of 4-NP, LC–MS analysis yields a single very broad peak. Using an ESI, APs give exclusively $[M - H]^-$ ions with m/z 205 for OP and m/z 219 for NP as shown in Fig. 2.6.8.

Using an APCI, at higher voltages, using so-called in-source CID, the spectra show fragmentation that closely resembles that obtained by the MS–MS technique. APs give, in addition to the $[M - H]^-$ ions, fragment m/z 133, resulting from the loss of a C_5H_{12} (OP) and C_6H_{14} (NP) group [31]. A similar fragmentation pattern is obtained using a MS–MS [32]. Fragments m/z 147, 133, 119 and 93 result from the progressive fragmentation of the alkyl chain, whereas m/z 117, observed also by Pedersen and Lindholst [31], cannot be explained in such a straightforward way. Therefore, reaction channels m/z 205 → 133 (for OP) and m/z 219 → 133 (for NP) and precursor (parent) ion scan of m/z 133 can be used to monitor APs. However, the reaction is not specific, and it can also be used for APECs (Fig. 2.6.9).

2.6.3 Carboxylated degradation products (APECs and CAPECs)

Alkylphenoxy carboxylates (APE_nC) were detected in both the NI mode [1,18,33] and PI mode [21]. In the NI mode, using ESI, APECs give two types of ions, one corresponding to the deprotonated molecule $[M - H]^-$ and the other to deprotonated alkylphenols $[M - CH_2COOH]^-$ in the case of APE_1Cs and $[M - CH_2CH_2OCH_2COOH]^-$ for the APE_2Cs. The relative abundance of these two ions depends on the extraction voltage. Using a low voltage, the ESI source is capable of producing deprotonated molecular ions, and the spectra display only signals at m/z 277 and 263 corresponding to NPE_1C and OPE_1C and m/z 321 and 307 for NPE_2C and OPE_2C. At higher voltages, using so-called in-source CID, the spectra give fragmentation that closely resembles that obtained by the MS–MS technique [34]. Figure 2.6.9(a) shows a product

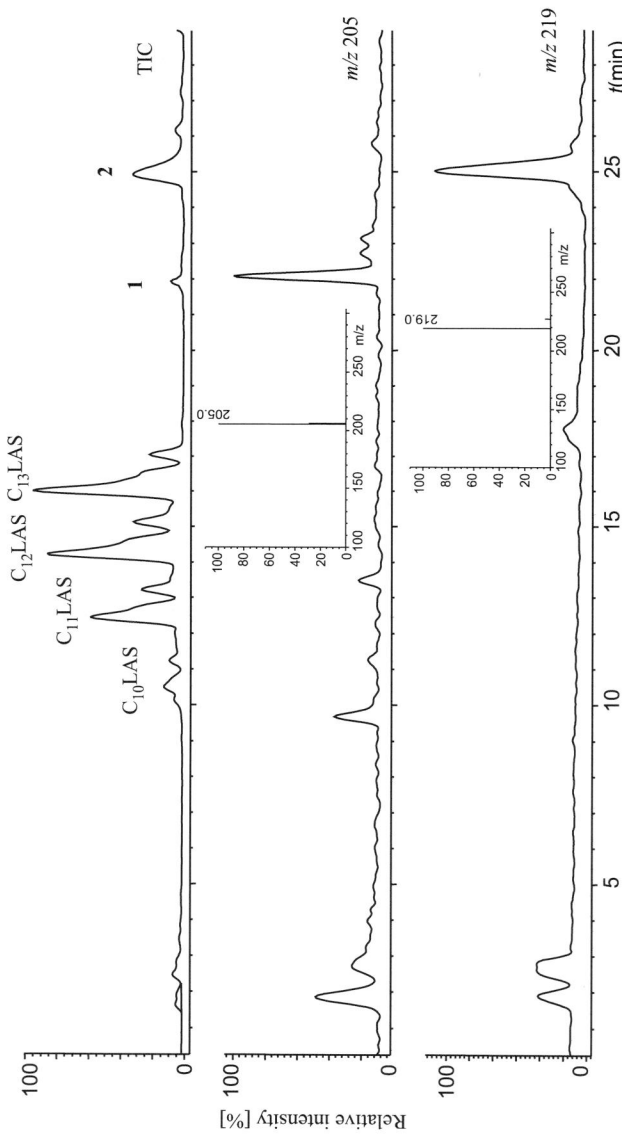

Fig. 2.6.8. Full-scan LC–ESI–MS chromatogram (upper trace) of river sediment (Anoia, downstream of WWTP) and extracted chromatogram for m/z 205 and 219 corresponding to OP and NP, respectively. Inset: mass spectra of OP and NP. Reprinted with permission from Ref. [40] © 2001 AOAC International.

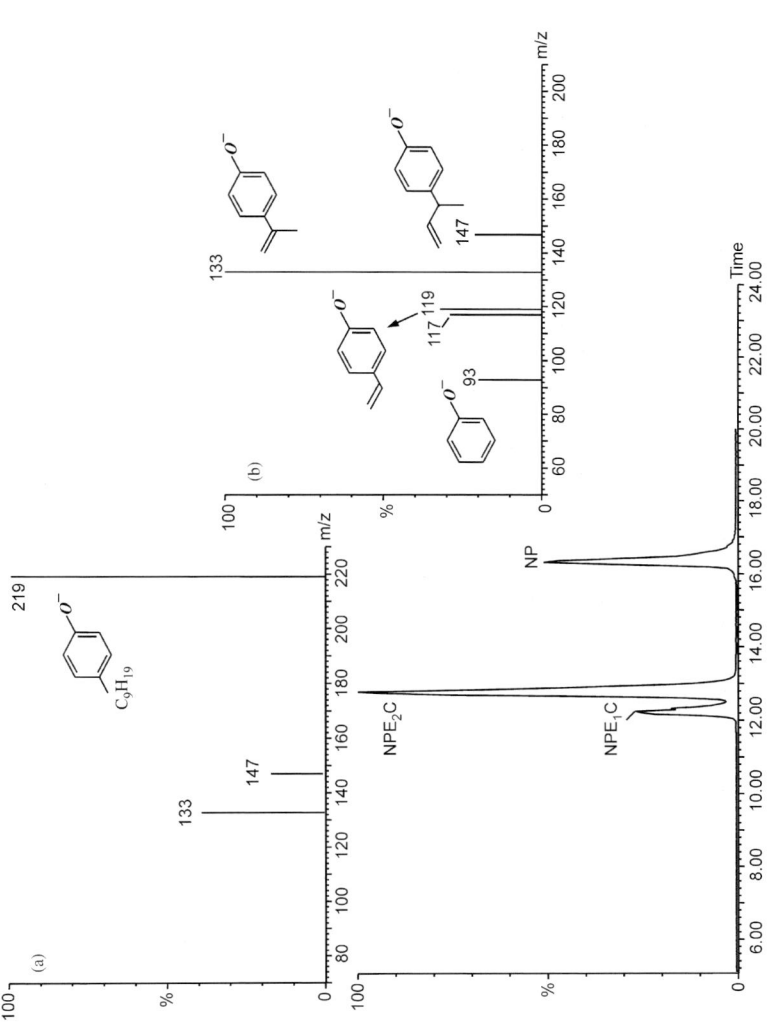

Fig. 2.6.9. MS–MS chromatogram (MRM channel m/z 219 → 133) of raw effluent (river water) treated in a Barcelona drinking water treatment plant. Insets: product ion scan of NPE_2C (A) and NP (B), obtained using argon as collision gas at collision energy of 40 eV. Reprinted with permission from Ref. [32] © 2002 Elsevier.

ion scan of m/z 321 for NPE_2C. Intense signals at m/z 219 and 205 are produced after the loss of the carboxylated (ethoxy) chain, while m/z 133 and 147 corresponded to the fragmentation of the alkyl chain, as described above for NP.

The short chain NPE_1C, which had been synthesised from branched NPs for standard comparison purposes, was also detected by APCI–FIA–MS in the positive and negative modes. In the presence of ammonium acetate, APCI under PI conditions resulted in $[M + NH_4]^+$ ions at m/z 296, whereas NI generated the prominent $[M - H]^-$ ion besides negatively charged acetate adduct ions $[M - H + acetate]^-$ and the dimeric ions $[2M - H]^-$ at m/z 277, 337 or 555, respectively [13]. NPE_2C gave $[M + NH_4]^+$ ion at m/z 340. This ion, examined by CID($+$), Labo generated alkyl fragments at m/z 57, 71 and 85, the $[C_9H_{19}]^+$ product ion at m/z 127 and the prominent ion at m/z 103, characteristic for carboxylated PEG chains as presented in the product ion spectrum and fragmentation scheme in Fig. 2.6.10.

Aerobic degradation in laboratory scale experiments was carried out by Di Corcia et al. [167]. They could prove the postulate, which resulted in either NPECs, alkyl chain carboxylated NPEO (CNPEO) or compounds carboxylated in both positions, in the alkyl chain and in the polyether chain (CNPEC). Additionally these compounds could be generated in the mechanical–biological WWTP [21]. Analysis of treated sewages showed that CAPECs were the dominant products of the A_9PEO biotransformation, accounting for 66% of all the metabolites leaving the WWTP [18].

The identity of the dicarboxylated breakdown products was confirmed by reversed-phase LC–ESI–MS [18] and LC–ESI–MS–MS [33]. Under NI conditions, at low cone voltage no CID process was possible and the spectrum displayed only signals tentatively assigned to the $[M - H]^-$ and $[MNa - 2H]^-$. However, with the cone voltage of 55 V, structural confirmation of these species was achieved by observing different fragment ions (Fig. 2.6.11). Typical total ion current (TIC) chromatogram and ion current profiles of CAPECs obtained using a C_{18} reversed-phase column (Altima, Alttech, Italy) and gradient elution with acetonitrile and water (both acidified with formic acid, 1 mmol L^{-1}) is shown in Fig. 2.6.12.

At low extraction voltages, the in-source CID process is greatly inhibited and the spectra display intense signals for the protonated molecular ions. By raising the extraction voltage, in-source CID spectra were obtained. Neutral losses of the carboxylated ethoxy chain and

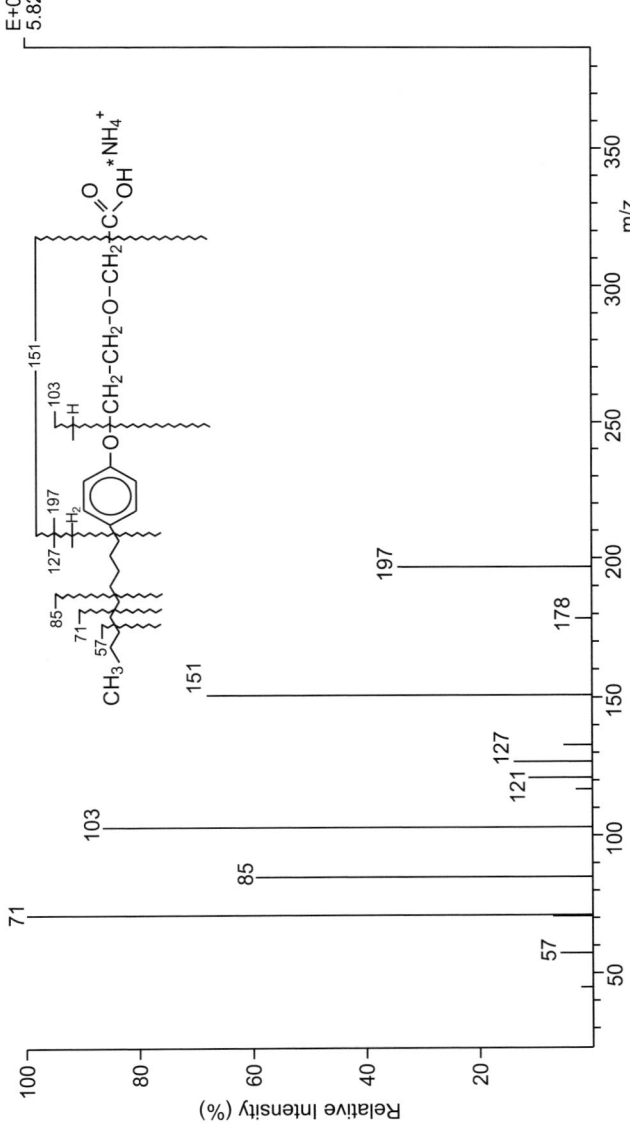

Fig. 2.6.10. APCI–FIA–MS–MS(+) (CID) daughter ion mass spectrum of selected $[M + NH_4]^+$ parent ion (m/z 340) of potential carboxylated non-ionic surfactant metabolite of precursor NPEO prepared by chemical synthesis; structure of short-chain NPEC: $C_9H_{19}–C_6H_4–O–(CH_2–CH_2–O)–CH_2–COOH$; fragmentation behaviour under CID presented in the inset [28].

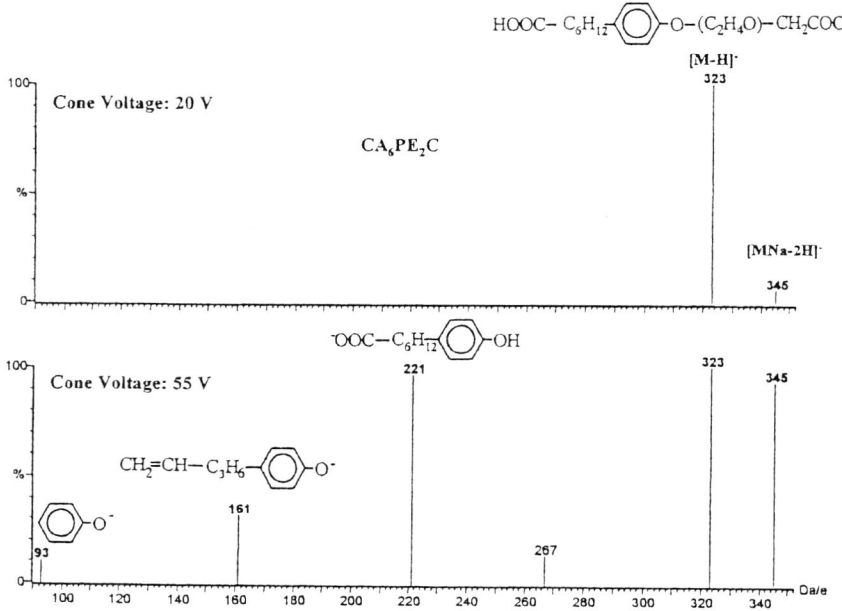

Fig. 2.6.11. NI mass spectra of the dicarboxylated (CA_6PE_2C) metabolite of NPEO taken at two different cone voltages. Reprinted with permission from Ref. [19] © 2000 American Chemical Society.

carboxylated alkyl chain, respectively, and methanol loss followed by formation of acylium ions, were found to be typical fragmentation patterns for methylated CAPECs. In Fig. 2.6.13 the daughter ion spectra of underivatised CNPECs (A) and methylated CNPECs (M) together with their fragmentation behaviour are presented. The alkyl chain branched $CNPE_1C$ compounds could be confirmed as extremely recalcitrant intermediates in the biochemical degradation process.

However, using LC–MS, under conditions giving solely molecular ions, identification of dicarboxylated compounds is difficult since CA_nPE_mCs have the same molecular mass as APECs having one ethoxy unit less and a shorter alkyl chain ($A_{n-1}PE_{m-1}C$). Moreover, since some compounds partially co-elute, the unequivocal assignment of the individual fragments can be accomplished only by using LC–MS–MS. Jonkers et al. [33] studied the aerobic biodegradation of NPEOs in a laboratory scale bioreactor. The identity of the CAPEC metabolites was

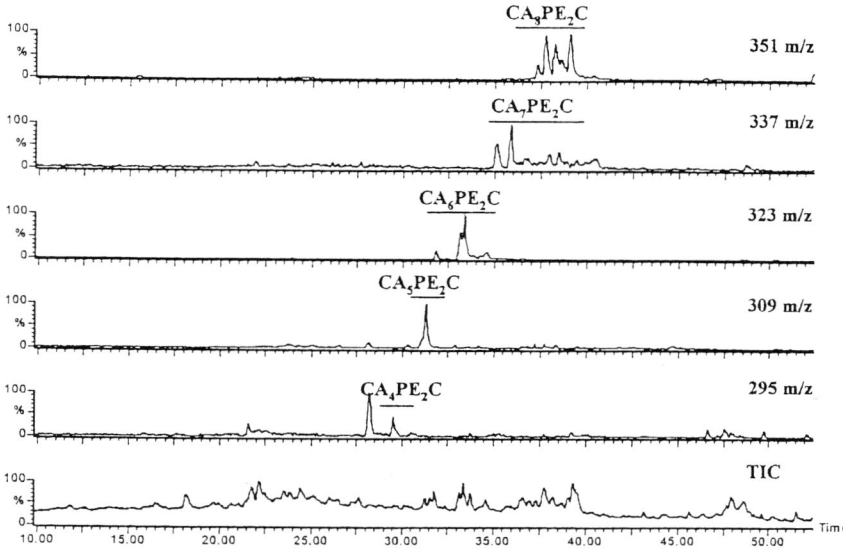

Fig. 2.6.12. TIC chromatogram (bottom trace) and extracted chromatograms for CAPECs obtained in NI mode. Reprinted with permission from Ref. [19] © 2000 by American Chemical Society.

confirmed by the fragmentation pattern obtained with LC–ES–MS–MS. Of 17 degradation products that were found to accumulate, nine were confirmed to be CAPEC metabolites. In Fig. 2.6.14 MS–MS fragmentation pattern is given for some of the compounds that were positively identified. In addition to the carboxy-alkylphenoxy fragment, typical fragments observed were the carboxy-alkylphenoxy fragment that has additionally lost CO_2^- or an acetic acid group, which, in the case of $CA_5PE_{1-2}C$, led to fragments of m/z 149 and 133.

2.6.4 Halogenated derivatives of alkylphenolic compounds

APEOs and their acidic and neutral metabolites can be halogenated to produce chlorinated and brominated products. The formation of these compounds has been reported during the chlorination processes at drinking water treatment plants [1,35,36] and after biological waste-water treatment [37].

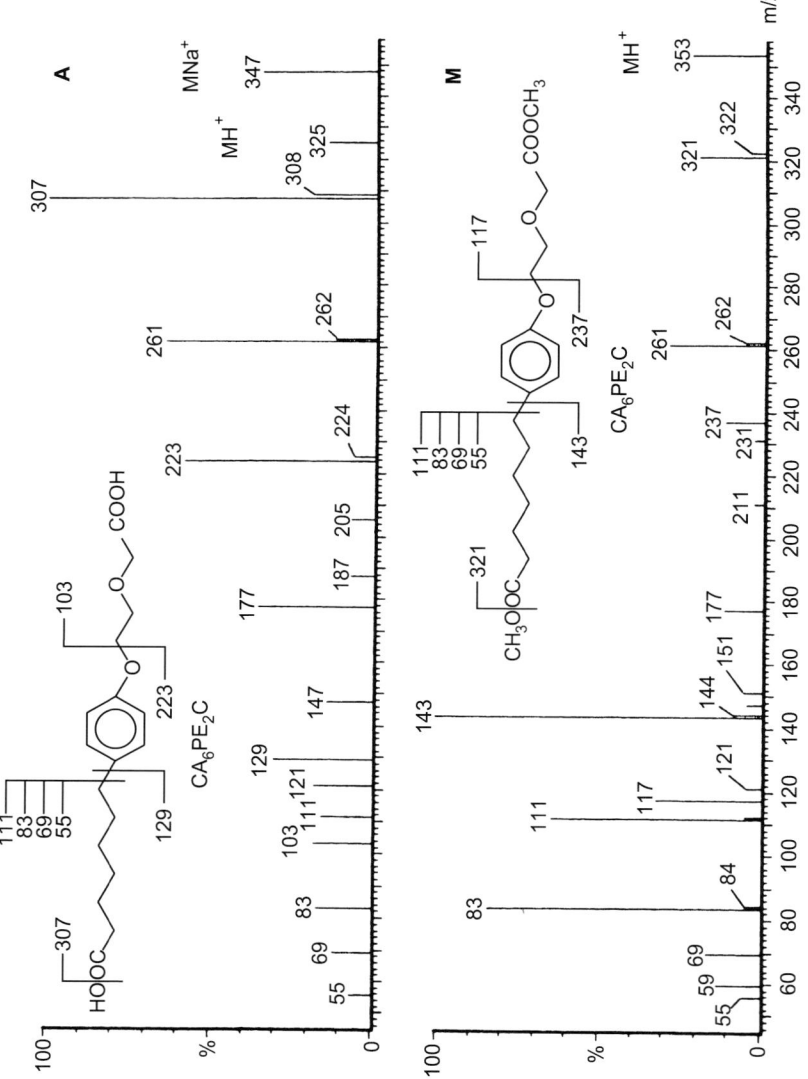

Using an ESI interface, halogenated APEOs like their non-halogenated analogues, show a great affinity for alkali metal ions, and they give exclusively evenly spaced sodium adduct peaks $[M + Na]^+$ with no further structurally significant fragmentation. The problem arises from the fact that the chlorinated derivatives ($ClAPE_nO$) have the same molecular mass and they gave the same ions as brominated compounds with one ethoxy group less ($BrAPE_{n-1}O$, respectively). However, they can be distinguished by their different isotopic profiles. The doublet signal in the mass spectrum of brominated compounds shows the contribution of bromine isotopes of $^{79}Br/^{81}Br = 100:98$, while the contribution of chlorine isotopes is $^{35}Cl/^{37}Cl = 100:33$. Therefore, chromatographic separation of these two groups of compounds is a pre-requisite of their quantitative determination. Halogenated NPEOs were detected in flocculation sludge from a Barcelona drinking water treatment plant (DWTP) (Figs. 2.6.15 and 2.6.17) [1]. Halogenated OPEOs were also identified. Like halogenated NPEOs, OPEO derivatives gave regularly spaced signals corresponding to $[M + Na]^+$ ions with m/z 395/397–571/573 corresponding to ClOPEOs, $n_{EO} = 3$–6 (assigned as ■ in Fig. 2.6.16(c)) and m/z 395/397–615/617 corresponding to BrOPEOs, $n_{EO} = 2$–7 (assigned as ■ in Fig. 2.6.16(d)).

Halogenated APs and halogenated APECs were analysed in the NI mode using an ESI interface [1,17]. Halogenated NPs (XAPs) gave a characteristic isotope doublet signal of the $[M - H]^-$ ions (m/z 297/299 for BrNP and m/z 253/255 for ClNP). Halogenated NPECs (XNPECs) gave two signals, one corresponding to a quasi-molecular ion and another to $[M - CH_2COOH]^-$ in the case of XNPE$_1$Cs or $[M - CH_2$ $CH_2OCH_2COOH]^-$ for XNPE$_2$Cs. The relative abundance and absolute intensity of these two ions, as compared with quasi-molecular ions, depends largely on the cone voltage. At higher values, the base peak, with high absolute intensity, for ClNPE$_1$C and ClNPE$_2$C is m/z 253/255 and for BrNPE$_1$C and BrNPE$_2$C m/z 297/299.

With gradient elution using methanol–water on a C_{18} reversed-phase column, APECs, APs, XAPs and XAPECs can be easily separated.

Fig. 2.6.13. Fragmentation behaviour of the acidic NPEO metabolites (CNPEC); (A) ESI–LC–MS(+) of underivatised compound and (M) di-methyl ester of HOOC(CH$_2$)$_6$–C$_6$H$_4$–O–(CH$_2$CH$_2$O)COOH: Source CID conditions, see Ref. [21]. Reprinted with permission from Ref. [21] © 1998 by American Chemical Society.

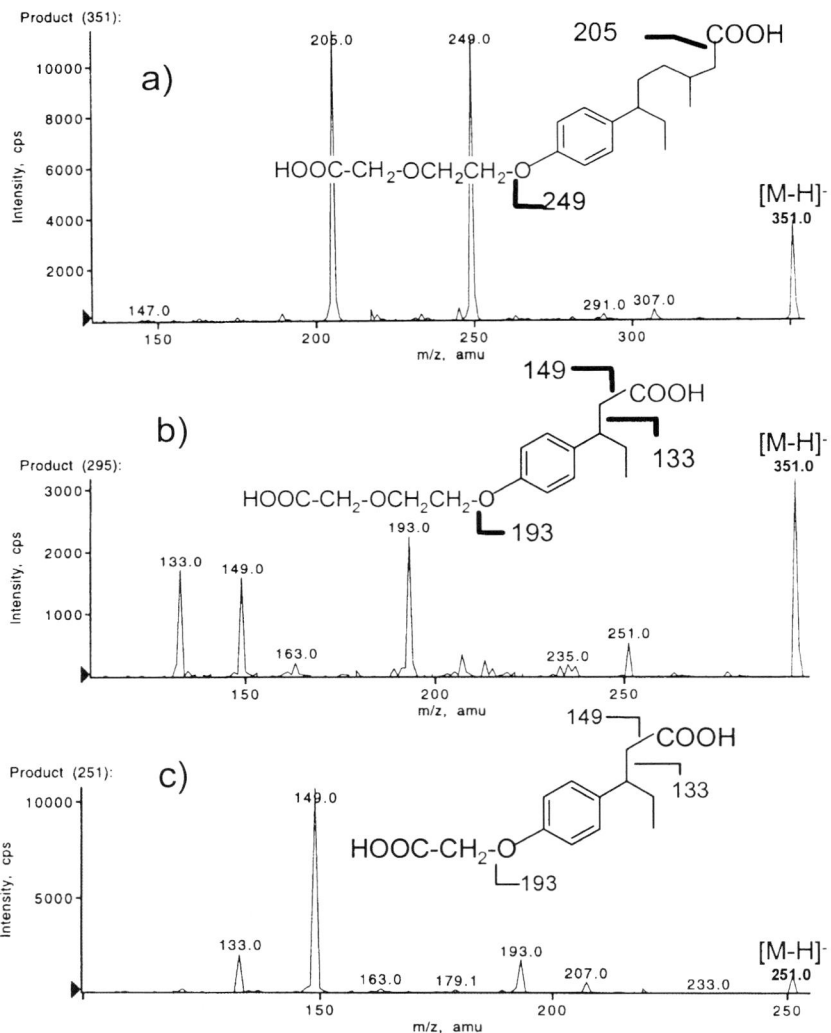

Fig. 2.6.14. MS–MS fragmentation patterns of some CAEC (NI mode): (a) CA$_9$PE$_2$C, (b) CA$_5$PE$_2$C, and (c) CA$_5$PE$_1$C. The exact branching of the alkyl chain is unknown; the alkyl isomer structures shown are chosen arbitrarily. Reprinted with permission from Ref. [33] © 2001 American Chemical Society.

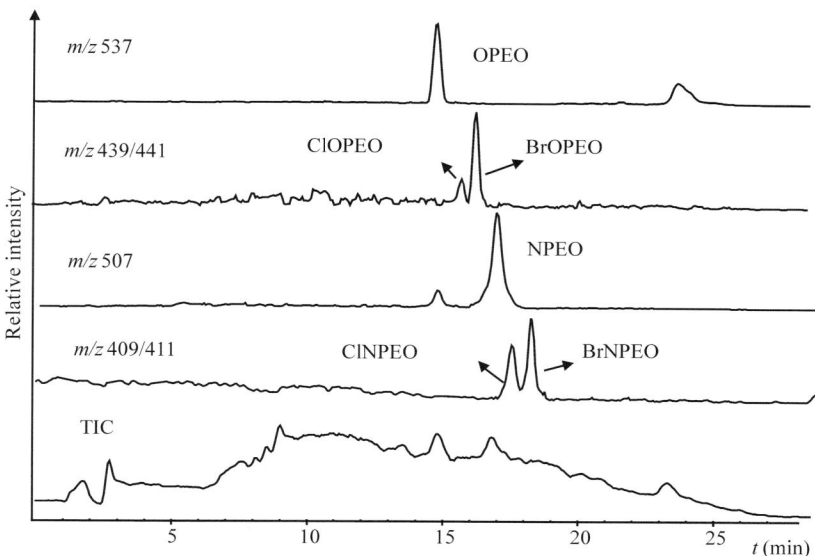

Fig. 2.6.15. Total ion LC–ESI–MS chromatogram (bottom trace) and reconstructed chromatograms of halogenated APEOs and APEOs, obtained in PI mode, found in sludge from a Barcelona drinking water treatment plant. Reprinted with permission from Ref. [1] © 2001 American Chemical Society.

However, when analysing real samples containing LAS, $ClNPE_1C$ co-elutes with $C_{11}LAS$, a compound having the same molecular weight ($M_W = 312$) and base ion m/z 311. LASs are the major surfactant class used in detergents throughout the world and high concentrations (up to several $mg\,L^{-1}$) are often found in environmental and wastewater samples. All attempts to separate $ClNPE_1C$ and $C_{11}LAS$ using gradient elution with standard mobile phases for reversed-phase separation (methanol/water or acetonitrile/water) failed and determination of $ClNPE_1C$ was achieved by monitoring a fragment ion at m/z 253/255, which also suffered the isobaric interferences in some real samples.

ESI–MS–MS permitted unambiguous identification and structure elucidation of compounds detected under NI conditions (halogenated NPECs and NPs), while for NPEOs, detected under positive ionisation conditions no fragmentation was obtained and these compounds were analysed using a single stage MS as $[M + Na]^+$ in selected ion

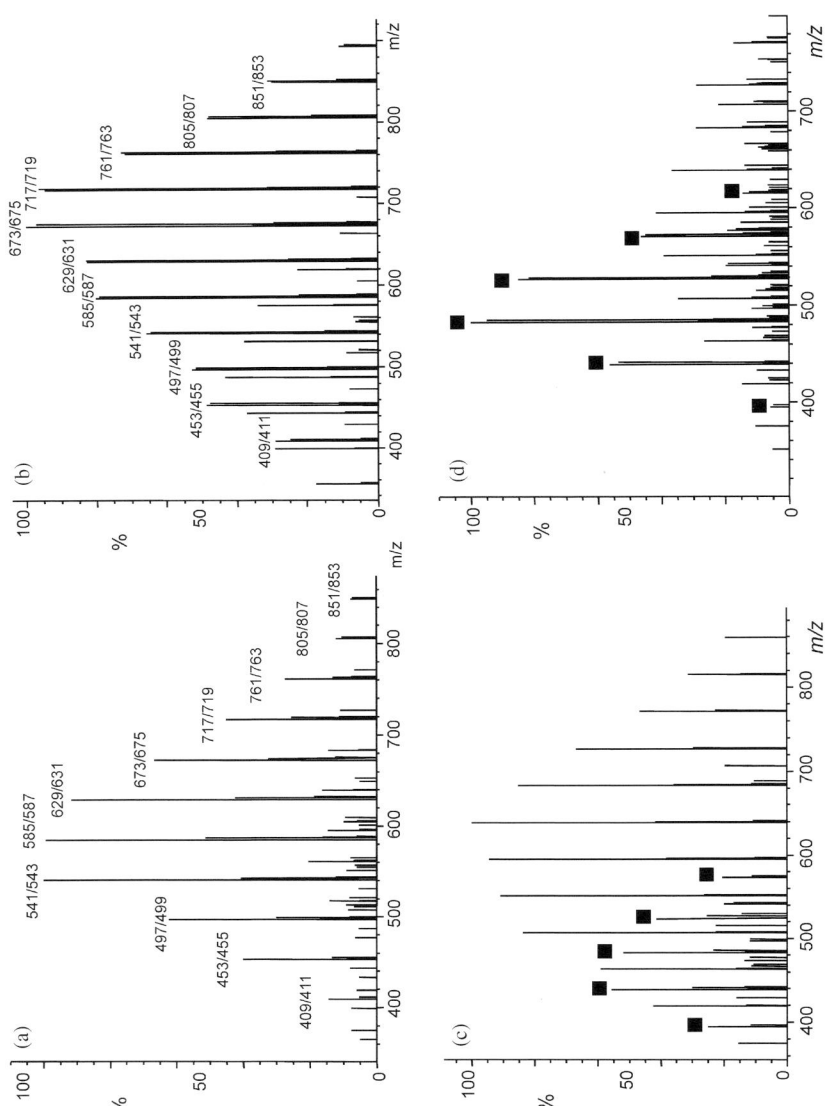

184

monitoring (SIM) mode [38,39]. With collision-induced dissociation (CID) under NI conditions, halogenated NPECs and NPs undergo fragmentation with few major pathways, depicted in Fig. 2.6.17. For halogenated NPE_1Cs and NPE_2Cs, the predominant reaction was loss of CH_2COOH and $CH_2CH_2OCH_2COOH$, respectively, which resulted in intense signals at m/z 253/255 for ClNPECs and m/z 297/299 for BrNPECs. Further fragmentation of $[ClNP]^-$ occurred primarily on the alkyl moiety leading to a sequential loss of m/z 14 (CH_2 group), with the most abundant fragments at m/z 167 for ^{35}Cl and m/z 169 for ^{37}Cl with the relative ratio of intensities of 3.03. A fragment corresponding to a chlorine ion was produced only when sufficient collision energy was applied. The intensity of this ion was not very pronounced, but nevertheless remained useful for the identification of chlorinated NP. The brominated compounds showed a markedly different fragmentation pathway. Even at low collision energy $[BrNP]^-$ (m/z 297/299) yielded intense signals at m/z 79 and 81 corresponding to the $[Br]^-$ (ratio of isotopes is 1.02), while the fragmentation of the side chain was suppressed and resulted just in a low-intensity fragment at m/z 211/213 produced after the loss of C_6H_{14}. Such a difference in the mechanism of fragmentation of chlorinated and brominated compounds is presumably the consequence of the lower energy of a Br–benzene bond compared with a Cl–benzene bond.

The detection limits of the LC–MS–MS method [38] fell down to 1–2 ng L^{-1} for the analysis of halogenated NPECs and NPs in water samples (after SPE pre-concentration) and to 0.5–1.5 ng g^{-1} in sludge samples (target compounds were extracted using pressurised liquid extraction). A specificity of MRM mode permitted unequivocal identification and quantification of isobaric target compounds (e.g. $BrNPE_1C$ and $ClNPE_2C$) and elimination of interference of co-eluting isobaric no-target compounds (e.g. $C_{11}LAS$), both having a base ion at m/z 311, observed using an LC–MS in SIM mode as shown in Fig. 2.6.18.

Fig. 2.6.16. ESI mass spectra of halogenated APEOs detected in the DWTP sludge (chromatogram shown in Fig. 2.6.17): (a) ClNPEO; (b) BrNPEO; (c) ClOPEO (assigned as ■); and (d) BrNPEO (assigned as ■). Reprinted with permission from Ref. [1] © 2001 American Chemical Society.

186

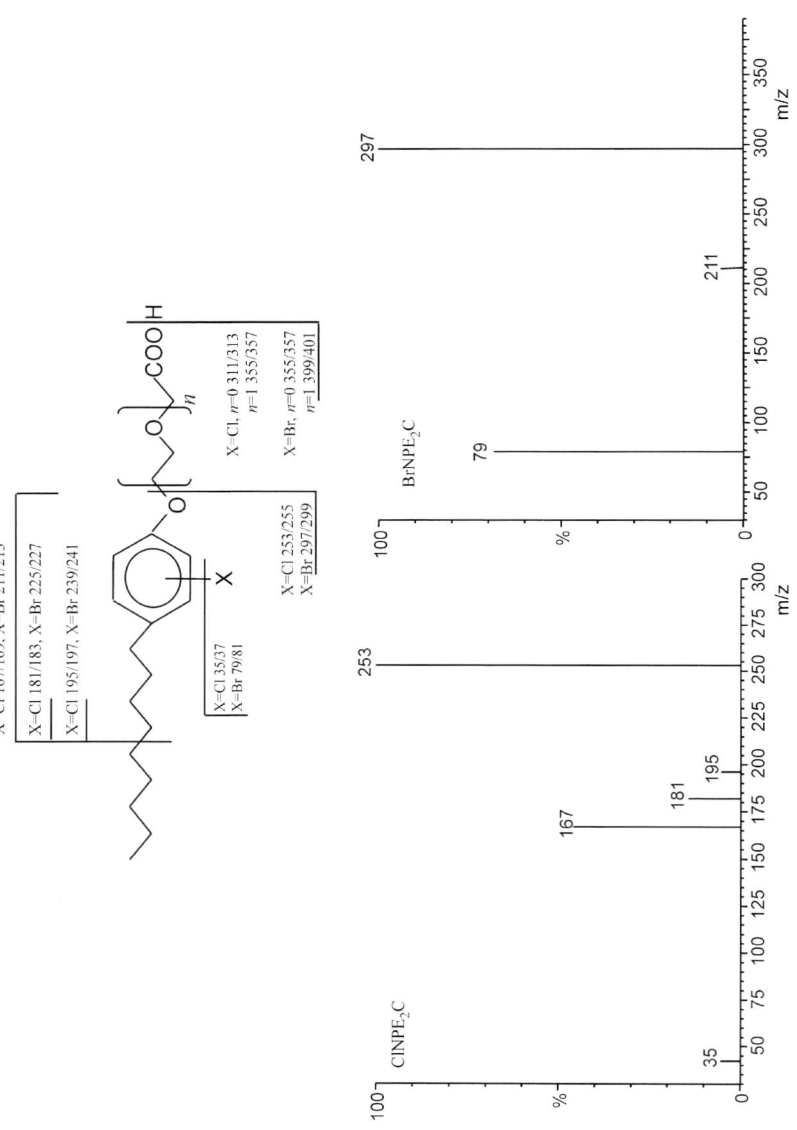

Fig. 2.6.17. CID spectra of (a) ClNPE$_2$C (precursor ion m/z 355) and (b) BrNPE$_2$C (precursor ion m/z 399) obtained at collision energy of 30 eV and the proposed fragmentation pattern. Reprinted with permission from Ref. [32] © 2002 Elsevier.

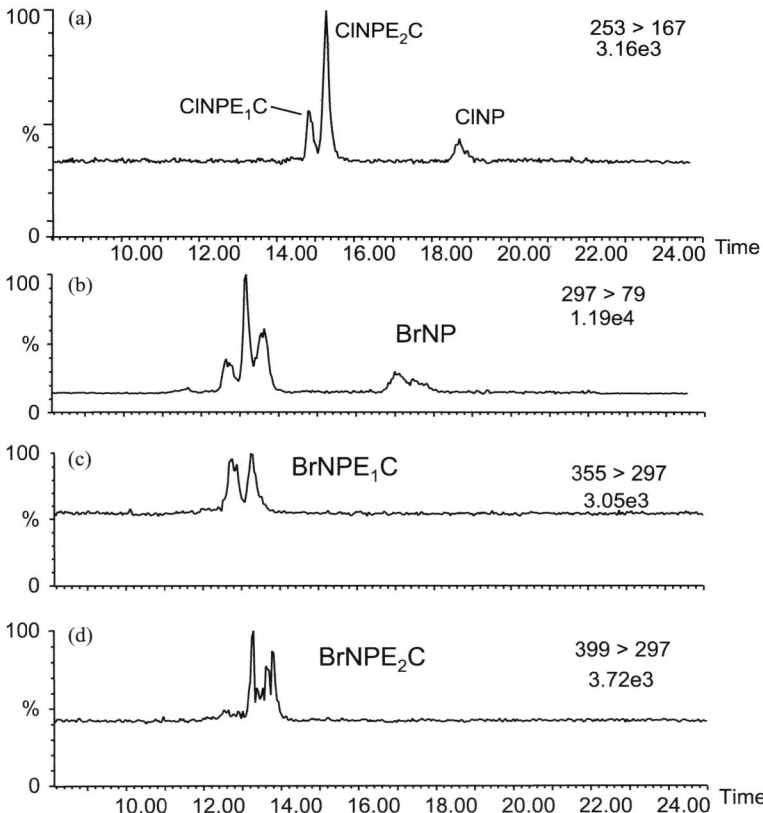

Fig. 2.6.18. LC–MS–MS analysis of halogenated alkylphenolic compounds in chlorinated river water (the Llobregat river, Spain). MRM mode: (a) $253 \rightarrow 167$ for detection of ^{35}Cl nonylphenolic compounds; (b) $297 \rightarrow 79$ for $^{79}BrNP$; (c) $355 \rightarrow 297$ for $^{79}BrNPE_1C$; (d) $399 \rightarrow 297$ for $^{79}BrNPE_2C$. Reprinted with permission from Ref. [39] © 2002 American Chemical Society.

REFERENCES

1   M. Petrovic, A. Diaz, F. Ventura and D. Barcelo, *Anal. Chem.*, 73 (2001) 5886.
2   H.Q. Li and H.F. Schroeder, *J. Chromatogr. A*, 889 (2000) 155.
3   H.F. Schroeder and K. Fytianos, *Chromatographia*, 50 (1999) 583.

4   M. Castillo, M.C. Alonso, J. Riu and D. Barcelo, *Environ. Sci. Technol.*, 33 (1999) 1300.

5   M. Castillo, E. Martinez, A. Ginebreda, L. Tirapu and D. Barcelo, *Analyst*, 125 (2000) 1733.

6   M. Castillo, G. Penuela and D. Barcelo, *Fresenius J. Anal. Chem.*, 369 (2001) 620.

7   M. Petrovic and D. Barcelo, *Anal. Chem.*, 72 (2000) 4560.

8   M. Castillo, F. Ventura and D. Barcelo, *Waste Mgmt*, 19 (1999) 101.

9   M. Petrovic and D. Barcelo, *Fresenius J. Anal. Chem.*, 368 (2000) 676.

10 S.D. Scullion, M.R. Clench, M. Cooke and A.E. Ashcroft, *J. Chromatogr. A*, 733 (1996) 207.

11 P. Jandera, M. Holčapek and G. Teodoridis, *J. Chromatogr. A*, 813 (1998) 299.

12 T. Yamagishi, S. Hashimoto and A. Otsuki, *Environ. Toxicol. Chem.*, 17 (1998) 670.

13 H.Fr. Schröder and F. Ventura, In: D. Barceló (Ed.), *Techniques and Instrumentation in Analytical Chemistry*, Sample Handling and Trace Analysis of Pollutants—Techniques, Applications and Quality Assurance, Vol. 21., Elsevier, Amsterdam, 2000, p. 828.

14 D.Y. Shang, M.G. Ikonomou and R.W. Macdonald, *J. Chromatogr. A*, 849 (1999) 467.

15 M. Takino, S. Daishima and K. Yamaguchi, *J. Chromatogr. A*, 904 (2000) 65.

16 D.Y. Shang, R.W. Macdonald and M.G. Ikonomou, *Environ. Sci. Technol.*, 33 (1999) 1366.

17 P.L. Ferguson, C.R. Iden and B.J. Brownawell, *Anal. Chem.*, 72 (2000) 4322.

18 P.L. Ferguson, C.R. Iden and B.J. Brownawell, *J. Chromatogr. A*, 938 (2001) 79.

19 A. Di Corcia, R. Caballo, C. Crescenzi and M. Nazzari, *Environ. Sci. Technol.*, 34 (2000) 3914.

20 A. Di Corcia, C. Crescenzi, A. Marcomini and R. Samperi, *Environ. Sci. Technol.*, 32 (1998) 711.

21 A. Di Corcia, A. Constantino, C. Crescenzi, E. Marinoni and R. Samperi, *Environ. Sci. Technol.*, 32 (1998) 2401.

22 C. Crescenzi, A. Di Corcia, R. Samperi and A. Marcomini, *Anal. Chem.*, 67 (1995) 1797.

23 B. Shao, J.Y. Hu and M. Yang, *J. Chromatogr. A*, 950 (2002) 167.

24 M. Spiteller and G. Spiteller, *Massenspektrensammlung von Lösungsmitteln, Verunreinigungen, Säulen-belegungsmaterialien und einfachen aliphatischen Verbindungen*, Springer, Wien, 1973, p. 69.

25 H.Fr. Schröder, *J. Chromatogr. A*, 777 (1997) 127.

26 K.B. Sherrard, P.J. Marriott, M.J. McCormick, R. Colton and G. Smith, *Anal. Chem.*, 66 (1994) 3394.

27 J.B. Plomley, P.W. Crozier, Proc. of 46th ASMS Conf. on Mass Spectrometrics and Allied Topics, Orlando, FL, 1998, p. 369.
28 R. Scheding and H. Fr. Schröder, Chromatographia, in preparation.
29 F. Houde, C. DeBlois and D. Berryman, *J. Chromatogr. A*, (2003), in press.
30 M. Petrovic and D. Barcelo, *J. Mass Spectrom.*, 36 (2001) 1173.
31 S.N. Pedersen and C. Lindholst, *J. Chromatogr. A*, 864 (1999) 17.
32 M. Petrovic, E. Eljarrat, M.J. Lopez de Alda and D. Barcelo, *J. Chromatogr. A*, 974 (2002) 23.
33 N. Jonkers, T.P. Knepper and P. De Voogt, *Environ. Sci. Technol.*, 35 (2001) 335.
34 C. Hao, T.R. Croley, R.E. March, B.G. Koenig and C.D. Metcalfe, *J. Mass Spectrom.*, 35 (2000) 818.
35 M. Reinhard, N. Goodman and K.E. Mortelmans, *Environ. Sci. Technol.*, 16 (1982) 351.
36 F. Ventura, J. Caixach, A. Figueras, I. Espalader, D. Fraisse and J. Rivera, *Water Res.*, 23 (1989) 1191.
37 F. Ventura, A. Figueras, J. Caixach, I. Espadaler, J. Romero, J. Guardiola and J. Rivera, *Water Res.*, 22 (1988) 1211.
38 M. Petrovic, A. Diaz, F. Ventura and D. Barcelo, *J. Am. Soc. Mass Spectrom.* (in press).
39 M. Petrovic and D. Barceló, In: I. Ferrer and E.M. Thurman (Eds), *Mass spectrometry, LC/MS/MS and TOF/MS: Analysis of Emerging contaminants*, ACS Symposium, Vol. 850., Oxford University Press and the American Chemical Society, Washington, (in press).
40 M. Petrovic and D. Barceló, *J. AOAC Int.*, 89 (2001) 1079.

2.7 ATMOSPHERIC PRESSURE IONISATION MASS SPECTROMETRY—IV. NON-IONIC SURFACTANTS: LC–MS OF ALKYL POLYGLUCOSIDES AND ALKYL GLUCAMIDES

Peter Eichhorn

2.7.1 Introduction

Alkyl polyglucosides (APG) and alkyl glucamides (AG) are non-ionic surfactants produced on the basis of renewable feedstocks such as glucose and fatty alcohols, which are derived from starch and palm oil, respectively.

2.7.2 Alkyl polyglucosides

APG consists of a complex mixture of a variety of homologues and isomers, including stereoisomers, binding isomers and ring isomers within the glucose moiety (Fig. 2.7.1). Using reversed phase-high performance liquid chromatography (RP-HPLC) with a C_8 column followed by electrospray ionisation–mass spectrometry (ESI–MS), separation and identification of individual components in a technical blend of APG were accomplished [1]. The detection was possible in both positive and negative ionisation modes. Figure 2.7.2(a) shows a typical LC–ESI–MS chromatogram from a standard mixture of APG acquired in negative ion mode scanning a mass range from m/z 200 to 600. The complex peak pattern obtained was resolved by extracting the masses of the molecular ions $[M - H]^-$ of the constituents (Table 2.7.1). In Fig. 2.7.2(b) and (c), the extracted ion chromatogram (XIC) of alkyl monoglucosides (C_8, C_{10} and C_{12}) and alkyl diglucosides (C_8, C_{10} and C_{12}), respectively, are depicted.

The order of elution depended on the length of the alkyl chain as well as on the number of glucose units. Increasing chain length resulted in increased retention time. On the other hand, the introduction of a second glucose moiety led to weaker retention on the C_8 column owing to higher hydrophilicity of the molecule. C_8-, C_{10}- and C_{12}-monoglucosides were entirely separated according to the alkyl chain length (Fig. 2.7.2(b)). Moreover, a separation with regard to the ring isomerism

T.P. Knepper, D. Barceló and P. de Voogt (Eds)

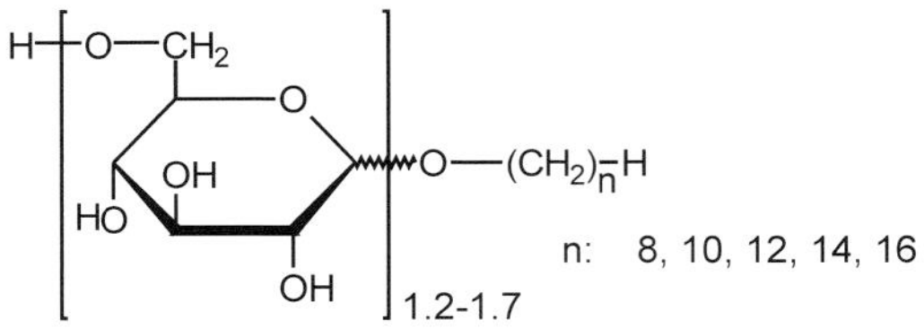

Fig. 2.7.1. Simplified structure of APG.

of the glucose moiety, i.e. pyranosidic or furanosidic form, was achieved. Peaks arising from stereoisomeric α- and β-alkyl monoglucosides were partially resolved (indices a and b). As for alkyl diglucosides (Fig. 2.7.2(c)) the peak pattern of a specific alkyl homologue was much more complex when compared with the corresponding monoglucoside. This was due to the higher number of possible isomers including ring isomers, stereoisomers and binding isomers. In the latter case, 1,4- or 1,6-glucosidal bonds might link two glucose rings.

With the selected ion masses besides the assigned monoglucosides, a series of signals with low intensity occurred, which are denoted as $\mathbf{7}^{*}$, $\mathbf{8}^{*}$ and $\mathbf{9}^{*}$ in Fig. 2.7.2(b). These signals originated from fragment ions of diglucosides, which were formed by the loss of one glucose unit ($\Delta\,m/z$ 162), resulting in fragments with the same m/z value as the corresponding monoglucosides with an identical chain length (see below).

The spectra of C_8-monoglucoside registered under positive and negative ionisation modes are displayed in Fig. 2.7.3 [1]. In $(+)$-ionisation mode, the molecule ion peaks of both the sodium (m/z 310) and ammonium adducts (m/z 315) were detected, whereas in $(-)$-ionisation mode the base peak was assigned exclusively to $[M - H]^-$ (m/z 291). In both instances, characteristic fragmentation of the parent compound occurred tracing back to the cleavage of the glucosidal bond. Under positive ionisation, three fragments of the carbohydrate unit were detected at m/z 163, 145 and 127 corresponding to the fragment ions $[Gluc + H-H_2O]^+$, $[Gluc + H-(H_2O)_2]^+$, and $[Gluc + H-(H_2O)_3]^+$, respectively. In negative ionisation mode, the fragment $[Gluc-H_2O-H]^-$ with m/z 161 was detected. When ESI–MS was operated in positive ion mode, formation of sodiated and ammoniated molecular ions was strongly dependent on the cone voltage. Ramping this parameter from 0 to $+100$ V displayed a rapid decrease in the intensity of $[M + NH_4]^+$, while the intensity of $[M + Na]^+$ rose until it reached a maximum at $+35$ V.

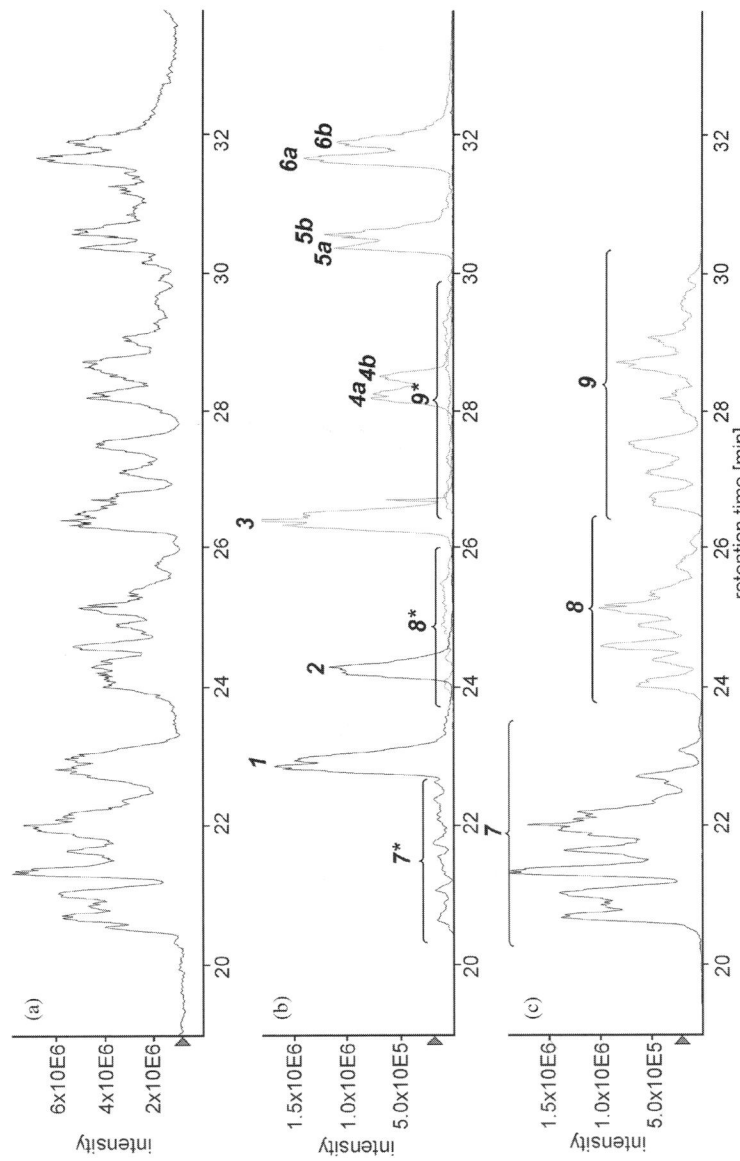

Fig. 2.7.2. (−)LC−ESI−MS chromatograms of a standard APG solution. (a) Total ion current trace from m/z 200 to 600. (b) XIC of C_8-, C_{10}-, and C_{12}-monoglucoside, and (c) XIC of C_8-, C_{10}-, and C_{12}-diglucoside. Peak numbering as in Table 2.7.1; indices a and b denote different stereoisomeric forms. (Separation on a RP-C_8 column with a water/acetonitrile gradient) (Reprinted from [1], Copyright 1999, with permission from Elsevier).

193

TABLE 2.7.1

Ion masses (m/z) of different molecular and adduct ions of alkyl mono- and diglucosides (as assigned in Figs. 2.7.2 and 2.7.5) (Reprinted from [1]. Copyright 1999, with permission from Elsevier)

Compound	Peak no.	Ion (m/z)		
		$[M - H]^-$	$[M + NH_4]^+$	$[M + Na]^+$
C_8-monoglucoside	1,2	291	310	315
C_{10}-monoglucoside	3,4	319	338	343
C_{12}-monoglucoside	5,6	347	366	371
C_8-diglucoside	7	453	472	477
C_{10}-diglucoside	8	481	500	505
C_{12}-diglucoside	9	509	528	533

In (+)-ionisation mode, a distinct selectivity of monoglucosides with respect to the formation of adduct ions was observed. Depending on the ring form (glucopyranoside or glucofuranoside) and the stereoisomerism (α- or β-alkyl monoglucoside), a different affinity towards Na$^+$ and

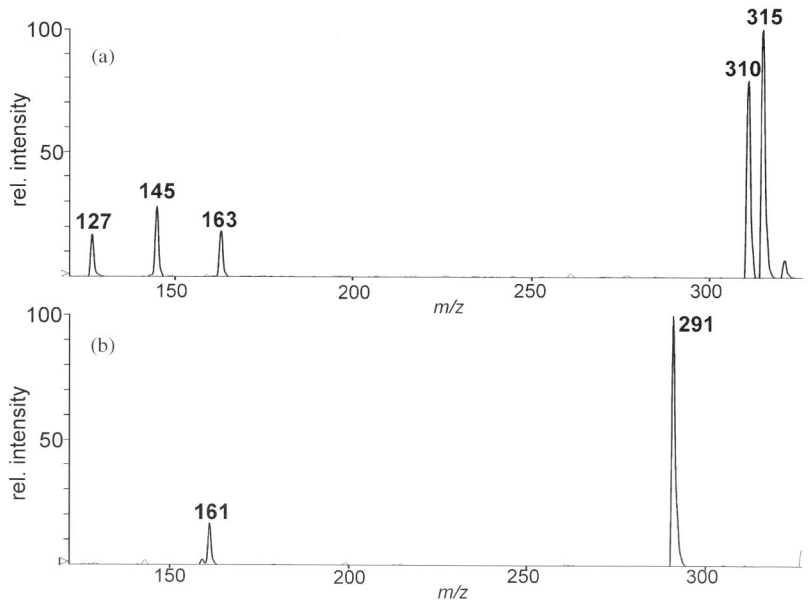

Fig. 2.7.3. LC–ESI–MS spectra of C_8-glucopyranoside under (a) positive and (b) negative ionisation. (Reprinted from [1], Copyright 1999, with permission from Elsevier).

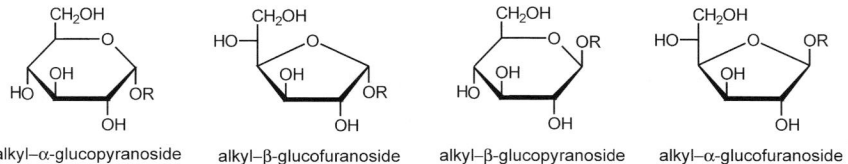

alkyl–α-glucopyranoside   alkyl–β-glucofuranoside   alkyl–β-glucopyranoside   alkyl–α-glucofuranoside

Fig. 2.7.4. Structures of ring and stereoisomers of alkyl monoglucoside.

NH_4^+ became apparent. The molecular structures of the four possible isomers of the alkyl monoglucosides are shown in Fig. 2.7.4.

The selected ion monitoring-chromatograms of C_8-, C_{10}-, and C_{12}-monoglucoside—each time the window corresponds to the sum of $[M + Na]^+$ and $[M + NH_4]^+$—are displayed in Fig. 2.7.5(a) (masses of the adduct ions as listed in Table 2.7.1) [1]. The XIC showing the adduct ions $[M + Na]^+$ and $[M + NH_4]^+$, respectively, are given in (b) and (c). While the alkyl glucofuranosides exhibited a higher tendency to form

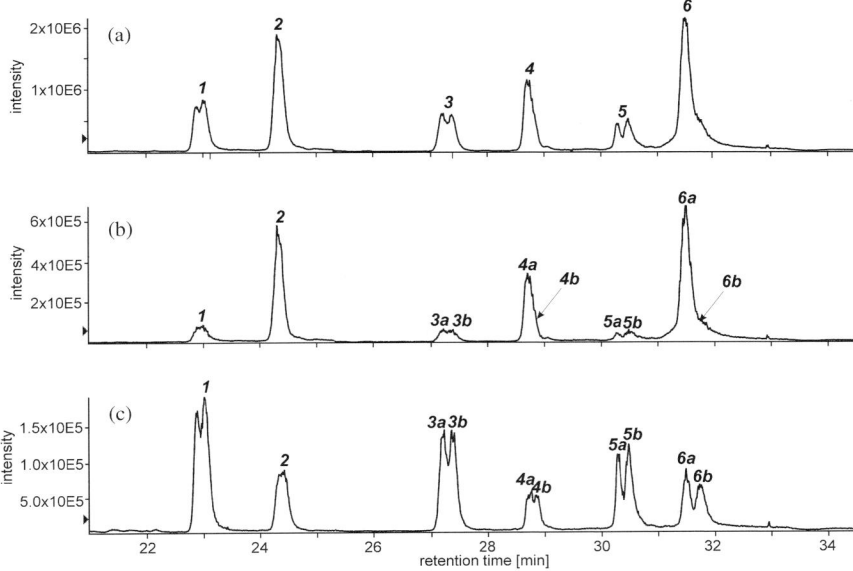

Fig. 2.7.5. (+)-LC–ESI–MS chromatograms of standard APG solution. (a) Sum of XIC of $[M + Na]^+$ and $[M + NH_4]^+$ of C_8-, C_{10}-, and C_{12}-monoglucoside. (b) XIC of $[M + Na]^+$, and (c) XIC of $[M + NH_4]^+$. Peak numbering as in Table 2.7.1; a denotes α-glucoside, b denotes β-glucoside (Reprinted from [1], Copyright 1999, with permission from Elsevier).

the sodium adduct, the alkyl glucopyranosides were preferably detected as the ammonium adduct ion. Also the α- and β-isomers of the glucofuranosides manifested a much higher selectivity for Na^+ and NH_4^+ compared with the two corresponding glucopyranosides. These observations were explained by specific interactions of the cation with the carbohydrate moiety. Since Na^+ and NH_4^+ have different ion radii (Na^+: 0.95 Å; NH_4^+: 1.43 Å) their coordination with the glucose moiety—most likely via oxygen atoms—depended on the spatial availability, which was determined by the constitution and configuration of the glucose part.

With regard to quantitative measurements of APG surfactants in, e.g. environmental samples, the authors stressed that it was of crucial importance to promote the formation of the desired molecular (or adduct) ion in order to obtain reproducible mass spectra. If tuning of the ESI interface parameters did not suffice to yield abundant ions of the selected species, acquisitions of the mass spectrometric detector after negative ionisation in conjunction with appropriate selection of the mobile phase composition were used as an alternative despite the lower sensitivity in this mode [1,2].

Negative ion mode was also applied for the atmospheric pressure chemical ionisation–mass spectrometric analysis (APCI–MS) of APG, solid-phase extracted from technical mixtures, as all components exhibited abundant deprotonated molecular ions [3]. The separation on a RP-C_8 column produced single peaks of C_8- and C_{10}-alkyl mono-glucoside with co-eluting stereoisomers of the respective homologue. From the ion traces of C_8- and C_{10}-diglucosides (ion masses as in Table 2.7.1), which consisted of a large number of incompletely resolved peaks, the two prominent signals of the decyl homologue (m/z 481) were selected for further structural elucidation performing APCI–MS/MS experiments. The daughter ion spectra of these dominating components produced only fragments resulting from the carbohydrate moiety of APG (Fig. 2.7.6). The ion at m/z 319 corresponded to the loss of one glucose ring ($\Delta m/z$ 162), while the series of fragments in the mass range from about m/z 60 to 200 were assigned to the carbohydrate part. Only the fragment ion with m/z 185 seemed to be characteristic of analyte A, since this ion could not be found in the spectral pattern of analyte B. But this subtle difference did not allow elucidation of the stereochemistry of alkyl diglucosides. The stereochemical characterisation of compound A, however, was finally achieved through application of online LC–^{1}H-NMR, identifying substance A as C_{10}-$\alpha(1 \to 6)$ isomaltoside.

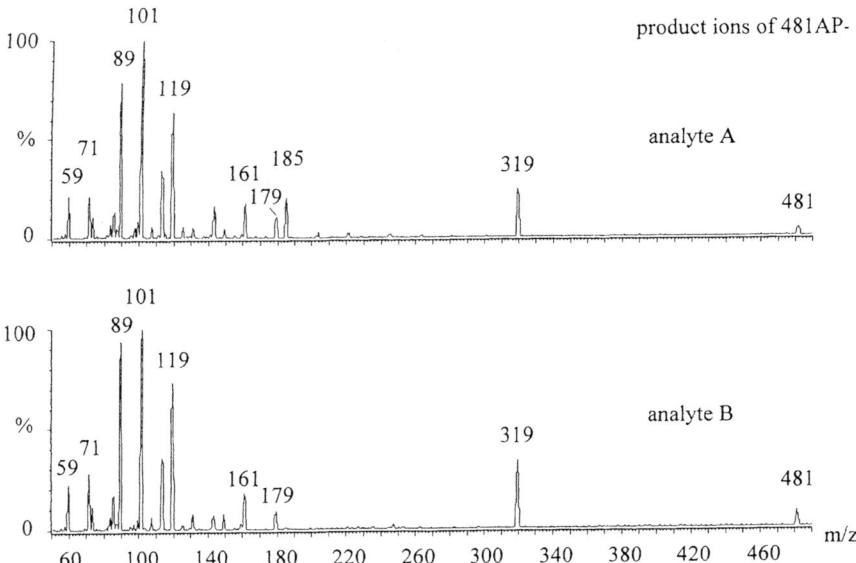

Fig. 2.7.6. (−)-LC–APCI–MS–MS daughter ion spectra of two C_{10}-alkyldiglucosides (parent ion: m/z 481) (Reprinted with permission from [3], Copyright 2000, American Chemical Society).

In the (−)-flow injection-ESI–MS analysis of concentrated APG standard solutions amended with 1 mM sodium acetate, Klaffke and colleagues [4] not only observed the sodiated adduct ions of the C-even monoglucosides C_8 to C_{14}, but also a series of clusters ions of the type $[2M + Na]^+$. Upon infusion of a technical blend containing C_8- and C_{10}-APG, the acquired spectrum gave peaks at m/z 607, 635 and 663 assigned as $[2M_{C_8} + Na]^+$, $[M_{C_8} + M_{C_{10}} + Na]^+$ and $[2M_{C_{10}} + Na]^+$, respectively. In the mass spectrum of an APG mixture consisting of C_{12}- and C_{14}-APG, analogous clusters were detected at m/z 719, 747 and 775. The formation of clusters was explained as a concentration-dependent effect, which means that with increasing concentration of the analyte in the ESI interface more cluster formation occurred. As no doubly sodiated ions of diglucosides—which a priori offer two glucose rings for the oxygen atom-based association of the cations—nor multiple charged ions of the monoglucosides appeared in the spectra, it was suggested that only the glucosidic oxygen connecting the sugar and

197

the alkyl chain could form the ion pair bond with the sodium ion. Due to the formation of this ion pair the oxygen had to be re-stabilised, which occurred via electron shift. The mechanism of this charge stabilisation in the molecule is shown in Fig. 2.7.7.

A comparison between FIA–ESI–MS and FIA–APCI–MS for the detection of APG in commercial blends was performed by Schröder [5]. In the (+)-FIA–ESI spectrum, mono- to tetraglucosides in the form of their $[M + Na]^+$ ions were detected, while under both positive and negative FIA–APCI conditions only the signals of the mono- and diglucosides were found. An excess of ammonium acetate for ionisation support suppressed the generation of $[M + Na]^+$ ions. The presence of acetate anions under negative ionisation mode using the ESI or APCI interface resulted in the formation of $[M + CH_3COO]^-$ adduct ions of C_{12}- and C_{14}-mono- and diglucoside with m/z 407, 435, 569 and 597, respectively. The ionisation efficiencies for these two homologues were found to be $(+)$-ESI $= (+)$-APCI $> (-)$-ESI $= (-)$-APCI.

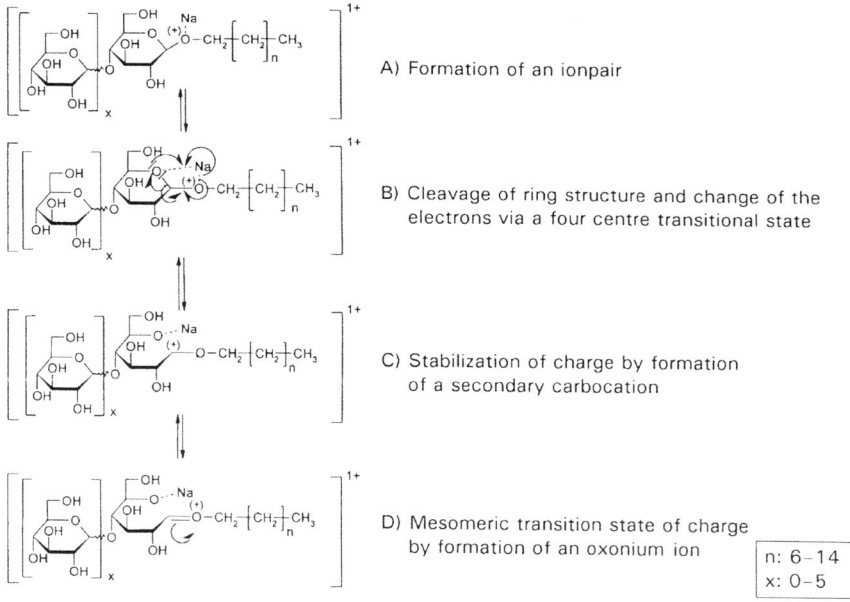

A) Formation of an ionpair

B) Cleavage of ring structure and change of the electrons via a four centre transitional state

C) Stabilization of charge by formation of a secondary carbocation

D) Mesomeric transition state of charge by formation of an oxonium ion

n: 6–14
x: 0–5

Fig. 2.7.7. Stabilisation of charge in sodiated alkyl diglucoside (from [4] with permission from author).

198

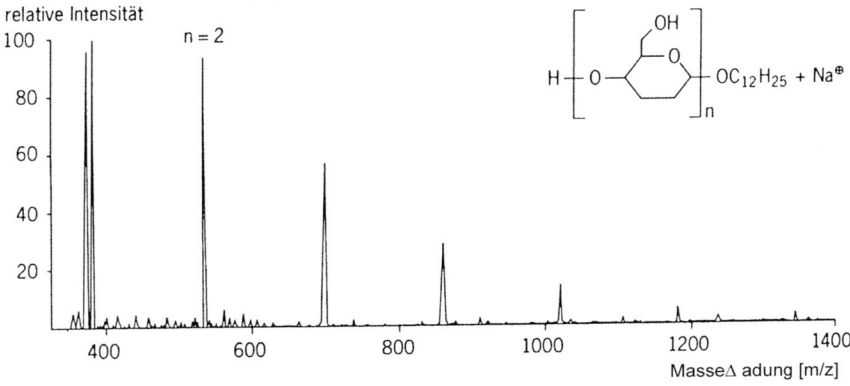

Fig. 2.7.8. MALDI-TOF-MS mass spectrum of C_{12}-alkyl polyglucoside (from [7]).

A rapid technique for the identification of surfactants in consumer products by ESI–MS was proposed by Ogura and co-workers [6]. After a simple preparation procedure, infusion of the sample, which was prepared in a water/methanol mixture (50:50) containing 10 mM ammonium acetate, allowed assignment of the $[M + NH_4]^+$ ions of C_{10}- and C_{12}-mono- and -diglucoside in the mass spectrum (ion masses as in Table 2.7.1). The approach even permitted quantitative analysis when deuterated internal standards were used.

For the quick characterisation of polydisperse surfactants with relative high molecular weight distributions matrix-assisted laser desorption/ionisation (MALDI)-time of flight (TOF)-MS represented an interesting alternative since low mass compounds did not interfere with the mass spectrometric detection of the compounds of interest. For example, the mass spectrum of C_{12}-APG (Fig. 2.7.8) exhibited equally spaced signals with $\Delta m/z$ 162 corresponding to sodiated adduct ions of the mono- (m/z 371) to heptaglucosides (m/z 1343) [7].

2.7.3 Alkyl glucamides

Mass spectrometric detection of AG (Fig. 2.7.9) was achievable in an ESI interface in positive as well as negative ionisation modes [8]. Figure 2.7.10 displays the spectra of C_{12}-AG, which are characterised by several molecular ions and fragments. The peak assignment with the corresponding masses of all ions is listed for the C_{12}- and C_{14}-homologues in

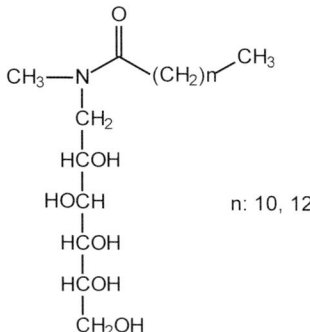

n: 10, 12

Fig. 2.7.9. Structure of alkyl glucamides.

Table 2.7.2. In (+)-ionisation mode, apart from the $[M + H]^+$ ion, the sodiated and potassiated ions could be detected. Furthermore, the fragment ion $[M + H-H_2O]^+$ arising from the loss of one water molecule out of the carbohydrate moiety was formed. No ammonium adduct ion was observed although ammonia was used for pH adjustment of the eluent. In (+)-ionisation mode two fragments were detected, both containing the carbohydrate part of the molecule, $[Gluc-NH_2-CH_3]^+$ and $[(Gluc-H_2O)-NH_2-CH_3]^+$.

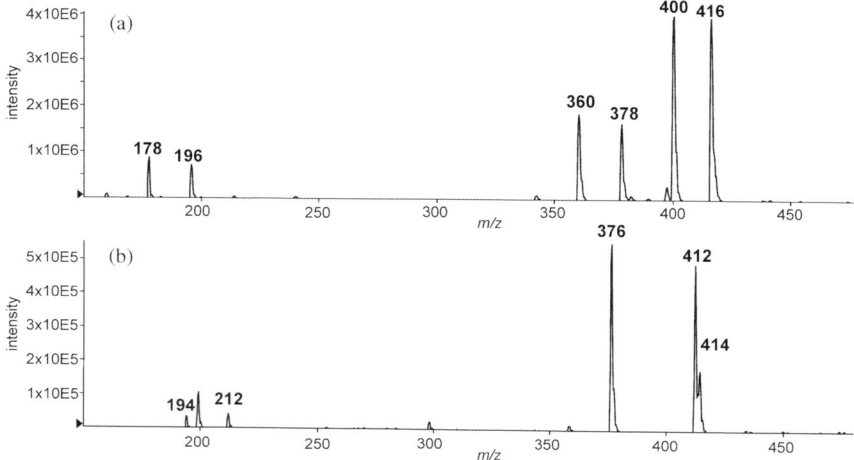

Fig. 2.7.10. LC–ESI–MS spectra of C_{12}-glucamide under (a) positive ionisation and (b) negative ionisation (from Ref. [8]).

TABLE 2.7.2

Molecular ions and fragments of C_{12}- and C_{14}-AG under positive and negative ionisation (from [8]. 2001. © John Wiley & Sons Limited. Reproduced with permission)

Positive ionisation			Negative ionisation		
Ion/fragment	C_{12}-AG (m/z)	C_{14}-AG[a] (m/z)	Ion/fragment	C_{12}-AG (m/z)	C_{14}-AG[a] (m/z)
$[M + H]^+$	378	406	$[M - H]^-$	376	404
$[M + Na]^+$	400	428	$[M + {}^{35}Cl]^-$	412	440
$[M + K]^+$	416	444			
$[M + H-H_2O]^+$	360	388			
$[Gluc-NH_2-CH_3]^+$	196	196	$[CH_3-N-CO-R]^{-\,b}$	212	240
$[(Gluc-H_2O)-NH_2-CH_3]^+$	178	178	$[Gluc-N-CH_3]^-$	194	194

[a]Spectra not shown.
[b]R: alkyl chain.

In $(-)$-ion mode besides the $[M - H]^-$ the chloride adduct also appeared in the MS spectra with a similar intensity, but optimisation of the ESI interface parameters allowed the formation to significantly shift towards the $[M - H]^-$ ion, while the chloride adduct was largely suppressed. The molecular ion yielded two fragments, one including the alkyl chain $[CH_3-N-CO-R]^-$, the other the sugar part $[Gluc-N-CH_3]^-$.

Despite the fact that the sensitivity of $(+)$-ionisation mode was roughly one order of magnitude higher, the $(-)$-ion mode was chosen for quantitative measurements in wastewater samples [2]. On one hand, the intensity ratio of the four possible quantitation ions $[M + H-H_2O]^+$, $[M + H]^+$, $[M + Na]^+$ and $[M + K]^+$ was not reproducible, probably due to varying contents of the cations, Na^+ and K^+, in the LC eluent and the sample. On the other hand, it was preferable to refer to a qualifier ion including the alkyl chain, i.e. $[CH_3-N-CO-R]^-$, since this provided greater confirmation for identification of the homologues.

Studies on an industrial blend of AG applying FIA–MS analysis demonstrated that the non-ionic surfactant was ionised under both $(+/-)$-ESI and APCI conditions [5]. In negative ionisation, the APCI produced deprotonated molecular ions, whereas the ESI produced acetate adduct ions.

Investigations were carried out in a laboratory fixed-bed bioreactor in order to identify aerobic degradation products of AG [8]. A metabolite postulated to originate from the oxidative breakdown of the alkyl chain

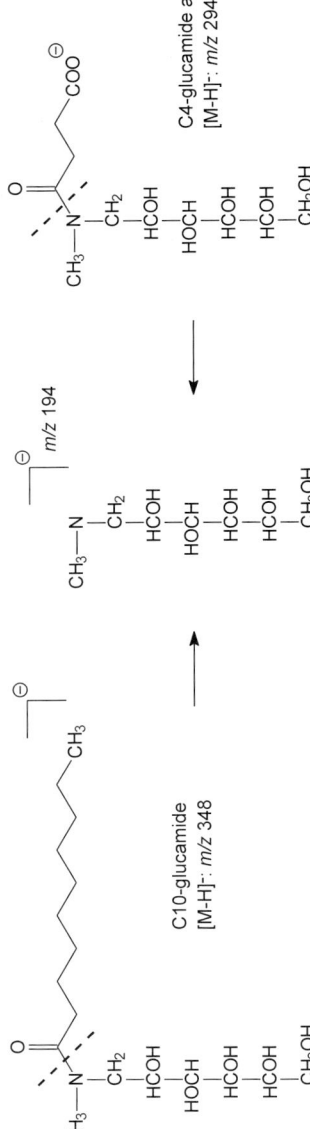

Fig. 2.7.11. Formation of fragment ion m/z 194 out of C_{10}-AG and C_4-glucamide acid (from Ref. [8]).

of C_{10}-AG (initial ω-oxidation followed by three β-oxidations) was detected after ion-pair RP-HPLC separation. Applying $(-)$-ESI–MS, this species, named C_4-glucamide acid, yielded a molecular ion $[M - H]^-$ at m/z 294 and a fragment ion at m/z 194. The latter ion, which likewise appeared in the spectrum of C_{12}-AG (see Table 2.7.2), was formed by heterolytic cleavage of the amide bond according to Fig. 2.7.11. Ramping of the orifice voltage allowed unequivocal assignment of the fragment to the parent ion.

REFERENCES

1 P. Eichhorn and T.P. Knepper, *J. Chromatogr. A*, 854 (1999) 221.
2 P. Eichhorn, M. Petrovic, D. Barceló and T.P. Knepper, *Vom Wasser*, 95 (2000) 245.
3 P. Billian, W. Hock, R. Doetzer, H.-J. Stan and W. Dreher, *Anal. Chem.*, 72 (2000) 4973.
4 S. Klaffke, T. Neubert and L.W. Kroh, *Tenside Surf. Det.*, 36 (1999) 178.
5 H.Fr. Schröder and F. Ventura, In: D. Barceló (Ed.), *Techniques and Instrumentation in Analytical Chemistry*, Sample Handling and Trace Analysis of Pollutants—Techniques, Applications and Quality Assurance, Vol. 21., Elsevier, Amsterdam, 2000, p. 828.
6 I. Ogura, D.L. DuVal, S. Kawakami and K. Miyajima, *J. Am. Oil Chem. Soc.*, 73 (1999) 137.
7 C. Hammes, J. Meister and M. Berchter, *Henkel-Referate*, 37 (1997) Henkel KGaA Düsseldorf.
8 P. Eichhorn and T.P. Knepper, *J. Mass Spectrom.*, 35 (2000) 468.

2.8 ATMOSPHERIC PRESSURE IONISATION MASS
SPECTROMETRY—V. NON-IONIC SURFACTANTS: FLOW
INJECTION ANALYSIS—MASS SPECTROMETRY AND
LIQUID CHROMATOGRAPHY—MASS SPECTROMETRY
OF ORGANOSILICONE SURFACTANTS

Lea S. Bonnington, William Henderson and Jerzy A. Zabkiewicz

2.8.1 Introduction

Industrially, silicone surfactants are used in a variety of processes
including foam, textile, concrete and thermoplastic production, and
applications include use as foam stabilisers, defoamers, emulsifiers,
dispersants, wetters, adhesives, lubricants and release agents [1]. The
ability of silicone surfactants to also function in organic media creates a
unique niche for their use, such as in polyurethane foam manufacture
and as additives to paints and oil-based formulations, whilst the ability
to lower surface tension in aqueous solutions provides useful super-
wetting properties. The low biological risk associated with these
compounds has also led to their use in cosmetics and personal care
products [2].

A broad range of silicone surfactants are commercially available,
representing all of the structural classes—anionic, non-ionic, cationic,
and amphoteric. The silicone moiety is lyophobic, i.e. lacking an affinity
for a medium, and surfactant properties are achieved by substitution of
lyophilic groups to this backbone. The most common functionalities
used are polyethylene glycols; however, a broad range exist, as shown in
Table 2.8.1 [2,3].

Organosilicone surfactants can be classified according to the position
of the lyophilic moiety with respect to the Si—O—Si polymer. The
abbreviations M, D, T and Q, referring to the number of oxygen atoms
bonded to the silicon, are used to define the silicone portion as shown in
Table 2.8.2. Insertion of the lyophilic moiety (A) into the silicone
polymer itself leads to monofunctional, i.e. M(D′A)M, and polyfunctional linear, i.e. M(D′A)$_x$D$_y$M, type surfactants. Polyfunctional linear
versions are also described as rake, comb- or graft-copolymers.
Substitution at the ends of the silicone polymer yields surfactants

Comprehensive Analytical Chemistry XL   205
T.P. Knepper, D. Barceló and P. de Voogt (Eds)

TABLE 2.8.1

Examples of different types and substituents of silicone surfactants [2,3]

Structural class	Lyophilic moiety
Anionic	Carboxylates, phosphate esters, sulfates, sulfonates, sulfosuccinates, thiosulfates
Non-ionic	Polyethers, alkanolamides, alkyls, alkylethoxylates, amines, benzyls, carbohydrates, esters, perfluoroalkyls
Cationic	Alkyl-, amidoimidazoline- and carboxy-quaternary ammonium salts
Amphoteric	Betaines, phosphobetaines, sulfobetaines

classified as difunctional, i.e. $(D'A)-D_x-(D'A)$ and polyfunctional branched, i.e. $(D'A)-D_x-(D'R)_y-D_z-(D'A)$ [1]. The difunctional derivatives are also known as ABA (block) copolymers, $\alpha-\omega$ or bolaform surfactants [3]. $(D'A)-D_x$ (AB-type copolymers) and cyclic siloxane surfactants, i.e. $D_3(D'A)$, have also been described [3]. The use of hydrophobic alkyl substituents, i.e. $A = (CH_2)_nCH_3$, is also well established, and yields surfactants appropriate for use in non-aqueous media, e.g. in oil-based agrochemical formulations [4]. The trisiloxane surfactants used in agrochemical formulations (the main focus of this chapter) fall into the monofunctional linear category, whilst the varieties used in polyurethane foam manufacture possess polyfunctional branched structures.

The polyether-modified structures are the most important silicone surfactants currently in use, examples of which are shown in Fig. 2.8.1. These are known by many names, including silicone polyethers, polyethermethylsiloxanes (PEMSs), dimethicone copolyols,

TABLE 2.8.2

Silicone nomenclature

Notation	Abbreviation	Structure
Mono	M	$R_3SiO_{0.5}$
Di	D	$R_2SiO_{(0.5)2}$
Tri	T	$RSiO_{(0.5)3}$
Quaternary	Q	$SiO_{(0.5)4}$

The nomenclature usually assumes $R = Me$, and substitution of methyl groups is indicated by a superscript suffix, e.g. D^{Ph} represents the unit $Me(Ph)SiO_{(0.5)2}$, or with the substituted group following a prime notation and enclosed in brackets, e.g. $M(D'A)M$.

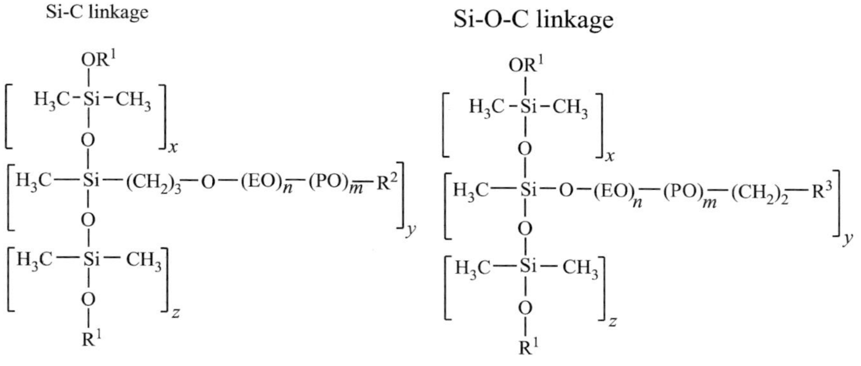

Si-C linkage   Si-O-C linkage

$$x = 0 - 50; \quad y = 1 - 100; \quad n = 3 - 30; \quad m = 0, 3 - 30;$$

$R^1 = Si(CH_3)_3$, alkylpolyether; $R^2 = H, CH_3, COCH_3$; $R^3 = OH, OCH_3$

Fig. 2.8.1. Typical structures of polyether-modified silicone surfactants.

silicone glycol copolymers, siloxylated polyethers, polyalkyleneoxide silicone copolymers, and silicone polyoxyalkylene copolymers, and can range from 30 to 80% polyoxyalkyene content by weight [5]. The alkyleneoxide distribution is polydisperse, and can consist of both ethoxylate (EO) and propoxylate (PO) moieties, in both random and block arrangements. Those used as foam promoters, such as those in polyurethane foam manufacture (non-aqueous systems), are usually randomly distributed whilst those used in antifoam applications (aqueous and hot non-aqueous systems) are generally block copolymers [5]. Structures with both Si–C and Si–O–C linkages are used (Fig. 2.8.1), with the trisiloxane surfactants based on the former structure and those used in polyurethane foam manufacture belonging to the latter.

2.8.2   Surfactants as agrochemical adjuvants

Surfactants are often used in agrochemical formulations as adjuvants, e.g. as wetting agents to improve the physico-chemical characteristics of the solution and to increase the uptake of active ingredients (e.g. pesticides) [6–8]. The surfactants can be included in pesticide products and/or added to the tank mix prior to use; however, they rarely exceed 1% of the total applied spray. Examples of typical surfactants used in agrochemical formulations are shown in Table 2.8.3 [9].

TABLE 2.8.3

Surfactants used as adjuvants in crop protection [9]

Structural class	Surfactants used
Anionic	Alkylbenzene sulfonates, alkylsulfates and ethoxysulfates, sulfosuccinates, phosphate esters, taurates, alkylnapthalene sulfonates, lignosulfonates
Non-ionic	Alkylphenol ethoxylates, long chain alkanol ethoxylates, long chain alkylamine ethoxylates, sorbitan esters and ethoxylates, castor oil ethoxylates, EO/PO copolymers, acrylic copolymers, polysiloxane-polyether copolymers, fluorosilicones
Cationic	Quaternary ammonium salts
Amphoteric	n-Alkyl betaines

The total world consumption of surfactants as agricultural adjuvants was estimated to be 60,000 t in 1993 [10], and this figure is expected to continue to increase as more concerted efforts are made to reduce the amounts of crop protection chemicals used. Indeed, whilst figures in 1995 showed the worldwide adjuvant market valued at only 3% that of the corresponding agrochemicals, the increase in use per annum was 5% as compared with 1.5% per annum for the agrochemicals [9].

Pesticides are classified as non-systemic or systemic depending on whether they act within or on the surface of the plant. Non-systemic pesticides often require complete and uniform foliage coverage for maximum effectiveness, and thus surfactant adjuvants can be added as spray modifiers to improve the wetting/spreading properties of the formulations. Fungicides and insecticides commonly fall into this category, whilst herbicides are more often systemic. Surfactants can aid foliar absorption and/or the biological activity of systemic pesticides but pesticide uptake and activity is a complex process influenced by a large number and range of metabolic processes in the plant. In most cases, the mechanisms of uptake and the enhancing role that surfactants play in these processes, are not well understood. The surfactant additive may enhance uptake at the leaf surface or at sites within the leaf [11]. Thus, there are many physical properties that can be influenced by surfactants including spray droplet spreading and drying, hygroscopicity and alterations to the permeability of the cuticular wax or leaf cuticle [12].

The use of organosilicone surfactants in agrochemical formulations has become increasingly widespread [13]. These products are known to enhance spray coverage and in many cases increase the foliar uptake of

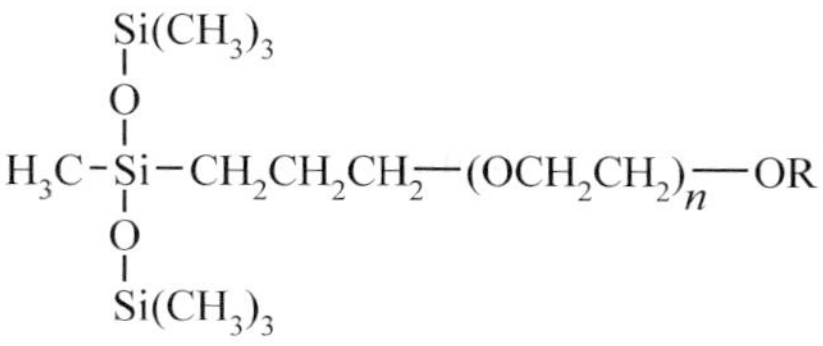

$$\begin{array}{c} \mathrm{Si(CH_3)_3} \\ | \\ \mathrm{O} \\ | \\ \mathrm{H_3C-Si-CH_2CH_2CH_2-(OCH_2CH_2)}_n\mathrm{-OR} \\ | \\ \mathrm{O} \\ | \\ \mathrm{Si(CH_3)_3} \end{array}$$

n = 3 - 16, Average n ~ 7.5; R = CH$_3$ (**1**), R = H (**2**), R = COCH$_3$

Fig. 2.8.2. Structure of the trisiloxane surfactants commonly used as agrochemical adjuvants.

agrochemicals, with the spreading ability and reduction of surface tension the most significant attributes as compared with conventional surfactants [14–16]. Such properties can provide rapid and efficient pesticidal uptake [17,18] and also render rainfastness to the formulation [19], desirable both economically and environmentally. The trisiloxane monofunctional analogues shown in Fig. 2.8.2 are the optimal structures for this form of silicone surfactant application, and as such are the focus of this chapter.

2.8.3 Trisiloxane surfactants

The usefulness of organosilicone surfactants as adjuvants in agrochemical formulations was first reported in 1973 [20], but the first commercial product was not introduced until 1985 (Silwet L-77, Union Carbide). Since then, their use has increased and many others have been brought to the market [4,21]. The trisiloxane methyl-capped ethoxylate surfactant Silwet L-77 (**1**, Fig. 2.8.2, R = CH$_3$), with the general formula $[(CH_3)_3SiO]_2-(CH_3)Si-(CH_2)_3-(OCH_2CH_2)_n-OCH_3$ ($n = 3-16$, ave. $n \sim 7.5$), is still the most widely used organosilicone surfactant for agrochemical applications. The commercial formulation of Silwet L-77 consists mainly of the trisiloxy polyethoxylate monomethyl ethers $[M_2D-C_3-O-(EO)_n-Me]$, but also contains some of the uncapped polyethoxylate analogue $[M_2D-C_3-O-(EO)_n-H]$ and some more polar constituents.

2.8.4 Analysis

Various methods have been applied to the analysis of silicones in general, and these have been well summarised elsewhere [22]. In particular, size exclusion chromatography (SEC) [23], nuclear magnetic resonance

(NMR) [24], and Fourier transform-infrared (FT-IR) spectroscopy [25] are commonly applied methods. Analysis using mass spectrometric (MS) techniques has been achieved with gas chromatography–mass spectrometry (GC–MS), with chemical ionisation (CI) often more informative than conventional electron impact (EI) ionisation [26]. For the qualitative and quantitative characterisation of silicone polyether copolymers in particular, SEC, NMR, and FT-IR have also been demonstrated as useful and informative methods [22] and the application of high-temperature GC and inductively coupled plasma-atomic emission spectroscopy (ICP-AES) is also described [5].

Whilst these methods are informative for the characterisation of synthetic mixtures, the information gained and the nature of these techniques precludes their use in routine quantitative analysis of environmental samples, which requires methods amenable to the direct introduction of aqueous samples and in particular selective and sensitive detection. Conventionally, online separation techniques coupled to mass spectrometric detection are used for this, namely gas (GC) and liquid chromatography (LC). As a technique for agrochemical and environmental analyses, high performance liquid chromatography (HPLC) coupled to atmospheric pressure ionisation–mass spectrometry (API–MS) is extremely attractive, with the ability to analyse relatively polar compounds and provide detection to very low levels.

Reliable quantitative determinations of organosilicone surfactants have been reported using online HPLC with light scattering mass detection (HPLC-LSD) [27,28]. These methods established that it was possible to quantify surfactant uptake into plants without the need for radio-labelled chemicals, previously the only method that was sufficiently sensitive. However, the detection limits of this method still require relatively large numbers of replications, tedious extraction and separation methods, and in the case of agrochemical formulations, necessitates the agrochemical to be analysed in parallel using radio-tracer methodology. The non-selectiveness of the method also limits its application to environmental samples. GC–MS methods have also been applied, with some success, but the high molecular weight of the larger oligomers precluded their observation, even following derivatisation [27]. Preliminary results of quantitation of trisiloxane surfactants and selected hydrolytic degradation products by HPLC-ICP-AES has been reported [5]. More recently, API–MS methods have been described for the qualitative and quantitative analysis of trisiloxane surfactants in agrochemical formulations [29–31], and under hydrolytic degradation conditions [29].

Atmospheric pressure ionisation–mass spectrometric characterisation of silicone surfactants

In general, non-ionic silicone compounds are not readily observed by API–MS methods as silicones show only very weak, if any, H-bonding capacity. This is because the dative $d_\pi-p_\pi$ nature of the Si–O bond reduces the electron density on the O atom, rendering it less basic than usual [32]. However, modified silicone compounds containing groups capable of H-bonding, such as polyether and amine moieties, will form adducts with H^+, NH_4^+ or alkali metal ions present in the analytical solvent and thus be detectable by API methods [33,34], as is observed for polar non-ionic molecules [35]. Furthermore, other studies have shown the chelating effect of polyether chains to be unaffected by the introduction of a Si atom at the beginning of the chain [36].

Non-ionic silicone surfactants such as $M_2D-C_3-O-(EO)_n-CH_3$ (**1**), are thus detectable by API–MS methods as adducts with NH_4^+, Na^+ and K^+ cations as shown in Fig. 2.8.3 [29]. Minor $[M + H]^+$ adducts for $M_2D-C_3-O-(EO)_n-H$ (**2**) have been observed, although, in the case of $M_2D-C_3-O-(EO)_n-CH_3$ these were not detected even with the addition of acidic media. Only singly charged species of $M_2D-C_3-O-(EO)_n-Me$ ($n \approx 3-16$) were observed for all cone voltages [29], as is consistent with that expected for the polyethylene glycol (PEG) oligomeric distribution [37,38].

Detection of $M_2D-C_3-O-(EO)_n-CH_3$ was possible by both positive ion mode atmospheric pressure chemical ionisation (APCI) and electrospray ionisation (ESI) MS methods, with good response down to absolute injections of 0.1 ng. However, ionisation in the negative ion mode was negligible at all concentrations analysed, as the polyether-modified structure has no sites capable of adducting with anions, nor has it any moieties capable of cleavage to yield anionic species.

In the absence of any added salts, the distribution of adducts formed was found to be variable by ESI–MS. In general, however, the lower molecular weight surfactant oligomers showed a preference for the Na^+ adduct, whilst the higher oligomers showed an increase in preference for the K^+ adduct. This is consistent with the metal-ion complexing abilities described for polyether derivatives [39,40]. Simplification of the spectra was achieved with the addition of salt solutions to the mobile phase, as shown for NH_4Cl in Fig. 2.8.4 [29]. With the addition of salt

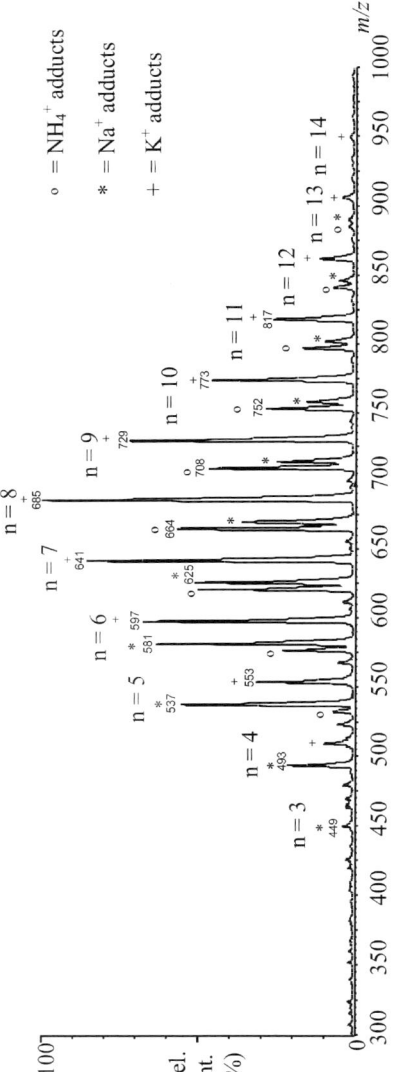

Fig. 2.8.3. FIA–ESI–MS(+) spectrum of the organosilicone surfactant M_2D–C_3–O–$(EO)_n$–CH_3.

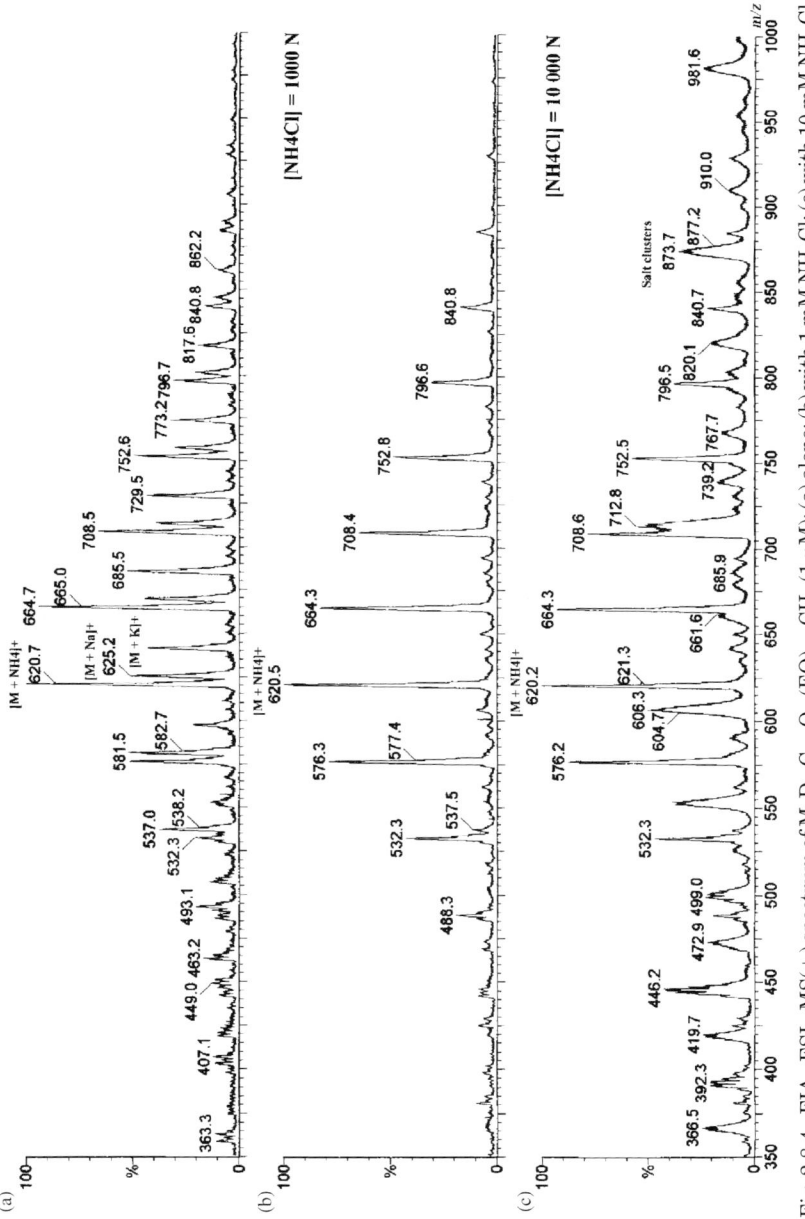

Fig. 2.8.4. FIA–ESI–MS(+) spectrum of $M_2D-C_3-O-(EO)_n-CH_3$ (1 µM): (a) alone; (b) with 1 mM NH_4Cl; (c) with 10 mM NH_4Cl.

213

solutions at a thousand-fold molar excess of cation to surfactant, the corresponding cation adduct dominated the spectrum, for any of the NH_4^+, Na^+, or K^+ salt solutions investigated (Fig. 2.8.4(b)). However, at higher salt concentrations, salt clusters also form, affecting analyte response and detector efficiency, as demonstrated in Fig. 2.8.4(c).

The use of the volatile salt $CH_3CO_2NH_4$ (1 mM) for controlling adduct formation showed similar improvements to the spectra (data not shown), whilst its relative volatility minimised the negative aspects of salt addition in ESI–MS (i.e. detector saturation and precipitation on the sampling orifice) [29].

In the absence of any added salts, the APCI–MS spectra were dominated by the Na^+ adducts, as shown in Fig. 2.8.5. The NH_4^+ and K^+ adducts were present at lower intensities, the latter especially for the higher molecular weight analogues. Addition of $CH_3CO_2NH_4$ did not simplify the adduct formation to $[M + NH_4]^+$ species as observed in ESI–MS and the best results for APCI–MS analysis were obtained without addition of any salt solutions. Application of this method to determinations of $M_2D–C_3–O–(EO)_n–Me$ recovery from solid substrates was achieved, using triethylene glycol monohexyl ether $[C_6(EO)_3]$ as the internal standard (Fig. 2.8.5) [29].

Flow injection analysis (FIA) ESI–MS and APCI–MS spectra for an EO/PO polyether modified silicone surfactant (PEMS) used as a personal care product have been obtained in positive and negative ionisation modes with the positive ionisation mode yielding the best results [41]. The spectra obtained in both modes were highly complicated, and thus no assignment was given. Significant differences in the ionisation results were obtained from the two interfaces, with those ions observed in the ESI–MS spectrum appearing in the lower

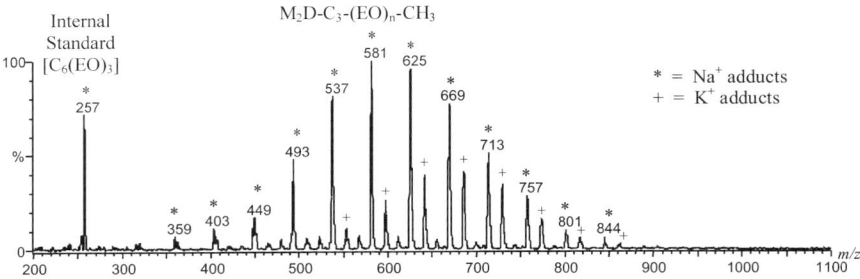

Fig. 2.8.5. FIA–APCI–MS(+) spectra of $M_2D–C_3–O–(EO)_n–Me$ obtained in recovery studies from montmorillonite clay.

mass range and thus likely to be carrying multiple charges, although no further investigation was performed to confirm this.

In general, characteristic fragmentation patterns for this class of compounds by API–MS methods have not been observed. Increasing the cone voltage did not yield any characteristic fragment ions for $M_2D-C_3-O-(EO)_n-Me$ by ESI–MS, APCI–MS or Fourier transform ion cyclotron resonance (FTICR)–MS [29]. Collision induced dissociation (CID) has been reported for some surfactants by API–MS, but only for protonated species and only for branched alcohol ethoxylates, which are non-ionic surfactants where the polyethoxylate chain is attached to a secondary carbon [42,43]. The conditions used to achieve this CID did not provide fragmentation of the $M_2D-C_3-O-(EO)_n-Me$ molecules. Similarly, fast atom bombardment (FAB) MS–MS fragmentation of PEG compounds (50 eV) was only observed for protonated species [44]. However, no detectable fragments were obtained for $M_2D-C_3-O-(EO)_n-H$ (2), for which $[M + H]^+$ adducts were observed by API–MS, with either increasing cone or skimmer lens offset voltages [29]. Characteristic fragments to aid differentiation between silicone surfactants with Si–C and Si–O–C linkages (for structures see Fig. 2.8.1) by FIA–MS–MS on a triple quadrupole were also not observed [41]. Information obtained from the daughter ion spectra (MS–MS) of EO modified polyether siloxanes following HPLC separation was also limited. Fragmentation for both $[M + H]^+$ and $[M + X]^+$ ($X = Na^+$ and K^+) ions of linear and cyclic PEG derivatives has been described under high energy CID conditions (~ 20 keV) [45], and the use of xenon as the collision gas also reportedly increases the CID efficiency of $[M + Na]^+$ ions [44]. Although not investigated, these may also be possible methods for the analysis of the silicone surfactant fragmentation by API–MS.

Quantitative analysis by atmospheric pressure ionisation–mass spectrometry
API–MS methods have been successfully applied to the quantification of $M_2D-C_3-O-(EO)_n-Me$, with reliable and reproducible results obtained after online HPLC separation [29,30]. The method was used to quantify recoveries of the surfactant from the surface of plant foliage and from solid substrates under controlled laboratory conditions. Extension of the method to environmental samples has not been investigated. The entire linear dynamic range for HPLC–APCI–MS was not determined, but linearity was observed within the required

application levels $(0.07-1.93$ ng $\mu L^{-1})$. The lowest concentration measured was easily within the range of detection for the APCI–MS method, although it was below the limit of detection by the HPLC-LSD method, even with a 50-fold higher injection volume (500 μL).

An example of the standard curve data obtained with HPLC–APCI–MS and HPLC-LSD is presented in Table 2.8.4, for comparison. In addition to the lower sensitivity, the reproducibility and correlation of the HPLC-LSD data was far inferior to that obtained by HPLC–APCI–MS.

Results obtained for the application of HPLC–APCI–MS to the quantification of $M_2D-C_3-O-(EO)_n-CH_3$ recoveries from *Chenopodium album* plant foliage are shown in Table 2.8.5, as compared with HPLC-LSD analysis [29]. The improvements in the sensitivity and reproducibility were obtained with the use of HPLC–APCI–MS as the analytical method and the HPLC–APCI–MS method also enabled detection of the $n = 3$ $M_2D-C_3-O-(EO)_n-CH_3$ molecule.

HPLC conditions resulting in co-elution of all the $M_2D-C_3-O-(EO)_n-Me$ oligomers were used in the quantitative analyses. This was found to simplify the quantitation (data processing), enhance the detection levels, and reduce analysis time down to 9 min. This method

TABLE 2.8.4

Relative response of $M_2D-C_3-O-(EO)_n-Me^a$ as determined by HPLC-LSD and HPLC–APCI–MS

HPLC-LSD				HPLC–APCI–MS			
$M_2D-C_3-O-(EO)_n-CH_3$ mass injected (ng)[b]	Relative response[c]	SD	CV (%)	$M_2D-C_3-O-(EO)_n-CH_3$ mass injected (ng)[d]	Relative response[c]	SD	CV (%)
35	ND	–	–	0.7	0.21	0.07	31
105	ND	–	–	2.1	0.74	0.13	17
170	0.40	0.08	21	3.4	1.37	0.40	29
240	0.57	0.06	10	4.8	2.04	0.31	15
345	0.91	0.08	9	6.9	2.73	0.27	10
485	1.18	0.17	15	9.7	3.85	0.03	1
690	1.89	0.19	10	13.8	4.87	0.41	9
965	2.88	0.28	10	19.3	6.89	0.31	4

SD = standard deviation; CV = coefficient of variation; ND = not detected.
[a]With Triton X-45 as the internal standard.
[b]10 μL injection volume.
[c]Response $(M_2D-C_3-O-(EO)_n-CH_3/$Triton X-45). Average of two injections.
[d]500 μL injection volume.

TABLE 2.8.5

Comparison of percentage uptake of $M_2D-C_3-O-(EO)_n-Me$ oligomers into *Chenopodium album* after 5 min as determined by HPLC-LSD and HPLC–APCI–MS

Surfactant solution	HPLC-LSD %uptake	HPLC–APCI–MS %uptake
$M_2D-C_3-O-(EO)_3-Me$	ND	22 ± 14
R^2 (regression)[a]	–	0.963 (linear)
$M_2D-C_3-O-(EO)_6-Me$	37 ± 11	34 ± 5
R^2 (regression)	0.974 (linear)	0.991 (power)
$M_2D-C_3-O-(EO)_9-Me$	37 ± 11	43 ± 2
R^2 (regression)	0.972 (linear)	0.997 (linear)
Commercial $M_2D-C_3-O-(EO)_n-Me$	34 ± 9	35 ± 6
R^2 (regression)	0.971 (linear)	0.983 (linear)
Purified $M_2D-C_3-O-(EO)_n-Me$	29 ± 4	31 ± 8
R^2 (regression)	0.981 (linear)	0.981 (linear)

ND = not detected.
[a]Least-squares fit trendline.

was not found to adversely affect the quantitative HPLC–APCI–MS results obtained, as also successfully adopted elsewhere [46]. Analysis was conducted in total ion mode and thus accounted for all the oligomers in the mixture, and also removed the need to control adduct formation.

FIA–APCI–MS has also been used to quantify recoveries of $M_2D-C_3-O-(EO)_n-Me$ from plant foliage and substrates with an example of the results obtained shown in Table 2.8.6, along with comparative HPLC-LSD results [29,31].

The FIA–APCI–MS data showed improved detection limits, reproducibility and correlation of standards over the HPLC-LSD results. However, extension of the FIA–MS method to other systems yielded erroneous quantitative data, attributed to signal response discrimination through competitive ionisation effects, and as such incorporation of online HPLC separation methods was determined to be the preferred API–MS method for quantitative analyses [29].

2.8.5 By-products

The commercial formulation of $M_2D-C_3-O-(EO)_n-Me$ is marketed as a trisiloxane ethoxylate product with a maximum of 20% alkoxide by-products.[1] The synthetic reaction is always performed with an excess

---

[1] http://www.loveland.co.uk/adjuvants.

TABLE 2.8.6

Comparison of percentage uptake of $M_2D-C_3-O-(EO)_n-Me$ into *Chenopodium album* foliage over time as determined by HPLC-LSD and FIA–APCI–MS[a]

	Time (min)	%Uptake (SD)	
		[HPLC-LSD]	[FIA–APCI–MS]
	10	58 (13)	64 (9)
	30	49 (16)	72 (10)
	240	ND	80 (4)
Regression		Power	Power
Least-squares fit[b]		$y = 0.0009x^{1.0498}$	$y = 2.373x^{0.7503}$
R^2		0.9249	0.9896

SD = standard deviation; ND = not detected.
[a]SIR, 2 channels selected [$(EO)_6$ oligomers of $M_2D-C_3-O-(EO)_n-Me$ and Agral-90].
[b]y = relative response ($M_2D-C_3-O-(EO)_n-Me$/Agral-90), x = [$M_2D-C_3-O-(EO)_n-Me$]/[Agral-90].

of the allyloxypolyalkyleneoxide starting material, to ensure complete consumption of the Si–H functionality, resulting in the excess polar constituents. Due to the nature of the synthetic procedure some variation among batches is expected. Quantitation of a commercial blend using RP-C_{18} HPLC-RI and HPLC-LSD methods demonstrated that the trisiloxane surfactants $M_2D-C_3-O-(EO)_n-Me$ (**1**) and $M_2D-C_3-O-(EO)_n-H$ (**2**) constituted 75% of the commercial mixture, with the former averaging 70%, and the remaining 25% constituting early eluting polar fractions [29].

A comparison of the HPLC–ESI–MS chromatograms obtained for commercial formulations of the trisiloxane surfactants, $M_2D-C_3-O-(EO)_n-Me$ and $M_2D-C_3-O-(EO)_n-H$ are presented in Fig. 2.8.6 [29]. Additional to the $M_2D-C_3-O-(EO)_n-Me$ (**1**) structures, some polar material and the uncapped analogue (**2**) can also be observed in the $M_2D-C_3-O-(EO)_n-Me$ product trace.

The more polar components have been characterised by FTICR–MS and HPLC–ESI–MS, and include excess PEG [$CH_3O(EO)_nH$ and $HO(EO)_nH$] starting materials, confirmation of which was obtained by comparison with commercial standards [29]. ESI–MS, FTICR–MS and ^1H NMR data analysis also indicated the presence of the Si–O–C by-product **3**, abbreviated to $MD^{OH}-O-(EO)_n-CH_2CH=CH_2$, shown in Fig. 2.8.7. The formation of **3** can occur through side reactions of the

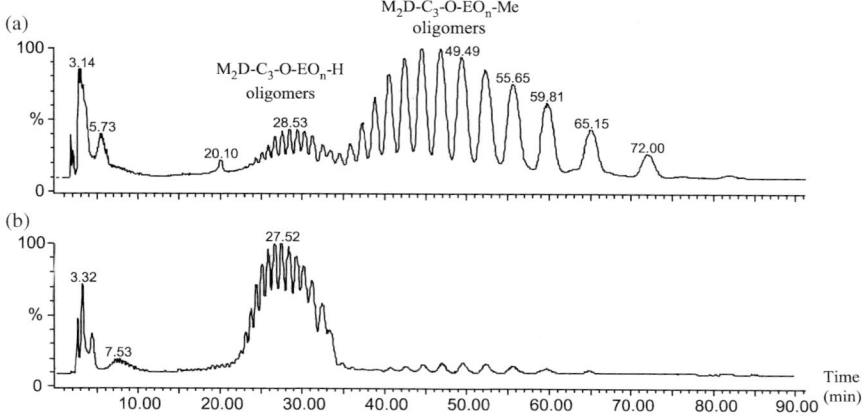

Fig. 2.8.6. RP-C_{18} HPLC–ESI–MS(+) chromatograms of: (a) $M_2D–C_3–O–(EO)_n–Me$; (b) $M_2D–C_3–O–(EO)_n–H$.

hydrosilylation procedure [47], followed by hydrolysis of a terminal M group.

Two higher molecular weight ion series (m/z 562–1134 and m/z 934–1154) were also observed in the ESI–MS and FTICR–MS spectra, corresponding with highest agreement to linear dimer structures

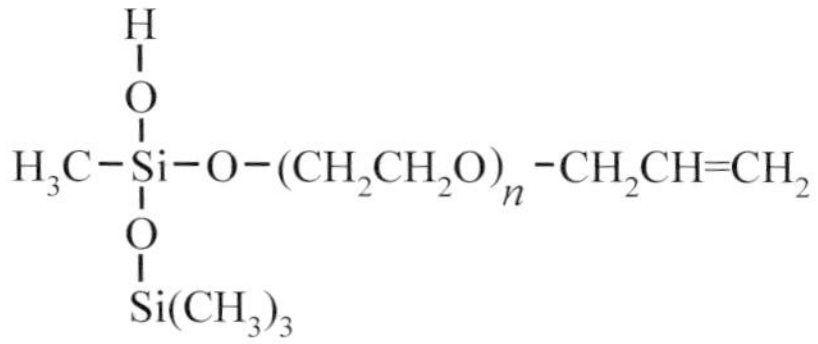

$$H_3C-\underset{\underset{Si(CH_3)_3}{|}}{\overset{\overset{H}{|}}{\overset{\overset{O}{|}}{\underset{|}{Si}}}}-O-(CH_2CH_2O)_n-CH_2CH{=}CH_2$$

3

$$H_3C-\underset{\underset{O}{|}}{\overset{\overset{Si(CH_3)_3}{|}}{\overset{\overset{O}{|}}{Si}}}-CH_2CH_2CH_2-OH$$
$$H_3C-\underset{\underset{Si(CH_3)_3}{|}}{\overset{\overset{O}{|}}{Si}}-CH_2CH_2CH_2-(OCH_2CH_2)_n-OMe$$

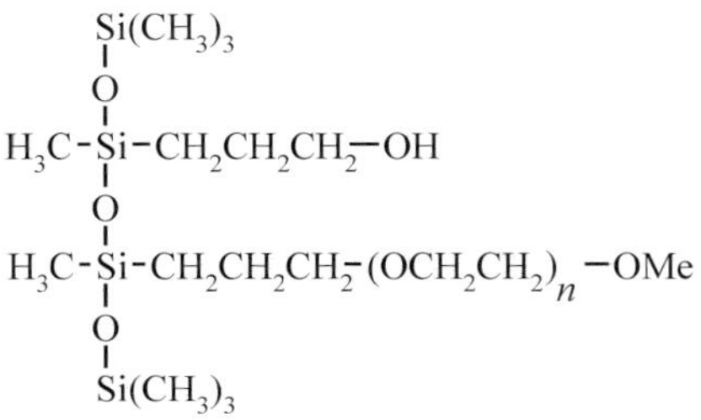

4

$$H_3C-\underset{\underset{O}{|}}{\overset{\overset{Si(CH_3)_3}{|}}{\overset{\overset{O}{|}}{Si}}}-CH_2CH_2CH_2-(OCH_2CH_2)_n-OMe$$
$$H_3C-\underset{\underset{Si(CH_3)_3}{|}}{\overset{\overset{O}{|}}{Si}}-CH_2CH_2CH_2-(OCH_2CH_2)_n-OMe$$

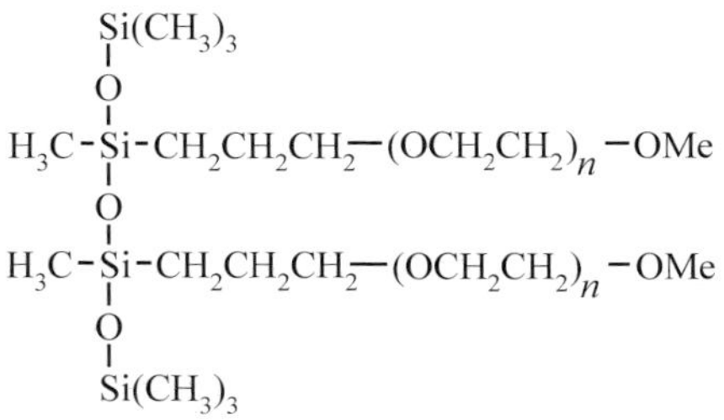

5

Fig. 2.8.7. Proposed by-products of $M_2D–C_3–O–(EO)_n–Me$ synthesis.

(compounds **4** and **5**, Fig. 2.8.7). The former series eluted with the PEG constituents and can be assigned to the linear dimer **4** ($n = 3$–16), whilst the latter, for which the linear dimer **5** ($n_{\text{TOTAL}} = 11$–16; $n_{\text{EO chain\#}} = 5.5$–8) is most likely, co-eluted with the M_2D–C_3–O–$(EO)_n$–H (**2**) oligomers.

An industrial blend of ethylene oxide (EO) PEMS marketed as a personal care product was examined by positive ion FIA–APCI–MS and LC–APCI–MS–MS (Fig. 2.8.8) [41]. The FIA–APCI–MS spectrum without LC separation (Fig. 2.8.8(a)) is dominated by ions corresponding to unreacted PEG (m/z 520, 564, 608, 652,...), whilst the ions corresponding to the PEMS (m/z 516, 560, 604, 648,...) could only be clearly observed following LC separation (Fig. 2.8.8(b)). Comparison of the TIC chromatograms of PEMS and PEG (Fig. 2.8.8(c) and (h)) demonstrates the dominance of the PEG by-products in the commercial formulation. It is unclear whether the observed relative intensities are representative of the actual amounts or of the different ionisation efficiencies, due to the confidential nature of the product composition. However, the spectra indicate a trisiloxane surfactant structure of that shown in Fig. 2.8.2 (R = Ac) and FIA–MS analysis of another commercial formulation of this product showed good spectra dominated by the silicone surfactants [48], indicating that the PEG by-product composition can vary significantly in commercially available PEMS formulations.

2.8.6 Degradation products

API–MS methods have also been applied to qualitative and quantitative determinations of the degradation of silicone surfactants [29]. Comparison of the APCI–MS spectrum of M_2D–C_3–O–$(EO)_n$–Me with that following 24 h equilibration in the presence of halloysite clay is shown in Fig. 2.8.9.

Fig. 2.8.8. APCI–MS and HPLC–APCI–MS of a PEMS and PEG. (a) APCI–FIA–MS(+) overview spectrum of a commercial blend of PEMS showing the precursor PEGs (starting at m/z 300,...; $\Delta m/z$ 44 u) from chemical synthesis; (b) Averaged APCI–LC–MS(+) mass spectrum of PEMS (RT 10.0–13.5 min) from the chromatogram shown in (c); (c) APCI–LC–MS(+) chromatogram of a commercial blend of PEMS acquired in total ion current (TIC) mode (d)–(g) selected mass traces of $[M + NH_4]^+$ ions of PEMS (m/z 516, 560, 604 and 648) and PEG homologues (m/z 520, 564, 608 and 652); (h) APCI–LC–MS(+) chromatogram of PEG in TIC mode [41].

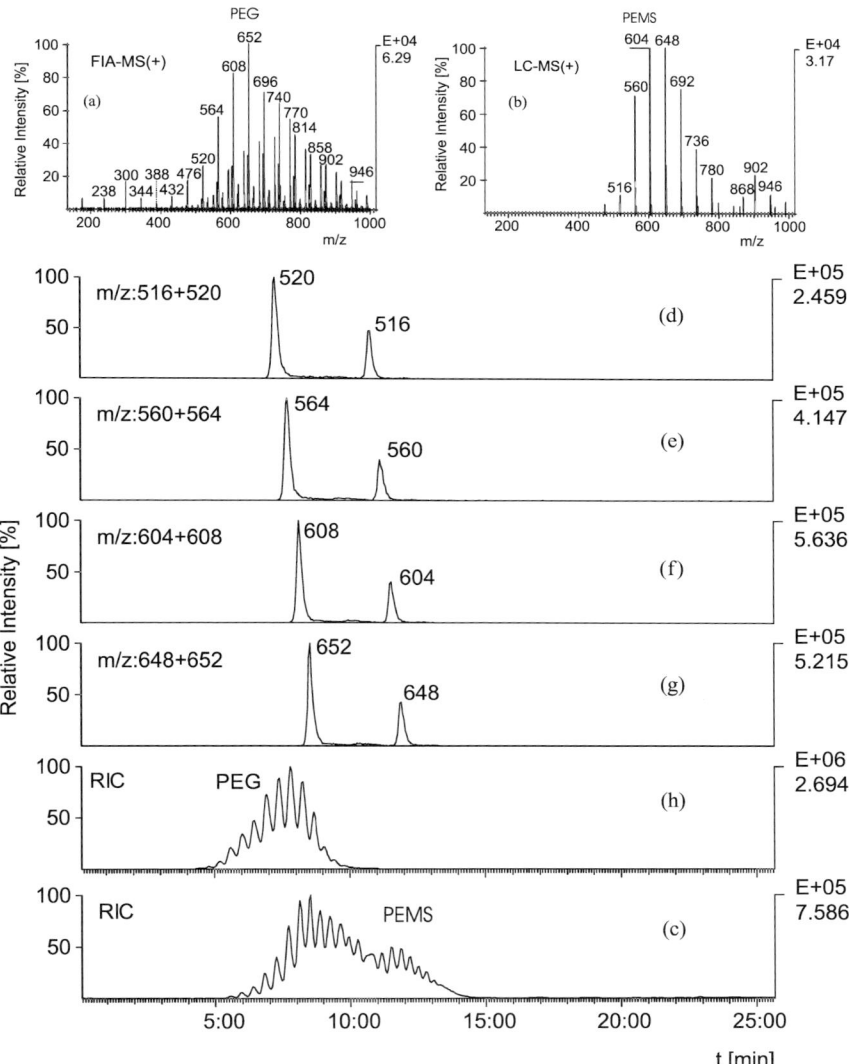

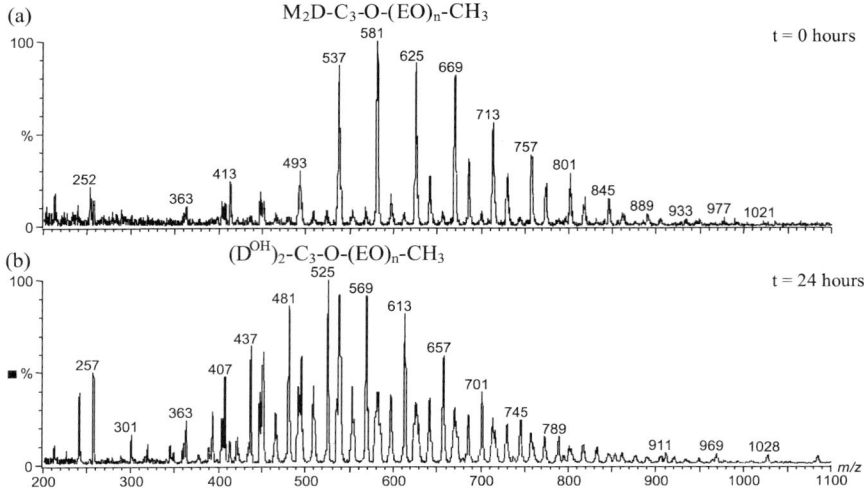

Fig. 2.8.9. FIA–APCI–MS(+) spectra of $M_2D-C_3-O-(EO)_n-Me$ after (a) 0 h and (b) 24 h in the presence of halloysite clay.

Significant changes in the spectra were observed, with the major ions following exposure to the clay substrate corresponding to the disilanol hydrolysis product, i.e. $(HO)_2-Si(CH_3)-C_3-O-(EO)_n-CH_3$. Further discussion of the degradation of these compounds is addressed in Chapter 5.5.

API–MS methods also enabled differentiation between the synthetic by-products and degradation products of an $M_2D-C_3-O-(EO)_n-Me$ commercial formulation as shown in Fig. 2.8.10. $HO(EO)_nH$, $M_2D-C_3-O-(EO)_n-H$ (2) and $D^{(R)2}-O-(EO)_n-CH_2CH=CH_2$ (R = OH or CH_3) compounds observed in the degradation mixture of commercially obtained $M_2D-C_3-O-(EO)_n-Me$ were confirmed as synthetic by-products, rather than degradation products, with the use of purified trisiloxane starting materials [29].

2.8.7 Conclusion

API–MS techniques (ESI and APCI) have been shown to be informative methods for the qualitative and quantitative analysis of organosilicone surfactants. The use of HPLC–API–MS has enabled improvements in

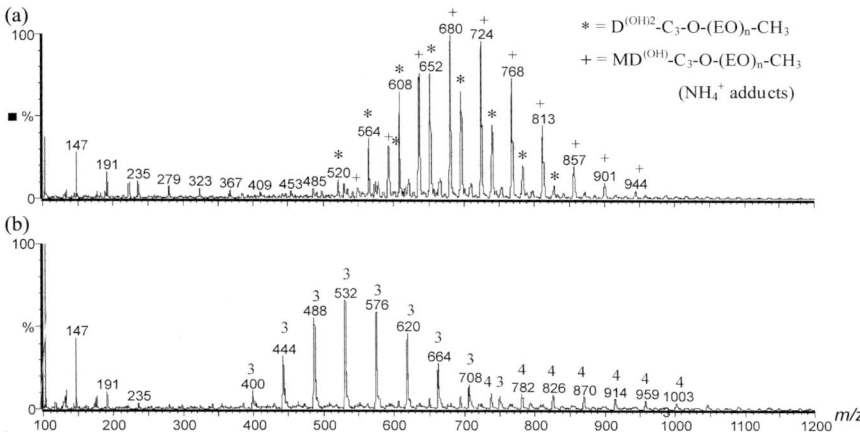

Fig. 2.8.10. FIA–ESI–MS(+) spectra of: (a) polar hydrolysis products of $M_2D-C_3-O-(EO)_n-Me$ and (b) polar components of the commercial formulation (for structures **3** and **4** refer to Fig. 2.8.7).

detection limits and reproducibility for the quantitation of these compounds, as used in agrochemical applications. Application of the technique to the detection of by-products and degradation products of the organosilicone surfactants has provided new information regarding their composition. However, the application of the method to environmental samples has not been investigated, and is an area to be addressed in future research.

Acknowledgements. This research was funded through the Public Good Science Fund (PGSF) of New Zealand (New Zealand Forest Research Institute) and forms part of a D.Phil. study by Dr L.S. Bonnington. Financial assistance from the New Zealand Federation of University Women (NZFUW), the New Zealand Vice Chancellors Committee (NZVCC), Waikato University and the New Zealand Lotteries Commission are also gratefully acknowledged. We are also thankful to Dr G. Willett of the University of New South Wales for kindly providing access to FTICR–MS, Dr G. Policello, Dr P. Stevens and Dr M. Costello of Witco Corporation for generous samples and information, and Alison Forster of Forest Research for scientific assistance.

REFERENCES

1 G.L.F. Schmidt, In: D.R. Karsa (Ed.), *Industrial Applications of Surfactants*, The Royal Society of Chemistry, London, 1987, pp. 24–32.

2 D.T. Floyd, In: R.M. Hill (Ed.), *Silicone Surfactants*, Surfactant Science Series, Vol. 86., Marcel Dekker Inc., New York, 1999, Ch. 7, pp. 181–207.

3 R.M. Hill, In: R.M. Hill (Ed.), *Silicone Surfactants*, Surfactant Science Series, Vol. 86, Marcel Dekker Inc., New York, 1999, Ch. 1, pp. 1–47.

4 G.A. Policello, P.J.G. Stevens, R.E. Gaskin, B.H. Rohitha and G.F. McLaren, In: R.E. Gaskin (Ed.), NZFRI Bulletin No. 193, *Proc. 4th International Symposium on Adjuvants for Agrochemicals*, 1995, pp. 303–307.

5 D.E. Powell and J.C. Carpenter, In: G. Chandra (Ed.), *Organosilicon Materials*, The Handbook of Environmental Chemistry, Vol. 3H., Springer-Verlag, Berlin, Heidelberg, 1997, Ch. 8, pp. 225–239.

6 R.C. Kirkwood, *Pestic. Sci.*, 38 (1993) 93.

7 D. Stock and P.J. Holloway, *Pestic. Sci.*, 38 (1995) 165.

8 P.J. Holloway and D. Stock, Industrial Applications of Surfactants II, In: D.R. Karsa (Ed.), *Special Publication No. 77*, Royal Soc., Cambridge, 1990, pp. 303–337.

9 L. Copping, *Adjuvants and additives in crop protection*, Agrow Report, Richmond, Surrey, UK, 2000.

10 L.M. Rogiers, In: R.E. Gaskin (Ed.), NZFRI Bulletin No. 193, *Proc. 4th International Symposium on Adjuvants for Agrochemicals*, 1995, pp. 1–10.

11 R.C. Kirkwood, *Pestic. Sci.*, 55 (1999) 69.

12 H. de Reuiter, E. Meinen and M.A.M. Verbeck, In: C.L. Foy (Ed.), *Adjuvants for Agrichemicals*, CRC Press, Florida, 1992, Ch. 8, pp. 109–116.

13 M. Knoche, *Weed Res.*, 34 (1994) 221.

14 G.A. Policello, P.J. Stevens, W.A. Forster and G.J. Murphy, In: F.R. Hall, P.D. Berger and H.M. Collins (Eds.), *Pesticide Formulations and Application Systems*, Vol. 14, ASTM Publishers, Philadelphia, 1995, pp. 313–317.

15 R.E. Gaskin, In: R.E. Gaskin (Ed.), NZFRI Bulletin No. 193, *Proc. 4th International Symposium on Adjuvants for Agrochemicals*, 1995, pp. 243–248.

16 J. Schönherr and M.J. Bukovac, *Plant Physiol.*, 49 (1972) 813.

17 D.S. Murphy, G.A. Policello, E.D. Goddard and P.J. Stevens, In: B.N. Devisetty, D.G. Chasin and P.D. Berger (Eds.), *Pesticide Formulations and Application Systems*, Vol. 12, ASTM Publishers, Philadelphia, 1993, pp. 45–56.

18   P.J.G. Stevens, R.E. Gaskin, S.-O. Hong and J.A. Zabkiewicz, *Pestic. Sci.*, 33 (1991) 371.

19   D. Penner, R. Burow and F.C. Roggenbuck, In: R.M. Hill (Ed.), *Silicone Surfactants*, Surfactant Science Series, Vol. 86, Marcel Dekker Inc., New York, 1999, Ch. 9, pp. 241–258.

20   L.L. Jansen, *Weed Sci.*, 21 (1973) 130.

21   P.J.G. Stevens, *Pestic. Sci.*, 38 (1993) 103.

22   A. Lee Smith (Ed.), *The Analytical Chemistry of Silicones*, John Wiley and Sons Ltd, New York, 1991.

23   R.D. Steinmeyer and M.A. Becker, In: A. Lee Smith (Ed.), *The Analytical Chemistry of Silicones*, Chromatographic methods, John Wiley and Sons Ltd, New York, 1991, Ch. 10, pp. 255–304.

24   R.B. Taylor, B. Parbhoo and D.M. Fillmore, In: A. Lee Smith (Ed.), *The Analytical Chemistry of Silicones*, Nuclear magnetic spectroscopy, John Wiley and Sons Ltd, New York, 1991, Ch. 12, pp. 347–419.

25   D. Lipp and A.L. Smith, In: A. Lee Smith (Ed.), *The Analytical Chemistry of Silicones*, Infrared, Raman, near-infrared and ultraviolet spectroscopy, John Wiley and Sons Ltd, New York, 1991, Ch. 11, pp. 305–346.

26   A. Ballistreri, D. Garozzo and G. Montaudo, *Macromolecules*, 17 (1984) 1312.

27   J.A. Zabkiewicz, P.J.G. Stevens, W.A. Forster and K.D. Steele, *Pestic. Sci.*, 38 (1993) 135.

28   W.A. Forster, K.D. Steele and J.A. Zabkiewicz, In: R.E. Gaskin (Ed.), NZFRI Bulletin No. 193, *Proc. 4th International Symposium on Adjuvants for Agrochemicals*, 1995, p. 267.

29   L.S. Bonnington, Analysis of Organosilicone Surfactants and their Degradation Products, PhD Thesis, University of Waikato, Hamilton, New Zealand, 2001.

30   J.A. Zabkiewicz, W.A. Forster, L.S. Bonnington and W. Henderson, *Proc. 6th International Symposium on Adjuvants for Agrochemicals*, Amsterdam, August 13–17, 2001.

31   J.A. Zabkiewicz, L.S. Bonnington, W.A. Forster and W. Henderson, In: P.M. McMullan (Ed.), *Proc. 5th International Symposium on Adjuvants for Agrochemicals*, Chem. Prod. Coop., 1998, pp. 85–92.

32   M.G. Voronkov, V.P. Milesshkevich and Y.A. Yuzhelevskii, *The Siloxane Bond: Physical Properties and Chemical Transformations*, Consultants Bureau, New York, 1978.

33   T. Kemmitt and W. Henderson, *J. Chem. Soc., Perkin Trans.*, 1 (1997) 729.

34   T. Kemmitt and W. Henderson, *Aust. J. Chem.*, 51 (1998) 1031.

35   T. Covey, Biochemical and Biotechnological Applications of Electrospray Ionisation Mass Spectrometry, In: A.P. Synder (Ed.), ACS Symposium Series 619, ACS, Washington, DC, 1996, pp. 21–59.

36   U. Siemeling, *Polyhedron*, 16 (1997) 1513.

37 J.B. Fenn, M. Mann, C.K. Meng, S.F. Wong and C.M. Whitehouse, *Science*, 246 (1989) 64.

38 T. Nohmi and J.B. Fenn, *Anal. Chem.*, 114 (1992) 3241.

39 T. Okada, *Anal. Chem.*, 62 (1990) 327.

40 R. Colton, S. Mitchell and J.C. Traeger, *Inorg. Chim. Acta*, 231 (1995) 87.

41 R. Scheding and H.Fr. Schröder, Technical University of Aachen, Germany, unpublished results.

42 A. Di Corcia, C. Crescenzi, A. Marcomini and R. Samperi, *Environ. Sci. Technol.*, 32 (1998) 711.

43 K.B. Sherrard, P.J. Marriott, M.J. McCormick, R. Colton and G. Smith, *Anal. Chem.*, 66 (1994) 3394.

44 R.P. Lattimer, H. Muenster and H. Budzikiewicz, *Int. J. Mass Spectrom. Ion Process.*, 90 (1989) 119.

45 T.L. Selby, C. Wesdemiotis and R.P. Lattimer, *J. Am. Mass Spectrom.*, 5 (1994) 1081.

46 C. Crescenzi, A. Di Corcia, R. Samperi and A. Marcomini, *Anal. Chem.*, 67 (1995) 1797.

47 C.J. Herzig, In: N. Auner and J. Weis (Eds.), *Organosilicon Chemistry: From Molecules to Materials*, VCH, Verlagsgesellschaft mbH, Weinheim, 1994, pp. 253–260.

48 L.S. Bonnington, W. Henderson, J.A. Zabkiewicz, University of Waikato, Hamilton/Forest Research, Rotorua, New Zealand, unpublished results.

2.9 ATMOSPHERIC PRESSURE IONISATION MASS SPECTROMETRY—VI. NON-IONIC SURFACTANTS LC–MS OF OTHER NON-IONIC SURFACTANTS

Horst Fr. Schröder

2.9.1 Introduction

The term 'non-ionic surfactants' refers mainly to alkyl and alkylaryl derivatives coupled with polyethylene and polypropylene glycol moieties as shown in Fig. 2.9.1. Their main use now and in the foreseeable future will be in the area of detergent and personal care formulations in combination with other surfactants, e.g. anionic surfactants. But traditional non-ionic polyglycol surfactants and newly developed non-ionics such as fluorinated surfactants and gemini surfactants (Surfynols) as well as non-ethoxylated sugar esters and alkylglucamides have begun to find applications in new areas such as environmental protection systems, paper processing and thin films. In parallel, an enormous growth was observed in fundamental and applied research on non-ionic surfactants covering chemical and physico-chemical analyses, organic and physical chemistry and biochemistry of non-ionic surfactants, which is reflected in fundamental books [1–4] and in research papers.

Surfactants are produced on very large or medium technical scales. Their analysis by manufacturers in products and their formulations sometimes may be complicated because of the great variety of surfactants [5]. After use as directed in aqueous systems they were discharged mainly with wastewaters. Their analysis in environmental samples then becomes quite difficult because analysis must be performed at trace concentrations with limited sample amounts after essential matrix-dependent pre-concentrations steps. In addition, homologues and isomers that exist for many surfactants, besides metabolites which are generated in biochemical processes, complicate their specific determination [6].

Some surfactants were found to be hardly degradable in the biological wastewater treatment process. Therefore, non-ionic surfactants are observed not only in wastewater and surface water but also in drinking water [7,8] and other environmental samples. In addition, they could be

Comprehensive Analytical Chemistry XL
T.P. Knepper, D. Barceló and P. de Voogt (Eds)

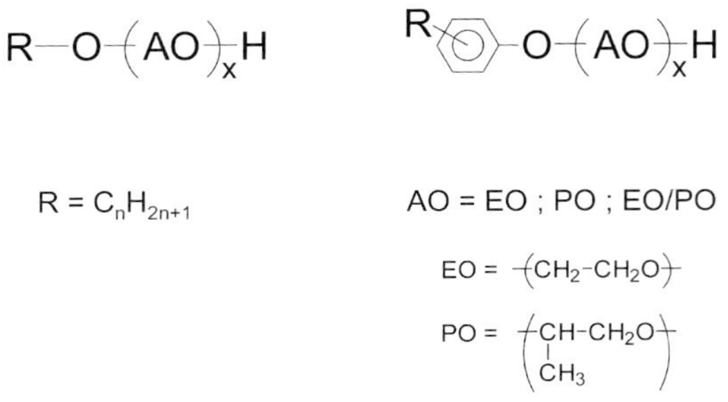

$$R{-}O{-}(AO)_x{-}H \qquad R{-}\bigcirc{-}O{-}(AO)_x{-}H$$

$R = C_nH_{2n+1}$ $\qquad$ AO = EO ; PO ; EO/PO

EO = $-(CH_2-CH_2O)-$

PO = $\left(-CH-CH_2O-\atop CH_3\right)$

Fig. 2.9.1. The general structure of non-ionic alkylpolyglycolethers (AE) (I) and alkyl aryl polyglycol ethers (APEO) (II).

detected in food stuff and personal care products. However, they are also found in laboratory equipment like glass fibre filters [9,10] or in syringe filters [11].

Non-specific sum parameter analysis [12,13], which is still used today, failed [14,15] in the analyses of some of these compounds. Chromatographic methods in combination with non-substance specific detectors, e.g. colorimetric and photometric [5] or with substance specific detectors such as IR (infrared spectroscopy), NMR (nuclear magnetic resonance spectroscopy) or MS (mass spectrometry), are applied increasingly nowadays.

MS techniques have met this need in the analysis of involatile, polar surfactants after coupling techniques of liquid chromatographic methods with MS became available. Different types of interfaces for off-line and on-line coupling of liquid chromatography (LC) and MS in the analyses of surfactants had been in use [7,16] while the methods applied at present were performed predominately with soft-ionising atmospheric pressure ionisation (API) interfaces [16–19].

Several types of surfactants and especially the most common non-ionic surfactants are analysed by soft ionisation MS techniques in the positive mode in a great variety of isomeric and homologue compounds in industrial blends, formulations of household cleaning agents, agrochemicals, etc. as well as in environmental samples. The screening approach by flow injection analysis coupled to MS (FIA–MS) without any chromatographic separation (by-passing the analytical column and using mixture analysis [20,21]) allows us to recognise the ethoxylate compounds because of their characteristic equidistant signals with $\Delta m/z$ 44 or 58 in their overview spectra. Besides surfactants based on

228

ethoxylate polyalcohols, polyamines, fatty acid polypropyleneglycol esters, fatty acid polypropylene glycol amides and ethoxylate modified siloxanes also belong to this group of compounds.

For a quick pre-screening, the FIA–MS approach with pattern recognition can be used, but if unknown samples are examined, the characteristic pattern of equispaced signals observed in the FIA–MS spectrum should be judged with caution. The presence of ionic surfactants that probably contain the same structural elements—polyethylene (PEG) or polypropylene glycol (PPG)—in the molecule must be taken into consideration, despite the fact that most of them cannot be ionised in their original form in positive mode. Moreover, ionisation of these ionic compounds occur in a destructive ionisation process resulting in alkyl or arylpolyglycolether ions, which then imagines the presence of non-ionics.

The most important anionic alkyl ether sulfate (AES; $C_nH_{2n+1}-(OCH_2CH_2)_m-OSO_3H$) surfactants belong to this type of anionic compounds with these characteristic structural elements. Other compounds with the ethoxylate structural element in their molecule exhibiting this behaviour are alkyl ether sulfonates ($C_nH_{2n+1}-(OCH_2CH_2)_m-SO_3H$), alkyl ether carboxylic acids ($R-(O-CH_2-CH_2)_m-CH_2-COOH$; $R = $ alkyl; $C_nH_{2n+1}-$), alkylphenolether carboxylic acids ($R = $ alkylaryl; $C_nH_{2n+1}-C_6H_4-$) or alkylphenolether sulfates, sulfonates and phosphates. All of them belong to the group of anionic surfactants. The presence of these compounds can be verified in the negative ionisation mode, more or less characteristic for most of the anionics. Cationic surfactants like fatty acid polyglycol amines ($R-N^{\oplus}H((CH_2-CH_2-OH)_{x,y})_2\ X^-$) may also exhibit equally spaced signals in the FIA–MS spectra and therefore should be taken into account if the pattern recognition approach is chosen.

Positive ionisation is the method of choice for the detection of all non-ionic surfactants generating molecular $[M + H]^+$ or ammonium adduct ions ($[M + NH_4]^+$) in the presence of ammonium acetate. Often $[M + Na]^+$ ions were also observed; however, an excess of ammonium acetate will suppress their generation. While sensitivity in this mode is very high and can be improved by an excess of ammonium acetate to suppress sodium or potassium adduct ions, selectivity compared with negative ionisation for anionics is low.

Variations in alkyl- or ethoxylate chain lengths, the differences in the degree of alkyl chain branching as well as mixing of ethyleneglycol (EO) and propyleneglycol ether (PO) moieties during the large-scale production of alkyl- and EO/PO-polyglycolethers were chosen to improve the adaptation to special requirements in applications. The substitution of hydrogen atoms against fluorine in the lipophilic part of the surfactant molecules changed performance and applicability of these compounds to quite a large extent. In parallel, all these modifications in the molecules may lead to very complex patterns of signals in the FIA–MS spectra even in blends and formulations. The extracts from untreated industrial and municipal wastewater, however, normally exceed these spectra in complexity. But potential complexity of these mixtures will not be expressed completely by the patterns of signals that belong to defined m/z ratios because multiple different structured types of surfactants with the same molecular weight—isomeric or isobaric compounds—may be hidden behind one peak in the FIA–MS spectrum characterised by one specific m/z ratio.

Here, the mixture analytical FIA–MS–MS approach reached its limitation to identify compounds. Hence, LC separations prior to MS analysis are essential to separate compounds with the same m/z ratio but with different structures. The behaviour in the LC separation will be influenced by characteristic parameters of the surfactant such as linear or strongly branched alkyl chain, the type, the number and the mixture of glycolether groups—PEG and/or PPG—and the ethoxylate chains. The retardation on SPE materials applied for extraction and/or concentration also depends on these properties and can therefore be used for an appropriate pre-separation of non-ionic surfactants in complex environmental samples as well as in industrial blends and household detergent formulations. A sequential selective elution from SPE cartridges using solvents or their mixtures can improve this pre-separation and saves time in the later LC separation [22].

2.9.2 Analyses of non-ionic ethoxylate surfactants

Alcohol ethoxylates (AE)
An industrial blend of AE surfactants with the general formula ($C_nH_{2n+1}O-(CH_2-CH_2-O)_xH$; $n = 12$, 14, 16 and 18) was examined using APCI–FIA–MS(+) for screening purposes (see Fig. 2.9.2(a)). According to the number of glycol units and the number of alkyl chain links, a series of homologue ammonium adduct ions ($[M + NH_4]^+$) equally spaced either with $\Delta m/z$ 44 ($-CH_2-CH_2-O-$) or $\Delta m/z$ 28

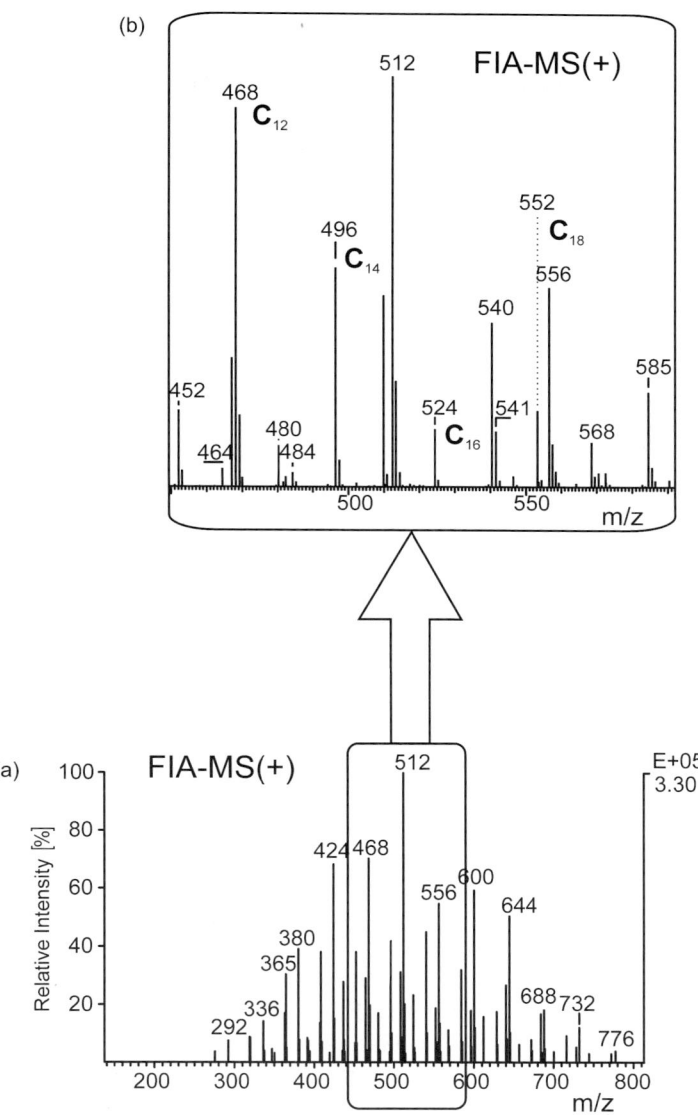

Fig. 2.9.2. (a) APCI–FIA–MS(+) overview spectrum of AE blend ($C_nH_{2n+1}O-(CH_2-CH_2-O)_x-H$); (b) expanded detail of APCI–FIA–MS(+) spectrum as in (a) [23].

($-CH_2-CH_2-$), respectively, could be observed in the expanded overview spectrum (see Fig. 2.9.2(b)) [23].

The separation of this blend performed on a RP-C_{18} column resulted in an excellent separation as shown in Fig. 2.9.3(a)–(d). The separation resulted in four signals according to the different numbers of alkyl chain links as recognisable in the reconstructed ion chromatogram (RIC) (Fig. 2.9.3(d)). Hidden behind each signal, the AE compounds with a varying number of glycol units could be detected as shown with peak 1 containing AE surfactants ($C_nH_{2n+1}O-(CH_2-CH_2-O)_xH$) with $n = 12$ and $x = 2$–13 ionised predominantly as $[M + NH_4]^+$ ions (m/z 292–776) or to a smaller extent as $[M + H]^+$ ions (m/z 275–495) (cf. Fig. 2.9.3(e)) [23,24].

Jandera et al. [25] studied the chromatographic behaviour of a technical blend of AE surfactants ($C_nH_{2n+1}-O(CH_2CH_2O)_mH$; $n = 12$, 14, 16, 18). Normal-phase (NP) and reversed-phase (RP) conditions were applied and atmospheric pressure chemical ionisation (APCI) combined with MS(+) was used to monitor LC separation. Cluster ions such as $[M + H]^+$, $[M + Na]^+$ and $[M + K]^+$ ions could be recorded. Different materials used as stationary phases and solvents and their mixtures applied as mobile phases were compared. LC under RP conditions was able to separate according to the alkyl chain lengths, whereas the ethoxylate chain length was not determinative for separation. The influence of the different organic solvents during separation was examined in both RP- and NP-modes. NP separation allows a differentiation according to the number of polyglycolethers in the homologues [25].

Overview spectra of a mixture of primary alcohol ethoxylate blend (PAE) and the isomeric secondary alcohol ethoxylate blend (SAE) used in wool scouring were recorded by Sherrard et al. [26]. The non-ionic PAE surfactant blend ($C_nH_{2n+1}-O(CH_2CH_2O)_mH$; $n = 6$ or 8; $m = 9$–14) and the SAE blend ($CH_3-(CH_2)_n-CH(O(CH_2CH_2O)_mH)-(CH_2)_x-CH_3$; $n + x = 9$–11; $m = 9$) were detected qualitatively by electrospray ionisation (ESI) in FIA–MS(+) mode. The FIA–MS spectra of the blends showed a Gaussian partition (statistical normal distribution) in molecular weight recognisable by their peak shape. Because of the impurities present in the blend, the surfactants were ionised not only as $[M + H]^+$ but also as $[M + Na]^+$ or $[M + K]^+$ with equidistant signals ($\Delta m/z$ 44). The number of different adduct ions other than $[M + H]^+$ and that generated by sodium and potassium led to a quite complex and therefore confusing diversity of signals because

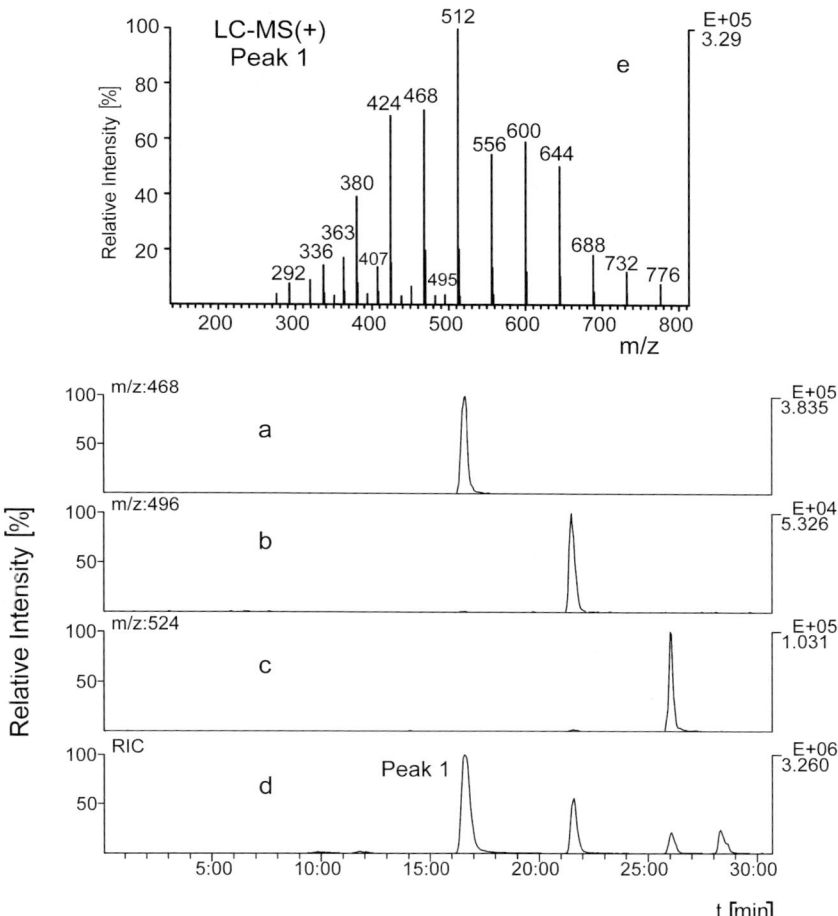

Fig. 2.9.3. APCI–LC–MS(+) separation of AE blend as in Fig. 2.9.2; (a)–(c) selected mass traces of AE blend, (d) RIC and (e) averaged mass spectrum of peak 1 in (d) presenting the AE homologues of $C_{12}H_{25}O–(CH_2–CH_2–O)_x–H$ ionised as $[M + NH_4]^+$ (m/z 292–776) or $[M + H]^+$ ions (m/z 275–495) [23].

the signals of the surfactant homologues arising from sodium clustering shifted by $\Delta m/z$ 22; potassium clustering led to a shifting of $\Delta m/z$ 38 compared with the $[M + H]^+$ ions. The addition of an excess of NaCl into the solutions suppressed all other clustering reactions and led to $[M + Na]^+$ ions resulting in a defined molecular weight information.

Collisionally induced dissociation (CID, which is also termed collisionally activated dissociation (CAD)) was performed by increased skimmer voltages to generate fragments. When CID was applied to the mixture in the FIA–MS mode, a large number of fragment ions like $[(HO(CH_2CH_2O)_mH + H]^+$ ($m = 3$–14; $m/z = 151$–635) was observed. All ions originated from the PEG moieties while fragments of the alkyl chain were not observable [26]. In addition, photocatalytic degradation by ultraviolet (UV) radiation in combination with titanium dioxide (TiO_2) particles of SAE was also performed with these surfactants. The surfactants were treated over a maximum period of 77 h. Then positive FIA–MS was performed, which resulted in ions comparable with those of CID of untreated compounds. The authors conclude that the carbon–oxygen bond in the SAE molecules can easily be cleaved, resulting in the more stable secondary carbon-centered cation [26]. AE of PAE type with a distribution of homologues in the FIA–MS(+) spectra comparable with the SAE ions, however, were fragmented by CID, resulting in product ions that were observed to be different from those of SAE surfactants [26].

Standards of AE (C_nH_{2n+1}–$O(CH_2CH_2O)_mH$) with different alkyl and ethoxylate chain lengths besides a cyclohexane derivative, ethoxylated with a maximum of two ether units were determined using APCI–FIA–MS(+) for screening purposes [27]. From this mixture of non-ionic surfactants of the ethoxylate type, molecular ions were observed, all ionised with equidistant signals corresponding to their homologues ($\Delta m/z$ 44). Fragmentation resulting in alkyl and ethoxylate fragments took place under these conditions besides ionisation as $[M + H]^+$ ions and sodium adduct ions ($[M + Na]^+$). Qualitative as well as quantitative results from these examinations were used for the detection and identification of compounds in tannery wastewaters. These non-ionic surfactants of the AE and alkylphenol ethoxylate (APEO) type were determined qualitatively and quantitatively besides PEG as their biochemical degradation products.

Matrix assisted laser desorption ionisation (MALDI) and ESI–MS spectra of non-ionic surfactant blends of AE obtained after positive ionisation were compared [28]. Both the ionisation procedures, which produce $[M + Na]^+$ ion clusters, were very useful for this purpose, but the ESI spectra generated were more complex, whereas MALDI ionisation led to simpler spectra that can be interpreted more easily [28].

Laser desorption (LD) Fourier transform ion cyclotron resonance MS (FTICR-MS) in the positive mode was applied for the analysis of different types of non-ionic surfactants. All non-ionic surfactants with MW < 2000 'gave good spectra with little fragmentation showing nice Gaussian peak shapes' [29]. The main disadvantage is the fragmentation of high mass surfactants. This will be minimised in future by the application of MALDI.

The AE blend Brij 35 with the general formula $C_nH_{2n+1}-O(CH_2CH_2O)_mH$ was analysed by MALDI MS prior to use for biochemistry research. Separation results of thin-layer (TLC) and RP-LC of these surfactants were compared [30]. Brij 35, as a mixture of C_{12} and C_{14} homologues ($m = 15-39$), was detected qualitatively as $[M + Na]^+$ and $[M + K]^+$ ions and quantitatively after TLC and RP-LC separation.

Non-ionic surfactants of a commercial washing powder were separated by supercritical fluid chromatography (SFC) and determined by APCI–MS. The constituents were first extracted by supercritical fluid extraction (SFE) using CO_2 with or without methanol as a modifier. Variations of the conditions resulted in a selective extraction of the analytes, which could be determined without further purification. Six groups of surfactants were observed, four of which are alkyl-polyethoxylates. The presence of APEO could be excluded by identification recording SFC-FTIR (Fourier transform infrared) spectra [31].

An AE surfactant mixture was found in a syringe filter eluate. Equidistant ions ($\Delta m/z$ 44) were detected qualitatively in the FIA–MS mode. The compounds originated from a single-use syringe filter used for sample filtration. The ESI interface produced $[M + NH_4]^+$ ions. The parent ion at m/z 528 in the ESI–FIA–MS(+) spectrum was submitted to CID and was characterised by ESI–FIA–MS–MS(+) as AE surfactant homologue with the general formula $C_{10}H_{21}-O(CH_2CH_2O)_mH$ ($m = 8$) [11].

When APCI–FIA–MS(+) was applied for screening purposes to determine the standards of non-ionic surfactants of the ethoxylate type in different industrial wastewaters, source-CID effects could be observed. Fragmentation of the AE homologues ($C_nH_{2n+1}-O(CH_2CH_2O)_mH$), differing in alkyl and ethoxylate chain lengths took place. Under the conditions applied, alkyl as well as ethoxylate chain fragments were generated and allowed the identification of the AE compounds [27]. Fragment ions at m/z 141, 169 and 271 were attributed to the aliphatic chains of decylic, lauric and tridecylic ethoxylates. Product ions between m/z 107 and 327, all equally spaced ($\Delta m/z$ 44), were consistent with the ethoxylate fragment ions originating from the AE surfactants [27,32].

The non-ionic surfactant mixture Brij 78 is often used as a biological detergent and purification reagent. This mixture of AE homologues was examined by applying MS for detection and MS–MS for identification. Mixture analysis by API–FIA generated characteristic product ions at m/z 341 and 385 ($[CH_3(CH_2)_{17}(OCH_2CH_2)_n]^+$) [33]. MS–MS performed under CID conditions resulted in additional product ions that enabled the authors to identify these compounds by mixture analysis (FIA–MS–MS(+)).

In a screening approach, non-ionic surfactants were monitored in the form of their $[M + NH_4]^+$ ions, equally spaced with $\Delta m/z$ 44 and identified by FIA–MS–MS(+) in combination with APCI or ESI interface [34,35]. C_{18}-SPE was performed prior to selective elution by diethyl ether [35]. Ions of the non-ionics of AE type at m/z 350–570 ($\Delta m/z$ 44) were identified as surfactants with the general formula $C_{13}H_{27}$–$O(CH_2CH_2O)_mH$ ($m = 3$–7). The complexity of the mixture confirmed the results using the diagnostic parent scans m/z 89 for aliphatic non-ionic surfactants of ethoxylate type necessary [35].

APCI–FIA- and APCI–LC–MS(+) analyses were performed to characterise non-ionic surfactants in complex mixtures. Product ion and parent ion spectra were recorded to confirm the characterised non-ionics. MS–MS(+) results proved that AE compounds (C_nH_{2n+1}–$O(CH_2CH_2O)_mH$) with different alkyl and ethoxylate chain lengths were present in all samples [35,36].

APCI and ESI–FIA–MS–MS examinations in the positive and negative mode were performed. The product ions of the $[M + NH_4]^+$ parent ion m/z 394 from the feed extract allowed an unequivocal identification of the homologue as ion of the AE molecule with the formula $C_{13}H_{27}$–O–$(CH_2$–CH_2–$O)_4$–H. The product ions m/z 45, 89, 133 and 177, known as ethoxylate fragments of the PEG chain, and the alkyl fragments m/z 57, 71, 85, 99, 113 and 127 were also characteristic [7] for this AE type, which is presented with its fragmentation scheme in Fig. 2.9.4 [22].

One of the most observed degradation pathways of non-ionic surfactants of ethoxylate type in the biochemical wastewater treatment process is the bond scission between the lipophilic alkyl chain and the hydrophilic ethoxylate moieties. The resulting ethoxylate compounds, PEG or PPG, are highly polar and are not quite easy to degrade, therefore often they can be observed in wastewater discharges. So, APCI–FIA–MS(+) product ion spectra of selected $[M + NH_4]^+$ ions, which were under suspicion as PEG (general formula: HO–$(CH_2$–CH_2–$O)_xH$)

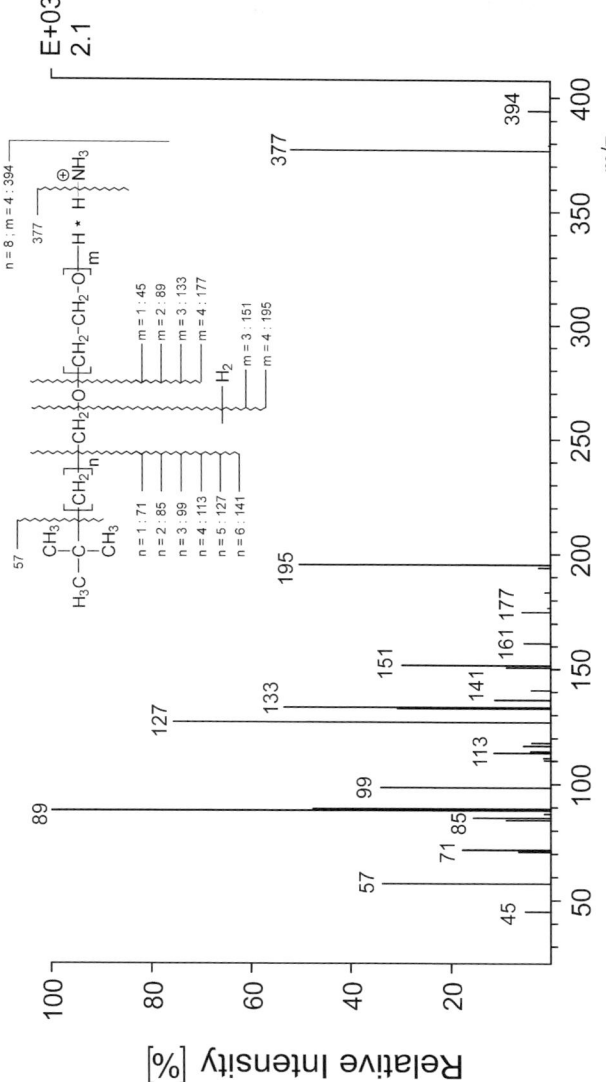

Fig. 2.9.4. APCI–FIA–MS–MS(+) (CID) product ion mass spectrum of parent ion with m/z 394 from C_{18}-SPE, identified as non-ionic surfactant of AE type (C_nH_{2n+1}–O–(CH_2–CH_2–O)$_m$–H); (inset) fragmentation scheme under CID conditions [22].

were generated. CID-spectra together with the characteristic fragmentation scheme of PEG molecules are presented in Fig. 2.9.5 [34].

When an industrial blend of C_{13}-AE surfactants $C_nH_{2n+1}O-(CH_2-CH_2-O)_xH$ was biodegraded with different biocoenoses in lab scale reactors, different degradation pathways were observed applying biocoenoses of different WWTPs. Besides a mixture of precursor compounds reduced in concentration, metabolites were also observed by APCI and ESI–MS(+) as $[M + NH_4]^+$ ions, which were then confirmed by MS–MS. The AE blend $C_nH_{2n+1}O-(CH_2-CH_2-O)_xH$ ($n = 13$; $x = 1-16$) with equally spaced ions ($\Delta m/z$ 44) at m/z 262, 306, 350,...,922 was degraded, resulting in a series of homologue compounds with (1) carbonylic groups in the terminal ethoxylate position ($C_nH_{2n+1}O-(CH_2-CH_2-O)_{x-1}-CH_2-C(H)O$; $x = 1-13$; m/z 304, 348, 392,...,832), (2) or was carboxylated in the terminal position ($C_nH_{2n+1}O-(CH_2-CH_2-O)_{x-1}-CH_2-COOH$; $x = 2-6$; m/z 364, 408, 452, 496 and 540) [37]. However, the intramolecular scission process was seen as the dominating reaction pathway of these AE compounds. This first led in one reactor to ions at m/z 300, 344, 388, 432,...,696, characteristic of PEG ($HO-(CH_2-CH_2-O)_x-H$; $x = 7-15$), before these primary degradation products were degraded further, resulting in

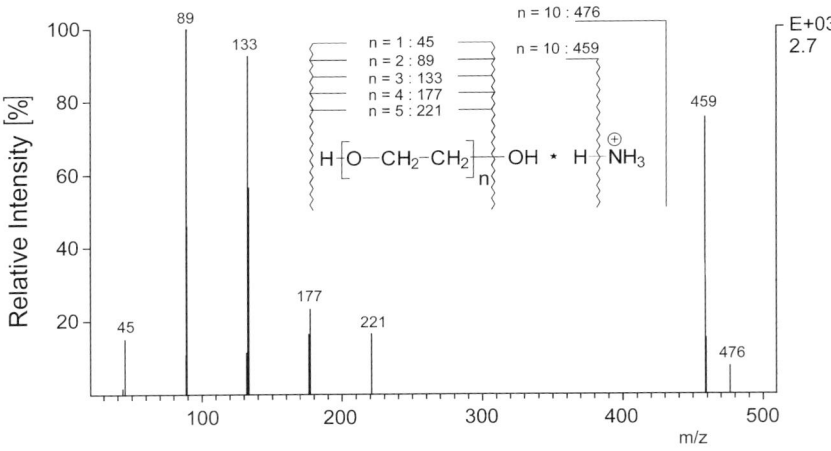

Fig. 2.9.5. APCI–FIA–MS–MS(+) (CID) product ion mass spectrum of selected $[M + NH_4]^+$ parent ion (m/z 476) of non-ionic surfactant metabolite identified as PEG homologue (general formula: $HO-(CH_2-CH_2-O)_x-H$); fragmentation behaviour under CID presented in the inset [35].

carboxylated PEG with the general formula $HO-(CH_2-CH_2-O)_x-$
CH_2-COOH observed as $\Delta m/z + 14$ $[M + NH_4]^+$ ions at m/z 314, 358,
402,...,622 ($x = 5-12$) applying APCI–FIA–MS(+). Identity was
confirmed by MS–MS [37].

ESI–LC–MS was used to analyse a mixture of 2-butyl branched C_{12}-
AE containing an average of five PEG units after submitting the
mixture to biodegradation. SPE applying Carbograph 4 cartridges was
used for the extraction of the compounds in samples from biodegradation. Neutral and acidic analytes or metabolites could also be
concentrated. Selective elution was performed applying neutral organic
solvents and acidified organics. After solvent removal, ESI–LC–MS
and ESI–LC–MS–MS spectra in the in-source CID mode were
recorded. After addition of a decyl-based ethoxylate internal standard,
neutral and acidic metabolites were confirmed by mass and MS–MS
spectra and mass chromatograms. AE were shortened in ethoxylate
chain lengths and/or oxidised resulting in carbonylic or carboxylic
compounds during the biodegradation process while, in parallel, highly
polar intermediates could be confirmed. Several alternative biodegradation pathways were observed [38].

Surface water samples often contain surfactants and their metabolites. After C_{18}-SPE combined with selective elution [7,9,10] the
metabolites, PEG and PPG, were observed in the ether fraction (PPG) or
in the combined methanol–water and methanol (PEG) fractions,
respectively. They could be ionised in the form of their $[M + NH_4]^+$
ions applying ESI–FIA–MS(+) in combination with ammonium
acetate for ionisation support. ESI–LC–MS(+) resulted in an excellent
separation of both metabolites, as presented in the total ion current
(TIC) trace in Fig. 2.9.6(7) together with selected mass traces of PEG
(m/z 300, 344 and 388) and PPG (m/z 442, 500, 558) ($\Delta m/z$ 44 and 58) in
Fig. 2.9.6(1)–(6) [36].

Mono- and di-carboxylate polyethylene glycol (MCPEG and DCPEG)
and PEG besides APEO and nonylphenol ethoxylates (NPEO) were
found in wastewaters that were discharged to surface waters without
any treatment. These compounds were identified after SPE and liquid
chromatographic separation followed by ion-spray mass spectrometric
determination (ESI–LC–MS) in the 'in-source-CID' mode. With an
elevated cone voltage of 80 V, examination was performed resulting in
fragment ions that permitted the confirmation of MCPEGs, DCPEGs
and PEGs [39].

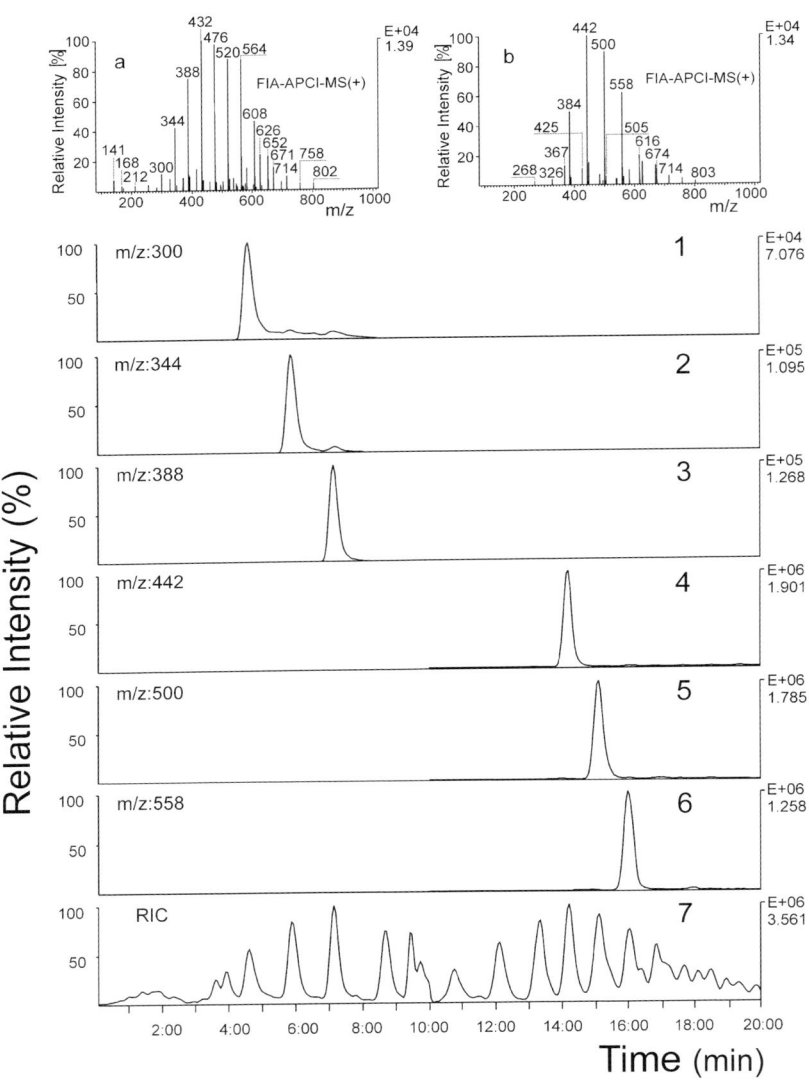

Fig. 2.9.6. (Inset) ESI–FIA–MS(+) overview spectrum of (a) PEG homologues and (b) PPG homologues contained in wastewater effluent SPE extract (7) ESI–LC–MS(+) RIC and selected mass traces of (1)–(3) PEG and (4)–(6) PPG homologues from mixture of (a) and (b); C_{18}-SPE with selective elution, compounds ionised as $[M + NH_4]^+$ ions [36].

A synthetically produced mixture of DCPEG homologues used for calibration purposes in the analysis of negatively ionised compounds by FIA and LC–MS was examined by APCI and ESI in the FIA and LC–MS($+/-$) mode. All FIA–MS spectra were highly complex because of the mixture ionised and in addition the ESI–FIA–MS spectrum contained double-charged ions—recognisable by ions equally spaced with $\Delta m/z$ 22 besides the single-charged ions that were also equally spaced, however, with $\Delta m/z$ 44 as shown in Fig. 2.9.7(a). As observed with the poor separation by LC homologues and isomeric compounds as products of an untargeted synthesis could not be separated from each other, which did not facilitate MS–MS examinations and identification of compounds. For recognition of complexity, mass traces of homologues with $\Delta m/z$ 14 spaced were presented in Fig. 2.9.7. For identification purposes, selection of parent ions before CID in the LC–MS–MS($+$) mode was performed. Identification by CID provided the information that the mixture produced from precursor PEG compounds by an unselective oxidation process consisted of five dominating series of homologues of compounds different in their structural formulae [25]. First, attempts to identify by interpretation of the product ion spectra of the $[M + NH_4]^+$ precursor ions at m/z 592, 606 and 620 were made. The product ion spectra of the selected parent ions are presented in Fig. 2.9.8(a)–(c) together with a general fragmentation scheme amenable to DCPEG [24].

Alkylpolypropyleneglycolethers
The thermal stability of the ions generated from the non-ionic surfactant mixture was examined using alcohol propoxylates (AP) (alkyl polypropylene glycolether). ESI and APCI in the FIA–MS mode was applied for the ionisation of the AP with the general formula $C_nH_{2n+1}-O(CH(CH_3)CH_2O)_mH$, where $n = 7$ and m varied from 2 to 10. An industrial blend containing the compound mixture presented with its general structural formula in Fig. 2.9.9 was studied under positive ESI and APCI ionisation conditions. The blend was spiked into wastewater before it was concentrated by C_{18}-SPE. The cartridges were eluted using sequential selective elution. The first eluting solvent mixture (hexane/ether; 1:1, v/v) [40] contained the compound mixture of AP compounds. For ionisation support in the FIA–MS mode, ammonium acetate was added to the mobile phase. In contrast to results obtained by TSP($+$) ionisation [7,10,40],

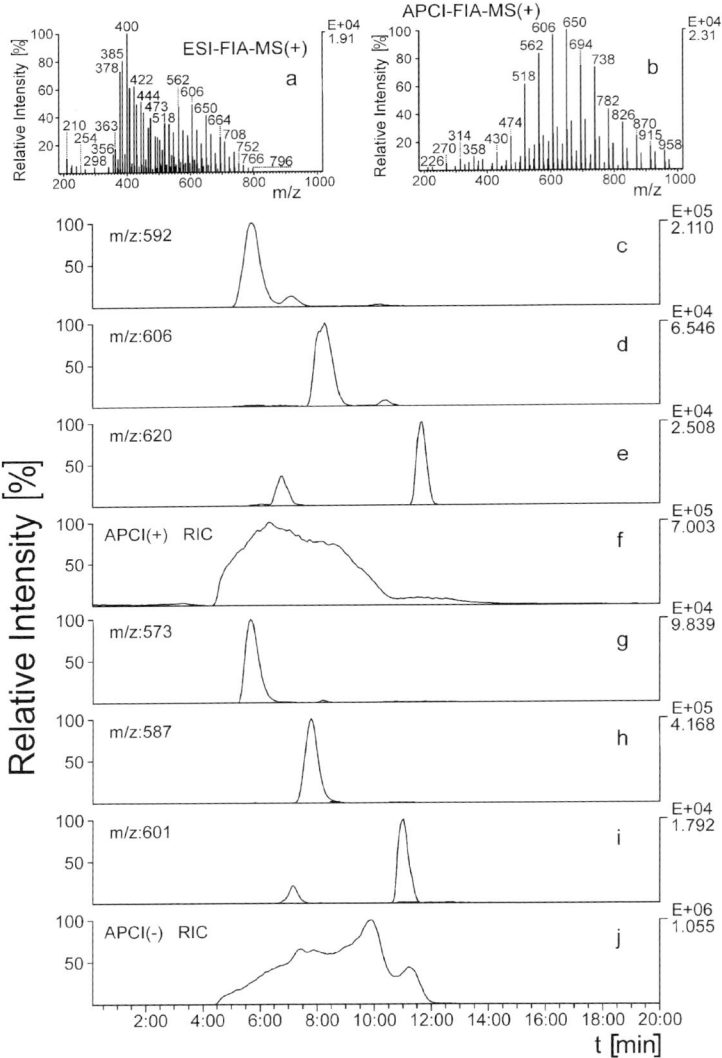

Fig. 2.9.7. (a) ESI–FIA–MS(+) and (b) APCI–FIA–MS(+) overview spectra of synthetically produced mixture of di-carboxylated PEG homologues; (f) APCI–LC–MS(+) and (j) APCI–LC–MS(−) RICs of mixture as in (a,b); (c–e) selected mass traces of di-carboxylated PEG homologues under positive and (g–i) negative ionisation. Gradient elution separated by RP-C_{18} column [24].

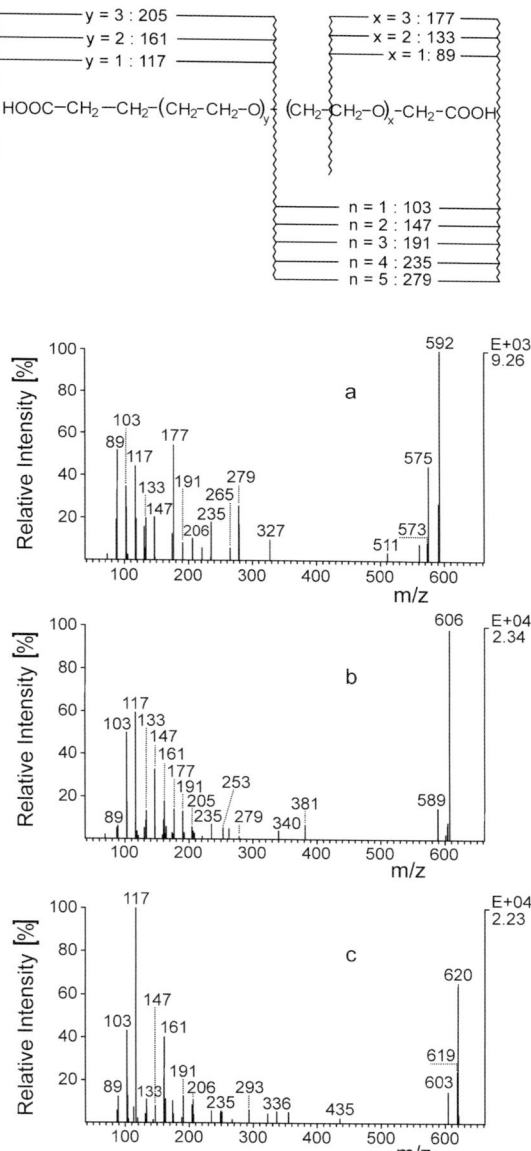

Fig. 2.9.8. (a)–(c) APCI–FIA–MS–MS(+) (CID) product ion mass spectra of selected $[M + NH_4]^+$ parent ions (m/z 592, 606 and 620) of synthetically produced carboxylated PEG homologues; generalised fragmentation behaviour under CID presented in the inset [24].

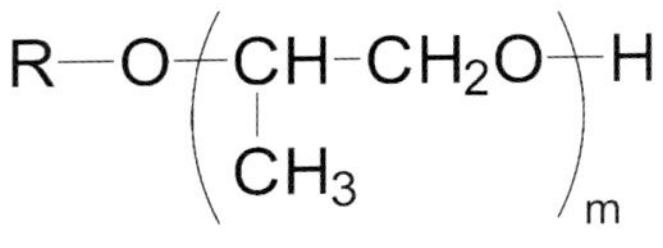

$$R-O+CH-CH_2O+H$$
$$CH_3$$
$$m$$

$$R = C_nH_{2n+1}$$

Fig. 2.9.9. General structural formula of alkyl polypropylene glycol ethers.

FIA-ESI(+) was not successful [16]. Besides $[M + NH_4]^+$ ions ($\bullet$) with equidistant signals ($\Delta m/z$ 58) at m/z 250–946 and $[M + H]^+$ ions ($\star$) at 233–581, $[M + NH_4]^+$ ions of PPG ($\blacktriangle$) at 384–964 ($\Delta m/z$) 58 can also be observed by APCI–FIA–MS(+) (cf. Fig. 2.9.10(a)). Moreover, fragment ions of PPG accentuated by $\downarrow$ in Fig. 2.9.10(a) appeared as dominant ions at 117 and 175 using a standard ion source temperature of 400°C. Reduction in temperature down to 200°C resulted in an overview spectrum more similar to TSI-ionisation [16], which now consisted only of $[M + NH_4]^+$ ($\bullet$) and $[M + H]^+$ ions ($\star$) of the non-ionic AP surfactant homologues (cf. Fig. 2.9.10(b)).

Separation of the C_7-AP blend, as presented with its APCI–FIA–MS(+) overview spectrum in Fig. 2.9.11 was also performed. Signals of the surfactant homologues differing by $\Delta m/z$ 58 could be recognised under optimised ionisation conditions in the overview spectrum. APCI–LC–MS(+) then recorded on a RP-C_{18} LC column [40] resulted in an excellent separation shown in Fig. 2.9.12. Mass spectra of selected peaks at m/z 250 and 656 in insets 1 and 2, respectively, prove the excellent separation efficiency of this method, showing either the $[M + NH_4]^+$ ion of the non-ionic surfactant $C_7H_{15}-O(CH(CH_3)CH_2O)_mH$ with $m = 2$ at m/z 250 and its $[M + H]^+$ ion at m/z 233 or the homologue with $m = 9$ with m/z 656 [16].

The fragmentation behaviour of an industrial blend of AP compounds with the general formula $(C_nH_{2n+1}-O-(CH(CH_3)-CH_2-O)_x-H)$ was observed under API conditions. It was found to be marginally different from the results generated by TSI-MS–MS(+) [7,9]. This mixture of non-ionic surfactants was examined by APCI–FIA–MS–MS(+) resulting in product ion spectra that contained alkyl fragments at m/z 57, 85 and 99 as well as fragments of the PPG chain at m/z 59, 117, 175 and 233. The product ion spectrum, together with the scheme of the

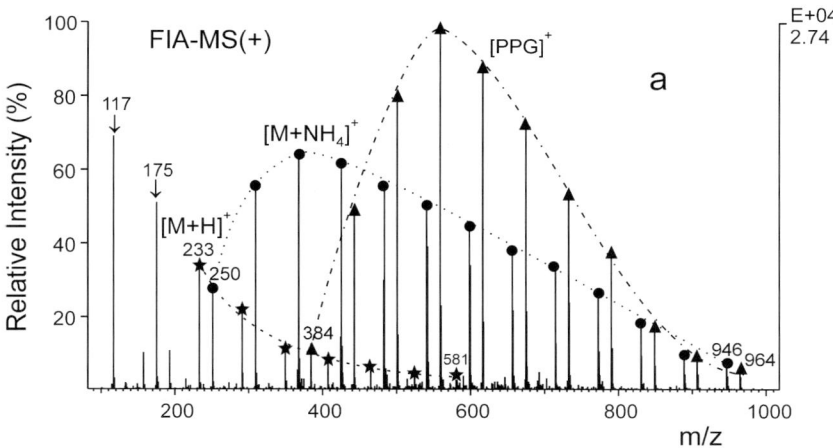

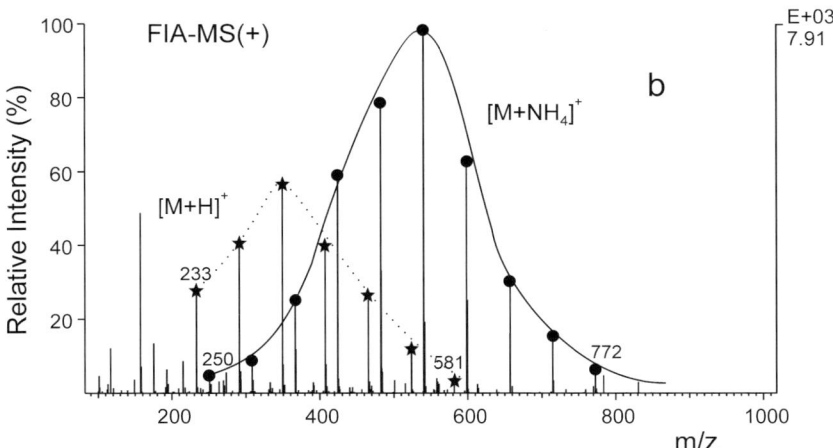

Fig. 2.9.10. Interdependences of temperature and ions examined in the APCI–FIA–MS(+) process. Overview spectra of AP blend ionised at source temperatures ((a) 400°C and (b) 200°C) resulting in $[M + NH_4]^+$ and $[M + H]^+$ ions and $[M + NH_4]^+$ ions of PPG fragments [16].

fragmentation behaviour of the $C_7H_{15}-O-(CH(CH_3)-CH_2-O)_8-H$ homologue is presented in Fig. 2.9.13 [24].

As described with AE biodegradation, ethoxylates such as PEG are the most prominent degradation products that are also observable

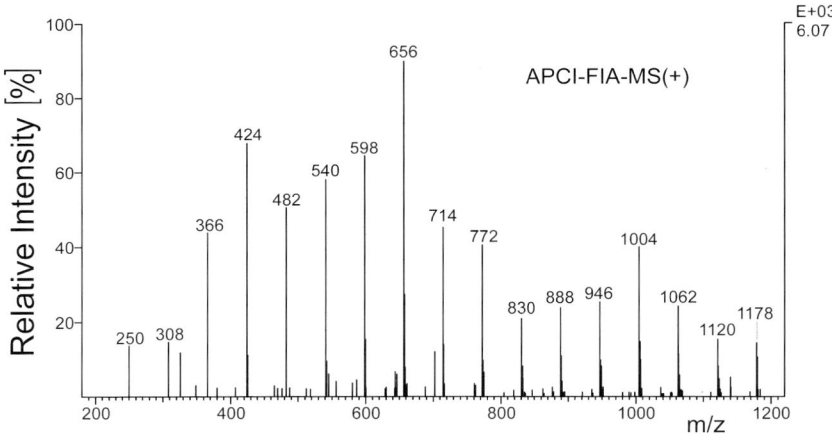

Fig. 2.9.11. APCI–FIA–MS(+) overview spectrum of AP blend (C_nH_{2n+1}–O(CH(CH$_3$)-CH$_2$O)$_m$–H) containing $\Delta\, m/z$ 58 spaced ions [16].

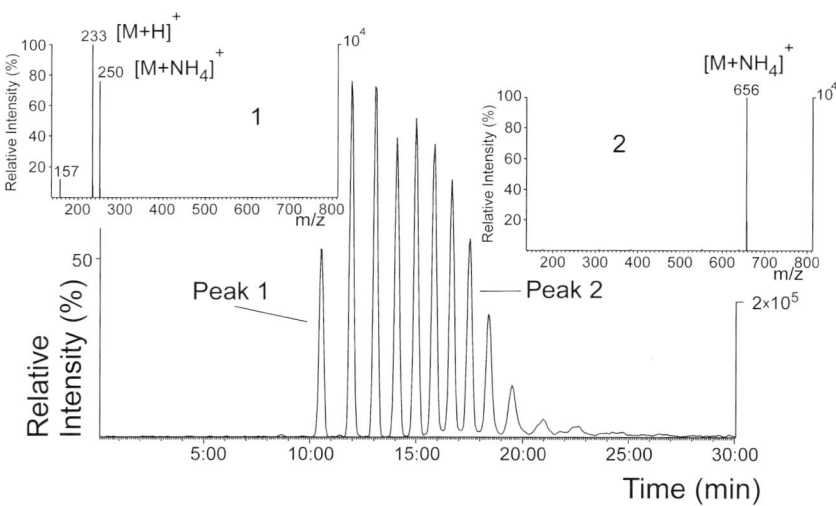

Fig. 2.9.12. (bottom): APCI–LC–MS(+) RIC with insets of averaged mass spectra of selected signals. (top): Separation efficiency obtained by averaged signals is presented in insets 1 and 2 [16].

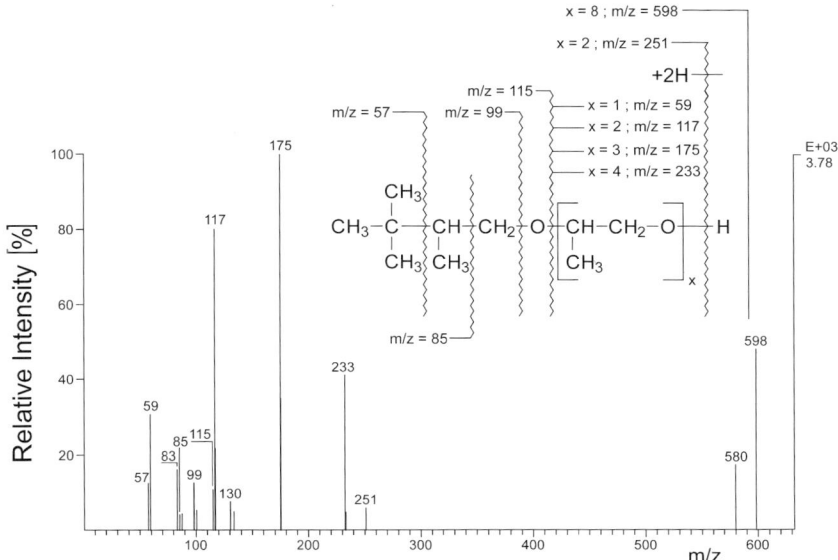

Fig. 2.9.13. APCI–FIA–MS–MS(+) (CID) product ion mass spectrum of unknown parent ion with m/z 598, identified as non-ionic surfactant of AP type (C_nH_{2n+1}–O–(CH(CH$_3$)–CH$_2$–O)$_m$–H); (inset) fragmentation scheme under CID conditions [24].

with the degradation of AP surfactants. So PPG will be generated in the biochemical wastewater treatment process by bond scission between the lipophilic alkyl chain and the hydrophilic ethoxylate moiety. The resulting ethoxylate compounds, PPG, are less polar than PEG but more stabile in the biochemical degradation process. In water samples of the Saale river (Germany) and Thessaloniki (Greece) wastewater treatment plant (WWTP), PEG and PPG as metabolites of ethoxylate-surfactants could be confirmed successfully by APCI–FIA–MS–MS(+). The APCI–FIA product ion spectrum of selected $[M + NH_4]^+$ ions of PPG (general formula: HO–(CH$_2$–CH(CH$_3$)–O)$_x$H) together with its characteristic fragmentation scheme is presented in Fig. 2.9.14 [22,34].

A PPG biodegradation product as metabolite of PPG generated by metabolisation of AP could be observed and verified by MS–MS(+) when WWTP effluents were examined by APCI–FIA–MS(+) and selected compounds were identified by FIA–MS–MS(+). In the SPE effluent extracts, two ions at m/z 266 and 324 were observed, which could be confirmed as homologues because of their identical fragmentation pattern [22]. Compounds could be characterised as secondary metabolites of the biochemical degradation product 'PPG'. Applying the diagnostic parent ion scan m/z 115 [35] to the mixture containing these compounds, ions at m/z 266 and 324 can be observed [22]. By analogy with alkyl polyethylene glycol ethers [40], PPG had been degraded resulting in short chain aldehyde homologues with the general formula $HO-(CH(CH_3)-CH_2-O)_x-CH(CH_3)-CHO$ ($x = 3$ or 4). These metabolites could be ionised as $[M + NH_4]^+$ ions at m/z 266 or 324 when APCI–FIA–MS(+) was applied. CID of the ion at m/z 266 led to fragments that are presented as product ions and fragmentation behaviour in Fig. 2.9.15 [22].

Mixed EO/PO compounds
AE and AP surfactants are synthesised by polymerisation of monomeric or short chain oligomeric glycol moieties. If mixtures of ethylene and propylene glycols, however, were applied for the synthesis, so called EO/PO block polymers in the form of complex mixtures with the general formula $R-O-(EO)_m-(PO)_n-H$ and the structure as shown with their

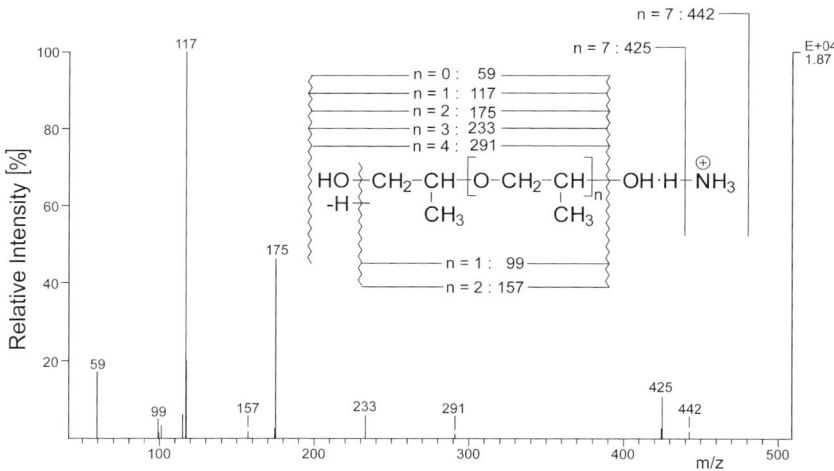

Fig. 2.9.14. APCI–FIA–MS–MS(+) (CID) product ion mass spectrum of selected $[M + NH_4]^+$ parent ion (m/z 442) of non-ionic surfactant metabolite observed in C_{18}-SPE eluates; PPG homologue (general formula: $HO-(CH_2-CH(CH_3)-O)_x-H$); fragmentation behaviour under CID presented in the inset [35].

248

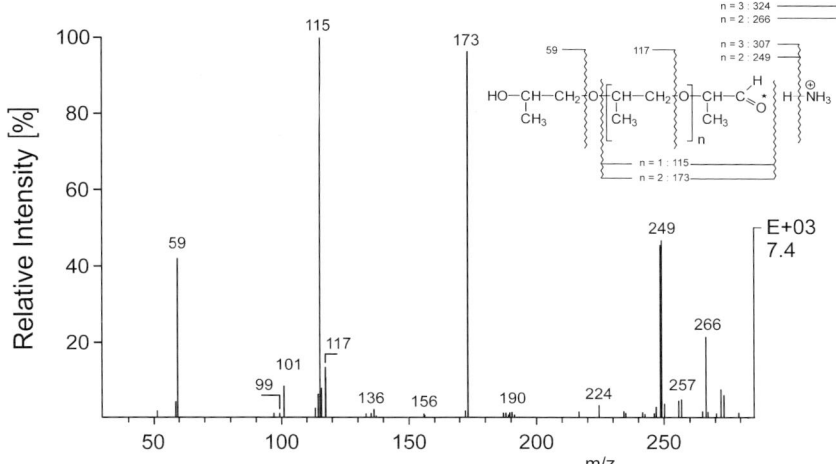

Fig. 2.9.15. APCI–FIA–MS–MSMS–MS(+) (CID) product ion mass spectrum of selected $[M + NH_4]^+$ parent ion (m/z 266); compound was identified as metabolite (aldehyde compound) of degradation product of non-ionic surfactant compound PPG (general formula: $HO–(CH(CH_3)–CH_2–O)_x–CH(CH_3)–CHO$ ($x = 3$); fragmentation behaviour under CID presented in the inset [22].

structural formula in Fig. 2.9.16, will be generated [4]. Because of the varying number of m and n and the variation in the number of carbon atoms in the alkyl chains, the number of isomers and homologues generated gives rise to the APCI–FIA–MS(+) spectrum which is presented in Fig. 2.9.17 [16]. Obviously the identification of these EO/PO-block polymers can be considered as very difficult. Even the detection of these surfactants is quite complicated because they cannot be monitored by optical systems because of a missing chromophore in the molecules. These compounds are hardly degradable in biochemical degradation processes and therefore accumulation cannot be excluded. Their acute toxicity against water organisms was found to be low [14].

As 'separation' in FIA–MS by-passing the analytical column shows separation because of different m/z ratios, isomeric (isobaric)

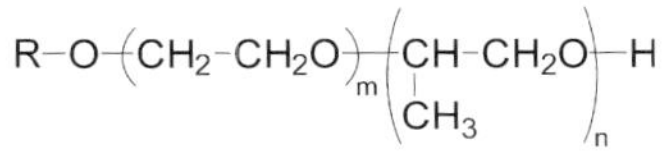

$$R–O–(CH_2–CH_2O)_m–(CH–CH_2O)_n–H$$
$$\qquad\qquad\qquad\qquad CH_3$$

Fig. 2.9.16. General structural formula of alkyl polyethylene–polypropylene glycolethers (EO/PO compounds).

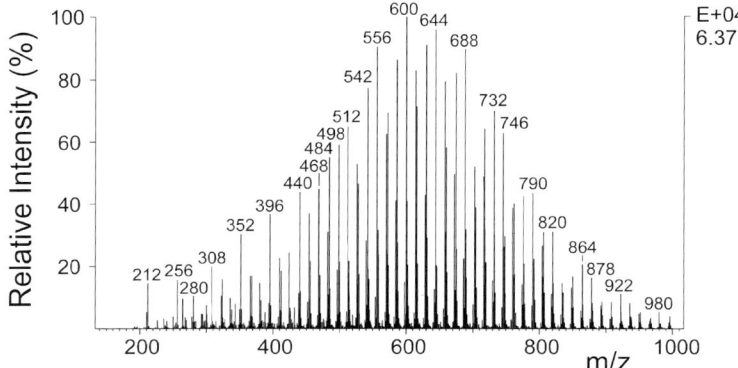

Fig. 2.9.17. APCI–FIA–MS(+) overview spectrum of fatty acid EO/PO polyglycolether blend ionised as $[M + NH_4]^+$ ions [16].

compounds with the identical molar mass cannot be differentiated. Here, only the separation on analytical columns combined with MS detection (LC–MS) may help to differentiate between isomers with a variation in the number of alkyl chain links and/or different EO/PO compositions as presented in Table 2.9.1.

The RIC of LC separation performed in a very time consuming manner using RP-C_{18} was not very convincing, whereas mass trace analysis showed a separation [16]. As an example of the results obtained

TABLE 2.9.1

Possible combinations in the molecules of the EO/PO surfactant blend with the general formula $C_nH_{2n+1}O-(CH_2-CH(CH_3)-O)_y-(CH_2-CH_2-O)_x-H$ ionised by FIA–MS(+) and presented in Fig. 2.9.17; $(-CH_2-CH(CH_3)-O-) = PPG$; $(-CH_2-CH_2-O-) = PEG$. Selection according to constant m/z ratio or constant number of alkyl chain links

$[M + NH_4]^+$ (m/z)	n $(R = C_nH_{2n+1})$	x $((PEG)_x)$	y $((PPG)_y)$
600	9	6	3
600	10	7	2
600	11	8	1
600	12	9	0
n $(R = C_nH_{2n+1})$	$[M + NH_4]^+$ (m/z)	x $((PEG)_x)$	y $((PPG)_y)$
9	586	7	2
9	600	6	3
9	614	5	4
9	628	4	5

by LC separation, only the $[M + NH_4]^+$ mass traces of four pure alternatively EO ($\Delta m/z$ 44) or PO ($\Delta m/z$ 58) containing homologue compounds were plotted together with the total ion mass trace in Fig. 2.9.18(a)–(e). In all cases presented in the selected mass traces the alkyl chain was fixed consisting of a chain that contained 10 carbon atoms ($R = C_{10}H_{21}-$). The LC method applied was able to separate the compounds roughly according to the increasing PO-number, as the APCI(+)–TIC proves. Mass trace analysis of $R-O-(PO)_y-H$ with m/z 292 and 350 ($y = 2$ or 3) confirms the separation of compounds containing PO-units, however, the EO-compounds with m/z 264 and 308 ($x = 2$ or 3) are hidden behind only one signal showing how the drift that increased EO numbers reduce retention time as also reported in the literature [22].

A special type of surfactant compounds, so-called H-active compounds consisting of condensation products of EO/PO polymers with com-

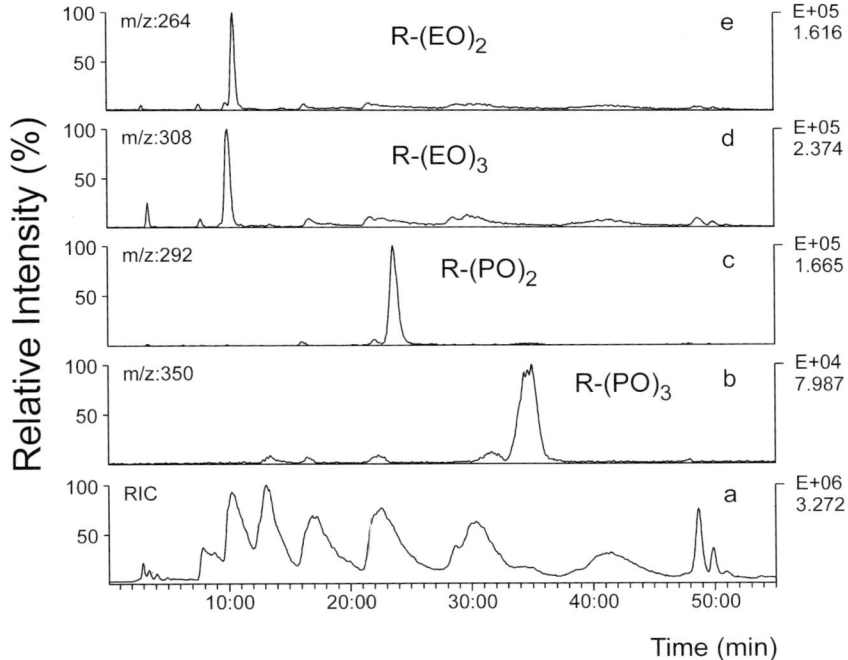

Fig. 2.9.18. (a) APCI–LC–MS(+) RIC of fatty acid EO/PO polyglycolether blend as in Fig. 2.9.17(b)–(e) selected mass traces as in (a) [16].

pounds having acidic protons, was screened applying APCI–FIA–MS(+), resulting in a complex mixture of compounds ionised in the form of [M + NH$_4$]$^+$ ions (cf. Fig. 2.9.19). The LC separation performed on a RP-C$_{18}$ column using methanol/water in combination with gradient

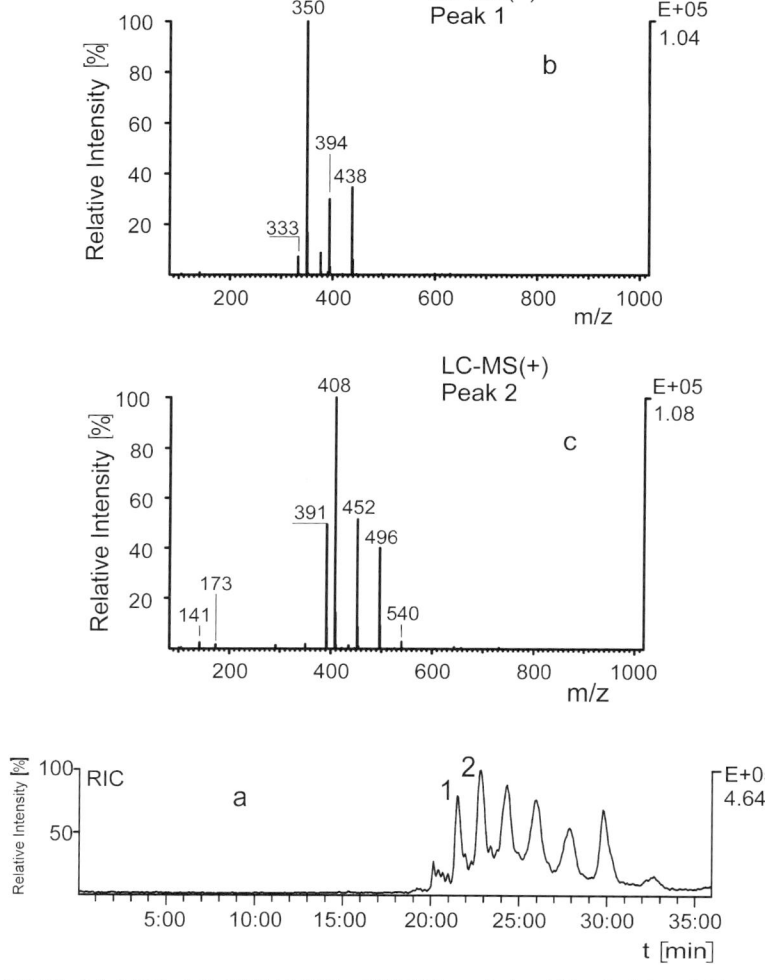

Fig. 2.9.19. (a) APCI–LC–MS(+) RIC of EO/PO surfactant blend; (b) and (c) averaged LC–MS(+) spectra of peaks 1 and 2 as in (a). Gradient elution separated on RP-C$_{18}$ column [24].

elution and APCI–MS(+) determination led to the chromatogram shown as RIC (Fig. 2.9.19(a) and selected mass spectra in Fig. 2.9.19(b) and (c)). The mass trace analysis of ions equally spaced with $\Delta m/z$ 58 proved that the surfactant compounds were separated according to the number of propylene oxide units in the molecules. Homologues compound ions with $\Delta m/z$ 44 also equally spaced, were hidden behind these Δ 58 spaced signals as presented in the MS spectra of peak 1 (b) and 2 (c). Signals differing by $\Delta m/z$ 17 could be identified as $[M + H]^+$ ions [24].

An industrial blend of the EO/PO block polymer with the general formula $C_nH_{2n+1}-O-(EO)_x-(PO)_y-H$ was examined under APCI–FIA–MS–MS(+) conditions. A proposition for the fragmentation behaviour observed under CID conditions is presented in the inset of the product ion spectrum shown in Fig. 2.9.20. The statistical variations of the EO and PO units in the ethoxylate chain of the parent ions of isomers with m/z 598 under CID conditions resulted in a product ion spectrum that contained, in addition to the alkyl fragments at m/z 57, 71, 85, 99, 113 and 141, PPG fragments at m/z 59, 117 and 175 and PEG fragments at m/z 89, 133 and 177. Mixed EO/PO ions at m/z 191 (EO$_3$–PO$_1$) and m/z 249 (EO$_3$–PO$_2$) from EO/PO-chain could also be observed. Furthermore fragments containing the alcohol moiety $C_{10}H_{21}-O-$ combined with ethoxylate chain units were generated which could be recorded as ions at m/z 199 (PO$_1$), 257 (PO$_2$) and 301 (PO$_2$–EO$_1$) [24]. The information about the sequence in which these mixed EO/PO moieties were linked together has not yet been elucidated.

Methylethers of mixed EO/PO compounds
Biodegradation experiments performed with a non-ionic surfactant blend as shown in Fig. 2.9.21 proved that a slight chemical modification—substitution of H by a methyl group—in the easily degradable AE molecules, leads to a persistent surfactant mixture. Methylation of the non-ionic EO/PO surfactant blend with the formula $C_nH_{2n+1}O-$ $(CH_2-CH_2-O)_x-(CH_2-CH(CH_3)-O)_y-H$ resulted in the methyl ether blend $C_nH_{2n+1}O-(CH_2-CH_2-O)_x-(CH_2-CH(CH_3)-O)_y-CH_3$ persistent against aerobic treatment over a period of more than 6 weeks [37]. This surfactant mixture was found to be highly toxic for water organisms [14].

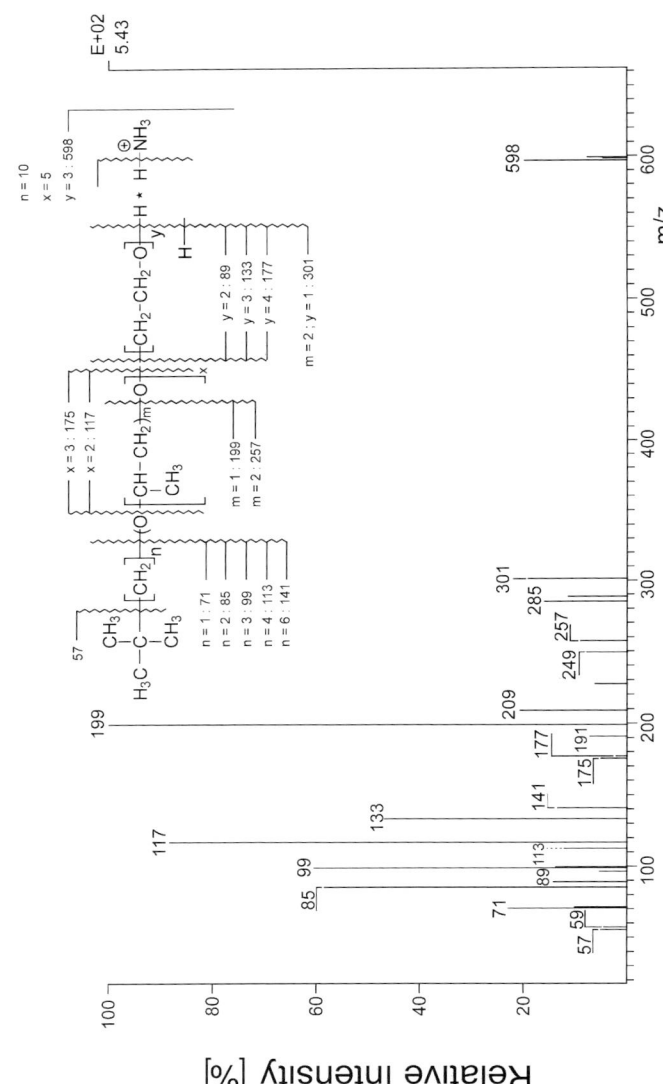

Fig. 2.9.20. APCI–FIA–MS–MS(+) (CID) product ion mass spectrum of $[M + NH_4]^+$ parent ion of EO/PO surfactant blend $(C_nH_{2n+1}–O–(EO)_x–(PO)_y–H)$. Selected parent compound ion: m/z 598; (inset) fragmentation scheme under CID conditions [24].

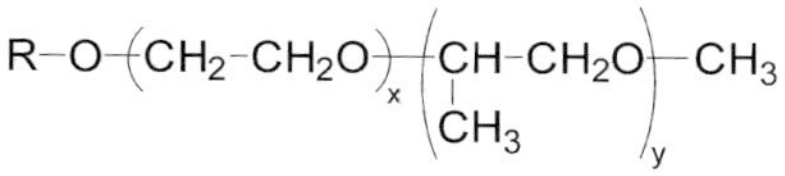

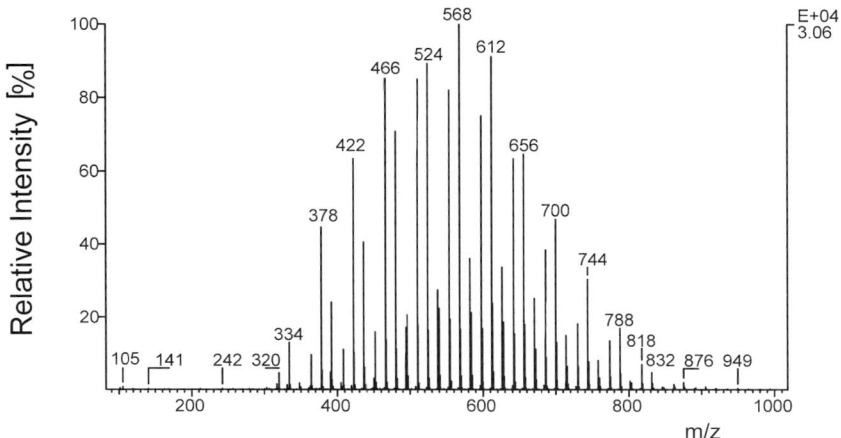

$$R-O-(CH_2-CH_2O)_x-\left(\begin{array}{c}CH-CH_2O\\ |\\ CH_3\end{array}\right)_y-CH_3$$

Fig. 2.9.21. General structural formula of methylated alkyl polyethylene–polypropylene glycol ethers (methyl ethers of EO/PO compounds).

The compounds are shown with their APCI–FIA–MS(+) spectrum in Fig. 2.9.22 ionised in the form of $[M + NH_4]^+$ ions. The compounds that had been examined have the general structural formula $C_nH_{2n+1}O-(CH_2-CH_2-O)_x-(CH_2-CH(CH_3)-O)_y-CH_3$. The compound with $n = 12$; $x = 3$ and $y = 2$ and other compounds that can be composed as presented in Table 2.9.2 will be ionised as $[M + NH_4]^+$ ions at m/z 466.

The possible combinations of PEG and PPG in the ethoxylate chain of the surfactant molecules are demonstrated in Table 2.9.2 according to the FIA–MS results in Fig. 2.9.22. The complexity of the FIA spectra is recognisable, but the complexity of possible isomers by variation in the composition of the molecules—C_n, PEG, can only be recognised in Table 2.9.2.

The structure implicates biochemical persistence, which was confirmed by FIA–MS and FIA–MS–MS(+) applying the pattern recognition and the CID approach. LC separation of this complex mixture of isomers and homologues under ion-pairing conditions, however, failed. APCI–LC–MS(+) resulted in broad, unresolved signals [37].

Fig. 2.9.22. APCI–FIA–MS(+) overview spectrum of methylated EO/PO polyglycolether blend ionised in the form of $[M + NH_4]^+$ ions [37].

TABLE 2.9.2

Possible combinations in the molecules of the methylated EO/PO surfactant blend with the general formula $C_nH_{2n+1}O-(CH_2-CH(CH_3)-O)_y-(CH_2-CH_2-O)_x-CH_3$ ionised by FIA–MS(+) and presented in Fig. 2.9.22; $(-CH_2-CH(CH_3)-O-)$ = PPG; $(-CH_2-CH_2-O-)$ = PEG). Selection according to constant m/z ratio or constant x/y ratio

$[M + NH_4]^+$ (m/z)	n $(R = C_nH_{2n+1})$	x $((PEG)_x)$	y $((PPG)_y)$
466	9	0	5
466	10	1	4
466	11	2	3
466	12	3	2
466	13	4	1
466	14	5	0

x $((PEG)_x)$	y $((PPG)_y)$	n $(R = C_nH_{2n+1})$	$[M + NH_4]^+$ (m/z)
3	2	9	424
3	2	10	438
3	2	11	452
3	2	12	466
3	2	13	470
3	2	14	494
3	2	15	508

The homologues of the methylated non-ionic EO/PO surfactant blend were ionised as $[M + NH_4]^+$ ions. A mixture of these isomeric compounds, which could not be defined by their structure because separation was impossible, was ionised with its $[M + NH_4]^+$ ion at m/z 568. The mixture of different ions hidden behind this defined m/z ratio was submitted to fragmentation by the application of APCI–FIA–MS–MS(+). The product ion spectrum of the selected isomer as shown with its structure in Fig. 2.9.23 is presented together with the interpretation of the fragmentation behaviour of the isomer. One of the main difficulties that complicated the determination of the structure was that one EO unit in the ethoxylate chain in combination with an additional methylene group in the alkyl chain is equivalent to one PO unit in the ethoxylate chain (cf. table of structural combinations). The overview spectrum of the blend was complex because of this variation in homologues and isomers. The product ion spectrum was also complex, because product ions obtained by FIA from isomers with different EO/PO sequences could be observed complicating the spectrum. The statistical variations of the EO and PO units in the ethoxylate chain of the parent ions of isomers with m/z 568 under CID

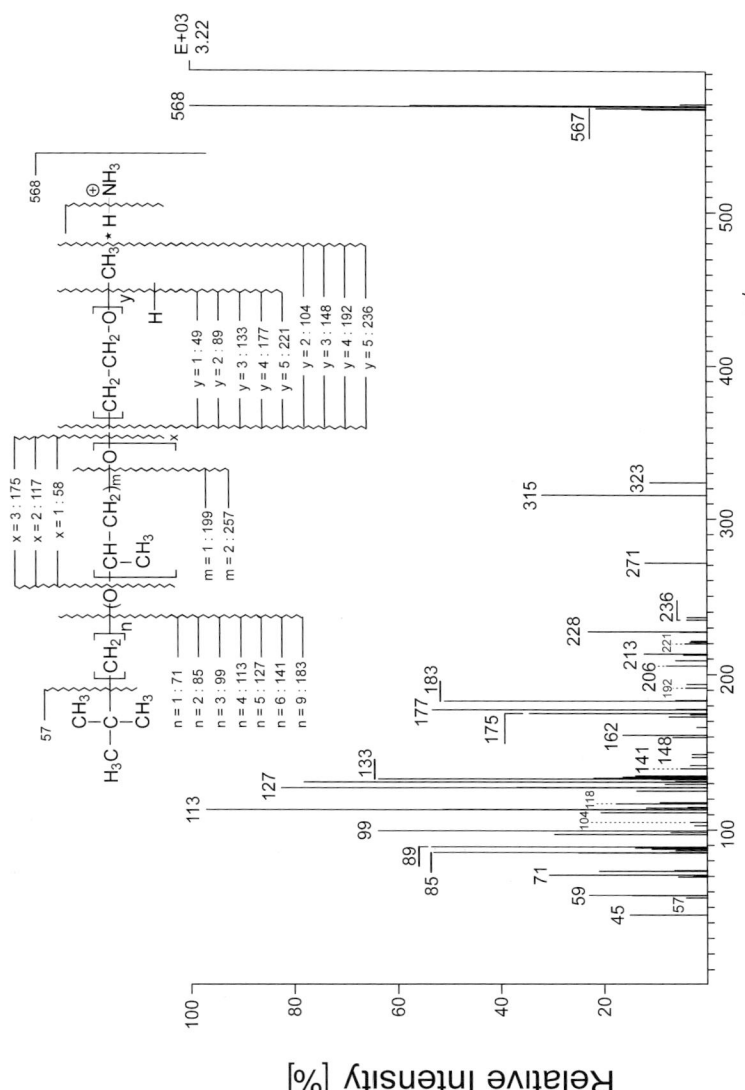

Fig. 2.9.23. APCI–FIA–MS–MS(+) (CID) product ion mass spectrum of $[M + NH_4]^+$ ion of methylated EO/PO surfactant compound $(C_nH_{2n+1}O-(CH_2-CH(CH_3)-O)_y-(CH_2-CH_2-O)_x-CH_3)$ with m/z 568; (inset) fragmentation scheme under CID conditions [24].

conditions resulted in a product ion spectrum that contained on one hand the alkyl fragments at m/z 57, 71, 85, 99, 113, 127, 141 and 183, and on the other hand PPG fragments at m/z 59, 117 and 175 and PEG fragments at m/z 89, 133 and 177. Besides these fragments, methylated (EO_x)-fragments at 104, 148, 192 and 236 and methylated mixed (EO_x-PO)-fragments were also found at m/z 162 $(EO_2-PO_1-CH_3)$ and m/z 206 $(EO_3-PO_1-CH_3)$ [24]. The information about the sequence, how these mixed EO−PO moieties were linked together, will be the job of experiments using trap MS^n experiments following up fragmentation.

Fatty acid polyglycol esters
ESI−LC−MS(+) as well as ESI−FIA−MS(+) [41] was used for the investigation of polyethoxylated fatty acid esters shown with their structural formula in Fig. 2.9.24. Impurities originating from the technical scale synthesis process could be observed in the overview spectrum. For this purpose, non-ionic surfactants of this type were ionised in the FIA−MS mode resulting in equal-spaced signals with $\Delta m/z$ 44 because of the ethoxylate units contained in the molecules. Both techniques provide detailed characterisation of the constituents of these surfactant blends [42]. Using FIA and LC−MS, PEG ($HO(CH_2-CH_2O)_mH$) could be found in varying quantities besides fatty acid polyglycol esters. LC provided a means of solving the problems with interfering molecular ions, PEG and non-ionic surfactants, observed by FIA−MS in the overview spectra obtained in quality assurance process. Depending on the application of the blends, precursor PEG may cause various problems [42].

An industrial blend of polyethoxylated fatty acid esters was examined by APCI−FIA−MS(+) and −LC−MS(+). The screening approach by FIA−MS confirmed that the industrial blend contained small amounts of impurities from synthesis as recognisable from Fig. 2.9.25.

The separation of this mixture performed on a RP-C_8 column applying a gradient elution led to a separation of the surfactant mixture. Two peaks can be recognised in the TIC trace in Fig. 2.9.25(d) consisting of a

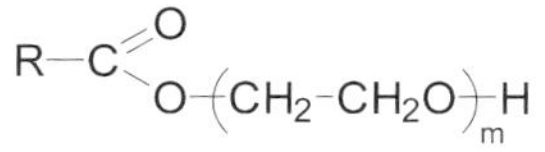

$$R-C\overset{\displaystyle O}{\underset{\displaystyle O-(CH_2-CH_2O)_m-H}{\Big\langle}}$$

Fig. 2.9.24. General structural formula of fatty acid polyethyleneglycolesters.

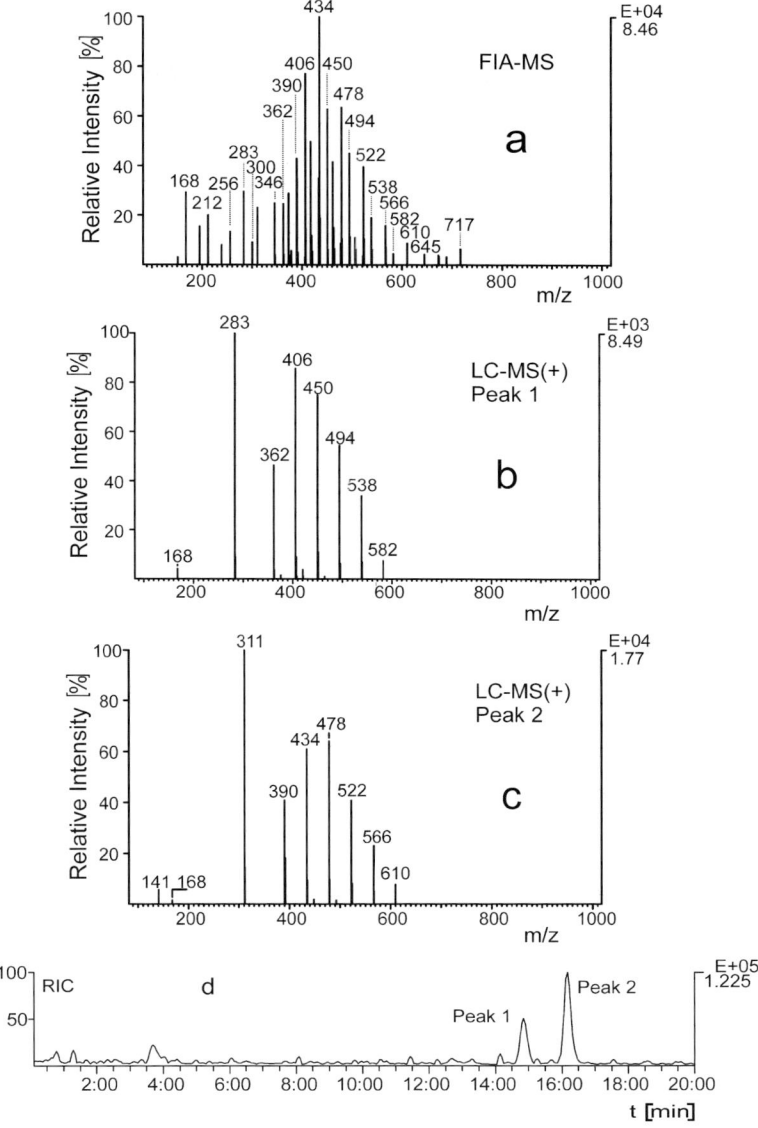

Fig. 2.9.25. (a) ESI–FIA–MS(+) overview spectrum of polyethoxylated fatty acid ester blend (R–C(O)O(CH$_2$–CH$_2$–O)$_x$H) ionised in the form of [M + NH$_4$]$^+$ ions; (d) ESI–LC–MS(+) RIC and (b) and (c) averaged mass spectra of peaks 1 and 2 as in (c). Gradient elution separated on RP-C$_{18}$ column [24].

series of homologues with equally spaced ions ($\Delta m/z$ 44) as shown in Fig. 2.9.25(b) and (c). The surfactants were analysed as $[M + NH_4]^+$ ions. The adduct ions at m/z 362, 406,...,538 and 582 in (b) represent the compounds with the general formula $C_nH_{2n+1}-C(O)O(CH_2-CH_2-O)_xH$ with $n = 15$ and $x = 1-6$, whereas the ions at m/z 390, 434,...,566 and 610 in (c) were generated from homologues with $n = 17$ and $x = 1-6$. The prominent ions at m/z 283 and 311 hidden in the peaks of the polyglycol esters were recognised as fragment ions [24].

An industrial blend of a non-ionic surfactant mixture of fatty acid polyglycolesters with the general formula $C_nH_{2n+1}-C(O)-O-(CH_2-CH_2-O)_m-H$ was examined by ESI–FIA–MS–MS(+) in the product ion mode (Fig. 2.9.26). MS–MS applied to the $[M + NH_4]^+$ parent ion at m/z 406 led to the identification of the homologue compound $C_{15}H_{31}-C(O)-O-(CH_2-CH_2-O)_3-H$, which belonged to a series of ions, all equally spaced with $\Delta m/z$ 44 because of their different PEG chain lengths. Only one single product ion at m/z 283 was obtained from the parent ion at m/z 406 as well as from the other homologue ions ($\Delta m/z$ 44) by scission of the PEG chain and ring closure resulting in the cyclic ion as shown in the inset of Fig. 2.9.26. Vice versa, the parent ion scan of 283 was used to prove the presence of fatty acid polyglycolesters in an environmental water sample, which contained

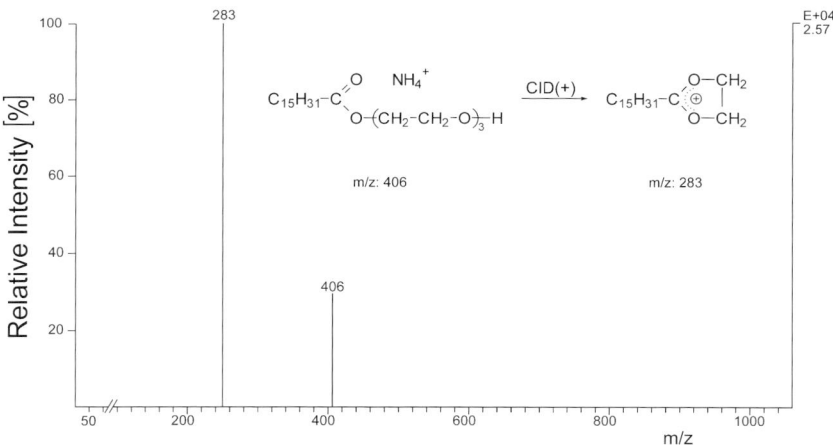

Fig. 2.9.26. ESI–FIA–MS–MS(+) (CID) product ion mass spectrum of $[M + NH_4]^+$ parent ion of fatty acid polyglycolester surfactant blend ($C_nH_{2n+1}-C(O)-O-(CH_2-CH_2-O)_m-H$); selected ion under CID: m/z 406; fragmentation behaviour in the inset [24].

non-ionic surfactants. Ions of fatty acid polyglycolesters at m/z 362, 406, 450, 494, 538 and 582 could be observed under these conditions in Fig. 2.9.27. Fatty acid polyglycolesters with different PEG chain lengths ($m = 2-7$) ionised as parent ions were present in the solution [16].

Results of CID–MS–MS enabled the authors [33] to identify surfactants of fatty acid polyglycolester type in Tween 40 and Myrj 40 formulations together with AE surfactant homologues. These blends were applied in biochemistry as purification reagents. Product ion spectra were generated by mixture analysis (FIA–MS–MS(+)). First MS for detection and then MS–MS for identification was performed using API–FIA. The characteristic product ion at m/z 283 ($[CH_3(CH_2)_{14}CO(CH_2CH_2O)_n]^+$) could be observed proving the explanation of the fragmentation behaviour as shown in the inset of Fig. 2.9.26 [16].

Fatty acid and unsaturated fatty acid mono- and diethanolamides
Fatty acid mono- (FAMA) and diethanol amides (FADA) with the general formula ($C_nH_{2n+1}-C(O)N(H)_x(CH_2-CH_2-OH)_{2-x}$; $x = 1$ or 0) are presented with their general structural formulae in Fig. 2.9.28(a) and (b). These surfactants have found a widespread application in household and personal care formulations because of their quite good

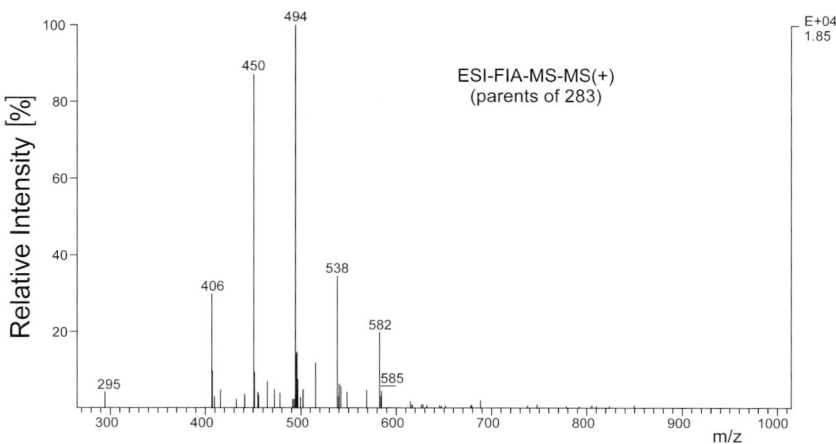

Fig. 2.9.27. ESI–FIA–MS–MS(+) (CID) parent ion mass spectrum of product ion at m/z 283 of C_{18}-SPE extract presenting fatty acid polyglycolester characteristic ions at m/z 406, 450, 494, 538 and 582, all equally spaced with $\Delta\, m/z$ 44 [16].

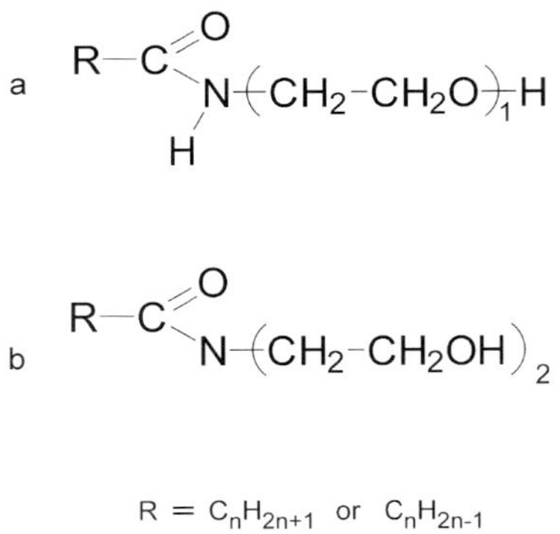

Fig. 2.9.28. General structural formulae of saturated and unsaturated FAMA or FADA compounds, respectively.

performance and their easy degradability in the biological sewage treatment process [7]. As nitrogen-containing compounds, the ions of the surfactant homologues were generated by APCI–FIA–MS(+) and ESI–FIA–MS(+) with identical patterns of homologues in the form of their $[M + H]^+$ ions.

The FAMA blend ($x = 1$) as presented in the APCI–FIA(+) spectrum in Fig. 2.9.29(a) contained two dominating compounds with ions at m/z 244 ($n = 11$) and 272 ($n = 13$) besides the C_9 and C_{15} homologues, although, in very low quantities (m/z 216 and 300). As presented in Fig. 2.9.29(c)–(f) besides these, homologue compounds with $n = 9$ and 15, also recognisable in the FIA–MS(+) spectrum, could be observed in the form of their mass traces at m/z 216 (c) or 300 (f), respectively, if APCI–LC–MS with a separation on a RP-C_{18} column was applied. The TIC is shown in Fig. 2.9.29(b) [24]. Dimeric ions ($[2M + H]^+$) could also be observed in the FIA–MS overview spectrum (cf. Fig. 2.9.29(a)) consisting of either two C_{11} (m/z 487), one C_{11} and one C_{13} (m/z 515) or two C_{13} homologues (m/z 543).

The LC separation of an industrial blend of FADA homologues (C_nH_{2n+1}–$C(O)N(H)_x(CH_2$–CH_2–$OH)_{2-x}$) as shown in the overview spectrum in Fig. 2.9.30(a) ($x = 1$) was performed on an RP-C_{18} column and is presented in Fig. 2.9.30(b)–(e). $[M + H]^+$ ions at m/z 232 ($n = 7$), 260 ($n = 9$), 288 ($n = 11$), 316 ($n = 13$), 344 ($n = 15$) and 372 ($n = 17$) were obtained in the ESI–LC–MS(+) mode [16]. The separation efficiency obtained under gradient elution conditions was checked and was demonstrated by the averaged mass spectrum [16]. Extraction

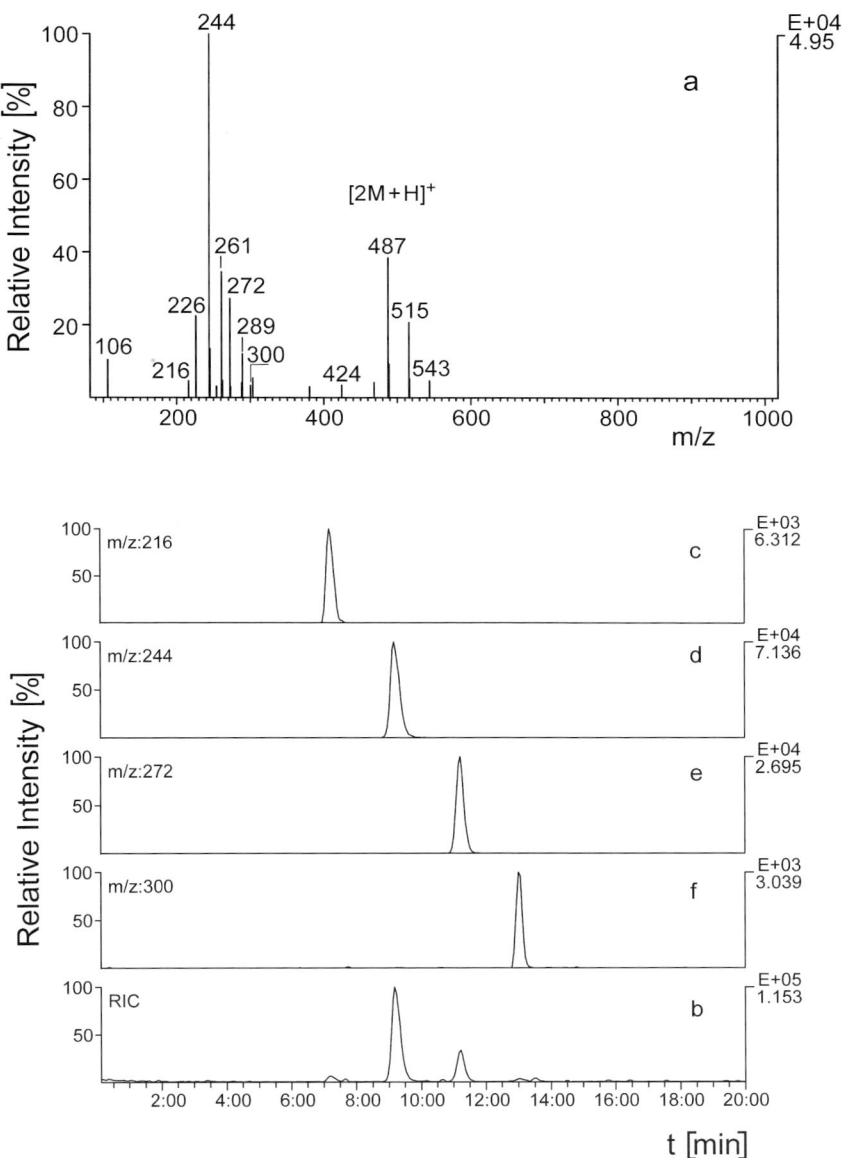

Fig. 2.9.29. (a) APCI–FIA–MS(+) overview spectrum of FAMA blend; (b) APCI–LC–MS(+) RIC of FAMA blend (C_nH_{2n+1}–C(O)NH(CH$_2$–CH$_2$–OH)) and selected mass traces (c)–(f) ($m/z = 216$, 244, 272 and 300; $\Delta\,m/z$ 28) [24].

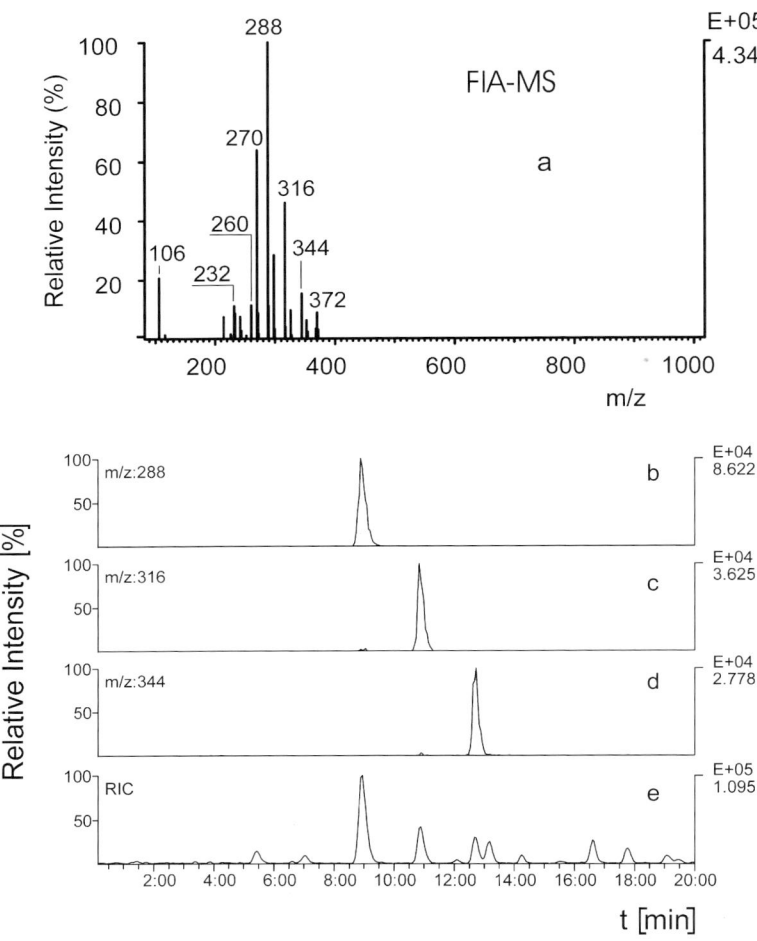

Fig. 2.9.30. (a) ESI–FIA–MS(+) overview spectrum of fatty acid diethanolamide blend $[M + H]^+$ ions; (e) ESI–LC–MS(+) RIC and (b)–(d) selected mass traces of $m/z = 288$, 316 and 344; $\Delta m/z$ 28. Gradient elution separated on RP-C_{18} column [16].

and concentration of this mixture from environmental samples was possible by C_{18}-SPE with methanol elution after preceding sequential elution steps applying hexane/ether, ether and methanol/water as reported in the literature [9].

The fragmentation behaviour of FAMA and FADA, however, differs quite obviously. While FAMA compounds under ESI–FIA–MS–MS(+) conditions resulted in alkyl fragments, these ions were not observed with FADA surfactants. The FAMA compounds with the general formula $C_nH_{2n+1}-C(O)-N(H)-CH_2-CH_2-OH$ ($n = 9$, 11, 13, 15) could be ionised with parent ions at m/z 216, 244, 272 and 300. CID then resulted in alkyl fragments at m/z 57, 71, 85 and 113 [16]. The same behaviour—loss of 61 under MS–MS conditions (m/z 244 → 183) as observed with FADA—could also be confirmed with FAMA. The fragment ion at m/z 62 was the protonated monoethanolamine ($[H_3N^{\oplus}-CH_2-CH_2OH]$). As fragmentation by abstraction of water (loss of 18), which was observed with diethanolamides here was not possible, the neutral NH_3 and the NH_4^+ ions were abstracted from the parent ion. ESI–FIA–MS–MS(+) applied to monoethanolamide blend resulted in the fragmentation behaviour shown in the inset of the product ion spectrum in Fig. 2.9.31.

FADA compounds with the general formula $C_nH_{2n+1}-C(O)-N(CH_2-CH_2-OH)_m$; ($n = 9$, 11, 13, 15; $m = 2$) are present in numerous detergent formulations. They could be easily identified by their characteristic product ion spectra obtained by ESI–FIA–MS–MS(+) as also observed under TSP–MS–MS(+) conditions. Product ion spectra contain fragments with m/z 70, 88, 106 and 227 [7,9,43] as presented in Fig. 2.9.32. The elimination of the neutral 61 ($[H_2N-CH_2-CH_2OH]$) was the most important reaction generating the fragment ion m/z 227 if $n = 11$. This reaction was observed in CID processes as a gas-phase rearrangement [16,44], as shown in the inset of Fig. 2.9.32.

The MS^n analysis using ESI was applied for the determination of an unknown surfactant compound contained in an extract of a shampoo formulation [44]. MS^n leading to sequential product ions helped to identify the constituents. The MS^4 experiments together with other spectral observations confirmed the hypothesis that the unknown compound was a N-(2-aminoethyl) fatty acid amide with the general formula $R-C(O)-NH(CH_2-CH_2-N)R'R''$. An authentic sample of the proposed laury ampho mono acetate (LAMA) ($R' = -CH_2-CH_2-OH$ and $R'' = -CH_2-CH_2-COOH$) that was available led to the same $[M + H]^+$ parent ion at m/z 345. The fragmentation that could be observed under ESI–FIA–MS–MS(+) conditions resulted in an intensive examination of amide surfactants. However, only two of them—lauryl diethanol amide ($[M + H]^+$: m/z 288), a non-ionic surfactant and laurylamido-β-propyl betaine ($[M + H]^+$: m/z 343)—

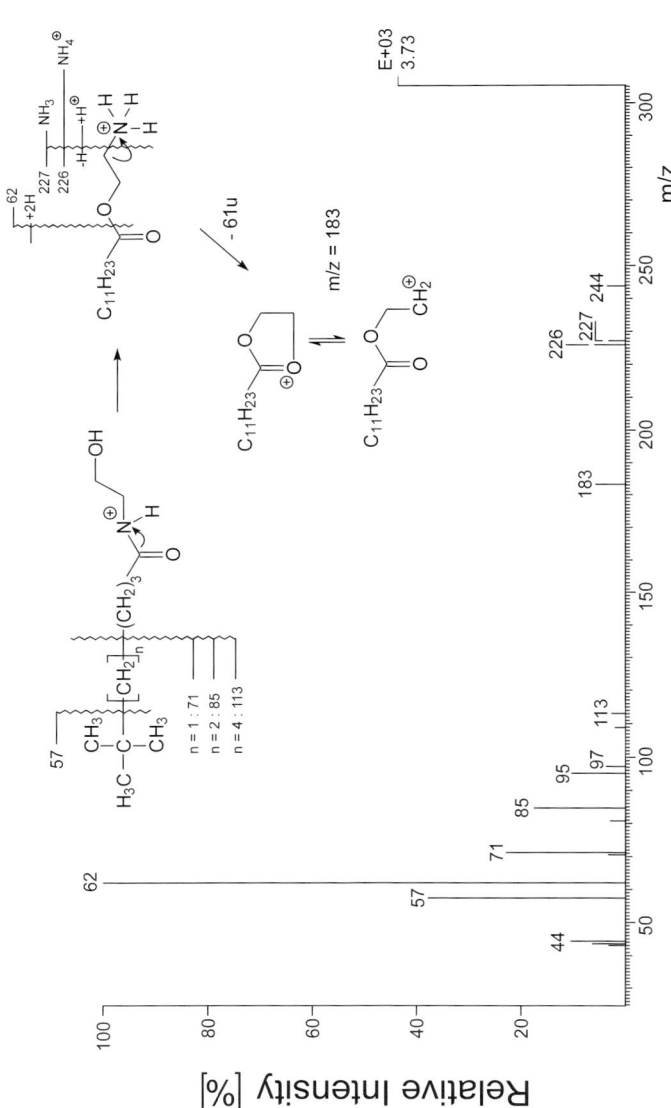

Fig. 2.9.31. ESI–FIA–MS–MS(+) (CID) product ion mass spectrum of parent ion with m/z 244 of FAMA homologue (C_nH_{2n+1}–C(O)–N(H)–CH$_2$–CH$_2$–OH; $n = 11$); (inset) fragmentation scheme observed under CID conditions [24].

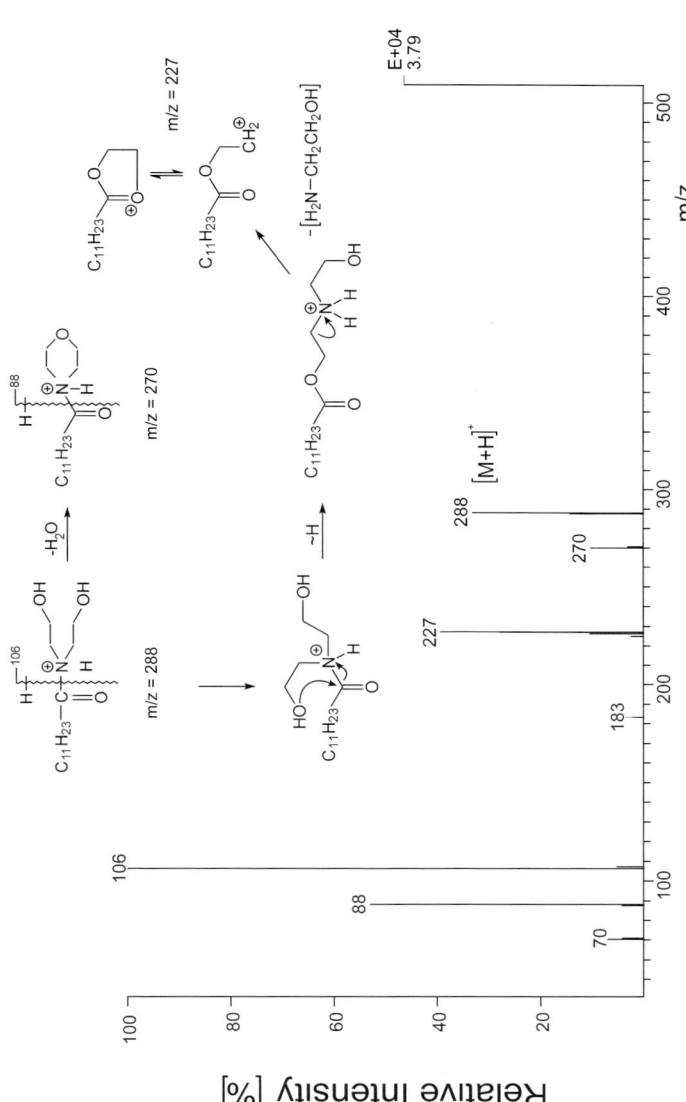

Fig. 2.9.32. ESI–FIA–MS–MS(+) (CID) product ion mass spectrum of parent ion with m/z 288; FADA ion C_nH_{2n+1}–C(O)–N(CH$_2$–CH$_2$–OH)$_m$; $n = 11$, $m = 2$; (inset) fragmentation scheme observed under CID conditions [16].

showed a similar behaviour to LAMA [9,16]. MS–MS examinations of surfactants that had been performed on a TSQ but not on an ion trap had led to the explanation that a gas-phase rearrangement prior to the fragmentation had happened or, alternatively an isomer had been present. The later on MS^3 experiments by in-source CID and MS–MS in the second quadrupole of a triple quadrupole instrument resulted in fragment ions, which confirmed the results obtained by ion trap experiments [44].

Both saturated and unsaturated fatty acids were used for the synthesis of FADA. The unsaturated compounds have the general formula $(C_nH_{2n-1}-C(O)N(CH_2-CH_2-OH)_2)$. The commercial blend of the oleic acid diethanolamide $(CH_3-C_7H_{14}-CH=CH-C_7H_{14}-C(O)-N(CH_2-CH_2-OH)_2)$ could be ionised as the $[M + H]^+$ ion at m/z 370 applying ESI–FIA–MS(+). This unsaturated FADA containing oleic acid as the lipophilic moiety exhibits only one single ion because of the high purity of precursor compounds used for the diethanolamide synthesis [16].

A commercial blend of this FADA derivative was examined by MS–MS. The application of ESI–FIA–MS(+) to this very pure blend resulted in a single $[M + H]^+$ ion at m/z 370. Then ESI–FIA–MS–MS(+) was performed and the diethanolamide derivative of this unsaturated fatty acid showed a comparable behaviour to that observed with saturated diethanolamides [16], i.e. fragments with m/z 88, 106, 352 and $(370-61 = 309)$ as shown in the CID spectrum in Fig. 2.9.33 were generated, which can then be explained as presented in the inset [24].

Fatty acid amido polyethoxylates
In addition to the FAMA compounds described before, fatty acid amido polyethoxylates $(C_nH_{2n+1}-C(O)N(H)(CH_2-CH_2-O)_xH;\ x > 1)$ were also applied as non-ionic surfactants contained in special formulations. These compounds are shown with their general structural formula in Fig. 2.9.34. Like FAMA, the molecules contain amino nitrogen and therefore could be easily ionised in the presence as well as absence of ammonium acetate by APCI or ESI in the positive mode in the form of $[M + H]^+$ ions. Comparable spectra in both API techniques were observed under APCI or ESI ionisation.

A complex mixture of homologue compounds was observed from an industrial blend ionised in the positive FIA–APCI mode (Fig. 2.9.35). The ions were generated either by the homologues differing by the

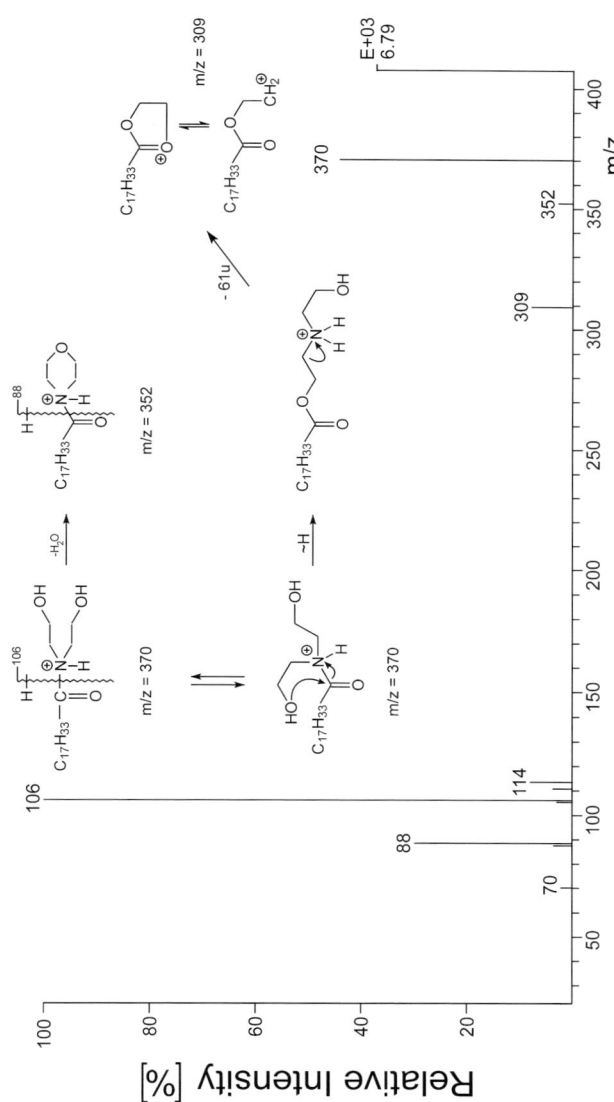

Fig. 2.9.33. ESI–FIA–MS–MS(+) (CID) product ion mass spectrum of parent ion with m/z 370 of industrial blend of unsaturated FADA; surfactant compound could be identified as diethanolamide derivative of oleic acid ($CH_3-C_7H_{14}-CH=CH-C_7H_{14}-C(O)-N(CH_2-CH_2-OH)_2$); (inset) fragmentation scheme observed under CID conditions [16].

269

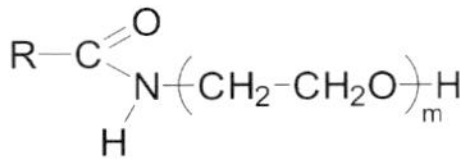

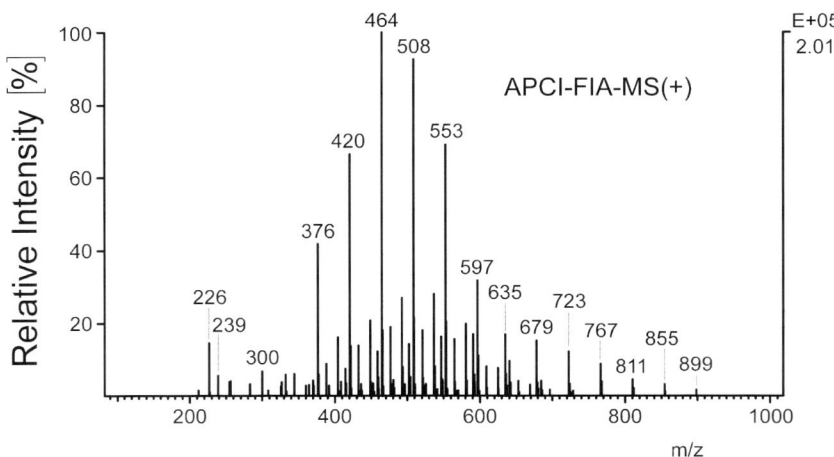

Fig. 2.9.34. General structural formula of fatty acid amido polyethoxylates.

number of alkyl chain links or according to the number of PEG units in the molecules. This blend of fatty acid amido polyethoxylates examined contained polyethoxylates with the general formula $C_nH_{2n+1}-$ $C(O)N(H)(CH_2-CH_2-O)_xH$ with C_7, C_9, C_{11}, C_{13}, C_{15}, C_{17} and C_{19} alkyl moieties. This was confirmed by APCI–LC–MS(+) presented as TIC (Fig. 2.9.36(f)) and in $\Delta m/z$ 28 spaced mass traces of C_9, C_{11}, C_{13} and C_{15}-homologues (Fig. 2.9.36(b)–(e)). The C_7, C_9, C_{11}, C_{13}, C_{15}, C_{17} and C_{19} homologues with $x = 6$ EO units could be ionised with ions at 408, 436, 464, 492, 520, 548 and 602 amu. Behind every signal in the mass traces, a series of compounds with different EO numbers were hidden, which is presented as an averaged mass spectrum for peak 1 in Fig. 2.9.36(a). Negative ionisation was also possible, however, as expected with reduced sensitivity, resulting in $[M + \text{acetate}]^-$ ions of the most prominent ions of peak 1 [24].

Fatty acid amido polyethoxylates ($C_nH_{2n+1}-C(O)N(H)(CH_2-CH_2-O)_xH$; $x > 1$) may consist of a large number of homologue compounds

Fig. 2.9.35. APCI–FIA–MS(+) overview spectrum of fatty acid amido polyethoxylate blend ($C_nH_{2n+1}-C(O)N(H)(CH_2-CH_2-O)_xH$; $x > 1$) [16].

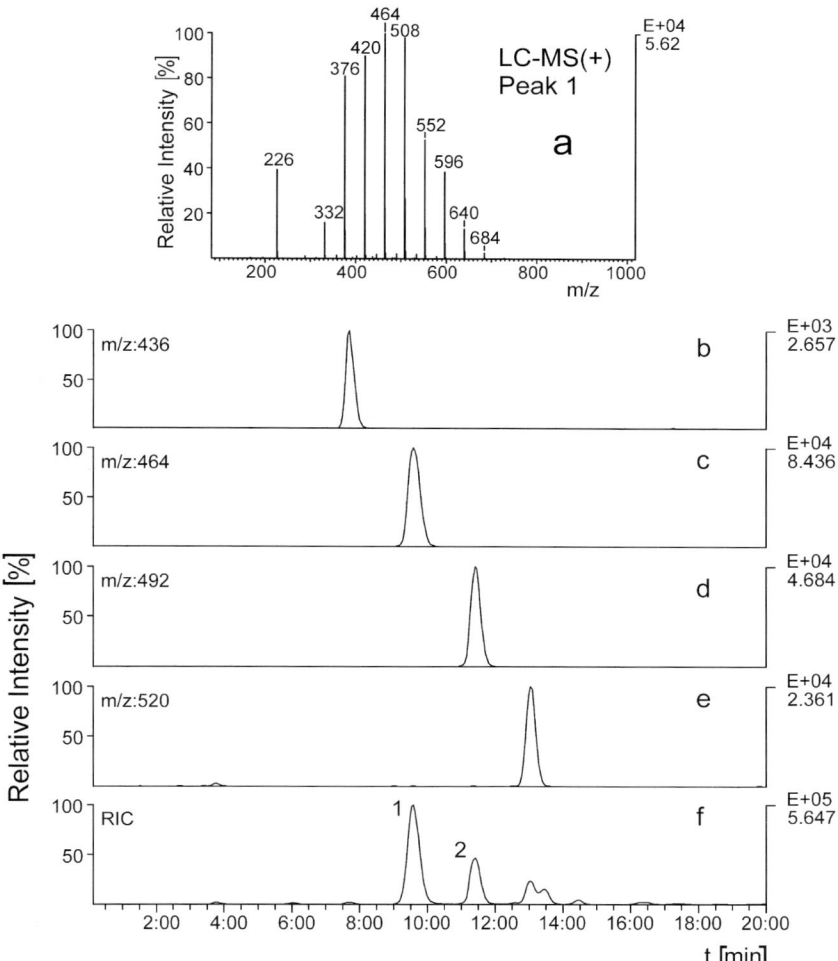

Fig. 2.9.36. (bottom) ESI–LC–MS(+) RIC and selected mass traces (b)–(e) ($m/z = 436$, 464, 492 and 520; $\Delta m/z$ 28) of fatty acid amido polyethoxylate blend as in Fig. 2.9.35 and (a) averaged ESI–MS(+) spectrum of peak 1 as in (f). Gradient elution separated on RP-C$_{18}$ column [16].

that differ from the FAMA compounds because of their hydrophilic chain of a varying number of ethoxylate units. As nitrogen-containing compounds $[M + H]^+$ ions were generated from these homologue molecules. For characterisation purposes, the $[M + H]^+$ ion of the homologue $C_{11}H_{23}-C(O)N(H)(CH_2-CH_2-O)_6H$ was submitted to APCI–FIA–MS–MS(+) resulting in one single fragment ion at m/z 226, which under CID conditions arose in a similar way by an intramolecular fission as described before with ethanolamides. The product ion spectrum combined with the scheme of fragmentation is shown in Fig. 2.9.37 [24].

Polyethoxylated sorbitan derivatives
Ethoxylated sorbitan ester surfactant mixtures like Tween 20 (cf. Fig. 2.9.38) were often used in biochemical applications. Detergents of this type were analysed by MALDI MS. The aim was to compare the separation results of TLC and RP-LC and to detect impurities within these ethoxylated sorbitan esters [30]. Tween 20, the ethoxylated sorbitan carboxylate was ionised resulting in $[M + Na]^+$ and $[M + K]^+$ ions. The Tween 20 isomeric and homologue molecules contained a varying number of ethoxylate units. The number of EO units ($-CH_2CH_2O-$) was determined from 18 to 34 resulting in $\Delta m/z$ 44 equally spaced signals [30].

A mixture of sorbitan derivatives named Tween 81 specified as ethoxylated sorbitan esters containing oleic acid, was examined by APCI–FIA–MS and –LC–MS in the positive and negative modes. APCI–MS ionisation was supported by the addition of ammonium acetate resulting in equal spaced ($\Delta m/z$ 44) $[M + NH_4]^+$ ions which, in parallel, suppressed $[M + Na]^+$ ions. The FIA–MS(+) spectrum contained ions with m/z between 358 and 974 while negative ionisation led to a series of ions from 399 to 971, $\Delta m/z$ 44 equally spaced, too.

Results of LC separation in the positive and negative modes are presented in Fig. 2.9.39. The TIC traces confirmed the presence of some impurities while the series of homologues hidden under signals 1 and 2, which were shown as insets in the positively or negatively ionised TIC, respectively, as ions of pure sorbitan derivatives, present as homologues with different numbers of polyglycolethers. Positive ionisation resulted in $[M + NH_4]^+$ ions but in parallel loss of H_2O (-18) was observed. Negative ionisation generated the homologue sorbitan derivatives in the form of $[M - H]^-$ ions [24].

The ethoxylated sorbitan ester Tween 81 was examined by APCI–FIA–MS–MS(+). The compound was ionised by APCI–FIA–MS(+) under addition of ammonium acetate resulting in ions with $\Delta m/z$ 44 equally spaced signals. The APCI–FIA-MS–MS(+) examination of the

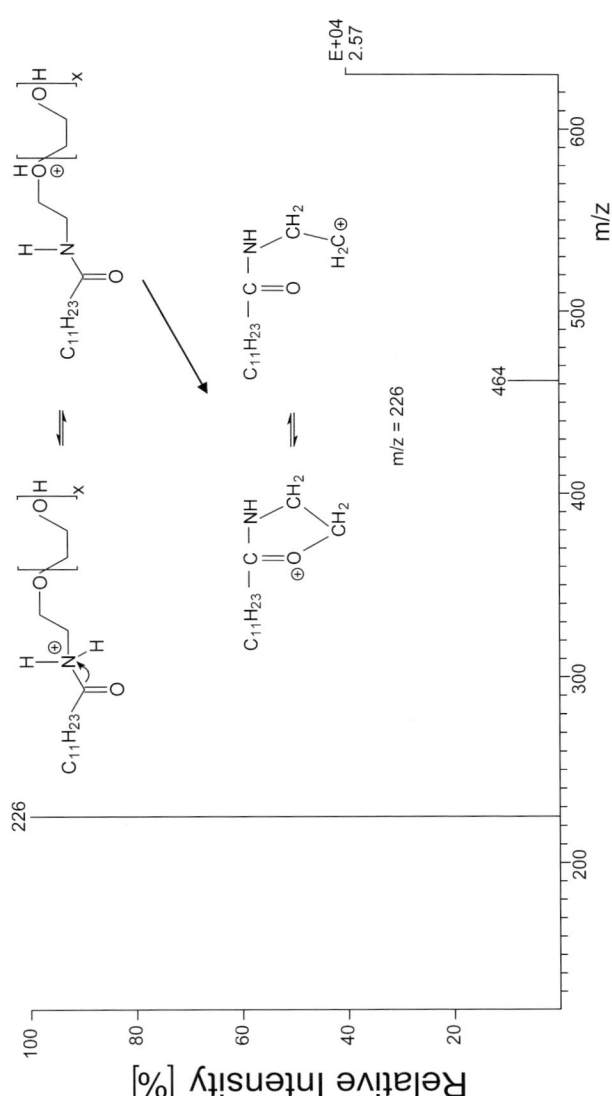

Fig. 2.9.37. APCI–FIA–MS–MS(+) (CID) product ion mass spectrum of parent ion with m/z 464 ($n = 11, x = 6$) of industrial blend of fatty acid amido polyethoxylate surfactants (C_nH_{2n+1}–C(O)N(H)(CH$_2$–CH$_2$–O)$_x$H; $x > 1$); (inset) fragmentation scheme observed under CID conditions [24].

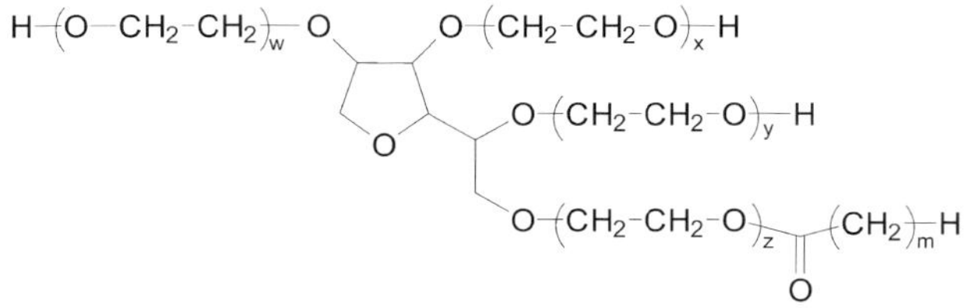

Fig. 2.9.38. General structural formula of polyethoxylated sorbitan esters.

selected homologue parent ion observed with m/z 578 resulted in a product ion spectrum that contained the characteristic fragment ions of ethoxylated sorbitan ester derivatives, which were equally spaced by $\Delta m/z$ 44. All product ion spectra recorded from different homologues contained these Δ 44 equally spaced ions because of their differing PEG content. So, these fragment ions consisting of PEG units could be observed at m/z 89, 133, 177 while three other series of Δ 44 spaced fragments could be observed at m/z 111, 155, 199, 243 and 287 or at 279, 323 and 367 or at 129, 173, 217, 261 and 305, respectively. These fragment ions arose from the destruction of the sorbitan sub-structure containing a series of PEG units. These sub-structure fragments have yet to be explained and it was impossible to make a reasonable proposal for the sorbitan ester fragmentation behaviour [24].

Dimeric non-ionic surfactants (gemini surfactants)
For the examination of a new type of surface active compounds, the so-called gemini surfactants with an improved surface activity [45,46], different mass spectrometric techniques, e.g. fast atom bombardment (FAB), time-of-flight secondary ion mass spectrometry (ToF-SIMS), MALDI, ESI and field desorption (FD) [47,48] were used. Molecular weight information was obtained from a selection of polyethoxylated compounds based on the 2,4,7,9-tetramethyl-5-decyne-4,7-diol (Surfynol) (see Fig. 2.9.40). These polyethoxylated decyne diols can be estimated as non-ionic gemini surfactants that were applied on the one hand because of their good surface activity and on the other hand because of their herbicidic effects. By applying APCI or ESI techniques to ionise such compounds in a different manner with API interfaces, comparable effects in the ionisation efficiency of Surfynols could be observed [16]. The blends Surfynol 420 and 440 produced similar results in their molecular weight (MW) distribution using the different MS

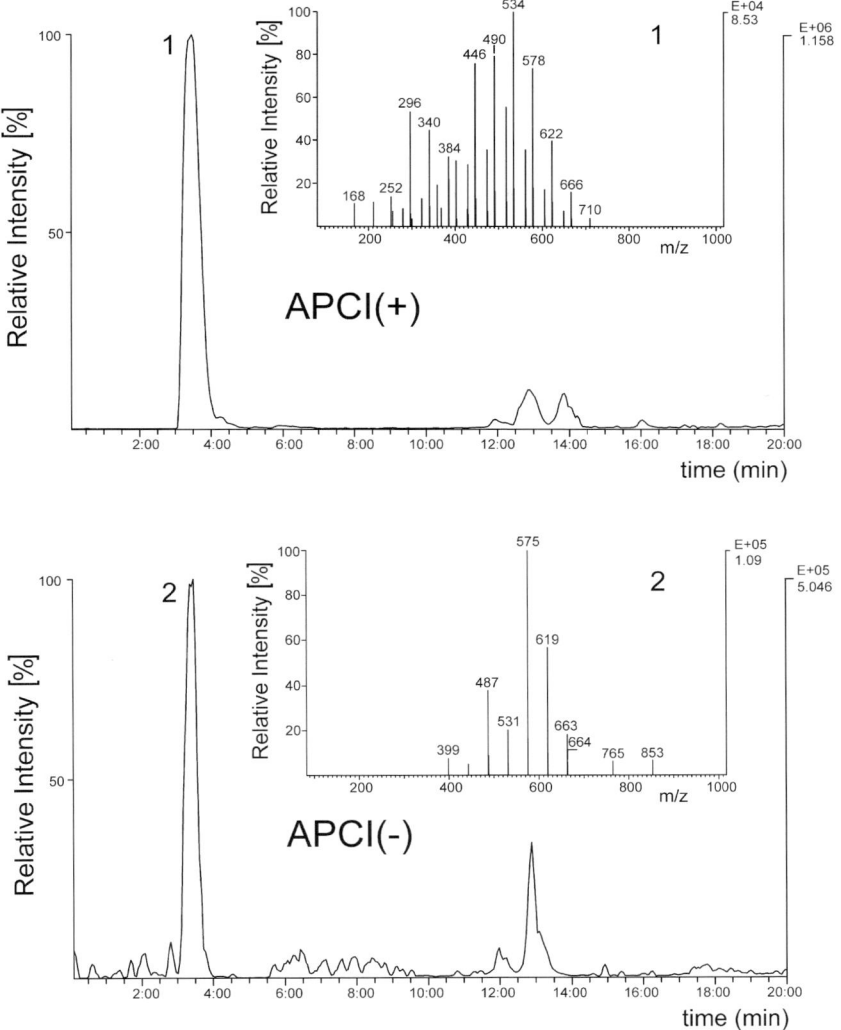

Fig. 2.9.39. (top) APCI–LC–MS(+) RIC of polyethoxylated sorbitan derivative and (inset) averaged mass spectrum under peak 1; (bottom) APCI–LC–MS(−) reconstruction chromatogram as before and (inset) averaged mass spectrum under peak 2. Gradient elution separated on RP-C$_{18}$ column [24].

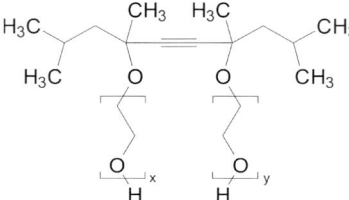

Fig. 2.9.40. General structural formula of polyethoxylated non-ionic gemini surfactant (2,4,7,9-tetramethyl-5-decyne-4,7-polyol).

techniques, whereas the blends 465 and 485 showed variations. Although the MS techniques applied are providing MW data, little was known about their accuracy [47].

Surfynol 440 with the structural formula as shown in Fig. 2.9.40 was examined by APCI–FIA–MS(+), LC–MS(+) and FIA–MS–MS(+) [16]. The addition of ammonium acetate under APCI–FIA–MS(+) conditions led to $[M + NH_4]^+$ ions equally spaced by $\Delta m/z$ 44 (cf. Fig. 2.9.41) because of the variation in the number of PEG units in the hydrophilic moieties. This mixture of homologues could be ionised with ions at m/z 332, 376,…,816, 860 representing PEG chain numbers of $x + y = 2–14$. LC separation was performed on an RP-C$_{18}$ column with a

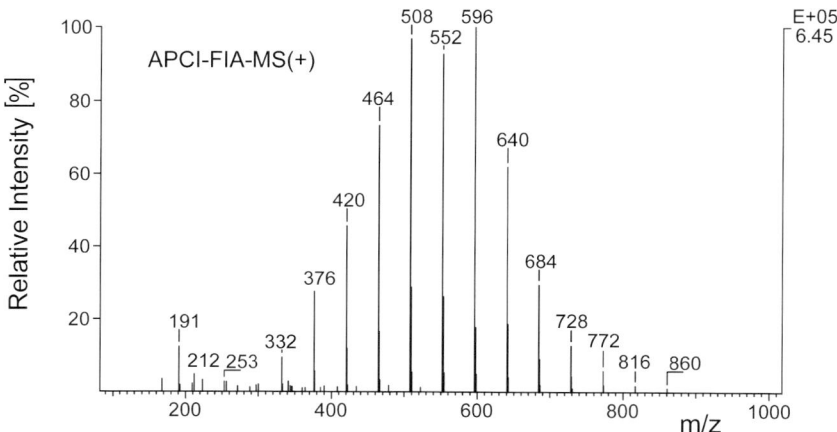

Fig. 2.9.41. APCI–FIA–MS(+) overview spectrum of gemini surfactant blend Surfynol (2,4,7,9-tetramethyl-5-decyne-4,7-polyol) [16].

gradient elution using methanol/water. This compound as well as other Surfynols (Surfynol 440, 465, 485) has no chromophoric group and therefore made UV detection impossible. LC separations of a mixture of these gemini surfactant homologues differing by their PEG content failed under RP-C_{18} conditions as also observed with AE compounds with different numbers of PEG units.

The 2,4,7,9-tetramethyl-5-decyne-4,7-diol mixture which belongs to the new group of polyethoxylated non-ionic gemini surfactants was examined by APCI–FIA–MS–MS(+). The $[M + NH_4]^+$ homologue ion at m/z 640 (number of PEG units: $x + y = 9$) of the Surfynol 440 blend was fragmented in CID experiments on a TSQ. The product ion spectrum of this homologue is presented in Fig. 2.9.42.

A fragmentation scheme cannot yet be proposed and makes MS^n experiments on a trap necessary to get more information about the product ion sequence. Besides PEG chain fragments at m/z 89, 133, 221 and 264, fragment ions equally spaced by $\Delta 44$ arising from alkyl abstraction reactions could also be observed at m/z 107, 151, 195, 239, 283, 327, 371 and 415. The prominent ion at m/z 191 is generated from abstraction of both PEG chains resulting in an alkyl fragment containing conjugated double bounds [24].

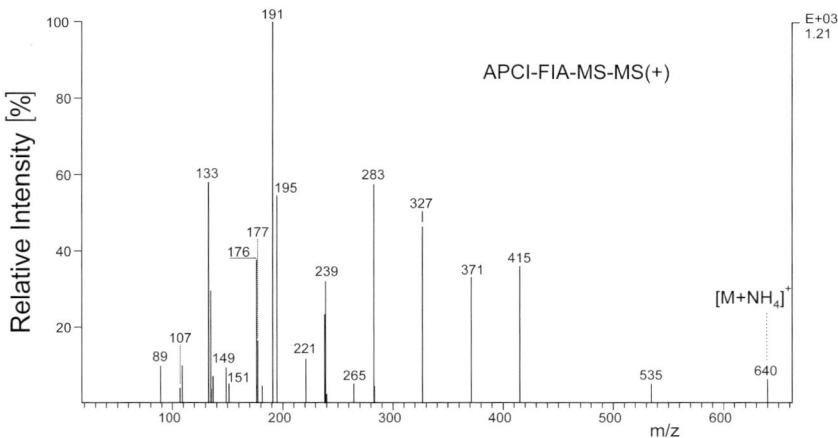

Fig. 2.9.42. APCI–FIA–MS–MS(+) (CID) product ion mass spectrum of $[M + NH_4]^+$ parent ion of 2,4,7,9-tetramethyl-5-decyne-4,7-polyol blend at m/z 640 (number of PEG units: $x + y = 9$) [24].

Fluorinated alkylpolyglycolethers

Besides gemini surfactants, fluorinated surfactants are also unique in their surface activity. These effects have made fluorinated surfactants excellent auxiliary agents in a large number of industrial processes where non-fluorinated surfactants must fail [49]. The non-ionic, partly fluorinated surfactants with the general formula $C_nF_{2n+1}-CH_2-CH_2-O-(CH_2-CH_2-O)_{m-1}-H$, which are identical to the perfluorinated surfactant compounds $C_nF_{2n+1}-(CH_2-CH_2-O)_m-H$, are presented in Fig. 2.9.43 with their structural formula resembling AE compounds, but partly fluorinated in the alkyl chain. Chemical and physical properties of these compounds first cited in the literature in 1950 [50] allow the application in combination with acidic, caustic, hot or even highly oxidising reagents. The physico-chemical and physical stability of the non-ionic ethoxylate compounds $C_nF_{2n+1}-CH_2-CH_2-O-(CH_2-CH_2-O)_{m-1}-H$ was examined under chemical oxidation conditions proving an extreme stability under hydrogen peroxide/sulfuric acid oxidation or even in the hydrogen/oxygen flame Wickbold torch [51].

The properties, behaviour under biochemical degradation conditions in WWTPs as well as the extraction, concentration and ionisation efficiency of the fluorine-containing surfactants were described in the literature [49,51].

The ESI–FIA–MS(+) spectrum of a non-ionic compound mixture is presented in Fig. 2.9.44. Comparing ESI and APCI, differences in ionisation efficiency and in the ions generated could be observed [16]. While in the ESI–FIA–MS(+) ionisation mode, a series of $[M + H]^+$ ions at m/z 497–717 and $[M + NH_4]^+$ ions at m/z 470–866 were generated, both equally spaced by $\Delta m/z$ 44, APCI–FIA–MS(+) mainly resulted in $[M + NH_4]^+$ ions. It was observed that APCI in the same blend ionised the compounds with longer PEG chains, i.e. the more polar compounds. This is in contradiction to the behaviour of non-fluorinated AE compounds, which with an increasing number of PEG chain links became more polar and therefore easily ionisable by ESI. This result contradicted our own experiences [16] and also those reported by

$$R-O\overbrace{(CH_2-CH_2O)}^{}_{m-1}H$$

$$R = CF_3(CF_2)_x-CH_2-CH_2-$$

Fig. 2.9.43. General structural formula of partly fluorinated non-ionic AE compounds.

278

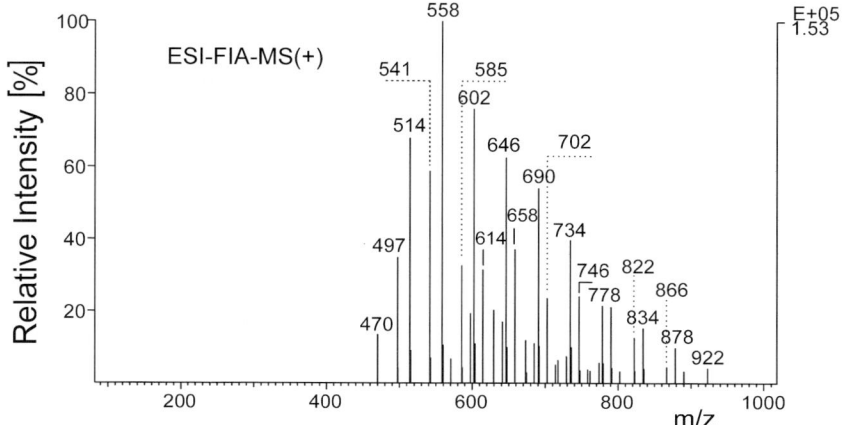

Fig. 2.9.44. ESI–FIA–MS(+) overview spectra of partly fluorinated non-ionic surfactant blend with the general formula $C_nF_{2n+1}-(CH_2-CH_2-O)_m-H$ identical with $C_nF_{2n+1}-CH_2-CH_2-O-(CH_2-CH_2-O)_{m-1}-H$ [16].

Castillo et al. [27]. This behaviour may be an effect of the electronic partition in this type of molecules because of the fluorine substitution in the alkyl chain, which is responsible for the unique surface activity.

The separation of the AE surfactant mixture 'partly fluorinated' in the alkyl chain applying a LC–MS on a RP-C_{18} column and APCI(+) ionisation is presented in Fig. 2.9.45(a)–(c) [24]. The good separation according to the different CF_2-chain lengths resulted in four groups of $[M + NH_4]^+$ ions at m/z 514–910 ($n = 6$; $m = 4$–13) Fig. 2.9.45(a), 614–966 ($n = 8$; $m = 4$–12) Fig. 2.9.45(b), 714–846 ($n = 10$; $m = 4$–7) and 814–902 ($n = 12$; $m = 4$–6). The averaged mass spectra of peaks 1 and 2 are presented in Fig. 2.9.45(a) and (b). The APCI results were comparable with the results also reported from TSP–LC–MS(+) examinations [7,10].

Under RP-C_{18} column separation conditions, no differentiation between conventional AE and fluorinated AE in retardation behaviour could be observed. The application of a perfluorinated RP-C_8 column to a mixture of standards of non-ionic fluorinated surfactants and the extracts of wastewater sludges spiked with these compounds using realistic concentrations, however, resulted in good separation of AE and fluorinated AE (cf. mass traces and TIC in Fig. 2.9.46(a), (b) or (e) and Fig. 2.9.46(c), (d) or (e), respectively) or matrix and fluorinated AE surfactants (Fig. 2.9.46(f), (g) or (h), respectively) [52].

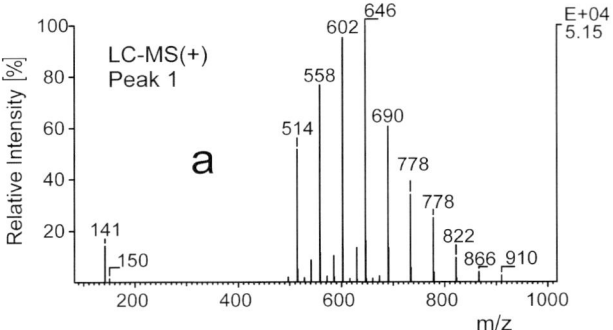

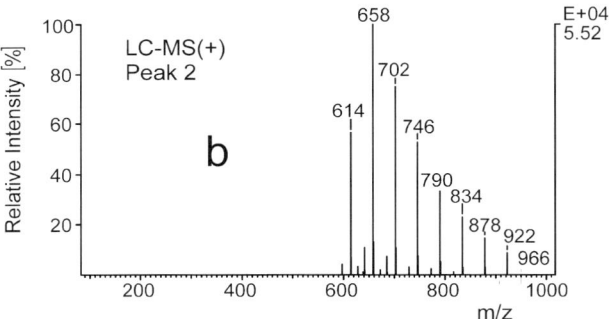

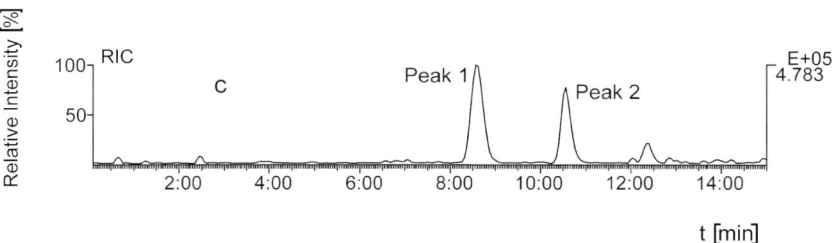

Fig. 2.9.45. (c) APCI–LC–MS(+) RIC of partly fluorinated non-ionic surfactant blend $(C_nF_{2n+1}-(CH_2-CH_2-O)_m-H)$; (a) and (b) averaged mass spectra of peak 1 and peak 2 as in (c). Gradient elution separated on RP-C_{18} column [24].

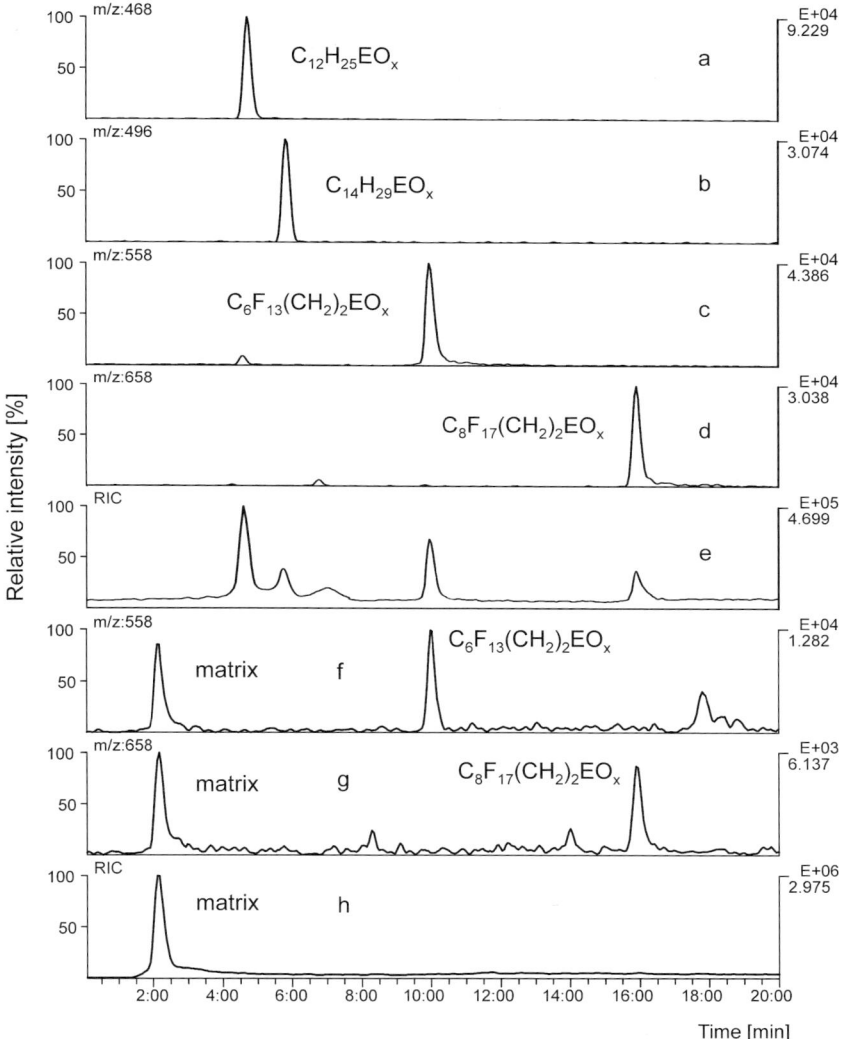

Fig. 2.9.46. (e) APCI–LC–MS(+) RIC of a mixture of standards containing a conventional AE blend (C_{12} and C_{14} homologues) and a fluorinated non-ionic surfactant blend (C_nF_{2n+1}–$(CH_2$–CH_2–$O)_m$–H; $n = 6$ and 8); (a)–(d) selected mass traces of conventional C_{12} and C_{14} AE compounds or C_6 and C_8 fluorinated AE compounds; (h) APCI–LC–MS(+) RIC of wastewater sludge extract containing non-ionic fluorinated surfactants; (f) and (g) selected mass traces of C_6 and C_8 fluorinated AE compounds extracted from sewage sludge. Gradient elution separated on perfluorinated RP-C_8 column [52].

Overview spectra of the non-ionic fluorinated polyglycolether surfactant blend (C_nF_{2n+1}–$(CH_2$–CH_2–$O)_m$–H) obtained by ESI–FIA–MS(+) and APCI–FIA–MS(+) examinations showed differences with the spectra observed under TSI–FIA–MS(+) conditions [7,10,16,51]; however, they were comparable. When FIA–MS–MS(+/−) was performed, fragmentation behaviour of the homologue C_6F_{13}–$(CH_2$–CH_2–$O)_9$–H ionised from the $[M + NH_4]^+$ ion at m/z 734 by APCI-, ESI- [16] or TSI [7] were quite similar. The product ion spectrum is presented in Fig. 2.9.47 together with the fragmentation scheme [19]. Polyglycol ether fragments consisting of 1–4 EO units (m/z 45, 89, 133 and 177) could be observed while fragments of the fluorinated alkyl chain were not generated under positive ionisation conditions. The electronic configuration of the fluorinated alkyl chain is the reason that fluorinated alkyl compounds under positive CID conditions are so stable [16]. Negative ionisation, however, will result in alkyl chain fragmentation [24].

Despite the fact that physico-chemical and chemical degradations were not possible, the isolation of persistent metabolites of the C_nF_{2n+1}–$(CH_2$–CH_2–$O)_m$–H compound generated by β and ω oxidations of the terminal PEG unit of the non-ionic blend was reported, but environmental data about this type of compound are still quite rare [49]. TSI(+) ionisation results of the industrial blend Fluowet OTN have been reported in the literature [7,51]. Actual data of non-ionic fluorinated surfactants were applied using ESI- and APCI–FIA–MS(+) and –MS–MS(+), which reported the biodegradation of the non-ionic partly fluorinated alkyl ethoxylate compounds C_nF_{2n+1}–$(CH_2$–CH_2–$O)_x$–H in a lab-scale wastewater treatment process.

Acidic metabolites with the general formula C_nF_{2n+1}–$(CH_2$–CH_2–$O)_m$–CH_2–COOH besides PEG as metabolite of non-ionics [7,10,51] could be isolated by C_{18}-SPE. APCI–FIA–MS–MS(+) was used for confirmation of these metabolites generated with the same pattern as found under TSI–CID conditions on a TSQ instrument [7]. Besides these metabolites, the precursor fluorine-containing surfactants (C_nF_{2n+1}–$(CH_2$–CH_2–$O)_x$–H), now shortened in the PEG chains, were still present. The APCI–FIA–MS–MS(+) spectrum of the selected ion at m/z 528 together with its fragmentation scheme is presented in Fig. 2.9.48(a) [16]. The negatively recorded CID spectrum in Fig. 2.9.48(b) contained non-identified fragment ions as well as the characteristic fluorinated alkyl chain ion at m/z 119 ($[CF_3$–$CF_2]^-$) and

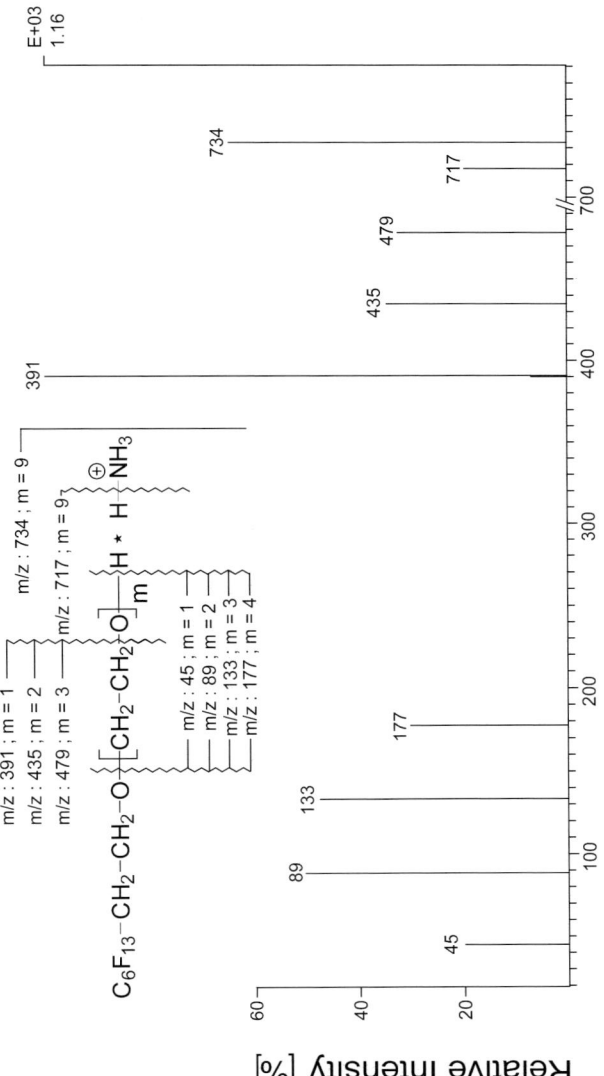

Fig. 2.9.47. ESI–FIA–MS–MS(+) product ion mass spectrum of $[M + NH_4]^+$ parent ion of non-ionic fluorinated polyglycolether surfactant blend homologue at m/z 734 (C_nF_{2n+1}–(CH_2–CH_2–O)$_m$–H; $n = 6$; $m = 10$; fragmentation behaviour in the inset [24].

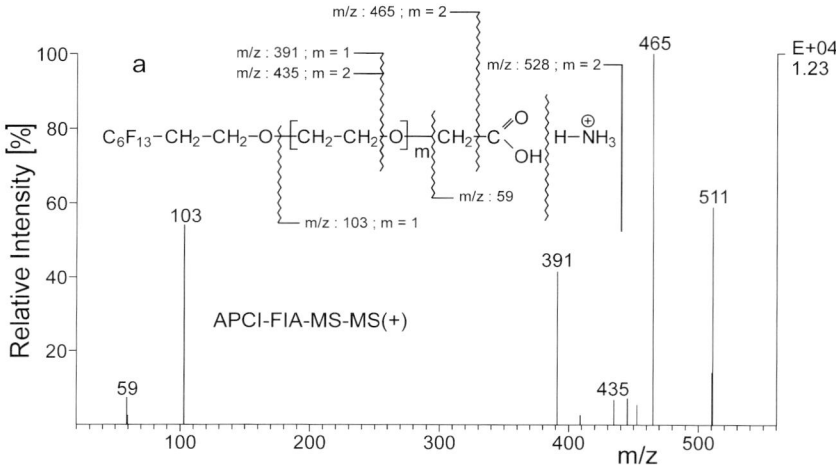

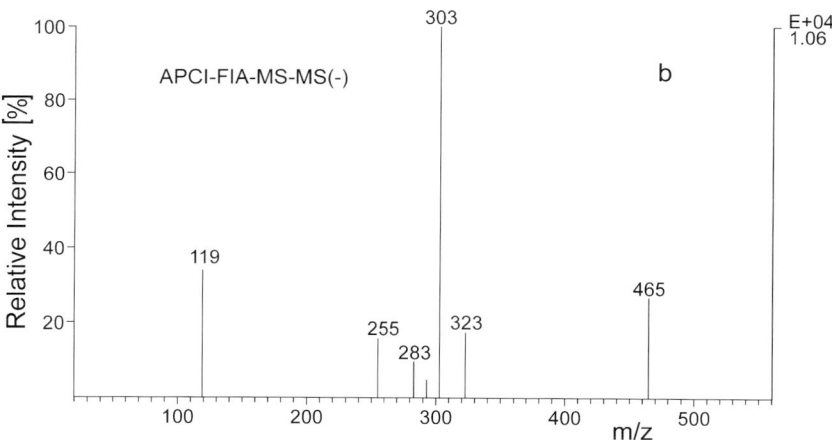

Fig. 2.9.48. (a) APCI–FIA–MS–MS(+) (CID) product ion mass spectrum of selected $[M + NH_4]^+$ parent ion (m/z 528) of fluorinated non-ionic surfactant metabolite identified as PEG-chain carboxylated fluorine containing non-ionic surfactant homologue (general formula: C_nF_{2n+1}–$(CH_2$–CH_2–$O)_m$–CH_2–$COOH$); fragmentation behaviour under CID(+) presented in the inset; (b) APCI–FIA–MS–MS(−) (CID(−)) product ion mass spectrum of $[M − H]^-$ parent ion (m/z 465) of compound as in (a) [37].

the ion at m/z 465 as a result of a CO_2 loss from the negative parent ion at m/z 509 [24].

2.9.3 Conclusion

The prediction that LC–MS will become a powerful tool in the detection, identification and quantification of polar compounds such as surfactants in environmental analysis as well as in industrial blends and household formulations has proven to be true. This technique is increasingly applied in substance-specific determination of surfactants performed as routine methods. From this it becomes obvious that no other analytical approach at that time was able to provide as much information about surfactants in blends and environmental samples as that obtainable with MS and MS–MS coupled with liquid insertion interfaces.

This development of surfactant analysis by MS started with FAB [8] and TSI [7] during the late 1980s and early 1990s. Nowadays these techniques have been replaced by the API methods, ESI or APCI [16], and MALDI, all soft-ionising techniques that ionised the compounds in the form of molecular ions or adduct ions. So identification makes fragmentation essential and therefore product ions will be generated either by source-CID or by MS–MS and MS^n in multiple stage instruments or by ion trap spectrometers, respectively. Identification of non-ionic surfactants is supported by their product ion spectra, generated in positive mode, often more informative because of an increased number of characteristic fragment ions than observed in negatively generated product ion spectra. The use of product ion spectra in computer-aided library searches makes identification easier [35,53–55]; however, the lack of commercially available standards for the generation of product ion spectra and the missing compatibility between MS instruments of different producers or even between different instruments of the same producer is an important obstacle in the identification. This problem with missing standards of anthropogenic surfactants or their metabolites, however, would prevent identification and quantification. But even with the availability of product ion libraries, a reliable quantification will only be possible with standards, either labelled with stabile isotopes or unlabelled, applying standard addition.

REFERENCES

1 M.J. Schick, *Nonionic Surfactants*, Surfactant Science Series, Vol. 1, Marcel Dekker, New York, USA, 1967.
2 J. Cross, *Nonionic Surfactants: Chemical Analysis*, Surfactant Science Series, Vol. 19, Marcel Dekker, New York, USA, 1987.
3 M.J. Schick, *Nonionic Surfactants: Physical Chemistry*, Surfactant Science Series, Vol. 23, Marcel Dekker, New York, USA, 1987.
4 V.M. Nace, *Nonionic Surfactants: Polyoxyalkylene Block Copolymers*, Surfactant Science Series, Vol. 63, Marcel Dekker, New York, USA, 1996.
5 T.M. Schmitt, *Analysis of Surfactants*, Surfactant Science Series, Vol. 63, Marcel Dekker, New York, USA, 1992.
6 M.J. Schwuger, *Detergents in the Environment*, Surfactant Science Series, Vol. 65, Marcel Dekker, New York, USA, 1996.
7 H.Fr. Schröder, In: D. Barceló (Ed.), *Applications of LC/MS in Environmental Chemistry*, Vol. 59, Elsevier, Amsterdam, The Netherlands, 1996, p. 263.
8 Ventura, In: D. Barceló (Ed.), *Environmental Analysis: Techniques, Applications and Quality Assurance*, Vol. 13, Elsevier, Amsterdam, The Netherlands, 1993, p. 481.
9 H.Fr. Schröder, *J. Chromatogr.*, 643 (1993) 145.
10 H.Fr. Schröder, *J. Chromatogr.*, 647 (1993) 219.
11 A. Raffaelli, S. Pucci, L. Campolmi and P. Salvadori, Proc. of 44th ASMS Conf. on Mass Spectrometry and Allied Topics, Portland, USA, 1996, p. 163.
12 L.S. Clesceri, A.E. Greenberg and A.D. Eaton (Eds.), *Standard Methods for the Examination of Water and Wastewater*, 20th edn., American Public Health Association, Washington, DC, 1998, pp. 5–47 (5540 C, Anionic Surfactants as MBAS) and pp. 5–49 (5540 D, Nonionic Surfactants as CTAS).
13 Z. Liu, S. Laha and R.G. Luthy, *Water Sci. Technol.*, 23 (1991) 475.
14 H.-Q. Li and H.Fr. Schröder, *Water Sci. Technol.*, 42 (2000) 391.
15 H.Fr. Schröder, *Vom Wasser*, 79 (1992) 193.
16 H.Fr. Schröder and F. Ventura, In: D. Barceló (Ed.), *Techniques and instrumentation in analytical chemistry*, Sample Handling and Trace Analysis of Pollutants—Techniques, Applications and Quality Assurance, Vol. 21, Elsevier, Amsterdam, The Netherlands, 2000, p. 828.
17 A. DiCorcia, *J. Chromatogr. A*, 794 (1998) 165.
18 A. Marcomini and M. Zanette, *J. Chromatogr. A*, 733 (1996) 193.
19 A.T. Kiewiet and P. de Voogt, *J. Chromatogr. A*, 733 (1996) 185.
20 J. Johnson and R. Yost, *Anal. Chem.*, 57 (1985) 758 A.
21 H. Schwarz, *Nachr. Chem. Tech. Lab.*, 29 (1981) 687.
22 H.Fr. Schröder and K. Fytianos, *Chromatographia*, 50 (1999) 583.

23 H.Fr. Schröder, *Vom Wasser*, 73 (1989) 111.
24 R. Scheding and H.Fr. Schröder, Chromatographia, in preparation.
25 P. Jandera, M. Holcapek and G. Theodoridis, *J. Chromatogr. A*, 813 (1998) 299.
26 K.B. Sherrard, P.J. Marriott, M.J. McCormick, R. Colton and G. Smith, *Anal. Chem.*, 66 (1994) 3394.
27 M. Castillo, F. Ventura and D. Barceló, *Waste Mgmt*, 19 (1999) 101.
28 B.Y. Giang and L. Chang, Proc. of 45th ASMS Conf. on Mass Spectrometry and Allied Topics, Palm Springs, FL, USA, 1997, p. 1268.
29 J. Via, D.E. Doster and R.A. Ludicky, Proc. of 44th ASMS Conf. on Mass Spectrometry and Allied Topics, Portland, USA, 1996, p. 1275.
30 G.A. Cumme, E. Blume, R. Bublitz, H. Hoppe and A. Horn, *J. Chromatogr. A*, 791 (1997) 245.
31 R.H. Auerbach, K. Dost, D.C. Jones and G. Davidson, *Analyst*, 124 (1999) 1501.
32 M. Castillo and D. Barceló, In: D. Barceló (Ed.), *Techniques and instrumentation in analytical chemistry*, Sample Handling and Trace Analysis of Pollutants—Techniques, Applications and Quality Assurance, Vol. 21, Elsevier, Amsterdam, The Netherlands, 2000, p. 537.
33 H. Wu, M. Winkler and H. Aboleneen, Proc. of 46th ASMS Conf. on Mass Spectrometry and Allied Topics, Orlando, FL, USA, 1998, p. 1063.
34 H.Fr. Schröder, Abschlußbericht Verbundforschungsvorhaben 02 WT 9358; Bundesminister für Bildung und Wissenschaft, Forschung und Technologie, Bonn, 1996.
35 H.Fr. Schröder, *J. Chromatogr. A*, 777 (1997) 127.
36 H.Fr. Schröder, In: M. Dohmann (Ed.), Gewässerschutz, Wasser, Abwasser, GWA 166, Gesellschaft zur Förderung der Siedlungswasserwirtschaft an der RWTH Aachen e.V., Aachen, 1997.
37 H.Fr. Schröder, *J. Chromatogr. A*, 926 (2001) 127.
38 T. Reemtsma, *J. Chromatogr. A*, 733 (1996) 473.
39 Priority surfactants and their toxic metabolites in waste effluent discharges: an integrated study (pristine), Progress report No. 1, Feb. 1st, 1998–July 31st, 1998.
40 H.Fr. Schröder, In: DVGW Deutscher Verein des Gas- und Wasserfaches (Eds.), DVGW-Schriftenreihe Wasser, No. 108, Wirtschafts- und Verlagsgesellschaft Gas und Wasser mbH, Bonn, 1990, pp. 121–144.
41 L.A. Ashfield, T.G. Pottinger and J.P. Sumpter, *Environ. Toxicol. Chem.*, 17 (1998) 679.
42 E. Elizondo, D.L. Present and S.B.H. Bach, Proc. of 45th ASMS Conf. on Mass Spectrometry and Allied Topics, Palm Springs, FL, USA, 1997, p. 1269.
43 H.Fr. Schröder, *Vom Wasser*, 81 (1993) 299.

44 R.J. Strife and M.M. Ketcha, Proc. of 44th ASMS Conf. on Mass Spectrometry and Allied Topics, Portland, USA, 1996, p. 1204.

45 F.M. Menger and C.A. Littau, *J. Am. Chem. Soc.*, 113 (1991) 1451.

46 R. Zana, In: K. Holmberg (Ed.), *Dimeric Surfactants, Novel Surfactants*, Surfactant Science Series, Vol. 74, Marcel Dekker, New York, USA, 1998, p. 241.

47 D.M. Parees, S.D. Hanton, P.A. Cornelio Clark and D.A. Willcox, *J. Am. Soc. Mass Spectrom.*, 9 (1998) 282.

48 Erratum to [47] Mass Spectrom., 9 (1998) 651.

49 E. Kissa, In: M.J. Schick and F.M. Fowkes (Eds.), *Fluorinated surfactants: synthesis, properties, applications*, Surfactant Science Series, Vol. 50, Marcel Dekker, New York, USA, 1994.

50 H.G. Bryce, In: J.H. Simons (Ed.), *Fluorine Chemistry*, Vol. 5, Academic Press, New York, USA, 1950, p. 370.

51 H.Fr. Schröder, *Vom Wasser*, 78 (1992) 211.

52 H.Fr. Schröder, Determination of Fluorinated Surfactants and their Metabolites in Sewage Sludge Samples by LC/MS and MS^n after Accelerated Solvent Extraction and Separation on Fluorine Modified RP-Phases, *J. Chromatogr. A*, in press.

53 H.Fr. Schröder, *Water Sci. Tech.*, 34 (1996) 21.

54 R.J. Strife and M.M. Ketcha, Proc. of 44th ASMS Conf. on Mass Spectrometry and Allied Topics, Portland, USA, 1996, p. 1203.

55 J.M. Halket, In: A. Kornmüller, T. Reemtsma (Eds.), Mass spectral libraries for LC–MS, Schriftenreihe 'Biologische Abwasserreinigung' 2. Kolloquium im Sonderforschungsbereich 193 der DFG an der Technischen Universität Berlin 'Anwendung der LC–MS in der Wasseranalytik' Berlin, Germany, 2001, Vol. 16, pp. 173–192, ISBN 3 7983 1852 2.

2.10 ATMOSPHERIC PRESSURE IONISATION MASS SPECTROMETRY—VII. ANIONIC SURFACTANTS: LC–MS OF ALKYLBENZENE SULFONATES AND RELATED COMPOUNDS

Peter Eichhorn

2.10.1 Introduction

Linear alkylbenzene sulfonate (LAS) is the surfactant with the highest worldwide production volume apart from soap. This class of surfactants finds wide application because of its excellent detergent properties, versatility, compatibility with all types of detergent ingredients and best cost/performance ratio among surfactants. In formulations, the alkyl chain length generally ranges from C_{10} to C_{13} with an average distribution of approximately 11.4–11.7. Isomer configurations range from the 2-phenyl position to the 7-phenyl position depending on the alkyl chain length.

2.10.2 Alkylbenzene sulfonates

Owing to the anionic character of LAS, an electrospray ionisation (ESI) interface operated in negative ion mode is particularly attractive for the mass spectrometric detection of this surfactant type. Consequently, a great part of the atmospheric pressure ionisation–mass spectrometry (API–MS) work on LAS is devoted to the application of $(-)$-ESI–MS.

In one study, however, atmospheric pressure chemical ionisation (APCI)–MS was applied for the simultaneous determination of LAS and octylphenol ethoxylates (OPEO) in surface waters after pre-concentration by solid-phase extraction (SPE) on C_{18} cartridges [1]. In the chromatogram from a C_1-reversed phase (RP) column, peaks arising from both the anionic LAS and the non-ionic OPEO were detected after positive ionisation, while in negative ionisation mode, OPEO were discriminated and only the anionic surfactant was observed. Surprisingly, the relative sensitivity for detection of LAS was approximately five times higher in positive ion mode, which led the authors to the conclusion that this ionisation mode was desirable for quantitative work.

Comprehensive Analytical Chemistry XL 289
T.P. Knepper, D. Barceló and P. de Voogt (Eds)

As for OPEO, the dominant pseudo-molecular species were sodium adducts of the type $[M - H_{(n-1)} + Na_n]^+$. It was assumed that this multiple adduction was due to an obviously sodium-rich sample. In the full scan spectra of C_{13}-LAS in turn, only the molecular ions $[M + 2Na]^+$ (m/z 385) in positive and $[M - H]^-$ (m/z 339) in negative ion mode were assigned, but the ions at m/z 747 and 701, corresponding to the clusters $[2M - 2H + 3Na]^+$ and $[4M - 4H + 2Na]^{2-}$, were ignored. The formation of these sodium-associated clusters, however, reflected the high LAS and sodium levels in the samples. The latter strongly promoted the abundant formation of the sodium adduct $[M - H + 2Na]^+$ whose detection suggested that APCI in positive ion mode was more favourable than negative ionisation.

LAS analysis by ESI–MS with negative ionisation led to mass spectra with the four $[M - H]^-$ ions of m/z 297, 311, 325, and 339 resulting from the LAS homologues C_{10}–C_{13}, respectively. With increasing cone voltage, additional fragment ions at m/z 183 and 80 were obtained, which were assigned to styrene-4-sulfonate and $[SO_3]^-$, respectively (Fig. 2.10.1). As no side chain-specific fragment ions were produced under API conditions, a mass spectrometric distinction of positional isomers of LAS could not be achieved. If this is required, gas chromatography (GC)–MS with electron impact or chemical ionisation has to be applied after conversion of the sulfonated surfactant into volatile derivatives [2,3].

For the identification of surfactants in influent and effluent samples from German and Greek wastewater treatment plants, Schröder and co-workers [4,5] employed an analytical approach comprising SPE, selective elution and screening analysis by flow-injection analysis (FIA)–MS(MS) in combination with liquid chromatography–(tandem)

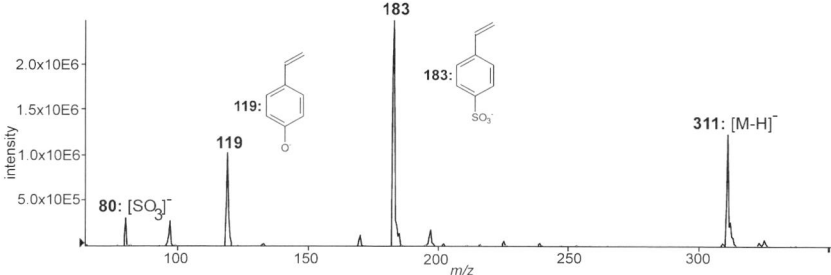

Fig. 2.10.1. $(-)$-ESI–MS mass spectrum of C_{11}-LAS.

mass spectrometry (LC–MS(MS)) for confirmation purposes. Thereby, a series of surfactants including LAS, nonylphenol ethoxylates (NPEOs) and alcohol ethoxylates (AEs) were unambiguously identified. For the (−)-FIA–MS–MS analysis of LAS, the four molecular ions $[M - H]^-$ were selected and the daughter ion m/z 183 was screened. Tandem mass spectrometry after RP-C_{18} separation was also performed to determine LAS concentrations in marine and estuarine sediment samples [6].

The use of solid-phase micro-extraction (SPME) for the qualitative and quantitative determination of LAS in wastewater samples was investigated by Ceglarek et al. [7]. When examining the effect of salt addition on the extraction efficiency, NaCl, commonly used in SPME to improve extraction yields, turned out to be unsuitable because of the formation of $[(NaCl)_n Cl]^-$ clusters in the ESI–MS (prior to injection LAS was desorbed from the fibre by methanol/water (50:50)), the formation of which were assumed to be responsible for the quantitative suppression of the LAS signals. These quenching effects were excluded when using ammonium acetate instead of NaCl.

The mass spectrometric distinction between LAS and its branched congener alkylbenzene sulfonate (ABS)—still in use in several countries—was feasible thanks to their different fragmentation behaviour. It was reported [8] that under the soft ionisation provided by fast atom bombardment (FAB), ABS formed fragments with a double bond in the benzylic position (m/z 197) (Fig. 2.10.2). The use of negative ESI likewise yielded this ion typical of ABS [9]. Formation of the fragments m/z 197 and 183, respectively, permitted the distinction between not fully chromatographically separated ABS and LAS, occurring in surface waters from the Philippines where the hard ABS surfactant still finds application in certain laundry detergent formulations [10]. Though not present in today's commercial ABS mixtures, shorter chain ABS with 4–7 carbons, corresponding to molecular ions $[M - H]^-$ with m/z 213, 227, 241 and 255, were detected in ground water samples collected in the same investigated area and identified as persistent contaminants [11].

Commercial LAS blends usually contain some sulfonated impurities, mainly dialkyltetralin sulfonates (DATS) with a similar distribution of aliphatic carbons than that of LAS (C_{10}–C_{13}). The (−)-ESI–MS spectra of DATS display a series of equally spaced signals with $\Delta m/z$ 14, shown for C_{11}-DATS in Fig. 2.10.3, arising from the cleavage of the alkyl

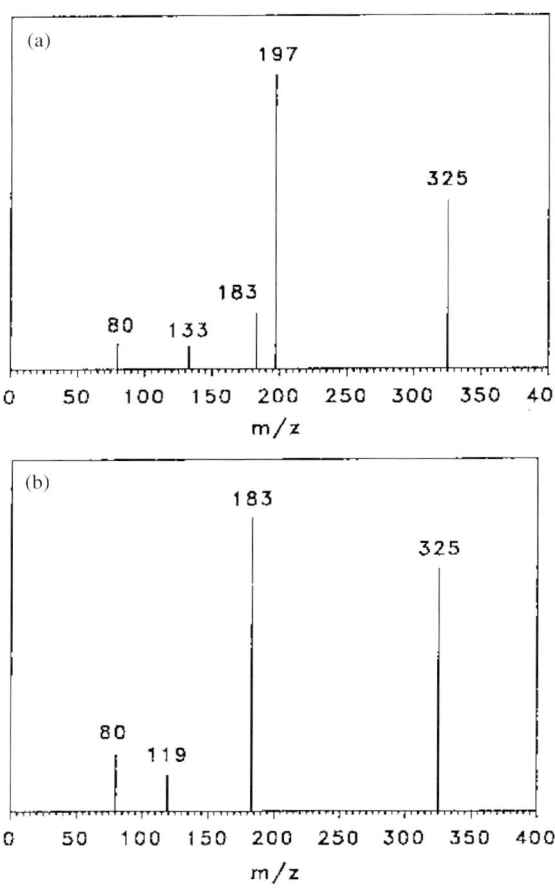

Fig. 2.10.2. FAB daughter ion spectrum obtained from the CID of the molecular ion of C_{12}-ABS at m/z 325 (a) and of C_{12}-LAS (b) Reprinted with permission from [18]. Copyright 1992. American Chemical Society).

substituents. The diagnostic fragment ion m/z 209 corresponds to sulfonated dihydronaphthalene.

2.10.3 Carboxylated degradation products

The aerobic biodegradation of LAS that occurs in wastewater treatment plants and surface waters leads to the formation of carboxylated

292

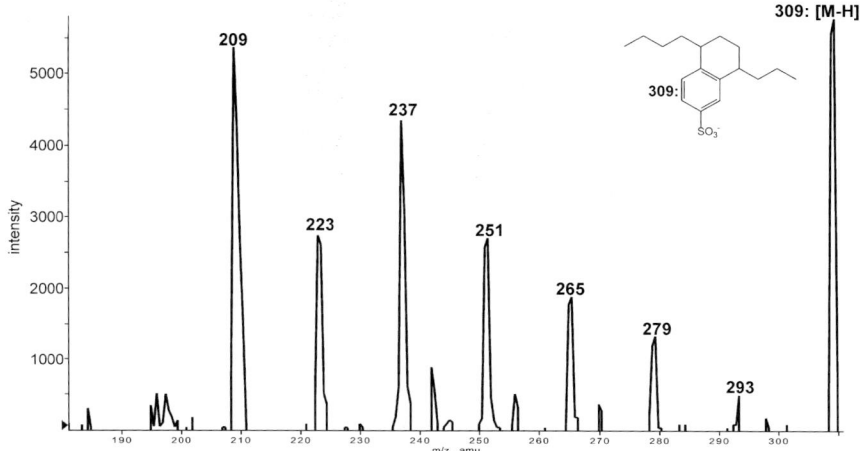

Fig. 2.10.3. (−)-LC−ES−MS spectrum of C_{11}-DATS. The distribution of the alkyl chain length is arbitrarily chosen.

intermediates of which sulfophenyl carboxylates (SPC) are major components. It is generally recognised that these polar intermediates are formed through ω-oxidation of LAS followed by successive oxidative shortening of the alkyl chain by two-carbon units, termed as β-oxidation [12,17], which may result in a complex mixture of alkyl homologues $(C_3 – C_{13})$, each of which is made up of phenyl isomers.

For sufficient retention of these very polar sulfonated carboxylates on RP columns, the addition of an ion-pairing (IP) agent such as tetraethylammonium acetate (TEAA) to the LC buffer was compulsory [13]. To maintain the compatibility of the eluent with the MS interface, the use of such an involatile cationic additive entailed a cation exchanger to be placed between the column and the interface [13]. Alternatively, equimolar amounts (5 mM) of acetic acid and triethyl-amine, which form the volatile IP agent triethylammonium, were added to the mobile phase in order to improve the retardation of very polar SPC [14]. While the first approach with TEAA was effective in retaining even the very short-chain C_3- and C_4-SPC (Fig. 2.10.4), the weaker IP agent triethylammonium notably increased the retention of C_5-SPC and higher, whereas C_4-SPC elutes almost with the dead volume of the LC (Fig. 2.10.5). Addition of commonly used ammonium acetate as buffer component led to the co-elution of the short-chain SPC ($\leq C_6$) [13].

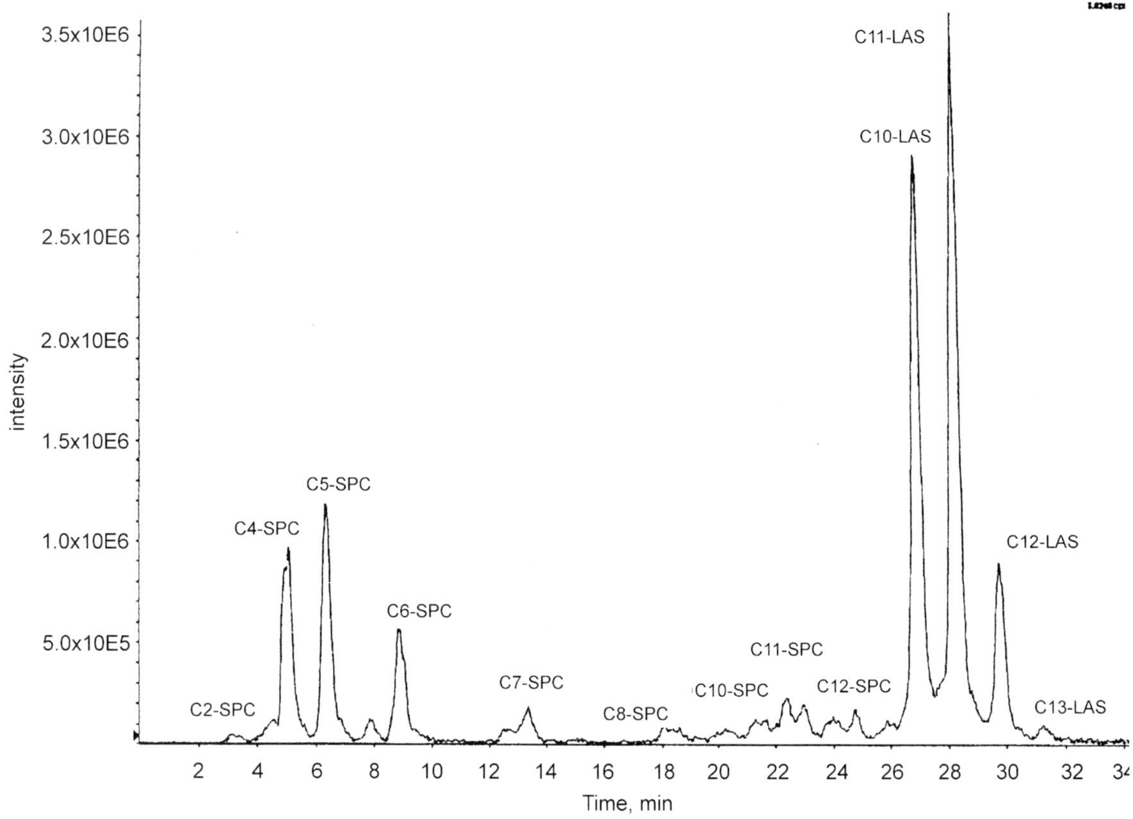

Fig. 2.10.4. (−)-LC–ESI–MS chromatogram of LAS and SPC separation using water/acetonitrile containing 5 mM tetraethylammonium acetate (from Ref. [13] with permission from author).

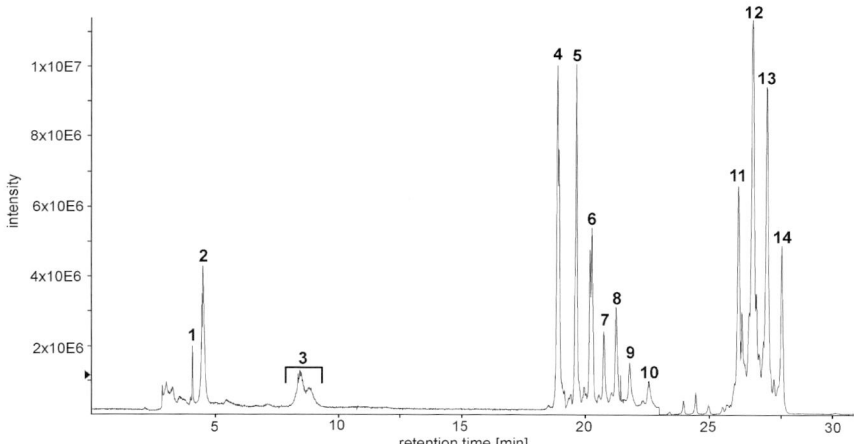

Fig. 2.10.5. (–)-LC–ESI–MS chromatogram of LAS and SPC separation using water/ acetonitrile containing 5 mM acetic acid/triethylamine. Peak numbering: (1) C_4-SPC, (2) C_5-SPC, (3) C_6-SPC, (4) C_7-SPC, (5) C_8-SPC, (6) C_9-SPC, (7) C_{10}-SPC, (8) C_{11}-SPC, (9) C_{12}-SPC, (10) C_{13}-SPC, (11) C_{10}-LAS, (12) C_{11}-LAS, (13) C_{12}-LAS, (14) C_{13}-LAS (from Ref. [22]).

In general, the presence of IP agents in the eluant entering the electrospray interface exerts adverse effects on the process of ion formation [15], thus the ionisation efficiency diminishes. In this respect, the efficacy of a cation exchanger is demonstrated in Fig. 2.10.6 comparing the extracted ion chromatogram (XIC) (molecular ion: m/z 285) of C_7-SPC (constituted of several phenyl isomers) without and with the use of the device. The overall peak intensity increased by roughly an order of magnitude after removal of cationic species. This, however, occurred at the expense of peak resolution since interaction of the analytes with the exchanger material caused appreciable peak broadening.

To avoid LC conditions involving the use of IP agents—as reported above—an alternative route was reported by Di Corcia et al. [16] for the separation and detection of breakdown products generated out of commercial LAS in a degradation test. This method was based on the conversion of the carboxylic groups into methyl esters. After this modification, the addition of 0.2 M ammonium acetate to the mobile phase sufficed to give sharp peaks. In addition, a 10-fold enhancement of the sensitivity of the ESI–MS detector was obtained.

296

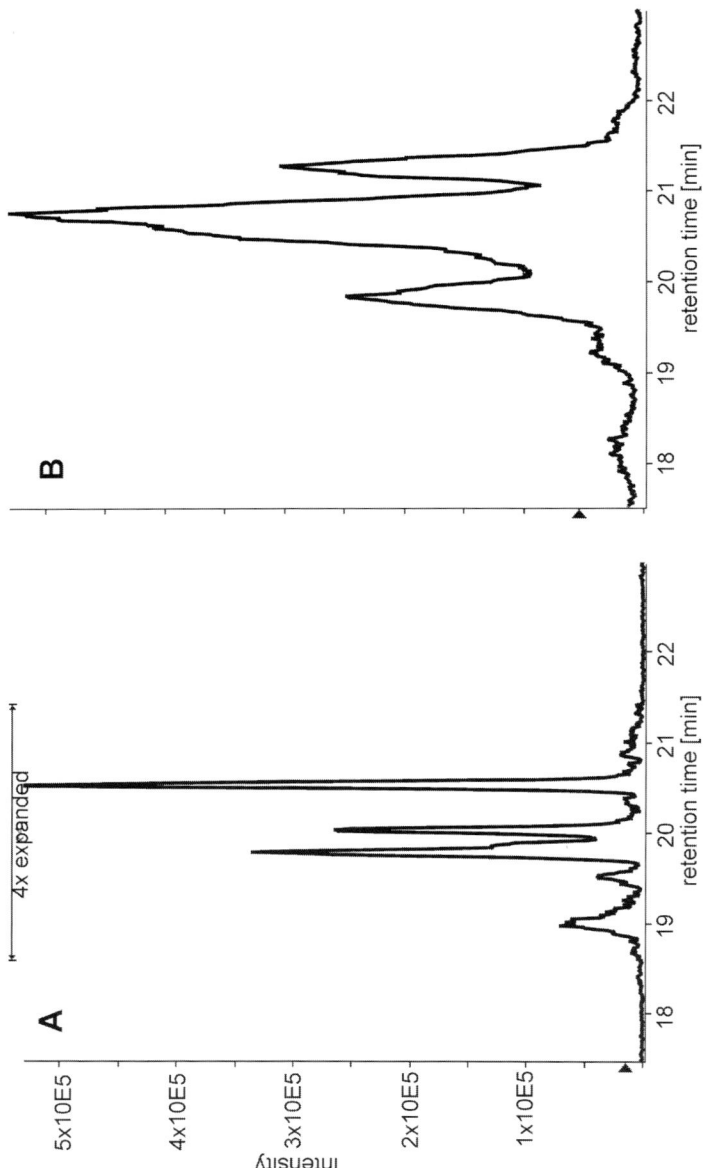

Fig. 2.10.6. (−)-LC−ESI−MS extracted ion chromatogram of C$_7$-SPC (m/z 285), (a) without suppressor, (b) with suppressor (from Ref. [22]).

In samples from a fixed-bed bioreactor, transient intermediates of the aerobic degradation of C_{12}-LAS were detected and identified by means of LC–ESI–MS and tandem MS using negative ionisation [17]. The in-source collision-induced dissociation (CID) spectra obtained from the single quadrupole instrument were almost identical to the daughter ion spectrum (DIS) of the $[M - H]^-$ anions generated by triple-quad MS (Fig. 2.10.10). The authors could confirm the identity of the SPC homologues from C_4 to C_{12} (Table 2.10.1); the C-odd SPC detected at comparably lower levels originated from impurities of C-odd LAS in the starting material [17]. In detail, there was a loss of m/z 60 corresponding to the elimination of acetic acid for the short-chain SPC ($\leq C_7$) (Fig. 2.10.7) or a loss of water, m/z 18, for the SPC with a chain length $>C_7$. The spectra of individual phenyl isomers of a set of SPC homologues differed only slightly, making it difficult to ascribe a definite structure, i.e. the position of the phenyl ring and the location of the carboxylic group.

A significant move forward in the elucidation of the breakdown mechanism of LAS/SPC was accomplished by structural identification of α,β-unsaturated SPC (SPC-2H), which were intermediates in the β-oxidation of the alkyl chain [17]. The SPC-2H eluting in the total ion current chromatogram under RP conditions adjacent to the SPC signals with a slightly shorter retention time exhibited a similar fragmentation pattern resulting from a deprotonated molecular ion being m/z 2 lower than the ones of the SPC (Fig. 2.10.8; Table 2.10.1). There was always a loss of m/z 44, corresponding to the elimination of CO_2, as well as formation of the characteristic fragment ions at m/z 211, 209, and 195 corresponding to olefinic 4-benzenesulfonate, and m/z 145 resulting from olefinic 4-benzene-phenolate. The fragment ion dominating the mass spectrum acquired from the SPC species, m/z 183 (4-styrenesulfonate), could not be found in the spectra obtained from the SPC-2H species $<C_7$-SPC since the double bond already present in the precursor molecule excluded this fragmentation step.

At low cone voltages, the spectra of the methylated SPC showed molecular ions $[M - H]^-$ with m/z of 14 units higher than the corresponding underivatised compounds, e.g. m/z 299 of C_7-SPC (Fig. 2.10.9(a)) [16]. Under in-source CID conditions, an abundant ion signal at m/z 183 was obtained from this homologue, specific of the entire group of LAS and SPC in negative ionisation mode. In the same samples, the authors could detect another series, signals

TABLE 2.10.1

Observed deprotonated molecular ions $[M - H]^-$ and characteristic fragmentation pattern of SPC and SPC-2H.(Reprinted with permission from [17]. Copyright SETAC, Pensacola, Florida, USA.)

Alkyl chain length (C_n)	SPC (m/z)		SPC-2H (m/z)		SPdC (m/z)	
	$[M - H]^-$	Fragment ions	$[M - H]^-$	Fragment ions	$[M - H]^-$	Fragment ions
4	243	183				
5	257	183, 197	255	145, 195, 209, 211		
6	271	183, 211	269	145, 195, 209, 211, 225	301[a]	183, 239,241
7	285	183, 225	283	145, 183, 195, 209, 211, 239	315[a]	183, 253, 255
8	299	183, 281	297	145, 183, 195, 209, 211, 253	329[a]	183, 267, 269
9	313	183, 295	311	145, 183, 195, 209, 211, 267	343[a]	183, 281, 283
10	327	183, 309	325	145, 183, 195, 209, 211, 281		
11	341[a]	183, 323	339	145, 183, 195, 209, 211, 295		
12	355	183, 337	353	145, 183, 195, 209, 211, 309		

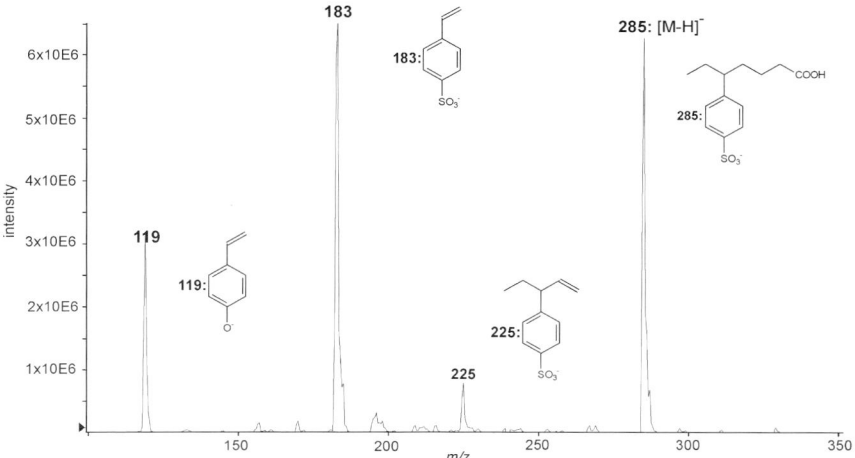

Fig. 2.10.7. ($-$)-LC–ESI–MS spectrum of C_7-SPC. The position of the phenyl ring in the ions m/z 285 and 225 is arbitrarily chosen (Reprinted with permission from [17]. Copyright SETAC, Pensacola, Florida, USA.).

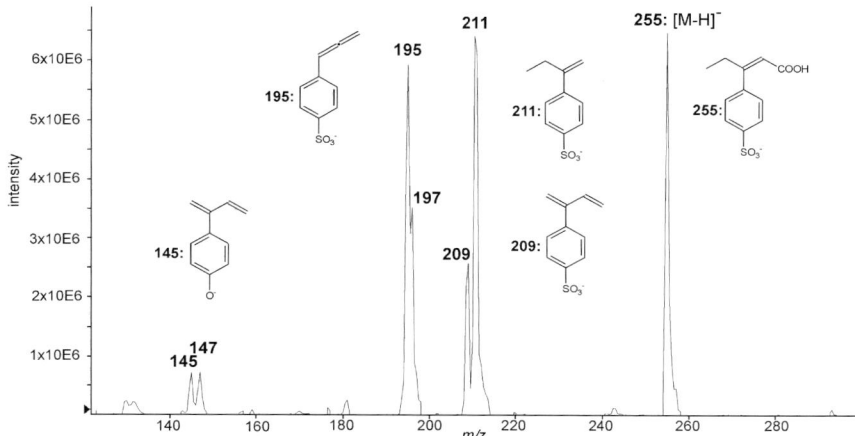

Fig. 2.10.8. $(-)$-LC–ESI–MS spectrum of C_5-SPC-2H. The position of the phenyl ring in the ions m/z 255 and 211 is arbitrarily chosen (Reprinted with permission from [17]. Copyright SETAC, Pensacola, Florida, USA.).

equally spaced of m/z 14, which were postulated to belong to dicarboxylated SPC (SPdC), such as, e.g. m/z 357 of the C_8-homologue. The CID spectrum of this dimethylated metabolite (Fig. 2.10.9(b)) gave, apart from the ion m/z 183, a fragment ion at m/z 285 corresponding to the loss of a methyl acetate moiety of the compound believed to carry the 4-sulfophenyl group in the 3-position of the alkyl chain.

In contrast to this, distinct fragmentation patterns of SPdC were observed in $(-)$-ESI–MS upon direct analysis of these species produced through LAS degradation by a pure culture [17,18]. At elevated cone voltages besides the deprotonated molecular ions (m/z 301, 315, 329 and 343 for C_6-, C_7-, C_8- and C_9-SPdC, respectively) and the fragment ion m/z 183, there were always losses of m/z 60, corresponding to the elimination of acetic acid, as well as m/z 62, corresponding to the simultaneous elimination of CO_2 from one carboxylic group together with H_2O from the second carboxylic group at the opposite end of the alkyl chain (Fig. 2.10.10). Since the latter fragmentation is only possible for the SPdC, it may serve as a fingerprint in the total ion chromatogram. The spectra from the individual SPdC isomers could not be distinguished from each other.

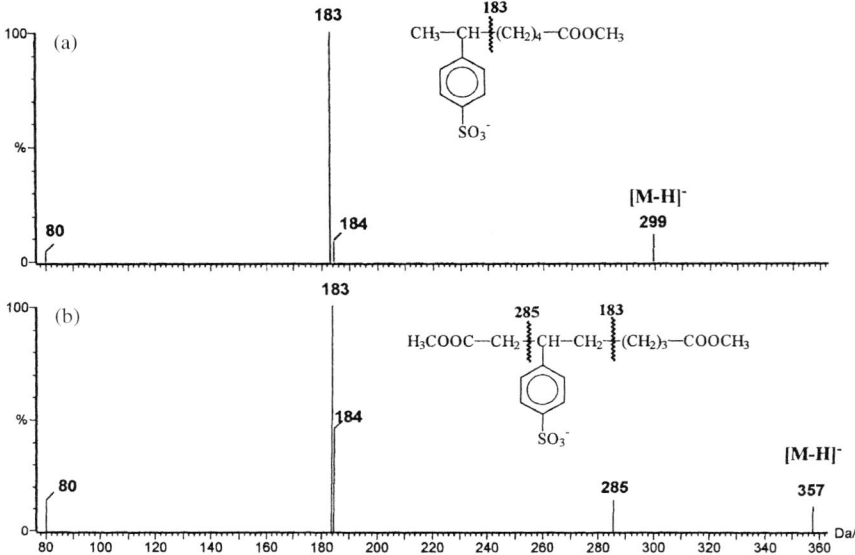

Fig. 2.10.9. (−)-LC−ESI−MS spectra of methylated C$_7$-SPC (a) and dimethylated C$_8$-SPdC (b) (Reprinted with permission from [16]. Copyright 1999. American Chemical Society).

Besides SPC and SPdC, Di Corcia et al. [16] could identify single and double-carboxylated DATS (DATSC and DATSdC) in samples from the test liquor after SPE and methylation in acidic methanol [16]. In contrast to the products originating from the biodegradation of LAS, the CID ESI−MS spectra (Fig. 2.10.11) of methylated DATS metabolites did not show the ion m/z 183, but a series of fragments characterising the length of the two alkyl side chains on the tetralinsulfonate moiety as well as the position of the carboxylic group (for DATSC). The cleavage of both side chains of methylated DATSC or DATSdC produced the diagnostic ions at m/z 223 and 209. Additionally, fragments assigned to the loss of one side chain were observed. For instance, the C$_8$-DATSdC having a methyl and a propyl substitution on the aliphatic six-membered ring produced fragment ions at m/z 297 and 267 formed after cleavage of one of the methylalkanoate moieties.

The derivatisation-LC−ESI−MS method was applied in a further study to determine LAS and DATS along with SPC, SPdC, DATSC and DATSdC in aqueous samples from sewage treatment plants and rivers [19].

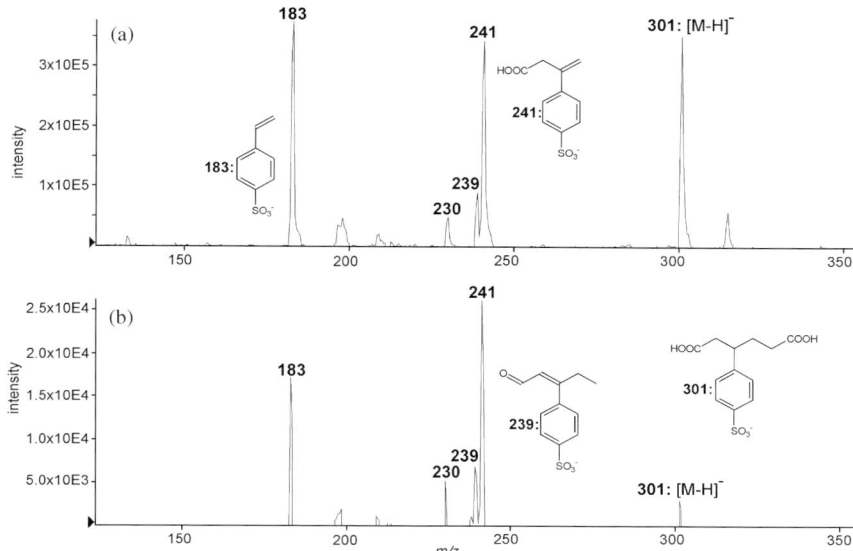

Fig. 2.10.10. $(-)$-LC–ESI–MS (a) and MS–MS spectra (b) of C_6-SPdC. The position of the phenyl ring in the ions m/z 301, 241 and 239 is arbitrarily chosen (Reprinted with permission from [17]. Copyright SETAC, Pensacola, Florida, USA.).

The importance of a good chromatographic separation prior to the mass spectrometric analysis has to be underlined in face of the large number of possible surfactant-derived isobaric compounds to be encountered during the microbial degradation of LAS containing traces of DATS. For example, C_{11}-LAS and methylated C_8-DATSC yield the same molecular ion $[M - H]^-$ at m/z 311; identical masses of the $[M - H]^-$ ions (m/z 355) are calculated for methylated C_{11}-SPC and dimethylated C_8-DATSdC.

With respect to quantification of SPC in environmental samples, one has to deal with the scarcity of available reference compounds. Of the four SPC homologues, C_6, C_8, C_{10} and C_{11}, analysed in coastal waters after SPE [20], the response factor of C_6-SPC in the ESI–MS was found to be about three times lower than those of the longer chain species. This was attributed to the much higher water content in the eluent at the time of elution of C_6-SPC, which resulted in a notable reduction of the ionisation efficiency. This fact was taken into account [17] by quantifying individual SPC based on the responses of the two

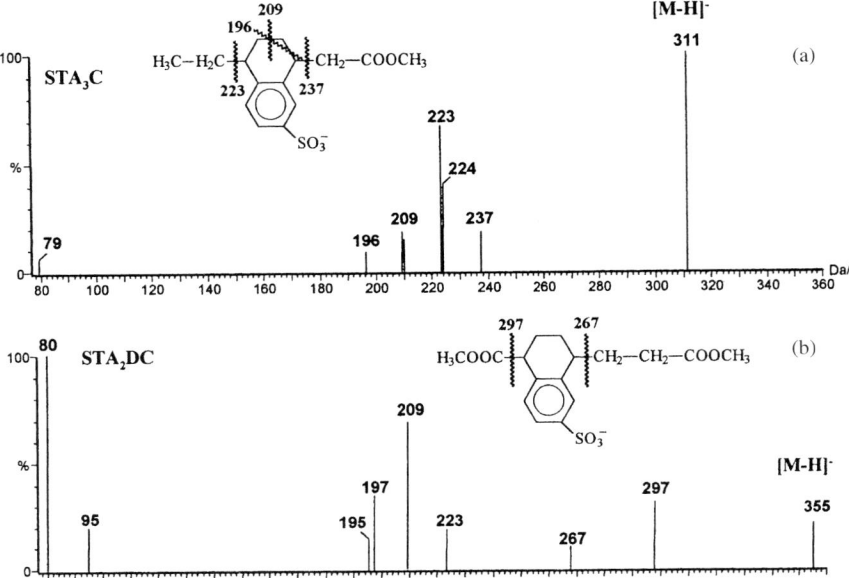

Fig. 2.10.11. (−)-LC−ESI−MS spectra of methylated C_8-DATSC (a) and dimethylated C_8-DATSdC (b) (Reprinted with permission from [16]. Copyright 1999. American Chemical Society).

homologues C_5-SPC and C_{11}-SPC. According to the chromatographic behaviour shown in Fig. 2.10.5, the early eluting SPC (C_4–C_6), generated out of a commercial LAS mixture in a laboratory degradation experiment, were quantified using C_5-SPC, while the quantitation of those species eluting closely to C_{11}-SPC, i.e. C_7-SPC and higher, was performed relating to this longer chain homologues.

In a study of the distribution of LAS and SPC at a salt marsh heavily impacted by untreated urban effluent [14], SPC with chain length from C_7 to C_{13} were examined after RP-C_{18} chromatography by (−)-ESI–LC−MS. As the authors only had short-chain SPC (C_2–C_5) available, they opted to quantify the longer chain species by measuring their peak areas by means of C_{10}-LAS homologue. The SPC data were expressed as equivalent C_{10}-LAS. For the mass spectrometric confirmation of the SPC identity, an elevated voltage was applied to the extraction cone yielding an abundant diagnostic fragment ion m/z 183.

The non-linear ABS had been proved to be quite resistant against microbial degradation in the aquatic environment [21], but degradative products analogous to SPC could be identified by LC–ESI–MS–MS [22]. As with ABS, these alkyl chain-branched carboxylated compounds (bSPC) also yielded the characteristic fragment m/z 197. After chromatographic separation, the MS–MS analysis of Philippine surface waters in multiple reaction monitoring mode demonstrated that the long-chain bSPC ($C_{10}–C_{12}$) accounted for approximately 70% of the total SPC amount, but of the $<C_8$ homologues only linear SPC were detectable [10].

For the characterisation of the biodegradation intermediates of C_{12}-LAS, metabolised in pure culture by an α-proteobacterium, Cook and co-workers [23] used matrix-assisted laser desorption/ionisation (MALDI)-time of flight (TOF)-MS as a complementary tool to HPLC with diode array detection and ^{1}H-nuclear magnetic resonance. The dominating signal in the spectrum at m/z 271 and 293 were assigned to the ions $[M - H]^-$ and $[M - 2H + Na]^-$ of C_6-SPC. Of minor intensity were the ions with m/z 285 and 299, interpreted to be the deprotonated molecular ions of C_7- and C_8-SPC, respectively.

REFERENCES

1 S.D. Scullion, M.R. Clench, M. Cooke and A.E. Ashcroft, *J. Chromatogr. A*, 773 (1996) 207.

2 M.J.-F. Suter, R. Reiser and W. Giger, *J. Mass Spectrom.*, 31 (1996) 357.

3 W.-H. Ding, J.-H. Lo and S.-H. Tzing, *J. Chromatogr. A*, 818 (1998) 270.

4 H.Fr. Schröder and K. Fytianos, *Chromatographia*, 50 (1999) 583.

5 H.-Q. Li, F. Jiku and H.Fr. Schröder, *J. Chromatogr. A*, 889 (2000) 155.

6 K. Bester, N. Theobald and H.Fr. Schröder, *J. Chromatogr.*, 45 (2001) 817.

7 U. Ceglarek, J. Efer, A. Schreiber, E. Zwanziger and W. Engewald, *Fresenius J. Anal. Chem.*, 365 (1999) 674.

8 A. Borgerding and R.A. Hites, *Anal. Chem.*, 64 (1992) 1449.

9 H.Fr. Schröder, *J. Chromatogr. A*, 777 (1997) 127.

10 P. Eichhorn, M.E. Flavier, M.L. Paje and T.P. Knepper, *Sci. Total Environ.*, 269 (2001) 75.

11 T.P. Knepper and P. Eichhorn (unpublished results).

12 R.D. Swisher, *Surfactant Degradation*, Marcel Dekker, New York, 1987.

13 T.P. Knepper and M. Kruse, *Tenside Surf. Det.*, 37 (2000) 41.

14 E. González-Mazo, M. Honing, D. Barceló and A. Gómez-Parra, *Environ. Sci. Technol.*, 31 (1997) 504.

15 D. Temesi and B. Law, *LC–GC Int.*, 3 (1999) 175.

16 A. Di Corcia, F. Casassa, C. Crescenzi, A. Marcomini and R. Samperi, *Environ. Sci. Technol.*, 33 (1999) 4112.

17 P. Eichhorn and T.P. Knepper, *Environ. Toxicol. Chem.*, 21 (2002) 1.

18 W. Dong, P. Eichhorn, S. Radajewski, D. Schleheck, K. Denger, T.P. Knepper, J.C. Murrell and A.M. Cook.

19 A. Di Corcia, L. Capuani, F. Casassa, A. Marcomini and R. Samperi, *Environ. Sci. Technol.*, 33 (1999) 4119.

20 J. Riu, E. González-Mazo, A. Gomez-Parra and D. Barceló, *Chromatographia*, 50 (1999) 275.

21 H.L. Webster and J. Halliday, *Analyst*, 84 (1959) 552.

22 P. Eichhorn, PhD Thesis, University of Mainz, Germany, 2001, http://deposit.ddb.de/cgi-bin/dokserv?idn = 962727733.

23 D. Schleheck, W. Dong, K. Denger, E. Heinzle and A.M. Cook, *Appl. Environ. Microbiol.*, 66 (2000) 1911.

2.11 ATMOSPHERIC PRESSURE IONISATION MASS SPECTROMETRY—VIII. ANIONIC SURFACTANTS: LC–MS OF OTHER ANIONIC SURFACTANTS

Horst Fr. Schröder

2.11.1 Introduction

At the end of the 1990s statistics show that the non-ionic surfactants achieved the highest growth in production rates world-wide, though anionic surfactants (anionics) maintained the dominant position in the surfactant market. Today they are produced in a larger variety by the petrochemical industry than all other types of surfactants. Their production spectrum covers alkyl sulfates (ASs), secondary alkane sulfonates (SASs) and aryl sulfonates and carboxylates via derivatives of partly fluorinated or perfluorinated alkyl surfactants to compounds with an alkylpolyglycolether substructure combined with an anionic moiety such as alkylether sulfates (AESs), phosphates, phosphonates or carboxylates.

In addition, aromatic species alkylphenol ethoxylate (APEO) derivatives of carboxylates, sulfonates, sulfates and phosphates also belong to the large group of anionic surfactants. Only some of them, e.g. the branched alkylbenzene sulfonates (ABSs) in the 1970s or the linear alkylbenzene sulfonates (LASs), reported in Chapter 2.10, today are produced in quantities of $\geq 100,000$-tons per year. Some of them with very low production rates, e.g. the fluorinated compounds, also termed as super-surfactants [1], can be used for special applications [1,2] because of their properties such as stability against heat, acidic or caustic conditions in parallel to their surface activity. Included in this group of anionic surfactants are also some novel surfactants with outstanding properties, e.g. the gemini or dimeric surfactants [3], which under identical critical micelle concentration (cmc) exhibit an improved surface activity compared with the other monomeric conventional surfactants. Some silicon surfactants with their 'unique feature to be insoluble in water and oils' [4] as well as some cleavable surfactants should be mentioned [5], which in a controlled way break down into non-surface active products—they also belong to the large group of anionics.

Comprehensive Analytical Chemistry XL
T.P. Knepper, D. Barceló and P. de Voogt (Eds)
© 2003 Elsevier Science B.V. All rights reserved

In parallel to the development of new types of surfactants, the established determination methods lost their specificity. So photometric and titrimetric methods applied to industrial blends or detergent formulations were less sensitive and selective or even insensitive compared with methods that were applied in combination with separation techniques prior to detection. The established methods, together with the more current analytical techniques, were reviewed in detail and analytical strategies were provided by Schmitt [6]. Compared with product analyses, the qualitative and moreover the quantitative determination of surfactants in environmental samples is more complicated, as already mentioned in the chapter on atmospheric pressure ionisation mass spectrometric (API–MS) analysis of non-ionics, and will be reported on in selected chapters of this book. Here, the substance-group-specific (non-specific sum parameter) methylene-blue method is still used nowadays for screening and as a standard determination method for anionic surfactants [7,8]. Compared with determination methods applied for the other types of surfactants, this method can be considered the most reliable substance-group-specific method [9,10] applied to environmental samples. Nevertheless, results of quantitative determination can be subjected to interferences arising from other compounds of similar structure. This leads to false positive as well as negative results and consequently to alternative determination methods. The more promising analytical techniques in surfactant analyses are based on molecular spectroscopic methods such as infrared (IR) and Raman spectroscopy [11] or nuclear magnetic resonance spectroscopy (NMR) [12]. The most promising methods, however, combined without or with a chromatographic pre-separation such as liquid chromatography (LC), supercritical fluid chromatography (SFC) or capillary zone electrophoresis (CZE), are all interfaced to MS [13] or gas chromatography (GC) coupled to MS (GC–MS) after derivatisation of the analytes.

These types of surfactants have the largest share of world-wide total production [14] and the highest application rates in household, trade and industrial processes. As these compounds were handled mainly in aqueous systems, they were consequently discharged with the waste-water. After a more or less efficient wastewater treatment (WWT) process, which results in an elimination, i.e. degradation by wastewater biocoenosis and/or adsorption of the surfactants to the sludge, these polar compounds and their metabolites either reach the environment by wastewater discharges or are adsorbed to wastewater sludge, respectively. Elimination by adsorptive effects in the WWT process leads to the problem shifting in the case of sludge that is not incinerated but used as fertiliser in agriculture for soil amelioration. Nevertheless, most of these anionic surfactants can easily be eliminated in the WWT process by degradation, mineralisation or adsorption to the sludge and therefore can only be found in the inflows of biological wastewater treatment plants (WWTP). Other hardly degradable surfactants [15], however, were not only found in feeds and effluents of WWTP but also in surface [16,17], ground waters or even in drinking water after soil filtration processes and activated carbon treatment [18]. Moreover, in marine ecosystems [19–21], along with different types of their biochemical degradation products, surfactants were observed [22–24], too.

Compared with pollutants such as pesticides only a few results—but an increasing number—dealing with the application of API–MS methods for the determination of anionic surfactants had been reported in the literature until now and will be cited later on. This is in accordance with the problems arising in the substance-specific analysis of polar compounds contained in complex mixtures. Quite a few methods concerning this topic were reviewed [22,25,26] some years ago and the number of papers dealing with the analyses of anionic surfactants in the form of blends, formulations and moreover in environmental samples is also limited. So substance-specific MS analysis of anionic surfactants was applied in combination with desorption methods [27–39], in thermospray (TSI) studies [10,18,25, 40–44] or using API techniques [22,23,45–55]. Problems with identification, however, were observed in the analyses of so-called primary degradation products and metabolites, as observed especially for LAS [22,24,56], as reported in another chapter of this book, or with hardly degradable metabolites of anionic APEO derivatives [15].

The qualitative determination of anionic surfactants in environmental samples such as water extracts by flow injection analysis coupled with MS (FIA–MS) applying a screening approach in the negative ionisation mode sometimes may be very effective. Using atmospheric pressure chemical ionisation (APCI) and electrospray ionisation (ESI), coupled with FIA or LC in combination with MS, anionic surfactants are either predominantly or sometimes exclusively ionised in the negative mode. Therefore, overview spectra obtained by FIA–MS(−) often are very clear and free from disturbing 'matrix' components that are ionisable only in the positive mode. However, the advantage of clear overview spectra conversely leads to tandem mass spectrometry (MS–MS) product ion spectra with a quite reduced number of product ions compared with spectra generated in the positive MS–MS mode and therefore are less informative in the identification of surfactants [25]. Despite this disadvantage of a poor fragmentation in the negative ionisation mode, the MS determination with the high selectivity for detection or characterisation of anionic surfactants has become the method of choice.

This present chapter covers the MS analyses of most types of anionics applied with the exception of LAS or ABS, which are separately discussed in Chapter 2.10.

2.11.2 Analyses of anionic surfactants

Alkane and alkene sulfonates
Alkane- and alkene sulfonates presented with their general structural formulas in Fig. 2.11.1 belong to the group of anionic surfactants. In contrast to aromatic sulfonates, the aliphatic linear alkane sulfonates (general formula: $C_nH_{2n+1}-SO_3H$; structural formula see Fig. 2.11.1(Ia)) and secondary alkane sulfonates (SASs; structural formula see Fig. 2.11.1(II)) with the modified formula $CH_3-(CH_2)_n-CH(SO_3^-)-(CH_2)_x-CH_3$ ($n + x = 10-15$) are detected nearly exclusively in the inflows of WWTPs [22,41,44]. These compounds are mainly constituents of surfactant formulations applied as household cleaning agents. A mixture of isomeric and homologue compounds was detected by ESI–FIA–MS(−) in the negative APCI mode in an industrial blend as $[M − H]^-$ ions at m/z 263, 277, 291, 305, 319 and 333 equally spaced with $\Delta m/z$ 14 because of a variation in the number of alkyl chain links [47]. The same peak shapes were recognised under ESI(−) conditions in

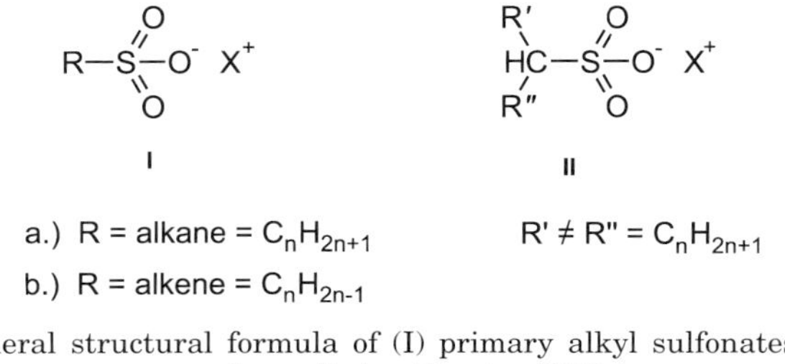

a.) R = alkane = C_nH_{2n+1} R' ≠ R" = C_nH_{2n+1}
b.) R = alkene = C_nH_{2n-1}

Fig. 2.11.1. General structural formula of (I) primary alkyl sulfonates (PAS) and (II) secondary alkyl sulfonates (SASs).

solutions containing SAS in low concentrations. An increase in SAS concentrations, however, resulted in monomeric SAS molecular $[M - H]^-$ ions at m/z 263, 277, 291, 305, 319 and 333 and their dimeric, mono charged counterparts $[2M - H]^-$ at m/z 555, 569, 583, 597 and 611 observed under elevated concentrations with a ratio of 8:2 [22] when APCI or ESI–FIA–MS(−) ionisation was applied (Fig. 2.11.2(a)).

The RP-C_{18} LC separation of this mixture of aliphatic linear alkane sulfonates (C_nH_{2n+1}–SO_3^-) and SAS from an industrial blend in combination with APCI–LC–MS(−) detection is presented as total ion mass trace (Fig. 2.11.2(f)) together with selected mass traces (m/z 277, 291, 305 and 319 for ($n + x = 11$–14)) in Fig. 2.11.2(b)–(e), respectively. The resolved mass traces proved the presence of large number of isomers of every SAS homologue in this blend. This complexity is generated because of the linear isomer precursor and the mixture of branched alkyl precursor compounds applied to chemical synthesis [22]. In parallel to elution behaviour observed in GC the branched isomers of alkylsulfates in LC separation were expected to elute first.

LC separation applying ion chromatography in combination with ion spray mass spectrometric detection was applied for the examination of a synthetic mixture of alkyl sulfonates (C_nH_{2n+1}–SO_3^-; $n = 8$) and AS with different alkyl chain lengths in the selected ion monitoring (SIM) ESI–MS(−) mode [53]. Selected ion current profiles provided the separation of the compounds. The ionic matrix constituents of the eluent were removed by a suppressor module prior to MS detection to improve the signal to noise (S/N) ratio.

Alkene sulfonates as olefinic compounds with the general formula C_nH_{2n-1}–SO_3^- (cf. Fig. 2.11.1(Ib)) were studied by API–MS(−) methods. The industrial blend contained two homologues with $C_{12}H_{23}$ and $C_{14}H_{27}$ alkyl chains that were ionised in the negative mode resulting in the form of $[M - H]^-$ ions at m/z 275 and 303 as presented in the APCI–FIA–MS(−) spectrum in Fig. 2.11.3. RP-C_{18} column chromatography performed with this unsaturated surfactant mixture confirmed comparable variations in structure as observed with the SAS blend [22].

The application of collision-induced dissociation (CID(−)) to n-alkane- or linear (primary) alkyl sulfonates (PASs; C_nH_{2n+1}–SO_3^-) or their isomers, the SASs with the modified formula CH_3–$(CH_2)_n$–$CH(SO_3^-)$–$(CH_2)_x$–CH_3, and alkene sulfonates using FIA–MS–MS(−) conditions leads to only one, not quite specific, fragment ion at m/z 80,

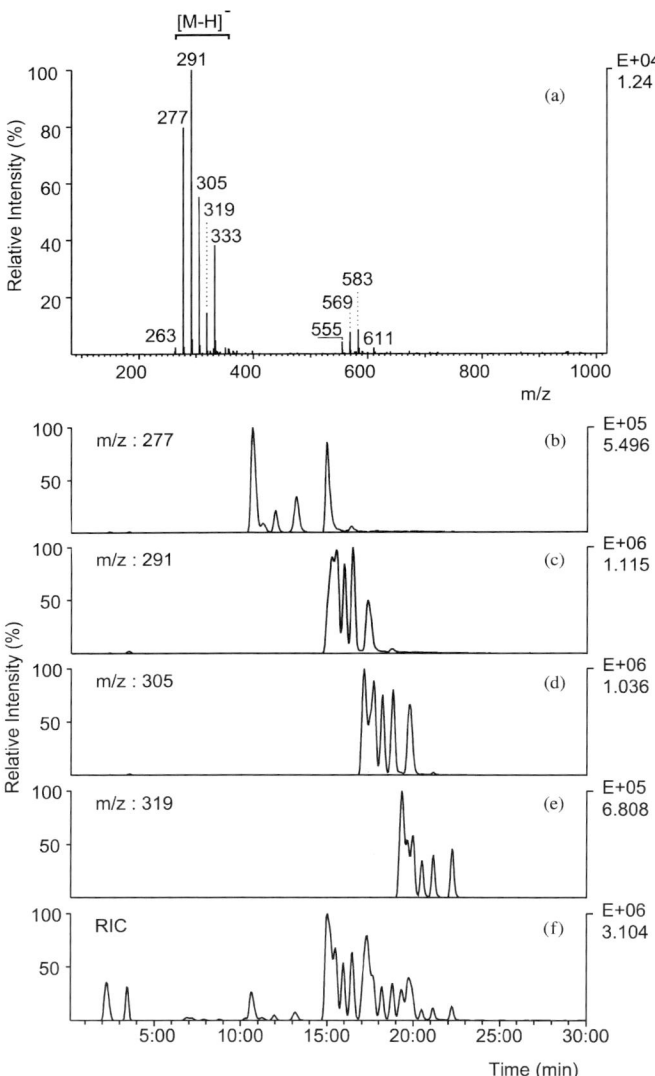

Fig. 2.11.2. (a) ESI–FIA–MS(−) overview spectrum of SAS (CH_3–$(CH_2)_n$–$CH(SO_3H)$–$(CH_2)_x$–CH_3) presenting $[M − H]^-$ and $[2M − H]^-$ ions; (f) APCI–LC–MS(−) total ion current chromatogram and (b)–(e) selected mass traces of SAS blends as in (a); gradient elution separated on RP-C_{18} column [22,61].

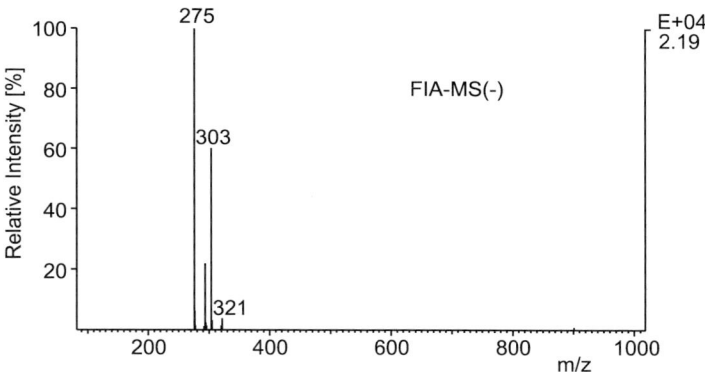

Fig. 2.11.3. APCI–FIA–MS(−) overview spectrum of an alkene sulfonate blend (general formula C_nH_{2n-1}–SO_3H) [22].

i.e. irrespective of the type of surfactant—alkane sulfonates or alkene sulfonates—or other sulfonates ESI or APCI–CID(−) at any collision energy, collision gas pressure or extraction voltage lead to only this product ion at m/z 80 ($[SO_3]^-$). As alkyl or alkenyl moieties do not result in negative ions, the same result was observed by generating negative product ions from the unsaturated alkene homologues. The parent ion scan of m/z 80 used for the confirmation of all types of saturated and unsaturated sulfonates in environmental samples and vice versa here is a diagnostic method for the recognition of these compounds [22,49] but does not allow any differentiation between sulfonate surfactants and other sulfonates.

Alkylsulfates

An industrial blend of ASs presented with its general structural formula in Fig. 2.11.4 was examined to elaborate a method for their quantification in trace amounts. ASs (C_nH_{2n+1}–O–SO_3^-) only were applied as surfactants in special applications. Moreover, these compounds are the basis of synthesis in the production of AES and therefore the trace analysis of AS in AES is an important task to estimate impurities. Applying APCI–FIA–MS(−), the overview spectrum in Fig. 2.11.5(a) containing $[M - H]^-$ ions at m/z 265 and 293 was obtained. The mixture then was separated by RP-C_{18} and recorded by ESI–MS(−). The same $[M - H]^-$ ions could be recognised that belonged to

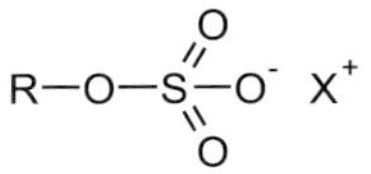

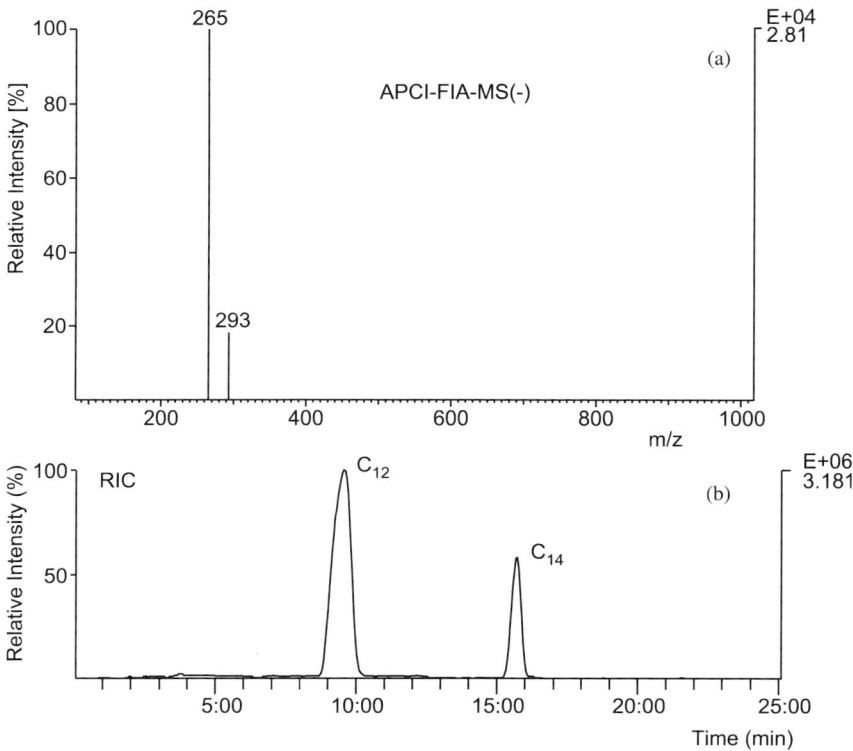

Fig. 2.11.4. General structural formula of alkyl sulfates (ASs).

the C_{12} and C_{14} AS homologues (Fig. 2.11.5(b)). These compounds resulted in $[M - H]^-$ ions at m/z 265 and 293, as observed by APCI, but in addition adduct ions at m/z 363 and 391 can be observed in LC separation in mass trace analysis. These ions at m/z 363 and 391 were

Fig. 2.11.5. (a) APCI–FIA–MS($-$) overview spectrum of an AS blend (C_nH_{2n+1}–O–SO_3H); (b) ESI–LC–MS($-$) total ion current chromatogram and selected mass traces of $[M - H]^-$ ions of C_{12}- and C_{14}-AS homologues as in (a); gradient elution separated on a RP-C_{18} column [22].

sulfuric acid adduct ions of the alkylsulfates ($[M + 97]^-$), which could be identified by LC–ESI–MS–MS($-$) and resulted in two product ions, the $[M - H]^-$ ions at m/z 265 and 293, besides the product ion at m/z 97 ($[HSO_4]^-$). LC–ESI–MS–MS($-$) of parent ions at m/z 265 and 293, however, resulted in a product ion at m/z 97 while the parent ions were no longer observed [22].

Conboy et al. [53] studied the synthetic mixtures of ASs (C_nH_{2n+1}–O–SO_3^-; $n = 8, 10, 12, 14$ and 18). These mixtures were separated by LC applying ion chromatography (IC) and compounds were detected in the SIM mode using ESI($-$) ionisation (ion spray) with MS detection. The MS–MS data of ASs that were presented contained both negatively charged ions, a $[SO_3]^-$ ion at m/z 80 with low intensity, and the prominent $[HSO_4]^-$ ion at m/z 97 [53].

The development of an easy-to-handle method for the qualitative and quantitative determination of surfactants in consumer products was the goal for applying ESI in the FIA–MS($+/-$) mode by direct infusion into the mass spectrometer. In this way C_{12}, C_{14}, C_{16} and C_{18} ASs could be determined besides other anionics (LASs, alkylcarboxylates), non-ionics (alkyl polyglucosides (APGs)) and cationics (quats and ester-quats). The methods applied for concentration and determination (MS–MS) helped to identify the compounds and in addition deuterated internal standards were applied for confirmation [57].

As a by-product of AES analysis, ASs were separated by LC-RP-C_{18} [52], combined with ion spray (ESI($-$)) ionisation and then this method was used for monitoring individual species of the ASs (C_nH_{2n+1}–O–SO_3^-) and AESs by ESI–LC–MS. In addition, RP-C_2 material had been applied for SPE of the pollutants prior to MS analysis.

Octadecyl sulfate sodium salt ($C_{18}H_{37}$–O–SO_3^- Na^+) was examined by laser desorption (LD) Fourier transform ion cyclotron resonance mass spectrometry (FTICR–MS) in the negative mode resulting in $[M - H]^-$ ions. Little fragmentation was observed under these conditions [28].

Alkylether sulfates (AESs)

AESs with the general structural formula C_nH_{2n+1}–O–$(CH_2$–CH_2–O$)_x$–SO_3^- (cf. Fig. 2.11.6) were determined qualitatively using ESI–LC–MS($-$). TSI analysis of these compounds in the negative mode was only possible with reduced sensitivity [22,25], whereas positive ionisation led to the cleavage of the SO_3 moiety, imagining

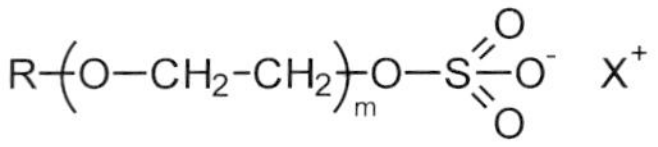

$$R-(O-CH_2-CH_2)_m-O-S\overset{\displaystyle O}{\underset{\displaystyle O}{\lessgtr}}O^- \quad X^+$$

Fig. 2.11.6. General structural formula of alkylether sulfates (AESs; alkylpolyglycol-ethersulfates).

non-ionic surfactants of the polyglycolether type (C_nH_{2n+1}–O–(CH_2–CH_2–O)$_x$–H). API methods, however, were able to ionise these compounds in their original form [22].

FIA–MS overview spectra in the APCI($-$) and ESI($-$) mode proved that the AES blend examined, contained two mixtures of homologues with C_{12} and C_{14} alkyl chains (Fig. 2.11.7(a)). FIA–MS($-$) spectra confirmed that, in addition to AES compounds, the ASs precursor compounds with $n = 12$ and 14 and $x = 0$ (m/z 265 and 293) were also present. In the AES molecules, the C_{12} and C_{14} alkyl chains were coupled with ethoxy chains that could be ionised to a quite different extent. While APCI($-$) ionised AESs up to 14 or 9 polyglycol ether units for the C_{12} and C_{14} homologues, respectively, ESI($-$) ionised AESs up to 6 or 4 PEG units, all equally spaced by $\Delta m/z$ 44 (cf. Table 2.11.1).

Positive ionisation in the API mode such as FIA–ESI–MS($+$) applying ammonium acetate resulted in a complex mixture of ions that were not completely identified but intact molecules of AES in the form of [M + NH$_4$]$^+$ ions could be observed. In the positive APCI mode using ammonium acetate for ionisation support, however, the AESs were ionised in a destructive way and could be observed as [M $-$ SO$_3$ + NH$_4$]$^+$ ions, i.e. as ammonium adduct ions of alkylethoxylates (C_nH_{2n+1}–O–(CH_2–CH_2–O)$_x$–H) [22] as presented in Fig. 2.11.8 starting with m/z 292 (C_{12}) or 320 (C_{14}) (cf. Table 2.11.1). The homologue ions observed under these conditions showed the widest range of homologues because sensitivity for AE ionisation in the positive mode is higher than for AES in the negative mode.

For comparison of the different ionisation methods and detection modes, the results obtained as FIA overview spectra are presented in Figs. 2.11.7 and 2.11.8. Reconstructed ion chromatograms (RIC) of APCI and ESI combined with selected mass traces of all LC separations and, in parallel, the selected standardised mass traces of the C_{12} and C_{14} homologues containing three ethoxy chain links recorded in the negative mode are presented in Fig. 2.11.9. These results again demonstrate the quite large variation in the ionisation efficiency of

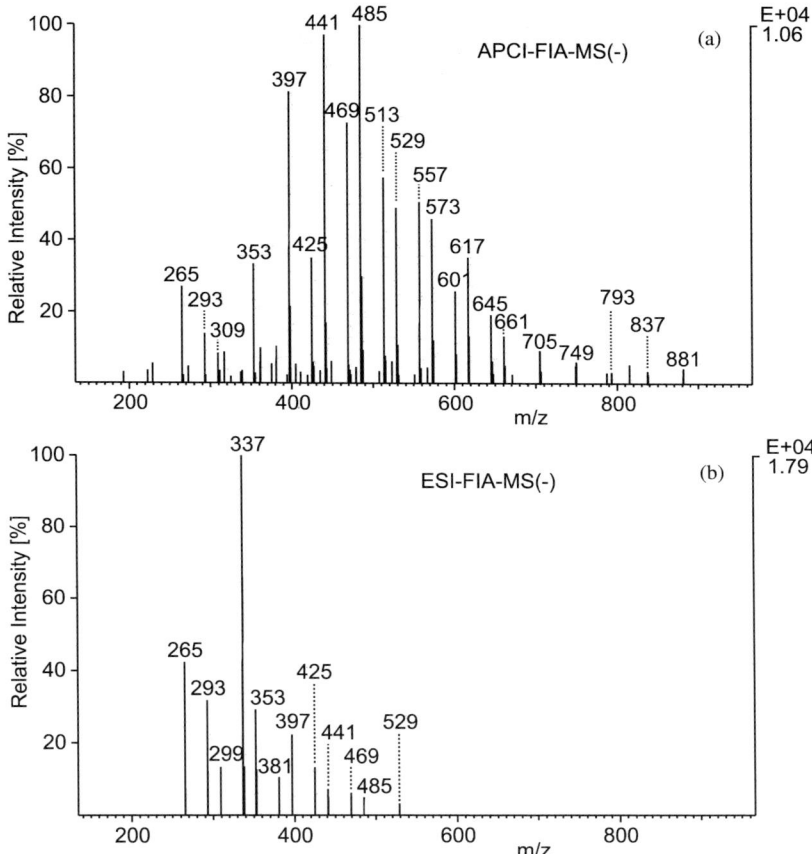

Fig. 2.11.7. (a) APCI–FIA–MS(−) and (b) ESI–FIA–MS(−) overview spectra of an AES blend $(C_nH_{2n-1}-(O-CH_2-CH_2)_n-OSO_3H)$ for visual comparison of ionisation results obtained with the application of different API interfaces in negative ionisation mode [61].

different API interface types as observed before in FIA–MS mode. Under LC conditions, ESI(−) was found to be the most effective in the ionisation of AES homologues containing the whole range of ethoxy chain links ($x = 1$–9) (cf. Fig. 2.11.9(12)). APCI–LC–MS(+) resulted in a total ion current chromatogram (TIC) (cf. Fig. 2.11.9) that started with signals of compounds containing three or more EO units in the ions ($x \geq 3$). This comparison of the LC separations allows the recognition of

TABLE 2.11.1

Range of ionised AES (C_nH_{2n+1}–O–$(CH_2$–CH_2–O$)_x$–SO$_3^-$; $n = 12$ or 14) homologues contained in an industrial blend applying ESI or APCI interfaces in negative or positive mode, respectively, as presented in Figs. 2.11.7 and 2.11.8

n (R = C_nH_{2n+1})	x –$(CH_2CH_2O)_x$–	(m/z)		
		ESI($-$) $[M - H]^-$	APCI($-$) $[M - H]^-$	APCI($+$) $[M - SO_3 + NH_4]^+$
12	0	265	265	n.d.
12	1	299	299	n.d.
12	2	353	353	292
12	3	397	397	336
12	4	441	441	380
12	5	485	485	424
12	6	529	529	468
12	7	n.d.	573	512
12	8	n.d.	617	556
12	9	n.d.	661	600
12	10	n.d.	705	644
12	11	n.d.	749	688
12	12	n.d.	793	732
12	13	n.d.	837	n.d.
12	14	n.d.	881	n.d.
14	0	293	293	n.d.
14	1	337	337	n.d.
14	2	381	381	320
14	3	425	425	364
14	4	469	469	408
14	5	n.d.	513	452
14	6	n.d.	557	496
14	7	n.d.	601	540
14	8	n.d.	645	584
14	9	n.d.	628	628
14	10	n.d.	n.d.	672
14	11	n.d.	n.d.	716

n.d. = not detected.

compounds belonging to the signals with identical retention times (RT), i.e. separation can be observed as AES compounds, ions generated by APCI($+$), however, imagine alkylpolyglycolethers because of the loss of SO$_3$ moiety. In contrast to this, the presence of ammonium acetate under ESI($+$) conditions led to $[M + NH_4]^+$ ions (Fig. 2.11.9). A tremendous loss of sensitivity, however, could be observed in the negative APCI mode (Fig. 2.11.9) [22].

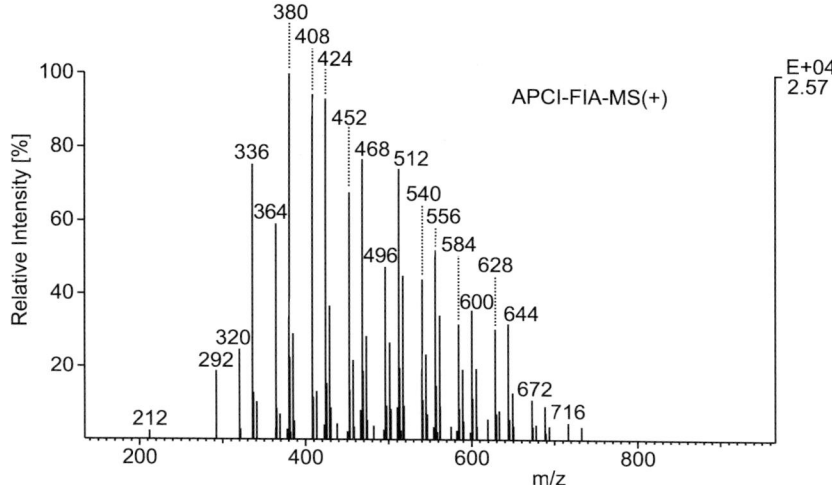

Fig. 2.11.8. APCI–FIA–MS(+) overview spectrum of an AES blend as in Fig. 2.11.7 (C_nH_{2n-1}–(O–CH_2–CH_2)$_n$–OSO_3H) confirming the destructive ionisation results obtained under positive APCI ionisation [61].

Laser desorption FT–MS in the negative mode was used for the determination of laureth ethoxy sulfate ($C_{12}H_{25}$–O–(CH_2–CH_2–O)$_x$–SO_3^-) besides ASs, SASs, PASs and dihexyl sulfosuccinate. All anionic surfactants with the exception of laureth ether sulfate gave one-peak mass spectra originating from the $[M - H]^-$ ions of the sulfate or sulfonated compounds. Laureth ethoxy sulfate gave a series of equally spaced ions ($\Delta m/z$ 44 u) because of a varying number of ethoxy units (PEG) in the molecules of the homologues ionised. Ionisation under the conditions applied took place with little fragmentation [28].

AES type surfactants were examined qualitatively in the ESI–FIA–MS mode but, from the result, the presence of AES seemed to be doubtful. After a LC separation presenting the TIC in the form of a 'contour plot' the identification of AES surfactants in parallel with an amphoteric surfactant mixture of alkylamido propyl betaine type was possible [50].

The sodium salts of propoxy/ethoxy ethersulfates (general structure: C_nH_{2n+1}–O–(CH(CH_3)–CH_2–O)$_x$–(CH_2–CH_2–O)$_y$–SO_3^- Na^+) were examined by FIA–MS, producing complex spectra because of overlapping of the surfactant homologues synthesised by a more or less

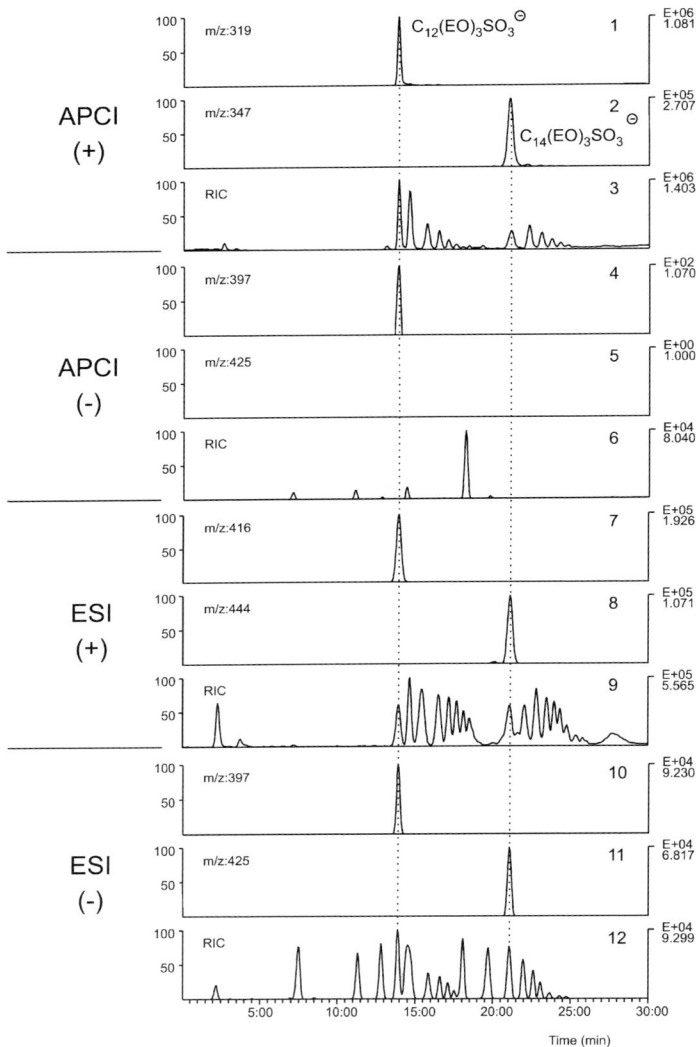

Fig. 2.11.9. APCI- (3, 6) and ESI–LC–MS (9, 12) reconstructed ion current chromatograms (RIC) and selected mass traces of the C_{12} (1, 4, 7 and 10) and C_{14} (2, 5, 8 and 11) homologues containing three ethoxy chain links of AES blend as in Fig. 2.11.7, ionised in the positive or negative mode, respectively; gradient elution separated on RP-C_{18} column [22].

untargeted condensation of the precursor compounds, glycols and alcohols, in synthesis [58]. A homologue mixture with various alkyl chain lengths and a varying number of ethoxylate (EO) and propoxylate (PO) groups in the hydrophilic chain were investigated by ESI–FIA–MS(+/−) and APCI–FIA–MS(+) [59]. When qualitative and quantitative information was desired, ESI(+) produced $[M + Na]^+$ ions from ASs, whereas the EO/PO-alkylether sulfates (EO/PO-AESs) were ionised as $[M + Na]^+$ and $[M + NaSO_3 + H + Na]^+$ ions, which made the interpretation of the spectra more difficult. In contrast, ESI(−) ionisation resulted in one series of $[M − H]^-$ ions enabling qualitative and semi-quantitative analysis. APCI(+) produced no molecular ions for alkylsulfate. For the EO/PO-AESs, the $[M − SO_3 + H]^+$ ions were generated predominantly. LC separation was found to be necessary to get more information about constituents [59].

When AESs of an industrial blend were examined by ESI–MS–MS(−), all AES homologues under standard CID conditions (collision energy: +20–50 eV; collision gas$_{Argon}$ pressure: 1.5 mTorr) resulted in a very simple product ion spectrum that contained the parent ion and only one product ion: m/z 97 ($[HSO_4]^-$) (cf. Fig. 2.11.10). Vice versa, the parent ion scan mode applying m/z 97 helped to recognise AES in blends, formulations and environmental samples [49,60]; however, interferences with sulfate compounds other than AES cannot be excluded. The same fragmentation behaviour was found if APCI–FIA–MS–MS(−) was applied; however, the sensitivity was reduced compared with ESI–FIA–MS–MS(−).

While negative ionisation helped to confirm the presence of AES in the original form resulting in $[HSO_4]^-$ product ions, CID in the positive mode using APCI or ESI first led to the abstraction of SO_3 in the ionisation process imaging the presence of alkylethoxylates. This procedure performed sequentially with AES after a preceding negative ionisation, however, then provides the information of the alkyl chain length and its branching via its AE product spectrum. ESI–LC–MS–MS(+) was applied to characterise the degree of branching in the AES molecule with the molar mass of 442 contained in an industrial blend. This homologue was characterised as $C_{12}H_{25}–O–(CH_2–CH_2–O)_4–SO_3^-$, which could be ionised as the $[M − SO_3 + NH_4]^+$ ion at m/z 380 before it was submitted to ESI–LC–MS–MS(+) examinations. Positive CID experiments led to alkyl ($\Delta m/z$ 14) as well as ethoxy fragments ($\Delta m/z$ 44), which were observed at m/z 71 and 85 or 45, 89 and 133 from the

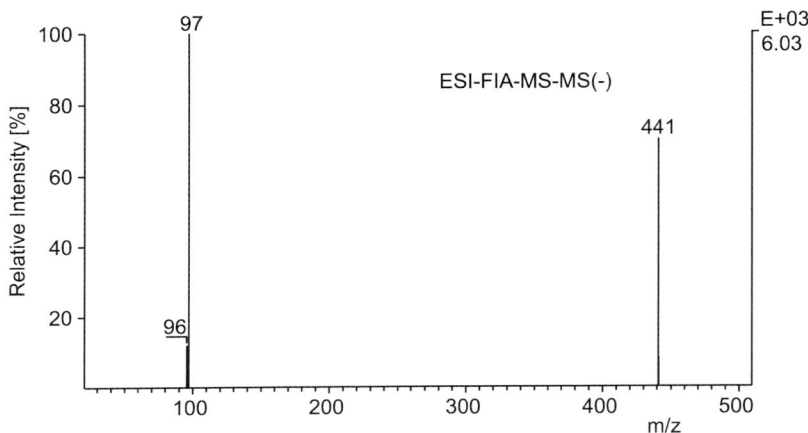

Fig. 2.11.10. ESI–FIA–MS–MS (CID) negative product ion mass spectrum of [M − H]⁻ parent ion at m/z 441 of anionic AES surfactant blend (AES; C_nH_{2n+1}–O–$(CH_2$–CH_2–O$)_x$–SO₃H, $n = 12$, $x = 4$) [22,60,61].

[M − SO₃ + NH₄]⁺ ions, respectively. These results proved a medium degree of branching in the alkyl chain because fragments at m/z 57 and 99 were missing [22,61].

Alkylether carboxylates

Because of their excellent biodegradability, alkylether carboxylates (AECs) (general structure: C_nH_{2n+1}–O–$(CH_2$–CH_2–O$)_n$–CH_2–COO⁻) are used more frequently today in households for cleaning purposes. From the structure shown in Fig. 2.11.11, mass spectra as observed in the ionisation of AES or non-ionic surfactants of the ethoxy type with equally spaced signals ($\Delta m/z$ 44) can be expected [22].

The behaviour of AEC under API–MS conditions in both positive and negative ionisation modes was the same as that observed with AES. Under APCI and ESI–FIA–MS(−), ionisation spectra of this commercial blend of AEC contained the deprotonated molecular [M − H]⁻ ions as presented in the overview spectrum in Fig. 2.11.12. The signals observed represent the equally spaced ions of the C_8H_{17}–O–$(CH_2$–CH_2–O$)_n$–CH_2–COO⁻ homologues ($n = 5$–17; m/z 407–891, equally spaced by $\Delta m/z$ 44) [61]. The same mixture ionised in the positive APCI mode, however, contained a more complex pattern of signals consisting

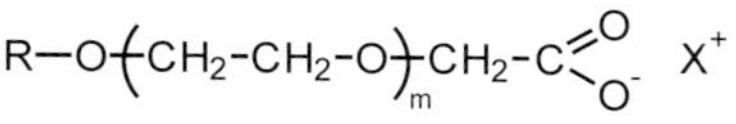

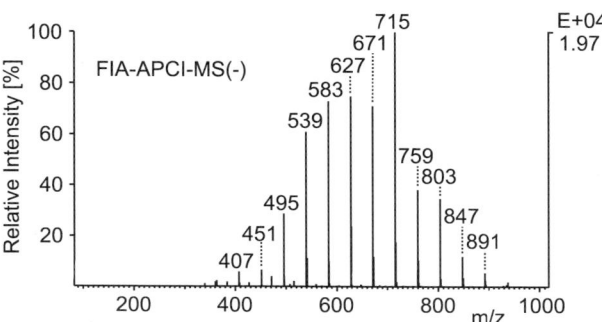

Fig. 2.11.11. General structural formula of alkylether carboxylates (AECs; alkylethoxy carboxylates).

of adduct ions ($[M + Na]^+$, $[M + NH_4]^+$), $[M - CO_2 + NH_4]^+$ ions of the decarboxylated AEC precursor and in addition $[M + NH_4]^+$ ions of the precursor AE, which dominate the spectrum, as also observed in the RIC under positive ionisation (cf. Fig. 2.11.13(b)) [22]. RP-C$_{18}$ LC separations in combination with APCI(+/−) ionisation of this industrial blend containing traces of the synthesis precursor AE compounds were performed. The total ion current traces recorded in the negative as well as positive APCI modes in the presence of ammonium acetate are presented in Fig. 2.11.13(d) and (b), respectively. Precursor compounds (AE) from synthesis could be confirmed as impurities in the commercial blend under positive ionisation while negative ionisation proved its selectivity. Positively recorded mass trace at m/z 558 (a) represented the ion of the decarboxylated compound (AEC-CO$_2$) ($[C_8H_{17}-O-(CH_2-CH_2-O)_9-CH_2-H*NH_4]^+$) which, however, was separated as AEC (molar mass 584; $C_8H_{17}-O-(CH_2-CH_2-O)_9-CH_2-COOH$) prior to ionisation combined with its decarboxylation. The small signal for AEC (RT 2.5−4 min) compared with the large AE signal (RT = 8−10 min) in the RIC(+) Fig. 2.11.13(b) demonstrates the insensitivity of these compounds to MS detection performed in the positive mode, whereas AE compounds exhibit an over proportional response. The negatively

Fig. 2.11.12. APCI–FIA–MS(−) overview spectrum of an AEC blend ($C_nH_{2n+1}-O-(CH_2-CH_2-O)_x-CH_2-COO^-$ H$^+$; $n = 8$) [61].

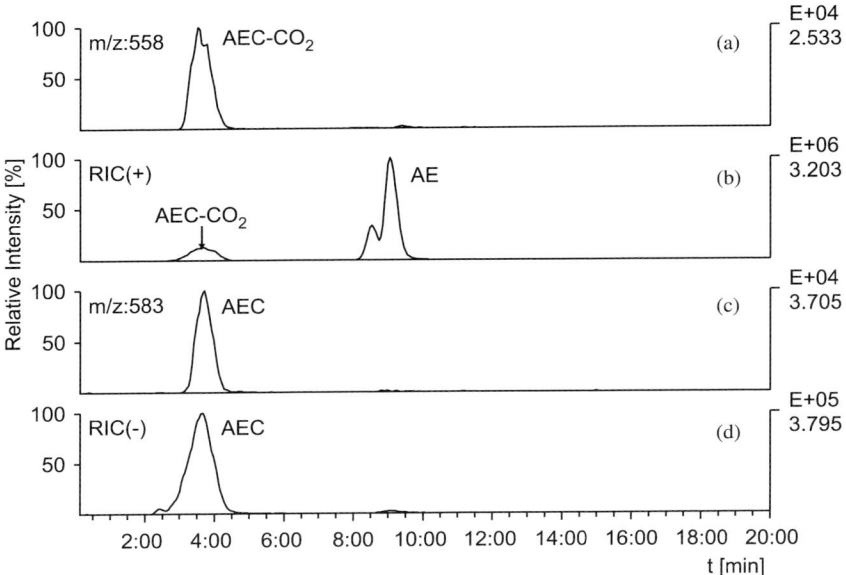

Fig. 2.11.13. APCI–LC–MS(+/−) RICs and selected mass traces of an AEC blend as in Fig. 2.11.12; gradient elution separated by RP-C_{18} column [61].

recorded RIC in Fig. 2.11.13(d), however, contained the $[M − H]^-$ ions, while in the mass trace at m/z 583 the negatively charged $C_8H_{17}–O–(CH_2–CH_2–O)_9–CH_2–COO^-$ ion could be observed [61].

For identification of AEC, two different strategies for substance-specific identification are amenable in API–FIA–MS–MS [22]. The application of CID in the positive mode led to a destructive ionisation that resulted in the loss of CO_2, as also observed with other anionic AE derivatives. Under these conditions, the precursor parent ions from a commercial blend subjected to API-CID(+) resulted in a characteristic pattern of product ions containing alkyl as well as ethoxy fragments (m/z 57, 71 and 113 or 89, 133 and 177) slightly different from MS–MS(+) spectra observed with conventional AE compounds. Nevertheless, the presence of AE compounds was imagined because of decarboxylation of AEC. Therefore, the analysis of the mixture of AECs was performed in the ESI–LC–MS–MS mode because of interferences of AE in FIA mode. The ESI–LC–MS–MS(+) spectrum of the decarboxylated parent ion at m/z 558 ($[C_8H_{17}–O–(CH_2–CH_2–O)_9–CH_2–H^*NH_4]^+$) as also observed in LC separation is presented together with the fragmentation scheme under positive ionisation in Fig. 2.11.14(a) [61]. The proposed fragmentation confirmed the decarboxylation under ESI–LC–MS–MS(+).

Negative CID applying APCI or ESI was performed successfully in the FIA–MS–MS mode. Though the spectra could be recorded with lower sensitivity, selectivity, however, could be improved considerably. The product ion spectrum of the respective $[M - H]^-$ parent ion of the non-decarboxylated $C_8(EO)_9$ homologue ion at m/z 583 contained a very specific product ion with m/z 101 as shown in Fig. 2.11.14(b). The selectivity of this product ion could be confirmed in the parent ion mode for the confirmation of these compounds in blends, formulations and environmental samples examined [22].

Alkylarylether derivatives—alkylarylether sulfates, -sulfonates, -phosphates and di-alkyl arylether-carboxylates
Besides the APEO compounds used as non-ionic surfactants, anionic derivatives of APEO were also produced as sulfonates, sulfates, phosphates and carboxylates and were used for special applications. The compounds are presented in Fig. 2.11.15(I)–(IV) with their structural formulae. Their general formulae are: I (di-alkylphenolether carboxylates $(C_nH_{2n+1})_2$–C_6H_3–O–$(CH_2$–CH_2–O$)_x$–CH_2–CO_2^-; di-APEC); II (alkylphenolether-sulfonates: C_nH_{2n+1}–C_6H_4–O–$(CH_2$–CH_2–O$)_x$–CH_2–SO_3^-; APEO-SO$_3$); III (alkylphenolether-sulfates: C_nH_{2n+1}–C_6H_4–O–$(CH_2$–CH_2–O$)_x$–SO_3^-; APEO-SO$_4$); IV (alkylphenolether-phosphates: C_nH_{2n+1}–C_6H_4–O–$(CH_2$–CH_2–O$)_x$–P(OH)$_2$O$^-$; APEO-PO$_4$). These anionic surfactant compounds were built up from the basic structural element 'APEO'. In their chemical syntheses, APEOs are coupled with –COO$^-$, –SO$_3^-$, –O–SO$_3^-$ or –O–P(OH)$_2$O$^-$ moieties, respectively. All alkyl chains of the compounds examined contained nine carbon atoms and the carboxylates were bis-(nonyl)-phenolethoxylate compounds. These mixtures were studied as industrial blends [22].

In the negative FIA–MS spectra, the sulfate, phosphate and carboxylate of these anionic alkylphenol ether derivatives in parallel to the aliphatic ethoxy surfactants show equally spaced signals with $\Delta m/z$ 44. However, according to the ionisation method applied—APCI or ESI in the negative or positive mode—and also that observed in the ionisation process of AES, equally spaced signals came either from the anionic compounds themselves or from the alkylphenol ethers

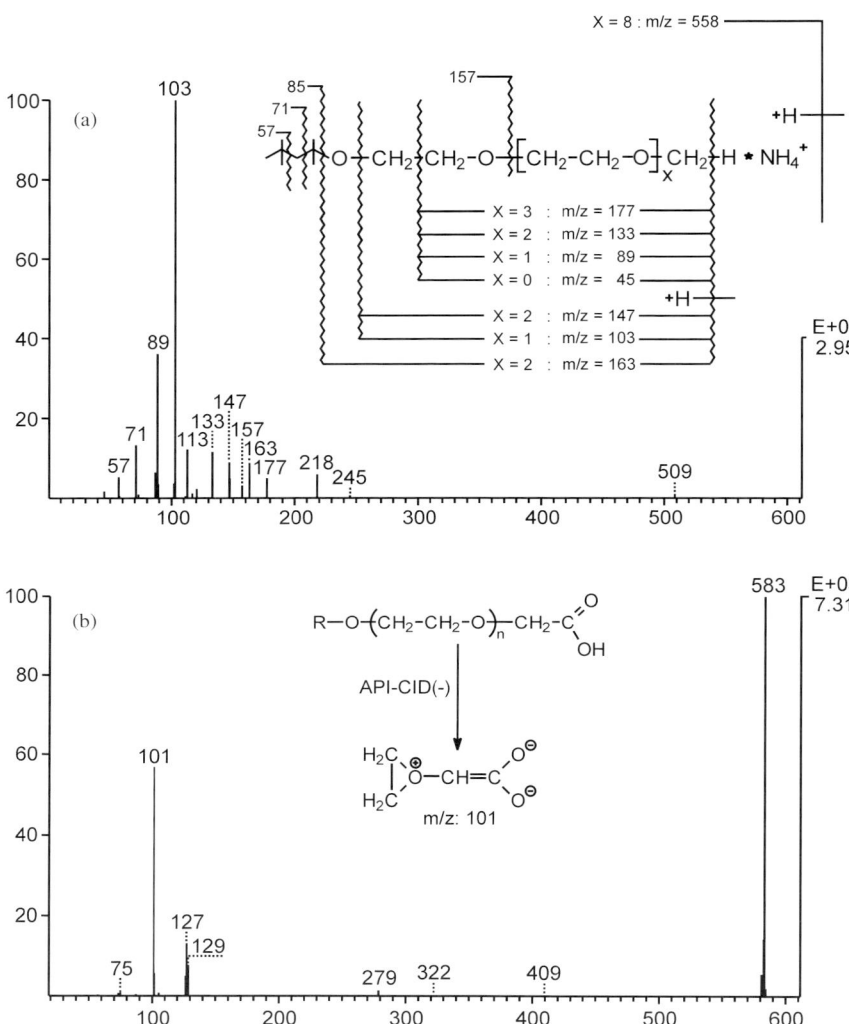

Fig. 2.11.14. (a) ESI–LC–MS–MS(+) product ion mass spectrum of the decarboxylated parent ion of AEC surfactant blend resulting in an ion at m/z 558 ([C_8H_{17}–O–(CH_2–CH_2–O)$_9$–CH_2–H*NH_4]$^+$) [61]; (b) ESI–FIA–MS–MS(−) CID spectrum of respective [M − H]$^-$ parent ion at m/z 583 as in (a); insets: fragmentation scheme or fragmentation behaviour under (+/−) ionisation [22].

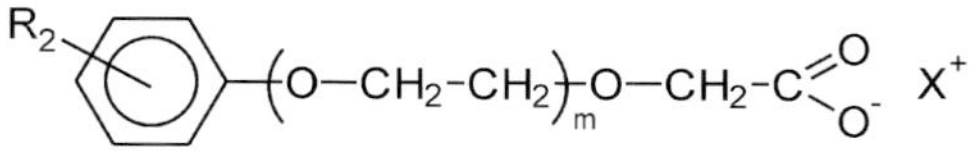

Fig. 2.11.15. General structural formulae of (I) di-nonylphenolethoxy carboxylates (di-NPEC; $C_9H_{19})_2-C_6H_3-O-(CH_2-CH_2-O)_x-CH_2-CO_2^-$), (II) nonylphenoldiethoxysulfonates (NP(EO)$_2$-SO$_3$; $C_9H_{19}-C_6H_4-O-(CH_2-CH_2-O)_2-CH_2-SO_3$), (III) nonylphenolethoxy sulfates (NPEO-SO$_4$; $C_9H_{19}-C_6H_4-O-(CH_2-CH_2-O)_x-SO_3H$), and (IV) nonylphenolethoxy phosphate (NPEO-PO$_4$; $C_9H_{19}-C_6H_4-O-(CH_2-CH_2-O)_x-P(OH)_2OH$).

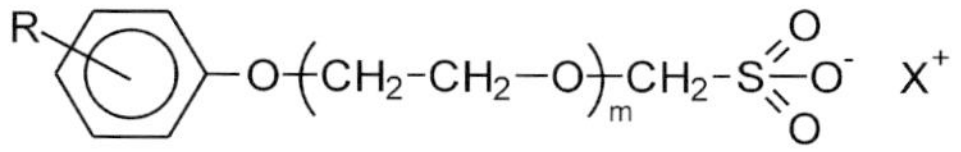

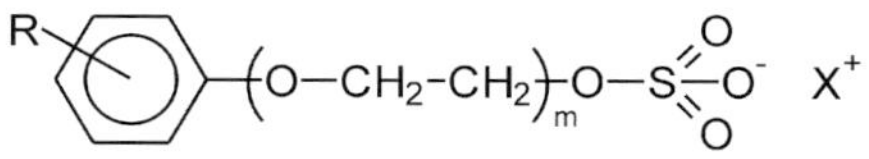

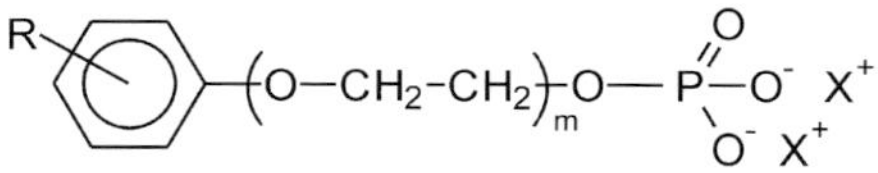

generated in a destructive ionisation process. As also observed with the AES surfactants, bond cleavage between ethoxy chain and acid groups in the compounds containing the functional groups $-O-SO_3^-$, $-O-PO(OH)O^-$ and $-CH_2-CO_2^-$ here also took place during ionisation. While negative ESI–FIA–MS of these compounds happened without any decomposition to $[M - H]^-$ ions and therefore led to spectra that could be easily interpreted, APCI or ESI ionisation in the positive mode, however, led to APEO ions and complex mixtures of adduct ions.

A mixture of an industrial blend of di-nonylphenolethoxy carboxylates (di-NPECs; $(C_nH_{2n+1})_2-C_6H_3-O-(CH_2-CH_2-O)_x-CH_2-COO^-$; (Fig. 2.11.15(I) R = $(C_9H_{19})_2$), which contained nonylphenolethoxylates (NPEO) as impurities from synthesis, was first analysed by APCI–FIA–MS(+/−) followed by a separation on a RP-C$_{18}$ column in the presence of ammonium ions for ion-pairing purposes and for ionisation support.

The FIA–MS(−) overview spectrum is presented in Fig. 2.11.16. It shows the di-NPEC homologue ions all equally spaced by $\Delta m/z$ 44. The ion starting at m/z 535 contains three ethoxylate units while the homologue at m/z 1019 represents a molecule with 14 EO units. In the APCI–LC–MS(+/−) mode, RICs and the selected mass traces at m/z 760 or 799 represent the di-NPEC homologue with nine EO units in the form of $[M - CH_2 - CO_2 + NH_4]^+$ or $[M - H]^-$ ions as shown in Fig. 2.11.17. Under positive ionisation conditions, the detection of the NPEO homologues ionised as $[M + NH_4]^+$ ions are favoured in the reconstructed ion mass trace (b) [22,61].

While APEO cannot be ionised successfully under negative conditions and consequently for identification MS–MS(−) is not informative, the identification of all anionic APEO derivatives is possible in the negative ionisation mode. For some derivatives, negative as well as positive ionisation can be applied. The loss of the anionic moiety, however, must be taken into account if ionisation is performed in the positive mode. Di-NPEC surfactant homologues submitted to negative CID resulted in the prominent di-alkyl-phenolate ion at m/z 345 (Fig. 2.11.18(a) as shown with the homologue m/z 799 under CID conditions). Therefore, the application of the parent ion scan of m/z 345 in the negative ESI and APCI–FIA–MS–MS mode is very specific for the detection of all anionic derivatives of di-NPEO comparable in

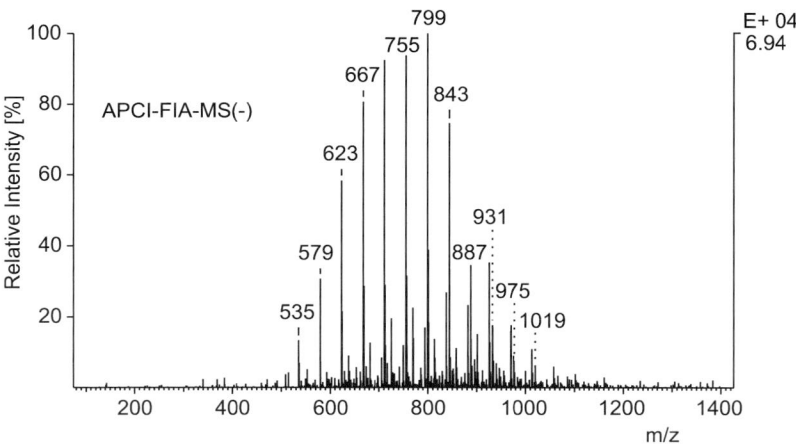

Fig. 2.11.16. APCI–FIA–MS(−) overview spectra of di-NPEC (C_9H_{19})$_2$–C_6H_3–O–(CH_2–CH_2–O)$_x$–CH_2–CO_2^- H$^+$) blend [22,61].

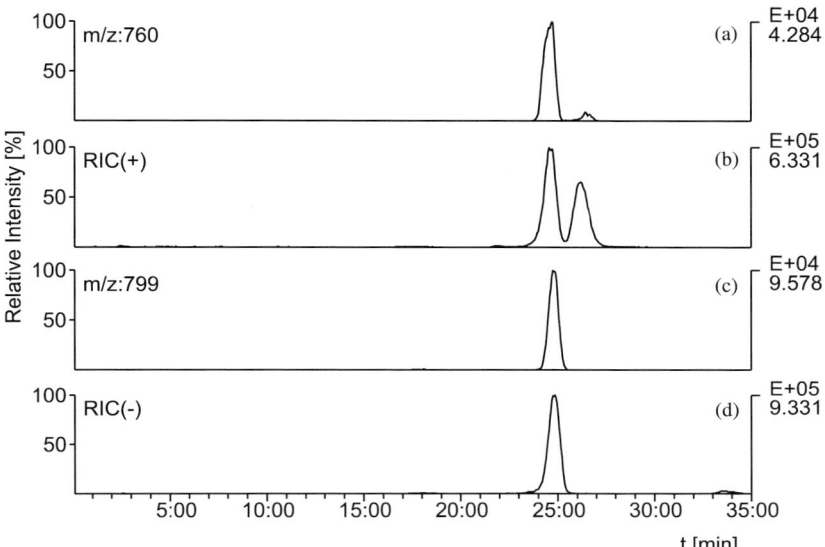

Fig. 2.11.17. APCI–LC–MS(+/–) reconstructed ion current chromatograms (RIC) and selected mass traces (di-NPEC homologue with nine EO units) of di-NPEC separation of compound as in Fig. 2.11.16 [22,61].

the selectivity of the positive parent scan 291 or 277 for NPEO and octylphenol ethoxylates (OPEO), respectively. The application of this parent scan m/z 345 in the negative ESI and APCI–FIA–MS–MS mode to water extracts in the presence of di-nonylphenol ethercarboxylates (di-NPEOC) (($(C_9H_{19})_2$–C_6H_3–O–$(CH_2$–CH_2–O$)_m$–CH_2–CO_2^-) led to the spectrum of all homologues, as shown in Fig. 2.11.18(b) [22].

The product ion spectra generated from the aromatic di-NPEC homologues ionised in the positive mode, however, suffered from the destructive ionisation of anionics, as also observed in the ionisation process of the aliphatic alkylether sulfates [22].

The commercially available nonylphenol diethoxysulfonate NP(EO)$_2$–SO$_3$ (Triton X-200, formula: C_9H_{19}–C_6H_4–O–$(CH_2$–CH_2–O$)_2$–CH_2–SO_3^-, and shown with its structural formula in Fig. 2.11.15(II) ($m = 2$) was studied using API–FIA–MS. Negative APCI-mode spectra contained the signal of one compound that could be ionised in the form of a $[M - H]^-$ ion at m/z 401. Positive ionisation of NP(EO)$_2$–SO$_3$ resulted in a $[M + NH_4]^+$ ion at m/z 420 besides impurities with ions at m/z 356

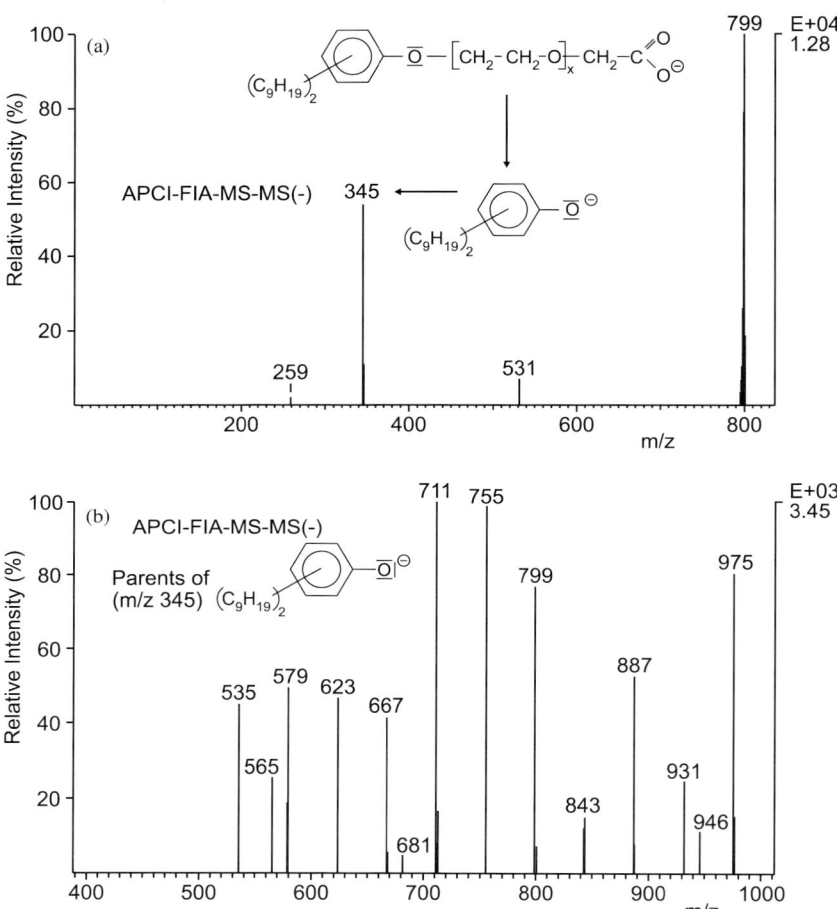

Fig. 2.11.18. (a) APCI–FIA–MS–MS(−) (CID) product ion mass spectrum of $[M − H]^{-}$ parent ion at m/z 799 of di-NPEC (($C_9H_{19})_2$–C_6H_3–O–$(CH_2$–CH_2–$O)_m$–CH_2–CO_2^{-}) with fragmentation behaviour in the inset; (b) APCI–FIA–MS–MS(−) (CID) parent ion mass spectrum of product ion at m/z 345 for confirmation of di-NPEC in an environmental sample extract [22].

and 312 not observed under negative ionisation conditions. This behaviour could be confirmed by comparison of APCI–LC–MS(+/−) total ion mass traces and selected mass traces in Fig. 2.11.19. Under these conditions, it became obvious that the sulfonate Triton X-200 was not a mixture of homologues but a defined molecule type with the formula

$C_9H_{19}-C_6H_4-O-(CH_2-CH_2-O)_2-CH_2-SO_3^-$ ($NP(EO)_2-;SO_3$). In contrast to sulfate, carboxylate and phosphate homologues no bond cleavage could be observed in APCI(+) mode. The sulfonate could also be ionised in the positive mode resulting in a $[M + NH_4]^+$ ion at m/z 420 [15].

Triton X-200 is known to be hardly biologically degradable. To confirm this behaviour, the compound characterised as $NPEO-CH_2-SO_3$ containing two EO units in the ethoxy chain ($NP(EO)_2-SO_3$) was submitted to a LC separation. RT comparison in APCI–LC–MS(+/−) experiments and APCI–FIA–MS–MS in the product ion mode were examined (Fig. 2.11.20(a)) in positive and negative modes [15]. In the positive CID spectrum of $NP(EO)_2-SO_3$, the parent ion 420 disappeared and the product ion m/z 291 (cf. inset in Fig. 2.11.20(a)) known as the characteristic NPEO product ion was observed as base peak [22]. Application of CID furthermore led to alkyl fragments (m/z 57, 71, and 113) and ions at m/z 165 and 121 resulting from the abstraction of the alkyl chain of ion 291 or from the abstraction of the alkyl chain and one

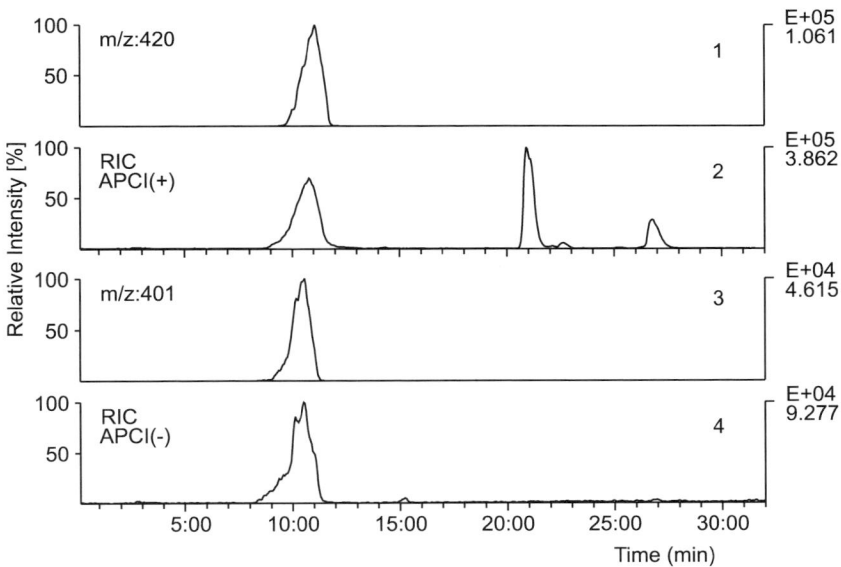

Fig. 2.11.19. APCI–LC–MS(+) (2) and APCI–LC–MS(−) (4) reconstructed ion current chromatograms (RIC) and (1) (m/z 420), (3) (m/z 401) selected mass traces of nonylphenol diethoxy sulfonate blend Triton X-200 ($NP(EO)_2-SO_3$); gradient elution separated by RP-C_{18} column [22,61].

glycol unit, respectively. Fragmentation of the corresponding negative parent ion m/z 401 by triple quad CID resulted in negative ions (see Fig. 2.11.20(b) as far as possible identified and found to be characteristic for this type of alkyl-phenoxy sulfonates [22].

The APEO-SO$_4$ (C_nH_{2n+1}–C_6H_4–O–$(CH_2$–CH_2–O)$_x$–SO$_3^-$; $n = 9$), with their general structural formula as shown in Fig. 2.11.15(III), were studied by APCI–FIA- and APCI–LC–MS in the positive as well as negative modes. The ions observed in the positive mode were [M – SO$_3$ + NH$_4$]$^+$ ions with m/z ratios at 326–942 equally spaced with $\Delta \times m/z$ 44 generated by cleavage of the EO–SO$_3$-bond during the destructive ionisation process of the precursor NPEO-SO$_4$ [15] according to the number of EO units. Negative ionisation of the NPEO-SO$_4$ blend resulted in [M – H]$^-$ ions of the NP(EO)$_x$–SO$_4$ ($x =$ 1–15) at m/z 343 up to 959 as shown in the FIA overview spectrum in Fig. 2.11.21(a). The LC separation of a standard mixture resulting in a broad signal covering all the different homologues was recorded in the APCI–MS(–) mode and is presented in the form of the total ion mass trace and the selected mass traces m/z 695 and 871 in Fig. 2.11.21(b) and (c) [15].

While under negative CID conditions, no structural information could be obtained by CID from NPEO-SO$_4$ compounds on a MS–MS instrument despite the variation of CID parameters, although the behaviour of the homologue ion m/z 590 (equivalent to the negative ion 667) under positive CID conditions was comparable with NPEO product ion spectra because of the destructive ionisation process of the precursor NPEO-SO$_4$ [15].

NPEO-SO$_4$ is one of the rare anionic surfactant compounds on which aerobic biodegradation monitoring has been performed, where metabolites could be observed by API–MS. Using FIA–MS, however, differentiation of precursor compounds and metabolites was impossible. Both compounds showed the same molar masses but could be recognised because of their quite different RTs in RP-LC [15]. MS^n CID performed by trap confirmed a fragmentation behaviour of the metabolite quite different from precursor NPEO-SO$_4$ compounds, whose structure is not yet clear.

An industrial blend of NPEO-PO$_4$ homologues (C_9H_{19}–C_6H_4–O– $(CH_2$–CH_2–O)$_x$–P(OH)$_2$O$^-$) with the general structural formula as shown in Fig. 2.11.15(IV) was examined by APCI–FIA–MS in the positive and negative modes. Results are presented in Fig. 2.11.22. The positive APCI–FIA–MS spectrum (a) contained the polyglycolether

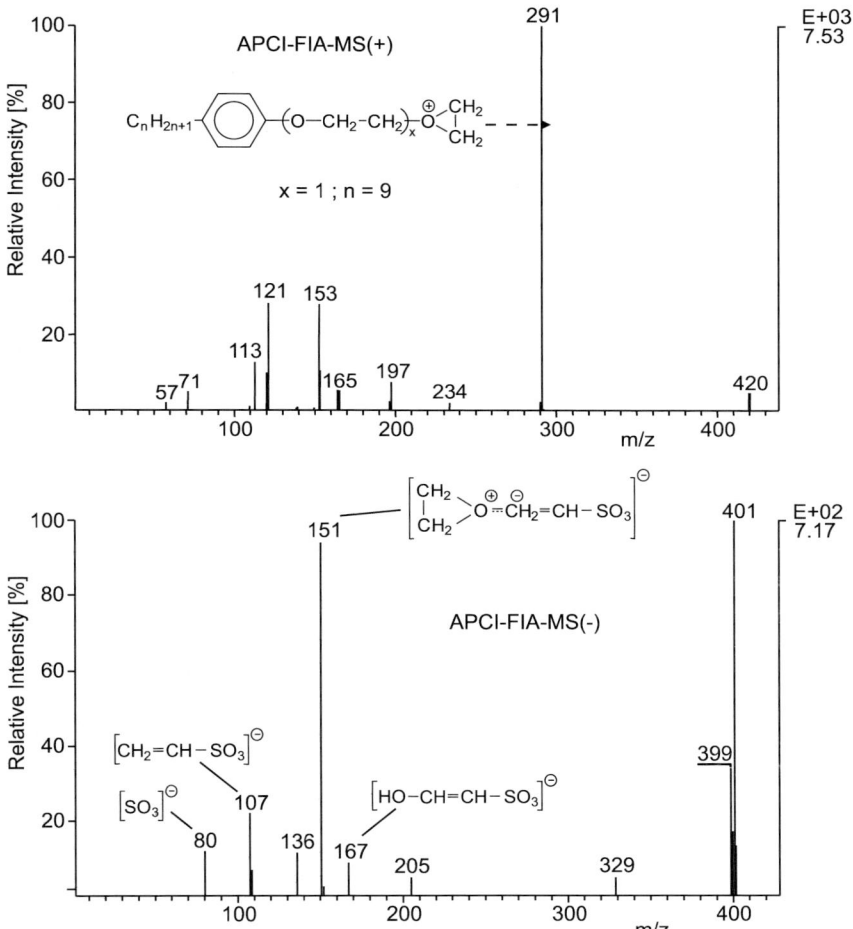

Fig. 2.11.20. (a) APCI–FIA–MS–MS(+) (CID) product ion mass spectrum of a $[M + NH_4]^+$ parent ion of $NP(EO)_2$–SO_3 (C_9H_{19}–C_6H_4–O–$(CH_2$–CH_2–$O)_2$–CH_2–SO_3H) at m/z 420; (b) APCI–FIA–MS–MS(−) (CID) product ion mass spectrum and fragments of compound as in (a) with the $[M − H]^-$ parent ion at m/z 401 [22].

homologues, generated by an abstraction of the $(P(OH)_2O^-)$ moiety. The ions observed in the positive mode (a) were $[M − PO_3H + NH_4]^+$ ions with m/z ratios at 370–942 ($x = 3$–16). $[M − H]^-$ ions of this anionic derivative were generated in negative mode. These ions were equally

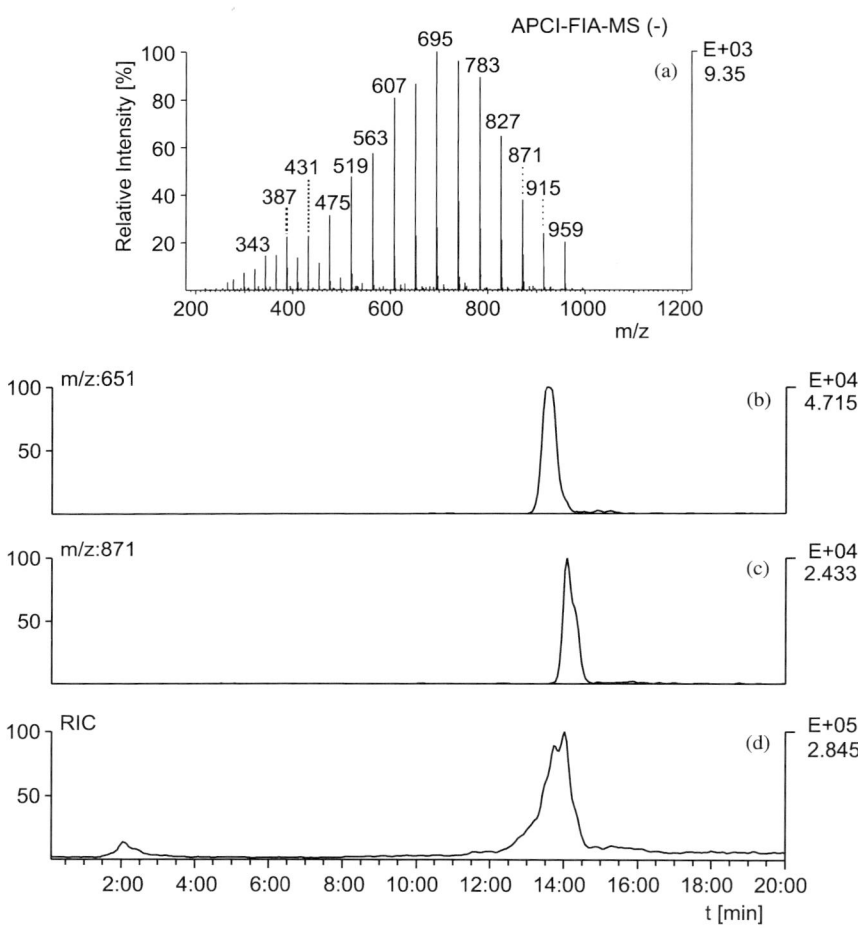

Fig. 2.11.21. (a) APCI–FIA–MS(−) and (d) APCI–LC–MS(−) reconstructed ion current chromatogram (RIC) and (b) and (c) selected mass traces of NPEO-SO$_4$ (C$_9$H$_{19}$–C$_6$H$_4$–O–(CH$_2$–CH$_2$–O)$_x$–SO$_3$H); column separation performed by RP-C$_{18}$ gradient elution [15].

spaced with $\Delta m/z$ 44, too (m/z 431–915 containing 3–14 CH$_2$–CH$_2$–O-units) [61].

The APCI-RP-C$_{18}$-LC-MS(+/−) separation of this mixture consisted of the anionic nonylphenol derivatives besides synthetic by-products to the total ion current traces and selected mass traces as presented in Fig. 2.11.23. By-products from synthesis could be confirmed as

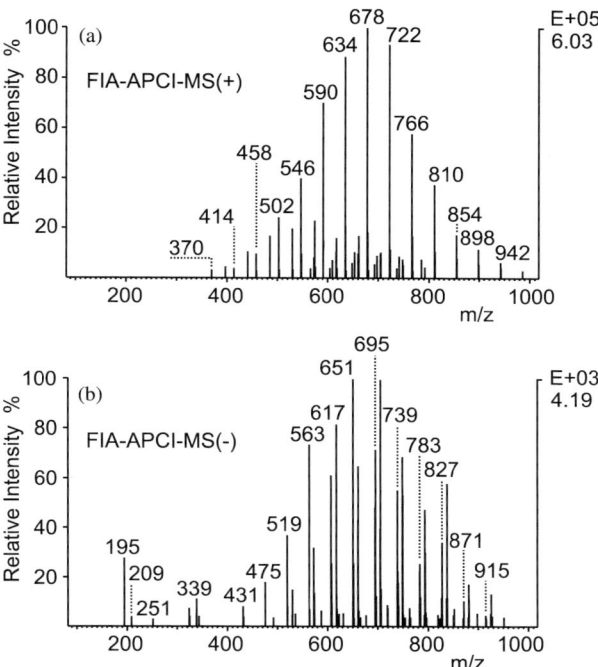

Fig. 2.11.22. (a) APCI–FIA–MS(+) and (b) APCI–FIA–MS(−) overview spectra of NPEO-PO$_4$ (C$_9$H$_{19}$–C$_6$H$_4$–O–(CH$_2$–CH$_2$–O)$_x$–P(OH)$_2$OH (cf. Table 2.11.2) [61].

impurities in the commercial blend under positive ionisation while the NPEO-PO$_4$ compounds were ionised as [M − PO$_3$H + NH$_4$]$^+$ adduct ions of NPEO-compounds (a) because of the loss of the (P(OH)$_2$O$^−$) moiety, as also observed under FIA(+) conditions. Negative ionisation resulted in the [M − H]$^−$ ions of the phosphates, presented with the mass trace (c) m/z 739 of the C$_9$H$_{19}$–C$_6$H$_4$–O–(CH$_2$–CH$_2$–O)$_{10}$–P(OH)$_2$O$^−$ homologue ion equivalent to the [M − PO$_3$H + NH$_4$]$^+$ ion at m/z 678 [61].

The selected [M − H]$^−$ ion at m/z 651 representing the C$_9$H$_{19}$–C$_6$H$_4$–O–(CH$_2$–CH$_2$–O)$_8$–P(OH)$_2$O$^−$ homologue of the NPEO phosphate was submitted to APCI–FIA–MS–MS(−). Two fragment ions, one at m/z 219 ([C$_9$H$_{19}$–C$_6$H$_4$–O]$^−$) and the other fragment ion at m/z 79 ([PO$_3$]$^−$) could be observed. These ions were also found if MS–MS(−) was applied to the other anionic NPEO derivatives [61].

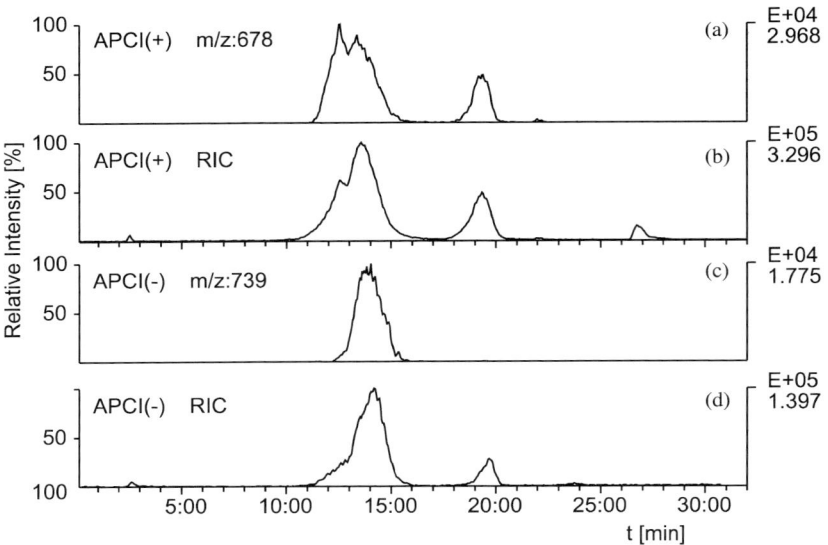

Fig. 2.11.23. (b) APCI–LC–MS(+) and (d) APCI–LC–MS(−) reconstructed ion current chromatograms (RICs) and (a) and (c) selected mass traces of NPEO-PO$_4$ as in Fig. 2.11.22, $x = 10$. Gradient elution separated by RP-C$_{18}$ column [61].

Fluorsulfonic and fluorcarboxylic surfactants

Partly fluorinated or perfluorinated sulfonic and carboxylic acids are compounds with excellent surface activity combined with an extreme stability against chemical or physico-chemical attacks as also described for non-ionic fluorine containing surfactant compounds. The anionic surfactants are shown with their structural formulae in Fig. 2.11.24(I)–(III). A selected ESI–FIA–MS(−) spectrum of a partly fluorinated surfactant (CF_3–$(CF_2)_4$–$(CH_2)_8$–SO_3H, m/z 461) is presented in Fig. 2.11.25. The FIA spectrum also contains ions of the by-products CF_3–$(CF_2)_2$–$(CH_2)_8$–SO_3H (m/z 361) and CF_3–$(CF_2)_3$–$(CH_2)_8$–SO_3H (m/z 411).

A mixture of these fluorinated surfactants was separated according to the method applied by Popenoe et al. [52]. For ionisation and detection, ESI–MS in the negative mode was applied. The result of LC separation is presented in the RIC in Fig. 2.11.26(a). Some of the separated compounds in the TIC and mass traces could be identified as perfluoro alkylsulfonates and perfluoro alkylcarboxylates such as

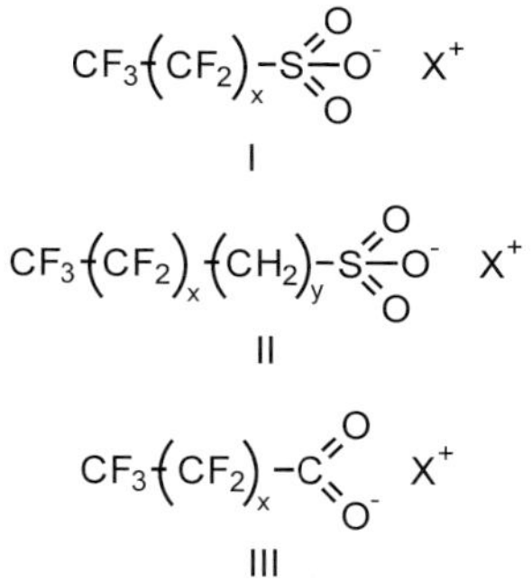

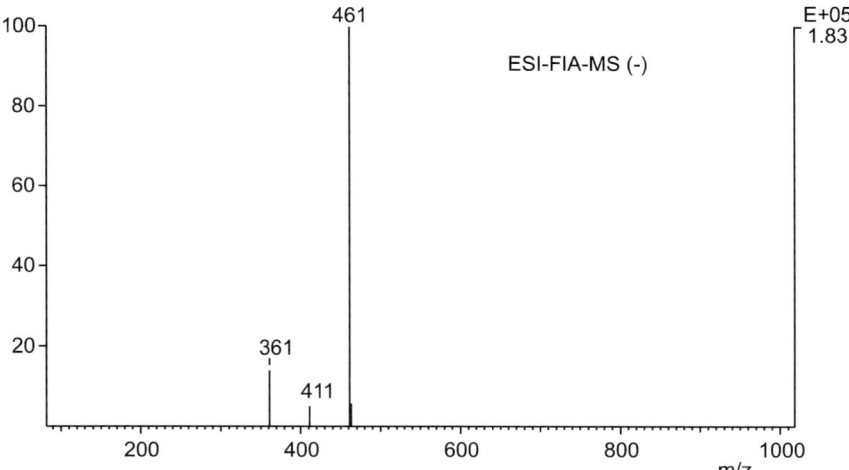

Fig. 2.11.24. General structural formulae of (I) perfluorinated alkylsulfonic acid, (II) partly fluorinated alkylsulfonic acid, and (III) perfluorinated carboxylic acid.

perfluorohexane sulfonic acid (PFHxS, $CF_3-(CF_2)_5-SO_3H$; m/z 399) (b)), perfluorooctanoic acid (PFOA; $CF_3-(CF_2)_6-COOH$; m/z 413) (c)), perfluorooctane sulfonic acid (PFOS, $CF_3-(CF_2)_7-SO_3H$; m/z 499) (e)) and perfluorodecane sulfonic acid ($CF_3-(CF_2)_9-SO_3H$; m/z 599) (g)) or partly fluorinated sulfonic acids ($CF_3-(CF_2)_4-(CH_2)_8-SO_3H$; m/z 461) while some compounds reproduced with their mass traces

Fig. 2.11.25. ESI–FIA–MS(−) spectrum of a partly fluorinated PFOS surfactant blend ($CF_3-(CF_2)_4-(CH_2)_8-SO_3H$, m/z 461) containing by-product ions of $CF_3-(CF_2)_2-(CH_2)_8-SO_3H$ (m/z 361) and $CF_3-(CF_2)_3-(CH_2)_8-SO_3H$ (m/z 411) [61].

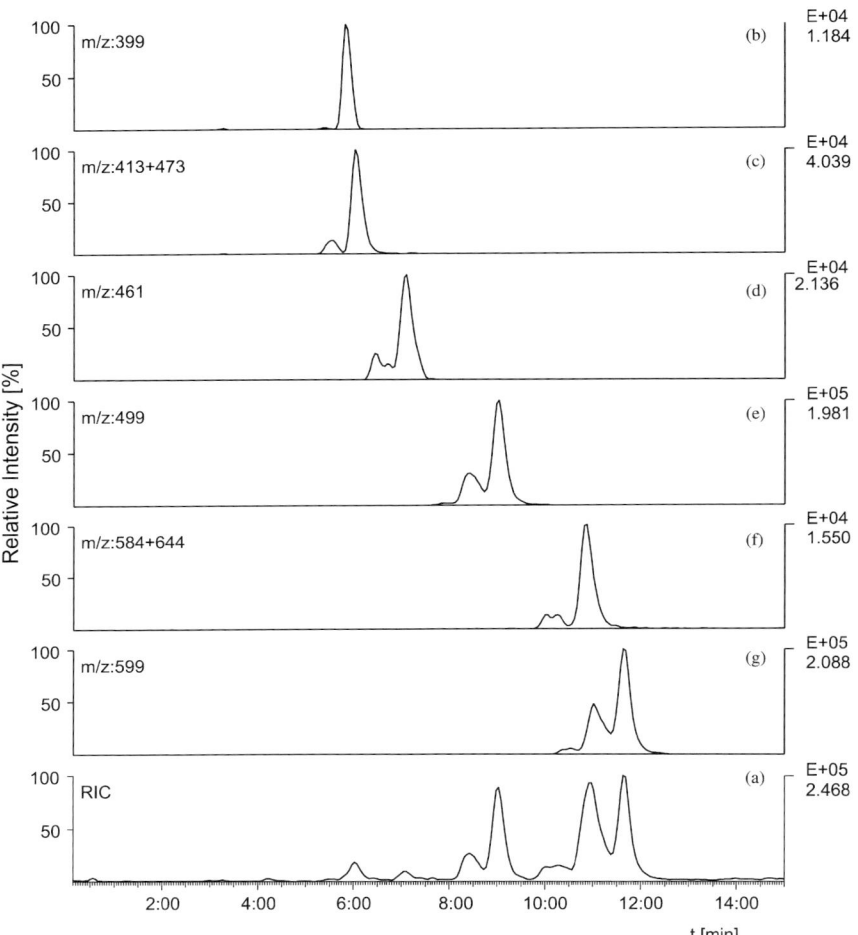

Fig. 2.11.26. (a) ESI–LC–MS(–) reconstructed ion current chromatogram (RIC) and selected mass traces of (b) CF_3–$(CF_2)_5$–SO_3H (PFHxS), (c) CF_3–$(CF_2)_6$–COOH (PFOA), (d) CF_3–$(CF_2)_4$–$(CH_2)_8$–SO_3H, (e) CF_3–$(CF_2)_7$–SO_3H (PFOS), (g) CF_3–$(CF_2)_9$–SO_3H of a mixture of partly fluorinated or perfluorinated sulfonic and carboxylic acids; gradient elution separated by RP-C_{18} column [61].

remained unidentified, e.g. compound in mass trace (f). The presence of isomers of these compounds in the mixtures could be recognised because of the peak splitting observed in the mass traces during LC separation [61,62].

Two independent analytical methods—LC–MS–MS and ^{19}F-NMR— for the determination of perfluorinated anionic surfactants in environmental water samples were presented. Perfluorinated alkanesulfonates and perfluorocarboxylates were determined qualitatively and quantitatively because of an accidental release of perfluorosurfactant contaminated fire-fighting foam [55]. C_{18}-SPE was applied for concentration of the compounds from water samples. Methanol was used for elution prior to ESI–LC–MS(−) analysis. The negatively recorded LC–MS–MS TIC for the determination of PFOS, PFHxS, PFOA, perfluoroheptanoic acid (PFHpA), perfluorododecanoic acid (PFDoA; internal standard) in water samples was presented [55].

Selected components of the mixture of industrial blends of partly fluorinated or perfluorinated sulfonic and carboxylic acids were examined by CID using MS–MS on a TSQ instrument. For ionisation and detection, ESI–MS–MS in the negative mode was applied. Predominantly fluorine-containing negatively charged product ions were observed. Results of the performed FIA–MS–MS(−) are presented in Figs. 2.11.27 and 2.11.28. The behaviour of the perfluorinated carboxylic acid (PFOA; $CF_3-(CF_2)_6-COOH$; m/z 413) under CID conditions is shown in the inset (Fig. 2.11.27), demonstrating that the alkyl chain was destroyed yielding $[CF_3-(CF_2)_x]^-$ ions at m/z 119, 169 and 219 ($x = 1-3$). In contrast to Moody et al. [55], who used the fragment ion 369 for quantification, we could not observe this product ion under the CID conditions applied on a triple quad instrument [61].

The negatively recorded product ion spectrum of PFOS ($CF_3-(CF_2)_7-SO_3H$; m/z 499) is presented in Fig. 2.11.28. Here, besides alkyl chain fragments, the fragment ions $[SO_3]^-$, $[SO_3F]^-$ and perfluoro alkyl-SO_3 fragment ions could be observed as shown in the fragmentation scheme reproduced in the inset [61].

Fluorinated phosphinic and phosphonic acid derivatives
Perfluoro derivatives of alkyl phosphonic acid $C_nF_{2n+1}-P(O)(OH)_2$ and alkyl phosphinic acid $C_nF_{2n+1}(C_mF_{2m+1})-P(O)OH$ ($n = m$ or $n \neq m$) shown with their general structural formulae in Fig. 2.11.29(I) and (II) were examined by negative ESI- and APCI–FIA–MS. These anionic surfactant compounds contained perfluoro alkyl chains [2,22,25]. By analogy with their behaviour in the TSI–FIA–MS(−) process [25], the phosphonic acid formed $[M - H]^-$ ions at m/z 399 and 499

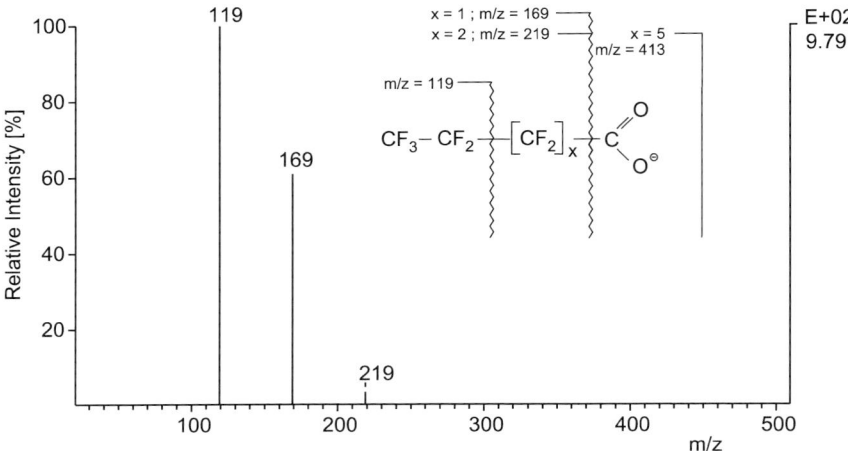

Fig. 2.11.27. ESI–FIA–MS–MS(–)(CID) product ion mass spectrum of $[M - H]^-$ parent ion at m/z 413 of PFOA (CF_3–$(CF_2)_6$–COOH); fragmentation scheme in the inset [61].

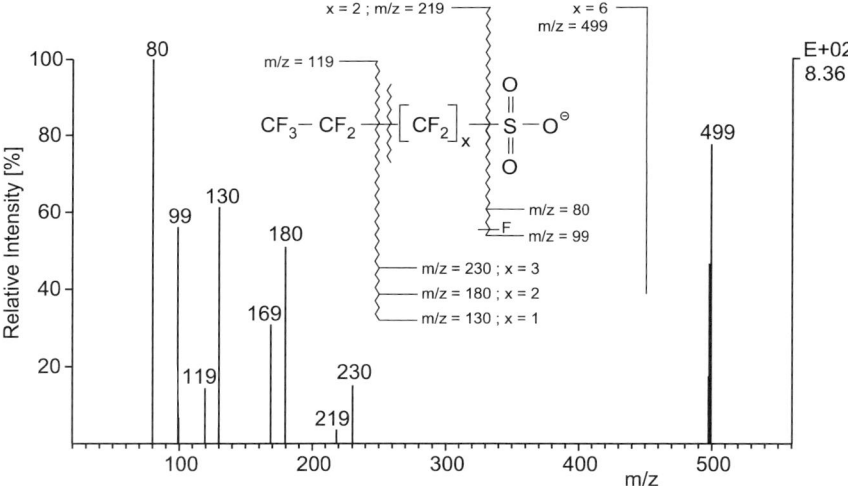

Fig. 2.11.28. ESI–FIA–MS–MS(–)(CID) product ion mass spectrum of $[M - H]^-$ parent ion at m/z 499 of PFOS (CF_3–$(CF_2)_7$–SO_3H); fragmentation scheme reproduced in the inset [61].

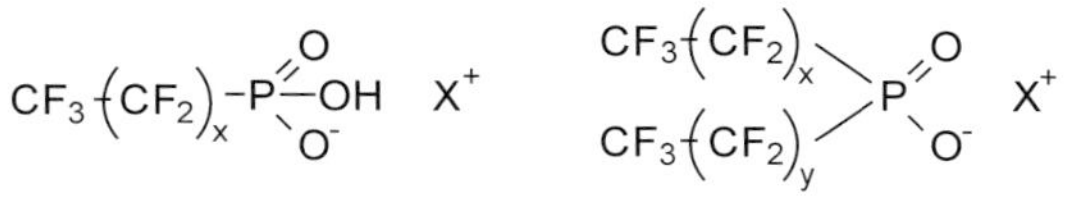

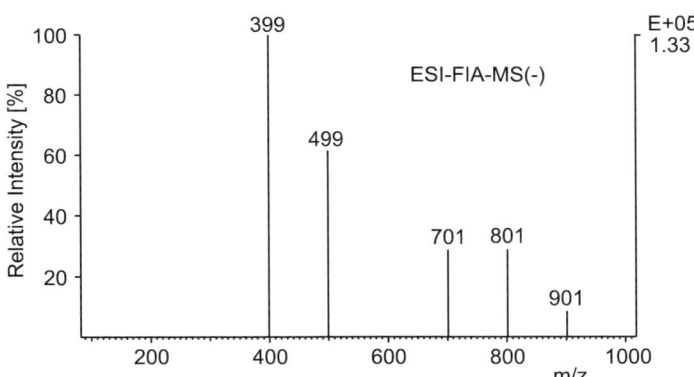

Fig. 2.11.29. General structural formulae of (I) perfluoroalkyl phosphonic acid C_nF_{2n+1}–P(O)(OH)$_2$ and (II) perfluoroalkyl phosphinic acid $C_nF_{2n+1}(C_mF_{2m+1})$–P(O)OH.

representing a perfluoroalkyl chain with $n = 6$ or 8 CF$_2$ chain links (cf. Fig. 2.11.30). The signals appearing at m/z 799 and 899 under TSI($-$) conditions, however, were not observed under ESI and APCI ionisations. Signals at m/z 701, 801 and 901 belonging to the phosphinic acid ions could be observed applying ESI- or APCI–FIA–MS($-$) (Fig. 2.11.30). The equally spaced signals with $\Delta m/z$ 100 could be explained by the different chain lengths of the fluorinated alkyl groups, which are always represented by m/z 100 for every (CF$_2$)$_2$ unit. In addition to these molecular [M $-$ H]$^-$ ions of the phosphinic acid, the APCI–FIA–MS($-$) provides ions with a m/z ratio of 682, 782 and 882. These ions, which appear in LC separation together with the ions at m/z 701, 801 and 901 and differ to the phosphinic acid with $\Delta m/z$ 19, seem to be fragments generated in the APCI($-$) ionisation process. Positive ionisation of these compounds was impossible with both API techniques, as also observed with TSI [25]. The quite different RTs observed

Fig. 2.11.30. ESI–FIA–MS($-$) overview spectrum of a mixture of perfluoro derivatives of alkyl phosphonic acid (C_nF_{2n+1}–P(O)(OH)$_2$) and alkyl phosphinic acid (C_nF_{2n+1} (C_mF_{2m+1})–P(O)OH) [22].

for the phosphonic and phosphinic acid compounds were in agreement with the number and length of the perfluoro alkyl chains that determined the lipophilicity and therefore the RT of the compounds [22]. Therefore, LC–MS in the APCI(−) mode resulted in an excellent separation of the perfluorinated alkylphosphinic and alkylphosphonic acids. These two different components could be observed in the mixture and a differentiation of their homologues was possible, as demonstrated in Fig. 2.11.31, which shows the mass traces (1–3) and the RIC (4) of the fluorinated alkyl phosphonic acid and alkyl phosphinic acid homologues [22,61].

The mixture of anionic fluorinated alkylphosphinic and -phosphonic acid surfactants with perfluorinated alkyl moieties was examined using ESI–FIA–MS–MS(−). For CID in the negative mode, precursor ions at m/z 399, 499 and 599 were selected [22]. From all precursor ions only one product ion at m/z 79 ($[PO_3]^-$) besides the $[M − H]^-$ ion was

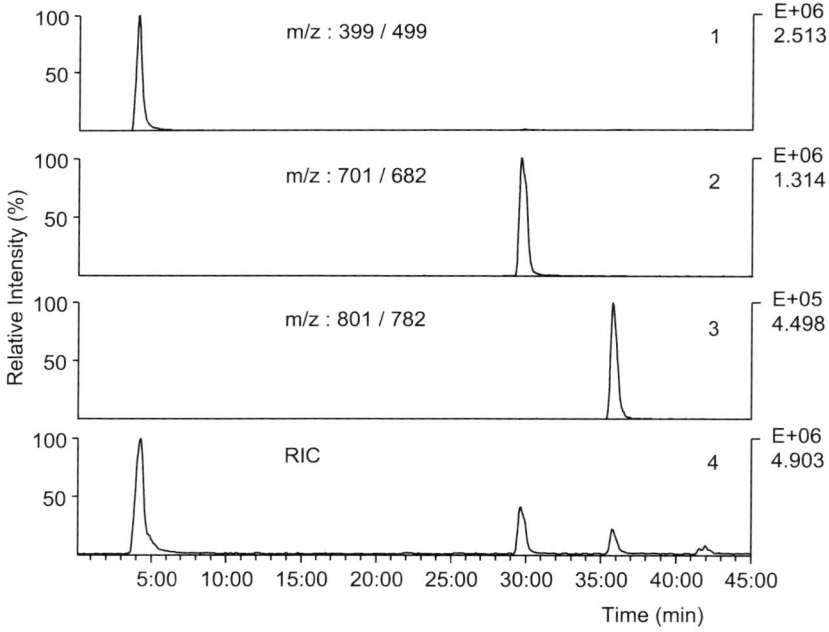

Fig. 2.11.31. APCI–LC–MS(−) total ion current chromatogram (TIC) (4) and selected mass traces (1)–(3) of compounds as in the overview spectrum in Fig. 2.11.30; gradient elution separated by RP-C_{18} column [22,61].

observed, as presented in Fig. 2.11.32. These product ions originated from the homologue phosphonic acids ($C_xF_{2x+1}-P(O)(OH)_2$). Under TSI–MS–MS($-$) ionisation conditions the same negative product ion at m/z 79 with high intensity could be observed [22,25,42].

The second dominating component of the surfactant blend was phosphinic acid with the general formula $C_xF_{2x+1}(C_yF_{2y+1})-P(O)OH$. The phosphinic acid molecules contained two perfluoro alkyl chains [25]. The negative ion at 801 ($x+y = 14$) of the fluorinated phosphinic acid was submitted to CID [22]. According to the collision energy applied, the fragment ions with m/z 401 (60 eV) or 401 and 501 (80 eV), respectively, were produced. These product ions generated by MS–MS were the result of an abstraction of only one of the two perfluorinated alkyl chains present in the molecules. As the perfluorinated alkyl chains either contained six or eight CF_2 chain links representing an abstraction of one alkyl chain from the precursor ion at m/z 801 combined with a fluorine, transfer resulted in product ions at m/z 401 and 501 as shown in the inset of Fig. 2.11.32. A product ion at m/z 79 ($[PO_3]^-$) as observed with phosphonic acids for this compound class cannot be generated because phosphinic acids contain only a PO_2 moiety not stable as an ion [61].

Sulfosuccinates

Sulfosuccinates, as presented with their general structural formula in Fig. 2.11.33, are applied as surfactants for personal hygiene because of their hypoallergenic features. The sodium salt of the sulfosuccinate blend with the formula $ROOC-CH-(SO_3^-)-CH_2-COOR$ Na^+ ($R = C_8H_{17}$) was examined by APCI–FIA–MS in the positive and negative modes. The addition of an excess of ammonium acetate under FIA–APCI–MS($+$) conditions resulted in $[M - NH_4]^+$ ions with m/z 440 while $[M - H]^-$ ions with m/z 421, however, were observed in the negative APCI–FIA–MS mode (Fig. 2.11.34(a)). Only one type of ion could be observed in this industrial blend by FIA–MS. This purity could also be confirmed by APCI–LC–MS($-$), as shown in the total ion current trace (cf. Fig. 2.11.34(b)), which is presented in combination with the averaged mass spectrum under the signal in the inset of Fig. 2.11.34(b) [22].

Dihexyl sulfosuccinate sodium salt ($C_6H_{11}OOC-CH-(SO_3^-)-CH_2-COOC_6H_{11}$ Na^+) was examined by LD FTICR–MS in the negative mode

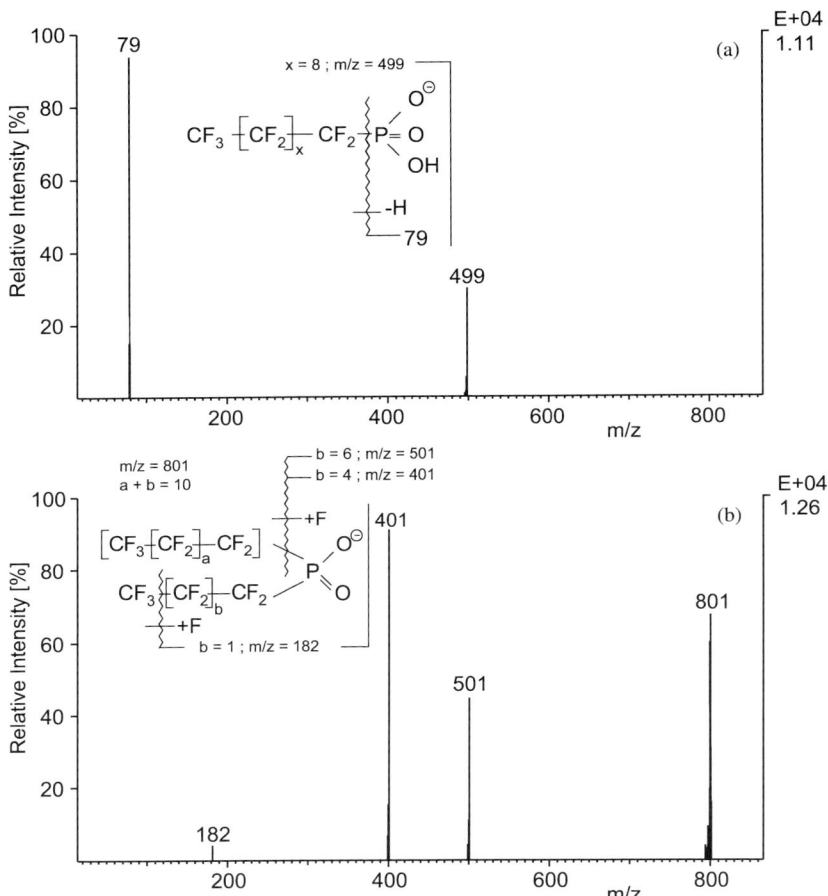

Fig. 2.11.32. (a) ESI–FIA–MS–MS(−) (CID) product ion mass spectrum of the $[M − H]^-$ parent ion at m/z 499 of a phosphonic acid homologue $(C_xF_{2x+1}-P(O)(OH)_2;\ x = 8)$; (b) ESI–FIA–MS–MS(−) (CID) product ion mass spectrum of the $[M − H]^-$ parent ion at m/z 801 of a phosphinic acid $(C_xF_{2x+1}(C_yF_{2y+1})-P(O)OH;\ x + y = 14)$ homologue; fragmentation schemes presented in the insets [61].

resulting in $[M − H]^-$ ions. No fragments could be observed under these conditions [28].

Under FIA–ESI–MS–MS(−) conditions, the $[M − H]^-$ ion of the sulfosuccinate anion $(ROOC–CH–(SO_3^-)–CH_2–COOR\ (R = C_8H_{17}))$ at m/z 421 resulted in only one product ion at m/z 81 $([HSO_3]^-)$ [22].

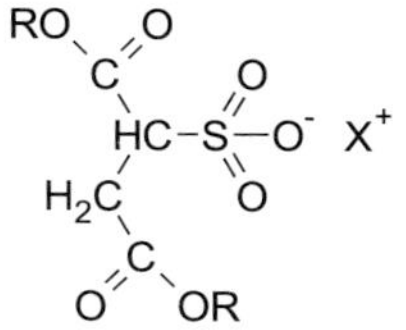

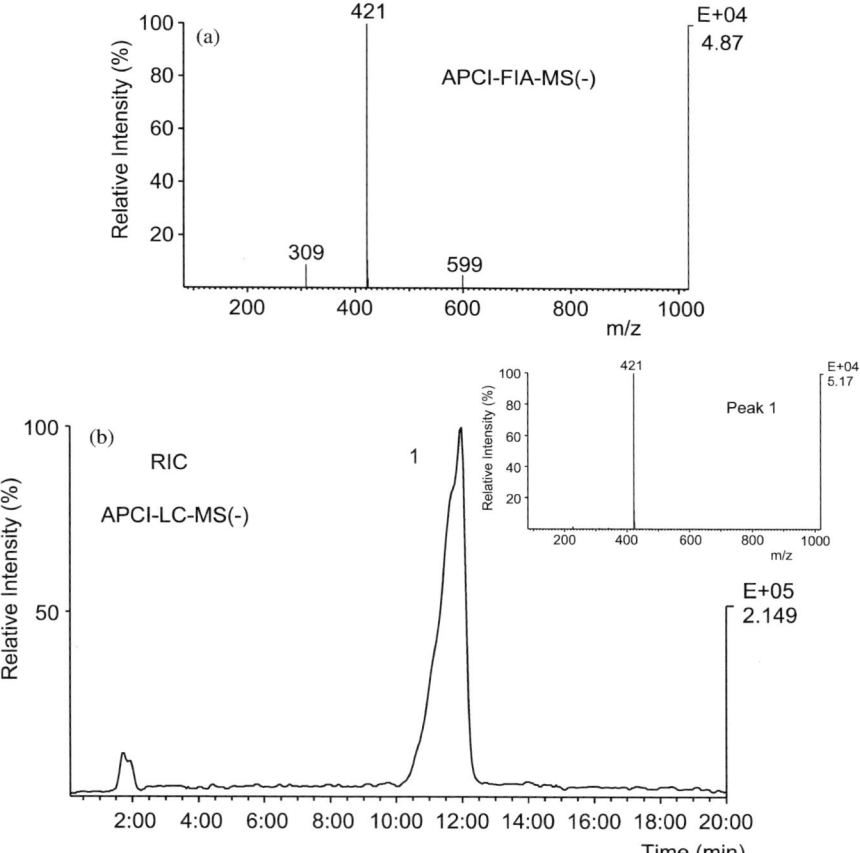

Fig. 2.11.33. General structural formula of sulfosuccinates.

Fig. 2.11.34. (a) APCI–FIA–MS(−) overview spectrum of sulfosuccinate blend (ROOC–CH–(SO$_3^-$)–CH$_2$–COOR Na$^+$; R = C$_8$H$_{17}$) containing the [M − H]$^-$ ion at m/z 421; APCI–LC–MS(−) RIC of blend as in (a) with averaged MS spectrum of peak 1 in the inset [22,61].

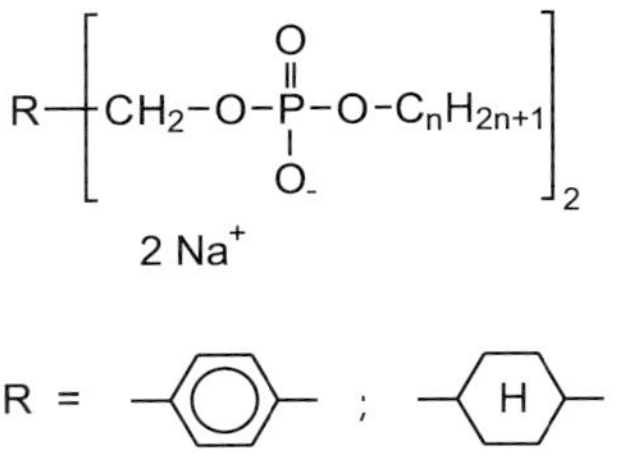

R $-$ $\left[CH_2-O-\overset{\overset{O}{\|}}{\underset{\underset{O_-}{|}}{P}}-O-C_nH_{2n+1} \right]_2$

2 Na$^+$

R = ⟨○⟩ ; ⟨H⟩

1,4 - isomers

Fig. 2.11.35. General structural formula of anionic gemini surfactants (R = spacer).

Dimeric anionic surfactants (gemini surfactants)

As already described in the chapter on 'Non-ionics' a new type of surfactant, the so-called gemini surfactants in the form of non-ionic, anionic and cationic compounds with an improved performance in surface activity [3,63] and biodegradability [64] is under research. Two anionic gemini surfactants (Fig. 2.11.35) were examined to optimise their substance-specific detection by MS methods and to evaluate their biodegradability under aerobic conditions. The anionic compounds were sodium salts of $C_{16}H_{33}$- or $C_{12}H_{25}$-alkylphosphoric acid derivatives (di-sodium-(p-phenylenedi-methylene)bis(hexadecylphosphate); general formula: $C_{16}H_{33}-OPO^-(O_2)-CH_2-C_6H_4-CH_2-OPO^-(O_2)-C_{16}H_{33}^*$ $2Na^{\oplus}$ and di-sodium-(1,4-cyclohexylenedimethylene)-bis(dodecylphosphate); general formula: $C_{12}H_{25}-OPO^-(O_2)-CH_2-C_6H_{10}-CH_2-OPO^-(O_2)-C_{12}H_{25}^*2Na^{\oplus}$), which were coupled together by an aromatic or alicyclic spacer group, respectively, as shown in Fig. 2.11.35.

The application of ESI–FIA–MS($-$) to both surfactants led to the overview spectra as presented in Fig. 2.11.36(a) and (b). The FIA spectra of the gemini surfactants contain the mono-deprotonated molecular ions ($[M - H]^-$) of the mono-protonated compounds with m/z ratios of 745 (a) or 639 (b) besides the di-deprotonated ($[M - 2H]^{2-}$) molecular ions at m/z 372 (a) or 319 (b), respectively [64].

From the separation of a mixture of both compounds applying RP-C_{18} and ESI–MS($-$), it was recognisable from mass trace analysis that the C_{12} derivative eluted first as shown in Fig. 2.11.36(c) and (d) [64]. The signals in the reconstructed ion current trace (e) contained the same ions at m/z 745 or 639 and at m/z 372 or 319 as also observed under ESI–FIA–MS($-$) conditions.

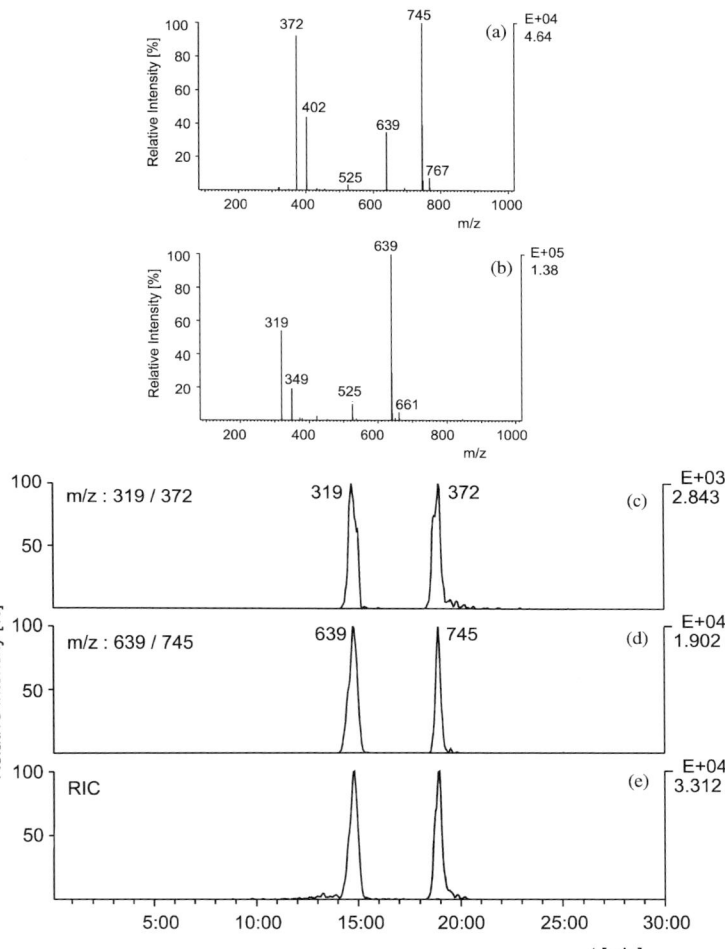

Fig. 2.11.36. (a) ESI–FIA–MS(−) overview spectrum of di-sodium-(p-phenylenedimethy-lene)bis-hexadecylphosphate ($C_{16}H_{33}$–$OPO^-(O_2)$–CH_2–C_6H_4–CH_2–$OPO^-(O_2)$–$C_{16}H_{33}$* $2Na^{\oplus}$) and (b) ESI–FIA–MS(−) overview spectrum of di-sodium-(1,4-cyclohexylene-dimethylene)bis-dodecylphosphate ($C_{12}H_{25}$–$OPO^-(O_2)$–CH_2–C_6H_{10}–CH_2–$OPO^-(O_2)$–$C_{12}H_{25}$*$2Na^{\oplus}$); (e) ESI–LC–MS(−) reconstructed ion current chromatogram (RIC) of anionic gemini surfactants as in (a) and (b); (c) and (d) selected mass traces of mono- and di-protonated gemini surfactants as in (a) and (b); gradient elution separated by RP-C_{18} column [64].

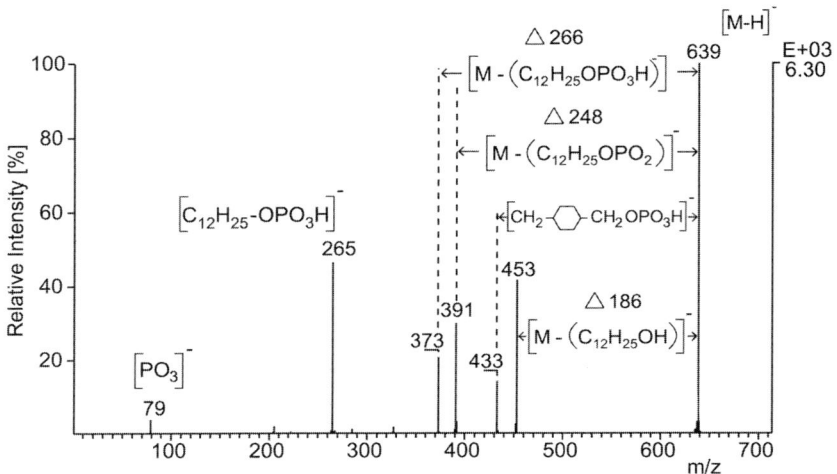

Fig. 2.11.37. ESI–FIA–MS–MS(−)(CID) product ion mass spectrum of [M − H]⁻ gemini surfactant parent ion at m/z 639 of di-protonated (1,4-cyclohexylene-dimethylene)-bis(dodecylphosphate)) anion; fragmentation behaviour in the inset [64].

The di-sodium-(1,4-cyclohexylenedimethylene)bis(dodecylphosphate; $C_{12}H_{25}-OPO^-(O_2)-CH_2-C_6H_{10}-CH_2-OPO^-(O_2) - C_{12}H_{25}*2Na^\oplus$) which was examined in biodegradation experiments was submitted to CID on a MS–MS instrument. ESI–FIA–MS–MS(−) was applied for the identification and confirmation of the compound in its original form. The CID spectrum generated from the [M − H]⁻ ion (M = $C_{12}H_{25}-$ $OP(OH)(O_2)-CH_2-C_6H_{10}-CH_2-OP(O^-)(O_2)-C_{12}H_{25}$) at m/z 639 contained characteristic fragments that allowed its identification, as shown in the fragmentation scheme in the inset of Fig. 2.11.37 [64].

2.11.3 Conclusion

In the analysis of surfactants by MS in combination with interfacing techniques that were able to handle liquids, the FIA–MS and LC–MS methods performed by the API technique have become a powerful tool. No other analytical approach at that time was able to provide as much information as that obtainable with MS and MS–MS. The analysis of surfactants, even in complex mixtures on the one hand, is facilitated

346

by soft ionisation techniques such as ESI or APCI resulting in molecular ions or adduct ions separated by their m/z ratio essential in FIA–MS. On the other hand, the small quantities of analytes needed for the different steps in identification and quantification by LC–MS and MS–MS allow separation methods that are easily handled on-line, whereas spectroscopic methods such as NMR have to apply stop-flow techniques [65].

While fast atom bombardment (FAB) [66] and TSI [25] built up the basis for a substance-specific analysis of the low-volatile surfactants within the late 1980s and early 1990s, these techniques nowadays have been replaced successfully by the API methods [22], ESI and APCI, and matrix assisted laser desorption ionisation (MALDI). In the analyses of anionic surfactants, the negative ionisation mode can be applied in FIA–MS and LC–MS providing a more selective determination for these types of compounds than other analytical approaches. Application of positive ionisation to anionics of ethoxylate type compounds led to the abstraction of the anionic moiety in the molecule while the alkyl or alkylaryl ethoxylate moiety is ionised in the form of AE or APEO ions. Identification of most anionic surfactants by MS–MS was observed to be more complicated than the identification of non-ionic surfactants. Product ion spectra often suffer from a reduced number of negative product ions and, in addition, product ions that are observed are less characteristic than positively generated product ions of non-ionics. The most important obstacle in the identification and quantification of surfactants and their metabolites, however, is the lack of commercially available standards. The problems with identification will be aggravated by an absence of universally applicable product ion libraries.

REFERENCES

1   E. Kissa, In: M.J. Schick and F.M. Fowkes (Eds.), *Fluorinated Surfactants: Synthesis, Properties, Applications*, Surfactant Science Series, Vol. 50, Marcel Dekker, New York, USA, 1994.
2   H.Fr. Schröder, *Vom Wasser*, 78 (1992) 211.

3   R. Zana, In: K. Holmberg (Ed.), *Dimeric Surfactants, Novel Surfactants*, Surfactant Science Series, Vol. 74, Marcel Dekker, New York, USA, 1998, p. 241.

4   I. Schlachter and G. Feldmann-Krane, In: K. Holmberg (Ed.), *Silicone Surfactants, Novel Surfactants*, Surfactant Science Series, Vol. 74, Marcel Dekker, New York, USA, 1998, p. 201.

5   K. Holmberg, In: K. Holmberg (Ed.), *Cleavable Surfactants, Novel Surfactants*, Surfactant Science Series, Vol. 74, Marcel Dekker, New York, USA, 1998, p. 333.

6   T.M. Schmitt, In: T.M. Schmitt (Ed.), *Analysis of Surfactants*, Surfactant Science Series, Vol. 63, Marcel Dekker, New York, USA, 1992.

7   L.S. Clesceri, A.E. Greenberg and A.D. Eaton (Eds.), *Standard Methods for the Examination of Water and Wastewater*, 20th Edn, American Public Health Association, Washington, DC, 1998, pp. 5–47, (5540 C, Anionic Surfactants as MBAS) and pp. 5–49 (5540 D, Non-ionic Surfactants as CTAS).

8   German Standard Methods for the Examination of Water, Waste Water and Sludge; General Measures of Effects and Substances (Group H). (a) Determination of Anionic Surfactants by Measurement of the Methylene Blue Index MBAS (H 24), (b) Determination of Bismut Active Substances (H 23-2), (c) Determination of the Disulfine Blue Active Substances (H 20), WILEY/VCH, Weinheim, Germany, and Beuth, Berlin, Germany, 1998.

9   H.-Q. Li and H.Fr. Schröder, *Water Sci. Technol.*, 42 (2000) 391.

10 H.Fr. Schröder, *Vom Wasser*, 79 (1992) 193.

11 T.M. Hancewicz, In: J. Cross (Ed.), *Molecular Spectroscopy I—Infrared and Raman Spectroscopy, Anionic Surfactants, Analytical Chemistry*, Surfactant Science Series, Vol. 73, Marcel Dekker, New York, USA, 1998, p. 125.

12 L.C.M. Van Gorkom and A. Jensen, In: J. Cross (Ed.), *Molecular Spectroscopy II—Nuclear Magnetic Resonance Spectroscopy, Anionic Surfactants, Analytical Chemistry*, Surfactant Science Series, Vol. 73, Marcel Dekker, New York, USA, 1998, p. 169.

13 H.T. Kalinoski, In: J. Cross (Ed.), *Mass Spectrometry of Anionic Surfactants, Anionic Surfactants, Analytical Chemistry*, Surfactant Science Series, Vol. 73, Marcel Dekker, New York, USA, 1998, p. 209.

14 J.L. Berna, A. Moreno and C. Bengoechea, *J. Surf. Det.*, 1 (1998) 263.

15 H.Fr. Schröder, *J. Chromatogr. A*, 926 (2001) 127.

16 H.Fr. Schröder, *Water Sci. Technol.*, 25 (1992) 241.

17 H.Fr. Schröder, *J. Chromatogr. A*, 777 (1997) 127.

18 H.Fr. Schröder, *J. Chromatogr. A*, 712 (1995) 123.

19 K. Bester, W. Giger, C. Schaffner and H.Fr. Schröder, Proc. GDCh-Tagung Umwelt und Chemie, Karlsruhe, Germany, 1998.

20 K. Bester, N. Theobald and H.Fr. Schröder, *Chemosphere*, 45 (2001) 817.

21 R. Kvestac, S. Terzic and M. Ahel, *Mar. Chem.*, 46 (1994) 89.

22 H.Fr. Schröder and F. Ventura, In: D. Barceló (Ed.), *Techniques and Instrumentation in Analytical Chemistry*, Sample Handling and Trace Analysis of Pollutants—Techniques, Applications and Quality Assurance, Vol. 21, Elsevier, Amsterdam, The Netherlands, 2000, p. 828.

23 E. González-Mazo, M. Honig, D. Barceló and A. Gómez-Parra, *Environ. Sci. Technol.*, 31 (1997) 504.

24 J. Riu, E. Gonzalez-Mazo, A. Gomez-Parra and D. Barceló, *Chromatographia*, 50 (1999) 275.

25 H.Fr. Schröder, In: D. Barceló (Ed.), *Applications of LC/MS in Environmental Chemistry*, Vol. 59, Elsevier, Amsterdam, The Netherlands, 1996, p. 263.

26 T. Reemtsma, *J. Chromatogr. A*, 733 (1996) 473.

27 E. Stephanou, M. Reinhard and H.A. Ball, *Biomed. Environ. Mass Spectrom.*, 15 (1988) 275.

28 J. Via, D.E. Doster and R.A. Ludicky, Proc. 44th ASMS Conf. on Mass Spectrometry and Allied Topics, Portland, OR, USA, 1996, p. 1275.

29 E. Schneider, K. Levsen, P. Daehling and F.W. Röllgen, *Fresenius' Z. Anal. Chem.*, 316 (1983) 488.

30 P.A. Lyon, W.L. Stebbings, F.W. Crow, K.B. Tomer, K.L. Lippstreu and M.L. Gross, *Anal. Chem.*, 56 (1984) 8.

31 M.J.G. Righton and C.D. Watts, Water Research Centre, Report ER 1194-M, Marlow, Bucks, UK, 1986.

32 R.M. Facino, M. Carini, P. Minghetti, G. Moneti, E. Arlandini and S. Melis, *Biomed. Environ. Mass Spectrom.*, 18 (1989) 673.

33 R. Large and H.J. Knof, *J. Chem. Soc., Chem. Commun.*, (1974) 935.

34 H.R. Schulten and D. Kümmler, *Fresenius' Z. Anal. Chem.*, 278 (1976) 13.

35 H. Shiraishi, A. Otsuki and K. Fuwa, *Bull. Chem. Soc. Jpn*, 52 (1979) 2903.

36 P. Dähling, F.W. Roellgen, J.J. Zwinselman, R.H. Fokkens and N.M.M. Nibbering, *Fresenius' Z. Anal. Chem.*, 312 (1982) 335.

37 R.K. Weber, K. Levsen, G.J. Louter, A.J.H. Boerboom and J. Haverkamp, *Anal. Chem.*, 54 (1982) 1458.

38 K. Levsen, E. Schneider, F.W. Röllgen, P. Dähling, A.J.H. Boerboom, P.G. Kistemaker and S.A. McLuckey, Comm. Eur. Communities (1984) EUR 8518, *Anal. Org. Micropollut. Water*, 132.

39 P. Zhu and K. Su, *Org. Mass Spectrom.*, 25 (1990) 260.

40 H.Fr. Schröder, *Vom Wasser*, 81 (1993) 299.

41 H.Fr. Schröder, *J. Chromatogr.*, 647 (1993) 219.

42 H.Fr. Schröder, *Vom Wasser*, 77 (1991) 277.

43 H.Fr. Schröder, *Korresp. Abwasser*, 39 (1992) 387.

44 C. Wetter and H.Fr. Schröder, (Eds), Abschlussbericht Forschungsvorhaben 02-WT 8733; Bundesminister für Forschung und Technologie, Bonn, Germany, 1992.

45 H.Fr. Schröder, In: M. Dohmann (Ed.), Gewässerschutz, Wasser, Abwasser, GWA 166, Gesellschaft zur Förderung der Siedlungswasserwirtschaft an der RWTH Aachen e.V., Aachen, Germany, 1997, ISBN 3 932590 43 0.

46 H.Fr. Schröder, Abschlußbericht Verbundforschungsvorhaben 02 WT 9358; Bundesminister für Bildung und Wissenschaft, Forschung und Technologie, Bonn, 1996.

47 J. Efer, U. Ceglarek, T. Anspach, W. Engewald and W. Pihan, *gwf Wasser, Abwasser*, 138 (1997) 577.

48 H.Fr. Schröder and K. Fytianos, *Chromatographia*, 50 (1999) 583.

49 H.Fr. Schröder, *J. Chromatogr. A*, 777 (1997) 127.

50 A. Raffaelli, S. Pucci, L. Campolmi and P. Salvadori, Proc. 44th ASMS Conf. on Mass Spectrometry and Allied Topics, Portland, OR, USA, 1996, p. 163.

51 N.J. Jensen, K.B. Tomer and M.L. Gross, *J. Am. Chem. Soc.*, 107 (1985) 1863.

52 D.D. Popenoe, S.J. Morris III, P.S. Horn and K.T. Norwood, *Anal. Chem.*, 66 (1994) 1620.

53 J.J. Conboy, J.D. Henion, M.W. Martin and J.A. Zweigenbaum, *Anal. Chem.*, 62 (1990) 800.

54 R.M. Facino, M. Carini, G. Depta, P. Bernardi and B. Casetta, *J. Am. Oil Chem. Soc.*, 72 (1995) 1.

55 C.A. Moody, W.C. Kwan, J.W. Martin, D.C.G. Muir and S.A. Mabury, *Anal. Chem.*, 73 (2001) 2200.

56 M.L. Trehy, W.E. Gledhill, J.P. Mieure and J.E. Adamove, *Environ. Toxicol. Chem.*, 15 (1996) 233.

57 I. Ogura, D.L. DuVal, S. Kawakami and K. Miyajima, *J. Am. Oil Chem. Soc.*, 73 (1996) 137.

58 X. Domingo, In: H.W. Stache (Ed.), *Alcohol and Alcohol Ether Sulfates, Anionic Surfactants*, Surfactant Science Series, Vol. 56, Marcel Dekker, New York, USA, 1996, p. 223.

59 B.N. Jewett, L. Ramaley and J.C.T. Kwak, Proc. 40th ASMS Conf. on Mass Spectrometry and Allied Topics, Palm Springs, FL, USA, 1997, p. 1270.

60 H.Fr. Schröder, *Water Sci. Technol.*, 34 (1996) 21.

61 R. Scheding and H.Fr. Schröder, Chromatographia, in preparation.

62 H.Fr. Schröder, Determination of Fluorinated Surfactants and their Metabolites in Sewage Sludge Samples by Liquid Chromatography-Mass Spectrometry and -Tandem Mass Spectrometry after Pressurised Solvent Extraction and Separation on Fluorine Modified RP-Phases, *J. Chromatogr. A*, in press.

63 F.M. Menger and C.A. Littau, *J. Am. Chem. Soc.*, 113 (1991) 1451.

64 H.Fr. Schröder, In: A. Kornmüller and T. Reemtsma, (Eds.), Geminitenside Schriftenreihe "Biologische Abwasserreinigung" 2, Kolloquium im Sonderforschungsbereich 193 der DFG an der Technischen Universität Berlin

"Anwendung der LC–MS in der Wasseranalytik" Berlin, Germany, 2001, Vol. 16, pp. 223–247, ISBN 3 7983 1852 2.

65 A. Preiss, U. Sänger, N. Karfich, K. Levsen and C. Mugge, *Anal. Chem.*, 72 (2000) 992.

66 F. Ventura, In: D. Barceló (Ed.), *Environmental Analysis: Techniques, Applications and Quality Assurance*, Vol. 13, Elsevier, Amsterdam, The Netherlands, 1993, p. 481.

2.12 ATMOSPHERIC PRESSURE IONISATION MASS SPECTROMETRY—IX. LC–MS ANALYSES OF CATIONIC SURFACTANTS: METHODS AND APPLICATIONS

Imma Ferrer, Horst Fr. Schröder and Edward T. Furlong

2.12.1 Introduction

Cationic surfactants represent a broad family of commercial compounds that are widely used in households and industrial enterprises as active ingredients in fabric softeners, antistatic agents in laundry detergents, and disinfectants. Conventional cationic compounds with surface-active properties are all typically quaternary amines and have at least one or two alkyl chains containing a combined total of 8–22 carbon atoms. The other alkyl groups are mostly short-chain substituents, such as methyl or benzyl groups. Compounds with alkyl fractions of $C_8–C_{14}$ are biocidally active and are preferred as disinfectants. Because of their positive charge located at the quaternary nitrogen atom, cationic surfactants have a strong affinity for negatively charged surfaces, to which their effectiveness as a fabric softener or as a disinfectant is attributed.

Until recently, the chemical compounds most commonly used for fabric conditioning were the dialkyl dimethyl quaternary ammonium compounds. For example, until 1991, ditallow dimethyl ammonium chloride (DTDMAC) was the most-used agent; this compound has been in use since the 1960s as a rinse-added fabric-softening active ingredient in the United States and Europe. However, in the last few years, it has been replaced by new groups of cationic surfactants, such as the esterquats or quaternary ammonium compounds that are different in structure. These compounds are expected to be more rapidly and completely degraded in the environment [1–3]. In Fig. 2.12.1, the chemical structures of some of the most widely used cationic surfactants are shown together with some quite exotic compounds. The consumption of cationic surfactants was estimated to exceed 250 000 t in 1986 [4] and in 1987 was 80 000 t in the US and 72 000 t in Europe [5].

Several techniques have been developed for the trace analysis of cationic surfactants. Most of the methodologies are based on high-performance liquid chromatography (HPLC) techniques, because most of the commercial cationic surfactants are produced as homologous

Comprehensive Analytical Chemistry XL
T.P. Knepper, D. Barceló and P. de Voogt (Eds)
© 2003 Elsevier Science B.V. All rights reserved

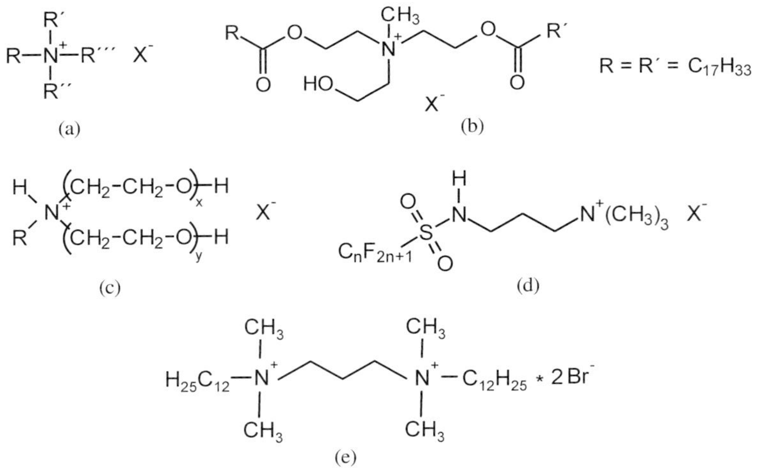

Fig. 2.12.1. Chemical structure of different classes of cationic surfactants: (a) quaternary ammonium surfactants (quats); (b) dialkylcarboxyethyl hydroxyethyl methyl ammonium surfactants (esterquats); (c) alkyl polyglycol amine surfactants; (d) quaternary perfluoro-alkyl ammonium surfactants; (e) N, N, N′, N′-tetramethyl-N, N′-didodecyle-1,3-propane-diyle-diammonium dibromide (cationic gemini surfactant). R = alkyl or benzyl group.

mixtures and are strong polar compounds. The majority of the cationic surfactants are not detectable with conventional detection techniques, such as ultraviolet (UV) absorption or fluorescence (FL) because any chromophores and fluorophores are either weak or absent. Some methods have been developed by using HPLC with conductivity detection [6,7], evaporative light-scattering detection (LSD) [8], and post-column ion-pair formation with phase separation and UV detection [9]. However, in all these techniques, the separation of the homologues is difficult because a complete chromatographic separation for the linear alkyl cationic surfactants, such as DTDMAC, is only possible using ion exchange columns. A more specific method for the determination of individual cationic surfactant homologues is mass (MS) and tandem mass spectrometry (MS–MS) because molecular weight information is provided and unequivocal structural identification of each homologue can be obtained.

Numerous chromatographic methods coupled with MS have been developed for quaternary ammonium compounds. HPLC and fast atom bombardment mass spectrometry (FAB-MS) techniques have been applied to analyse concentrations of alkyldimethylbenzyl ammonium homologues in a variety of matrices, such as in pharmaceutical formulation and ophthalmic products [10–14]. Capillary electrophoresis (CE) also has been applied for the determination of these compounds [15,16]. This technique has the advantage of high resolution and efficiency, low consumption of solvent, and rapid method development. Gas chromatography (GC) has also been used for qualitative determination of long-chain cationic surfactants by converting them to the corresponding tertiary amines by thermal decomposition in the injection port [17,18]. GC coupled with MS (GC–MS) recently has been applied for the detection of benzalkonium chloride homologues in river water and sewage effluent [19] after derivatisation of the surfactants with potassium *tert*-butoxide.

To date, the environmental distribution, fate and behaviour of cationic surfactants have not been investigated extensively, although the likely source of cationic surfactants is wastewater discharge. For example, during the textile washing process, a fabric softener is applied in the rinse cycle, where the active ingredient adsorbs onto fabrics and makes textiles soft. During this washing process, the cationic surfactants are transferred to the aqueous phase by the excess of anionic surfactants and thus are discharged into the aquatic environment through wastewaters, mainly as ion-pair complexes [7,20]. Thus, these compounds are expected to be found in trace amounts in environmental water samples. Few papers reporting environmental concentrations of cationic surfactants have been published until now [12,21–32] because of the recent availability of appropriate analytical methods and difficulties in analysis.

2.12.2 Methods of analysis

As mentioned above, the most commonly used method for the analysis of cationic surfactants has been HPLC coupled with conductometric, UV, or fluorescence detectors, the latter typically utilizing post-column ion-pair formation for enhanced sensitivity. Analysis by GC is only possible for cationic compounds after a derivatisation step [33] because of the ionic character of this compound. However, structural information might be lost.

In the last few years, MS has been used to detect cationic surfactants because the fixed ionic charge of these molecules makes them suitable for the analysis by atmospheric pressure ionisation (API) MS. Radke et al. [28] reported a method for DTDMAC and two of its most important substitution products by using HPLC on-line coupled with an electrospray (ESI) interface to a mass spectrometer. A study reporting the use of liquid chromatography (LC) combined with MS–MS (LC–MS–MS) has been described for the analysis of benzalkonium chloride on skin [34]. The specific methods used for each type of cationic surfactant will be explored in detail in the following sections.

Quaternary alkyl ammonium compounds
The homologous quaternary ammonium surfactants (so-called 'quats') are characterised by their general formula $[RR'N^{\oplus}R''R''']\,X^-$ as shown in Fig. 2.12.1(a). Factors affecting the ionisation efficiency of these quats were examined, using flow injection analysis (FIA) coupled with MS (FIA–MS) in the positive ESI mode [35]. These compounds were applied as powerful counterions in ion-pair chromatography. Compounds with a varying number of alkyl chain links were studied. The results obtained demonstrated that quats containing only alkyl groups give the best sensitivity under these conditions. FAB and thermospray ionisation (TSI), however, gave rise to dealkylation reactions, whereas ESI spectra showed one single peak representing $[M]^+$ ions. Basic research results regarding the ionisation efficiency observed with these compounds were reported [35].

Synthetic mixtures of industrially important quats were examined by applying ion chromatography (IC) with selected resins for separation purposes. ESI in the positive mode was applied for detection [36]. The mixture was separated successfully under ion-pairing conditions by applying methane sulfonic acid (CH_3SO_3H). The industrial mixtures contained four symmetrical tetraalkylammonium compounds varying in the alkyl groups starting at C_3H_7 (propyl) and ending at C_6H_{13} (hexyl). Ion current profiles reveal the chromatographic integrity and signal response by the instrument used. The ionic constituents that were applied for ion-pairing in the separation process were removed by a suppressor device to improve the signal-to-noise (S/N) ratio prior to MS detection. Results of the separation of the quats, ionised as $[M]^+$ ions, were presented [36,37]. MS–MS data of quats also were elaborated and presented [36].

The determination of alkyltrimethyl ammonium surfactant compounds with the general formula $RN^{\oplus}(CH_3)_3\,X^-$ were performed using continuous flow (CF) FAB (CF-FAB) and ESI. Dodecyl-, tetradecyl- and hexadecyl compounds dissolved in water were examined to compare the performance of both MS ionisation techniques. CF-FAB and FIA–ESI were applied in the selected ion monitoring mode (SIM) and in the selected reaction monitoring mode (SRM). Quantitative results were shown and discussed [38].

The homologue quaternary ammonium surfactant mixture with the general formula $RR'N^{\oplus}(CH_3)_2X^-$ where $R = C_nH_{2n+1}-$ ($n = 12, 14$ and 16), $R' = $ benzyl ($C_6H_5-CH_2-$) and $X^- = $ acetate, was examined by FIA–ESI–MS(+) resulting in $[M]^+$ ions at m/z 304 (C_{12}), 332 (C_{14}) and 360 (C_{16}). FIA–MS(+) studies using atmospheric pressure chemical ionisation (APCI) applied to the same commercial blend gave rise to dealkylations or Hoffmann-type elimination reactions, as also reported for other quats in the literature [35]. Ions detected under these conditions proved either an elimination of CH_2 or the cleavage of the benzyl moiety resulting in ions at m/z 290, 318 and 346 $[M - (CH_2)]^+$ or 214, 242 and 256 $[M - (C_6H_5-CH_2-)]^+$, respectively [39].

A mixture of seven different quaternary ammonium compounds was examined by LC–MS applying the methane sulfonic acid (CH_3SO_3H) ion-pairing reagent. ESI–LC–MS in the positive mode was applied for detection. The tetraalkylammonium compounds of the mixture were ionised in the form of their $[M]^+$ ions at m/z 228 $[C_{12}H_{25}N^{\oplus}(CH_3)_3]$, 270 $[C_{16}H_{33}N^{\oplus}H(CH_3)_2]$, 298 $[C_{18}H_{37}N^{\oplus}H(CH_3)_2]$, 326 $[C_{20}H_{41}N^{\oplus}H(CH_3)_2]$, 304 $[C_{12}H_{25}N^{\oplus}(CH_3)_2-CH_2-C_6H_5]$, 332 $[C_{14}H_{29}N^{\oplus}(CH_3)_2-CH_2-C_6H_5]$, and 360 $[C_{16}H_{33}N^{\oplus}(CH_3)_2-CH_2-C_6H_5]$. LC separation resulted in the total ion current (TIC) chromatogram in Fig. 2.12.2(a) and the selected mass traces (b)–(h) [39]. Each of these compounds contained one linear alkyl group, which varied from C_{12} to C_{20} and two methyl groups directly bonded to the central nitrogen. In addition, an H, a CH_3 group or a benzyl moiety ($-CH_2-C_6H_5$) were attached to the central nitrogen atom of the molecules.

The determination of alkyltrimethyl ammonium compounds with the general formula $RN^{\oplus}(CH_3)_3X^-$ surfactant were performed using CF-FAB and ESI. The compounds examined were dodecyl-, tetradecyl-, and hexadecyl-trimethyl ammonium compounds dissolved in water. The product ion spectra of dodecyltrimethyl ammonium compound and its methyl-deuterated homologue were presented. The product ion at m/z 60 $[(CH_3)_3NH]^+$ of the non-deuterated compound at m/z 228 was the only ion that shifted after deuteration to m/z 69 [38].

Recently, the authors developed a novel methodology for the determination of alkyl (C_{12}, C_{14} and C_{16}) dimethylbenzyl ammonium

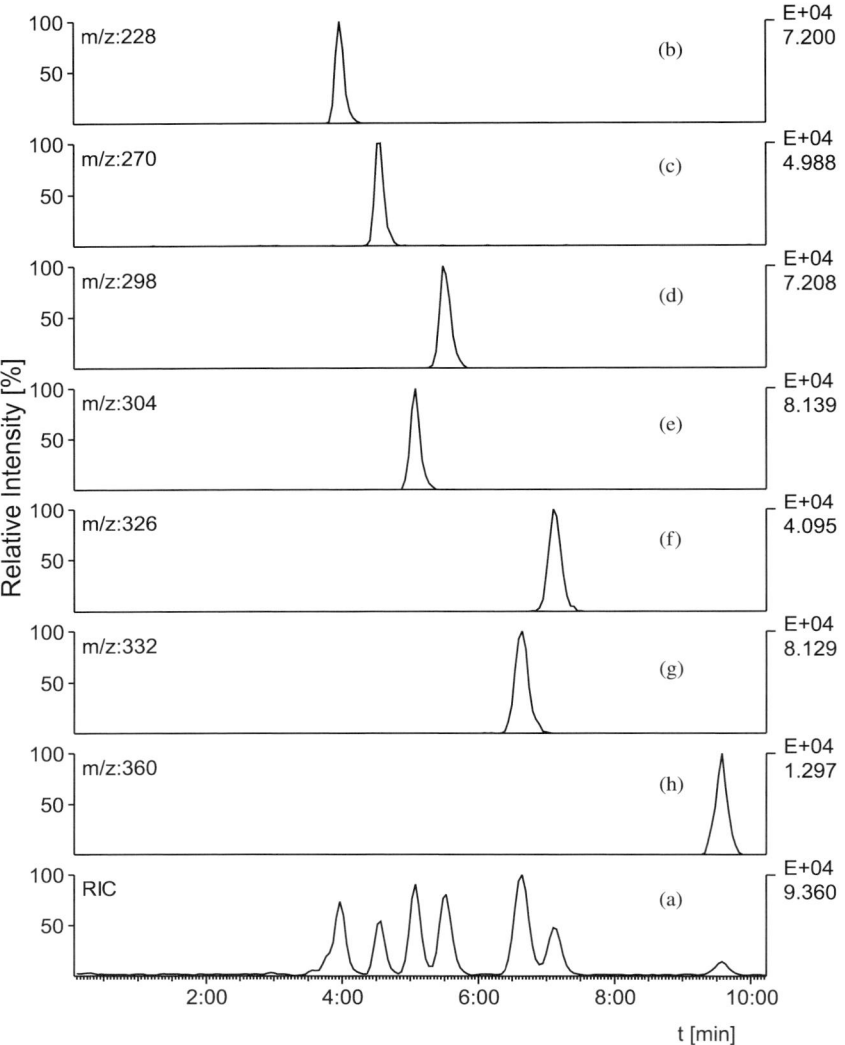

Fig. 2.12.2. (a) ESI−LC−MS(+) TIC chromatogram and selected mass traces of a mixture of seven different quats; (b) $C_{12}H_{25}N^{\oplus}(CH_3)_3$; (c) $C_{16}H_{33}N^{\oplus}H(CH_3)_2$; (d) $C_{18}H_{37}N^{\oplus}H(CH_3)_2$; (e) $C_{12}H_{25}N^{\oplus}(CH_3)_2-CH_2-C_6H_5$; (f) $C_{20}H_{41}N^{\oplus}H(CH_3)_2$; (g) $C_{14}H_{29}N^{\oplus}(CH_3)_2-CH_2-C_6H_5$; (h) $C_{16}H_{33}N^{\oplus}(CH_3)_2-CH_2-C_6H_5$; isocratic separation on a PLRP column under ion-pairing conditions applying methane sulfonic acid (CH_3SO_3H) [39].

chloride (also known as benzalkonium chloride or BAC) in water samples [23]. BAC is a quaternary ammonium salt with antiseptic properties and uses similar to other cationic surfactants. BAC is manufactured from coconut oil, which results in a distribution of C_{12}, C_{14} and C_{16} alkyl chain homologues (see Fig. 2.12.1(a)). These quaternary ammonium surfactants act as detergents by disrupting cell membranes and thus are bacteriostatic or bacteriocidal, depending on the concentrations used. As a result, the primary use of BAC is as an active ingredient in cleaning and disinfection products. The percentage of the homologues varies from one product to another, with a typical total BAC concentration of about 0.1–0.8%. Other reported uses of BAC include preservatives and antiseptics in medical formulations, in health care solutions (ophthalmologic, nasal and skin), fungicides, spermicides, and virucides.

The method developed was based on solid-phase extraction (SPE) by using polymeric cartridges, followed by HPLC ion trap mass spectrometry (IT-MS) and MS–MS detection, equipped with an ESI interface in positive ion mode. LC separation was achieved for three BAC homologues by using a C_{18} column and a gradient of acetonitrile/10 mM aqueous ammonium formate. Total method recoveries were higher than 71% in different water matrices. Total method recoveries, obtained after the pre-concentration of 50 mL of distilled water, tap water, and wastewater sample, spiked at 0.1625 μg L^{-1} with BAC onto polymeric (PLRP-s) and Hysphere cartridges, are listed in Table 2.12.1. Recoveries in excess of 80% were obtained for both homologues from distilled and tap-water matrices. For wastewater samples, extraction recoveries averaged 74.5% for both SPE phases because of the complexity of the sample matrix.

The main ions observed by LC–MS were at mass-to-charge ratios (m/z) of 304, 332, and 360, which correspond to the molecular ions of the C_{12}, C_{14}, and C_{16} alkyl BAC, respectively. The unequivocal structural identification of these compounds in water samples was determined by LC–MS–MS after isolation and subsequent fragmentation of each molecular ion. Fig. 2.12.3 shows the mass spectra of both homologues at a fragmentor voltage of 120 V, as well as the proposed fragmentation interpretation. In both cases, molecular ions and two main fragment ions are produced. For the C_{12} homologue, the base peak is the molecular ion, at m/z 304, and the peak at m/z 212 results from the cleavage of the carbon–nitrogen bond between the toluyl substituent and the quaternary amine. The m/z 91 ion is a stable tropylium ion

TABLE 2.12.1

Mean percent total method recoveries and relative standard deviations ($n = 3$) of benzalkonium homologues obtained after the percolation of 50 mL of deionised water, tap water and wastewater samples spiked at 0.1625 μg L^{-1} with benzalkonium chloride through Hysphere and PLRP-s cartridges

Compound	DI water	Tap water	Wastewater
Total method recoveries in Hysphere cartridges (%)			
$C_{12}BAC$	93 (1)	90 (2)	78 (3)
$C_{14}BAC$	91 (5)	83 (8)	71 (6)
Total method recoveries in PLRP-s cartridges (%)			
$C_{12}BAC$	90 (3)	87 (2)	74 (4)
$C_{14}BAC$	89 (7)	84 (5)	75 (7)

typically formed by fragmentation of benzylalkyl compounds. Similarly, for the C_{14} homologue, we obtained the molecular ion at m/z 332, a secondary fragment at m/z 240, and the tropylium ion at m/z 91. Both compounds have the same loss of the benzyl group and formation of the tropylium ion. Similar results were reported in a previous work [30] and are shown in Fig. 2.12.4. It is important to note that the latter work [30] was carried out using an APCI interface, whereas the results shown in Fig. 2.12.3 were done with an ESI interface [23]. A previous work [40] showed that an APCI interface is not suitable for cationic compounds because of the fixed ionic charge in the molecule. The results in Fig. 2.12.4 could be caused by the ESI process occurring at the same time in ACPI interfaces (especially in the older instrument models).

Because of the ionic character of cationic surfactants, they would be expected to associate predominantly with solids rather than with water, thus being present at higher concentrations in sediments compared with water samples. Thus, a study to determine the applicability of accelerated solvent extraction (ASE) to the extraction and recovery of BAC homologues from sediment was performed [41]. BAC was extracted successfully from sediment samples using a new methodology based on ASE followed by an on-line cleanup step, detection and identification by LC–MS or MS–MS using an ESI interface operated in the positive ion mode. This methodology combines the high efficiency of extraction provided by a pressurised fluid and the high sensitivity offered by the IT-MS–MS. Moreover, the on-line cleanup step included in the methodology provides average reproducibility of 12%, with few sample manipulations, improving the analytical

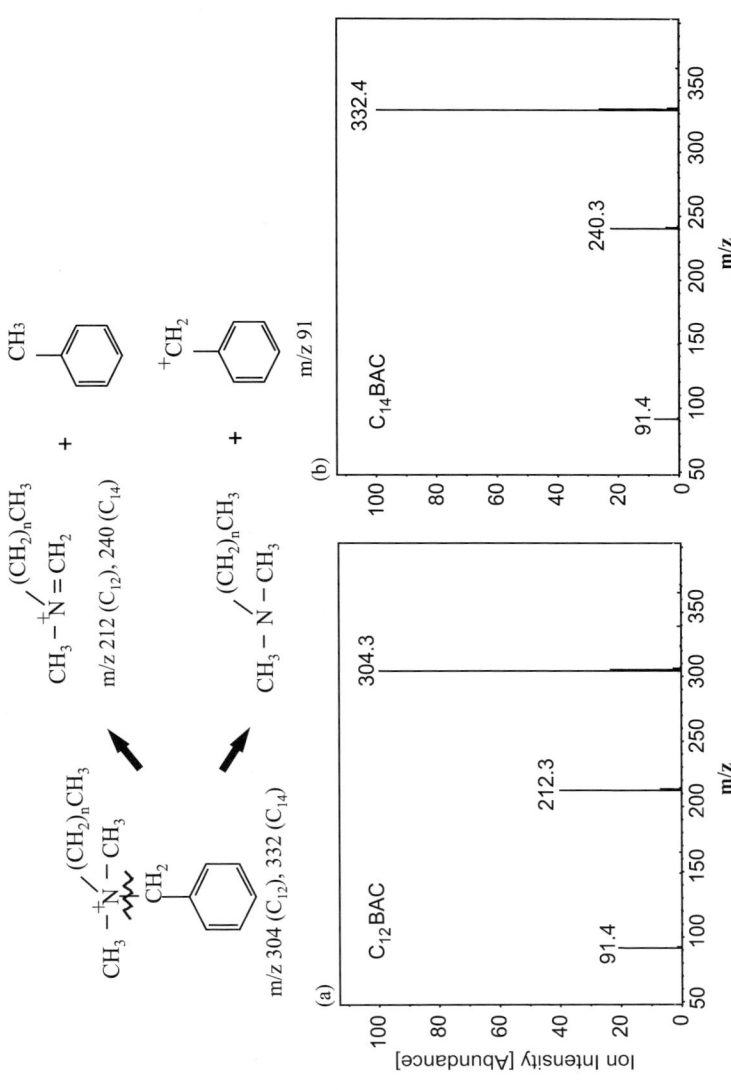

Fig. 2.12.3. Mass spectra and proposed fragmentation patterns for (a) C_{12} and (b) C_{14} benzalkonium chloride homologues at a capillary exit voltage of 120 V [23].

361

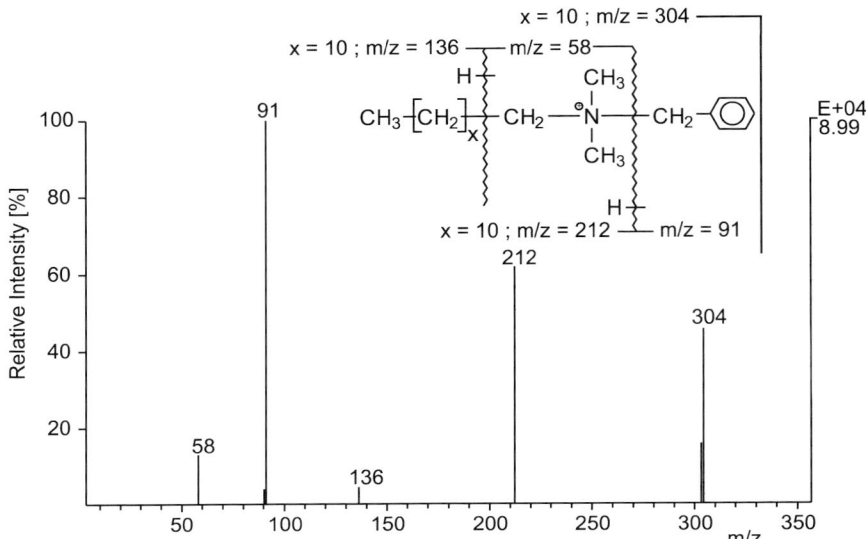

Fig. 2.12.4. FIA–APCI–MS–MS(+) CID product ion mass spectrum of selected $[M]^+$ parent ion of cationic quat surfactant blend $RR'N^{\oplus}(CH_3)_2X^-$ at m/z 304 characterised as alkyl di-methylbenzyl ammonium acetate; fragmentation pattern presented in the inset [37].

performance of the analyses. The effect of solvent type and ASE operational variables, such as temperature and pressure, were evaluated. After optimisation, a mixture of acetonitrile/water (6:4) was found to be most efficient to extract BAC from the sediments. The optimised conditions for the extraction of BAC from sediment samples are listed in Table 2.12.2. The recoveries of extraction from the sediment samples ranged from 95 to 105% for C_{12} and C_{14} homologues. The methodology developed in this work provides detection limits in the low nanogram-per-kilogram range.

TABLE 2.12.2

Conditions used for ASE

Amount of sediment	10 g
Solvent	ACN/H$_2$O (6:4)
Temperature	120°C
Pressure	1500 psi
Number of cycles	3
Solvent flush	40%

Quaternary carboxyalkyl ammonium compounds
Esterquats, quaternary carboxyalkyl ammonium compounds, have been developed as substitutes for the quats, especially DTDMACs which are applied as softeners in household products because they are resistant to degradation [42] and toxic [43,44]. Today, the esterquats are preferentially used (see Fig. 2.12.1(b)). As previously noted, the detection of all types of esterquats in environmental samples is difficult because of their tendency to adsorb at surfaces. Even in the inflows of wastewater treatment plants (WWTP) it is extremely difficult to find these compounds, which are reduced in toxicity [45] compared with the DTDMACs. One reason is that these carboxyalkyl ammonium compounds are not very stable in the environment. A second reason is that these compounds are not stable during the ionisation process.

The industrial blend of esterquats, with the general formula $(R(CO)OCH_2CH_2)_2-N^{\oplus}(CH_3)CH_2CH_2OH$ X^- where $R =$ mixture of alkyl (C_{13}, C_{15} and $C_{17} =$ tallowyl) or unsaturated alkenyl ($C_{17}H_{33} =$ oleyl) moieties, was examined using APCI- and FIA–ESI–MS(+). Both API techniques showed low sensitivity, but in addition the application of FIA–APCI–MS(+) for the examination of these esterquats resulted in quite complex spectra. Large numbers of fragment ions were more commonly observed than molecular ions. Ionisation of the di-hydrogenated tallowethyl hydroxyethyl ammonium methane sulfate compound (general formula $(R(CO)OCH_2CH_2)_2-N^{\oplus}(CH_3)CH_2CH_2OH$ $CH_3OSO_3^-$) by FIA–ESI–MS(+) resulted in $[M]^+$ ions of the compounds with $R = C_{17}H_{33}$ and of compounds containing a mixture of $C_{17}H_{33}$ and $C_{15}H_{29}$ moieties as an impurity of the blend shown in Fig. 2.12.5. Besides these small molecular ions at m/z 692 and 664, the fragment ions from a loss of R(CO) under standard ionisation conditions could be observed at m/z 400 and 428 [37]. Both ions resulted from the abstraction of one $C_{17}H_{33}COO$ moiety from the $[M]^+$ ions of the compounds containing two identical moieties with $R = C_{17}H_{33}$ or the other substituent being a $C_{15}H_{31}$ moiety.

The results of ion-pairing separation of this compound according to the method of Conboy et al. [36] with MS detection in the ESI–LC–MS(+) mode was not promising. Even at a concentration of 50 mg L^{-1} of this esterquat in the standard solution, the S/N ratio was poor [39]. The base peak was the ion at m/z 428, representative of the fragment ions $(R(CO)OCH_2CH_2)-N^{\oplus}(CH_3)(CH_2CH_2OH)_2$ with $R = C_{17}H_{33}$.

ESI could be applied in the FIA–MS $(+/-)$ mode to achieve a quick and reliable method to detect and identify surfactants in consumer

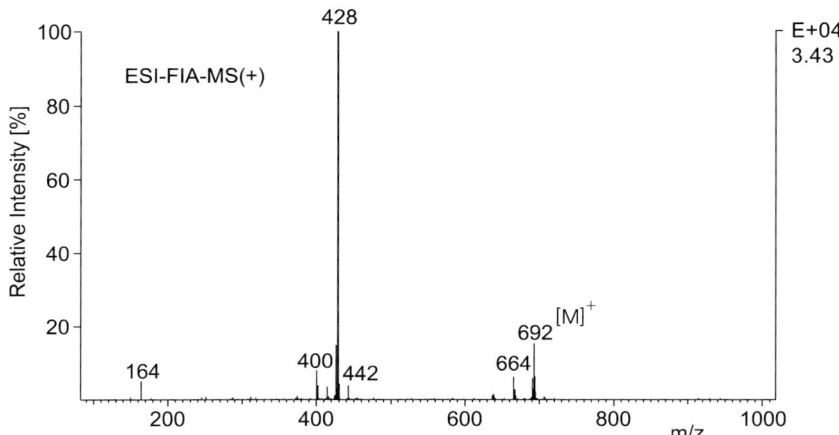

Fig. 2.12.5. FIA–ESI–MS(+) overview spectrum of industrial blend of an esterquat compound $(R(CO)OCH_2CH_2)_2–N^{\oplus}(CH_3)CH_2CH_2OH \ X^-$ $(R = C_{17}H_{33}$ (oleyl)) [37].

products. FIA–MS allowed qualitative and quantitative determination and identification of esterquats and quats in commercial fabric softeners. FIA overview spectra of compounds with the general formulae $(R(CO)–NH–CH_2CH_2CH_2)–NH^{\oplus}(CH_3)–CH_2CH_2OH \ X^-$ and $(R'(CO)–NH–CH_2CH_2CH_2)–NH^{\oplus}(CH_3)–CH_2CH_2–OC(O)R'' \ X^-$ were presented [46].

To establish a product ion library from which parent ions could be identified by reverse interpretation of product ion data, the commercially available esterquat surfactant blend $C_{13}H_{27}C(O)N(H)CH_2CH_2$ $N(CH_2CH_2OH)CH_2CH(OH)CH_2N^{\oplus}RR'R''$ $(R = R' = R'' = CH_3$; m/z 430) was examined by FIA–ESI–MS^n in the positive mode using IT-MS [47]. The map of ions generated in the different steps under MS^n $(n = 1–5)$ was presented. For the explanation of the fragmentation behaviour, an H· transfer from the alkyl chain to the nitrogen and the loss of trimethylamine was proposed. The results of MS^n experiments and the 'explosion of information' available from these experiments were discussed. The differences in the fragmentation behaviour observed with the application of a triple stage quadrupole spectrometer or an IT instrument were discussed [47].

APCI- and FIA–ESI–MS–MS(+) were applied for the examination of esterquats used in the household as textile softener $(R(CO)OCH_2\,CH_2)_2–N^{\oplus}(CH_3)CH_2CH_2OH \ X^-$ where R = tallowyl or oleyl moieties.

The mixture used for the study was an industrial blend. The first reaction observed for the ion of the di-hydrogenated tallowethyl hydroxyethyl ammonium methane sulfate compound (m/z 692; general formula $(R(CO)OCH_2CH_2)_2-N^{\oplus}(CH_3)CH_2CH_2(OH)-CH_3OSO_3^-$ where $R = C_{17}H_{33}$) under FIA–ESI–MS–MS($+$) conditions was an elimination of an RCO moiety, resulting in the fragment ion at m/z 428, which then was further fragmented. The dioleyl compound under very soft FIA–ESI–MS($+$) conditions (low extraction voltage) resulted in the ion at m/z 692 together with a prominent ion at m/z 428. If the blend was submitted to 'normal' MS conditions, the ion at m/z 428 could be observed, which under MS–MS resulted in a product ion spectrum containing only one fragment ion with m/z 309. This ion has the composition $[C_{17}H_{33}(-CO)OCH_2CH_2]^+$ [37]. This product ion spectrum with the fragmentation pattern in the inset is shown in Fig. 2.12.6 [39].

With the application of FIA–ESI–MS in the positive and negative modes, the purpose of a study for a quick and reliable detection and identification method of surfactants in consumer products could be reached. FIA–MS, bypassing the analytical column, and performing mixture analysis FIA–MS–MS in the positive mode allowed identification of esterquats and quats in commercial fabric softeners. Parent ions were structured as follows: $(R(CO)-NH-CH_2CH_2CH_2)-NH^{\oplus}(CH_3)-CH_2CH_2OH$ X$^-$; $(R'(CO)-NH-CH_2CH_2CH_2)-NH^{\oplus}(CH_3)-CH_2CH_2-OC(O)R''$ X$^-$; $R'(CO)-OCH_2CH_2-NH^{\oplus}(CH_3)-CH_2CH_2CH_2-NH-C(O)R''$ X$^-$ ($R' = C_{15}H_{31}$ and $R'' = C_{15}H_{31}$). The parent ion m/z 637 cleaved on the side of the tertiary amine, resulting in a loss of an electron from the alkyl chain, generating a positively charged fragment, whereas the dialkyl tertiary amine was eliminated by neutral loss [46].

Quaternary fluorinated alkyl ammonium compounds
The fluorine-containing cationic surfactants of quat type with the general formula $C_nF_{2n+1}-SO_2-NH-CH_2-CH_2-CH_2-N^{\oplus}(CH_3)_3$ X$^-$ ($n = 8$) (Fig. 2.12.1(d)) were examined by FIA–MS using APCI and ESI in the positive and negative modes. The APCI($+/-$) ionisation resulted in a dealkylation at the nitrogen with ions at m/z 585 or 583, respectively. The alkyl chain of this compound contained the moiety C_8F_{17}. The ions generated under APCI conditions were characterised as dealkylation products—m/z 585 [M $-$ CH$_2$]$^+$ or m/z 583 [M $-$ H$-$CH$_3$]$^-$—as reported in the literature [35,37].

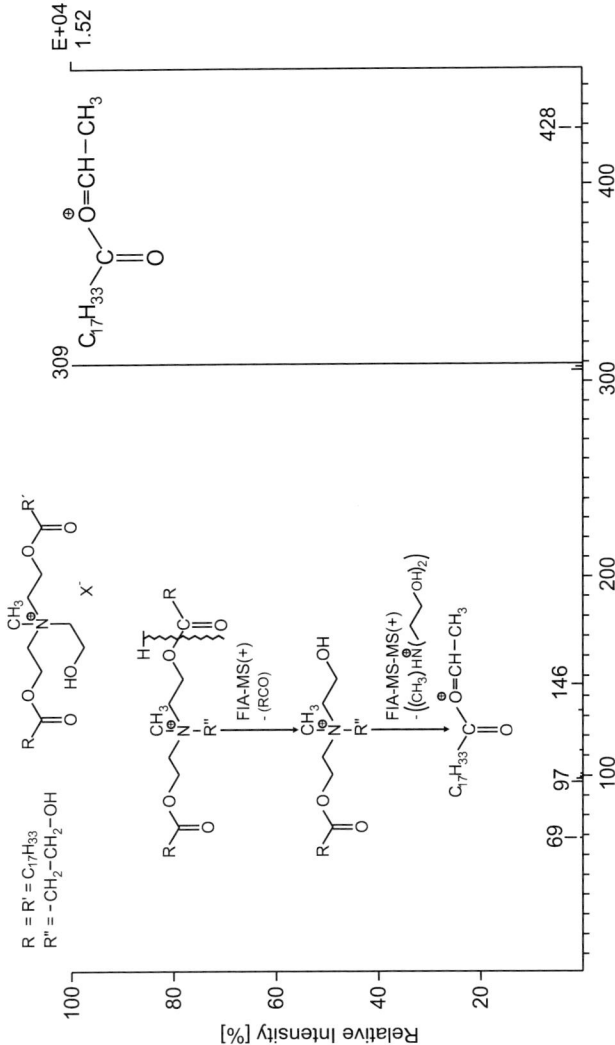

Fig. 2.12.6. Identification of esterquat compounds: FIA–APCI–MS–MS(+) (CID) product ion mass spectrum of selected [M – RCO]⁺ base peak ion of cationic surfactant blend of di-hydrogenated tallowethyl hydroxyethyl ammonium methane sulfate type (m/z 692; general formula (R(CO)OCH₂CH₂)₂–N⁺(CH₃)–CH₂CH₂(OH)CH₃OSO₃⁻); fragmentation behaviour under CID presented in the inset [39].

ESI(+), however, mainly produced the [M]$^+$ ions at m/z 599 as well as a smaller number of dealkylated ions [M − CH$_2$]$^+$ at 585 (cf. Fig. 2.12.7), whereas ESI(−) predominantly produced the m/z 583 [M − H–CH$_3$]$^-$ ion, which was also found under APCI(−) conditions. This compound was the first type of cationic surfactants that could be ionised in the negative mode, even though it contained ammonium nitrogen [37].

APCI collision induced dissociation (CID) experiments performed with this compound in the negative mode first led to a dealkylation resulting in the m/z 583 [M − H–CH$_3$]$^-$ ion, which then was destroyed, resulting in CF$_3$–(CF$_2$)$_x$ fragments at m/z 119, 169, 219, and 269 as shown in the inset of Fig. 2.12.8. Once more, we should note that APCI is not suitable for the analysis of charged molecules [40]. This fact might also explain the unusual fragmentation and no observation of molecular ions in the spectra. Positive FIA–ESI–MS–MS by CID applied to the [M]$^+$ ion at m/z 599 of fluorine-containing cationic surfactant C$_n$F$_{2n+1}$–SO$_2$–NH–CH$_2$–CH$_2$–CH$_2$–N$^{\oplus}$(CH$_3$)$_3$ X$^-$ resulted in fragments at m/z 60 [(CH$_3$)$_3$NH]$^+$, 72, 88, and 116 [M − (NH–CH$_2$–CH$_2$–CH$_2$–N(CH$_3$)$_3$]$^+$. The dealkylation step [35] of the [M]$^+$ ion prior to CID resulting in an ion at m/z 585 led to the parallel product ions ($-\Delta \times$ m/z 14) at 46 and 58. In addition, a fragment ion at m/z 85 was most abundant under these conditions [37].

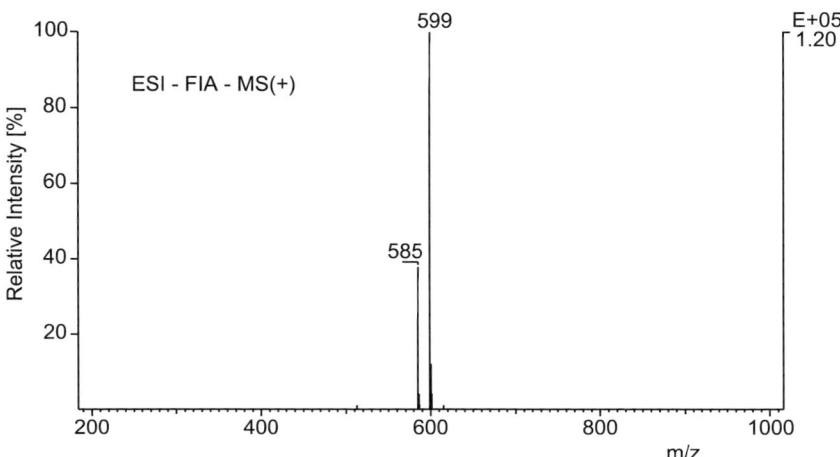

Fig. 2.12.7. FIA–ESI–MS(+) overview spectrum of fluorinated cationic surfactant of quat type C$_n$F$_{2n+1}$–SO$_2$–NH–CH$_2$–CH$_2$–CH$_2$–N$^{\oplus}$(CH$_3$)$_3$ X$^-$ [37].

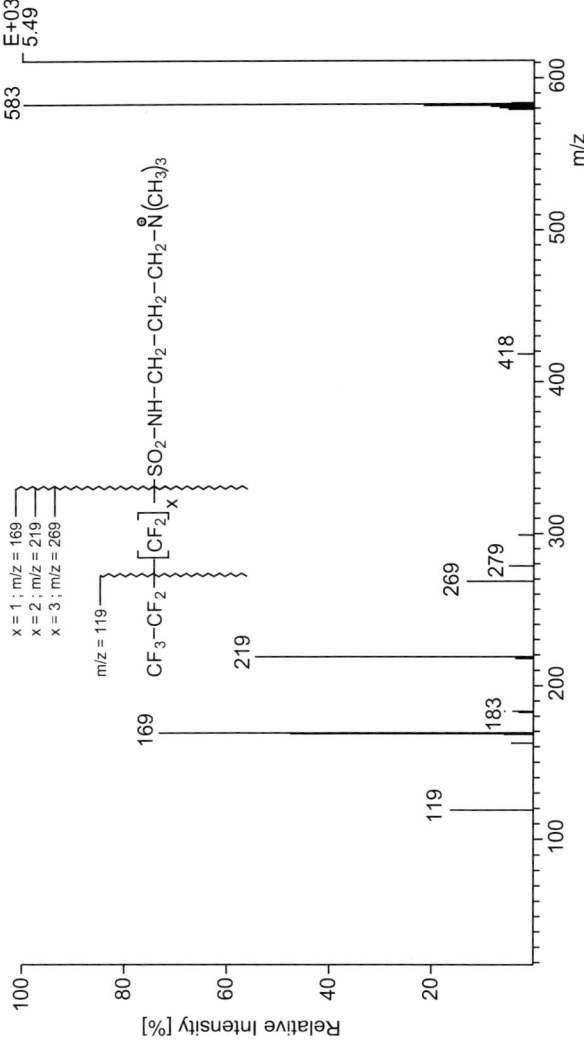

Fig. 2.12.8. FIA–APCI–MS–MS(−) (CID) product ion mass spectrum of a selected [M − H]⁻ parent ion (*m/z* 583) of cationic fluorinated quaternary alkyl ammonium surfactant blend (general formula: C_nF_{2n+1}–SO_2–NH–CH_2–CH_2–CH_2–$N^{\oplus}(CH_3)_3$ X⁻ [22]); fragmentation behaviour under CID shown in the inset [39].

Dimeric cationic surfactants (gemini surfactants)

The so-called gemini surfactants also are cationic compounds. This new type of surfactant, with a dramatically improved performance in surface activity [48,49], was examined by mass spectrometry. One cationic gemini surfactant with the general formula $C_{12}H_{25}-N^{\oplus}(CH_3)_2-CH_2-CH_2-CH_2-N^{\oplus}(CH_3)_2-C_{12}H_{25}*2Br^-$ and the systematic name N,N,N',N'-tetramethyl-N,N'-didodecyle-1,3-propanediyle-diammonium dibromide, as shown with its structural formula in Fig. 2.12.1(e), was examined by MS methods after submission to a biodegradation experiment [50].

The use of APCI interface was not very successful because complex spectra were generated instead of molecular ions. The application of FIA–ESI–MS(+) using methanol/water (1:1, v/v) with 0.5% of acetic acid, however, led to a simple overview spectrum, as shown in Fig. 2.12.9(a). The ions observed at m/z 547 and 549 are the $[M - Br]^+$ ions containing one bromine atom with its characteristic isotopic pattern of $^{79}Br/^{81}Br$ isotopes. The base peak at m/z 234 originated from the double-charged $[M - 2Br]^{2+}$ ion. Ionisation in the presence of the ion-pairing reagent, methane sulfonic acid essential for LC separation, resulted in the FIA–ESI–MS(+) spectrum, as shown in Fig. 2.12.9(b). The ion at m/z 563 is the $[M^{2+} + CH_3-SO_3^-]^+$ adduct ion, whereas the ion at m/z 659 originated from the adduct ion $[M^{2+} + CH_3-SO_3^- + (CH_3-SO_3H)]^+$. LC separation of this cationic gemini surfactant could be performed successfully [39] under ion-pairing conditions on a PLRP column [36]. MS detection in the ESI(+) was necessary because of missing chromophores in the molecule.

This Gemini cationic surfactant could not be recovered from samples of biodegradation experiments performed in a lab-scale aerobic bioreactor [50] either because of complete biodegradation (mineralisation) or because of irreversible adsorption at the surface of the biodegradation device.

In CID experiments using FIA–ESI–MS–MS(+), cationic gemini surfactant fragmented as shown in Fig. 2.12.10. While the fragmentation of the $[M]^+$ ion resulted in unspecific product ions. Fragmentation of the base peak ion at m/z 234, the double charged $[M - 2Br]^{2+}$ compound, resulted in a plausible product ion spectrum presented together with its fragmentation pattern in Fig. 2.12.10. Besides alkyl fragments at 57 and 71 generated from the $C_{12}H_{25}$ alkyl moieties of the surfactant ions, fragment ions that contain nitrogen $(CH_2)=N^{\oplus}$

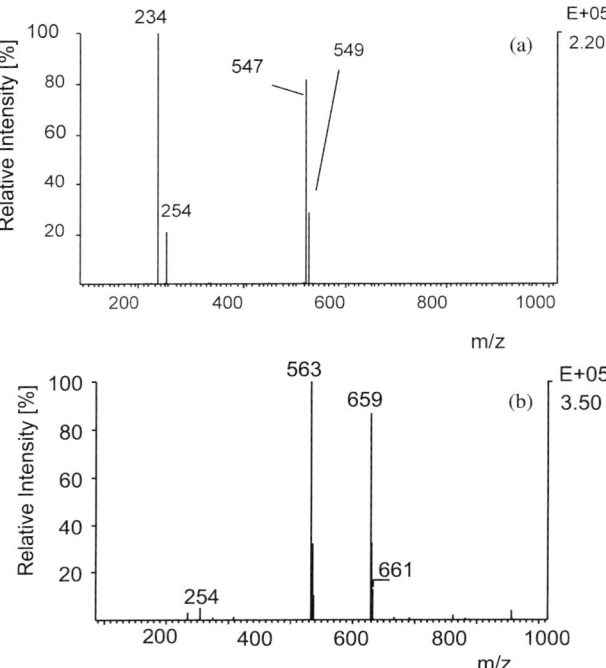

Fig. 2.12.9. (a) FIA–ESI–MS(+) overview spectrum of cationic gemini surfactant N,N,N',N'-tetramethyl-N,N'-didodecyle-1,3-propane-diyle-diammonium dibromide $C_{12}H_{25}-N^{\oplus}(CH_3)_2-CH_2-CH_2-CH_2-N^{\oplus}(CH_3)_2-C_{12}H_{25}*2Br^-$; (b) FIA–ESI–MS(+) mass spectrum of cationic gemini surfactant as in (a) in the presence of methane sulfonic acid (CH_3SO_3H) [50].

$(CH_3)-CH_2-CH_2-CH_3$ (m/z 86); $(CH_2)=N^{\oplus}(CH_3)_2$ (m/z 58) could be observed here [50].

2.12.3 Applications of LC–MS for the detection of cationic surfactants in the environment

Detections of cationic surfactants in commercial products and environmental samples
Cationic surfactant blends of quats were studied by FIA–ESI–MS(+). The general formula of the compounds was $(R)_nN^{\oplus}(CH_3)_{4-n}$ with R = $C_{12}-C_{22}$. These compounds applied in products for personal

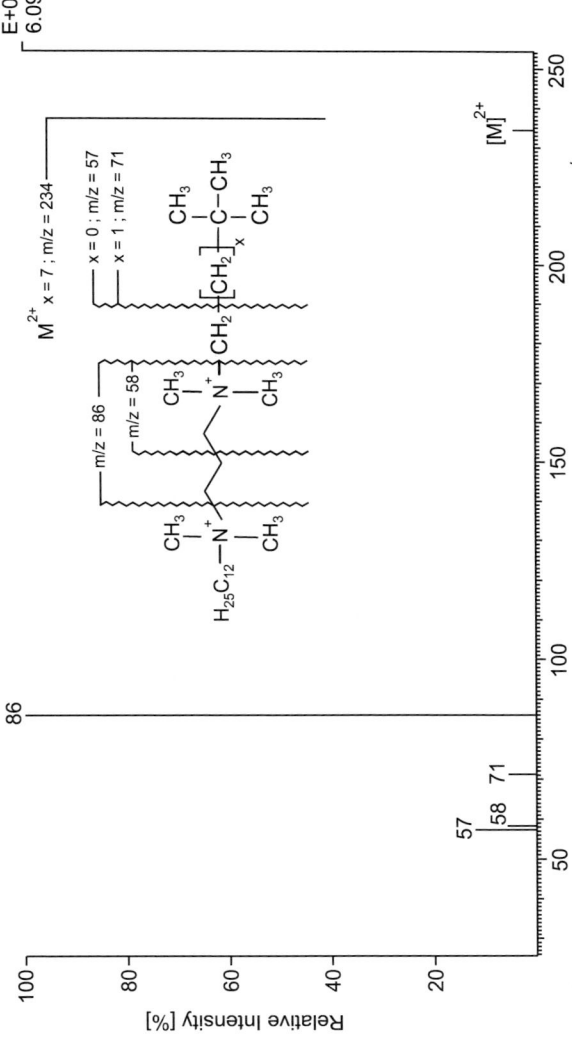

Fig. 2.12.10. FIA–APCI–MS–MS(+) (CID) product ion mass spectrum of $[M - 2Br]^{2+}$ parent ion (m/z 234) of cationic gemini surfactant; general formula: $C_{12}H_{25}-N^{+}(CH_3)_2-CH_2-CH_2-CH_2-N^{\oplus}(CH_3)_2-C_{12}H_{25}$; fragmentation behaviour under CID shown in the inset [50].

hygiene were analysed according to their chain length. Qualitative and quantitative determinations were made for commercial products and the amount of quats deposited on hair coming into contact with these compounds. MS–MS was applied for identification [51].

Quaternary ammonium cationic surfactants, such as DTDMAC, were determined in digested sludge by using supercritical fluid extraction (SFE) and FIA–ESI–MS(+) after separation by normal phase LC. Standard compounds—commercially available DTDMAC— were used to check the results. The DTDMAC mixture examined showed ions at m/z 467, 495, 523, 551, and 579 all equally spaced by $\Delta\, m/z$ 28 $(-CH_2-CH_2-)$ resulting from the ionisation of compounds like $RR'N^{\oplus}(CH_3)_2\, X^-$ $(R = I \neq R')$ as shown in Fig. 2.12.11(a) and (b) [22].

One of the first results dealing with MS–MS investigations of quats was published by Conboy et al. [36]. Industrially important quats were separated by IC and determined by ion spray MS in the ESI(+) mode. Full scan IC–MS–MS spectra were presented from tetraalkyl ammonium compounds examined. They varied in the alkyl groups, starting at C_3H_7 and ending at C_6H_{13}.

The FIA–ESI–MS(+) mode was found to be useful in meeting analysis needs for quats applied in products for personal hygiene. API combined with an ion trap instrument was applied. Industrial blends of quaternary ammonium surfactants with the general formula $(R)_nN^{\oplus}(CH_3)_{4-n}$, where R could be an alkyl chain of 12–22 or more carbon atoms, were studied by FIA–ESI–MS–MS(+) for identification. These compounds were determined according to their chain length distribution [51]. Fragmentation on an ion-trap instrument in the MS^2 and MS^3 by CID almost exclusively resulted in sequential losses of intact long alkyl chains. A proton transfer from alkyl chain to the nitrogen resulted in alkenes. The comparison of the behaviour of these compounds under CID conditions using IT-MS was described as similar to that observed on a triple quadrupole instrument [51].

MS–MS was applied to characterise alkyl quats. The blend of cetyl-dimethyl-ammonium bromide was examined by FIA–APCI–MS–MS(+). Besides the nitrogen-containing fragment at m/z 46, mainly alkyl chain fragments from the C_{16} chain could be observed at m/z 43, 57, 71, and 85. The APCI(+) product ion spectrum is presented in Fig. 2.12.12. The inset contains the pattern of the CID fragmentation behaviour observed on a triple quadrupole instrument [39].

The qualitative and quantitative surfactant contents of a WWTP discharge, surface water, and foam resulting from an overflow drop

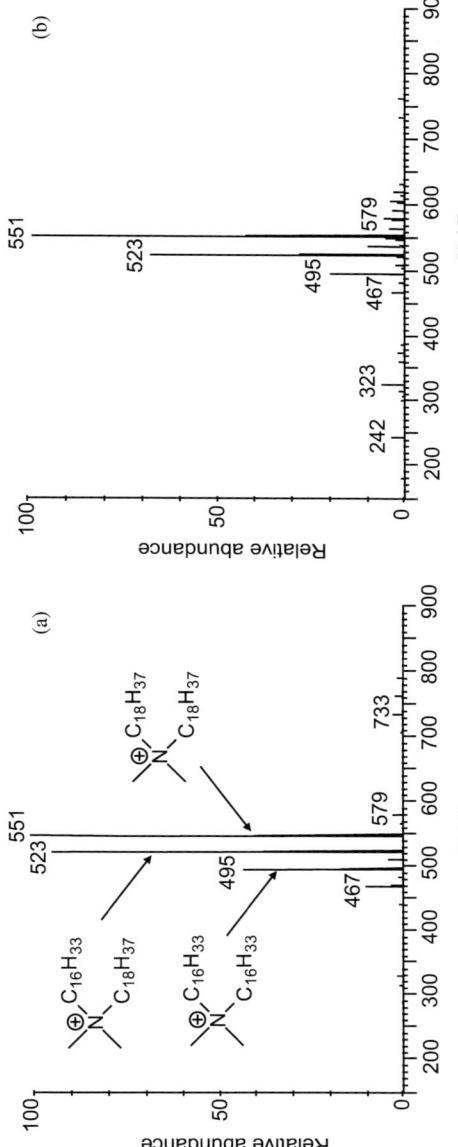

Fig. 2.12.11. FIA–ESI–MS(+) overview spectra of (a) SFE extract of wastewater sludge and (b) of an industrial blend of quats. Reproduced with permission from Ref. [22] © 1996 by American Chemical Society.

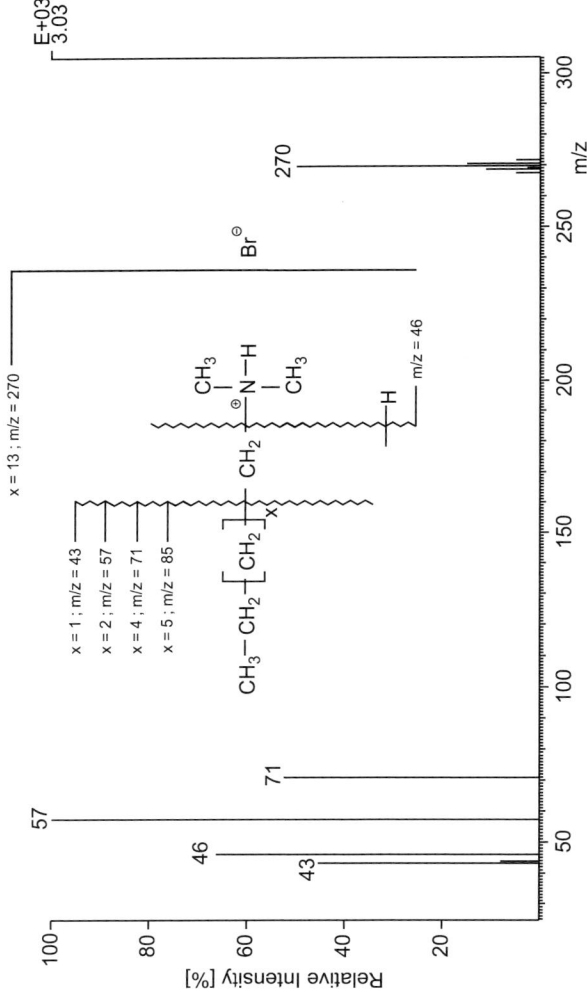

Fig. 2.12.12. FIA–APCI–MS–MS(+) (CID) product ion mass spectrum of selected $[M]^+$ parent ion of cationic quaternary alkyl ammonium surfactant blend quats; general formula $(R)_n N^{\oplus}(CH_3)_{4-n}$ at m/z 270 together with its fragmentation pattern in the inset [39].

were screened and determined by FIA–APCI–MS(+/−) in environmental samples. Surface-water samples of a tributary of the Elbe River (Germany) were examined for cationics by applying FIA–APCI–MS(+) [29]. The cationic surfactants of fatty acid ethoxy amine type with the general formula $R-N^{\oplus}H((CH_2-CH_2-O)_{x,y}H)_2 \ X^-$ (Fig. 2.12.1(c)) could be determined qualitatively. For concentration purposes, SPE with RP-C_{18} in combination with selective elution was used [52,53]. After a sequential selective elution [30–32] with hexane/ether, ether, methanol/water and methanol, the pollutants in the fractions were ionised and the pattern of equal-spaced signals ($\Delta \, m/z$ 44) provides an excellent indicator of the selective elution efficiency of the solvents applied [52, 53]. The ethoxy amines appear as $[M + H]^+$ ions. A co-elution effect, resulting from the high surface activity of the alkyl ethoxy amines, could be observed under methanol/water elution conditions. The ethoxy amines with short ethoxylate chains could be observed in the FIA–APCI–MS(+) spectra of the methanol/water eluate. The alkyl ethoxy amines with long ether chains and elevated m/z ratio dominated the FIA–APCI–MS(+) spectrum of the methanol fraction (Fig. 2.12.13), which also contains impurities of PEG as precursor compounds of chemical synthesis [37,52,53]. The ion masses observed for the cationic surfactants of fatty acid polyglycol amine type are listed in Table 2.12.3.

As expected, LC separation of the dichloromethane/acetone SPE eluate in the RP-mode, presented as FIA–APCI–MS(+) in Fig. 2.12.13, was impossible because the alkyl ethoxy amines as cationic surfactants could not be eluted under conventional RP-separation conditions [37, 53]. The use of methane sulfonic acid for ion-pairing resulted in the separation of the compounds in the methanol eluate as shown as TIC (d) and selected ion trace masses (m/z 504 (a), 670 (b), and 802 (c)) in Fig. 2.12.14. Here, the short-chain ethoxy amines were eluted later than the more polar long-chain homologues [39].

The cationic surfactant mixture, which was also observed by FIA–MS(+) in the Saale River (Germany) but could not be separated under RP-C_{18} conditions, was classed as a cationic surfactant mixture of fatty acid ethoxy amine type with the general formula $R-N^{\oplus}H((CH_2-CH_2-OH)_x)-(CH_2-CH_2-OH)_y X^-$ by FIA–MS–MS(+) [29]. The CID spectrum of the parent ion at m/z 538 generated by FIA–MS–MS(+) resulted in a series of equally spaced product ions ($\Delta \, m/z$ 44) starting with 212 and ending at 520. Besides these product ions, alkyl- and ethoxylate fragment ions with low intensity were observed at 57, 71

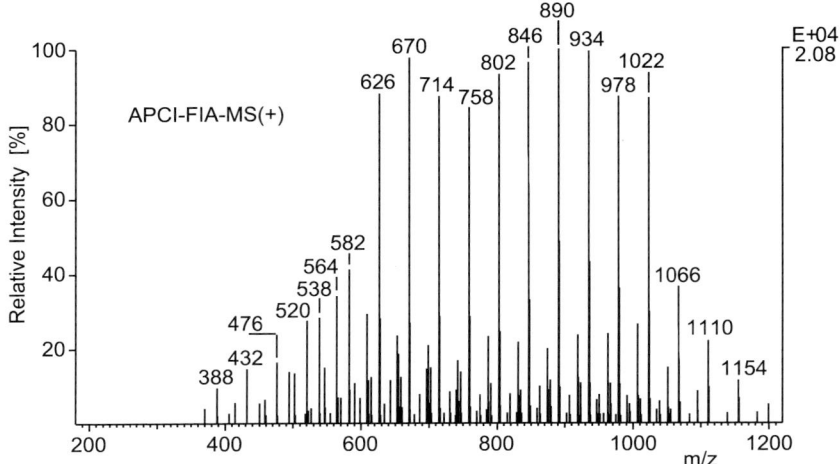

Fig. 2.12.13. FIA–APCI–MS(+) overview spectrum of methanol eluate of C_{18}-SPE containing fatty acid polyglycol amines $(R-N^{\oplus}H((CH_2-CH_2-O)_{x,y}H)_2\ X^-))$ [37].

and 85 or 45 and 89, respectively. The product ion spectrum of m/z 538 together with the fragmentation pattern is shown in Fig. 2.12.15 [37]. For confirmation of the results, the CID was examined for two industrial blends characterised as 'alkylamine oxalkylates' by their data sheets. The generation of parent ions in the FIA–APCI–MS–MS(+) mode also presented the ions equally spaced with $\Delta\,m/z$ 44 starting at m/z 212 as observable in the environmental water sample [37].

TABLE 2.12.3

Ion masses observed in FIA–APCI–MS(+) mode

$R-N^{\oplus}H((CH_2-CH_2-O)_{x,y}H)_2\ X^-$ $R = C_{12}H_{25}$		$HO-(CH_2-CH_2-O)_xH$	
$x+y$	$m/z\ [M]^+$	x	$m/z\ [M+NH_4]^+$
8	538	8	388
9	582	9	432
10	626	10	476
11	670	11	520
12	714	12	564
20	1066	20	916

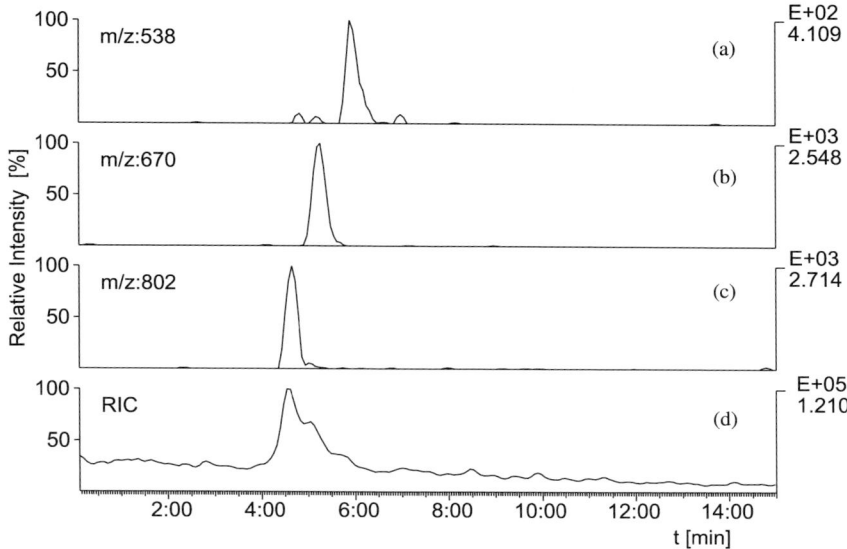

Fig. 2.12.14. (a–c) Selected mass traces and (D) TIC of ESI–LC–MS(+) analysis of polyglycol amine surfactant mixture $R-N^{\oplus}H((CH_2-CH_2-O)_{x,y}H)_2X^-$. Isocratic separation conditions applying a PLRP column using the ion-pairing reagent methane sulfonic acid (CH_3SO_3H) [39].

Environmental detections of benzalkonium chlorides

BAC homologue concentrations in wastewater samples collected from various WWTPs across the US were determined by on-line SPE–LC–ESI–MS (Table 2.12.4) [23,41]. Concentration levels of BAC detected in effluents of WWTPs reached maximum levels of 36.6 µg L^{-1} and in surface-water samples collected downstream from different WWTP discharges detected concentrations ranged from 1.2 to 2.4 µg L^{-1}, thus indicating its input and persistence through the wastewater treatment process.

The structural confirmation of the BAC homologues in wastewater samples was possible by fragmentation of the main product ions using MS–MS. The presence of the C_{16} homologue could be confirmed in three different samples by MS–MS experiments, although it was not possible to quantify it because of lack of adequate signal in the standard (C_{16} homologue accounts for only 17% of the formulation in the technical blend). However, the detection of the ion at m/z 268, which corresponds to the C_{16}-methylmethaneamine cation, could confirm the presence of

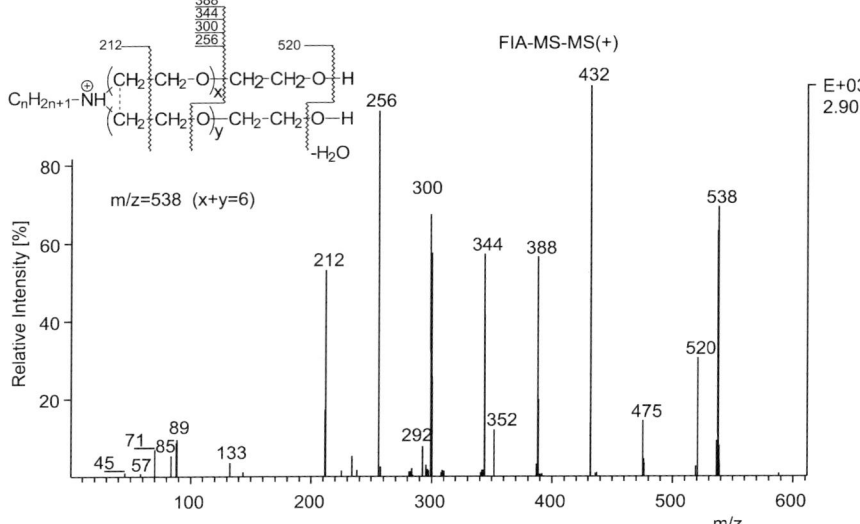

Fig. 2.12.15. FIA–APCI–MS–MS(+) (CID) product ion mass spectrum of cationic surfactant compound (*m/z* 538) fatty acid polyglycol amine type observed in the Saale river, Germany; (general formula: $R-N^{\oplus}H((CH_2-CH_2-OH)_x)-(CH_2-CH_2-OH)_y \; X^-$; fragmentation behaviour of $[M]^+$ parent ion at *m/z* 538 under CID conditions is presented in the inset [37].

this compound in the water sample. In most of the samples, especially in the WWTP effluents, the concentrations for the C_{14} homologue are higher than the corresponding concentrations for the C_{12} homologue, which is the inverse of the composition of the technical BAC mixture, where the C_{12} homologue is the major compound. The low concentrations of C_{12} and C_{14} homologues observed in rivers suggests that their concentrations may be driven distribution water and other sources in the hydrologic system, particularly soil and sediment.

Few data have been published on cationic surfactants in aquatic sediments. Levels from 3 to 67 mg kg^{-1} were determined in sediment samples from Rapid Creek [54] and 6–69 mg kg^{-1} concentrations were found in sediment samples from Japan [4]. DTDMAC was reported in digested sewage sludge at levels ranging from 0.15 to 5.87 g kg^{-1} [22]. Also, cationic surfactants were determined in urban coastal environments and they were reported as markers [18,55]. In our work, concentrations of BAC ranged from 23 to 206 μg kg^{-1} in sediment samples from different rivers near to wastewater treatment plants,

TABLE 2.12.4

Concentrations ($\mu g\ L^{-1}$) of benzalkonium homologues in wastewater samples, analysed by on-line SPE–LC–ESI–MS under positive mode of operation (taken from Ref. [23])

Sample location	$C_{12}BAC$, m/z 304	$C_{14}BAC$, m/z 332	$C_{16}BAC$, m/z 360
Wastewater			
Blue River WWTP (MO)	5.35	36.6	
Blue River WWTP (MO)	3.19	23.5	Yes
Johnson County WWTP (MO)	2.74	16.6	
Blue River WWTP (KS)[a]	4.17	7.66	Yes
Johnson City WWTP (MO)[a]	5.82	6.28	Yes
Surface water			
Indian Creek at 103rd St. (MO)	<LOD[b]	<LOD	
Cobb County Intake (GA)	<LOD	<LOD	
Indian Creek, Kansas City (KS)[a]	1.34	2.38	Yes
Hohokus Brook at Hohokus (NJ)	<LOD	<LOD	
Columbia Wetlands Units 1 (MO)	<LOD	<LOD	
City of Roswell water treatment plant (GA)	<LOD	<LOD	
Blue River near Kansas City (MO)	1.22	<LOD	

[a]Sample extracts from liquid/liquid extraction; For all other sites a sample volume of 50 mL was preconcentrated by on-line SPE. The relative standard deviation of triplicate analyses varied between 5 and 10%.
[b]LOD = less than detection limit.

suggesting that the high affinity of BAC for soil and sediments will result in preferential concentration of BAC on solids, compared with water or other environmental compartments. On a mass-to-mass basis, concentrations of BAC were much higher than those concentrations found in wastewater samples [23]. This result reflects the high affinity of this compound for solids and preferential partitioning to sediment in the aquatic environment. Higher concentrations were found for $C_{14}BAC$ and particularly for $C_{16}BAC$, which was present only as a trace in wastewater samples [23]. This observation is likely caused by hydrophobicity increasing with BAC/homologue alkyl chain length, resulting in increased partitioning of higher homologues onto sediment. Chromatograms of a sediment sample taken from the Missouri site (MO), US are shown in Fig. 2.12.16.

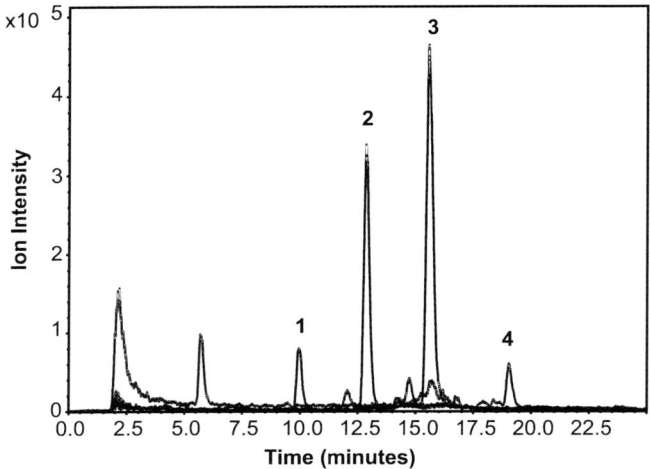

Fig. 2.12.16. Extracted ion chromatograms (C_{12} homologue: $m/z = 304$, C_{14} homologue: $m/z = 332$, C_{16} homologue: $m/z = 360$, and C_{18} homologue: $m/z = 388$) from a sediment sample extracted with ASE and analysed by SPE–LC–ESI–MS with positive ion mode of operation under full-scan conditions. Peak numbers: $1 = C_{12}$BAC, $2 = C_{14}$BAC, $3 = C_{16}$BAC, $4 = C_{18}$BAC [41].

2.12.4 Conclusion

Various classes of cationic surfactants, including quats, esterquats, alkyl ethoxy amines, quaternary perfluoroalkyl ammoniums and gemini surfactants have been analysed extensively with LC–MS and LC–MS–MS techniques, and their spectra have been fully character-ised. Different ionisation methods have been applied for the detection of such surfactants, including API techniques (APCI and ESI) in negative and positive modes of operation. In addition, detailed examples regarding MS–MS fragmentation of these compounds have been reported and presented in this chapter.

The widespread use of cationic surfactants as detergents and disinfectants in many different home- and industrial-cleaning products suggests its likely presence in the aquatic environment, such as in surface water from WWTP. Environmental detections of cationic surfactants have been reported at the submicrogram-per-litre level in water and sediment matrices. The presence of surfactants in surface water also suggests potential toxicity for organisms, such as fish and related species. The acute toxicity of BAC to fish has been estimated ($LC_0 = 0.5-4$ mg L^{-1}, $LC_{100} = 2-5$ mg L^{-1}), and toxicity to *Daphnia magna* is found at even lower concentrations [56,57].

REFERENCES

1   S.T. Giolando, R.A. Rapaport, R.J. Larson, T.W. Federle, M. Stalmans and P. Masscheleyn, *Chemosphere*, 30 (1995) 1067.
2   R. Puchta, P. Krings and P. Sandkuhler, *Tenside Surf. Det.*, 30 (1993) 186.
3   J. Waters, H.H. Kleiser, M.J. How, M.D. Barratt, R.R. Birch, R.J. Fletcher, S.D. Haigh, S.G. Hales, S.J. Marshall and T.C. Pestell, *Tenside Surf. Det.*, 28 (1991) 460.
4   R.S. Boethling, In: J. Cross and E.J. Singer (Eds.), *Cationic Surfactants*, Vol. 53, Marcel Dekker, New York, 1994, p. 95.
5   R.S. Boethling and D.G. Lynch, In: N.T. de Oude (Ed.), *The Handbook of Environmental Chemistry*, Vol. 3, Springer-Verlag, Berlin, 1992, p. 144.
6   V.T. Wee and J.M. Kennedy, *Anal. Chem.*, 54 (1982) 1631.
7   P. Gerike, H. Klotz and J.G.A. Kooijman, *J. Water Res.*, 28 (1994) 147.
8   A.J. Wilkes, G. Walraven and J.-M. Talbot, *J. Am. Oil Chem. Soc.*, 69 (1992) 609.
9   C. de Ruiter, J.C. Hefkens, U.A.Th. Brinkmann, R.W. Frei, M. Evers, E. Matthijs and J.A. Meijer, *Int. J. Environ. Anal. Chem.*, 31 (1987) 325.
10 S.L. Abidi, *J. Chromatogr.*, 324 (1985) 209.
11 G. Bleau and M.J. Desaulniers, *J. Chromatogr.*, 487 (1989) 221.
12 K. Kummerer, A. Eitel, U. Braun, P. Hubner, F. Daschner, G. Mascart, M. Milandri, F. Reinthaler and J. Verhoef, *J. Chromatogr. A*, 774 (1997) 281.
13 M. Mambagiotti-Alberti, S. Pinzauti, G. Moneti, G. Agati, V. Giannelli, S.A. Coran and F.F. Vincieri, *J. Pharm. Biomed. Anal.*, 2 (1984) 409.
14 S. Pinzauti, M. Mambagiotti-Alberti, G. Moneti, E. La Porta, S.A. Coran, F.F. Vincieri and P. Gratteri, *J. Pharm. Biomed. Anal.*, 7 (1989) 1611.
15 C.E. Lin, W.C. Chiou and W.C.J. Lin, *J. Chromatogr. A*, 723 (1996) 189.
16 E. Piera, P. Erra and M.R.J. Infante, *J. Chromatogr. A*, 757 (1997) 275.
17 L.D. Metcalfe, *J. Am. Oil Chem. Soc.*, 61 (1984) 363.
18 M. Valls and J.M. Bayona, *Fresenius J. Anal. Chem.*, 339 (1991) 212.
19 D. Wang-Hsien and L. Ying-Hsiao, *Anal. Chem.*, 73 (2001) 36.
20 E. Matthijs, P. Gerike, H. Klotz, J.G.A. Kooijman, H.G. Karber and J. Waters, *Removal and Mass Balance of the Cationic Fabric Softener Di(hydrogenated)tallow Dimethyl Ammonium Chloride in Activated Sludge Sewage Treatment Plants*, European Association of Surfactants Manufacturers (AIS/CESIO), Brussels, Belgium, 1992.
21 P. Fernandez, M. Valls, J.M. Bayona and J. Albaiges, *Environ. Sci. Technol.*, 25 (1991) 547.

22 P. Fernandez, A.C. Alder, M.J.-F. Suter and W. Giger, *Anal. Chem.*, 68 (1996) 921.

23 I. Ferrer and E.T. Furlong, *Environ. Sci. Technol.*, 35 (2001) 2583.

24 L. Elrod Jr., T.G. Golich and J.A. Morley, *J. Chromatogr.*, 625 (1992) 362.

25 V.T. Wee, *Water Res.*, 18 (1984) 223.

26 G. Ambrus, L.T. Takahashi and P.A. Marty, *J. Pharm. Sci.*, 76 (1987) 174.

27 M.C. Prieto-Blanco, P. Lopez-Mahia and D. Prada-Rodriguez, *J. Chromatogr. Sci.*, 37 (1999) 295.

28 M. Radke, T. Behrends, J. Forster and R. Herrmann, *Anal. Chem.*, 71 (1999) 5362.

29 H.Fr. Schröder, *J. Chromatogr. A*, 777 (1997) 127.

30 H.Fr. Schröder, In: D. Barceló (Ed.), *Applications of LC/MS in Environmental Chemistry*, Vol. 59, Elsevier, Amsterdam, 1996, p. 263.

31 H.Fr. Schröder, *J. Chromatogr.*, 643 (1993) 145.

32 H.Fr. Schröder, *J. Chromatogr.*, 647 (1993) 219.

33 S. Suzuki, Y. Nakamura, M. Kaneko, K. Mori and Y.J. Watanabe, *Chromatographia*, 463 (1989) 188.

34 S. Kawakami, R.H. Callicott and N. Zhang, *Analyst*, 3 (1998) 489.

35 A. Raffaelli and A.P. Bruins, *Rapid Commun. Mass Spectrom.*, 5 (1991) 269.

36 J.J. Conboy, J.D. Henion, M.W. Martin and J.A. Zweigenbaum, *Anal. Chem.*, 62 (1990) 800.

37 H.Fr. Schröder and F. Ventura, In: D. Barceló (Ed.), *Techniques and Instrumentation in Analytical Chemistry*, Sample Handling and Trace Analysis of Pollutants—Techniques, Applications and Quality Assurance, Vol. 21, Elsevier, Amsterdam, 2000, p. 828.

38 S.A. Coran, M. Bambagiotti-Alberti, V. Giannellini, G. Moneti, G. Pieraccini and A. Raffaelli, *Rapid Commun. Mass Spectrom.*, 12 (1998) 281.

39 R. Scheding and H.Fr. Schröder, Chromatographia, in preparation.

40 E.M. Thurman, I. Ferrer and D. Barceló, *Anal. Chem.*, 73 (2001) 5441.

41 I. Ferrer and E.T. Furlong, *Anal. Chem.*, 74 (2002) 1275.

42 R.D. Swisher, *Surfactant Biodegradation*, Marcel Dekker, New York, 1986.

43 P. Rudolph, *Aquatic Toxicity Data Base*, Federal Environmental Agency, Berlin, 1989.

44 M.A. Lewis, *Water Res.*, 25 (1991) 101.

45 H.-Q. Li and H.Fr. Schröder, *Water Sci. Technol.*, 42 (2000) 391.

46 I. Ogura, D.L. DuVal, S. Kawakami and K. Miyajima, *J. Am. Oil Chem. Soc.*, 73 (1996) 137.

47 R.J. Strife and M.M. Ketcha, *Proc. 44th ASMS Conf. on Mass Spectrometry and Allied Topics*, Portland, 1996, p. 1203.

48 F.M. Menger and C.A. Littau, *J. Am. Chem. Soc.*, 113 (1991) 1451.

49 R. Zana, Dimeric Surfactants, In: K. Holmberg (Ed.), *Novel Surfactants*, Surfactant Science Series, Vol. 74, Marcel Dekker, New York, 1998, p. 241.

50 H.Fr. Schröder, In: A. Kornmüller, and T. Reemtsma (Eds.), *Geminitenside Schriftenreihe Biologische Abwasserreinigung 2. Kolloquium im Sonderforschungsbereich 193 der DFG an der Technischen Universität Berlin,* Anwendung der LC–MS in der Wasseranalytik, Berlin, Vol. 16, 2001, p. 223, ISBN 3 7983 1852 2.

51 B.H. Solka and G.W. Haas, *Proc. 45th ASMS Conf. on Mass Spectrometry and Allied Topics,* Palm Springs, FL, 1997, p. 549.

52 H.Fr. Schröder, in M. Dohmann, Ed., Gewässerschutz, Wasser, Abwasser, GWA 166, Gesellschaft zur Förderung der Siedlungswasserwirtschaft an der RWTH Aachen e. V., Aachen, 1997..

53 H.Fr. Schröder, Abschlußbericht Verbundforschungsvorhaben 02 WT 9358; Bundesminister für Bildung und Wissenschaft, Forschung und Technologie, Bonn, 1996.

54 M.A. Lewis and V.T. Wee, *Environ. Toxicol. Chem.*, 2 (1983) 105.

55 N. Chalaux, J.M. Bayona and M.I. Venkatesan, *Mar. Pollut. Bull.*, 24 (1992) 403.

56 Technical Information Sheet, Preventol R 50 and Preventol R 80, Bayer, Leverkusen, August 1995.

57 Sanchez Leal, J.J. Gonzalez, K.L.E. Kaiser, V.S. Palabrica, F. Comelles and M.T. Garcia, *Acta Hydrochim. Hydrobiol.*, 22 (1994) 13.

2.13 ATMOSPHERIC PRESSURE IONISATION MASS SPECTROMETRY—X. LC—MS OF AMPHOTERIC SURFACTANTS

Peter Eichhorn

2.13.1 Cocamidopropyl betaines

The components of a technical cocamidopropyl betaine (CAPB; Fig. 2.13.1) mixture were separated under reversed phase conditions in the order of increasing length of the alkyl chain (Fig. 2.13.2) [1]. Since the hydrophobic moiety of the surfactant molecule is derived from coconut oil, the two homologues C_{12} and C_{14} form the major constituents according to the distribution in the natural raw product containing approximately 49% of C_{12}- and 19% of C_{14}-fatty acid [2].

Mass spectrometric detection after electrospray ionisation (ESI) was achieved in positive and negative ion modes [1]. In (+)-ionisation mode, the mass spectrum was characterised by a series of adduct ions, cluster ions as well as doubly charged monomers. In Fig. 2.13.3, the mass spectrum of C_{12}-CAPB is shown and the assigned ions are listed in Table 2.13.1. Apart from the protonated molecular ion $[M + H]^+$ at m/z 343, the sodiated $[M + Na]^+$ and the potassiated adduct ion $[M + K]^+$ were also detected at m/z 365 and m/z 381, respectively. Furthermore, dimer and, to a minor degree, trimer clusters were observed (not shown in the spectrum) as their protonated, sodiated and potassiated adducts $[2M + H]^+$, $[2M + Na]^+$ and $[2M + K]^+$, and $[3M + H]^+$, $[3M + Na]^+$ and $[3M + K]^+$, respectively (m/z values are given in Table 2.13.1). Higher clusters of C_{12}-CAPB could not be investigated because of the limited mass range of the quadrupole MS used (up to m/z 1200), but the quadromers of the C_{12}-homologue were expected to be formed since the corresponding species of C_8-CAPB were observed (e.g. m/z 1145 of $[4M + H]^+$).

In addition, adducts of the surfactant molecule with two cations leading to doubly charged species were detected (m/z 183, 191, 199.5 and 211; Table 2.13.1). While the ammonium ions arose from ammonia added to the eluent for pH adjustment, the Na^+ and K^+ ions were assumed to originate from impurities in the solvents and the technical surfactant used. Cleavage of the bond between the methylene group

Comprehensive Analytical Chemistry XL
T.P. Knepper, D. Barceló and P. de Voogt (Eds)

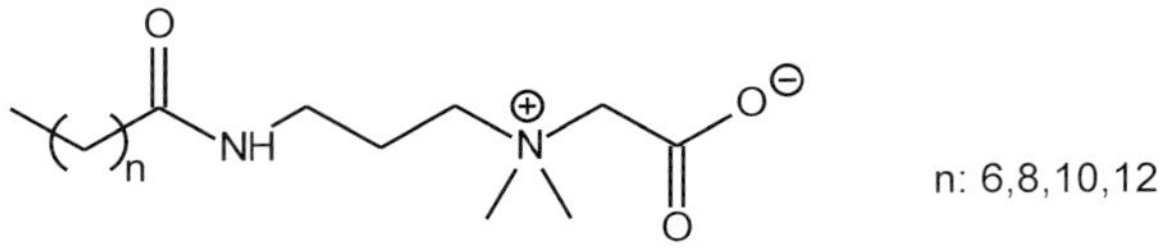

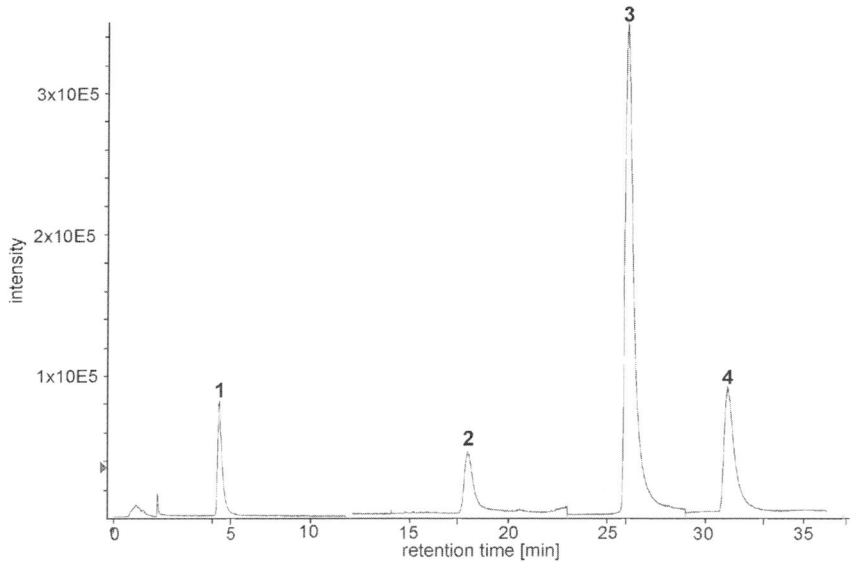

Fig. 2.13.1. General structure of cocamidopropyl betaine.

and the quaternary nitrogen atom released the stable fragment $[M-H-(N(CH_3)_2-CH_2-COO)]^+$ with m/z 240.

A further ion detected at m/z 362 was assigned as the dimer $[2M + H + K]^{2+}$. Since it was the only doubly charged dimer observed, it seemed that the combination of two C_{12}-CAPB molecules with one proton and one potassium cation yielded a particularly stable species.

It was pointed out that the ratios between the protonated, sodiated, and potassiated forms of the molecular ions, the clusters and the doubly charged ions exhibited relatively high day-to-day variations. Obviously, changing contents of competing cations derived from the sample as well as from impurities in the solvents interfered with the formation of a desired species.

Fig. 2.13.2. (−)-LC−ESI−MS chromatogram of a CAPB standard. Peak numbering: (1) C_8-CAPB, (2) C_{10}-CAPB, (3) C_{12}-CAPB, (4) C_{14}-CAPB (from [1]. 2001. © John Wiley & Sons Limited. Reproduced with permission).

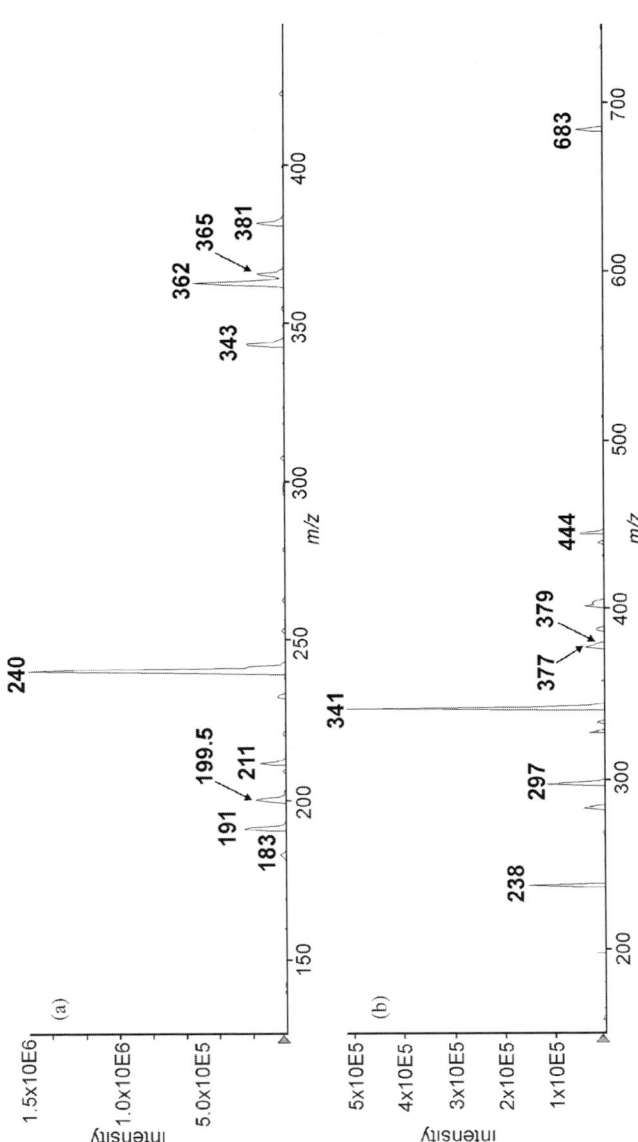

Fig. 2.13.3. LC−ESI−MS spectra of C_{12}-CAPB: (a) positive ionisation and (b) negative ionisation (from [1]. 2001. © John Wiley & Sons Limited. Reproduced with permission).

387

TABLE 2.13.1

Ions and fragments of C_{12}-CAPB detected in positive and negative ionisation modes (from [1]. 2001. © John Wiley & Sons Limited. Reproduced with permission).

Positive ionisation		Negative ionisation	
Ion/fragment	m/z	Ion/fragment	m/z
$[M + H]^+$	343	$[M - H]^-$	341
$[M + Na]^+$	365	$[M + {}^{35}Cl]^-$	377
$[M + K]^+$	381	$[M + {}^{37}Cl]^-$	379
$[2M + H]^+$	685[a]	$[2M - H]^-$	683
$[2M + Na]^+$	707[a]		
$[2M + K]^+$	723[a]		
$[3M + H]^+$	1027[a]	$[3M - H]^-$	1025[a]
$[3M + Na]^+$	1049[a]		
$[3M + K]^+$	1065[a]		
$[M + H + Na]^{2+}$	183		
$[M + H + K]^{2+}$	191		
$[M + NH_4 + K]^{2+}$	199.5		
$[M + 2K]^{2+}$	211		
$[2M + H + K]^{2+}$	362		
$[M + H-(N(CH_3)_2-CH_2-COO)]^+$	240[b]	$[M - H-(N(CH_3)_2-CH_2-COO)]^-$	238
		$[M - H-CO_2]^-$	297
		$[2M-(R-CO-NH-(CH_2)_3)]^{-c}$	444

[a]Not shown in selected spectrum window.
[b]Base peak.
[c]R: alkyl chain.

In Fig. 2.13.3, the mass spectrum of C_{12}-CAPB acquired in negative ionisation is displayed and the assigned ions are summarised in Table 2.13.1 [1]. While the most abundant ion corresponded to the molecular ion $[M-H]^-$ (m/z 341), the intensity of the adduct ion $[M + {}^{35}Cl]^-$ at m/z 377 was about 10-fold less. In analogy to the behaviour in $(+)$-ion mode, clusters other than those of monomeric ions of the amphoteric surfactant were also observed. The ion m/z 683 was assigned to the dimer $[2M - H]^-$, and the trimer $[3M - H]^-$ was detected at m/z 1025. The loss of CO_2 from $[M - H]^-$ yielded the fragment ion m/z 297. A further fragment detected at m/z 238 corresponded to $[M - H-(N(CH_3)_2-CH_2-COO)]^-$. It was emphasised that the same fragmentation step occurred under positive ionisation (see above). As in this instance, where the fragment carried two protons more, the ion was detected at two mass units higher (m/z 240).

An unusual ion was observed at m/z 444. It arose from the dimeric species $[2M - H]^-$ after losing the fragment $[R-CO-NH-(CH_2)_3]$.

388

Apparently a strong electrostatic interaction between the hydrophilic head groups of the two surfactant molecules—quaternary nitrogen atom of a first molecule interacted with an oxygen atom of the carboxylic group of a second molecule and vice versa—resulted in a pronounced stability of the dimeric form (see also cluster formation under (+)-ion mode), out of which the fragment $[R-CO-NH-(CH_2)_3]$ was released without destroying the arrangement of the charged moieties.

Comparing the overall ion intensity of negative and positive ionisation it was marked that the latter was about one order of magnitude more sensitive.

To study the effect of different concentrations (0.1, 1, 10, 100 mg L^{-1}) and orifice voltages (OR: -20, -50 and -80 V) on the formation of dimer and trimer CAPB-clusters and on the extent of fragmentation in negative ion mode, the OR was switched during the same run at intervals of 0.2 s [1]. This provided a rapid means for the comparison of ion intensities. The dependence of the area ratios of monomeric to dimeric and to trimeric C_{12}-CAPB on the concentration and on the OR is shown in Fig. 2.13.4. At an OR of -20 V the area ratio was most affected by the concentration. A predominance of the monomeric species could

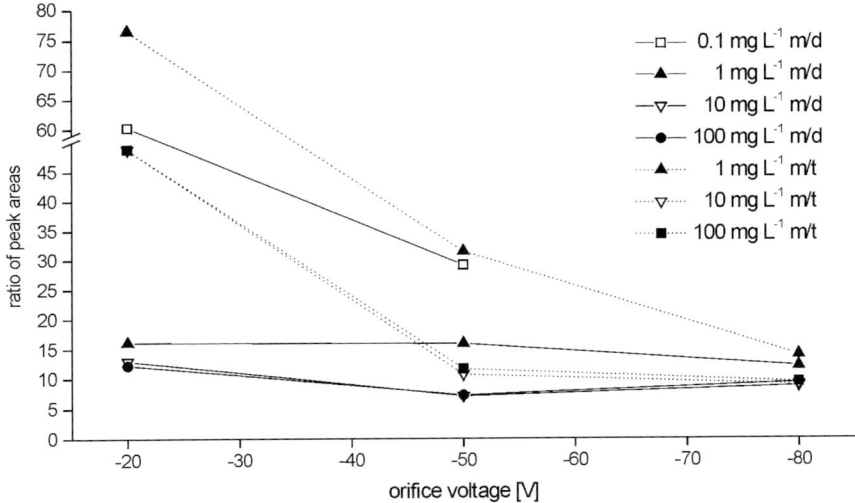

Fig. 2.13.4. Dependence of peak area ratio of monomer $[M - H]^-$ to dimer $[2M - H]^-$ (m/d) and to trimer $[M - H]^-$ (m/t) on OR and C_{12}-CAPB concentration (from [1]. 2001. © John Wiley & Sons Limited. Reproduced with permission).

be clearly noticed at low analyte concentrations. With increasing concentration of C_{12}-CAPB from 0.1 to 100 mg L^{-1}, cluster formation became more likely and hence the ratios monomer/dimer (m/d) and monomer/trimer (m/t) decreased. It might have been expected that elevating the voltage from -20 to -50 V resulted in a higher ratio since declustering was promoted. However, such a trend could not be extracted from the diagram representing only area ratios. The absolute intensity of the cluster ions, especially the trimers, was lowered, but the signal intensities of the monomer decreased overproportionally. A further increase of OR to -80 V led to minor changes in the ratios affecting virtually only the m/t-ratio at 1 mg L^{-1}. The $[2M - H]^-$ ion was no more detectable at 0.1 mg L^{-1} when switching from -50 to -80 V. Besides the trimer was not observed at any voltage running the lowest test concentration.

The dependence of absolute signal intensity of $[M - H]^-$, $[2M - H]^-$ and their two fragments $[M - H-(N(CH_3)_2-CH_2-COO)]^-$ and $[2M - (R-CO-NH-(CH_2)_3)]^-$, respectively, on the OR was also investigated in Ref. [1] (Fig. 2.13.5). The molecular ion $[M - H]^-$ was

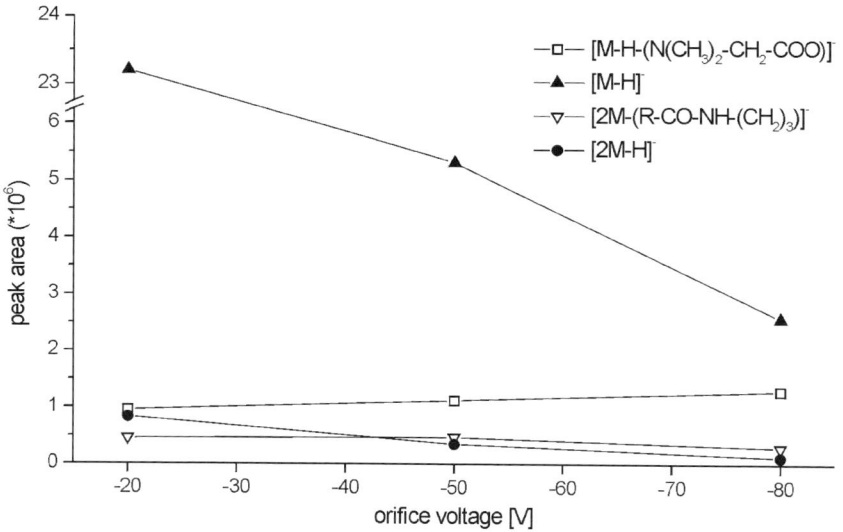

Fig. 2.13.5. Absolute peak intensities of $[M - H]^-$ and $[2M - H]^-$, and the two fragments (C_{12}-CAPB concentration: 1 mg L^{-1}) (from [1]. 2001. © John Wiley & Sons Limited. Reproduced with permission).

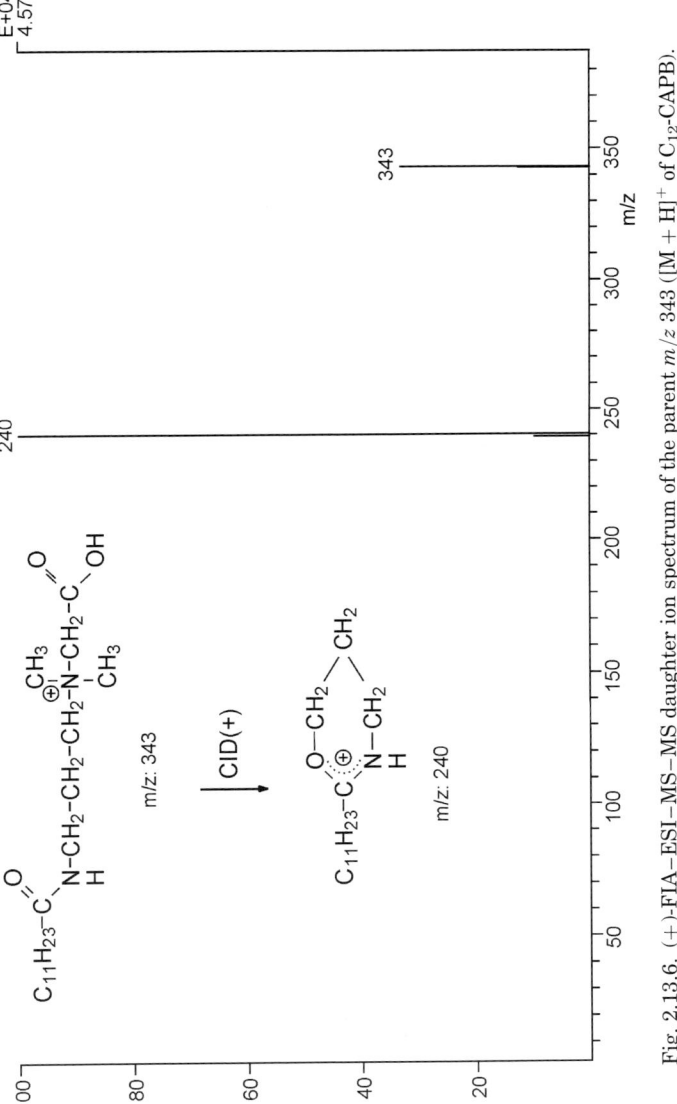

Fig. 2.13.6. (+)-FIA–ESI–MS–MS daughter ion spectrum of the parent m/z 343 ([M + H]$^+$ of C$_{12}$-CAPB).

by far the most abundant ion at -20 V. While its intensity dropped by more than a factor of four when the OR was set to -50 V, the relative decrease of $[2M - H]^-$ was less pronounced. At the same time, the intensities of both fragments slightly increased. This was a similar behaviour than the one described above where the intensity of the monomer decreased overproportionally compared with that of dimer and trimer clusters. The elevation of OR to -80 V finally led to a further reduction in $[M - H]^-$ intensity paralleled by a more intense signal of its fragment. In contrast with this, a loss of peak intensity of the dimer and its corresponding fragment was registered.

Taking into consideration the irreproducible adduct formation observed under the positive ion mode and the fact that the molecular ion $[M - H]^-$ was the most prominent ion in negative ionisation at low concentrations (≤ 1 mg L^{-1}) prevailing over dimers and trimers, the latter ion mode was considered a better choice for quantitative analysis even if it exhibited a somewhat lower sensitivity. This disadvantage, however, was partly outweighed by the fact that $(-)$-ion mode was less prone to interferences from co-eluting compounds generally present in real samples [1].

$(+)$-ESI-flow injection analysis-MS/MS was applied for the examination of CAPB in industrial blends [3,4]. Submission of the protonated molecular ion of C$_{12}$-CAPB (m/z 343) resulted in the abstraction of m/z 103 corresponding to the loss of the amphoteric moiety (HN(CH$_3$)$_2-$CH$_2-$COO). The single dominating fragment with m/z 240 (for identity refer to Table 2.13.1) was believed to be generated due to thermodynamical reasons, which favoured the formation of this ion as it had the ability to build up a cyclic structure allowing the delocalisation of the positive charge as shown in the inset of the daughter ion spectrum (Fig. 2.13.6).

REFERENCES

1 P. Eichhorn and T.P. Knepper, *J. Mass Spectrom.*, 36 (2001) 677.
2 R.A. Reck, *J. Am. Oil Chem. Soc.*, 62 (1985) 355.
3 H.Fr. Schröder and F. Ventura, In: D. Barceló (Ed.), *Techniques and Instrumentation in Analytical Chemistry*, Sample Handling and Trace Analysis of Pollutants—Techniques, Applications and Quality Assurance, Vol. 21, Elsevier, Amsterdam, 2000, p. 828.
4 R. Scheding and H.Fr. Schröder, Chromatographia, in preparation.

Chapter 3

Sample handling

3.1 SAMPLING AND SAMPLE TREATMENT FOR SURFACTANT ANALYSIS IN WATER

Eduardo González-Mazo, Thomas P. Knepper,
Francesc Ventura and Pim de Voogt

3.1.1 Introduction

The complex composition of aqueous environmental sample matrices, especially sewage and marine water samples, and the low concentrations in which the surfactants are generally found, have made it necessary to perform an initial stage of concentration and purification of the analytes prior to its analysis. Traditionally, such steps were carried out off-line with procedures based on liquid–liquid extraction (LLE), sublation or steam distillation, followed by chromatographic clean-up steps.

In the past two decades quite a few new techniques have emerged for the treatment of aqueous samples prior to organic analysis. Perhaps the most important development is that of solid-phase extraction (SPE), which has successfully replaced many off-line steps. This technique can be considered to have introduced a genuine new era in sample handling [1]. The many varieties in which the technique is available and can be applied have made it the key step in handling of aqueous samples. Among the successful varieties are solid-phase microextraction (SPME), matrix solid-phase dispersion, disk extraction and immuno-sorbent extraction. Several reviews covering these topics have appeared in the literature in the past decade (see e.g. Refs. [2,3] for nonylphenol

Comprehensive Analytical Chemistry XL
T.P. Knepper, D. Barceló and P. de Voogt (Eds)

(NP) and nonylphenol ethoxylates (NPEOs), [4] for alcohol ethoxylates (AEs), and [5] for linear alkylbenzene sulfonates (LASs) and sulfopheny carboxylates (SPCs)).

The present chapter provides an overview of methods currently used for extracting and concentrating surfactants from aqueous matrices. Some of the techniques will be addressed only briefly, because they are discussed elsewhere in this book more extensively.

3.1.2 Sampling

Due to the physico-chemical properties of surfactants, it is difficult to obtain samples that are fully representative of sampling sites [6]. The intrinsic nature of surfactants results in them tending to concentrate or adsorb at interfaces, and this affects not only the way they disperse, but also in which environmental compartments they accumulate. This means that the vertical distribution of surfactants is not homogeneous and depends on the degree of physical mixing occurring in the water column. For example, in the case of the distribution of an anionic surfactant in the marine medium (in a stagnant zone), the highest concentrations of LAS homologues were detected in the surface microlayer (around one order of magnitude greater than that found at depths of more than 20 cm) [7–9]. Figure 3.1.1 shows a typical vertical profile of a surfactant-type molecule.

For the NPEO vertical differentiation in the water column has also been demonstrated (see Fig. 6.4.1). The compounds showed a tendency to accumulate at the two-phase boundaries of air/fresh water and fresh water/saline water (the halocline), with concentration maxima observed at depths of 0 and 2 m [10], respectively.

Accumulation of surfactants in the surface microlayer may also occur locally during sampling as a result of sampling vessel discharges. High concentrations of alkylphenol ethoxylates (APEO) in seawater sampled with buckets were observed when compared with a submerged (depth 2 m) automatic torpedo sampler [11]. The oligomer profile observed in the high level samples corresponded to that of bilge and wastewater discharged by the ship.

Consequently, for an accurate quantification of surfactants in the water column, sampling must be performed by taking surface micro-layer samples (at depths between 0 and 3–5 mm), using a surface sampler, and at various greater depths with Ruttner or similar bottles.

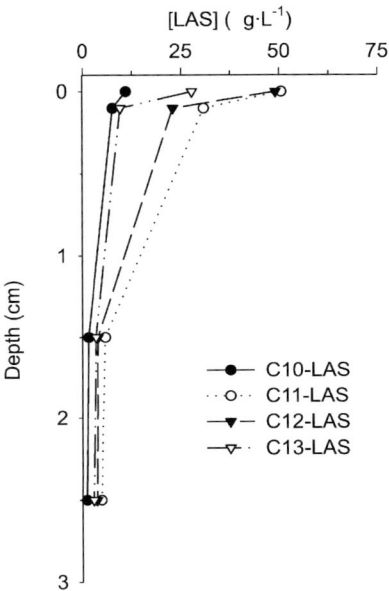

Fig. 3.1.1. Vertical profile of LAS homologues in the water column in the Besaya Estuary (north of the Iberian Peninsula, Spain) (Figure taken from Ref. [8]).

As for the many surfactant metabolites, their distribution is often more homogeneous due to the loss of surfactant properties by the molecules. This would render the sampling procedure much simpler.

Also, for an accurate quantification in the water column, it must be ensured that an effective desorption of the surfactants in the suspended solids is performed, before or during the pretreatment of the samples. The suspended solids fraction can significantly contribute to the total concentrations of surfactants in the water sampled, depending on the suspended matter concentration in the water. de Voogt et al. [12] showed that in wastewater treatment plant (WWTP) influents, suspended solids can contribute up to 50% of the total NP, and 20% of the total NPEO concentration in the water, respectively. Di Corcia et al. [13] proposed that a specific volume of methanol should be added to the sample before the extraction stage to encourage the desorption of the surfactants. Another procedure they suggested is to filter the sample and analyse the LAS in two phases. However, e.g. marine waters, this involves processing large volumes of water to obtain a sufficient quantity of solids from which to extract the surfactant and the risk of contamination would thus be greater.

In order to avoid further biodegradation of surfactants after sampling or during storage, various substances have been used for the preservation of samples. So far, mainly formaldehyde and salts of copper(II) and mercury(II) [7,14] have been applied, and formaldehyde has been found to be the most effective of these, although still not sufficient for long term storage. This topic is discussed in more detail in Chapter 4.3.

Another aspect to take into account is that surfactants are often ingredients in the cleaning products used in chemical laboratories. The sampling (as discussed above) and sample handling to which the samples are subjected always carries with it the risk of contamination. It is therefore necessary to process sufficient numbers of blank samples together with the real samples. Numerous samples of oceanic seawater and fossil seawater (taken from wells) have shown traces of LAS (around 1 ppb). Therefore, environmental concentrations found at similar levels should be regarded with caution.

3.1.3 Sample enrichment and clean-up steps

Prior to the quantitative determination of surfactants from aqueous environmental samples, a preconcentration step is needed to enrich low amounts of target analytes and to remove interfering matrix components.

Initial attempts
The extraction of an anionic surfactant from river water has been performed by ion-pair LLE with the methylene blue cation [15–17]; a recovery in river water of 96.7% and a detection limit of 0.05 mg L^{-1} have been reported.

The solvent sublation procedure of Wickbold [18] is another method that has been used for the analysis of LAS present in seawater [19,20]. The solvent sublation technique (gaseous stripping into organic solvent, often ethyl acetate) has also been used to isolate and concentrate non-ionic surfactants, e.g. AEs and APEO in aqueous samples [21,22]. The co-extracted interferences can be eliminated by cation/anion ion-exchange and alumina chromatography [23,24].

For samples of diverse origins, such as river water and sediment, raw and treated municipal wastewater and activated and digested sewage sludge, a simultaneous distillation and solvent extraction has been employed to isolate the NP, $NPEO_1$ and $NPEO_2$ [14]. Recoveries were higher than 80% and the estimated detection limit in water samples was 0.5 $\mu g\, L^{-1}$. For $APEO_n$ ($n \geq 3$) the solvent sublation technique was used [21]. For the determination of NPEO in liquid cleaning products, the analytes were batch-extracted by stirring with hexane–acetone (75:25) [25]. For the quantification of non-ionic surfactants (AE) in raw and settled influent and effluent of municipal sewage treatment plants, the separate extraction of liquid and solid phases and chromatographic clean-up have been performed [26]. For the liquid phase, solvent sublation was chosen since it enables large volumes of sample to be used. Analytical recoveries of standard additions were >80%. The detection limit of the procedure for effluent samples was 3.0 $\mu g\, L^{-1}$.

With the aim of minimising the time taken in the preconcentration, and extending the chemical analysis of surfactants to more complex aqueous matrices in which very low detection limits are required, preconcentration techniques using solid–liquid extraction with various adsorbent materials, such as XAD [27] and anionic exchange resin [28] have been developed.

Recent developments for recovery improvement
Nowadays, SPE has been shown to be a very powerful and robust alternative to the traditional extraction methods stated above [29,30]. It offers the advantages of very low solvent consumption, speed, and ease of handling, and combines the concentration and (partial) clean-up into one on-line step. Various types of commercially available stationary phases enable the application of distinct separation mechanisms, thereby increasing the selectivity of the adsorbent to the analytes.

To date, typical SPE materials are based on silica gel or highly cross-linked styrene-divinylbenzene (PS-DVB). The former is functionalised with distinct chemical groups to yield various sorbents with non-polar or polar characteristics. Non-polar materials are modified with alkyl groups of different chain length (C18, C8, C2), while polar sorbents have cyano-, amino-, or diol-bonded groups. Ion-exchange phases have either anionic or cationic functional groups.

Up to now C18 and C8 phases have been most frequently employed [25,31–34]. However, both phases have a high retention power and therefore a great many substances may interfere in the analysis. In these cases it is necessary to use a second cartridge as a clean-up stage. The hydrophobic and ionic nature of the surfactant molecule has consolidated the use of minicolumns of the hydrophobic (C_{18}, C_8, C_2) and strong anionic exchange (SAX) types [6,7,25,35–38]. Minicolumns that act as both reverse and anionic exchange phases at the same time, such as that of graphitised carbon black (GCB), have also been employed. The applications have been very diverse, taking into account the many different matrices studied (Table 3.1.1; [39–42]).

The main drawback of all silica-based materials arises from their limited pH stability. Reversed-phase (RP) polymeric sorbents such as PS-DVB offer several advantages: this material can be used under extreme pH conditions, has higher specific surface areas and shows greater retention for polar compounds. Chemical modification of the polymeric resin yields highly selective stationary SPE phases [43]. Another alternative sorbent is non-porous GCB with positively charged active centres on the surface, which offers the singular feature of behaving as a non-specific RP sorbent and anion exchanger [44]. By exploiting this feature, the separation of acidic compounds from co-extracted base/neutral ones can be easily achieved by differential elution. In Table 3.1.1, a selection of analytical methods is compiled, which are based on SPE for the preconcentration of a range of surfactant classes from wastewater, surface, and drinking water samples.

Verification of the proper choice of the enrichment method can be done through recovery studies under varying conditions of the most important parameters, such as amount and quality of the enrichment material, pH of the sample, as well as the amount and quality of analyte and matrix to be enriched. These parameters not only differ based on the chemical behaviour of the analyte, but are also strongly dependent on the matrix to be analysed.

Quite often the most crucial parameter for SPE of compounds for which no enrichment techniques for environmental samples can be found in the literature is the selection of the proper SPE material. For example, during extraction of the two non-ionic surfactants, alkyl glucamides (AGs) and alkyl polyglucosides (APGs) from a spiked WWTP effluent ($5\ \mu g\ L^{-1}$) using RP-C18 cartridges, it was found that this material was unsuitable for quantitative enrichment (recovery $<30\%$). Satisfactory results were obtained using PS-DVB material (Lichrolut EN) [45] (Table 3.1.2).

TABLE 3.1.1

Selected enrichment and isolation methods of surfactants in wastewater (WW), coastal water (CW), estuarine water (EW), surface water (SW), ground water (GW) and drinking water (DW) matrices with adjacent LC-separation; detection by MS, UV and FL

Compound	Matrix	Isolation	Reference
LAS	WW, SW	GCB	[92]
LAS	CW	C18 + SAX	[84]
LAS	CW	C18 + SAX	[85]
LAS	WW, SW	C2 + SAX	[40]
LAS	SW	C8	[93]
LAS	WW, SW	GCB	[41,42]
LAS	GW	Cationic/anionic + XAD-8	[94]
LAS	CW	C8 + C8/SAX	[19]
LAS		C18 + SAX	[82]
LAS, NPEO	WW	GCB	[13]
LAS, NPEO	SW	C18 (+SAX)	[95]
LAS, NPEO, AE	WW	C18 + PS-DVB	[96]
LAS, NPEO, AE	WW	C18 + PS-DVB	[70]
LAS, NPEO, AE	WW	C18	[97]
LAS, NPEO, AE	WW	C18	[98]
LAS, AES, AE, CDEA	WW	C18	[99]
NPEO	SW	PS-DVB	[100]
NPEO	SW	–	[101]
NPEO	EW	C18	[63]
NPEO	EW	C18	[102]
NPEO	DW	LLE	[103]
NPEO	DW	C18	[11]
NPEO, AE	WW	C18	[104]
NPEO, AE	WW	C18 + PS-DVB	[105]
NPEO, AE	WW	C18 + PS-DVB	[106]
NPEO, AE	WW	C18 + PS-DVB	[107]
NPEO, AE	WW, SW	GCB	[71]
NPEO, AE	WW, DW	C18	[108]
AE	WW	C18	[109]
AE	WW	C8	[73]
AE	WW	C2	[110]
AE	WW, SW	C8	[72]
AE	SW	C18; PS-DVB	[111]
AE, AES	WW, SW	C2	[112]
AE, AES	WW	C2	[113]
PFAS, PFAC	SW	C18	[114]
CDEA	WW	C18	[115]
AG	WW	C18	[116]
APG	WW	C18	[117]
Esterquats	WW	C2 + clean-up	[118]
Esterquats	WW	IP-LLE	[119]

TABLE 3.1.2

Recoveries of AG and APG in secondary effluents (in triplicate)

AG		APG		
C12	C14	C8	C10	C12
89 ± 3	91 ± 3	79 ± 7	82 ± 4	82 ± 5

Recently techniques have been developed to further reduce the sample volume required. Perhaps the most promising technique is SPME. This method consists of utilising high capacity adsorbents that are brought into contact directly with the sample ($<$ 10 mL) or indirectly through a headspace above the sample.

3.1.4 Applications

Non-ionic surfactants
The occurrence of APECs in water, wastewater, sediments or sludges is well documented. The sample handling methodologies have been gradually changed from solvent sublation, steam distillation and LLE to SPE with cartridges or disks, whereas their analysis by GC-MS is now complemented by the new information obtained by using LC-MS. The methods described in the literature include extraction of sewage effluent samples with chloroform or gaseous stripping and further analysis by GC-MS of methylated extracts [46] or by LC-UV [47,48]; adsorption on GAC or XAD-2 then Soxhlet with dichloromethane and analysis by GC-MS or FAB-MS of methylated extracts from river and drinking water samples [49–51]; SPE of WWTP effluents and river water with C_{18} cartridges and LC-FL analysis [52] or GC-MS of methylated extracts [53]; SPE employing SAX disks and analysis of methylated esters by GC-MS of sewage and paper mill effluents [38]; C_{18}-SPE, silica chromatography clean-up and analysis by LC-FL of effluents, surface waters and drinking water [11]; C_{18}-SPE silica chromatography clean-up and analysis of estuarine and marine water by LC-ESI-MS [54]; SPE of WWTP effluents using GCB cartridges and analysis by GC-MS as propylated [55] or butylated esters [56]; LC-FL [13]; LC-APCI-MS [57], or LC-ESI-MS [58–60].

In addition, brominated derivatives of alkylphenols (BrNPnEO and BrNPnEC) can be formed during chlorine disinfection of water if bromide is present. The concentration methods used to analyse these halogenated derivatives include adsorption on XAD-8 resins and further GC-MS analysis [61]; LLE of acidified sewage samples with diethyl ether [62]; adsorption on carbon (GAC) columns and analysis by GC-MS and FAB mass spectrometry [50]; SPE extraction using a C18 cartridge for water samples and a high temperature continuous flow sonication extraction apparatus for sediments and further LC-ESI-MS analysis [63], or SPE extraction of residues from sludges using Lichrolut-EN cartridges and LC-ESI-MS [64] (see Fig. 3.1.2).

The chromatographic determination of AEs in the environment has been extensively reviewed [3,4]. AEs have been extracted from the aqueous matrix, among others, by solvent sublation [65], LLE [66], SPE with Amberlite XAD-2 [67], cartridges [68–73] (i.e. C_1, C_{18}, GCB, C_8) or C_{18} disks [74] and matrix solid-phase dispersion for bioconcentration studies in fish samples [75].

The preconcentration of NP, OP, NPEO, and NPEC [36,76] and the simultaneous enrichment of LAS and NPEO, as well as NP, is feasible

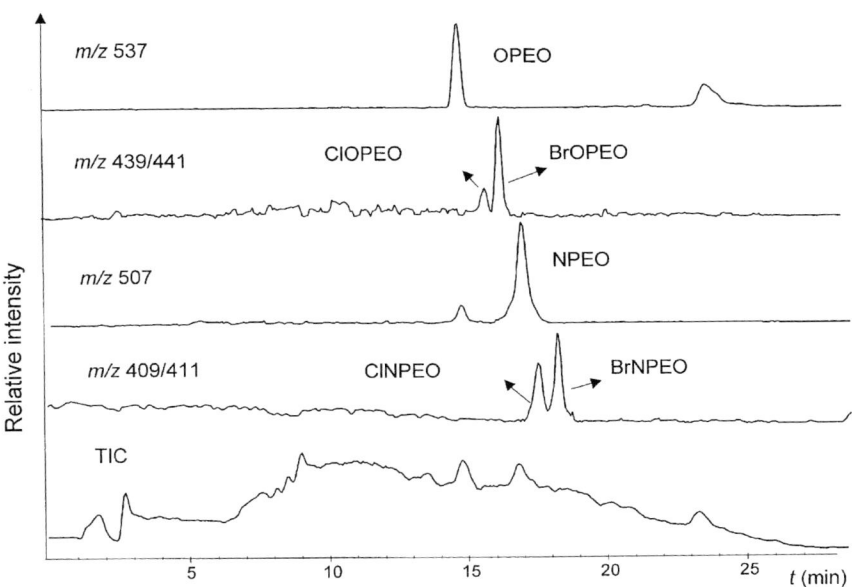

Fig. 3.1.2. Total-ion LC-ESI-MS chromatogram (bottom trace) and reconstructed chromatograms of APEOs and halogenated APEOs in positive ionisation mode of sludge from a Barcelona drinking water plant [64] (Figure taken from Ref. [64]).

by employing an octadecylsilica minicolumn [39]. The analytes were desorbed from the cartridge with 5 mL of acetone. A procedure for the determination of all NPEO oligomers enriched from marine waters is presented [25,60]. Subsequently the extraction of LAS and NPEO as well as their related biointermediates from the influents and effluents of WWTPs, river water, and drinking water were performed using a GCB cartridge [13]. The presence of positively charged active centres on the GCB surface allows the fractionation of the complex mixture of the analytes by differential elution, according to differences in their acid strengths. Strong anion exchange has also been used for the extraction of NPEC to NPE$_4$C, in paper mill effluents, sewage effluents and river waters [38].

An analytical procedure that quantifies the total AE concentration resolved by alkyl chain length for various environmental matrices (influent, effluent, and river water) was developed by Di Corcia et al. [41]. The method utilises a reverse-phase column to extract and concentrate AE from surface waters and wastewaters and utilises strong anionic and cationic exchange columns to remove potential interferences. Samples are passed through the RP extraction column (C_1). AE and potential anionic and cationic interferences are eluted from the C_1 column and passed directly through the SAX and SCX. The SAX and SCX columns retain anionic and cationic materials while non-ionic AE are not retained. Recovery of AE from influent, treatment plant effluent, and river water is quantitative (65–102%) over a range of concentrations for all matrices.

Boyd-Boland and Pawliszyn [77] pioneered the SPME analysis of APEOs by SPME-HPLC using normal-phase gradient elution with detection by UV absorbance at 220 nm. The Carbowax-template resin (CW-TR) and Carbowax-divinylbenzene (CW-DVB) fibres allowed the analysis of APEO with a linear range of 0.1–100 mg L^{-1}. The former coating produced the best agreement between the distribution of ethoxymers before and after extraction. This CW-TR fibre provided a limit of detection for individual AP ethoxamers at the low ppb level. The determination of NP in water by SPME-GC (FID) was accomplished by Chee et al. [78] using a polydimethylsiloxane (PDMS) fibre. The linear range was between 1 and 15 mg L^{-1} with an estimated detection limit of 0.1 mg L^{-1}.

On the other hand, the recent developments of new coatings have expanded the range of compounds analysed by SPME. Thus, the determination of metabolites of NPEO$_n$ ($n = 0$–2) and their halogenated derivatives ($BrNPEO_n$) formed by chlorination practices in water treatment plants has been performed by SPME-GC-MS (SIR) using a DVB/CAR/PDMS fibre [79]. The optimised conditions that allow detection limits in the low $\mu g\, L^{-1}$ level were obtained immersing the fibre for 1 h at room temperature and adding 100 μL of methanol to the sample (40 mL). The simultaneous determination of short chain metabolites of NPEO and their acidic metabolites ($NPEO_n$, $n = 0-2$ and NPE_nC, $n = 0-1$) has been accomplished by an in-sample derivatisation headspace (HS) SPME method. The analytical procedure involves derivatisation of both NPEOs and NPECs to their methyl esters–ethers, respectively, using dimethyl sulfate/NaOH and further HS-SPME-GC-MS determination [80]. Parameters affecting both derivatisation efficiency and HS-SPME procedure, such as the selection of the fibre coating, derivatisation-extraction time, temperature and ionic strength were optimised. Among the six studied fibres exposed to the headspace for 60 min, the CW/DVB fibre was the most suitable for the simultaneous determination of NPEOs and NPECs (Fig. 3.1.3). The effect of sample temperature on the derivatisation was examined and showed increases in the response for all compounds until 60°C (Fig. 3.1.4), which was considered to be the optimum temperature. Fig. 3.1.5 shows, as an example, the GC-MS-SIR profiles of the derivatised metabolites from river water entering Barcelona's drinking water treatment plant.

Aranda and Burk [81] have established an SPME-HPLC-FL method and on-line derivatisation to determine AEs (Brij 56) in water. The surfactant was extracted with a PDMS-DVB fibre and pre-column derivatisation with 1-naphthoyl chloride in the presence of 4-(dimethyl-amino)pyridine as catalyst. The method has a limit of detection of 0.1 mg L^{-1}.

Anionic surfactants
González-Mazo and Gómez-Parra [82] have put forward a scheme for purification and concentration by SPE, designed specifically for samples of marine origin. This procedure shows good selectivity for LAS and a high recovery percentage over a wide range of concentrations, but the recovery of SPC is low, especially those with short chains. Research has been directed towards the development of concentration and purification methods that would allow as many as possible of both the surfactants and intermediates to be analysed. To deal with the wide range of polarities of these compounds, sequential extraction methods

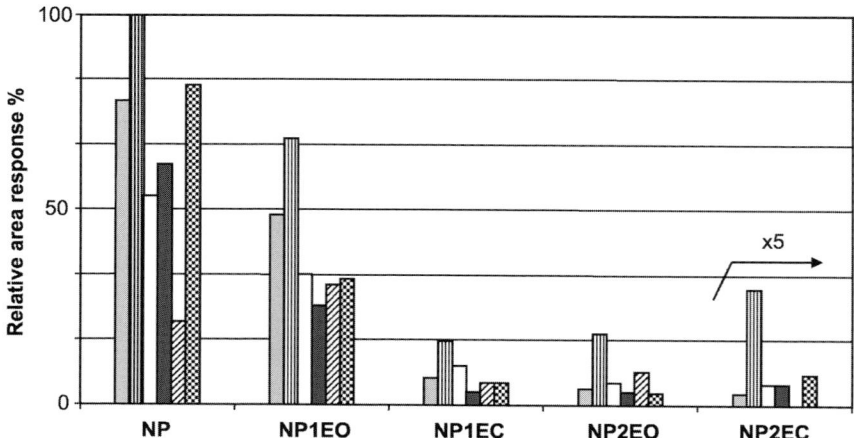

Fig. 3.1.3. Extraction efficiency of six commercial SPME fibres. Milli-Q water containing NP (33 µg L^{-1}), NPEO$_1$, NPEO$_2$ (83 µg L^{-1} each), and 165 µg L^{-1} NPE$_1$C and NPE$_2$C; DMS 200 µL; NaOH 5 M, 1 mL; sodium chloride (7.5 g); extraction time, 60 min; extraction temperature, 65°C; and stirring rate 1200 rpm. Compound identification: (grey shaded) DVB-CAR-PDMS; (double vertical lines) CW-DVB 65 µm; (0) PA; (9) PDMS 1000 µm; (hatched) PDMS 7 µm; (checkerboard) PDMS-DVB (Figure taken from Ref. [80]).

have been devised using XAD-8 resin [94], octadecylsilica [52] and GCB [44]. Field et al. [94] have determined LAS, DATS, APEO and their degradation intermediates using different eluents and varying the pH. Specifically, the separation of the LAS and the SPC of the final fraction

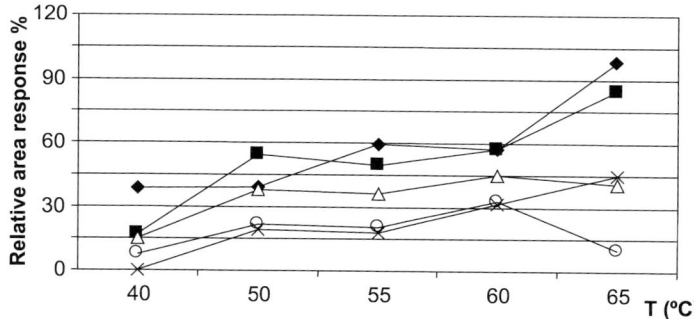

Fig. 3.1.4. Temperature profiles for in-sample derivatisation HS-SPME of NPEOs–NPECs using the CW-DVB fibre. Conditions as in Fig. 3.1.3. Compound identification: (◆) NP; (■) NPEO$_1$; (△) NPE$_1$C; (×) NPEO$_2$; and (○) NPE$_2$C compounds; the relative area response is magnified by a factor of three (Figure taken from Ref. [80]).

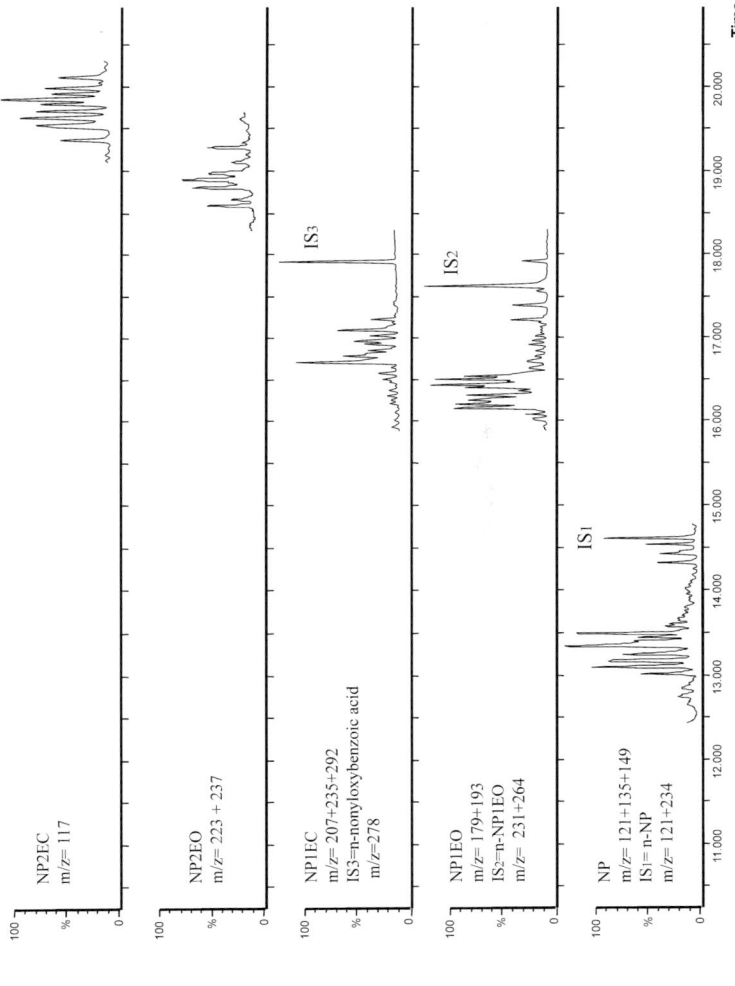

Fig. 3.1.5. HS-SPME-GC-MS single-ion chromatograms of derivatised compounds from river water entering Sant Joan Despí water treatment plant (Figure taken from Ref. [80]).

obtained is achieved by eluting the SPC with alkaline solution (pH 12, 0.01N NaCl), since the homologues of LAS are retained absorbed onto the resin under these conditions. Di Corcia et al. [42] have adjusted the sample (10–15 mL) to an acid pH (pH 3) with HCl, in order to encourage the retention of the dicarboxylic SPC (SPDC), and have then eluted sequentially the material retained with three fractions of greater to less polarity (in a solution of variable composition, of methanol and methylene chloride).

Crescenzi et al. [44] have obtained high recoveries for C3, C4 and C5-SPC using GCB and large volumes of sample (0.5–4 L). Sarrazin et al. [83,86] increased the ionic strength of the medium and adjusted the pH of the solution to 1.5 or 1.0 to encourage the retention of the more polar compounds (sulfobenzoic acids and SPC) on a C_{18} column. The same analytical procedure has been employed by León et al. [6] but including also an anionic SAX type minicolumn. This second stage was necessary in order to obtain chromatograms free of interference that would enable an accurate detection and quantification of environmental concentrations of the LAS and SPC simultaneously in complex samples. This method presented high recoveries even for samples with low concentrations (Table 3.1.3), and for sample volumes until 500 mL. The recoveries decreased slightly at higher concentrations (above that detected in the environment) due probably to the saturation of the minicolumn (500 mg). The detection limit obtained with this method is lower than 1 μg L^{-1} for all LAS and monocarboxylic SPC homologues, thus allowing an accurate quantification of each analyte.

The matrix can be an important factor in method efficiency. For example, in earlier studies regarding simultaneous analysis of LAS and SPC from water samples on a single RP-C18 extraction cartridge, the samples were acidified to pH 3 or 1.5 prior to enrichment [84,85] and [6,86], respectively. Although the adjustment to 1.5 provided good retention of all SPC homologues including the very polar short chain SPC (< C6), the higher value of 3.0 yielding poorer recoveries of these species can be the best choice since humic acids, occurring in natural freshwater settings, are co-extracted to increasing extents by lowering the pH of the sample [87]. Subsequent elution of the cartridges with methanol did not only redissolve the compounds of interest, but also released the trapped humic substances. Their presence in the extract used for LC analysis without further clean-up should be kept at a minimum level, since the negligible volatility of these interferences resulted in precipitation in the ion source and on the skimmer of the MS

TABLE 3.1.3

LAS and SPC recovery percentages and standard deviation ($n = 3$) obtained for the proposed method by León et al. [6] and different concentrations of each homologue (10, 50, 100, 200 and 450 $\mu g\ L^{-1}$)

Compound	Concentration ($\mu g\ L^{-1}$)				
	10	50	100	200	450
C2-SPC	3 ± 1	2 ± 0	3 ± 1	1 ± 1	1 ± 1
C3-SPC	20 ± 1	15 ± 2	20 ± 2	11 ± 2	5 ± 1
C4-SPC	84 ± 2	89 ± 5	92 ± 3	48 ± 8	20 ± 1
C5-SPC	100 ± 3	108 ± 2	105 ± 1	84 ± 4	80 ± 1
C6-SPC	103 ± 3	109 ± 3	102 ± 3	91 ± 1	84 ± 2
C8-SPC	105 ± 5	103 ± 2	104 ± 1	83 ± 5	77 ± 2
C9-SPC	104 ± 7	135 ± 2	108 ± 9	102 ± 7	97 ± 5
C10-SPC	95 ± 5	103 ± 3	101 ± 2	91 ± 3	75 ± 3
C11-SPC	101 ± 6	99 ± 1	96 ± 2	89 ± 3	77 ± 2
C12-SPC	99 ± 8	95 ± 3	90 ± 5	87 ± 4	79 ± 4
C13-SPC	108 ± 7	94 ± 4	79 ± 5	79 ± 6	73 ± 1
C10-LAS	91 ± 7	91 ± 7	69 ± 5	65 ± 4	55 ± 3
C11-LAS	102 ± 10	71 ± 5	65 ± 5	64 ± 4	59 ± 2
C12-LAS	108 ± 5	57 ± 3	64 ± 3	60 ± 2	56 ± 1
C13-LAS	72 ± 7	35 ± 3	47 ± 1	48 ± 3	45 ± 4

thereby gradually reducing the instrumental sensitivity [87]. Moreover, they might affect the ionisation process (see Chapter 2).

Recovery studies should always be performed in the matrix to be analysed. Thus, the recovery of LAS, for example, was performed by spiking a secondary effluent of a municipal WWTP, adjusted to pH 3.0, with 100 $\mu g\ L^{-1}$ of a commercial mixture of the anionic surfactant [88]. Recoveries of the four alkyl homologues are listed in Table 3.1.4. With regard to the degradation products of LAS, the spiking experiments were carried out using a mixture containing all investigated alkyl homologues with a representative distribution. This was prepared from the liquor of a fixed bed bioreactor (FBBR; see Chapter 5.1) amended with an LAS mixture. After finishing the degradation experiment, a volume of the test medium was withdrawn, concentrated by lyophilisation, and reconstituted with methanol. This solution was spiked into groundwater ($n = 3$) corresponding to an SPC concentration of approximately 0.1 $\mu g\ L^{-1}$. SPC were quantified (see Table 3.1.4) via a calibration made from the methanolic spiking solution. Methods for analysis of surfactants in groundwater are discussed in more detail in Chapter 6.7.

TABLE 3.1.4

Recoveries of SPCs in ground water and LAS in a secondary effluent (in triplicate)

SPC							LAS			
C6	C7	C8	C9	C10	C11	C12	C10	C11	C12	C13
37 ± 6	54 ± 2	50 ± 3	49 ± 3	60 ± 4	54 ± 8	68 ± 7	85 ± 3	85 ± 5	84 ± 4	82 ± 4

But even taking into account all possible parameters that can be verified during SPE extraction, there is no guarantee for an efficient (> 80%?) and reproducible recovery.

Ceglarek et al. [89] used the CW-TR fibre to analyse LAS in influent and effluent wastewater samples of a WWTP applying SPME. The optimised conditions included the addition of 0.5 g mL^{-1} of ammonium acetate to 3 mL of sample, extraction by immersion of the fibre (2 h) and static desorption (15 min) in isopropanol/methanol (1:1). The extracted LAS were analysed by SPME-HPLC-FD and LC-ESI-MS. The former was not suitable for quantifying LAS because of its limited extraction efficiency, whereas LC-ESI-MS showed a linear range from 0.5 to 100 µg L^{-1}, with detection limits of 0.5 µg L^{-1} for each individual homologue of LAS. The CW-TPR fibre also extracted alkylether sulfates (AESs) but not under optimised conditions.

Cuzzola et al. have studied, using ionspray and electrospray MS and SPME-GC-MS, the Fenton oxidation products of surfactants such as lauryl sulfate [90] and AES [91]. The oxidation leads to the formation of products with hydroxyl and epoxide groups because of insertion of oxygen atoms or aldehydes [90] or terminal ethoxylic moieties [91] derived from the loss of the hydrophilic sulfate group.

Amphoteric surfactants
To our knowledge, until today no SPE method exists for the extraction of amphoteric surfactants, such as CAPB, from aquatic matrices. A series of attempts were mounted to extract amphoteric surfactants from a fortified secondary effluent at various pH values (pH 3–7) on cartridges containing either RP-C18, Lichrolut EN or anion exchange material or combinations thereof [88]. In all instances, insufficient extraction efficiencies with high standard deviations were obtained. From these findings, it was concluded that CAPB was lost during sample preparation through adsorption onto surfaces of the extraction bottles. Seemingly, the partial cationic character of the charged head group of CAPB led to strong adsorptive forces with the silicate-based glassware. To circumvent the encountered problems, further efforts were made by adding methanol as an organic modifier to the sample (5 and 10%) as well as using polyethylene containers instead of glass vessels. None of these experiments gave acceptable recoveries with satisfactory standard deviations.

Applications for cationic surfactants are discussed in Chapter 2.12.

3.1.5 Conclusion

Modern methods of sample handling for determination of surfactants in aqueous samples are practically all based on SPE and modifications thereof. Substantial reductions in analysis time, solvent consumption, sample volume required, and number of off-line steps have thus been achieved. This has not only increased the analysts' capacity and analysis price per sample, but also decreased the risk of both analyte loss and contamination during sample handling. Whether or not this has indeed resulted in an increased quality of analytical results still needs to be validated through, e.g. intercalibration exercises. This aspect is discussed in more detail in Chapter 4.

REFERENCES

1   M.W.F. Nielen, R.W. Frei and U.A.T. Brinkman, *Journal of Chromatography Library*, Vol. 39A, Elsevier, Amsterdam, 1988, p. 5.
2   B. Thiele, K. Gunther and M.J. Schwuger, *Chem. Rev.*, 97 (1997) 3247.
3   A.T. Kiewiet and P. de Voogt, *J. Chromatogr. A*, 733 (1996) 185.
4   A. Marcomini and M. Zanette, *J. Chromatogr. A*, 733 (1996) 193.
5   M.L. Trehy, W.E. Gledhill, J.P. Mieure, J.E. Adamove, A.M. Nielsen, H. Perkins and W.S. Eckhoff, *Environ. Toxicol. Chem.*, 15 (1996) 233.
6   V.M. León, E. González-Mazo and A. Gómez-Parra, *J. Chromatogr. A*, 889 (2000) 211.
7   E. González-Mazo, J.M. Forja Pajares and A. Gómez-Parra, *Environ. Sci. Technol.*, 28 (1998) 1636.
8   V.M. León, M. Sáez, E. González-Mazo and A. Gómez-Parra, *Sci. Total Environ.*, 288 (2002) 215.
9   S. Terzic and M. Ahel, *Mar. Poll. Bull.*, 28 (1994) 735.
10 R. Kveštak and M. Ahel, *Ecotox. Environ. Saf.*, 28 (1994) 25.
11 P. de Voogt, O. Kwast, R. Hendriks, and N. Jonkers, In: D. Vethaak, et al. (Eds.), *Estrogens and Xeno-estrogens in the Aquatic Environment of The Netherlands, RIKZ-RIZA, The Hague*, 2002, p. 96.
12 P. de Voogt, O. Kwast, R. Hendriks and N. Jonkers, *Analusis*, 28 (2000) 776.
13 A. Di Corcia, R. Samperi and A. Marcomini, *Environ. Sci. Technol.*, 28 (1994) 850.
14 M. Ahel and W. Giger, *Anal. Chem.*, 57 (1985) 1577.
15 S. Takano, N. Yagi and K. Kunihiro, *Yukagaku*, 24 (1975) 389.

16 S. Takano, C. Takasaki, K. Kunihiro and M. Yamanaka, *Yukagaku*, 25 (1976) 31.

17 H. Hon-nami and T. Hanya, *J. Chromatogr.*, 161 (1978) 205.

18 R. Wickbold, *Tenside Surf Det.*, 8 (1971) 61.

19 E. Matthijs and M. Stalmans, *Tenside Surf Det.*, 30 (1993) 29.

20 S. Terzic, D. Hrsak and M. Ahel, *Mar. Poll. Bull.*, 24 (1992) 199.

21 M. Ahel and W. Giger, *Anal. Chem.*, 57 (1985) 2584.

22 E. Stephanou and W. Giger, *Environ. Sci. Technol.*, 16 (1982) 800.

23 M.C. Allen and D.E. Linder, *J. Am. Oil Chem. Soc.*, 58 (1981) 950.

24 D. Brown, H. De Henau, J.T. Garrigan, P. Gerike, M. Holt, E. Keck, E. Kunkel, E. Matthijs, J. Waters and R.J. Watkinson, *Tenside Surf Det.*, 23 (1986) 190.

25 A. Marcomini, S. Stelluto and B. Pavoni, *Int. J. Environ. Anal. Chem.*, 35 (1989) 207.

26 A. Kiewiet, J.M.D. van der Steen and J. Parsons, *Anal. Chem.*, 67 (1995) 4409.

27 P.W. Taylor and G. Nickless, *J. Chromatogr.*, 178 (1979) 259.

28 T. Saito, K. Higashi and K. Hagiwara, *Z. Anal. Chem.*, 313 (1982) 21.

29 M.C. Bruzzoniti, C. Sarzanini and E. Mentasti, *J. Chromatogr. A*, 902 (2000) 289.

30 V. Lopez-Avila, *Crit. Rev. Anal. Chem.*, 29 (1999) 195.

31 A. Marcomini and W. Giger, *Anal. Chem.*, 59 (1987) 1709.

32 M. Kikuchi, A. Tokai and T. Yoshida, *Water. Res.*, 20 (1986) 643.

33 I. Fujita, Y. Ozasa, T. Tobino and T. Sugimura, *Chem. Pharm. Bull.*, 38 (1990) 1425.

34 A. Marcomini, A.D. Corcia, R. Samperi and S. Capri, *J. Chromatogr.*, 644 (1993) 59.

35 A. Marcomini, S. Capri and W. Giger, *J. Chromatogr.*, 403 (1987) 243.

36 E. Kubeck and C.G. Naylor, *J. Am. Oil. Chem. Soc.*, 67 (1990) 400.

37 H.-B. Lee, J. Weng, T.E. Peart and R.J. Maguire, *Water Qual. Res. J. Canada*, 33 (1998) 19.

38 J.A. Field and R.L. Reed, *Environ. Sci. Technol.*, 30 (1996) 3544.

39 H. De Henau, E. Matthijs and W.D. Hopping, *Int. J. Environ. Anal. Chem.*, 26 (1986) 279.

40 M.A. Castles, B.L. Moore and S.R. Ward, *Anal. Chem.*, 61 (1989) 2534.

41 A. Di Corcia, M. Marchetti, R. Samperi and A. Marcomini, *Anal. Chem.*, 63 (1991) 1179.

42 A. Di Corcia, S. Marchese and R. Samperi, *J. Chromatogr.*, 642 (1993) 163.

43 M.E. León-González and L.V. Pérez-Arribas, *J. Chromatogr. A*, 902 (2000) 3.

44 C. Crescenzi, A. Di Corcia, G. Passariello, M. Samperi and M.I.T. Carou, *J. Chromatogr. A*, 733 (1996) 41.

45 P. Eichhorn and T.P. Knepper, *J. Chromatogr. A*, 854 (1999) 221.

46   M. Ahel, T. Conrad and W. Giger, *Environ. Sci. Technol.*, 21 (1987) 697.

47   M. Ahel, W. Giger and M. Koch, *Water Res.*, 28 (1994) 1131.

48   M. Ahel, C. Schaffner and W. Giger, *Water Res.*, 30 (1996) 37.

49   F. Ventura, A. Figueras, J. Caixach, I. Espadaler, J. Romero, J. Guardiola and J. Rivera, *Water Res.*, 22 (1988) 1211.

50   J. Rivera, D. Fraisse, F. Ventura, J. Caixach and A. Figueras, *Fres. J. Anal. Chem.*, 327 (1987) 377.

51   F. Ventura, J. Caixach, A. Figueras, D. Fraisse and J. Rivera, *Water Res.*, 23 (1989) 1191.

52   A. Marcomini, A. Di Corcia, R. Samperi and S. Capri, *J. Chromatogr. A*, 644 (1993) 59.

53   H.-B. Lee, J. Weng, T.E. Peart and R.J. Maguire, *Water Qual. Res. J. Canada*, 33 (1998) 19.

54   N. Jonkers, R.W.P.M. Laane and P. de Voogt, *Environ. Sci. Technol.*, (2002) in press.

55   W.H. Ding, S.H. Tzing and J.H. Li, *Chemosphere*, 38 (1999) 2957.

56   W.H. Ding and C.T. Chen, *J. Chromatogr. A*, 862 (1999) 113.

57   E.T. Furlong and P.M. Gates, 45th ASMS Conf. on Mass Spectom. and Allied Topics, Palm Springs, 1997, p. 942.

58   A. Di Corcia, A. Constantino, C. Crescenzi, E. Marinoni and R. Samperi, *Environ. Sci. Technol.*, 32 (1998) 2401.

59   A. Di Corcia, R. Cavallo, C. Crescenzi and M. Nazzari, *Environ. Sci. Technol.*, 34 (2000) 3914.

60   A. Marcomini, G. Pojana, A. Sfriso and J.M. Quiroga Alonso, *Environ. Toxicol. Chem.*, 19 (2000) 2000.

61   M. Reinhard, N. Goodman and K. Mortelmans, *Environ. Sci. Technol.*, 16 (1982) 351.

62   H.A. Ball, M. Reinhard and P.L. McCarthy, *Environ. Sci. Technol.*, 23 (1989) 951.

63   P. Lee Ferguson, C.R. Iden and B.J. Brownawell, *Anal. Chem.*, 72 (2000) 4322.

64   M. Petrovic, A. Diaz, F. Ventura and D. Barceló, *Anal. Chem.*, 73 (2001) 5886.

65   A. Kiewiet, J.M.D. van der Steen and J.R. Parsons, *Anal. Chem.*, 67 (1995) 4409.

66   K. Inaba, *Int. J. Environ. Anal. Chem.*, 31 (1987) 63.

67   T.M. Schmitt, M.C. Allen, D.K. Brain, K.F. Guin, D.E. Lemmel and Q.W. Osburn, *J. Am. Oil Chem. Soc.*, 67 (1990) 103.

68   N.J. Fendinger, W.M. Begley, D.C. Macavoy and W.S. Eckhoff, *Environ. Sci. Technol.*, 29 (1995) 856.

69   M. Castillo, M.C. Alonso, J. Riu and D. Barceló, *Environ. Sci. Technol.*, 33 (1999) 1300.

70   M. Castillo and D. Barceló, *Anal. Chem.*, 71 (1999) 3769.

71 C. Crescenzi, A. Di Corcia, R. Samperi and A. Marcomini, *Anal. Chem.*, 67 (1995) 1797.

72 K.A. Evans, S.T. Dubey, L. Kravetz, I. Dzidic, J. Gumulka, R. Mueller and J.R. Stork, *Anal. Chem.*, 66 (1994) 699.

73 K.A. Evans, S.T. Dubey, L. Kravetz, S.W. Evetts, I. Dzidic and C.C. Dooyema, *J. Am. Oil Chem. Soc.*, 74 (1997) 765.

74 S.T. Dubey, L. Kravetz and J.P. Salanitro, *J. Am. Oil Chem. Soc.*, 72 (1995) 23.

75 J. Tolls, M. Haller and D.T.H.M. Sijm, *J. Chromatogr. A*, 839 (1999) 109.

76 M.A. Blackbur and M.J. Waldock, *Water Res.*, 29 (1995) 1623.

77 A. Boyd-Boland and J. Pawliszyn, *Anal. Chem.*, 68 (1996) 1521.

78 K.K. Chee, M.K. Wong and H.K. Lee, *J. Microcol. Sep.*, 8 (1996) 131.

79 A. Diaz, F. Ventura and M.T. Galceran, *J. Chromatogr. A*, 963 (2002) 159.

80 A. Diaz, F. Ventura and M.T. Galceran, *Anal. Chem.*, 74 (2002) 3869.

81 R. Aranda and R.C. Burk, *J. Chromatogr.*, 829 (1998) 401.

82 E. González-Mazo and A. Gómez-Parra, *Trends Anal. Chem.*, 15 (1996) 375.

83 L. Sarrazin, W. Wafo and P. Rebouillon, *J. Liq. Chromatogr.*, 22 (1999) 2511.

84 E. González-Mazo, M. Honing, D. Barceló and A. Gómez-Parra, *Environ. Sci. Technol.*, 31 (1997) 504.

85 J. Riu, E. González-Mazo, A. Gómez-Parra and D. Barceló, *Chromatographia*, 50 (1999) 275.

86 L. Sarrazin, A. Arnoux and P.J. Rebouillon, *J. Chromatogr. A*, 760 (1997) 285.

87 R.J.C.A. Steen, A.C. Hogenboom, P.E.G. Leonards, R.A.L. Peerboom, W.P. Cofino and U.A.T. Brinkman, *J. Chromatogr. A*, 857 (1999) 157.

88 P. Eichhorn, PhD Thesis. University of Mainz, Germany, 2001.

89 U. Ceglarek, J. Efer, A. Schreiber, E. Zwanziger and W. Engewald, *Fres. J. Anal. Chem.*, 365 (1999) 674.

90 A. Cuzzola, A. Raffaelli, A. Saba, S. Pucci and P. Salvadori, *Rapid Commun. Mass Spectrom.*, 13 (1999) 2140.

91 A. Cuzzola, A. Raffaelli, A. Saba and P. Salvadori, *Rapid Commun. Mass Spectrom.*, 14 (2000) 834.

92 A. Di Corcia, L. Capuani, F. Casassa, A. Marcomini and R. Samperi, *Environ. Sci. Technol.*, 33 (1999) 4119.

93 M.L. Trehy, W.E. Gledhill and G. Orth, *Anal. Chem.*, 62 (1990) 2581.

94 J.A. Field, J.A. Leenheer, K.A. Thorn, L. Barber Jr., C. Rostad, D.L. Macalady and S.R. Daniel, *J. Contam. Hydrol.*, 9 (1992) 55.

95 S.D. Scullion, M.R. Clench, M. Cooke and A.E. Ashcroft, *J. Chromatogr.*, 733 (1996) 207.

96 M. Castillo, M.C. Alonso, J. Riu and D. Barceló, *Environ. Sci. Technol.*, 33 (1999) 1300.

97   M. Petrovic and D. Barceló, *Fres. J. Anal. Chem.*, 368 (2000) 676.
98   H.-Q. Li, F. Jiku and H.F. Schröder, *J. Chromatogr. A*, 889 (2000) 155.
99   H.F. Schröder, *Vom Wasser*, 79 (1992) 193.
100 K. Maruyama, M. Yuan and A. Otsuki, *Environ. Sci. Technol.*, 34 (2000) 343.
101 M. Takino, S. Daishima and K. Yamaguchi, *J. Chromatogr. A*, 904 (2000) 65.
102 P.L. Ferguson, C.R. Iden and B.J. Brownawell, *Environ. Sci. Technol.*, 35 (2001) 2428.
103 L.B. Clark, R.T. Rosen, T.G. Hartman, J.B. Louis, I.H. Suffet, R.L. Lippincott and J.D. Rosen, *Int. J. Air Water Pollut.*, 47 (1992) 167.
104 H.F. Schröder and K. Fytianos, *Chromatographia*, 50 (1999) 583.
105 M. Castillo, E. Martínez, A. Ginebreda, L. Tirapu and D. Barceló, *Analyst*, 125 (2000) 1733.
106 M. Castillo, M.C. Alonso, J. Riu, M. Reinke, G. Klöter, H. Dizer, B. Fischer, P.D. Hansen and D. Barceló, *Anal. Chim. Acta*, 426 (2001) 265.
107 M. Farré, M.-J. García, L. Tirapu, A. Ginebreda and D. Barceló, *Anal. Chim. Acta*, 427 (2001) 181.
108 H.F. Schröder, *J. Chromatogr.*, 643 (1993) 145.
109 H.F. Schröder, *Vom Wasser*, 73 (1989) 111.
110 J.C. Dunphy, D.G. Pessler, S.W. Morrall, K.A. Evans, D.A. Robaugh, G. Fujimoto and A. Negahban, *Environ. Sci. Technol.*, 35 (2001) 1223.
111 H.F. Schröder, *Fres. J. Anal. Chem.*, 353 (1995) 93.
112 D.D. Popenoe, S.J. Morris, P.S. Horn and K.T. Norwood, *Anal. Chem.*, 66 (1994) 1620.
113 E. Matthijs, M.S. Holt, A. Kiewiet and G.B.J. Rijs, *Environ. Toxicol. Chem.*, 18 (1999) 2634.
114 C.A. Moody, W.C. Kwan, J.W. Martin, D.C.G. Muir and S.A. Mabury, *Anal. Chem.*, 73 (2001) 2200.
115 H.F. Schröder, *Vom Wasser*, 81 (1993) 299.
116 P. Eichhorn and T.P. Knepper, *J. Mass Spectrom.*, 35 (2000) 468.
117 P. Eichhorn and T.P. Knepper, *J. Chromatogr. A*, 854 (1999) 221.
118 J. Waters, K.S. Lee, V. Perchard, M. Flanagan and P. Clarke, *Tenside Surf Det.*, 37 (2000) 161.
119 M. Radke, T. Behrends, J. Förster and R. Herrmann, *Anal. Chem.*, 71 (1999) 5362.

3.2 METHODS FOR THE SAMPLE HANDLING OF NON-IONIC SURFACTANTS IN SLUDGES AND SEDIMENTS

Niels Jonkers and Pim de Voogt

3.2.1 Introduction

Strategies for sample handling strongly depend on the nature of the analytes and matrices involved, and on the concentration levels of the analytes [1]. In addition, the method of detection will set specific requirements to the sample treatment procedures applied, in order that the final extract is compatible with the method's ways of sample introduction.

As the nature of surfactants varies from relatively apolar, non-ionic to essentially highly polar or ionic species, with, in addition, varying specific characteristics such as persistence towards (bio)chemical or physical degradation, the sample treatment may also vary considerably.

The environmental levels that we will consider here range from minute (i.e. ppt–ppb levels in, e.g. marine sediments) to relatively high (i.e. 10–5000 ppm, as in sewage sludge) levels.

3.2.2 Methods for the sample handling of non-ionic surfactants in sludges and sediments

Since the beginning of the 1980s, the presence of surfactants in solid matrices has slowly received more attention. The publication by Giger and coworkers in 1984 [2] on the anaerobic formation in sewage sludge of nonylphenol (NP) out of alkylphenol ethoxylates (APEOs) particularly stressed the importance of this environmental compartment. However, environmental data of surfactants in sediments and sludges remains scarcer than data on aqueous concentrations, undoubtedly because sediment samples are more difficult to handle.

A variety of sample handling techniques is available, which can be divided into four types: Soxhlet extraction, steam distillation, sonication and accelerated solvent extraction (ASE) [3].

In recent years, a revolution has taken place in the field of sediment extraction, with the introduction of ASE techniques. While in their

Comprehensive Analytical Chemistry XL
T.P. Knepper, D. Barceló and P. de Voogt (Eds)

review on alkylphenol ethoxylates from 1997, Thiele et al. [4] mentioned Soxhlet and steam distillation as the main techniques for extraction of APEO from solid matrices, publications since that time show that those classical techniques have been replaced by ASE or sonication techniques, which are more efficient, and less time and solvent consuming [3].

In this chapter, all four types of sediment and sludge sample handling techniques for non-ionic surfactants will be discussed and compared. Most of the studies published on non-ionics focus on APEOs and their degradation products, viz. the alkylphenols, but some extraction methods for alcohol ethoxylates (AEOs) and coconut diethanol amides will also be discussed.

Performances of the methods described can be found in Table 3.2.1.

General aspects
In most cases, samples are first freeze-dried, ground, sieved and stored at − 20°C. Air or oven drying of samples is also performed occasionally. However, it has been reported that freeze-drying gives more consistent recoveries than air drying [5]. Especially for the relatively volatile NP, oven drying is not recommended [6]. Conservation agents, such as formaldehyde, are usually purposely avoided, because of the risk of sample contamination. In a small-scale laboratory experiment, con-servation of solid residues from wastewater was compared when stored without formaldehyde at − 20°C or with formaldehyde at +4°C. No significant differences were found between the two regimens for NP and NPEO concentrations after 3 and 6 weeks of storage [7].

The surface-active properties of surfactants could give rise to sorption of analytes to glassware and equipment. However, the literature shows that when a comparison is made between the use of silanised glassware and careful rinsing of normal glassware with methanol during sample treatment, the latter (cheaper) solution is sufficient [5].

Matrix effects can influence significantly the extraction efficiency and signal intensity. For heavily contaminated samples such as sewage sludge, this problem is particularly relevant, and therefore the use of internal standards is essential in these applications. Internal standards applied in the extraction procedures for non-ionic surfactants include: perinaphtenone [8], 4-fluoro-4'-hydroxyl-benzophenone [5] and 4-bro-mophenyl acetic acid [9]. More appropriate are internal standards from the same compound class: butylphenol [10], heptylphenol [11],

416

TABLE 3.2.1

Performance of sample handling techniques for the extraction of non-ionic surfactants from solid matrices

Compounds	Matrix	Extraction technique	Recovery (%)	Relative standard deviation (%)	Limit of detection (ng g^{-1})[a]	Reference
4-NP	Recycled paper sludge	Soxhlet dichloromethane, silica clean-up		9.6–35		[24]
4tOP	Recycled paper sludge	Soxhlet dichloromethane, silica clean-up		9.6–35		[24]
NP	Marine sediment	Soxhlet hexane/isopropyl alcohol cyanopropyl silica SPE clean-up	84	3.2	4	[5]
NPEO$_{1-9}$	Marine sediment	Soxhlet hexane/isopropyl alcohol cyanopropyl silica SPE clean-up	53–103	3.2–27	2–10	[5]
NP	Estuarine sediment	Soxhlet hexane NH$_2$-SPE clean-up	88	5	4	[27]
NPEO$_1$	Estuarine sediment	Soxhlet hexane NH$_2$-SPE clean-up	90	7	4	[27]
NPEO$_2$	Estuarine sediment	Soxhlet hexane NH$_2$-SPE clean-up	89	8	4	[27]
NP	Sewage sludge and sediment	Soxhlet methanol/diethyl ether	71	9		[23]
OP	Sewage sludge and sediment	Soxhlet methanol/diethyl ether	70	21		[23]
NP	River and estuarine sediment	Soxhlet dichloromethane silica clean-up	108	4	15 (= LOQ)	[25]
OP	River and estuarine sediment	Soxhlet dichloromethane silica clean-up	110	4	3 (= LOQ)	[25]
NPEO$_1$	River and estuarine sediment	Soxhlet dichloromethane silica clean-up	91	7	40 (= LOQ)	[25]
NPEO	Marine sediment	Soxhlet methanol C$_{18}$-SPE clean-up	84		0.5–4.3	[16]
NP	Marine sediment	Soxhlet methanol C$_{18}$-SPE clean-up	104		0.5	[16]
NP	River sediment	Soxhlet methanol alumina clean-up	79	24		[49]
NPEO	River sediment	Soxhlet methanol alumina clean-up	82	12		[49]
alkylphenols	River sediment	Soxhlet methanol alumina clean-up	85	6		[48]
NPEO$_1$	Freshwater sediment	Soxhlet dichloromethane	98		10	[26]
NPEO$_2$	Freshwater sediment	Soxhlet dichloromethane	100		2.1	[26]
AEO	Sludge	Soxtec methanol C$_{18}$-SPE clean-up	85–92	11–15	10	[18]
NP	Sludge	Soxtec methanol C$_{18}$-SPE clean-up	72	16	1	[18]
OP	Sludge	Soxtec methanol C$_{18}$-SPE clean-up	74	16	1	[18]
NPEO	Sludge	Soxtec methanol C$_{18}$-SPE clean-up	71	12	20	[18]
OPEO	Sludge	Soxtec methanol C$_{18}$-SPE clean-up	69	12	20	[18]
NP	Sewage sludge	Steam distillation	82	16		[34]
NP	Marine sediment	Ultrasonication methanol/dichloromethane, C$_{18}$-SPE clean-up	81	10	10	[39]
OP	Marine sediment	Ultrasonication methanol/dichloromethane, C$_{18}$-SPE clean-up	75	8	10	[39]
NPEO	Marine sediment	Ultrasonication methanol/dichloromethane, C$_{18}$-SPE clean-up	95	9	10	[39]

(continued on next page)

TABLE 3.2.1 (continued)

Compounds	Matrix	Extraction technique	Recovery (%)	Relative standard deviation (%)	Limit of detection (ng g^{-1})[a]	Reference
NPE$_1$C	Marine sediment	Ultrasonication methanol/dichloromethane, C$_{18}$-SPE clean-up	83	12	5	[39]
CDEA	Marine sediment	Ultrasonication methanol/dichloromethane, C$_{18}$-SPE clean-up	92	7	0.5–1	[39]
NP	Estuarine sediment	Sonication methanol aminopropyl silica clean-up	82.8	5.4	21.5	[13]
NPEO$_1$	Estuarine sediment	Sonication methanol aminopropyl silica clean-up	93.5	6.7	37	[13]
NPEO$_{>1}$	Estuarine sediment	Sonication methanol aminopropyl silica clean-up	84.3	5.9	0.78–5	[13]
alkylphenols	River sediment	Ultrasonication methanol alumina clean-up	82	5		[48]
NP	Marine sediment	Sonication hexane/acetone cyanopropyl silica SPE clean-up	88	4.1		[5]
NPEO1-9	Marine sediment	Sonication hexane/acetone cyanopropyl silica SPE clean-up	45–92			[5]
NP	Land applied sewage sludge	Enhanced solvent extraction dichloromethane SEC clean-up	126			[8]
NPEO$_1$	Land applied sewage sludge	Enhanced solvent extraction dichloromethane SEC clean-up	78			[8]
NPEO$_2$	Land applied sewage sludge	Enhanced solvent extraction dichloromethane SEC clean-up	108			[8]
NP	River sediment	Pressurised liquid extraction methanol alumina clean-up	85	6–33		[49]
NPEO	River sediment	Pressurised liquid extraction methanol alumina clean-up	87	6–33		[49]
alkylphenols	River sediment	Pressurised liquid extraction methanol alumina clean-up	75–111	2–7		[48]
NPEO$_{1-17}$	Sewage sludge	SFE, water modifier	85–105	2–10		[43]
NPEO	River sediment	SFE, methanol modifier	85			[28]
OPEO	River sediment	SFE, methanol modifier	65			[28]
NP	River sediment	SFE, methanol modifier	90			[28]
NP	Sewage sludge	SFE silica clean-up	96–98	4.5–5.3		[41]
NP	Freshwater sediment	SFE silica clean-up			46	[26]
OP	Freshwater sediment	SFE silica clean-up			1	[26]
NPE$_{1-4}$C	Sewage sludge	Subcritical SFE, ethanol/water SAX clean-up	100		200–2000 (= LOQ)	[9]

[a]LOQ, limit of quantitation.

decylphenol monoethoxylate [11], *n*-octylphenol [12] or *n*-nonylphenol monoethoxylate [12] (in high performance liquid chromatography (HPLC) the linear alkylphenols and linear APEO are easily baseline separated from the branched isomers, and they do not occur in environmental samples). Currently, the optimal (but most expensive) selection of internal standards are the (linear) NPs and shortchain NPEO labeled with ^{13}C in the phenolic ring, which have recently become available from Cambridge Isotope Laboratories [13]. Ferguson and coworkers [14] used specially synthesised branched ^{13}C-labeled NP and NPEO internal standards, having exactly the same retention time as the analytes studied, thereby correcting for matrix effects in the best way. Unfortunately, in most studies no internal standards are used at all.

Soxhlet extraction
Although the Soxhlet extraction technique is termed a 'classical method', it is still applied regularly for the extraction of non-ionic surfactants from solid matrices. The advantages of this technique are its robustness (virtually any solid matrix can be extracted) and the fact that no expensive equipment is needed.

Many extraction solvents have been used, of which the more polar protic solvents are found to perform best. While Soxhlet recoveries for NPEO are reported at 70% using pentane [13], and 80% using acetone/hexane [13], solvents like propanol or methanol yield higher recoveries, especially for the higher ethoxylates.

The most commonly used solvent for the extraction of both alkylphenols and APEO is methanol [15–19]. If sodium hydroxide (20%, w/w) is added to the methanol, recoveries may increase by 25% [16,20]. Nonylphenoxy carboxylates (NPECs) can also be extracted using Soxhlet with methanol, but recoveries are often below 50% [19]. Only one method similar to Soxhlet has been reported for the extraction of AEOs, in which a Soxtec$^{©}$ system with methanol is used [18].

Other solvents that have been used are isopropanol [21], dichloromethane/methanol [22], methanol/diethyl ether (10:1) with hydrochloric acid [23] for alkylphenols, dichloromethane [24–26], and hexane [27] for alkylphenols and short-chain NPEO, and hexane/isopropyl alcohol for NP and $NPEO_{1-9}$ [5].

As Soxhlet extraction is not a very specific technique, interferences will be present in the raw extract, and in most cases, a clean-up step is necessary. Several clean-up types have been reported, including adsorption chromatography using silica [21,24,25], alumina [16,22], and several kinds of SPE clean-up using C18 cartridges [16–18], dual column-LC with Lichrosorb© ADS restricted access material [6], cyanopropyl silica cartridges [5], NH$_2$-cartridges [27] or ENV + polymer [23]. In some cases, activated copper powder is added to the clean-up column to remove elemental sulfur [5,28].

Further steps may include derivatisation, necessary when using GC methods for the final detection [29]. Derivatisation has been performed with pentafluorobenzyl bromide [22,30], acetic anhydride [24], or *N,O*-bis(trimethylsilyl)trifluoroacetamide [25] (see also Chapter 2.1). A liquid–liquid separation with *n*-hexane and an alkaline aqueous solution proved to be unsuccessful as a clean-up step for alkylphenols, as the compounds ended up in both phases [23].

Steam distillation
For several decades, steam distillation has been applied successfully for the extraction of non-ionic surfactants from sediments and sludges, but in recent years the interest in this method has declined. In only one recent publication has steam distillation been applied [31].

In this technique sediment or sludge is suspended in water, which is refluxed for 3–5 h, while the distillate is percolated through a small amount of extraction solvent (1–5 mL), which is in most cases cyclo-hexane [2,31–34]. Jobst [35] used isooctane as the extraction solvent.

Clean-up after steam distillation is usually performed by alumina [31–33] or silica [2]. Sweetman reported a method without any clean-up at all [34]. The steam distillation method has only been applied to the alkylphenols and short-chain ethoxylates, because of the poor volatility of longer chain ethoxylates.

Sonication
One of the new techniques that has emerged in recent years is sonication extraction. A short overview of the application of this method to alkylphenols and APEO has been given recently by Petrovic et al. [3]. The reduction in extraction time and solvent use for sonication compared with Soxhlet extraction are substantial, and recoveries are comparable or better.

As with Soxhlet, the preferred extraction solvent is methanol, which has been used for the extraction of alkylphenols, halogenated NPs and NPEO [13,14,36], and AEOs [37]. Other solvents that have been used are ethyl acetate for alkylphenols and short-chain NPEO [10], hexane/acetone (60:40) for NP and $NPEO_{1-9}$ [5] and methanol/ dichloromethane (7:3) for alkylphenols, NPEO, AEO and coconut diethanol amides [38,39]. Usually 1–2 g of dried sample is treated, and two or three extraction cycles of 15–30 min are performed, using 15–60 mL of solvent per cycle. In some cases, the extraction tube is cooled on an ice bath [10] or heated to 50–65°C [13,37]. Ferguson et al. [37] developed an elegant extraction device, in which the sediment was packed into a column, and ultrasonic extraction was performed under solvent flow at elevated temperatures. With a methanol flow of 0.5 mL min^{-1}, the extraction was completed in 7–10 min. Clean-up steps after sonication extraction that have been applied are C_{18}-SPE [10,38,39], cyanopropyl silica SPE [5] or alumina [37]. An extensive clean-up procedure was reported by Ferguson and coworkers [13,36], in which first a normal phase clean-up is performed using aminopropyl silica SPE, and then a reversed phase semipreparative C_{18}-HPLC clean-up follows. However, this method should only be used for low NPEO oligomers, as the long-chain NPEOs are lost in the normal phase clean-up step. In a subsequent publication by the same authors, the normal phase clean-up was omitted, as samples from the vicinity of a wastewater treatment plant were analysed, in which the long-chain NPEO can be expected [14].

For the analysis of AEO sediment extracts on HPLC-fluorescence (FL), it was necessary to derivatise using naphthoil chloride and N-methylimidazole in order to incorporate a chromophore into the analytes [37].

Supercritical fluid extraction/accelerated solvent extraction
The current trend in the development of solid sample handling techniques is the rapid increase of interest in new techniques such as supercritical fluid extraction (SFE) and ASE. Many names have been given to more or less similar techniques, including pressurised liquid (or fluid) extraction (PLE) and hot (subcritical) water extraction. Although a general name for all these techniques does not yet exist, they share one feature in common. Pressure is artificially imposed to achieve temperatures higher than the normal boiling point of the extraction solvent. These temperatures lead to higher solubilities and diffusion, lower viscosities and in the end improved extraction recoveries [40]. After the extraction, the analyte is collected either by deposition onto a solid-phase trap (e.g. octadecyl silica [41]) or into a collection solvent.

For SFE techniques, CO_2 is always used as the supercritical fluid, while for ASE several extraction solvents are possible.

The first publications on SFE of APEO were discussed in a review on analytical methods for APEO [42]. For the determination of alkylphenols in sewage sludge and sediment, a SFE technique was optimised, using CO_2 at 80°C and 351 atm, at a flow rate of 2 mL min^{-1}. Extraction times were 15 min static and 10 min dynamic, with a sample intake of 0.1–1 g [41]. In this method, in situ acetylation of the alkylphenols using acetic anhydride was performed. The extract was washed with an aqueous K_2CO_3 solution to remove co-extracted acetic acid, and cleaned up using 5% deactivated silica.

Later, this method was extended to $NPEO_{1-17}$ and the NPEC metabolites [43]. In that method, water is added as a static modifier, which turned out to be critical for obtaining good recoveries for the long-chain oligomers. For $NPEO_1$ and $NPEO_2$, the addition of water as modifier increased recoveries with 8 and 22%, respectively. NPE_1C and NPE_2C were methylated off-line using BF_3 in methanol [43].

The same SFE method, but without the modifier, was used for alkylphenols in freshwater sediments, at a temperature of 110°C [26,44, 45]. Between all extractions, a blank dynamic extraction step was performed to avoid carry-over. In this method, $NPEO_1$ and $NPEO_2$ were extracted separately using Soxhlet, instead of using SFE with modifier.

A second SFE-based method for the extraction of NPEC from solid matrices is called subcritical (hot) water extraction [9]. The method was applied for $NPE_{1-4}C$ in 0.25 g sludge samples. Subcritical water is used as the extraction solvent, with 30% ethanol as the modifier, yielding quantitative recoveries. Other tested extraction solvent compositions were CO_2, hot water, methanol-modified hot water, none of which were judged satisfactory.

When a modifier of 20% ethanol is used, quantitative recoveries are still obtained for NPE_1C and NPE_2C, but recoveries of NPE_3C and NPE_4C drop to 70 and 55%, respectively. The decreasing recovery with increasing number of ethoxylate groups indicates that interactions between the ethoxylate groups and the soil matrix, rather than the hydrophobic interactions between the NP group and soil, control the NPEC recovery from soil.

The extraction time applied was 20 min at 75°C and 150 atm and a flow rate of 1.5–2 mL min^{-1}. The clean-up step consisted of a SAX Empore© disk, which was simultaneously eluted and methylated by methyl iodide in acetonitrile [9].

Kreisselmeier and Dürbeck [28] extracted alkylphenols, NPEO and OPEO by SFE using methanol as a modifier. The extraction pressure had to be set as high as possible (>450 atm), with a dynamic extraction time of 30 min. Neither an elevation of the temperature from 100 to 150°C nor an extension of the static extraction time improved the recovery. Comparable results were even obtained at 30°C while omitting the static extraction step. The modifier proved again to be quite important, as the recoveries increased by 20% with the addition of 27.5% methanol. Problems arose with the extraction of aged samples, for which the extractability was reduced by 40% for NPEO and NP, and 22% for OPEO.

An ASE method was optimised for alkylphenols and APEO as well as the anionic LAS surfactants [11]. However, acceptable recoveries were only obtained for LAS and the alkylphenols. Several modifiers were tested, all of which gave similar results for NP and NPEO. For LAS, methanol was clearly the best modifier. Tests without a modifier showed a clear drop in recovery for all compounds. Finally, 27.5% methanol was chosen as the modifier in CO_2. Extractions were performed at 100°C and 150 atm in a 10 min static and 5 min dynamic step. Heise and Litz [46] obtained improved (70–100%) recoveries by using a very similar ASE procedure with an even shorter extraction time of 10 min static.

Shang et al. [5] used ASE for the extraction of NP and NPEO from estuarine sediments. A sample of 15–25 g was extracted three times using hexane/acetone at 100°C and 103 atm. This was followed by a clean-up step using CN-SPE. A blank sample was extracted between all samples to avoid contamination. Hexane/acetone was also used in the ASE method for alkylphenols and NPEO by Heemken et al. [12]. Extraction conditions for samples of 0.5–1 g were 100°C, 150 atm, with a static extraction step of 15 min, and a rinse step with 20 mL solvent. After a clean-up by HPLC, the analytes were derivatised with heptafluorobutyric acid anhydride for GC analysis.

ASE using dichloromethane has been applied to extract alkylphenols and short-chain NPEO from sediment [8,47]. Samples of 2–5 g were extracted in two cycles of 30 mL at 100°C and 69 atm. Clean-up was performed using size exclusion chromatography to remove high molecular weight lipids, and then using normal phase HPLC.

PLE using methanol was applied for extraction of NP [48,49] and NPEO [49] from sediments. With 5–6 g sample, PLE was performed at 100–150°C, and 100–150 atm, with two static mode cycles [49] or one static and three dynamic cycles with subsequent elution [48]. Although the two procedures differed in some details, the recoveries were comparable.

Petrovic and coworkers [6] used PLE applying MeOH:acetone (1:1) for the extraction of APEOs, APECs, APs and their halogenated derivatives from sediment samples. In combination with column-switching LC, using a restricted access material precolumn, hours of sample treatment were saved.

Other methods
Two examples of other extraction methods of APEO from solid matrices that have been studied include solid-phase microextraction (SPME) [50] and microwave extraction [46].

For the SPME method a sludge sample was suspended in water, and extracted for 60 min at room temperature using an SPME fibre with Carbowax coating. During HPLC analysis, OPEO was detected in the extract but not NPEO, while from SFE extraction it was known that both OPEO and NPEO were present. This suggests that some matrix interferences are occurring that should be investigated further [50].

The microwave extraction was performed with 25 mL methanol at 150°C, 10 atm and 20 min extraction time. Large variations in recovery were obtained between 31 and 85% for different sediments [46].

3.2.3 Comparative studies

Several studies compared different extraction methods for alkylphenols and APEO from solid matrices. The reference method in all cases is Soxhlet, as this has been the standard extraction method for several decades. Although the obtained recoveries are not always better for the newly developed methods compared with Soxhlet, the new methods all share the advantages of shorter extraction times and lower solvent use.

Lee et al. [43] compared the extraction of NPEO with SFE (water as modifier) to Soxhlet with acetone/hexane, Soxhlet with dichloromethane and steam distillation. The SFE method gave recoveries around 10% higher than the acetone/hexane Soxhlet. SFE performed much better than Soxhlet with dichloromethane or steam distillation, as in those methods the higher oligomers were not detected in the extract at all.

In another study, PLE with methanol was compared with Soxhlet with methanol [49]. Recoveries and relative standard deviations were comparable for both the methods, confirming that Soxhlet can easily be substituted by PLE in already developed methods, without other substantial changes in the analytical procedure.

Ding and Fann [48] compared both PLE and sonication with Soxhlet extraction. The extraction solvent was methanol for all three methods. PLE proved to be the favourable method, with 20% higher recoveries than Soxhlet and sonication, and equal relative standard deviations for all methods.

Two studies compared sonication with Soxhlet extraction with pentane [13], acetone/hexane [13] and hexane/ispropyl alcohol (70:30) [5]. The first two Soxhlet extraction solvents yielded lower recoveries than sonication, but the most polar solvent hexane/isopropyl alcohol gave recoveries and relative standard deviations for NP and $NPEO_{1-19}$ comparable with sonication with hexane/acetone [5].

An important finding was reported by Heise and Litz [46] who investigated the recoveries of cationic, anionic and non-ionic surfactants (NPEO) from different solid matrices, such as sludge, soil and sediment. Three extraction methods were tested: Soxhlet, ASE and microwave extraction, all using methanol as the extraction solvent. All methods gave the same results for all three surfactant types: although relative standard deviations were between 3 and 12%, recoveries varied widely, i.e. from 29 to 116% among the different matrices. For equal matrices, the three methods give comparable results. These results showed that matrix effects have a much higher influence on recoveries than the extraction method that is used, and that all current extraction methods are quite sensitive to these effects.

3.2.4 Conclusion

It is quite clear that in the near future extraction of non-ionic surfactants from solid matrices will be based on SFE, PLE and sonication methods. The initial costs for SFE/PLE equipment will be earned back quickly by the reduction of man-hours and solvent consumption per sample. For sonication methods, not even this financial barrier is present. More emphasis must be laid on quality assurance of the extraction, clean-up and detection methodologies. The recently available ^{13}C-labeled standards will help to account for matrix effects that have been shown to occur, and thus improve the reliability of measurements of non-ionic surfactants.

REFERENCES

1   M.C. Hennion, In: D. Barceló (Ed.), *Sample Handling and Trace Analysis of Pollutants*, Elsevier, Amsterdam, 2000, pp. 3–71.
2   W. Giger, P.H. Brunner and C. Schaffner, *Science*, 225 (1984) 623.
3   M. Petrovic, E. Eljarrat, M.J. López de Alda and D. Barceló, *Trends Anal. Chem.*, 20 (2001) 637.
4   B. Thiele, K. Günther and J. Schwuger, *Chem. Rev.*, 97 (1997) 3247.
5   D.Y. Shang, M.G. Ikonomou and R.W. Macdonald, *J. Chromatogr. A*, 849 (1999) 467.
6   M. Petrovic, S. Lacorte, P. Viana and D. Barceló, *J. Chromatogr. A*, 959 (2002) 15.
7   P. de Voogt and F.W.M. van der Wielen, *Analysis of AP and APE*, Report. University of Amsterdam, Amsterdam (in Dutch), 1998.
8   M.J. La Guardia, R.C. Hale, E. Harvey and T.M. Mainor, *Environ. Sci. Technol.*, 35 (2001) 4798.
9   J.A. Field and R.L. Reed, *Environ. Sci. Technol.*, 33 (1999) 2782.
10 M.A. Blackburn, S.J. Kirby and M.J. Waldock, *Mar. Poll. Bull.*, 38 (1999) 109.
11 A. Kreisselmeier and H.W. Dürbeck, *Fres. J. Anal. Chem.*, 354 (1996) 921.
12 O.P. Heemken, H. Reincke, B. Stachel and N. Theobald, *Chemosphere*, 45 (2001) 245.
13 P.L. Ferguson, C.R. Iden and B.J. Brownawell, *Anal. Chem.*, 72 (2000) 4322.
14 P.L. Ferguson, C.R. Iden and B.J. Brownawell, *J. Chromatogr. A*, 938 (2001) 79.
15 A. Marcomini and W. Giger, *Anal. Chem.*, 59 (1987) 1709.
16 P. de Voogt, K. de Beer and F. van der Wielen, *Trends Anal. Chem.*, 16 (1997) 584.
17 K. Bester, N. Theobald and H.F. Schröder, *Chemosphere*, 45 (2001) 817.
18 S. Chiron, E. Sauvard and R. Jeannot, *Analusis*, 28 (2000) 535.
19 N. Jonkers, R.W.P.M. Laane and P. de Voogt, *Environ. Sci. Technol.*, 37 (2003), 321–327.
20 A. Marcomini, F. Cecchi and S. Friso, *Environ. Technol.*, 12 (1991) 1047.
21 G.A. Jungclaus, V. Lopez-Avila and R.A. Hites, *Environ. Sci. Technol.*, 12 (1978) 88.
22 N. Chalaux, J.M. Bayona and J. Albaigés, *J. Chromatogr. A*, 686 (1994) 275.
23 U. Bolz, H. Hagenmaier and W. Körner, *Environ. Poll.*, 115 (2001) 291.

24 M. Hawrelak, E. Bennett and C. Metcalfe, *Chemosphere*, 39 (1999) 745.
25 T. Isobe, H. Nishiyama, A. Nakashima and H. Takada, *Environ. Sci. Technol.*, 35 (2001) 1041.
26 E.R. Bennett and C.D. Metcalfe, *Environ. Toxicol. Chem.*, 19 (2000) 784.
27 A. Marcomini, G. Pojana, A. Sfriso and J.M. Quiroga, Alonso, *Environ. Toxicol. Chem.*, 19 (2000) 2000.
28 A. Kreisselmeier and H.W. Dürbeck, *J. Chromatogr. A*, 775 (1997) 187.
29 A.T. Kiewiet and P. de Voogt, *J. Chromatogr. A*, 733 (1996) 185.
30 C. Wahlberg, L. Renberg and U. Wideqvist, *Chemosphere*, 20 (1990) 179.
31 C.M. Lye, C.L.J. Frid, M.E. Gill, D.W. Cooper and D.M. Jones, *Environ. Sci. Technol.*, 33 (1999) 1009.
32 M. Ahel and W. Giger, *Anal. Chem.*, 57 (1985) 1577.
33 M. Ahel, W. Giger and M. Koch, *Water Res.*, 28 (1994) 1131.
34 A.J. Sweetman, *Water Res.*, 28 (1994) 343.
35 H. Jobst, *Fres. Z. Anal. Chem.*, 328 (1987) 644.
36 P.L. Ferguson, C.R. Iden and B.J. Brownawell, *Environ. Sci. Technol.*, 35 (2001) 2428.
37 L. Cavalli, G. Cassani, L. Vigano, S. Pravettoni, G. Nucci, M. Lazzarin and A. Zatta, *Tenside Surfact. Det.*, 37 (2000) 282.
38 M. Petrovic, A.R. Fernández-Alba, F. Borrull, R.M. Marce, E. González-Mazo and D. Barceló, *Environ. Toxicol. Chem.*, 21 (2002) 37.
39 M. Petrovic and D. Barceló, *J. Assoc. Offic. Anal. Chem.*, 84 (2001) 1074.
40 T.L. Chester, J.D. Pinkston and D.E. Raynie, *Anal. Chem.*, 70 (1998) 301R.
41 H.B. Lee and T.E. Peart, *Anal. Chem.*, 67 (1995) 1976.
42 H.B. Lee, *Water Qual. Res. J. Canada*, 34 (1999) 3.
43 H.B. Lee, T.E. Peart, D.T. Bennie and R.J. Maguire, *J. Chromatogr. A*, 785 (1997) 385.
44 E.R. Bennett and C.D. Metcalfe, *Environ. Toxicol. Chem.*, 17 (1998) 1230.
45 D.T. Bennie, C.A. Sullivan, H.B. Lee, T.E. Peart and R.J. Maguire, *Sci. Total Environ.*, 193 (1997) 263.
46 S. Heise and N. Litz, *Tenside Surfact. Det.*, 36 (1999) 185.
47 R.C. Hale, C.L. Smith, P.O. De Fur, E. Harvey, E.O. Bush, M.J. La Guardia and G.G. Vadas, *Environ. Toxicol. Chem.*, 19 (2000) 946.
48 W.H. Ding and J.C.H. Fann, *J. Chromatogr. A*, 866 (2000) 79.
49 S. Valsecchi, S. Polesello and S. Cavalli, *J. Chromatogr. A*, 925 (2001) 297.
50 A.A. Boyd-Boland and J.B. Pawliszyn, *Anal. Chem.*, 68 (1996) 1521.

3.3 SAMPLE HANDLING FOR THE DETERMINATION OF SURFACTANTS IN BIOTA

Pim de Voogt, Monica Sáez and Eduardo González-Mazo

3.3.1 Introduction

Organic surfactants tend to accumulate in biological membranes due to their lipophilic alkyl chains and their affinity for interfaces. Through this characteristic, organisms are capable of concentrating surfactants. Surfactants of all classes are readily taken up across gills [1]. Therefore, it is important to dispose of techniques that make it possible to extract and quantify the analyte in biological tissues and media.

Despite the increasing interest over the last decade in non-ionic surfactants resulting from the possible weak estrogenic behaviour of some representatives (notably octylphenol (OP), nonylphenol (NP) and nonylphenol ethoxylates with one or two ethoxylate groups, $NPEO_{1,2}$), surprisingly few reports are available on the levels of non-ionics in biota. Quite a few reports have been published on residue tissue levels and formation of metabolites in rodents. In these studies usually ^{14}C-labeled OP, NP or alkylphenol ethoxylates (APEOs) are used, and detection is based on measuring radioactivity often without any treatment of the samples. Method reviews of analytical procedures for APEO and alkylphenols (APs) including the determination in biota have been published [2–4].

There is also little information about biota sample treatment for determining anionic surfactants. Most of the literature available on anionics corresponds to LAS.

Pretreatment of biological samples for surfactant analysis is usually straightforward and includes either homogenising with anhydrous sodium sulphate or freeze-drying. Below, the sample treatment methods for extraction and clean-up of non-ionic and anionic surfactants in biota that have been encountered in the literature are reviewed.

3.3.2 Classical methods

The majority of methods published in the literature for extraction of biological samples comprises classical methods such as solid/liquid

Comprehensive Analytical Chemistry XL
T.P. Knepper, D. Barceló and P. de Voogt (Eds)

extraction (including Soxhlet extraction) or liquid/liquid (L/L) extraction. After such extractions, samples are either directly injected into the chromatographic system, or further cleaned by partitioning or conventional column adsorption chromatography with, e.g. Florisil or alumina columns.

Non-ionics

Rodents. Danzo et al. [5] used a simple ethyl acetate extraction with NaCl added in a vortex tube to determine NP concentrations in serum of exposed guinea pigs with GC-MS. Doerge et al. [6] analysed rat serum from exposure experiments for total NP content (i.e. conjugated and non-conjugated forms). Phase-transfer catalysis conditions were used to extract NP and the internal standard, 4-*t*-octylphenol, from serum into ethyl acetate and simultaneously derivatise the phenolic hydroxyl group using pentafluorobenzylbromide. The resulting PFB-NP derivative was analysed after an LC separation by using electron capture negative ion APCI-MS detection. Before derivatisation, the mixture was vortex mixed and precipitated proteins removed by centrifugation.

Fish. Initial reports to determine NP in fish referred to the non-specific analysis of ^{14}C-labeled compounds administered in exposure experiments. Granmo and Kollberg [7] used a solubiliser employing pancreatic enzymes to extract ^{14}C-labeled NP from exposed cod and analysed it by liquid scintillation counting.

Ferreira-Leach and Hill [8,9] applied radioactively labeled OP in a bioaccumulation experiment with rainbow trout. Soluble *t*-OP residues were extracted from each tissue by repeated vortexing with MeOH and centrifugation until the radioactivity fell to background levels (usually four to six extractions). The supernatants were combined and an aliquot taken for determination of radioactivity by liquid scintillation counting. For high-performance liquid chromatography (HPLC) analysis, purification of the OP-containing extracts was performed on OASIS SPE cartridges using MeOH elution.

McLeese et al. [10] studied the accumulation of AP in aquatic fauna including salmon. Fish were mixed with anhydrous sodium sulphate and Soxhlet extracted with ethyl acetate. Lipids were removed by GPC on Biobeads using cyclohexane/dichloromethane (1:1, v/v). The final analysis was performed with GC-FID. The recovery of p-NP was reported as 86%.

Ahel et al. [11] used steam distillation/solvent extraction and determination by normal phase HPLC for the analysis of NP, NPEO$_1$ and NPEO$_2$ in freshwater fish.

Snyder et al. [12] and Keith et al. [13] used steam distillation for the extraction of NP and NPEO$_{1-3}$ from fish tissues. Normal phase HLPC was used as a subsequent clean-up step prior to final analysis by GC-MSD using selected ion monitoring (SIM).

Thibaut et al. [14] published a procedure for determination of NP and NP transformation products in snails, duckweed and trout liver and viscera. Samples were homogenised in MeOH, and either directly chromatographed on HPLC (liver, duckweed), or subjected to a further clean-up using L/L partitioning with methanol, chloroform and acetonitrile/isooctane, successively.

Arukwe et al. [15] dosed Atlantic salmon with radioactive labeled NP. Fish were frozen immediately until gall bladder, skin, kidney, gill, liver, muscle, fat, remaining carcass and viscera were sampled and analysed. Tissue radioactivity was analysed by liquid scintillation counting after combustion of aliquots in an oxidiser apparatus. Metabolites (biliary and urinary) were separated by radio-HPLC.

Tsuda et al. [16] determined NP and OP in bioaccumulation experiments with freshwater fish. The analytes were extracted with acetonitrile and the lipid in the sample extract was eliminated by partitioning between hexane and acetonitrile. After Florisil PR column clean-up, the sample extract was analysed by GC-MS in the SIM mode.

Liber et al. [17] determined NP from bluegill sunfish exposed in littoral enclosures after blending and extracting with hexane, Florisil clean-up, eluting with 5% acetone in hexane and detection with GC-FID.

Blackburn et al. [18] collected a series of fish samples taken from marine and freshwater for food quality assurance purposes. The marine fish contained no detectable levels of APEO residues (0.05–0.1 μg g^{-1} NP). Samples were extracted with DCM in a Soxhlet extractor. Samples were concentrated and further purified on alumina columns using hexane to elute lipids and 1:1 hexane/acetone to collect the analytes. Samples of marine fish were further cleaned up with a second alumina column. APs were collected by combining the elutriates from a first elution with 10% acetic acid in hexane, and a subsequent hexane elution.

Invertebrates. In a sediment exposure experiment using invertebrates, for extraction of contaminants (including NP), invertebrates were added to microvials and extracted with methanol/iso-octane [19]. Tissues were pulverised and the iso-octane layer was removed. Tissues were re-extracted ethyl acetate. The ethyl acetate layer was removed and added to the iso-octane extract.

Wahlberg et al. [20] determined NPEO in Blue mussels by GC-ECD after derivatisation of the phenols with PFBCl. To this end, samples were homogenised in acetone/hexane (5:2), followed by extraction with hexane/diethylether (9:1). Lipids were removed by L/L extraction with acetonitrile in 0.1 M NaOH followed by L/L extraction in TMP/sulphuric acid. Recoveries of 93, 34, 65 and 100% were reported for NP, $NPEO_1$, $NPEO_2$ and $NPEO_3$, respectively.

Granmo et al. [21] used the method reported above for cod [7] for the extraction of ^{14}C-NP from Blue mussels and shrimp.

Anionics

In contrast with the variety of methods of extraction of LAS from other solid samples (such as sediment, suspended solid, sludge) and water samples (see corresponding section), there are relatively few reports of attempts to extract LAS from biological samples [22–28]. Direct HPLC measurement of accumulation of LAS in rat liver lysosomes has been reported [29].

In the early years of analysis data on the bioconcentration of LAS in biota were obtained using radiolabeled compounds and LSC, not distinguishing between parent compounds and metabolites [1,27,30]. These methods were non-specific and the data produced were unreliable due to overestimation. Therefore, it was important to develop accurate and specific analytical techniques in order to isolate and quantify these separately.

Kimerle [27] reviewed the ecotoxicology of LAS focusing on the results rather than on the method of analysis, for which the author referred to the review undertaken by Painter and Zabel [30], alluding only to two papers on biota sample preparation. Litz et al. [31] determined the concentration of LAS in rye grass by Azure A active substances (AzAAS). AzAAS is a non-specific colorimetric method, which has not been used as frequently as MBAS (see Chapter 3.1). Briefly, it consists of the formation of an ion association complex with a dyed solution of Azure A (cationic). The complex formed is solvent-extractable and is separated from unreacted dye prior to colour measurement.

Fish. Solid biota samples, mainly fish, should be quickly killed by liquid N_2 [28] or cervical dislocation [32] and kept at low temperatures ($-20°C$). Some authors preferred desiccation of the sample at high temperature (70°C) [33,34] or lyophylisation [28]. The extraction and isolation steps would be combined when using lyophylisation and homogenisation, followed by a Soxhlet extraction, usually with MeOH, and a subsequent solid-phase extraction (SPE) clean-up, prior to the quantification.

In 1986, Kikuchi et al. [22] reviewed the methods available for the analysis of LAS in marine samples and concluded these were not sufficiently specific or sensitive, and too time-consuming. They then developed a specific LAS extraction method, in which the fish tissue was homogenised and extracted three times with MeOH (this isolation step with MeOH has been demonstrated to improve selectivity [30,35]). This extract was then diluted in distilled water to a percentage below 10% of MeOH, passed through a C_{18} minicolumn and eluted with MeOH prior to its analysis by HPLC with FLD under an isocratic regime. The percentage recovery was determined using ^{14}C-LAS. The method required a series of four minicolumns to obtain recoveries from 72 to 98% in 50 g (w/w) of fish tissue spiked with 0.02 $\mu g\, g^{-1}$ (w/w). The detection limit of the method was 0.3 $\mu g\, g^{-1}$, even though LAS was not detected in environmental samples.

An off-line Soxhlet-SPE sequence was developed by Sáez et al. [28] for marine fish and clams and is described in the next section.

Invertebrates. Microbiota samples (microalgae or microorganisms), similarly to suspended solids, are usually dried, ground and extracted with MeOH prior to analysis. Caution should be taken in order to obtain a representative volume of sample and preconditioned material in order to minimise the losses due to sorption onto the walls of the containers and analytical equipment [35–37]. One inconvenience arises when it is not possible to distinguish between suspended solids and microorganisms. Regarding microalgae, the determination of the chlorophyll-a could give an estimate of the biomass.

Marin et al. [33] determined the LAS concentration in the faeces of the marine mussel *Mytilus galloprovincialis* Lmk. The sample was pipetted, dried at 70°C and treated in the same way as sediment samples, referring to Matthijs and De Henau [34], and measured by HPLC with spectroscopic detection.

In order to obtain a further purification and to determine LAS and SPC simultaneously from marine organism samples, Sáez et al. [28] developed a method consisting of a sequential Soxhlet extraction followed by an SPE step. Wet fish or soft tissue of clams were homogenised with an Ultra-turrax, lyophilised and ground to a powder, and sequentially extracted with Soxhlet extraction equipment using the following procedure: 9 h with Hexane to remove the lipids and fats that could interfere, and 6 h with MeOH. This was shown to be the best Soxhlet protocol for high recoveries, after testing different polar solvents. The methanolic extract was evaporated and redissolved in distilled water, acidified with HCl to a pH of 3. The SPE step consisted of a C_{18} minicolumn, eluted with MeOH, followed by a SAX strong anionic exchanger on which the analytes were eluted with 2 M HCl in MeOH. This method is based on the sequence described by González-Mazo et al. [38], in which the method recommended by Castles et al. [39] has been modified. The eluted material was dried by heating under a nitrogen flow and redissolved in the same mobile phase used in the HPLC analysis. Because the single C_{11}-homologues for LAS and SPC were used for optimisation of the method, the analyses were performed under isocratic conditions. The method showed a detection limit of 15 ng g^{-1} (wet basis) for undecylbenzene sulphonate and 30 ng g^{-1} for sulpho-phenylundecanoic acid; recoveries were higher than 80% for both the compounds. This method has been applied successfully to the extraction and isolation of C_{11}-LAS and C_{11}-SPC from several marine fish species and clam tissues. Here, the lyophylisation of the samples and the performance of a sequential extraction step allowed the employment of just one C_{18} minicolumn to obtain very good recoveries. Another advantage of this method is the possibility of analysing LAS and their degradation intermediates (SPCs) together, a very important issue when determining biodegradable compounds such as LAS.

3.3.3 Pressurised fluid, microwave and ultrasonic extraction

Non-ionics
Datta et al. [40] used pressurised fluid extraction to extract ground fish tissue, and the resulting extract was purified on aminopropyl silica (APS) extraction cartridges. With no further sample preparation, NP and its ethoxylates, up to nonylphenol pentaethoxylate, were quantified using normal phase (APS Hypersil) HPLC with fluorescence detection.

Pedersen and Lindholst [41] used microwave-assisted solvent extraction followed by SPE to determine octylphenol in liver and muscle of Rainbow trout.

Anionics

Ou et al. [42] used methanol-ultrasonic extraction followed by clean-up with aluminium oxide, and enrichment with a C-18 SPE column for the determination of LAS in plant tissues by HPLC. Both efficiency and accuracy of the overall method were high, with a mean recovery of 89% (84–93% for LAS concentrations ranging from 1 to 100 mg kg^{-1}) and a repeatability of 3% relative standard deviation for six replicate analyses. With a 2 g sample for analysis, LAS levels of 0.5 mg kg^{-1} in plants could be detected with the proposed method.

3.3.4 Matrix solid-phase dispersion

In matrix solid-phase dispersion (MSPD) the sample is mixed with a suitable powdered solid-phase until a homogeneous dry, free flowing powder is obtained with the sample dispersed over the entire material. A wide variety of solid-phase materials can be used, but for the non-ionic surfactants usually a reversed-phase C18 type of sorbent is applied. The mixture is subsequently (usually dry) packed into a glass column. Next, the analytes of interest are eluted with a suitable solvent or solvent mixture. The competition between reversed-phase hydrophobic chains in the dispersed solid-phase and the solvents results in separation of lipids from analytes. Separation of analytes and interfering substances can also be achieved if polarity differences are present. The MSPD technique has been proven to be successful for a variety of matrices and a wide range of compounds [43], thanks to its sequential extraction; matrices analysed include fish tissues [44,45] as well as other diverse materials [46,47].

Non-ionics

Tolls et al. [48] performed experiments with fathead minnows to determine the bioconcentration of alcohol ethoxylates in fish. Fish were homogenised in a mortar adding 1 mL of MeOH and 5 g of octadecylsilica. The dry dispersed powder was transferred to a syringe barrel and eluted. Four different fractions were obtained, using hexane, ethyl acetate, ethyl acetate/MeOH (1:1), and MeOH, respectively. The fractions were subjected to further clean-up with alumina. To that end two different methods were compared, and turned out to be equally effective.

In the first method [49], the extract is reconstituted in hexane/DCM and transferred to the column. The column is then eluted first with hexane/DCM (1:1), which is discarded, and then with DCM/MeOH (100:1). This method consumes relatively large solvent volumes (typically 100 mL). The second method [50] makes use of two prepacked alumina columns. The extract is reconstituted in hexane/DCM, and eluted with DCM/MeOH (95:5). The elutriate is reconstituted in cyclohexane/DCM, and transferred to the second prepacked column. The column is washed first with cyclohexane/DCM (1:1) and the analytes are next eluted with DCM/MeOH (95:5). This procedure uses typical volumes of 10–20 mL of solvent.

In an earlier experiment with radioactive labeled $C_{13}EO_8$, the same elution scheme had been used by Tolls and Sijm [51] to analyse tissue residues for radioactivity (see Section 3.3.3—anionics). It was then shown that less than 5% radioactivity remained in the MSPD column with the applied elution profile.

Zhao et al. [52] developed a modified MSPD method with sequential clean-up over alumina to isolate and purify APEO and AP in biological tissues. The dispersed material was packed in the barrel of a 25 or 50 mL glass syringe, on top of a layer of slightly deactivated alumina. An elution profile with methanol or acetonitrile and then DCM was applied for the separation of analytes from lipids. MeOH was found to yield better results than acetonitrile. The solvents were eluted using the stainless steel plunger of the syringe. The MeOH fraction was found to contain all AP and APEO. Alumina and silica were evaluated as sequential clean-up materials. Alumina was found to be more efficient in removing interferences. Recoveries from the sequential MSPD/alumina column amounted to >90% for OP, NP and NPEO. The technique enables the reliable determination of these compounds at the ppb level.

This method has been applied successfully to zebrafish [53], trout and zebra mussels [52], and blue mussels, bream and flounder [54].

Anionics
Tolls et al. [23,51] used MSPD with ion-pair liquid–liquid partitioning for the extraction and isolation of radiolabeled LAS from a freshwater

fish. For the simultaneous determination of LAS and SPC from the fish tissue, they combined MSPD off-line with graphitised carbon black (GCB) SPE. The sample was ground in a mortar with octadecylsilica powder to obtain a paste, which was then dried and powdered. The SPE column, a syringe packed with the powder (C_{18} and fish), was fractionally eluted with different solvents depending on whether only LAS on its own, or LAS and SPC together, had to be isolated. For LAS isolation, the extracts were eluted with hexane, EtOAc, EtOAc/MeOH (1:1, v/v). The latter fraction was evaporated, resuspended in MeOH and partitioned between the organic phase CH_2Cl_2/MeOH (3:1) and the aqueous phase (25 mM TBA-OH and 0.7 M NaOH). This was repeated and both the organic phases were dried, redissolved in MeOH and analysed by HPLC. For the simultaneous isolation of LAS and SPC together, five solvents (Hexane, EtOAc, EtOAc/MeOH (1:1), MeOH, and MeOH/H_2O (1:1)) were employed sequentially. The first two fractions were discarded, the others were dried and resuspended in MeOH and buffered with 0.1 M H_3PO_4. They were kept at 4°C overnight and centrifuged at 1000g in order to precipitate the proteins. The pellet and the third fraction (EtOAc/CH_3OH (1:1)) were treated with ion-pair L/L extraction in order to extract the portion of LAS that is associated with the pellet. The supernatants of the other fractions were individually diluted in distilled water and extracted with GCB SPE cartridges. LAS and SPC were eluted with CH_2CL_2/MeOH/NEt_3 (8:2:1), and neutralised with trifluoroacetic acid, evaporated and redissolved in MeOH. Before their analysis by HPLC, a solution of 5 mM of TBA-HSO_4, 10 mM PO_4^{3-} at pH 6, one of the components of the mobile phase mixture employed, was added. Recoveries of the method exceeded 70% and the limit of quantitation was approximately 0.2 mg kg^{-1}.

3.3.5 Conclusions and recommendations

Despite an increased interest in their environmental fate and significance during the last decade, relatively few reports have been published in the literature on analytical methods for the sample handling prior to determination of surfactants in biological media. The methodologies that have been published are confined to anionics (LAS) and non-ionics (AEO, AP and APEO). Classical methods including Soxhlet and L/L extraction are still often used, despite being laborious and time and solvent consuming. Steam distillation may limit

TABLE 3.3.1

Performance of sample handling techniques for the extraction of non-ionic and anionic surfactants from biological matrices

Compounds	Matrix	Extraction technique	Recovery (%)	Relative standard deviation (%)	Limit of detection (ng g^{-1})	Reference
OP, NP	Fish Zebra mussel	MSPD, alumina (sequential)	91 (OP) 108 (OP)	12 10	10	[52]
NPEO	Fish Zebra mussel	MSPD, MeOH, alumina (sequential)	100 101	5 1	30	[52]
NP, NPEO$_{1-3}$	Mussel	Acetone/hexane + hexane/diethylether, LLE	93 (NP), 34(NPEO$_1$), 65 (NPEO$_2$), 100 (NPEO$_3$)			[20]
C$_{13}$EO	Fish	MSPD, four solvents, alumina, DCM/MeOH	>70	10–25	80	[48]
C$_n$EO$_m$, n = 12–14, m = 8,11,14	Fish	MSPD, four solvents, alumina, DCM/MeOH	75–95	3–25	80 (C$_{13}$EO$_8$), 800 (C$_{14}$EO$_{14}$)	[48]
NP	Fish	Soxhlet DCM, alumina, hexane	107	10–20	100	[18]
NPEO$_{1,2}$	Fish	Soxhlet DCM, alumina, hexane		32	100	[18]
LAS	Fish	MeOH, C18-SPE	72–98		30	[22]
LAS	Plant stem, leaves	US-MeOH, alumina, C18-SPE	89	3	500	[42]
LAS (C$_{10-13}$), LAS + SPC (simultaneously)	Fish	MSPD, L/L extraction, MSPD + SPE (carbon)	>70, 84–88 (LAS), 54–65 (SPC)	3, 3 (LAS), 5 (SPC)	200, 600 (LAS), 2000 (SPC)	[23]
C$_{11}$LAS	Fish Clam	Soxhlet hexane, MeOH, C18 + SAX-SPE	87–104 80–88	1–7 2–4	15	[27]
C$_{11}$SPC	Clam	Soxhlet hexane, MeOH, C18 + SAX-SPE	81–100	2	30	[27]

coextraction of lipids as in classical extraction, thus reducing the time necessary to separate lipids from analytes. From the more recent methods in particular, the MSPD methodology appears promising, as isolation from the matrix and purification are accomplished within one sequential step, thus significantly limiting time and solvent use. Table 3.3.1 provides an overview of sample handling techniques and their performances.

From the available literature it becomes clear that method evaluation studies do not surpass the level of within-laboratory performances. Although several of these (see Table 3.3.1) reveal satisfactory levels of quality and environmentally relevant limits of detection, a genuine quality assurance of these methods is still lacking. There are no reports of interlaboratory studies and certified reference materials for surfactants are not available on the market yet. It can therefore be concluded that there remains much to be done in the field of improving and evaluating quality of analytical measurements of surfactants in biota.

REFERENCES

1   J. Tolls, P. Kloepper-Sams and D.T.H.M. Sijm, *Chemosphere*, 29 (1994) 693.
2   S.S. Talmage, *Environmental and Human Safety of Major Surfactants: Alcohol Ethoxylates and Alkylphenol Ethoxylates*, Soap and Detergent Association, Lewis, Boca Raton, FL, 1994.
3   B. Thiele, K. Guenther and M.J. Schwuger, *Chem. Rev.*, 97 (1997) 3247.
4   C.A. Staples, J. Weeks, J.F. Hall and C.G. Naylor, *Environ. Toxicol. Chem.*, 17 (1998) 2470.
5   B.J. Danzo, H.W. Shappell, A. Banerjee and D.L. Hachey, *Reprod. Toxicol.*, 16 (2002) 29.
6   D.R. Doerge, N.C. Twaddle, M.I. Churchwell, H.C. Chang, R.R. Newbold and K.B. Delclos, *Reprod. Toxicol.*, 16 (2002) 45.
7   A. Granmo and S. Kollberg, *Water Res.*, 10 (1976) 189.
8   A.M.R. Ferreira-Leach and E.M. Hill, *Analusis*, 28 (2000) 789.
9   A.M.R. Ferreira-Leach and E.M. Hill, *Mar. Environ. Res.*, 51 (2001) 75.
10 D.W. McLeese, V. Zitko, D.B. Sergeant, L. Burridge and C.D. Metcalfe, *Chemosphere*, 10 (1981) 723.
11 M. Ahel, J. McEvoy and W. Giger, *Environ. Pollut.*, 79 (1993) 243.
12 S.A. Snyder, T.L. Keith, C.G. Naylor, C.A. Staples and J.P. Giesy, *Environ. Toxicol. Chem.*, 20 (2001) 1870.

13   T.L. Keith, S.A. Snyder, C.G. Naylor, C.A. Staples, C. Summer, K. Kannan and J.P. Giesy, *Environ. Sci.Technol.*, 35 (2001) 10.

14   R. Thibaut, A. Jumel, L. Debrauwer, E. Rathahao, L. Lagadic and J.P. Cravedi, *Analusis*, 28 (2000) 793.

15   A. Arukwe, A. Goksoyr, R. Thibaut and J.P. Cravedi, *Mar. Environ. Res.*, 50 (2000) 141.

16   T. Tsuda, A. Takino, M. Kojima, H. Harada and K. Muraki, *J. Chromatogr. B*, 723 (1999) 273.

17   K. Liber, J.A. Gangl, T.D. Corry, L.J. Heinis and F.S. Stay, *Environ. Toxicol. Chem.*, 18 (1999) 394.

18   M.A. Blackburn, S.J. Kirby and M.J. Waldock, *Mar. Poll. Bull.*, 38 (1999) 109.

19   A.A. Fay, B.J. Brownawell, A.A. Elskus and A.E. McElroy, *Environ. Toxicol. Chem.*, 19 (2000) 1028.

20   C. Wahlberg, L. Renberg and U. Widequist, *Chemosphere*, 20 (1990) 179.

21   A. Granmo, R. Ekelund, K. Magnusson and M. Berggren, *Environ. Pollut.*, 59 (1989) 115.

22   M. Kikuchi, A. Tokai and T. Yoshida, *Water Res.*, 20 (1986) 643.

23   J. Tolls, M. Haller and D.T.H.M. Sijm, *Anal. Chem.*, 71 (1999) 5242.

24   Z. Yong-Yuan, T. Yu-Yun and F. Korte, In: A. Yediler (Ed.), *Cooperative Research Projects of the Institute of Hydrobiology of Academia Sinica, Wuhan, P.R. China and the Institute of Ecological Chemistry of the Gesellschaft für Strahlen-und Umweltforschung mbH*, Munich, Germany, 1981–1989.

25   J. Tolls, M.P. Lehmann and D.T.H.M. Sijm, *Environ. Toxicol. Chem.*, 19 (2000) 2394.

26   J. Tolls, M. Haller, W. Seinen and D.T.H.M. Sijm, *Environ. Sci. Technol.*, 34 (2000) 304.

27   R.A. Kimerle, *Tens. Surf. Det.*, 26 (1989) 169.

28   M. Sáez, V.M. León, A. Gómez-Parra and E. Gónzález-Mazo, *J. Chromatogr. A*, 889 (2000) 99.

29   M. Bragadin, G. Perin, S. Raccanelli and S. Manente, *Environ. Toxicol. Chem.*, 15 (1996) 1749.

30   H.A. Painter and T.F. Zabel, *Review of the Environmental Safety of LAS*, Water Research Center, Henley Road, Medmenham, UK, CO 1659-M/1/EV 8658, (1988).

31   N. Litz, H.W. Doering, M. Thiele and H.P. Blume, *Ecotoxicol. Environ. Saf.*, 14 (1987) 103.

32   J. Tolls, M. Haller, I. de Graaf, M.A.T.C. Thijssen and D.T.H.M. Sijm, *Environ. Sci. Technol.*, 31 (1997) 3426.

33   M.G. Marin, L. Pivotti, G. Campesan, M. Turchetto and L. Tallandini, *Water Res.*, 28 (1994) 85.

34   E. Matthijs and H. De Henau, *Tenside Deterg.*, 24 (1987) 193.

35 J. Waters, In: M.R. Porter (Ed.), *Recent Developments in the Analysis of Surfactants*. Critical Reports On Applied Chemistry, Vol. 32, 1991.

36 J. Waters and W. Kupfer, *Anal. Chim. Soc.*, 85 (1982) 241.

37 Q.W. Osburn, *J. Am. Oil Chem. Soc.*, 59 (1982) 453.

38 E. González-Mazo, J.M. Quiroga, D. Sales and A. Gómez-Parra, *Toxicol. Environ. Chem.*, 59 (1997) 77.

39 M.A. Castles, B.L. Moore and S.R. Ward, *Anal. Chem.*, 61 (1989) 2534.

40 S. Datta, J.E. Loyo-Rosales and C.P. Rice, *J. Agric. Food Chem.*, 50 (2002) 1350.

41 S.N. Pedersen and C. Lindholst, *J. Chromatogr. A*, 864 (1999) 17.

42 Z.Q. Ou, L.Q. Jia, H.Y. Jin, A. Yediler, T.H. Sun and A. Kettrup, *Chromatographia*, 44 (1997) 417.

43 S.A. Barker, *J. Chromatogr. A*, 885 (2000) 115.

44 H.M. Lott and S.A. Barker, *Environ. Monit. Assess.*, 28 (1993) 109.

45 E. Iosifidou, P. Shearan and M. O'Keeffe, *Analyst*, 119 (1994) 2227.

46 S.A. Barker, A.R. Long and C.R. Short, *J. Chromatogr. A*, 475 (1989) 353.

47 E. Viana, J.C. Moltó and G. Font, *J. Chromatogr. A*, 754 (1996) 437.

48 J. Tolls, M. Haller and D.T.H.M. Sijm, *J. Chromatogr. A*, 839 (1999) 109.

49 A.T. Kiewiet, J.M.D. van der Steen and J.R. Parsons, *Anal. Chem.*, 67 (1995) 4409.

50 G. Cassani, L. Cavalli, M. Lazzarin and G. Nucci, Analytica Conference 98, Munich, Germany, 1998, p. 253.

51 J. Tolls and D.T.H.M. Sijm, *Environ. Toxicol. Chem.*, 18 (1999) 2689.

52 M. Zhao, F. van der Wielen and P. de Voogt, *J. Chromatogr. A*, 837 (1999) 129.

53 J. Legler, L.M. Zeinstra, F. Schuitemaker, P.H. Lanser, J. Bogerd, A. Brouwer, A.D. Vethaak, P. de Voogt, A.J. Murk and B. van den Burg, *Environ. Sci. Technol.*, 36 (2002) 4410.

54 A.D. Vethaak, G.B.J. Rijs, S.M. Schrap, H. Ruiter, A. Gerritsen and J. Lahr, *Estrogens and xeno-estrogens in the aquatic environment of The Netherlands*, Ministry of Transport, Public Works and Water Management, The Hague, Netherlands, 2002, pp. 1–292.

Chapter 4

Quantification and quality assurance in surfactant analysis

4.1 INTRODUCTION

Pim de Voogt and Thomas P. Knepper

The quantitative environmental analysis of surfactants, such as alcohol ethoxylates, alkylphenol ethoxylates (APEOs) and linear alkylbenzene sulfonates (LASs), is complicated by the presence of a multitude of isomers and oligomers in the source mixtures (see Chapter 2). This issue bears many similarities to the quantitation problems that have occurred with halogenated aromatic compound mixtures, e.g. poly-chlorinated biphenyls (PCBs) [1].

In the case of PCBs, the solutions to the quantitation problem have been:

- to synthesise all possible single congeners instead of quantifying against commercial mixtures;
- to synthesise isotope-labelled reference compounds;
- to require the use of high resolution capillary gas chromatography (GC) columns for unequivocal identification;
- to organise many interlaboratory exercises; and
- to produce certified or standard reference materials.

Comprehensive Analytical Chemistry XL
T.P. Knepper, D. Barceló and P. de Voogt (Eds)

Compared with the tremendous achievements in improvement of the quality of quantitative PCB analysis, surprisingly little attention has been paid in the scientific literature to the quality assurance of quantitative analysis of surfactants. Only a few 'single' standards are available for the quantitation of surfactants such as APEOs and LAS, and even fewer for the qualitative and quantitative analysis of their (bio)degradation products, which is well documented in Chapter 4.2. For LAS and their primary biodegradation products, the sulfophenyl carboxylates (SPCs) only the technical blends of LAS are commercially available, which differ in their homologue as well as phenylisomer composition. Some pure homologues can only be obtained from some industrial producers. In the case of SPC, there are no commercial products available at all. With respect to nonylphenol ethoxylates (NPEOs) only mixtures consisting, for a given oligomer (e.g. $NPEO_1$, $NPEO_2$), of several possible isomers of the alkyl chain, and some of the polar degradation products, such as nonylphenoxy monocarboxylate (NPE_1C), can be obtained.

Most of the quantitative analysis of surfactants is, therefore, still based on calibration with commercially available technical mixtures. Chapter 4.3 discusses the advances and limitations of this state-of-the-art technique for quantitation in liquid chromatographic mass spectrometry.

Stability of calibrants and analytes is another frequently over-looked aspect of quality assurance, which is particularly relevant to surfactants. This aspect is discussed in Chapter 4.4. Very few intercalibration studies have been performed for the surfactant types of analytes (cf. Chapter 4.5). Currently, no certified reference material is available for surfactants. The European Commission has recently tendered for production of a reference material with certified surfactant concentrations [2]. We can conclude that quality assurance in quantitative surfactant analysis is still in its infancy when compared to analysis of PCB or chlorinated dioxins. Notwithstanding this, several important achievements have been made during recent years regarding improvement of the accuracy and reliability of qualitative analysis of surfactants, which will be the subject of the following chapters.

REFERENCES

1   P. de Voogt, D.E. Wells, L. Reutergårdh and U.A.Th. Brinkman, *Int. J. Environ. Anal. Chem.*, 40 (1990) 1.
2   European Commission, Dedicated call GROW/DC5MTI, 10/01 Topic V.27 Organic Components in Sludge, Brussels, 16 October 2001.

4.2 REFERENCE COMPOUNDS IN QUANTIFICATION OF SURFACTANTS, THEIR METABOLITES AND REACTION BY-PRODUCTS

Francesc Ventura and Thomas P. Knepper

4.2.1 Introduction

As mentioned in Chapters 1, 5 and 6, surfactants are degraded during wastewater treatment and in the soil leading to a series of extremely complex metabolites. Furthermore during water treatments, such as chlorination, these surfactants can undergo additional chemical modifications. To obtain a better knowledge of their environmental and health effects, accurate determination of recalcitrant metabolites of different types of surfactants in various aquatic compartments is needed, and this requires the use of reference standards. Owing to their complex chemical nature, such standards are, in general, commercially unavailable. Consequently, there is a need for the synthesis of reference compounds. In this chapter the availability and synthesis of reference compounds for surfactant analysis is reviewed.

4.2.2 Non-ionic surfactants

The quantification of short-chain metabolites from nonylphenol ethoxylates (NPEOs) has been carried out in different ways because of the lack of individual standards. Thus, several authors have employed commercial mixtures with low ethoxylation degree and assuming similar molar response factors, such as Marlophen 83 with an average of 3.5 mol of ethylene oxide (EO) [1–5], Imbentin-N/7A, a mixture of $NPEO_1$ and $NPEO_2$ (75:25) [6–11] or a mixture of miscellaneous origin [12,13], to quantify $NPEO_1$ and $NPEO_2$.

In contrast, other authors have preferred to synthesise the individual metabolites for a more confident determination (see Fig. 4.2.1). Table 4.2.1 shows the methods described in the literature to synthesise the non-commercial metabolites of NPEO. Thus, $NPEO_1$ and $NPEO_2$ can be synthesised, according to Shang et al. [14], by reacting nonylphenol (NP) with 2-chloroethanol and 2-(2-chloroethoxy) ethanol in the presence of potassium hydroxide and dimethyl sulfoxide. These compounds

Comprehensive Analytical Chemistry XL
T.P. Knepper, D. Barceló and P. de Voogt (Eds)

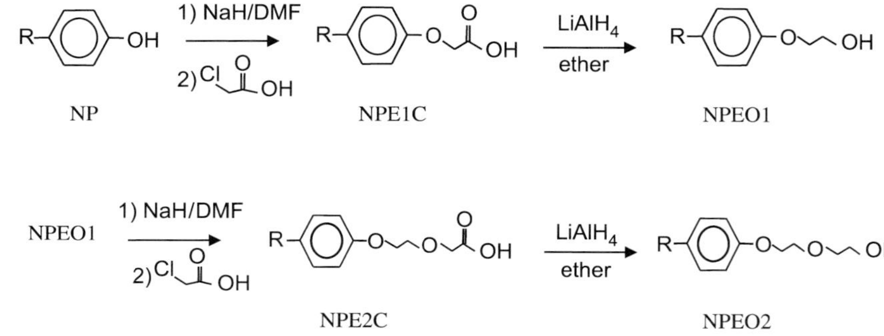

Fig. 4.2.1. Scheme for the synthesis of NPEO and NPEC.

were isolated by using a liquid chromatographic (LC) separation method reported by Wahlberg et al. [15]. Ferguson et al. [16] synthesised the mono-, di- and triethoxylated 4-n-NP to be used as internal standards from 4-n-NP and 2-chloroethanol according to the procedure described by Mansfield and Locke [17], whereas they employed commercially available stably isotope-labelled $^{13}C_6$ n-NP, mono-, di- and triethoxylates as surrogate standards. An alternative approach to synthesise mono- and diethoxylated NP has been carried out by Díaz and Ventura [18] by reduction of the corresponding carboxylates with lithium aluminium hydride.

Ejlertsson et al. [13] used labelled ^{14}C-NPEO$_{1-2}$ to monitor the anaerobic degradation of both compounds chosen as model compounds in digestor sludges, landfilled municipal solid wastes and landfilled sludges. The ^{14}C-NP previously synthesised according to the method of Ekelund et al. [19] was ethoxylated by adding sodium methoxide in methanol, evaporating the methanol and adding a solution of EO in toluene, which rendered the target compounds after heating to 140°C for 18 h.

For the quantification of alkylphenol ethoxylates (APEOs), several related compounds have been used as internal standard, e.g. for recovery determination, decylphenol monoethoxylate [20]. Octylphenol nonaethoxylate (t-OPEO9) was synthesised by reacting octylphenol (OP) with 1,2-bis(2-chloroethoxy)ethane to give the chloro derivative, followed by reaction with the sodium salt of hexaethylene glycol [21,22]. Another approach used the synthetic standard 1-(4′-methoxyphenyl)-hexan-1-ol as a surrogate to monitor the efficiency of the extraction for

TABLE 4.2.1

Review of methods for the synthesis of several alkylphenol metabolites

Raw material	Synthesis	Metabolite	Reference
OP	 $\xrightarrow[\substack{80\text{-}90^\circ C \\ Na\ met}]{atm\ N_2}$ OPNa $\xrightarrow{\triangle}$	OPEO$_1$	[17]
NP	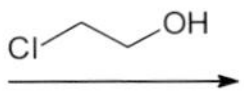 Cl~~OH $\xrightarrow{KOH/DMSO}$	NPEO$_1$	[14]
NP	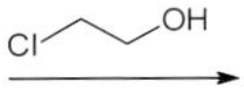 Cl~~OH $\xrightarrow{NaOH/H_2O}$		[16]
NPE$_1$C	$\xrightarrow[ether,\ rt]{LiALH_4}$		[18]
NPE$_2$C	$\xrightarrow[ether,\ rt]{LiALH_4}$	NPEO$_2$	[18]
NP	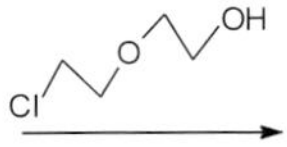 Cl~O~OH $\xrightarrow{KOH/DMSO}$		[14]
NP	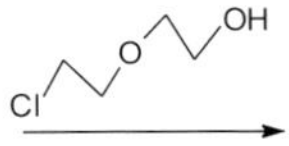 Cl~O~OH $\xrightarrow{NaOH/H_2O}$		[16]
NP	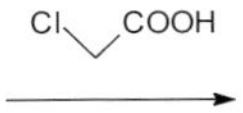 Cl~COOH $\xrightarrow{70^\circ C\ NaOH/H_2O}$	NPE$_1$C	[27]
NPEO$_1$	$\xrightarrow[ox.\ react\ Jones]{Cr_2O_7^{2+}/H_2SO_4}$		[3,6,9,12,24–27,94]
NP	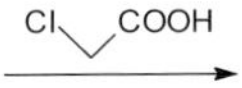 Cl~COOH $\xrightarrow{NaH/DMF}$		[18]
NPEO$_1$	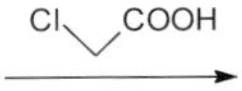 Cl~COOH $\xrightarrow{60^\circ C\ NaH/DMF}$	NPE$_2$C	[18]
NP	$\xrightarrow[CHCl_3]{SO_2Cl_2}$	ClNP	[16]

(continued on next page)

449

TABLE 4.2.1 (continued)

Raw material	Synthesis	Metabolite	Reference
NP	Br_2 — H_2O or AcOH	BrNP	[26,32]
NP	Br_2 — 1,2-diCl ethane		[16]
OPE_1C	Br_2 / H_2O — rt, 10 min	$BrOPE_1C$	[26,39]
$OPEO_n$ ($n = 0-3$)	1) $Br_2 / AcOH$ — 2) $Cr_2O_7^{2+} / H_2SO_4$	$BrOPE_nC$ ($n = 0-3$)	[32]
BrNP	Cl—COOH — NaH / DMF 8 h. 60°C	$BrNPE_1C$	[18]
OPE_1C	Br_2 — H_2O or AcOH rt	$BrOPE_1C$	[26,39]

the determination of APEO in effluents by liquid chromatography with fluorescence detection [23].

OPE_1C and also NPE_1C and NPE_2C have all been synthesised from OP and a commercial mixture of $NPEO_n$ ($n = 1, 2$), by respectively, oxidation with Jones reagent and further purification [3,6,9,12,24–27]. Another approach involves the reaction of NP with chloroacetic acid in alkaline aqueous solution or in a sodium hydride–dimethyl sulfoxide solution to obtain NPE_1C [18,28]. Accordingly, NPE_2C can also be obtained individually by reacting $NPEO_1$ with chloroacetic acid [18]. Other authors have employed the same method as described before, but using commercial mixtures of low degree of ethoxylation to obtain the corresponding mixture of NPE_nC [6,29].

The situation is less promising with respect to dicarboxylated metabolites (CAPEC). Thus, reports on the presence of CAPECs in environmental samples have to deal with semiquantitative results because of the lack of any standard (see Chapter 5.1). For example Ding et al. [30] concentrated water samples from a tertiary effluent by rotary evaporation to dryness and, after derivatisation to propylesters, CAPECs were analysed by gas chromatography mass spectrometry

(GC-MS). The authors assumed that the response factor of the internal standard (anthracene-d_{12}) and the metabolites found in the extracts were the same. Ding and Chen [31] analysed carboxylates and dicarboxylated NP residues in polluted rivers receiving effluents from wastewater treatment plants (WWTPs) by concentration through a graphitised carbon black (GCB) cartridge and direct derivatisation in the GC injector-port using a large volume direct sample introduction device with tetraalkyl ammonium salts. Quantification was performed by assuming that the response factors for the $OPEC_1$ and CAPECs were equal. DiCorcia et al. [7] carried out degradation experiments on a laboratory scale. Compounds were concentrated by solid-phase extraction (SPE) using Carbograph-4 and further analysed by liquid chromatography electrospray (LC-ESI)-MS using a C_{18} reversed-phase column. Identification of CAPECs was achieved by comparison of the collision-induced dissociation (CID) spectra from both underivatised metabolites and their methyl esters. Quantification was performed by assuming that the molar response of the $APEO_1 + APEO_2$ mixture did not differ significantly from those of species bearing a carboxylic group in the alkyl chain and/or the ethoxylate chain. In a further survey of WWTP also carried out by DiCorcia et al. [24] using the same analytical method described before [7], molar response factors of the different CAPEC homologues were calculated by analysing a known aliquot of a biodegradation solution performed under the same conditions used to analyse extracts of the WWTP effluents. These relative response factors were then used to quantify the dicarboxylated species.

Halogenated NP derivatives such as $BrNPEO_n$ or $BrNPEC_n$ can be formed during chlorine disinfection in the presence of elevated bromide concentrations and have been identified in chlorinated wastewater effluents [32], reclaimed water [33], tap water [34], sediments [16] and sludges [35]. Several methods related to the synthesis of these compounds are described in the literature. Thus, ClNP was prepared by Ferguson et al. [16] by chlorination of NP using sulfuryl chloride according to the method developed by Stokker et al. [36] or like Stephanou et al. [26], who synthesised ClOP by adding an aqueous chlorine solution to octylphenol. BrNP has been synthesised by using elemental bromine [26,37,38]. Brominated acidic metabolites of OP were prepared by bromination of OP and further oxidation of the synthesised BrOP to the corresponding carboxylate by oxidation with Jones reagent [32] or by reacting the brominated AP with chloroacetic acid in the presence of sodium hydride and dimethylformamide [18].

Alternatively, bromination of the corresponding octylphenol carboxylate has also been described as an efficient way to synthesise brominated alkylphenoxy carboxylates [26,39]. Because of the lack of standards the presence of brominated carboxyalkylphenoxy ethoxy carboxylates (BrCAPEC) in a tertiary-treated wastewater effluent has been reported on a semiquantitative basis [30].

Alcohol ethoxylates (AEs) of general formula $C_nH_{2n+1}-(OCH_2CH_2)_m-OH$ ($n = 12-18$, $m = 1-23$) are rapidly biodegraded in water. The proposed mechanisms [40,41] for their aerobic biodegradation include: (a) the central cleavage of the molecule, leading to the formation of polyethylene glycols (PEGs) and aliphatic alcohols; (b) the ω,β-oxidation of the terminal carbon of the alkyl chain; and (c) the hydrolytic shortening of the terminal carbon of the polyethoxylated chain (see Chapter 5.1).

Marcomini et al. [42] studied the aerobic biodegradation mechanisms of different commercial AE blends (linear, oxo- and multibranched alkyl chains) by analysing the metabolites under the same standardised conditions applying liquid chromatography mass spectrometry LC-MS.

DiCorcia et al. [43] have fully characterised the biodegradation intermediates of branched alcohol ethoxylates (BAEs), which have a single branching in position two with respect to the ether bridge and account for ca. 25% of the commercial AEs in domestic detergents [44]. The model compound, 2-butyl-branched $A_{12}5E$, was submitted to biodegradation and samples from this bioassay solution were enriched by SPE (Carbograph 4) and further analysed by LC-ESI-MS using CID spectra to elucidate the metabolite structures (see Chapter 5.1).

Quantitation of AEs and their metabolites in the environment has been carried out in different ways. Thus, commercial AE blends (i.e. Dobanol 25-9, Neodol 25-9, Dehydrol LT7) have been used to quantitatively determine their presence in water by supercritical fluid chromatography [45]; by thermospray LC-MS [46] using a linear $C_{11}9EO$ as internal standard; by LC-ESI-MS of underivatised AEs [47,48], the last-named using a fully deuterated alcohol $[C_{13}D_{27}O(CH_2CH_2O)_nH]$ as internal standard, or derivatised with 2-fluoro-*N*-methylpyridinium *p*-toluenesulfonate [49]; by LC-Atmospheric Pressure Chemical Ionization (APCI)-MS [50,51]; or capillary electrophoresis [52]. Other approaches involve the reaction of AEs with hydrogen bromide to form the corresponding alkylbromide derivatives followed by GC-MS [45,53] or to derivatise, among others, with phenyl isocyanate followed by HPLC with UV detection [54,55], with 1-naphthoyl chloride [56,57] or with 3,5-dinitrobenzoyl chloride followed by HPLC [58].

Several pure homologues of AEs have been synthesised by reacting 1-bromo-[1-^{14}C]tridecane with tri-, hexa-, or nonaethylene glycol [59] to study their sorption on sediments.

AEs labelled with ^{14}C in the ethoxylate chain or separately at the α-carbon in the alkyl chain were used for biodegradation studies [60]. The two pure chain length AEs synthesised were $C_{12}EO9$ and $C_{16}EO3$ to cover the low to middle range of ethoxylate chain lengths and the middle to high range of alkyl chain lengths. The $C_{16}EO3$ was prepared by condensing hexadecanol with [^{14}C]ethylene oxide in the presence of sodium metal and using diglyme as solvent. The alkyl-labelled [^{14}C]$C_{16}EO3$ was prepared by reacting C_1-labelled [^{14}C]hexadecanol with a previously synthesised E_3 ethoxylate chain. The two $C_{12}EO9$-labelled compounds were synthesised in a manner similar to that described before by using labelled and unlabelled dodecanol, an E_9 ethoxylate chain and [^{14}C]ethylene oxide. Steber and Wierich [40,61] have used two commercially available labelled AEs (^{14}C$_{18}$EO7 and C_{18}^{14}EO7) and a synthesised carboxylated [^{14}C]PEG to study the biodegradation pathways of AEs in microbial biocenoses of sewage treatment plants. The carboxylated [^{14}C]PEG 400 (ca. 10% MCPEG and 90% DCPEG) was prepared by adding [^{14}C]PEG 400 to a chromic/sulfuric solution, followed by extraction of the acids from a saturated magnesium sulfate solution into chloroform. Custom-synthesised labelled compounds (^{14}C$_{13}$EO3 and ^{14}C$_{13}$EO9) have been used to optimise extraction conditions of AEs in natural waters [55] and (^{14}C$_{13}$EO8 and ^{14}C$_{13}$EO4) to monitor the presence of AEs in fish [57].

Carboxylated AEs (AECs) can be prepared by oxidation of the corresponding AE with Jones reagent followed by extraction from a saturated magnesium sulfate solution with dichloromethane [43,62]. Very few papers deal with the determination of AECs in environmental samples and their presence has been qualitatively identified in raw and treated water [63], whereas their quantification by using commercial blends (i.e. Lialet 123/6.5-COOH) was assessed to study the aerobic biodegradation of AEs [42,64]. The compounds were SPE-enriched on GCB and analysed as 9-chloromethylanthracene derivatives by HPLC with fluorescence (FL) detection. AECs identified as biodegradation intermediates of branched AEs [43] were quantified by assuming that there were no differences in response for different configurations of the alkyl chain, using molar responses of different linear C_{12}ECs relative to the internal standard ($C_{10}EO6$). Alkyl chain carboxylated AEs (CAEs) and dicarboxylated AEs (CAECs) also identified in the same experiment were quantified by assuming that in compounds bearing, in addition to an E-chain, one or more carboxylic groups, the molar absorbance (UV at 210 nm) is dependent upon the presence of the latter group. Thus, the concentrations of the different CAEs and CAECs were estimated by assigning to each class of acids the respective molar absorbances obtained from model mono-, di- and tricarboxylated compounds (i.e. hexanoic, adipic and 1,3,6-cyclohexane tricarboxylic acids). Therefore, molar responses of the different compounds of interest relative to the internal standard with the ESI-MS detector were calculated. Another approach followed for the determination of CAEs in the absence of adequate standards, is the assumption that their response factor is double the value determined for AECs [64].

Studies on PEG and carboxylated metabolites (MCPEG and DCPEG) levels in the environment are relatively scarce because of the lack of simple and efficient analytical methods to detect them at trace levels. Marcomini and Pojana [65] have determined MCPEGs and DCPEGs in a biodegradation bench test and sewage by SPE enrichment on GCB followed by LC-FL detection of 9-chloromethylanthracene derivatives. Crescenzi et al. [66] extracted sewage effluents, river water, seawater and groundwater by using a GCB (Carbograph-4) cartridge to analyse carboxylated PEGs, which were isolated from neutral ones by differential elution with a mixture of dichloromethane/methanol (80:20, v/v) acidified with HCl and made it possible to convert into their methyl esters. Both extracts were analysed by LC-ESI-MS in the positive mode. Quantification of neutral and acidic PEGs in environmental samples was performed by an external standard quantification procedure using PEG 400 and dicarboxylated PEG 600 calibration curves. For neutral PEGs, the molar responses of each pure PEG, which increased steadily from $n_{EO} = 4$ to $n_{EO} = 8$, were taken into account. No significant changes were observed for $n_{EO} > 8$, giving the same response factor of PEG 8 for PEGs > 8. With respect to the dicarboxylated species, it was found that their molar response factor was not influenced by the ethoxy chain length whereas, due to the lack of individual monocarboxylated PEGs, their quantification was achieved by assigning the same response factor as that used for dicarboxylated species. The method is suitable for species having ethoxy chain lengths ≥ 4 with recoveries higher than 83% and limits of quantification of 0.1–0.3 ng/l if the selected ion-monitoring mode is used [66]. A commercially available dicarboxylated

PEG oligomer (3,6,9-trioxaundecanedioic acid) was used to quantify carboxylated PEGs generated under standardised aerobic biodegradation conditions [42]. DCPEGs were also identified and quantified by assuming the same response factor as that for PEG in influents and effluents of Spanish WWTPs with secondary biological treatment [67].

Alkylglucamides (AGs) are efficiently removed by biodegradation in wastewater treatment plants, but data on influent and effluent concentrations for these compounds is scarce [68]. Eichhorn and Knepper [69] developed an SPE method using Lichrolut-EN cartridges followed by LC-ESI-MS to analyse AGs in wastewaters and to study the biodegradability and metabolic pathway of a C_{10}-glucamide (see Chapter 5.1). The postulated mechanism (see Fig. 5.1.18), includes an ω-oxidation of the alkyl chain followed by subsequent β-oxidations and was confirmed by LC-ESI-MS. For the proper identification and quantification of a degradation intermediate, namely C_4-glucamide acid (see Fig. 4.2.2), it was indispensable to be able to synthesise this compound as it was not available from chemical suppliers. The organic synthesis was performed by Díaz (unpublished results) reacting N-methyl glucamine in pyridine with succinic anhydride at room temperature for 5 h followed by recrystallisation with ethanol [18] according to the scheme given in Fig. 4.2.2.

A series of attempts to discover optimum reaction conditions was made in view of preventing formation of by-products arising from the reaction of the acyl halogenide with different functional groups within the N-methyl glucamine molecule. Nonetheless, a subsequent purification step was necessary to remove by-products and unreacted starting material. This turned out to be complicated due to similar solubilities of both the

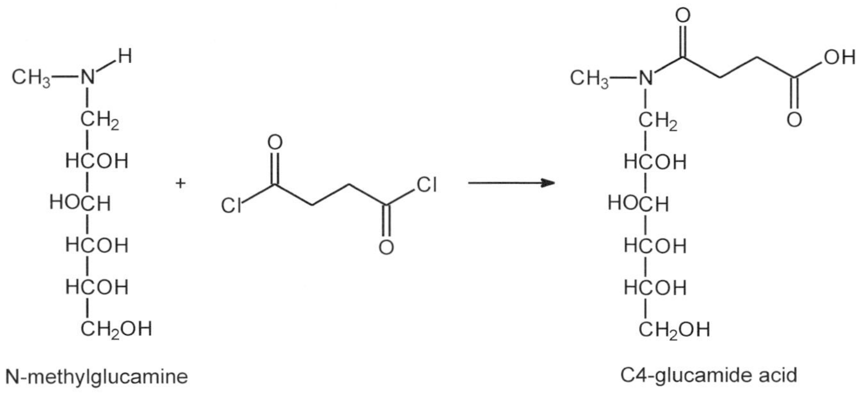

N-methylglucamine   C4-glucamide acid

Fig. 4.2.2. Synthetic route for C_4-glucamide acid.

educts and the product as well as of reaction side-products. In order to pursue the progress and the quality of the synthetic work, ^{1}H- and ^{13}C-NMR measurements were completed by Díaz and Ventura [18]. Additionally, LC-ESI-MS was applied within this work to check for the purity of the obtained products and to confirm the identity of the components.

As shown in Chapter 2.4.3.2, it was feasible to separate C_4-glucamide acid from C_{10}-AG by Reversed phase (RP)-HPLC owing to very different polarities of both compounds. Regarding the chromatography of a mixture containing the very polar species involved in the synthesis of C_4-glucamide acid, the use of RP-LC did not offer a reasonable separation. C_4-glucamide acid, succinic acid and a by-product, identified as N-methyl glucamide whose secondary amine group and terminal hydroxy group of the sugar moiety were esterified, were nearly co-eluting under the given conditions yielding rather broad peaks (Fig. 4.2.3(A)). Instead, a very good separation was achieved on an anion exchange (AX) column benefiting from the fact that the three compounds are carboxylic acids, i.e. they interact distinctly with the cationic exchanger material. The column used (Metrohm Dual 2), developed for specific ion chromatography of small inorganic anions, allowed a baseline separation. C_4-glucamide acid carrying a single carboxylate group was eluted first, whereas theoduct and the succinic acid, both having two negatively charged groups, were significantly strongly retained (Fig. 4.2.3(B)).

The operation of the AX column required aqueous alkaline eluents containing $Na_2CO_3/NaHCO_3$. Such an eluent was utterly incompatible for a direct coupling of the LC column with the ESI-MS interface since the inorganic salts crystallised instantly upon entering the ion source. This problem was overcome by placing a cation exchanger before the interface.

Commercially, *trisiloxane ethoxylates*, such as those used as adjuvants in agrochemical applications (see Chapter 5.5), are manufactured by the hydrosilylation of 1,1,1,3,5,5,5-heptamethyltrisiloxane (MD^HM) with an allyl-capped polyalkyleneoxide, in the presence of a platinum catalyst, according to the procedure described by Bailey and Snyder [70]. Two methods have also been reported for the synthesis of pure oligomers of this surfactant type, both utilising monodisperse oligoethoxylate monomethyl ether starting materials [71,72]. The procedure shown in Fig. 4.2.4 was used to yield the $n = 3$, 6 and 9 oligomers, with the $n = 6$ and 9 oligoethylene glycols produced by

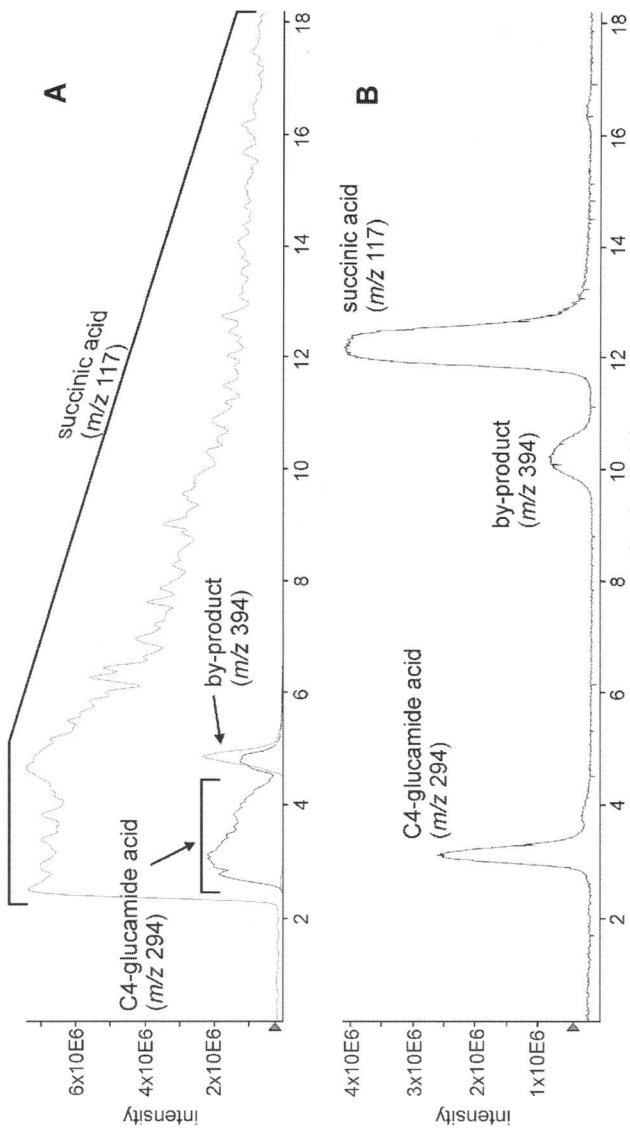

Fig. 4.2.3. LC-ESI-MS chromatograms—of a reaction mixture of C_4-glucamide acid on a C_8-RP column (A) and anion exchanger column (B). LC conditions: (A) 125×2.1 mm, 4 µm, C_8-bonded silica with multistep gradient profile, water–CH_3CN, pH 7.9 at 0.2 mL min^{-1}; (B) 150×3.1 mm, anion exchange column with a gradient profile, $Na_2CO_3/NaHCO_3$ (3 min isocratic $1.3/2.0$ mM, within 12 min to $13/20$ mM) at 0.5 mL min^{-1}, split 1:1.

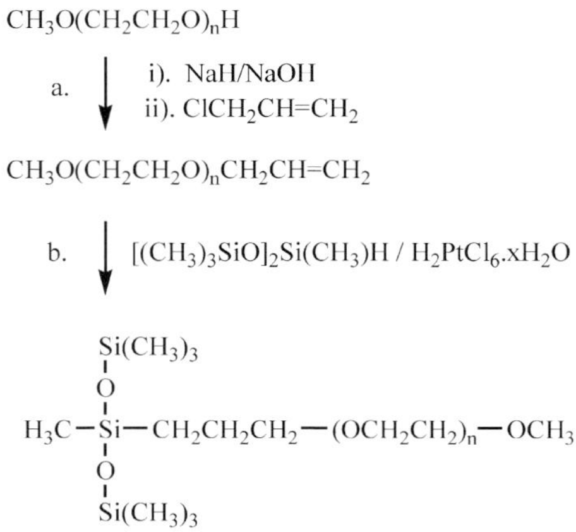

$$CH_3O(CH_2CH_2O)_nH$$

a. $\quad$ i). NaH/NaOH
$\quad\quad$ ii). ClCH$_2$CH=CH$_2$

$$CH_3O(CH_2CH_2O)_nCH_2CH=CH_2$$

b. $\quad$ [(CH$_3$)$_3$SiO]$_2$Si(CH$_3$)H / H$_2$PtCl$_6$.xH$_2$O

$$\begin{array}{c} Si(CH_3)_3 \\ | \\ O \\ | \\ H_3C-Si-CH_2CH_2CH_2-(OCH_2CH_2)_n-OCH_3 \\ | \\ O \\ | \\ Si(CH_3)_3 \end{array}$$

Fig. 4.2.4. Reaction sequence used to yield monodisperse trisiloxane alkylethoxylate surfactants ($n = 3$, 6 and 9).

etherifications and chlorinations of smaller oligomers [71]. The resulting trisiloxane alkylethoxylate products were purified by chromatographic methods and characterised by GC-MS and API-MS. RP C$_{18}$ HPLC-APCI-MS gave purities of greater than 95, 96 and 90% for the $n = 3$, 6 and 9 oligomers, respectively, whilst GC-MS indicated 100% purities for all with only the peak corresponding to the desired oligomer quantitatively observed in the chromatograms. The use of these monodisperse oligomers to analyse the degradation processes of the commercially used surfactant product is described in later sections.

Wagner et al. used the reaction procedure shown in Fig. 4.2.5, and variations thereof, to yield the monodisperse oligoethoxylate monomethyl ether oligomers ($n = 3$–9). The trisiloxanes were then produced by hydrosilylation as in step b of Fig. 4.2.4 [72]. Distillation procedures were used to purify the intermediates and resulting trisiloxane alkylethoxylate products, and structural characterisation was performed by GC-MS and NMR. Purities for the $n = 3$–9 oligomers of $\gg 99$, $\gg 99$, 99, 97.5, 96, 95 and 90%, respectively, as determined by GC-MS, were reported.

4.2.3 Anionic surfactants

Commercial linear alkylbenzene sulfonates (LAS) contain mixtures of several alkyl chain homologues each with a variable number of

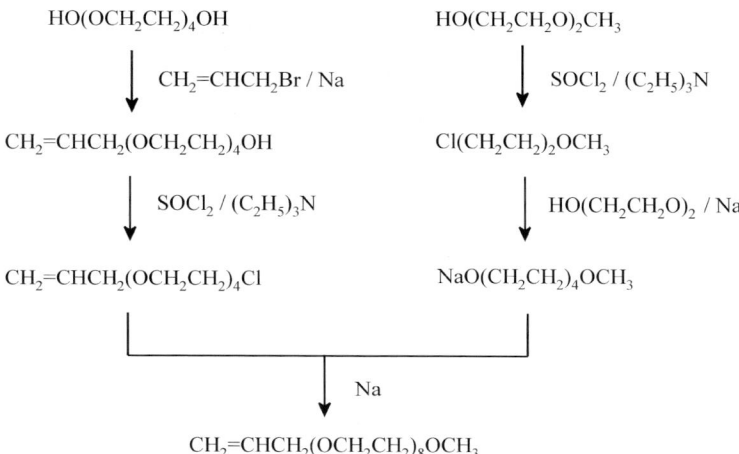

Fig. 4.2.5. Reaction sequence adopted to yield the (EO)$_8$ ethoxylate monomethyl ether.

phenyl-positional isomers ranging from C$_{10}$ to C$_{14}$. Most of the LAS isomers are chiral and they exist as racemic mixtures in commercial products. The biodegradation pathway for LAS under aerobic conditions yields mostly chiral sulfophenyl carboxylates (SPCs) (see Chapter 5.1). The existence of sulfophenyl dicarboxylates (SPDCs), resulting from attack on both methyl groups on the alkyl chain, has been observed [73,74]. Recently, it has also been reported that degradation of LAS studied on a laboratory-scale bioreactor yielded additionally α,β-unsaturated SPC intermediates, which would be an indicator of β-oxidation [75].

Commercial LAS also contain coproducts such as dialkyltetralin sulfonates (DATS), which are formed during alkylation of benzene with olefins and subsequent sulfonation (70 isomers), and iso-LAS, which have a single methyl branching attached either to the non-benzylic (type I) or benzylic carbon (type II) of the alkyl chain (see Fig. 4.2.6). Concentrations of DATS ranged from <1–10% depending on the process [76,77] but at present the most used LAS in Europe has relative DATS concentrations of <0.5% [78], while relative concentrations of iso-LAS account for 3–6% of the LAS formulations [79]. A laboratory aerobic biodegradation experiment of LAS and coproducts has proved that analogues of LAS, both DATS and iso-LAS, undergo primary biodegradation leading to the formation of carboxylic intermediates and, to a lesser extent, of dicarboxylated species. A significant difference

459

460

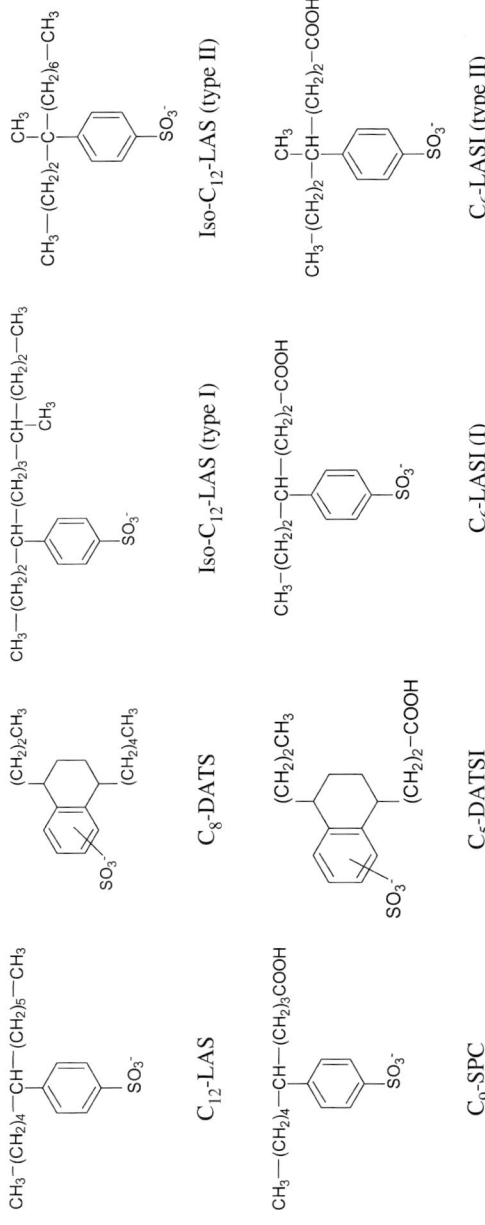

Fig. 4.2.6. Top sequence. Examples of structures of LAS and coproducts (DATS and *iso*-LAS). Bottom sequence. Examples of metabolites derived from LAS and coproducts.

is that the intermediates from coproducts are more resistant to ultimate biodegradation probably derived from the two-ring structure of DATS and the methyl branching close to or on the benzylic carbon [74]. Figure 4.2.6 shows examples of structures and acronyms of LAS, their coproducts and metabolites.

SPE employing either GCB [6,25,73–75,80,81], C_8 [77], C_{18} [82–87] or a combination of C_{18}+ SAX [88–91] cartridges is the most used method to determine SPC in environmental samples accompanied by detection with HPLC-FL [6,25,90], LC-MS [73–75,82,83,86], GC-MS with derivatisation [77,80,92] or capillary electrophoresis [81]. For biota samples, matrix SPE [85] or Soxhlet [91] followed by clean-up with SPE have been used to determine SPC. SPE employing GCB cartridges and further analysis by LC-MS has been used for the determination of SPDCs [73,74].

SPCs are not commercially available, but different methods of synthesising them have been carried out. Thus, SPCs can be obtained by sulfonation and subsequent neutralisation of the corresponding commercially available phenylcarboxylic acid. The conditions reported by Taylor and Nickless [93] to synthesise a series ranging from C_2 to C_6 and C_8 SPCs have been generally followed. Briefly, concentrated sulfuric acid is added to the phenylcarboxylic acid and reacted at 60°C with gentle stirring for 24 h. The mixture is diluted with water, washed with ether and neutralised with NaOH solution. The supernatant solution, once sodium sulfate is precipitated, is evaporated to dryness and the white solid obtained is purified by recrystallisation with methanol–ether. This method has been adopted, among others, by Marcomini et al. [27,94], who synthesised a series of C_2–C_5 SPCs as well as two sulfophenyldicarboxylates (C_3 and C_5), and Field et al. [92], who synthesised C_3, C_6, C_8 and C_{10} SPCs, or C_4 SPCs [81,95]. Other authors [6,25,73,74,80,89–91,96] cited the same synthetic procedure as described by Marcomini et al. [94]. Another approach to the synthesis of SPCs, which involves the isolation of SPCs in the form of calcium salts, was developed by Sarrazin et al. [84]. In this procedure, the aqueous phase is progressively neutralised with calcium carbonate instead of sodium hydroxide as described before [87], filtered to eliminate the calcium sulfate formed, and the recovered liquid, once evaporated to dryness, is recrystallised in boiling water giving yields ranging from 50 to 85%. The synthetic pathway to prepare C_9, C_{12} and C_{13} SPCs, where corresponding phenylcarboxylic acids are commercially unavailable, can be carried out in three different steps [97] (Fig. 4.2.7). Initially,

461

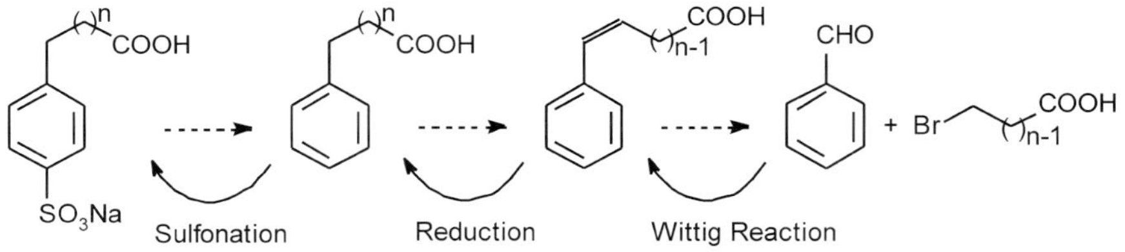

Fig. 4.2.7. Retrosynthetic scheme explored for the preparation of SPC (taken from Ref. [97] with permission).

a Wittig reaction [98] was performed with benzaldehyde and the corresponding yields of bromoalkylacids of different chain lengths generated with tributylphosphine and NaH. The unsaturated phenylalkylcarboxylates obtained with around 50% yield were reduced by using H_2 and Pd/C as catalyst in MeOH. Finally, sulfonation can be carried out by following the methods cited above [84,93,94]. The overall yield of the procedure was estimated at around 20–30% depending on the length of the alkyl chain.

The p-toluenesulfonic acid [99] and the C_3-SPC isomer have been used as internal standards, the latter to study the presence of SPCs in fish [85]. At present, there are no standards of α,β-unsaturated SPCs available commercially.

DATS and *iso*-LAS are minor compounds present in LAS formulations with varying proportions depending on the manufacturing process. However, the widespread use of LAS implies that significant amounts of these coproducts can reach the environment. It has been demonstrated [73,79,95,100] that both coproducts analogous to LAS undergo primary biodegradation leading to the formation of monocarboxylic intermediates (DATSI and *iso*-LASI) and dicarboxylated species to a lesser extent. These metabolites have been identified in groundwater [92], river water [24] and WWTP [24] even with concentration values for DATSI 3.2 times higher than for SPC [77]. Enrichment methods for DATSI and LASI include evaporation to dryness in water samples and methanol extraction of sediments with further clean-up by SPE (C_8 cartridge) and further analysis by GC-MS of their trifluoroethyl derivatives after condensation with trifluoroethanol of sulfonyl and acyl chlorides obtained by reaction of DATS with PCl_5 [77]. SPE using GCB cartridges and LC-MS analysis [74] or ion exchange and fractionation by solvent extraction and adsorption chromatography and further analysis of the fractions by ^{13}C-NMR and GC-MS of trifluoroethyl derivatives have also been used [92].

The 1-butyl-4-methyltetralinsulfonate (C_9-DATS) internal standard [77,101] and the C_6–C_{10} [73,74,77,102] or C_{10}–C_{14} [101] DATS homologous blends have been used to estimate the concentration of DATS in environmental samples. Other authors have estimated DATS by using C_8-LAS quantitation curves [92]. Biodegradation studies in laboratory simulations of activated sludge treatment were carried out by using synthesised ^{14}C-ring-labelled C_8-DATS and *iso*-LAS as model compounds to observe the formation of DATSI and *iso*-LASI [100]. DATSI concentrations have been estimated in groundwater by assuming the same response as an SPC standard [92] or assuming that DATSI and *iso*-LASI gave the same molar response factors as C_9-LAS [73,74].

4.2.4 Cationic surfactants

The environmentally most relevant compound by far is ditallow dimethylammonium chloride (DTDMAC), also called distearyldimethyl ammonium chloride (DSDMAC). Quantification of DTDMAC has been carried out by HPLC with conductivity detection [103–112]; by combination of thin layer chromatography (TLC) or chromatography on aluminium oxide and infrared spectroscopy [113]; by a gradient HPLC method applying post-column ion-pair formation and UV or FL detection [107,110,114,115]; by microbore HPLC-ESI-MS [116]; or by Fast Atom Bombardment (FAB)-MS [117–120]. GC has also been used for the qualitative determination of DTDMAC by converting them into the corresponding tertiary amines by decomposition in the injection port (80% at 400°C) [121,122]. Commercially available didecyldimethyl ammonium chloride and synthesised dioctyldimethyl ammonium iodide have been used as surrogate standards for the determination of DTDMAC [20,115].

Labelled $[^2H_3]$ditallowdimethyl ammonium iodide ($[^2H_3]$DTDMAI) was synthesised to be employed as an internal standard for the quantitative determination of trace levels of DTDMAC in sewage and river water samples by using FAB-MS [117]. The synthesis of ($[^2H_3]$ DTDMAI) involved the reaction of commercial ditallowmethyl amine with CD_3I in 1-propanol.

Other less used cationic surfactants, such as benzalkonium chloride (BAC), have been determined by solid-phase extraction followed by HPLC with post-column ion-pairing and FD detection in effluents from several European hospitals [123], whereas dihexyldimethyl ammonium and benzylhexadecyldimethyl ammonium chlorides have also been determined in dried sewage sludge, soil and sediment by comparing extraction by Soxhlet (methanol), accelerated solvent extraction and microwave-assisted extraction followed by HPLC with post-column derivatisation [124]. Several dialkyldimethyl ammonium chloride surfactants and other miscellaneous cationic surfactants (imidazolinium, alkyltrimethyl ammonium and alkyldimethylbenzyl ammonium compounds) were determined with extraction of an ion-pair from acid into chloroform and alumina clean-up for sewage water samples, and extraction using methanolic hydrochloric acid after centrifugation and further clean-up for activated sludges followed by HPLC with conductivity detection [109]. In all cases quantification was carried out with commercial standards.

Labelled [^{2}H$_3$]dodecyltrimethyl ammonium iodide ([^{2}H$_3$]C$_{12}$TMAI) was synthesised as an internal standard for the quantitative determination of dodecyltrimethyl ammonium chloride (C$_{12}$TMAC) in sewage and river water samples by using FAB-MS [117]. The synthesis of [^{2}H$_3$]C$_{12}$TMAI involved the reaction of dimethylamino dodecane with deuterated methyl iodide (CD$_3$I) in methanol.

Determination of two esterquats used as substitution products of DTDMAC, such as diethylester dimethylammonium chloride (DEEDMAC) and diesterquaternary (DEQ) (Fig. 4.2.8) in sewage water samples was carried out by the same ion-pair extraction procedure for the analysis of DTDMAC reported elsewhere [103,111] followed by microbore HPLC-ESI-MS analysis [116] and quantification employing commercial blends.

Simms et al. [119] employed an integrated approach to determine the mechanism and kinetics of surfactant biodegradation by FAB-MS and liquid scintillation counting. Two compounds, the esterquat N-octadecyl-N-[palmytoyloxyethyl]-N,N-dimethyl ammonium chloride and N-(2-hydroxyethyl)-N,N-dimethyloctadecyl ammonium chloride,

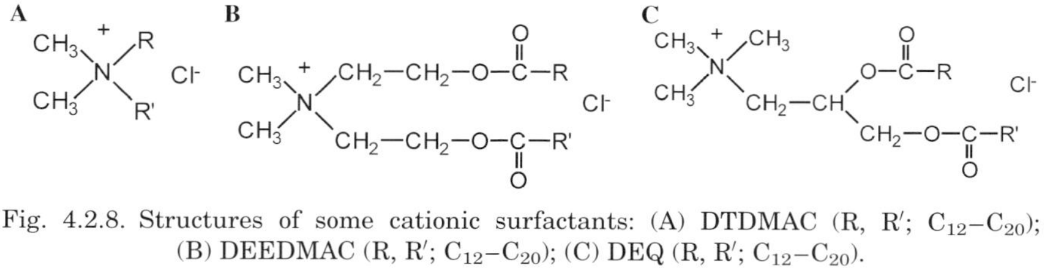

Fig. 4.2.8. Structures of some cationic surfactants: (A) DTDMAC (R, R'; C$_{12}$-C$_{20}$); (B) DEEDMAC (R, R'; C$_{12}$-C$_{20}$); (C) DEQ (R, R'; C$_{12}$-C$_{20}$).

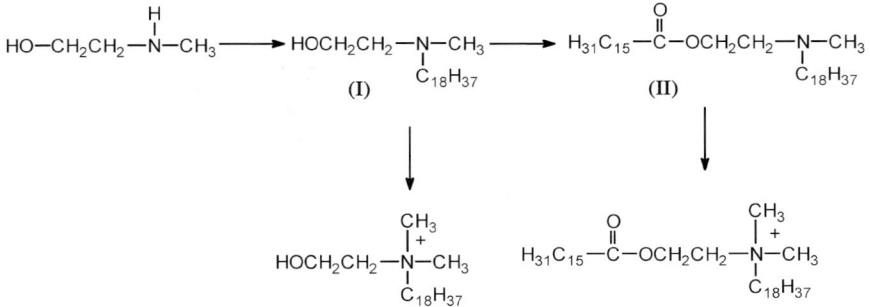

Fig. 4.2.9. Synthesis of cationic surfactants (taken from Ref. [119] with permission).

were studied. Figure 4.2.9 depicts the synthetic scheme of both compounds.

Labelled [^{13}C] and [^{14}C]-N-octadecyl-N-[palmytoyloxyethyl]-N,N-dimethyl ammonium iodide were prepared by reaction of (II) with ^{13}CH$_3$I and ^{14}CH$_3$I, respectively. Labelled [^{13}C] and [^{14}C]-N-(2-hydroxyethyl)-N,N-dimethyloctadecylammonium iodide were prepared by reaction of (I) with ^{13}CH$_3$I and ^{14}CH$_3$I, respectively.

On the other hand, the widespread occurrence of trialkylamines [$R_1R_2NCH_3$], where $R_1 = CH_3$ or C_nH_{2n+1} and $R_2 = C_nH_{2n+1}$ for $n = 14-18$, in sludges, coastal water and sediments [120,121,125–128], open seawater [129] and biota [128] has led to consideration of these compounds as indicators of urban sewage contamination of coastal areas. Trialkylamines (TAMs) come from trace impurities in quaternary ammonium salts used as fabric softeners in household laundry detergents, which were estimated to be 0.05% in DTDMAC [125] and exhibit an homologous distribution of odd–even carbon number alkyl derivatives, where the two long-chain substituents (C_{16} and C_{18}) are the predominant components. The methods employed to isolate TAMs included liquid–liquid for wastewater, and Soxhlet extraction (dichloromethane/methanol) and further fractionation by column chromatography followed by GC-NPD and GC-MS for sediments [120]. Quantification of TAMs was performed by using N,N-dimethyloctadecylamine (commercial), N,N-dioctadecylmethylamine, and trioctadecylamine as external standards. The two last-named compounds were synthesised as follows: N,N-dioctadecylmethylamine was obtained from dioctadecylamine by reductive alkylation with formic acid and formaldehyde in ethanol according to a previously described

465

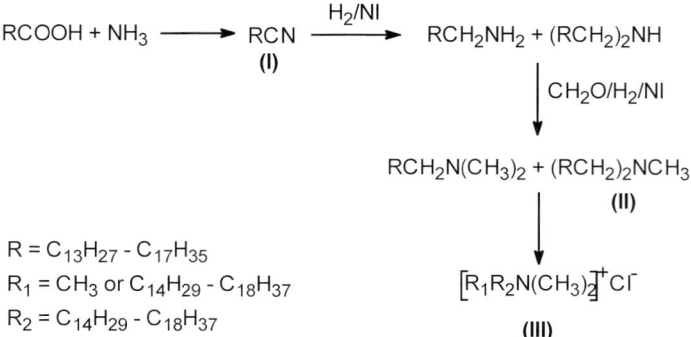

Fig. 4.2.10. Synthesis of trimethylalkyl ammonium salts.

method [130], whereas trioctadecylamine was prepared from the
N,N-dioctadecylamine and octadecyliodide [131], which were obtained
from n-octadecanol and PI_3 [132].

Another class of widely distributed contaminants in urban coastal
environments [120,121,127] related to cationic surfactants are the long-
chain alkylnitriles (LANs) (RCN, where $R = C_{13}–C_{17}$ saturated and
unsaturated), which are intermediates in their production and have
also been identified as impurities in cationic surfactant formulations.
LANs come from the reaction of fatty acids with ammonia, in a
combined liquid phase–vapor phase process, to form the corresponding
fatty nitriles (I). LANs are then converted by hydrogenation into
primary or secondary amines, depending on the reaction conditions.
Reductive alkylation of the amines with formaldehyde yields TAMs (II),
which are quaternised by exhaustive alkylation with methyl chloride to
obtain the final di- or trimethylalkyl ammonium salts (III) [133]
(Fig. 4.2.10). The same methodological approach described for TAMs
was employed for LANs and its quantitation performed by using
heptadecylnitrile as standard [120,121,128]. This compound was
synthesised by reaction of hexadecylchloride and potassium cyanide
according to the method previously described elsewhere [134].

4.2.5   Conclusion

For a quantification an addition of stable isotope-labelled analytes,
which exhibit the same chromatographic and ionisation behaviour as
their unlabelled analogues present in the sample, and which can be

distinguished from the latter with the aid of the MS detector separating the different ion masses, is the most reliable method. Not only for surfactants, but much more for their metabolites, their availability is very scarce. Until recently even the availability of unlabelled reference compounds was quite limited. Nowadays, after recognition of the importance of the quite often more recalcitrant metabolites, such as, e.g. the NPECs, synthetic pathways have been developed in order to have such compounds available for their quality assured quantification.

REFERENCES

1   E. Stephanou and W. Giger, *Environ. Sci. Technol.*, 16 (1982) 800.
2   M. Ahel and W. Giger, *Anal. Chem.*, 57 (1985) 1577.
3   M. Ahel, T. Conrad and W. Giger, *Environ. Sci. Technol.*, 21 (1987) 697.
4   M. Ahel, W. Giger and M. Koch, *Water Res.*, 28 (1994) 1131.
5   C.M. Lye, C.L.J. Frid, M.E. Gill, D.W. Cooper and D.M. Jones, *Environ. Sci. Technol.*, 33 (1999) 1009.
6   A. DiCorcia, R. Samperi and A. Marcomini, *Environ. Sci. Technol.*, 28 (1994) 850.
7   A. DiCorcia, A. Constantino, E. Marinoni, C. Crescenzi and R. Samperi, *Environ. Sci. Technol.*, 32 (1998) 2401.
8   A. Marcomini, P.D. Capel, Th. Lichtensteiger, P.H. Brunner and W. Giger, *J. Environ. Qual.*, 15 (1989) 523.
9   A. Marcomini, S. Bussetti, A. Sfriso, S. Capri, T. LaNoce and A. Liberatori, In: G. Angeletti and A. Björseth (Ed.), *Organic Micropollutants in the Aquatic Environment*, Kluwer Academic Press, Dordrecht, The Netherlands, 1991, p. 294.
10 A. Marcomini, F. Cecchi and A. Sfriso, *Environ. Technol.*, 12 (1991) 1047.
11 M. Ahel, W. Giger, E. Molnar and S. Ibric, *Croatica Chem. Acta*, 73 (2000) 209.
12 H.-B. Lee, T.E. Peart and D.T. Bennie, *J. Chromatogr. A*, 785 (1997) 385.
13 J. Ejlertsson, M.-L. Nilsson, H. Kylin, A. Bergman, L. Karlson, M. Öquist and B.H. Svensson, *Environ. Sci. Technol.*, 33 (1999) 301.
14 D.Y. Shang, M.G. Ikonomou and R.W. Macdonald, *J. Chromatogr. A*, 849 (1999) 467.
15 C. Wahlberg, L. Renberg and U. Wideqvist, *Chemosphere*, 20 (1990) 179.
16 P.L. Ferguson, C.R. Iden and B.J. Brownawell, *Anal. Chem.*, 72 (2000) 4322.
17 R.C. Mansfield and J.E. Locke, *J. Am. Oil Chem. Soc.*, 41 (1964) 267.
18 A. Díaz and F. Ventura, Pristine European Project ENV4-CT-0494, 2000.

19 R. Ekelund, A. Bergman, A. Granmo and M. Berggren, *Environ. Pollut. A*, 64 (1990) 107.

20 A. Kreisselmeier, M. Schoester and G. Kloster, *Fresenius J. Anal. Chem.*, 353 (1995) 109.

21 N.M.A. Ibrahim and B.B. Wheals, *Analyst*, 121 (1996) 239.

22 N.M.A. Ibrahim and B.B. Wheals, *J. Chromatogr. A*, 731 (1996) 171.

23 L.G. Mackay, M.Y. Croft, D.S. Selby and R.J. Wells, *J. AOAC Int.*, 80 (1997) 401.

24 A. DiCorcia, R. Cavallo, C. Crescenzi and M. Nazzari, *Environ. Sci. Technol.*, 34 (2000) 3914.

25 A. Marcomini, G. Pojana, A. Sfriso and J.M. Quiroga-Alonso, *Environ. Sci. Technol.*, 19 (2000) 2000.

26 E. Stephanou, M. Reinhard and H.A. Ball, *Biomed. Environ. Mass Spectrom.*, 15 (1988) 275.

27 A. Marcomini, C. Tortato, S. Capri and A. Liberatori, *Annali di Chimica*, 83 (1993) 461.

28 K. Yoshimura, *J. Am. Oil Chem. Soc.*, 63 (1986) 1590.

29 W.-H. Ding, S.-H. Tzing and J.-H. Lo, *Chemosphere*, 38 (1999) 2597.

30 W.-H. Ding, Y. Fujita, R. Aeschimann and M. Reinhard, *Fresenius J. Anal. Chem.*, 354 (1996) 48.

31 W.-H. Ding and C.-T. Chen, *J. Chromatogr. A*, 862 (1999) 113.

32 M. Reinhard, N. Goodman and K.E. Mortelmans, *Environ. Sci. Technol.*, 16 (1982) 351.

33 Y. Fujita, W.H. Ding and M. Reinhard, *Water Environ. Res.*, 68 (1996) 867.

34 J. Rivera, D. Fraisse, F. Ventura, J. Caixach and A. Figueras, *Fresenius J. Anal. Chem.*, 327 (1987) 377.

35 M. Petrovic, A. Díaz, F. Ventura and D. Barceló, *Anal. Chem.*, 73 (2001) 5886.

36 G.E. Stokker, A.A. Deana, S.J. Desolms, E.M. Schultz, R.L. Smith, E.J. Cragoe, J.E. Baer, C.T. Ludden, H.F. Russo, A. Scriabine, C.S. Sweet and L.S. Watson, *J. Med. Chem.*, 23 (1980) 1414.

37 H. Kammerer, K. Eberle, V. Bohmer and M. Grossmann, *Makromol. Chem.*, 176 (1975) 3295.

38 H.B. Ball and M. Reinhard, In: R.L. Jolley, R.J. Bull, W.P. Davis, S. Katz, M.H. Roberts and V.A. Jacobs (Eds.), *Water Chlorination*, Lewis, Chelsea, MI, 1984, p. 1505.

39 Y. Fujita and M. Reinhard, *Environ. Sci. Technol.*, 31 (1997) 1518.

40 J. Steber and P. Wierich, *Appl. Environ. Microbiol.*, 49 (1985) 530.

41 P. Schöberl, K.J. Bock and M. Hubert, *Tenside Deterg.*, 25 (1988) 86.

42 A. Marcomini, M. Zanette and G. Pojana, *Environ. Toxicol. Chem.*, 19 (2000) 549.

43 A. DiCorcia, C. Crescenzi and A. Marcomini, *Environ. Sci. Technol.*, 32 (1998) 711.

44   B. Fell, *Tenside Deterg.*, 28 (1991) 385.
45   F.I. Onuska and K.A. Terry, *J. High Resol. Chrom. Commun.*, 11 (1988) 874.
46   K.A. Evans, S.T. Dubey, L. Kravetz, I. Dzidic, J. Gumukla, R. Mueller and J.R. Stork, *Anal. Chem.*, 66 (1994) 699.
47   C. Crescenzi, A. DiCorcia, R. Samperi and A. Marcomini, *Anal. Chem.*, 67 (1995) 1797.
48   K.A. Evans, S.T. Dubey, L. Kravetz and S.W. Evetts, *JAOCS*, 74 (1997) 765.
49   J.C. Dunphy, D.G. Pessler and S.W. Morrall, *Environ. Sci. Technol.*, 35 (2001) 1223.
50   M. Castillo, M.C. Alonso, J. Riu and D. Barceló, *Environ. Sci. Technol.*, 33 (1999) 1300.
51   M. Castillo and D. Barceló, *Anal. Chem.*, 71 (1999) 3769.
52   K. Heinig, C. Vogt and G. Werner, *Anal. Chem.*, 70 (1998) 1885.
53   N.J. Fendinger, W.M. Begley and D.C. Mcavoy, *Environ. Sci. Technol.*, 29 (1995) 856.
54   A.T. Kiewet, J.M.D. van der Steen and J.R. Parsons, *Anal. Chem.*, 67 (1995) 4409.
55   E. Matthijs, M.S. Holt, A. Kiewiet and G.B.J. Rijs, *Environ. Toxicol. Chem.*, 18 (1999) 2634.
56   S.T. Dubey, L. Kravetz and J.P. Salanitro, *J. Am. Oil Chem. Soc.*, 72 (1995) 23.
57   J. Tolls, M. Haller and H.M. Sijm, *J. Chromatogr. A*, 839 (1999) 109.
58   C. Sung, M. Baird and H.A. Anderson, *J. Chromatogr. A*, 771 (1997) 145.
59   B.J. Brownawell, H. Chen, W. Zhang and J.C. Westall, *Environ. Sci. Technol.*, 31 (1997) 1735.
60   R.J. Larson and L.M. Games, *Environ. Sci. Technol.*, 15 (1981) 1488.
61   J. Steber and P. Wierich, *Water Res.*, 21 (1987) 661.
62   W.-H. Ding, Y. Fujita and M. Reinhard, *Rapid Commun. Mass Spectrom.*, 8 (1994) 1016.
63   F. Ventura, D. Fraisse, J. Caixach and J. Rivera, *Anal. Chem.*, 63 (1991) 2095.
64   A. Marcomini, G. Pojana, C. Carrer, L. Cavalli, G. Cassani and M. Lazzarin, *Environ. Toxicol. Chem.*, 19 (2000) 555.
65   A. Marcomini and G. Pojana, *Riv. Ital. Sostanze Grasse*, 75 (1998) 35.
66   C. Crescenzi, A.D. Corcia and A. Marcomini, *Environ. Sci. Technol.*, 31 (1997) 2679.
67   M. Castillo, E. Martinez, A. Ginebreda, L. Tirapu and D. Barceló, *Analyst*, 125 (2000) 1733.
68   G. Glod, P.M. Campbell, T.C.J. Feijtel, P.H. Masschelein, E. Matthijs and A. Rottiers, Annual Meeting of the German Chemical Society, Würzburg, Germany, 1999.

69 P. Eichhorn and T.P. Knepper, *J. Mass Spectrom.*, 35 (2000) 468.

70 D.L. Bailey and N.Y. Snyder, US Patent, 3299112, 1967.

71 L.S. Bonnington, Analysis of Organosilicone Surfactants, their Degradation Products. PhD Thesis. University of Waikato, Hamilton, New Zealand, 2001.

72 R. Wagner, Y. Wu, G. Czichocki, H.V. Berlespech, B. Weiland, F. Rexin and L. Perepelittchenko, *Appl. Organometal. Chem.*, 13 (1999) 611.

73 A. DiCorcia, F. Casassa, C. Crescenzi, A. Marcomini and R. Samperi, *Environ. Sci. Technol.*, 33 (1999) 4112.

74 A. DiCorcia, L. Capuani, F. Casassa, A. Marcomini and R. Samperi, *Environ. Sci. Technol.*, 33 (1999) 4119.

75 P. Eichhorn and T.P. Knepper, *Environ. Toxicol. Chem.*, 21 (2002) 1.

76 A. Moreno, J. Bravo and J.L. Berna, *J. Am. Oil Chem. Soc.*, 65 (1988) 1000.

77 M.L. Trehy, W.E. Gledhill, J.P. Mieure, J.E. Adamove, A.M. Nielsen, H.O. Perkins and W.S. Eckhoff, *Environ. Toxicol. Chem.*, 15 (1996) 233.

78 A. Moreno, Personal communication.

79 P. Kölbener, U. Baumann, A.M. Cook and T. Leisinger, *Environ. Toxicol. Chem.*, 14 (1995) 571.

80 W.H. Ding, J.H. Lo and S.H. Tzing, *J. Chromatogr. A*, 818 (1998) 270.

81 C. Kanz, M. Nölke, T. Fleischmann, H.P.E. Kohler and W. Giger, *Anal. Chem.*, 70 (1998) 913.

82 M. Petrovic and D. Barceló, *Anal. Chem.*, 72 (2000) 4560.

83 P. Eichhorn, M. Petrovic, D. Barceló and T.P. Knepper, *Vom Wasser*, 95 (2000) 245.

84 L. Sarrazin, A. Arnoux and P. Rebouillon, *J. Chromatogr. A*, 760 (1997) 285.

85 E. González-Mazo, A. Gomez-Parra and D. Barceló, *Anal. Chem.*, 71 (1999) 5242.

86 P. Eichhorn, M.E. Flavier, M.L. Paje and T.P. Knepper, *Sci. Total Environ.*, 269 (2001) 75.

87 P. Eichhorn, T.P. Knepper, A. Díaz and F. Ventura, *Water Res.*, 36 (2002) 2179.

88 J. Riu, E. González-Mazo and A.G.-P.y.D. Barceló, *Chromatographia*, 50 (1999) 275.

89 E. González-Mazo, M. Honing, D. Barceló and A. Gómez-Parra, *Environ. Sci. Technol.*, 31 (1997) 504.

90 V. León, E. González-Mazo and A. Gómez-Parra, *J. Chromatogr. A*, 889 (2000) 211.

91 M. Saez, V.M. León, A. Gómez-Parra and E. González-Mazo, *J. Chromatogr. A*, 889 (2000) 99.

92 J.A. Field, J.A. Leenheer, K.A. Thorn, L.B. Barber, C. Rostad, D.L. Macalady and R. Daniel, *J. Contam. Hydrol.*, 9 (1992) 55.

93 P.W. Taylor and G. Nickless, *J. Chromatogr.*, 178 (1979) 259.

94   A. Marcomini, A. DiCorcia, R. Samperi and S. Capri, *J. Chromatogr.*, 644 (1993) 59.
95   P. Kölbener, U. Baumann, T. Leisinger and M. Cook, *Environ. Toxicol. Chem.*, 14 (1995) 561.
96   V.M. León, E. González-Mazo, J.M. Forja-Pajares and A. Gómez-Parra, *Environ. Toxicol. Chem.*, 20 (2001) 2171.
97   M.P. Marco, Personal communication.
98   R. Greenwald, M. Chaykosky and E.J. Corey, *J. Org. Chem.*, 28 (1963) 1128.
99   P. Rovellini, N. Cortesi and E. Fideli, *Riv. Ital. Sostanze Grasse*, 72 (1995) 381.
100   A.M. Nielsen, L.N. Britton, C.E. Beall, T.P. McCormick and G.L. Russell, *Environ. Sci. Technol.*, 31 (1997) 3397.
101   M.L. Trehy, W.E. Gledhill and R.G. Orth, *Anal. Chem.*, 62 (1990) 2581.
102   C. Crescenzi, A. DiCorcia, E. Marchiori, R. Samperi and A. Marcomini, *Water Res.*, 30 (1996) 722.
103   V.T. Wee, *Anal. Chem.*, 54 (1982) 1631.
104   V.T. Wee, *Water Res.*, 18 (1984) 223.
105   E. Matthijs and H.D. Henau, *Vom Wasser*, 69 (1987) 73.
106   H. Klotz, *Tenside Deterg.*, 24 (1987) 370.
107   L. Nitschke, R. Müller, G. Metzner and L. Huber, *Fresenius J. Anal. Chem.*, 342 (1992) 711.
108   M. Emmrich and K. Levsen, *Vom Wasser*, 75 (1990) 343.
109   K. Levsen, M. Emmrich and S. Behnert, *Fresenius J. Anal. Chem.*, 346 (1993) 732.
110   C.D. Ruiter and J.C.H.F. Hefkens, *Int. J. Environ. Anal. Chem.*, 31 (1987) 325.
111   P. Gerike, H. Klotz, J.G.A. Kooijman, E. Matthijs and J. Waters, *Water Res.*, 28 (1994) 147.
112   D.C. McAvoy, C.E. White, B.L. Moore and R.A. Rappaport, *Environ. Toxicol. Chem.*, 13 (1994) 213.
113   H. Hellmann, *Tenside Deterg.*, 28 (1991) 111.
114   S.M. Gort, E.A. Hogendoorn, R.A. Baumann and P. Van Zoonen, *Int. J. Environ. Anal. Chem.*, 53 (1993) 289.
115   P. Fernández, A.C. Alder, M.J.F. Suter and W. Giger, *Anal. Chem.*, 68 (1996) 921.
116   M. Radke, T. Behrends and R.H.J. Förster, *Anal. Chem.*, 71 (1999) 5362.
117   J.R. Simms, T. Keough, S.R. Ward, B.L. Moore and M.M. Bandurraga, *Anal. Chem.*, 60 (1988) 2613.
118   F. Ventura, J. Caixach J. Romero, I. Espader J. Rivera, *Water Sci Technol.*, 25 (1992) 257.
119   J.R. Simms, D.A. Woods and T.K.D.R. Walley, *Anal. Chem.*, 64 (1992) 2951.
120   M. Valls, J.M. Bayona and J. Albaigés, *Int. J. Environ. Anal. Chem.*, 39 (1990) 329.

121 P. Fernández, M. Valls, J.M. Bayona and J. Albaigés, *Environ. Sci. Technol.*, 25 (1991) 547.

122 G. Grossi and R. Vece, *J. Gas Chromatogr.*, 3 (1965) 170.

123 K. Kümmerer, A. Eitel, U. Braun, F.D.P. Hubner, G. Mascart, M. Milandri, F. Reinthaler and J. Verhoef, *J. Chromatogr. A*, 774 (1997) 281.

124 S. Heise and N. Litz, *Tenside Surf. Deterg.*, 36 (1999) 185.

125 M. Valls, J.M. Bayona and J. Albaigés, *Nature*, 337 (1989) 722.

126 M. Valls and J.M. Bayona, *Fresenius J. Anal. Chem.*, 339 (1991) 212.

127 N. Chalaux, J.M. Bayona, M.I. Venkatesan and J. Albaiges, *Marine Pollut. Bull.*, 24 (1992) 403.

128 C. Maldonado, J. Dachs and J.M. Bayona, *Environ. Sci. Technol.*, 33 (1999) 3290.

129 M. Valls, P. Fernández and J.M. Bayona, *Chemosphere*, 19 (1989) 1819.

130 A.W. Ralston, D.N. Eggenberger and P.L. DuBrow, *J. Am. Chem. Soc.*, 70 (1948) 977.

131 A.W. Ralston, C.W. Hoerr and P.L. DuBrow, *J. Org. Synth.*, 9 (1944) 259.

132 W.W. Hartmann, J.R. Byers and J.B. Dickers, *Org. Synth.*, 15 (1935) 29.

133 E. Jungerman, In: D. Swern (Ed.), *Bayleys Industrial Oil and Fat Products*, Wiley, New York, 1979, p. 587.

134 R.A. Smiley and C. Arnold, *J. Org. Chem.*, 25 (1960) 257.

4.3 ADVANTAGES AND LIMITATIONS IN SURFACTANT QUANTIFICATION BY LIQUID CHROMATOGRAPHY-MASS SPECTROMETRY

Pim de Voogt, Mira Petrovic, Niels Jonkers and Thomas P. Knepper

4.3.1 Introduction

For quantitative analysis of organic compounds in general by means of liquid chromatography-electrospray ionisation mass spectrometry (LC-ESI-MS), one should be aware of two major factors, which may strongly impact on the outcomes. These are directly associated with the process of ion generation in the interface.

- Formation of distinct adducts with ions originating from the buffer, the sample and/or the introduction system (e.g. H^+, Na^+, K^+, Cl^-, CH_3COO^-) and the formation of cluster and multiply charged ions [1].
- Ionisation suppression caused by co-elution of other analytes or natural matrix components or due to the presence of ion pairing (IP) agents or high salt concentrations in the mobile-phase buffer used [2–4].

The adduct formation can be largely controlled and directed into the formation of a single selected species by adequate choice of the ionisation mode (possibly at the expense of sensitivity), the eluent composition (buffer addition, pH adjustment, type of organic modifier) and by optimisation of the ion source parameters influencing the stability of individual (adduct) ions. In contrast to the variations in adduct or cluster formation, which principally can be diagnosed by recording more than one (adduct) ion in SIM mode, the occurrence of ion suppression requires more careful diagnosis.

A quite reliable method involves the addition of stable isotope-labelled analytes, which exhibit the same chromatographic and ionisation behaviour as unlabelled analogues present in the sample, and which can therefore be distinguished from the latter with the aid of the MS detector separating the different ion masses. However, whereas for many environmental pollutants such as pesticides, pharmaceuticals,

T.P. Knepper, D. Barceló and P. de Voogt (Eds)

or polycyclic aromatic hydrocarbons, a large pool of isotope-labelled compounds is provided by chemical suppliers, the availability of such reference standards for surfactants, as discussed in Chapter 4.2, remains scarce. This is not surprising in light of the difficulties in obtaining even unlabelled defined compositions of commercial surfactants comprising up to several tens to hundreds of homologues and isomers, such as nonlyphenol ethoxylates (NPEOs).

In this instance, the standard addition of a surfactant to the real sample is an appropriate alternative, i.e. along with the analysis of the original sample, where a second sample is amended with a known concentration of the pure standard material, preferably at a similar level, and prepared for analysis alongside the authentic sample. Comparison of the analytical results provides information on both the extraction efficiencies and possible ion suppressions in the interface.

4.3.2 Quantitative LC-MS analysis of alkylphenol ethoxylates (APEOs)

Ion adduct formation in atmospheric pressure ionisation (API)
For sensitive quantification in LC-MS analysis of non-ionic surfactants, selection of suitable masses for ion monitoring is important. The non-ionic surfactants easily form adducts with alkaline and other impurities present in, e.g. solvents. This may result in highly complicated mass spectra, such as shown in Fig. 4.3.1(A) (obtained with an atmospheric pressure chemical ionisation (APCI) interface) and Fig. 4.3.2 (obtained with an ESI interface).

Other ions that can be formed include methanol adducts (e.g. $[NPEO_1 + MeOH + H]^+$, which is observed in LC-ESI-MS for the monoethoxylate only, but not for higher oligomers [5]), water clusters ($[M + (H_2O)_n + H]^+$, (often in APCI), dimeric complexes $[(APEO_n)_2 + M]^+$, which have been observed for mono-, di-, tri- and tetraethoxylates of nonylphenol (NP) [6], and doubly charged ions, such as disodium adducts [7]. Disodiated adducts of higher odd-numbered ethoxymers may interfere with singly charged ions of less ethoxylated APEO, in particular, in RP-HPLC because of co-elution [7].

Matrix-induced signal suppression in LC-MS has been investigated and discussed by several authors [8,9]. In general, an increase in analyte response factor with increasing electrolyte concentrations in the mobile-phase is observed for electrolyte concentrations below

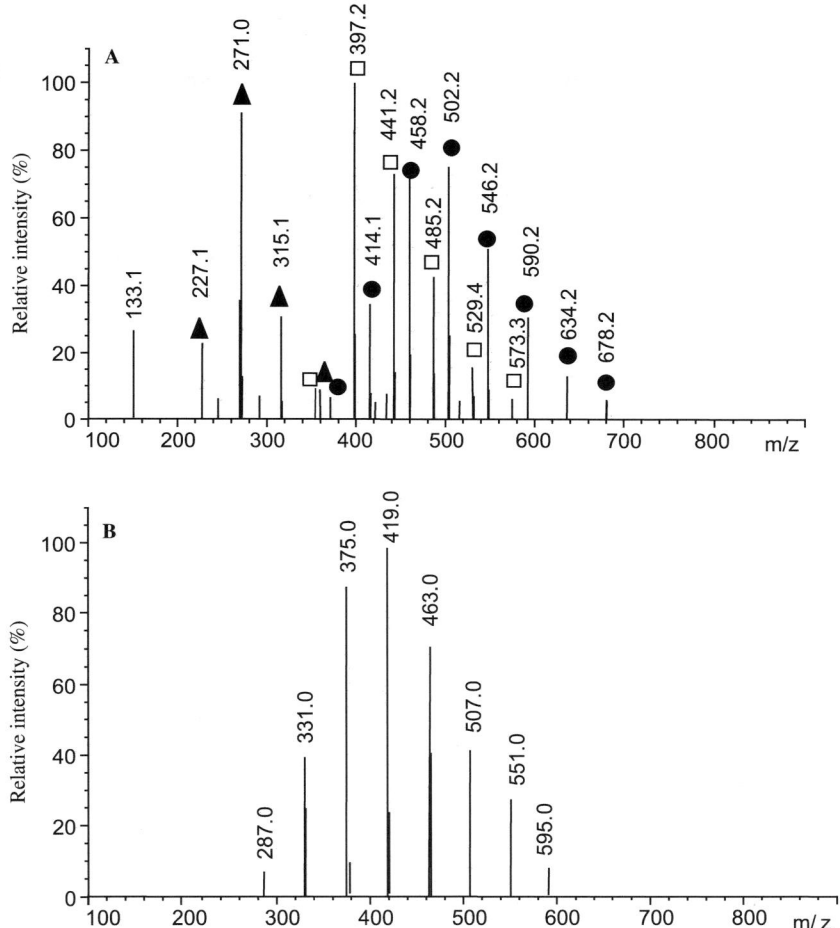

Fig. 4.3.1. Mass spectra of NPEO oligomeric mixture with an average of four EO units. (A) APCI interface ($\square$: $[M + H]^+$, $\blacktriangle$: $[M - C_9H_{18}]^+$, $\bullet$: $[M + NH_4]^+$); (B) ESI interface.

10^{-3} M [8]. Ferguson et al. [10] have shown that matrix effects may result in severe ion suppression in APEO analysis (this phenomenon occurs particularly in negative mode electrospray ionisation). They also demonstrated that the effect may be accounted for appropriately when using internal and surrogate standards, which closely mimic the ionisation behaviour of the analytes. In Ferguson's study, specially

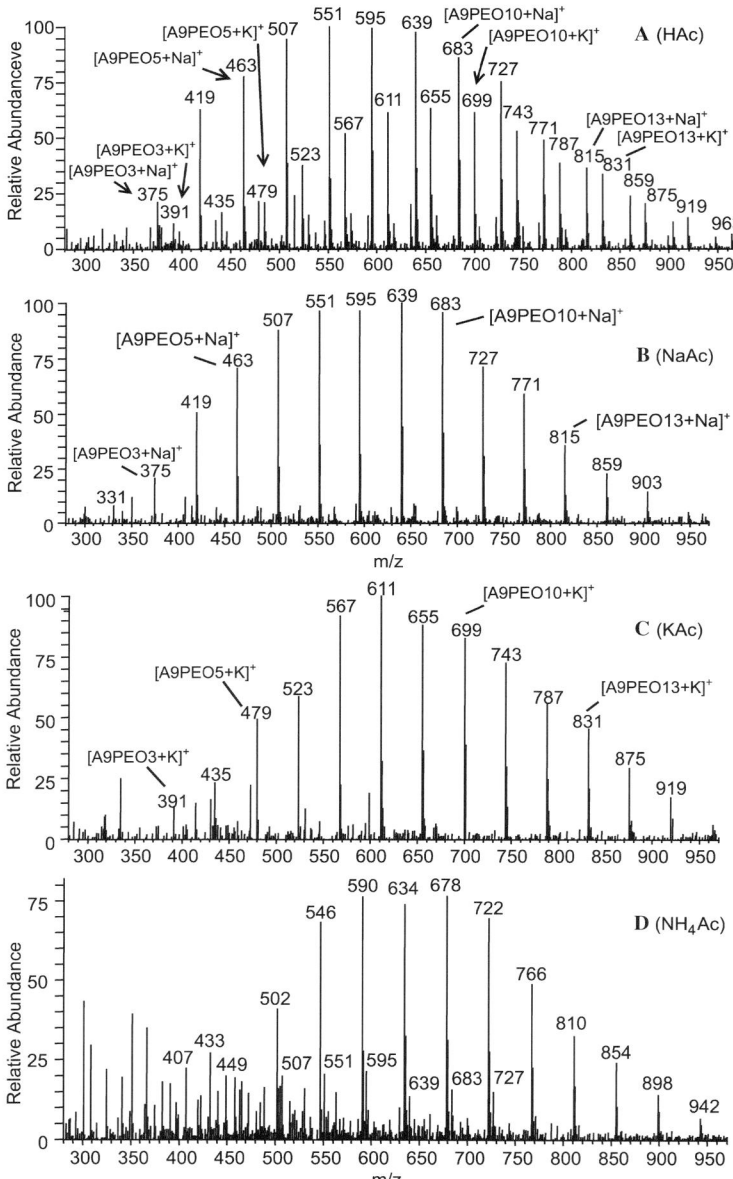

Fig. 4.3.2. Influence of the use of different buffers in the mobile phase on the formation of adduct formation in LC-ESI-MS of APEO mixtures.

synthesised branched ^{13}C-labelled NP and NPEO were used, which correct ideally for matrix effects, as they have exactly the same retention time as the investigated compounds.

The performance of atmospheric pressure interfaces appears to vary widely from instrument to instrument. A variety of interface designs are available from the various manufacturers. Optimisation of operating parameters, such as cone voltage, temperature, and mobile-phase composition is always necessary prior to actual analysis of samples. A given optimised set of parameters is likely to change with changing matrices, and may also vary with local conditions, such as the alkaline metal content of water or tubing used.

Differences between APCI and ESI can be found in adduct ion formation, as well as in sensitivity. Using an APCI interface the typical spectrum of APEO reveals a characteristic pattern showing several series of equidistant signals with mass differences of 44 Da. The main series corresponds to $[M + H]^+$ ions (assigned as $\square$ in Fig. 4.3.1(A)), the second series corresponds to $[M - C_9H_{18}]^+$ (assigned as $\blacktriangle$) and the third one to adduct molecular ions $[M + NH_4]^+$ (assigned as $\bullet$). Additionally, an indiscriminate number of other adduct ions $[M + Na]^+$ and $[M + K]^+$ and clusters $[M + (H_2O)_nH]^+$ were formed. Therefore, the ionisation of APEO molecules is dispersed among many molecular adduct ions and the abundance of each ion highly variable, depending on operating parameters and concentration of NH_4^+, Na^+, and K^+ ions, which are difficult to control due to the ubiquity of these ions that originate from the sample itself or from the solvents used.

Using the same instrument with an ESI interface NPEOs and OPEOs show a great affinity for alkali metal ions. They gave exclusively, evenly spaced sodium adducts $[M + Na]^+$, due to the ubiquity of sodium in the solvents and surfaces (Fig. 4.3.1(B)). In another type of instrument applying ESI, however, a relatively complicated spectrum was obtained (Fig. 4.3.2(A)), with similarities to the APCI spectrum of the first instrument.

Although sodium adducts were preferably formed in both instruments, even in the absence of added electrolyte, in order to avoid the possible reduction in ionisation due to insufficient concentration of metal ions in solution, it is recommended that sample extracts are fortified with sodium ions prior to injection. However, the addition of higher concentrations can induce system instability and analyte suppression. Therefore, the concentration of added electrolyte should compromise enhancing the adduct formation and signal intensity with negative effects.

Shang et al. [7] studied the effect of different additives (NaAc, NaOH, NaCl, NH$_4$Ac) on analyte signal intensity and they found that the relative intensity of NPEO adduct ions may be enhanced by all additives, but NaAc produced the most abundant adduct ions for the entire ethoxylate series with good reproducibility. Additionally, the intensity of adducts, especially for mono- and diethoxylates was found to depend on reaction time prior to LC-ESI-MS analysis and concentration of NaAc. They recommended 0.5 mM NaAc for normal-phase separation with solvent system toluene−MeOH−water. In reversed-phase systems the highest abundance of sodium adducts for NPEOs ($n_{EO} = 1-10$) was observed at concentrations higher than 10 μM, while any further increase in concentration had very low influence on signal intensity [10,11].

In a similar study with ESI the influence of different buffers was studied [12]. In the presence of acetic acid (HAc) only in the MeOH/H$_2$O mobile-phase, a mass spectrum resulted with ion adducts of Na and K appearing as the most abundant ones. However, minor peaks could also be observed in the mass spectrum resulting from ammonium adducts (Fig. 4.3.2(A)). The respective ions could be suppressed or enhanced by changing the nature of the buffer used in the mobile-phase. For example, when a potassium buffer was used, sodium and ammonium adducts were suppressed, and the spectrum became less complicated with primarily the potassium adduct ion being visible (Fig. 4.3.2(C)). In addition, the signal-to-noise ratio improved by about a factor of 1.5−2. Similarly, sodium or ammonium acetate buffers enhanced the sodium and ammonium adduct ions, meanwhile suppressing other adducts (Fig. 4.3.2(B) and (D), respectively).

Therefore, the way to ensure reproducible adduct formation is to use mobile-phase additives (e.g. ammonium acetate or formate, formic, acetic or trifluoroacetic acid (in APCI), ammonium hydroxide, etc.). Their application in the mobile phase can be an effective way to improve the intensity of the MS signal and LC-MS signal correlation between matrix and standard samples. However, it is observed that some additives like trifluoroacetic acid or some ion-pairing agents (triethyl-amine) may play a role in ionisation suppression [3]. In addition, high concentrations of involatile buffers will cause precipitation on, and eventually blocking of, the MS entrance cone, leading to a fast decrease of sensitivity. For the involatile NaAc buffer, it is advisable to maintain concentrations well below 0.5 mM. An effective way of keeping the MS clean is using a programmable six-way valve to divert the LC flow to the waste at time intervals in the analysis during which buffer concentrations are high, and no analytes elute, such as during column equilibration. This feature is available as standard in some LC-MS systems, but can easily be built into any system.

Response factors in MS
The issue of response factors in quantitative ethoxymer LC-MS analysis has been discussed in detail by Crescenzi et al. [13] for the alcohol ethoxylates, and by Ferguson et al. [14] for the APEOs. Crescenzi et al. argued that (i) the ability of molecular species to form adducts with cations would be affected by the nature and the number of polar sites present in the molecule, and (ii) the stability of the complexes formed would increase with increasing number of EO units in the molecules. The latter observation has been confirmed for NPEOs [10]. Longer EO chains possess a higher affinity for alkali metal ions [15]. Competitive ionisation suppression of the less ethoxylated oligomers by highly ethoxylated AEO and APEO is also believed to occur [10,13], but probably not in field situations where the higher ethoxymers tend to be relatively less concentrated [10,16]. Other mechanisms may also account for the observed adduct ion suppression of the lower ethoxylated oligomers, e.g. saturation of the electrospray ion current or modification of ion evaporation processes from small droplets [10].

Ferguson et al. [14] showed that the cone voltage applied in the ESI source also greatly influences the molar responses of ethoxymers. Optimum cone voltages increased with increasing EO chain length (from 35 eV for $NPEO_1$ to 105 eV for $NPEO_{15}$). A maximum will appear for each ethoxymer, since at increasingly higher cone voltages collision-induced dissociation will occur. At a given cone voltage, initially there is an exponential increase in sensitivity with increasing number of EO units in the chain until about six EO units. A further increase in number of EO units resulted in a slight decrease of the response, which was explained by the formation of doubly charged adducts as well as a decrease in transmission of higher mass ions by the quadrupole mass analyser [14].

Quantitative analysis
Crescenzi et al. [13] estimated that a reversed-phase LC-ESI-MS method using an NPEO mixture with an average of 10 EO units as calibrated standard solution, underestimated NPEO concentration in

wastewater treatment plant (WWTP) effluents by a factor ranging between 1.2 and 2.5. The underestimation was explained by the fact that the oligomeric distribution of a biodegraded NPEO mixture was substantially different from that in the standard solution, and by the low sensitivity of the ESI-MS detector for compounds containing one and two EO units [17].

Figure 4.3.3 shows ESI mass spectra of NPEOs in a commercial blend Findet 9Q/22, with an average of 10 EO units (Fig. 4.3.3(A)), in WWTP influent (Fig. 4.3.3(B)) and in an effluent sample (Fig. 4.3.3(C)). The mass spectra of NPEOs in influent is characterised by an oligomer distribution typical of the mixture that was slightly altered by biological degradation, due to the in-sewer degradation. Besides a maximum at NPEOs with nine EO units (typical commercial mixture), the abundance of lower oligomers ($n_{EO} = 2-5$), which are considered stable metabolic products, increased. Biologically treated wastewater contained only traces of oligomers with $n_{EO} = 5-12$, and the maximum is centred at lower oligomers ($n_{EO} = 1-4$). Consequently, in the absence of pure NPEOs with EO chains of discrete lengths, these samples cannot be quantified using the same oligomeric mixture as a standard solution.

As shown in Figure 4.3.4, the results obtained using different calibrants and different ionisation sources varied substantially. In this study, compounds used as a calibrant were laboratory-synthesised APEOs with a low degree of ethoxylation ($APEO_1$ and $APEO_2$) and commercial blends containing an average of three, six and 10 EO units per molecule. When calibrated with $NPEO_{10}$ (oligomeric mixture with 10 EO units as an average) the concentration of NPEOs in WWTP effluent obtained using an ESI was 40% less than when calibrated with the oligomeric mixture with an average of 3 and a range of 1–6 EO units (cf. Fig. 4.3.4(A)). In contrast, in WWTP influent and in water spiked with a commercial blend of $NPEO_{10}$, the results obtained using the calibrants that have significantly different oligomeric distribution ($NPEO_3$ or $NPEO_6$) were underestimated by a factor of 1.5–1.7 compared to calibration with $NPEO_{10}$. In APCI, a similar trend was observed (cf. Fig. 4.3.4(B)). However, the accuracy of the APCI-MS method differed significantly when measuring NPEO concentrations in

Fig. 4.3.3. ESI mass spectra of: (A) a commercial blend Findet 9Q/22 (NPEO with an average of 10 EO units); (B) sewage treatment plant influent; and (C) sewage treatment plant effluent. Reprinted with permission from Ref. [17]$^{©}$ 2001 John Wiley & Sons, Ltd.

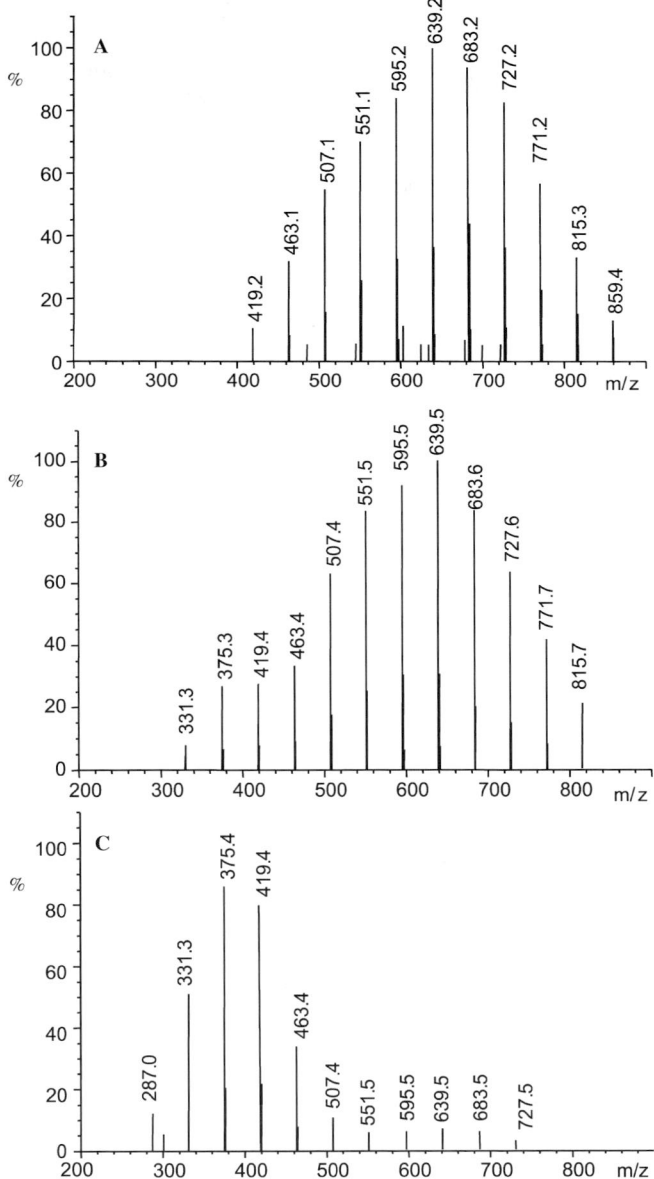

effluent samples. Concentrations measured with an APCI (using NP_3EO as a calibrant) were 25% lower than those obtained with an ESI (compare Fig. 4.3.4(A) and (B)), because, as discussed above, APCI does not give an appreciable ion signal for $NPEO_1$ and $NPEO_2$. To overcome the problem of underestimation of lower oligomers it is recommended first, to estimate the oligomeric distribution of each sample (or group of samples, e.g. influent, effluent, surface waters, sediment, sludge, etc.) in full scan mode and then to use a calibrant with an oligomeric distribution as similar as possible.

Besides the differing response of $NPEO_1$ and $NPEO_2$ compared to higher ethoxymers, another reason for quantifying these two ethoxymers separately is because of their differing toxicological effects. As only the two short-chain NPEOs are endocrine disruptors, it is desirable to report the short-chain NPEO ethoxymer concentrations separately when discussing results from monitoring campaigns. For environmental risk evaluations, information on total concentrations of APEO only is insufficient.

By far the most often analysed non-ionic surfactant metabolite, and indeed the most environmentally relevant, is the endocrine disruptor NP. For NP, commercial standards are cheap, and most of the quantification problems mentioned above for NPEO appear to be absent, as NP is usually considered to be a pure substance. However, NP in fact consists of at least 10 alkyl isomers (see Chapter 2.1), which can only be separated using GC analysis. Of these isomers, probably only a few are endocrine disruptors, and therefore it would be desirable to quantify all NP isomers separately. Not much work has been done in this field, but gaining more insight into these specific problems is expected to be the focus of research in the coming years. A recent example of a study on the separate environmental fates of NP isomers can be found in Ref. [18], in which 10 NP isomers were quantified separately, using GC-MS. The first report of an HPLC-FL method in which alkyl isomers of NP and NPEO were separated into 12 alkyl isomers using a graphite carbon column has recently appeared [19]. These results show promise for obtaining pure NP isomers in the future using preparative HPLC.

For several other non-ionic surfactant metabolites, standards are either commercially available, such as for dicarboxylated polyethylene glycols (DCPEGs) [20], or can be synthesised relatively easily, such as for nonylphenoxy acetic acids (NPECs) [21]. For quantification of metabolites for which standards are absent, the most similar available

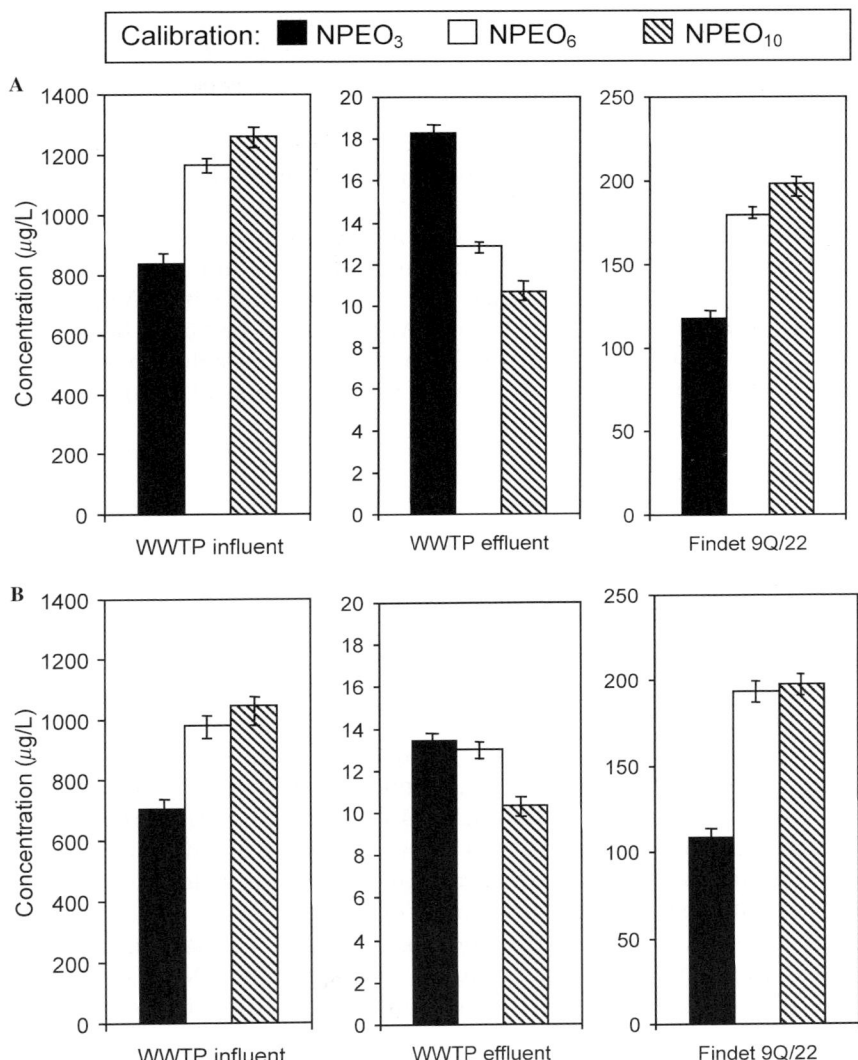

Fig. 4.3.4. Effect of composition of a technical mixture used for calibration in the quantification of APEO concentrations of influents and effluents from a wastewater treatment plant. (A) LC-ESI-MS; (B) LC-APCI-MS.

standard is used, assuming an equal molar response in the MS. This assumption is expected to be valid for, e.g. the quantification of long-chain ethoxymers ($NPE_{>2}C$) using NPE_2C as a reference [22]. However, a considerable error may be introduced when using the same standard for the quantification of the APEO metabolites carboxy alkylphenoxy acetic acids (CAPEC) [22]. Another approach for quantifying CAPECs was followed by DiCorcia et al. who used a solution of undetermined amounts of CAPECs, obtained by submitting an APEO mixture to biodegradation. CAPECs in this solution were quantified by UV detection, using an $NPEO_{1,2}$ mixture, assuming that with UV detection the molar response of NPEO and CAPEC would be equal, while this would not be the case with an MS detector. This solution was then diluted and used for quantification of the experimental samples [23].

Detection limits

The comparison of instrumental detection limits of analytes injected into the MS clearly indicates that the ESI interface offered better sensitivity and specificity for a higher range of oligomers than APCI does. It was found that monoethoxylate cannot be detected using an APCI interface, while diethoxylates can be detected only at very high concentrations, far higher than levels found in environmental and wastewater samples. In contrast, ESI permits detection of the full range of APEO oligomers. It must be emphasised again that in order to obtain a high sensitivity especially for $APEO_1$ and $APEO_2$, a careful optimisation of operating parameters is necessary, including proper selection of ion adducts.

As mentioned above, it is important to note that the formation of adducts, fragmentation and overall sensitivity of MS detection highly depends on interface design and can vary significantly depending on the instrument used. For example, Shang et al. [7] reported detection limits for normal-phase LC-ESI-MS analysis of marine sediment in the range of 0.8–4 ng (injected) depending on the individual NPEO oligomer, while Crescenzi et al. [13] reported limits of quantification ($S/N = 10$) for reversed-phase LC-ESI-MS analysis of WWTP influents, effluents, river water and drinking water of 2, 0.07, 0.007 and 0.0007 μg L^{-1} for preconcentration of 10, 100, 1000 and 4000 mL, respectively. The lowest detection limits were reported by Ferguson et al. [10]. Using SPE (concentration factor 2000) and isocratic elution using a narrow-bore C_8 column and ESI-MS detection; they reported method detection limits (MDLs, at $S/N = 3$) in the range of 0.2–1.0 ng L^{-1} for APEOs and APs in water samples. In sediments, the MDLs obtained were between 0.9 and 5.0 ng g^{-1} for NPEO$_{2-15}$, whereas for NP and NPEO$_1$ these values were 21 and 37 ng g^{-1}, respectively [14].

4.3.3 Quantitative LC-MS analysis of LAS and SPC

The ESI-MS is nowadays the most commonly used LC-MS coupling device for the analysis of LAS and SPC. But also for the analysis of these anionic analytes the most serious drawbacks of this ionisation technique are matrix effects. A crucial role is played by mobile phases containing relatively high salt concentrations or ion-pair reagents that lead to signal instability or even, in some cases, to plugging of the orifice plate. For successful removal of alkaline salts, a suppressor was incorporated between the LC column and the mass spectrometer [24]. With this set-up it was possible to reduce a spiked sodium concentration of 15 mg L^{-1} in the HPLC eluent before the suppressor by more than 99.8% at the entry into the MS.

A further common problem the analyst faces when integrating SPC into analytical procedures for the determination of LAS is the scarcity of available reference compounds, thereby complicating their determination. Therefore, the identification of the analytes has to be performed by comparison of retention time and absolute peak area ratio between the deprotonated molecular ion and the fragment ion, relative to the ratio obtained from the authentic standard ($\pm 20\%$). Retention times of SPC, for which no standards were available, can be determined once by mass spectrometric identification in full-scan mode.

The concentration of each analyte in the sample (extract) can be calculated by measuring the peak area obtained by selecting ion signals for the parent ion and comparing the peak area with that obtained from standard solutions. Depending on the sensitivity of the analyte, quantitative analyses should be done using 5- or 6-point calibrations in the range from 5 to 1000 or 50 to 1000 μg L^{-1} constructing linear calibration graphs ($r \geq 0.9991$). Calibrations should be performed daily for all analytes measured. The set of calibration solutions has to be run at the beginning of each sample series as well as at the end and if needed within the series. By this, gradual losses of sensitivity due to contamination of the MS instrument by precipitation of inorganic salts or non-volatile matrix components will be taken into account. The

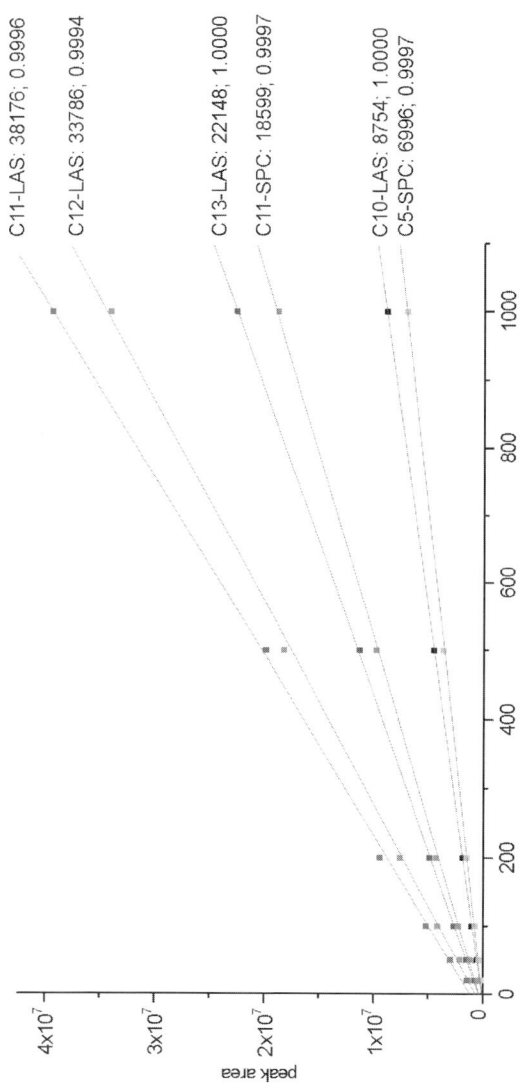

Fig. 4.3.5. Calibration graphs of LAS and SPC. For each compound the slope of the fitted linear regression and the correlation coefficient are given behind the substance name. The concentrations of the LAS homologues represent the sum of the concentration of the four components, i.e. for comparison of the ionisation efficiencies the value of the slope has to be multiplied by the relative percentage of each homologue in the commercial mixture.

mean of two calibrations bracketing a sample series is applied for quantification. Typical graphs of C_{10}- to C_{13}-LAS are shown in Fig. 4.3.5.

Due to the rather narrow linear range of the MS detector and owing to the strongly varying concentrations of the target analytes in environmental samples, dilutions have to be prepared in most instances.

The quantification of individual SPC can be performed based on the responses of the individual homologues C_5-SPC and C_{11}-SPC. By taking into account that an increasing portion of the organic solvent in the LC eluent enhances the ionisation efficiency [25,26] (Fig. 4.3.5), the early eluting SPC, $\leq C_6$, should be quantified using, e.g. C_5-SPC, while the quantification of those species eluting closely to C_{11}-SPC, i.e. C_7-SPC and higher, should be performed relating to this longer chain homologue.

For quality assurance, repeatability studies need to be performed on the LC-ESI-MS by consecutive injections of the same sample. An example is shown in Fig. 4.3.6 for the analyses of four SPC homologues in a sample from a biodegradation experiment.

Except for one outlier, the results varied within a range of $\pm 10\%$ of the mean value. The variability was high for the short-chain C_6-SPC. A positive correlation with slight shifts in retention times could not be established for any analyte (C_6: 8.1 $\pm$ 0.20 min; C_8: 19.5 $\pm$ 0.16 min; C_{10}: 20.7 $\pm$ 0.15 min; C_{12}: 21.9 $\pm$ 0.19 min). Overall, a tendency of decreasing sensitivity was observed during this experiment attributable to the contamination of the mass spectrometer with matrix components (discernible in the crystallisation of inorganic salts on the curtain plate). As pointed out earlier, this effect can be taken into consideration by regularly placing entire calibration series within a sample sequence.

4.3.4 Quantitative analysis of APEO with the fluorescence detector in comparison with LC-MS

Quantitative analysis of AP/APEO by HPLC-FL can be performed with external standard solutions of mixtures of AP or APEO. Initially quantification of oligomeric mixtures was based on the elaborate procedure of normal-phase analysis with subsequent quantification of all oligomeric peaks [27]. Kiewiet et al. [28] have described the general principle of quantification of ethoxymers in reversed-phase LC with spectroscopic detection in detail using the example of derivatised alcohol ethoxylates. Based on this method the quantitative analysis of

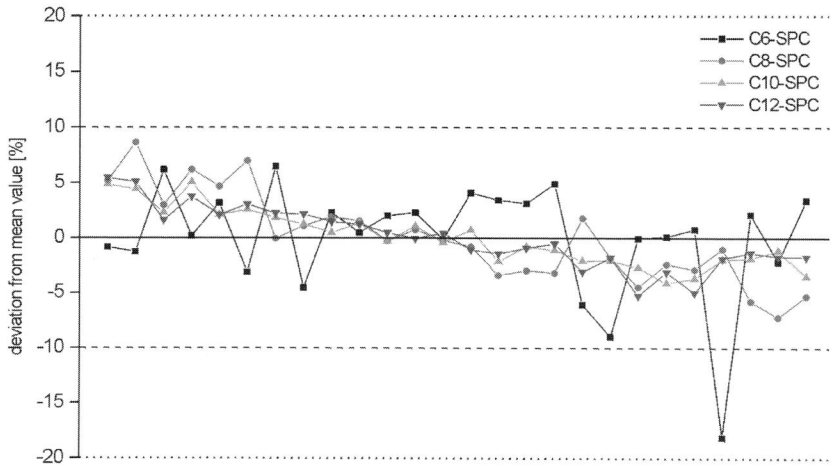

Fig. 4.3.6. Repeatability of peak area of four SPC homologues, determined from 26 consecutive LC separations of 10 μL injections of a biodegradation sample. Total duration: 19 h.

APEO with LC-FL was further elaborated by de Voogt et al. [16]. Quantification of the APEO was based on the assumption that the average EO chain length in sediment samples was close to the EO chain length of the standard mixtures. Whether this assumption is valid can be evaluated by normal-phase HPLC results (see Chapter 2.2 and Section 4.3.3). If the average chain length in actual samples does not deviate much from that in calibration mixtures, taking into account differences in molar responses, the sum of ethoxylates found in NP-HPLC should be close to the RP-HPLC result. Contrary to ESI-MS, the response factor in fluorescence detection does not vary substantially with EO chain length, whereas in UV at higher chain lengths absorptivity decreases [29].

As an example, Table 4.3.1 presents the results calculated from reversed-phase and normal-phase HPLC analysis, respectively, of samples taken from an industrial WWTP [16]. Concentrations were expressed in microgram per gram of material as received. In the water sample, reversed-phase HPLC analysis showed the presence of NP and NPEO (Fig. 4.3.7(A)) with levels of 0.008 and 0.383 μg g^{-1} (sum of NPEO), respectively. Normal-phase analysis of NPEO oligomers

TABLE 4.3.1

Concentrations of NP and NPE calculated from both normal- (NP) and reversed-phase (RP) HPLC-FL analysis of wastewater and sludge samples from an industrial waste water plant [16]

Compound	Water		Sludge[a]		Analytical limit of detection (μg g^{-1} sample)
	RP-analysis (μg g^{-1} water)	NP-analysis (μg g^{-1} water)	RP-analysis (μg g^{-1} sludge)	NP-analysis (μg g^{-1} sludge)	
NP	0.008		6.822 (0.004)		0.5
NPEO$_3$				5.13 (2.40)	1.6
NPEO$_4$				1.47 (0.64)	1.8
NPEO$_5$				1.71 (0.65)	2.0
NPEO$_6$				1.51 (0.48)	2.2
NPEO$_7$		0.016		1.78 (0.43)	2.5
NPEO$_8$		0.029		2.25 (0.52)	2.7
NPEO$_9$		0.045		2.55 (0.50)	2.9
NPEO$_{10}$		0.057		3.20 (0.56)	3.1
NPEO$_{11}$		0.064		3.23 (0.39)	3.3
NPEO$_{12}$		0.061		3.06 (0.54)	3.5
NPEO$_{13}$		0.045		2.75 (0.54)	3.7
NPEO$_{14}$		0.025		2.40 (0.58)	3.9
NPEO$_{15}$		0.013		2.39 (0.38)	4.1
NPEO$_{16}$		0.005		2.04 (0.52)	4.3
NPEO sum	0.383	0.360	40.16 (0.88)	35.48 (6.32)	0.5

[a]Standard deviation (based on duplicate analysis) are given in parentheses.

showed the presence of oligomers ranging from 7 to 16 EO units (cf. chromatogram (C) in Fig. 4.3.7). The total sum of NPEOs (0.36 μg g^{-1}) calculated from normal-phase HPLC was in good agreement with the concentration quantified by reversed-phase analysis. Similarly, the concentrations in the sewage sludge were compared and found to be 40.2 μg g^{-1} (reversed-phase) and 35.5 μg g^{-1} (normal-phase), respectively. This finding confirms that the assumption of equal molar absorptivity for ethoxymers in normal-phase appears to hold.

The principle outlined above, i.e. combination of normal- and reversed-phase analysis, with quantification based on the peaks obtained in reversed-phase LC and using the normal-phase runs to identify the appropriate (i.e. most closely matching) calibrant mixture, has been applied in analysis of marine sediments. The samples were

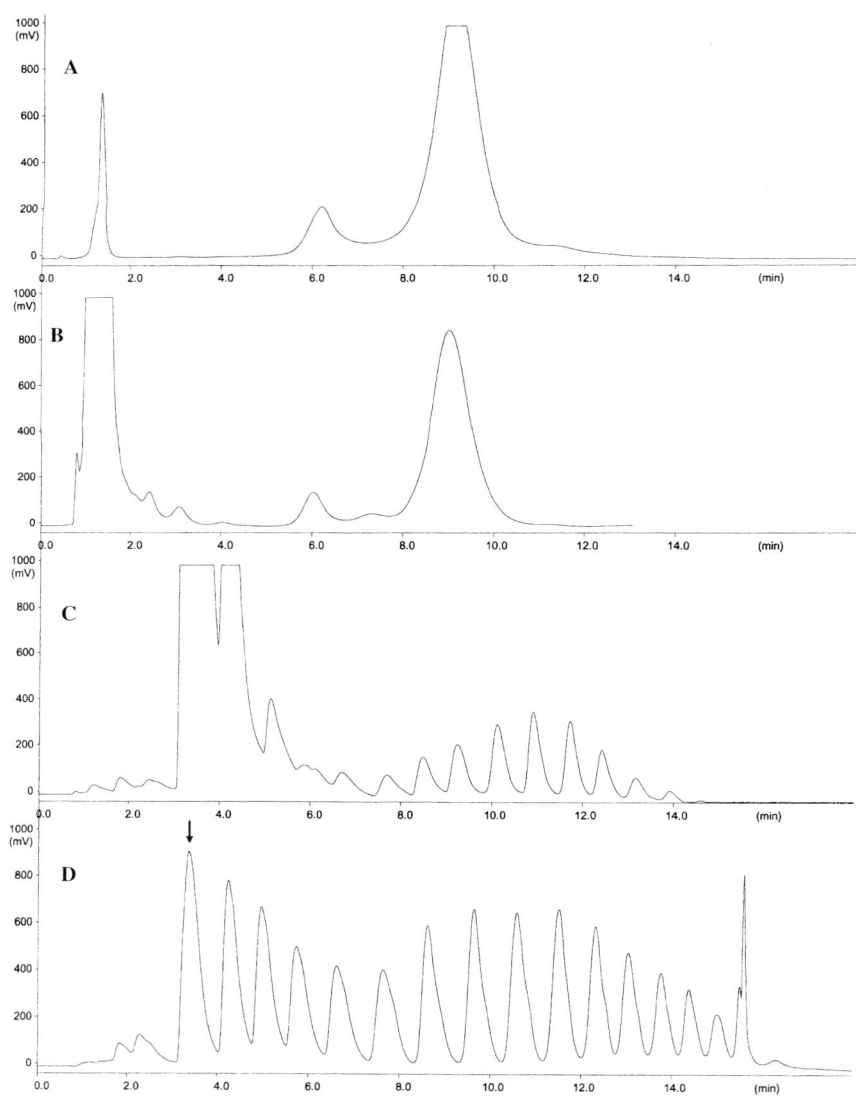

Fig. 4.3.7. Reversed-phase HPLC (A + B) and normal-phase HPLC (C + D) chromatograms of wastewater (A + C) and sludge (B + D) taken from an industrial wastewater treatment plant. Detector: fluorescence detector. Chromatograms used for calculation of the results shown in Table 4.3.1.

analysed using fluorescence detection with an excitation wavelength of 225 nm and an emission wavelength of 301 nm. Two different HPLC columns were used to acquire the maximum amount of information. A normal-phase (Hypersil$^©$ 3 NH$_2$) column was used for identification and selection of the standard that most closely resembled the sample. A reversed-phase (Lichrospher$^©$ C$_{18}$) column was used for quantification of the total amount of AP/APEO. NPEO was quantified using the closest matching commercial NPEO mixture as a quantification standard. Results thus obtained for sediments sampled in 1995 and 1999 from the Dutch coastal zone of the North Sea varied between 10 and 50 ng NPEO g^{-1} (dry weight) [16,30].

In another study, samples from the Dutch coastal zone of the North Sea were collected in 1999 and 2000 and analysed by LC-ESI-MS [31]. At locations corresponding to those of the 1995/1999 studies, levels (10–30 ng NPEO g^{-1}) were observed in close agreement with the ones reported earlier. It can be concluded that, provided careful evaluation of which commercial mixture will be used as a calibrant in LC-FL, the result thus calculated can be similar to those obtained by LC-MS.

It is obvious that due to the complexity of APEO mixtures, precise quantification of environmental samples with LC-FL is very difficult. In such samples, each of the isomers originally present within one oligomeric group may have undergone different changes as a result of processes such as weathering and (bio)degradation, which can be expected to be structure dependent. In addition, the various oligomeric groups may have been degraded (i.e. shortening of the EO chain) to a different extent. Hence the actual samples taken from the environment usually contain mixtures with a composition different from that of the available standards. The total NPEO concentration can be quantified satisfactorily with the closest matching commercial standard because the fluorescence response factors do not vary much with ethoxymer chain length. More accurate analysis can only be accomplished when: (i) individual isomer and oligomer standards of sufficient purity become available through synthesis; and (ii) analytical techniques have more resolving power. As to the latter, LC-MS methods are described extensively in Chapter 2 of this book, whereas high-temperature GC-MS, or hyphenated-normal- and reversed-phase LC-FL are techniques that need further investigation.

In HPLC of APEO and AP, retention time shifts relative to those in standard mixtures can easily occur when the composition of the sample mixture differs from the standard. To give an example, in normal-phase

HPLC-FL the chromatograms of a standard mixture provide a single common peak for all APEOs with, for example, eight EO units (APEO$_8$). However, if the sample contains a mixture of octylphenol (OPEO$_8$) and nonylphenol (NPEO$_8$) octaethoxylates, with the octyl compounds more abundantly present, while the standard contains nonyl compounds only, a shift in retention times may be observed. A similar phenomenon can occur in RP-HPLC (where, in principle, separation of OP, NP, OPEO and NPEO is achieved, see Fig. 2.2.4 in Chapter 2.2) if the ethoxylate composition of, say, the NPEO peak (containing all ethoxylate oligomers containing a nonyl moiety) differs from that of a standard. In practice, these phenomena can sometimes hamper the proper identification of peaks in NP-HPLC, thereby impeding a quantitative assessment. As a consequence, identification and quantification of non-ionic surfactants by HPLC must be based on concomitant reversed-phase and normal-phase HPLC analysis.

4.3.5 Conclusion

The quantification of surfactants in environmental samples needs further development, particularly in so far as quality assurance of the analysis is concerned. Since the majority of the individual isomers and oligomers involved are not yet available as standards, quantification has to be based in part on external standards of commercially available mixtures. As this holds for both LC-FL and LC-MS analyses, both suffer from this shortcoming. Yet, samples analysed with both the methods have shown good agreement of resulting data.

For the quantification of LAS and their metabolites in LC-MS, in particular the respective position of the phenyl ring at the alkane or the alkylcarboxylate chain influences the retention time in RP-HPLC significantly. This might also hamper the proper allotment and thus quantification of the individual species too.

For the quantification of ethoxylated non-ionic surfactants in LC-MS, the effect of changing response factors with changes in EO chain length requires particular attention. The urgent need for proper internal standards has recently been addressed with the ^{13}C-labelled NP and NPEO compounds that have become available. These will help improve the quality of LC-MS quantification of ethoxylated non-ionic surfactants significantly.

REFERENCES

1   A.P. Bruins, *J. Chromatogr. A*, 794 (1998) 345.
2   P.L. Ferguson, C.R. Iden and B.J. Brownawell, *Anal. Chem.*, 72 (2000) 4322.
3   R. King, R. Bonfiglio, C. Fernández-Metzler, C. Miller-Stein and T. Olah, *J. Am. Soc. Mass Spectrom.*, 11 (2000) 942.
4   R.J.C.A. Steen, A.C. Hogenboom, P.E.G. Leonards, R.A.L. Peerboom, W.P. Cofino and U.A.T. Brinkman, *J. Chromatogr. A*, 857 (1999) 157.
5   N. Jonkers, R.W.P.M. Laane and P. de Voogt, *Environ. Sci. Technol.*, 37 (2002) 321.
6   H. Moriwaki, T. Nakano, S. Tsunoi and M. Tanaka, *Rapid. Commun. Mass Spectrom.*, 15 (2001) 2208.
7   D.Y. Shang, M.G. Ikonomou and R.W. Macdonald, *J. Chromatogr. A*, 849 (1999) 467.
8   T.L. Constantopoulos, G.S. Jackson and C.G. Enke, *J. Am. Soc. Mass Spectrom.*, 10 (1999) 625.
9   B.K. Choi, D.M. Hercules and A.I. Gusev, *J. Chromatogr. A*, 907 (2001) 337.
10 P.L. Ferguson, C.R. Iden and B.J. Brownawell, *Anal. Chem.*, 72 (2000) 4322.
11 M. Petrovic, F. Ventura, A. Díaz and D. Barceló, *Anal. Chem.*, 73 (2001) 5886.
12 N. Jonkers and P. de Voogt, unpublished results.
13 C. Crescenzi, A. DiCorcia, R. Samperi and A. Marcomini, *Anal. Chem.*, 67 (1995) 1797.
14 P.L. Ferguson, C.R. Iden and B.J. Brownawell, *J. Chromatogr. A*, 938 (2001) 79.
15 T. Okada, *Macromolecules*, 23 (1990) 4216.
16 P. de Voogt, K. de Beer and F. van der Wielen, *Trends Anal. Chem.*, 16 (1997) 584.
17 M. Petrovic and D. Barceló, *J. Mass Spectrom.*, 36 (2001) 1173.
18 T. Isobe, H. Nishiyama, A. Nakashima and H. Takada, *Environ. Sci. Technol.*, 35 (2001) 1041.
19 J.L. Gundersen, *J. Chromatogr. A*, 914 (2001) 161.
20 C. Crescenzi, A. DiCorcia, A. Marcomini and R. Samperi, *Environ. Sci. Technol.*, 31 (1997) 2679.
21 A. Marcomini, C. Tortato, S. Capri and A. Liberatori, *Ann. Chim.*, 83 (1993) 461.
22 N. Jonkers, T.P. Knepper and P. de Voogt, *Environ. Sci. Technol.*, 35 (2001) 335.
23 A. Di Corcia, R. Cavallo, C. Crescenzi and M. Nazzari, *Environ. Sci. Technol.*, 34 (2000) 3914.
24 T.P. Knepper and M. Kruse, *Tenside Surf. Deterg.*, 37 (2000) 41.
25 M.G. Ikonomou, A.T. Blades and P. Kebarle, *Anal. Chem.*, 63 (1991) 1989.

26   C. Molina, M. Honing and D. Barceló, *Anal. Chem.*, 66 (1994) 4444.

27   M. Ahel and W. Giger, *Anal. Chem.*, 57 (1985) 1577.

28   A.T. Kiewiet, J.M.D. van der Steen and J.R. Parsons, *Anal. Chem.*, 67 (1996) 4409.

29   C.S. Zhou, A. Bahr and G. Schwedt, *Anal. Chim. Acta*, 236 (1990) 273.

30   P. de Voogt, O. Kwast, R. Hendriks and N. Jonkers, In: A.D. Vethaak, et al. (Eds.), *Estrogens and Xeno-Estrogens in the Aquatic Environment in The Netherlands*, RIKZ/RIZA, The Hague, 2002.

31   N. Jonkers, R.W.P.M. Laane and P. de Voogt, Proceedings 12th Annual Meeting SETAC Europe, Vienna, 2003, p. 185.

4.4 STABILITY OF SURFACTANTS IN POST-SAMPLING STORAGE

Mira Petrovic and Damia Barceló

4.4.1 Introduction

Environmental monitoring programmes often require the analysis of numerous samplfes that cannot be analysed immediately after sampling. The stability of target compounds in water samples during their transport and storage is the key issue as regards quality assurance parameters. Loss of sample integrity for some compounds during transportation, because of hydrolysis and adverse shipping conditions (high temperature and/or humidity), may bring into question the reliability of the results obtained. The half-life of surfactants at low $\mu g \, L^{-1}$ levels is very much dependent on storage conditions (temperature, pH, exposure to light, use of preservatives) and on the matrix (concentration of particulate matter, total organic carbon (TOC), concentration of other contaminants and nutrients). In some instances, the degradation products of surfactants are more stable than the parent compounds, so analysis should be aimed at the breakdown products. For some compounds, degradation can be rapid and samples should be analysed as soon as possible after collection. However, government agencies and private laboratories carrying out environmental monitoring often do not possess the instrumentation needed to analyse surfactants and samples are routinely sent to other laboratories for analysis.

During the last two decades a large amount of environmental data for ionic and non-ionic surfactants in various matrices has been generated. However, one of the doubts regarding the quality of analytical data is attributed to the stability of target compounds. In approximately 30–40% of papers dealing with surfactant analysis, the storage conditions and the time interval between sampling and analysis are not specified or the vague phrases, such as 'stored at 4°C until analysis' are used. To improve the quality assurance of analytical protocols for different groups of non-ionic and ionic surfactants several comprehensive stability studies were performed comparing standard and alternative preservation methods [1–4]. Standard preservation

Comprehensive Analytical Chemistry XL

T.P. Knepper, D. Barceló and P. de Voogt (Eds)

495

techniques employ the addition of an additive to prevent microbiological degradation and storing at 4°C. Preservation agents used in surfactant analysis are formaldehyde, $HgCl_2$, sulphuric acid and sodium azide. However, these so-called standard or conventional methods are not always appropriate and stability studies showed that significant losses of some compounds occur when the sample is kept in a water matrix, even preserved with the chosen additive. Alternative methods include the freeze-drying and the use of solid phase extraction (SPE) materials (disks, cartridges and disposable pre-columns) for the stabilisation and storage of surfactants pre-concentrated from water samples.

4.4.2 Stability in a water matrix

Non-ionic surfactants
The experimental evaluation [4] of the stability of non-ionic surfactants (nonylphenol ethoxylates, NPEOs, and alcohol ethoxylates, AEOs) during sample storage showed that aqueous samples can be stored at 4°C without addition of any preservative only for a short time (a maximum of 5 days). The most often used preservative is formaline (1–8% (v/v) of 37% solution of formaldehyde in water).

Ahel and Giger [1] determined that after adding formaldehyde solution (20 g L^{-1}) to a biologically treated municipal wastewater, the concentrations of $NPEO_1$, $NPEO_2$ and nonylphenol (NP) remained constant over a storage period of 12 days at 4°C. At room temperature and without preservation significant degradation occurred. To prevent biological degradation of non-ionic surfactants DiCorcia et al. [5] recommended addition of formaline (1%) to the sample and its subsequent freezing. To overcome problems associated with the storage and transport in frozen conditions, de Voogt et al. [6] recommended a higher formaldehyde concentration (3%) and storage at 4°C. Marcomini et al. [7] used an even higher concentration of formaldehyde (5%, v/v) for the stabilisation of NPEOs and nonylphenol carboxylates (NPECs) in seawater from the Venice lagoon. Naylor et al. [3] studied the stability of NPEOs and NP in river water preserved with or without 1% formaline and stored at 4°C. After 2 weeks, there was a clear deterioration of water samples kept without added preservative. The losses of $NP_{3-17}EO$, NP_2EO, NP_1EO and NP after 4 weeks were 70, 63, 49 and 37%, respectively. When river water was preserved with 1% formaline the losses were 25, 42, 7 and 15%, respectively. An additional problem, when using formaldehyde as a preservative additive, is incompatibility with some analytes (formaldehyde reacts with glycols to form cyclic acetals) [8] and with some derivatisation agents, e.g. phenyl isocyanate used for the derivatisation of AEOs [9]. Szymanski and Lukaszewski [10] showed that: (i) raw and treated sewage samples preserved with 1% formaline are insufficiently stable for long-term maintenance of non-ionic surfactant concentration; (ii) raw ethyl acetate extracts of sewage samples demonstrate limited long-term stability of non-ionic surfactant concentration; and (iii) dried ethyl acetate extracts of sewage samples mixed with chloroform (1:2) exhibit stable non-ionic surfactant concentration over a minimum of 28 months. This manner of sewage sample preparation may be recommended for long-term storage.

Another pre-treatment often used to prevent biological degradation of organic contaminants in water is acidification with sulfuric acid to pH < 3 [4]. The change in concentrations of NPEO, AEOs and PEGs in spiked groundwater (500 μg L^{-1}) and in wastewater treatment plant (WWTP) influent preserved by addition of formaline (1 and 3%, v/v) and by acidification to pH 3 with sulfuric acid and kept for 30 days at 4°C are shown in Table 4.4.1.

Acidification to pH 3 was a more effective preservation method than addition of formaldehyde. The losses of alcohol ethoxylates were less than 5% for C$_{10}$ and C$_{12}$ and up to 17% for CEO$_{14}$. When the groundwater

TABLE 4.4.1

Change in concentration (%) of surfactants in spiked groundwater (500 μg L^{-1}) and in the influent of WWTP Igualada (Catalonia, NE Spain) kept for 30 days at 4°C under different preservation conditions [4]

Compound	Ground water		Wastewater[a]		
	pH = 3	Formaline 3%	pH = 3	Formaline	
				1%	3%
NPEO	0	0	− 45	− 85	− 45
C$_{10}$EO	− 4	− 4	89	105	93
C$_{12}$EO	− 3	− 27	− 74	− 80	− 18
C$_{14}$EO	− 18	− 40	− 87	− 91	− 89
PEG	7	8	20	29	16

[a]Initial concentrations: NPEO, 719 μg L^{-1}; C$_{10}$EO, 150 μg L^{-1}; C$_{12}$EO, 368 μg L^{-1}; C$_{14}$EO, 209 μg L^{-1}; PEG, 3420 μg L^{-1}.

sample was preserved with formaldehyde, losses of AEOs were up to 40%. No loss of NPEO in groundwater matrix was observed. In more complex wastewater samples, the serious changes of sample occurred during the period of time studied, indicating that conventional methods of preservation and storage cannot prevent changes: significant losses of NPEO occurred. The mass spectra of NPEOs show disappearance of the highest and enrichment in the lower oligomers in an NPEO oligomeric mixture. The shortening of the polyethoxy chain led to the formation of biointermediates containing 3–5 fewer ethoxy units than the parent compound. The disappearance of the highest AEO homologues ($n = 12$–18) in wastewater stored for 30 days and enrichment of the $C_{10}EO$ was observed. The losses of C_{12}–C_{14} were up to 88% when the sample was preserved by acid, almost 90% when the sample was preserved with 1% of formaldehyde and from 17 to 86% when the concentration of formaldehyde was increased to 3%. The enrichment of the C_{10} homologue, up to 105%, was observed. The C_7 and C_8 homologues were also detected. In contrast to NPEOs, only negligible changes in the oligomer distribution were observed, leading to the conclusion that shortening of alkyl-chain and central fission of the molecule prevail over shortening of the ethoxy chain. As a result of central fission of AEOs, an increase in concentration of PEGs, up to 28%, was determined. Additionally, different acidic forms of polyethylene glycols, such as mono-carboxylated (MCPEGs) and dicarboxylated (DCPEGs) were detected.

Kubeck and Naylor [2] tested four chemical preservatives (formaline, sulfite, bisulfite and a mixture of formaline/sulfite) by analysing NPEOs in river water over a period of 4 weeks. The sulfite- and bisulfite-preserved samples deteriorated drastically beginning after the first week, dropping 60–80% in NPEO concentration after four weeks. Formaline maintained the NPEO concentration within 10% of the original through at least two weeks. Formaline/sulfite mixture showed 25% loss after four weeks.

Another preservative used in the analysis of non-ionic surfactants is sodium azide (NaN_3), normally used at a concentration of 100 mg L^{-1}. Kiewiet et al. [11] tested much higher concentrations and found that 0.01 M (650 mg L^{-1}) sodium azide could not prevent substantial losses of the alcohol ethoxylates, probably due to the fact that sodium azide is only an inhibitor for aerobic degradation. However, in closed bottles, anaerobic degradation processes could play an important role as well. They observed substantial losses (17–81%) of non-ionic surfactants during transport and seven days storage of the wastewater samples.

Mercuric chloride (HgCl$_2$) is another preservative used in surfactants analysis. However, the evaluation of its effectiveness and practicability has still to be established. HgCl$_2$ (10 mg L^{-1}) is recommended by the USEPA for the preservation of polar pesticides in water samples. Crescenzi et al. [12] determined that the addition of HgCl$_2$ reached the goal of preserving carbamate pesticide. However, negative side-effects overwhelmed the beneficial ones, because up to 100% loss of 14 of the 34 tested pesticides was observed. A possible explanation is an increase of the rate of hydrolysis induced by Hg(II), the chelate-forming metal, as observed for chlorpyriphos.

Ionic surfactants
Addition of formaldehyde up to a concentration of 8% has been found to prevent biodegradation of linear alkylbenzene sulphonates (LAS) in wastewater and environmental samples. The stability study conducted in the authors' laboratory [4] showed that when preserved with formaldehyde (3%) the losses of LAS in a groundwater matrix were up to 14% after 30 days of storage at 4°C. When kept without any preservation, the losses of LAS in the groundwater matrix ranged from 23 to 57% depending on the length of the alkyl chain. However, in a wastewater matrix, due to the complexity of such samples, including high content of inorganic salts and total organic carbon (TOC), LAS suffered more degradation (Table 4.4.2), and when preserved with 3%

TABLE 4.4.2

Change in concentration (%) of LAS in spiked groundwater (total concentration 500 µg L^{-1}) and in the influent of WWTP Igualada (Catalonia, NE Spain) kept for 30 days at 4°C under different preservation conditions [4]

LAS	Ground water			Wastewater[a]		
	Without preservation	pH = 3	Formaline, 3%	pH = 3	Formaline	
					1%	3%
C$_{10}$	− 23	− 17	− 11	− 55	− 38	− 14
C$_{11}$	− 28	− 13	− 11	− 70	− 53	− 23
C$_{12}$	− 38	− 13	− 14	− 73	− 47	− 30
C$_{13}$	− 57	− 14	− 13	− 78	− 43	− 29

[a]Initial concentrations: C$_{10}$LAS, 208 µg L^{-1}; C$_{11}$LAS, 904 µg L^{-1}; C$_{12}$LAS, 257 µg L^{-1}; C$_{13}$LAS, 123 µg L^{-1}.

formaldehyde the losses were up to 30%. It was observed that degradation affects longer alkyl chain homologues more than shorter ones causing the shift in homologue distribution. The results show significant differences between distribution of LAS homologues in samples kept under different storage and preservation conditions, yielding a lower LAS average molecular weight after storage.

4.4.3 Stability on SPE cartridges

An alternative to storage of the original water matrix is enrichment and temporary storage of organic contaminants on solid phase extraction (SPE) material (disposable cartridges, columns or disks). One of the advantages of using disposable SPE columns or disks is the storage space, since one cartridge usually replaces bottles of 500–1000 mL. The easy shipping of cartridges to the central laboratory for final analyses is another advantage, making it unnecessary to perform the analysis immediately after sampling. Stability studies performed for pesticides showed that, for example, triazine herbicides pre-concentrated on C-18 disks were found to be stable for up to three months when stored at $-20°C$ [13,14], some organophosphorus pesticides pre-concentrated on disposable SPE pre-columns for up to 8 months [15], and phenolic compounds for up to two months [16]. Recently, similar stability studies have also been undertaken for non-ionic (NPEOs and AEOs) and ionic (LAS) surfactants [4].

All of these studies showed that temporary storage of surfactants in an SPE matrix is more efficient than storage of water samples. However, the handling and storage of SPE cartridges, columns or disks after pre-concentration require particular attention. It is suggested that the presence of residual water in the SPE matrix can cause losses of analytes due to the hydrolysis of compounds during storage. To avoid this negative effect it is necessary to remove traces of water through desiccation. Senesman et al. [17] studied the effects of different desiccation methods (freeze-drying, vacuum desiccation, desiccation with $CaSO_4$) and concluded that freeze-drying is probably the most appropriate method for fast removal of residual water from SPE C-18 disks. However, potential adverse effects such as loss of volatile compounds and excessive chromatographic noise from undesirable chemical interferences may overshadow any increase in analyte stability. Therefore, depending on the vapour pressure of analytes, an alternative desiccation treatment should be applied. Standard procedures used are air-drying (e.g. using a Baker SPE apparatus, J.T. Baker, Deventer, The Netherlands connected to a vacuum system) or drying under N_2. DiCorcia et al. [18] recommended the washing of cartridges with methanol before storage to minimise the amount of residual water remaining in the cartridge. However, in this instance the maximum volume of methanol that can be passed through the cartridge, depending on the sorbent volume and analytes studied, should be carefully optimised.

Another important step is slow defreezing of SPE cartridges at room temperature before elution. SPE cartridges (or columns and disks) should be removed from storage at 4°C and −20°C and kept at room temperature for a period varying from 1–2 to 6–8 h depending on the storage conditions.

The change in concentration of LAS and non-ionic surfactants (NPEOs and AEOs) pre-concentrated on LiChrolut C-18 SPE cartridges from spiked groundwater and stored at room temperature and −20°C, respectively, for 90 days, are summarised in Table 4.4.3. Complete recovery of NPEO, $C_{10}EO$ and $C_{12}EO$ was observed on cartridges stored at −20°C during the period of time studied. The study showed only small losses of $C_{14}EO$, up to a maximum of 15%. Slightly higher losses occurred during 90 days at 4°C (up to a maximum of 16% for $C_{14}EO$, and less than 10% for the remaining compounds studied). After 30 days, losses were negligible for all compounds. At room temperature, after 7 and 14 days, respectively, there were no serious losses of compounds studied. After 30 days some degradation was noticed and small losses of all compounds occurred (up to 15%).

The further results obtained also indicate that storage on SPE material is a better preservation procedure for LAS extracted from ground- or wastewater samples than storage in a water matrix at 4°C (Table 4.4.3). Evaluation of stability of LAS pre-concentrated from groundwater on LiChrolut C-18 disposable cartridges showed that the complete recovery of C_{10}–$C_{12}LAS$ was observed on cartridges stored at −20°C for the 90 days. Only $C_{13}LAS$ was partially lost after 90 days (maximum 15%). Slightly higher losses occurred during 90 days at 4°C. At room temperature, after 7 and 14 days, respectively, only negligible losses occurred. After 30 days some degradation was noticed and small losses, of up to 17%, were observed.

The data for stability of non-ionic surfactants pre-concentrated from wastewater are shown in Fig. 4.4.1. At −20°C (Fig. 4.4.1(B)) the losses

TABLE 4.4.3

Change in concentration (%) of LAS and non-ionic surfactants in spiked groundwater pre-concentrated on SPE cartridges (RSD values of three replicates are given in parentheses) [4]

Compound	Storage conditions														
	Room temperature					4°C					−20°C				
	7 days	14 days	30 days	60 days	90 days	7 days	14 days	30 days	60 days	90 days	7 days	14 days	30 days	60 days	90 days
NPEO	0[a] (6)	0 (5)	0 (7)	−8 (4)	−12 (7)	0 (5)	0 (5)	0 (7)	0 (6)	0 (7)	0 (5)	0 (5)	0 (4)	0 (6)	0 (8)
$C_{10}EO$	0 (5)	0 (6)	0 (5)	−13 (8)	−19 (10)	0 (2)	0 (6)	0 (4)	0 (6)	−6 (4)	0 (6)	0 (3)	0 (5)	0 (9)	0 (13)
$C_{12}EO$	0 (4)	0 (8)	−5 (6)	−16 (9)	−22 (8)	0 (3)	0 (5)	0 (5)	−5 (3)	−8 (5)	0 (3)	0 (8)	0 (5)	0 (5)	−5 (5)
$C_{13}EO$	0 (4)	−6 (4)	−15 (10)	−20 (13)	−25 (11)	0 (1)	0 (5)	−4 (3)	−8 (3)	−16 (8)	0 (4)	0 (3)	0 (4)	−6 (5)	−14 (5)
$C_{10}LAS$	0 (4)	0 (5)	0 (6)	−10 (8)	−9 (6)	0 (4)	0 (7)	0 (8)	0 (6)	0 (10)	0 (2)	0 (2)	0 (6)	0 (7)	0 (7)
$C_{11}LAS$	0 (2)	0 (5)	0 (7)	−10 (6)	−15 (7)	0 (3)	0 (3)	0 (5)	0 (5)	0 (5)	0 (3)	0 (4)	0 (4)	0 (5)	0 (5)
$C_{12}LAS$	0 (5)	0 (4)	0 (5)	−9 (6)	−16 (11)	0 (3)	0 (5)	−5 (4)	−6 (4)	−9 (6)	0 (3)	0 (6)	0 (3)	0 (3)	−5 (4)
$C_{13}LAS$	−8 (4)	−16 (7)	−17 (10)	−25 (10)	−29 (9)	0 (4)	−7 (4)	−14 (8)	−18 (10)	−23 (10)	0 (4)	0 (4)	−11 (5)	−11 (6)	−15 (8)

[a]Values less than RSD are expressed as 0.

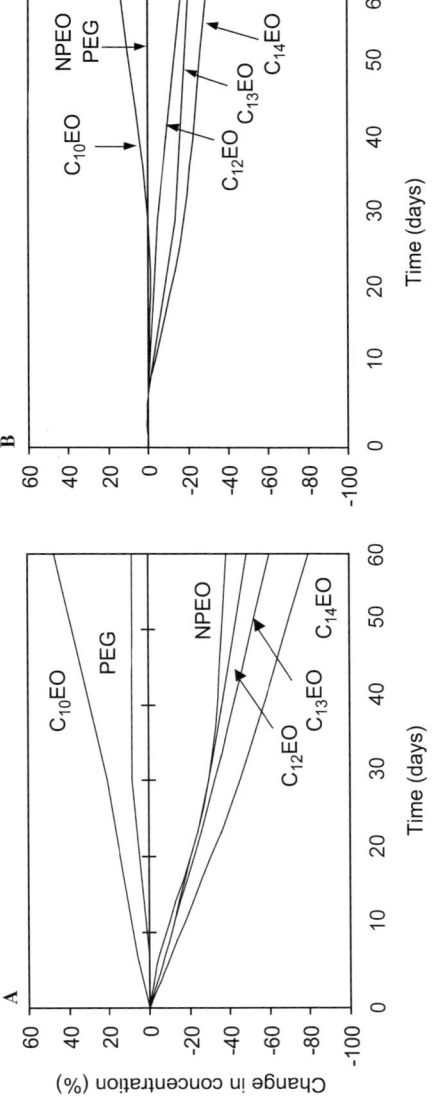

Fig. 4.4.1. Change in concentration (%) of non-ionic surfactants in the wastewater influent of WWTP Igualada pre-concentrated on LiChrolut C-18 cartridges kept at room temperature (A) and at −20°C (B) [4].

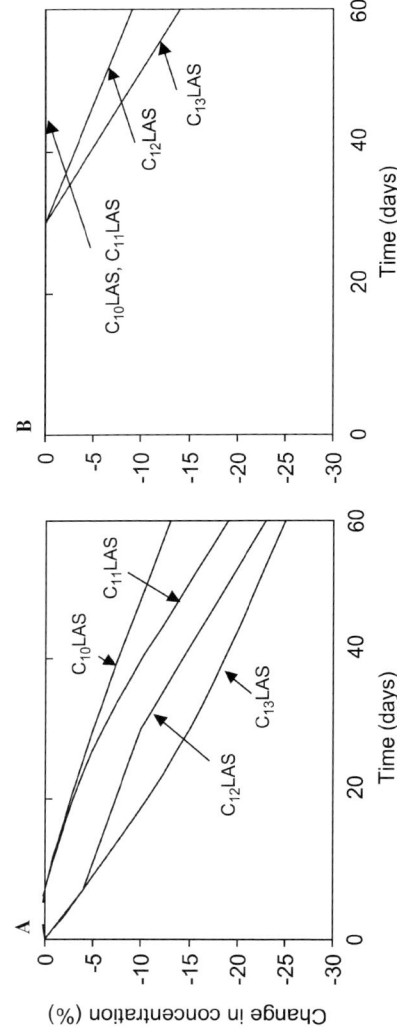

Fig. 4.4.2. Change in concentration (%) of LASs in the wastewater influent of WWTP Igualada pre-concentrated on LiChrolut C-18 cartridges kept at room temperature (A) and at − 20°C (B) [4].

of AEOs were from 17 to 29%, showing an increase in concentration of $C_{10}EO$, as observed previously in a water matrix. NPEOs and PEGs suffered no losses, and were completely recovered from cartridges stored for 60 days at $-20°C$.

At room temperature (Fig. 4.4.1(A)), after 7 days' storage, the changes in concentrations measured were less than 11%, which can be considered an acceptable value. A one-week period usually permits transport of samples to the central laboratory or shipping of samples by mail without any special requirements. However, longer storage of SPE cartridges at room temperature, led to the significant change of this complex sample. The loss of NPEOs was 39%. The least stable of the surfactants evaluated were $C_{12}-C_{14}$ alcohol polyethoxylates with losses ranging from 49 to 80%. The shortening of the ethoxy chain of NPEOs is observed, is as the increase in concentration of short alkyl-chain AEOs and PEGs. The transformation products identified were MCPEGs and DCPEGs, which means that the degradation processes that occur on SPE material are equivalent to those that occur in a water matrix.

When pre-concentrated from wastewater the losses of LAS ranged from 13 to 25% after 60 days of storage at room temperature (Fig. 4.4.2(A)). After 7 days, losses were lower than 5%, which allows manipulation and transportation of cartridges without any special requirements. When kept at $-20°C$ (Fig. 4.4.2(B)) the losses after 60 days were lower than 15% for C_{12} and $C_{13}LAS$, while C_{10} and C_{11} homologues were completely recovered.

4.4.4 Conclusion and recommendations

Stability of surfactants in a water matrix, even using different preservation agents, is poor and serious, quantitative and qualitative changes in sample integrity occur if the storage exceeds 7 days. The most suitable preservation additive is formaldehyde (minimum 3%) for non-ionic surfactants and LAS, and acidification to pH < 3 for benzene and naphthalene sulphonates. However, storage for longer than 7 days is not recommended.

The use of SPE cartridges for the stabilisation of non-ionic and ionic surfactants is more suitable than storing water samples under conventional conditions. A great advantage of SPE cartridges is easy shipping and the reduction in the storage space. All non-ionic surfactants studied (APEOs, AEOs, CDEAs) and LAS showed significantly better stability on SPE material than in a water matrix.

Therefore, as a rule, the sample should be pre-concentrated on SPE cartridges immediately after sampling or, as a maximum, within 1–2 days. In order to avoid the hydrolysis of analytes pre-concentrated in the SPE matrix it is necessary to remove the traces of residual water by desiccation or methanol washing. The dry cartridges can be shipped at room temperature if the duration of transport does not exceed 7 days. Otherwise, transport and storage should be accomplished at 4°C. Storage at −20°C is feasible for a longer period (up to two months), especially for samples of less complex matrix (groundwater).

REFERENCES

1   M. Ahel and W. Giger, *Anal. Chem.*, 57 (1985) 1577.
2   E. Kubeck and C.G. Naylor, *J. Am. Oil Chem. Soc.*, 67 (1990) 400.
3   C.G. Naylor, J.P. Mieure, W.J. Adams, J.A. Weeks, F.J. Castaldi, L.D. Ogle and R.R. Romano, *J. Am. Oil Chem. Soc.*, 69 (1992) 695.
4   M. Petrović and D. Barceló, *Fresenius J. Anal. Chem.*, 368 (2000) 676.
5   A. DiCorcia, R. Samperi and A. Marcomini, *Environ. Sci. Technol.*, 28 (1994) 850.
6   P. de Voogt, K. de Beer and F. van der Wielen, *Trends Anal. Chem.*, 16 (1997) 584.
7   A. Marcomini, G. Pojana, A. Sfriso and J.-M. Quiroga Alonso, *Environ. Toxicol. Chem.*, 19 (2000) 2000.
8   C. Crescenzi, A. DiCorcia, A. Marcomini and R. Samperi, *Environ. Sci. Technol.*, 31 (1997) 2679.
9   E. Matthijs, M.S. Holt, A. Kiewiet and G.B.J.H. Rijs, *Environ. Toxicol. Chem.*, 18 (1999) 2634.
10 A. Szymanski and Z. Lukaszewski, *Water Res.*, 34 (2000) 3635.
11 A.T. Kiewiet, J.M.D. van der Steen and J.R. Parsons, *Anal. Chem.*, 67 (1995) 4409.
12 C. Crescenzi, A. DiCorcia, M.D. Madbouly and R. Samperi, *Environ. Sci. Technol.*, 29 (1995) 2185.
13 E. Martínez and D. Barceló, *Chromatographia*, 42 (1996) 72.
14 D. Barceló, S. Chiron, S. Lacorte, E. Martínez, J.S. Salau and M.C. Hennion, *Trends Anal. Chem.*, 13 (1994) 352.
15 S. Lacorte, N. Ehresmann and D. Barceló, *Environ. Sci. Technol.*, 29 (1995) 2834.

16   M. Castillo, D. Puig and D. Barceló, *J. Chromatogr. A*, 778 (1997) 301.
17   S.A. Senesman, T.L. Lavy and J.D. Mattice, *Anal. Chem.*, 67 (1995) 3064.
18   A. DiCorcia, F. Casassa, C. Crescenzi, A. Marcomini and R. Samperi, *Environ. Sci. Technol.*, 33 (1999) 4112.

4.5 INTERLABORATORY STUDIES FOR THE DETERMINATION OF SURFACTANTS

Pim de Voogt, Mira Petrovic and Thomas P. Knepper

4.5.1 Introduction

The ability to provide accurate and reliable data is central to the role of analytical chemists, not only in areas like the development and manufacture of drugs, food control or drinking water analysis, but also in the field of environmental chemistry, where there is an increasing need for certified laboratories (ISO 9000 standards). The quality of analytical data is a key factor in successfully identifying and monitoring contamination of environmental compartments. In this context, a large collection of methods applied to the routine analysis of prime environmental pollutants has been developed and validated, and adapted in nationally or internationally harmonised protocols (DIN, EPA). Information on method performance generally provides data on *specificity, accuracy, precision (repeatability* and *reproducibility), limit of detection, sensitivity, applicability* and *practicability*, as appropriate.

In research laboratories, where the objectives are to develop and establish new analytical methods, standardised procedures may commonly not be applied. To prove the reliability of the methods and to guarantee the quality of data generated, characterisation of these methods through the aforementioned performance parameters represents an appropriate means. Additionally, there are some basic requirements which also have to be obeyed in research laboratories in order to guarantee the desired quality of the obtained data: for example, bulk material for an already optimised solid phase extraction (SPE) enrichment should always be from the same source; temperature-related shifts in chromatographic retention times should be avoided by maintaining constant temperature; the stability of reference compounds has to be continuously controlled by comparing previously used calibration series with freshly prepared ones; obtained analytical results have to be verified for their plausibility with the reproducibility checked by standard addition experiments in the matrix investigated; the repeatability of the mass spectrometric results and the reliability of

Comprehensive Analytical Chemistry XL   509
T.P. Knepper, D. Barceló and P. de Voogt (Eds)
© 2003 Elsevier Science B.V. All rights reserved

the autosampler have to be checked by consecutive injections of identical samples (see Chapter 4.4).

Given the possibility of participating in interlaboratory studies, a laboratory's own methods should be subjected to such trials. Taking this all into account, interlaboratory studies provide an important tool to assess and improve laboratory performance [1,2].

Although recognised as of high importance [3], until recently, very few studies for the determination of surfactants have been performed. Gerike et al. [4] tested a high-pressure liquid chromatography (HPLC)–conductometric detection procedure for the analysis of ditallow dimethylammonium chloride (DTDMAC) in sludges, sediments, soil and aqueous environmental samples and demonstrated its applicability during the course of interlaboratory exercises. The procedure gave accurate and reproducible results (i.e. within laboratory standard deviations of typically $\leq 7\%$) for environmental samples and high recoveries of standard additions, generally $\geq 90\%$, from all matrices. Naylor et al. [5] explicitly addressed the quality of measurements of nonylphenol ethoxylates (NPEOs) in samples of river water. Castillo et al. [6] are the only workers to report an interlaboratory study on a variety of surfactants using liquid chromatography mass spectrometric (LC-MS) techniques.

The present chapter discusses the results of a series of three small-scale interlaboratory exercises performed within the framework of the European Union-funded PRISTINE project (ENV4-CT97-0494), employing HPLC-fluorescence (FL) and LC-MS techniques for the determination of ionic and non-ionic surfactants.

4.5.2 Study set-up

Interlaboratory study I
Based on their use in industrial and domestic applications and abundance in environmental samples, six surfactants were selected by the participants of interlaboratory study I. Two different standard mixture solutions (*standards I1* and *I2*) were distributed to all participants for the preparation of the calibration graphs. *Standard I1* contained Arkopal® N100 (NPEO), Marlamid DF 1218 (coconut fatty acid diethanolamide, CDEA) and Marlon A 350 (linear alkylbenzene sulfonate, LAS) at 1000 mg L^{-1} for each compound, while *standard I2* contained Rewopol NOS 5 (nonylphenol ethoxylate sulfate, NPEO-SO$_4$),

Genapol C 050 (alcohol ethoxylate, AEO) and Marlon PS 65 (secondary alkane sulfonate, SAS) at $1000 \, \text{mg L}^{-1}$ for each compound. These mixtures were also used for the preparation of the spiked wastewaters (effluent from WWTP Aachen-Soers, Germany, sampled on January 7, 1999).

Interlaboratory study II

For interlaboratory study II, each participant received three cartridges spiked with water matrices containing different levels of surfactants, and three solutions of analytes. The cartridges were prepared as follows:

Cartridge II1. Distilled water was spiked with $1 \, \text{mg L}^{-1}$ of NPEO (Arkopal® N100) and LAS (a commercial mixture containing $C_{10}-C_{13}$ homologues) (February 2000) and pre-concentrated on C-18 LiChrolut cartridges (Merck, Darmstadt, Germany) using an 'Automated Sample Preparation with Extraction Columns' system (ASPEC XL) from Gilson (Villiers-le-Bel, France).

Cartridge II2. Influent from the WWTP Igualada (Catalonia, NE Spain) was taken into Pyrex borosilicate amber glass bottles and preserved immediately after sampling (February 2000) with addition of formaldehyde (3%, v/v). The sample was filtered through a 0.45 μm membrane filter and pre-concentrated (50 mL) on C_{18} cartridges within 24 h.

Cartridge II3: Filtered wastewater, which was also used for cartridge II2, was spiked with 1 ppm of $NPEO_{10}$ and 1 ppm of LAS C_{10-13}, and then pre-concentrated on C_{18} cartridges using the procedure described above.

Each participant received nine ready-to-elute cartridges corresponding to three replicates for each sample, a spiking solution containing $1000 \, \text{mg L}^{-1}$ of NPEO and LAS (total concentration), respectively, and standards of NPEO and a commercial mixture of LAS with the proportional composition of different homologues as follows: C_{10}, 13.8%; C_{11}, 36.6%; C_{12}, 29.2%; and C_{13}, 19.1%.

Interlaboratory study III

This interlaboratory study was conducted in the same manner as interlaboratory study II, but at lower concentration levels for both the spiked and real samples. Primary effluent from WWTP Igualada (sampled on September 9, 2000) was used as sample III2, and sample III3 was made by spiking sample III2 (spiking level of III3: 200 ppb of NPEO$_{10}$, and 200 ppb of LAS (C_{10-13})).

4.5.3 Results

Interlaboratory study I
Seven laboratories participated in the interlaboratory evaluation within the framework of the PRISTINE, SANDRINE and INEXsPORT European Union Projects [6]. The results obtained for the analysis of diverse classes of surfactants by different analytical methods are listed in Table 4.5.1. The analytical strategies were based on LC coupled to either MS or FL detection in all cases with the exception of one laboratory using a test tube ELISA kit. Samples were spiked with the surfactants NPEO, CDEA, LAS, AEO, NPEO-SO$_4$ and SAS.

Comparison of the different techniques employed was realised by considering the results provided for NPEO, which was spiked at a concentration of 1.8 mg L^{-1} and by being the only compound determined in all of the laboratories. With an RSD of only 2.3%, the results obtained by LC-MS, even when utilising different ionisation techniques [6], were very satisfactory (Table 4.5.1). The results obtained by LC-FL (*laboratories 2 and 6*) highlighted the fact that they were the most uncertain ones and the lowest for the determination of NPEO, indicating an underestimation of the NPEO concentration. For *laboratory 2* this is due to the fact that no specific wavelengths for the determination of NPEO were used. More accurate LC-FL results were obtained in the case of *laboratory 6* which used optimal wavelengths for the determination of NPEO and also by *laboratories 2 and 7* for LAS analysis. Although *laboratory 4* did not use chromatographic techniques, results obtained for the determination of NPEO were comparable to those of the other laboratories. This ELISA determination performed by *laboratory 4* gave the highest value (1.99 mg L^{-1}) for the determined NPEO concentrations. This is a common fact when applying ELISA methods to environmental analysis since cross-reacting substances may interfere with and give additional signals to the biological assay.

Comparison of the three techniques (LC-MS, LC-FL and ELISA) used to determine NPEO leads to the conclusion that LC-MS is the most

TABLE 4.5.1

Results obtained for the analyses of laboratory study I

Surfactant	Quantified values (mg L^{-1})								RSD (%)	
	Spiked level	Lab 1, LC-MS	Lab 2, LC-FL	Lab 3, LC-MS	Lab 4, ELISA	Lab 5, LC-MS	Lab 6, LC-MS	Lab 7, LC-FL	Lab 1–7	LC-MS
NPEO	1.8	1.61	1.22	1.70	1.99	1.89	1.75, 1.53[a]	NA	6.4	2.3
CDEA	1.2	0.56	NA	1.20	NA	0.55	1.00	NA	10.6	10.6
LAS	1.2	0.88	0.95	1.11	NA	1.11	1.52	1.02	5.1	7.1
NPEO-SO$_4$	2.0	2.58	1.66	1.82	NA	NA	2.09	NA	16.2	14.8
AEO	1.8	1.32	NA	1.77	NA	NA	2.38	NA	28.3	28.3
SAS	1.6	2.51	NA	1.70	NA	NA	1.72	NA	21.3	21.3

NA: not analysed.
[a]LC-FL.

513

accurate and precise. In addition, the technique enables unequivocal identification of samples through MS spectra and is thus reliable in determining NPEO at the submicrogram to microgram per litre level in wastewater. Between-laboratory precision values were calculated and given as the reproducibility relative standard deviation (RSD_R). This was determined from the reproducibility standard deviation (S_R) and the average concentration at a particular concentration level. When data from all laboratories were pooled, the RSD_R values ranged from 1.80 to 32.8% for the determination of target analytes. The most accurate result corresponded to that provided for LAS. These types of surfactants are widely used and routinely determined, demonstrating the fact that the more experienced the analyst is, the lower the deviation from the target value. The accuracy of the results obtained by LC-MS determination of LAS was similar to that obtained by the well-established technique of LC-FL.

Overall, it has to be commented that although very different methodologies and analytical conditions were used for the analysis of a complex mixture of compounds, acceptable values of reproducibility were obtained in this interlaboratory study. Considering the principles established by *Aquacheck*, a well-known organisation distributing interlaboratory exercises, all the results obtained were within an acceptable range except for those corresponding to AEO for which a double-flagged error was obtained (28.3%).

LC-MS techniques led to lower or equal deviations from target values when compared with well-established LC-FL techniques as mentioned previously (see results for NPEO, NPEO-SO_4 and LAS in Table 4.5.1). The between-laboratory precision for LC-MS results varied between 2.3 and 28.3% (in terms of RSD), indicating the good reproducibility of LC-MS results when considering that four different instrumentations were used. The most inaccurate LC-MS result corresponded to that obtained for AEO. This was due to the fact that this product corresponds to a mixture of different homologues (from C_{12} to C_{18}) and ethoxymers (average $n_{EO} = 5$) the spectrum of which is the most complex (compared to the other surfactants analysed). Relative abundances of high m/z ions in the spectra corresponding to ethoxylated species can suffer some variations from analysis-to-analysis owing to different ionisations promoted by small differences in the mobile phase, nebulisation, temperature, nitrogen flow, etc. Consequently, quantification based on high m/z ions did lead to the most variable results in the present study. On the other hand, the most accurate result corresponded to that obtained for LAS. The lowest LAS concentration was obtained by laboratory 1 which used the quantification of the m/z ion corresponding to C_{12}LAS. *Laboratories 3, 5 and 6* used all the m/z ions common to each LAS homologue and *laboratories 2 and 7*, using fluorescence detection, also used all four chromatographic peaks for LAS quantification, thus explaining why *laboratory 1* results are somewhat lower compared to the others [6]. The differences observed in the quantification of CDEA could only be explained on the basis of different sample storage times—and thus possible degradation—of the compound, and is thus one of other factors demanding the need for further interlaboratory studies.

In general terms, the variation from laboratory to laboratory (between-laboratory) was greater than that attributed to the analytical error displayed within laboratories (intralaboratory). There are many reasons for the interlaboratory variation that can be attributed to operational parameters such as mobile phase flowrate, mobile phase and buffer composition, vaporiser temperature, tip temperature and source temperature.

Interlaboratory study II
The results of interlaboratory study II are presented in Fig. 4.5.1. Five sets of results were obtained for the LAS exercise, and four sets for the NPEO exercise. For LAS, the within-laboratory variability ranged between 2 and 8% (RSD) for sample II1 (distilled water spiked with $1 \, \text{mg} \, \text{L}^{-1}$ LAS), 1 and 13% for sample II2 (wastewater influent), and 3 and 8% for sample II3 (sample II2 spiked with $1 \, \text{mg} \, \text{L}^{-1}$ LAS). Between-laboratory variations (calculated from the mean of laboratory means, MOLM) amounted to RSDs of 15, 30 and 30% for samples II1, II2 and II3, respectively. The LAS values reported were in the range of $700-1100 \, \mu\text{g} \, \text{L}^{-1}$ in sample II1, $1100-1800 \, \mu\text{g} \, \text{L}^{-1}$ in sample II2 and $1900-3000 \, \mu\text{g} \, \text{L}^{-1}$ in sample II3, indicating that even in the matrix wastewater influent, the spiked concentration of $1 \, \text{mg} \, \text{L}^{-1}$ LAS could be almost quantitatively determined by all laboratories.

Considering the type of analytes, the above results are quite satisfactory, although it has to be taken into account that the samples contained relatively high concentrations, which would intrinsically decrease variabilities. However, the matrix was very complex (untreated wastewater), which also resulted in high concentrations of

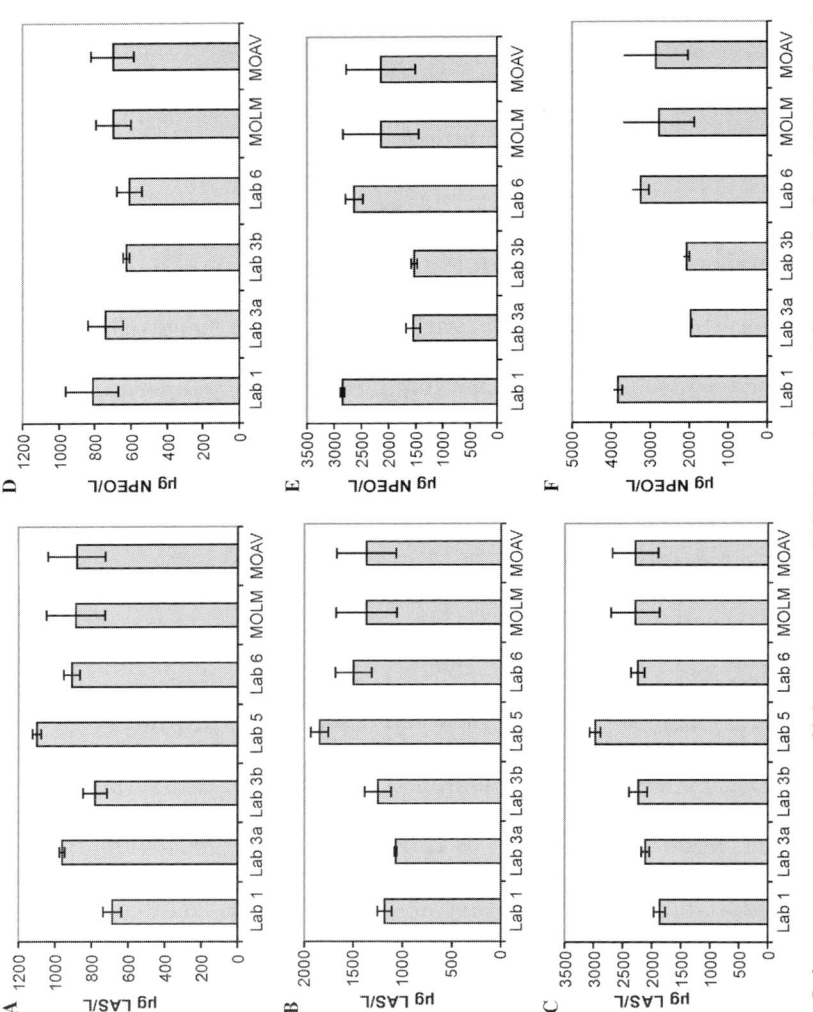

Fig. 4.5.1. Laboratory means, mean of laboratory means (MOLM) and mean of all reported values (MOAV) from interlaboratory study II. (A) LAS, cartridge II1; (B) LAS, cartridge II2; (C) LAS, cartridge II3; (D) NPEO, cartridge II1; (E) NPEO, cartridge II2; (F) NPEO, cartridge II3.

interfering compounds and severe ion suppression. Since the between-laboratory variability is somewhat higher than the within-laboratory RSD, there may be an indication of systematic differences between the individual laboratories. However, apart from *laboratory 5*, which invariably reported the highest concentrations, no clear laboratory (i.e. methodological) trends could be observed. The results of the analysis of NPEO show a good-to-excellent agreement for sample II1 (spiked distilled water), with comparable within- and between-laboratory RSDs. Hence, no systematic differences occurred between the methodologies used for this sample. However, for samples II2 and II3, *laboratory 3* (using flow injection analysis FIA-MS) showed systematically lower values than *laboratories 1 and 6*, which both used LC-ESI-MS. Within-laboratory variabilities (based on three independent replicate analyses) were relatively low (2–8%), with the exception of *laboratory 1* in sample II1 (RSD, 15%) showing that repeatability for this type of analysis is satisfactory. Again the NPEO spiked at a concentration of $1 \, mg \, L^{-1}$ to the matrix wastewater influent could almost be quantitatively determined by *laboratories 1 and 6*, whereas FIA-MS failed (*laboratory 3*). In order to study these effects more thoroughly, especially the influence of ion suppression in LC-MS, as well as the quality assurance at lower concentrations normally occurring in the matrix wastewater effluent, a third interlaboratory study was conducted.

Interlaboratory study III
Six sets of results from five laboratories were obtained for the analyses of NPEO in three cartridges. All laboratories used MS for quantitative analysis, except *laboratory 5*, which used LC-FL. *Laboratory 1* used an LC with an APCI interface; *laboratories 2 and 4* used LC-ESI-MS; and *laboratory 3* used FIA-MS analysis. All laboratories performed three independent replicate analyses, i.e. analysed three replicate cartridges of each type of spike.

In general, the results obtained in this trial showed relatively poor agreement when compared to results achieved at higher concentrations. Cartridge III1 (Fig. 4.5.2(A)) provided the best results with a between-laboratory variation amounting to 29%. The differences between the highest ($190 \, ng \, mL^{-1}$) and lowest ($80 \, ng \, mL^{-1}$) observed mean values of NPEO, which was spiked at a concentration of $200 \, ng \, mL^{-1}$ into distilled water, amounted to a factor of two. In general, within-laboratory variance was acceptable, with an RSD between 5 and 10%.

Results for cartridges III2 and III3 (Fig. 4.5.2(B) and (C)) showed large between-laboratory variabilities, with RSDs of 65 and 49%, respectively, whereas within-laboratory RSDs were 5–10%.

The trends that were observed showed errors that were most likely of a systematic nature, because these could be grouped according to the interface type used in LC-MS. Both ESI methods yielded results in reasonable agreement, as well as in accordance with the LC-FL method. Thus, in sample III2 (wastewater effluent), the average values determined by these three laboratories were 530 and 580 ng mL^{-1}. In sample III3, corresponding to sample III2 spiked with 200 ng mL^{-1} NPEO, an almost 100% recovery could be observed, thus giving satisfactory results for these laboratories. However, the APCI interface invariably showed lower results than those obtained with ESI-MS. This phenomenon has been confirmed in other studies and was discussed in Chapter 4.3. Flow injection analysis led to significantly lower results in cartridges III1 and III2. The systematic deviation from other MS methods is confirmed by the agreement between the two types of quantification used within the FIA method, i.e. using either MID or daughter ions.

Additionally, the LAS interlaboratory study III resulted in data sets from five laboratories. For three of these, individual C_nLAS data were made available. One participant did not comply with the interlaboratory protocol and their results could not be included in the evaluation. *Laboratory 3* only supplied total LAS results (discussed below), while *laboratory 5* supplied individual C_nLAS data which was too late for inclusion in the final evaluation. However, their total LAS results have been included. The results of interlaboratory study III for LAS are shown in Fig. 4.5.3 for the single congeners (C_{10}, C_{11}, C_{12} and C_{13}LAS), grouped according to cartridges. In Fig. 4.5.4 the total LAS (sum of C_{10}–C_{13}LAS) results for each laboratory are shown.

Invariably, results obtained for the separate LAS congeners showed that the data obtained with LC-APCI-MS (*laboratory 1*) were again lower than those obtained with LC-ESI-MS (*laboratories 2 and 4*), varying between 10% and more than 50% less than the mean of the two higher values. For C_{13}LAS the value obtained with APCI was only 16% of the mean of the two values obtained with an ESI-MS interface. Between the two laboratories using ESI-MS, a

518

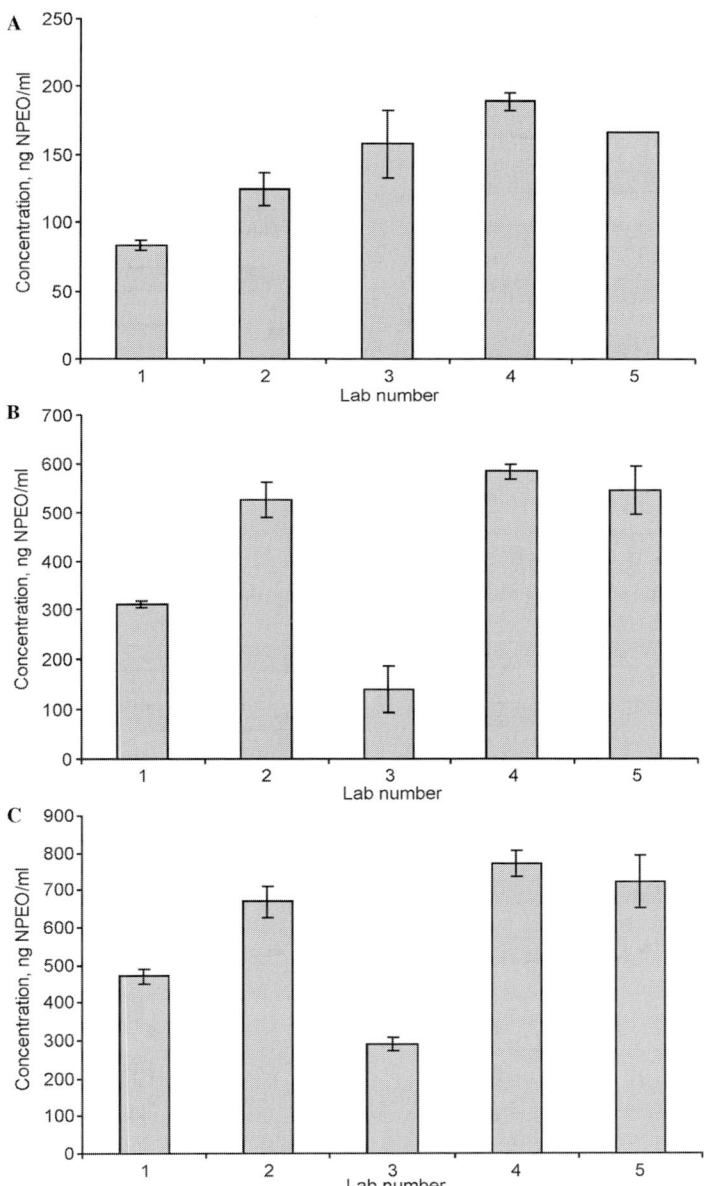

Fig. 4.5.2. Results of interlaboratory study III for NPEO analyses. (A) Cartridge III1; (B) cartridge III2; (C) cartridge III3.

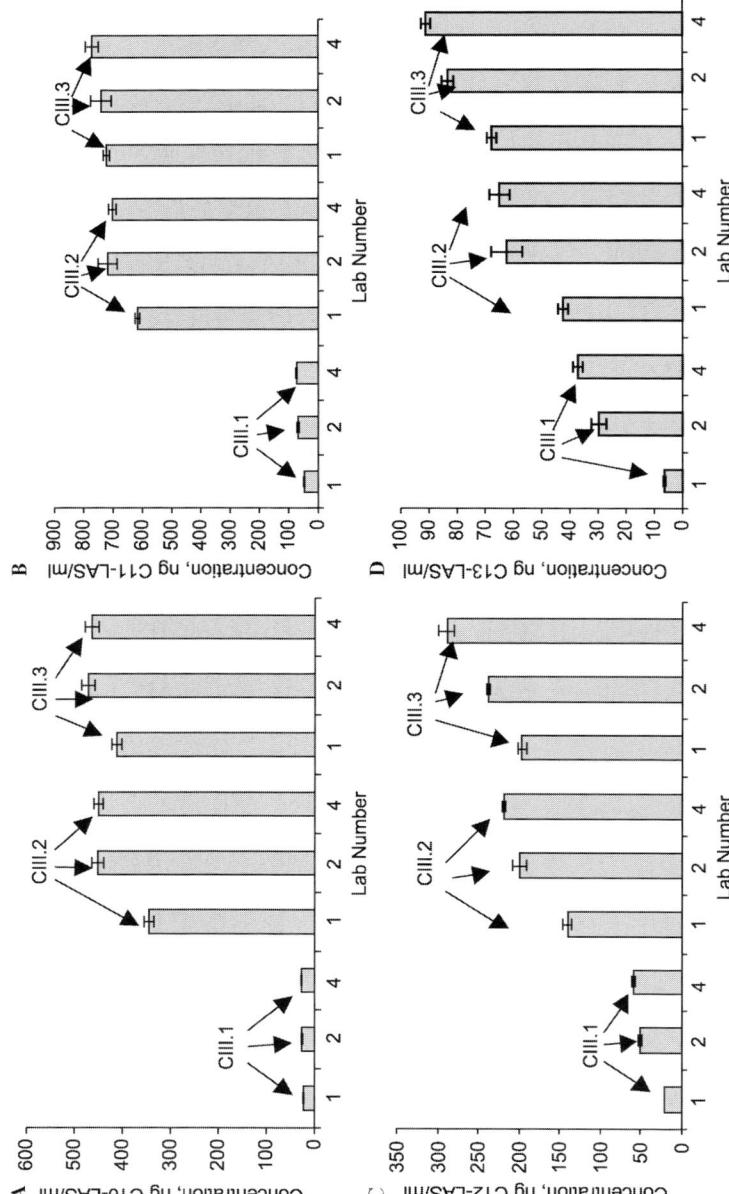

Fig. 4.5.3. Results of interlaboratory study III for individual LAS compounds (C_{10}–C_{13}LAS) grouped according to cartridges III1–3. (A) C_{10}LAS; (B) C_{11}LAS; (C) C_{12}LAS; (D) C_{13}LAS.

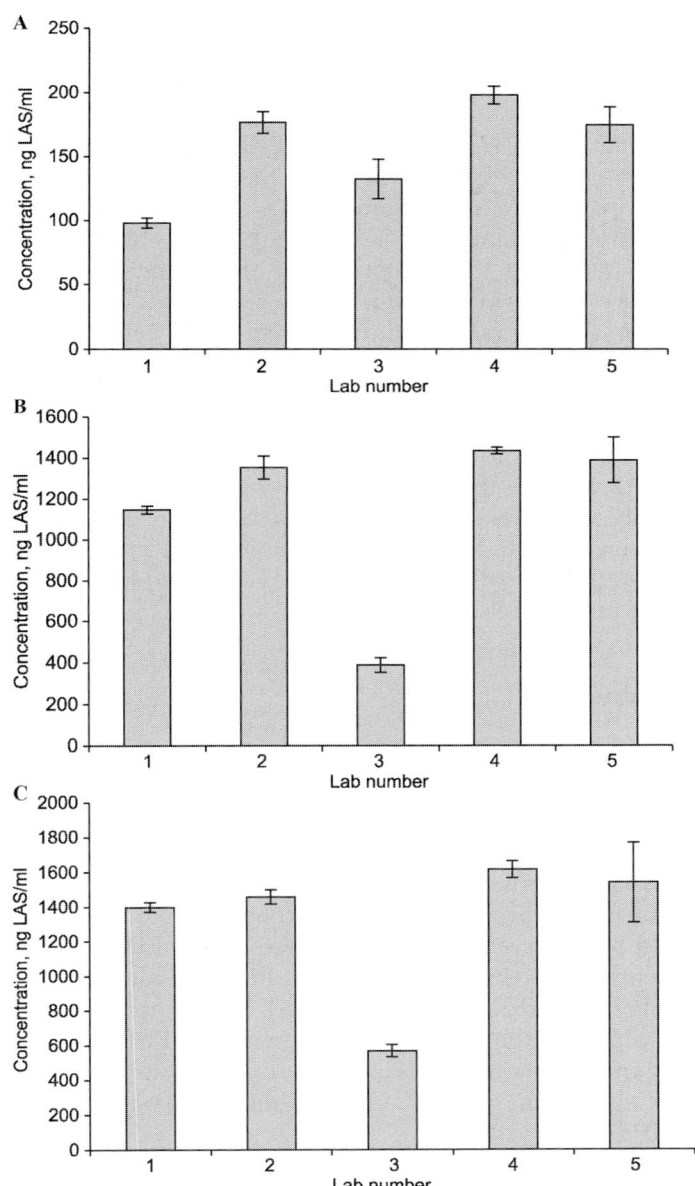

Fig. 4.5.4. Results of interlaboratory study III for total LAS analyses. (A) Cartridge III1; (B) cartridge III2; (C) cartridge III3.

reasonably to good agreement was found consistently (with differences between laboratory means of 1–15%). Within-laboratory variability was excellent (based on three replicates, RSDs between 1 and 5% were observed).

The systematic difference between APCI and ESI is also demonstrated in the summed LAS results (Fig. 4.5.4). Similar to the trends observed for NPEO, in the summed LAS results, the data obtained by FIA-MS are systematically lower than those obtained with ESI-MS or APCI-MS (with the exception, for the latter, of cartridge III1 results, where APCI-MS (mean value 98 ng mL^{-1}) and FIA-MS (mean value 137 ng mL^{-1}) data agree more closely, but still unsatisfactorily).

The data obtained by LC-ESI-MS are in good agreement with data reported from LC-FL measurements. For example, from the 200 ng mL^{-1} spiked into distilled water, the mean recovery found by these laboratories was 92 ± 13%. In the wastewater effluent (sample III2) the LAS concentrations determined by these three laboratories varied between 1350 and 1440 ng mL^{-1} and in the spiked wastewater sample III3 again, the spiked 200 ng mL^{-1} LAS could be quantified with a recovery of up to 100%.

4.5.4 Conclusions

The results of the interlaboratory exercises involving LC-APCI-MS, LC-ESI-MS, FIA-MS and LC-FL show that in general repeatability (as reflected by within-laboratory RSD) is at a satisfactory level (i.e. RSD around 10%) for LAS and NPEO.

However, large between-laboratory variabilities occurred that are indicative of poor reproducibility of methods. In so far as systematic trends could be observed, FIA-MS resulted more often in significantly lower concentrations than LC-MS or LC-FL methods. Within the LC-MS methods, the APCI interface yielded quantitative results which were invariably less than those obtained with LC-ESI-MS. This aspect is discussed in more detail in Chapter 4.3. LC-FL data were generally in fair-to-good agreement with LC-ESI-MS data. This held true for both LAS and NPEO analyses.

It can be concluded that further work is necessary to improve method reproducibility. The lack of reference materials and isotope-labelled standards to control the quality assurance of analytical methods is a serious drawback in this area. Synthesis of ^{13}C-labelled LAS compounds and development of reference materials for LAS and NPEO (as discussed in Chapter 4.2) are immediate research needs to improve method reliability in surfactant analysis.

REFERENCES

1 P. Quevauviller, W. Cofino, R. Cornelis, P.P. Fernandez, R. Morabito and H. Van der Sloot, *Int. J. Environ. Anal. Chem.*, 67 (1997) 173.
2 P. de Voogt, J. Hinschberger, E.A. Maier, B. Griepink, H. Muntau and J. Jacob, *Fresenius. J. Anal. Chem.*, 356 (1996) 41.
3 Z. Lukaszewski, *Resolved and Unresolved Questions of Analysis of Surfactants in the Aquatic Environment; Adsorption and Its Applications in Industry and Environmental Protection*, Vol II: Applications in Environmental Protection, Vol. 120, Elsevier Science, Amsterdam, 1999, pp. 135–176.
4 P. Gerike, H. Klotz, J.G.A. Kooijman, E. Matthijs and J. Waters, *Water Res.*, 28 (1994) 147.
5 C.G. Naylor, J.P. Mieure, W.J. Adams, J.A. Weeks, F.J. Castaldi, L.D. Ogle and R.R. Romano, *J. Am. Oil Chem. Soc.*, 69 (1992) 695.
6 M. Castillo, J. Riu, F. Ventura, R. Boleda, R. Scheding, H.F. Schröder, C. Nistor, J. Emneus, P. Eichhorn, T.P. Knepper, C.C.A. Jonkers, P. de Voogt, E. Gonzalez-Mazo, V.M. Leon and D. Barceló, *J. Chromatogr. A*, 889 (2000) 195.

Chapter 5

Environmental processes

5.1 AEROBIC BIODEGRADATION OF SURFACTANTS

Thomas P. Knepper, Peter Eichhorn and Lea S. Bonnington

5.1.1 Introduction

Regulations mandating a certain degree of assessment of the biode-gradability of surfactants were first released in Germany shortly after the low degradability of the branched alkylbenzene sulfonates was realised (ABSs; see Chapter 1), and have since also been established in all Member States of the European Union [1–3]. The general fate of surfactants after use is disposal into the sewage system and thereby redirection to wastewater treatment plants (WWTPs), where they are subsequently submitted to biodegradation. In the initial step of microbial action, known as the primary degradation, minor alterations in the chemical structure of the molecule occur, often resulting in the loss of the surface-active properties. This is usually paralleled by a decrease in the aquatic toxicity (see Chapter 7) since, for the majority of the substances, the surface activity itself is often responsible for the adverse interactions with biological surfaces, e.g. cell membranes.

To reduce the potential risk of environmentally harmful surfac-tants, requirements of a minimum primary degradation amounting to 80% for anionic and non-ionic surfactants was stipulated as far back as 1977 [4]. However, within this early regulation no restraints were included regarding cationic or amphoteric surfactants as these did not hold a significant market share when the laws came into force.

Comprehensive Analytical Chemistry XL
T.P. Knepper, D. Barceló and P. de Voogt (Eds)

Methods monitoring the primary degradation of anionic and non-ionic surfactants traditionally used a group-specific analysis based on the formation of a surfactant complex, which was subsequently quantified by photometric or titrimetric methods [5,6]. Anionic surfactants were complexed with the cationic dye methylene blue and extracted with chloroform prior to photometric analysis. The major anionic surfactants, such as linear alkylbenzene sulfonates (LASs), alkyl sulfates (ASs) and alkylether sulfates (AESs), give a positive response as methylene blue active substances (MBASs), but the detection can be adversely affected by many interferences [7,8]. Similar drawbacks are encountered with the analysis of non-ionic surfactants, which are determined as bismuth iodide active substances (BiASs). Furthermore, the positive complexes MBAS and BiAS are usually formed only for intact surfactant molecules, i.e. degradation intermediates do not contribute to the overall value. The fate of surfactants, however, is not sufficiently described if the primary degradation is the only parameter taken into account. Knowledge about the ultimate transformation is also of considerable importance, as has become particularly evident in the case of the nonylphenol ethoxylates (NPEO; see Chapter 7.4). For this purpose the total degradation of a compound, which can occur either by mineralisation or assimilation, is measured as a substance-independent sum parameter such as evolution of CO_2, oxygen consumption or removal of the dissolved organic carbon (DOC).

A series of international standard methodologies has been established by the Organization of Economic Co-operation and Development (OECD) to assess the biodegradability of surfactants [9]. These tiered-based tests distinguish among 'ready biodegradability', 'inherent biodegradability' and 'simulation biodegradability'. They differ mainly in the comparability with real environmental conditions, the concentration of the test compound and the test duration, the used inoculum, the measured analytical parameter and the pass level. In the first stage the surfactant is submitted to the ready biodegradability tests (OECD 301A to F). If it meets the pass level, i.e. 60% CO_2 formation or 70% DOC removal of the theoretically possible amounts within 28 days, it is concluded that with high likelihood the examined substance is (ultimately) degraded under environmental conditions. In instances of failure to reach this designated pass level, a test of inherent bio-degradability under ideal conditions is thus required (OECD 302A to C). A positive result in this test defines the compound as inherently biodegradable, i.e. it has the potential to be degraded. The third tier has to be initiated if the pass level in the former two tests has not been attained. In this instance the biodegradability under WWTP conditions is tested (OECD 303A and 304A).

5.1.2 Aerobic biodegradability tests for surfactants

For the study of the aerobic biodegradability of surfactants in aqueous media, in addition to the standardised OECD methods employed for the testing of organic compounds [9], a large variety of assays has been developed. These can differ in instrumental and technical set-up, measured parameters, type and origin of test medium and inoculum used, test duration, and substrate concentration [10]. Analogous to OECD methods, their design allows determination of surfactant degradability by measuring DOC removal, CO_2 formation and chemical oxygen demand (COD) or biological oxygen demand (BOD). However, examples where they are used to investigate the metabolic pathway of numerous surfactants are also plentiful [11–19].

A rather simple set-up, termed the river die-away test, was introduced by Swisher [10] to assess the primary biodegradability of surfactants in rivers and was later employed in several studies working with river, estuarine and seawater [20–24]. More sophisticated test systems for simulating the elimination in surface waters were developed by Schöberl et al. [25] and Boeije et al. [26] based on a staircase model. The use of trickling filters permitted determination of the degree of bioelimination and mineralisation [27–29]. In addition, degradation intermediates could be easily recovered at amounts required for further analytical purposes. Gerike and Jasiak [30,31] modified the coupled-units test (OECD 303A), normally applied to the simulation of aerobic wastewater treatment, to accumulate recalcitrant metabolites in the test liquor. The reproducibility of this concept was later demonstrated by Guhl and Ranke [32].

A biologically active fixed-bed bioreactor (FBBR) operated under aerobic conditions can be employed for investigating both the primary degradation of the targeted surfactants and their breakdown pathway [33–40]. The FBBR represents a modification of so-called test filters widely utilised in German chemical industries and waterworks to study poorly degradable compounds in waste and surface water, respectively. Use of the FBBR benefits from its simple experimental set-up, ease of operation and the ability to provide aqueous samples apt for direct injection into the electrospray ionisation mass spectrometer (ESI-MS). Moreover, wide experience has been gained with the FBBR as a model system in terms of assessing the biodegradability of structurally very distinct organic compounds, identifying degradative products and following their formation and decline. Among the tested substances in addition to surfactants described in the literature, are aliphatic amines, aromatic sulfonates, sulfonamides, organic acids and phosphates [41–45]. In these studies, the test concentrations of the spiked analytes were chosen depending on the objectives of the studies and the analytical methods applied. Low concentrations at environmentally relevant levels ($10\,\mu g\,L^{-1}$) were used for examining primary biodegradability, whereas higher amounts of the parent compounds ($10\,mg\,L^{-1}$) were selected when the aim was to identify the unknown metabolites, which often occur only at traces in the test liquor. In this way, the need for sample enrichment procedures was minimised.

In order to investigate the metabolism of surfactants while detecting simultaneously the parent compounds and the degradation intermediates, which presumably exhibit higher polarity than the spiked surfactant, the FBBR was amended with the test compounds at concentrations in the $mg\,L^{-1}$ range. In particular, unfiltered river water (10 L) was continuously pumped in a closed loop from a storage tank upwards into a glass column packed with native, porous sintered glass beads as carrier material for the immobilisation of microorganisms. In most cases, after finishing a single experimental set-up, the biofilm was removed from the FBBR in order to avoid adaptation processes.

5.1.3 Aerobic biodegradation of anionic surfactants

Linear alkylbenzene sulfonates
LAS generally consists of a mixture of homologues and isomers (see also Chapter 1). Individual components are classified by the length of the alkyl substituent and the position of attachment of the sulfophenyl ring (Fig. 5.1.1(a)). In formulations, the alkyl-chain length may vary from C_{10} to C_{13} with an average distribution of 11.4–11.7 and the phenyl can range from the 2- to 7-carbon depending on the alkyl-chain length. Commercial blends usually contain some sulfonated impurities ($<1\%$), namely dialkyltetralin sulfonates (DATSs), ABS, and single

methyl-branched LAS (iso-LAS) (Fig. 5.1.1(b)). The level of these impurities depends on the synthetic route used (see Chapter 1).

As a consequence of the vast amounts of LAS entering the aquatic environment via wastewater discharges, their environmental fate and distribution have been the subject of thorough investigations for more than 30 years. Hence, an enormous body of literature exists regarding the analytical methods used for the determination of LAS in various environmental compartments (see Chapter 2, specifically sections 2.1–2.3, 2.4.2, 2.4.4.1), investigations on their biodegradability under both laboratory and field conditions [23,46–52] and assessment of their toxic potential towards aquatic organisms (Chapter 7 and Refs. [53–56]). In essence, LAS are one of the best-studied organic substances produced on an industrial scale with respect to environmental behaviour.

The gap in understanding the metabolic pathway of LAS and the occurrence and fate of the aerobic degradative products in the environment was bridged utilising an FBBR [34]. The major degradation intermediates are sulfophenyl carboxylates (SPCs; Fig. 5.1.1(c)), which have been observed in wastewaters and in the environment (see Chapter 6, specifically sections 6.1, 6.2.2, 6.3.2 and 6.5). SPC originating from commercial LAS constitute even more complex mixtures than the parent compound, since the initial attack on the alkyl chain can take place on either side of the phenyl substituent such that a wider range of alkyl homologues ranging from C_4 to C_{13}-SPC can be produced.

Standardised aerobic biodegradation methods. Since the introduction of LAS more than 30 years ago, an extensive database has been developed supporting their aerobic biodegradability. A large fraction of these data were gathered from standard OECD laboratory screening tests used routinely in Europe and North America to determine the biodegradation potential of organic substances [9]. Evidence for mineralisation was provided from non-specific gross biodegradation parameters: O_2 uptake, organic carbon removal and CO_2 formation [57–62]. Two major factors, however, significantly impact on the environmental relevance of such screening test results and strongly limit their extrapolation to 'real world' settings. In particular (a) standard screening tests do not accurately simulate the physical, chemical and biological conditions found in environmental systems and (b) the tests ignore several compartments such as sediments, soils, and groundwater, estuaries

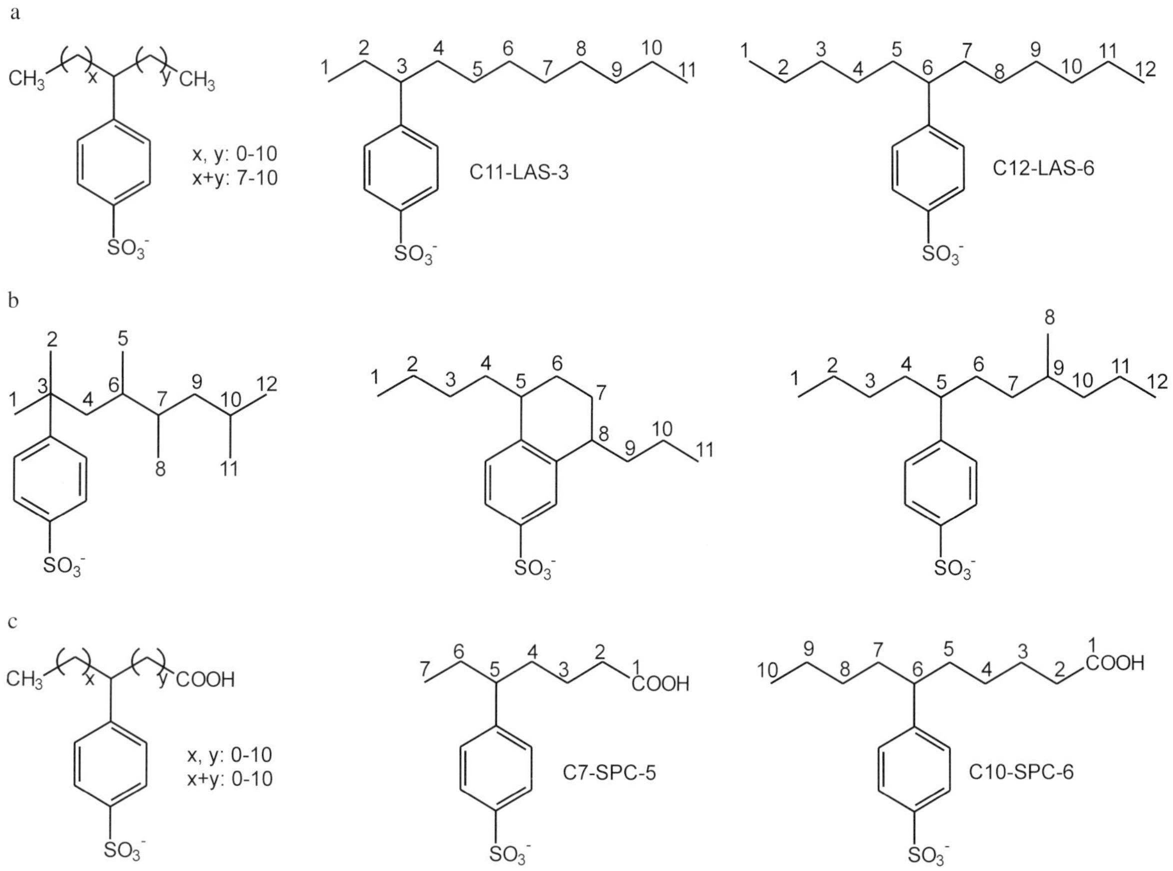
a
x, y: 0-10
x+y: 7-10
C11-LAS-3
C12-LAS-6
b
c
x, y: 0-10
x+y: 0-10
C7-SPC-5
C10-SPC-6

and oceans. These may be important with regard to the actual biodegradation of surfactants in the environment (see Chapter 6).

To overcome these limitations many experiments have been performed to simulate natural environmental settings using both laboratory and field microcosms. These studies were conducted on sediments [63,64], soils [65–67], groundwaters [68,69], surface waters [25,70], estuarine waters [71,72], and seawater [23,73–75]. In most instances a reasonable rate of LAS primary degradation was observed which was dependent on the history of exposure among other parameters. In environments with continuous exposure to LAS residues, e.g. coastal areas with direct discharges of domestic sewage, the microbial communities were rather well adapted to the substrate and thus advanced levels of degradation were observed. The very low kinetics of primary LAS degradation in sediments [63], in contrast, are most likely due to oxygen limitation.

A detailed description of the aerobic degradation of LAS was provided by Swisher [10] and Schöberl [76]. The majority of the laboratory studies published have indicated that metabolism starts with oxygenation of one of the terminal methyl groups of the alkyl chain and the conversion of the alcohol to a carboxylic group (ω-oxidation), releasing an SPC with the same number of carbon atoms in the side-chain as in the original LAS isomer (Fig. 5.1.2) [77,78].

The conversion rates of individual components in a commercial LAS mixture are dependent on the molecular structure. For example, the length of the alkyl chain is positively correlated with the primary degradation rate, and as such, isomers with the phenyl substituted at central positions are degraded more slowly than other isomers [79,80]. Both effects are a direct consequence of the enzymatic attack on the hydrophobic moiety. The relation between surfactant structure and the biodegradation has been termed as *Swisher's distance principle* which, in summary, describes that an increased distance between

---

Fig. 5.1.1. (a) General structure of LAS (left) and structures of two components (middle and right) illustrating the nomenclature used to identify individual species; (b) possible structures of branched C_{12}-alkylbenzene sulfonate (ABS) (left), C_{11}-dialkyltetralin sulfonate (DATS) (middle), and C_{12}-single methyl-branched linear alkylbenzene sulfonate (*iso*-LAS) (right); (c) general structure of sulfophenyl carboxylates (SPCs) (left) and the structure of two possible components deriving from LAS biodegradation (middle and right) illustrating the nomenclature used to identify individual species.

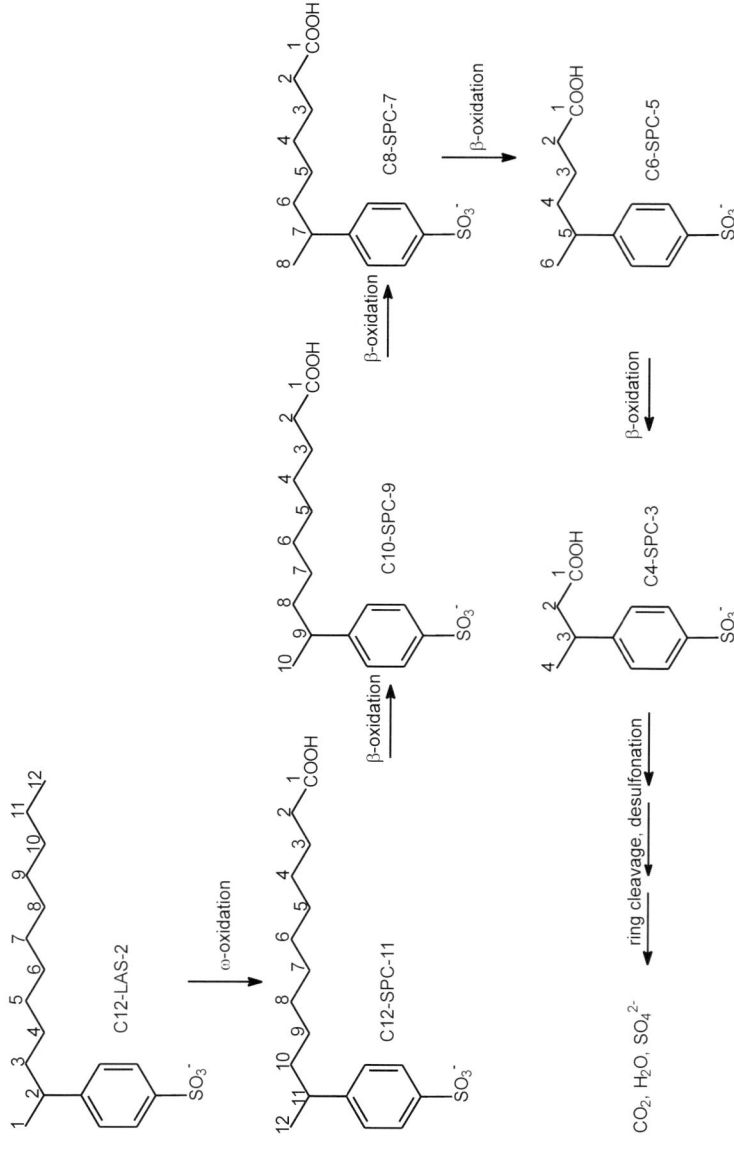

Fig. 5.1.2. Aerobic degradation pathway of LAS shown for C_{12}-LAS-2.

the xenobiotic arylsulfonate moiety and the far end of the hydrophobic group increases the speed of primary degradation [10].

It is generally accepted that ω-oxidation of LAS is followed by successive oxidative shortening of the alkyl chain by two-carbon units, termed β-oxidation (Fig. 5.1.2) [77,81]. The resulting very short-chain SPCs are further broken down by ring cleavage, which is considered the rate-determining step within the total process, and desulfonation then occurs to result in complete degradation.

Although β-oxidation is certainly the most important mechanism for the destruction of the alkyl chain in LAS, some indication had been given that removal of a single carbon, i.e. α-oxidation, may also occur to a minor extent. This alternative route has been proposed in order to explain the isolation of C_5-SPC as a by-product in the degradation of C_{12}-LAS [82] and the detection of a multitude of both C-even and C-odd SPC intermediates observed during the degradation of C_{11}-LAS [83].

According to the scheme shown in Fig. 5.1.2, for the pure C_{12} homologue as the starting compound, C-even-SPC should have been formed exclusively. Direct support for the scheme was provided by Hrsak and Begonja [84] and Dong et al. [85], who used pure cultures of bacteria, which attacked C_{12}-LAS-2 and C_{12}-LAS-3, respectively. The SPCs formed were subject to chain shortening in pure or mixed cultures. The ring cleavage mechanism of an SPC, via 4-sulfocatechol, was established [86] and this was suspected to be common to the general degradation of LAS, because mixed cultures which degrade LAS contain high levels of 4-sulfocatechol-1,2-dioxygenase [85].

LAS congeners are usually racemic mixtures of optically active compounds [10], and their transient degradative intermediates are also usually optically active [87]. Most of these intermediates are transients, but some can have long half-lives [88]. Some LAS congeners are evidently subject to attack at both methyl groups on the alkyl chain as sulfophenyl dicarboxylates (SPdCs) are also detected during the degradation of LAS [88].

However, the observation of SPC and SPdC during the degradation of LAS still does not prove either ω-oxygenation or β-oxidation of SPC as the sole significant mechanism, as Schleheck et al. [89] realised when they isolated the first heterotrophic bacterium able to attack commercial LAS. They could not distinguish between ω-oxygenation and sub-terminal oxygenation with subsequent Baeyer–Villiger oxygenation and hydrolysis to yield an SPC. Furthermore, they were working with a pure culture and not with an environmentally realistic microbial mixture.

Aerobic biodegradation of LAS utilising an FBBR. In order to resolve the oxidative mechanisms of LAS, Eichhorn and Knepper [34] used the following experimental approach:

(a) the material subjected to the biodegradation study used was a pure LAS homologue ideally with very low amounts of structurally related synthetic by-products in the test medium, which could disturb the chemical analysis of LAS, and for which their corresponding intermediates could also complicate the determination of SPC species [90];
(b) the experimental set-up was based on a mixed microbial community from a natural setting other than isolated cultures, as this guaranteed a higher environmental relevance of the outcomes;
(c) the applied analytical protocol included a sample preparation method allowing an accurate quantification of all individual degradation products.

In the first preliminary investigations to examine the formation of individual SPC species, a pure C_{12}-LAS homologue was added to the FBBR at $100 \ \text{mg} \ \text{L}^{-1}$ (about $290 \ \mu\text{mol} \ \text{L}^{-1}$). This elevated concentration, found in the aquatic environment only under extreme discharge conditions, was designed to yield sufficient amounts of intermediates for analysis, without the need for any sample enrichment, which could result in analyte discrimination. In most studies performed to determine the ultimate biodegradability of LAS (standard OECD tests) lower concentrations—typically in the range between 1 and $20 \ \text{mg} \ \text{L}^{-1}$—were chosen to be closer to the levels found in environmental compartments and also to rule out adverse effects on the biocoenosis caused by the surface-active properties of the compound. In comparison to higher organisms, bacterial populations are less sensitive to surfactants: Painter [91] reported that LAS did not show inhibition of biodegrading microorganisms at a concentration of $100 \ \text{mg} \ \text{L}^{-1}$ using the OECD-activated sludge respiration inhibition method. Performing the same test procedure, Verge and Moreno [55] determined the toxicity EC_{50} (3 h) of single LAS homologues to $500-1200 \ \text{mg} \ \text{L}^{-1}$. A similar test concentration of $100 \ \text{mg} \ \text{L}^{-1}$ LAS used in a trickling filter also proved to cause no inhibitory problem [28,92].

The samples collected from the FBBR were analysed by LC–ESI–MS for the occurrence of SPC [34] (see also Chapter 2.4.4.1). The C-even-SPC were by far the most prominent intermediates (Fig. 5.1.3), whereas the C-odd-SPC were found at much lower amounts (data not shown).

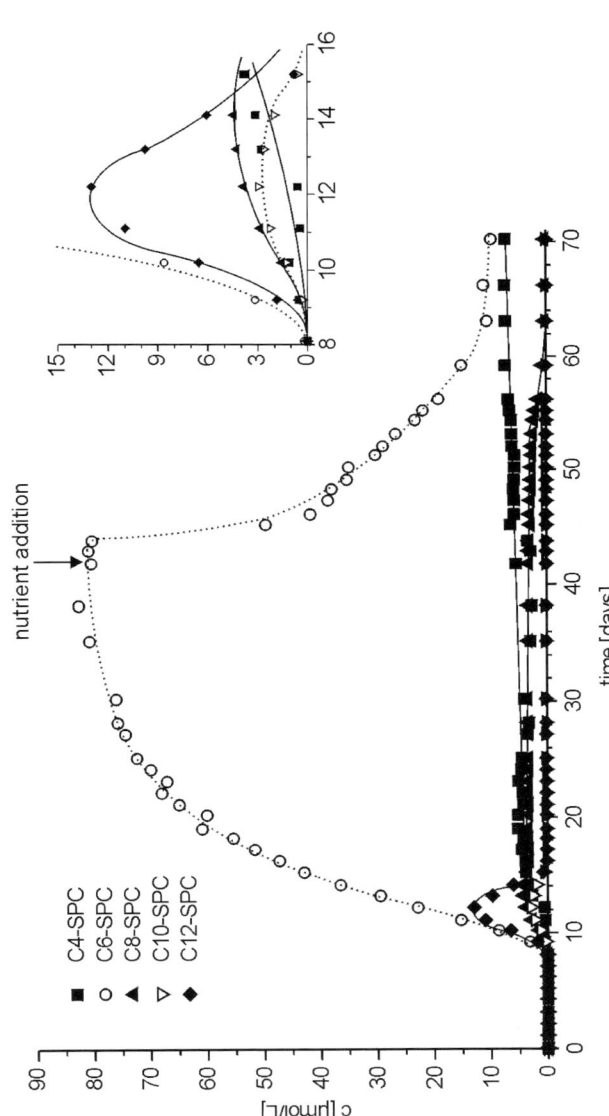

Fig. 5.1.3. Concentration profile of C-even-SPC intermediates formed during biodegradation of C_{12}-LAS on FBBR without initial nutrient supply. The inset shows the evolution between days 8 and 16.

In the inset of Fig. 5.1.3 it can be seen that the SPC formation begins in earnest around day 8 or 9 resulting in a build-up of C-even-SPC. The concentrations of the two long-chain SPC, C_{12} and C_{10}, reach a maximum on day 12, and it seems that they are then rapidly converted into the shorter-chain homologues. In all cases, C_6-SPC is the most abundant homologue while its direct precursor, C_8-SPC, and its successor, C_4-SPC, are detected at substantially lower levels. The increase in C_6-SPC is almost linear from the onset of formation until about day 19 when its formation slows down finally reaching a steady level at approximately 80 μmol L^{-1}.

This behaviour could be attributed to the lack of nutrients in the natural test medium, which were necessary for the metabolism of degrading microorganisms, as addition of a nutrient solution on day 42 led to restoration of the microbial activity [34]. After a short delay of 2 days, the biodegradation of SPC continued. As can be seen in Fig. 5.1.3 the concentration of C_6-SPC started to decrease exponentially on day 44. In parallel, the level of C_4-SPC increased slightly. Finally, the concentrations of all species reached relatively to constant levels. With respect to C_8-SPC, the level remained unchanged until the degradation of C_6-SPC decreased. This was attributed to the higher degradation rate of the latter compound.

Proof of the β-oxidation mechanism was demonstrated in the LC–ESI–MS chromatogram of the FBBR sample taken on day 8, shown in Fig. 5.1.4 [34]. The C-even-SPC were identified as the major degradation intermediates. Under the applied chromatographic conditions several isomeric SPC, e.g. four phenyl isomers of C_6-SPC, could be separated (Fig. 5.1.4).

As mentioned previously some SPC material was not further degraded. Identification could be achieved through closer analysis of the isomeric pattern of a single SPC homologue, for instance C_6-SPC, whose constituents are very well separated. In Fig. 5.1.5 the change in the three most abundant isomers, arbitrarily termed a, b and c, is plotted semi-quantitatively for the run time between day 35 and the termination of the experiment on day 69 (Fig. 5.1.3). Prior to the addition of the nutrient solution (day 42) the isomers showed constant levels, but after the addition a very distinct degradation of the single isomers could be observed. Once the microorganisms started to metabolise C_6-SPC, isomer a disappeared within 2 days, whilst the degradation of b was somewhat slower. On day 63 this latter compound was no longer detectable. The level of isomer c, on the other hand, was

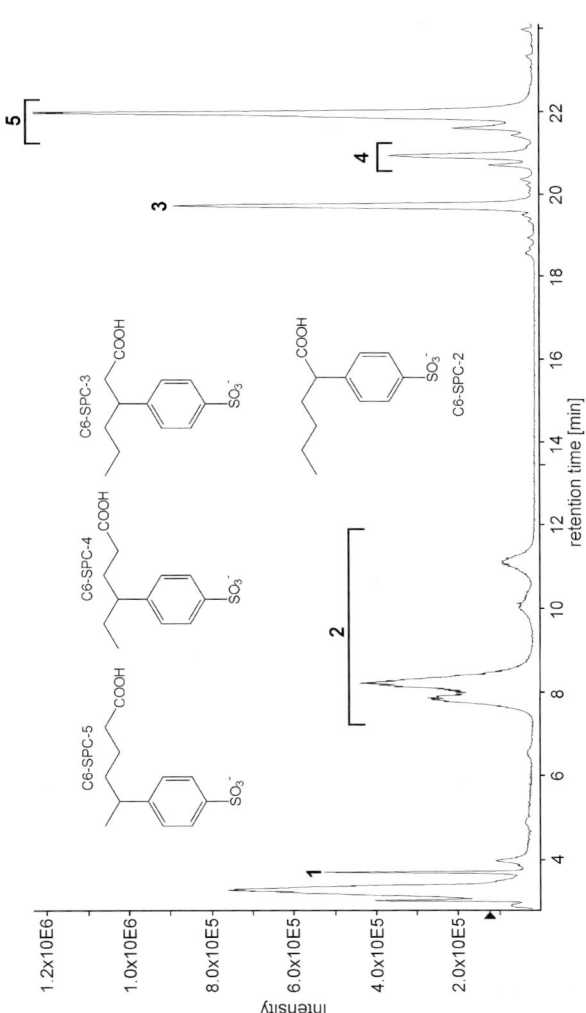

Fig. 5.1.4. (−)-LC−ESI−MS chromatogram of an FBBR degradation experiment on LAS taken after 8 days. Peak numbering: (1) C_4-SPC, (2) C_6-SPC, (3) C_8-SPC, (4) C_{10}-SPC, (5) C_{12}-SPC. Only the time window where the SPC are eluting is shown with the C_{12}-LAS eluting at 27.5 min under the selected conditions.

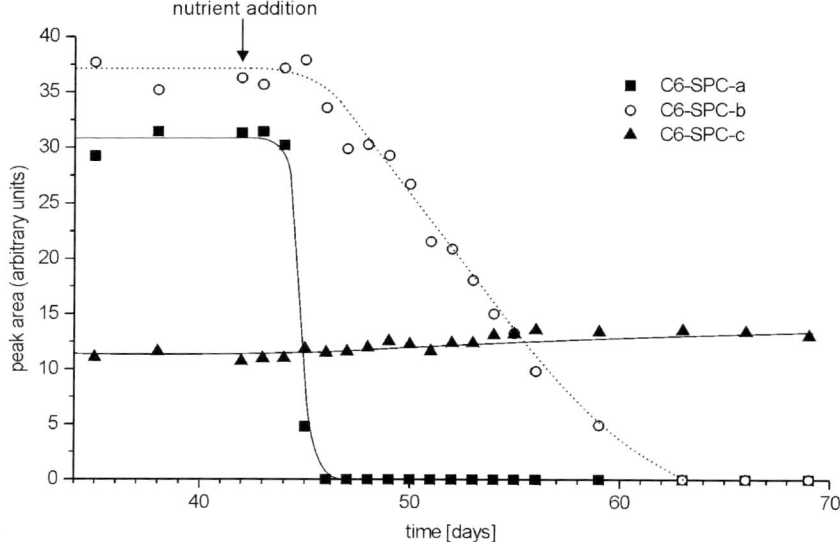

Fig. 5.1.5. Profile of major C_6-SPC isomers, with addition of nutrients on day 42 (Of the four isomers separated in Fig. 5.1.4, the first, second and fourth peaks were integrated).

unaffected by the nutrient addition. Its level remained constant throughout, indicating a high resistance to further metabolisation. Structural assignment of the species c to one of the structures given in Fig. 5.1.4 can be made tentative on a mechanistic basis. C_6-SPC-c might correspond to C_6-SPC-5 since, in this isomer, the proximity of the sulfophenyl ring would make the attack more difficult, thereby hindering the next β-oxidation step [10]. For an unequivocal elucidation it would be necessary to synthesise all possible C_6-SPC isomers.

With respect to the detection of C-odd-SPC in the present test solution, their occurrence can be explained by the fact that the C_{12}-LAS used contained an impurity of some 0.5% of C-odd-LAS. The concentration profile of the C-odd SPC is displayed in Fig. 5.1.6. The long-chain C_{11} SPC is not detectable, while the levels of C_9- and C_7-SPC are below 0.3 μmol L^{-1}. The change in C_5- and C_7-SPC concentrations over time qualitatively resembles that of C_6- and C_8-SPC, respectively. C_5-SPC is accumulated in the test liquor up to about 5 μmol L^{-1}, but upon addition of the nutrient solution, the concentration drops rapidly. This is reduced to a certain level, but is not completely eliminated, with

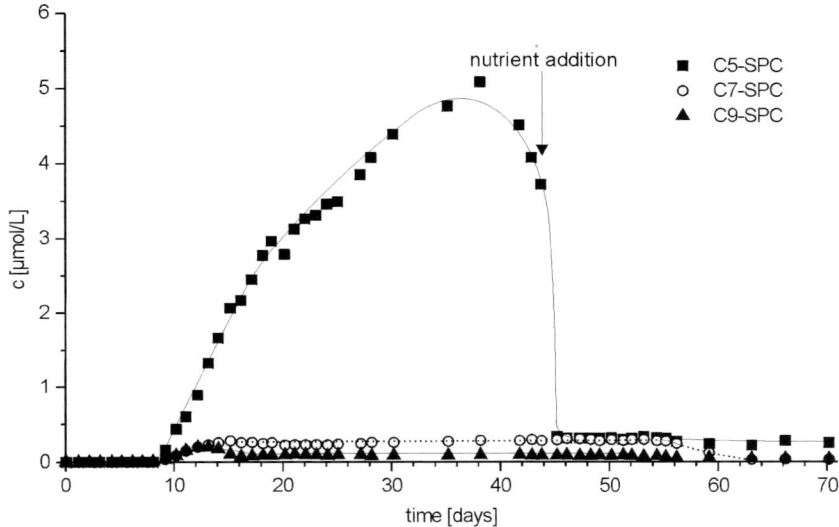

Fig. 5.1.6. Concentration profile of C-odd-SPC intermediates formed during biodegradation of C_{12}-LAS on FBBR without initial nutrient supply.

a recalcitrant fraction remaining at a constant level. The degradation of C_7-SPC in contrast is retarded, as in the case of C_8-SPC, with a notable decrease beginning only after day 58.

The results of the first LAS degradation study, conducted at a high test concentration of 100 mg L^{-1}, proved additional nutrient supply was necessary to assure optimum growth conditions of the bacteria involved. Thus, in a subsequent experiment, addition of a nutritive solution at the beginning of the assay was investigated with the aim of enhancing the degradation process. A second aspect, which was investigated under these optimised circumstances, was the removal of the parent compound, C_{12}-LAS. For this purpose, it was imperative to distinguish between primary degradation and losses due to adsorption. Although adsorption had not posed problems in earlier studies using the same equipment—even when assaying relatively non-polar organic compounds [93]—it was speculated that the inherent property of surfactants to adhere to surfaces might influence the observed outcome. For this reason, two experiments were carried out in parallel in the test devices using identical initial concentrations of C_{12}-LAS (100 mg L^{-1}). To trace the sorption behaviour, the test medium in one experiment was supplemented with 5% formaldehyde prior to the spiking of the surfactant, in order to inhibit microbial activity, while the conditions for the other experiment were maintained as previously.

The eluate from the FBBR inhibited with formaldehyde shows an initial rapid loss of LAS followed by a steady loss until day 7, after which the concentration of LAS remains stable at about half the value in the reservoir (Fig. 5.1.7). No degradative products were detected in this experiment, and therefore the loss can be attributed to sorption. However, it is important to note that precipitation of magnesium and calcium salts can also contribute to the loss of the anionic surfactant [94].

The concentration of C_{12}-LAS in the eluate from the uninhibited FBBR is similar to that in the sorption experiment until day 7, after which the compound disappeared within 4 days (Fig. 5.1.7). No degradative intermediates were detected before day 7, which can be attributed to acclimation of the microorganisms to the substrate. Thereafter, degradative intermediates could be observed (data not shown), with the C-even-C_{12}-, C_{10}-, C_8-, C_6- and C_4-SPC being most prominent. The formation of C_{12}-SPC can occur only if ω-oxygenation occurs. Varying concentrations of the C-even-SPC were observed during

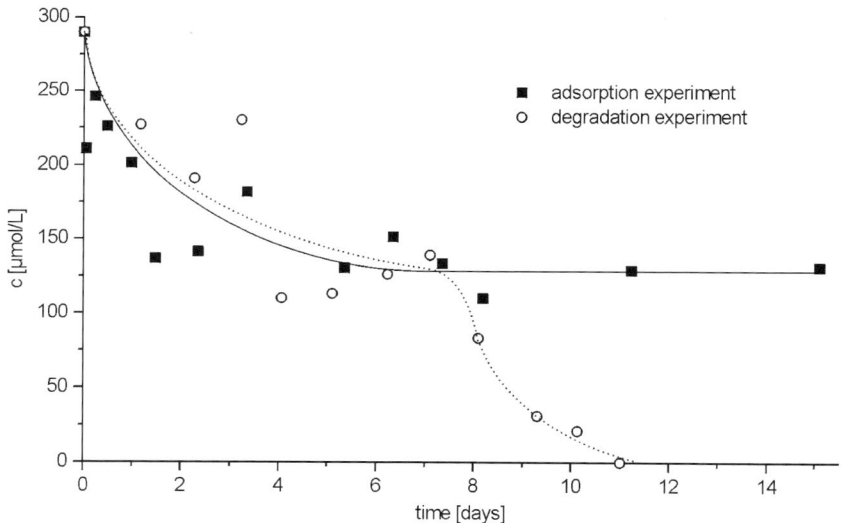

Fig. 5.1.7. Concentration profile of C_{12}-LAS on FBBR during adsorption and degradation experiments at a surfactant concentration of 100 mg L^{-1}.

TABLE 5.1.1

Molar concentrations of the LAS intermediates, SPC and SPC-2H, at curve maximum and at the end-point of the experiment (spiking concentration: 290 μmol LAS L^{-1})

Alkyl-chain length (C_n)	SPC (μmol L^{-1})		SPC-2H (μmol L^{-1})		c (μmol L^{-1})	
	c_{max}	c_{final}	c_{max}	c_{final}	SPdC	Desulfonated metabolites
4	3.6	ND	ND	ND	ND	ND
5	1.7	0.6	ND	ND	ND	ND
6	37	4.8	0.67	0.03	ND	ND
7	0.6	0.04	ND	ND	ND	ND
8	5.6	0.1	0.25	0.16	ND	ND
9	0.1	0.04	ND	ND	ND	ND
10	0.4	0.06	0.07	0.03	ND	ND
11	ND	ND	ND	ND	ND	ND
12	0.3	0.02	0.03	0.01	ND	ND

ND: not detected.

the experiment. The C_{12}- or C_{10}-SPC never exceeded 0.4 μmol L^{-1}. The C_8- or C_4-SPC reached a maximum concentration about 10-fold higher, whereas the C_6-SPC rose to 37 μmol L^{-1} (Table 5.1.1). The concentrations towards the end of the experiment were much lower, usually well below 0.1 μmol L^{-1}. These variations over the course of the experiment presumably reflect changes in the microbial population and their different affinities for the surfactants as a carbon source. C-odd-SPC (C_9, C_7, and C_5) were also observed in the experiment (Table 5.1.1). They presumably arise from C_{11}-LAS, present as an impurity in the C_{12}-LAS. With the exception of a transient accumulation of C_5-SPC, they were never observed above 1 μmol L^{-1}. The intermediate build-up of about 1.8 μmol L^{-1} C_5-SPC on day 9 was due to the excretion of one isomer, in particular, which was well separated under the chosen chromatographic conditions (not shown). Seemingly, after production of this isomer into the test solution, it was taken up again into the cells for further metabolisation. By the next day the isomer had disappeared almost quantitatively, while at least one other isomer remained in the test liquor at a relative constant level.

In conclusion, it has been observed that: (a) the kinetics of the degradation of individual SPC homologues decreases with shortening of the oxidised side-chain, i.e. higher concentrations of metastable SPC with shorter alkyl chains are detected; and (b) within a set of phenyl isomers for each SPC homologue, some species show increased resistance to further degradation. Hence, an accumulation in the test liquor occurs, which is dominated by the shorter-chain SPC.

Observation (a) has also been described in other biodegradation assays conducted under controlled conditions using defined microbial communities and LAS homologues. Divo and Cardini [95] examined the metabolism of single C_{10}- to C_{13}-LAS homologues using an isolated microorganism belonging to the genus *Pseudomonas*. Depending on the identity of the starting compound, the degradation intermediates had between four and seven carbon atoms in the side-chain. Another bacterial strain belonging to the proteobacteria group was isolated from a trickling filter by Schleheck et al. [89] and investigated for its capacity to degrade C_{12}-LAS-3. Mass spectrometric analysis by matrix-assisted laser desorption/ionisation time-of-flight mass spectrometry (MALDI-ToF-MS) identified C_6-SPC-4 as the principal degradative product. Hrsak [96] made use of a mixed methanotrophic–heterotrophic culture to degrade a C_{11}-LAS mixture consisting of all five possible positional isomers. The major intermediates were C_5- and C_7-SPC identified by comparison of high performance liquid chromatography (HPLC) retention times with authentic standards. A commercial LAS mixture was subjected to biodegradation in river water which was used both as a source of organisms and as the medium [97]. A series of transient metabolites were formed in the mixture and assigned as C_4- to C_7-SPC. Cavalli et al. [11] tested the biodegradability of commercial LAS with a modified OECD 301E procedure over a period of 80 days. The LAS concentration was restored every 4 days with fresh substrate. At the end of the test, four SPC corresponding approximately to 85% of the total SPC content were structurally identified by nuclear magnetic resonance (NMR) as isomers of C_6-SPC-3 and C_7-SPC-3, while two SPdC (C_5 and C_7) were likewise recovered.

With respect to observation (b) above, the pronounced stability of some degradation intermediates towards further breakdown might be due to steric inhibition, in which the proximity of the sulfophenyl ring impedes further β-oxidation. This has already been discussed above for the four possible isomers of C_6-SPC.

In the literature the persistency of LAS breakdown products has been addressed in several studies. The residual fraction was primarily attributed to metabolites of manufacturing by-products in technical LAS blends used for the assays. Kölbener et al. [92] kinetically examined the degradation of commercial LAS in a trickling filter. Of the LAS-carbon initially added, 15% remained as DOC in the eluate of which roughly 50% was identified by HPLC-ultraviolet (UV) and infrared (IR) spectroscopy as carboxylated DATS and SPC. In a subsequent study it was shown that the source of the remaining refractory organic carbon (3–14%) was carboxylated DATS and partially degraded ABS [28]. The same authors identified a series of recalcitrant arylsulfonates originating from biodegradation of pure DATS and ABS [90]. The significance of resistant metabolites of LAS by-products was also claimed by DiCorcia et al. [88], who identified along with regular SPC, SPdC and mono- and dicarboxylated DATS using mass spectrometry.

With regard to the findings of the studies conducted by Eichhorn and Knepper [34], an interference with monocarboxylated DATS could be ruled out as the molecular ions of these intermediates have m/z two units lower than that of SPC with the same number of aliphatic carbon atoms.[1] Moreover, they do not produce the fragment m/z 183, which was always detected along with the $[M - H]^-$ of SPC.[2]

According to manufacturing information, 2.9% of nonlinear LAS was contained in the C_{12}-LAS assayed [98]. It seems most reasonable that this impurity is made up of iso-LAS, containing a methyl branch located on one of the carbon atoms of the alkyl chain (for the structure refer to Fig. 5.1.1). Although β-oxidation is not prevented by the presence of a tertiary carbon, the biodegradation rates characteristically show slow initial phases [99]. Hereafter, chain shortening may proceed, cleaving the resulting propionate moiety, in place of the acetate moiety of the non-capped analogues. It has been suggested that this specific step requires alterations to the enzymatic system of the cells [10].

Support for this hypothesis on iso-LAS is given through the findings of Nielsen et al. [100] in a degradation study carried out with radiolabelled substrate, in which it was reported that most of the iso-LAS isomers underwent ultimate biodegradation (79–90%), although some of their carbon (10–20%) was released as water-soluble intermediates. Furthermore, Kölbener et al. [90] could identify structures

---

[1] The mass spectrometric identification of LAS metabolites is given in Chapter 2.4.4.1.

[2] Contrary to LAS and SPC yielding the typical fragment m/z 183, corresponding to 4-styrenesulfonate, DATS and their carboxylated intermediates form the diagnostic ion m/z 209. This can be assigned as the fragment carrying the intact tetralin structure with its two six-membered rings from which the hydrocarbon chains have been cleaved.

mass spectrometrically corresponding to *iso*-C$_5$-SPC, which formed part of the organic residue in LAS biotransformation. In studies on commercial mixtures of LAS, DiCorcia et al. [88] monitored the formation of different SPC species, and indicated that *iso*-SPC exhibited slower degradation rates than the later eluting SPC.

The relevance of the stable compounds originating from LAS breakdown and the elucidation of their possible structures is further discussed in Chapter 2.4.4.1. The environmental significance of all SPC intermediates stemming from commercial LAS degradation is described in Chapter 6.

Novel metabolites from the β-oxidation of sulfophenyl carboxylates. Besides the evidence from the repetitive cycles of β-oxidation in C$_{12}$-LAS biotransformation, further support for β-oxidation could be gathered if other intermediates occurring via this pathway were detectable. β-Oxidation, though known for decades, is still the subject of active research [101,102]. The class of compound that has been identified to date, albeit in mammalian systems, is the α,β-unsaturated carboxylate.

Indications for the formation of analogous species in microbial metabolism of LAS were found by Knepper and Kruse [33] during biotransformation of commercial LAS surfactant on an FBBR. However, the low concentrations of the tentative metabolites in the test liquor, which eluted under the applied reversed-phase (RP)-HPLC conditions somewhat earlier than the normal SPC, did not permit acquisition of full-scan mass spectra as was needed for unequivocal identification. Further evidence for the formation of the intermediate with a double bond in the alkanoate moiety was reported by Bird [103]. During biodegradation of C$_{11}$-LAS by a bacterial strain, a new UV adsorption band centred near 260 nm was detected, which was assumed to result from a double bond, although a definite confirmation could not be provided.

A further successful attempt to detect α,β-unsaturated SPC (SPC-2H) during the degradation of LAS was achieved by Eichhorn and Knepper [34] (Fig. 5.1.8). An SPC-2H has m/z of two mass units below that of the corresponding SPC, and compounds of this type have been described in the degradation of impurities in LAS, the monocarboxylated DATS [88,104]. However, a definite distinction between the known DATS carboxylates [88] and the putative SPC-2H could be provided by LC–ESI–MS analysis, which enabled structural assignment via

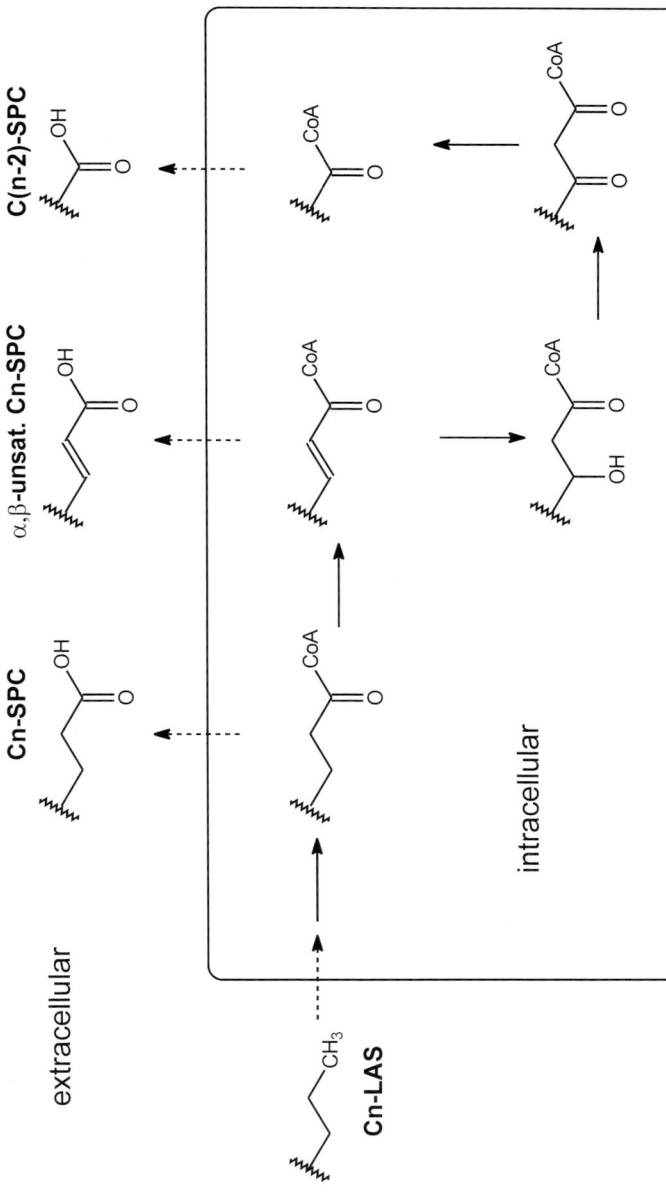

Fig. 5.1.8. Presumed major pathway for intracellular aerobic biodegradation of LAS (as acetyl-CoA derivatives) to SPC via ω-oxidation followed by successive oxidative shortening of the alkyl chain by two carbon units (β-oxidation). Intermediates of β-oxidation, such as SPC-2H formed by enzymatic dehydrogenation, are also transported out of the cell after cleavage of the CoA ester. Modified from Ref. [34].

the recorded mass spectra (see Chapter 2.4.4.1). The C-even-SPC-2H compounds were present at about 1–10% of the concentrations of the corresponding SPC (Table 5.1.1). No C-odd-SPC-2H were detected, presumably due to the low levels of C-odd-LAS and C-odd-SPC present. Four SPC-2H were detected, namely the C_{12}, C_{10}, C_8 and C_6 homologues. These correspond to the β-oxidation products of the C-even-SPC, thereby directly supporting β-oxidation as a mechanism for all relevant SPC. The route of metabolism of LAS observed in the FBBR was thus ω-oxidation followed by β-oxidations.

The fact that no C_4-SPC-2H were detected, i.e. that the β-oxidation cycle was terminated at C_4-SPC, is indicative of a different mechanism for further steps in the degradation process, with desulfonation or ring cleavage beginning to dominate (see Fig. 5.1.2). Schulz et al. [86] grew enrichment cultures on C_4-SPC-2 yielding a pure culture of a bacterium capable of metabolising sequentially the enantiomeric forms of C_4-SPC-2. 4-Sulfocatechol was identified as an intermediate, which subsequently underwent *ortho*-cleavage. Desulfonation of the same compound was accomplished by another organism, *Pseudomonas putida*, yielding the 4-hydroxyphenyl derivative [105]. A lack of adequate enzymes for the biodegradation of C_4-SPC-3, originating from the biotransformation of C_{12}-LAS-2 in fish, resulted in an accumulation to about 70% of the molar concentration of the parent compound [106].

Overall, there is no need to suggest alternative pathways at this point, as the currently accepted theory adequately explains all the observations. There was no experimental evidence to support sub-terminal oxygenation in the FBBR experiments. The better-known attack on both methyl groups of LAS to give SPdC was also not observed.

Apart from SPC and SPC-2H, no other biodegradation products occurring in the degradation pathway as depicted in Fig. 5.1.8, nor the previously reported SPdC, were found in this study. This may in part be due to the polarity and/or stability of compounds such as CoA-esters, which makes it almost impossible for them to be excreted out of the cell. Moreover, no desulfonated metabolites were detected in gas chroma-tography (GC)–MS analyses of the FBBR samples. Such intermediates must be formed in the sequence of (ultimate) biodegradation [10]. The lack of their detection in the work of Eichhorn and Knepper [34], (Table 5.1.1) might be explained by the fact that desulfonation is likely to occur at a later stage of biodegradation [107,108].

Isomer pattern of sulfophenyl carboxylates. To illustrate the different fate of individual SPC isomers in the FBBR experiment, which has been discussed in the preceding section for C_6-SPC, LC–ESI–MS chromatograms from the samples taken on day 8, 11, 18 and 26 following spiking of C_{12}-LAS are compared in Fig. 5.1.9. As can be clearly seen, there are at least four different, sufficiently separated C_6-SPC species, occurring in the sample taken at day 8 (Fig. 5.1.9, left-hand side). The individual signals, assigned $a–d$ according to the order of elution, all correspond to different isomers of C_6-SPC, which all exhibit different biodegradation rates and final fates. In the sample taken after 11 days, in which the relative amount of the C_6-SPC species is the highest from the samples analysed, the C_6-SPC isomer a is predominant followed by isomer d, whereas the isomers b and c are less abundant. After 18 days the amount of isomer d is increased relative to isomer a and after 26 days the apparently recalcitrant isomer d dominates the mixture. In Fig. 5.1.9 (right-hand side) the biodegradation behaviour is shown for the different C_8-SPC species. In general, a similar behaviour to the C_6-SPC isomers is observed, except that after 8 days almost only isomer b is present, and after 26 days isomer a is dominant with lesser amounts of b and c.

These observations are consistent with earlier observations, indicating that those SPC isomers with a greater resistance to further degradation stem from $(C_{12}-)$LAS in which the phenyl ring is attached to central positions of the alkyl chain, or those which are oxidised on the shorter side of the alkyl moiety [10]. In the course of microbial degradation, subsequent shortening of the alkyl chain via β-oxidation becomes hindered in close proximity to the aromatic ring.

Commercial linear alkylbenzene sulfonate—effect of adaptation. A third degradation experiment was performed in order to confirm the build-up of recalcitrant SPC species by assaying a commercial LAS mixture instead of a pure LAS homologue. The use of an LAS blend yielded results of higher relevance to the aquatic environment, which is generally exposed to complex mixtures of the anionic surfactant. Furthermore, information was obtained on the impact of acclimation on the qualitative and quantitative formation of SPC intermediates as well as on the behaviour of recalcitrant phenyl isomers. For this purpose, the river water in the FBBR was spiked with LAS at $10\ \text{mg L}^1$ as those degradative products, which were detectable only at high substrate concentrations (SPC-2H), were not

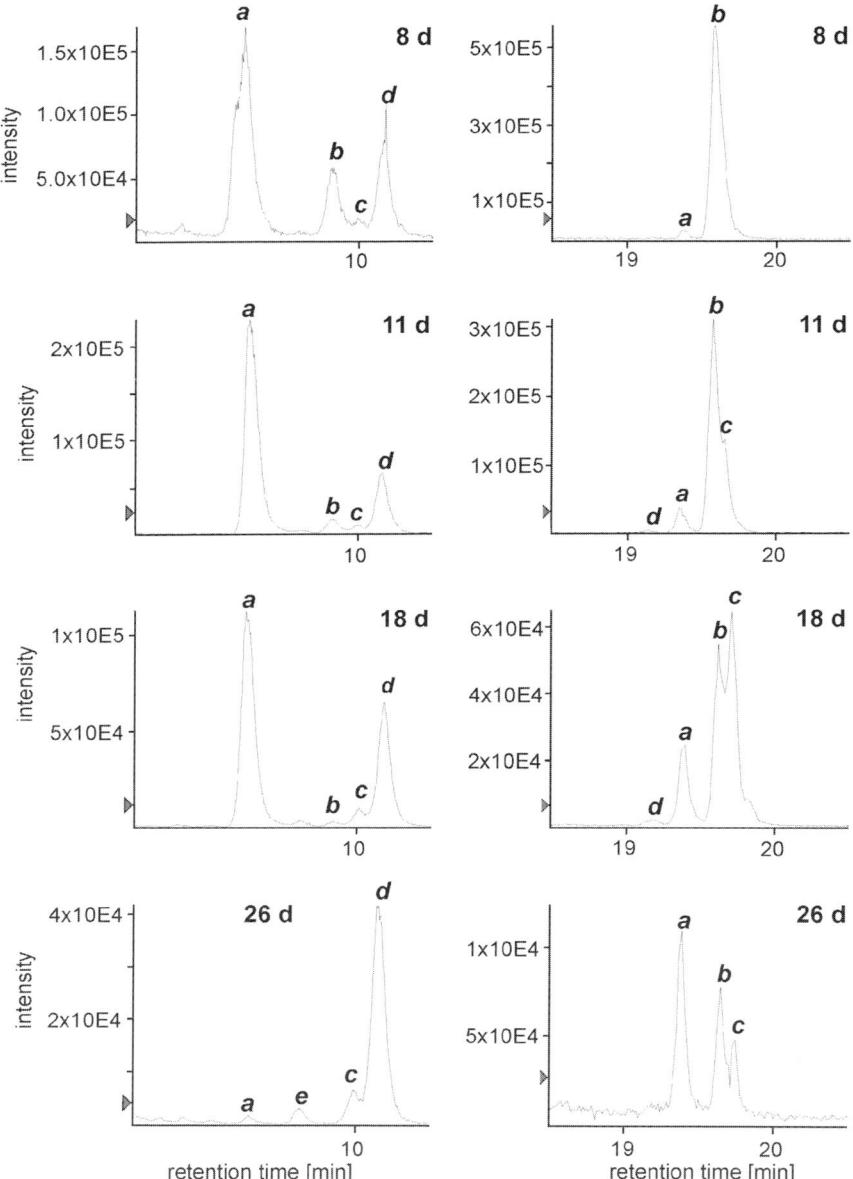

investigated. The FBBR samples were directly subjected to LC–ESI–MS analysis without further treatment and SPC were semi-quantitatively determined via absolute peak area. The dependence of LAS primary degradation on the alkyl-chain length and position of phenyl attachment was not monitored as this has been the subject of many thorough investigations since the mid-1960s [21,49,72,77, 109,110].

In Fig. 5.1.10 the evolution of C-even (a) and C-odd-SPC (b) over-time from the LAS-spiked FBBR is plotted. Their formation starts at day 5 with the entire range of examined SPC observed. The curves show a maximum at day 7 with C_7-, C_8-, and C_9-SPC prevailing. However, it should be noted that in quantitative terms the short-chain homologues C_4–C_6 are underestimated, as their ionisation efficiency is lower than that of the longer-chain SPC. A relatively constant level of all SPC compounds was achieved on day 15 and maintained until the end of the experiment on day 30 (not all data shown).

In order to investigate the acclimatisation of the microorganisms, the reservoir tank was emptied and the entire unit rinsed twice with water, before filling again with freshly acquired Rhine river water. The system was then respiked with 10 mg L^{-1} of the same LAS substrate. It can be seen in Fig. 5.1.11 (a) and (b) that after the first day, SPCs are already present in the test liquor. On the apex at day 2, the absolute intensities of single SPC are similar to those observed in the previously described assay at day 7, and C_7-, C_8- and C_9-SPC are the most abundant homologues. After day 9, no substantial changes in detected amounts were found, with the final homologue distribution greatly resembling that of the first degradation test.

In summary, acclimatisation of the microorganisms, which had settled on the glass beads of the fixed-bed during the first run, resulted in a more rapid degradation of LAS, thus leading to the formation of SPC without an acclimatisation delay. This ability is of particular significance in natural waters since it may prevent accumulation of LAS residues in receiving streams.

Fig. 5.1.9. $(-)$-LC–ESI–MS extracted ion chromatograms of C_6-SPC (m/z 271; left-hand side) and C_8-SPC (m/z 299, right-hand side) from samples of FBBR taken after 8, 11, 18 and 26 days after spiking of C_{12}-LAS. Assignment of the individual recalcitrant species according to order of elution and formation $(a-e)$. Modified from Ref. [34].

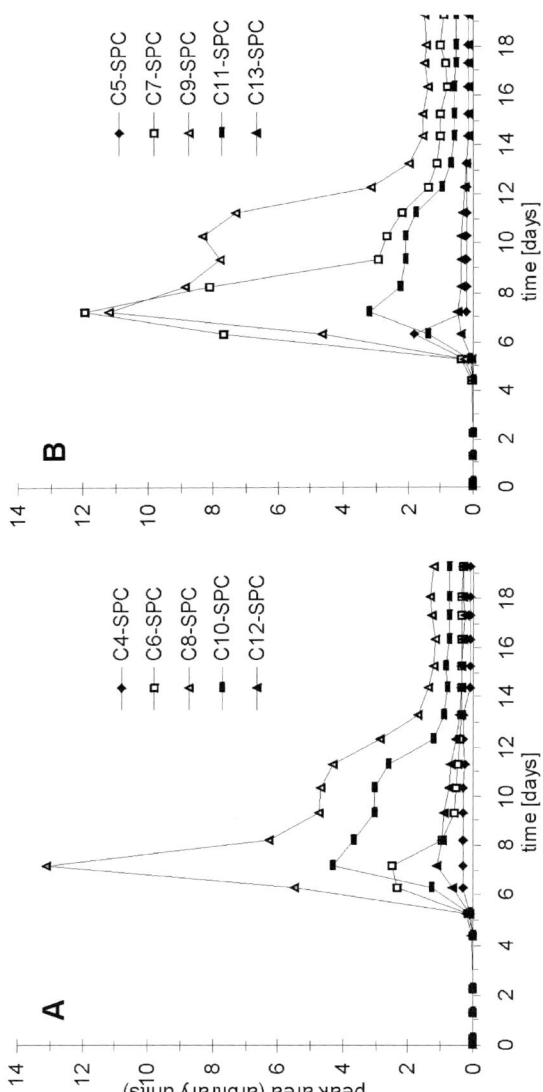

Fig. 5.1.10. Profile of (a) C-even and (b) C-odd-SPC intermediates formed during biodegradation of commercial LAS on FBBR after first spiking.

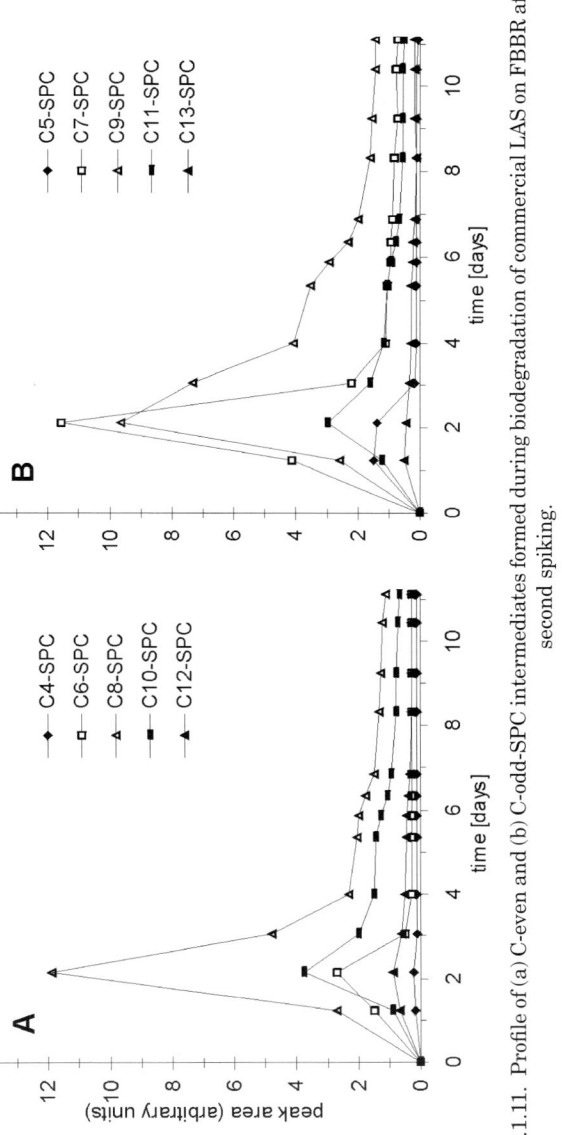

Fig. 5.1.11. Profile of (a) C-even and (b) C-odd-SPC intermediates formed during biodegradation of commercial LAS on FBBR after second spiking.

However, observations made in the previously described acclimatisation experiment indicated that the ultimate fate of SPC is not affected by the time of exposure to the LAS intermediates. It seems that the microorganisms are not capable of complete degradation of all isomers. It is possible though that the microbial community established on the fixed-bed is not capable of producing the enzymes needed for total conversion. It is a generally accepted view that LAS is predominantly biodegraded by a consortium of bacteria [89]. Different lines may be responsible for different functions with, e.g., a single member of the community inducing the terminal oxidation of the alkyl chain and its shortening by β-oxidations, and a second organism completing the degradation via biologically mediated ring cleavage. Consequently, a mixture of organisms may be requested with the omission of any member of the consortium resulting in incomplete LAS mineralisation.

Relevance to the environment. To provide evidence that the recalcitrance of some SPC species in the FBBR liquor was not restricted by limitations in the microbial community, keeping in mind that it was sourced from surface water with a known history of LAS exposure, a comparison with the fate of SPC in a natural environment was sought. For this purpose, the isomer patterns of a selected homologue detected in a steady-state FBBR sample was compared to that found in an extract from the river Rhine [34].

In Fig. 5.1.12, the extracted total ion chromatograms (XIC) of C_7-SPC in the enriched river water sample (a) and the FBBR sample (b) are shown. In (a) at least six isomers $(a-f)$ can be distinguished at retention times between 18 and 21 min. Comparison of this chromatogram with that of the enriched river sample shows several similarities for the different isomers. In the river water, six individual isomers $(a-f)$ could also be assigned, with these occurring at the same retention times and at almost the same ratio of abundance. These results prove unambiguously that the recalcitrance of the SPC recovered from the FBBR was not the result of a limitation in the experimental set-up. Regarding the identity of the persistent C_7-SPC constituents in the FBBR liquor and surface water samples, it might be assumed that they comprise those positional isomers in which the sulfophenyl substituent impedes a subsequent β-oxidation, i.e. namely C_7-SPC-2 and C_7-SPC-3. This supposition, however, is not sufficient to explain the high number of signals detected, with six evident, that are shown in Fig. 5.1.12.

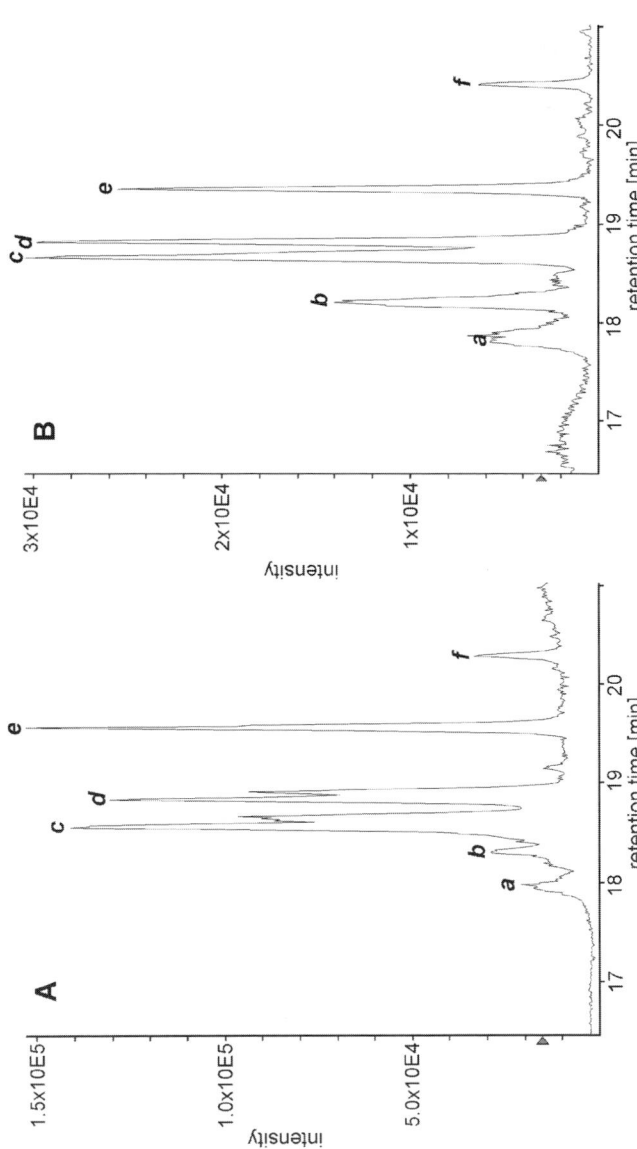

Fig. 5.1.12. (−)-LC–ESI–MS extracted ion chromatograms of C_7-SPC (m/z 285) from (A) an enriched river water sample, and (B) an FBBR sample taken 20 days (constant concentrations) after spiking with commercial LAS. The individual isomers (a–f) are assigned according to elution order. Modified from Ref. [34].

A possible source of these additional metabolites could also arise from the presence of *iso*-LAS (an impurity in the LAS tested), as observed in the C_{12}-LAS experiment described previously. Degradation of *iso*-C_{11}-LAS or *iso*-C_{13}-LAS could theoretically liberate a total set of 12 *iso*-C_7-SPC isomers. In Fig. 5.1.13, these are arbitrarily termed as *a* through *l*, noting that additionally each isomer can also comprise four diastereomers, except for *c*, *f* and *i*, which can only have two enantiomers. It can be speculated that the species that most slowly degraded, or those being virtually recalcitrant, will be those where both the sulfophenyl ring and the methyl branch are close to the carboxylate group, i.e. *a−d*, and to a lesser extent *e−h*. As such the isomers *i−l* can be assumed to be the most easily converted *iso*-C_5-SPC derivatives.

Cavalli et al. [11] claimed that *iso*-LAS was amenable to biodegradation without producing any evident accumulation of recalcitrant products. The authors suggested that the contrasting results in a previous study by Kölbener et al. [90], who reported persistent *iso*-SPC compounds, arose from the unbalanced microbial community of the particular biodegradation system used. But our own findings described here demonstrate good accordance between the lab-scale experiment and the conditions in real-world settings, and indicate that linear SPC cannot exclusively account for the persistent SPC fraction.

Overall, it can be concluded that the aerobic biodegradation of LAS under laboratory conditions results in the formation of some stable SPC metabolites and that residues with a very similar composition, defined by peak pattern and retention time, occur in surface waters (see Chapter 6.2.2).

Other anionic surfactants

Secondary alkane sulfonates. Secondary alkane sulfonates (SASs) have been shown to be readily biodegradable under aerobic conditions [10, 111], with ω-oxidation and oxidative desulfonation the significant steps in the conversion [112]. As such, in most aquatic environmental compartments ultimate degradation is considered likely. However, a tendency to adsorb to particulate matter may cause their removal from such environments prior to complete conversion, i.e. in WWTP clean-up processes, thereby preventing the degradative process from occurring [114,113]. The fate of SAS in sludge removed from WWTPs will thus be dependent on the extent of aerobic digestion in

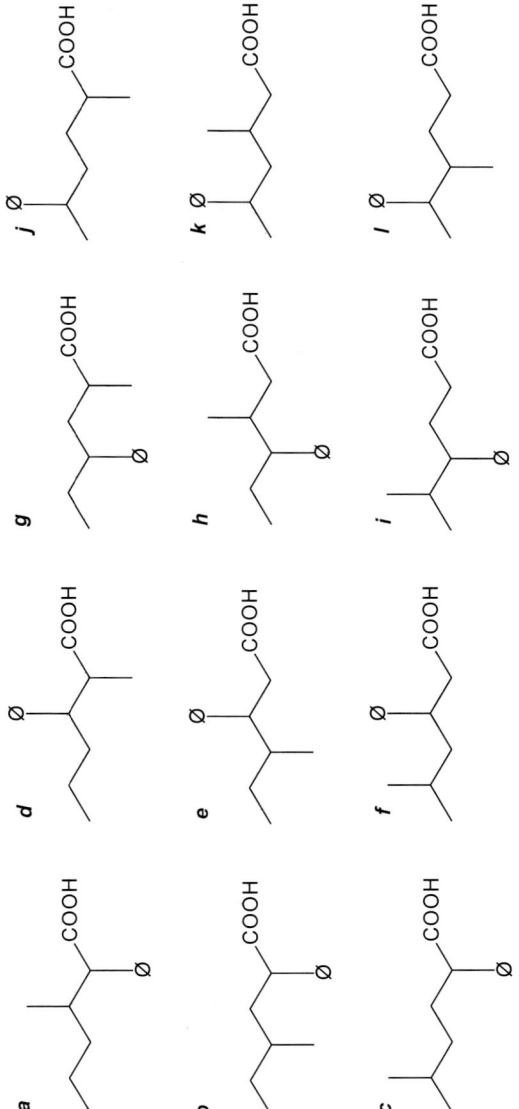

Fig. 5.1.13. Possible structures of *iso*-C$_7$-SPC. The '∅' symbolises the sulfophenyl group.

the plant, and the aerobic nature of the subsequent compartment to which it is transferred, with degradation under anaerobic conditions limited.

Alcohol sulfonates. The aerobic degradation of AS is rapid and extensive [10,111,114,115], as much as it has been speculated that a broad range of microorganisms must be capable of catalysing the process [10]. The degradation is thought to proceed by enzymatic cleavage of the sulfate ester bond, followed by subsequent oxidation steps [10,114], although none of the steps requires molecular oxygen [114]. The products obtained are all amenable to ultimate biodegradation, and no recalcitrant metabolites have been indicated. Adsorption to sludge does not pose any persistence threats as anaerobic digestion is also extensive for this surfactant class [10,116]. Furthermore, biodegradability of linear and branched derivatives has been shown to be equally rapid and extensive [116] and available data indicate that all AS homologues are readily degradable irrespective of chain length [117]. Removal of AS isomers of $>99\%$ has been described in several studies using a variety of detection methods including MBAS [118], ^{14}C-radiolabelling [116] and LC−MS [119].

Alcohol ether sulfates. Ready aerobic biodegradation of AESs has been described [113], with ω/β-oxidation and cleavage of the sulfate and ether bonds attributed to the process [10]. However, molecular oxygen is not necessary for the two latter steps, and primary and ultimate degradation has been described under both aerobic and anaerobic conditions [114].

Fatty acid esters. Fatty acid esters (FESs) are readily degraded in aerobic environments [120], by ω- and β-oxidation steps, followed by desulfonation [122], such that extensive mineralisation has been described. Persistence from aquatic sources can only be envisaged in cases where adsorption leads to removal of the surfactant from the aerobic conditions, with anaerobic degradation not observed in any studies described to date [122]. Aerobic degradation in sludge-amended soils has, however, been described as rapid [122].

5.1.4 Aerobic biodegradation of non-ionic surfactants

Nonylphenol ethoxylates

The biodegradation of NPEO was generally believed to start with a shortening of the ethoxylate chain, leading to short-chain NPEO containing one or two ethoxylate units (NPEO$_1$ and NPEO$_2$). Further transformation proceeds via oxidation of the ethoxylate chain, producing mainly alkylphenoxy ethoxy acetic acid (NPE$_2$C) [121] and alkylphenoxy acetic acid (NPE$_1$C) (Fig. 5.1.14(a)) [122]. In a only few studies NPE$_3$C or longer-chain carboxylates have been detected in environmental samples (see Chapter 6.2.1). Little has been reported the further degradation of NPEC. In a semi-quantitative biodegradation study by DiCorcia et al., the formation of doubly carboxylated metabolites is described, with both alkyl and ethoxylate chain oxidised (CAPEC) [123]. Ding et al. detected CAPEC in river and wastewater [124,125].

It was generally assumed that the endocrine disruptor nonylphenol (NP) is the most persistent metabolite of NPEO. However, experimental data on the formation of the metabolite NP from NPEO are surprisingly scarce, and mostly under anaerobic conditions NP has been reported to be formed [126,127]. Only one article reports a slight increase in NP concentration from the aerobic degradation of NPEO during the composting of wool scour effluent sludge [128].

The purpose of a study performed by Jonkers et al. [40] was to elucidate the aerobic biodegradation route of NPEO spiked at a concentration of 10 mg L^{-1} to river water similar to the experimental set-up described for LAS in Chapter 5.1.3.1. LC–ESI–MS and LC–ESI–MS–MS after solid-phase enrichment were applied to clarify identity and routes of formation of metabolites. For NPEO a relatively fast primary degradation of >99% degradation was observed after 4 days. Contrary to the generally proposed degradation pathway of EO-chain shortening, it could be shown that the initiating step of the degradation is ω-carboxylation of the individual ethoxylate chains: NPEC metabolites with long carboxylated EO chains were identified (Fig. 5.1.14(b)). Further degradation proceeds gradually into short-chain carboxylated EO with the most-abundant species being NPE$_2$C (Fig. 5.1.15(c)). It could be shown that the oxidation of the nonyl chain proceeded concomitantly with this degradation, leading to metabolites having both a carboxylated ethoxylate and alkyl chain of varying lengths (CAPEC; Fig. 5.1.15). The identity of the CAPEC metabolites

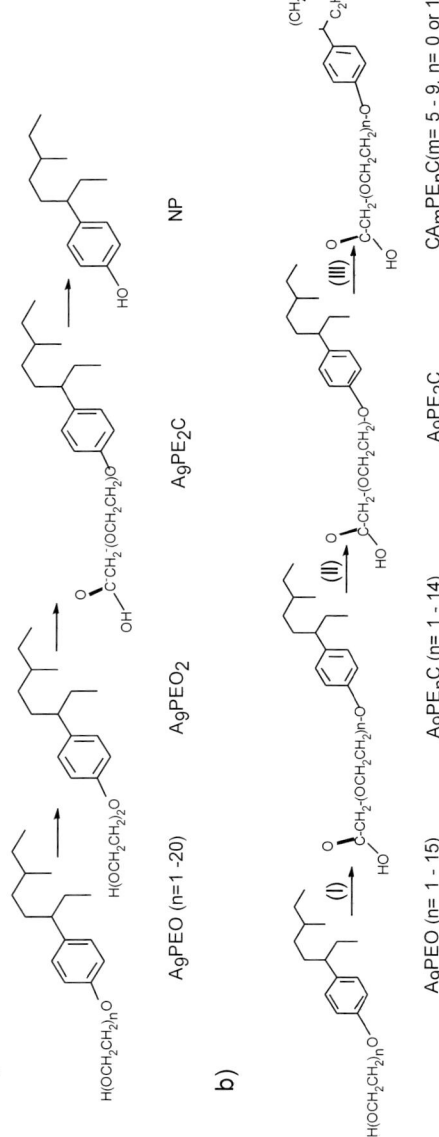

a)

A$_9$PEO (n=1-20) A$_9$PEO$_2$ A$_9$PE$_2$C NP

b)

A$_9$PEO (n= 1 - 15) A$_9$PE$_n$C (n= 1 - 14) A$_9$PE$_2$C CA$_m$PE$_n$C(m= 5 - 9, n= 0 or 1)

Fig. 5.1.14. (a) Generally proposed aerobic biodegradation pathway of NPEO. [122] (b) Newly proposed aerobic biodegradation pathway [40]. With permission from Env. Sci Technol, © by the American Chemical Society, 2001.

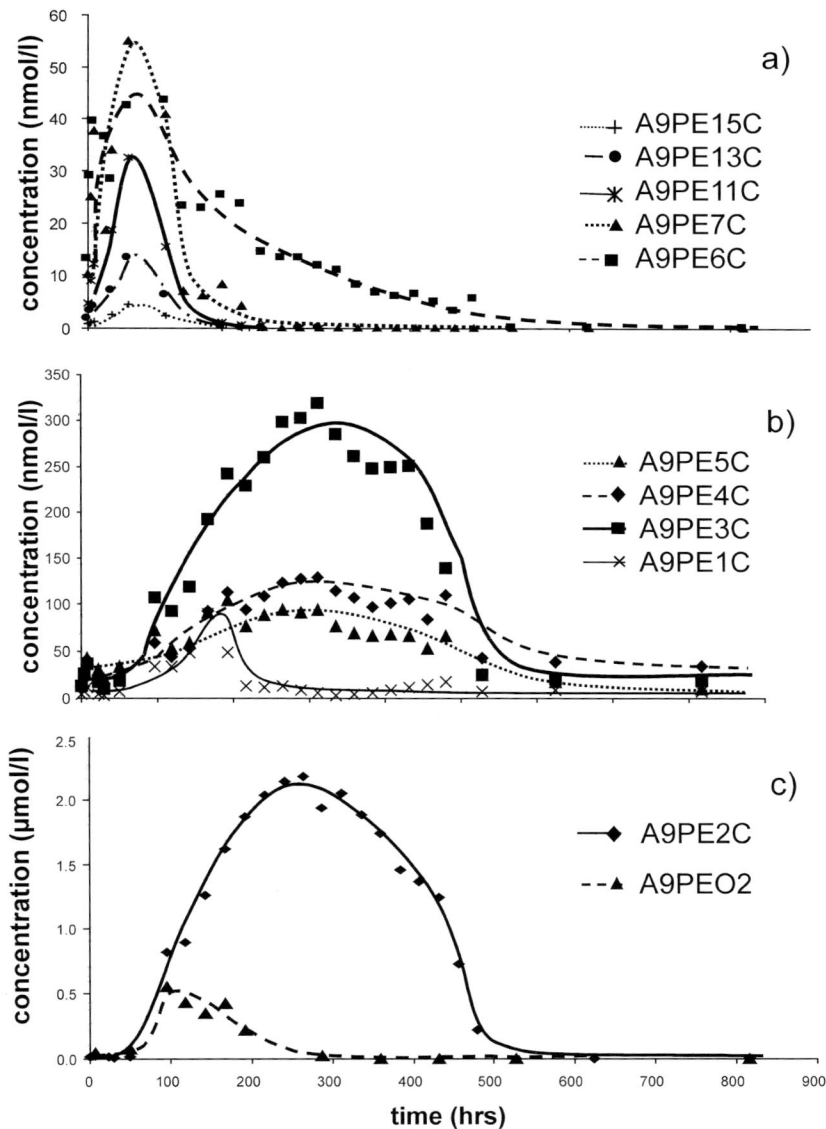

Fig. 5.1.15. Follow-up of NPEC (A_9PEC) and NPEO (A_9PEO_2) formation during the bioreactor experiment spiked with A_9PEO_{10}: (a) Long-chain A_9PEC; (b) mid-chain A_9PEC; (c) A_9PE_2C and A_9PEO_2 (taken from Ref. [40]). With permission from Env. Sci Technol, © by the American Chemical Society, 2001.

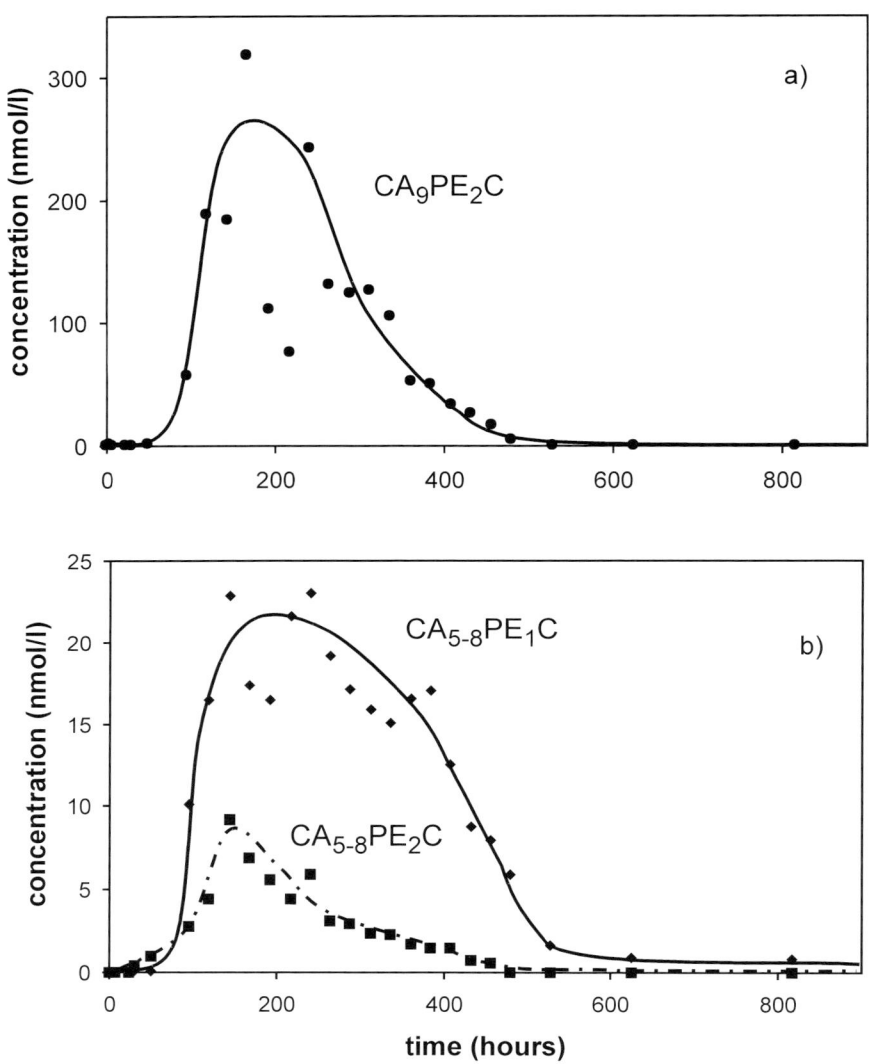

Fig. 5.1.16. Formation and subsequent degradation of CAPEC in the FBBR: (a) CA_9PE_2C; (b) $CA_{5-8}PE_1C$ and $CA_{5-8}PE_2C$ (nmol L^{-1}) (taken from Ref. [40]). With permission from Env. Sci Technol, © by the American Chemical Society, 2001.

was confirmed by the fragmentation pattern obtained with LC–ESI–MS–MS [40]. Both NPEC and CAPEC metabolites were still present in the bioreactor after 31 days (Fig. 5.1.16). In the aerobic degradation pathway, $NPEO_2$ is formed only to a minor extent and is even further degraded over several days; NP was not found as a metabolite in this study.

Alkyl glucamides
As already pointed out in the chapter describing the analytical methods (Chapter 2.4.3.2), results on biodegradation of alkyl glucamides (AGs) are also very scarce. Stalmans et al. [129] applied standardised degradation tests using a batch-activated sludge system and the modified Sturm test, and thereby indicated a high mineralisation rate. In both the tests the DOC removal exceeded 98% and primary degradation was greater than 99%.

Eichhorn and Knepper [38] investigated the biodegradability and the metabolic pathway of C_{10}-AG on an FBBR spiked at a concentration of 1 mg L^{-1}. This homologue was used as it was advantageous to work with solitary substances in order to obtain unambiguous results, despite the fact that the higher homologues C_{12}- and C_{14}-AG were predominant in formulations. Analogous behaviour with respect to the breakdown mechanism, however, could be expected for the higher homologues.

The strategy pursued to elucidate the metabolism of AG was to postulate a mechanism based on the known degradation pathway of the structurally related surfactant LAS. This exclusively proceeds via ω-oxidation of the alkyl chain, resulting in the formation of carboxylic acids, which are then further broken down through β-oxidations as described previously. The corresponding pathway relevant to AG is shown in Fig. 5.1.17. From the putative carboxylic acids that are theoretically possible, only the C_4-glucamide acid could be detected (Fig. 5.1.18) [38].[3]

The lack of detection of the C_6, C_8, and C_{10} carboxylic acids can potentially be attributed to their short lifetime in solution or to the possibility that they are not released from the cells into the bulk. It is also likely that they are rapidly further degraded to shorter homologues,

---

[3] Due to its simplicity, the term 'glucamide acid' was used in this work to describe the new metabolite.

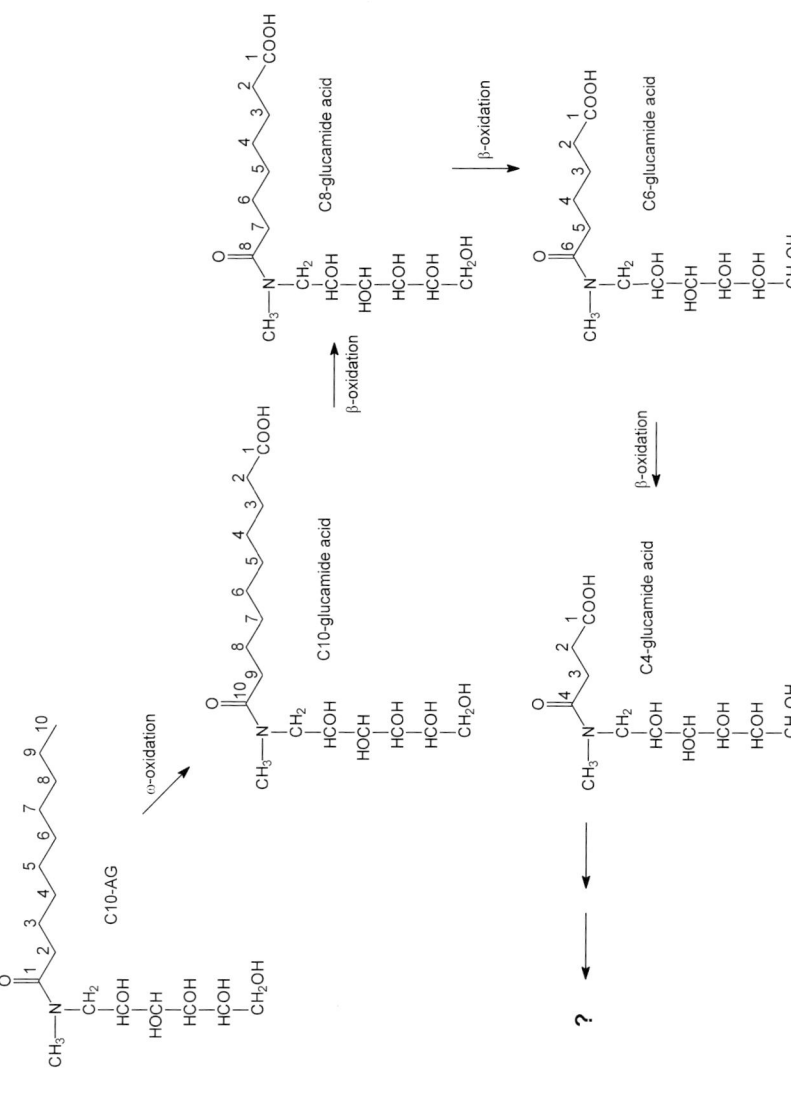

Fig. 5.1.17. Postulated aerobic breakdown pathway of C_{10}-AG (taken from Ref. [38]). With permission from J. Mass Spec., © by Wiley, 2000.

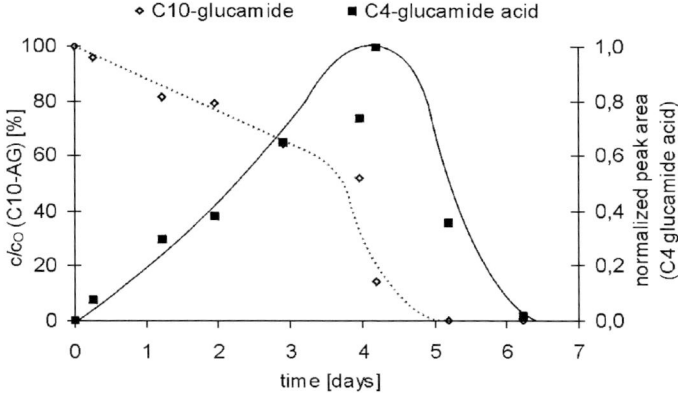

Fig. 5.1.18. Degradation curve of C_{10}-AG and formation of metabolite C_4-glucamide acid.
© by Wiley, 2000.

with similar observations made for long-chain SPC, which undergoes rapid conversion into shorter-chain species during LAS metabolism. The relative longevity of the C_4-glucamide acid can be explained by the fact that in contrast to the preceding steps, a subsequent β-oxidation would not lead to a regular carboxylic acid, but to an α-keto acid, which is expected not to occur.

It is expected that the final conversion of C_4-glucamide acid occurs as an intracellular process. This assumption is supported by the findings of Stalmans et al. [131], in which high DOC removal and CO_2 formation in standardised biodegradation tests was described.

Alkyl polyglucosides
Alkyl polyglucosides (APGs) consist of a complex mixture of a variety of homologues and isomers, including stereoisomers, binding isomers and ring isomers within the glucose moiety (see Chapter 1). In order to assess the aerobic and anaerobic biodegradability of APG, a series of standard laboratory tests with activated sludge were carried out by Steber et al. [130] using both the discontinuous screening tests and continuous test systems. Evaluation of the results of sum parameters analysis (DOC removal, TOC removal, COD and BOD) allowed classification of APG as 'readily biodegradable' according to OECD definitions. Ultimate biodegradation was proved by the coupled-units test (OECD 303A) giving a DOC removal of 89% and even the anaerobic

563

biodegradability testing (ECETOC screening test) indicated complete transformation into CO_2 and CH_4 amounting to 84% theoretical gas production. The findings of Steber et al. [133] were confirmed in two other studies, working likewise with OECD standard tests [131,132]. However, all these test methods were exclusively based on sum parameters, and hence they did not yield any information regarding the metabolic pathway of APG.

Eichhorn and Knepper [39], however, applied more specific methods to investigate the aerobic biodegradation of 1 mg L^{-1} of each C_8-, C_{10}-, and C_{12}-β-monoglucoside spiked into river water, utilising the FBBR technique and MS detection. In this assay a cocktail of identical concentrations of the three homologues was added to investigate the effect of different alkyl-chain lengths. In Fig. 5.1.19 relative concentrations of each homologue are plotted versus time. The biodegradation curve showed a sigmoidal-shaped course, which can be subdivided into: (a) a lag phase, where the acclimation of the microorganisms to the test substances was required (0–13 h); (b) a degradation phase, in which the compounds are metabolised (13–21 h) at increasing speed; (c) a plateau

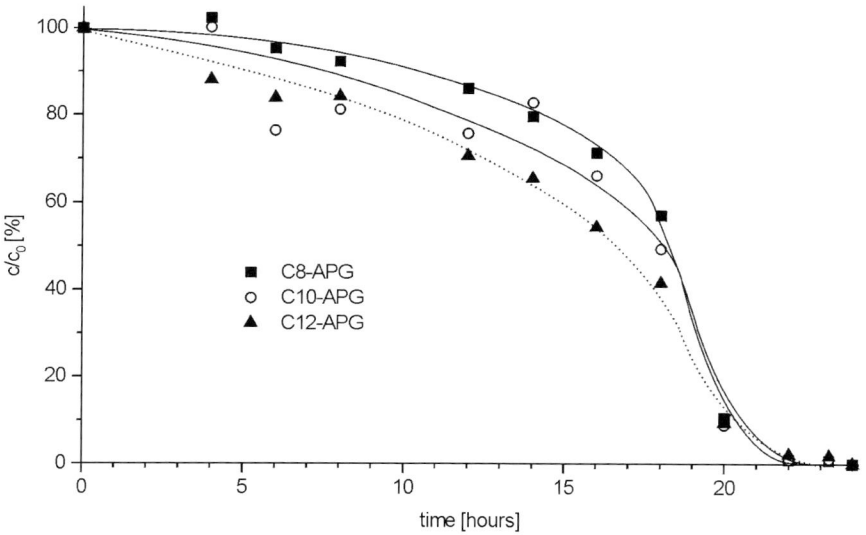

Fig. 5.1.19. Degradation curve of C_8-, C_{10}- and C_{12}-β-glucopyranoside (taken from Ref. [39]). © by Elsevier, 1999.

phase, in which degradation is nearly complete (after 1 day). Over the first period, a slight dependence of the adaptation rate on the alkyl-chain length was observed, with faster degradation observed with increasing chain length. However, almost no difference was evident in the final degradation status.

The FBBR samples were also investigated for other possible metabolites. A breakdown pathway of the APG was postulated according to the metabolism of LAS and AG, which are degraded by ω-oxidation of the alkyl chain resulting in carboxylic acids (compare with Fig. 5.1.2). These are in turn further degraded by β-oxidation releasing C_2 units. The corresponding mechanism for APG is shown in Fig. 5.1.20 (Pathway I). The search for the putative 'polyglucoside alcanoic acids' was not successful. An alternative degradation mechanism has thus

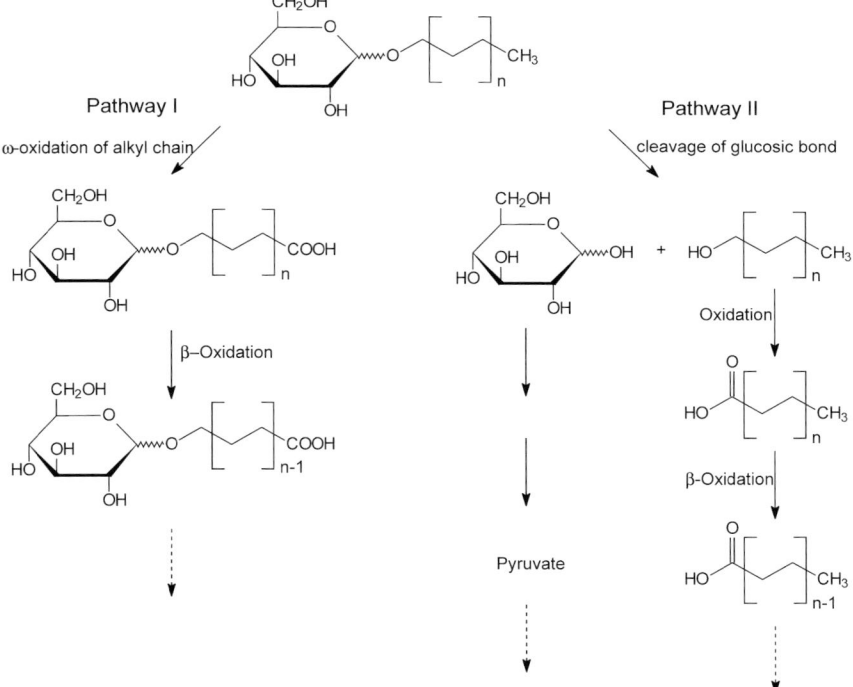

Fig. 5.1.20. Possible degradation pathways of alkyl glucopyranosides; $n = 8, 10, 12, 14, 16$ (taken from Ref. [39]). © by Elsevier, 1999.

been proposed (Fig. 5.1.20, Pathway II), which involves the cleavage of the glucosidal bond, which leads to glucose and the fatty alcohol intermediates. The generated glucose is rapidly further metabolised via pyruvate and the fatty alcohol oxidised to the corresponding acid, which can subsequently undergo classical fatty acid degradation. The lack of detection of any fatty acids, which were also investigated by $(-)$-LC–ESI–MS, was attributed to their rapid intracellular metabolism.

The hypothetic Pathway II is supported by the results of Kroh et al. [133], who investigated the ability of carbohydrolases α-glucosidase, β-glucosidase and isomaltase to break the glucosidal bond between fatty alcohol and carbohydrate. The model compounds C_8-α- and C_8-β-glucopyranoside used were completely hydrolysed by a mixture of the three enzymes. However, a technical APG blend, which also contained higher oligomers, was not quantitatively cleaved, presumably due to the incapability of the employed enzymes to break the linkages between the sugar moieties.

Alcohol ethoxylates
Alcohol ethoxylates (AEs) have been developed as an environmental friendly replacement of APEO and as such the data available regarding their biodegradability are relatively extensive. The linear derivatives exhibit good degradation [134]; however, the branched analogues are converted to a much lesser extent [135]. The process is considered to be by cleavage of the hydrophobe–hydrophile bond, and subsequent ω/β-oxidation of the hydrophobe [136]. The proposed mechanisms for their aerobic biodegradation is shown in Fig. 5.1.21 [137] including: (a) the central cleavage of the molecule, leading to the formation of polyethylene glycols (PEGs) and aliphatic alcohols; (b) the ω,β-oxidation of the terminal carbon of the alkyl chain; (c) the hydrolytic shortening of the terminal carbon of the polyethoxylated chain. PEGs and aliphatic alcohols are degraded independently. PEG biodegradation proceeds by successive depolymerisation of the ethoxy chain via non-oxidative and oxidative cleavage of the C_2 units. This process leads to the formation of shorter-chain neutral PEGs, mono- and dicarboxylated metabolites of PEG (MCPEGs and DCPEGs), alcohol polyethoxylates carboxylated on the polyethoxy chain (AECs) or in the alkyl chain (CAEs), or alcohol polyethoxylates carboxylated on both ends (CAECs). Taking into account the dependence on the aquatic system in which PEGs are found, their half-lives can vary from weeks to months.

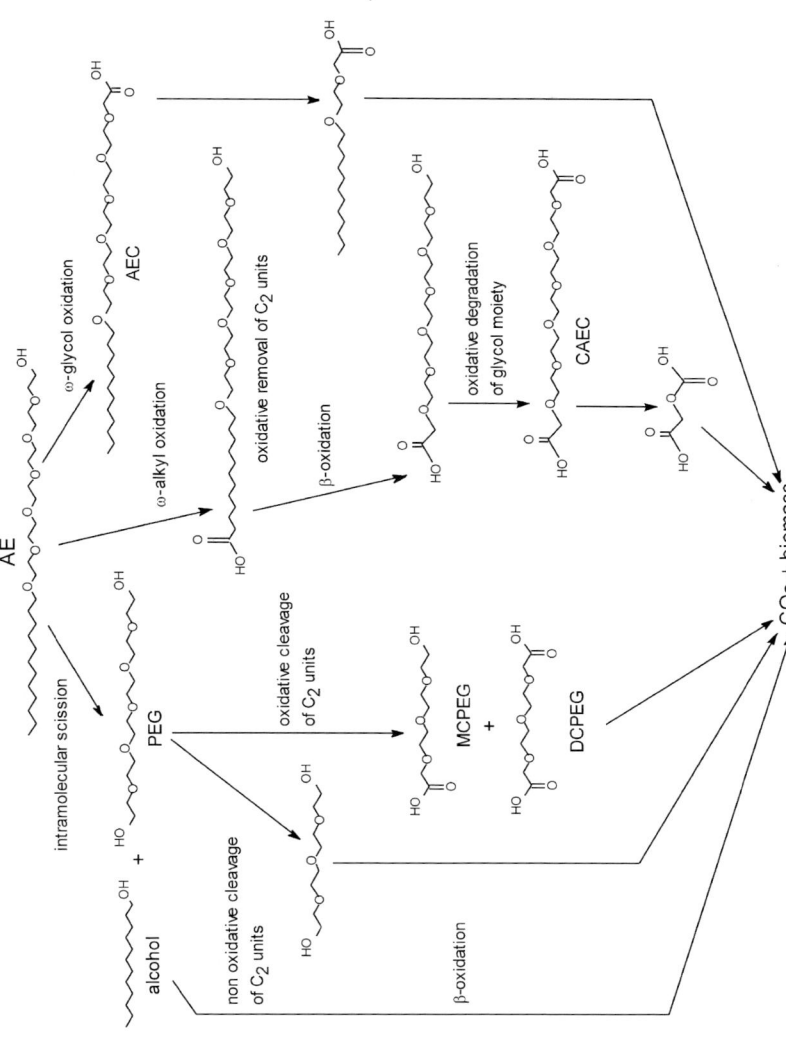

Fig. 5.1.21. Proposed biodegradation pathways for aliphatic alcohol polyethoxylates under aerobic conditions (taken from Ref. [137]). With permission from Analusis., © by EDP Sciences, 1997.

5.1.5 Cocamidopropyl betaines

Only limited data have been published on the degradability of betaines, such as cocamidopropyl betaine (CAPB, Fig. 5.1.22), and the limited information available is somewhat contradictory. Swisher reported that total degradability was rather poor (45–58%) [10], whilst Brunner et al. [138] obtained mixed results with a series of the amphoteric surfactants in an extended OECD 302B test and a laboratory trickling test filter. While the two compounds cocoampho diacetate and cocoampho dipropionate were mineralised only to a minor extent, CAPB and cocoampho acetate proved to be nearly totally degradable.

Eichhorn and Knepper [36] studied the aerobic biodegradation of four CAPB homologues (C_8–C_{12}) at a concentration of 10 mg L^{-1} on an FBBR, with detection of the CAPB and potential metabolites performed by LC–ESI–MS (see Chapter 2).

The concentration profile of the analysed CAPB homologues revealed initial losses (Fig. 5.1.23) attributable to adsorption onto the glass surfaces of the system employed—most likely onto the porous glass beads used as the carrier material of the fixed-bed. A correlation between the alkyl-chain length and the tendency to adsorption was observed, with approximately 70% of the most hydrophobic C_{14}-CAPB component lost within the first hours of the experiment. The dissolved concentration of C_{12}-CAPB dropped to about 80% of the initial value and the least hydrophobic homologues C_{10}- and C_8-CAPB exhibited only minor losses in the range of 5–10%. After these initial losses, relatively stable levels were achieved, which were maintained for differing lengths of time thereafter. This period was interpreted as the acclimation phase of the microbial population. A substantial decline of C_{12}-CAPB caused by biodegradation was observed after about 26 h, whereas the analogous point of C_{14}-CAPB could not be defined clearly, although after 33 h both compounds had almost completely disappeared. The concentration decline of the two shorter alkyl-chain homologues, C_8 and C_{10}, exhibited a longer lag phase after spiking for 33 h. The rate of biodegradation

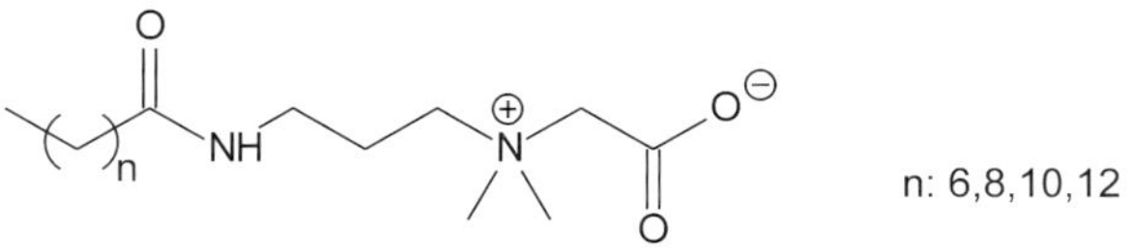

n: 6,8,10,12

Fig. 5.1.22. General structure of cocamidopropyl betaine.

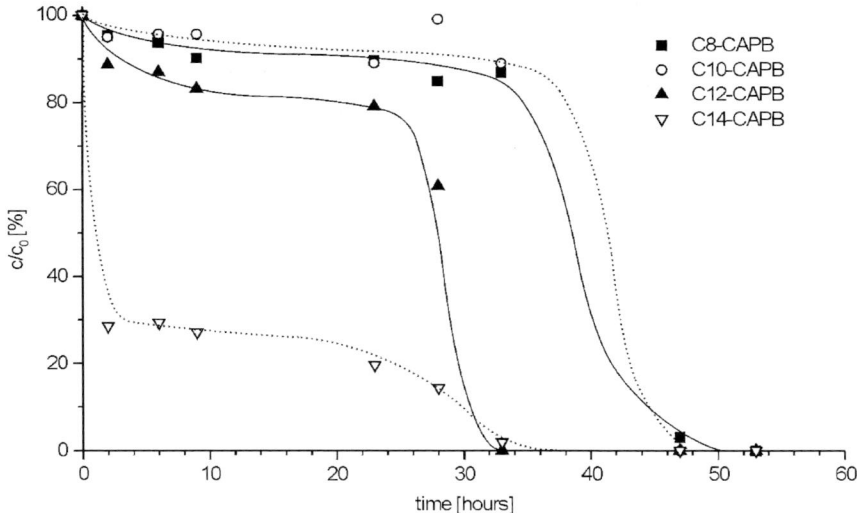

Fig. 5.1.23. Profile of primary degradation of C_8- to C_{14}-CAPB on FBBR at a test concentration of 10 mg L^{-1} (superposed by adsorption) (Modified from Ref. [36]).

thereafter, however, was rapid, with their concentrations decreased by more than 95% in the samples taken after 48 h. In the 53 h samples, none of the analytes was detectable (Fig. 5.1.23).

In conclusion, the degradation rates of the four homologues can be correlated to the length of the hydrocarbon chain, with higher alkyl homologues, i.e. the more lipophilic ones degraded faster. Such a relationship has likewise been described for alkyl homologues of LAS (*Swisher's distance principle*). The initial step in LAS conversion is the ω-oxidation of the terminal methyl group of the alkyl chain, thus resulting in the formation of carboxylic acids. Subsequently, the breakdown proceeds through β-oxidations, i.e. via successive shortening of the alkyl chain by C_2-units to form lower carboxylated homologues.

Accordingly, FBBR samples from CAPB spiking experiments were thus investigated for the presence of analogous intermediates to those of LAS. Acquisition of $(-)$-mode ionisation full-scan chromatograms ranging from m/z 120–400, as well as the direct search for the theoretical masses in SIM mode, did not yield positive results on CAPB with carboxylated alkyl-chain lengths of C_{14} to C_4 (m/z 391, 363, 335,

307, 279 and 251). Searches for any other metabolites potentially formed from the amphoteric surfactant were also unsuccessful. It can be presumed that the intact surfactant molecules were taken up by the cells and rapidly degraded intracellularly without the release of any metabolite. This assumption is supported by the results of Brunner et al. [138], who demonstrated high ultimate degradation of CAPB.

REFERENCES

1 EEC, Council Directive of 22 November 1973 on the approximation of the laws of the Member States relating to detergents (73/405/EEC), Official Journal of the European Communities, No. L 347/53.

2 EEC, Council Directive of 31 March 1982 amending Directive 73/405/EEC on the approximation of the laws of the Member States relating to methods of testing the biodegradability of non-ionic surfactants (82/242/EEC), Official Journal of the European Communities, No. L 109/1.

3 EEC, Council Directive of 31 March 1982 amending Directive 73/405/EEC on the approximation of the laws of the Member States relating to methods of testing the biodegradability of anionic surfactants (82/243/EEC), Official Journal of the European Communities, No. L 109/18.

4 Verordnung über die Abbaubarkeit anionischer und nichtionischer grenzflächenaktiver Stoffe in Wasch- und Reinigungsmitteln vom 30.01.1977, BGBl, 1, 244.

5 DIN German Standard Method for the Examination of Water, Waste Water and Sludge; General Measures of Effects and Substances (Group H). (A) Determination of Methylene Blue Active Substances (H23-1), (B) Determination of Bismuth Active Substances (H23-2), (C) Determination of the Disulfine Blue Active Substances (H20). VCH, Weinheim, Germany, 1992.

6 Anionic surfactants as MBAS, Standards Methods for the Examination of Water and Wastewater, Vol. 512B. American Public Health Association, 1985.

7 Nederlandse Vereniging van Zeepfabrikanten NVZ, Environmental data review of soap, Bongerts, Kuyer and Huiswaard, consulting engineers, Delft, The Netherlands, 1994.

8 L.H.M. Vollebregt and J. Westra, Environmental effects of surfactants, Onderzoeks- en Adviescentrum Chemie Arbeid Milieu, University of Amsterdam, 1998.

9   OECD, Guidelines for testing of chemicals. Organisation for Economic Co-operation and Development (OECD), Paris, 1993.

10 R.D. Swisher, *Surfactants and Biodegradation*. Marcel Dekker, New York, NY, USA, 1987.

11 L. Cavalli, G. Cassani, M. Lazzarin, C. Maraschin, G. Nucci and L. Valtorta, *Tenside Surf. Deterg.*, 33 (1996) 393.

12 A. DiCorcia, F. Casassa, C. Crescenzi, A. Marcomini and R. Samperi, *Environ. Sci. Technol.*, 33 (1999) 4112.

13 A. Marcomini, G. Pojana, C. Carrer, L. Cavalli, G. Cassani and M. Lazzarin, *Environ. Toxicol. Chem.*, 19 (2000) 555.

14 A. DiCorcia, C. Crescenzi, A. Marcomini and R. Samperi, *Environ. Sci. Technol.*, 32 (1998) 711.

15 A. Moreno, J. Ferrer, J. Bravo, J.L. Berna and L. Cavalli, *Tenside Surf. Deterg.*, 35 (1998) 375.

16 A. Remde and R. Debus, *Chemosphere*, 32 (1996) 1563.

17 C.A. Staples, J.B. Williams, R.L. Blessing and P.T. Varineau, *Chemosphere*, 38 (1999) 2029.

18 C. Crescenzi, A. DiCorcia, A. Marcomini and R. Samperi, *Environ. Sci. Technol.*, 31 (1997) 2679.

19 M. Huber, U. Meyer and P. Rys, *Environ. Sci. Technol.*, 34 (2000) 1737.

20 M.A. Manzano, J.A. Perales, D. Sales and J.M. Quiroga, *Water Res.*, 33 (1999) 2593.

21 J.A. Perales, M.A. Manzano, D. Sales and J.M. Quiroga, *Bull. Environ. Contam. Toxicol.*, 63 (1999) 94.

22 T.L. Potter, K. Simmons, J. Wu, M. Sanchez-Olvera and E. Calabrese, *Environ. Sci. Technol.*, 33 (1999) 113.

23 D. Sales, J.M. Quiroga and A. Gómez-Parra, *Bull. Environ. Contam. Toxicol.*, 39 (1987) 385.

24 S. Terzic, D. Hrsak and M. Ahel, *Water Res.*, 26 (1992) 585.

25 P. Schöberl, W. Guhl, N. Scholz and K. Taeger, *Tenside Surf. Deterg.*, 35 (1998) 279.

26 G.M. Boeije, D.R. Schowanek and P.A. Vanrolleghem, *Water Res.*, 34 (2000) 1479.

27 U. Baumann, M. Benz, E. Pletscher, K. Breuker and R. Zenobi, *Tenside Surf. Deterg.*, 36 (1999) 288.

28 P. Kölbener, U. Baumann, T. Leisinger and A.M. Cook, *Environ. Toxicol. Chem.*, 14 (1995) 571.

29 J. Mampel, T. Hitzler, A. Ritter and A.M. Cook, *Environ. Toxicol. Chem.*, 17 (1998) 1960.

30 P. Gerike and W. Jasiak, *Tenside Surf. Deterg.*, 22 (1985) 305.

31 P. Gerike and W. Jasiak, *Tenside Surf. Deterg.*, 23 (1986) 300.

32 W. Guhl and A. Ranke, *Tenside Surf. Deterg.*, 30 (1993) 6.

33 T.P. Knepper and M. Kruse, *Tenside Surf. Deterg.*, 37 (2000) 41.

34 P. Eichhorn and T.P. Knepper, *Environ. Toxicol. Chem.*, 21 (2002) 1.

35 S. Trimpin, P. Eichhorn, H.-J. Räder, K. Müllen and T.P. Knepper, *J. Chromatogr. A*, (2001).

36 P. Eichhorn and T.P. Knepper, *J. Mass Spectrom.*, 36 (2001) 677.

37 T.P. Knepper, P. Eichhorn, J. Müller, F. Karrenbrock and F. Sacher, *Proc. 48 ASMS*, (2000).

38 P. Eichhorn and T.P. Knepper, *J. Mass Spectrom.*, 35 (2000) 468.

39 P. Eichhorn and T.P. Knepper, *J. Chromatogr. A*, 854 (1999) 221.

40 N. Jonkers, T.P. Knepper and P. De Voogt, *Environ. Sci. Technol.*, 35 (2001) 335.

41 T.P. Knepper, J. Mueller, T. Wulff and A. Maes, *J. Chromatogr. A*, 889 (2000) 245.

42 F. Karrenbrock, T.P. Knepper, F. Sacher and K. Lindner, *Vom Wasser*, 92 (1999) 361.

43 T.P. Knepper, F. Kirschhöfer, I. Lichter, A. Maes and R.-D. Wilken, *Environ. Sci. Technol.*, 33 (1999) 945.

44 T.P. Knepper, F. Karrenbrock, F. Sacher and K. Lindner, *Proceedings of the Int. Riverbank Filtration Conference*, Düsseldorf, Germany, 2–4/11/2000, IARW- Schriftenreihe, Rheinthemen, Vol. 4, 2001, p. 291.

45 T.P. Knepper, L. Weber, F.T. Lange and B. de Groot, *ARW-Jahresbericht*, 1998 (1999) 143.

46 J.L. Berna, J. Ferrer, A. Moreno, D. Prats and F.R. Bevia, *Tenside Surf. Deterg.*, 26 (1989) 101.

47 H. Brüschweiler, H. Felber and F. Schwager, *Tenside Surf. Deterg.*, 28 (1991) 348.

48 D.B. Knaebel, T.W. Federle and J.R. Vestal, *Environ. Toxicol. Chem.*, 9 (1990) 981.

49 R.J. Larson, *Environ. Sci. Technol.*, 24 (1990) 1241.

50 R.J. Larson and A.G. Payne, *Appl. Environ. Microbiol.*, 41 (1981) 621.

51 D. Prats, M. Rodríguez, J.M. Llamas, M.A. Muela, J. de Ferrer, A. Moreno and J.L. Berna, *Fifth World Surfactant Congress*. Firenze, Italy, 2000.

52 P. Schöberl and K.J. Bock, *Tenside Surf. Deterg.*, 17 (1980) 262.

53 E. Aidar, T.C.S. Sigaud-Kutner, L. Nishihara, K.P. Schinke, M.C.C. Braga, R.E. Farah and M.B.B. Kutner, *Marine Environ. Res.*, 43 (1997) 55.

54 V.A. Shcherbakove, K.S. Laurinavichius and V.K. Akimenko, *Chemosphere*, 39 (1999) 1861.

55 C. Verge and A. Moreno, *Tenside Surf. Deterg.*, 33 (1996) 323.

56 C. Verge and A. Moreno, *Tenside Surf. Deterg.*, 37 (2000) 172.

57 W.K. Fischer, P. Gerike and W. Holtmann, *Water Res.*, 9 (1975) 1131.

58 W.K. Fischer, P. Gerike and W. Holtmann, *Water Res.*, 9 (1975) 1137.

59 P. Gerike and W.K. Fischer, *Ecotoxicol. Environ. Safety*, 3 (1979) 157.

60 P. Gerike, W.K. Fischer and W. Holtmann, *Water Res.*, 14 (1980) 753.

61 R.J. Larson, *Appl. Environ. Microbiol.*, 38 (1979) 1153.

62 G. Pecenik, G. Borgovoni, P. Castro and I. Gagliardi, *First World Surfactant Congress*, 1 (1984) 121.

63 T.W. Federle and B.S. Schwab, *Water Res.*, 26 (1992) 123.

64 R.J. Larson, T.M. Rothgeb, R.J. Shimp, T.E. Ward and R.M. Ventullo, *J. Am. Oil Chem. Soc.*, 70 (1993) 645.

65 U. Branner, M. Mygind and C. Jorgensen, *Environ. Toxicol. Chem.*, 18 (1999) 1772.

66 K. Figge and P. Schöberl, *Tenside Surf. Deterg.*, 26 (1989) 122.

67 N. Litz, H.W. Doering, M. Thiele and H.-P. Blume, *Ecotoxicol. Environ. Safety*, 14 (1987) 103.

68 C.J. Krueger, L.B. Barber, D.W. Metge and J.A. Field, *Environ. Sci. Technol.*, 32 (1998) 1134.

69 R.J. Larson, T.W. Federle, R.J. Shimp and R.M. Ventullo, *Tenside Surf. Deterg.*, 26 (1989) 116.

70 J.A. Perales, M.A. Manzano, D. Sales and J.M. Quiroga, *Int. Biodeterg. Biodegr.*, 43 (1999) 155.

71 R.J. Shimp, *Tenside Surf. Deterg.*, 26 (1989) 390.

72 S. Terzic, D. Hrsak and M. Ahel, *Water Res.*, 26 (1992) 585.

73 J.M. Quiroga, J.A. Perales, L.I. Romero and D. Sales, *Chemosphere*, 39 (1999) 1957.

74 J.M. Quiroga and D. Sales, *Bull. Environ. Contam. Toxicol.*, 44 (1990) 851.

75 J. Vives-Rego, R. López-Amorós, T. Guindulain, M.T. García, J. Comas and J. Sánchez-Leal, *J. Surf. Deterg.*, 3 (2000) 303.

76 P. Schöberl, *Tenside Surf. Deterg.*, 26 (1989) 86.

77 R.L. Huddleston and R.C. Allred, *Dev. Ind. Microbiol.*, 4 (1963) 24.

78 W.J. Payne and V.E. Feisal, *Appl. Microbiol.*, 11 (1963) 339.

79 E.A. Setzkorn, R.L. Huddleston and R.C. Allred, *J. Am. Oil Chem. Soc.*, 41 (1964) 826.

80 R.D. Swisher, *J. Water Poll. Control Fed.*, 35 (1963) 877.

81 R.D. Swisher, *J. Water Poll. Control Fed.*, 35 (1963) 1557.

82 G. Baggi, D. Catelani, A. Colombi, E. Galli and V. Treccani, *Ann. Microbiol. Enzymol.*, 24 (1974) 317.

83 C.R. Eggert, R.G. Kaley and W.E. Gledhill, In: A.W. Bourquin and P.H. Pritchard (Eds.), *Proc. of the Workshop: Microbial Degradation of Pollutants in Marine Environments*, 1979, U.S. EPA Report No. 600/9-79-012 451.

84 D. Hrsak and A. Begonja, *J. Appl. Microbiol,* submitted for publication (since three weeks) 85 (1998) 448.

85 W. Dong, S. Radajewski, P. Eichhorn, K. Denger, T. P. Knepper, J.C. Murrell and A.M. Cook, *Appl. Environ. Microbiol.*, submitted for publication.

86 S. Schulz, W. Dong, U. Groth and A.M. Cook, *Appl. Environ. Microbiol.*, 66 (2000) 1905.

87   C. Kanz, M. Nölke, H.-P. Fleischmann, E. Kohler and W. Giger, *Anal. Chem.*, 70 (1998) 913.

88   A. DiCorcia, F. Casassa, C. Crescenzi, A. Marcomini and R. Samperi, *Environ. Sci. Technol.*, 33 (1999) 4112.

89   D. Schleheck, W. Dong, K. Denger, E. Heinzle and A.M. Cook, *Appl. Environ. Microbiol.*, 66 (2000) 1911.

90   P. Kölbener, A. Ritter, F. Corradini, U. Baumann and A.M. Cook, *Tenside Surf. Deterg.*, 33 (1996) 149.

91   H.A. Painter, *Environmental Effects of Chemicals Report DOE 1237—M/1*, Water Research Center, Medmenham, UK, 1986.

92   P. Kölbener, U. Baumann, T. Leisinger and A.M. Cook, *Environ. Toxicol. Chem.*, 14 (1995) 561.

93   T.P. Knepper, F. Sacher, F.T. Lange, H.J. Brauch, F. Karrenbrock, O. Roerden and K. Lindner, *Waste Management*, 19 (1999) 77.

94   ECOSOL (1999) LAS facts and figures European Centre of Studies on LAB-LAS, Report.

95   C. Divo and G. Cardini, *Tenside Surf. Deterg.*, 17 (1980) 30.

96   D. Hrsak, *Environ. Pollut.*, 89 (1995) 285.

97   P.W. Taylor and G. Nickless, *J. Chromatogr.*, 178 (1979) 259.

98   A. Moreno, Petresa, Spain, Personal communication.

99   A.J. Willets, *A. van Leeuwenhoek*, 40 (1974) 561.

100   A.M. Nielsen, L.N. Britton, C.E. Beall, T.P. McCormick and G.L. Russell, *Environ. Sci. Technol.*, 31 (1997) 3397.

101   S. Eaton, K. Bartlett and M. Pourfarzam, *Biochem. J.*, 320 (1996) 345.

102   R.J.A. Wanders, R. Vreken, M.E.J. de Boer, F.A. Wijburg, A.H. van Gennip and L. Ijlst, *J. Inherit. Metab. Dis.*, 22 (1999) 442.

103   J.A. Bird, (1972) Biodegradation of alkylbenzenesulfonates and related compounds, PhD Thesis, University. Newcastle upon Tyne, UK.

104   J.A. Field, J.A. Leenheer, K.A. Thorn, L.B. Barber, C. Rostad, D.L. Macalady and S.R. Daniel, *J. Contam. Hydrol.*, 9 (1992) 55.

105   M.A. Kertesz, P. Kölbener, H. Stockinger, S. Beil and A.M. Cook, *Appl. Environ. Microbiol.*, 60 (1994) 2296.

106   J. Tolls, M.P. Lehmann and D.T.H.M. Sijm, *Environ. Toxicol. Chem.*, 19 (2000) 2394.

107   T.C. Cordon, E.W. Maurer and A.J. Stirton, *J. Am. Oil Chem. Soc.*, 47 (1970) 203.

108   P.W. Taylor and G. Nickless, *J. Chromatogr.*, 178 (1979) 259.

109   A.H. Mann and V.W. Reid, *J. Am. Oil Chem. Soc.*, 48 (1971) 588.

110   A. Moreno, L. Cohen and J.L. Berna, *Tenside Surf. Deterg.*, 25 (1988) 216.

111   P. Schölberl, K.J. Bock and L. Huber, *Tenside Surf. Deterg.*, 25 (1988) 86.

112   M.J. Scott and M.N. Jones, *Biochim. Biophys. Acta*, 1508 (2000) 235.

113   A.M. Bruce, J.D. Swanwick and R.A. Owensworth, *J. Proc. Inst. Sew. Purif. Pt.*, 5 (1966) 661.

114   J. Steber, P. Gode and W. Guhl, *Soap Cosmet. Chem. Spec.*, 64 (1988) 44.
115   O.R.T. Thomas and G.F. White, *Biotechnol. Appl. Biochem.*, 11 (1989) 318.
116   N.S. Battersby, L. Kravetz and J.P. Salanitro, *Proceedings of the Fifth CESIO World Surfactant Congress, Firenze*, (2000) 1397–1407.
117   S. Hirzel, BUA Report 189, August 1996.
118   W. Fischer and P. Gericke, *Water Res.*, 9 (1975) 1137.
119   M.L. Cano, J.P. Salanitro, K.A. Evans, A. Sherren and L. Kravetz, *92nd AOCS Annual Meeting and Expo, Minneapolis*, 2001, May 13.
120   J. Steber and P. Weirich, *Tenside Surf. Deterg.*, 2 (1989) 406.
121   S.S. Talmage, *Environmental and Human Safety of Major Surfactants: Alcohol Ethoxylates and Alkylphenol Ethoxylates*, A report to the Soap and Detergent Association, Lewis Publishers, Boca Raton, FL, 1994.
122   B. Thiele, K. Günther and M. Schwuger, *Chem. Rev.*, 97 (1997) 3247.
123   A. DiCorcia, A. Costantino, C. Crescenzi, E. Marinoni and R. Samperi, *Environ. Sci. Technol.*, 32 (1998) 2401.
124   W.H. Ding, Y. Fujita, R. Aeschimann and M. Reinhard, *Fresenius J. Anal. Chem.*, 354 (1996) 48.
125   W.H. Ding and C.T. Chen, *J. Chromatogr. A*, 824 (1998) 79.
126   W. Giger and P.H. Brunner, *Science*, 225 (1984) 623.
127   J. Ejlertsson, M. Nilsson, H. Kylin, A. Bergman, L. Karlson, M. Öquist and B. Svensson, *Environ. Sci. Technol.*, 33 (1999) 301.
128   W. Jones and D. Westmoreland, *Environ. Sci. Technol.*, 32 (1998) 2623.
129   M. Stalmans, E. Matthijs, E. Weeg and S. Morris, *SÖFW J.*, 13 (1993) 10.
130   J. Steber, W. Guhl, N. Stelter and F.R. Schröder, *Tenside Surf. Deterg.*, 32 (1995) 515.
131   M.T. Garcia, L. Ribosa, E. Campos and J. Sanchez Leal, *Chemosphere*, 35 (1997) 545.
132   T. Madsen, G. Petersen, C. Seiero and J. Torslov, *J. Am. Oil Chem. Soc.*, 73 (1996) 929.
133   L.W. Kroh, T. Neubart, E. Raabe and H. Waldhoff, *Tenside Surf. Deterg.*, 36 (1999) 19.
134   D.B. Knaebel, T.W. Federle and J.R. Vestal, *Environ. Toxicol. Chem.*, 9 (1990) 981.
135   L. Kravetz, J.P. Salanitro, P.B. Dorn and K.F. Guin, *J. Am. Oil Chem. Soc.*, 68 (1991) 610.
136   T. Balson and M.S.B. Felix, In: D.R. Karsa and M.R. Porter (Eds.), *Biodegradability of Surfactants*. Blackie Academic & Professional, Glasgow, UK, 1995, p. 204.
137   A. Marcomini and G. Pojana, *Analusis*, 25 (1997) 35.
138   C. Brunner, U. Baumann, E. Pletscher and M. Eugster, *Tenside Surf. Deterg.*, 37 (2000) 276.

5.2 ANAEROBIC BIODEGRADATION OF SURFACTANTS

Eduardo González-Mazo, V.M. León, Jose Luis Berna and Abelardo Gómez-Parra

5.2.1 Introduction

The major part of the biosphere is aerobic and consequently priority has been given to the study and assessment of biodegradability under aerobic conditions. Nevertheless, there are environmental compartments that can be permanently (e.g. anaerobic digesters) or temporarily anaerobic (e.g. river sediments and soils) and surfactants do reach these. The majority of surfactants entering the environment is exposed to and degraded under aerobic conditions. This is the predominant mechanism of removal even in cases of absence of wastewater treatment practices (direct discharge) and it is estimated that less than 20% of the total surfactant mass will potentially reach anaerobic environmental compartments [1]. Only in a few cases, however, will the presence of surfactants in these compartments be permanent. The presence of surfactants in anaerobic zones is not exclusively due to the lack of anaerobic degradation. Physico-chemical factors such as adsorption or precipitation play an important role as well as the poor bioavailability of surfactant derivatives (chemical speciation) in these situations.

While O_2 serves as the electron acceptor in aerobic biodegradation processes forming H_2O as the final product, degradation in anaerobic systems depends on alternative electron acceptors such as sulfate, nitrate or carbonate, which yield, ultimately, hydrogen sulfide (H_2S), molecular nitrogen (N_2) and/or ammonia (NH_3) and methane (CH_4), respectively.

Anaerobic biodegradation is a multistep process in which different bacterial groups take part. It involves hydrolysis of polymeric substances like proteins or carbohydrates to monomers and the subsequent decomposition to soluble acids, alcohols, molecular hydrogen (H_2) and carbon dioxide (CO_2). Depending on the prevailing environmental conditions, the final steps of ultimate anaerobic biodegradation are performed by denitrifying, sulfate-reducing or methanogenic bacteria. These reactions are dependent on the availability of organic and

Comprehensive Analytical Chemistry XL
T.P. Knepper, D. Barceló and P. de Voogt (Eds)

inorganic substrates, the redox potential of the environment and the types of bacteria present [2].

The available information about surfactant degradation under anaerobic conditions is restricted to anionic and non-ionic surfactants. Anaerobic biodegradation is strongly dependent on the chemical structure of the compound, the presence of a sufficient amount of anaerobically degrading microorganisms and a fulfilment of their growth requirements.

5.2.2 Mineralisation under anaerobic conditions

Screening tests
Most of the test methods used to assess the anaerobic biodegradability of organic chemicals have been screening level tests based upon measuring the increase in gas volume or pressure resulting from the conversion of a test material to carbon dioxide (CO_2) and methane (CH_4), such as the ECETOC-28 test [3,4].

The anaerobic degradability of anionic surfactants with $S-O_x$ as the anion varies significantly. Sulfonated anionic surfactants such as linear alkybenzene sulfonates (LAS), secondary alkane sulfonates (SAS), alpha olefin sulfonates (AOS) and methylester sulfonates (MES) are usually classified as poorly degradable under laboratory screening tests [5–7]. However, anionic surfactants with a sulfate group such as alkylether sulfate (AES) and alkyl sulfate (AS) are readily degradable under anaerobic conditions [8]. Concerning non-ionic surfactants, those containing straight-chain alcohol ethoxylates (AE) can be transformed completely under anaerobic conditions, whereas branched AE are often more difficult for the organisms to degrade [8,9].

These test methods are known to have significant drawbacks. The diluted sludge inoculum corresponds to about 10% or less of the real digester sludge concentration. In addition, for analytical reasons, the test compound concentration is usually in the range of 20–100 mg of organic carbon L^{-1}, which is significantly higher than the concentrations usually found in digesters. Therefore, in some cases, inhibitory effects are to be expected and also have been observed in these screening tests, which is particularly problematic for chemicals to which methanogens are especially susceptible, such as surface-active agents [10]. From these facts it is understandable that anaerobic screening tests are more stringent than the test systems simulating realistic environmental conditions. As in aerobic screening tests, it can be concluded that a positive test result indicates good biodegradation under real environmental conditions, whereas negative result is not necessarily proof of recalcitrance.

An additional aspect of the environmental relevance of anaerobic screening test results also has to be kept in view. These tests determine the ultimate biodegradation of a chemical by the measurement of the production of the final gaseous products. Therefore the bacterial transformation of the parent chemical (primary biodegradation) by anaerobic bacteria is not reflected by these data. Thus, even a poor degradation result in such a screening test does not necessarily indicate anaerobic persistence of the parent compound. In other words, if the ultimate degradation of a substance under these test conditions is significant it can be concluded that the extent of primary biodegradation of this chemical is virtually 100% since already a significant proportion of the carbon atoms of every molecule have been mineralised. Obviously, this assumption is only valid for pure compounds.

Simulation tests

These test systems, also known as anaerobic mineralisation tests, assesses the mineralisation of ^{14}C-radiolabelled chemicals to CO_2 and CH_4 under anaerobic (methanogenic) conditions. The inoculum is usually obtained from an active digester but in principle the system can also be used to simulate other anaerobic compartments, such as septic tanks or sediments. The system follows the formation of CO_2 and CH_4 separately over time. The underlying principle of this method is that the headspace of the anaerobic vessel is continuously purged with N_2, which is passed through a series of base traps, to capture first the $^{14}CO_2$ evolved. The effluent gas is mixed subsequently with oxygen and passed through a combustion tube (CuO at 800°C) to convert the $^{14}CH_4$ into $^{14}CO_2$. The latter is trapped in a second series of base traps. The test can be run in a batch, or fed-batch mode (with addition of fresh sludge/test material at regular intervals). The technical details of this system are provided in Steber and Wierich [11] and Nuck and Federle [12].

Nuck and Federle [12] used this test system to evaluate the mineralisation of anionic surfactants such as AES and AS in anaerobic digester sludge. Those compounds have been shown to degrade anaerobically (total yields of $^{14}CH_4$ and $^{14}CO_2$ exceeded 80%), as have been previously shown in screening tests [8]. For non-ionic surfactants like nonylphenol ethoxylates (NPEOs), only partial degradation such as conversion to nonylphenol diethoxylate (NPEO$_2$), nonylphenol monoethoxylate (NPEO$_1$) and NP was observed using [U–^{14}C]-NPEO$_{1-2}$ [13]. NP is the most recalcitrant of the intermediates formed and was not further degraded, thus agreeing with earlier reports [14,15] suggesting that the phenol ring structure remained intact during the period of incubation.

Using both the methods, indirect (pressure measurements in the headspace of the vials) and direct (quantifying the concentration of surfactant), anaerobic sludge showed potential for degrading LAS and NPEO [16]. These authors found inocula showing capability for the conversion of LAS under anaerobic conditions, namely by an inoculum obtained from lake sediments. In addition, inocula that were found in aerobic environments such as compost and activated sludge from a wastewater treatment plant, showed capability of anaerobic degradation of LAS. No bacterial strain or a microbial enrichment capable of degrading LAS by utilising LAS carbon under anaerobic conditions has been isolated [17]. The same authors reported anaerobic biodegradation of NPEO$_2$ and NPEO$_1$, whereas NP was not degraded. All the compounds were tested at three concentrations of 20, 100 and 200 mg L^{-1}, and no inhibition effect was observed.

The use of ^{14}C materials permits the direct measurement of the mineralisation of LAS under anaerobic conditions, but not the monitoring of the various intermediates formed during the process, and their quantification.

5.2.3 Surfactants in anaerobic environmental compartments

Surfactants such as LAS and NPEO have been found in compartments with low oxygen content, such as anaerobic sludge digesters or anaerobic continental and marine sediments [14,15,18–25]. One of the possible causes of this persistence is the inhibition of the anaerobic digestion [17,26,27]. Battersby and Wilson [27] observed inhibitory effects of NP at 50 mg C L^{-1} on methane formation in a survey of the anaerobic biodegradation potential of organic chemicals in digesting sludge, an effect that was also observed lately by Ejlertsson et al. [13]. Some authors have found a critical LAS concentration, between 190 and 284 mg L^{-1} and 292 and 492 mg L^{-1} where methanogenesis from acetate and propionate consumption, respectively, are completely inhibited [17]. In experiments performed either at 200 mg L^{-1} of LAS [16], or at concentrations below 50 ppm [28], which have been detected in environmental compartments, no inhibiting effect of LAS on anaerobic bacteria has been observed.

A considerable portion of the scientific literature on biodegradation under anaerobic conditions of surfactants refers to LAS due to its widespread usage in detergent preparations. Some authors suggest that anaerobic biodegradation of LAS can only take place if the surfactant has been previously exposed to aerobic conditions [29,30]. The process is less extensive than aerobic degradation (Fig. 5.2.1), and does not occur if the conditions are strictly anaerobic. According to these authors, the ω-oxidation can only occur in the presence of oxygen while the β-oxidation can take place under anaerobic conditions [31].

Other steps such as desulfonation may occur under anaerobic conditions [32], although some authors have shown resistance of the aromatic ring cleavage in the natural medium [33,34]. Other authors,

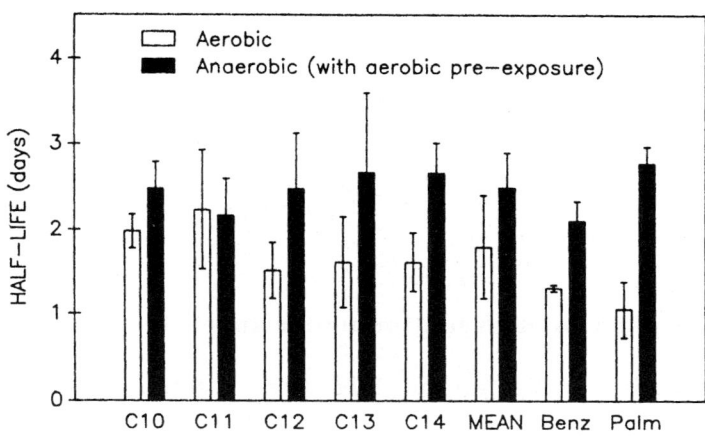

Fig. 5.2.1. Half-lives for aerobic and anaerobic biodegradation of LAS homologues in batch activated-sludge systems. The error bars represent two standard deviations. Benz: benzoate, Palm: palmitate. (Figure taken from Ref. [30].)

however, have confirmed the biodegradability of both benzene and alkylbenzenes under anaerobic enrichments [35,36] as well as in anaerobic sediments of estuaries and salt marshes [37]. The most recent studies indicate that, under appropriate conditions, benzene is oxidised into CO_2 utilising sulfate or Fe(III) as electron acceptors. For example, a rapid mineralisation of this compound under sulfate reduction conditions has been detected in continental and marine sediments, and in aquifers [38–40]. Kazumi et al. [41] have shown that the sulfate reduction agents in estuarine sediments are capable of degrading 80% of the benzene present. Recently, anaerobic degradation of a long-chain alkane in the absence of sulfate-reducing bacteria was reported [42].

As mentioned before, the presence of surfactants in anaerobic compartments cannot be separated from their physico-chemical characteristics and in fact surfactants which degrade extensively in the laboratory under anaerobic conditions, e.g. soap, are also found in considerable concentrations in anaerobic compartments. Due to their hydrophobic character surfactants are strongly sorbed to sludges and therefore a large amount of the load of these compounds into a sewage treatment plant (reportedly 20–50%) is associated with suspended solids [43,44]. The relevance of the presence of surfactants in the environment should be assessed, therefore, on the basis of their potential impact on the structure and function of the various compartments. In most cases, ionic surfactants are present as insoluble salts and therefore their potential impact is negligible as reflected in the lack of known negative impacts.

Conditions prevailing in the environment, however, are extremely difficult to simulate in the laboratory, in particular those occurring under strict anaerobic zones. Chemical speciation of the test substance in such environments is difficult to assess and consequently to reproduce in laboratory tests.

5.2.4 Primary biodegradation under anaerobic conditions

Recent studies using various laboratory techniques and specific analytical methods such as liquid chromatography (LC) or gas chromatography (GC) have reported considerable degrees of disappearance of the parent test substance in particular when LAS [45–48] or $NPEO_1$ and $NPEO_2$ [13,16] were used. Primary biodegradation between

40 and 85% after 250 days, and between 14 and 23% for C_{12}-LAS during sewage sludge digestion, have been reported (Fig. 5.2.2). The above-referred laboratory techniques used are of a different nature. The most widely used have been batch tests using specific analysis instead of bio-gas determination, continuous stirred tank reactors (CSTR) [49] and up-flow anaerobic sludge blanket (UASB) [50] reactors. When using UASB reactors in particular substantial removal degrees of LAS have been reported (>80%) and it is shown that benzene sulfonate and benzaldehyde are formed. These are the first identified metabolites of LAS degradation under anaerobic conditions. According to Shcherbakova et al. [51], these metabolites are virtually non-toxic to methanogenic bacteria. The formation of benzene sulfonate during anaerobic degradation of LAS is presumably caused by β-oxidation of the alkyl chain as known from the alkyl sulfates.

The anaerobic biodegradability and toxicity on anaerobic bacteria of cationic surfactants such as ditallow dimethylammonium chloride (DTDMAC) and two esterquats have been investigated in a recent study [52]. For the esterquats studied, high biodegradation levels were obtained and no toxic effects on anaerobic bacteria were observed even

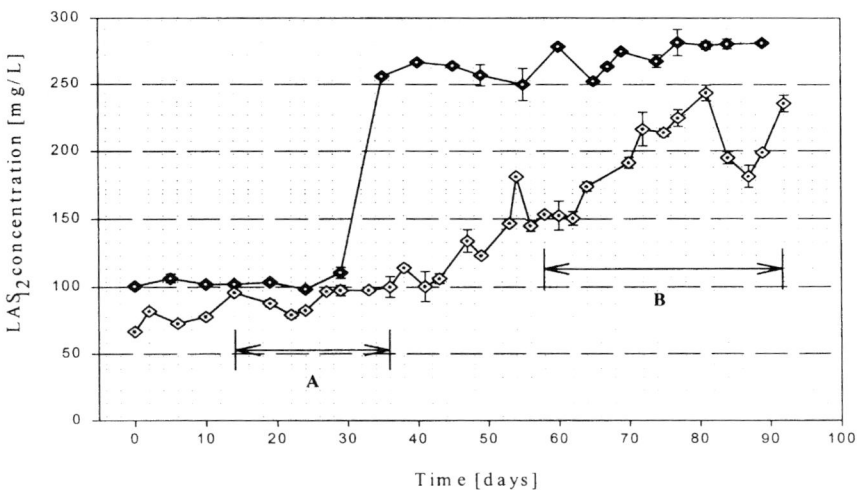

Fig. 5.2.2. Continuous stirred tank reactors experiment (different test periods: A and B). Reactor fed with sludge spiked with C12-LAS, the concentration of which in the influent (filled symbols) was increased approximately from 100 to 268 $mg\,L^{-1}$; the effluent concentration (empty diamonds). (Figure taken from Ref. [47].)

at the highest concentrations tested, 100 and 200 mg C L^{-1}, respectively. On the contrary, DTDMAC was not degraded under the same test conditions and no inhibitory effects on biogas production were detected for this surfactant at concentrations < 100 mg C L^{-1}.

To our best knowledge there are no studies on the anaerobic biodegradation of amphoteric surfactants hitherto published.

5.2.5 LAS and SPC distribution in sub-oxic and anoxic compartments

The environmental relevance of anaerobic degradability as a property for surfactants is emphasised by the amount that can accumulate and potentially cause effects in anaerobic environmental compartments. The most important final recipient of surfactants not degraded through various processes is the marine environment (sediments).

In underground waters, LAS is mineralised in the oxic zone, and degradation intermediates have not been detected. However, in sub-oxic layers, the process is much slower and here sulfophenyl carboxylates (SPC) have been detected [7,53]. Even in this layer, a preferential biodegradation of the homologues of the longer-chain LAS and of the external isomers takes place. Krueger et al. [53] observed that around 20% of LAS was ω-oxidised with oxygen of concentration 1 mg L^{-1} in underground waters, and Miura and Nishizawa [54] observed biodegradation of LAS adsorbed on the sediments of rivers and lakes with concentrations of dissolved oxygen of less than 1 mg L^{-1}.

The distribution of LAS in continental sediments has been studied [55], and the vertical profiles of LAS concentrations with depth in several lake sediments have been established [56,57]. In Swiss lakes, the concentration of LAS increases with depth and this is due to the efficiency of the wastewater treatment plants. Amano et al. [56] have, however, observed a decrease in the concentration of LAS with depth, and detected seasonal variations in the profile of LAS in the uppermost surface layer.

However, as far as we know, the distribution of LAS biodegradation intermediates according to depth in the sediment column has been determined only in marine sediments [58]. This study was performed in a saltmarsh channel (Sancti Petri Channel, Cádiz Bay, Spain), receiving untreated urban wastewater effluents. In this zone the benthic organisms are very scarce [59], and the capacity for irrigation of the sediment is also very low [60]; consequently bioturbation is virtually absent. In these conditions, diffusion is the mechanism controlling the vertical transport of the chemical species dissolved in the interstitial water.

At two stations with different degrees of exposure to the effluent, the vertical profiles of the total SPC concentration (taking the sum of SPC between 6 and 13 carbon atoms of the aliphatic chain) show a maximum at sediment depths between 10 and 14 cm (Fig. 5.2.3), with values as high as 2100 μg L^{-1} in one sampling point. Because the predominant mechanism of transport in the interstitial water in the sampling area is diffusion [61], the SPC found should have been produced at this depth, where the conditions are strictly anoxic ($E_h = -380$ mV). The long-chain SPCs are produced in the initial stages of the degradation of LAS. Hence, their appearance at considerable depth (10–14 cm) suggests that their rate of degradation under anoxic conditions must be very slow.

Figure 5.2.4 shows the variation of the concentration of LAS and long-chain SPCs with depth, with reference to the wet sediment (sediment + interstitial water), versus the variation of NO_3^- and SO_4^{2-}, in the interstitial water. A very sharp decrease in the concentration of LAS and low concentrations of SPC are observed in the oxic zone, the lower

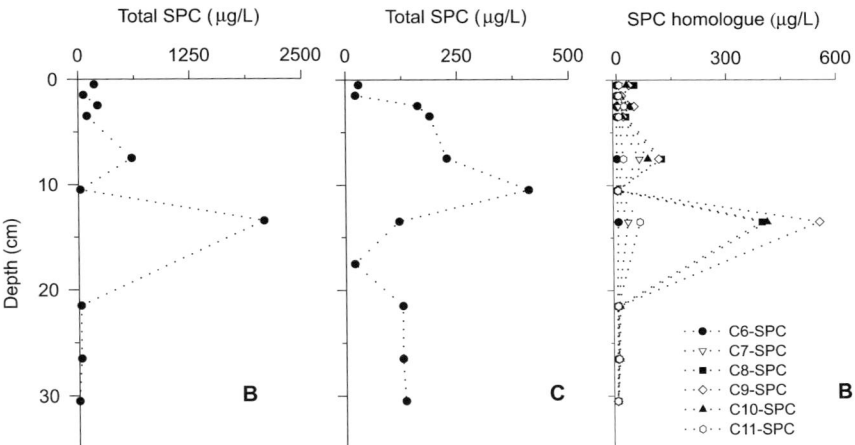

Fig. 5.2.3. Vertical profiles of total SPC in interstitial water found at stations with different exposition grade to non-treated wastewater effluent, 0.1 km (B) and 3 km (C). (Figure taken from Ref. [58].)

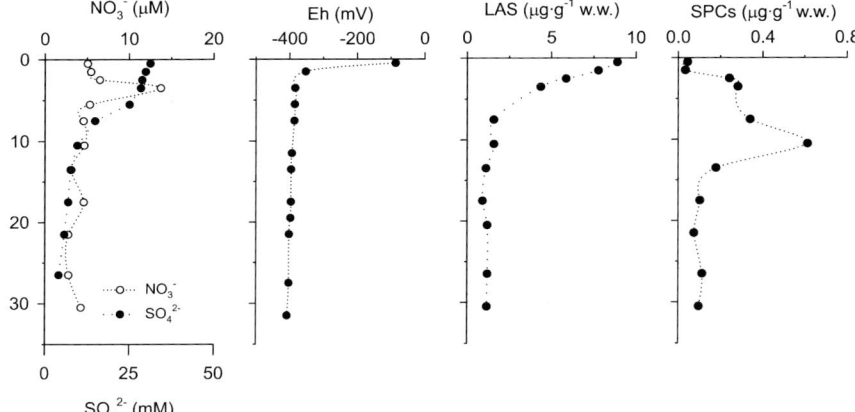

Fig. 5.2.4. Vertical profiles of nitrate and sulfate concentrations in interstitial water, and total LAS and SPC concentrations in sediment (wet weight) for station C. (Figure taken from Ref. [58].)

limit ($\cong 3$ cm) of which could be estimated from the maximum concentration of NO_3^- [62]; this indicates that the successive β-oxidations are relatively rapid in an oxic medium [17].

At greater depths, sulfate reduction prevails as the main anaerobic route for the degradation of organic matter in the studied area [63]. At this depth (4–8 cm), the vertical gradients of the concentration of LAS and long-chain SPC are smaller. However, Klump and Martens [64,65] observed that sulfate reduction is not an efficient metabolic route for the degradation of organic molecules of complex structure. The sulfate concentration is low (< 10 mM) at the depths (10–14 cm) at which the maximum concentration of long-chain SPC have been found. Other anaerobic mechanisms of organic matter degradation (denitrification, Fe and Mn reduction and fermentation) could also be involved in the LAS degradation process.

5.2.6 Conclusion

The environmental relevance of anaerobic degradability as a property for surfactants is driven by the amount that can accumulate and potentially cause effects in anaerobic environmental compartments such as continental or marine sediments. Due to their hydrophobic character surfactants are strongly sorbed and therefore a large quantity of the load of these compounds in sewage treatment plants (reportedly 20–50%) is associated with suspended solids and thus escapes aerobic treatment processes.

In comparison with aerobic biodegradability, relatively few test methods have been published for assessing the anaerobic biodegradability of surfactants. Most test methods used have been screening level tests that determine the ultimate biodegradation of a chemical by measurement of the production of the final gaseous products, i.e. methane and carbon dioxide. Sulfonated anionic surfactants are usually classified as poorly degradable under laboratory screening tests. However, anionic surfactants with a sulfate group are readily degradable under anaerobic conditions. With respect to non-ionic surfactants, those containing straight-chain AE can be transformed completely under anaerobic conditions, whereas branched AE are often more difficult for the organisms to degrade. However, a poor degradation result in such a screening test does not necessarily indicate anaerobic persistence of the parent compound.

The anaerobic mineralisation tests assess the mineralisation of ^{14}C-radiolabelled chemicals to CO_2 and CH_4 under anaerobic (methanogenic) conditions and bacterial transformation of the parent chemical (primary biodegradation) is shown. Anionic surfactants such as AES, AS, LAS and non-ionic $NPEO_{1-2}$ have been shown to degrade anaerobically under mineralisation tests. Recent studies using various laboratory techniques and specific analytical methods such as LC or GC have reported considerable degrees of disappearance of the parent test substance in particular when LAS or $NPEO_1$ and $NPEO_2$ were used. Furthermore, in sub-oxic and anoxic compartments (marine sediments) degradation intermediates (SPC) of LAS have been detected.

REFERENCES

1 Anaerobic Biodegradation of Surfactants, review of scientific information. ERASM, Brussels, 1999.
2 A.J.B. Zehnder and W. Stumm, In: A.J.B. Zehnder (Ed.), *Biology of Anaerobic Microorganisms*, Wiley-Liss, 1988, p. 1.
3 ECETOC, Evaluation of anaerobic biodegradation. Technical Report 28, 1988.

4 EPA, EPA 40 CFR Ch. 1 (7-1-88 Edition) 796.3140: Anaerobic biodegradability of organic chemicals, 1988, p. 568.

5 S. Wagener and B. Schink, *Water. Res.*, 21 (1987) 615.

6 S. Ito, S. Naito and T. Unemoto, *Yukagaku*, 37 (1988) 1006.

7 J.A. Field, L.B. Barber, E.M. Thurman, B.L. Moore, D.L. Lawrence and D.A. Peake, *Environ. Sci. Technol.*, 26 (1992) 1140.

8 J.P. Salanitro and L.A. Diaz, *Chemosphere*, 30 (1995) 813.

9 L. Kravetz, J.P. Salanitro, P.B. Dorn and K.F. Guin, *J. Am. Oil. Chem. Soc.*, 68 (1991) 610.

10 J.B. Jones, B. Bowers and T.C. Stadtman, *J. Bacteriol.*, 130 (1977) 1357.

11 J. Steber and P. Wierich, *Water Res.*, 21 (1987) 661.

12 B.A. Nuck and T.W. Federle, *Environ. Sci. Technol.*, 30 (1996) 3597.

13 J. Ejlertsson, M.-L. Nilsson, H. Kylin, A. Bergman, L. Karlson, M. Öquist and B.H. Svensson, *Environ. Sci. Technol.*, 33 (1999) 301.

14 A. Marcomini, P.D. Capel, T. Lichtensteiger, P.H. Brunner and W. Giger, *J. Environ. Qual.*, 18 (1989) 523.

15 W. Giger, O.H. Brunner and C. Schaffner, *Science*, 225 (1984) 623.

16 I. Angelidaki, A.S. Mogensen and B.K. Ahring, *Biodegradation*, 11 (2000) 377.

17 H.N. Gavala and B.K. Ahring, *Biodegradation*, 13 (2002) 201.

18 A.M. Bruce, J.D. Swanwick and R.A. Ownsworth, *J. Proc. Inst. Sew. Depurif.*, (1966) 427.

19 K. Oba, Y. Yoshida and S. Tomiyama, *Yukagaku*, 16 (1967) 517.

20 W.E. Gledhill, *Tox. Assessment*, 2 (1987) 89.

21 W. Janicke and G. Hilge, *Tenside Surfact.*, 16 (1979) 472.

22 W. Giger, A.C. Alder, P.H. Brunner, A. Marcomini and H. Siegrist, *Tenside Surfact. Det.*, 26 (1989) 95.

23 R.A. Rapaport and W.S. Eckhoff, *Environ. Toxicol. Chem.*, 9 (1990) 1245.

24 T.W. Federle and B.S. Schwab, *Water Res.*, 26 (1992) 123.

25 E. González-Mazo, J.M. Forja and A. Gómez-Parra, *Environ. Sci. Technol.*, 32 (1998) 1636.

26 R.D. Swisher, *Surfactant Science Series*, Vol. 18, Marcel Dekker, New York, 1987.

27 N.S. Battersby and V. Wilson, *Appl. Environ. Microbiol.*, 55 (1989) 433.

28 J.L. Sanz, M. Rodríguez, J. Ferrer, A. Moreno and J.L. Berna, *The Cler Review, Research Technology Public Policy*, 6 (2000) 26.

29 T.E. Ward, *7th Annual Meeting of the Society of Environmental Toxicology and Chemistry*, Arlington, Virginia, 1986.

30 R.J. Larson, T.M. Rothgeb, R.J. Shimp, T.E. Ward and R.M. Ventullo, *J. Am. Oil. Chem. Soc.*, 70 (1993) 645.

31 P. Schöberl, *Tenside Surfact. Deterg.*, 26 (1989) 86.

32 K. Denger and A.M. Cook, *J. Appl. Microbiol.*, (1999) 86.

33 M. Reinhard, N.L. Goodman and J.F. Barker, *Environ. Sci. Technol.*, 18 (1984) 953.

34 J.R. Barbaro, J.F. Barker, L.A. Lemon and C.I. Mayfield, *J. Contam. Hydrol.*, 11 (1992) 245.

35 B.H. Wilson, G.B. Smith and J.F. Rees, *Environ. Sci. Technol.*, 20 (1986) 997.

36 D. Grbic-Galic and T.M. Vogel, *Appl. Environ. Microbiol.*, 53 (1987) 254.

37 D.M. Ward, R.M. Atlas, P.D. Boehn and J.A. Calder, *Ambio*, 9 (1980) 277.

38 E.A. Edwards and D. Grbic-Galic, *Appl. Environ. Microbiol.*, 58 (1992) 2663.

39 D.R. Lovley, J.D. Coates, J.C. Woodward and E.J.P. Phillips, *Appl. Environ. Microbiol.*, 61 (1995) 953.

40 J.D. Coates, R.T. Anderson and D.R. Lovley, *Appl. Environ. Microbiol.*, 62 (1996) 1099.

41 J. Kazumi, M.E. Caldwell, J.M. Suflita, D.R. Lovley and L.Y. Young, *Environ. Sci. Technol.*, 31 (1997) 813.

42 R.T. Anderson and D.R. Lovely, *Nature*, 404 (2000) 722.

43 M.S. Greiner and E. Six, *Tenside Surfact. Deterg.*, 34 (1997) 250.

44 D.C. McAvoy, S.D. Dyer and N.J. Fendinger, *Environ. Toxicol. Chem.*, 17 (1998) 1705.

45 J.L. Berna, J. Ferrer, A. Moreno, D. Prats and F. Ruiz Bevia, *Tenside Surf. Deterg.*, 26 (1989) 101.

46 J.L. Sanz, M. Rodríguez, R. Amils, J.L. Berna, J. Ferrer and A. Moreno, *Riv. Ital. Sost. Gras.*, 76 (1999) 307.

47 I. Angelidaki, F. Haagensen and B.K. Ahring, *The Cler Review, Research Technology Public Policy*, 6 (2000) 32.

48 D. Prats, M. Rodríguez, J.M. Llamas, M.A. De la Muela, J. Ferrer, A. Moreno and J.L. Berna, *The Cler Review, Research Technology Public Policy*, 6 (2000) 46.

49 A.S. Mongensen, F. Haagensen and B.K. Ahring, *Environ. Toxicol. Chem.*, in press.

50 A.S. Mogensen and B.K. Ahring, *Biotechnol. Bioengng.*, 77 (2002) 483.

51 V.A. Shcherbakova, K.S. Laurinavichius and V.K. Akimenko, *Chemosphere*, 39 (1999) 1861.

52 M.T. Garcia, E. Campos, J. Sanchez-Leal and I. Ribosa, *Chemosphere*, 41 (2000) 705.

53 C.J. Krueger, K.M. Radakovich, T.E. Sawyer, L.B. Barber, R.L. Smith and J.A. Field, *Environ. Sci. Technol.*, 32 (1998) 3954.

54 K. Miura and H. Nishizawa, *Yukagaku*, 31 (1982) 367.

55 H. Takada and N. Ogura, *Estuar. Mar. Chem.*, 37 (1992) 257.

56 K. Amano, T. Fukushima and O. Nakasugi, *Hydrobiologia*, 235/236 (1992) 491.

57 R. Reiser, H. Toljander, A. Albrecht and W. Giger, In: R.P. Eganhouse (Ed.), ACS SYMPOSIUM SERIES 671, American Chemical Society, 1997, p. 196.

58 V.M. León, E. González-Mazo, J.M. Forja and A. Gómez-Parra, *Environ. Tox. Chem.*, 20 (2001) 2171.

59 T.A. Del Valls, J.M. Forja and A. Gómez-Parra, *Environ. Toxicol. Chem.*, 17 (1998) 1073.

60 J.M. Forja, J. Blasco and A. Gómez-Parra, *Estuar. Coast Shelf Sci.*, 39 (1994) 417.

61 A. Lerman, *Ann. Rev. Earth Planet. Sci.*, 6 (1978) 281.

62 W.M. Smethie, *Deep-Sea Res.*, 34 (1987) 983.

63 A. Gómez-Parra and J.M. Forja, *Hydrobiologia*, 252 (1993) 23.

64 J.V. Klump and C.S. Martens, In: E.J. Carpenter and D.G. Capone (Eds.), *Nitrogen in the Marine Environment*, Academic Press, New York, 1983, p. 411.

65 J.V. Klump and C.S. Martens, *Geochim. Cosmochim. Acta*, 51 (1987) 1161.

5.3 BIODEGRADATION OF LINEAR ALKYLBENZENE SULFONATES IN THE MARINE ENVIRONMENT

Victor M. León and Eduardo González-Mazo

5.3.1 Introduction

Although the marine environment can generally be considered the final destination of industrial and urban wastewater effluents, studies of biodegradation of linear alkylbenzene sulfonates (LAS) in this compartment have been scarce until recently [1–8]. The removal of LAS from the marine medium seems to be an efficient process, as shown by the low levels of LAS detected in samples of both water and sediment [9–11]. High values have only been found in zones close to the direct wastewater effluent discharge points of urban areas [11].

LAS degradation in seawater is slower than in other ecosystems [7,8,12] due to marine microorganisms being less capable of degrading xenobiotic compounds [13] and more sensitive to LAS than terrestrial microorganisms [14,15]. For this reason the presence of terrestrial halotolerant bacteria may play an important role in the biodegradation processes in coastal ecosystems [16]. The identification of LAS intermediates, the sulfophenyl carboxylates (SPC) of chain lengths ranging from 7 to 13 carbon atoms, in the marine medium [10], suggests that the degradation pathway is analogous to that found in freshwater (see Chapter 5.1). These authors have even detected C_{13}-SPC in interstitial marine water, a finding confirming that ω-oxidation is also the first step in the aerobic LAS biodegradation process. Other biodegradation intermediates may be generated at lower concentrations (sulfophenyl dicarboxylates or desulfonated phenyl carboxylates), as previous studies have detected in other systems [17–19].

The characterisation of LAS degradation in the marine environment requires laboratory experiments, although due to the special characteristics of this compartment (e.g. its high salinity and its normally oligotrophic status) and the numerous variables that affect it, divergent results may be obtained. Marine-specific bacterial communities cannot be cultivated as a whole in standard media due to the difficulty of reproducing original ecosystem conditions where they have been

Comprehensive Analytical Chemistry XL
T.P. Knepper, D. Barceló and P. de Voogt (Eds)
© 2003 Elsevier Science B.V. All rights reserved

sampled [20] and the tendency in recent years is to use natural seawater in biodegradation experiments [12,21].

5.3.2 Primary biodegradation and mineralisation

Generally, LAS biodegradation processes were previously monitored either as primary biodegradation (a decreasing LAS concentration) or as mineralisation (formation of biomass and/or CO_2). Both processes have usually been fitted to a first order kinetics, but this approximation is only valid if there has been no prior acclimatisation phase; an initial lag phase is usually present in biodegradation tests [4]. Nevertheless, first order kinetic constants may be used for special cases or for a preliminary characterisation of the process.

From a few studies available for the marine medium, it has been observed that the extent of LAS primary biodegradation varies with the initial concentration of the compound and with the temperature [4,7,22], and generally reaches more than 95% in estuarine and marine systems.

However, precise fits require the use of at least second order models, which include three characteristic phases in the biodegradation process [4]: (i) an acclimatisation or lag phase; (ii) an exponential phase with most of the effective degradation; and (iii) a final phase in which the microorganism population is stable and some portion of the substrate could persist (see Fig. 5.3.1(a) [21]).

Primary biodegradation intermediates (SPC), some of them termed 'key intermediates' of LAS degradation [22], have received less attention, as a consequence of their standards being unavailable, and because until relatively recently no suitable method of recovery was available.

The intermediates generated during the biodegradation of C_{11}-LAS and C_{12}-LAS are shown in Fig. 5.3.1(b) [21]. Further biodegradation tests were carried out by adding C_{11}-SPC and 5-phenyl-C_5-SPC to the marine medium (Fig. 5.3.2(a) and (b)). These biodegradation tests were carried out by adding to the marine medium a natural inoculum at environmentally representative concentrations (1 ppm) of LAS at 25°C. In general it can be concluded that, under these conditions, LAS biodegradation is a very fast process and the lag phase is short, especially for SPCs, as previously found by Hrsak and Grbic-Galic [23]. For this reason, intermediates with longer carboxylic chains (C_9-C_{12} SPC) generated by ω-oxidation and initial β-oxidations have only been

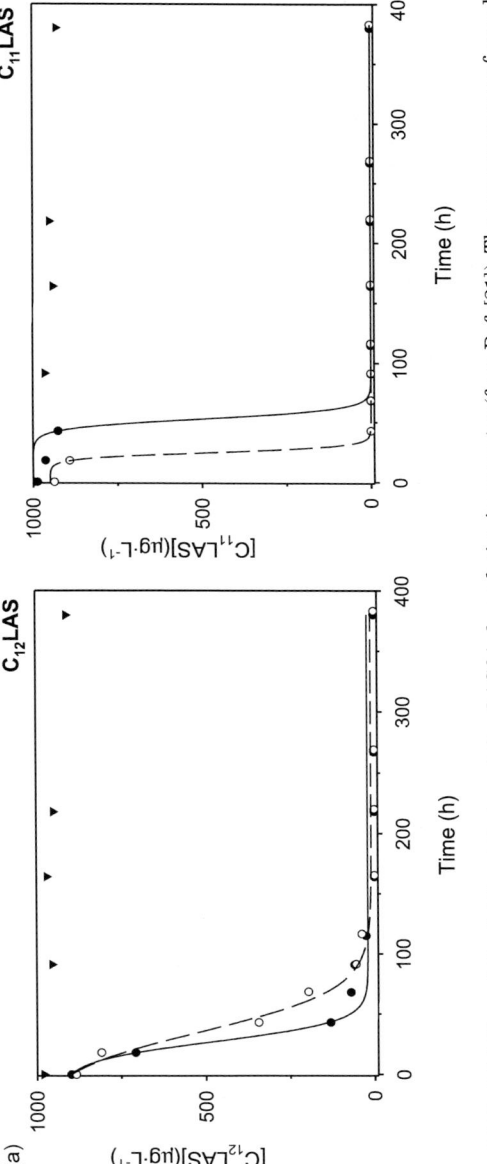

Fig. 5.3.1. (a) 3-Phenyl-C_{11}-LAS and 2-phenyl-C_{12}-LAS biodegradation in seawater (from Ref. [21]). The assays were performed by duplicate (circles) and an abiotic control (4% formaldehyde) (triangles). (b) Main intermediates generated from 3-phenyl-(ϕ)-C_{11}-LAS and 2-phenyl-(ϕ)-C_{12}-LAS (upper graphs) and mixtures of several isomers (lower graphs) in seawater with an initial concentration of 1 ppm (from Ref. [21]).

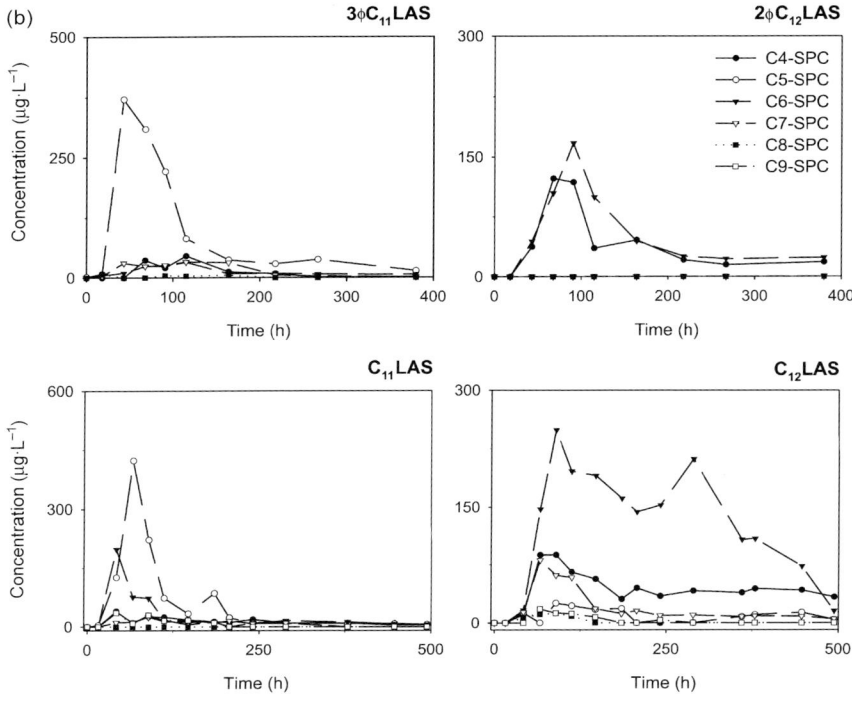

Fig. 5.3.1 (*continued*)

detected in the advanced exponential phase (Figs. 5.3.1(b) and 5.3.2(a)) due to their high reactivity. The first SPCs detected have been those of shorter chain lengths, probably generated from LAS external isomers that are more readily biodegradable. Subsequently the process is faster and extends to all the isomers resulting in a wide spectrum of SPC intermediates, with the SPC of C_5- to C_7-chain length being the most abundant (Fig. 5.3.3). Therefore the oxidation of these intermediates must be the slowest step, possibly involving ring cleavage [24]. C_{11}-LAS and C_{11}-SPC have been shown to generate the same intermediates (Figs. 5.3.1(b) and 5.3.2(a)), as previously revealed by Hrsak and Grbic-Galic [23]. The biodegradation pathway is the same for both compounds, following the oxidation of C_{11}-LAS at the terminal methyl group.

β-Oxidation is the main route responsible for the shortening of the carboxylic chain, and consequently the reagents with an odd number of carbon atoms generate preferentially SPCs with an odd-numbered chain, while the SPC with an even-numbered carbon chain result from corresponding even-numbered intermediates. However lower concentrations of intermediates generated by α-oxidation have also been detected by various authors already cited [24–26], as shown in Fig. 5.3.1(b) (lower graphs). In fact the existence of this oxidation

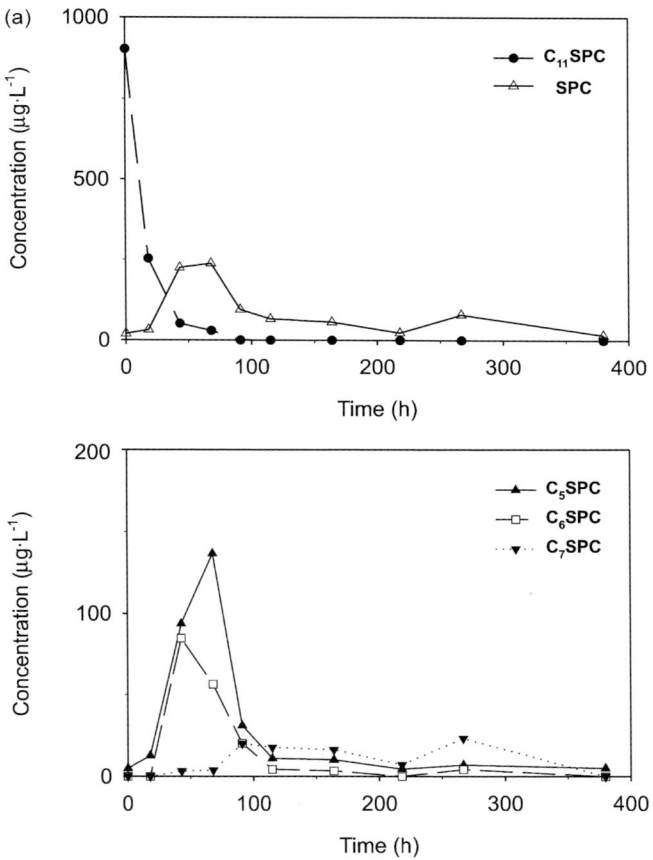

Fig. 5.3.2. (a) C_{11}-SPC biodegradation in seawater and total SPC concentration generated (upper graph) and main primary biodegradation intermediates (lower graph) (from Ref. [21]). (b) 5-Phenyl-(ϕ)-C_5-SPC biodegradation in seawater and main primary biodegradation intermediates generated (lower graph) (from Ref. [21]).

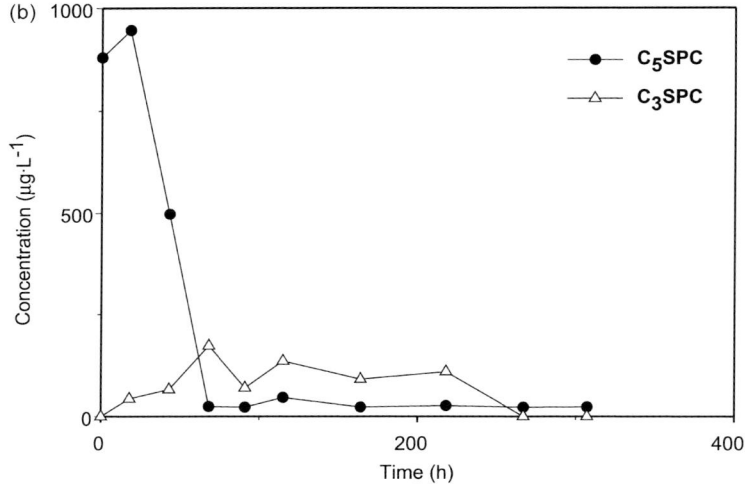

Fig. 5.3.2 (*continued*)

reaction has also been detected in the alkylbenzene sulfonate branch, taking place more slowly than β-oxidation [27].

During the degradation of LAS, SPCs may also be generated from the more internal isomers of LAS, which are those presenting two relatively long alkyl chains. These intermediates constitute a minority fraction in the set of intermediates, in comparison with the monocarboxylates, as has been described by DiCorcia et al. [17].

LAS mineralisation has been determined in estuaries [28] at low concentrations ($20 \, \mu g \, L^{-1}$) reaching 40–60% of the original concentration after 50 days. As the final biodegradation of SPC (ring cleavage) is the limiting step in LAS mineralisation, the disappearance of SPC indicates that LAS biodegradation has been completed. During C_{11}-LAS and C_{12}-LAS biodegradation in seawater at 25°C, it has been found that the corresponding SPCs have completely biodegraded [21].

5.3.3 Parameters affecting biodegradation

Both the extent and the kinetics of LAS biodegradation depend on the characteristics of the compound (e.g. isomer and homologue distribution, and the initial concentration), the properties of the medium

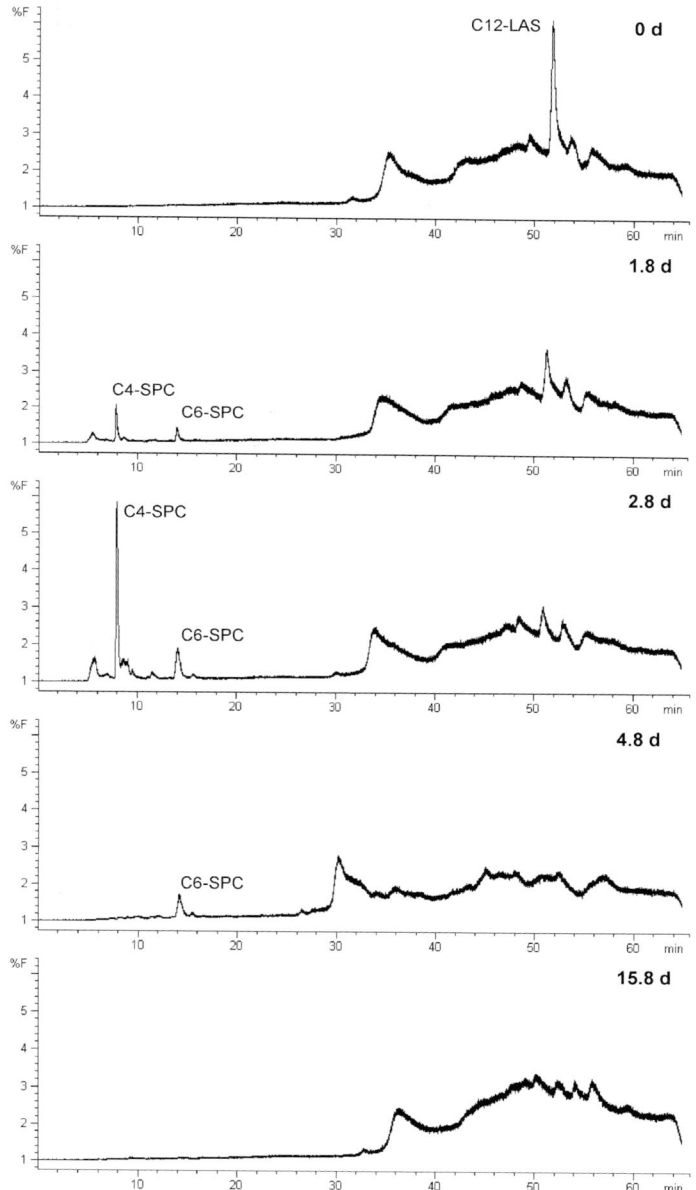

Fig. 5.3.3. HPLC chromatograms obtained during a C_{12}-LAS biodegradation test, showing the typical sequence of SPC detected in this process (from Ref. [21]).

(suspended solids concentration, dissolved oxygen, temperature, pH, and nutrient concentration) and its microbiological aspects (e.g. density or acclimatisation of the microbiota).

Initial LAS concentration, and isomer and homologue distribution
Tests conducted with concentrations found in the natural environment (1 ppm) present percentages of primary biodegradation in excess of 99%, although when the initial concentration tested is higher (10–20 ppm), the extent is less (see Table 5.3.1). The rate of degradation also decreases in line with the initial concentration due to the toxic effect on the microbiota [29]. In Table 5.3.1 it can be observed that the average life-times increase in line with the increase in the initial concentration of LAS.

From commercial LAS, the homologues with longer alkyl chains are preferentially biodegraded in coastal ecosystems [7,8,10], since these are the most easily oxidised [22]. However when the degradation of different homologues (C_{11}-LAS and C_{12}-LAS) has been studied separately under identical conditions, the rate and extent of biodegradation do not show significant differences (Figs. 5.3.1(a) and 5.3.4). In Fig. 5.3.4, the evolutions of the concentration of C_{11}-LAS, C_{12}-LAS and total SPC generated, over time, are shown. LAS primary biodegradation in marine water at 25°C is completed relatively rapidly in both cases (6 days), but the removal of SPC needs at least 21 days.

TABLE 5.3.1

Main characteristics of biodegradation experiments and first order kinetic parameters (degraded LAS(%) $= A\, e^{-kt}$) obtained for LAS primary biodegradation in different estuarine and coastal media at different initial concentrations and for different homologues ($t = 20$–25°C)

Compound	Initial concentration (ppm)	$t_{1/2}$ (d)[a]	A (%)[b]	Reference
LAS	20	6–9	>70 (10 ds)	[3]
LAS	10	6–7	>95	[4]
LAS	1.0	4.4	97–99	[7]
C_{10} to C_{13}-LAS	20.0		26–89	[8]
C_{12}-LAS	1.0	1.8 ± 0.2	99.8	[21]
C_{11}-LAS	1.0	1.6 ± 0.3	99.5	[21]

[a]Half-life time ($t_{1/2}$) calculated as $t_{1/2} = \ln 2/k$.
[b]Extension of primary degradation (A).

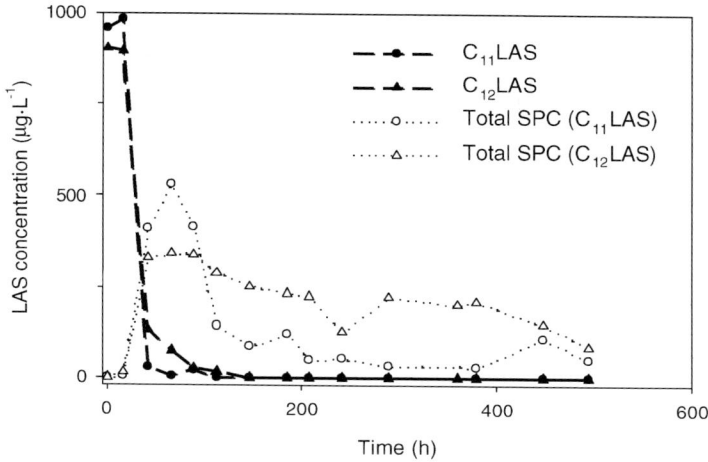

Fig. 5.3.4. Biodegradation of C_{11}-LAS and C_{12}-LAS and formation of total SPC at 25°C with natural inoculum (from Ref. [21]).

The biodegradation of pure external isomers is faster than that of isomer mixtures due to the steric impediment [7,22]. The faster biodegradation rate for internal isomers of different LAS homologues is shown for the freshwater layer in a stratified estuary in Fig. 5.3.5 (from Terzic et al. [7]).

5.3.4 Properties of the medium

Bacterial community and acclimatisation / lag phase
LAS mineralisation takes place by bacterial associations in which one strain uses the metabolic products of others [16,30,31]. Basically, these processes involve three kinds of microorganisms: metabolic organisms responsible for ω-oxidation, organisms that produce the shortening of the carboxylic chain and specialist organisms involved in desulfonation and/or ring cleavage [31,32]. Specifically in the marine environment Sigoillot and Nguyen [16] found that complete C_{12}-LAS mineralisation was produced by four types of microorganisms. However, a more complex community is necessary to degrade all the isomers of the commercial mixture [33].

LAS degradation is faster and more extensive when the microorganisms have previously been exposed to this compound [28]. The duration

599

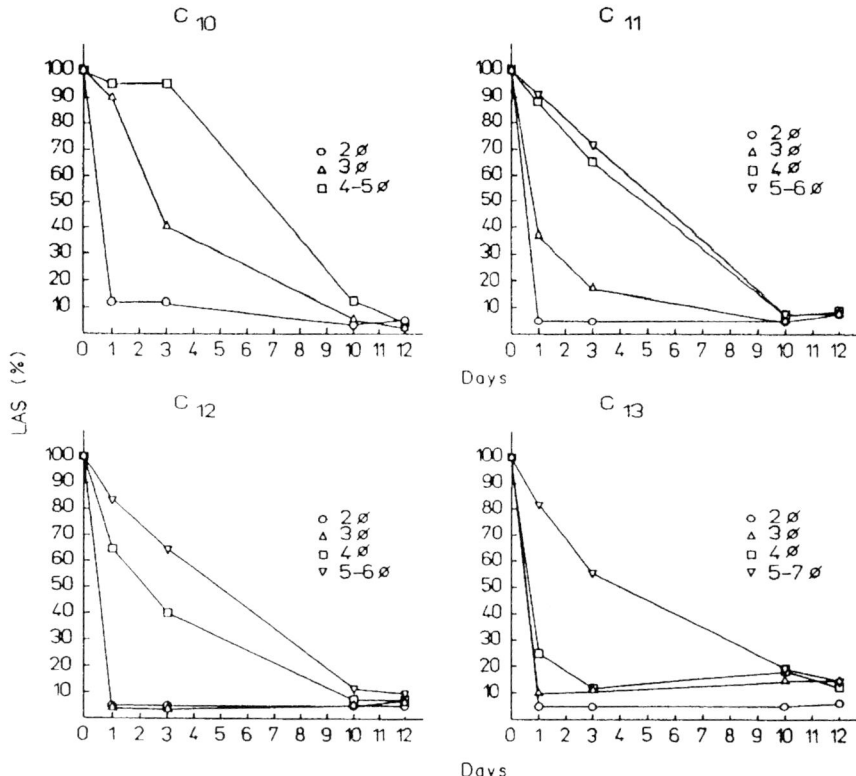

Fig. 5.3.5. Die-away curves of the individual isomers of each LAS homologue ($C_{10}-C_{13}$) achieved by mixed bacterial culture originating from the freshwater layer of the estuary at 23°C. Initial concentration of LAS was 1 mg L^{-1} (from Ref. [8]).

of the lag phase increases when the conditions of the medium are adverse, and in general this phase is longer in seawater than in freshwaters. During the lag phase, bacterial activity decreases due to the presence of the surfactant and the biodegradation process starts when the bacterial biomass and activity are at a maximum [34]. The bacterio-plankton undergoes changes during the LAS biodegradation [35] because only a small percentage of bacteria are capable of degrading LAS [28,29]. The rate and extent of LAS biodegradation are correlated with the density of bacteria present in the medium [29], but more recent research has shown a closer correlation with bacterial

600

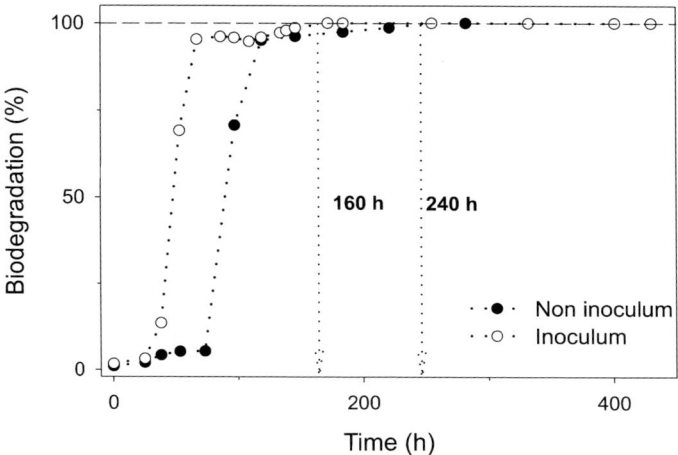

Fig. 5.3.6. C_{12}-LAS biodegradation percentage, with and without inoculum, at 25°C (from Ref. [21]).

activity, since only a proportion of the total population is able to perform the biodegradation process [34].

Dodecylbenzene sulfonate biodegradation (percentage with and without inoculation) is shown in Fig. 5.3.6 [21]. This experiment has been performed at environmentally representative concentrations (1 μg mL^{-1}). The extent of the biodegradation processes (>98% in all cases) has not been significantly modified by the addition of inoculum adapted to LAS (Fig. 5.3.6). The presence of inoculum (4 mL L^{-1}) has reduced the lag phase for biodegradation of the LAS homologue and consequently the time needed to complete the biodegradation, due to the higher density of bacteria available. The use of natural inoculum is appropriate because neither the intermediates generated nor their proportions have been significantly modified, and this reduces the duration of the test.

Temperature
The degradative bacterial activity is increased with water temperature [2,4,5,7,22], as it increases all the metabolic processes. Consequently, an increase in this factor encourages both the extent and the kinetics of the process (see Table 5.3.2). In assays performed at high concentrations

601

TABLE 5.3.2

Main characteristics of biodegradation experiments and first order kinetic parameters (degraded LAS(%) = $A\,e^{-kt}$) obtained for LAS primary biodegradation in different estuarine and coastal media at different temperatures

Compound	Initial concentration (ppm)	T (°C)	$t_{1/2}$ (d)[a]	A (%)[b]	Reference
LAS	10	5–15	>20	<50	[4]
LAS	10	20–25	6–7	>95	[4]
C_{12}-LAS	1.0	25 ± 0.2	2.5 ± 0.2	99.5	[21]
C_{12}-LAS	1.0	5 ± 0.2	9.6 ± 3.2	99.2	[21]

[a]Half-life time ($t_{1/2}$) calculated as $t_{1/2} = \ln 2/k$.
[b]Extension of primary degradation (A).

(20 ppm), Quiroga-Alonso et al. [5] found that LAS primary biodegradation in seawater did not take place at low temperatures (5–10°C) whereas at 25°C it reached more than 90% after 21 days. However, recent studies have shown that primary biodegradation of C_{11}-LAS and C_{12}-LAS at 5 and 25°C were more than 99% for environmentally representative concentrations (1 ppm), similar to the results obtained by Terzic et al. [7] in estuarine water.

Taking into account such environmentally representative experiments, it could be deduced that the extent of the biodegradation should not be affected by temperature, although this parameter clearly influences the rate at which this process takes place. In this respect, the lag phase is increased and oxidation of the LAS homologues (C_{11}–C_{12}) needs a longer time to develop at lower temperatures (see Fig. 5.3.7). The experiments performed previously at high concentrations and low temperatures may be affected predominantly by toxic effects, rather than by the particular temperature used.

There can be no doubt that there is an important seasonal factor in LAS biodegradation in coastal areas, as other authors have indicated previously [4,7]. Temperature variations are only relevant in shallow coastal systems. In these areas, LAS will probably be efficiently removed from the water column in summer time, whereas degradation could be less extensive during winter months due to the lower kinetics of biodegradation under these conditions, which permits access to sediments of a higher proportion of the LAS present.

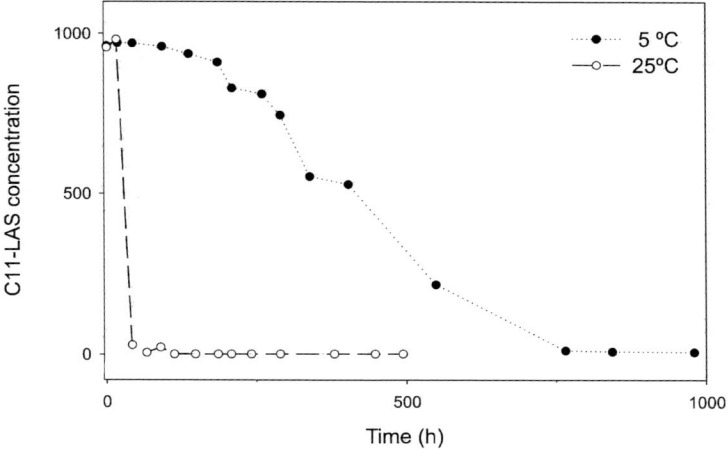

Fig. 5.3.7. C_{11}-LAS biodegradation in inoculated natural marine water at 5 and 25°C, representing the extreme temperatures of seawater in winter and summer time, respectively (from Ref. [21]).

Other physico-chemical conditions

The presence of suspended solid materials increases the extent of LAS biodegradation [13,28], but the rate of the process remains invariable. The influence of the particulate material is due specifically to the increased density of the microbiota associated with sediments. However, suspended solids may also reduce the bioavailability of LAS as a result of its sorption onto preferential sites (e.g. clays, humic acids), although this is a secondary effect due to the reversibility of the sorption process. Salinity does not affect LAS degradation directly, but could also reduce LAS bioavailability by reducing the solubility of this molecule [5]. Another relevant factor to be taken into account is that biodegradation processes in the marine environment could be limited by the concentration of nutrients, especially of phosphorus and nitrogen [34].

5.3.5 Conclusion

The removal of LAS from the marine medium is, in general, a slow, but efficient process. ω-Oxidation is the first step in the LAS biodegradation process, resulting in SPCs. β-Oxidation is the main route responsible for the shortening of the carboxylic chain. Temperature and characteristics of the microbiota present in the environmental compartment are important factors in the rate and extent of degradation and mineralisation. At high environmental temperatures ($\geq 25°C$) complete disappearance of LAS is observed much more rapidly in marine water, i.e. within 7 days, whereas the further degradation of LAS degradation products (SPC) takes longer (3 weeks). The extent of biodegradation of LAS is not affected by temperature, however. Factors such as the suspended solids concentration, presence of humic acids, and nutrient limitation on the other hand may limit the bioavailability and thereby the extent of biodegradation.

REFERENCES

1   H. Hon-nami and T. Hanya, *Water Res.*, 41 (1980) 1251.
2   M. Devescopi, D. Hrask and D. Fuks, *Rapp. Commun. Int. Mer. Medit.*, 30 (1986) 31.
3   J. Vives-Rego, M.D. Vaqué, J. Sánchez-Leal and J.L. Parra, *Tenside Surfact. Deterg.*, 24 (1987) 20.
4   D. Sales, J.M. Quiroga-Alonso and A. Gómez-Parra, *Environ. Toxicol. Chem.*, 39 (1987) 385.
5   J.M. Quiroga-Alonso, D. Sales and A. Gómez-Parra, *Water Res.*, 23 (1989) 801.
6   M. Stalmans, E. Matthijs and N.T. de Oude, *Water Sci. Technol.*, 24 (1991) 115.
7   S. Terzic, D. Hrsak and M. Ahel, *Water Res.*, 26 (1992) 585.
8   S. Terzic, D. Hrsak and M. Ahel, *Mar. Pollut. Bull.*, 24 (1992) 199.
9   H. Takada, N. Ogura and R. Ishiwatari, *Environ. Sci. Technol.*, 26 (1992) 2517.
10 E. González-Mazo, M. Honing, D. Barceló and A. Gómez-Parra, *Environ. Sci. Technol.*, 31 (1997) 504.
11 E. González-Mazo, J.M. Forja and A. Gómez-Parra, *Environ. Sci. Technol.*, 32 (1998) 1636.
12 J. Vives-Rego, R. López-Amorós, T. Guindulain, M.T. García, J. Comas and J. Sánchez-Leal, *J. Surfact. Deterg.*, 3 (2000) 303.
13 R.J. Larson, T.M. Rothgeb, R.J. Shimp, T.E. Ward and R.M. Ventullo, *J. Am. Oil Chem. Soc.*, 70 (1993) 645.
14 M. Bressan, R. Brunetti, S. Casellato, G.C. Fava, L. Tosoni, M. Turchetto and G.C. Campesan, *Tenside Surfact. Deterg.*, 26 (1989) 148.
15 M.A. Lewis, *Water Res.*, 25 (1991) 101.

16 J.C. Sigoillot and M.H. Nguyen, *Appl. Environ. Microbiol.*, April (1992) 1308.

17 A. DiCorcia, F. Casassa, C. Crescenzi, A. Marcomini and R. Samperi, *Environ. Sci. Technol.*, 33 (1999) 4112.

18 A. DiCorcia, L. Capuani, F. Casassa, A. Marcomini and R. Samperi, *Environ. Sci. Technol.*, 33 (1999) 4119.

19 L. Sarrazin, W. Wafo and P. Rebouillon, *J. Liquid Chromatogr. Rel. Technol.*, 22 (1999) 2511.

20 I. Barcina, P.I. Lebaron and J. Vives-Rego, *FEMS Microbiol. Ecol.*, 23 (1997) 1.

21 V.M. León, PhD Thesis, Cádiz University, Spain, July 2001.

22 R.D. Swisher, *Surfactant biodegradation. Chemical structure and primary biodegradation*, Surfactant Science Series, Vol. 18, Marcel Dekker, New York, 1987, p. 415.

23 D. Hrsak and D. Grbic-Galic, *J. Appl. Bacteriol.*, 78 (1995) 487.

24 P. Schöberl, *Tenside Surfact. Deterg.*, 26 (1989) 86.

25 R.D. Swisher, *J. WPCF*, (1963) 877.

26 R.B. Cain, A.J. Willets, J.A. Bird, In: A.W. Waters and E.H. Hueck van der Plas (Eds.), *Biodegradation of Materials*, Vol. II, Applied Science Publishers, London, 1972, p. 136.

27 L. Cavalli, G. Cassani and C. Maraschin, *The CLER Review*, 2 (1996) 4.

28 R.J. Shimp, *Tenside Surfact. Deterg.*, 26 (1989) 390.

29 J.M. Quiroga-Alonso and D. Sales, *Tenside Surfact. Deterg.*, 28 (1991) 200.

30 L. Jiménez, A. Breen, N. Thomas, T.W. Federle and G.S. Sayler, *Appl. Environ. Microbiol.*, 75 (1991) 1566.

31 D. Hrsak and A. Begonja, *J. Appl. Microbiol.*, 85 (1998) 448.

32 D. Schleheck, W. Dong, K. Denger, E. Heinzle and A.M. Cook, *Appl. Environ. Microbiol.*, 66 (2000) 1911.

33 R.D. Swisher, *Yakagaku*, 21 (1972) 130.

34 J. Vives-Rego, J. Sánchez-Leal, J. Martínez and J. González, *Microbios*, 69 (1992) 187.

35 R. López-Amorós, J. Comas, M.T. García and J. Vives-Rego, *FEMS Microbiol. Ecol.*, 27 (1998) 33.

5.4 SURFACTANT SORPTION ON NATURAL SEDIMENTS

Eduardo González-Mazo and Victor M. León

5.4.1 Introduction

Sorption plays a significant role in the environmental fate and effects of compounds released into the aquatic environment, largely determining their distribution between different environmental compartments. Apart from affecting the mobility, and therefore the potential of a surfactant to reach groundwater and surface water, sorption can affect its toxicity and biodegradation by influencing bioavailability. This process is especially relevant for surfactants, since their molecular structure presents a pronounced tendency to sorb onto interfaces.

In general, potential sorption mechanisms on the surface, of dissolved particulate material [1] include: chemical reactions on the surfaces (hydrolysis, complexation, ligand exchange and hydrogen bonding), electrical interactions (electrostatic and polarization) on surfaces, and interactions with a solvent (hydrophobic repulsion). Surfactant sorption is related to the chemical potential of surfactant molecules in solution and to the nature of the solid. The surface activity and the structure of sorbed films are greatly influenced by the chemical structure of the hydrophobic and hydrophilic groups of surfactants. The hydrophobic groups tend to sorb preferentially onto the organic matter and sorption usually increases with the hydrophobicity [2–5]. The nature of the hydrophilic group (anionic, cationic or non-ionic) determines which specific interactions of surfactant molecules could take place with any specific sorbent.

Cationic surfactants are usually sorbed strongly onto sediments due to favourable electrostatic interactions with the predominantly negatively charged surfaces of natural materials. The sorption of cationic surfactants depends on the cation exchange capacity (CEC) of the sorbent. The primary adsorption mechanism is cation exchange [6], but this process has received little attention in respect of soils and sediments [7]. As for anionic and non-ionic surfactants, these tend to be less strongly, but still extensively, sorbed, but with a certain amount remaining in the aqueous phase. Anionic sorption depends on the length of the alkylic chain and the isomer position [3,8], and the process

Comprehensive Analytical Chemistry XL
T.P. Knepper, D. Barceló and P. de Voogt (Eds)

is then controlled mainly by hydrophobic mechanisms. For non-ionic surfactants, sorption depends on their molecular structure and, in particular, on the organic carbon content of the sediment [6].

The comments given above provided a simplified overview: there are several other parameters that affect sorption (including physico-chemical conditions, the chemical nature and properties of the sorbent, and concentration) and must be taken into account, making more difficult an adequate characterisation and modelling of this process. For example, in the literature, linear alkylbenzene sulfonates (LAS) sorption has usually been explained by hydrophobic mechanisms, but at low concentrations this process is predominantly electrostatic [9,10]. Also other compounds (including non-surfactants) are present in the environment at the same time, thus increasing the complexity of sorption processes. Somasundaran and Huang [11] detected synergism and competition between the anionic LAS and non-ionic surfactants (alcohol ethoxylates, AEs), and cationic and non-ionic surfactants (nonylphenol, NP) in sorption onto alumina, depending upon the ratio of the two surfactants in the mixture.

5.4.2 Field measurement of sorption coefficients

Some authors have measured partition coefficients of surfactants in the natural medium, expressed as the ratio between particulate material concentration and water concentration (Table 5.4.1; see also Chapter 6.4). Considerable variability is observed in these data mainly due to differences in the sedimentary matrices (e.g. grain size, suspended material concentration, ionic charges) and different characteristics of the medium (e.g. salinity, pH). Partition coefficients have usually been normalised by organic carbon content [12], but even in these cases the values for different materials and media cannot be taken as comparable. Thus other processes that take place simultaneously with sorption in the environment (mainly biodegradation in the water column and precipitation) directly affect the LAS distribution between the particulate and dissolved phases. In this respect, considering that LAS is a mixture of different homologues each with a distinct hydrophobicity, and each with a different composition in the two phases, partition coefficients (for total LAS) may not be useful for chemically rigorous analysis. These coefficients are affected by changes

TABLE 5.4.1

Partition coefficients, expressed as the ratio between concentration in particulate material and concentration in water, in different systems

Compound	Compartment	K_d (L kg^{-1})	Reference
LAS	River sediment	100–2600	[50]
	River sediment	1000–5900	[38]
	River sediment and suspended solids	500–14 000	[13]
	Estuarine suspended solids	500–7000	[51]
	Marine sediment	100–2216	[52]
	Marine suspended solids	3500–13 000	[53]
	Marine sediment	50–3500	[54]
NP	Estuarine suspended solids	23 000–64 000	[55]
NPEO$_1$		9200–127 000	
NPEO$_2$		6400–81 000	
NPEO$_{3-15}$		18 000–62 000	
NP	Estuarine suspended solids	16 000–66 000	[56]
OP	Estuarine suspended solids	22 000	[57]
NP		36 000	
NPEO$_1$		42 000	
NPEO$_2$		22 000	
NPEO$_3$		11 000	
NPEO	Estuarine sediment	700–33 000	[58]
NP	River sediment	<26.4–4625	[59]
NPEO$_1$		<38.3–292	
NPEC	Estuarine suspended solids	700–4800	[55]

in homologue composition both due to biodegradation and to the nature of the particles and cannot be used to characterise the sorption mechanism of LAS.

Despite these reservations, environmental distribution values may be considered valid for the sorption process, to a first approximation. On this basis, it can be concluded that detected environmental partition coefficients show the clear affinity of surfactants to particulate material. The affinity is higher for cationic surfactants than for other surfactants, as shown by the high partition coefficient values (Table 5.4.1). Partition coefficients are also higher for the water column than for sediments (Table 5.4.1), and it is difficult to offer an explanation for this, bearing in mind the many factors affecting the partition coefficient in both natural water and sediment.

In Fig. 5.4.1, a clear difference can be observed in the ratios between the concentration in dissolved and adsorbed phases, comparing the water column and the sediment of Lake Teganuma [13].

The partition coefficients for different LAS homologues (Table 5.4.2) are higher in the marine environment due to the higher ionic strength that promotes sorption of anionic surfactants [14] and an increase in the partition coefficient with the alkylic chain length has been observed (Table 5.4.2). The evolution of the concentration of the various homologues of LAS in solids in suspension (cf. Fig. 5.4.2) is similar to that found in water and, in the process of adsorption, an increase can be observed in line with the chain length, as commented on previously.

5.4.3 Characterisation in laboratory experiments

Numerous laboratory sorption studies have been conducted for the most common surfactants: non-ionics, such as AE and alkylphenol ethoxylates (APEOs); anionics such as LAS, secondary alkane sulfonates (SASs) and sodium dodecylsulfates (SDS); and on different natural sorbents [3,8,15–17]. Until now, cationic and amphoteric surfactants have received less study than the other types, probably because they represent only 5 and 2%, respectively, of the total surfactant consumption in Western Europe (1998) [18].

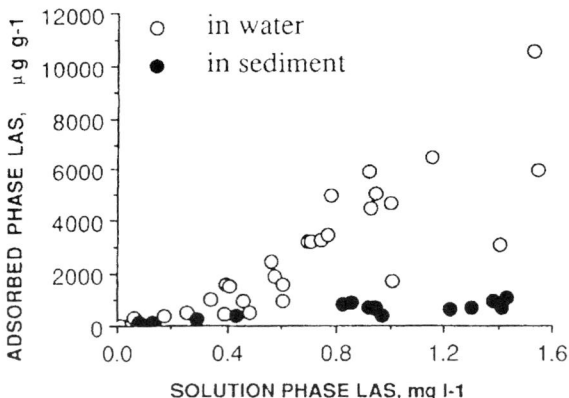

Fig. 5.4.1. Distribution of LAS between particulate and dissolved phases in sediments and water columns from Lake Teganuma, Japan (from Ref. [13]).

TABLE 5.4.2

Partition coefficients for LAS homologues in river and marine-suspended solids

Compound	Compartment	K_d (L kg^{-1})	Reference
C$_{10}$-LAS	Riverine-suspended solids	112–183	[60]
C$_{11}$-LAS		198–514	
C$_{12}$-LAS		406–1618	
C$_{13}$-LAS		1217–6296	
C$_{10}$-LAS	Marine-suspended solids	3400 ± 800	[21]
C$_{11}$-LAS		4750 ± 1200	
C$_{12}$-LAS		6570 ± 1900	
C$_{13}$-LAS		12 200 ± 3800	

Laboratory sorption experiments are used to determine the distribution of a compound between the solid and liquid phases in the absence of other processes involved (e.g. biodegradation, precipitation) under controlled conditions. Sorption experiments are usually performed by adding particulate matter, free of surfactants, to solutions with different known surfactant concentrations at constant temperature (sorption isotherms). The experiments performed at high concentrations of

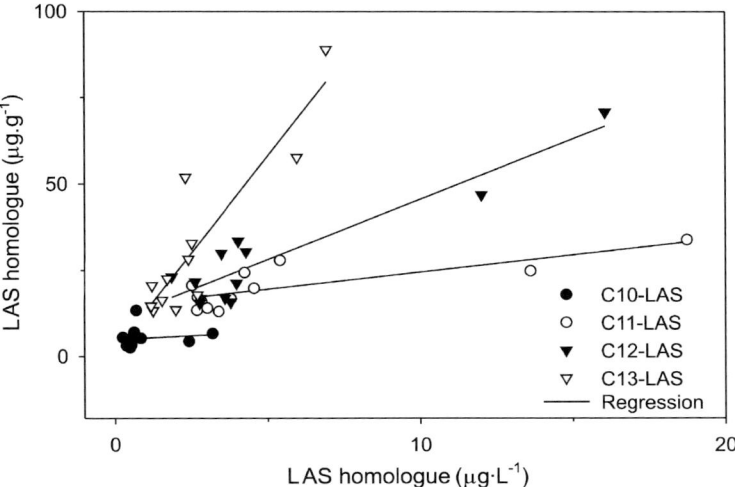

Fig. 5.4.2. LAS homologue concentrations in marine-suspended solids (vertical scale), as against their concentrations in water (Cádiz Bay, Spain; horizontal scale) (from Ref. [21]).

surfactant enable sorption to be evaluated using a small volume of sample [3,10,17]. However, when environmentally representative concentrations are used (low concentrations), it is difficult to obtain representative samples for analysis since surfactants tend to concentrate or sorb at interfaces, and conditions of foaming must be avoided, or at least the foam should be allowed to subside before sampling. Two methods are usually employed to counter these problems: surfactants are labelled with [14]C [19,20] or large volumes are treated by SPE followed by a specific detection method [14,21].

In order to estimate the time necessary to reach the stationary state, the process should be characterised kinetically; this involves frequent sampling at short time intervals. The kinetics of the sorption process depend on surfactant and sorbent; this process is slower when it takes place on a heterogeneous surface than on a homogeneous one [22]. In relation to the nature of the surfactant, sorption is very fast for cationics [7], probably faster than for other types of surfactant, whereas experiments with non-ionics need longer times to reach the stationary state. The equilibrium time for surfactant sorption experiments is usually less than 6 h (Table 5.4.3), in order for biodegradation not to occur. However, recent studies for AE [17,23] and LAS [24] have shown that, during this period of time, up to 80–90% of the sorption process takes place, but that 12 h are necessary for the stationary state to be reached (see Fig. 5.4.3). Primary biodegradation of LAS intermediates has not been detected in sediment assays [30] due to the prior sterilisation of the water and sediment used, and the performance of microalgae assays by adding formaldehyde to avoid bacterial proliferation [31].

The partition coefficient is determined by measuring the surfactant concentration in the dissolved and particulate phases when the steady state has been reached. The separation of the two phases is performed by either filtration or centrifugation. There have been many studies to characterise the sorption of surfactants onto minerals and textiles [8,25,26], but relatively few have characterised sorption into environmental compartments [2,3,15,17,19,20,27,28], particularly for marine environments [14].

The surfactant partition coefficient in natural sediments determined in laboratory experiments ranges from 1.9 to 2100 L kg^{-1} for non-ionic, 780 to 2900 L kg^{-1} for anionic (LAS) and 3800 to 2.5×10^5 L kg^{-1} for cationic surfactants (see Table 5.4.4). The partition coefficients of cationic surfactants are one or two orders of magnitude higher than for

TABLE 5.4.3

Equilibrium times of sorption experiments for different surfactants in different environmental matrices

Sorbent	Compound	Equilibrium time (h)	Reference
Sediment	AE	2–4 (12)	[17]
Sediment	AE	12	[19]
Sediment	AE	48	[23]
River sediment	LAS	4	[3]
River sediment	LAS	<3	[8]
Sediment	LAS	4	[20]
Sediment	LAS	0.3	[61]
Marine sediment	LAS	4	[14]
Marine sediment	C_{11}-LAS	12	[30]
		4 (80%)	
Algae residue	LAS	10	[27]
Marine microalgae	LAS and SPC	12–24	[31]
Marine sediment	C_{11}-SPC	24	[30]

other types of surfactants, similar to behaviour observed in the natural environment. There is a considerable variability in the distribution of surfactants, depending on the compound, the physico-chemical conditions and the nature of the sorbent, with more intense sorption corresponding to cationic surfactants. The environmental and laboratory

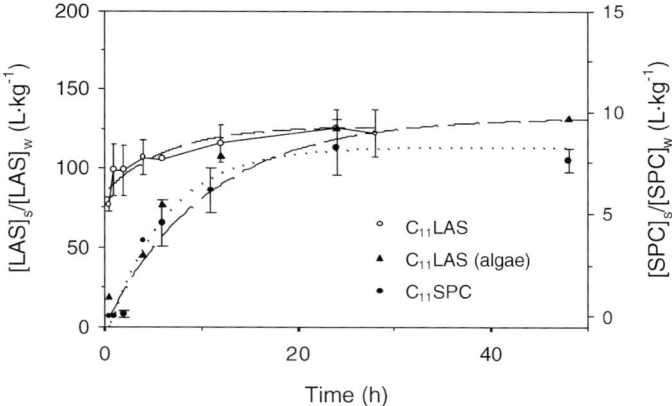

Fig. 5.4.3. Sorption curve of C_{11}-LAS and C_{11}-SPC in sediment, and C_{11}-LAS in microalgae. In the figure the actual partition coefficient of microalgae was multiplied by a factor of 0.1.

TABLE 5.4.4

Partition coefficients of different surfactants, determined in the laboratory, for various environmental matrices

Compound	Sorbent	K_d (L kg^{-1})	Reference
LAS	River sediment	2900 ± 1000	[3]
	River sediment	780 ± 130	[38]
	River sediment	6115–7416	[13]
TMAC	River sediment	45–423	[62]
STAC	Sediment	225 677	[7]
	River sediment	65 724	[7]
CTAB	Sediment	71 051	[7]
	River sediment	17 337	[7]
DSDMAC	Sediment	12 489	[7]
	River sediment	3833	[7]
AE	Sediment	350–2100	[17]
	Sediment	1.9–73	[19]
NPEO$_{3-10}$	Sediment	1460–450	[63]

CTAB, cetyl trimethyl ammonium bromide; STAC, stearyl trimethyl ammonium chloride; DSDMAC, distearyl dimethyl chloride.

partition coefficients of surfactants (see Tables 5.4.1 and 5.4.4, respectively) have shown surfactants to present greater affinity to the solid phase. In general, higher values and wider variability have been detected for environmental coefficients, due to other processes that take place simultaneously in the media (such as biodegradation, precipitation), which have a significant effect on the sorption equilibrium, as has been observed previously by other authors [29].

It has been reported for both anionic [14,20,28] and non-ionic surfactants [2,15,23] that sorption increases with the number of carbon atoms in the hydrophobic chain (Table 5.4.5). The sorption coefficient of LAS in activated sludges [22] increases by 2.8 times with each methylene group, and a similar variation has been observed for river and marine sediments (Table 5.4.5). The partition coefficients obtained for the marine medium are slightly higher than those for river sediment, as a consequence of the higher ionic strength of seawater [14].

Other factors of the molecular structure favour sorption, such as isomeric position for anionic, or the number of ethylene oxides in non-ionic surfactants (see Table 5.4.5). There is an increased sorption in

TABLE 5.4.5

LAS and AE homologue partition coefficients, expressed as the ratio between concentrations in particulate and dissolved phases, in various systems

	Activated sludges [24]	River sediments [8]	Marine sediments [14]	Sediment [34]		Sediment [23]
C_{10}-LAS	220	41	96	15	$A_{10}E_3$	41
C_{11}-LAS	1000	100	377	–	$A_{10}E_5$	48
C_{12}-LAS	3070	330	640	77	$A_{10}E_8$	126
C_{13}-LAS	9330	990	1112	–	$A_{12}E_8$	1230
C_{14}-LAS	–	2950	–	709	$A_{14}E_8$	3548
					$A_{16}E_8$	6166

the cases of external isomers [8,13,20] (greater distance between the phenyl linkage and the terminal carbon of the alkylic chain), and of alkyl ethoxylates with a higher number of ethylene oxides in the chain [6,17] (see Fig. 5.4.4, adapted from the results presented by Brownawell et al. [6]), suggesting non-polar mechanisms for sorption. However, cationic sorption is predominantly conducted by electrostatic mechanisms.

Recently, new sorption experiments have been conducted with marine sediments [30] and microalgae [31]. Specifically these experiments have

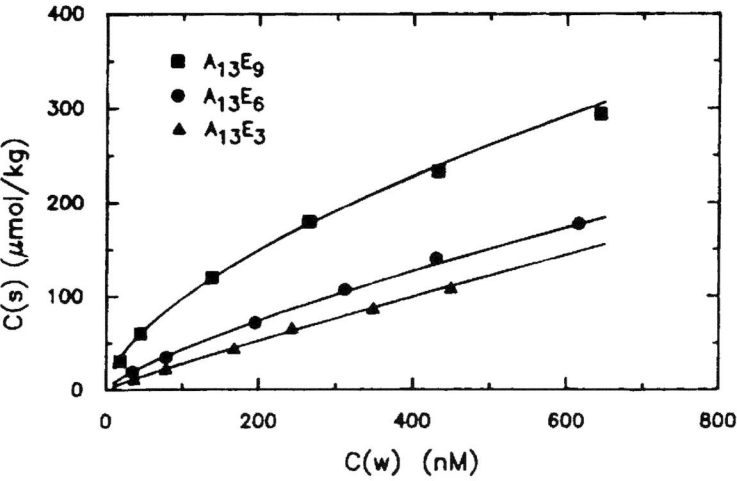

Fig. 5.4.4. Sorption isotherms of AE homologues on sediment (EPA-12), showing the fits obtained with the Freundlich isotherm (from Brownawell et al. [6]).

used polypropylene centrifuge tubes, continuously shaken and maintained at constant temperature in a culture chamber. Sorption of LAS and some of their biodegradation intermediates (sulfophenyl carboxylates, SPCs) have been characterised, including control tests (without sorbent) in every assay. In this case, two sorption surfaces must be taken into account, the plastic and the test material, either sediment $(10\ g\ L^{-1})$ or microalgae $(4\ g\ L^{-1})$. The dissolved phase was recovered after intense centrifugation, pre-concentrated by solid-phase extraction and subsequently analysed by HPLC (using the specific LAS and SPC method proposed by León et al. [32]).

Two different partition coefficients are shown in Table 5.4.6, one corresponding to the sediment tested and the other to the complete system comprising plastic and sediment; from these results, the plastic surface can be considered not significantly different from that of the sediment. The carboxylic group increases molecule polarity thus reducing the presence of SPC in the sorbed phase (lower partition coefficients than LAS homologues, Table 5.4.6. The chain length (alkylic or carboxylic) is correlated positively with the partition coefficient, with the effect of the presence of the carboxylic group of the molecule being more significant than the additional methylene groups. Sorption onto the tube walls is relevant for LAS homologues but not for SPC due to their higher polarity.

Ionic surfactants also participate in specific chemical and electrostatic interactions with sorbents that do not occur with non-ionic compounds. In fact, recent research has shown that hydrophobic

TABLE 5.4.6

LAS and SPC $(C_{10}-C_{13})$ partition coefficients (K_d), obtained for a marine sediment [30] and for the complete sorption system (sediment + plastic), at 15°C and 400 $\mu g\ L^{-1}$

Homologue	Partition coefficient (L kg^{-1})	
	K_d (sediment)	K_d (sediment ± plastic)
C$_{10}$-SPC	1 ± 1	1 ± 1
C$_{11}$-SPC	3 ± 2	3 ± 2
C$_{12}$-SPC	4 ± 6	4 ± 6
C$_{13}$-SPC	7 ± 4	7 ± 4
C$_{10}$-LAS	12 ± 5	40 ± 11
C$_{11}$-LAS	62 ± 7	118 ± 9
C$_{12}$-LAS	116 ± 13	218 ± 20
C$_{13}$-LAS	137 ± 38	469 ± 77

sorption takes place after an initial sorption on ionic particle sites by electrostatic strengths [10]. However, this sorption of ionic surfactants must be conducted predominantly by hydrophobic mechanisms, as numerous previous studies have revealed [8,13,14].

5.4.4 Sorption isotherms

Sorption isotherms show the relationship between the concentration of sorbate and the mass sorbed at constant temperature. The surfactant isotherm can be divided into four regions (see Fig. 5.4.5) [5,11]: in the first section, at low concentrations, where this curve has a slope close to unity (region I), the sorption occurs by electrostatic interactions with monomers on the surface; the sharp increase of adsorption taking place in the second region (II) is due to surfactant association at the surface through lateral interactions of the hydrocarbon chains (bidimensional aggregates on the sorbed phase); the slope decreases in the third region (III); and finally, at a concentration higher than CMC, the sorption is asymptotic (region IV). These distinctive regions of the curve have been detected for SDS [33] and LAS [34]. Surfactant concentrations typically found in the natural environment give isotherms not only in region I, but also in region II of the curve.

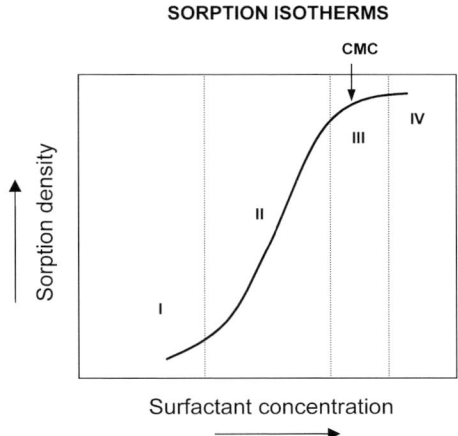

Fig. 5.4.5. Schematic diagram of the typical surfactant sorption behaviour [5,11] related to surfactant concentration at constant temperature.

Nevertheless, surfactant sorption isotherms on natural surfaces (sediments and biota) are generally non-linear, even at very low concentrations. Their behaviour may be explained by a Freundlich isotherm, which is adequate for anionic [3,8,14,20,30], cationic [7] and non-ionic surfactants [2,4,15,17] sorbed onto solids with hetero-geneous surfaces. Recently, the 'virial-electrostatic' isotherm has been proposed to explain anionic surfactant sorption; this is of special interest since it can be interpreted on a mechanistic basis [20]. The virial equation is similar to a linear isotherm with an exponential factor, i.e. with a correction for the deviation caused by the heterogeneity of the surface or the energy of sorption.

5.4.5 Influence of physico-chemical conditions on sorption processes

Apart from the structural properties of the molecule, surfactant sorption is affected by physico-chemical parameters, such as tempera-ture, pH, organic carbon content of sediment, suspended solids concentration, ionic strength, etc. The most relevant of these are discussed next.

Temperature
Higher temperatures increase the sorption kinetics, as generally occurs in other chemical reactions, and may also affect the extension of the process, although this aspect of the subject has received little attention until now. The isotherms of C_{11}-LAS sorption on marine sediments at different temperatures (15, 20 and 25°C) are shown in Fig. 5.4.6 [30]. At lower temperatures, a slightly higher proportion of the LAS is associated with the suspended sediment. This effect could be explained by the fact that LAS sorption is an exothermic reaction, and is thus more likely to take place at lower temperatures. This change with temperature detected enables the thermodynamic parameters of C_{11}-LAS sorption to be determined by using sorption coefficients at different temperatures and applying the procedure proposed by Biggar and Cheng [35]. The sorption enthalpy estimated for this process ($\Delta H_{\text{sorption}} = -114$ KJ mol^{-1}) indicates that C_{11}-LAS is sorbed by chemisorption, and the negative free energy value (-24.0 kJ mol^{-1}) indicates that this is a spontaneous process [30].

618

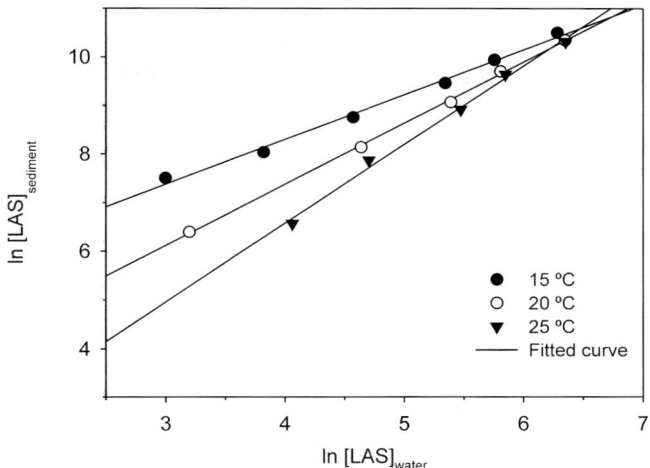

Fig. 5.4.6. C_{11}-LAS sorption isotherms on marine sediments at different temperatures.

Particulate matter concentration, organic content and granulometry
For the different types of surfactant, the effect of solids concentration on sorption is unclear. While some studies have shown that the concentration of solids has no effect on the extent of sorption [19,36], in others, it has been detected that the partition coefficient decreases with particle concentration [17,27,37,38].

Sorption of surfactants onto the surfaces of microorganisms and suspended solids may be relevant in the environment, since their specific surface is relatively high. In the marine ecosystem, suspended solids and phytoplankton are the abiotic and biotic environmental compartments, respectively, that constitute the greater part of suspended material. These are the compartments where the dispersion of a chemical occurs preferentially in the water column. LAS and SPC partition coefficients are slightly higher on microalgae than on sediments [30], probably due to their higher specific surface or preferential sorption sites compared with sediments (i.e. greater organic carbon content). LAS partition coefficients on microalgae [30] are similar for C_{11} and C_{12} homologues (727 $\pm$ 195 and 710 $\pm$ 11 L kg^{-1}, respectively) and higher than for C_{11}-SPC (28 $\pm$ 2 L kg^{-1}).

The role of organic carbon associated with sediments is still unclear. Many studies have shown that there is a good correlation between the organic carbon content and LAS sorption [2,3,20,36,38], but at environmentally relevant concentrations (in the ppb range) sorption mechanisms do not seem to be hydrophobic [8,10]. But the good correlation between the clay content of sediments and sorption is generally accepted, for both anionic [3,8,10] and non-ionic [17] surfactants. Figure 5.4.7 shows virial sorption isotherms of C_{12}-LAS on different sediments (EPA-13, EPA-12, EPA-16 and EPA-25) obtained by Westall et al. [20]. Increased sorption was found for soils with higher contents of organic carbon (3.04, 2.33, 1.20 and 0.76%, respectively) and clay (52.6, 35.4, 39.0 and 20.5%, respectively).

The most notable correlation found for the organic carbon partition coefficient (K_{oc}) of hydrophobic chemicals is that obtained with the octanol/water partition coefficient (K_{ow}). Recently, a single correlation has been proposed for all these compounds based on previous data ($\log(K_{oc}/K_{ow})$ versus $\log K_{ow}$); the specific approximation proposed was $K_{oc} = 0.35 K_{ow}$, but this could vary by a factor of 2.5 in either direction [39].

Ionic strength and pH
Ionic strength and pH are closely related, and their influence on the sorption process may be explained by electrostatic or chemical interactions. The ionic strength of the medium is correlated positively

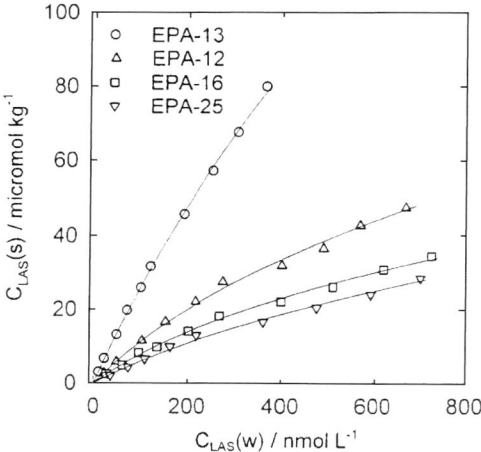

Fig. 5.4.7. Sorption isotherms of C_{12}-LAS on different sediments. The lines were calculated with the virial equation (from Westall et al. [20]).

with anionic surfactant sorption [6,14] and negatively with the non-ionic surfactant sorption [40]. Ionic strength and pH of the solution have only a small effect on the sorption of AE, but their effect increases with the number of ethoxy groups, showing that sorption presumably takes place through the mechanism of a hydrogen bond [6]. The effect of ionic strength is more relevant in estuarine areas, since in these systems it presents a wide variability (0.01–0.7 M). The sorbent surface is positively charged at low pH, which makes anionic sorption easier [6,20,26,27,28,41] and non-ionic and cationic sorption more difficult.

Ionic charge and precipitation
The presence of ions on the sorbent surface with an opposite charge to that of the ionic surfactants under consideration tends to promote the sorption process, due to the existence of specific interactions [20,26,41], although this effect has not always been detected [3]. The dissolved ionic surfactants may compete with ions of the same charge in the sorption process [26], or may precipitate with ions having the opposite charge. Surfactant precipitation is usually regarded as part of the sorption process, due to the difficulty of differentiating between sorption and precipitation. LAS precipitation with Ca^{2+} is especially relevant [20,42,43] as the calcium salt has low solubility, particularly in estuarine and marine environments. LAS precipitation has also been detected with other cations such as aluminium [11] or magnesium compounds [44].

Surfactant presence in a medium may increase the mobility of other compounds, and this property is used to treat contaminated soils and sludges; however, the sorbed contaminants may also be redissolved.

5.4.6 Desorption and reversibility

The reversibility of sorption has considerable environmental relevance, because it permits the recovery of ecosystems after exposure to contamination, and makes possible the re-dissolution of chemicals that have been sorbed onto particulate material. For this reason, this aspect has been studied extensively, with results showing that sorption onto natural sediments is a reversible process for non-ionic [17] and anionic surfactants [8,20,30,45,46]. This desorption process, however, has not been confirmed by other authors [3,14].

TABLE 5.4.7

Percentage of C_{11}-LAS desorbed at 5°C and different concentrations [30]

Concentration ($\mu g\ L^{-1}$)	Desorption (%)
50	0
100	9.4
200	21.5 ± 14.3
400	49.2 ± 19.3
600	51.3 ± 21.5
1000	61.9 ± 5.8

Desorption experiments are usually conducted as a continuation of sorption tests, by adding water without surfactant [10,30]. C_{11}-LAS desorption percentages in marine sediments at 5°C are shown in Table 5.4.7. Those sediments with higher LAS concentrations undergo a more extensive desorption, more than 50% of the initial sorbed surfactant for concentrations being higher than 200 $\mu g\ L^{-1}$. The affinity of LAS for active sediment sites tends to decrease with surfactant concentration, due to the most-favoured sites having already been occupied. As with other organic chemicals, LAS desorption from soils and sediments may occur in two stages: a rapid stage (lasting hours or days) followed by a stage of much slower release (during months or years) [47,48]. Consequently, LAS desorption is fast and reversible for the weakly sorbed fractions, but other fractions probably require longer times to be desorbed, if this happens at all. Nevertheless, the desorption of SPC is complete, given the relatively weak interactions between this molecule and the sorbent surface. Slow desorption has been studied for other compounds and it seems to be controlled by slow diffusion processes in the sediment aggregates [49] with the organic matter being mainly responsible for the process [48]. Studies of PCBs and PAHs have corroborated the slow diffusion process that depends on diffusion across pores in organic matter or through pores in solids covered by organic matter [48].

5.4.7 Conclusion

Sorption in the environment encompasses adsorption/absorption as well as desorption. The former depends on both physico-chemical properties of the compound and environmental conditions, such as pH, temperature and ionic strength of the medium. The latter is characterised by at least two phases, a relatively fast and a slow phase. Sorption is very important for the bioavailability of the surfactant under consideration. In the aqueous environment all surfactants have relatively high affinities for sorption to sediment particles. Cationic surfactants adsorb stronger to sediments than other surfactants due to the predominantly negatively charged surfaces of natural materials.

REFERENCES

1 J. Westall, In: W. Stumm (Ed.), *Aquatic Surface Chemistry*, John Wiley and Sons, New York, 1987, p. 3.
2 K. Urano, M. Saito and C. Murata, *Chemosphere*, 13 (1984) 293.
3 E. Matthijs and H. De Henau, *Tenside Surf. Det.*, 22 (1985) 299.
4 C. Palmer, D.A. Sabatini and J.H. Harwell, In: D.A. Sabatini and R.C. Knox (Eds.), *Transport and Remediation of Subsurface Contaminants*, ACS Symposium Series 491, American Chemical Society, Washington, DC, 1992.
5 T. Rheinlander, T.E. Klumpp and M.J. Schwuger, *J. Disper. Sci. Technol.*, 19 (1998) 379.
6 B.J. Brownawell, H. Chen, J.M. Collier and J.C. Westall, *Environ. Sci. Technol.*, 24 (1990) 1234.
7 R.J. Larson and R.D. Vashon, *Dev. Ind. Microbiol.*, 24 (1983) 425.
8 V.C. Hand and G.K. Williams, *Environ. Sci. Technol.*, 21 (1987) 370.
9 T. Suhara, H. Fukui, M. Yamaguchi and F. Suzuki, *Colloids Surf. A*, 119 (1996) 15.
10 Z. Ou, A. Yediler, Y. He, L. Jia, A. Kettrup and T. Sun, *Chemosphere*, 32 (1996) 827.
11 P. Somasundaran and L. Huang, *Polish J. Chem.*, 71 (1997) 568.
12 S.W. Karickhoff, D.S. Brown and T.A. Scott, *Water Res.*, 13 (1979) 241.
13 K. Amano and T.J. Fukushima, *Environ. Sci. Health*, 28 (1993) 683.
14 J.A. Rubio, E. González-Mazo and A. Gómez-Parra, *Mar. Chem.*, 54 (1996) 171.
15 Z. Liu, D.A. Edwards and R.G. Luthy, *Water Res.*, 26 (1992) 1337.
16 M. Ahel and W. Giger, *Chemosphere*, 26 (1993) 1471.
17 M.L. Cano and P.B. Dorn, *Environ. Toxicol. Chem.*, 15 (1996) 684.
18 J.L. Berna, N. Battersby, L. Cavalli, R. Fletcher, A. Guldner, D. Chowanek and J. Steber, *Proceedings of the Fifth World Surfactant Congress CESIO*, Firenze, 2000.

19   B.J. Brownawell, H. Chen, W. Zhang and J.C. Westall, *Environ. Sci. Technol.*, 31 (1997) 1735.

20   J.C. Westall, H. Chen, W. Zhang and B.J. Brownawell, *Environ. Sci. Technol.*, 33 (1999) 3110.

21   V.M. León, M. Sáez, E. González-Mazo and A. Gómez-Parra, *Sci. Total Environ.*, 288 (2002) 215.

22   W. Stum, *Chemistry of the Solid–Water Interface. Processes at the Mineral-water and Particle Water Interface in Natural Systems*, Wiley/Interscience Publisher, John Wiley & Sons Inc, New York, 1992, p. 87.

23   A.T. Kiewiet, K.G.M. de Beer, J.R. Parsons and H.A.J. Govers, *Chemosphere*, 32 (1996) 675.

24   L.M. Games, In: K.L. Dickson, A.W. Maki and J. Carins (Eds.), *Modelling the Fate of Chemicals in the Aquatic Environment*, Ann Arbor Science, Ann Arbor, MI, 1982, p. 325.

25   A. Beveridge and W.F. Pickering, *Water Res.*, 17 (1983) 215.

26   K. Dao, A. Bee and C. Treiner, *J. Colloid Interface Sci.*, 204 (1998) 61.

27   D.M. Di Toro, L.J. Dodge and V.C. Hand, *Environ. Sci. Technol.*, 24 (1990) 1013.

28   N.A. Fernández, E. Chacin, E. Gutierrez, N. Alastre, B. Llamoza and C.F. Foster, *Biores. Technol.*, 54 (1995) 111.

29   C.T. Chiou, S.T. McGroddy and D.E. Kile, *Environ. Sci. Technol.*, 32 (1998) 264.

30   V.M. León, (2001) PhD Thesis. Cádiz University, Cádiz.

31   M. Sáez, (2002) PhD Thesis. Cádiz University, Cádiz.

32   V.M. León, E. González-Mazo and A. Gómez-Parra, *J. Chromatogr. A*, 889 (2000) 211.

33   P. Chandar, P. Somasundaran and N.J. Turro, *J. Colloid Interface Sci.*, 117 (1987) 31.

34   B. Li and E. Ruckenstein, *Langmuir*, 12 (1996) 5052.

35   J.W. Biggar and M.W. Cheng, *Soil Sci. Soc. Am. Proc.*, 37 (1973) 863.

36   D.C. McAvoy, C.E. White, B.L. Moore and R.A. Rapaport, *Environ. Toxicol. Chem.*, 13 (1994) 213.

37   V.C. Hand, R.A. Rapaport and R.H. Wendt, *Environ. Toxicol. Chem.*, 9 (1990) 467.

38   V.C. Hand, R.A. Rapaport and C.A. Pittinger, *Chemosphere*, 21 (1990) 741.

39   R. Seth, D. Mackay and J. Muncke, *Environ. Sci. Technol.*, 33 (1999) 2390.

40   T. van den Boomgaard, T.F. Tadros and J. Lyklema, *J. Colloids Surf.*, 116 (1987) 8.

41   K. Inoue, K. Kaneko and M. Yoshida, *Soil Sci. Plant Nutr.*, 24 (1978) 91.

42   K.L. Matheson, M.F. Cox and D.L. Smith, *J. Am. Oil Chem. Soc.*, 62 (1985) 1391.

43   J.L. Berna, J. Ferrer, A. Moreno, D. Prats and F. Ruiz Bevia, *Tenside Surf. Deterg.*, 26 (1989) 101.

44 L. Cohen, R. Vergara, A. Moreno and J.L. Berna, *J. Am. Oil Chem. Soc.*, 70 (1993) 723.
45 W.A. House and I.S. Farr, *Colloids Surf.*, 40 (1989) 167.
46 P. Siracusa and P. Somasundaran, *Colloids Surf.*, 26 (1987) 55.
47 S.C. Wu and P.M. Gschwend, *Environ. Sci. Technol.*, 20 (1986) 717.
48 G. Cornelissen, P.C.M. van Noort and H.A.J. Govers, *Environ. Sci. Technol.*, 32 (1998) 3124.
49 J.J. Pignatello and B. Xing, *Environ. Sci. Technol.*, 30 (1996) 1.
50 C.F. Tabor and L.B. Barber, *Environ. Sci. Technol.*, 30 (1998) 161.
51 S. Terzic and M. Ahel, *Mar. Pollut. Bull.*, 28 (1994) 735.
52 E. González-Mazo, J.M. Quiroga, D. Sales and A. Gómez-Parra, *Toxicol. Environ. Chem.*, 59 (1997) 77.
53 E. González-Mazo, J.M. Forja and A. Gómez-Parra, *Environ. Sci. Technol.*, 32 (1998) 1636.
54 V.M. León, E. González-Mazo, J.M. Forja and A. Gómez-Parra, *Environ. Toxicol. Chem.*, 20 (2001) 2171.
55 N. Jonkers, R.W.P.M. Laane and P. de Voogt, *Environ. Sci. Technol.*, (2002), in press.
56 D.A. Van Ry, J. Dachs, C.L. Gigliotti, P.A. Brunciak, E.D. Nelson and S.J. Eisenreich, *Environ. Sci. Technol.*, 34 (2000) 2410.
57 P.L. Ferguson, C.R. Iden and B.J. Brownawell, *Environ. Sci. Technol.*, 35 (2001) 2428.
58 A. Marcomini, B. Pavoni, A. Sfriso and A. Orio, *Mar. Chem.*, 29 (1990) 307.
59 C.G. Naylor, J.P. Mieure, W.J. Adams, J.A. Weeks, F.J. Castaldi, L.D. Ogle and R.R. Romano, *J. Am. Oil Chem. Soc.*, 69 (1992) 695.
60 P. Schöberl, H. Klotz and R. Spilker, *Tenside Surf. Deterg.*, 33 (1996) 329.
61 J.R. Marchesi, W.A. House, G.R. White, N.J. Russell and I.S. Farr, *Colloids Surf.*, 53 (1991) 163.
62 C.A. Pittinger, V.C. Hand, J.A. Masters and L.F. Davidson, In: W.J. Adams, G.A. Chapman and W.G. Landis (Eds.), *Aquatic Toxicology and Hazard Assessment*, ASTM STP 971, Vol. 10. American Society for Testing Materials, Philadelphia, 1988, p. 138.
63 D.M. John, W.A. House and G.F. White, *Environ. Toxicol. Chem.*, 19 (2000) 293.

5.5 FATE OF ORGANOSILICONE SURFACTANTS

Lea S. Bonnington, William Henderson and Jerzy A. Zabkiewicz

5.5.1 Introduction

Ecotoxicological effects have been demonstrated for a number of surfactants or their metabolites, including some still currently in use, such as the nonylphenol ethoxylates [1], and as such there is a necessity to find more environmentally acceptable alternatives. Whilst the silicones are not the major surfactant type in use to date, the efficient properties and indications of low environmental persistence and toxicity demonstrate their potential for widespread use [2–4]. Relatively little is known about these new, rapidly emerging surfactants and the purpose of this chapter is thus to collate the available data, present new data, and identify the future research required in this area in order to evaluate the environmental relevance of this class of surfactants.

In general organosilicone materials are considered to be of minimal concern, and thus there is relatively little environmental regulation worldwide [5]. Consequently, limited information regarding the environmental loading and occurrence of the silicone surfactants is available, despite the existence of silicone environmental health and safety councils in several countries.[1] The most recent data available were published in 1993, where the estimated annual use of the polyethermethylsiloxanes (PEMSs), the most widely used silicone surfactants, was reported to be 18 300 tons in the USA [6]. The resultant annual environmental loading was reported to be 10 240 metric tonnes, with the remaining 44% of that produced (8100 metric tonnes) being used as intermediates in manufacturing processes and thus not entering the environment in surfactant form. Entry into the environment by dispersion over time and into the air compartment was described as negligible (< 10 metric tonnes annually). A summary of the methods of disposal, and annual amounts used for the main PEMS surfactant applications, is presented in Table 5.5.1 [6].

---

[1] USA: Silicone Environmental Health and Safety Council (SEHSC); Europe: Centre Européen des Silicones (CES); Japan: Silicone Industry Association of Japan (SIAJ); Global: Global Silicones Council (GSC).

Comprehensive Analytical Chemistry XL
T.P. Knepper, D. Barceló and P. de Voogt (Eds)

Worldwide use of silicone surfactants in polyurethane foam production was 30 000 metric tonnes in 1995 [7]. Assuming analogous use profiles for the USA, this would infer a worldwide annual use of 70 000 metric tonnes for all silicone surfactants. Whilst this figure is not large compared with total world surfactant consumption ($<1\%$ of the 10.2 million metric tonnes reported in 1996) [8], an average annual growth rate of approximately 7% for the silicone industry in general [9], indicates that the use of these compounds may also increase.

Silicone surfactants—especially the trisiloxane surfactants—have demonstrated useful properties as additives (adjuvants) in agricultural formulations (outlined in Chapter 2.8). This application results in the direct introduction of surfactants into the environment and the analysis of their impact is thus of considerable importance. However, the relatively low amounts of surfactants used in agrochemical formulations ($\leq 1\%$) and the much lower ecotoxicological risk than that found for corresponding pesticides, has meant that these compounds are currently not considered to be of great concern, and as such are not subject to the legislative controls that are in place for pesticides and surfactant use in other applications. As our general environmental awareness improves, this attitude is changing and we are beginning to

TABLE 5.5.1

Application, annual usage and method of disposal for industrial PEMS surfactants in the USA for 1993 [6]

Application	Function	Annual amount used (metric tonnes)	Method of disposal
Polyurethane foam	Surfactant	8100 (44%)	Site limited[a]
Polyurethane foam	Surfactant	7210 (39%)	Landfill/incineration/ recycling
Plastics moulding	Mould release		
Cosmetic products	Surfactant, dispersant, aesthetics	2690 (15%)	Wastewater
Textile industry	Antifoam, feel, durability		
Agricultural adjuvants	Surfactant, wetting/spreading	340 (2%)	Soil

[a]Industrial intermediate—not released into the environment.

acknowledge that any introduction of synthetic products into the natural environment should be monitored. The development of sensitive and selective methods, such as atmospheric pressure chemical ionisation (APCI) and electrospray (ESI) mass spectrometric (MS) techniques, has contributed significantly to our ability to achieve this. Atmospheric pressure ionisation (API) MS has been demonstrated as a viable technique for the analysis of organosilicone surfactants (refer to Chapter 2.8), and as such, could also be used to follow the degradation process for the first time [10], the results of which will be summarised in following sections.

5.5.2 Behaviour in environmental media

Due to the instability of the $Si-O-Si$ backbone, the primary degradation of silicone surfactants under hydrolytic conditions is readily achieved [3,7,11–13]. This is evidenced primarily by a loss of surface-activity, which is environmentally significant, as the detrimental effects caused by surfactants in soil systems are often due to alterations in surface- active properties [14]. The amphiphilic nature of surfactant molecules causes alterations to soil physics, chemistry and biological interactions and thus rapid primary degradation of surfactants is highly desirable. Consequently, the hydrolytic instability of the trisiloxane molecules appears ecotoxicologically favourable. However, the ecotoxicity and persistence of the resulting degradation products must also be determined in order to truly evaluate environmental effects, with complete degradation or inactivation being the ideal for any chemical contaminant.

As shown in Table 5.5.1, 15% of the silicone surfactants annually used were disposed of via wastewater treatment plants [6], but no studies have addressed their fate or persistence in this environmental compartment. Due to the hydrolytic instability and tendency for sorption to surfaces, it is generally thought that limited persistence of the parent molecule in aqueous systems should occur. Consequently more attention has been focused on interactions with solid media such as that resulting from direct application as agricultural adjuvants, and in re-use of sludge. Increased water solubility for the degradation products of trisiloxane surfactants has, however, been observed [10,12,15], demonstrating the need to also monitor the

629

degradation products of silicone surfactants in aqueous systems to truly ascertain removal from wastewater systems.

Information regarding the behaviour of silicone surfactants in environmental media is limited with all investigations reported to date confined to laboratory conditions. In general ^{14}C-labelling methods and changes in surface tension and spreading ability have been used to determine degradation [15–17]. However, these results need to be interpreted carefully due to the non-specific nature of the methods. The ^{14}C-labelling method does not discriminate between the parent molecule and degradation products thereof, and surface tension and spreading ability may also be influenced by the degradation products formed.

An investigation into the behaviour of the trisiloxane surfactant, $M_2D-C_3-O-(EO)_n-COCH_3$,[2] on five soils showed low mobility of the surfactant over a 24 hour period, as determined by ^{14}C-radiolabelling methods in conjunction with adsorption/desorption isotherms and TLC methods [17]. The results indicated both clay and organic matter content influence adsorptive behaviour. This work was further extended to soil columns, where ^{14}C-labelling was used to determine the distribution of the surfactant in the gas, liquid and solid phases of the system over 10 weeks [15]. Over this time period, more than 50% of the radiolabel was recovered in the leachate. This was interpreted as indicating a relatively high mobility of the surfactant. However, although not interpreted in this manner by the authors, it seems logical that the results probably more accurately reflect the behaviour of the degradation products. Indeed, the authors noted that the mass recovery of THF-soluble fractions from the aqueous leachates was significantly less than that obtainable for the parent molecule, indicating the formation of more polar water-soluble molecules, as consistent with observations for the degradation products of other trisiloxane surfactants [10,12].

The behaviour of $M_2D-C_3-O-(EO)_n-CH_3$ (average $n = 7.5$, 1) on a silt loam and silty clay loam soil has been investigated, with surface tension and spreading ability used as the indicators of degradation/recovery [16]. The results obtained showed that the 'inactivation' of the surfactant in soil far exceeded any likely loading through

---

[2] Abbreviation adopted for $[Si(CH_3)_3-O]_2-Si(CH_3)-CH_2CH_2CH_2-O-(CH_2CH_2O)_n-COCH_3$, Average $n = 7.5$, where $M = (CH_3)_3-Si-O_{(0.5)}$, $D = O_{(0.5)}-Si(CH_3)-O_{(0.5)}$ and $EO = CH_2CH_2O$ (refer to Chapter 2.8).

agricultural practices. Adsorption to soil particulates was considered to have contributed significantly to the effective removal of the trisiloxane surfactant from solution in both slurry and soil column determinations, and an increase in degradation was described with more extreme pH values for both solid and aqueous media.

The recovery of $M_2D-C_3-O-(EO)_n-CH_3$ after exposure to various solid media has been investigated by API-MS, high performance liquid chromatography light scattering mass detection (HPLC–LSD) and HPLC–APCI-MS methods [10]. Recoveries with extraction immediately following application were determined (surfactant concentration: 0.1%, surfactant/solid: 10 mg g^{-1}) with complete recoveries obtained on all media other than the clays illite and montmorillonite (Table 5.5.2) [10].

The structural integrity of $M_2D-C_3-O-(EO)_n-CH_3$ was maintained for all solid media investigated, as confirmed by qualitative analysis of the flow injection analysis (FIA) API-MS spectra (example shown in Fig. 2.8.10a). Intercalation of the surfactant solution between the clay layers was considered as the likely cause of the reduced recovery on montmorillonite (74%), according to experimental observations and

TABLE 5.5.2

Recovery of $M_2D-C_3-O-(EO)_n-CH_3$ with immediate extraction from various solid media

Media	% Recovery (SD)[a]	
	HPLC–LSD	HPLC–APCI-MS
Aqueous	96 (1)	103 (12)
Sand	100z (2)	85 (7)
Pumice	109 (10)	90 (3)
Goethite	104 (1)	104 (14)
Fe_2O_3	98 (1)	88 (8)
Halloysite	90 (2)	97 (4)
Kaolinite	96 (6)	94 (7)
Illite	78 (5)	77 (9)
Montmorillonite	77 (6)	74 (11)[b]
Soil[c]	101 (8)	85 (15)
TiO_2	111 (8)	104 (2)
Al_2O_3	103 (–)[d]	94 (4)
$Al(OH)_3$	103 (3)	91 (0.1)

[a]Average of two replicates. SD, standard deviation.
[b]Flow injection analysis (FIA) APCI-MS data.
[c]Soil type: Te Kowhai; Composition: 49% volcanic glass, 47% halloysite [28].
[d]One replicate only.

consistent with known sorption of organic [18], polyethylene glycol (PEG) [19] and surfactant [20] compounds to this clay. The reduced recovery on illite was attributed to surface interactions, with sorption of PEG-300 to external surfaces of illite previously described [19]. Results from analyte-solid studies obtained elsewhere have also demonstrated stronger interactions for surfactants [20,21] and pesticides [18,22] with montmorillonite and illite as compared with other media.

The behaviour of $M_2D-C_3-O-(EO)_n-CH_3$ over time (4 weeks) has been monitored by FIA–APCI-MS in the presence of $Al(OH)_3$, $CaCO_3$ (calcite), $FeO(OH)$ (goethite), Fe_2O_3 (hematite), halloysite, illite, kaolinite, sand, pumice, talc and TiO_2 (anatase), and provides some useful comparative information regarding silicone surfactant behaviour in the presence of various solid media [10]. In general, the results indicated a dependence of parent molecule recovery on pH, with lower recoveries obtained with more extreme pH values (i.e. halloysite and sand, pH 3.7 and 4.8, respectively), consistent with the known pH instability of trisiloxanes under aqueous conditions [3,11,12,16]. In particular, the loss of the parent molecule was most rapid in the presence of the clay, halloysite, and is consistent with other reports of acceleration of silicone hydrolysis in the presence of acid clays [23–25]. Comparison of the recovery of $M_2D-C_3-O-(EO)_n-CH_3$ in the presence of halloysite, kaolinite and illite clays (0.1%, 10 mg g^{-1}) by FIA–APCI-MS is presented in Fig. 5.5.1 [10].

The lowest recoveries were observed in the presence of halloysite. Whilst this could be due to either sorption or degradation, or a combination of the two mechanisms, contributions due to hydrolytic degradation on the strongly acidic clay (pH 3.7) can be considered significant, as illustrated by the changes to the ESI–MS spectra (Fig. 2.8.9). The ability of halloysite to intercalate a range of organic liquids including mono-, di- and triethylene glycols has been described [26], but larger molecules were not investigated. Complete recovery of the $M_2D-C_3-O-(EO)_n-CH_3$ trisiloxane surfactant molecule was obtained with extraction performed immediately following application (Table 5.5.2), indicating that intercalation does not occur readily and thus, due to the low supernatant pH, degradation can be considered the dominant mechanism. The reduction in recovery of $M_2D-C_3-O-(EO)_n-CH_3$ in the presence of illite was more rapid than for kaolinite despite comparable acidities for the two media. Both clays are non-swelling and intercalation is not likely [19,27], but illite carries an overall negative charge whereas kaolinite is a neutral-type lattice. Adsorption

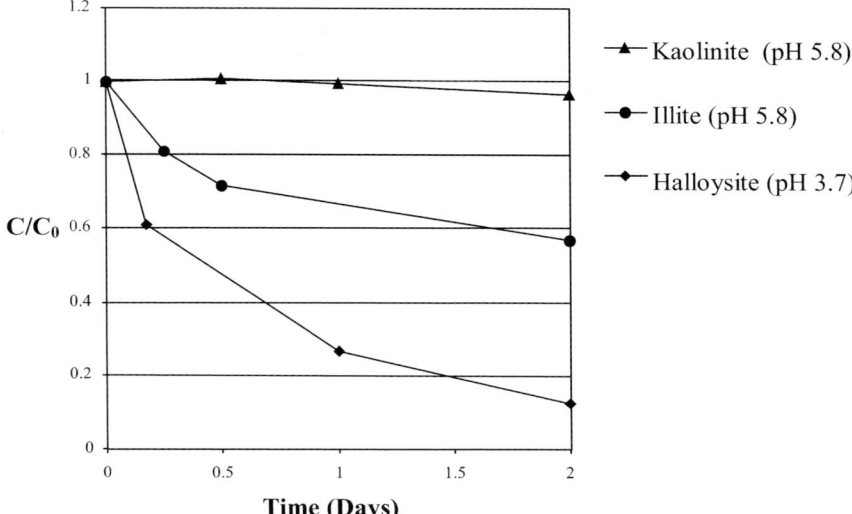

Fig. 5.5.1. Recovery of $M_2D-C_3-O-(EO)_n-CH_3$ over time from various clays as determined by FIA–APCI-MS (average of three replicates).

phenomena/surface interactions are thus the most likely cause for the reduced recovery on illite, consistent with the control recovery results (Table 5.5.2). In contrast, recovery in the presence of talc was high over the four-week analysis period (data not shown). Talc is a non-swelling, non-charged clay, for which a relatively neutral supernatant pH was observed (pH 7.7), indicating the importance of pH, intercalation and surface charge factors in the sorption/degradation process.

The percentage recoveries of $M_2D-C_3-O-(EO)_n-CH_3$ over time in the presence of sand, soil and the clays kaolinite, halloysite and montmorillonite (0.1%, 10 mg g^{-1}) as determined by HPLC–APCI–MS methods are shown in Fig. 5.5.2 [10]. Validation of the data was obtained by parallel HPLC–LSD analysis (data not shown).

The time (days) to 50% loss were: sand (17) > kaolinite (4) > soil (3) > halloysite (1) > montmorillonite(< 0.1). The rate was most rapid for the solid media exhibiting extreme pH values (halloysite, montmorillonite), and highest for the intercalating clay, montmorillonite. The observed recovery from the soil sample (Te Kowhai) was intermediate to

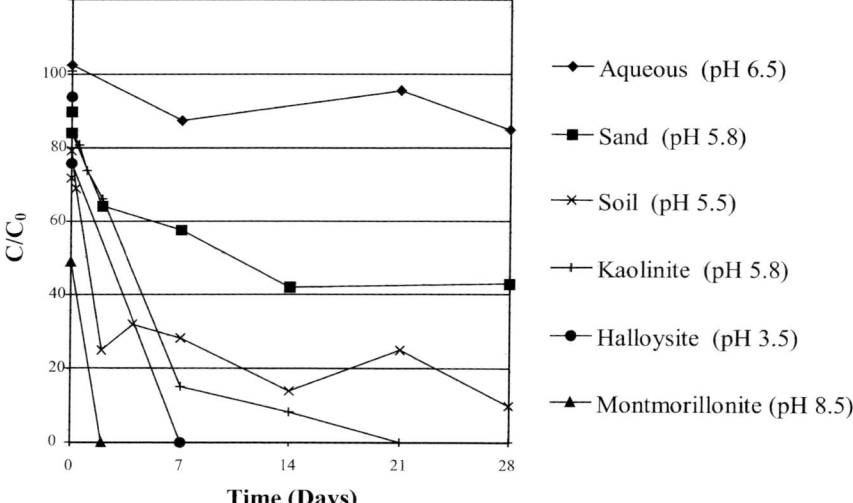

Fig. 5.5.2. Recovery of $M_2D-C_3-O-(EO)_n-CH_3$ over time in various media by HPLC–APCI-MS.

that of sand and halloysite consistent with the described composition (49% volcanic glass, 47% halloysite) [28].

Recoveries of $M_2D-C_3-O-(EO)_n-CH_3$ solutions (0.1%, 10 mg g^{-1}) over time under aqueous and acidified aqueous conditions, and in the presence of alumina, sand and silica have also been investigated by HPLC–LSD [10]. Again the recovery showed a dependence on the supernatant pH (Table 5.5.3), with alumina (pH 7.2) > aqueous (pH 5.8) > sand (pH 5.8) > silica (pH 4.9) > 0.01% HCO$_2$H (pH 3.3).

A summary of the rates to 50% recovery for the trisiloxane surfactant, $M_2D-C_3-O-(EO)_n-CH_3$ in various media, as determined by FIA–ACPI-MS, HPLC–LSD and HPLC–APCI-MS methods, is presented in Table 5.5.3 [10].

These results indicate limited persistence of $M_2D-C_3-O-(EO)_n-CH_3$ in the parent form in typical soil and aqueous environments. In general, the rates of loss observed in the presence of solid media show a dependence on pH as is observed in aqueous media, with lower recoveries in acidic and alkaline pH media, and in the presence of charged and intercalating clay minerals.

TABLE 5.5.3

Time to 50% loss in recovery of $M_2D-C_3-O-(EO)_n-CH_3$ in various media

Media	pH	Time (days)
Calcite	7.8	≫28
Talc	7.7	≫28
Al(OH)$_3$	6.7	≫28
Pumice	6.6	32[a,b]
Hematite	7.8	32[a,b]
Goethite	7.8	25[a]
Alumina	7.2	22[b,c]
TiO$_2$	7.0	20[a]
Aqueous	5.8	19[b,c]
Sand	5.8	14[b,c,d]
Kaolinite	5.8	4[c,d]
Soil[e]	5.5	3[c,d]
Illite	5.9	2.5[a,b]
Silica	4.9	0.6[c]
Halloysite	3.5	0.4[a]
Montmorillonite	8.5	0.1[c,d]
Acidified aqueous (HCO$_2$H)	3.3	0.1[c]

[a]APCI-MS.
[b]Obtained by extrapolation.
[c]HPLC–LSD data.
[d]HPLC–APCI-MS data.
[e]Soil type: Te Kowhai; Composition: 49% volcanic glass, 47% halloysite [28]; ND, not determined (≫28 days).

5.5.3 Product identification

Limited information is available regarding the nature of the degradation products formed from silicone surfactants, although characteristic aspects of silicone chemistry (i.e. influence of chain length and ring size on compound stability) [23,29] can be used to aid prediction of product formation and distribution.

The potential sites of cleavage in the hydrolytic degradation of the trisiloxane surfactant, $M_2D-C_3-O-(EO)_n-CH_3$ (**1**) are illustrated in Fig. 5.5.3. The Si–O bond (*c*) is a likely site of cleavage according to the chemistry of silicones and the relative instability of this bond to hydrolysis [23].

Cleavage between the hydrophobic head group and the hydrophilic tail group (*a*) would generate the products **3** and **4**, shown in Fig. 5.5.3, whilst cleavage at the terminal methyl group (*b*) would give rise to

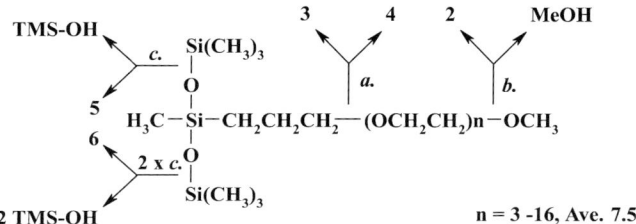

Fig. 5.5.3. Potential sites of cleavage (→) in the abiotic degradation of the trisiloxane alkylethoxylate surfactant, $M_2D-C_3-O-(EO)_n-CH_3$ (**1**); formation of compounds **2**–**6**.

$M_2D-C_3-O-(EO)_n-H$ (**2**)[3] type products. A combination of hydrolysis at (a) and (b) would also yield $HO-(EO)_n-H$ type products. If hydrolysis at (b) did occur then shorter chain analogues of $M_2D-C_3-O-(EO)_n-H$, $HO-(EO)_n-H$ and $HO-(EO)_n-CH_3$ would also be expected due to the comparable nature of the bonds being cleaved. Hydrolysis of the $Si-O$ bond (c) will lead to mono- and di-silanol intermediates, (**5** and **6**), with the concomitant generation of $(CH_3)_3Si-OH$ (TMS–OH).

Subsequent condensation of the organofunctionalised silanol intermediates (**3**, **5**, **6**) will yield a broad range of linear (**7**) and cyclic (**8**) products as shown in Fig. 5.5.4, whilst the condensation of two molecules of TMS–OH to yield $(TMS)_2-O$ is expected to be rapid.

Taking into account this ability of silanols to rearrange into a variety of lengths and structures, then further compounded by the polydisperse EO composition of the trisiloxane surfactants, the potential products of their degradation are numerous. To date, monitoring of the hydrolytic degradation has mostly been by non-selective methods, such as surface tension and spreading [3,11,12]. However, more sensitive and specific analytical techniques are necessary to obtain conclusive structural information. In order to achieve this, the abiotic degradation of the trisiloxane surfactant $M_2D-C_3-O-(EO)_n-CH_3$, under hydrolytic conditions, has been investigated using API-MS [ESI-, APCI- and Fourier transform ion cyclotron resonance (FTICR)], HPLC, nuclear magnetic resonance (NMR) and gas chromatographic (GC-) MS methods [10].

---

[3] Abbreviation for $[Si(CH_3)_3-O]_2-Si(CH_3)-CH_2CH_2CH_2-O-(CH_2CH_2O)_n-H$, Average $n = 7.5$ (refer to Chapter 2.8).

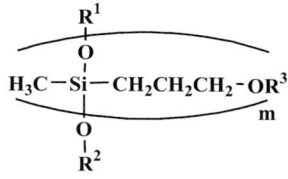
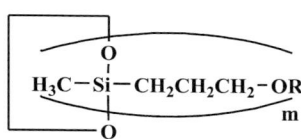

5, m =1; R^1 = H, R^2 = Si(CH$_3$)$_3$; R^3 = (EO)$_n$–CH$_3$

6, m =1; R^1, R^2 = H; R^3 = (EO)$_n$–CH$_3$

7, m ≥ 1, R^1, R^2 = H, Si(CH$_3$)$_3$; R^3 = H, (EO)$_n$–CH$_3$ **8**, m ≥ 3, R = H, (EO)$_n$–CH$_3$

Fig. 5.5.4. Proposed products of M$_2$D–C$_3$–O–(EO)$_n$–CH$_3$ (**1**) abiotic degradation.

FIA–API-MS analysis of acid-degraded M$_2$D–C$_3$–O–(EO)$_n$–CH$_3$ (**1**) was conducted following organic/aqueous solvent partitioning according to the scheme shown in Fig. 5.5.5 [10]. The results obtained from several FIA–ESI-MS analyses of abiotically degraded M$_2$D–C$_3$–O–(EO)$_n$–CH$_3$ demonstrated that the ions observed can be classified into six major series [10]. These, abbreviated to *Series A, B, C, D, E*, and *F*, are summarised in Table 5.5.4, along with possible assignments.

Examples of the FIA–ESI-MS spectra obtained for the water- and organic-soluble fractions of selected degradation mixtures are shown in Figs. 5.5.6 and 5.5.7, respectively. Assignment (*Series A, B, C, D* and *F* as per Table 5.5.4) was aided by the addition of salt solutions (1 mM CH$_3$CO$_2$Na), thereby simplifying the spectra to the Na$^+$ adducts.

It is important to note that the ion series observed by the API-MS method may not be representative of all the products present, not the quantities thereof. Cleavage of the ethoxylate chain removes the capacity of silicone surfactants to be ionised and therefore detected by these methods. As such, for example, the cleaved silicone head group [M(D'CH$_2$CH$_2$CH$_2$OH)M, **3**] was never observed by API-MS. The nature of the API-MS process is such that competition between analytes for ionisation occurs, and as such compounds with higher surface activity and EO content can be expected to dominate in the resulting spectra. Suppression effects may thus preclude observation rather than confirm absence. As a consequence, the use of additional techniques such as FTICR-MS, GC-MS, HPLC and NMR to provide complementary data was also necessary. Furthermore, the high number of possible structures for each ion series observed, rendered it difficult to assign structures with confidence. Consequently simplified M$_2$D–C$_3$–O–(EO)$_n$–R

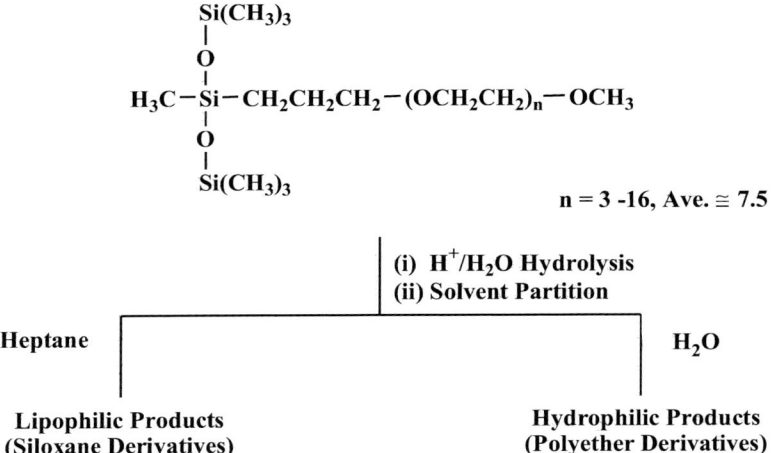

$n = 3$ -16, Ave. $\cong$ 7.5

(i) H^+/H_2O Hydrolysis
(ii) Solvent Partition

Heptane H_2O

Lipophilic Products   Hydrophilic Products
(Siloxane Derivatives)   (Polyether Derivatives)

Fig. 5.5.5. Hydrolysis and solvent extraction of the trisiloxane surfactant $M_2D-C_3-O-(EO)_n-CH_3$.

analyte mixtures[4] were also used to aid in the assignments. These also enabled differentiation between synthetic by-products and degradation products as discussed in Chapter 2.8 (see Fig. 2.8.10) [10].

Application of ESI-MS, FTICR-MS, HPLC, HPLC–ESI-MS and NMR techniques confirmed formation of $CH_3O(EO)_nH$ (**4**), $MD^{(OH)}-C_3-O-(EO)_n-R$ (**5**) and $D^{(OH)2}-C_3-O-(EO)_n-R$ (**6**) as degradation products. Comparison with commercially available $CH_3O(EO)_nH$ standards also provided unequivocal confirmation of $CH_3O(EO)_nH$ as hydrolysis products of $M_2D-C_3-O-(EO)_n-CH_3$. Formation of a range of linear and cyclic products both with and without the EO chains, and terminal TMS groups intact, were also indicated. In general, the more silylated analogues of a structural series partitioned preferentially into the heptane-soluble fraction, and higher EO content derivatives were more commonly observed in the water-soluble fraction. Typically the products formed followed that for expected silanol/siloxane stabilities,[5] with longer EO chain products more stable as low condensation polymers

---

[4] Obtained by synthetic procedures or chromatographic separation of the commercial product.
[5] The stability of alkylsilanols and cyclic siloxanes is known to increase with increased chain length of the alkyl substituent and of the siloxane backbones [23,29].

TABLE 5.5.4

Ions observed (FIA–ESI-MS) and possible structures for hydrolytically degraded $M_2D–C_3–O–(EO)_n–CH_3$ [a]

Series	Ions observed ($\Delta m/z = 44$) [b]	Molecular mass ($\Delta MW = 44$)	Possible structures	Compound type
A	$[M + H]^+$: 121–649 $[M + NH_4]^+$: 314–358 $[M + Na]^+$: 187–803 $[M + K]^+$: 335–731	120–780	$HO(EO)_nCH_3$	4
B	$[M + H]^+$: 311–443 $[M + NH_4]^+$: 328–944 $[M + Na]^+$: 333–1257	310–1234	$D^{(OH)2}–O–(EO)_n–CH_2CH=CH_2$ [c] $M(D^{OH})–C_3–O–(EO)_n–CH_3$ Linear silicones: $R^1 = H; R^2 = TMS; R^3 = (EO)_n–CH_3$ Linear silicones: $R^1, R^2 = TMS; m\text{-}2[R^3 = (EO)_n–CH_3]; 2[R^3 = H]$ Cyclic tetramer, $m = 4$: $R^1 = (CH_2)_3–O–(EO)_n–CH_3; R^2, R^3, R^4 = (CH_2)_3–OH$	5 7 7 8
C	$[M + H]^+$: 327–591 $[M + NH_4]^+$: 520–960 $[M + Na]^+$: 305–965	282–1030	$D^{(OH)2}–C_3–O–(EO)_n–CH_3$ Linear silicones: $R^1, R^2 = H; R^3 = (EO)_n–CH_3$ Linear trimer, $m = 3$: $R^1 = H; R^2 = TMS; R^3 = (EO)_n–CH_3; R^4, R^5 = H$ $HO(EO)_nH$ [b]	6 7 7
D	$[M + K]^+$: 409–1069 $[M + Na]^+$: 893–1070	870–1047	Linear trimer, $m = 3$: $R^1, R^2 = H; R^3 = (EO)_n–CH_3; R^4, R^5 = H$ Linear tetramer, $m = 4$: $R^1, R^2 = H; R^3, R^4 = (EO)_n–CH_3; R^5, R^6 = H$ Linear pentamer, $m = 5$: $R^1, R^2 = H; R^3, R^4, R^5 = (EO)_n–CH_3; R^6, R^7 = H$ Linear pentamer, $m = 5$: $R^1 = H; R^2 = TMS;$ $R^3 = (EO)_n–CH_3; R^4, R^5, R^6, R^7 = H$	7 7 7 7

(continued on next page)

639

TABLE 5.5.4 (continued)

Series	Ions observed ($\Delta m/z = 44$)[b]	Molecular mass ($\Delta MW = 44$)	Possible structures	Compound type
E	$[M+H]^+$: 633–809	456–984	Linear dimer, $m = 2$: R^1, R^2 = TMS; $R^3 = (EO)_n$–CH_3; R^4 = H	**7**
	$[M+Na]^+$: 479–1007		Linear trimer, $m = 3$: R^1, R^2 = TMS; R^3, $R^4 = (EO)_n$–CH_3; R^5 = H	**7**
			Cyclic trimer, $m = 3$: $R^1 = (CH_2)_3$–O–$(EO)_n$–CH_3; R^2, $R^3 = (CH_2)_3$–OH	**8**
			Cyclic tetramer, $m = 4$: R^1, $R^2 = (CH_2)_3$–O–$(EO)_n$–CH_3; R^3, R^4 ($CH_2)_3$–OH	**8**
			$M(D^{OH})$–O–$(EO)_n$–$CH_2CH=CH_2^c$	
F	$[M+H]^+$: 647–735, 823–955	294–954, 1571–1879	Linear silicones: R^1, R^2 = TMS; $R^3 = (EO)_n$–CH_3	**8**
	$[M+NH_4]^+$: 400–796		Cyclic trimer, $m = 3$: R^1, $R^2 = (CH_2)_3$–O–$(EO)_n$–CH_3; $R^3 = (CH_2)_3$–OH	**8**
	$[M+Na]^+$: 317–977, 1594–1902		Cyclic tetramer, $m = 4$: R^1, R^2, $R^3 = (CH_2)_3$–O–$(EO)_n$–CH_3; $R^4 = (CH_2)_3$–OH	**8**

[a]Results of several acid and alkaline degradations: $[M_2D$–C_3–O–$(EO)_n$–$CH_3]$ = 50, 4, 2, 0.2, 0.02%; [HCl] = 2, 0.2, 0.02 mol L^{-1}; [NaOH] = 0.2 mol L^{-1}; degradation time = 1, 2, 3, 4 days.
[b]Mass of CH_2CH_2O monomer.
[c]Synthetic by-products (refer to Chapter 2.8).

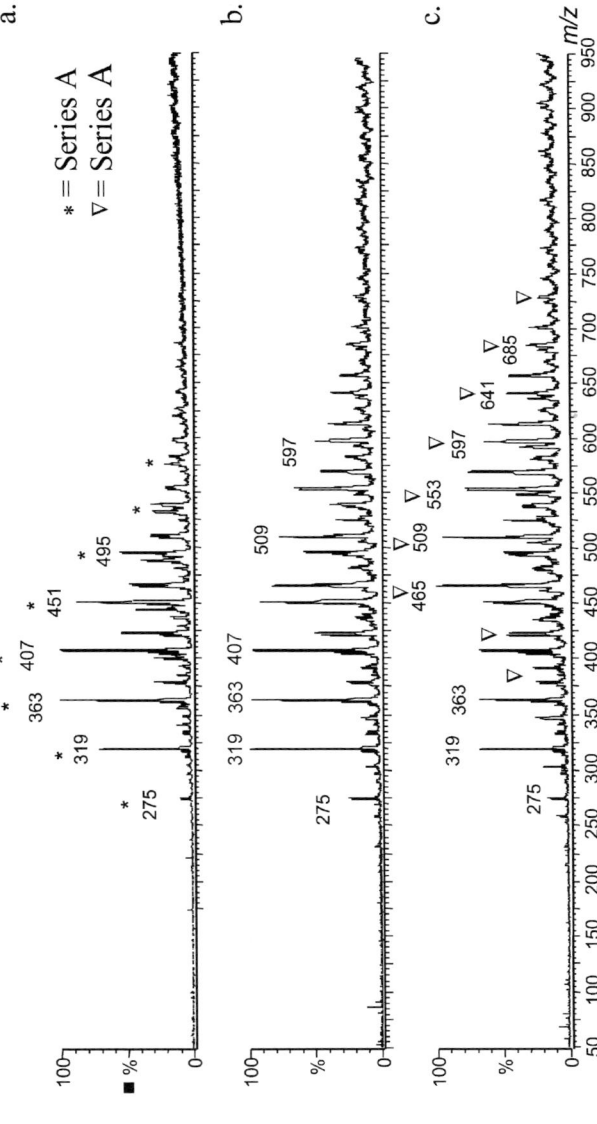

Fig. 5.5.6. FIA–ESI-MS spectra of water-soluble fractions of M_2D–C_3–O–$(EO)_n$–CH_3 degraded with: a, 2 M; b, 0.2 M; and c, 0.02 M HCl. Series A and B are as per Table 5.5.4 (Na^+ adducts).

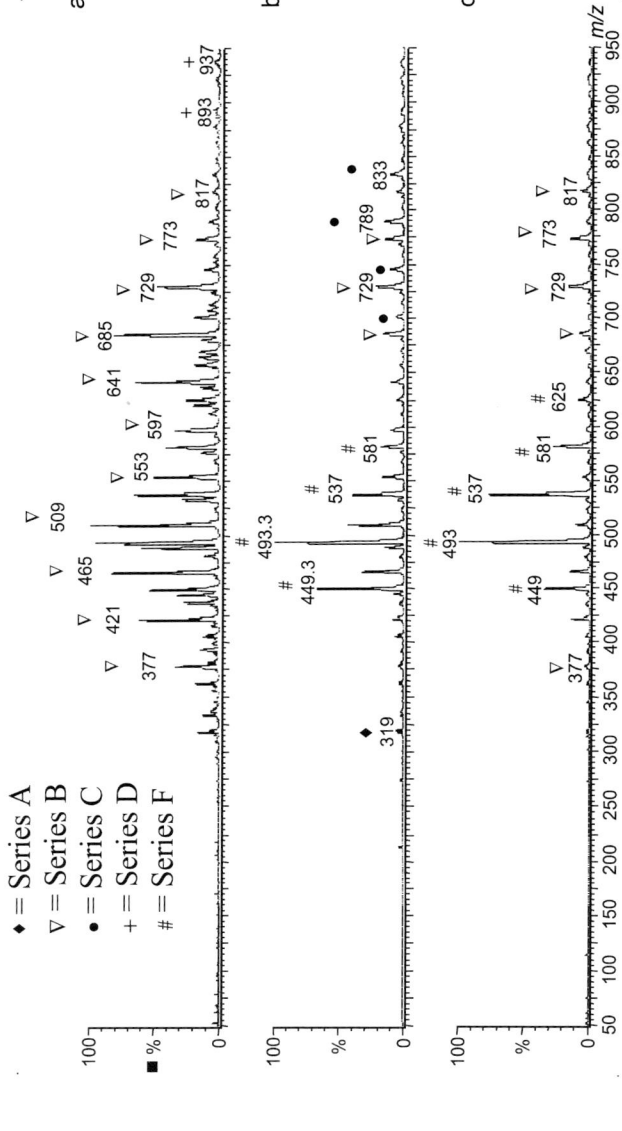

Fig. 5.5.7. FIA–ESI-MS spectra of organic-soluble fractions of M_2D–C_3–O–$(EO)_n$–CH_3 degraded with: a, 2 M; b, 0.2 M; and c, 0.02 M HCl. Series A, B, C, D and F are as per Table 5.5.4 (Na^+ adducts).

and silanols, and short EO chains more commonly observed in cyclic products [10].

GC-MS (electron impact, EI), ^{1}H and ^{13}C-NMR analysis indicated the cyclic trimer with all the EO chains cleaved [8; $m = 3$; R = H] as the major component of the heptane-soluble fraction of acid-degraded $M_2D-C_3-O-(EO)_n-CH_3$. The ^{1}H and ^{13}C spectra of the water-soluble fraction were indicative of linear disilanol structures, i.e. $D^{(OH)2}-C_3-O-(EO)_n-R$, **6**, or other linear disilanols (**7**; R^1, R^2 = H), and ^{29}Si-NMR studies indicated formation of cyclic and linear structures [10].

Determination of the recovered mass of the products following degradation and solvent partitioning showed a reduction from the starting mass, under all conditions investigated (varying concentrations, pH etc.) [10]. In theory mass recoveries should increase after hydrolysis due to addition of H_2O. However, the formation of volatile silicone compounds, as is well documented in the literature [25,30–33], would lead to mass loss. Analysis of the headspace of degraded samples of $M_2D-C_3-O-(EO)_n-CH_3$ by GC-MS demonstrated that the mass loss was due to the production of volatile species [10]. The most significant ions in the spectra (m/z 221) were assigned to the structure [{Si(CH$_3$)$_3$–O}$_2$–Si(CH$_3$)]$^+$ ([M$_2$D]$^+$) consistent with expected EI-MS fragmentation patterns [34–36].[6] The observation of this fragment ion indicates M_2D^R type volatile compounds are produced in the degradation of $M_2D-C_3-O-(EO)_n-CH_3$. The spectra of the other peaks observed in the chromatogram were also characteristic of silicone compounds, confirming the formation of various volatile silicone products, and thus providing an explanation for the observed mass loss in the degradation of the trisiloxane surfactant.

5.5.4 Stability

Initial degradation of silicone surfactants is thought to occur predominantly as hydrolyis of the silicone Si–O moiety, and this is known to be enhanced at acidic or alkaline pH values [3,11,12]. At neutral pH the surfactants are generally considered stable, but residual acidity or basicity of glassware is also known to catalyse their hydrolytic degradation [7]. Silicone polyether surfactants possessing the Si–O–C

---

[6] In general, the larger R substituents on the Si atom are cleaved preferentially by EI-MS. Fragmentation occurs readily and as such the parent molecules are often not observed.

structure, such as those used in polyurethane foam manufacture, are generally much less hydrolytically stable than the Si–C analogues. The trisiloxane surfactants however show increased hydrolytic instability over other Si–C surfactants, attributed to the relatively high water solubility, lack of micelle formation, and high proportion of terminal TMS groups [7,37]. The solubility and lack of micelle formation enable access to hydrolytic sites and kinetically the equilibrium of the hydrolytic reaction is driven by the high proportion of terminal TMS groups (M) due to the formation and subsequent transfer of M–M [7]. The hydrolytic instability of the trisiloxane surfactants, whilst favourable for environmental reasons, can limit their use in certain agrochemical formulations [11], and as such several reports have addressed variations in structure in order to improve their stabilities for this application. Limited improvements to the stability of the Si–O–Si substructure was achieved with variation of the alkyl spacer group and hydrophobic moeities of M_2D^R trisiloxane surfactants [38]. However, hydrolytic stability was reported for $[(CH_3)_3SiO]_2-(CH_3)Si-(CH_2)_3-O-(PO)_{1.5}-(EO)_8-(PO)_{1.5}-H$ [39]. Improved hydrolytic stability is also described for higher MW (>1000) silicone polyether copolymers, although these surfactants do not exhibit as efficient wetting properties [12]. Silane surfactants of the type $(CH_3)_3Si-(CH_2)_x-R$ and $[(CH_3)_3Si(CH_2)]_2-Si(CH_3)-X$ also show much higher stabilities than the silicone analogues [40–42], with comparable efficacy as adjuvants reported [40].

5.5.5 Quantitation

To date, mostly only non-selective methods have been used to monitor the degradation process of the silicone surfactants. Complete primary degradation of trisiloxane surfactants has been observed within 12 h at pH 3, with loss of the parent compound by GC used as the index [12]. However, at pH 6–8, stability over 40 days, as determined by surface tension [12], and over 2 years at high concentration with buffering to pH 7, as determined by spreading [3], has been reported. The use of HPLC online with inductively coupled plasma atomic emission spectroscopy (ICP–AES) to quantify the hydrolysis of aqueous trisiloxane dispersions through detection of the degradation product $(CH_3)_3Si(OH)$ has been described, although no experimental data were published [43]. Quantitative HPLC–LSD studies on the recoveries of

$M_2D-C_3-O-(EO)_n-CH_3$ (0.2% solutions) at varying pH values (pH 5.8, 3.2, 2.8), indicated half-life values of 19 days (by extrapolation), 2 h, and <1 h, respectively [10].

FIA–APCI-MS has been applied to the analysis of the degradation and rates of formation of selected degradation products of $M_2D-C_3-O-(EO)_n-CH_3$ in the presence of halloysite clay (0.1%, 10 mg g^{-1}) as shown in Fig. 5.5.8 [10].

The reduced recovery of $M_2D-C_3-O-(EO)_n-CH_3$ observed in the presence of halloysite was rapid and paralleled by the concomitant production of silanol monomer (**5** and **6**) and free PEG (**4**) products thereof. The persistence of these products over the duration of the experiment shows the need for further monitoring studies. The application of more informative online separation techniques such as HPLC–API-MS would also improve the quality of data obtainable. Other degradation products were not monitored, but the subsequent reduction of the response of the monomers is indicative of further transformation or immobilisation.

The rates and products for $M_2D-C_3-O-(EO)_n-CH_3$ hydrolysis as observed by preliminary FIA-MS analyses showed a dependence on pH,

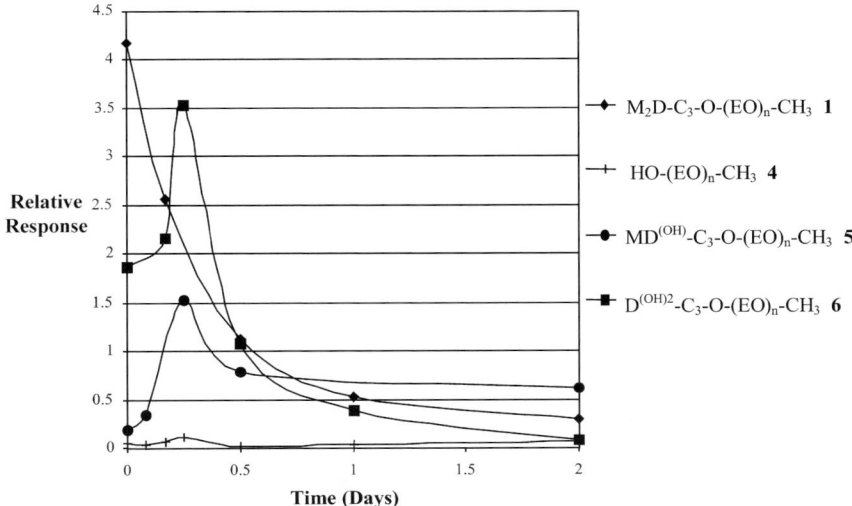

Fig. 5.5.8. Behaviour of $M_2D-C_3-O-(EO)_n-CH_3$ and selected degradation products on halloysite clay (pH 3.7) over two days in aqueous suspension by FIA–APCI-MS.

siloxane concentration and time [10]. Higher stability with higher siloxane concentration [12], and the influence of silanol concentration on rates and product formation as determined by spread areas, has also been described [3].

5.5.6 Ultimate fate

Environmental concentrations of polyether-modified silicone surfactants have not been measured, but considering that these materials account for 15% of the total organosilicone released to wastewater treatment and assuming sorption to particulate matter, it has been estimated that concentrations in sediments could be up to 4 ppm [43]. No experimental data exist regarding the ultimate environmental fate of silicone surfactants, and thus inferences can only be made according to that of structurally analogous compounds. The primary degradation of the trisiloxane surfactants to yield products including $(CH_3)_3SiOH$, $(HO)_2-Si(CH_3)-C_3-O-(EO)_n-CH_3$, $HO-(EO)_n-CH_3$ has been demonstrated, and it follows that the environmental impact may thus be as for the silicone and PEG constituents. The biodegradation [44–48] and environmental analysis [49] of PEG compounds are well documented elsewhere, and will not be elaborated on further here. Several studies have also been made on the environmental occurrence of poly(dimethyl)siloxane polymers [33,50] and in particular on their degradation on soils [30,51–54] and clay minerals [23–25].

Fate of silicones
Degradation of siloxanes is thought to proceed first via hydrolytic production of silanols. Low molecular weight water-soluble poly(dimethyl)siloxanes have been shown to be readily degraded photolytically in aqueous solution, with silicate as the final degradation product [55]. The degradation of high molecular weight silicones under aqueous conditions occurs less readily [33], and first requires depolymerisation to water-soluble silanols. Abiotic hydrolysis in contact with solid media is well documented, the major product of which appears to be the monomeric diol $[(CH_3)_2Si(OH)_2]$ [25,31–33,50–54]. The monomeric silanols are thought to be oxidised by photolytic demethylation in water, microbially in soil, and sunlight-induced hydroxyl radicals in the atmosphere following volatilisation. The results indicate that the ultimate degradation products are CO_2 and $Si(OH)_4$. Both occur

naturally in the environment, and the latter should precipitate to form soil minerals. Volatilisation of methylated silanol derivatives [i.e. $(CH_3)_2Si(OH)_2$, $(CH_3)_3SiOH$] [31–33], low molecular weight cyclic (i.e. D_3, D_4, D_5) [25,30,56] and linear TMS-endblocked MD_nM (i.e. TMS–O– TMS) [30] siloxanes and other degradation products [10] has also been described, and is considered a major elimination pathway for silicones from dry soil surfaces [4]. Volatile methyl siloxanes have also been detected in the offgases of landfills and wastewater treatment plants [57].

Fate of trisiloxane metabolites

The ultimate fate of higher alkyl silanols such as those produced in trisiloxane surfactant degradation, for example CH_3–$Si(OH)_2$–CH_2-CH_2CH_2OH, has not been described, and is an area requiring further investigation. The mechanisms described above for the degradation of the methyl siloxanes may or may not be applied to higher alkylated versions. The Si–C bond is not susceptible to hydrolysis [7], and as such the abiotic elimination processes are not likely to occur.

Biodegradation. The compound used in the biodegradation studies of silicone derivatives was the simple silanediol, $(CH_3)_2Si(OH)_2$ [58] and it is difficult to predict whether this elimination pathway will be applicable to the trisiloxane metabolites. The biodegradation of ethoxysilanes has also been described [4], but the ability of these microorganisms or others, to degrade the alkylated silanols formed from the silicone surfactants remains to be determined. Although studies are few, biological degradation of polyether-modified silicone surfactants on activated sludge has been reported as very limited to non-existent [43]. However, enhancement of the biodegradation of the monomeric dimethylsilanediol with the addition of the C_3 and C_4 compounds, acetone, 2-propanol and 2,3-butanediol has been observed [58], and may indicate that the soil-originating microorganisms (fungal and bacterial) responsible may also be capable of metabolising higher alkyl silanols.

Sorption. Any reduction in analyte recovery in the presence of solid media may be a reflection of losses through sorption or actual degradation, and in reality it is difficult to quantify degradation due to the unknown extent of the sorption processes. Sorption can be by intercalation (absorption), and/or electrostatic attraction and covalent bonding (adsorption), for which the latter requires degradation of the silicone to a silanol. Condensation of silanols with free hydroxyl groups (including other silanol functionalities) is a well-known and broadly utilized aspect of silicone chemistry with many of the applications of silicone compounds involving covalent bonding through this moiety as either linking agents or coatings [4]. The solid media that are modified are numerous, indicating that silylation of a broad range of substrates in the environment can also be expected. Cross-linking of halloysite clay to glass in the presence of the trisiloxane surfactant $M_2D-C_3-O-(EO)_n-CH_3$, indicated silanol formation from the surfactant, with confirmation obtained by mass recovery determinations, elemental analysis of the clay, and API-MS spectra of the supernatant [10]. The covalent binding of silanols to silicate substrates [23], soils [31–33,59, 60], sludge [43] and humic acids [4] has been described. However, little is known about the environmental fate of chemically bound silanols, and is an area for further research in this field.

Concentration effects

The low application rates of the trisiloxane surfactants in agricultural formulations are important positive factors with respect to subsequent degradation in the environment. Increased degradation rates have been observed with lower concentrations [12], and the condensation of the silanol degradation products is also concentration dependent [4]. At high concentrations in land applications, the covalent bonding of silicone surfactant derivatives could result in aggregation of silicate minerals, inhibiting drainage and/or affecting the sorption capacity of the soil. However, it is important to note that at the application rates used for agricultural adjuvants, this is not likely. Agrochemical formulations, generally comprising the adjuvants at 0.5%, are typically applied at rates of $200 \, \text{L ha}^{-1}$. This would result in a loading of only $1 \, \text{mg m}^{-2}$ of soil for the silicone surfactants, and as such detrimental cumulative effects should be minimal.

Moisture effects

Moisture levels are also expected to influence the degradation process of the silicone surfactants. The degradation of surfactants on soils and peat is known to increase with soil moisture content [61], and this will also be expected for the primary degradation (hydrolysis) of silicone surfactants. Further studies of these compounds under different moisture levels should be conducted to fully determine degradation rates in the presence of solid media. The drying/rewetting cycles occurring in natural soil environments should, however, be conducive to silicone surfactant degradation [33] enabling hydrolysis and elimination through evaporation of volatile products.

Inhibition of polydimethylsiloxane hydrolysis on soils [52,60] and clays [25] by high moisture levels has been described, which is in general attributed to their low water solubility. This is not expected to be applicable to the more soluble surfactant analogues, however, which show increased water solubility with degradation [10,12,15], and which are used at concentrations much lower than are required for the formation of hydrophobic degradation products.

5.5.7 Conclusion

The hydrolytic degradation of organosilicone surfactants and derivatives can occur readily. The range of products potentially formed are numerous as a result of the ability of the silicone moiety to rearrange into a variety of lengths and structures, and compounded by the large number of components constituting the commercial formulations. Unequivocal assignment of the degradation product mixture is thus complicated, although application of API-MS, FTICR-MS, GC-MS, HPLC, HPLC–API-MS and NMR techniques as described here have provided new insights into this process. However, due to the highly complicated degradation pathways, the application of MS-MS and online HPLC separation methods is recommended.

The results indicate that persistence of organosilicone surfactants in the parent molecule form will be limited on typical soil media and in aqueous environments. Reduced recovery was considered to be a result of abiotic degradation and/or strong sorption processes. Losses were most significant on solid media exhibiting extreme pH values and were also enhanced in the presence of clay substrates. Studies on clays indicated that pH, potential for intercalation and surface charges are important factors in the removal process.

The available data summarised in this review support previous conclusions that trisiloxane silicone surfactants should be relatively benign in the natural environment. Primary degradation is rapid and ultimate degradation to naturally occurring compounds, i.e. CO_2, H_2O and $Si(OH)_4$, is indicated, though not yet demonstrated unequivocally.

In particular, the low application rates used for these compounds in agricultural formulations are an important positive factor which should contribute significantly to the elimination process, and to the minimisation of any potential negative accumulative effects.

ACKNOWLEDGEMENTS

Financial support from the Public Good Science Fund (PGSF) of New Zealand, New Zealand Federation of University Women (NZFUW), New Zealand Vice Chancellors Committee (NZVCC), Waikato University and the New Zealand Lotteries Commission are gratefully acknowledged. Thanks also to Dr G. Willett of the University of New South Wales, Australia, for kindly providing access to FTICR-MS and to Dr G. Policello, Dr P. Stevens and Dr M. Costello of Witco Corporation for generous samples and information.

REFERENCES

1 J.A. Leeks, W.J. Adams, P.D. Guiney, J.F. Hall and C.G. Naylor, *Proc. 4th World Surfactants Congress*, Barcelona, Vol. 3, 1996, p. 276.
2 D. Coupland, J.A. Zabkiewicz and F.J. Ede, *Ann. Appl. Biol.*, 115 (1989) 147.
3 G.A. Policello, P.J. Stevens, W.A. Forster and G.J. Murphy, In: F.R. Hall, P.D. Berger, and H.M. Collins (Eds.), *Pesticide Formulations and Application Systems*, Vol. 14, ASTM publishers, 1995, p. 313.
4 J.L. Spivack, E.R. Pohl and P. Kochs, In: G. Chandra (Ed.), *Organosilicon Materials*, The Handbook of Environmental Chemistry, Vol. 3H, Springer-Verlag, Berlin Heidelberg, 1997, pp. 105, Chapter 5.
5 J.A. Hatcher, G.S. Slater, D. Wischer, C. Stevens, Y. Miyakawa, G. Chandra, L.D. Maxim and T. Sawano, In: G. Chandra (Ed.), *Organosilicon Materials*, The Handbook of Environmental Chemistry, Vol. 3H, Springer-Verlag, Berlin, Heidelberg, 1997, Chapters 9–12.
6 R.B. Allen, P. Kochs and G. Chandra, In: G. Chandra (Ed.), *Organosilicon Materials*, The Handbook of Environmental Chemistry, Vol. 3H, Springer-Verlag, Berlin Heidelberg, 1997, pp. 1–25, Chapter 1.
7 R.M. Hill, In: R.M. Hill (Ed.), *Silicone Surfactants*, Surfactant Science Series, Vol. 86, Marcel Dekker, New York, 1999, pp. 1–47, Chapter 1.
8 D.R. Karsa, *Chem. Ind.*, 9 (1998) 685.
9 G. Chandra, L.D. Maxim and T. Sawano, In: G. Chandra (Ed.), *Organosilicon Materials*, The Handbook of Environmental Chemistry, Vol. 3H, Springer-Verlag, Berlin, Heidelberg, 1997, pp. 295, Chapter 12.

10 L.S. Bonnington, Analysis of Organosilicone Surfactants and their Degradation Products, PhD Thesis, University of Waikato, Hamilton, New Zealand, 2001.

11 M. Knoche, *Weed. Res.*, 34 (1994) 221.

12 M. Knoche, H. Tamura and M.J. Bukovac, *J. Agric. Food Chem.*, 39 (1991) 202.

13 P.J.G. Stevens, *Pestic. Sci.*, 38 (1993) 103.

14 G. Kuhnt, *Environ. Toxicol. Chem.*, 12 (1993) 1813.

15 E.F.C. Griessbach, A. Copin, R. Deleu and P. Dreze, *Sci. Total Environ.*, 221 (1998) 159.

16 P.J.G. Stevens, In: R.E. Gaskin (Ed.), *Proc. 4th Int. Symp. on Adjuvants for Agrochemicals*, 1995, pp. 345–349, Melbourne, NZ FRI Bulletin No. 193.

17 E.F.C. Griessbach, A. Copin, R. Deleu and P. Dreze, *Sci. Total Environ.*, 201 (1997) 89.

18 J.B. Weber and H.D. Coble, *J. Agr. Food Chem.*, 16 (1968) 475.

19 B.K.G. Theng, *Developments in Soil Science –Formation and Properties of Clay-Polymer Complexes*, Vol. 9, Elsevier, New York, 1979.

20 D.B. Knaebel, T.W. Federle, D.C. McAvoy and J.R. Vestal, *Environ. Toxicol. Chem.*, 15 (1996) 1865.

21 D.B. Knaebel, T.W. Federle, D.C. McAvoy and J.R. Vestal, *Appl. Environ. Microbiol.*, 60 (1994) 4500.

22 A. Torrents and S. Jayasundera, *Chemosphere*, 35 (1997) 1549.

23 M.G. Voronkov, V.P. Milesshkevich and Y.A. Yuzhelevskii, *The Siloxane Bond: Physical Properties and Chemical Transformations*, Consultants Bureau, New York, 1978.

24 S. Xu, R.G. Lehmann, J.R. Miller and G. Chandra, *Environ. Sci. Technol.*, 32 (1998) 1199.

25 S. Xu, *Environ. Sci. Technol.*, 32 (1998) 3162.

26 B.K.G. Theng, *The Chemistry of Clay-Organic Reactions*, Adam Hilger Ltd, London, 1974.

27 B. Velde, *Developments in Sedimentology—Clay Minerals*, Vol. 40, Elsevier, New York, 1985.

28 P.L. Singleton, DSIR Land Resources Scientific Report No. 5, 1991.

29 T.C. Kendrick, B. Parbhoo, J.W. White, In: S. Patai, and Z. Rappoport (Eds.), *The Silicon–Heteroatom Bond*, John Wiley and Sons Ltd, 1991, pp. 67–150, Chapter 3.

30 R.R. Buch and D.N. Ingebrigtson, *Environ. Sci. Technol.*, 13 (1979) 676.

31 R.G. Lehmann and J.R. Miller, *Environ. Toxicol. Chem.*, 15 (1996) 1455.

32 R.G. Lehmann, S. Varaprath and C.L. Frye, *Environ. Toxicol. Chem.*, 13 (1994) 1753.

33 C. Stevens, *J. Inorg. Biochem.*, 69 (1998) 203.

34 U. Just, F. Mellor and F. Keidel, *J. Chromatogr. A*, 683 (1994) 105.

35 A. Ballistreri, D. Garozzo and G. Montaudo, *Macromol.*, 17 (1984) 1312.

36 J.A. Moore, In: A. Lee Smith (Ed.), *The Analytical Chemistry of Silicones*, John Wiley and Sons Ltd, 1991, pp. 421–470, Chapter 12.

37 B. Gruning and G. Koerner, *Tenside Surf. Det.*, 26 (1989) 312.

38 K.D. Klein, W. Knott, G. Koerner, In: N. Auner, and J. Weis (Eds.), *Organosilicon Chemistry II: From Molecules to Materials*, VCH, 1996, pp. 613–618.

39 G. Feldmann-Krane, W. Hoehner, D. Schaefer and S.S. Dietmar, DE Patent, 4 (1993) 317 605.

40 K.D. Klein, S. Wilkowski and J. Selby, *Proc. 4th Int. Symp. Adj. for Agrochem.*, FRI Bulletin No. 193, 1995, pp. 27–31.

41 K.D. Klein, D. Schaefer and P. Lersch, *Tenside Surf. Det.*, 31 (1994) 115.

42 A.R.L. Colas, F.A.D. Renauld and G.C. Sawicki, GB Patent, 8 (1988) 819–567.

43 D.E. Powell, J.C. Carpenter, In: G. Chandra (Ed.), *Organosilicon Materials*, The Handbook of Environmental Chemistry, Vol. 3H, Springer-Verlag, Berlin, Heidelberg, 1997, pp. 225–239, Chapter 8.

44 G.F. White, N.J. Russell and E.C. Tidswell, *Microbiol. Rev.*, 60 (1996) 216.

45 D.P. Cox, *Adv. Appl. Microbiol.*, 23 (1978) 173.

46 J.R. Haines and M. Alexander, *Appl. Microbiol.*, 29 (1975) 621.

47 T. Balson, M.S.B. Felix, In: D.R. Karsa, and M.R. Porter (Eds.), *Biodegradable Surfactants*, Blackie, Glasgow, UK, 1995, p. 243.

48 S.J. Gonsio and R.J. West, *Environ. Toxicol. Chem.*, 14 (1995) 1273.

49 T. Schmitt, *Analysis of Surfactants*, Surfactant Science Series, Vol. 96, Marcel Dekker Inc, New York, 2001, Chapter 18, pp. 562–582.

50 N.J. Fendinger, D.C. McAvoy, W.S. Eckhoff and B.B. Price, *Environ. Sci. Technol.*, 31 (1997) 1555.

51 J.C. Carpenter, J.A. Cella and S.B. Dorn, *Environ. Sci. Technol.*, 29 (1995) 864.

52 R.G. Lehmann, J.R. Miller, S. Xu, U.B. Singh and C.F. Reece, *Environ. Sci. Technol.*, 32 (1998) 1260.

53 S. Varaprath and R.G. Lehmann, *J. Environ. Polym. Degrad.*, 5 (1997) 17.

54 R.G. Lehmann, *Environ. Toxicol. Chem.*, 12 (1993) 1851.

55 C. Anderson, K. Hochgeschwender, H. Weidemann and R. Wilmes, *Chemosphere*, 16 (1987) 2567.

56 J.F. Hobson, R. Atkinson, W.P. Carter, In: G. Chandra (Ed.), *Organosilicon Materials*, The Handbook of Environmental Chemistry, Vol. 3H, Springer-Verlag, Berlin, Heidelberg, 1997, pp. 137–179, Chapter 6.

57 J.C. Carpenter, R. Gerhards, In: G. Chandra (Ed.), *Organosilicon Materials*, The Handbook of Environmental Chemistry, Vol. 3H, Springer-Verlag, Berlin, Heidelberg, 1997, pp. 27–51, Chapter 2.

58 C.L. Sabourin, J.C. Carpenter, T.K. Leib and J.L. Spivack, *Appl. Environ. Microbiol.*, 62 (1996) 4352.

59 R.G. Lehmann, S. Varaprath, R.B. Annelin and J.L. Arndt, *Environ. Toxicol. Chem.*, 14 (1995) 1299.
60 G. Lehmann, S. Varaprath and C.L. Frye, *Environ. Toxicol. Chem.*, 13 (1994) 1061.
61 N. Valoras, J. Letey, J.P. Martin and J. Osborn, *Soil Sci. Soc. Am. J.*, 40 (1976) 6.

Chapter 6

Occurrence of surfactants in the environment

6.1 CONCENTRATIONS OF SURFACTANTS IN WASTEWATER TREATMENT PLANTS

Mira Petrovic and Damia Barceló

6.1.1 Introduction

Currently, under optimised conditions, more than 90–95% of surfactants can be eliminated by conventional biological wastewater treatment. Even if such high elimination rates are achieved, the principal problem is the formation of recalcitrant metabolites out of the parent surfactants. Some compounds, e.g. alkylphenol ethoxylates (APEOs), are eliminated incompletely, forming metabolites that are more resistant to further degradation and more toxic than parent compounds. Discharges of sewage effluent into rivers and lakes and the disposal of sludge onto soil are likely to cause diffuse contamination in the aquatic environment. In order to reduce the concentration of surfactants and metabolites in wastewater effluents and to increase and complete their elimination in treatment processes, new biochemical and physical procedures are presently being investigated. However, the majority of the wastewater treatment plants (WWTPs) in Europe still employ physicochemical and biological treatment, while additional treatment practice occurs only in Scandinavian countries.

There is no complete knowledge about the degradation pathway of the main surfactants and this is especially so of some less used or new

Comprehensive Analytical Chemistry XL
T.P. Knepper, D. Barceló and P. de Voogt (Eds)

surfactants. As a consequence, very little data about their occurrence in wastewaters and elimination in WWTP are available. One reason for this is the very difficult analysis of the complex sewage water matrix. On the other hand, some surfactants, such as linear alkylbenzene sulfonates (LASs) and APEOs, have been frequently and systematically analysed during the past two decades.

In this chapter, we have summarised the available data for major surfactants and their metabolites in WWTPs. Since the use of some poorly biodegradable surfactants has been reduced considerably and new surfactants fulfilling requirements for environmental acceptability are now introduced into daily practice, the present chapter lists mainly data generated in the last 5–6 years.

6.1.2 Ionic surfactants

Anionic surfactants

Linear alkylbezene sulfates, their co-products and metabolites. Several extensive monitoring programmes were conducted with the objective to determine the concentrations of LAS in raw and treated wastewaters and removal during sewage treatment. An overview of the concentrations of LAS found in raw WWTP influents, secondary effluents and digested sludges in Europe and the United States is given in Table 6.1.1. The reported concentrations in raw influent ranged from 0.5 to more than 15 mg L^{-1}. The removal efficiency was generally higher than 95% (average removal range was 98.5–99.9%) and the corresponding effluent values were from <5 to 1500 µg L^{-1}. The results of five national pilot studies on LAS [1] showed that the differences observed in the main operating characteristics of the WWTP (i.e. treatment type, plant size, sludge retention time, hydraulic retention time and temperature) were not found to greatly influence the removal of LAS.

The homologue distribution shift of LAS during in-sewer degradation and wastewater treatment is repeatedly reported. In raw WWTP influent the relative concentration of the higher homologues C$_{12}$-LAS and C$_{13}$-LAS is lower, as compared to the original distribution in detergent formulations. This is likely due to a partial removal of these species during transportation of the wastewater in the sewage system, which also may explain the low levels detected in some influents, since the longer alkyl-chain homologues exhibit both higher degradation rate and adsorption tendency due to higher lipophilicity. The greater

TABLE 6.1.1

Reported concentrations of LAS (SPC value is given in parentheses) in WWTP influent, effluent and sludge

Country	Influent (μg L^{-1})	Effluent (μg L^{-1})	Sludge (mg kg^{-1})	Reference
Spain	2020–2170 (6–19)	10–91 (49–158)	–	[8]
	988–1309	136–197	–	[36]
	–	–	2060–11 700	[37]
	1423	17.8	–	[38]
Switzerland	–	–	2900–11 900	[39]
Italy	4600	68	6000	[1]
	3140–8400 (35–304)	13–115 (80–260)	–	[5]
	3400–10 700	21–290	–	[6]
		(220–1150)	–	[9]
Germany	1900–8000	65–115		[40,41]
	–	–	16–11 900	[42,43]
The Netherlands	3400–8900	19–71	3400–5930	[10,44]
UK	15 100	10	8300[a] 300[b]	[1]
	1850–5580	40–1090	–	[45]
Portugal	–	–	7700	[37]
USA	1800–7700	<1–1500	3100–15 600	[7,25,46]

[a]Primary sludge.
[b]Activated sludge.

proportion of LAS in wastewater is associated with the suspended solids, representing nearly 60% of the incoming LAS [2]. The homologue distribution of dissolved LAS has a lower molecular weight distribution than adsorbed or precipitated LAS. The average chain length of dissolved LAS in wastewater entering WWTP is 11.3 (measured in three treatment plants in Spain), whereas that of the adsorbed/precipitated LAS is 12.0. The in-sewer removal of LAS is estimated to be 10–68% depending on the length of the sewer, travel time and the degree of microbiological activity in the sewer [3]. In WWTP, a substantial percentage (15–35%) of LAS is physically removed by precipitation during the primary settling step. It is not exposed to any biological treatments as this precipitate (primary sludge) follows anaerobic digestion processes together with the bulk of the sludge

657

produced. LAS concentration in anaerobically digested sludge was found to be much higher than in aerobically stabilised sludge. Prat et al. [2] found that the LAS concentrations in non-digested sludge from three WWTP in Spain were between 8.4 and 14.0 mg g^{-1} (average 12.6 mg g^{-1}), whereas after anaerobic digestion they ranged from 12.1 to 18.8 mg g^{-1} (average 15.8 mg g^{-1}). During the aerobic fermentation process, the LAS content dropped to 6.0 mg g^{-1}.

Dialkyltetralin sulfonates (DATS) are alicyclic by-products of commercial LAS. They represent 0.4–14% of the corresponding LAS [4]. There have been only a few studies on the occurrence of DATS in WWTP. The levels of DATS in raw influent of four WWTP located in the area of Rome, Italy, ranged from 310 to 1200 µg L^{-1} [5,6]. In treated water, concentrations ranged from 6 to 106 µg L^{-1} with an average removal efficiency of 95%. These values are congruent with data which formed a US survey [7]. In all the studies, an increase of the DATS-to-LAS concentration ratio in the final effluent was found, due to the fact that DATS are significantly more resistant to biodegradation than LAS. Removal of DATS in an activated sludge treatment was found to be 95%, while in trickling filters the removal efficiency was only 63%.

The degradation of LAS, as it occurs in the environment and WWTP, yields sulfophenyl carboxylates (SPC), mostly with 6–10 carbon atoms. The active processes of biodegradation occurring in the sewer system are confirmed by the detection of SPC in the influent samples. Three WWTP in Spain were monitored for the occurrence of LAS and their degradation products [8]. SPC$_6$ to SPC$_{12}$ were detected in raw influents in total concentrations of 6–19 µg L^{-1}. Higher levels between 49 and 158 µg L^{-1} were found in the corresponding effluents after microbial formation from LAS. The most abundant species were short- to mid-chain intermediates (SPC$_7$–SPC$_9$). DiCorcia and Samperi [9] determined significantly higher levels of SPC (calculated as a sum of the various isomers and homologues of SPC coming from both LAS and DATS) in treated sewage samples leaving the WWTP at Ostia, Italy. Concentrations ranged from 220 to 1150 µg L^{-1}. In another study, monocarboxylated and dicarboxylated metabolites of LAS and DATS were monitored separately in three activated-sludge WWTP from the area of Rome [5]. The concentration of mono- and dicarboxylated SPC in influents ranged from 35 to 304 µg L^{-1} and from 7 to 60 µg L^{-1}, respectively. In effluent, concentrations ranged from 80 to 260 µg L^{-1} (mono-carboxylated SPC) and from 20 to 106 µg L^{-1} (dicarboxylated SPC).

The total concentration of carboxylated metabolite of DATS ranged from 9 to 115 μg L^{-1} in influents and from 152 to 390 μg L^{-1} in treated wastewater.

Alkyl ether sulfates (AES) and alkyl sulfates (AS). Another important group of anionic surfactants are AES, which are used in laundry and cleaning detergents as well as in cosmetic products. The typical commercial AES have an alkyl-chain length distribution of 12–18 C units and an ethoxylene distribution from 0 to 12, with an average of three ethylene oxide units. Concentrations of AES in WWTP influents and effluents are summarised in Table 6.1.2. The maximal reported concentration in the raw municipal wastewater was 6000 μg L^{-1}, while in treated water it ranged from 3 to 15 μg L^{-1} (a value of 164 μg L^{-1} was measured in an overloaded trickling filter plant). Eichhorn et al. [8] studied the change in AES concentration in a WWTP that discontinuously receives heavily contaminated wastewater from a company formulating personal care products, such as shower gels, shampoos and foam baths. Whereas in normal domestic sewage the levels of AES, showing a daytime-dependent profile influenced by the washing and cleaning activities of the population, vary between 0.27 and 4.6 mg L^{-1}, a very strong increase with concentration can be observed when the influent is polluted with industrial wastewater. The AES in the grab samples taken from the WWTP influent during the discharge time were found to be 46 and 298 mg L^{-1}, respectively.

The commercial AES mixture contains up to 20% of non-ethoxylated compounds and alkyl sulfates (AS). AS is also applied as a main

TABLE 6.1.2

Concentrations of alcohol ether sulfates (AES) in WWTP influent and effluent

Country	Total AES (μg L^{-1})		References
	Influent	Effluent	
USA	105–2180	4–164	[25]
The Netherlands	1200–6000	3–11.5	[3,10]
Germany	270–4600[a]	–	[8]
	46 000–298 000[b]		

[a]Domestic wastewater.
[b]Discontinuous load of industrial wastewater (surfactant formulating company).

surfactant in laundry and cleaning formulations. This widespread use is reflected by their ubiquity in wastewaters entering sewage treatment works. Feijtel et al. [10] determined range and average WWTP removal and concentrations of AS in influent and effluent of seven WWTPs in The Netherlands. In influent AS, concentrations ranged from 100 to 1300 µg L^{-1} and in effluent from 1.2 to 12.1 µg L^{-1}, with an average removal of 99.2%. The average alkyl-chain length was 12.3. These levels are in good agreement with the values found in German WWTP [11], where elimination rates were higher than 99.9% and AS daytime-dependent influent concentrations ranged from < 20 to 620 µg L^{-1}, with peak values in the evening at 21:00 h.

Cationic surfactants

Cationic surfactants are used widely as fabric softeners, anti-static agents, antiseptic components and as the main ingredients in rinse conditioner products. However, there is little data on their occurrence in wastewater and removal during wastewater treatment. The concentrations of some cationic surfactants and their biodegradation intermediates, found in WWTP influents and effluents, are listed in Table 6.1.3.

TABLE 6.1.3

Concentrations of cationic surfactants in WWTP influent and effluent

Country	Surfactant	Influent (µg L^{-1})	Effluent (µg L^{-1})	Reference
UK	DEQ diol quat	49–132[a]	< 1.2–6.1	[12]
	MTEA	6–17[a]	< 1.5–19	
	DEDMA	99–263[a]	< 2.7–11	
The Netherlands	DEQ diol quat	240–325[b]	0.7–2.2	[12]
	MTEA	216–272[a]	0.5–3.9	
	DEDMA	51–64[a]	< 0.1–3	
Germany	C$_{18}$ monoalkyl quaternaries of DEQ and DEEDMAC	2.4 ± 0.1	0.02 ± 0.01	[13]
	Sum of diester homologues (C$_{16/16}$, C$_{16/18}$, C$_{18/18}$)	140 ± 4.3	0.40 ± 0.04	

[a]Predicted concentration in a raw sewage for non-esterified quaternary intermediates calculated from per capita use, population served and prevailing plant flows.
[b]Measured.

Waters et al. [12] monitored the cationic degradation intermediates in six WWTP in UK and The Netherlands. This study focuses on 2,3-dihydroxypropanetrimethyl ammonium chloride (Lever's diester quat, DEQ diol quat), methyltrihydroxyethyl ammonium chloride (MTEA) and dimethyldihydroxyethyl ammonium chloride (DMDEA), formed as transitory species during the biodegradation of the ditallow esters of Lever's diester quat (DEQ), tri(hydroxyethyl)methyl ammonium methosulfate (Esterquat-EQ) and di(hydroxyethyl)dimethyl ammonium chloride (DEEDMAC). The limited monitoring for DEQ in The Netherlands indicated high removal through sewage treatment (>99%). The removals for all three cationic degradation intermediates were >95%. Radke et al. [13] also reported very efficient removal of DEQ and DEEDMAC (>99%) in the WWTP, Bayreuth, Germany.

6.1.3 Non-ionic surfactants

Alkylphenol ethoxylates (APEO) and their degradation products
APEOs are among the most widely used surfactants, although their environmental acceptability is strongly disputed. The widespread occurrence of APEO-derived compounds in treated wastewaters raises concerns about their impact on the environment following disposal of effluents into aquatic system. Studies have found that their neutral and acidic metabolites are more toxic than the parent compounds and possess the ability to mimic natural hormones by interacting with the estrogen receptor (see Chapter 7.3). These findings have raised public concern and a voluntary ban on APEO use in household cleaning products began in 1995, and restrictions on industrial cleaning applications in 2000 [14]. In Western Europe and the USA, the APEOs in household detergents have been completely replaced by alcohol ethoxylates (AEO). As a result, the general trend of declining APEO concentrations was observed, especially in the Scandinavian countries, Denmark, The Netherlands, Switzerland and UK. For example, concentrations measured in Swiss effluents prior to the voluntary ban on NPEO surfactants in laundry detergent formulations ranged from <10 to 200 μg L^{-1} (sum of NP, NPEO$_1$ and NPEO$_2$), while present-day levels (data from 1998) show a 10-fold decrease [15]. However, mainly because of its lower price, APEOs are still being used in substantial amounts in institutional and industrial applications.

De Voogt et al. [16] and Belfroid et al. [17] compared the levels of APEOs in domestic and industrial wastewaters in The Netherlands. The average concentration of APEOs, in 12 samples of domestic wastewater, collected in 1999, was 32.3 µg L^{-1}, while in three samples of industrial wastewater the average concentration reached 7600 µg L^{-1}.

Wastewater-derived alkylphenolic compounds have been studied extensively. The concentrations of nonylphenol ethoxylates (NPEOs), as the strongly prevalent sub-group of APEOs, determined in the influents of WWTPs (Table 6.1.4), varied widely among various WWTPs from <30 to 1035 µg L^{-1}. However, values can go up to 22 500 µg L^{-1} in industrial wastewaters (especially from tannery, textile, pulp and paper industry). Levels of octylphenol ethoxylates (OPEOs) are significantly lower, comprising approximately 5–15% of total APEOs in WWTP influents, which is congruent with their lower commercial use.

In 40 WWTP all over Japan, primary effluents were found to contain mainly NPEOs with longer ethoxy units ($n_{EO} = 4$–8), while NPECs accounted for <5% (mol mol^{-1}) of total nonylphenolic compounds [18]. The average removal of nonylphenolic compounds was approximately 60% after biological treatment and 70% after full treatment (almost all studied WWTPs have a final disinfection process).

NPEO metabolites, nonylphenol (NP) and nonylphenoxy carboxylates (NPECs) could be already detected in WWTP influents at levels up to 40 µg L^{-1} (high values up to 250 µg L^{-1} are detected in industrial wastewaters). The presence of NP and NPECs in the influents cannot only be attributed to the metabolism of NPEOs, but also their entry in WWTP resulting from application in other fields, has to be taken into account. For example, NP is also used as an ingredient of pesticide formulations and NP$_1$EC as a corrosion-inhibiting agent. A comprehensive monitoring of NP in effluents from eight WWTP with activated sludge secondary treatment in the USA detected concentrations from 1.2 to 23 µg L^{-1} [19], while the maximum concentration found in 40 WWTP in Japan did not exceed 1.7 µg L^{-1} [18].

As for the elimination efficiency of NPEO during wastewater treatment, values from 81 to 99.5% were calculated. However, the removal of parent NPEOs led to the formation of transformation products that are much more resistant to microbial degradation and overall elimination of nonylphenolic compounds during sewage treatment is very low. Ahel et al. [20] estimated that approximately 60–65% of all nonylphenolic compounds introduced to WWTP are discharged into the environment: 19% in the form of carboxylated

TABLE 6.1.4

Concentration ranges ($\mu g\,L^{-1}$) of NPEOs and their metabolites detected in influents and effluents of WWTPs

Country	NPEO Influent	NPEO Effluent	NPEC Influent	NPEC Effluent	NP Influent	NP Effluent	Reference
Spain	33–820	<0.2–49	<0.2–14[a]	13–113	<0.5–22	<0.5–21	[8,29,32]
	27–2120[b]	10–24	<0.4–219	13–33	17–251[b]	15–225[b]	[33,38]
	140	2	–	–	58	0.65	[47]
Switzerland	96–430	<0.1–35	–	4–16	–	<0.1–3.8	[20,21,48,49]
Italy	–	–	–	–	2–40	0.7–4	[9]
	29–145	1.7–6.6	–	0.6–15	–	–	[22]
	127–221	2.2–4.1	–	–	–	–	[27]
Germany	120–270	24–48	–	1.7–5.4	–	0.3–1	[40]
The Netherlands	<LOD–125[c]	<LOD–2.2[c]	–	–	<LOD–19[c]	<LOD–1.5[c]	[16,17]
	50–22.500[d]				<LOD–40[d]		
United Kingdom	–	1–60	–	–	–	<0.2–5.4	[50]
	–	45 ± 16[e]	–	–	–	3 ± 0.85	[51]
Denmark	30	<0.5	–	–	–	<0.5	[52]
	25	2.4	–	–	–	–	[53]
Belgium	–	–	–	–	6 (122)[f]	<1.0	[54]
Sweden	–	0.5–27[g]	–	–	–	0.5–3	[55]

(continued on next page)

TABLE 6.1.4 (continued)

Country	NPEO		NPEC		NP		Reference
	Influent	Effluent	Influent	Effluent	Influent	Effluent	
USA	–	<1.0–330	–	–	–	–	[56]
				142–272[h]		<1.0–33	[57]
	–	160–460	–	–	–	11–23	[58]
							[59]
Canada						0.8–15	[60]
Japan	–	<0.1–60 ($n_{EO} = 1–3$) <0.1–245 ($n_{EO} = 4–18$)	–	<0.1–1119		<0.1–1.7	[18,61]

WWTP receiving mainly industrial wastewater, data 1999.

[a]NPE$_1$C.
[b]WWTP receiving >60% of industrial wastewater (mainly textile).
[c]WWTP receiving domestic wastewater, data 1999.
[d]Industrial WWTP, data 1999.
[e]NPEO$_1$.
[f]Effluent of a textile plant.
[g]NPEO$_1$ and NPEO$_2$.
[h]NPE$_{1–4}$C.

derivatives, 11% as lipophilic NP_1EO and NP_2EO, 25% as NP and 8% as untransformed NPEO. Furthermore, they also estimated that 92–96% of NP is discharged from the WWTP via digested sludge and only 4–8% via secondary effluents. However, in some WWTPs, especially those receiving substantial contributions from the textile or tannery industries, very high concentrations of NP were measured in both sewage sludge and secondary effluents (up to 400 $\mu g\,L^{-1}$).

The oligomeric distribution of NPEOs in primary effluent is characterised by a typical mixture that was slightly altered by biological degradation. Besides a maximum at NPEOs with 8–10 EO units (typical commercial mixture), another maximum appears at lower oligomers ($n_{EO} = 1$–3), which are considered as stable metabolic products. Ahel et al. [21] found that the elimination of higher oligomers was almost complete in an activated sludge treatment, and higher oligomers ($n_{EO} > 8$) were not present in the secondary effluents. Biologically treated wastewater contained only traces of oligomers with $n_{EO} = 3 - 8$, and the maximum concentration centred at more lipophilic NP_2EO.

More polar NPECs are discharged mainly via secondary effluents and concentrations reported were approximately 2–10 times higher than in primary effluents. DiCorcia et al. [22] determined the concentration of carboxylated alkylphenoxy carboxylates (CAPECs) in treated effluents of five major activated sludge WWTP in Rome. Unexpectedly, contrary to the general belief that NPECs are refractory metabolites, the results showed that CAPECs were the dominant products of the NPEO biotransformation. The concentrations were as follows: 2.5–24 $\mu g\,L^{-1}$ (CA_8PEC), 0.3–2.0 $\mu g\,L^{-1}$ (CA_7PEC), 2.3–16 $\mu g\,L^{-1}$ (CA_6PEC), 0.3–2.1 $\mu g\,L^{-1}$ (CA_5PEC), 0.1–1.4 $\mu g\,L^{-1}$ (CA_4PEC) and 0.1–0.6 $\mu g\,L^{-1}$ (CA_3PEC). By averaging data relative to the five WWTP over 4 months, relative abundances of NPEO ($n_{EO} = 1$ and 2), NPECs and CAPECs were found to be 10 ± 2, 24 ± 5 and $66 \pm 7\%$, respectively.

Fujita et al. [18] studied formation of halogenated (chlorinated and brominated) NPEOs and NPECs during wastewater treatment. Halogenated derivatives were found to be produced during the disinfection processes by chlorination accounting for up to 10% of total nonylphenolic compounds. They were found in 25 of 40 WWTPs with concentrations up to 6.5 $\mu g\,L^{-1}$ in secondary effluent and 52.4 $\mu g\,L^{-1}$ in final effluent. Of all halogenated compounds, BrNPECs ($n_{EO} = 1$–2) were found to be the most abundant.

High sorption coefficients for lipophilic nonylphenolic compounds (NP, $NPEO_1$ and $NPEO_2$) strongly influence their mass flows during sewage treatment. The high portion of surfactants is sorbed on the solid fraction of a WWTP sample and it should be added to the water-dissolved fraction or analysed separately. De Voogt et al. [16] determined the contributions of the solid and aqueous fractions to the total concentration of NP and NPEOs in WWTP samples. The results showed that a significant portion of NP remains in the solid phase and the contribution of NP in the water fraction of influent, effluent and primary sludge was 52, 86 and 9%, and of NPEOs 82, 86 and 11%, respectively.

An overview of the concentrations of nonylphenolic compounds in digested sludge is shown in Table 6.1.5. The NPEO metabolites were found with concentrations ranging from lower $mg\,kg^{-1}$ values to lower $g\,kg^{-1}$ values. A new draft proposal limits the concentrations of total nonylphenolic compounds (sum of NP, $NPEO_1$ and $NPEO_2$) in sludge disposed on soil, e.g. as fertiliser, at $50\,mg\,kg^{-1}$. To date, significantly higher concentrations were found in a majority of WWTP sludges analysed, limiting disposal options and implying an environmental risk

TABLE 6.1.5

Reported concentration ranges $(mg\,kg^{-1})$ of NPEOs and their metabolites detected in digested sewage sludge

Country	NPEO	NPEC	NP	Reference
Spain	17–86	<10–48[a]	14–74	[33]
	21–135	<0.1–14[a]	172–600	[37]
Switzerland	–	–	80–500	[23]
Germany	21	–	2.5–300	[37]
	–	–	2.5–3.7	[62]
The Netherlands	0.7–2390[b]	–	<LOD–125[b]	[16,17]
	<LOD–1410[c]		<LOD–2480[c]	
Portugal	2.1	10	234	[37]
Denmark	43.6[d]	–	–	[53]
UK	–	–	255–823	[63]
USA	–	27–113[e]	368	[24,64]
Canada	41–564	<dl–65[f]	137–470	[60,65]

[a]NP_1EC.
[b]Domestic WW(TP), data 1997.
[c]Industrial WW(TP), data 1997.
[d]Primary sludge.
[e]$NP_{1-4}EC$.
[f]$NP_1EC + NP_2EC$.

because of diffuse pollution that is likely to occur when such sludge is applied to soil.

Adsorption of NP onto primary and secondary sewage sludge, and its formation from NPEOs during anaerobic stabilisation of sludge, results in extremely high concentrations of NP in anaerobically digested sludge. The high levels of NP (more than 1 g kg^{-1}) are consistently found in anaerobically stabilised sludge. Giger et al. [23] reported that a 15-fold increase of NP concentration during anaerobic digestion results in extremely high concentrations of NP in anaerobically digested sewage sludge ($450-2530 \text{ mg kg}^{-1}$, mean 1010 mg kg^{-1}). The concentrations in activated sewage sludge, mixed primary and secondary sludge, and aerobically stabilised sludge were substantially lower, suggesting that the formation of NP is favoured under mesophilic anaerobic conditions. Field and Reed [24] studied the occurrence of NPECs in anaerobically digested municipal and industrial sludges. $NPE_{1-4}C$ concentrations ranged from 27 to 113 mg kg^{-1} with NPE_2C as the most abundant oligomer and *ortho*-to-*para* isomer ratios ≥ 1, which indicated the depletion of *para* NPEC isomers relative to *ortho* isomers during anaerobic sludge treatment. In contrast, sludge that had not undergone anaerobic treatment contained only *para* isomers.

Alcohol ethoxylates

AEOs are widely accepted as environmentally safe surfactants, and today they are often used as an alternative to APEOs. The most commonly used AEOs contain a mixture of even−odd linear and partially α-methylated-branched alkyl chains containing between 10 and 18 C units and an ethoxylate chain of between 2 and 20 ethylene oxide units.

AEOs were detected in influent waters at concentrations of $125-3600 \text{ μg L}^{-1}$ (Table 6.1.6), whereas concentrations up to 509 μg L^{-1} were found for effluent samples. The linear and α-methyl-branched AEOs easily pass ready biodegradability tests. McAvoy et al. [25] found that the overall average removal of AEOs by trickling filter treatment was $89.9 \pm 6.4\%$ and in activated sludge treatment was $97.0 \pm 4.4\%$. No apparent change in alkyl-chain length distribution was observed during primary and secondary treatment (average chain length was 13.4 ± 0.1 C units in raw influent and 13.6 ± 0.2 C units in final effluent). Several studies [8,26,27] also reported elimination efficiency ranging from 98 to 99.9%, irrespective of the type of sewage treatment, operating conditions or influent concentration.

TABLE 6.1.6

Reported concentration ranges of AEOs in WWTP influent, effluent and digested sludge

Country	A_nEO ($n = 10-18$)			Reference
	Influent (μg L^{-1})	Effluent (μg L^{-1})	Sludge (mg kg^{-1})	
Spain	275–451	5–12	–	[8,29,32]
	90–418	<0.2–300	–	[28,33]
	–	–	42–190	[37]
Italy	242–570	<0.1–2.4		[27]
	1300	28		[66]
Germany	–	–	13	[37]
	2150–3800[a]	190–380		[40]
The Netherlands	1600–4700	2.2–13	–	[3,10,26,67]
Portugal	–	–	150	[37]
Sweden	125–168	–	–	[68]
USA	685–3600	8–509	–	[25]

[a]$n = 12-15$.

AEOs, especially those with a long alkyl chain, showed great tendency to adsorb onto particulate matter. Kiewiet et al. [26] determined that the greatest proportion (>50%) of AEOs (fractions $C_{12}-C_{15}$ and $C_{12}-C_{18}$) is associated with the solid phase. This also resulted in high concentrations of AEOs in digested sludge (up to 190 mg kg^{-1}).

Polyethylene glycols (PEG) and their carboxylic derivatives
PEGs are widely used in different household products, including shampoos, liquid soaps, toothpaste, etc. PEGs can also be taken as effective markers of industrial activity. Most commercial formulations have very large amounts of PEG originating from the starting materials or formed as by-products during the manufacturing processes. Since it is too expensive to separate them from the final product, PEGs can be found at high concentrations in raw influents.

Several studies report concentrations of PEG in WWTP. In Spanish WWTPs concentrations in influents (PEG > 4) ranged from 310 to 3720 μg L^{-1} [8,28,29], whereas an Italian study reported levels of 85–270 μg L^{-1} [30]. The elimination efficiency ranged from 81 to 98%. The discrepancy between different WWTPs in the removal of PEGs may partly be a result of differences in operational conditions during treatment processes, and also, the formation of PEGs as primary degradation products from AEOs contributing to the apparently low removal of PEGs in some treatment plants. The initial biodegradation of linear alkyl-chain AEOs occurs by cleavage at the ether bridge between the alkyl group and the ethylene oxide moiety leading to the formation of PEGs and aliphatic alcohols. PEG biodegradation continues by successive oxidative and non-oxidative cleavage of C_2 units and formation of shorter chain neutral and acidic PEGs. Analyses of influent and effluent at WWTP employing activated sludge treatment showed that even high molecular mass PEGs were very efficiently removed from wastewater. However, as a consequence of biodegradation processes occurring in the sewer and WWTPs, relatively high concentrations of acidic metabolites were found in raw and treated wastewaters.

The analysis of tannery and textile wastewaters revealed the presence of monocarboxylated polyethylene glycols (MCPEGs) with concentrations of 820–940 µg L^{-1}, and dicarboxylated metabolites (DCPEGs) with concentrations up to 370 µg L^{-1} [28]. In WWTP influents these acidic species were determined at levels of 22–85 µg L^{-1} (MCPEG) and 10–100 µg L^{-1} (DCPEG) [30]. In effluents, MCPEGs and DCPEGs were detected at concentrations of 0.5–7.7 and <0.2–5.8 µg L^{-1}, respectively.

Non-ionic surfactants containing an amide group
This group of non-ionic surfactants comprises different sub-groups, such as polyethoxylated amides, fatty acid diethanol amides, aminosugars and dialkyl disugar amides (gemini surfactants).

Fatty acid diethanolamides (FADAs) are used in certain household textile washing and hand dish-washing formulations. The presence of coconut diethanolamides (CDEAs) with an alkyl chain between 7 and 15 C units has been reported in municipal wastewaters in Germany at levels of 111–124 µg L^{-1} [31] in influent and 14 µg L^{-1} in effluent indicating approximately 90% elimination, while the monitoring of CDEAs in a Swedish WWTP indicated significantly lower removal (about 40%). Other studies [8,29,32,33] reported levels up to 470 µg L^{-1} in raw influents (Table 6.1.7). Very low concentrations found in effluents indicate efficient elimination (>96%), whereas nothing is known about the formation of persistent metabolites.

Aminosugars are the basic compounds of a new class of highly biodegradable detergents based on 're-growing resources'. Alkyl glucamides (AGs) (e.g. $C_{12}/_{14}$-N-methyl glucamides) are one of the main

TABLE 6.1.7

Reported concentrations of non-ionic surfactants containing an amide group in WWTP samples

Country	Surfactant	Influent (μg L^{-1})	Effluent (μg L^{-1})	Sludge (mg kg^{-1})	Reference
Spain	CDEA	273–472	3–11	–	[8,29,32]
		4.1–460	–	–	[33]
		–	–	4.5–25	[37]
	AG	26–45	<0.1–0.2	–	[8]
Sweden	CDEA	<1–40	<1–12	–	[68]
Germany	CDEA	112–124	14	–	[31]
		–	–	11	[37]

representatives of aminosugars. They are predominantly used as co-surfactants and, therefore, their application volume is quite low. However, AGs are detected in Spanish WWTP influents with concentrations ranging from 26 to 45 μg L^{-1} [8]. In the effluent samples, only AG reaches a maximum concentration of 0.2 μg L^{-1} in one WWTP, while in two other WWTPs levels were below 0.1 μg L^{-1} (limit of detection). These data are in the same range as data obtained from German WWTPs [34,35].

Alkyl polyglucosides (APG)
APGs are non-ionic surfactants produced from renewable feedstock, such as glucose and fatty alcohols derived from starch and palm oil. Although produced in significantly lower amounts than other non-ionic surfactants (80 000 t in 1997), APGs play an important role as co-surfactants in personal care products and laundry detergents. They were found in three WWTPs from the Catalonia region of Spain with concentrations ranging from 7 to 13 μg L^{-1} in influents, while in effluent samples they were not detected [8], which corresponds to an overall removal efficiency reaching 100%.

REFERENCES

1 J. Waters and T.C.J. Feijtel, *Chemosphere*, 30 (1995) 1939.
2 D. Prat, F. Ruiz, B. Vazquez, D. Zarzo, J.L. Berna and A. Moreno, *Environ. Toxicol. Chem.*, 12 (1993) 1599.

3   E. Matthijs, M.S. Holt, A. Kiewiet and G.B.J. Rijs, *Environ. Toxicol. Chem.*, 18 (1999) 2634.

4   A. Moreno, J. Bravo and J.L. Berna, *J. Am. Oil Chem. Soc.*, 65 (1988) 1000.

5   A. DiCorcia, L. Capuani, F. Casassa, A. Marcomini and R. Samperi, *Environ. Sci. Technol.*, 33 (1999) 4119.

6   C. Crescenzi, A. DiCorcia, E. Marchiori, R. Samperi and A. Marcomini, *Water Res.*, 30 (1996) 722.

7   M.L. Trehy, W.E. Gledhill, J.P. Mieure, J.E. Adamove, A.M. Nielsen, H.O. Perkins and W.S. Eckhoff, *Environ. Toxicol. Chem.*, 15 (1996) 233.

8   P. Eichhorn, M. Petrović, D. Barceló and T.P. Knepper, *Vom Wasser*, 95 (2000) 245.

9   A. DiCorcia and R. Samperi, *Environ. Sci. Technol.*, 28 (1994) 850.

10 T.C. Feijtel, J. Struijs and E. Matthijs, *Environ. Toxicol. Chem.*, 18 (1999) 2645.

11 H.Fr. Schröder, M. Schmitt and U. Reichensperger, *Waste Mgmt.*, 19 (1999) 125.

12 J. Waters, K.S. Lee, V. Perchard, M. Flanagan and P. Clarke, *Tenside Surf. Det.*, 37 (2000) 161.

13 M. Radke, T. Berhrends, J. Foster and R. Herrmann, *Anal. Chem.*, 71 (1999) 5362.

14 R. Renner, *Environ. Sci. Technol.*, 31 (1997) 316A.

15 M. Ahel and W. Giger, *Am. Chem. Soc. Natl. Meeting Extended Abst.*, 38 (1998) 276.

16 P. de Voogt, O. Kwast, R. Hendriks and C.C.A. Jonkers, *Analusis*, 28 (2000) 776.

17 A.C. Belfroid, P. de Voogt, E.G. van der Velde, G.B.J. Rijs, G.J. Schafer and A.D. Vethaak, In: A.D. Vethaak, B. van der Burg and A. Brouwer (Eds.), *Endocrine-disrupting Compounds: Wildlife and Human Health Risks*, RIZK, The Hague, 2000, pp. 31–37.

18 M. Fujita, M. Ike, K. Mori, H. Kaku, Y. Sakaguchi, M. Asano, H. Maki and T. Nishihara, *Water Sci. Technol.*, 42 (2000) 23.

19 L.B. Barber, G.K. Brown and S.D. Zaugg, *ACS Symp. Ser.*, 747 (2000) 97.

20 M. Ahel, W. Giger and M. Koch, *Water Res.*, 28 (1994) 1131.

21 M. Ahel, W. Giger, E. Molnar and S. Ibrić, *Croat. Chem. Acta*, 73 (2000) 209.

22 A. DiCorcia, R. Cavallo, C. Crescenzi and M. Nazzari, *Environ. Sci. Technol.*, 34 (2000) 3914.

23 W. Giger, P.H. Brunner and C. Schaffner, *Science*, 225 (1984) 623.

24 J.A. Field and R.L. Reed, *Environ. Sci. Technol.*, 33 (1999) 2782.

25 D.C. McAvoy, S.D. Dyer, N.J. Fendinger, W.S. Eckhoff, D.L. Lawrence and W.M. Begely, *Environ. Toxicol. Chem.*, 17 (1998) 1705.

26 A.T. Kiewiet, J.M.D. van der Steen and J.R. Parsons, *Anal. Chem.*, 67 (1995) 4409.

27 C. Crescenzi, A. DiCorcia and R. Samperi, *Anal. Chem.*, 67 (1995) 1797.

28 M. Castillo, M.C. Alonso, J. Riu and D. Barceló, *Environ. Sci. Technol.*, 33 (1999) 1300.

29 D. Barceló, J. Dachs, and S. Alcock (Eds.), *BIOSET: Biosensors for Evaluation of the Performance of Wastewater Treatment Works*, Final Report, 2000.

30 C. Crescenzi, A. DiCorcia, A. Marcomini and R. Samperi, *Environ. Sci. Technol.*, 31 (1997) 2679.

31 H.Fr. Schroeder, *Vom Wasser*, 81 (1993) 299.

32 M. Petrović and D. Barceló, *Fresenius J. Anal. Chem.*, 368 (2000) 676.

33 M. Castillo, E. Martinez, A. Ginebreda, L. Tirapu and D. Barceló, *Analyst*, 125 (2000) 1733.

34 P. Eichhorn and T.P. Knepper, *J. Chromatogr. A*, 854 (1999) 221.

35 P. Eichhorn and T.P. Knepper, *J. Mass Spectrom.*, 35 (2000) 468.

36 J. Riu, P. Eichhorn, J.A. Guerrero, Th.P. Knepper and D. Barceló, *J. Chromatogr. A*, 889 (2000) 221.

37 M. Petrović and D. Barceló, *Anal. Chem.*, 72 (2000) 4560.

38 M. Castillo, M.C. Alonso, J. Riu, M. Reinke, G. Klöter, H. Dizer, B. Fisher, P.D. Hansen and D. Barceló, *Anal. Chim. Acta*, 426 (2001) 265.

39 J. McEvoy and W. Giger, *Naturwissenschaften*, 72 (1985) 429.

40 H-Q. Li, F. Jiku and H.Fr. Schroeder, *J. Chromatogr. A*, 889 (2000) 155.

41 P. Spengler, W. Körner and J.W. Metzger, *Vom Wasser*, 93 (1999) 141.

42 T. Kuchler and W. Schnaak, *Chemosphere*, 35 (1997) 153.

43 W. Schnaak, Th. Küchler, M. Kujawa, K.P. Henscherl, D. Suesenbach and R. Donau, *Chemosphere*, 35 (1997) 5.

44 T.C. Feijtel, E. Matthijs, A. Rottiers, G.B.J. Rijs, A. Kiewiet and A. de Nijs, *Chemosphere*, 30 (1995) 1053.

45 M.S. Holt, K.K. Fox, M. Burford, M. Daniel and H. Bickland, *Sci. Total Environ.*, 210/211 (1998) 255.

46 D.C. McAvoy, W.S. Eckhoff and R.A. Rapaport, *Environ. Toxicol. Chem.*, 12 (1993) 977.

47 C. Planas, J.M. Guadayol, M. Doguet, A. Escalas, J. Rivera and J. Caixach, *Water Res.*, 36 (2002) 982.

48 M. Ahel, W. Giger, E. Molnar, S. Ibrić, C. Ruprecht and C. Schaffner, *Div. Environ. Chem. Preprints Extended Abst.*, 38 (1998) 267.

49 M. Ahel, E. Molnar, S. Ibric and W. Giger, *Water Sci. Technol.*, 42 (2000) 15.

50 T.P. Rodgers-Gray, S. Jobling, S. Morris, C. Kelly, S. Kirby, A. Janbakhsh, J.E. Harries, M.J. Waldock, J.P. Sumpter and C.R. Tyler, *Environ. Sci. Technol.*, 34 (2000) 1521.

51 C.M. Lye, C.L.J. Frid, M.E. Gill, D.W. Cooper and D.M. Jones, *Environ. Sci. Technol.*, 33 (1999) 1009.

52 B.N. Jacobsen and T. Guildal, *Water Sci. Technol.*, 42 (2000) 315.

53 A. Cohen, K. Klint, S. Bowadt, P. Persson and J.A. Jönsson, *J. Chromatogr. A*, 927 (2001) 103.

54 T. Tanghe, G. Devriese and W. Verstraete, *J. Environ. Qual.*, 28 (1999) 702.
55 N. Paxéus, *Water Res.*, 30 (1996) 1115.
56 S.A. Snyder, T.L. Keith, D.A. Verbrugge, E.M. Snyder, T.S. Gross, K. Kannan and J.P. Giesy, *Environ. Sci. Technol.*, 33 (1999) 2814.
57 R.C. Hale, C.L. Smith, P.O. de Fur, E. Harvey, E.O. Bush, M.J. La Guardia and G.G. Vadas, *Environ. Toxicol. Chem.*, 19 (2000) 946.
58 J.A. Field and R.L. Reed, *Environ. Sci. Technol.*, 30 (1996) 3544.
59 J.R. Todorov, A.A. Elskus, D. Schlenk, P.L. Ferguson, B.J. Brownawell and A.E. McElroy, *Mar. Environ Res.*, 54 (2002) 691.
60 H.B. Lee and T.E. Peart, *Anal. Chem.*, 67 (1995) 1976.
61 T. Isobe, H. Nishiyama, A. Nakashima and H. Takada, *Environ. Sci. Technol.*, 35 (2001) 1041.
62 U. Bolz, H. Hagenmaier and W. Körner, *Environ. Pollut.*, 115 (2001) 291.
63 A.J. Sweetman, *Water Res.*, 28 (1994) 343.
64 E.R. Bennett and C.D. Metcalfe, *Environ. Toxicol. Chem.*, 17 (1998) 1230.
65 H.B. Lee, T.E. Peart, D.T. Bennie and R.J. Maguire, *J. Chromatogr. A*, 785 (1997) 385.
66 M. Zanette, A. Marcomini, E. Marchiori and R. Samperi, *J. Chromatogr. A*, 756 (1996) 159.
67 A.T. Kiewiert, J.R. Parsons and H.A.J. Govers, *Chemosphere*, 34 (1997) 1795.
68 N. Paxeus and H.Fr. Schröder, *Fourth World Surfactant Congr.*, 4 (1996) 463.

6.2   OCCURRENCE OF SURFACTANTS IN SURFACE WATERS AND FRESHWATER SEDIMENTS—I. ALKYLPHENOL ETHOXYLATES AND THEIR DEGRADATION PRODUCTS

Thomas P. Knepper, Mira Petrovic and Pim de Voogt

The monitoring studies conducted to assess the removal of surfactants during wastewater treatment have illustrated the capacity of treatment plants to effectively reduce the surfactant loads in sewage, thereby reducing significantly the proportion of these compounds that reach surface waters (see Chapter 6.1). In the face of the large-scale usage of surfactants consumed daily and subsequently discharged into domestic and industrial sewage water, this is of significant importance for the protection of aquatic life forms in receiving water bodies. Since alkylphenol ethoxylates (APEOs) form one of the most widely used classes of surfactants, it is obvious that, even after wastewater treatment, APEO and the degradation products formed during treatment are present in receiving rivers [1–6]. Additionally, adequate facilities for the treatment of wastewater are not always available in various regions, even in highly industrialised countries. Hence, surfactant residues in wastewaters are continuously emitted into rivers and oceans without appropriate removal treatment [7,8].

In this chapter, in addition to presenting a comprehensive summary of reviews already published about the presence of APEOs and their metabolites [9,10], which cover mainly nonylphenol (NP), NP monoethoxylate (NPEO$_1$) and NP diethoxylate (NPEO$_2$), we focus on describing a complete picture of APEO and so far known metabolites, including those less widely investigated, such as the NP ethoxycarboxylates (NPECs).

Furthermore, the impact of direct wastewater discharges to surface waters is addressed, for which discussion of investigations into the distribution and metabolisation of APEOs in a river in Brazil has been taken as a representative example of such situations.

Comprehensive Analytical Chemistry XL
T.P. Knepper, D. Barceló and P. de Voogt (Eds)

Review of literature data
In a review of monitoring campaigns describing the occurrence of NPEO, NP and octylphenols (OPs) in Canadian surface waters and sediments, values were compared with those obtained worldwide [10]. The latter included a total of 12 different studies conducted in Switzerland, UK, USA, Spain, and Egypt (Table 6.2.1). All recorded concentrations were roughly in the same range with values of NPEO up to 70 $\mu g\,L^{-1}$ in surface waters and 45 $\mu g\,g^{-1}$ in sediments. NP concentrations were reported to be as high as 180 $\mu g\,L^{-1}$ in surface waters and 72 $\mu g\,g^{-1}$ in sediments.

A more detailed set of data for NPEO derivatives, such as $NPEO_1$, $NPEO_2$, $NPEO_3$, NP and OP, analysed in surface water and sediments of various countries, has been put together by Ying et al. in another review paper (see Table 6.2.2) [9]. The detected levels of the analysed APEO derivatives in surface water were below 1 $\mu g\,L^{-1}$, but values as high as 644 $\mu g\,L^{-1}$ could also be detected, for example in surface waters taken from Spain. In sediments, the highest concentrations detected were NP values of up to 14 $\mu g\,g^{-1}$.

Occurrence of NPEO and NP in freshwater environments
of The Netherlands
Freshwater environments in The Netherlands have been locations of a recent large survey on estrogenic substances, including APEO

TABLE 6.2.1

Concentration ranges of AP, NPEO and NPEC in freshwaters and sediments in Canada and the World (Data from overview by Servos et al., 2000 [23])

	NPEO	OP	NP	NPEC
Canada[a]				
Surface waters	0.02–18	<0.003–0.6	<0.02–4.3	0.44–4.3
($\mu g\,L^{-1}$)	($n = 55$)	($n = 95$)	($n = 95$)[b]	($n = 37$)
Sediments	<0.02–45	<0.01–24	<0.02–72	n.d.a.
($\mu g\,g^{-1}$)	($n = 14$)	($n = 52$)	($n = 58$)	
World [c]				
Surface waters	<0.06–70	<LOD–3	<0.11–180	<0.1–45
($\mu g\,L^{-1}$)	($N = 7$)	($N = 2$)	($N = 9$)	($N = 3$)
Sediments ($\mu g\,g^{-1}$)	<0.002–9	n.d.a.	<0.003–69	n.d.a.
	($N = 3$)		($N = 6$)	

[a]n, number of samples.
[b]Servos et al. [23] quote one study (Carey et al., 1981) with levels of 11–2600 $\mu g\,L^{-1}$.
[c]N, number of studies included; n.d.a., no data available.

TABLE 6.2.2

Concentrations of AP(EO)s in surface waters and sediments in various countries[a] (Data from overview by Ying et al., 2002 [9])

Country (n)	NPEO$_1$	NPEO$_2$	NPEO$_3$	NP	OP	References
Surface water in μg L^{-1}						
Canada (38)	<LOD–7.8 (0.11)	<LOD–10 (<LOD)		<LOD–0.92 (<LOD)	<LOD–0.084 (<LOD)	[3]
UK (69)				<0.2–22 (<0.2)		[4]
Switzerland (181)	<LOD–0.35	<0.6–32 (<0.6)		<LOD–0.48		[24]
Spain (6)		<LOD–2.55		<LOD–644 (51)		[25]
Japan 1 (24)	0.04–0.81			0.05–1.08	0.01–0.18	[26]
Japan 2 (48)				0.11–3.08	<LOD–0.08	[27]
Japan 3 (256) (summer)				<LOD–1.9 (0.25)		[6]
Japan 4 (261) (autumn)				<LOD–3.0 (0.15)		[6]
USA 1 (14)			<LOD–17.8 (6.97)	<LOD–1.19 (1.52)	<LOD–0.081 (0.017)	[28]
USA 2 (5)				12–95 (48)		[29]
USA 3 (22)	0.06–0.33 (0.145)		0.03–0.33 (0.153)	0.08–0.42 (0.2)	0.0016–0.007 (0.002)	[5]
USA 4 (30)	<0.06–0.60 (0.09)	<0.07–1.2 (0.1)	<1.6–14.9 (2)	<0.11–0.64 (0.12)		[2]
Germany (16)				0.007–0.134 (0.023)	0.008–0.054 (0.0038)	[30]
Taiwan (5)			2.8–25.7 (21.3)	1.8–10 (3)		[7]

(continued on next page)

TABLE 6.2.2 (*continued*)

Country (*n*)	NPEO$_1$	NPEO$_2$	NPEO$_3$	NP	OP	References
Sediments in μg g^{-1}						
UK		<0.0005–0.0092 (<0.0005)		<0.0001–0.015 (<0.0001)		[4]
USA 1	<0.0023–0.175 (0.018)			<0.0029–2.960 (0.162)		[2]
USA 2	0.0264–13.3 (2.617)	0.0161–3.580 (0.585)	<LOD–0.876 (0.171)	0.007–13.70 (2.107)	<LOD–0.045 (0.030)	[5]
Canada	<LOD–0.038 (0.007)	<LOD–0.006 (0.0012)		0.0001–0.072 (0.10.6)	<LOD–0.001.8 (0.00041)	[3]
Japan	0.010–3.470			0.030–13.000	0.003–0.670	[26]

^{a}Concentration ranges given, median in parentheses.

678

and alkylphenols (APs) [11]. Samples were collected during three sampling campaigns in spring, summer and autumn of 1999 from surface water, particulate matter and sediments.

As a general picture, surface waters did not contain detectable levels of APEOs (Table 6.2.3). In only a few surface water samples were levels above the limit of detection (LOD) found. In the Koudevaart canal, which receives the effluents of a municipal wastewater treatment plant (WWTP), NP was detected at a level of 0.7 μg L^{-1} in autumn. Levels between 1 and 3 μg L^{-1} were found in spring in the same canal (NPEO), in the brackish water of the Waddensea to which the Koudevaart canal is discharging (NPEO), and in another canal in a heavily industrialised area (NP). In summer levels of NPEO between 2 and 3 μg L^{-1} were detected again in the Koudevaart and the adjacent Waddensea, and also in the Rhine river at Lobith and Maassluis, in the river Meuse at Eijsden, Belfeld and Haringvliet, in the Scheldt near Schaar van Ouden Doel, and in the Ijmuiden harbour. Surface water levels of AP(E) in The Netherlands are thus similar to many locations elsewhere in the world (see Table 6.2.2), but lower than the values reported for Spain.

Particulate matter generally contained higher levels than sediments (see also Chapter 6.3). In general, NPEO and NP were found in most of the particulate matter and sediment samples at levels above the LOD, whereas levels of OPEO and OP were below the LOD (Table 6.2.3). In the river Rhine, concentrations in particulate matter increased going downstream from the German–Dutch border to the estuary (Fig. 6.2.1). In the Meuse, the opposite trend was observed, with levels decreasing from the Belgian–Dutch border to the estuary (Fig. 6.2.2). In the river Scheldt levels between 0.7 and 4.2 μg g^{-1} were found in samples taken at the Belgian–Dutch border.

In all three sampling periods, relatively high levels (2–2.2 μg NPEO g^{-1} and 1–4.1 μg NP g^{-1}) were found in particulate matter samples taken at Terneuzen in the canal that links the Belgian city of Ghent to the western Scheldt estuary.

Sediments were mainly sampled in autumn. Two samples taken in spring showed concentrations comparable with those found for the same locations in autumn. Significant levels of 0.3–2.8 μg g^{-1} dry weight (d.w.) of NPEO were found in the Dommel, a small tributary of the river Meuse, in samples taken downstream from the effluent discharge of a municipal WWTP, and in samples from the Rhine, Scheldt and Meuse rivers. Elevated levels of NP (0.5–3.8 μg g^{-1}) were

TABLE 6.2.3

Concentration ranges of AP(EO)s observed in freshwater environments in The Netherlands during the 1999 LOES survey (Data from de Voogt et al., 2002 [31])

	No. of samples	OP		OPEO		NP		NPEO	
		Concentration range	Median[a]	Concentration range	Median	Concentration range	Median	Concentration range	Median
Surface waters ($\mu g\ L^{-1}$)	86	<0.05–6.3	0.30 (8)	<0.16–17	11 (2)	<0.11–4.1	0.99[b] (9)	<0.18–87	1.5[b] (29)
Particulate matter ($\mu g\ g^{-1}$ d.w.)	54	<0.001–0.40	0.015 (5)	<0.002–1.7	0.8 (2)	<0.003–4.1	0.17 (39)	<0.005–22	0.31 (47)
Sediments ($\mu g\ g^{-1}$ d.w.)	25	<0.002–0.026	0.008 (3)	<0.034	–[c] (0)	<0.01–3.8	0.16 (21)	<0.01–2.8	0.11 (19)

[a]Median values have been calculated from samples with concentrations >LOD. The number of samples with concentrations >LOD is given in parentheses.
[b]Value is between LOD and LOQ.
[c]No median, all values <LOD.

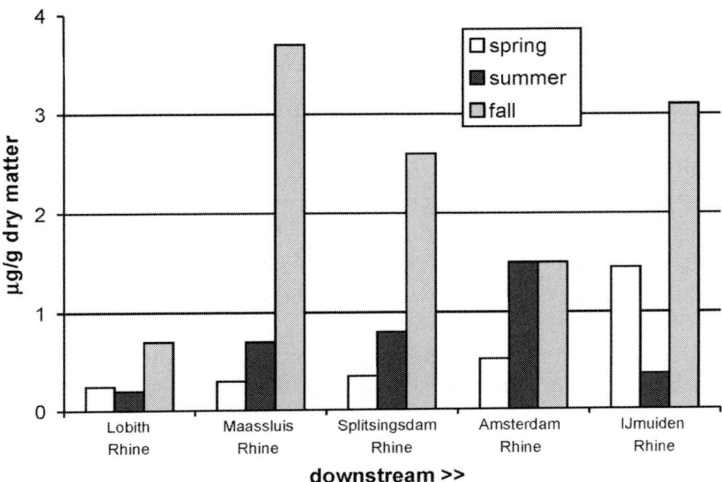

Fig. 6.2.1. Spatial concentration trend of NPEO in suspended matter from samples taken
from the river Rhine (source: Ref. [31]).

observed in sediments from the rivers Scheldt, Dommel, Meuse, and
from the North Sea Canal (which is fed by water from the river Rhine)
at Ijmuiden.

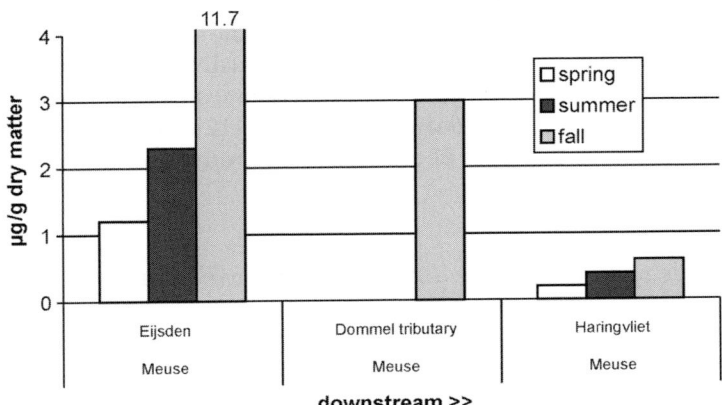

Fig. 6.2.2. Spatial concentration trend of NPEO in suspended matter from samples taken
from the river Meuse (source: Ref. [31]).

Sediment levels of AP(EO) in The Netherlands appear to be relatively low compared to samples from other areas of the world (see Tables 6.2.1 and 6.2.2).

Sediments appear to have relatively enriched levels of NP. The ratio of NP to NPEO in sediments is much higher than in samples of particulate matter (see also Chapter 6.3), and this ratio seems to increase from water < particulate matter < marine sediment < freshwater sediment. In 10 out of 12 freshwater sediments, and 5 out of 11 marine sediments, the NP/NPEO ratio was more than one. In particulate matter, this ratio was observed in fewer samples, namely in 7 out of 50 samples.

Sampling stations in this survey allow us to follow (by comparing riverine data) the fate of NPEO and NP from their entrance into The Netherlands until the point at which the estuaries are reached (corresponding to distances typically in the range of 150–300 km). The plots show that levels of NPEO (see Figs. 6.2.1 and 6.2.2) and NP in sediments and particulate matter are relatively high in the river Meuse, where it enters the territory of The Netherlands, compared to the river Rhine. However, as the Meuse discharges into relatively large estuaries, the concentrations appear to be 'diluted' significantly. On the contrary, the levels in sediments and particulate matter in the river Rhine increase from Lobith at the German border to Splitsingsdam or Ijmuiden, where the river flows into the North Sea. The main stream of the river Rhine, having passed through several highly industrialised areas, runs directly into the North Sea without much dilution and with relatively low residence time in the estuary. Due to the short residence time of the water, hardly any biodegradation occurs and concentrations are thus not favourably influenced to any significant degree by degradation or dilution [12]. Local diffuse sources also contribute to the lack of concentration reduction during passage to the sea.

Occurrence of NPEO, NP and NPEC in freshwater environments of The Netherlands and Germany
Two studies conducted between November 2000 and November 2001 in the neighbouring countries of The Netherlands (five locations at the rivers Rhine and Meuse) [13] and Germany (21 rivers and streams belonging to the river Rhine catchment area in Hesse) [14] allow for some comparison to be made. Dissolved aqueous concentrations

(Tables 6.2.4 and 6.2.6) and particulate matter concentrations (Table 6.2.5) of NP, NPEO and NPEC were analysed by liquid chromatography electrospray mass spectrometry (LC–ESI–MS) after enrichment by solid-phase extraction (SPE) and Soxhlet followed by SPE, respectively.

In samples of the river Rhine taken in July 2001, the concentrations of NPEO were invariably below 0.1 $\mu g\,L^{-1}$ whereas NP was present at levels between 0.17 and 0.30 $\mu g\,L^{-1}$. In two of the three samples taken in July, the NPEC concentrations were significantly higher than both NPEO and NP, with values up to 1.1 $\mu g\,L^{-1}$. In the samples taken in September from the river Rhine, NPEO concentrations were higher than during the sampling campaign in summer with a maximum value of $NPEO_{3-16}$ of 5.2 $\mu g\,L^{-1}$. NP concentrations were in the same range as in July, whereas from the NPECs only NPE_1C could be detected at concentrations between 0.06 and 0.12 $\mu g\,L^{-1}$. Comparing the NPEC concentrations from July and September, it appears that in September 2001 less aerobic degradation of the NPEOs discharged via WWTP into the river Rhine has occurred, which may correspond to the higher temperatures in July.

In the river Meuse the $NPEO_{total}$ and NP concentrations varied between 0.07 and 0.34 $\mu g\,L^{-1}$ and 0.13 and 0.38 $\mu g\,L^{-1}$, respectively. Thus they fall within the same range as those observed in the river Rhine. Detected levels of $NPE_{1-6}C$ ranged from less than the LOD up to 3.4 $\mu g\,L^{-1}$, with $NPE_{3-6}C$ being significantly higher than NPE_1C and NPE_2C. In the river Meuse, similar to the river Rhine, NPEC concentrations appeared to be higher in July than in September.

Particulate matter was also analysed during July and September 2001 at the same sampling points (Table 6.2.5). Significant levels of NPEO (0.18 – 5.2 $\mu g\,g^{-1}$ d.w.) and NP (0.75 – 4.9 $\mu g\,g^{-1}$ d.w.) were found in all samples of the rivers Rhine and Meuse. In the 30 samples investigated, from the NPECs only NPE_2C could be detected and this was found only in two samples. No seasonal differences could be observed for NP or NPEO.

In Table 6.2.6, an overview of the results of a screening on AP(EO) and metabolites in surface waters in Hesse, Germany conducted in November 2000 (21 samples) and November 2001 (5 samples) is presented. NP, OP and $NPEO_1$, $NPEO_2$ were not detected at all, whereas NPE_1C and NPE_2C were present in many samples with mean values obtained in 2000 of 0.23 and 0.39 $\mu g\,L^{-1}$, respectively. In general, values acquired in 2001 were lower owing to the higher flow

TABLE 6.2.4

Dissolved aqueous concentrations of NP, NPEO and NPEC in samples collected in 2001 from the rivers Meuse and Rhine, The Netherlands ($\mu g\,L^{-1}$)

Location; date	$NPEO_1$	$NPEO_2$	$NPEO_{3-16}$	$NPEO_{total}$	NP	NPE_1C	NPE_2C	$NPE_{3-6}C$	$NPE_{1-6}C$
Rhine A; July	<LOD	0.02	<LOD	0.02	0.30	0.44	0.23	0.45	1.1
Rhine B; July	0.04	0.03	<LOD	0.08	0.16	0.32	0.19	0.36	0.87
Rhine C; July	<LOD	0.01	0.03	0.06	0.17	0.09	<LOD	<LOD	0.09
Rhine A; Sept	<LOD	0.01	0.10	0.11	0.15	0.06	<LOD	<LOD	0.06
Rhine B; Sept	0.12	0.50	5.21	5.8	0.40	0.12	<LOD	<LOD	0.12
Rhine C; Sept	0.04	0.04	0.28	0.36	0.15	0.09	<LOD	<LOD	0.09
Meuse E [a]; July	0.10	0.09	0.16	0.34	0.38	0.94	0.80	1.62	3.4
Meuse D; Sept	<LOD	0.01	0.06	0.07	0.12	<LOD	<LOD	<LOD	<LOD
Meuse E; Sept	<LOD	0.06	0.25	0.32	0.17	0.18	0.17	0.32	0.68

[a]Internal standard not recovered, analysis not reliable for Meuse D; July.

TABLE 6.2.5

Concentrations of NP, NPEO and NPEC in particulate matter collected in 2001 from the rivers Meuse and Rhine, The Netherlands ($\mu g\,g^{-1}$ d.w.)

Location date	$NPEO_1$	$NPEO_2$	$NPEO_{3-16}$	$NPEO_{total}$	NP	NPE_1C	NPE_2C	$NPE_{3-6}C$
Rhine A; July	<LOD	0.012	0.17	0.18	1.9	<LOD	<LOD	<LOD
Rhine B; July	0.18	0.15	0.74	1.1	3.9	<LOD	<LOD	<LOD
Rhine C; July	0.23	0.17	0.89	1.3	2.2	<LOD	<LOD	<LOD
Rhine A; Sept	0.11	0.072	0.79	0.98	1.6	<LOD	<LOD	<LOD
Rhine B; Sept	<LOD	0.02	0.34	0.36	2.3	<LOD	0.49	<LOD
Rhine C; Sept	0.09	0.23	1.2	1.6	1.6	<LOD	<LOD	<LOD
Meuse D; July	2.2	1.4	1.6	5.2	2.1	<LOD	<LOD	<LOD
Meuse E; July	0.10	0.14	0.93	1.18	4.9	<LOD	<LOD	<LOD
Meuse D; Sept	<LOD	0.07	0.32	0.38	2.9	<LOD	1.5	<LOD
Meuse E; Sept	<LOD	0.04	0.80	0.84	0.75	<LOD	<LOD	<LOD

TABLE 6.2.6

Results of screening of APEO and metabolites in surface waters in Germany in November 2000 ($n = 21$) and November 2001 ($n = 5$) (data taken from Ref. [14])

Concentration ($\mu g\,L^{-1}$)	NPE$_1$C		NPE$_2$C		NP, OP, t-OP		NPEO$_1$, NPEO$_2$	
Year	2000	2001	2000	2001	2000	2001	2000	2001
Mean	0.23	0.14	0.39	<LOQ	<LOQ	<LOQ	<LOQ	<LOQ
Min.	<LOQ	<LOQ	<LOQ	<LOQ	<LOQ	<LOQ	<LOQ	<LOQ
Max.	0.48	0.28	1.20	0.22	<LOQ	<LOQ	<LOQ	<LOQ

LOQ: $0.06-0.08\ \mu g\,L^{-1}$ except NPEO$_1$ ($0.25-0.50\ \mu g\,L^{-1}$).

rate of the investigated rivers compared to 2000, thus resulting in lower concentrations of these permanently emitted pollutants.

Comparison of Tables 6.2.4 and 6.2.6 shows that dissolved concentrations of NP and NPEO are higher in the Rhine and Meuse rivers than in the Hessian river, which is probably caused by accumulation from diffuse and point sources along the rivers. On the contrary, levels of NPEC in Hessian streams are similar to those observed in the rivers Rhine and Meuse in The Netherlands. An explanation for this could be that degradation occurs relatively more readily in the smaller streams due to slower water velocities and a lower particle density in these waters.

Occurrence of NPEO, NP and NPEC in freshwater environments of Spain

The occurrence of APEO or their metabolites (AP and carboxylated derivatives) was also studied in two tributaries of the Llobregat river (northeastern Spain). The Llobregat river is one of the most important sources of drinking water for the Barcelona area with a population of 3.2 million people. Since 1960, the effluents from tannery, textile, pulp and paper industries, situated along the river basin, have discharged into the river and several comprehensive studies [15–17] have been performed in order to determine the levels of estrogenic alkylphenolic compounds (and steroid sex hormones) and to study the effects of WWTP effluents on the reproductive physiology of fish.

Similar to the studies conducted in The Netherlands and Germany, NPEOs and NP were found in both water and sediment samples, while more polar acidic metabolites (NPE$_1$C) were detected only in water samples (Figs. 6.2.3 and 6.2.4).

NP levels in receiving waters ranged from 0.5 to 15 μg L^{-1}, showing a clear increase in concentrations at sites downstream of WWTPs. The highest concentration (15 μg L^{-1}) was found in the Anoia river downstream of three WWTPs (Igualada, Piera and Calaf) (Fig. 6.2.3).

The concentrations of NPEOs in the Anoia river sampled at different points relative to the points of effluent discharge ranged from 11 μg L^{-1} (upstream) to 41 μg L^{-1} (downstream of discharges). NPEOs were far less abundant in the Cardener river, with concentrations ranging from 2.2 to 8.4 μg L^{-1} (Fig. 6.2.4). However, all these reported concentrations are still much higher than those reported for the rivers Rhine, Meuse and those investigated in Hesse, Germany.

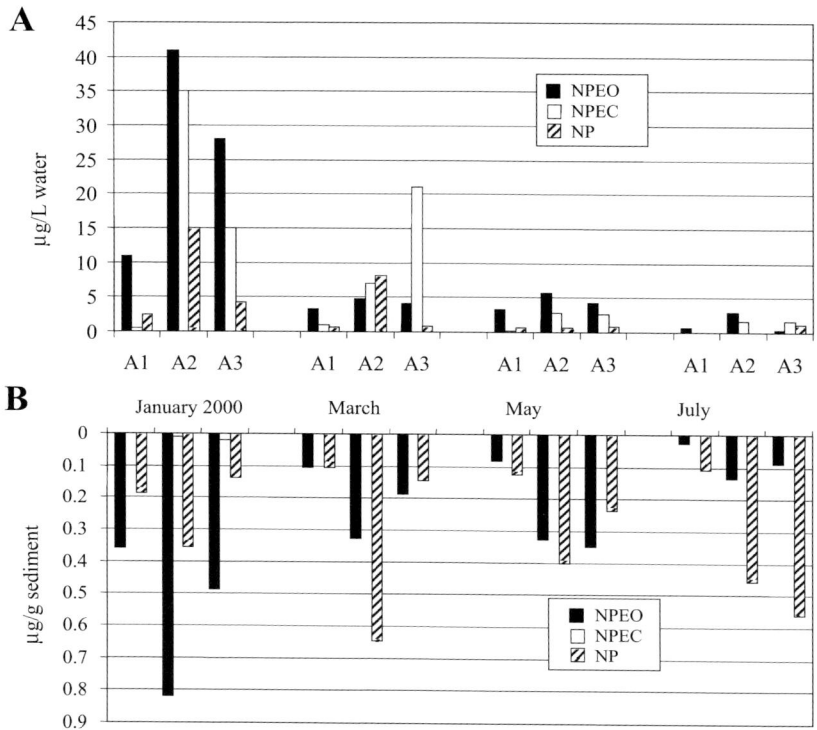

Fig. 6.2.3. Seasonal variation of alkylphenolic compounds in water (A) and sediment (B) samples from the Anoia river. Sampling points: A1 (1.5 km upstream of WWTP, Igualada); A2 (23 km downstream of WWTP); A3 (27 km downstream of WWTP).

The reason for that might not only be the high wastewater content of the investigated Spanish rivers, but also the continued intensive use of NPEO in Spain.

Again, the sediment appears to act as a sink for NPEOs and their non-polar degradation products. At sites situated downstream of the WWTP effluent discharge sediment concentrations of NPEOs and NP were up to 0.82 and 0.66 $\mu g\,g^{-1}$ in the Anoia river and 0.15 and 0.38 $\mu g\,g^{-1}$ in the Cardener river, respectively. OP was also found in samples located downstream of WWTP in concentrations ranging from 0.04 to 0.10 $\mu g\,g^{-1}$ in the Anoia river, and from 0.02 to 0.08 $\mu g\,g^{-1}$ in the Cardener river. The NPEO and NP values determined in

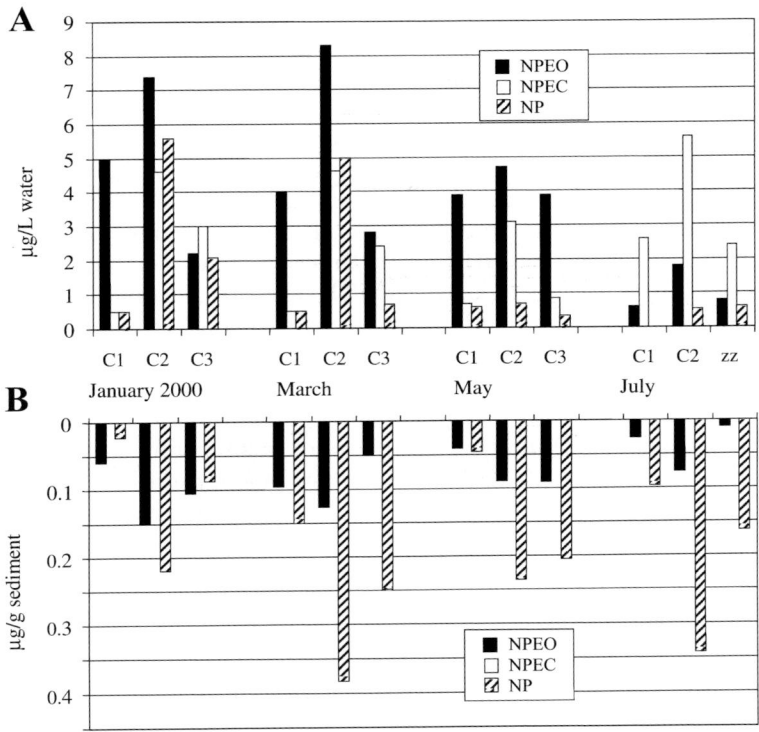

Fig. 6.2.4. Seasonal variation of alkylphenolic compounds in water (A) and sediment (B) samples from the Cardener river. Sampling points: C1 (5 km upstream of WWTP, Manresa); A2 (4 km downstream of WWTP); A3 (8 km downstream of WWTP).

the investigated Spanish sediments were thus in the range of those observed in many other countries.

However, an issue of concern is the relatively high level of alkylphenolic compounds in both water and sediment upstream of known sources of pollution (WWTPs). The concentrations of total alkylphenolic compounds (APEO, AP and APEC) ranged from 5 to 15 $\mu g \, L^{-1}$ in water and from 0.075 to 0.55 $\mu g \, g^{-1}$ in sediments, and may be attributed mainly to two origins. One is the use of sewage sludge in agriculture and the other is the uncontrolled discharge of industrial wastewaters in the upper course of the river.

During a seven-month period (from January to July 2000) of this study a seasonal fluctuation in concentrations of all target compounds was observed. The seasonal variation of concentrations of nonylphenolic compounds was especially marked in the Anoia river, where the total concentration of analytes in May and July was almost one order of magnitude lower than in January. This can be attributed to the increasing rainfall during May (36 L m^{-2}) and June 2000 (74 L m^{-2}) in the studied area which resulted in higher dilution of WWTP effluents in receiving waters. Indeed the pluviosity taking place at the various sampling periods, varied in approximately inverse order to the average concentrations measured. However, there is evidence that the biodegradation of NPEOs in both WWTP and in receiving water, is temperature dependent and similar seasonal fluctuation has already been observed in the Glatt river (Switzerland) [18]. Primary elimination and mineralisation of NPEOs decrease as the temperature decreases. A study by Manzano et al. [19] indicated a strong influence of temperature on the period of acclimation of microorganisms and on the rate of biodegradation of NPEOs [68% of primary biodegradation at 7°C (30% mineralisation) versus 96% at 25°C (70% mineralisation)].

Occurrence of NPEO and NPEC in freshwater environments of Brazil
Similar to the studies conducted by Eichhorn et al. [20], which are described in Chapter 6.3, the impact of direct discharge of wastewaters to surface waters was investigated by analysing the distribution and levels of NPEO and NPEC in a small Brazilian river located northeast of the city of Niterói (state of Rio de Janeiro, Brazil) (see also Fig. 6.3.1) [21]. In Brazil only up to 10% of the sewage water is treated. The concentrations of NPEO were determined by LC–ESI–MS, and NPEC by tandem MS after SPE at the six sampling sites shown in Fig. 6.2.5. After the passage of Cachoeiras de Macacu, the NPEO concentration increased by a factor of 6 from 100 to 600 ng L^{-1}. Reaching the town entrance of Japuíba, the level dropped again, down to 160 ng L^{-1}, which may be likely due to degradation of NPEO, similar to that reported for linear alkylbenzene sulfonates (LASs) in Chapter 6.2.2, or to adsorption onto sediments, which was not investigated during this study. At the three subsequent stations the NPEO concentrations averaged 160, 170 and 360 ng L^{-1}, respectively. Upstream from Cachoeiras de Macacu, NPE$_1$C and NPE$_2$C could not be detected. Downstream the levels were between 12 and 14 ng L^{-1} (NPE$_1$C) and 4–8 ng L^{-1} (NPE$_2$C).

The findings showed that microbial communities present in the river were capable of oxidising the relatively low concentrations of NPEO

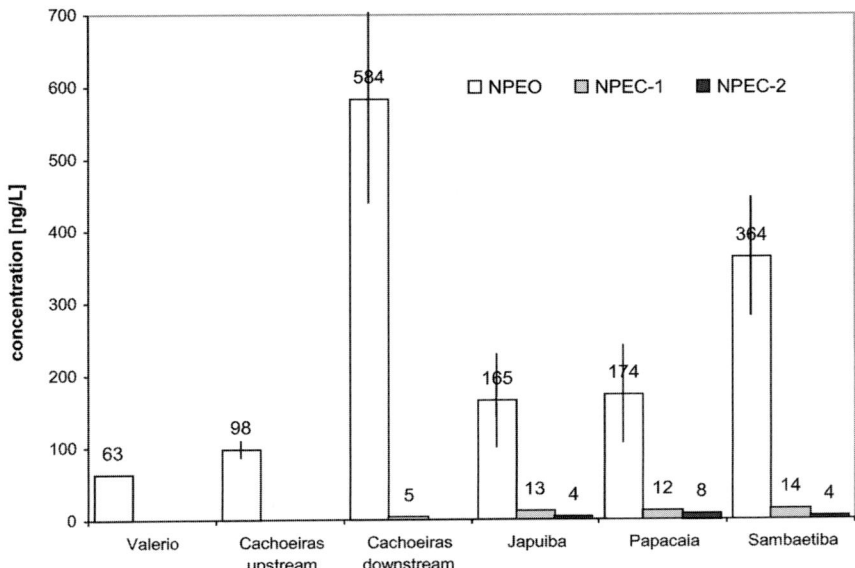

Fig. 6.2.5. Concentrations of NPEO (Σ EO_1–EO_{20}), NPE_1C and NPE_2C in samples from Rio Macacu, Brazil [21].

emitted into the river, thus yielding NPEC, but that the NPEC values reached (compared to NPEO) were much lower than those observed in European rivers (see earlier paragraphs).

Conclusion

Although, since 1995, the extensive use of NPEO in Germany and The Netherlands has been restricted on a voluntary basis [22], in particular in domestic applications, NPEO and its degradation products can still be found in freshwater environments. However, the reported concentrations of these pollutants in these countries nowadays are much less than data reported for other countries, such as the USA or Spain.

In recent monitoring campaigns conducted in Germany, Spain and The Netherlands, where NPEC was also included, the NPEC concentrations detected were almost always higher than NPEO and NP in the water phase. Thus it is clear that it is also necessary to include the NPECs, in particular $NPE_{1-6}C$, in the group of analytes to be monitored in further sampling campaigns.

In the sediments and suspended matter investigated, mainly NP and NPEO were found, reflecting the much stronger tendency of these substances to adsorb as compared to the more polar NPEC.

Furthermore, due to large variations both in concentrations and the ratio of the individual pollutants obtained, not only from the data acquired during the sampling campaigns discussed in this chapter, but also those described in Chapter 4, it is evident that in order to obtain statistically relevant data, sampling campaigns should at least span a period of several months.

REFERENCES

1   M. Ahel and W. Giger, *Anal. Chem.*, 57 (1985) 1577.
2   C.G. Naylor, J.P. Mieure, W.J. Adams, J.A. Weeks, F.J. Castaldi, L.D. Ogle and R.R. Romano, *J. Am. Oil Chem. Soc.*, 69 (1992) 695.
3   D.T. Bennie, C.A. Sullivan, H.B. Lee, T.E. Peart and R.J. Maguire, *Sci. Total Environ.*, 193 (1997) 263.
4   M.A. Blackburn, S.J. Kirby and M.J. Waldock, *Mar. Poll. Bull.*, 38 (1999) 109.
5   P.L. Ferguson, C.R. Iden and B.J. Brownawell, *Environ. Sci. Technol.*, 35 (2001) 2428.
6   A. Tabata, S. Kashiwa, Y. Ohnishi, H. Ishikawa, N. Miyamoto, M. Itoh and Y. Magara, *Water Sci. Technol.*, 43 (2001) 109.
7   W.H. Ding, S.H. Tzing and J.H. Lo, *Chemosphere*, 38 (1999) 2597.
8   E. González-Mazo, M. Honing, D. Barceló and A. Gómez-Parra, *Environ. Sci. Technol.*, 31 (1997) 504.
9   G.G. Ying, B. Williams and R. Kookana, *Environ. Int.*, 28 (2002) 215.
10 M.R. Servos, R.J. Maguire, D.T. Bennie, H.B. Lee, P.M. Cureton, N. Davidson, R. Sutcliffe and D.F.K. Rawn, Supporting document for nonylphenol and its ethoxylates, *National Water Research Institute 00-029*, Environment Canada, Burlington, 2000, pp. 1–217.
11 A.D. Vethaak, G.B.J. Rijs, S.M. Schrap, H. Ruiter, A. Gerritsen and J. Lahr, *Estrogens and xeno-estrogens in the aquatic environment of The Netherlands (LOES)*. RIKZ/RIZA 2002.001, The Hague, 2002, ISBN 9036954010.
12 N. Jonkers, R.W.P.M. Laane and P. de Voogt, *Environ. Sci. Technol*, 37 (2003) 321.
13 N. Jonkers, P. Serné and P. de Voogt, *Report to RIZA*, University of Amsterdam, Amsterdam, 2001.

14 Hessisches Landesamt fuer Umwelt und Geologie, Wiesbaden, Germany. Unpublished results 2000 and 2001.

15 M. Solé, M.J. López de Alda, M. Castillo, C. Porte, K. Ladegaard-Pedersen and D. Barceló, *Environ. Sci. Technol.*, 34 (2000) 5076.

16 N. Garcia-Reyero, E. Grau, M. Castillo, M.J. López de Alda, D. Barceló and B. Piña, *Environ. Toxicol. Chem.*, 20 (2001) 1152.

17 M. Petrovic, M. Solé, M.J. López de Alda and D. Barceló, *Environ. Toxicol. Chem.*, 21 (2002) 2146.

18 M. Ahel, W. Giger and C. Schaffner, *Water Res.*, 28 (1994) 1143.

19 M.A. Manzano, J.A. Perales, D. Sales and J.M. Quiroga, *Water Res.*, 33 (1999) 2593.

20 P. Eichhorn, S.V. Rodrigues, W. Baumann and T.P. Knepper, *Sci. Total Environ.*, 284 (2002) 123.

21 T. P. Knepper, P. Eichhorn, S.V. Rodrigues and W. Baumann, *J. Braz. Chem. Soc.*, submitted.

22 L.H.M. Vollebregt and J. Westra, *Milieu-effecten van tensiden, Chemiewinkel*, University of Amsterdam, Amsterdam (in Dutch) (1997).

23 M.R. Servos, R.J. Maguire, D.T. Bennie, H.-B. Lee, P.M. Cureton, N. Davidson and R. Sutcliffe, *Nonylphenol and its Ethoxylates: Supporting Document*, NWRI 00-029, Environment Canada, Burlington, 2000.

24 M. Ahel, E. Molnar, S. Ibric and W. Giger, *Water Sci. Technol.*, 42 (2000) 15.

25 M. Sole, M.J. López de Alda, M. Castillo, C. Porte, K. Ladegaard-Pedersen and K.D. Barceló, *Environ. Sci. Technol.*, 34 (2000) 5076.

26 T. Isobe, H. Nishiyama, A. Nakashima and H. Takada, *Environ. Sci. Technol.*, 35 (2001) 1041.

27 T. Tsuda, A. Takino, M. Kojima, H. Harada, K. Muraki and M. Tsuji, *Chemosphere*, 41 (2000) 757.

28 S.A. Snyder, T.L. Keith, D.A. Verbrugge, E.M. Snyder, T.S. Gross, K. Kannan and J.P. Giesy, *Environ. Sci. Technol.*, 33 (1999) 2814.

29 J. Dachs, D.A. Van Ry and S.J. Eisenreich, *Environ. Sci. Technol.*, 33 (1999) 2676.

30 H.M. Kuch and K. Ballschmiter, *Environ. Sci. Technol.*, 35 (2001) 3201.

31 P. de Voogt, O. Kwast, R. Hendriks and N. Jonkers, In: Vethaak et al. (Eds.), (2002), Ref. [11] Ch. 3, pp. 96–106.

6.3 OCCURRENCE OF SURFACTANTS IN SURFACE WATERS AND FRESHWATER SEDIMENTS—II. LINEAR ALKYLBENZENE SULFONATES AND THEIR CARBOXYLATED DEGRADATION PRODUCTS

Peter Eichhorn

6.3.1 Introduction

Surfactants and their biotransformation products enter surface waters primarily through discharges from wastewater treatment plants (WWTPs). Depending on their physicochemical properties, surface-active substances may partition between the dissolved phase and the solid phase through adsorption onto suspended particles and sediments [1,2]. Several environmental studies have been dedicated to the assessment of the contribution of surfactant residues in effluents to the total load of surfactants in receiving waters. This contribution reviews the relevant literature describing the presence of linear alkylbenzene sulfonates (LASs) and in particular of their degradation products in surface waters and sediments (Table 6.3.1).

6.3.2 Impact of discharges of treated sewage on surface water quality

Tabor and Barber examined the 2800 km reach of the Mississippi river for the occurrence and fate of LAS in waters and bottom sediments [3]. Dissolved LAS was concentrated on reversed-phase (RP)-C_{18} solid-phase extraction (SPE) cartridges, while sediment-bound LAS was extracted with methanol in three 24-h cycles on a rotating mixer. After derivatisation of the extracts to yield the trifluoroethyl esters, analyses were performed by gas chromatography mass spectrometry (GC–MS). Upon enrichment of 900 mL samples, detection limits of 0.1 μg L^{-1} were achieved. LAS was identified in 21% of the water samples at concentrations ranging from 0.1 to 28 μg L^{-1} and was detected at all of the Mississippi river sediment samples at concentrations ranging from 0.01 to 20 mg kg^{-1} with a mean value of 0.83 mg kg^{-1}. The concentrations of LAS sorbed onto sediment particles generally did not correlate with

Comprehensive Analytical Chemistry XL
T.P. Knepper, D. Barceló and P. de Voogt (Eds)

TABLE 6.3.1

Analysis of surfactants and degradation products in surface waters and sediments

Compound	Matrix	Extraction	Clean-up/treatment	Identification/ quantification	Concentration (μg L^{-1})	Concentration (mg kg^{-1})	Reference
LAS	River water; river sediment	RP-C$_{18}$ (l) rotating mixer (methanol) (s)	Derivatisation	GC–MS	0.1–28	0.01–20	[3]
LAS	River water	RP-C$_2$ reflux (methanol) (s)	SAX	HPLC–FL	<10–300	0.2–130	[4]
LAS	River water; river sediment			HPLC–FL	2.1–130	0.17–35	[5]
LAS	River water; river sediment	RP-C$_{18}$ (l) Soxhlet (methanol) (s)	SAX + RP-C$_{18}$ (s)	HPLC–FL	9–44	1–3	[6]
LAS	River water; suspended matter	RP-C$_{18}$ (l) Soxhlet (methanol) (s)	SAX + RP-C$_{18}$ (s)	HPLC–FL	0.9–517	2.8–21.8	[7]
LAS	River water; suspended matter	RP-C$_{18}$ (l) Soxhlet (methanol) (s)	SAX + RP-C$_{18}$ (s)	HPLC–FL	0.9–623	2–209	[8]
LAS	River water; river sediment	Sonication (methanol) (s)	RP-C$_{18}$	HPLC–FL		0.34–4.7	[10]

TABLE 6.3.1 (continued)

Compound	Matrix	Extraction	Clean-up/treatment	Identification/ quantification	Concentration ($\mu g\,L^{-1}$)	Concentration ($mg\,kg^{-1}$)	Reference
LAS, SPC, DATS, DATSC	River water	GCB (l)		LC–ESI–MS	1.5–36 (LAS) 0.84–154 (SPC) 0.16–33 (DATS) 1.9–216 (DATSC)		[11]
LAS, SPC, DATS, DATSC	River water; river sediment	RP-C$_8$ (l) (LAS, DATS) freeze-drying (l) (SPC, DATSC) sonication (methanol) (s)	RP-C$_8$ (s) (LAS, DATS) derivatisation	GC–MS	<1–94 (LAS) <1–53 (SPC) <1–23 (DATS) <1–159 (DATSC)	<0.2–300 (LAS) <1 (SPC) <0.2–4.2 (DATS) <1 (DATSC)	[12]
LAS	River sediment	Soxhlet (methanol) (s)	SAX + RP-C18 (s)	HPLC–FL		<0.1–3.3	[14]
LAS, SPC	River water	GCB	Derivatisation	GC–MS	12–135 (LAS) 0.3–3.1 (SPC)		[15]
LAS, SPC	River water	RP-C$_{18}$		LC–ESI–MS	12–155 (LAS) 1.7–14 (SPC)		[17]
LAS, ABS, SPC	River water	RP-C$_{18}$		LC–ESI–MS	1.7–88 (LAS) 1.0–51 (ABS)		[23]

697

those of dissolved LAS determined in the river water. Of 39 water samples collected at the same time and location as the sediment samples, LASs were detected only in eight samples indicating that after deposition of LAS-loaded particles, the surfactant is more stable in the sediment than in the water column.

McAvoy et al. [4] reported LAS concentrations in receiving waters and sediments above and below the outfall of US WWTPs. LASs were isolated from surface water samples by enrichment on RP-C_2 cartridges with subsequent clean-up on strong anion exchange (SAX) columns in order to remove non-ionic and cationic material. Sorbed LAS were extracted from sediments by refluxing the air-dried samples with methanol, followed by isolation onto a SAX column. The final extracts were analysed by high performance liquid chromatography (HPLC) with fluorescence (FL) detection providing limits of detection for the solution and the solid samples of 10 µg L^{-1} and 1 mg kg^{-1}, respectively. The background mean concentration of LAS above the outlets of the WWTPs was 18 µg L^{-1} with a range of less than the limit of detection (LOD) to 90 µg L^{-1}. The discharge of the effluents led to an increased LAS from activated sludge treatment plants (ASTPs) with mean concentrations of 1.1 mg kg^{-1} being determined, whereas sediments above trickling filter plants (TFPs) contained on average 0.5 mg kg^{-1}. Substantial differences in LAS concentrations in the river sediments were found downstream of the plants. The sediments below the ASTP were only slightly affected yielding mean LAS levels ranging between 1.0 and 2.3 mg kg^{-1}. The much less effective removal of the anionic surfactant during wastewater treatment in the TFP led to significantly higher residual concentrations in the effluents and consequently in the sediment samples collected below the outfalls. The average LAS concentrations ranged from 60 to 182 mg kg^{-1}.

A similar monitoring study was conducted by Waters and Feijtel [5] comparing LAS levels in waters and sediments above and below the discharge point of WWTPs in different European countries. In general, no significant additional contribution of the WWTP effluent to the LAS load of the river downstream of the discharge was observed. In the upstream samples from German, Dutch and Italian rivers surfactant concentrations ranging between 2.1 and 27 µg L^{-1} were determined, whereas below the outfalls between 2.1 and 11 µg L^{-1} were found. In the Spanish stream a concentration increase from 27 to 30 µg L^{-1} was determined. A more critical situation was identified in a small English river, which was quite polluted due to a number of unregulated discharges of untreated sewage into the upper course of the river. Hence, there was no apparent impact from the WWTP effluent: in the upstream samples 30–130 μg L^{-1} was determined, while in the downstream samples between 9 and 47 μg L^{-1} of LAS was found. In conclusion, the concentrations of LAS in the receiving waters were significantly lower than its mesocosm no effect concentration (NEC) of 100–350 μg L^{-1} for aquatic ecosystems. In the case of sediment samples, the concentration of LAS analysed in river sediments that were taken above the effluent outfalls were $\leq$1 mg kg^{-1} with two exceptions from The Netherlands where 12 and 35 mg kg^{-1} were found. Below the discharge slightly elevated concentrations were determined in the river sediment ranging from 0.49 to 5.3 mg kg^{-1}, which led the authors to conclude that LAS was effectively biodegraded in this environmental compartment.

A 7-day pilot study for LAS monitoring was carried out at the German river Isar by Schöberl et al. [6]. Surfactant levels were determined in water and sediment samples collected upstream and downstream of the outfall of the sewage treatment plant. Solid samples were Soxhlet extracted with methanol and the extracts were submitted to a two-step clean-up procedure on SPE cartridges (SAX followed by RP18). From the water samples, the surfactants were isolated on Megabond RP18 cartridges. Final quantification was achieved by means of RP–HPLC–FL analysis. Emission of the treated sewage (67 μg LAS L^{-1}) into the river resulted in a slight increase in LAS from 9 to 11 μg L^{-1} determined as the sum of dissolved and particle-bound LAS. A more pronounced effect of the discharge was observed in the sediments where 1 and 3 mg kg^{-1} were measured in the upstream and the downstream samples, respectively. Comparison of the mean homologue distribution in aqueous (11.5 carbon atoms) and solid samples (11.9 carbon atoms) revealed preferential adsorption of the higher homologues onto sediments.

The methodology previously developed was applied to analyse LAS in five German rivers serving as receiving waters for WWTP effluents [7]. Samples taken above the outfall of activated sludge plants contained between 1.5 and 6 μg L^{-1}, while downstream at distances of between 1.4 and 5 km from the plant concentrations in the range from 0.9 to 19 μg L^{-1} were measured. In contrast, a substantially higher degree of pollution was observed in a stream receiving discharges from a sewage plant that performed only mechanical treatment. The mean effluent concentration of 2260 μg L^{-1} led to a 10-fold increase in LAS in the water (background concentration: 63 μg L^{-1}). At 2.5 km below the plant outlet the concentration averaged 623 μg L^{-1} and at 11.5 km downstream as much as 517 μg L^{-1} was determined. Analysis of the suspended matter gave LAS contents of 0.2 mg kg^{-1} in the upstream sample and 1.1 mg kg^{-1} in the sample collected at the site located 2.5 km downstream the plant.

In a follow-up monitoring study, Schöberl and et al. [8] reported on the determination of the LAS fraction bound to suspended matter in comparison to the one in the dissolved phase. Particle sizes >60 μm were not considered for analysis as the specific contamination of individual particle fractions with anthropogenic compounds decreased with increasing particle size according to Ref. [9]. Hence, concentrations are highest in the clay fraction, markedly lower in the silt fraction and almost negligible in the sand fraction (>60 μm). These results showed that in the river Isar 0.1% (above WWTP outfall) and 0.3% LAS (below WWTP outfall), respectively, were associated with the suspended matter. The corresponding values for the river Wupper were 3.9 and 0.6%, whereas in the river Chemnitz, without activated sludge plants, 5.3 and 1.5% were found in the suspended matter, i.e. the adsorbed fraction was less than 6% of the total LAS. Analysis of the homologue distributions clearly showed that the longer-chain homologues, particularly C_{13}-LAS, were preferentially adsorbed to suspended matter. In the river Chemnitz, the mean carbon chain length of adsorptively bound LAS was 12.1, while in the dissolved phase the value was 11.3.

In a monitoring program conducted on the river Po by Cavalli et al. [10], sediment samples were collected on two occasions at 10 different sites along the river course of about 500 km. After sonication extraction with hot methanol and clean-up of the extracts on C_{18}-SPE cartridges, quantitative analyses were done on an HPLC equipped with a fluorescence detector (LOD: 0.05 mg kg^{-1}). The concentration range of LAS was 0.3–4.7 mg kg^{-1} testifying that a good biological recovery of the waters took place, despite the loading from the large catchment area. A comparison of LAS sediment concentrations—determined in Cavalli's study as well as those from previous programmes of other researchers—with toxicity data available for benthic organisms showed that the LAS levels were, in most instances, below the predicted no effect concentration (PNEC), which is placed at about 20 mg kg^{-1}.

In the study of DiCorcia et al. [11] investigating the impact of WWTP effluents on the concentrations of LAS-related compounds, the spectrum

700

of target compounds included—beside LAS and sulfophenyl carboxylates (SPCs)—the dialkyltetralin sulfonates (DATSs), by-products of LAS manufacturing, and their carboxylated biotransformation products (DATS(d)C). Analytes were extracted from 1000 mL samples by SPE on graphitised carbon black (GCB) cartridges and quantitatively determined by liquid chromatography negative electrospray ionisation mass spectrometry ((−)-LC–ESI–MS). Under full-scan conditions, detection limits of the method for LAS metabolites and co-products were set at $2 \, \text{ng} \, \text{L}^{-1}$. In samples collected 100 m upstream of the treatment plant, concentrations of LAS, SP(d)C, DATS and DATS(d)C were found to be 1.5, 0.84, 0,16 and 1.9 $\mu\text{g} \, \text{L}^{-1}$, respectively. Discharge of the secondary effluent containing 43, 184, 40 and 260 $\mu\text{g} \, \text{L}^{-1}$ of each sulfonated compound resulted in elevated concentrations in the samples taken 100 m downstream of the plant: $36 \, \mu\text{g} \, \text{L}^{-1}$ (LAS), $154 \, \mu\text{g} \, \text{L}^{-1}$ (SP(d)C), $33 \, \mu\text{g} \, \text{L}^{-1}$ (DATS) and $216 \, \mu\text{g} \, \text{L}^{-1}$ (DATS(d)C). For a more detailed description of the fate of SPC in the receiving waters, the authors distinguished between SPC with linear alkyl side-chains and those carrying one methyl branch in the carbon chain (termed iso-SPC). The latter species eluting under RP conditions before their linear ones largely predominated over SPC in the effluent samples showing a mean number of carbon atoms of 7.6. A slight shift of this value to 7.4 determined in samples taken 27 km downstream of the plant suggested that microorganisms were actively operating in the river water.

The studies conducted by Trehy et al. [12] on river waters and sediments considered the four surfactant-derived compounds LAS, SPC, DATS and DATSC. Concentrations were determined in the two compartments upstream and downstream of 10 US WWTPs using either trickling filter or activated sludge technologies. Unlike the parent compounds, which were extractable from river water on BondElut C_8-SPE cartridges, the carboxylated intermediates were not sufficiently retained by C_8 or C_{18} columns, even under acidic conditions, and therefore concentrated by freeze-drying of the aqueous samples. For the extraction of the target analytes from solid samples, dried sediments were sonicated with cold methanol; LAS and DATS were additionally isolated on C_8-SPE columns. All of the extracts evaporated to dryness were derivatised with PCl_5/trifluoroethanol to yield the trifluoroethyl esters. The high electron affinity of the trifluoroethyl group provided high sensitivity in negative chemical ionisation (CI) GC–MS. Concentrations of the degradation intermediates were calculated on the basis of the internal standard 1,2,3,4-tetrahydro-2-naphthoate sulfonate assuming that its response was equal in all cases to those of SPC and DATSC. On average the samples collected upstream of the plants contained 16 μg L^{-1} LAS and 9.3 μg L^{-1} SPC. The values below the plant outlet were 35 and 31 μg L^{-1}, respectively. In the river water, the mean level of DATS increased from 3.2 to 19 μg L^{-1}, while the concentration of their carboxylated degradation products rose from 19 to 55 μg L^{-1} after emission of treated sewage into the stream. Higher levels of DATSC in comparison to SPC were attributed to a slower biodegradation of the former, possibly due to the steric hindrance of the two-ringed structure. In the sediment samples taken upstream of the plants, the LAS concentrations ranged between <LOD (0.2 mg kg^{-1}) and 3.1 mg kg^{-1}. In the downstream samples, the concentrations were strongly associated with the type of treatment plant. The average LAS level below the activated sludge plants was 1.4 mg kg^{-1}, while a mean concentration of 99 mg kg^{-1} was determined in the sediments affected by emissions from TFPs. Apparently suspended solids were directly carried over from the trickling filter unit to the receiving water. The DATS concentrations in the sediments were comparatively low (<LOD to 4.2 mg kg^{-1}) possibly due to lower adsorption coefficients for DATS than for LAS. Concentrations of carboxylated intermediates in sediments were all estimated to be below the LOD of 1 mg kg^{-1}.

The lack of detection of LAS metabolites in sediments was also discussed by Sarrazin et al. [13] determining the biodegradability of the anionic surfactant in artificially contaminated sediments from a shallow pond. During the experiments, no LAS metabolites were observed. This was explained by the high solubility of these compounds, which may be quickly eliminated from the sedimentary layer by dissolution and diffusion into the pond water.

Dated sediment layers (1939–1991) of a small German river were studied for their LAS concentrations and the gathered data were related to political, economic and technical developments in Germany [14]. As expected the sediments corresponding to the period 1939–1945 contained no LAS, since at that time soap was the most important surfactant used for laundry washing. A significantly different picture was obtained in sediments from 1949 to 1951 when a particular LAS was used along with soap. The values determined in the two core samples amounted to 1.4 and 1.3 mg kg^{-1}, respectively. In the sediment horizons from 1954 to 1956 no LASs were detected, which is in accordance with the fact that the less expensive branched alkylbenzene sulfonate (ABS) had replaced the linear one. In the years from 1959 to

1961 the German detergent industry mounted efforts to develop a readily degradable ABS in order to substitute ABS, which had caused strong foaming in sewage treatment plants and in rivers owing to its poor microbial degradability. The LAS concentrations of 0.9 mg kg^{-1} were ascribed to a specific phenomenon of the investigated stream, which in the upper course had received discharges from surfactant-producing industries. In the sediment corresponding to the years 1964–1966 very low levels were found. A steady increase was recorded in the following decade detecting 0.9 mg kg^{-1} in the 1969–1971-sediment and about 1.5 mg kg^{-1} in the layer of 1974–1976, which was explained by an increase in detergent consumption. From 1979 to 1991 the sediment concentrations decreased linearly as a consequence of the progressive construction and later modernisation of sewage treatment plants. In summary, river sediments were recognised as a sink for the anaerobically non-degradable LAS, but the authors emphasised that the relevance of sediments to accumulate the anionic surfactant has been largely overestimated.

6.3.3 Impact of discharges of untreated sewage on surface water quality

Many of the preceding examples have demonstrated the great capacity of WWTPs to efficaciously reduce the surfactant loads in sewage, thereby preventing to a large extent these compounds reaching surface waters. However, the treatment of wastewater in adequate facilities is not available to various regions in less industrialised countries. Hence, surfactant residues in wastewaters are continuously discharged into rivers and oceans without previous treatment.

Ding et al. [15] presented preliminary results of a study on the concentrations of LAS and SPC in waters from a Taiwanese river receiving direct discharges of untreated municipal wastewater. The surfactant residues and their degradation products were concentrated on GCB cartridges and after elution converted into their trifluoroethyl esters. Quantitative analyses were accomplished by ion-trap GC–MS in electron impact and CI modes, which allowed calculation of isomer distributions of each LAS homologue. The total concentrations of the target analytes in the river water samples ranged from 12 to 135 μg L^{-1} for LAS and 0.3 to 3.1 μg L^{-1} for SPC. The concentrations of LAS and SPC isomers were determined semi-quantitatively by relating the CI-MS protonated molecular ions $[M + H]^+$ of individual LAS isomers to the internal standard, C_2-LAS, assuming identical response factors of the internal standard and the analytes. Calculations of the internal/external (I/E) isomer ratio of LAS, as proposed by Takada et al. [16], indicated that through natural processes in the river the external isomers were removed much faster than the internal isomers, this being in line with preferential elimination of these isomers through sorption and biodegradation.

The impact of direct wastewater discharges to surface waters was investigated by Eichhorn et al. [17] investigating the whereabouts of LAS and SPC in a small Brazilian river. In this country, only 10% of the urban population has their sewage treated. In rural areas this value could even be lower than 5%. Hence, surface waters were expected to be substantially affected by domestic wastewater. The objective was to investigate the fate of LAS and SPC in a river located northeast of the city of Niterói (state of Rio de Janeiro, Brazil) by monitoring the concentrations at several stations along the river course. The concentrations of LAS and SPC, determined by $(-)$-LC – ESI – MS after SPE on RP18 material, at the six sampling sites, are shown in Fig. 6.3.1. After the passage of Cachoeiras de Macacu, the LAS concentration increased by about a factor of 10 from 14 to 155 μg L^{-1}. Reaching the town entrance of Japuíba, the level dropped again to 12 μg L^{-1}, which was likely due to degradation of LAS in the river section between Cachoeiras de Macacu (downstream) and Japuíba. At the three following stations the LAS concentrations averaged 36, 38 and 31 μg L^{-1}, respectively.

The SPC levels in the Rio Macacu rose from 1.7 to 12 μg L^{-1} between the Cachoeiras de Macacu upstream and downstream sites correlating rather well with the concurrent rise of the parent compound LAS. In the next river section up to Japuíba a reduction of SPC was recognised. However, whereas the LAS value diminished by a factor of 14, the decrease in SPC was much less pronounced changing from 12 to 7.4 μg L^{-1}. A slight increase was observed at the sites of Papucaia (town centre) and Sambaetiba I and II ranging on average between 12 and 14 μg L^{-1}. This was paralleled by quite steady concentrations of the anionic surfactant at these locations. Although not significant, a certain trend of decreasing LAS and SPC levels might be identified in the river course moving from the sampling station Sambaetiba I to Sambaetiba II, which could be attributed to biodegradation of both investigated

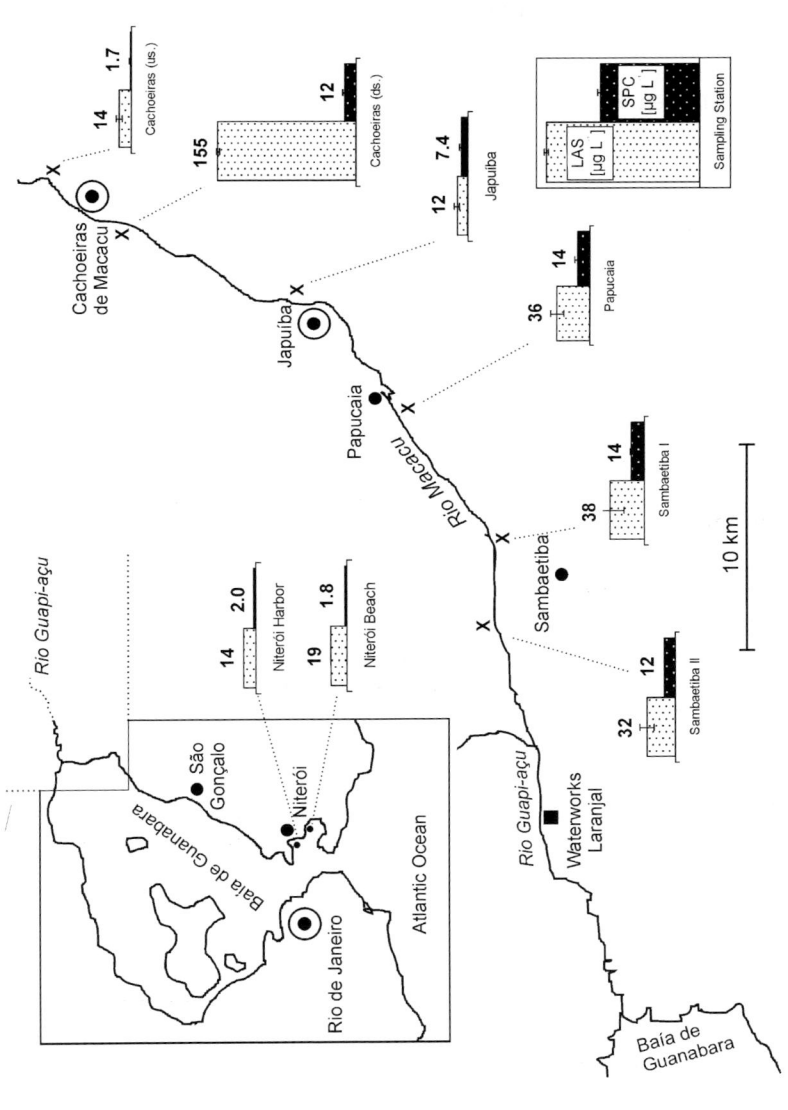

Fig. 6.3.1. Map of sampling sites at the Rio Macacu (Papucaia: lat. S 22°37′; long. W 42°45′) and the Baía de Guanabara, Brazil and the corresponding concentrations of LAS and SPC (from Ref. [17]).

sulfonates without addition of relevant amounts of urban wastewater in this section of the Rio Macacu.

In conclusion, the findings showed that microbial communities present in the river were qualified to oxidise LAS yielding long-chain SPC, which were subsequently further broken down to the shorter-chain homologues. Due to the continuous exposure to LAS via municipal wastewater, the microbial population has adapted to the surfactants by producing required enzymes for the aerobic degradation [18]. The self-purifying capacity of the water was impressively demonstrated between Cachoeiras de Macacu (downstream) and the next sampling site at Japuíba, where the surfactant level decreased by more than one order of magnitude. In the lower course of the Rio Macacu near Sambaetiba, relatively constant concentrations of LAS and SPC suggested that the degradation of both species had slowed. This might have been caused by insufficiently dissolved oxygen being essential for the initial attack of LAS or the presence of non-degradable and hence accumulated wastewater-borne contaminants, which negatively affected the microbial performance.

A further indicator giving hints on the progression of biodegradation of LAS and thereby the build-up of SPC was the peak pattern of phenyl isomers of individual SPC homologues. Since the degradation rate of a particular isomer depended on the position of the attachment of the phenyl ring on the oxidised alkyl chain [19], the peak distribution observed in the Macacu samples could be used as a marker and be compared to the one recorded from a surface water sample where a steady state in SPC breakdown was reached [17].

This is illustrated in Fig. 6.3.2 for the mass trace of C_7-SPC obtained from a Rio Macacu sample (A) and the German river Rhine (B) [17]. The relative intensities of the three major peaks, a, b and c, in the latter, were representative of a nearly stationary level. On the other hand, the predominance of peak a relative to b and c in Fig. 6.3.2(A) indicated that biodegradation of SPC was not yet completed. This finding was in accordance with the fact that the concentrations of LAS compared to those of SPC were several factors greater, i.e. SPC might still have been liberated by surfactant breakdown. However, the overall rate of conversion of the less rapidly degradable isomers seemed not to be rapid since very similar phenyl isomer patterns were observed in all the samples taken along the course of the Rio Macacu (data not shown).

With respect to the occurrence of LAS and SPC in the Baía de Guanabara, known to be heavily polluted with oil, land runoff

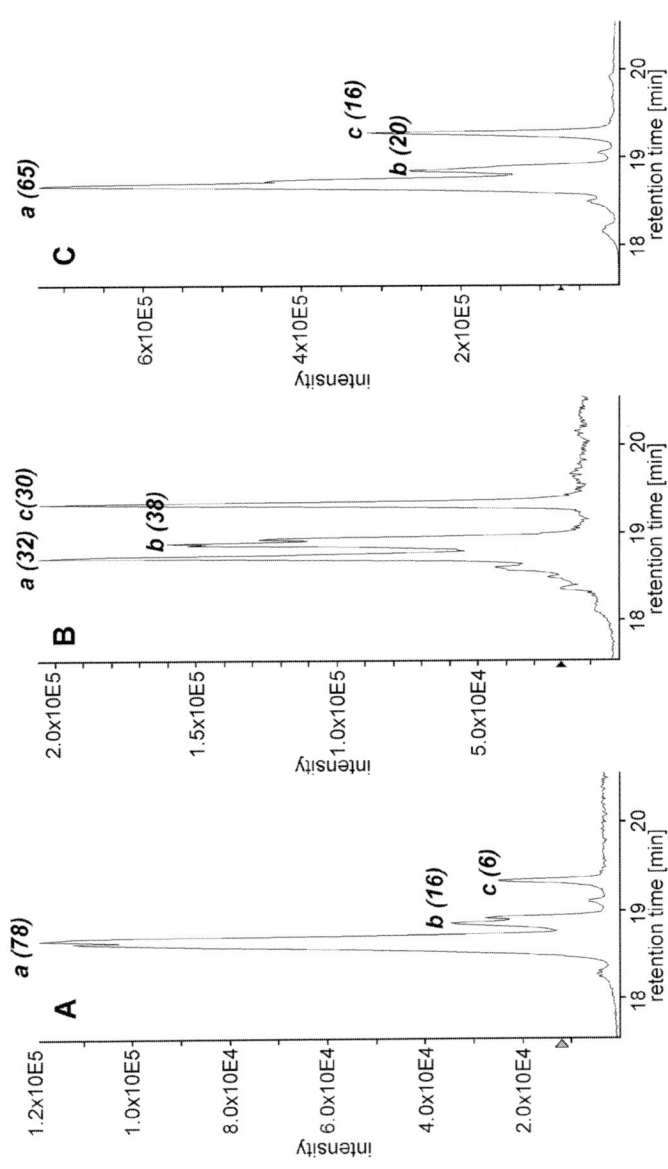

Fig. 6.3.2. (−)-LC–ESI–MS extracted ion chromatogram of C_7-SPC (m/z 285) in samples from (A) Rio Macacu, Brazil, (B) river Rhine, Germany, and (C) Baía de Guanabara, Brazil. Values in parentheses indicate relative peak area ($a + b + c = 100\%$) (from Ref. [17]).

and sewage [20,21], the surfactant levels found in the samples taken close to Niterói amounted to 14 and 2.0 $\mu g\,L^{-1}$ in the harbor, and 19 and 1.8 $\mu g\,L^{-1}$ near the beach (Fig. 6.3.1) [17]. At both locations the ratios LAS-to-SPC, which served as markers for discharges of untreated sewage waste [22], were about 10 and therefore lay within a similar range to those values observed in the Rio Macacu. It is worth mentioning that C_7-SPC in the water from the Baía de Guanabara exhibited a comparable pattern—with a slight shift in favour of b and c (Fig. 6.3.2(C))—to the one of the Rio Macacu being indicative of an unfinished biodegradation of the target analytes.

A second example on the fate of LAS and its degradation intermediates in surface waters was presented by Eichhorn et al. [23] describing the situation in the Philippines where no wastewater facilities existed at municipal level. Thus, the sewage water was discharged directly or via drainage systems to surface water bodies contributing to the continued deterioration of water quality. In the Philippines one encountered a particular situation that LAS was used along with its branched congener ABS, based on a tetramerised propylene-derived hydrophobic moiety instead of a paraffin-based hydrocarbon chain, as in LAS. The continuous usage of ABS, which was not voluntarily banned or legally restricted as in many other countries, was explainable by the availability of cheaper raw materials for ABS production and their favourable physical proper-ties in special formulations like synthetic laundry bars [24]. This form of application provided better performance and easier handling for manual laundry washing frequently done directly in surface waters. The resistance of ABS to biochemical degradation in aquatic environments—this had led to strong foam formation in WWTPs and sewage-polluted surface water in the 1950s in ABS-applying countries [25]—was the reason for their replacement.

Water samples were taken from three tributaries to the Laguna de Bay, Philippines (Fig. 6.3.3) [23]: Bucal stream and San Cristobal river in Calamba and San Pedro river, which all drained into Laguna de Bay

---

Fig. 6.3.3. Map of Laguna de Bay, Luzon/Philippines (lat. N 14°02′–14°05′, long. E 121°0′–121°5′) with sampling sites and concentrations. With the exception of the sites along the Pasig river (Manila Bay), where the values of the December 1999 and March 2000 batch are given separately, the values given for the other sampling locations represent mean values (from Ref. [22]).

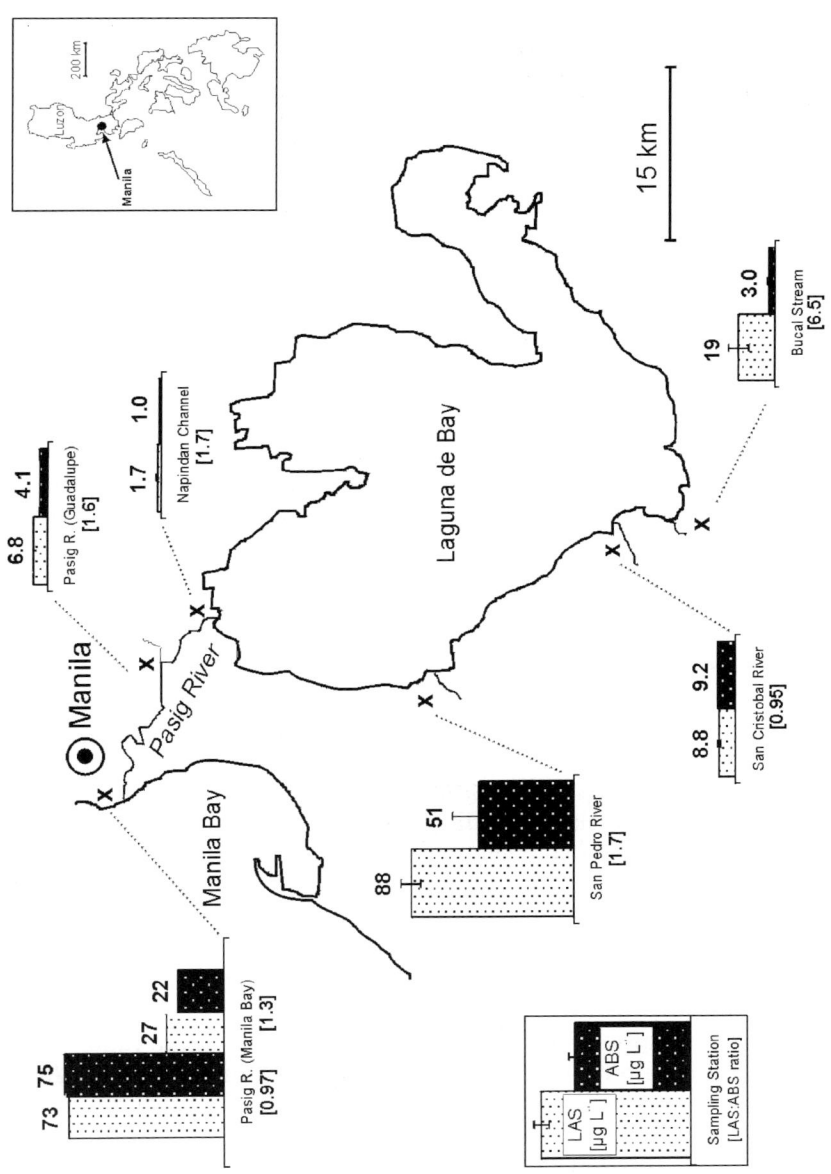

and three sites representing the outflow from the Napindan Channel to Guadalupe and the Pasig river, near the mouth of Manila Bay. These areas are part of metropolitan Manila and were therefore affected by the discharges of about 2 millions inhabitants.

In all of the investigated water streams, LAS as well as ABS were detectable in the low- to mid-concentration range in $\mu g \, L^{-1}$. The levels in the samples collected in December 1999 and March 2000 are given in Fig. 6.3.3. The calculated ratios of LAS-to-ABS concentration are presented in the same figure. The two series were found to be in good agreement in relation to the total concentration and LAS-to-ABS ratio. In detail, the following interpretations could be made: the sampling site at the Bucal stream located just about 10 m downstream of its spring had an embedded man-made 'traditional' basin, which was used for washing clothes. The LAS and ABS burdens were found to be rather low and since the source of surfactants, discharge was directly located upstream of the sampling point, no significant biodegradation could be expected. Hence, the LAS-to-ABS ratio of about 6 represented a fresh surfactant contamination resulting from washing activities.

In contrast, the San Pedro river was heavily contaminated by wastes originating from various human activities and industries and therefore it was not surprising that LAS and ABS concentrations of up to 90 $\mu g \, L^{-1}$ were found. The mean LAS-to-ABS ratio of 1.7 indicated that the surfactant discharges had occurred recently, but a partial degradation of LAS might already have taken place. In the third investigated tributary leading to Laguna de Bay, the San Cristobal river, the concentrations of LAS and ABS amounted to about 9 $\mu g \, L^{-1}$, but the LAS-to-ABS ratio was still lower than those found in the Bucal stream and the San Pedro river. This gave an indication that during transportation of the surfactants in the upper course of the San Cristobal river biodegradation of LAS had advanced to a certain extent.

The residues of LAS and ABS detected in the three investigated water streams entering Laguna de Bay could be further reduced by dilution, or biochemical degradation (mainly for LAS). Removal of the surfactants through sorption and precipitation had also been taken into account. This elimination route was particularly important for ABS since many of its derivatives were rather insoluble [26].

Low values (1–2 $\mu g \, L^{-1}$) of LAS and ABS were detected at the entrance of the Napindan Channel, which was the only outlet of Laguna de Bay towards Manila Bay [23]. Two other sampling points discharging into Manila Bay were located at the Pasig river. An increase in LAS and ABS values was observed at the Guadalupe site (between 4.1 and 6.8 μg L^{-1}) and their concentrations again rose to a level of up to 75 μg L^{-1} for ABS as it entered Manila Bay.

The significant increase in ABS and LAS concentrations between the outlet of Laguna de Bay at Napindan Channel and the mouth of Manila Bay (Pasig river) was explained by high emissions of surfactant-loaded wastewaters along the channel course. Approaching Manila Bay a rising number of industrial effluents together with an increasing number of people in the vicinity of the Pasig river, estimated to be more than 2 million, are responsible for the discharge of large amounts of surfactants. Since the average flow of the Pasig river was only 1.6 m^{3}s^{-1}, dissolved oxygen was also very low and rapid aerobic degradation of LAS was limited. On the other hand, the breakdown was likely to be negatively influenced by other industrial pollutants such as petroleum and heavy metals.

With continuing degradation of the set of LAS alkyl homologues, the peak pattern of the four components was shifted towards the shorter homologues [23]. This effect is shown explicitly in Fig. 6.3.4 for the samples taken in December 1999; comparable observations were made during the second sampling in March 2000 (data not shown). The LAS homologue pattern in the Bucal stream where washing-related surfactant discharges directly occurred upstream of the sampling point represented a rather unaffected pattern and corresponded quite well with a technical blend used in detergent formulations (C$_{10}$- to-C$_{13}$-LAS ratio of about 1). On the other side, a significant impact of degradation was stated for the contaminated samples from the San Cristobal river and the Pasig river mouthing to Manila Bay, where an increasing portion of C$_{10}$-LAS paralleled a decrease in C$_{13}$-LAS (C$_{10}$- to-C$_{13}$-LAS-ratio of about 10). A different situation arose for the poorly degradable ABS (Fig. 6.3.5). In this instance, similar homologue distributions of the branched species were observed for all samples underlining the fact that biodegradation was a negligible way of eliminating this compound. Referring to the above-discussed ratio of LAS-to-ABS, this parameter exhibited a good correlation with the degree of degradation described by the sampling site-specific alkyl homologue pattern. Furthermore, it could be perceived by comparing Fig. 6.3.4 (Bucal stream sample) and Fig. 6.3.5 that the distinct alkyl homologue distributions were characteristic for LAS and ABS, respectively, which was due to different raw materials used for the synthesis of alkylbenzenes.

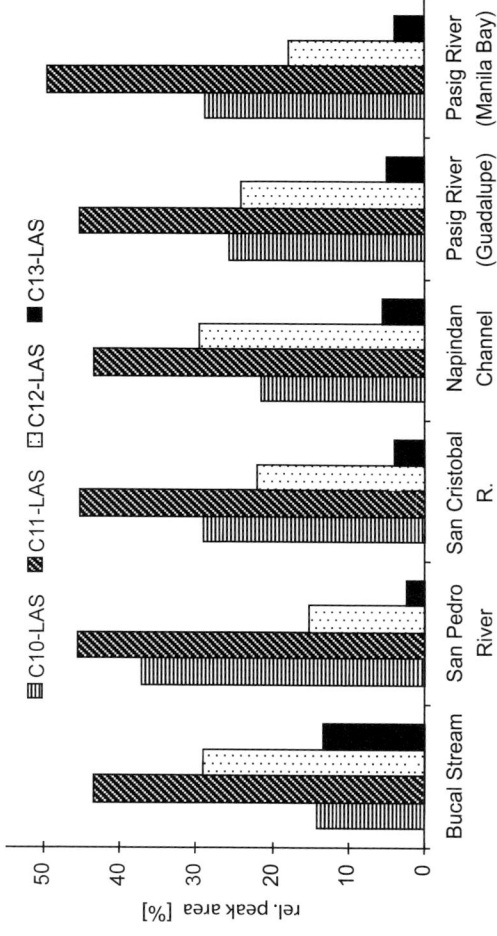

Fig. 6.3.4. Alkyl homologue distribution of LAS (sample batch from December 1999) (from Ref. [23]).

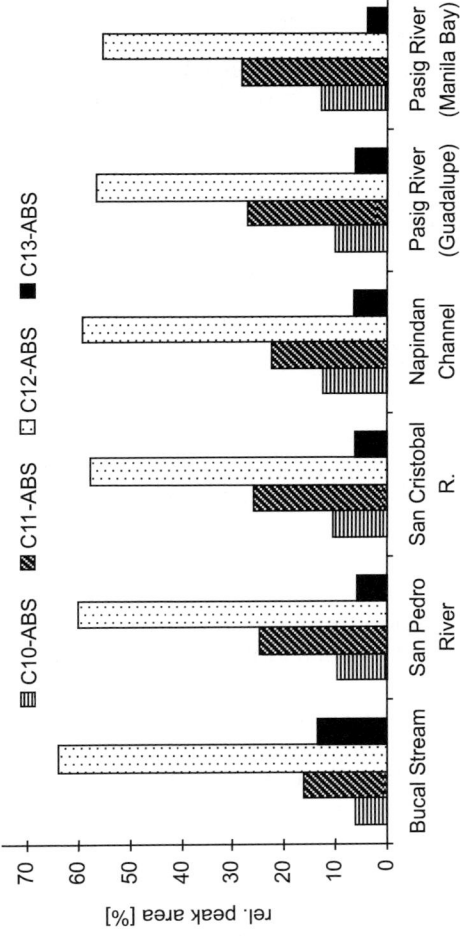

Fig. 6.3.5. Alkyl homologue distribution of ABS (sample batch from December 1999) (from Ref. [23]).

Regarding the occurrence of SPC in the surface waters (a typical chromatogram is shown in Fig. 6.3.6) a semi-quantitative analysis was performed by plotting the absolute peak area of the SPC homologues C_6 to C_{12}. Although the shorter-chain SPC were slightly underestimated due to lower extraction recoveries, the interpretation of relative burdens of SPC was possible through comparison of the peak areas. This is demonstrated in Fig. 6.3.7 for the first sample batch taken in December 1999. A similar homologue distribution was recorded at all sampling points showing a first relative maximum at C_9- or C_{10}-SPC and a second maximum at C_{12}-SPC. The detection of the former species stood in concordance with the fact that long-chain SPCs were rapidly broken down into lighter SPCs, but the rate of conversion decreased with progressive chain shortening resulting in an accumulation of mid-chain SPC. The second maximum at C_{12}-SPC was likely attributable to SPC originating from C_{12}-ABS as the most abundant homologue present in ABS-detergent formulations.

With respect to the formation of SPC out of ABS it had to be remembered that this surfactant was a mixture of hundreds of components, all minor in amount, with distinct constitutions within the alkyl group. The biological resistance of ABS was attributed to the presence of a quaternary carbon atom in the alkyl chain [27], which disabled the formation of a double bond as intermediate step of the β-oxidation cycle. A primary degradation, however, i.e. ω-oxidation of the alkyl chain, was not blocked by such a group. Hence, chain carboxylation of ABS was principally feasible and the breakdown might proceed as long as no quaternary carbon atom hindered the metabolism. Alternatively, the degradation might have been initiated by an open-ended chain if the other side was blocked by a terminal quaternary group.

Although no significant shift in the homologue pattern of ABS was observed in Fig. 6.3.7 [23], indicating no preferential removal of any homologue, the relative contribution of branched SPC homologues to the overall detected amounts of degradative products could be verified by LC–ESI–MS/MS analyses performing multiple reaction monitoring mode measurements of the C_8- to C_{12}-SPC parent ions [M − H]$^-$, yielding either the fragment m/z 183 for the linear species or m/z 197 for the branched congeners. In Fig. 6.3.8 the percentages of both species to the total peak area of an alkyl homologue are shown. Overall, the relative portions of the branched SPC decreased for the shorter-chain homologues. This was in line with the fact that steric hindrance became more probable with ongoing destruction ultimately impeding chain

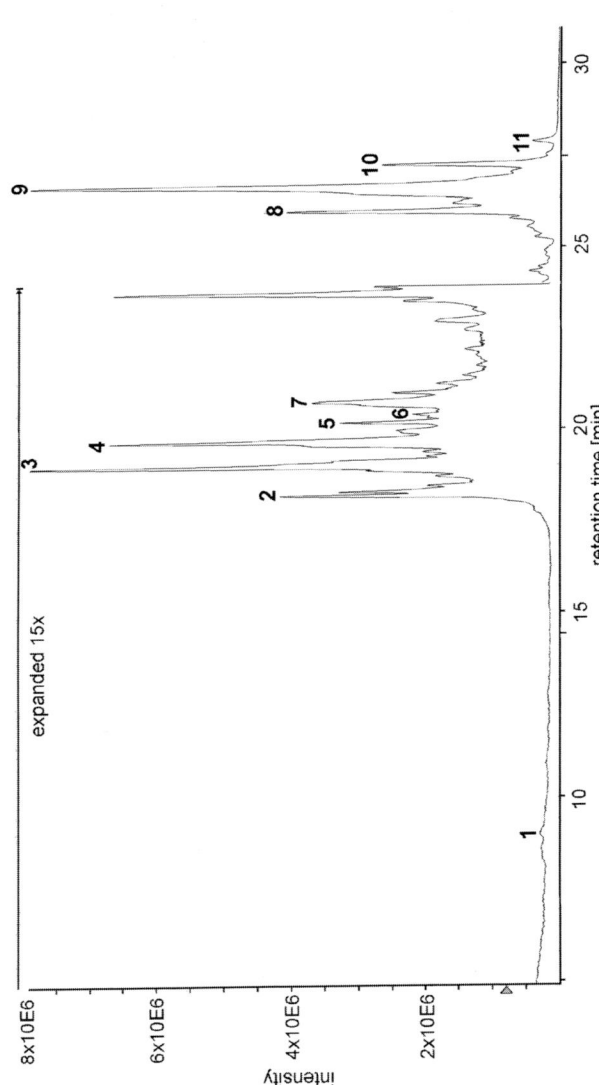

Fig. 6.3.6. (−)-LC−ESI−MS chromatogram of a sample from San Pedro river containing SPC and LASs. Peak numbering: (1) C_6-SPC, (2) C_7-SPC, (3) C_8-SPC, (4) C_9-SPC, (5) C_{10}-SPC, (6) C_{11}-SPC, (7) C_{12}-SPC, (8) C_{10}-benzene sulfonates, (9) C_{11}-benzene sulfonates, (10) C_{12}-benzene sulfonates, (11) C_{13}-benzene sulfonates (from Ref. [23]).

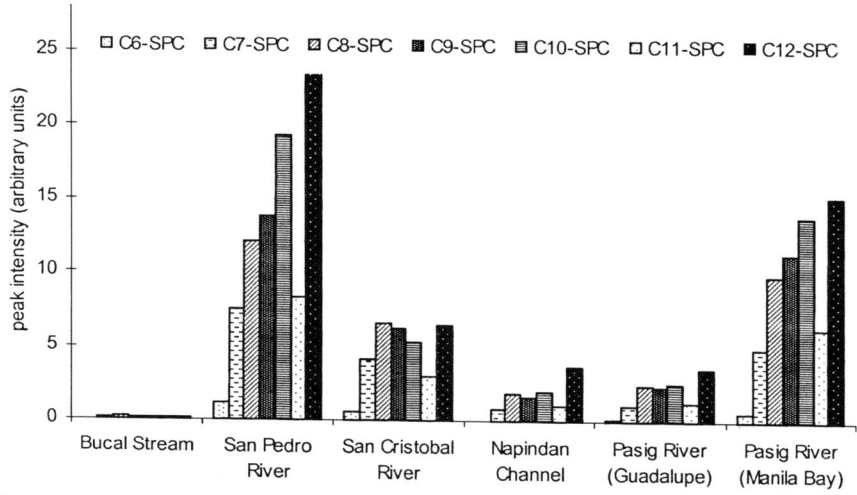

Fig. 6.3.7. Semi-quantitative analysis of C_6–C_{12}-SPC in investigated surface waters (sample batch from December 1999) (from Ref. [23]).

shortening via β-oxidation. The exceptional behaviour of C_{10}-SPC with its relatively high percentage of branched SPC could be assigned to the fact that it was likely to be formed by terminal oxidation of C_{10}-ABS

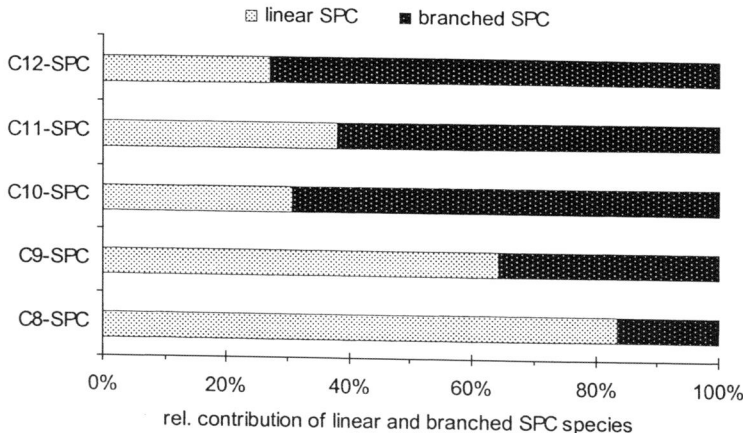

Fig. 6.3.8. Relative peak intensities of linear and branched SPC homologues determined by LC–ESI–MS/MS recording the ion transitions $[M - H]^- \rightarrow m/z$ 183 for linear SPC and $[M - H]^- \rightarrow m/z$ 197 for branched SPC.

as well as by β-oxidation of C_{12}-ABS, which was the most prominent alkyl homologue in the mixture (Fig. 6.3.8).

REFERENCES

1 J.C. Westall, H. Chen, W. Zhang and B.J. Brownawell, *Environ. Sci. Technol.*, 33 (1999) 3110.
2 E. Matthijs and H. de Henau, *Tenside Surf. Det.*, 22 (1985) 299.
3 C.F. Tabor and L.B. Barber, *Environ. Sci. Technol.*, 30 (1996) 161.
4 D.C. McAvoy, W.S. Eckhoff and R.A. Rapaport, *Environ. Toxicol. Chem.*, 12 (1997) 977.
5 J. Waters and T.C.J. Feijtel, *Chemosphere*, 30 (1995) 1939.
6 P. Schöberl, R. Spilker and L. Nitschke, *Tenside Surf. Det.*, 31 (1994) 243.
7 P. Schöberl, *Tenside Surf. Det.*, 32 (1995) 25.
8 P. Schöberl, H. Klotz and R. Spilker, *Tenside Surf. Det.*, 33 (1996) 329.
9 M. Tippner, *Z. Wasser Abwasserforsch.*, 25 (1992) 46.
10 L. Cavalli, G. Cassani, L. Vigano, S. Pravettoni, G. Nucci, M. Lazzarin and A. Zatta, *Tenside Surf. Det.*, 37 (2000) 282.
11 A. DiCorcia, L. Capuani, F. Casassa, A. Marcomini and R. Samperi, *Environ. Sci. Technol*, 33 (1999) 4119.
12 M.L. Trehy, W.E. Gledhill, J.P. Mieure, J.E. Adamove, A.M. Nielsen, H.O. Perkins and W.S. Eckhoff, *Environ. Toxicol. Chem.*, 15 (1996) 233.
13 L. Sarrazin, T. Schembri and P. Rebouillon, *Toxicol. Environ. Chem.*, 69 (1999) 487.
14 P. Schöberl and R. Spilker, *Tenside Surf. Det.*, 33 (1996) 400.
15 W.-T. Ding, S.-H. Tzing and H.-J. Lo, *Chemosphere*, 38 (1999) 2597.
16 H. Takada, N. Ogura and R. Ishiwatari, *Environ. Sci. Technol.*, 26 (1992) 2517.
17 P. Eichhorn, S.V. Rodrigues, W. Baumann and T.P. Knepper, *Sci. Total Environ.*, 284 (2002) 123.
18 R.J. Larson and A.G. Payne, *Appl. Environ. Microbiol.*, 41 (1981) 621.
19 R.D. Swisher, *Surfactant Biodegradation*, Marcel Dekker, New York, 1987.
20 R. Paranhos, S.M. Nascimento and L.M. Mayr, *Fresenius Environ. Bull.*, 4 (1995) 352.
21 G. Perin, R. Fabris, S. Manente, A.R. Wagener, C. Hamacher and S.A. Scotto, *Water Res.*, 31 (1997) 3017.
22 P. Eichhorn, PhD Thesis, University of Mainz, Germany (2001); http://deposit.ddb.de/cgi-bin/dokserv?idn = 962727733.
23 P. Eichhorn, M.E. Flavier, M.L. Paje and T.P. Knepper, *Sci. Total Environ.*, 269 (2001) 75.

24 D.L. Wharry, K.L. Matheson and J.L. Berna, *Chem. Age India*, 38 (1987) 87.
25 H.L. Webster and J. Halliday, *Analyst*, 84 (1959) 552.
26 R.D. Swisher, *Tenside*, 6 (1969) 135.
27 R.E. McKinney and J.M. Simmons, *Sewage Ind. Wastes*, 31 (1959) 549.

6.4 NON-IONIC SURFACTANTS IN MARINE AND ESTUARINE ENVIRONMENTS

Niels Jonkers and Pim de Voogt

6.4.1 Introduction

Whereas the occurrence of surfactants in wastewaters and freshwater has received substantial interest over the last decade, and thus become fairly well documented, the amount of data on surfactants in estuarine or marine waters remains quite limited. However, it has become clear that knowledge of the behaviour of surfactants in freshwater cannot be easily extrapolated to saline waters [1], as both microbial activities and partition coefficients may vary strongly between the two environments.

In this chapter the data available for non-ionic surfactants in estuarine and marine environments are reviewed and discussed. The available information is confined almost exclusively to the surfactants of the alkylphenol ethoxylates (APEO) type, and their metabolites. For other non-ionic surfactants there are hardly any data available for the marine environment. The APEO metabolites reviewed here include the short-chain ethoxylates A_9PEO_1 and A_9PEO_2, resulting from deethoxylation, as well as the nonylphenoxy ethoxy carboxylates (A_9PEC) resulting from oxidation. These are all formed during aerobic biodegradation processes. Also nonylphenol (NP) will be reviewed, which may be formed during anaerobic degradation.

The available information has been grouped according to the obvious aqueous compartments (dissolved; particulate, biota), and includes some relevant atmospheric data. In addition a division into geographical areas has been used when trying to group the data.

Biodegradation is discussed as a key phenomenon in the (differences between marine and freshwater) occurrence and fate of non-ionic surfactants. Finally conclusions are drawn with regard to the marine fate of these compounds.

Comprehensive Analytical Chemistry XL
T.P. Knepper, D. Barceló and P. de Voogt (Eds)

6.4.2 Non-ionic surfactants in saline waters

Alkylphenol derivatives in the Mediterranean Sea
The behaviour of lipophilic metabolites of A_9PEO in the Venice lagoon, Italy, was studied in detail by Marcomini et al. [2,3]. Salinity in this estuary is in the range of 25–35‰, and both industrial and domestic wastewater is discharged into the lagoon. Water samples were analysed by high performance liquid chromatography with fluorescence detection (HPLC–FL) [2].

In an early study, rather high APEO concentrations have been found: 0.20, 0.73, 1.1 and 17.5 µg L^{-1} for NP, A_9PEO_1, A_9PEO_2 and A_9PEO_{3-13}, respectively [3]. In a second study, A_9PEO concentrations in water were reported in the range of 0.6–4.5 µg L^{-1}, with average ethoxylate chain lengths of 5–10 units [2]. The highest values were systematically found near an industrial wastewater discharge. The sum of the lipophilic metabolites $NP + A_9PEO_1 + A_9PEO_2$ never exceeded 10% of total A_9PEO, reflecting an ineffective wastewater treatment.

In a later study by the same group, the carboxylated metabolites were also included in the analysis [4]. This time, water samples were taken in the Venice lagoon as well as in the Adriatic Sea and in a river reaching the estuary.

A_9PEO concentrations in the lagoon were found to range between 0.6 and 6.8 µg L^{-1}, while in the river this range was 14–62 µg L^{-1}. For A_9PEC, the concentration ranges were 0.3–6.2 and 7.3–225 µg L^{-1} in the lagoon and the river, respectively. Metabolites-to-surfactant ratios A_9PEC/A_9PEO amounted to 2–3 and 0.5 in the river and the lagoon, respectively. Therefore, it was suggested that the main biodegradation mechanisms are different for the estuary and the river. In the estuary, the hydrolytic biodegradation mechanism (hydrolytic shortening of the ethoxylate chain) is thought to prevail over the oxidative–hydrolytic mechanism (first an oxidation of the terminal hydroxylic group, then shortening of the ethoxylate chain), leading to a faster disappearance of A_9PEC from the estuary.

Spatial distributions of A_9PEO and A_9PEC were found to be rather homogeneous throughout the lagoon and the sea samples. However, elevated concentrations were present in samples taken near an industrial area and near a shipyard.

A seasonal dependence of water concentrations was observed, with winter concentrations of both surfactants and metabolites more than two times higher than those in summer. This was ascribed to increased biodegradation at temperatures $>20°C$. NP concentrations were below the detection limit in all water samples.

An extensive study of the behaviour of A_9PEO and their non-ionic metabolites was performed in the Krka estuary in Croatia, which revealed some interesting features involving the vertical stratification of the water column [5].

This stratified estuary has a depth of 40 m, with an upper fresh or brackish water layer of 0.2–4 m, depending on the river flow. The main source of pollution is untreated municipal wastewater, which is discharged into the estuary. Water samples were collected at different distances from these sewage outlets at two water depths: from the fresh and the marine water layers. Furthermore, at one location, a vertical profile of the water column was made, including a sample of the water surface micro layer. Total A_9PEO_n, and individual A_9PEO_1, A_9PEO_2 and NP concentrations were determined with normal phase HPLC–FL analysis.

In the wastewater, an average ethoxylate chain length of 10 was found, with total A_9PEO_n concentrations of 500 $\mu g\,L^{-1}$ [5]. Of these A_9PEO_n, 6–60% was sorbed onto suspended solids. One likely removal process of A_9PEO_n from the estuarine water column is therefore sedimentation of suspended matter. However, no sediments were analysed to test this hypothesis.

Water concentrations of A_9PEO_n were high at 1 m from the sewage outlet (8.5–89 $\mu g\,L^{-1}$), but decreased to below 1 $\mu g\,L^{-1}$ at 25 m distance, while further concentration decrease was slow. Concentration ranges of NP, A_9PEO_1 and A_9PEO_2 were <0.02–1.2, <0.02–0.44 and <0.020–0.3 $\mu g\,L^{-1}$, respectively [6]. The NP and A_9PEO_{1+2} concentrations represent 3–4% of the total surfactant-derived nonylphenolic compounds, which is considerably less than for situations where wastewater is treated before discharge into natural water (with values of usually more than 30% [7]).

The concentrations of both parent surfactant and lipophilic metabolites decreased sharply with increasing distance from the sewage outlet. At a distance of 100 m, the concentrations had all dropped to below 0.5 and 0.1 $\mu g\,L^{-1}$ for A_9PEO_n and metabolites, respectively. The decrease was mainly explained by an efficient dilution. The concentrations were significantly higher near the water surface, as a consequence of the wastewater plume spreading primarily into the upper freshwater layer, while leaving the saline layer less affected.

When going downstream in the estuary, a small increase in A_9PEO_1 and A_9PEO_2 concentrations was observed, indicating that biodegradation is likely to occur. However, the extent of biodegradation was not as significant as in some freshwater systems. It was concluded that biodegradation plays only a limited role in the fate of APEO in the estuary [5].

An interesting vertical profile of the metabolite concentrations was observed: the compounds showed a tendency to accumulate at the two-phase boundaries of air–freshwater and freshwater–saline water (the halocline). Thus, concentration maxima were observed at depths of 0 and 2 m (see Fig. 6.4.1) [6]. The observed distribution may result from either the physicochemical properties of these compounds (surface activity and hydrophobicity), or their formation at the interface due to increased biological activity. For the parent surfactants a similar but less pronounced vertical distribution pattern was observed (with maxima at 0 and 2 m of 17 and 9 $\mu g \, L^{-1}$, respectively) [5].

In seawater near Barcelona, an A_9PEO concentration of 0.85 $\mu g \, L^{-1}$ was found, using gas chromatography mass spectrometry (GC–MS) analysis. No A_9PEO were detected in the particulate phase of the samples [8].

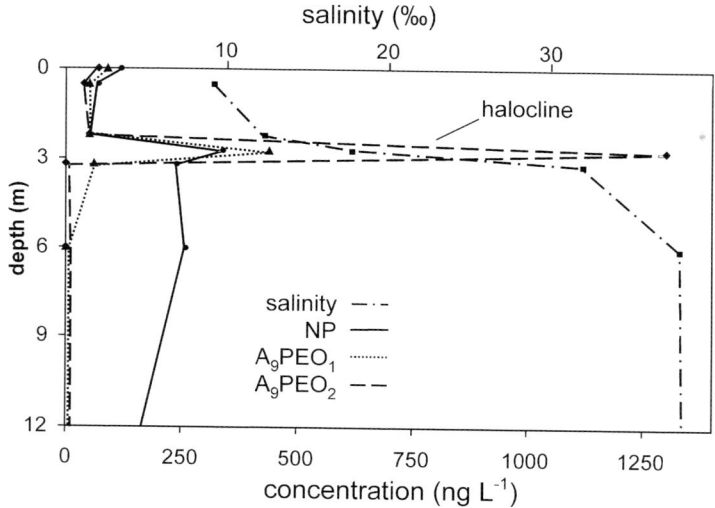

Fig. 6.4.1. Change in concentrations of non-ionic surfactants with depth in a haloclinic waterbody (adapted from Kveštak et al. [6]).

In a more recent study in the same area, A_9PEO water concentrations of $<0.2-4.8\ \mu g\ L^{-1}$ were found using LC–MS analysis. In the same study, other coastal areas of Spain showed maximum concentrations of $11\ \mu g\ L^{-1}$ at a site that receives large amounts of industrial and domestic wastewater. For NP a concentration range of $<0.15-4.1\ \mu g\ L^{-1}$ was reported, and A_9PE_1C were below detection limits $(0.19\ \mu g\ L^{-1})$ in all samples [9].

Very high APEO concentrations in coastal water were reported from Israel. At 10 locations along the coast, water samples were taken, all near the mouth of a river or stream. Samples were taken 2–3 m from the shoreline. With reversed phase HPLC, total APEO concentrations were determined. The levels varied from 4.2 to as much as $25\ \mu g\ L^{-1}$, while concentrations in the rivers themselves were in the range of $12.5-75\ \mu g\ L^{-1}$. In samples taken further offshore (between 50 and 150 km), concentrations had decreased to $<1.0-2.6\ \mu g\ L^{-1}$. In all samples, the higher oligomers (around 10 ethoxylate units) were most abundant. Unfortunately, as very little analytical details were reported, the validity of the results cannot be completely established [10].

Alkylphenol derivatives in the North Sea and UK coastal waters
Blackburn and Waldock [11] undertook a survey of water concentrations of alkylphenols (APs) in rivers and estuaries in England and Wales. Locations were selected to represent a wide range of possible AP inputs, from both agricultural and industrial sources. For all samples, both filtered and unfiltered water were treated, to determine both the dissolved and the total extractable APs by GC–MS.

River concentrations were generally in the range of $<0.2-12\ \mu g\ L^{-1}$ total extractable NP, with one exceptional river (the Aire river) containing $180\ \mu g\ L^{-1}$, located in an area of heavy textile industry. Estuarine levels were generally one order of magnitude less, with total extractable NP concentrations in the range of $<0.08-0.32\ \mu g\ L^{-1}$. In over 80% of the estuarine samples, total NP concentrations were below $0.1\ \mu g\ L^{-1}$. However, in the Tees estuary (unrelated to the Aire river) exceptionally high values were found, with a total extractable NP concentration of $5.2\ \mu g\ L^{-1}$ and an octylphenol (OP) concentration of $13\ \mu g\ L^{-1}$. This is the only estuarine site where OP was present above the detection limit of $0.1\ \mu g\ L^{-1}$. The values can be explained by the industrial activities taking place near the estuary: a road tanker cleaning facility, and the major UK production facility for A_9PEO.

A comparison between the dissolved and total concentrations revealed that at very high concentrations, the majority of NP is present in the particulate phase, while at lower concentrations ($<20\ \mu g\ L^{-1}$), the greater part appears to be in solution.

A subsequent study in the same areas also included biological samples [12]. Again, the Aire river contained the highest concentrations, with total extractable NP + A_9PEO_{1+2} concentrations of $15-76\ \mu g\ L^{-1}$. In many samples of this river, levels exceed the no observed effect concentration (NOEC) for vitellogenin induction in trout ($5-20\ \mu g\ L^{-1}$). However, in most samples of the other investigated rivers, levels were $<1\ \mu g\ L^{-1}$. OP levels were invariably below $0.5\ \mu g\ L^{-1}$.

In the UK estuaries, NP and APEO were only detected in two out of the nine investigated areas. The highest estuarine water concentrations were found in the Tees estuary, as was the case in the preceding study. The majority of these APEO were present in the particulate phase ($>75\%$).

Concentrations increased from <0.2 to $5.8\ \mu g\ L^{-1}$ for NP and from <0.6 to $76\ \mu g\ L^{-1}$ for A_9PEO_{1+2} when going downstream in the estuary, indicative of the input of APEO directly into the estuary by industrial activities in the area. These results demonstrate that although in most riverine and estuarine environments NP and A_9PEO_{1+2} concentrations are below the threshold to affect the investigated fish species, local concentrations in areas directly impacted by industrial effluents can reach concentrations far above NOEC levels.

A clear indication of APEO biodegradation in estuarine waters was found in the Scheldt estuary in The Netherlands [13]. In water samples with salinities ranging from 0.5 to 35‰, APEO and the carboxylated metabolites A_9PEC were analysed using LC−electrospray (ES)−MS. Upstream in the estuary, A_9PEO concentrations of $2.3\ \mu g\ L^{-1}$ were found, while the total A_9PEC concentration was much higher: $12\ \mu g\ L^{-1}$. When going downstream in the estuary, A_9PEO concentrations decreased rapidly, while the A_9PEC concentration decreased much more slowly, suggesting the formation of the latter out of the former compounds. A_9PEO oligomer distributions remained constant throughout the estuary. These results indicate that in contrast to the Venice lagoon, Italy [4], in the Scheldt estuary the oxidative-hydrolytic biodegradation mechanism predominates over the hydrolytic mechanism. At the mouth of the estuary, APEO and APEC concentrations were 0.04 and $0.09\ \mu g\ L^{-1}$, respectively. The NP concentrations decreased from 0.9 upstream to $0.04\ \mu g\ L^{-1}$ at the mouth of the estuary.

A survey of Dutch coastal waters was carried out in spring, summer and autumn of 1999 [14] using normal and reversed HPLC–FL. In general, dissolved concentrations in surface waters appeared to be below the limit of quantification ($\sim 0.5\ \mu g\ L^{-1}$). Interestingly, the exceptions were found in summer in the western Scheldt and the Rhine/Meuse estuaries, when levels ranged between 2 and 3 $\mu g\ L^{-1}$. At the same locations in these estuaries sampled in spring and autumn, concentrations were below quantification limits. Elevated summer concentrations may be caused by reduced water flows and by seasonal increase in detergent use.

Alkylphenol derivatives along the Atlantic coast, USA
A study by Ferguson et al. focused on the fate of short-chain A_9PEO and A_8PEO (1–3 ethoxylate units) and AP in the Jamaica Bay estuary in New York, USA [15]. This study also included halogenated APs (see Chapter 6.6), which can be formed during wastewater chlorination treatment. The largest input of freshwater into the bay is biologically treated wastewater, which has a residence time in the bay of approximately 35 days.

Water samples were analysed by reversed phase LC–ES–MS. Concentrations of 0.08–0.42 and 0.16–0.94 $\mu g\ L^{-1}$ were determined for NP and A_9PEO$_{1-3}$, respectively. Concentration levels of A_8PEO were lower by more than an order of magnitude, and halogenated APs were not detected in the water. The APEO metabolites showed a strong correlation with the sewage tracer silver, indicating a wastewater source of these compounds.

Although there is only a limited salinity gradient in the Jamaica Bay estuary (from 25 to 31‰), it was shown that the ratio A_9PEO$_{1-3}$/NP decreased with salinity, indicating that degradation occurs in the water (at a faster rate than possible volatilisation of NP). The total A_9PEO$_{0-3}$ concentration was, in general, higher at lower salinities. These trends were not observed for A_8PEO, and it was suggested that A_8PEO sources other than wastewater are present, possibly transporting A_8PEO into the bay from the marine side.

APEO concentrations in suspended matter samples were similar to the sediment concentrations. From these values, in situ organic carbon corrected suspended matter/water distribution coefficients ($\log K_{oc}$) were calculated. For the separate ethoxymers NP, A_9PEO$_1$, A_9PEO$_2$,

725

A$_9$PEO$_3$ and OP, log K_{oc} values of 5.4, 5.5, 5.2, 4.9 and 5.2, respectively, were calculated.

In another estuary near New York, namely the Hudson estuary, NP was analysed by GC–MS. Dissolved water concentrations ranged from 0.012 to 0.094 μg L^{-1}, while the concentrations of NP in the particulate phase of the water amounted to between 0.45 and 4.0 μg g^{-1} [16].

Other non-ionic surfactants in the Mediterranean Sea
Data on the marine occurrence of non-ionic surfactants other than APEO is very limited. Petrovic et al. reported alcohol ethoxylate (AE) concentrations along the coast of Spain of < 0.1–15 μg L^{-1} and coconut diethanol amide (CDA) concentrations of < 0.05–24 μg L^{-1}, the highest concentrations present at sites receiving high amounts of industrial and domestic wastewater [9].

The only study available on metabolites of AE was performed by Crescenzi et al. [17]. The initial biodegradation of AE occurs by cleavage at the ether bridge between the alkyl and ethoxylate chain, resulting in polyethylene glycols (PEG) and alcohols. In consecutive oxidation steps, the PEG chains are shortened and mono- and dicarboxylated metabolites (MCPEG and DCPEG, respectively) are formed.

The PEG, MCPEG and DCPEG metabolites were monitored in marine, estuarine, river and wastewater near and in the Italian river Tiber. Water samples were analysed by LC–ES–MS analysis. Suspended particulate matter (SPM) was also analysed, but none of the metabolites was detected in this matrix.

In the river water, PEG ranging from 6 to 20 ethoxylate units (average of 12 units) were found at 23 μg L^{-1}. The MCPEG (6–13 units) and DCPEG (4–20 units) were present at 1 and 7 μg L^{-1}, respectively. In the estuarine zone, total PEG concentrations of around 68 μg L^{-1} were found at a salinity of 1.3‰, decreasing to 0.5 μg L^{-1} at a salinity of 36‰. The average PEG chain length decreases slightly from 11 at salinity 1.3 to 8 at salinity 36‰. DCPEG are present in higher concentrations than MCPEG, and increasingly dominate the MCPEG at higher salinities. Total concentrations for MCPEG and DCPEG are 2.1 and 8.1 μg L^{-1}, respectively, at salinity 1.3, decreasing to < 0.1 and 0.7 at salinity 36‰.

The data suggest that in a saline medium the biodegradation of PEG continues and that the main biodegradation mechanism is simultaneous oxidation of both ends of the PEG chain, leading to the predominating presence of the dicarboxylated DCPEG metabolites.

An overview of non-ionic surfactant concentrations in saline waters can be found in Table 6.4.1.

6.4.3 Non-ionic surfactants in estuarine and marine sediments

The effect of salinity on sediment–water partition coefficients of surfactants is still a matter of scientific debate. Increase in salt concentration can lead to salting-out effects, increasing sorption coefficients, but alternatively, addition of certain electrolytes can decrease sorption of the polar part of the surfactant by blocking sorption sites in some fashion [18]. In a recent study, increasing salt concentrations were found to slightly increase sorption for the AE $A_{13}EO_3$, while for $A_{13}EO_9$ a small decrease was found [19]. Other parameters, like sediment properties [20] and the alkyl and the ethoxylate chain lengths of the surfactant [21], have a more significant influence on the sorption coefficient.

In Table 6.4.2, concentrations of non-ionic surfactants in estuarine and marine sediments are listed.

Mediterranean Sea
The environmental occurrence of the lower A_9PEO oligomers in several types of sediments was studied in the Venice lagoon [2]. A distinction was made between sediments covered with a layer of macroalgae (which are blooming massively in spring–summer), and sediments free of macroalgae.

A special device was used to collect artificially resuspended sediment. In this device, the sediment was placed in a tube with overlying water. An oscillating grid placed in the water above the sediment caused a steady state resuspension equilibrium, under the same conditions as typically generated by winds and currents in the lagoon. This overlying water was then filtered, and the resuspended material analysed.

Sediment concentrations of NP + A_9PEO_{1+2} varied from 0.018 to 0.129 $\mu g\,g^{-1}$ (dry weight). However, in the sediments covered with a layer of macroalgae biomass, 3–10 times higher concentrations were found. Concentrations in 10 samples of the macroalgae themselves were 0.25 $\pm$ 0.15 $\mu g\,g^{-1}$ dry weight.

TABLE 6.4.1

Reported concentrations of non-ionic surfactants and their metabolites in marine and estuarine waters

Sampling area	Compartment	Compounds analysed	Concentration (μg L^{-1})	Reference
Venice lagoon (Italy)	Estuarine water	NP	0.20	[3]
		A_9PEO_1	0.73	
		A_9PEO_2	1.1	
		A_9PEO_{3-13}	17.5	
Venice lagoon (Italy)	Estuarine water	A_9PEO_{1-14}	0.6–4.5	[2]
Venice lagoon (Italy)	Estuarine + river water	NP	<0.2	[4]
	Estuarine water	A_9PEO_{1+2}	0.6–6.8	
		$A_9PE_{1+2}C$	0.3–6.2	
	River water	A_9PEO_{1+2}	14–62	
		$A_9PE_{1+2}C$	7.3–225	
Barcelona (Spain)	Marine water	A_9PEO_{0-6}	0.85	[8]
	Marine suspended matter	A_9PEO_{0-6}	ND	
Krka estuary (Croatia)	Estuarine water	NP	<0.02–1.20	[6]
		A_9PEO_1	<0.02–0.44	
		A_9PEO_2	<0.020–1.30	
Krka estuary (Croatia)	Brackish estuarine water layer	A_9PEO_{3-18}	1.1–6	[5]
	Saline estuarine water layer	A_9PEO_{3-18}	0.1–0.7	
English estuaries	Estuarine water	NP	<0.08–5.2	[11]
		OP	<0.1–13	
English estuaries	Estuarine water	NP	<0.2–5.8	[12]
		A_9PEO_{1+2}	<0.6–76	
Hudson estuary (USA)	Estuarine water	NP	0.012–0.095	[32]
	Estuarine suspended matter	NP	0.0026–0.022	
Hudson estuary (USA)	Estuarine water, dissolved	NP	0.012–0.094	[16]
	Estuarine water, particulate phase	NP	0.0026–0.0039	
Jamaica Bay (USA)	Estuarine water	NP	0.077–0.416	[15]
		A_9PEO_{1-3}	0.158–0.942	
		OP	0.0016–0.0083	
		A_8PEO_{1-3}	0.0055–0.034	
		Halogenated NP	ND	
Israeli coast	Marine water	A_9PEO_{1-16}	<1.0–25	[10]

TABLE 6.4.1 *(continued)*

Sampling area	Compartment	Compounds analysed	Concentration (μg L^{-1})	Reference
Scheldt estuary (The Netherlands)	Estuarine water	A_9PEO_{1-12}	0.04–2.3	[13]
		$A_9PE_{1-6}C$	0.09–12	
		NP	0.04–0.9	
Scheldt (NL)	Estuarine water	$NPEO_{1-10}$	2.0–2.5	[14]
		NP	2.0	
Rhine/Meuse at Haringvliet (NL)	Estuarine water	$NPEO_{1-10}$	2.7	[14]
Rhine/Meuse at Maassluis (NL)	Estuarine water	$NPEO_{1-10}$	2.6	[14]
Tiber estuary (Italy)	Estuarine water	PEG_{4-21}	0.5–68	[17]
		$MCPEG_{4-18}$	<0.05–2.1	
		$DCPEG_{4-18}$	0.8–8.1	
Elbe (Germany)	Estuarine water	NP	0.0087–0.084	[39]
		A_9PEO_1	0.01–0.111	
		A_9PEO_2	0.0024–0.024	
Elbe & Weser estuaries (Germany)	Estuarine water	NP	0.033	[40]
		$A_9PEO_{>2}$	<0.005	
Coast of Spain	Coastal water	AE	<0.1–15	[9]
		CDA	<0.05–24	
		A_9PEO	<0.2–11	
		NP	<0.15–4.1	
		A_9PE_1C	<0.1	

ND, not detected.

A strong seasonal dependence of the lipophilic metabolites' concentration in resuspended material was observed: from an average of 5.8–6.7 to 0.5–1.0 μg g^{-1} dry weight in February and July, respectively. Concentrations of APEO in resuspended material were higher by one order of magnitude than in the sediment. These concentration ratios were also observed for polychlorinated biphenyls in the same sediment and resuspended matter, which are assumed to be hardly affected by biodegradation. Consequently, for APEO these concentration ratios were attributed to dilution during sedimentation rather than to biodegradation in the sediment.

Both the NP + A_9PEO_{1+2} concentrations in sediment with and without overlying macroalgae and the increased concentrations in resuspended material during the algal bloom showed the important role played by algae in the environmental fate of A_9PEO.

TABLE 6.4.2

Reported concentrations of non-ionic surfactants and their metabolites in marine and estuarine sediments

Sampling area	Compartment	Compounds analysed	Concentration ($\mu g\, g^{-1}$)	Reference
Venice lagoon (Italy)	Estuarine sediment	NP	$0.005-0.04^{a}$	[2]
		A_9PEO_1	$0.009-0.08^{a}$	
		A_9PEO_2	$0.003-0.02^{a}$	
	Resuspended estuarine sediment	NP	$<0.1-5.6^{a}$	
		A_9PEO_1	$0.2-6.6^{a}$	
		A_9PEO_2	$<0.1-1.5^{a}$	
California (USA)	Marine sediment	NP	<0.002	[30]
		A_9PEO_1	<0.002	
		A_9PEO_2	<0.002	
Barcelona (Spain)	Marine sediment	NP	$0.006-0.07$	[22]
Barcelona (Spain)	Marine sediment	A_9PEO_{0-6}	6.6	[8]
Nile estuary (Egypt)	Estuarine sediment	NP	$0.019-0.044$	[22]
English estuaries	Estuarine sediment	NP	$<0.1-1.7$	[12]
		A_9PEO_{1-2}	$<0.5-3.6$	
English estuaries	Estuarine sediment	OP	$0.002-0.340^{a}$	[24]
		NP	$0.03-9.05^{a}$	
		A_9PEO_1	$0.16-3.97^{a}$	
UK estuaries	Estuarine sediment	A_9PEO_{1-3}	$0.02-0.40^{a}$	[23]
German estuaries	Estuarine sediment	A_9PEO_{1-3}	$0.03-0.12^{a}$	[23]
S-Norway/W-Sweden estuaries	Estuarine sediment	A_9PEO_{1-3}	$0.03-0.15^{a}$	[23]

TABLE 6.4.2 (continued)

Sampling area	Compartment	Compounds analysed	Concentration ($\mu g\,g^{-1}$)	Reference
French/Belgian/Dutch North Sea coast	Estuarine sediment	A_9PEO_{1-3}	0.02–0.31[a]	[23]
Dutch coast	Estuarine sediment	A_9PEO_{1-10}	0.04–0.25[a]	[14]
Tokyo Bay (Japan)	Estuarine sediment	NP	0.12–0.64[a]	[25]
		OP	0.006–0.01[a]	
		A_9PEO_1	0.03[a]	
Strait of Georgia (Canada)	Estuarine sediment	A_9PEO_{1-19}	0.300–3.0	[28]
Jamaica Bay (USA)	Estuarine sediment	A_9PEO_{0-3}	0.05–30	[15]
		A_8PEO_{1-3}	<0.005–0.09	
		Halogenated NP	<0.001–0.027	
Scheldt estuary (The Netherlands)	Estuarine sediment	A_9PEO_{1-12}	<0.0003–0.240[a]	[13]
		NP	<0.0003–1.080[a]	
Elbe estuary (Germany)	Estuarine sediment	NP	0.370–0.480[a]	[39]
		A_9PEO_1	0.750–0.890[a]	
		A_9PEO_2	0.970–1.150[a]	
Coast of Spain	Marine and harbour sediment	AE	0.037–1.30	[9]
		CDA	0.030–2.70	
		A_9PEO	0.010–0.620	
		NP	<0.010–1.050	

[a]Calculated for dry weight.

731

In a later study in the Venice lagoon, a dated sediment core was analysed [4]. Seven separate layers were analysed, dating from 1972 back to 1910.

When analysing back to the 1950s, A_9PEO_1 and A_9PEO_2 decrease four- and three-fold, respectively, while NP decreases only two-fold. This is an indirect sign of the non-oxidative (anaerobic) biodegradation of A_9PEO_{1+2} to NP. From 1950 to 1920, surprisingly, an increase of NP was observed. This was ascribed to post-depositional vertical transport through the sediment layer caused by pore-water diffusion.

From a comparison of calculated sedimentary annual fluxes of A_9PEO (48 ng cm^{-2}), the area of the lagoon and the annual loading of A_9PEO into the lagoon, it was estimated that sedimentation contributes less than 0.1% to the total removal of A_9PEO from the estuary, and this process is therefore negligible compared to biodegradation.

Chalaux et al. investigated dumped sewage sludge and marine sediments near Barcelona, with GC–MS analysis [22]. In the sludge, NP concentrations were similar to the previous work: 20–350 μg g^{-1}. NP was found in the marine sediments in the range of 0.006–0.069 μg g^{-1}. In the same area, A_9PEO sediment concentrations of 6.6 μg g^{-1} were found, using GC–MS analysis [8].

More recently, Petrovic et al. [9] reported NP sediment concentrations of <0.01–0.45 μg g^{-1} at the coast of Barcelona, using LC–MS analysis. Other sites along the Spanish coast showed a maximum NP concentration of 1.05 μg g^{-1} at a site in the proximity of the outflow of partially untreated wastewater. AEs were detected between 0.037 and 1.30 μg g^{-1}, and curiously, the supposedly easily biodegradable CDAs were detected in all sediment samples, in the range of 0.03–2.71 mg g^{-1}.

Chalaux et al. reported NP concentrations in sediments from the Nile estuary of 0.044 μg g^{-1} NP [22].

North Sea and UK coastal waters

De Voogt et al. [23] analysed marine and estuarine sediments from 22 sites in northwestern Europe (extending from Ireland and France to Norway and Sweden) by HPLC–FL. NP, OP, A_9PEO and A_8PEO concentration ranges of 0.1–17, <LOD-2, 12–400 and 0.2–16 ng g^{-1} dry weight were found, respectively. The highest levels were found in the estuaries of the rivers Seine, Mersey, Rhine/Meuse, Weser and Elbe.

Of nine estuaries investigated in England by Blackburn et al., only the sediments from the highly industrialised Tees estuary contained detectable concentrations of NP and A_9PEO_{1+2} [12]. In this estuary, also the highest APEO metabolite concentrations in water were found. GC–MS analysis showed maximum sediment concentrations of 1.7 μg g^{-1} NP and 3.6 μg g^{-1} A_9PEO_{1+2}.

Lye et al. [24] found higher NP concentrations in sediment from the same Tees estuary. The maximum NP level found by GC–FID analysis was 9.05 μg g^{-1}, upstream of the estuary, and then a clear decreasing trend was observed down to 1.60 μg g^{-1} at the mouth of the estuary. A similar pattern was found for A_9PEO_1 (from 3.97 to 0.120 μg g^{-1} downstream) and OP (from 0.34 to 0.03 μg g^{-1}). This trend is somewhat surprising, as Blackburn et al. reported water concentrations of NP and A_9PEO_{1+2} increasing by factors of 30–126 when going downstream in the Tees estuary in the preceding study [12]. In another English estuary, concentration ranges of 0.03–0.080, 0.16–1.40 and 0.002–0.020 μg g^{-1} for NP, A_9PEO_1 and OP were found, respectively [24].

In general, the sediment levels reported by de Voogt et al. [23], Lye et al. [24] and Blackburn et al. [12] for UK estuaries are in reasonable agreement.

Jonkers et al. used LC–ES–MS to analyse sediments from the Scheldt estuary, The Netherlands [13]. Sediment concentrations ranged from <0.0003 to 0.242 and from <0.0003 to 1.080 μg g^{-1} dry weight for A_9PEO and NP, respectively. The highest concentrations were found near an industrial area and a commercial harbour. In most sediments, no A_9PEC were found.

In a Dutch survey conducted in 1999 [14], A_9PEO and NP were found to be present at detectable levels in most (42 out of 54) sediment and SPM samples taken, whereas A_8PEO and OP levels were below the limits of detection. Both suspended matter and bottom sediments were sampled. Figure 6.4.2 shows mean concentrations of A_9PEO and NP in SPM and sediments from freshwater and marine waters, normalised to organic carbon. These are the means calculated from all samples analysed in the Dutch survey, including freshwater sediments. Marine SPM generally contained higher levels of A_9PEO (from <0.01 to 3.5 μg g^{-1} dry wt.) than marine sediments (<0.01–0.3 μg g^{-1}). A similar trend was found for NP, with levels between <0.01 and 4 μg g^{-1} in SPM and <0.01–0.3 μg g^{-1} in sediments. For the A_9PEO the difference between SPM and sediment concentrations is much larger, however, than for the NP, as can be seen in Fig. 6.4.2. This is

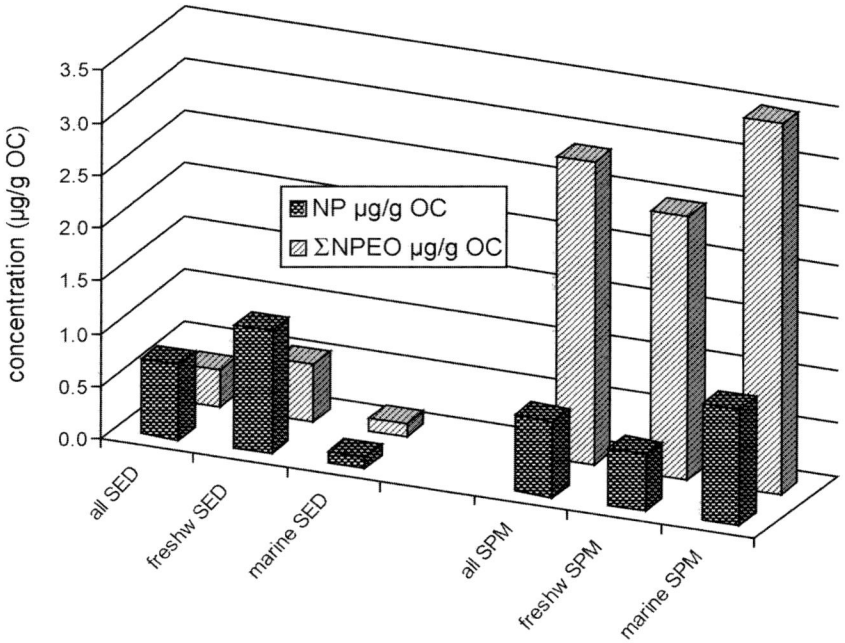

Fig. 6.4.2. Mean concentrations of ΣA_9PEO and NP in sediments and particulate matter collected in freshwater in The Netherlands and marine water from the Dutch coastal zone of the North Sea. For further explanation, see text (source: de Voogt et al. [14]).

probably due to the relatively rapid biodegradation of the A_9PEO sorbed to SPM.

Levels in SPM tended to decrease along the Dutch coast in a northerly direction with increasing distance from the main (Rhine/Meuse and Scheldt) estuaries. Locally in the western Scheldt estuary near the same industrial area mentioned in the preceding paragraph, levels in SPM up to 22 µg A_9PEO g^{-1} were reported.

A comparison of Rhine, Meuse and Scheldt estuaries revealed that in the Rhine estuary, with its relatively short residence times of the water, concentrations on estuarine SPM were similar to or higher (ratios estuarine/freshwater >1) than those upstream (i.e. on freshwater SPM from the river), whereas in Meuse and Scheldt estuaries with longer residence times of the water, particulate matter concentrations of A_9PEO are much lower (ratios <1) than in samples taken upstream (see Fig. 6.4.3). For NP this phenomenon is also

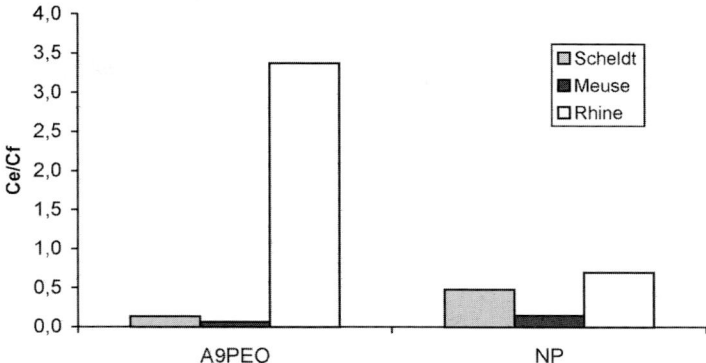

Fig. 6.4.3. Ratios of estuarine relative to freshwater concentrations on SPM of A_9PEO and NP observed in the Dutch rivers Scheldt, Meuse and Rhine. C_e, concentration on estuarine SPM; C_f, concentration on freshwater SPM. Each value represents the ratio of the means of three samples taken at salinity 0 and three samples taken at salinity $>30‰$ (source: de Voogt et al. [14]).

observed, although it is less pronounced (see Fig. 6.4.3). Probably, in estuaries with long residence times, biodegradation is likely to affect the APEO concentrations to a larger extent than in those estuaries with short residence times. In sediments from the same rivers a similar trend in ratios appeared to be present (data not shown) although the number of samples collected was much less in this case.

Pacific Ocean

Several studies have been performed on the fate of APEO in rivers flowing to Tokyo Bay, Japan, and in the bay itself [25,26]. Tokyo Bay receives vast amounts of treated and untreated domestic and industrial wastewater.

In the rivers, high concentrations of NP in sediment were observed [25] over relatively long stretches of about 10 km downstream from a sewage effluent discharge point (compared to elevated concentrations over a stretch of only 1 km length in a similar study conducted in the Detroit river [27]). Owing to the tidal current the horizontal mixing of surface sediments is possibly more extensive than in nontidal areas [25].

Concentrations in Tokyo Bay sediments were about one order of magnitude less than those in the river sediments: 0.12–0.64, 0.006–

0.01 and 0.03 μg g^{-1} dry weight for NP, OP and A$_9$PEO$_1$, respectively. A seaward concentration decrease in the estuary was observed [25].

In a dated sediment core from Tokyo Bay (going back to 1950), NP concentrations showed a clear maximum (0.53 μg g^{-1}) in the sediment layer deposited in the mid-1970s [25], around the time when legal regulations on industrial wastes were first implemented (although NP production has been ever increasing). The NP isomer composition was fairly constant in standard mixtures and all environmental compartments, including the layers of a dated sediment, indicating that the reactions occurring are not isomer selective.

In the Strait of Georgia, Canada, a study was performed on the fate of A$_9$PEO in marine sediments [28]. Box core sediment samples were taken in a region heavily impacted by two municipal wastewater effluents. The sewage only receives primary treatment, which is clearly reflected in the results of this study.

Normal phase LC–ES–MS was used for the analysis. In a previous work by the same authors, an interesting comparison was made between LC–FL, LC–UV and LC–ES–MS analysis of the same (spiked and unspiked) marine sediments [29]. The concentrations measured by LC–FL and LC–UV were consistently higher by about a factor of two than concentrations analysed by LC–ES–MS (however, with LC–UV analysis, for the unspiked sediments concentrations were all below the detection limit). This was attributed to the limited specificity of the LC–FL and UV techniques.

In the Strait of Georgia study, surface sediment concentrations in the vicinity of the effluent averaged 1.5 μg g^{-1}, with maximum concentrations of approximately 3 μg g^{-1}. At 20 km distance from the outfall, the concentration had dropped to 0.30 μg g^{-1} [28].

The ethoxymer distribution observed in the sediments was different from most other patterns reported in literature. Although A$_9$PEO$_1$ and A$_9$PEO$_2$ had the highest abundance, more than half of the total APEO had ethoxylate chains of more than 2 units. A second maximum was observed at around 9 ethoxylate units, which is the same maximum as for some commercial APEO products. The authors suggested that analyses that have not incorporated the higher oligomers in the past might have underestimated APEO concentrations by a factor of 2.

Some of the sediment cores were vertically sectioned and dated. Surprisingly, when going down in the sediments, no concentration decrease with depth was observed, and also the ethoxylate distribution remains the same. These results clearly show the lack of biodegradation once the A_9PEO have reached the marine sediment. Half-life values have been estimated to be more than 60 years. Although the observation that A_9PEO_1 and A_9PEO_2 exhibited a higher relative abundance than in commercial products points to some degradation, the lack of change in distribution in the sediment depth profiles indicates that these metabolites are formed prior to entering the sediments, e.g. in the municipal wastewater plant.

Despite this lack of degradation, the environmental significance of these compounds may be only minor, if preservation is accomplished by irreversible sorption to the sediments, which would render the A_9PEO non-bioavailable.

Surfactant markers were analysed by Chalaux et al. in marine sediments and sewage sludges [30] from California. The compounds investigated included APEO metabolites A_9PEO_1, A_9PEO_2 and NP, as well as linear alkylbenzenes (LAB) and trialkylamines (TAM), which are residues of anionic and cationic surfactants, respectively. Sediment cores were collected up to 60 km offshore.

With GC–FID analysis, NP, LAB and TAM were found in the sewage sludge in similar concentrations of 89–420 $\mu g\ g^{-1}$, while no $APEO_{1+2}$ were detected. In the marine sediments, LAB and TAM were detected, but all APEO metabolites were below the detection limit of 2 ng g^{-1}. The absence of NP in the sediments was attributed to its lower hydrophobicity compared to LAB and TAM.

Atlantic coast, USA

Ferguson et al. used reversed phase LC–ES–MS to analyse short-chain A_9PEO and A_8PEO (1–3 ethoxylate units), APs and halogenated APs in sediments from Jamaica Bay estuary in New York [15].

The concentration range of total A_9PEO_{0-3} observed in sediments was 0.05–30 $\mu g\ g^{-1}$, while A_8PEO were found at much lower concentrations of 0.0045–0.289 $\mu g\ g^{-1}$. This reflects the higher production and consumption of A_9PEO compared to A_8PEO.

The sediment concentrations were roughly correlated to the organic carbon content of the sediments, and the organic carbon normalised concentrations did not vary much throughout the basin. Only near one wastewater discharge were exceptionally high concentrations found.

The ethoxymers distribution appeared similar in all sediments, with A_9PEO_1 the dominant ethoxymer, and smaller but significant contributions of NP and A_9PEO_2. A_9PEO_3 represented less than 10% of the total concentration. For one sample, higher ethoxymers (4–15) were analysed, and these oligomers contributed only 8.5% to the total $\Sigma APEO$. As was reported previously [28] that APEO degradation in anaerobic environments is very slow, it is assumed that this APEO profile in the sediments (which are mostly anoxic, due to their high organic carbon content) is the result of degradation before the deposition, mainly during the biological wastewater treatment.

Chlorinated and brominated NPs were found in many sediment samples in the bay, with a maximum detected concentration of $0.027\ \mu g\ g^{-1}$ [15]. Their concentrations contributed less than 1% to the total A_9PEO metabolites, and therefore they are likely to be of little ecotoxicological concern.

Based on these monitoring results and some estimations of volatilisation and sediment burial rates, Ferguson et al. concluded that the estuarine fate of APEO metabolites in Jamaica Bay is determined mainly by sorption processes, degradation in the water, and advective transport out of the estuary [15].

6.4.4 Alkylphenol ethoxylates in estuarine and marine biota

Very few data are available for concentrations of non-ionic surfactants in biota. This is partly due to difficulties encountered in the analysis of these compounds in biological tissues [31]. The data reported in the literature for APEOs are shown in Table 6.4.3.

Bioconcentration factors (BCF) were determined in fish samples collected in the field as well as in experimentally exposed fish in a survey conducted in the UK [12]. The experimental BCF of NP was between 90 and 125, suggesting a moderate accumulation in rainbow trout muscle. Environmental BCF values for NP in fish muscle (for gudgeon, roach and chub) were between 10 and 50. For A_9PEO_{1+2}, a maximum BCF of 475 in chub liver was determined. A series of North Sea fish samples taken offshore contained no detectable APEO metabolites in liver or muscle tissue.

In another study on the Tees estuary, flounder tissue concentrations were reported in the range of $0.030–0.180\ \mu g\ A_9PEO\ g^{-1}$ wet weight, while A_9PEO_1 was not detected. In another estuary, NP was detected in three out of six fish tissues at levels between 0.005 and $0.055\ \mu g\ g^{-1}$ wet weight. A_9PEO_1 was observed in five out of six samples in the range of

TABLE 6.4.3

Reported concentrations of non-ionic surfactants and their metabolites in marine and estuarine biota

Sampling area	Compartment	Compounds analysed	Concentration (μg g^{-1})	Reference
Venice lagoon (Italy)	Macroalgae	NP + A_9PEO_{1+2}	0.25[a]	[2]
English estuaries	Marine fish	NP	<0.1	[12]
		A_9PEO_{1+2}	<0.5	
English estuaries	Estuarine fish	NP	0.005–0.180[b]	[24]
		A_9PEO_1	ND–0.240[b]	
Dutch coastal zone	Blue mussel	A_9PEO_{1+2}	<0.09[b]	[14]
	Flounder	A_9PEO_{1+2}	<0.04–0.5[b]	

ND, not detected.
[a]Calculated for dry weight.
[b]Calculated for wet weight.

0.190–0.240 μg g^{-1} wet weight. OP was found in only one fish sample (0.017 μg g^{-1}) [24].

In a Dutch survey, APEO and NP were determined in samples of blue mussel and flounder from the North Sea and adjacent coastal waters [14]. Levels were generally (in 85% of the 21 composite samples taken) below the limit of detection of 0.09 μg ΣAPEO g^{-1} (fresh weight), and maximum levels of 0.5 μg ΣAPEO g^{-1} and 0.12 μg NP g^{-1} in flounder were observed.

6.4.5 Alkylphenols in estuarine and marine atmospheres

The first scientists to investigate the coastal atmospheric presence of APs were Van Ry and Dachs, in a study conducted in the Hudson river estuary. GC–MS analyses showed that atmospheric NP isomer mixtures have a similar composition to technical mixtures, with relatively high total concentrations in the range of 0.0002–0.069 μg m^{-3} in the gas phase, and 0.0001–0.051 μg m^{-3} in the aerosol phase. These concentrations are higher than those of polycyclic aromatic hydrocarbons and up to two orders of magnitude higher than polychlorinated biphenyl concentrations in impacted urban-industrial areas [32].

Mean NP concentrations in the water phase were 0.048 μg L^{-1}, and the air–water exchange was quantified by calculating the NP fugacities in both phases. The ratio of the fugacities ($f_{water}/f_{gas} = C_{water} \times H/C_{gas} \times$

RT) was always higher than one, showing that a net volatilisation from estuarine waters is a source of NP to the atmosphere.

Considering that the reported NP water concentrations are below most values reported in the literature, the atmospheric presence of NP must be ubiquitous in coastal regions where APEO and NP are discharged into surface waters.

In a subsequent study [16] at the same three sampling sites, in the gas phase maximum concentrations of 0.081 and 0.0025 $\mu g\,m^{-3}$ were measured for NP and OP, respectively. Maximum aerosol concentrations amounted to 0.051 and 0.00063 $\mu g\,m^{-3}$, for NP and OP. Apparently, also in the atmosphere, OP is less abundant than NP.

The AP concentrations in air are temperature dependent: temperature explained 40–62% of the variability in the log gas phase NP and OP concentrations. Concentrations were significantly higher in summer than in winter. A study of the influence of wind direction on concentration showed that local sources of APs were more important than long-range atmospheric transport.

The similar gas phase NP concentrations for the three sites indicate that NP sources may be ubiquitous in the region. Suggested sources are volatilisation from the estuary, but also from land-applied sources, e.g. adjuvants in agricultural products.

To assess the relative importance of the volatilisation removal process of APs from estuarine water, Van Ry et al. constructed a box model to estimate the input and removal fluxes for the Hudson estuary. Inputs of NPs to the bay are advection by the Hudson river and air–water exchange (atmospheric deposition, absorption). Removal processes are advection out, volatilisation, sedimentation and biodegradation. Most of these processes could be estimated; only the biodegradation rate was obtained indirectly by closing the mass balance. The calculations reveal that volatilisation is the most important removal process from the estuary, accounting for 37% of the removal. Degradation and advection out of the estuary account for 24 and 29% of the total removal. However, the actual importance of degradation is quite uncertain, as no real environmental data were used to quantify this process. The residence time of NP in the Hudson estuary, as calculated from the box model, is 9 days, while the residence time of the water in the estuary is 35 days [16].

In Table 6.4.4, concentrations of APs in estuarine atmospheres reported in the literature are listed.

TABLE 6.4.4

Reported concentrations of non-ionic surfactants and their metabolites in marine and estuarine atmospheric samples

Sampling area	Compartment	Compounds analysed	Concentration (ng m^{-3})	Reference
Hudson estuary (USA)	Estuarine atmosphere, gas phase	NP	0.2–69	[32]
	Estuarine atmosphere, aerosol phase	NP	0.1–51	
Hudson estuary (USA)	Estuarine atmosphere, gas phase	NP	ND–81	[16]
		OP	ND–2.5	
	Estuarine atmosphere, aerosol phase	NP	0.02–51	
		OP	ND–0.63	

ND, not detected.

6.4.6 Biodegradation of non-ionic surfactants in estuarine and marine environments

An important process in which surfactants are removed from the aqueous environment is biodegradation. Whereas surfactant biodegradation in freshwater is in general quite fast, the degradation processes occur usually slower in saline waters [1,33] (see also Chapter 5.3). However, one study reported that the biodegradation of A$_9$PEO was faster in seawater than in freshwater: in 50 days, primary biodegradation was 33–36% in pond water and 95% in seawater [34].

Kveštak and Ahel investigated the biotransformation kinetics of A$_9$PEO$_n$ in the Krka estuary in Croatia [35]. Static die-away tests were performed with autochthonous bacterial cultures originating from the two compartments of the stratified estuary: the upper fresh/brackish water layer and the lower saline water layer. Experiments were performed at three different temperatures, and at two concentrations. Samples were taken daily and all separate ethoxylates (1–16) were quantified by normal phase HPLC–FL analysis. No other metabolites were analysed.

It was found that three factors affect the biotransformation rates of A$_9$PEO$_n$: the origin of the bacterial culture, temperature, and the initial concentration of A$_9$PEO$_n$. Biotransformation kinetics of mixed bacterial cultures from the brackish water layer was faster than those from the saline water layer. Rate constants (based on first order kinetics) for the brackish water cultures were 1.5–8.5 times larger than those for the saline water layer at 22.5°C. At lower temperatures, rate constants are smaller, and the difference between brackish and saline water is less pronounced. For A_9PEO the calculated half-lives are 23–69 days in winter (13°C) and 2.5–35 days in summer (22.5°C).

As it was shown before that the wastewater plume spreads mainly in the upper fresh/brackish layer of the estuary [6], the difference between biotransformation rates in the two water layers is explained by a better pre-adaptation of the brackish water bacteria to A_9PEO in their natural habitat, due to higher pre-exposures. It seems that the bacterial populations in these two physically very close habitats are quite different.

In addition to the origin of the bacterial cultures and the temperature, biotransformation kinetics was also influenced by the initial concentration of A_9PEO_n. At 0.1 mg L^{-1}, the lag phase is shorter and the transformation rate is higher than at 1 mg L^{-1}, suggesting an inhibitory effect on the biotransformation activity of estuarine bacteria. However, actual A_9PEO_n concentrations in estuaries are at least an order of magnitude below these experimental concentrations. Therefore, environmental concentrations should not be regarded as a limiting factor for an efficient biodegradation.

The transformation of the higher oligomers was faster than that for the lower oligomers. The lowest oligomers were actually formed during the experiment, with the strongest concentration increase for A_9PEO_2. Other metabolites were probably also formed, but were not detected with these analytical methods. No increase in NP concentration was found in these experiments.

Static die-away tests were performed by Potter et al. with an A_9PEO_{7-24} mixture in water from a vertically well-mixed estuary in Florida [36]. Lag times of 0–12 days were observed, and after 4–24 days, primary degradation was complete. These rates are similar to those reported by Kveštak and Ahel [6]. It is likely that the microorganisms in these experiments were pre-acclimated to biodegrade nonionic surfactants, as a municipal sewage treatment plant discharge is present a few kilometres upstream.

The first important metabolite that was formed was A_9PEO_2, which was further degraded after several tens of days, leading to an increase of A_9PE_2C. A_9PE_2C was found to be quite persistent, and was still present at the end of the experiment (after 183 days) at 23–76% of the initial APEO concentration. To a lesser extent

A_9PE_1C also increased during the experiment, while A_9PEO_1 was detected only in trace amounts. The completely deethoxylated NP was not detected at all.

To further investigate the recalcitrance of A_9PE_2C, the water at the end of the experiments was mixed with freshly collected estuarine water. In three of the five experiments, degradation of the residual A_9PE_2C started at day 20 and concentrations had dropped by 50% after 32 days.

In another study, the biodegradation rates of NP were determined in seawater from an unpolluted area in Sweden [37]. ^{14}C-labelled NP was used, and biodegradation was followed by measuring the labelled $^{14}CO_2$ that was formed. Initially the degradation of NP was slow, but after an adaptation period of 4 weeks, the degradation rate increased. A half-life of 58 days was observed.

When sediment was present in the flask, the initial degradation rate was already high (1.2% of NP degraded per day), and did not increase. The larger microbial community in the sediment can probably explain the initial rate, and the lack of increase is probably due to the richer supply of other carbon sources.

In separate experiments where sediment was present, nitrogen was bubbled through the water before starting the test, resulting in very low concentrations of oxygen. The degradation rate was reduced by half compared to an oxygen-rich experiment with sediment.

Nguyen and Sigoillot [38] investigated the bacterial communities degrading APEO in marine waters. A bacterial community was selected from coastal seawater in France, which is intermittently polluted by urban sewage. After 30 days of incubation with A_9PEO_{9-10}, the residual APEO concentration was 0.97%, as determined by liquid/liquid extraction with $CHCl_3$ and analysis by HPLC–UV. IR spectroscopy was used to determine the ethoxylate chain lengths in the extracts, and A_9PEO_2 was identified as the main metabolite. However, with more sophisticated analytical extraction techniques, other metabolites might have been found as well.

From the bacterial cultures, which were enriched on A_9PEO, 25 strains were isolated. With these isolated strains, further degradation experiments revealed that none of the pure cultures was able to degrade A_9PEO completely by itself. However, four strains were able to initiate the degradation of A_9PEO_{10}, giving A_9PEO_{3-5} as end products. These strains were of the *Pseudomonas* genus and of marine origin. Another strain degraded A_9PEO_5, and one degraded A_9PEO_3, both with A_9PEO_2

as end product. None of the strains could degrade A_9PEO_2, A_9PEO_1 or NP.

All the strains involved in APEO degradation were of marine origin (i.e. strains requiring sodium). It seems that in the marine environment, the role of freshwater strains originating from sewage is negligible.

6.4.7 Conclusion: fate of non-ionic surfactants in estuarine and marine environments

Several important processes influencing the fate of non-ionic surfactants and their metabolites in saline waters have been investigated in recent years. Although the knowledge of their estuarine and marine fate is far from complete, some general conclusions can be drawn for the APEOs.

Dissolved concentrations of non-ionic surfactant in estuaries are lower than those found in rivers, with reported differences of around one order of magnitude [4,11,25]. However, in some cases local sources in the estuary are the cause of high surfactant levels [9,11]. Concentrations of octylphenol ethoxylates usually amount to levels about one order of magnitude below those of the NP derivatives, reflecting the production volumes of both classes of compounds.

Surfactant loadings from estuaries to marine environments are in most cases very low.

In stratified estuaries, surfactants tend to accumulate at the phase boundaries air–water and freshwater–saline water, leading to a complex vertical distribution pattern in the water column [5,6].

Biodegradation processes are of high importance in saline waters, although degradation rates may be lower than in freshwater environments. The dissolved concentrations of the lower APEO oligomers, the carboxylated APEC, and the AE metabolites PEG, MCPEG and DCPEG are in many cases higher than those of the surfactant itself. A_9PE_2C and A_9PEO_2 are found as the most recalcitrant APEO metabolites [13,36]. Biodegradation is highly dependent on the residence time of the water in estuaries. The longer the residence time, the more biodegradation will occur.

Surfactants have been reported to accumulate in macroalgae. This may lead to higher APEO concentrations in the top layer of the sediment, when, due to decomposing algae, the sediment is covered with a layer of degrading biomass [2].

Isomer-selective reactions do not occur in the various environmental processes in estuaries, and therefore fate analysis of NP and APEO can be expressed in terms of sum of all the alkyl isomers [25].

Sorption of APEO to sediments does occur, and once the surfactants have entered the sediment, further degradation will be very slow [28]. In general, once buried in sediments, ethoxymer (APEO) and isomer (AP, APEO) distributions probably do not change substantially.

In particular, for the relatively volatile APs, volatilisation can be a significant process of removal from the water column [16,32]. A model study showed that volatilisation of NP was in fact a more important removal process than degradation and advection out of the estuary [16]. The estuarine fate of short-chain APEO metabolites in estuaries is determined mainly by sorption processes, degradation in the water, and advective transport out of the estuary [15].

In conclusion, the processes dominating the environmental fate of non-ionic surfactants in estuaries are biodegradation in the water column, sorption/sedimentation and volatilisation, together with the dynamics of the estuary, in particular water residence times.

REFERENCES

1   E. González-Mazo, M. Honing, D. Barceló and A. Gómez-Parra, *Environ. Sci. Technol.*, 31 (1997) 504.
2   A. Marcomini, B. Pavoni, A. Sfriso and A. Orio, *Mar. Chem.*, 29 (1990) 307.
3   A. Marcomini, S. Stelluto and B. Pavoni, *Int. J. Environ. Anal. Chem.*, 35 (1989) 207.
4   A. Marcomini, G. Pojana, A. Sfriso and J.M. Quiroga Alonso, *Environ. Toxicol. Chem.*, 19 (2000) 2000.
5   R. Kveštak, S. Terzić and M. Ahel, *Mar. Chem.*, 46 (1994) 89.
6   R. Kveštak and M. Ahel, *Ecotox. Environ. Saf.*, 28 (1994) 25.
7   W. Giger, M. Ahel, M. Koch, H.U. Laubscher, C. Schaffner and J. Schneider, *Water Sci. Technol.*, 19 (1987) 449.
8   M. Valls, M. Bayona and J. Albaigés, *Int. J. Environ. Anal. Chem.*, 39 (1990) 329.
9   M. Petrovic, A.R. Fernández-Alba, F. Borrull, R.M. Marce, E. González-Mazo and D. Barceló, *Environ. Toxicol. Chem.*, 21 (2002) 37.

10　U. Zoller and M. Hushan, *Water Sci. Technol.*, 43 (2001) 245.

11　M.A. Blackburn and M.J. Waldock, *Water Res.*, 29 (1995) 1623.

12　M.A. Blackburn, S.J. Kirby and M.J. Waldock, *Mar. Pollut. Bull.*, 38 (1999) 109.

13　N. Jonkers, R.W.P.M. Laane and P. de Voogt, *Environ. Sci. Technol.*, 37 (2003), 321–327.

14　P. de Voogt, O. Kwast, R. Hendriks, In: D. Vethaak et al. (Eds.), *Estrogens and Xeno-estrogens in the Aquatic Environment of The Netherlands*, RIZA/RIKZ report 2002.001, The Hague, 2002, ISBN 9036954010.

15　P.L. Ferguson, C.R. Iden and B.J. Brownawell, *Environ. Sci. Technol.*, 35 (2001) 2428.

16　D.A. Van Ry, J. Dachs, C.L. Gigliotti, P.A. Brunciak, E.D. Nelson and S.J. Eisenreich, *Environ. Sci. Technol.*, 34 (2000) 2410.

17　C. Crescenzi, A. DiCorcia, A. Marcomini and R. Samperi, *Environ. Sci. Technol.*, 31 (1997) 2679.

18　S.A. Savintseva, I.M. Sekisova, V.A. Kolosanova and A.F. Koretskii, *Kolloidn. Zh.*, 47 (1985) 901.

19　B.J. Brownawell, H. Chen, W. Zhang and J.C. Westall, *Environ. Sci. Technol.*, 31 (1997) 1735.

20　D.M. John, W.A. House and G.F. White, *Environ. Toxicol. Chem.*, 19 (2000) 293.

21　A.T. Kiewiet, K.G.M. de Beer, J.R. Parsons and H.A.J. Govers, *Chemosphere*, 32 (1996) 675.

22　N. Chalaux, J.M. Bayona and J. Albaigés, *J. Chromatogr. A*, 686 (1994) 275.

23　P. de Voogt, K. de Beer and F. van der Wielen, *Trends Anal. Chem.*, 16 (1997) 584.

24　C.M. Lye, C.L.J. Frid, M.E. Gill, D.W. Cooper and D.M. Jones, *Environ. Sci. Technol.*, 33 (1999) 1009.

25　T. Isobe, H. Nishiyama, A. Nakashima and H. Takada, *Environ. Sci. Technol.*, 35 (2001) 1041.

26　K. Maruyama, M. Yuan and A. Otsuki, *Environ. Sci. Technol.*, 34 (2000) 343.

27　E.R. Bennett and C.D. Metcalfe, *Environ. Toxicol. Chem.*, 19 (2000) 784.

28　D.Y. Shang, R.W. Macdonald and M.G. Ikonomou, *Environ. Sci. Technol.*, 33 (1999) 1366.

29　D.Y. Shang, M.G. Ikonomou and R.W. Macdonald, *J. Chromatogr. A*, 849 (1999) 467.

30　N. Chalaux, J.M. Bayona, M.I. Venkatesan and J. Albaigés, *Mar. Pollut. Bull.*, 24 (1992) 403.

31　M. Zhao, F.W.M. van der Wielen and P. de Voogt, *J. Chromatogr. A*, 837 (1999) 129.

32　J. Dachs, D.A. Van Ry and S.J. Eisenreich, *Environ. Sci. Technol.*, 33 (1999) 2676.

33 S. Terzić, D. Hrsak and M. Ahel, *Water Res.*, 26 (1992) 585.
34 P. Schöberl and H. Mann, *Arch. Fisch. Wiss.*, 27 (1976) 149.
35 R. Kveštak and M. Ahel, *Arch. Environ. Contam. Toxicol.*, 29 (1995) 551.
36 T.L. Potter, K. Simmons, J. Wu, M. Sanchez-Olvera, P. Kostecki and E. Calabrese, *Environ. Sci. Technol.*, 33 (1999) 113.
37 R. Ekelund, Å. Granmo, K. Magnusson, M. Berggren and Å. Bergman, *Environ. Pollut.*, 79 (1993) 59.
38 M.H. Nguyen and J.C. Sigoillot, *Biodegradation*, 7 (1997) 369.
39 O.P. Heemken, H. Reincke, B. Stachel and N. Theobald, *Chemosphere*, 45 (2001) 245.
40 K. Bester, N. Theobald and H.F. Schröder, *Chemosphere*, 45 (2001) 817.

6.5 ANIONIC SURFACTANTS IN MARINE AND ESTUARINE ENVIRONMENTS

Eduardo González-Mazo and Abelardo Gómez-Parra

6.5.1 Introduction

Littoral systems are spaces that are particularly severely affected by human activity, since a high proportion of the population lives in or near coastal zones, especially on estuaries (37% of the world's total population [1]). Therefore it is not surprising that coastal areas receive large quantities of surfactants originating from the discharge of urban wastewaters, both untreated and treated, together with those brought there by rivers. Other minor quantities of surfactants are received from surface water run-off, underground water and other sources, but these are less important.

Although a substantial body of data is available on the levels of linear alkylbenzene sulfonates (LASs) in rivers and estuaries, fewer studies have been conducted on their environmental behaviour, with reference to the mechanisms involved in their transport and to the reactivity they undergo. Studies of LAS in subterranean water and in the marine medium are scarce and have mainly been conducted in the last decade [2–6], coinciding with the development of new techniques of concentration/separation and analysis of LAS at ppb levels or less. Data on concentrations of sulfophenyl carboxylates (SPCs) are very scarce and the behaviour of these intermediates has hardly received any study. This chapter provides an overview of the current knowledge on behaviour of LAS and their degradation products in coastal environments.

6.5.2 Environmental processes relevant for LAS

The levels of LAS detected depend largely on whether the site at which samples are taken has any wastewater effluent points or treatment plants in the vicinity, and on the distance between the discharge point and the site. In turn, the influence of the main processes participating in the removal of LAS from the medium (biodegradation, dilution,

Comprehensive Analytical Chemistry XL   749
T.P. Knepper, D. Barceló and P. de Voogt (Eds)

precipitation and adsorption) is determined by the characteristics of the system into which the wastewaters are discharged (river, estuary, lake or sea).

The concentrations of LAS found in water and fluvial sediments show great variability (see Chapter 6.3). The values found in water (0–600 μg L^{-1}) [7–10] and in sediments (0–600 μg g^{-1}) [7,9,11,12] show the pronounced affinity of the compound for the solid phase. The partition coefficients of LAS observed in fluvial sediments range from 100 to 2600 L kg^{-1} [9]. The maximum concentrations have been detected close to urban centres in which untreated wastewater is discharged, and a rapid rate of decrease is observed as one moves downstream from these [9,13]. Some authors have found dilution to be the main factor responsible for decreasing concentrations along the course of a river towards the sea [9,13,14], but others consider biodegradation to be the most efficient process [9,15], while Rapaport and Eckhoff [16] hold adsorption onto solids in suspension to be a major factor for LAS removal from river water.

The detection of short-chain sulfophenylcarboxylic acids (SPCs) [2,9, 13,17,18] is clear proof of the primary biodegradation of LAS in the natural medium. In general the concentrations detected are small, one order of magnitude less than those found for LAS [2,9,13], with the exception of the river Osellino [18] where the total concentration of SPC found exceeded that of LAS. The highest concentrations are also observed near zones into which wastewater effluents are discharged [9, 18]. Both SPC and dialkyltetralines (DATSs) have also been detected in some samples of fluvial sediments [9].

The proportion of LAS adsorbed onto the solids in suspension is variable, and may constitute between 3 and 5% of the total [15,19]. Sedimentation takes place principally close to the discharge point, where high concentrations are consequently detected [3,9,20] and the rest is transported to more distant zones before being deposited [9]. Invariably, the proportion of long-chain homologues is greater in the solid phase (solids in suspension and sediment) than in water and greater than in commercial LAS [9,12].

The vertical distribution of LAS in the sediment column has been characterised in various lakes [21,22], where evidence has been found of its degradation in the top 5 cm, but not at greater depths where the conditions usually are anoxic. Amano et al. [23,24] have simulated the temporal variation of the concentration of LAS in the surface layer of sediments and have estimated the flux across the water–sediment interface. According to the model, in summer 85% of the LAS discharged is biodegraded and 9% is deposited on the sediment, whereas in winter, the degradation accounts for only 11%, sedimentation for 12%, and more than 75% reaches the lake water.

6.5.3 LAS in estuarine and marine systems

Estuaries are zones of great ecological and environmental relevance where water masses of different degrees of salinity mix together. The LAS discharged into rivers reach the marine medium via estuaries. The behaviour presented by LAS in estuaries is not as conservative as in the sea [4,18,25–27]. In estuaries, the decline in the concentration of LAS is sharper than would be expected due to the process of dilution [6]. In this respect, Takada and Ogura [25] describe estuaries as efficient barriers to the entry of LAS into the marine medium. The processes of adsorption and precipitation are more intensive than in the continental medium, and biodegradation is slower and less extensive than in fresh water [28]. Adsorption onto particulate matter is facilitated by the increase of the ionic strength [24,29], precipitation increases with the concentration of Ca^{2+} (10^{-2} M in seawater [30]) and the microorganisms of the marine medium are less active in biodegradation than those of freshwater.

The concentrations of LAS found present a high variability (Tables 6.5.1 and 6.5.2) as a result of the different processes facilitating their removal from the medium. Takada and Ogura [25] (see Table 6.5.1) characterise the distribution of LAS between water, solids in suspension and sediments along the length of the Tamagawa river estuary (Japan). The concentrations decrease towards the sea in all three phases, especially during the summer months when all the LAS reaching the estuary is degraded.

An increase has also been observed in the LAS concentration in the aqueous phase during the cold season, due to reduced microbial activity [3,18,23]. Therefore the flow of LAS towards the coastal zone will be strongly influenced by the seasonal variable as confirmed by the values obtained by Takada et al. [3], who found that for the Tamagawa estuary the flow of LAS was five times higher in winter than in summer. However, in long-term monitoring [6,31,32], the variations detected are principally a function of tidal influence and of the time of day, or the day of the week, when the LAS is introduced into the system [6,33].

TABLE 6.5.1

Levels of LAS in water (μg L^{-1}), in solids in suspension (μg g^{-1}) and in sediment (μg g^{-1}), in various estuarine systems

Phase	Estuary	[LAS]	Reference
Water	Tamagawa	8.1–444	[25]
	Krka	0.9–391	[40]
	Scheldt	<0.5–9.4	[4]
	Barbate	6.1–9.8	[5]
	Lagoon of Venice	1.3–256	[18]
	Suances	12–527	[27]
	Santoña	9–49	
	Palmones	14–38	
Solids in suspension	Tamagawa	3–704	[25]
Sediment	Various Japanese	4.82 ± 7.36	[41]
	Tamagawa	0.5–24	[25]
	Tamagawa	0.12–45.1	[3]
	Barbate	0.44 – 0.83	[42]
	Suances	0.02–0.5	[27]
	Santoña	0.36–1.18	
	Palmones	1.1–5.1	

TABLE 6.5.2

Levels of LAS in water (μg L^{-1}), solids in suspension (μg g^{-1}) and sediment (μg g^{-1}), in various coastal marine systems

Phase	Coastal zone	[LAS]	Reference
Water	Bay of Tokyo	<3–14	[43]
	Lagoon of Venice	1.2–296.5	[26]
	North Sea	<0.4–1.2	[4]
	Gulf of Thermaikos	10–60	[31]
	Bay of Cádiz (interior part)	3.5–30	[27,42]
	Bay of Cádiz (exterior part)	3.8–51.2	[42]
	Channel of Sancti Petri, Cádiz	2.7–1687	[42]
	Lagoon of Venice	1.4–2.4	[18]
Solids in suspension	Channel of Sancti Petri, Cádiz	14.4–5941.0	[6]
Sediment	Costa Alicante	0.1	[44]
	Bay of Tokyo	0.2 ± 0.2	[41]
	Bay of Tokyo	<0.1–0.4	[45]
	Channel of Sancti Petri, Cádiz	1.6–150	[6,34]
	Bay of Cádiz (interior part)	0.1–7	[35]
	Danish fjords	0.8–22	[46]
	Lagoon of Venice	0.1	[18]

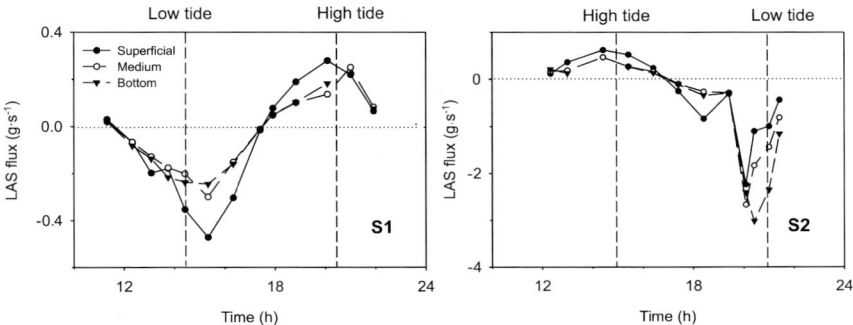

Fig. 6.5.1. LAS fluxes obtained for the three water depths studied (surface, middle and bottom), between the Sancti Petri Channel and Bay of Cádiz, Spain, during the periods of the two sampling campaigns S1 and S2 (taken from Ref. [27]).

Recently the influence of the phase of the tide has been demonstrated in a zone subjected to the discharge of effluent of urban wastewaters, and the consequent variation in the flow of LAS from the site of its discharge (the Sancti Petri Channel) to Cádiz Bay (Spain). Figure 6.5.1 shows the fluxes of LAS for the three depths studied (surface, middle and bottom), in two samplings at different tidal amplitudes. The largest flux of LAS occurs when there is a high tidal amplitude (S2). LASs originating from the wastewater discharge are exported from the Sancti Petri Channel to the bay when the tide is ebbing (negative flux). The input of LAS into the channel is much lower with the inflowing tide (positive direction of flux), therefore, a net export flux of -65.2 kg d^{-1} takes place towards the bay. The export is less (-12.0 kg d^{-1}) for a low tidal amplitude (S1).

It has been found that the vertical distribution in the water column of LAS is not homogeneous [6,27,40]; rather, it depends on the degree of physical mixing in the water column and on whether wastewaters are discharged at the surface level or at depth. In Fig. 6.5.2(a), the highest levels are found at 2.5 m depth (station SU6), as a result of wastewater being discharged into the estuary at depth. At a relatively short distance from this outlet (station SU4, Fig. 6.5.2(b)), LAS was found to be distributed more homogeneously, and in the more stagnant zone (station SU3, Fig. 6.5.2(c)) the distribution is inverse. At this latter point, the concentrations measured in the surface microlayer were up to one order of magnitude higher than the corresponding samples from

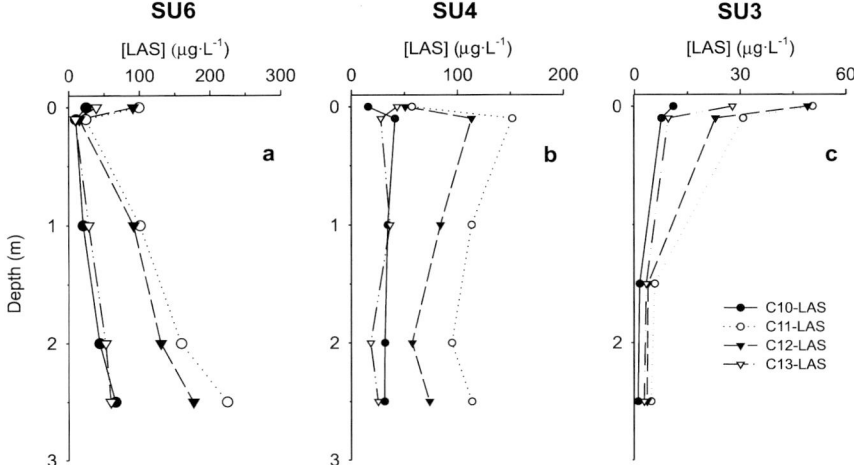

Fig. 6.5.2. Vertical profiles obtained for LAS homologue concentrations at several sampling points from the Besaya Estuary, Spain: (a) sampling site directly downstream from wastewater discharge point; (b) and (c) samples taken 15 km (b) and 50 km (c) downstream from discharge (from León et al. [27]).

10 cm depth in the water column, largely due to the characteristics of LAS as a surfactant (Fig. 6.5.2(c)).

The high coefficients of sediment–water partition detected, which varied between 500 and 10 000 L kg^{-1} in estuaries [27,40] and between 3000 and 13 000 L kg^{-1} in the marine medium [6,34] showed the great adsorption capacity of LAS onto particulate matter. As in continental systems, it has been detected that the more hydrophobic homologues are preferentially adsorbed onto the sediment [3,27,34,35,40].

In the proximity of a wastewater discharge point, the sediments contain a higher proportion of the longer-chain length homologues (C_{12} and C_{13}), with even C_{14}-LAS being detected [6,27,34,35,40]; the latter constitutes only $<0.5\%$ of commercial LAS. Progressively as the distance from the discharge point increases, the percentage of homologues of greater molecular weight decreases. In the Bay of Tokyo, the content of LAS in the sediments is very low (<0.01 μg g^{-1}) (Table 6.5.2), but in other coastal zones such as the Bay of Cádiz, the concentrations of LAS reach several tens of ppm [6,34]. This difference in the concentrations of LAS detected in the two bays is due to the different routes of access of LAS to the medium (e.g. the elimination of a large part of the LAS in the estuary before the water reaches the Bay of Tokyo, compared with the direct discharge into the Bay of Cádiz, without prior treatment of wastewaters). It is also due to the specific characteristics of each zone, principally with respect to the granulometry of the sediment and to the salinity of the water.

The proportion of LAS removed from the water column by adsorption depends on the water's content of solids in suspension, on the ionic strength, and on the hydrodynamics of the zone. In the Tamagawa estuary [3] and in the Bay of Cádiz [27] values of < 20% for the fraction of LAS adsorbed have been found. However, in other zones with higher contents of solids in suspension, higher percentages have been found (from 11 to 60%) [6,40]. The evolution of the concentration of the various homologues of LAS in solids in suspension is similar to that found in water (Fig. 6.5.3). An increase is seen in the process of adsorption onto the suspended solids, in line with the chain length: for each additional carbon atom in the alkyl chain, the adsorption is between 2.5 and 3.2 times greater.

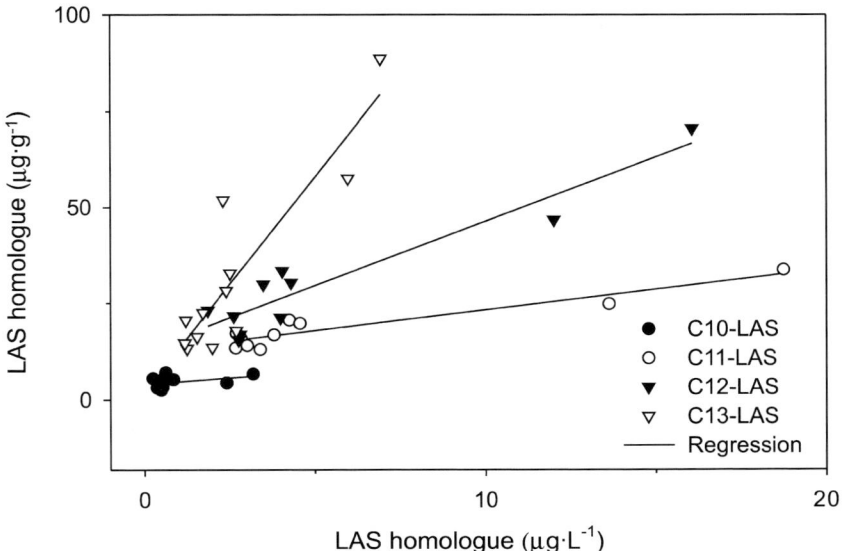

Fig. 6.5.3. Representation of the concentration of LAS in suspended solids against the concentration of LAS in water in the Bay of Cádiz, Spain (taken from Ref. [27]).

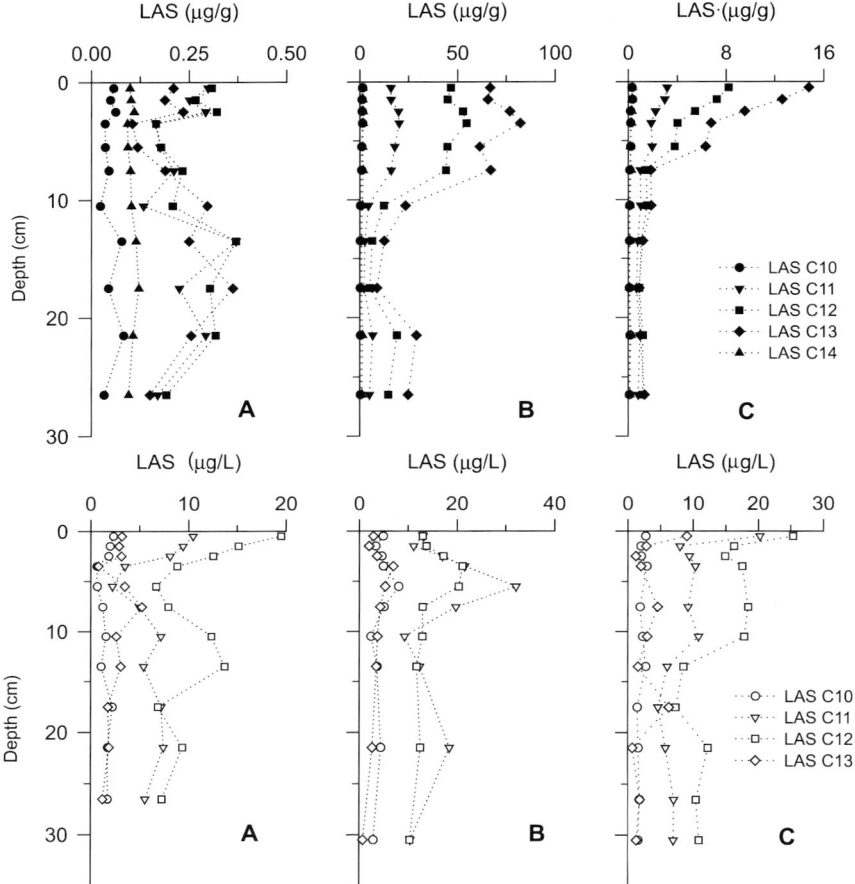

Fig. 6.5.4. Vertical profiles of LAS in sediment and interstitial waters from three stations situated at different distances from the non-treated wastewater effluent point (A: 12 km; B: 0.1 km; and C: 3 km (taken from Ref. [34])).

From the foregoing discussion, it will be appreciated that sediments constitute the final natural compartment for reception of LAS that have not been degraded. The vertical profiles of the concentrations of the LAS homologues in the sediment and interstitial water found for three sampling stations are shown in Fig. 6.5.4. There is a pronounced decrease in LAS concentration with depth, particularly in the first few centimetres, which may be related to greater discharges of effluent into the zone in recent years and to LAS degradation occurring in the sediment.

For stations B and C, where the LAS concentrations were higher than for A, the variation in total LAS concentration with sediment depth was determined by the homologues of 12 and 13 carbon atoms (Fig. 6.5.4). These homologues present a strong tendency to sorption and are readily biodegradable. In interstitial water, the vertical profile of the LAS concentration is similar to that observed for the sediment, particularly at stations B and C. The homologue-specific partition coefficient did not vary much with depth, because there is no appreciable variation in the composition of the sediment with depth [34].

Several specific studies have been made on the LAS transport processes in a variety of different estuarine and marine environmental compartments (e.g. LAS removal by biodegradation and/or adsorption), including the flux of LAS from freshwater to a coastal environment [3,6, 29,40]. However, a complete study of LAS behaviour requires knowledge of their primary biodegradation intermediates, SPC, that have been detected in different laboratory tests [15,36,37]. Their existence in fresh-water [2,13] and marine water [5] has been demonstrated recently.

6.5.4 SPC in estuarine and marine systems

León et al. [27] have characterised the distribution of LAS and SPC homologues in water, suspended solids, interstitial water and sediment of five Iberian estuaries and a salt marsh channel. Total SPC concentrations in water were low ($<50\mu g\,L^{-1}$) and at some stations SPC were not detected, probably due to the rapid LAS biodegradation kinetics. SPC homologues of between 3 and 12 carbon atoms were detected, although their concentrations were several orders of magnitude lower than those of the corresponding LAS. The highest SPC concentrations found were of medium carboxylic chain lengths (C_6-C_8) in water. In all the cases studied, the combined amount of these three homologues detected constituted more than 70% of the total SPC detected. These results are in agreement with those obtained previously by other authors [2,38] who have designated these homologues as 'key intermediates' since they are the most persistent in the LAS biodegradation process. In the more highly contaminated areas, the long-chain SPC (e.g. C_{11}-SPC) were detected at significant concentrations as a consequence of continuous LAS input to these zones.

In the solids in suspension, SPC homologues between C_5- and C_{13}-SPC were detected, with those of longer carboxylic chain predominating. Although variable, the percentages associated with the suspended solids were always $< 1\%$. SPC of between 4 and 13 atoms of carbon are present in the sediment. The presence of C_{13}-SPC demonstrates that ω-oxidation is the first stage in the biodegradation process, as previously indicated by González-Mazo et al. [5]. The highest concentrations of SPC homologues ($47-120$ μg kg^{-1}) corresponded to those of longer carboxylic chain (C_9-C_{11}), which present the greatest affinity for the solid phase. The heavier SPC have been previously detected in sediments of the same area by González-Mazo et al. [5], suggesting their higher tendency to sorb compared with the other SPC homologues. On the other hand, the adsorption of SPC homologues is less intense than that of LAS, and their distribution is a function of the length of the carboxylic chain (e.g. partition coefficients of 22 for C_6-SPC and 45 for C_7-SPC).

The LAS biodegradation pathways determined by Swisher [39], indicate that SPC are generated by successive ω and β-oxidations. Results from the Oka estuary (presented in Table 6.5.3) demonstrate this sequence very clearly: relatively close to the discharge outlet (UR7), the presence of all the monocarboxylic SPC was detected, but at increasing distances with longer time for the biodegradation process to take place, an increasing proportion of SPC of intermediate chain length (C_6-C_8) was found. Finally, these constituted 100% of the SPC detected at the mouth of the estuary (UR1).

The SPC are homogeneously distributed in the water column (as shown in Fig. 6.5.5), as a consequence of losing their surfactant character.

TABLE 6.5.3

Spatial distribution (%) of SPC homologues in water from the Oka estuary (taken from León et al. [27])

Station	Distance from seaward limit (km)	SPC homologue (%)					
		$<C_5$	C_5	C_6	C_7	C_8	$>C_8$
UR7	5.7	16.8	7.2	37.7	18.3	15.1	4.9
UR4	4.3	24.9	6.5	36.3	25.2	7.1	0
UR1	0.0	0	0	65.8	25.7	8.5	0

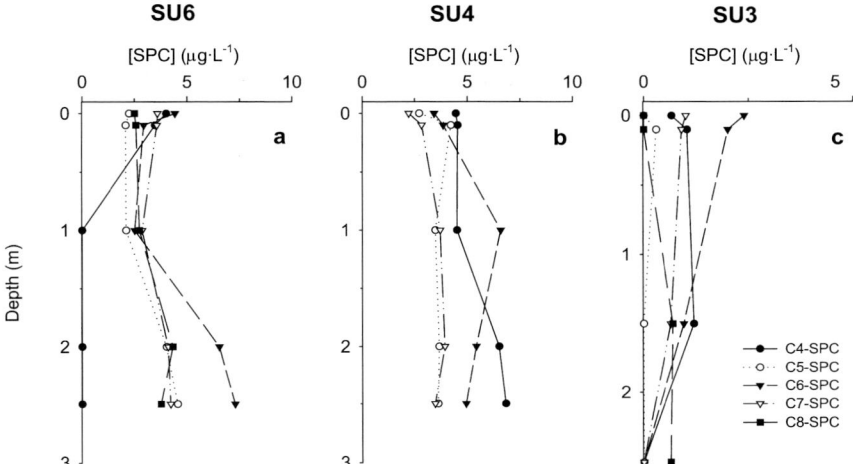

Fig. 6.5.5. Vertical profiles obtained for SPC homologue concentrations in several sampling points from the Besaya estuary (taken from Ref. [27]).

6.5.5 Conclusion

Spatial trends in LAS concentrations in marine and estuarine environments show clear relationships with the location of sampling points relative to point sources of wastewater discharges and a rapid rate of decrease is observed as one moves downstream from these. LAS tend to be degraded relatively rapidly in summer, whereas in winter a higher recalcitrance is observed. Sediments constitute the final sink of LAS that have not been degraded.

The proportion of long-chain LAS homologues is greater in the solid phase (solids in suspension and sediment) than in water and greater than in commercial LAS. For LAS clear vertical trends in distribution can be observed both in the water and sediment columns, with relatively enriched concentrations in the surface microlayer and sediment top layers.

SPCs of medium carboxylic chain lengths (C_6–C_8) are the most persistent surfactants and constitute more than 70% of the total SPC detected in water. In sediments, the longer-chain SPC (C_9–C_{11}) are the most abundant ones. The SPC metabolites are more evenly distributed in the water column than the parent LAS.

REFERENCES

1   J.E. Cohen, C. Small, A. Mellinger, J. Gallup and J. Sachs, *Science*, 278 (1997) 1211.
2   J.A. Field, J.A. Leenheer, K.A. Thorn, L. Barber Jr., C. Rostad, D.L. Macalady and S.R. Daniel, *J. Contam. Hydrol.*, 9 (1992) 55.
3   H. Takada, N. Ogura and R. Ishiwatari, *Environ. Sci. Technol.*, 26 (1992) 2517.
4   E. Matthijs and M. Stalmans, *Tenside Surfact. Det.*, 30 (1993) 29.
5   E. González-Mazo, M. Honing, D. Barceló and A. Gómez-Parra, *Environ. Sci. Technol.*, 31 (1997) 504.
6   E. González-Mazo, J.M. Forja and A. Gómez-Parra, *Environ. Sci. Technol.*, 32 (1998) 1636.
7   H. Takada and R. Ishiwatari, *Environ. Sci. Technol.*, 21 (1987) 875.
8   D.C. McAvoy, W.S. Eckhoff and R.A. Rapaport, *Environ. Toxicol. Chem.*, 12 (1993) 977.
9   C.F. Tabor and L.B. Barber, *Environ. Sci. Technol.*, 30 (1996) 161.
10 P. Schöberl, *Tenside Surfact. Det.*, 34 (1997) 233.
11 P. Schöberl and R. Spilker, *Tenside Surfact. Det.*, 33 (1996) 400.
12 L. Cavalli, G. Cassani, L. Vigano, S. Pravettoni, G. Nucci, M. Lazzarin and A. Zatta, *Tenside Surfact. Det.*, 37 (2000) 282.
13 W.H. Ding, S.H. Tzing and J.H. Lo, *Chemosphere*, 38 (1999) 2597.
14 J. Ferrer, A. Moreno, M.T. Vaquero and L. Comellas, *Tenside Surfact. Det.*, 34 (1997) 278.
15 H. Takada, K. Mutoh, N. Tomita, T. Miyadzu and N. Ogura, *Water Res.*, 28 (1994) 1953.
16 R.A. Rapaport and W.S. Eckhoff, *Environ. Toxicol. Chem.*, 9 (1990) 1245.
17 A. DiCorcia, L. Capuani, F. Casassa, A. Marcomini and R. Samperi, *Environ. Sci. Technol.*, 33 (1999) 4119.
18 A. Marcomini, G. Pjana, A. Sfriso and J.M. Quiroga, *Environ. Toxicol. Chem.*, 19 (2000) 2000.
19 P. Schöberl, H. Klotz and R. Spilker, *Tenside Surfact. Det.*, 33 (1996) 329.
20 M.L. Trehy, W.E. Gledhill and G. Orth, *Anal. Chem.*, 62 (1990) 2581.
21 K. Amano, T. Fukushima, K. Inaba and O. Nakasugi, *Jap. J. Water Poll. Res.*, 12 (1989) 38.
22 R. Reiser, H. Toljander, A. Albrecht and W. Giger, In: R.P. Eganhouse (Ed.), *American Chemical Society*, SYMPOSIUM SERIES 671, *Molecular Markers in Environmental Geochemistry*, 1997, p. 196.
23 K. Amano, T. Fukushima and O. Nakasugi, *Water Sci. Technol.*, 23 (1991) 497.
24 K. Amano, T. Fukushima and O. Nakasugi, *Hydrobiologia*, 235/236 (1992) 491.

25   H. Takada and N. Ogura, *Mar. Chem.*, 37 (1992) 257.

26   M. Stalmans, E. Matthijs and N.T. de Oude, *Water Sci. Technol.*, 24 (1991) 115.

27   V.M. León, M. Sáez, E. González-Mazo and A. Gómez-Parra, *Sci. Total Environ.*, 288 (2002) 215.

28   S. Terzic, D. Hrsak and M. Ahel, *Water Res.*, 26 (1992) 585.

29   E. González-Mazo and A. Gómez-Parra, *Trends Anal. Chem.*, 15 (1996) 375.

30   L. Cohen, A. Moreno and L. Berna, *J. Am. Oil Chem. Soc.*, 70 (1993) 79.

31   S. Kilikidis, A. Kamarianos, X. Karamanlis and U. Giannakou, *Fresenius Environ. Bull.*, 3 (1994) 95.

32   H.F. Schröder, *Tenside Surfact. Det.*, 32 (1995) 492.

33   J.M. Quiroga-Alonso, D. Sales-Márquez and A. Gómez-Parra, *Water Res.*, 23 (1989) 801.

34   V.M. León, E. González-Mazo, J.M. Forja and A. Gómez-Parra, *Environ. Toxicol. Chem.*, 20 (2001) 2171.

35   E. González-Mazo, J.M. Quiroga and A. Gómez-Parra, *Ciencias Mar.*, 25 (1999) 367.

36   W.E. Gledhill, M.L. Trehy and D.B. Carson, *Chemosphere*, 22 (1993) 873.

37   T.P. Knepper and M. Kruse, *Tenside Surfact. Det.*, 37 (2000) 41.

38   P.W. Taylor and G. Nickless, *J. Chromatogr.*, 403 (1987) 243.

39   R.D. Swisher, Surfactant biodegradation, *Chemical structure and primary biodegradation*, Surfactant Science Series, Vol. 18, Marcel Dekker, New York, 1987, p. 415.

40   S. Terzic and M. Ahel, *Mar. Pollut. Bull.*, 28 (1994) 735.

41   H. Takada and R. Ishiwatari, *Water Sci. Tech.*, 25 (1991) 437.

42   E. González-Mazo, J.M. Quiroga, D. Sales and A. Gómez-Parra, *Toxicol. Environ. Chem.*, 59 (1997) 77.

43   H. Hon-nami and T. Hanya, *Water Res.*, 41 (1980) 1251.

44   D. Prats, F. Ruiz, B. Vázquez, D. Zarzo, J.L. Berna and A. Moreno, *Environ. Toxicol. Chem.*, 12 (1993) 1599.

45   H. Takada, R. Ishiwatari and N. Ogura, *Estuar. Coastal Shelf Sci.*, 35 (1992) 141.

46   Lillebaelt Report. The Lillebaelt Cooperation, Denmark, December 1998.

6.6 SURFACTANTS IN DRINKING WATER: OCCURRENCE AND TREATMENT

Thomas P. Knepper and Pim de Voogt

6.6.1 Introduction

Drinking water intended for human consumption is the most protected foodstuff worldwide and thus many national and international regulations have been introduced to assure a safe drinking water supply. In general, drinking water must be of a quality that under no circumstances would present a hazard to human health. In many countries, such as Germany, standards for water resources that may be used for drinking water production are specified [1].

Raw water for drinking water production can be obtained from various sources including ground water, bank filtrate, rainwater, wastewaster and surface water resources depending on supply, demand and quality. In particular due to the scarcity of ground water supplies in many highly populated regions, the usage of river water for the production of potable water is common practice.

The production of drinking water from ground water, bank filtrate and surface water is influenced by the presence of anthropogenic, organic, polar persistent pollutants [2]. The problems associated with this group of pollutants, among them organic sulfonates, aromatic acids and phenols, are strongly related to their polarity, which enables them to bypass natural as well as man-made filtration steps [3]. Many of these polar compounds are also resistant to degradation and have thus been found in ground and drinking water. These poorly degradable and highly water soluble contaminants are difficult to remove even by state-of-the-art drinking water treatment processes, such as biodegradation and granular activated-charcoal filtration.

Because the quality of drinking water is strictly regulated, drinking water treatment steps are generally monitored very thoroughly. This has been especially the case for the removal of restricted anthropogenic compounds, e.g. pesticides from raw water. However, many other polar anthropogenic organic compounds and their metabolites, including surfactants, have not been subjected to regulation so far and thus the available monitoring data are quite scarce. Most of the analytical

Comprehensive Analytical Chemistry XL
T.P. Knepper, D. Barceló and P. de Voogt (Eds)

investigations related to surfactants in drinking water to date have, for obvious reasons, been directed to those compounds already detected in surface waters serving as raw water for drinking water production. Therefore, only data for nonylphenol ethoxylates (NPEO), linear alkybenzene sulfonates (LAS) and their known degradation products are available. Other surfactants, such as alkyl glucamides (AG), alkyl polyglucosides (APG), cationics or amphoterics have not been investigated because they have not been detected in the raw water.

In subsequent sections, an overview of the limited available data for non-ionic and anionic surfactants in drinking water will be given. A discussion of the efficiency of different production processes adopted in the preparation of drinking water to remove anionic surfactants will also be provided.

Generally, the presence of polar organic pollutants in raw water is already a significant factor determining the effort and costs involved in subsequent drinking water treatment. Furthermore, it is of no debate that this trend will continue in the future due to the increasing population density worldwide.

6.6.2 Routes of surfactants and their metabolites from surface to drinking waters

The mobility of slowly degradable compounds or persistent metabolites present in surface water or bank filtration-enriched ground water is of particular concern in the production of potable water. Certain surfactants, and especially their polar metabolites among others, have the potential to bypass technical purification units used, which may include flocculation, (active charcoal) filtration, ozonation or chlorination. As such, these compounds can reach drinking water destined for human consumption [4–6]. In most cases the origin of surfactant residues and their degradation intermediates in raw water is from wastewater treatment plant (WWTP) effluents (see Chapters 6.1 and 6.2) or direct emissions of wastewater, with the latter still common in many less developed countries.

In the production of drinking water from river water, the removal efficiencies of polar pollutants generally depends on their nature and on the type and arrangement of the natural and/or installed technical purification stages. In a series of investigations carried out in the 1990s, the behaviour of polar pollutants occurring in raw water during drinking water treatment, including, e.g. complexing agents [7], aromatic sulfonates [8], intermediates, metabolites and by-products from industrial syntheses, was studied [9,10]. Some of these compounds were found to be capable of bypassing all purification steps to some extent, and thus were ultimately detected in the finished drinking water.

Thus, especially since the development of appropriately specific and sensitive analytical methods, as discussed in Chapter 2, it seems only logical that some highly water-soluble surfactants and their even more polar metabolites have been positively detected in potable water. In the following sections, examples are given according to the different surfactant classes as well as the source of the raw water used.

6.6.3 Analysis of alkylphenol ethoxylates and degradation products in drinking water

In The Netherlands the drinking water supply is largely dependent on water from the rivers Meuse and Rhine. The quality assessment of these rivers is thus a long existing goal of RIWA (Association of River Waterworks, The Netherlands) and the associated water companies, with permanent monitoring for a variety of organic micropollutants in place [11,12].

During a sampling campaign carried out in 1999, alkylphenol ethoxylates (APEO) and degradation products, such as nonylphenol (NP), were also determined in raw, process and drinking waters for the first time [13]. Sampling locations were selected on the basis of RIWA's existing monitoring network. Samples from treatment facilities of the water companies were also taken at various stages of the purification process. The majority of the sites were sampled three times. Further details of sampling criteria and locations can be found in the RIWA report [13], and in Chapter 3.1. Table 6.6.1 presents the results of the analysis of samples of process and drinking water. For samples with levels below the limit of detection (LOD), the actual LOD is given. For samples with levels between LOD and limit of quantitation (LOQ), or above the LOQ, the actual value is given. The LOD is defined as three times, and the LOQ as ten times the standard deviation of the noise level of the sample. As the noise levels vary from sample to sample, so do the LOD and LOQ.

TABLE 6.6.1

Overview of results of determination of AP and APEO in process and drinking water sampled from the Meuse and Rhine rivers, NL

Location	Description by RIWA	Sampling	Concentrations ($\mu g\ L^{-1}$)			
			NPEO	OPEO	NP	OP
Meuse watershed (Process water)						
Petrusplaat	Finished product	Mar 99	< 0.88	< 0.54	< 0.51	< 0.19
		Jun 99	< 0.41	< 0.25	< 0.24	< 0.10
		Sep 99	< 0.30	< 0.16	< 0.16	< 0.07
Scheveningen	After extraction	Mar 99	1.53	< 0.24	< 0.22	< 0.08
		Jun 99	< 0.32	< 0.20	< 0.19	< 0.08
		Sep 99	< 0.38	< 0.20	< 0.21	< 0.09
Ouddorp	After infiltration	Jun 99	< 0.50	< 0.34	< 0.33	< 0.14
		Sep 99	< 0.36	< 0.19	< 0.20	< 0.08
Meuse watershed (Drinking water)						
Scheveningen	After infiltration	Mar 99	< 0.47	< 0.29	< 0.27	< 0.10
		Jun 99	< 0.57	< 0.35	< 0.33	< 0.14
		Sep 99	< 0.36	< 0.19	< 0.20	< 0.08
Ouddorp	Drinking water	Jun 99	< 0.48	< 0.29	< 0.28	< 0.12
		Sep 99	< 0.38	< 0.20	< 0.21	< 0.09
Braakman	Drinking water	Jun 99	< 0.37	< 0.22	< 0.21	< 0.09
		Sep 99	< 0.34	< 0.18	< 0.18	< 0.08
Kralingen	Drinking water	Jun 99	2.10	< 0.24	<*0.25	< 0.10
		Sep 99	< 0.36	< 0.19	< 0.20	< 0.08
Berenplaat	Drinking water	Jun 99	2.08	< 0.27	< 0.25	< 0.11
		Sep 99	< 0.51	< 0.27	<*0.5	< 0.12
Rhine watershed (Process water)						
Nieuwegein	Finished product	Mar 99	< 3.83	< 0.70	< 0.66	< 0.25
		Jun 99	< 0.37	< 0.22	< 0.21	< 0.09
		Sep 99	< 0.40	< 0.21	< 0.22	< 0.09

TABLE 6.6.1 (continued)

Location	Description by RIWA	Sampling	Concentrations (µg L^{-1})			
			NPEO	OPEO	NP	OP
Lekkerkerk	GLS PF 99	Mar 99	< 0.57	< 0.57	< 0.53	< 0.20
		Jun 99	<* 2.51	< 0.45	< 0.43	< 0.18
		Sep 99	< 0.33	< 0.17	< 0.18	< 0.07
Nieuw Lekkerland	GPU PE99B (De Put)	Mar 99	< 0.43	< 0.43	< 0.41	< 0.15
		Jun 99	1.59	< 0.29	< 0.28	< 0.12
		Sep 99	< 0.28	< 0.15	< 0.16	< 0.06
Leiduin	After infiltration	Mar 99	<* 1.01	< 0.57	< 0.54	< 0.20
		Jun 99	< 0.29	< 0.18	< 0.17	< 0.07
		Sep 99	< 0.36	< 0.19	< 0.20	< 0.08
Leiduin	After ozone treatment	Mar 99	< 0.62	< 0.54	< 0.51	< 0.19
		Jun 99	< 0.49	< 0.29	< 0.28	< 0.12
Leiduin	After GAC	Mar 99	< 0.86	< 0.67	< 0.63	< 0.23
Rhine watershed (Drinking water)						
Leiduin	Drinking water	Mar 99	<* 1.26	< 0.50	< 0.47	< 0.18
		Jun 99	< 0.32	< 0.19	< 0.18	< 0.08
		Sep 99	< 0.32	< 0.17	< 0.17	< 0.07
Andijk	Drinking water	Mar 99	<* 1.79	< 0.47	< 0.44	< 0.17
		Jun 99	< 0.60	< 0.36	< 0.34	< 0.15
		Sep 99	< 0.39	< 0.20	< 0.22	< 0.09
Twentekanaal WMO	Drinking water	Mar 99	< 0.71	< 0.42	< 0.39	< 0.15
		Jun 99	< 0.50	< 0.30	< 0.29	< 0.12
		Sep 99	< 0.47	< 0.25	< 0.26	< 0.11
Weesperkarspel	Drinking water	Jun 99	< 0.38	< 0.23	< 0.22	< 0.09
		Sep 99	< 0.40	< 0.21	< 0.22	< 0.09

Data from Refs. [13,14]: sampling and sample description from Ref. [13]; LOQ, if below LOD, the actual LOD is given (e.g. for calculation purposes); < is below LOD; <* is between LOD and LOQ. LOD is defined as 3 × SD in noise; LOQ is defined as 10 × SD in noise. Noise level varies from sample to sample, and consequently LODs differ.

From the compound groups analysed, only the NPEO were found above the LOQ. NPEO were found above the LOD in 5 out of 23 samples of process water; in these five, levels were ranging from 1.0 to 3.8 μg L^{-1}. In four out of 22 drinking water samples NPEO were found above the LOD, ranging from 1.2 to 2.1 μg L^{-1}. OP and OPEO were invariably below the LOD, and NP was also generally below the LOD with the exception of two samples of drinking water, where the levels found were between the LOD and LOQ (0.2–0.5 μg L^{-1}).

The values observed are consistent with the generally low levels of these compounds found in the rivers Rhine and Meuse. A clear distinction between the values found in the river Meuse and the Rhine river basin was not observed. Analysis of samples from the Rhine and Meuse basins in the same period has shown that in general levels were below the LOD, with occasional detectable values in the range of between 1.0 and 3.5 μg L^{-1} observed [13,14] (see Chapter 6.3). These findings have been confirmed in a recent study by Jonkers et al. [15], in which total dissolved APEO concentrations in the downstream area of the Rhine river of between 0.2 and 0.9 μg L^{-1}, and concentrations of NP averaging 0.1 μg L^{-1} with little variation, were reported.

In US drinking water, Sheldon and Hites [16] in the 1970s, reported that levels of NP were below LOD, whereas OP was found at a level of 0.01 μg L^{-1} in drinking water. Clark et al. [17] in 1992 reported levels of 0.08 (NPEO), 0.15 (NPEO$_2$) and 0.5 (NPEO$_{3-7}$) μg L^{-1}, respectively. OPEO$_2$ was found at a level of 0.002 μg L^{-1}. Clark et al. were also the first to describe, in addition to APEO, determinations of the carboxylate metabolites (NPEC) in drinking water. They reported values of 0.16 (NPEC$_2$), 0.06 (NPEC$_{3-7}$), 0.04 (OPEC$_2$), and 0.012 (OPEC$_{3-4}$) μg L^{-1}, respectively.

At Cape Cod, USA, a study was carried out by Rudel et al. [18] to investigate the impact of septic systems as a source of APEOs and their degradation products to ground water. In this study NP was detected in all sewage samples at concentrations above 1000 μg L^{-1}. In ground water downgradient of an infiltration bed for secondary treated effluent, NP and OP and ethoxylates were present at about 30 μg L^{-1}. NPE$_1$C and NP/OP(EO)$_4$ were detected in some drinking water wells at concentrations up to 33 μg L^{-1}.

Investigation of NPEC in European rivers used as raw waters in drinking water production revealed values in the lower μg L^{-1} range (see Chapter 6.3).

The Llobregat river in Spain is an extremely highly polluted river supplying water to the city of Barcelona and surrounding areas with a population of approximately 3.2 million. Despite a variety of treatment processes employed in the studied plant, satisfactory removal of organic compounds from water was not guaranteed. Guardiola et al. [19] reported maximum levels in Barcelona tapwater of 0.14 µg L^{-1} for NP, 1.1 µg L^{-1} for NPEO and 0.25 µg L^{-1} for NPEO$_2$. The same research group reported that during breakpoint chlorination of raw water taken from the Llobregat river, formation of brominated NP was also observed [20,21]. This was ascribed to high levels of bromide ions together with the presence of NP in the surface water. In the tapwater of Barcelona, monobrominated NP (BrNP) was found in later investigations.

In more recent in-depth investigations on this issue [22], a series of chlorinated and brominated by-products of alkylphenolic surfactants and their degradation intermediates formed during chlorine disinfection of raw water were identified using mass spectrometry (Table 6.6.2).

6.6.4 Behaviour of sulfophenyl carboxylates during drinking water production

Until recently the possible occurrence of very polar LAS intermediates, which are made up of some recalcitrant constituents, had not been investigated at all. In a study from Eichhorn et al. [23], two waterworks differing in both raw water quality and treatment technology were chosen for evaluation of their capacity to eliminate sulfophenylcarboxylates (SPC) from the water. A detailed scheme comparing all installed treatment steps of both waterworks is displayed in Fig. 6.6.1. The river Rhine and the Llobregat river, from where the two investigated waterworks withdraw their raw water, differ substantially in terms of hydrological conditions as well as in physical and chemical parameters of the water. The first waterworks is located on the Llobregat river, Spain (154 km) and is henceforth denoted as 'waterworks Llobregat'. As has been already described, this river is heavily polluted by discharges from urban and industrial activities and leaching from mining and agriculture [24,25]. The second waterworks is situated on the river Rhine, Germany (508 km) and is denoted as 'waterworks Rhine'. A particular characteristic process adopted at waterworks Rhine is the subsoil passage performed after granular activated-carbon (GAC) filtration. In contrast, the unique features of waterworks Llobregat

TABLE 6.6.2

Concentrations of halogenated NPEO and degradation products in $\mu g\ L^{-1}$ found in the Llobregat river and the drinking water treatment plant of Barcelona, Spain (modified from Ref. [22])

Sample	BrNPEC	BrNP	BrNPEO	ClNPEO	NPEO	OPEO	NPEC$_1$	NPEC$_2$	OPEC$_{1-2}$	NP	OP
Raw water	ND	ND	ND	ND	12.9	1.2	2.2	2.9	ND	0.45	ND
Preclorinated water	0.42 ($n_{EO} = 0–1$)	0.21	4.0	ND	8.9	ND	1.0	1.3	ND	0.06	ND
Treated water	ND	ND	ND	ND	ND	ND	ND	ND	ND	ND	ND

ND, not detected.

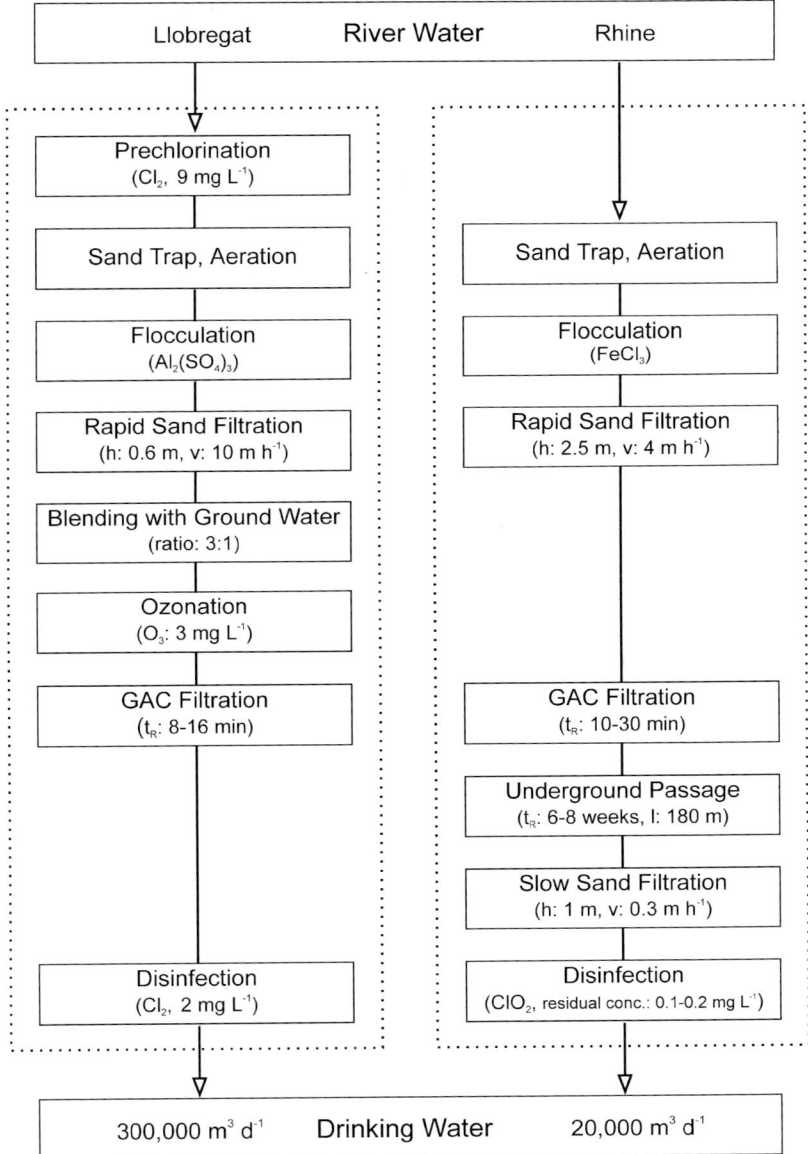

Fig. 6.6.1. Scheme of drinking water production plants on the Llobregat river, Spain (waterworks Llobregat) and the river Rhine, Germany (waterworks Rhine) (from Ref. [23] with permission from Elsevier).

are breakpoint chlorination of the raw water and ozonation prior to GAC filtration. Furthermore, the volumes of drinking water produced daily vary greatly, with $20\,000\,m^3$ for the German waterworks as compared to $300\,000\,m^3$ at the Spanish waterworks.

In the Llobregat river, the greater percentage of wastewater of both municipal and industrial origin is reflected by higher values of dissolved organic carbon (DOC) and conductivity [19]. It has to be noted that seasonal variations of the water quality are extremely pronounced in the Llobregat since this relatively small Mediterranean river is strongly influenced by meteorological conditions. During the dry periods, frequently occurring in summer, the overall flow of the river may decrease significantly, resulting in a worsening of the water quality due to the subsequent elevation in the contributing fraction of treated wastewater. In such cases foams, mainly caused by industrial wastes, are obtained in the raw waters, presenting certain obstacles in the later processing. Though it might be expected that higher ambient temperatures in the Spanish river augments the activity of microbial communities, thereby enhancing the self-purification capability of the water, it should be taken into consideration that in many instances biodegradation of surfactants requires the availability of dissolved oxygen whose solubility, however, decreases with rising water temperature. In contrast, pollutant concentrations in the river Rhine are seasonally influenced to a lesser degree, due to a more constant flow of this river.

Raw water samples taken from the river at the two waterworks contain traces of SPC [23]. The measured concentrations in the raw waters average 5.0 and 1.8 μg L^{-1} in the Llobregat and Rhine rivers, respectively, demonstrating the higher wastewater percentage in the former. This is corroborated through the LAS levels, which amount to 12 and 1.0 μg L^{-1}, respectively. Calculation of the ratios of LAS-to-SPC as an indicator for the progression of LAS breakdown, yields values of 2.4 for the Llobregat water and 0.6 for the Rhine water.

Regarding the distribution of the alkyl homologues analysed (C6 to C12) in the raw waters, the patterns are shown in the last row of Figs. 6.6.2 and 6.6.3 for the Llobregat and Rhine rivers, respectively [23]. It can be noticed that C7-SPC is the most prominent homologue in the Rhine water sample while the relative abundances of the longer-chain homologues C8- and C9-SPC are somewhat greater in the Llobregat water. Since the shorter-chain SPC are formed out of higher

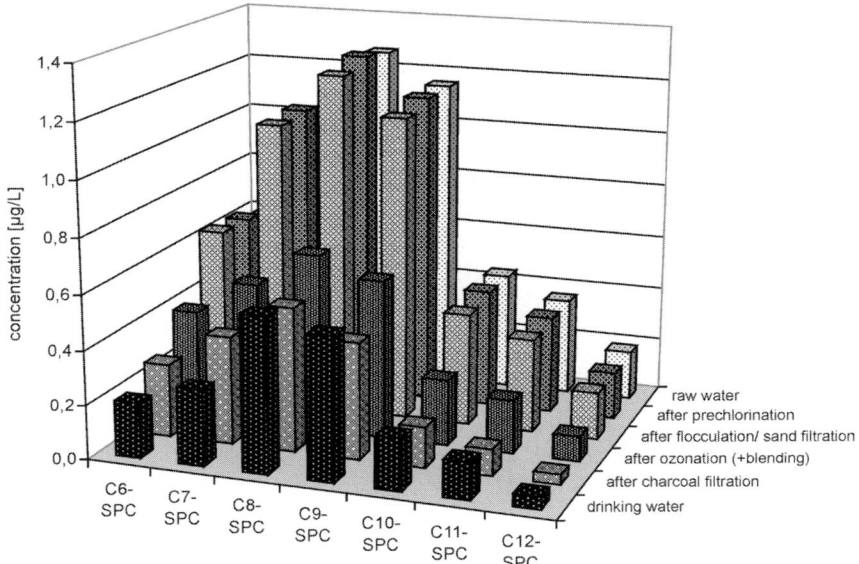

Fig. 6.6.2. Concentrations of individual SPC homologues in waters of different treatment stages at waterworks Llobregat (two sample batches) (from Ref. [23] with permission from Elsevier).

homologues, this observation indicates that SPC degradation/transformation in the Llobregat river occurs to a lesser extent.

Total concentrations of SPC during the different stages of water purification in waterworks Llobregat are given in Fig. 6.6.4, whilst the fate of single SPC homologues is shown in Fig. 6.6.2 [23]. As can be seen neither prechlorination nor flocculation followed by rapid sand filtration have an impact on the elimination. After ozonation and blending with ground water, an SPC reduction of about 50% is manifested, which is in part attributed to the mixing. The extent of the contribution of oxidative breakdown of the aromatic sulfonates to the decrease cannot be deduced from the analytical results. Also, no significant removal of SPC was observed during the following stage of GAC filtration in the waterworks Llobregat.

In field investigations it has been shown that the elimination efficiency of aromatic sulfonates is strongly dependent on the process conditions of the individual facility [26]. In a long-term study conducted over one year in a waterworks, where the treatment consisted of three

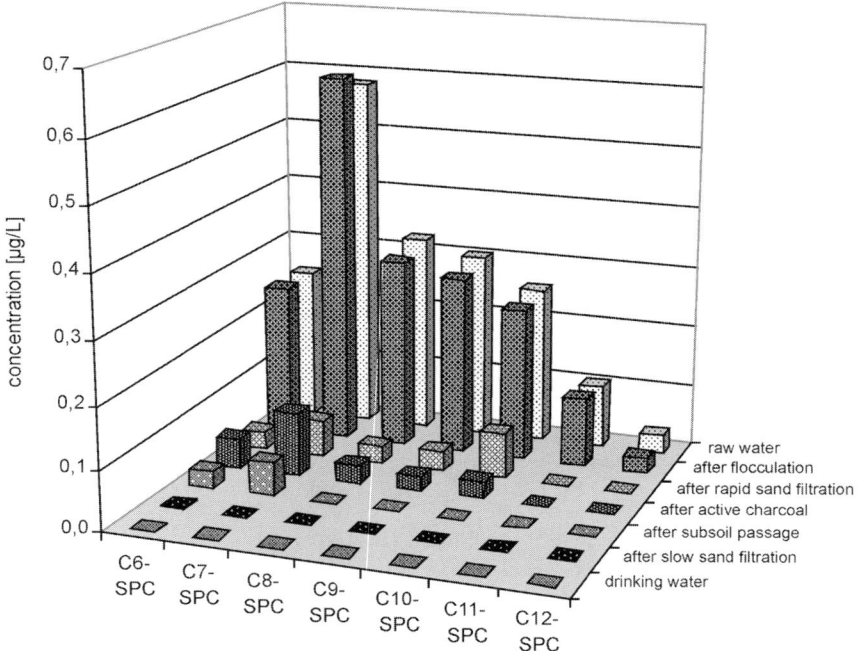

Fig. 6.6.3. Concentrations of individual SPC homologues in waters of different treatment stages at waterworks Rhine (three sample batches) (from Ref. [23] with permission from Elsevier).

main steps (iron removal, manganese removal, and a final filtration step by two parallel GAC adsorbers), the examined naphthalene sulfonates were not always efficiently removed. Directly after exchange of the activated carbon in the filter, no sulfonate was detectable. However, with increasing running time, i.e. loading of the filter material, the concentration observed steadily increased due to breakthrough on the GAC.

Early work by Rivera et al. [27] indicated that GAC filters installed at the waterworks Llobregat were capable of retaining different polar pollutants including LAS, NPEO and PEG, but most of them reached (to an unknown portion as the applied fast atom bombardment (FAB)–MS method only yielded qualitative results) the drinking water. Hence, further examinations covering at least one GAC cycle should be performed to obtain more information on the efficacy of GAC filtration

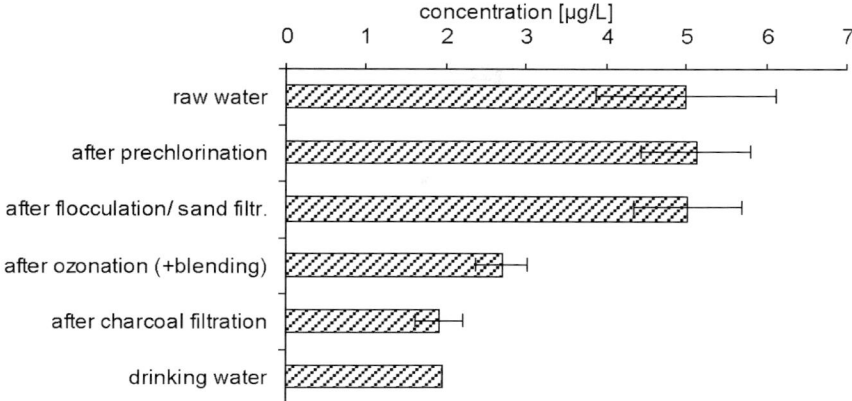

Fig. 6.6.4. Total concentration of SPC in waters of different treatment steps at waterworks Llobregat (two sample batches) (from Ref. [23] with permission from Elsevier).

to eliminate SPC. Furthermore, when performing lab-scale experiments on the adsorption capacity of single compounds, it should be taken into account that the outcomes will be strongly influenced by the nature of the carbon used, e.g. fresh or preloaded carbon [28].

In the final step of the water treatment process at the waterworks Llobregat, disinfection of the water with chlorine is performed; however, this also leads to no measurable removal of SPC, with resultant levels being around 2 μg L^{-1} in the finished drinking water [23].

A quite different picture is drawn at the waterworks Rhine, with a much better reduction of the SPC burden [23]. Figures. 6.6.3 and 6.6.5 show that flocculation with iron(III) chloride does not affect the presence of SPC, in accordance with the results obtained for the flocculation process at the waterworks Llobregat. In contrast, however, a marked reduction is observed during rapid sand filtration, where the SPC level drops from 1.8 down to 0.25 μg L^{-1}. It could be assumed that the specific elimination of SPC was brought about by microorganisms residing on the sand grains. Such an impact was not observed during the corresponding step in waterworks Llobregat, which might arise from differences in the contact times (about 10-fold less in the latter) and also by the fact that prechlorination of the raw water carried out in the Spanish waterworks most likely slows down biological processes (if not completely eliminates) in the subsequent treatment stages.

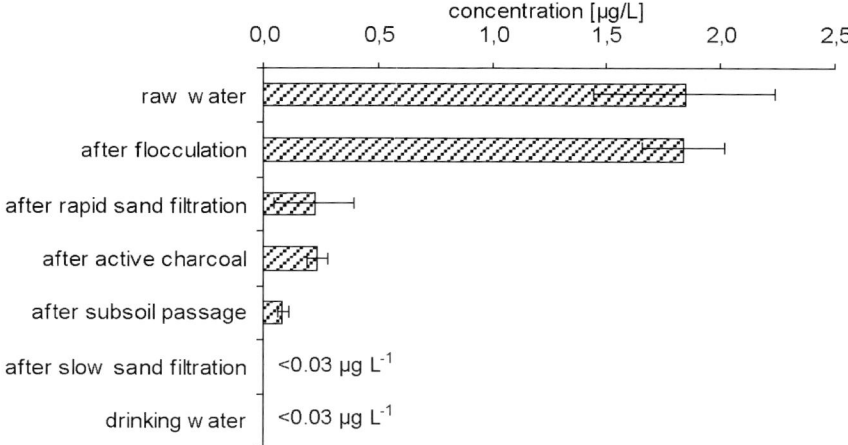

Fig. 6.6.5. Total concentration of SPC in waters of different treatment steps at waterworks Rhine (three sample batches) (from Ref. [23] with permission from Elsevier).

The next stage in the Rhine treatment, the GAC filtration, does not result in a measurable decrease of SPC [23]. As already discussed for the waterworks Llobregat, in this instance the actual state of the GAC, in terms of the extent of loading and breakthrough behaviour, might be a decisive factor in its ability to adsorb the polar sulfonates.

Following the subsequent subsoil passage of about 180 m length, the concentration of SPC amounts to $0.12 \ \mu g \ L^{-1}$. However, in order to interpret the cause of this apparent decrease (50%) accurately, the duration of the underground passage has also to be considered. The subsoil passage can take up to two months, and as such the samples taken before and after this infiltration process, even though sampling was performed in triplicate over a period of six weeks, do not necessarily correspond directly to each other. Hence, adsorptive removal of SPC during subsoil passage cannot be determined unequivocally from these results.

In degradation experiments of radiolabelled LAS in soil columns, Branner et al. [29] observed that microbial transformation products, believed to be SPC, were virtually not retained on the column at all. Conversely, the results from the waterworks Rhine study [23] show that the subsequent slow sand filtration leads to a nearly total elimination of the SPC homologues. The residues detected in the water after this step and after the final disinfection with chlorine dioxide were found to range well below the quantification limit of the applied analytical method (0.03 μg L^{-1}).

Comparing the capacities of both waterworks with regard to SPC elimination, it is evident that the polar pollutants exhibit a high potential to bypass the purification processes. Only the sand filtration steps, with their supposed biological activity, contribute significantly to reducing the SPC concentrations. This effect is very surprising in view of the known resistance to degradation of residues of these sulfonates in the river Rhine and in contaminated surface waters in general.

6.6.5 Possible formation of halogenated linear alkylbenzene sulfonates in waterworks Llobregat

The formation of halogenated NPs during drinking water disinfection in waterworks Llobregat has been described in preceding sections. Based on these observations, a study was conducted on the potential for formation of halogenated by-products of the anionic surfactant LAS, as likewise, LAS is present as an organic pollutant in the water entering the waterworks (at a mean level of 12 μg L^{-1}). Besides LC–ESI–MS analysis of the chlorinated raw water aimed at detection of the putative single chlorinated and brominated compounds, laboratory experiments were carried out with artificial raw water simulating the prechlorination process.

Neither in the raw nor in the test water were the hypothetical by-products, e.g. halogenated C12-LAS (i.e. ^{35}Cl-C12-LAS with [M–H]$^-$ of m/z 359 and ^{79}Br-C12-LAS with [M–H]$^-$ of 403), detected. This might be attributed to steric hindrance in substituting an aromatic hydrogen atom by a considerably larger halogen atom, due to the presence of the already bulky alkyl chain and sulfonate groups. Moreover, the latter sulfonate moiety acts as an electron-withdrawing substituent, which will thereby deactivate the aromatic ring with respect to electrophilic substitution.

6.6.6 Occurrence of surfactants in drinking waters analysed in countries with less developed WWT

As indicated by the studies on the two European waterworks, very polar SPC present at trace amounts in surface waters can reach treated waters to differing degrees. Therefore, a monitoring exercise was carried out in two towns on the east bank of the Baía de Guanabara, Brazil, namely Niterói and São Gonçalo. Both towns are supplied by the waterworks Laranjal, which is located 3 km downstream from the point where the Rio Macacu—investigated previously for its residues of LAS and SPC [30]—flows into the Rio Guapi-açu (see also Fig. 6.3.1). For comparative purposes, samples from the city of Rio de Janeiro, which receives its drinking water from another waterworks (Guandu at the Rio Guandu), were also analysed.

Results of the quantitative determination of SPC in tapwater samples from several sites including public buildings and private houses are compiled in Table 6.6.3 for two sampling periods. While in August 2000 the concentrations in Niterói averaged 3.3 μg L^{-1}, the mean level was markedly lower in the October samples, when 1.6 μg L^{-1} was determined. These differences are most likely due to differing SPC levels in the raw water withdrawn from the Rio Guapi-açu, which in turn varies depending on the hydrological status of the river and level of pollution of the domestic wastewaters. The almost identical value found in the October samples for both the Niterói and São Gonçalo sites is in accordance with the fact that both cities receive their drinking water from the same waterworks. In contrast, the drinking water from Rio de Janeiro, which is supplied by a waterworks situated in another river catchment area, had an SPC concentration of 3.7 μg L^{-1}.

During sampling, it was noticed by the authors that most buildings were equipped with reservoir tanks installed on the roofs to safeguard water supply in case of public shortages. These tanks were occasionally subjected to cleaning activities. To exclude the possibility that such direct actions represent a possible contamination source, through detergent residues such as LAS—out of which SPC might theoretically

TABLE 6.6.3

Concentrations of SPC in drinking water samples in Brazil

Sampling site	Sampling period	Concentration (μg L^{-1})
Niterói	August 2000	3.3 ± 0.4 (3)[a]
Niterói	October 2000	1.6 ± 0.2 (10)
São Gonçalo	October 2000	1.4 ± 0.2 (2)
Rio de Janeiro	October 2000	3.7 ± 0.7 (2)

[a]Number of samples analysed is in parentheses.

have been formed—in some instances, the water entering the cistern was sampled, in addition to that collected directly from the corresponding tap. However, no appreciable differences were observed.

Evidence for different raw water qualities and presumably also for distinct treatment stages in water processing at the two supplying waterworks Laranjal and Guandu, is given through comparison of the specific isomer pattern of individual SPC homologues detected in the water samples. In particular, the isomeric pattern of the individual SPC homologues provides a type of fingerprint which can be employed to elucidate the origin, determine level of contamination and extent of degradation and to assess the quality of the raw water resource used for the production of drinking water. The extracted (LC–ESI–MS) ion chromatograms for C7-SPC, which demonstrate the progression of SPC formation and breakdown, are displayed in Fig. 6.6.6 for the drinking water samples from Niterói (A) and Rio de Janeiro (B). The annotated peaks (a, b and c) are representative of different isomers of the C7 homologue, and the values in parentheses indicate the relative peak areas (where $a + b + c = 100\%$). These metabolites ($a-c$) are increasingly less readily degraded, respectively, and thus their relative distribution provides information regarding the extent of SPC transformation in the samples. The Niterói sample (Fig. 6.6.6(a)), which is representative of all samples from Niterói/São Gonçalo, shows a prevalence of peak a over b and c, which corresponds well with the pattern observed in the samples from the Rio Macacu and the Baía de Guanabara, additionally confirming the origin of Niterói drinking water. The sample from Rio de Janeiro (Fig. 6.6.6(b)), however, shows a relative homogeneous distribution of the three peaks resembling more the pattern found in the river Rhine (data not shown) and therewith indicative of a far more advanced biodegradation of SPC in the raw water used.

In studies carried out in November and December 2001 in Brazil [31], metabolites from NPEO were investigated in six different drinking water samples. Whereas in the three samples taken from Niteroi (see also Fig. 6.3.1) no analytes at all were detected, the levels in the drinking water from Rio de Janeiro, Brazil were in the range from 12 to 24 ng L^{-1} and below LOD to 6 ng L^{-1} for NPE_1C and NPE_2C, respectively. Thus the detected values are about 1000-fold less than the detected SPC values and also reflect the LAS/SPC to NPEO/NPEC concentration ratio found in Brazilian surface waters.

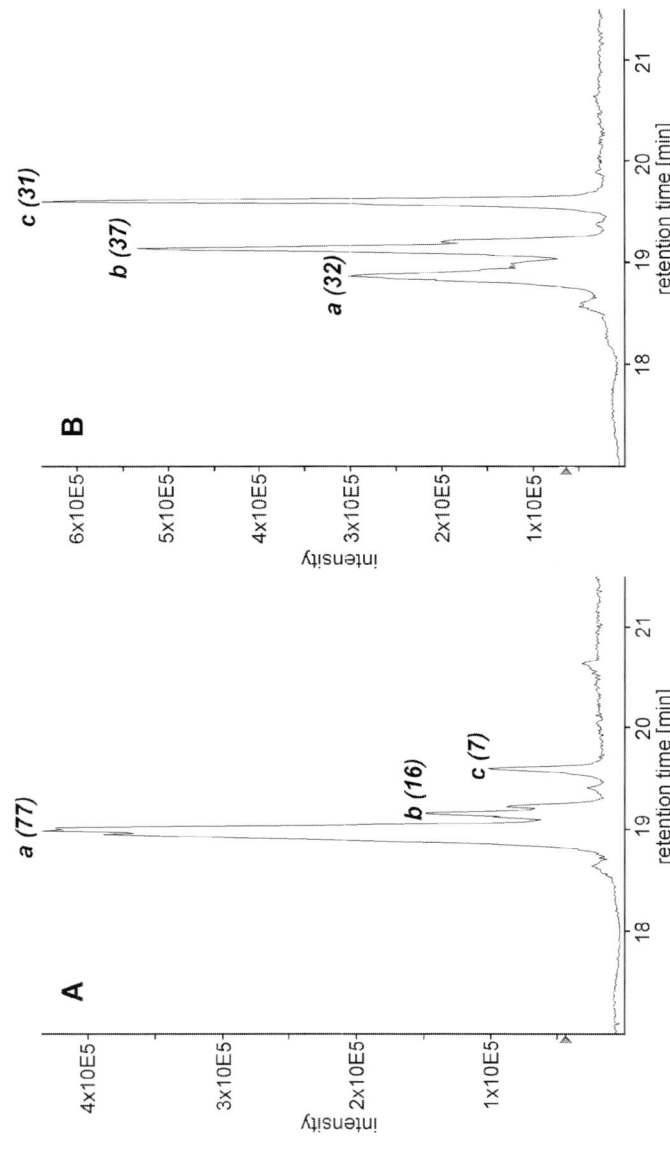

Fig. 6.6.6. (−)-LC–ESI–MS extracted ion chromatograms of C7-SPC (m/z 285) in drinking water samples from (a) Niterói and (b) Rio de Janeiro. Values in parentheses indicate relative peak area ($a + b + c = 100\%$) (from Ref. [30] with permission from Elsevier).

6.6.7 Risks arising from the presence of surfactants and their degradation products in drinking waters

Polar compounds

No toxicological data for SPC either on mammals or on humans are available. In contrast, the ecotoxicity to aquatic organisms (fish and *Daphnia*) has been investigated and showed that the LC_{50} was increased by a factor of 200–300 by the presence of a carboxyl group at the end of the chain, and another 10–20 times upon shortening the alkanoate chain from 11 to 5 [32]. This was attributed to the fact that the toxicity of LAS is directly related to its surface-active characteristics. Oxidation of the alkyl chain generating carboxylic acids leads to the loss of surface activity. This is in keeping with the findings of Berna et al. in investigations monitoring the toxicity towards *Daphnia magna* during biodegradation of various LAS blends [33]. The metabolites formed during this process were far less toxic than the parent molecule regardless of the mean molecular weight the type of LAS assayed. The residual material from trickling filter studies by Kölbener et al. showed no detectable impact on the growth of the algae *Selenastrum capricornutum* and the mobility of *Daphnia magna* [34,35].

Regarding the reported SPC concentrations of $2 \mu g L^{-1}$ in drinking water from the Llobregat waterworks described in Fig. 6.6.4, and assuming that a 60 kg body weight (BW) person consumes 2 L of water per day, this drinking water level converts to an SPC dose of $0.067 \mu g kg^{-1}$ BW day^{-1}. For LAS, a NOAEL for systemic toxicity has been identified as $69 mg kg^{-1}$ BD day^{-1} [36]. On the assumption that SPCs have equivalent toxicities to LAS, then the exposure level to SPCs is 1 million times less than the NOAEL. Based on the very large margin of exposure, it can be concluded that the presence of SPCs in drinking water poses little or no risk to consumers.

A second aspect that has to be evaluated is whether or not SPC are capable of disturbing the endocrine system of man, since a variety of xenobiotic substances with similar chemical properties have been shown to exert estrogenic effects on fish and mammals. Unlike several alkylphenol derivatives stemming from the degradation of APEO, a range of SPCs and a commercial LAS preparation proved not to be estrogenic in the recombinant estrogen yeast screen [37]. These findings were corroborated in a second in vitro test using the vitellogenin assay [37]. Neither C11-LAS nor the two SPC homologues C5 and C11 induced any positive response in the tests.

Nonpolar compounds

A screening study was conducted with samples from a Llobregat river WWTP in order to establish a better treatment for the removal of organic pollutants. Given that improvement of water quality is known to be achieved with the implementation of GAC filtration, the behaviour of surfactants (polyethoxylated nonylphenols, bromo-polyethoxylated nonylphenols) has been studied on seven different commercial GACs. These pollutants were efficiently removed using a bituminous coal-based GAC [38].

The applicability of GAC filtration for the removal of nonylphenol (NP) was evaluated using batch adsorption data [39]. The results showed that the sorption capacity of GAC for NP was at least 100 mg g^{-1} GAC. According to these data it can be concluded that a full-scale GAC filter, depending on the freshness of the GAC unit, will be sufficient to remove the environmentally relevant NP concentrations of 10 µg L^{-1}. Consequently, the existing GAC treatment technology in drinking water treatment should protect the consumer from the intake of the xeno-estrogenic micropollutant NP via drinking water. Investigations performed in our laboratories have indeed shown that NP was never present above the detection limit in all drinking waters analysed.

6.6.8 Conclusion

The absence of controlling regulation for surfactants in the past has meant little data are available concerning this compound class. In general only the NPEO, LAS and their known degradation products have been detected in surface waters serving as raw water for drinking water production, with other surfactants, such as AG, APG, cationics and amphoterics not detected. The introduction of legislation and development of LC–MS techniques appropriate to the detection of surfactants and their metabolites has improved the availability and quality of data obtained and added significantly to the knowledge base regarding these compounds in water samples.

From the available data it can be concluded that mainly the polar and recalcitrant metabolites, such as centred SPC isomers, resulting from LAS degradation, reach the drinking water plant at µg L^{-1} levels. From the viewpoint of toxicological effects, SPC-contaminated drinking water at these concentrations can be considered very unlikely to threaten the health of consumers.

Increase in demand and human activity worldwide has seen a reduction in the availability of pristine environments for use as drinking water resources. Raw water for drinking water production is thus obtained from a variety of sources, resulting in variable levels of contamination. Several methods for treatment, on laboratory and field scales and in practice in functioning waterworks, have been summarised here, and in particular rapid sand filtration, subsoil passage and ozonation have been demonstrated as important steps in the removal of undesirable xenobiotics.

The results described here demonstrate the importance of appropriate treatment and monitoring in actual drinking water processing plants, with attention to the specific requirements of the raw water matrix in use. In particular, the adverse effect of certain processes, namely pre-chlorination, which has been implicated in the inhibition of biodegradation in subsequent steps, and in the formation of alternative metabolites, is highlighted. Furthermore, the variable efficiency of GAC filtration in practice, emphasises the need for regular monitoring and quality control. The duration of specific process steps has also been shown to influence the efficacy of the technique, and should be addressed in application.

Comparison of the various results presented shows that the effect of different processing steps can be variable depending on the treatment process, the time and quality of the individual processing step and on the raw water matrix. As such, in the future implementation of drinking water treatment plants, the methods for treatment applied should be appropriate to the quality of the raw water resource. However, though sophisticated techniques may assist in diminishing to a large extent anthropogenic pollutants that have entered water purification plants, a sustainable drinking water production should be able to rely soley on natural purification steps in order to produce quality drinking water. A prerequisite for such drinking water production is an effective and adequate wastewater treatment, thus leading to below predicted no-effect concentrations of the residual levels of the surfactants and metabolites in receiving waters and thus also posing little risk to the aquatic environment. This applies especially to countries with less-developed WWT, whose residents should also be assured of a quality drinking water supply.

REFERENCES

1   DIN 2000, Zentrale Trinkwasserversorgung, Leitsätze für Anforderungen an Trinkwasser, Planung, Bau und Betrieb der Anlagen, Deutsche Normen, Berlin, November 1973.

2   I. Janssens, T. Tanghe and W. Verstraete, *Water Sci. Technol.*, 35 (1997) 13.

3   T.P. Knepper, F. Sacher, F.T. Lange, H.J. Brauch, F. Karrenbrock, O. Roerden and K. Lindner, *Waste Mgmt.*, 19 (1999) 77.

4   F. Ventura, D. Fraisse, J. Caixach and J. Rivera, *Anal. Chem.*, 63 (1991) 2095.

5   H.F. Schröder, *J. Chromatogr.*, 554 (1991) 251.

6   H.F. Schröder, *Water Sci. Technol.*, 25 (1992) 241.

7   H.-J. Brauch, F. Sacher, E. Denecke and T. Tacke, *GWF-Wasser/Abwasser*, 4 (2000) 141.

8   F.T. Lange, M. Wenz and H.-J. Brauch, *High Resolut. J. Chromatogr.*, 18 (1995) 243.

9   T.P. Knepper, F. Kirschhöfer, I. Lichter, A. Maes and R.-D. Wilken, *Environ. Sci. Technol.*, 33 (1999) 945.

10   T.P. Knepper, J. Müller, T. Wulff and A. Maes, *J. Chromatogr. A*, 889 (2000) 245.

11   RIWA, Organic micropollutants in Rhine and Meuse—Monitoring with HPLC/UV-fingerprint, Amsterdam, 2000.

12   RIWA, Inventory and toxicological evaluation of organic micropollutants— Revision 1999, Amsterdam, 2000.

13   R.T. Ghijsen and W. Hoogenboezem, *Endocrine Disrupting Compounds in the Rhine and Meuse Basin: Occurrence in Surface, Process and Drinking Water*, RIWA, Amsterdam, 2000.

14   P. de Voogt, R. Hendriks and O. Kwast, In: D. Vethaak (Ed.), *Estrogens and Xeno-estrogens in the Aquatic Environment of The Netherlands*, RIZA/ RIKZ report 2002.001, Lelystad/The Hague, 2002, ISBN 9036954010.

15   N. Jonkers, R.W.P.M. Laane and P. de Voogt, *Environ. Sci. Technol.*, (2003) accepted.

16   L.S. Sheldon and R.A. Hites, *Environ. Sci. Technol.*, 13 (1979) 574.

17   L.B. Clark, R.T. Rosen, T.G. Hartman, J.B. Louis, I.H. Suffet, R.L. Lippincott and J.D. Rosen, *Int. J. Environ. Anal. Chem.*, 47 (1992) 167.

18   R.A. Rudel, S.J. Melly, P.W. Geno, G. Sun and J.G. Brody, *Environ. Sci. Technol.*, 32 (1998) 861.

19   A. Guardiola, F. Ventura, L. Matia, J. Caixach and J. Rivera, *J. Chromatogr.*, 562 (1991) 481.

20   F. Ventura, A. Figueras, J. Caixach, I. Espalder, J. Romero, A. Guardiola and J. Rivera, *Water Res.*, 22 (1988) 1211.

21 F. Ventura, J. Caixach, A. Figueras, I. Espalder, D. Fraisse and J. Rivera, *Water Res.*, 23 (1989) 1191.

22 M. Petrovic, A. Diaz, F. Ventura and D. Barcelo, *Anal. Chem.*, 73 (2001) 5886.

23 P. Eichhorn, T.P. Knepper, F. Ventura and A. Diaz, *Water Res.*, 36 (2002) 2179.

24 F. Valero, R. Alcáraz, J.J. Rodríguez and M. Carnicero, *Eur. Water Mgmt.*, 2 (1998) 41.

25 J. Rivera, F. Ventura, J. Caixach, M. de Torres, A. Figueras and J. Guardiola, *Int. J. Environ. Anal. Chem.*, 29 (1987) 15.

26 F.T. Lange, M. Wenz and H.-J. Brauch, *Anal. Meth. Instr.*, 2 (1995) 277.

27 J. Rivera, D. Fraisse, F. Ventura, J. Caixach and A. Figueras, *Fres. J. Anal. Chem.*, 328 (1987) 577.

28 F. Sacher, F. Karrenbrock, T.P. Knepper and K. Lindner, *Vom Wasser*, 96 (2001) 173.

29 U. Branner, M. Mygind and C. Jorgensen, *Environ. Toxicol. Chem.*, 18 (1999) 1772.

30 P. Eichhorn, S.V. Rodrigues, W. Baumann and T.P. Knepper, *Sci. Total Environ.*, 284 (2002) 123.

31 T.P. Knepper, P. Eichhorn, S.V. Rodrigues and W. Baumann, *J. Braz. Chem. Soc.*, submitted.

32 R.D. Swisher, W.E. Gledhill, R.A. Kimerle and T.A. Taulli, *Surfact. Congress*, 74 (1976) 218.

33 J.L. Berna, A. Moreno and J.J. Ferrer, *Chem. Technol. Biotechnol.*, 50 (1991) 387.

34 P. Kölbener, U. Baumann, T. Leisinger and A.M. Cook, *Environ. Toxicol. Chem.*, 14 (1995) 561.

35 P. Kölbener, A. Ritter and U. Baumann, *Chimia*, 49 (1995) 204.

36 HERA, Human and Environmental Risk Assessment of LAS, CEFIC, Brussels, Belgium, 2002 (http/www.heraproject.com).

37 J.M. Navas, E. González-Mazo, A. Wenzel, A. Gómez-Parra and H. Segner, *Mar. Pollut. Bull.*, 38 (1999) 880.

38 F. Paune, J. Caixach, I. Espadaler, J. Om and J. Rivera, *Water Res.*, 32 (1998) 3313.

39 T. Tanghe and W. Verstraete, *Water, Air, Soil Pollut.*, 131 (2001) 61.

6.7 OCCURRENCE AND FATE OF SURFACTANTS IN SOIL, SUBSOIL, AND GROUNDWATER

Peter Eichhorn

6.7.1 Introduction

Surfactant residues do not only occur in the aquatic environment, into which they enter primarily through the discharge of sewage water, but they may also reach soil environments by various routes such as infiltration of or irrigation with (treated) wastewater, application of sewage sludge as organic fertiliser and soil conditioner or through the use of pesticide formulations containing surfactants as adjuvants. Since the late 1980s the amendment of soils with sludges, representing the dominating input into terrestrial compartments, has received considerable attention due to the high sludge concentrations of linear alkylbenzene sulfonates (LAS; Fig. 6.7.1) [1–7]. LAS adsorbed onto sludge or precipitated as insoluble Mg or Ca salts during wastewater treatment can persist in the subsequent anaerobic sludge treatment due to the poor biodegradability under these conditions. In anaerobically digested sludge the concentrations of LAS residues have been typically found in the range from 1000 to 30 000 mg kg^{-1} [8–10]. With the repetitive application of LAS-contaminated sewage sludge on to agricultural land and hence the introduction of considerable amounts of surface-active agents into soils, questions emerged regarding their whereabouts— issues such as accumulation over time or threat of groundwater contamination by leaching were addressed [11]—as well as possible adverse effects on soil organisms and plants. These concerns stimulated the research for investigating the fate and behaviour of LAS in soils.

A series of publications dedicated to soil-associated LAS has been authored or initiated by the surfactant-manufacturing industry. It was already in 1990 when Mieure et al. [12] presented a safety assessment for LAS for terrestrial plants and animals comparing measured LAS concentrations in soil environments with the lowest effect concentrations for typical organisms.[1] In 1998, de Wolf and Feijtel [13]

---

[1] Safety margins of 78–131 for fertilisation of agricultural fields with sewage sludge were calculated.

Comprehensive Analytical Chemistry XL
T.P. Knepper, D. Barceló and P. de Voogt (Eds)

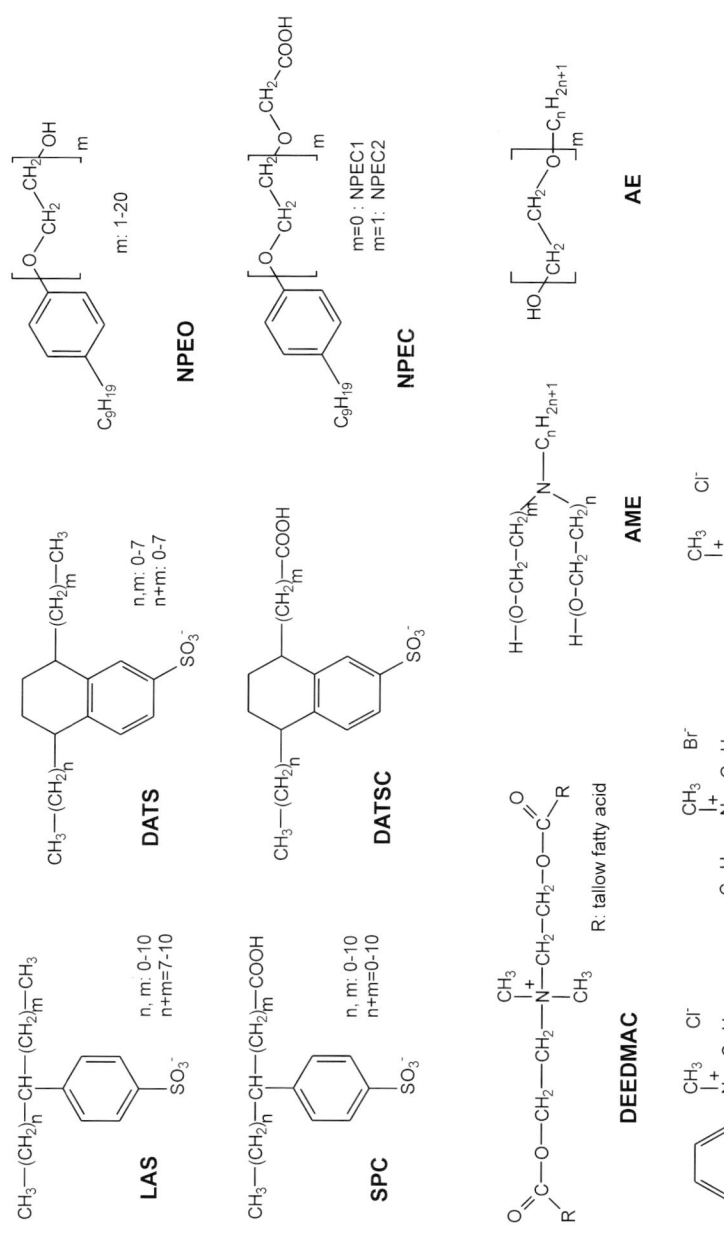

Fig. 6.7.1. Simplified structures of surfactants.

released an extended update of LAS risk assessment in soil reviewing processes that critically influence the fate of the anionic surfactant. Kloepper-Sams et al. [14] and more recently Jensen [15] reviewed the effects of surfactant, in particular LAS, in soil environments. Beside biological effects on soil organisms and plants, surfactants were also reported to alter soil physics and soil chemistry (review by Kuhnt [16]) and to interact with sludge-borne organic pollutants such as polycyclic aromatic hydrocarbons or chlorinated organics expressed as enhancement of solubilisation of hydrophobic contaminants and alterations of their adsorption and biodegradation behaviour (review by Haigh [17]).

Although the occurrence of the non-ionic surfactant nonylphenol ethoxylate (NPEO; Fig. 6.7.1) and its biodegradation intermediates in the aquatic environment has attracted large interest [18] because of the potential of some biotransformation products to mimic estrogenic effects, their fate and distribution in soil environments has been only investigated in a few cases.

In this contribution the current state of knowledge on the occurrence, fate and interactions of surfactants in terrestrial environments is summarised. It focuses mainly on LAS and NPEO for which most information has been gathered.

6.7.2 Occurrence of surfactants in sludge-amended soil and their biodegradability under laboratory conditions

Among the first analytical methods proposed for the determination of LAS in sludge-amended soil was the one developed by Matthijs and de Henau [1,19]. Samples of dried soil were extracted with methanol under reflux for 2 h. To remove soil matrix components the extracts were subjected to a clean-up procedure consisting of an anion exchange column (SAX) followed by a reversed-phase (RP)-C8 column. Quantitative measurements performed by RP-high performance liquid chromatography with ultraviolet detection (HPLC–UV) gave LAS levels in the range from 0.9 to 2.2 mg kg^{-1} with a mean value of 1.4 mg kg^{-1}. Of four soils analysed three were collected from lands with regular sludge application and one from a site used for sludge disposal.

Berna and co-workers [9] used the analytical method published in Ref. [1], after a slight modification, to determine LAS in sludge-amended soils on two different locations. On one field in a grape vine farm sludges were applied in a dried form twice a year. Directly after application an LAS level of 16 mg kg^{-1} was measured, which dropped to < 1 mg kg^{-1} by day 90. An average half-live ($t_{1/2}$) for LAS of 26 days was obtained. On the other field cultivated with vegetables, amendment of the land (without history) with sludge resulted in a surfactant concentration of 52 mg kg^{-1}. The results showed a rapid disappearance with a half-live of primary degradation of 33 days. From these findings it was concluded that sludge application at six-month intervals did not represent any risk of LAS accumulation in the soil.

The objective of the study carried out by de Ferrer and colleagues [20] was to determine LAS biodegradation in a land filling operation using sludges (15%) blended with soil (85%). The analytical protocol comprised extraction of the samples with alkaline methanol followed by HPLC–UV measurements. After sludge amendment the surfactant concentration was monitored over a period of 62 days showing a decrease from initially 155 to 17 mg kg^{-1}. The average half-life of primary degradation of the four alkyl homologues was about 20 days. In comparison to the lighter ones ($t_{1/2}$(C10): 13 days), the substantially lower conversion rate of the higher weight LAS components ($t_{1/2}$(C13): 21 days) was attributed to the stronger adsorption of these more hydrophobic homologues onto the soil.

In a programme of monitoring work, Holt et al. [2,21] determined the fate of LAS in sludge-amended soils collected from a total of 51 fields at 24 farms. Sample preparation followed the protocol developed by Matthijs and de Henau [1] using Soxhlet extraction. The authors achieved detection limits (LOD) of 0.4 mg kg^{-1} with the UV system and of 0.2 mg kg^{-1} with fluorescence (FL) analysis. Of the 42 farm sites that had received sludge prior to the year of sampling five contained LAS concentrations below the analytical detection limit, whilst the mean level for the other 37 fields was 0.7 mg kg^{-1}. A frequency distribution of the percentage LAS losses from soil showed that for 90% of the sites less than 5% of the estimated cumulative load added by sludge amendment remained in the soil. At those locations where sludge had been applied prior to sampling the concentration range was < 0.2–20 mg kg^{-1}.

Holt and Bernstein [3] conducted a second monitoring exercise to determine LAS along with linear alkyl benzene in sludge-amended soils. Analyses of the anionic surfactant followed the previously established method [2]. The agricultural lands sampled had all been surface spread with primary or anaerobically digested sludge. In seven out of nine soil samples collected from fields with sludge application in previous years, the mean LAS level was below

1 mg kg^{-1}. The other two soils contained 1.3 mg kg^{-1}. Comparison of the estimated cumulative LAS with the concentrations measured showed removals between 94.6 and 99.5%. Another set of soil samples was analysed originating from fields that had received sludge application within 6 months prior to sampling. The concentrations ranged from <0.2 to 8.1 mg kg^{-1} with a mean value of 3.3 mg kg^{-1}. Studies on the temporal evolution of the LAS concentration performed on two soils without previous history of sludge application showed on field one a decrease of the initially measured level of 117 mg kg^{-1} by 97.6% after a period of time of 103 days. On the other field with a concentration of 145 mg kg^{-1} immediately after sludge amendment, the surfactant level dropped to 15 mg kg^{-1} by day 22 and further decreased to the background level after 55 days. In all instances alkyl homologue-specific analyses by HPLC–UV revealed no preferential removal of any chain length.

Rapaport and Eckhoff [4] reported LAS levels in sludge-amended soils likewise employing the methodology of Matthijs and de Henau [1]. Shortly after application of anaerobically digested sludge three soil cores were taken from a cultivated land, which had regularly been treated with sludge over a period of 7 years. In the upper layer (0–7.5 cm) the LAS concentration averaged 28 mg kg^{-1} (range: 13–47 mg kg^{-1}). At a depth of 7.5–15 cm a mean level of 7 mg kg^{-1} was determined, whereas in the 15–22.5 cm-section 5 mg kg^{-1} was found in one core and <LOD in the other two cores. Based on sludge LAS content, application rate and ploughing depth, an LAS concentration of 55 mg kg^{-1} in the soil was calculated. The determined level of 28 mg kg^{-1} was considered satisfactory in view of the potential extent of environmental variation and biodegradation that might have occurred between sludge application and sampling. Indirect evidence of microbial degradation was indicated because measured concentrations were at least seven times lower than predicted levels based upon a cumulative load over the 7 years prior to sampling.

In a recent study, Carlsen and colleagues [22] investigated the occurrence of LAS in a series of soil samples from seven different locations with different histories. Ten centimetre subsamples of the 50 cm cores were extracted using a microwave technique and the analyses were carried out using HPLC–FL. With the exception of two sites the LAS concentrations were found to range from <LOD to 1.2 mg kg^{-1}. No substantial differences were noticed between undisturbed soils used for grazing for 50–100 years and soils being

moderately amended with sludge. However, on the field with heavy sludge amendment the surfactant concentration in the upper layer of $0-10$ cm reached 11.2 mg kg^{-1}. Slightly lower levels of approximately 9.5 mg kg^{-1} were determined in the three sections beneath, but in the $40-50$ cm horizon 7.1 mg kg^{-1} were still detected. The two top layers of a core taken from the same location were examined 2 years later showing LAS concentrations of 10.4 and 19.3 mg kg^{-1} in the $0-10$ and $10-20$ cm horizons, respectively. In the soil core taken in the run-off zone from sludge storage, LAS was detected in all layers at levels between 0.78 and 4.0 mg kg^{-1} with a maximum concentration in the $10-20$ cm section.

A method for the simultaneous determination of LAS, NPEO1 and NPEO2 in sludge-amended soils was presented by Marcomini and Giger [5]. Samples were extracted in a Soxhlet apparatus with methanol for 12 h. The extracts were evaporated to 5 mL and diluted with acetone/water to 20 mL in such a way to attain a solvent composition of methanol/acetone/water of 1:2:1. This caused precipitation of fine suspended material, which was removed by centrifugation prior to injection into the HPLC equipped with an FL detector. As an RP-C18 column gave no satisfactory results for the simultaneous and routine analysis of both surfactants due to long analysis time and broadening of the NPEO peak, the alkyl homologues of LAS were separated on a C8 RP column, while the ethoxymers of NPEO were resolved on an aminosilica column. Application of the methodology to a sample of sludge-amended soil gave concentrations for LAS, NPEO1 and NPEO2 of 15, 0.40 and 0.07 mg kg^{-1}, respectively.

The previously developed method was applied to determine the fate of the aromatic surfactants in sludge-treated soil at 19 time intervals during a one-year-period after sludge application [23,24]. The concentrations of LAS, NPEO1 and NPEO2, present in the freshly amended soil at 45, 1.1 and 0.1 mg kg^{-1}, respectively, decreased rapidly within 3 weeks to an extent of about 20% of the initial value, which was believed to be due to microbial degradation. After an additional 70 days of further, slow disappearance, constant levels during the following 130 days were observed. The final mean concentrations of the soil were 5, 0.12 and 0.01 mg kg^{-1} for LAS, NPEO1 and NPEO2, respectively. The residual levels reflected a certain persistency of the analytes in soil presumably because of their strong sorption to soil matter impeding further biodegradation. This assumption was supported by the temporal change of the LAS homologue distribution (C11- to C14-

LAS). The mean chain length of 12.1 in the applied sludge increased to 12.2 after a period of 23–31 days and was finally found to be 12.4 in the treated soil after 312–322 days.

Hawrelak and co-workers [25] determined lipophilic degradation products of APEO, including NPEO1, NPEO2 and alkylphenols, in samples collected from an agricultural field treated with recycled paper sludge. Prior to spreading on the field, the sludge had been deposited for 2 months on the field for stockpiling. Sludge samples were Soxhlet extracted with dichloromethane and aliquots of the extracts were analysed either directly by normal phase (NP)– HPLC–FL for the determination of the short-chain ethoxylates or after acetylation by gas chromatography–mass spectrometry (GC– MS) for nonylphenol and octylphenol (NP, OP). The concentrations of NPEO1 and NPEO2 in the sludge taken one day after deposition on the field were 1.21 and 0.39 mg kg^{-1}, respectively. The values declined during the following 2 months of field storage to 0.17 (NPEO1) and 0.14 mg kg^{-1} (NPEO2). In the sludge collected 6 weeks after spreading on the field, the residual concentrations amounted to 0.07 mg kg^{-1} NPEO1 and to 0.08 mg kg^{-1} NPEO2. The authors concluded that microbial degradation of NPEO, introduced into the terrestrial compartment through recycled paper sludge, substantially reduced the risk of environmental contamination.

Heise and Litz [26] investigated the extraction behaviour of surfactants (LAS, NPEO and cationics) from sand comparing Soxhlet extraction, accelerated solvent extraction (ASE) and microwave-assisted extraction. Fractionation of the three surfactant types anionic, non-ionic and cationic, was accomplished by column chromatography with aluminium oxide. Soxhlet extraction and ASE of spiked sand with methanol—stored during 7 days prior to extraction—gave similar recoveries for both LAS and NPEO with values between 88 and 116%. Less efficient extraction was achieved by microwave extraction (79% for NPEO).

The primary biodegradability of soap in sludge-amended soils was reported by Prats et al. [27]. The amendment of soil (27% sand, 34% lime, 39% clay) at a vineyard farm with anaerobically digested sewage sludge led to an increase of the soap concentration from the background level of 570 mg kg^{-1} (yearly treatment of the field) to 1470 mg kg^{-1} as determined after Soxhlet extraction, fluorescence labelling of the fatty acids with a coumarin derivative and HPLC–FL analysis. By day 150 after amendment, the value dropped to 470 mg kg^{-1} thereby reaching a similar level to the one before sludge application. The steady state concentration was assigned to the poor bioavailability of calcium and magnesium salts of the fatty acids precipitated during sewage treatment [28].

Comellas and co-workers [7] developed an analytical procedure to study the primary degradation of LAS in limy soils modified with anaerobically digested sewage sludge. As the use of basic methanol had proved to enhance extraction efficiencies due to reduced adsorption of LAS onto lipid organic matter, the authors compared Soxhlet extraction with NaOH added to the sample and extraction with alkaline methanol under reflux. The continuous basic methanolic treatment of the latter method was reported to allow extensive saponification of lipids thereby improving the extraction. Prior to the quantitative analyses by HPLC–UV/FL matrix components were removed from the extracts by column clean-up on RP-C18. Degradation experiments carried out with LAS contents in the sludge–soil mixtures of 1870, 5460 and 19 870 mg kg^{-1} showed that both the lag phases and the half-life times increased with higher initial LAS doses. At the lowest concentration tested the lag phase was 7.5 days, while the onset of biodegradation in the 19 870 mg kg^{-1} assay was observed on average after 33 days. The half-lives to 50% degradation ranged from 18 to 162 days.

Holt and colleagues conducted a series of field experiments investigating the disappearance of LAS with time [2,21]. On one field that had received a subsurface injection of 5 L m^{-2} of anaerobically digested sludge containing 12 mg L^{-1} LAS, the initial concentration of 66.4 mg kg^{-1} dropped to 4.5 mg kg^{-1} after 24 days and to <1.0 mg kg^{-1} within 49 days after application. Spraying of the same sludge to another field followed by immediate ploughing increased the background concentration from 0.3 to 2.6 mg kg^{-1}. By day 49 the LAS level had fallen to <0.3 mg kg^{-1}. On two further fields spread with digested (10 700 mg kg^{-1}) and activated sludge (800 mg kg^{-1}), respectively, the initial disappearance was very rapid. The concentrations decreased from 12.8 and 4.5 mg kg^{-1} to 1.0 and <0.2 mg kg^{-1}, respectively, within a period of time of 32 days. The authors observed no significant change in the homologue distribution over time suggesting that there was no preferential biodegradation of specific alkyl-chain length.

Litz and colleagues [29] investigated the behaviour of LAS in different soils under both field and laboratory conditions. The field experiments were carried out by spiking aqueous LAS solutions to plots (between 0.8

and $50 \, \mathrm{g \, m^{-2}}$), from which samples were drawn from different soil depths over a period of 8 weeks. LAS was rapidly degraded in loamy orthic Luvisol under arable land. Degradation reached approximately 50% after 5 days and nearly 80% after a further 7 days. A total precipitation of 86 mm within 2 weeks moved about 3% of the LAS into the 10–30 cm soil layer. In an acidic dystric cambisol under pine forest, LAS proved to be less mobile due to binding with organic matter in the humic litter. In cambisol irrigated with sewage water, LAS was primarily bound in the 0–5 cm horizon. Slight leaching to a depth of 10–30 cm was observed after application of 180 mm wastewater. In the laboratory tests iron oxide, humic content and pH value proved to be the decisive soil parameters influencing the sorption of LAS. Strong sorption was indicated for acidic samples with high iron levels and organic matter.

An extended study of the behaviour of LAS in sandy soils was conducted by Küchler and Schnaak [6] using field trials and lysimeter studies. In the field study two sewage sludges with LAS contents were applied under realistic amendment conditions to soil (91% sand, 0.9% C_{org}). During a period of one year core samples of 30 cm depth were taken, divided into subsamples of 10 cm and analysed by Soxhlet extraction/solid-phase extraction (SPE)/HPLC–FL to determine biodegradation and leaching within the investigated horizons. In the upper soil horizon of the first plot, amended with slightly contaminated sludge, the LAS concentration decreased from $1.1 \, \mathrm{mg \, kg^{-1}}$ in 18 days to below the LOD. At depths of 10–20 and 20–30 cm the surfactant was intermediately detected after 7 days at levels of 0.35 and $0.2 \, \mathrm{mg \, kg^{-1}}$, respectively, but after 4 weeks no LAS could be detected. In the second plot with an initial LAS concentration of $8.2 \, \mathrm{mg \, kg^{-1}}$ no LAS was detectable in the 0–10 cm layer after a period of 28 days. In the 10–20 cm horizon the surfactant reached a level of about $1 \, \mathrm{mg \, kg^{-1}}$ after 7 or 14 days and decreased again to below the LOD by day 28. At a depth of 20–30 cm, the concentrations were at any time <LOD. In the lysimeter studies, soil columns of 40 cm in height were amended with 152 and $2200 \, \mathrm{mg \, kg^{-1}}$ LAS and rained with artificial rainwater. LAS measurements were done for each soil horizon and the percolated water. Although intense percolation occurred, no LAS leached out of the 40 cm columns. The lysimeter experiments revealed a rapid biodegradation in the soil (77% sand, 1.1% C_{org}). Forty-two days after the beginning of the experiment, LAS was detectable in the soil horizon 0–5 cm in the lysimeter with an initial concentration of $2200 \, \mathrm{mg \, kg^{-1}}$. In this

lysimeter LAS was detected to a depth of 30–40 cm indicating relatively high mobility; however, 99% of the total LAS was calculated to be subject to biodegradation. In a leaching experiment with intensive raining (1333 mg kg^{-1} LAS), the homologue profile of the percolated water displayed a strong discrimination of the longer chain homologues due to stronger retention in the soil. In the water, C10-LAS accounted for about 55% of the total LAS concentration, while the percentage of C13-LAS was less than 3%. Increasing the LAS load to the soil by a factor of two resulted in a distribution shift towards the longer alkyl chain as a consequence of saturation of the sorption capacity of the organic soil substance. The authors summed up that the competition between leaching and microbial degradation determined the fate of LAS. The latter was more significant due to good aeration of the investigated sandy soils.

The primary biodegradation of a commercial LAS blend in a coarse sand soil without previous exposure to sewage sludge was studied by Elsgaard et al. [30] under aerobic and oxygen-limited conditions. Soil incubations were amended with either aqueous LAS solutions or LAS pre-adsorbed onto sludge resulting in surfactant contents between 8 and 488 mg kg^{-1}. After 2 and 8 weeks of incubation the LAS residues in the soil samples were extracted with alkaline methanol and the purified extracts analysed by HPLC–UV/FL. In the aerobic experiments with initial concentrations of up to 62 mg kg^{-1} the biodegradation of LAS was greater than 73% after 2 weeks and higher than 85% after 8 weeks. At a nominal concentration of 488 mg kg^{-1}, 7% of the LAS spiked in the sludge-bound form were recovered after 8 weeks, while 85% of the LAS amended as aqueous solution were not degraded. In soil samples incubated anaerobically[2] the extent of primary degradation was generally lower than under aerobic conditions ranging from 0 to 38% after a duration of 2 weeks. No clear correlation was apparent between the initial LAS concentration or the form of LAS amendment and the percentage of degradation.

Larson et al. [31] performed biodegradation assays of C13-LAS in surface soils encompassing a range of different soil types (silt loam, loam, sand, sandy loam, silty clay). No soil had been previously exposed to LAS surfactants. The rate and extent of mineralisation were rather independent of the soil type with half-lives in the range of 1.1–4.9 days. The onset of biodegradation was not preceded by a noticeable lag phase

---

[2] Used glass jars contained initial oxygen pool.

indicating ready adaptation of the native microbial communities to the xenobiotic substrate. In a further study the authors investigated the degradation of LAS in subsurface soil. To this aim, porous sand with no history of exposure was collected from the unsaturated zone of an unconfined aquifer (ca. 8 m) and amended with 0.2 mg kg^{-1} of an LAS mixture. After 0, 53 and 250 days subsamples were withdrawn and dosed with ^{14}C-LAS at a level of 0.8 mg kg^{-1}. A clear influence of adaptation at the beginning of degradation was perceived. In the unexposed soil the formation of ^{14}CO$_2$ started after a lag phase of about 20 days, while in the two other soils no adaptation phase was noticed. In addition, the authors collected soil and water samples in the immediate vicinity of an effluent plume from a septic tank as well as from sites at various distances (up to 20 m) from the effluent point. In these matrices the biodegradability of C13-LAS was studied separately for water and soil showing that depth-averaged half-lives were shorter and less variable in the first (mean value: 10.6 days) than in the latter (18.7 days). Whereas the half-lives of LAS degradation decreased with increasing distance from the discharge point, the opposite behaviour was observed for the soil samples.

In a second work Larson et al. [32] reported on results of two studies to characterise the kinetics of biodegradation of ^{14}C-labelled LAS homologues in sludge-amended soil. Both soils had received sludges during several years prior to the experiments. Although the rate and extent of mineralisation of the LAS homologues, amended at 25 mg kg^{-1}, were comparable in the two soils, a lag phase of about 2 weeks was noticeable in the soil with the reduced frequency of sludge loading probably indicating a less stable LAS degrading microbial population. For the five alkyl homologues C10 to C14 the mineralisation efficiencies ranged from 60 to 70% with half-lives of 16 and 21 days, respectively. No significant differences in $t_{1/2}$ were observed over the range of homologues. In summary, the findings provided evidence that thanks to the ubiquitous occurrence of LAS-degrading soil micro-organisms, an accumulation of LAS on soils as a consequence of repetitive sludge amendments did not occur.

Figge and Schöberl [33] reported on the degradation and fate of ^{14}C-LAS after introduction of digested sludge into the topsoils collected from two agricultural ecosystems. The first of the soil cores transferred into the laboratory test apparatus was a heavy, clay-like soil, while the second was classified as a loose, sandy soil. The initial concentrations of LAS in the top layer of the soils were 27.2 and 16.2 mg kg^{-1}, respectively. The distribution of radioactivity between the different compartments of each system, i.e. air, biomass, soil and percolated water, was determined after 76 days (first core) and 106 days (second core) showing that at least 64 and 72%, respectively, of the LAS incorporated with the digested sludge were totally mineralised to CO_2. The second largest fraction of radioactivity (27 and 18%) was recovered from the soil, mainly concentrated in the topsoils (cores were sectioned for analysis into 5 cm horizons). These radioactive residues were assigned as polar metabolites of LAS, which were firmly adsorbed onto soil particles. However, the identity of these secondary products was not further elucidated. The strong binding of these compounds to soil matter and therewith their reduced bioavailability was assumed to be a rate-limiting factor in the further degradation. No radioactivity could be detected in soil depths below 40 cm. Less than 1.5% of the radioactivity was detected in the percolated water—during the experiment the test systems were maintained under a defined standard climate—including mainly water-soluble carbonates and smaller amounts of mixed organic substances. The composition of the latter was further characterised by radio-thin layer chromatography indicating that these were dominated by biodegradation intermediates of LAS. On the basis of the $^{14}CO_2$ production over time LAS half-lives were calculated to be 26 days in the clay-like soil and 13 days in the loose, sandy soil. Seemingly, aerobic LAS degradation was influenced by the accessibility of atmospheric oxygen into the soil.

The mineralisation kinetics of ^{14}C-C12-LAS was studied by Dörfler et al. [34] in three freshly sampled soils of different physico-chemical properties. After an incubation phase of 63 days the total amounts of mineralised surfactant, dosed to the soils at 10 mg kg^{-1}, were rather independent on the soil type and varied between 64 and 66% $^{14}CO_2$. The kinetics could be represented best by a 3/2-order function with linear adaptation and three parameters, each of which describing a different part of the degradation curve (adaptation, exponential growth, asymptotic part). To assess the influence of the initial LAS concentration, a series of degradation experiments was performed amending the soils with 1 mg kg^{-1}. Tracing the CO_2 evolution showed that the mineralisation rates were not affected by the starting concentration. In contrast, investigations on the effect of soil temperature revealed a linear correlation between the amount of mineralised LAS and the soil temperature. Below 3°C the LAS degradation was negligible.

Branner and co-workers [35] investigated the ability of a microbial community to degrade LAS in soil columns under water-saturated conditions. This should simulate an infiltration situation where LAS was present in irrigation water. The surfactant was continuously added to the column complemented by three pulses of ^{14}C-labelled C12-LAS at intervals of 2 weeks. After a short adaptation phase, microbial degradation took place and mineralisation was observed. Determination of LAS in the column effluents by HPLC–UV indicated an almost quantitative primary degradation of the second and third pulse, but only a small fraction (9%) of the amount was measured as $^{14}CO_2$. Around 80% were recovered as unmineralised degradation intermediates, whose chemical identity was not further elucidated. The breakdown products were found to be much less sorptive than LAS itself and passed through the soil columns without considerable retardation.

The microbial mineralisation of ^{14}C-LAS and ^{14}C-alcohol ethoxylate (AE; Fig. 6.7.1) was examined by Knaebel et al. [36] in 11 soils differing in their physico-chemical properties and collected from distinct locations. With one exception, the soils had not previously been exposed to surfactants. After amendment of the soil samples with 0.05 mg kg^{-1} surfactant, the formation of $^{14}CO_2$ was monitored. Both LAS and AE were mineralised in every soil without a lag phase indicating that microbial communities indigenous to soil were capable of ultimately biodegrading the xenobiotic compounds even without previous exposure. Yields of $^{14}CO_2$ from LAS ranged from 16 to 70% (mean value: 49%) of added ^{14}C, whereas those from AE varied between 25 and 69% (46%). The authors emphasised that mineralisation was a conservative indicator of biodegradation as it did not provide information on the extent of ^{14}C incorporation into biomass or humus. Determination of the microbial activity, measured as the rate of ^{14}C-acetate incorporation into microbial lipids, revealed great variations as a function of soil, but no correlation was observed with the mineralisation rate representing half-lives from 3.7 to 1.1 days for LAS. As the kinetic parameters describing LAS and AE mineralisation were alike, it was suggested that similar microbial populations degraded both surfactants or that their biodegradation was controlled by the same environmental factors.

Federle and colleagues [37] examined the mineralisation of LAS and AE in two sandy soil profiles from the same area, one of which impacted by infiltrating wastewater from a laundromat. The mineralisation of ^{14}C-C13-LAS and ^{14}C-C12/14-AE in each profile was evaluated as a function of depth (groundwater table at 14 m below surface). The non-ionic surfactant was mineralised at every depth in both profiles with short lag phases in samples obtained below 2 m depth. Calculated half-lives ranged from 39 to 1.4 days. A drastic decrease in mineralisation extent was observed below 2 m. The mineralisation of the anionic surfactant displayed a similar behaviour characterised by little or no mineralisation in the vadose zone between 2 and 14 m in either profile. In the upper soil immediate onset of LAS mineralisation was noticed, but lag periods were apparent in the saturated zone. The marked discontinuity in the mineralisation went along with changes in microbial biomass and activity, but was not consistent with nitrogen and phosphorous contents. The reasons behind this phenomenon remained unclear. Regarding the influence of infiltrating wastewater, the degradation rates of neither LAS nor AE were enhanced in the unsaturated zone, but a small stimulation effect was noticed in the saturated zone. In a further experiment the surfactant mineralisation was studied in groundwater collected from wells upgradient and downgradient of the leaching field. In the water, free from LAS and AE traces as determined by methylene blue-active substances (MBAS) and cobalt thiocyanate-active substances (CTAS) procedures, respectively, LAS was not mineralised in any sample. The non-ionic AE showed instant onset of mineralisation in water from the downgradient well, but a lag period of 8 days in the water taken from the upgradient well. In summary, microbial populations occurring in groundwater exhibited less biodegradative activity but were capable of preventing an accumulation of LAS or AE.

The effect of mineral and organic soil constituents on the mineralisation of LAS, AE, stearyl trimethylammonium chloride (STAC) and sodium stearate (main soap component) in soils was studied by Knaebel and co-workers [38]. The four ^{14}C-labelled compounds were aseptically adsorbed to montmorillonite, kaolinite, illite, sand and humic acids and subsequently mixed with soil yielding surfactant concentrations of about 50 μg kg^{-1}. The CO_2 formation in the serum bottle respirometers was monitored over a period of 2 months indicating that the mineralisation extent was highest for LAS (49–75%). Somewhat lower amounts of produced CO_2 were reported for AE and the stearate ranging from 34–58% and 29–47%, respectively. The mineralisation extent of the cationic surfactant did not exceed 21% (kaolinite) and achieved only 7% in the montmorillonite-modified soil. Associating the mineral type with the mineralisation kinetics showed that sand

and kaolinite complexes exhibited the highest rates and extents of mineralisation, whereas in the assays with illite and montmorillonite-modified soils the total amounts of CO_2 measured were lower although initial mineralisation rates were comparable. Except for the degradation of AE, the yields of CO_2 were lowest in those soils where the surfactants had been introduced preabsorbed onto humics. The dissimilarities in the mineralisation behaviour were explained by different degrees of surfactant bioavailability associated with specific interactions of the test compounds with the soil minerals. Their structural and physico-chemical characteristics such as type and number of internal/external binding sites, ion exchange capacity and swelling capacity appeared to be decisive parameters in the surfactant—clay (sand) interactions, which were assumed to be of a polar or an ionic nature. Hydrophobic or in the same instances covalent bonding in turn was deemed responsible for the strong binding of the surfactants to humics. Overall, the studies demonstrated that the environmental form of a surfactant had a profound influence on its bioavailability and therefore on its biodegradability in soils.

The mineralisation of ^{14}C-labelled C12-LAS and NPEO2 was investigated in different activated sludge—soil mixtures and soils [39]. For comparison of the biodegradation potential in different soils, Gejlsbjerg and co-workers carried out experiments in a coarse, sandy soil (89% sand), a sandy soil (89.7% sand) and a more clayey soil (74.6% sand, 10.6% clay), all of which without previous exposure to sewage sludge. LAS and NPEO2 were rapidly mineralised in the predominantly aerobic sludge—soil mixtures, but mineralisation under anaerobic conditions was slow. The mineralisation of LAS in sludge—soil mixtures initially containing 36 and 7.5 mg kg^{-1} (water content of 40% corresponding to 100% aerobic conditions) amounted to about 77% after 2 months. At a water content of 80%, i.e. under partially anaerobic conditions, the value was substantially lower and found to be 19 and 39%, respectively. As for NPEO2, between 61 and 70% of the added ^{14}C was recovered as CO_2 from the biodegradation experiments with initial NPEO2 levels in the test mixture of 95 and 20 mg kg^{-1}, respectively. The mineralisation of NPEO2 was not affected by the soil type, whereas some differences were observed for LAS. The half-lives of LAS in the aerobic sludge—soil mixtures were calculated to be 7.0–8.5 days. With one exception of $t_{1/2} = 17$ days, the values for NPEO2 under identical conditions were 7.8–8.8 days. Comparable results in terms of mineralisation rate and half-lives were obtained in soils fortified directly with the model compounds without addition of sludge. In summary, the presence of oxygen was the key factor determining the mineralisation of LAS and NPEO2. It was anticipated that both surface-active agents were ultimately degradable under environmental conditions provided that disintegration of larger sludge lumps assured exposure of atmospheric oxygen to the compounds. The unmineralised surfactant residues were assumed to be strongly bound to the organic fraction in the sludge and soil. This was expected to reduce the bioavailability of these residues and hence the toxicity towards soil organisms.

One of the few studies on the biodegradation of cationic surfactants in sludge-amended soils was presented by Giolando et al. [40] as part of a comprehensive environmental safety database for the di-(tallow fatty acid) ester of di-2-hydroxyethyl dimethyl ammonium chloride (DEED-MAC; Fig. 6.7.1), which had been introduced as fabric softener to substitute the poorly degradable ditallow dimethyl ammonium chloride (DTDMAC; Fig. 6.7.1). The evolution of $^{14}CO_2$ production in the test system, dosed with 0.1 or 1 mg kg^{-1}, was monitored over a period of 108 days showing an extent of mineralisation in the range from 52 to 62% and half-lives of 18 days at both test concentrations. Somewhat shorter half-lives (14–15 days) and slightly higher CO_2 formation (64–72%) were obtained in the degradation studies carried out on the secondary hydrolysis product of DEEDMAC indicating that the ester hydrolysis was the rate-limiting step in the total breakdown of the cationic surfactant. Environmental concentrations of DEEDMAC in sludge-amended soils were predicted based on an assumed sludge concentration of 2000 mg kg^{-1}, an application rate of 2 ha^{-1} year^{-1} and homogeneous distribution in the upper 20 cm soil layer yielding a level immediately after amendment of 1.3 mg kg^{-1} and a steady state level of <0.001 mg kg^{-1}.

In summary, a large body of literature on the fate and distribution of LAS in (sludge-amended) soils concludes that biodegradation is the predominant removal mechanism of the anionic surfactant assuring that no accumulation occurs under normal environmental circumstances even after repeated field application of sludge over several years. LAS levels in surface soils were generally found in the low mg kg^{-1} range with exceptionally high concentrations of up to 155 mg kg^{-1} (Tables 6.7.1 and 6.7.2). Directly after sludge amendment the values rapidly dropped within a few weeks representing half-lives in the range of 7–33 days. Comparison of calculated accumulative loads with field data showed that LAS did not accumulate in soil even after several years of sludge application. Typical application rates of one or two treatments per year were estimated to represent no risk of accumulation. As for the potential of LAS to leach into deeper soil layers, a limited number of field studies [4,6,22] indicated that the sulfonate was only slightly transported into subsurface soil horizons. With one exception of heavy sludge amendment where low $mg\,kg^{-1}$ concentrations of LAS were detectable in the 40–50 cm layer, the analytical data of the two other studies showed weaker migration of LAS into the subsoil. Although not explicitly proved in field experiments, sandy soils with an a priori higher permeability appear to be no more amenable to leaching than more compact clayey or loamy soils. The better aeration of sandy soils promotes the primary biodegradation of LAS for which the presence of atmospheric oxygen is a prerequisite. Regarding the biodegradability of surfactants sorbed onto soil matter or present as precipitates of insoluble salts, contrasting explanations were presented: on the one hand it was argued that sorption positively affects the biodegradation, as it provides more time for the microbial community to adapt to the substrate and to degrade it; on the other hand the strongly adsorbed fraction is not bioavailable and therefore cannot be used by microorganisms.

Of the analytical procedures used for the determination of LAS in soils (Table 6.7.1), most methods rely on (Soxhlet) extraction with methanol, followed by clean-up on SPE cartridges (RP-C18 and/or SAX) and final quantitative measurements by HPLC–UV/FL. Applying this protocol, detection limits were achieved ranging between 0.05 and $5\,mg\,kg^{-1}$ depending on the matrix, the enrichment factor and the optical detection system employed.

No literature published could be found describing the metabolic pathway of LAS or the occurrence of biodegradation intermediates in sludge-amended soils. Although the many studies provided evidence that the anionic surfactant could be mineralised to a large extent in soils, no attempts have been made for the identification and structural elucidation of possible breakdown intermediates. As the occurrence of a series of sulfophenyl carboxylate (SPC) homologues has been reported in sewage water [41,42], river [43,44] and marine waters [45,46] and also in sewage-contaminated groundwater, it appears very likely that the biodegradation of LAS in sludge-amended soils follows the same metabolic route involving the intermediate formation of SPC [47]. Thus, field studies are necessary to confirm this assumption. Despite a low

TABLE 6.7.1

Analytical methods for the determination of surfactant in sludge-amended soils

Compound	Matrix	Extraction	Clean-up/treatment	Quantification	Concentration range (mg kg^{-1})	Reference
LAS	Sludge-amended soil	Methanol (reflux)	SAX + RP-C8	HPLC–UV	0.9–2.2	[1,19]
LAS	Sludge-amended soil	Soxhlet (methanol)	RP-C18 + SAX	HPLC–UV/FL	<1–52	[9]
LAS	Sludge-amended soil	Alkaline methanol		HPLC–UV	17–155	[20]
LAS	Sludge-amended soil	Soxhlet (methanol)	SAX + RP-C8/18	HPLC–UV/FL	<0.2–66	[2,21]
LAS	Sludge-amended soil	Soxhlet (methanol)	SAX	HPLC–UV/FL	<0.2–145	[3]
LAS	Sludge-amended soil	Methanol (reflux)	SAX + RP-C8	HPLC–UV	<5–47	[4]
LAS	Sludge-amended soil	Soxhlet	RP-C18	HPLC–FL	<0.05–8.2	[6]
LAS	Sludge-amended soil	Microwave extraction (acidic methanol)	SAX + RP-C8	HPLC–FL	<LOD[a]–19.3	[22]
LAS	Sludge-amended soil	Soxhlet (alkaline methanol); alkaline methanol (reflux)	RP-C18	HPLC–UV/FL		[7]
LAS	Soil	Alkaline methanol (shaking)	RP-C18	HPLC–UV/FL		[30]
LAS	Soil	Alkaline methanol	Complexation/extraction	Photometer		[29]
LAS, DTDMAC	Soil (below septic tank tile field)	Acidic methanol	RP-C2 + SAX (LAS)	HPLC–FL (LAS) FAB–MS (DTDMAC)	<1–20 (LAS) <1–60 (DTDMAC)	[78]
LAS, NPEO1 + 2, NP	Sludge-amended soil	Soxhlet (methanol)		HPLC–FL	15 (LAS) 0.4 (NPEO1) 0.07 (NPEO2)	[5]
LAS, NPEO1 + 2, NP	Sludge-amended soil	Soxhlet (methanol; LAS)		HPLC–FL	5–45 (LAS)	[23,24]

TABLE 6.7.1 (continued)

Compound	Matrix	Extraction	Clean-up/treatment	Quantification	Concentration range (mg kg^{-1})	Reference
		Soxhlet (hexane; NPEO1 + 2, NP)		NP–HPLC–FL (NPEO1 + 2, NP)	0.11–1.1 (NPEO1) 0.095–0.012 (NPEO2)	[26]
LAS, NPEO, DHDMAB[b], BDMHAC[c]	Soil	Soxhlet, ASE, microwave extraction (all methanol)	Alumina column; derivatisation (cationics)	HPLC–DAD (LAS, NPEO) NP–HPLC–FL (cationics)		
NPEO1 + 2, NP, OP	Sludge-amended soil	Soxhlet (dichloromethane)		NP–HPLC–FL GC–MS (NP, OP)	0.07–1.21 (NPEO1) 0.08–0.39 (NPEO2) 2.35–4.61 (NP) 0.05–0.18 (OP)	[25]
Soap	Sludge-amended soil	Soxhlet	Derivatisation	HPLC–FL	470–1470	[27]

[a]LOD: 0.29–14 mg kg^{-1}.
[b]DHDMAB: dihexadecyldimethyl ammonium bromide.
[c]BDMHAC: benzyldimethylhexadecyl ammonium chloride (for structures see Fig. 6.7.1).

TABLE 6.7.2

Elimination of surfactants in (sludge-amended) soil under field and laboratory conditions

Compound	Matrix	Field (F) or laboratory (L) study	Parameter	Initial field or test concentration (mg kg⁻¹)	Observation period (days)	Half-live (days)	¹⁴CO₂ formation, extent of prim. degrad. (p.d.) (%) or final field conc. (f.f.c) (mg kg⁻¹)	Reference
LAS	Sludge-amended soil	F		16–52	90	26–33	<1 (f.f.c.)	[9]
LAS	Sludge-amended soil	F		2.6–66.4	32–127	7–22	<0.2–4.5 (f.f.c.)	[2,21]
LAS	Sludge-amended soils	F		117–145	55–103			[3]
LAS	Landfilled sludge–soil mixture	F		155	62	13–21ᵃ	17 (f.f.c.)	[20]
LAS, NPEO1 + 2	Sludge-amended soil	F		45 (LAS)	104		5 (LAS)	[24]
				1.1 (NPEO1)			0.11 (NPEO1)	
				0.1 (NPEO2)			0.012 (NPEO2)	
Soap	Sludge-amended soil	F		1470	150		470 (f.f.c.)	[27]
LAS	Sludge-amended soil	L		1870–19 870	116	18–162		[7]
LAS	Soil	F, L	Different soils	0.8–50ᵇ	56		up to 80 (p.d.)	[29]
LAS	Soil	F, L		1.1–8.2	365	3–7	<LOD (f.f.c.)	[6]
LAS	Soil	L	Aerobic/partly oxygen-limited	8–488	56		0–85 (p.d.)	[30]
C13-LAS	Soil (below septic tank tile field)	L				12–32		[31]
¹⁴C-LAS	Soil	L	Different soils	0.05–250	10–32	1.1–4.9	60–70	[31]
¹⁴C-LAS	Subsurface soil	L		0.8	250	20	42–57	[31]
¹⁴C-LAS	Sludge-amended soil	L		25		16–21	60–70	[32]
¹⁴C-LAS	Sludge-amended soil	L	Different soils	16.2–27.2	76–180	13–26	64–72	[33]

TABLE 6.7.2 (continued)

Compound	Matrix	Field (F) or laboratory (L) study	Parameter	Initial field or test concentration (mg kg^{-1})	Observation period (days)	Half-live (days)	$^{14}CO_2$ formation, extent of prim. degrad. (p.d.) (%) or final field conc. (f.f.c) (mg kg^{-1})	Reference
^{14}C-C12-LAS	Soil	L	Different soils	1, 10	19–63		64–66	[34]
^{14}C-LAS	Soil	L	Water-saturated				<9	[35]
^{14}C-LAS; ^{14}C-AE	Soil	L	Different soils	0.05		1.1–3.7 (LAS); 1.1–9.9 (AE)	16–70 (LAS); 25–69 (AE)	[36]
^{14}C-C13-LAS; ^{14}C-C12/14-AE	Soil (below laundromat pond)	L		0.05	28–52	1.4–39 (AE)	3–52 (LAS); 21–44 (AE)	[37]
^{14}C-LAS; ^{14}C-AE; ^{14}C-STAC; ^{14}C-stearate	Soil	L	Soil mineral composition	0.05	60		49–75 (LAS); 34–58 (AE); 7–21 (STAC); 29–47 (stearate)	[38]
NPEO1 + 2	Soil (amended with paper sludge)	F		1.21 (NPEO1); 0.39 (NPEO2)			0.07 (NPEO1); 0.08 (NPEO2)	[25]
^{14}C-LAS, ^{14}C-NPEO2	Sludge-amended soil and soil	L	Different soils; water content	7.5–36 (LAS); 20–95 (NPEO2)	60	7.0–8.5 (LAS); 7.8–8.8 (NPEO2)	19–77 (LAS)	[39]
DEEDMAC	Sludge-amended soil	L		0.1, 1	108	18	52–62	[40]

^{a}Homologue-dependent.
^{b}In g m^{-2}.

ecotoxicological relevance of SPC, additional interest arises from the fact that some phenyl isomers of each alkyl homologue had proved to be quite resistant to ultimate biodegradation in the aquatic environment [48]. The high polarity and therefore mobility of these carboxylated compounds renders them particularly amenable to leach into deeper soil horizons possibly until reaching the aquifer.

In two laboratory experiments with radiolabelled LAS, degradation intermediates were detected by liquid scintillation counting, but the reported outcomes appear to be contrasting. In the studies conducted by Figge and Schöberl [33] radioactive residues were recovered from the topsoil—the soil columns were sporadically irrigated—and assigned as polar metabolites strongly bound to soil particles. In contrast to this, the adsorption experiments performed by Branner et al. [35] under water-saturated conditions indicated rapid leaching of a large fraction of incompletely degraded breakdown intermediates. The outcomes of the latter study would support the assumption that SPC are the major metabolites of LAS in soils, but the very strong binding of LAS intermediates observed in the first study do not match the predicted low adsorption tendency and high mobility of SPC in soils. As pointed out above, field studies are necessary to pursue the fate of LAS metabolites in sludge-amended soils.

Regarding the occurrence and behaviour of non-ionic NPEO in soils modified with sewage sludge, only a very limited number of studies has been published (Table 6.7.1) despite the environmental concern of NPEO-derived compounds, which is mainly related to their endocrine-disrupting effects. Although the results obtained by the group of Giger [23,24] in the late 1980s indicated a certain recalcitrance of NPEO1 and NPEO2 in agricultural soils amended with contaminated sludge, only a few efforts have been made since then to trace NPEO in soils. Moreover, nothing is known on the leaching potential of NPEO. Since biodegradation of polydisperse NPEO in sewage treatment plants is generally paralleled by a shift in the polyethoxylate distribution towards the lower ethoxymers [49], the more lipophilic fractions of mainly short-chain NPEO bound to sludge particles can be expected to exhibit a low potential to leach into deeper soil layers. In contrast, the carboxylated metabolites NPEC, which either might be directly introduced into the soil bound to sludge particles or formed out of NPEO after application, are rather hydrophilic and therefore likely to be more mobile in soils.

With respect to the future interest in the whereabouts of LAS and NPEO in sludge-amended soils, this can be expected to increase in the Member States of the European Union: on the one hand the amount of sewage sludge produced is predicted to increase due to the obligation to collect urban wastewater and subject it to secondary treatment [50], and on the other hand the European Commission has developed a draft directive [51] setting cut-off limits for the concentrations of inorganic and organic contaminants, among which are LAS and NPEO, in sewage sludge, which will be used on agricultural fields. In the context of these measures, monitoring exercises will be required to evaluate the impact of the legislative restrictions on surfactant levels in amended soils.

Apart from LAS and NPEO, other detergent-derived surfactants such as the widely used AE, alkyl ether sulfates, secondary alkane sulfonates or fatty acid esters have only been scarcely traced in sewage sludges or sludge-amended soils. Results from aerobic biodegradation tests conducted under laboratory conditions [52–54] and from monitoring programmes on wastewater treatment plants [55] indicated ready biodegradability and high elimination rates, respectively, but the knowledge on the fate of the sludge-associated surfactant fraction during anaerobic digestion is in many instances fragmentary [56–58]. Thus, residues of undegraded surfactants may pass through sewage treatment plants and finally reach soil compartments. Cationic surfactants having a positive charge with strong affinity for particle surfaces of sewage sludge are likewise candidates to find their routes into the terrestrial environment through sludge amendment [59]. In summary, monitoring work needs to be performed on amended sites to determine trace levels and to evaluate possible risks of surfactant accumulation.

The determination of such surfactant traces occurring in soils in the sub-mg kg^{-1} range will demand the employment of sophisticated analytical methods providing high sensitivity and selectivity. Surprisingly, liquid chromatographic separation coupled to mass spectrometric detection (LC–MS) has so far not been applied to the analysis of surfactants in soil samples (Table 6.7.1) although this technique has been successfully applied to sewage sludge [60–62] and sediment analysis [63,64]. In view of the sample preparation, ASE appears to be a very promising alternative to the established Soxhlet extraction as it combines reduction of solvent consumption and time saving with good extraction efficiencies [26]. The application of ASE to the extraction of sediments has already been proved to yield satisfactory results for LAS [65,66], NPEO [67–69] or cationic benzalkonium chloride [70].

6.7.3 Behaviour of surfactants in soil after infiltration from septic systems and sewage-infiltration ponds and from surface waters

Direct infiltration of (treated) wastewater through sand beds and infiltration of sewage water from laundry ponds or septic systems are among the major contamination sources of surfactants into subsurface and groundwaters. Most studies published on this issue were released by research groups from the USA where these treatment techniques have found some use (Table 6.7.3).

Thurman and colleagues [71] determined LAS and branched alkylbenzene sulfonates (ABS) in groundwater samples collected 500 and 3000 m downfield of sewage-infiltration ponds. The surfactants were concentrated on macroreticular resins (XAD-8), separated from non-ionic surfactants by passing through an anion exchange column and the extracts containing the anionics were titrated with a quaternary ammonium surfactant. In addition, on-site measurements of anionic surfactants were performed with an MBAS test kit. Distinction between isolated LAS and ABS was achieved through ^{13}C-nuclear magnetic resonance (NMR) yielding spectral differences mainly in the aliphatic region. Quantitative analysis of the groundwater samples showed that ABS was present in the 3000 m well (down-gradient) at 2300 μg L^{-1} therewith accounting almost quantitatively for the MBAS found (2500 μg L^{-1}). LAS was not detectable in the sample from this well, but was determined at 300 μg L^{-1} in the 500 m well (MBAS: 400 μg L^{-1}), where ABS in turn was absent. The detection of the alkyl-branched isomers in the 3000 m well samples gave indications that the groundwater contamination was older than 1965 (the last year of domestic ABS use), i.e. the poorly biodegradable compound had survived more than 20 years in the aquifer.

A field study was conducted by Larson et al. [31] to characterise the impact of effluent discharges on a sandy soil about 0.5 m below the surface. A 2.5 m thick unsaturated zone and a 3–4 m thick unconfined sand/gravel aquifer underlaid the tile field. LAS concentrations in the effluent plume decreased over a distance of 10 m from 10 000 to 30 μg L^{-1}. A further object of study was a laundromat pond exposed to LAS-containing sewage for more than 25 years. A clay layer separated the natural pond from the vadose zone made up of porous sand. Measurements of LAS levels as a function of soil depth beneath the pond showed a rapid decrease from about 220 mg kg^{-1} at 30 cm to

TABLE 6.7.3

Analytical methods used for the determination of surfactants and biodegradation products in subsurface and groundwaters

Source of contamination	Compound	Extraction	Clean-up/treatment	Identification/quantification	Reference
Infiltration of secondary effluent	LAS, DATS, SPC, DATSC	AX, LLE, XAD	Derivatisation	GC–MS, IR, 1H-NMR and 13C-NMR	[72]
Infiltration of secondary effluent	LAS, DATS	RP-C2 + SAX		HLPC–FL, FAB–MS, MBAS	[73]
Infiltration of secondary effluent	NPEO, NPEC1	LLE	Derivatisation (NPEO1 + 2, NPEC1)	NP–HPLC–UV (NPEO1-6); GC–MS (NPEO1 + 2, NPEC1)	[77]
Infiltration of secondary effluent from polishing pond	Non-ionics (NPEO)			CTAS	[75]
Tracer experiment in sewage-contaminated aquifer	LAS, SPC		Derivatisation	GC–MS	[74]
Infiltration from sewage ponds	LAS, ABS	XAD, AX		MBAS, 13C-NMR	[71]
Infiltration from septic system	NPEO1-16				[79]
Infiltration from septic tank	LAS, DTDMAC	RP-C2 + AX (LAS); LLE (DTDMAC)	RP-C2 + SAX (LAS)	HPLC–FL (LAS); HPLC-conductivity (DTDMAC)	[78]
Infiltration of from septic tank and laundromat pond	LAS				[31]
Infiltration from cesspools	LAS, DATS	GCB		HPLC–FL	[76]
Infiltration from surface water	Non-ionics (NPEO)			CTAS	[80]
Infiltration from surface water	NPEO1, NPEO2				[83]
Infiltration from surface water	NPEO1 + 2, NPEC1 + 2, NP	Steam distillation/solvent extraction; LLE (NPEC)	Silica column (NPEC); derivatisation (NPEC)	NP–HPLC–FL	[84]
Infiltration from surface water	NPEO1 + 2, NPEC			NP–HPLC–FL	[85]
Artificial ground water recharge with pre-treated surface water	SPC	SPE		LC–ES–MS	[86]
Artificial ground water recharge with tertiary effluent	APEC		Derivatisation	GC–MS	[82]
Various sources	Non-ionics (NPEO)			CTAS	[81]
Infiltration from fire-fighting training field	Perfluorinated surf.	LLE	Derivatisation	CTAS GC–MS	[87]
Infiltration from fire-fighting training field	Perfluorinated surf. (PFAA)	SAX	Derivatisation	GC–MS	[88]
Pesticide application	AE, AME	SPE		LC–ES–MS	[89]

approximately $20\,\mathrm{mg\,kg^{-1}}$ at $2.0\,\mathrm{m}$. At a depth of $3.6\,\mathrm{m}$ no LAS surfactant was detectable (LOD not specified). In the near groundwater LAS was likewise absent. Overall, the authors concluded from their outcomes that biodegradation was the principal factor limiting the accumulation of LAS in soil and groundwater compartments.

Field and co-workers [72] identified persistent anionic surfactant-derived chemicals in groundwater contaminated by rapid infiltration of secondary effluents through sand beds. The infiltration had caused a $3500\,\mathrm{m}$ effluent plume in a shallow water-table aquifer. The target analytes (for structures see Fig. 6.7.1) comprising LAS, SPC, dialkylte-tralin sulfonates (DATS) and their carboxylated metabolites (DATSC) were isolated from groundwater samples by ion exchange and the extracts obtained were subsequently purified by desalting and liquid–liquid extraction yielding one fraction containing LAS and DATS, the other SPC and DATSC. Qualitative analyses was performed by a combination of analytical tools including infrared spectroscopy, ^{1}H-NMR and ^{13}C-NMR. In addition, the two fractions were derivatised to convert the sulfonated (carboxylated) compounds into their trifluoro-ethylated esters making them amenable to GC–MS. An independently collected set of groundwater samples was analysed after SPE for LAS employing HPLC–FL and fast atom bombardment (FAB)–MS. The groundwater concentrations of LAS (C10 to C13) and SPC (C3 to C13) were estimated as 3 ± 1 and $8 \pm 8\,\mathrm{\mu g\,L^{-1}}$, respectively. The levels of DATS and DATSC, for which authentic standards were not available, were determined on the basis of C9-LAS and C4-SPC (assuming equivalent response factors) giving 5 ± 1 and $27 \pm 3\,\mathrm{\mu g\,L^{-1}}$, respectively. In comparison with the LAS concentration in the sewage effluent, the value decreased 97% until reaching the groundwater well located at about 500 m downgradient from the infiltration beds. The LAS removal was attributed to both sorption and biodegradation occurring during infiltration through the vadose zone, but not during transport in the groundwater. During the infiltration/groundwater transport a preferential removal of higher alkyl-chain homologues as well as of external phenyl isomers was detected. The LAS metabolites in turn showed similar homologue pattern in the sewage effluent and the groundwater, dominated by the mid-chain length C5 to C8. The higher ratio of DATS to LAS in groundwater (1.7) compared with the one determined in sewage effluent (0.2) indicated the preferential break-down of LAS in the subsurface soil. The survival of LAS and its derivatives in the groundwater was assumed to be due to unfavourable biogeochemical conditions (such as nutrient limitations, below threshold concentrations or decreased microbial populations). On the assumption that no retardation through adsorptive effects occurred during infiltration/groundwater transport, a residence time for LAS in the groundwater well was calculated to be between 2.7 and 4.6 years.

In a follow-up [73] conducted on the same contamination site the fate and transport of ABS and DATS were investigated. Groundwater samples collected from a longitudinal transect of wells along the effluent plume were analysed after concentration on SPE-C2 and clean-up on SAX cartridges by HPLC–FL and FAB–MS. In the groundwater samples LAS occurred at levels below the LOD established by HPLC–FL ($10–20\ \mu g\ L^{-1}$), but traces could be identified in selected samples applying FAB–MS. Determination of DATS residues in the groundwater samples gave levels likewise in the range $10–20\ \mu g\ L^{-1}$. Comparison of the DATS homologue distribution in sewage effluent and groundwater indicated a shift to the shorter alkyl chains likely attributable to increased partitioning or biodegradation of the more hydrophobic components during infiltration/groundwater transport. In the sample taken from the well located 3000 m downgradient from the infiltration point, $2100\ \mu g\ L^{-1}$ ABS was measured by HPLC–FL representing 90% of MBAS ($2330\ \mu g\ L^{-1}$). Its presence corresponded to a groundwater contamination that had occurred prior to 1965.

The goal of a further study [74] was to determine the transport and biodegradation behaviour of LAS under in situ conditions in a sewage-contaminated aquifer by conducting natural gradient tracer tests. A technical mixture of the anionic surfactant was dosed into two geochemically distinct zones of the sand and gravel aquifer: into the oxic zone characterised by high dissolved oxygen (about $8\ mg\ L^{-1}$) as well as into the transitions zone with only $1\ mg\ L^{-1}$ dissolved oxygen. At two monitoring wells situated at 4.6 and 9.4 m downgradient from the injection well, samples were withdrawn for chemical analysis of LAS and its biotransformation products. Breakthrough curves were recorded and associated with the unretained tracer bromide showing that LAS almost co-migrated with bromide in the transition zone, but it was retarded relative to inorganic anions in the oxic zone. Moreover, the composition of the breakthrough front was enriched by the shorter alkyl-chain homologues, which was attributed to sorptive processes in this zone since hydrophobic interactions of LAS with organic soil matter preferentially slowed the transport of the higher homologues. The nearly unretarded migration of the LAS homologues in the transition zone in turn correlated well with the low organic carbon content in the aquifer sediment ($<0.01\%$). As for the occurrence of biodegradation intermediates, namely SPC, no such compounds were detectable in the oxic zone (proved by GC–MS analysis of trifluoroethylated groundwater extracts). The absence of metabolites in conjunction with an unchanged LAS homologue/isomer pattern suggested that no biodegradation took place in the oxic zone. In contrast to the observations in the oxic zone, SPC could be identified in the transition zone. Their appearance went hand in hand with a decline of dissolved oxygen (down to 0.1 mg L^{-1}) associated with the oxidative conversion of LAS into the carboxylated metabolites. The total mass of metabolites determined was almost equal to that of the mass of total LAS lost. During the 40-day experiment the formed SPC appeared to be stable and were not completely mineralised. The differences in biodegradability of LAS in the two aquifer zones were assumed to be related to the exposure of indigenous microbial communities to the surfactant. Apparently, microorganisms in the oxic zone had not been previously exposed to sewage-contaminated groundwater, whereas the occurrence of LAS residues in the transition zone had led to a bacterial acclimation. However, once the dissolved oxygen in the water was depleted, both LAS and its biotransformation products seemed to persist.

Zoller [75] discussed the influence of activated sludge/aquifer treatments on the quality of reused groundwater. After secondary sewage treatment and intermittent flooding of the effluent in spreading basins, the water was infiltrated through the saturated zone to recharge groundwater supplies. In the surroundings of the recharge area recovery wells pumped water from the aquifer, which was used for agricultural irrigation. Despite an efficient removal of non-ionic surfactants during the wastewater treatment (including polishing ponds) reducing the level on average from 7500 to 300 µg L^{-1}, between 22 and 25 µg L^{-1} were still found in the reclaimed water. In view of the presence of hard non-ionic surfactants in reused water, which represented an issue of environmental and possibly health risk concern, the author stressed that the use of reclaimed effluents for irrigation and for aquifer recharge should be carried out with caution.

In an area where domestic wastewater was discharged through cesspools, Crescenzi et al. [76] could identify LAS and DATS in a groundwater sample. After preconcentration of the aqueous sample on a graphitised carbon black (GCB) cartridge, co-extracted basic, neutral and weakly acidic compounds were removed from the cartridge by a formic acid-containing solvent mixture. Subsequent elution with a basified organic solvent system liberated the target analytes. For the chromatographic separation of the structure-related sulfonates, differing only slightly in their retention behaviour, the use of an RP-C8 column was preferred over an RP-C18 column as the former provided co-elution of the isomers thereby facilitating peak assignment in the HPLC–FL chromatograms. In the groundwater samples taken from a 20 m depth-well, the LAS concentration amounted to 0.3 μg L^{-1}, while the DATS level totalled 0.9 μg L^{-1} (LOD 0.1 μg L^{-1}). Of the four DATS homologues determined (C10–C13) the lightest one was identified as the most abundant species. The change in the homologue pattern as compared to the one typical of wastewater was assumed to be due to a preferential removal of the longer chain homologues as a consequence of stronger sorption onto soil particles and/or faster biodegradation.

Rudel and co-workers [77] identified alkylphenols and other estrogenic phenolic compounds in groundwater impacted by infiltrations of secondary effluents and septage from septic tanks. The area was characterised by sand, gravel and boulders deposited by glaciers. The monitoring wells with depths between 12 and 24 m were located about 150–275 m downgradient of the infiltration beds. No wastewater had been discharged to these beds during the 6 months before sample collection. The analytical procedure comprised liquid–liquid extraction with dichloromethane and either NP–HPLC–UV determination (OP/NPEO1-6) or GC–MS analysis after silylation (NPEO1, NPEO2 and NPEC1). The sum concentration of OP/NPEO in the untreated wastewater and the septage samples ranged from 1350 to 11 000 μg L^{-1}. After treatment the water contained 5.5 μg L^{-1} of NPEO1, 0.8 μg L^{-1} of NPEO2 and 42 μg L^{-1} of the carboxylated metabolite NPEC1. In one of two samples collected the tetra- to hexaethoxylates were found at levels between 12 and 48 μg L^{-1}. In four groundwater samples analysed OP/NPEO2 was detected at levels between 14 and 38 μg L^{-1}, while traces of OP/NPEO3 and OP/NPEO4 occurred occasionally in the low μg L^{-1} range (4–5 μg L^{-1}, below the limit of quantification of the HPLC method). Due to the limited number of samples examined, no firm conclusions could be drawn in view of the source of contamination (treated effluent or septic systems), but the findings suggested that further studies were required to study the impact of septic tanks on groundwaters serving as drinking water supplies.

The sorption and transport of LAS and the cationic DTDMAC were determined in the upper soil horizons (78–97% sand, 0.03–0.5% C_{org}) and in the aquifer (95–99% sand, 0.04–0.50% C_{org}) below a septic tank tile field [78]. McAvoy and co-workers extracted the analytes from solid samples by refluxing with acidic methanol. For LAS determinations the evaporated extracts were diluted with water, concentrated by SPE (RP-C2 followed by SAX column) and analysed by HPLC–FL (LOD: 1 mg kg^{-1}). Analysis of the cationic surfactants was done by FAB–MS. Monitoring results showed that both LAS and DTDMAC were substantially removed during transport in the unsaturated zone. The LAS level dropped from 13 850 to 1640 µg L^{-1} within 0.5 m from the surface and was further reduced to < 10 µg L^{-1} before entering the groundwater at a depth of 2.5 m. The concentrations of dissolved DTDMAC decreased 25-fold from initially 4570 to 195 µg L^{-1} at 0.5 m and were below the LOD of 5 µg L^{-1} upon entering the groundwater. In the tile field gravel an LAS concentration of 21 mg kg^{-1} was determined, which decreased to non-detectable levels 5 cm beneath the gravel bed. Somewhat higher values were found for DTDMAC reflecting its higher sorption tendency: about 60 mg kg^{-1} in the field gravel, 10 mg kg^{-1} at 7.5 cm below the surface and less than 1 mg kg^{-1} at 10 cm depth. Measurements of sorption coefficients gave K_d values for DTDMAC being one order of magnitude higher than those for LAS. Experiments carried out on a series of core samples collected throughout the leachate plume indicated that sorption of LAS was controlled by organic carbon and clay content in the upper soil layers. No clear correlation between soil composition and sorption of DTDMAC could be established. With respect to the elimination of the cationic surfactant in the upper soil horizons, the stronger retention was believed to allow adequate time for biodegradation, as DTDMAC had been shown to degrade at slower rates when compared with LAS.

The fate of NPEO was examined in two household septic systems of different design and hydro-geological settings [79]. The first site (A) consisted of clay-dominated glacial till of low permeability, while the second site (B) was characterised by sand and gravel of high permeability. At both locations the groundwater table was 10 m below ground surface. After a one-year period of discharging NPEO-containing sewage to the septic tanks, the tank effluent from site A contained about 15 000 µg L^{-1} NPEO of which some 5000 µg L^{-1} corresponded to the two short-chain ethoxymers NPEO1 and NPEO2. In tank B effluent the sum concentration of NPEO1–2 was about 900 µg L^{-1}. The higher oligomers NPEO3-16 totalled $4000 \, \mu g \, L^{-1}$. Analysis of the septage leachate of site A, collected at the end of the first leach line, showed an NPEO reduction between 94.3 and 99.6%. NPEC levels were found in the range from 148 to $1340 \, \mu g \, L^{-1}$. In water samples taken from lysimeters beneath the leach field laterals, only traces of NPEO were detectable (<LOD to $2.7 \, \mu g \, L^{-1}$). Analysis of selected samples for NPEC gave values <LOD to $0.5 \, \mu g \, L^{-1}$. In all analysed groundwater samples NPEO-derived compounds were absent. The same held true for the occurrence of NPEC in groundwater from site B, which was a priori more permeable for infiltrating effluents. The non-ionic surfactant was occasionally detected at sub-$\mu g \, L^{-1}$ levels in waters from the observation wells, but no clear evidence was obtained that these originated from subsoil leaching (instead laboratory contamination was suspected).

The withdrawal of water from the aquifer in the close vicinity of surface waters may produce a hydraulic gradient, provoking the infiltration of surface water through the riverbank. Through biochemical and physical processes taking place during the subsoil passage, organic micro-pollutants can be removed from the water, which after reclaim might be used for drinking water production or irrigation of arable land (Table 6.7.3).

Zoller and co-workers [80] conducted a study, which aimed at determining the extent of which Israel's groundwater in the vicinity of sewage-polluted surface waters was affected by infiltration of NPEO-contaminated water. The wells with depths of 50–156 m were located at distances of between 40 and 300 m from the riverbank of two investigated rivers. Colorimetric analyses of non-ionic surfactants as CTAS in surface and adjacent groundwaters indicated their insufficient removal in the subsurface environment. In comparison with the levels determined in the streams ($1600-2600 \, \mu g \, L^{-1}$), the groundwaters contained between 120 and $780 \, \mu g \, L^{-1}$ non-ionic surfactant representing elimination rates of 92–54%. The data gathered revealed an inverse relationship between the distance of a given well from the stream and the surfactant concentration measured. The authors concluded that neither naturally occurring biodegradation nor adsorption processes onto soil were efficient enough to prevent contamination of groundwaters.

In a subsequent work Zoller [81] aimed at mapping the country's surface water, seawater and groundwater according to their non-ionic surfactant content. A further objective was to establish the connection between surface input and output into receiving surface water, seawater and groundwater as a function of the local physico-chemical and hydrogeological parameters. In water wells sampled during two campaigns the non-ionic surfactant concentrations ranged from 22 to 1800 μg L^{-1} ($n = 8$) confirming the findings of the previous study. It was suggested that surface and subsurface pollution by poorly biodegradable surfactants was not a specific problem, but a general contamination problem in Israel.

The occurrence of alkylphenol carboxylates (APEC) as an anthropogenic indicator of wastewater impact was determined by Fujita and Reinhard [82] in groundwater artificially recharged with tertiary treated effluents. Dry residues of evaporated water samples were derivatised with n-propanol/acetyl chloride yielding propyl esters, which were subjected to GC–MS analysis. Treated wastewater from two treatment trains applying different technologies contained sum concentrations of APEC, including their brominated analogues, of 5.1 and 0.5 μg L^{-1}. After blending of the effluents with local groundwater, taken from a depth of more than 260 m, at a ratio of 1:1:1 the water was directly injected into the aquifer. In a sample from a monitoring well located about 15 m downgradient of the infiltration site the APEC level amounted to 7.0 μg L^{-1}. Preliminary analytical data from the nearest production well (about 500 m downgradient) indicated the presence of APEC at an estimated level of 1–10 μg L^{-1}. The findings suggested that APEC persisted in groundwater during the 1–2 months travel from the injection location to the monitoring well and were apparently not eliminated during transport to the production well corresponding to a residence time of 2–3 years.

Schaffner and colleagues [83] presented preliminary results on the behaviour of short-chain NPEO during bank filtration of river water into groundwater being a first step in the treatment of river water for public water supplies. The water from the small heavily polluted stream (Glatt river), which upstream received discharges from several municipal wastewater treatment plants, infiltrated into a quaternary fluvioglacial valley fill aquifer. The concentrations of NPEO1 and NPEO2 were monitored during one year in samples ($n = 17$) from observation wells located at various distances from the riverbank (2.5–14 m). The levels of NPEO1 and NPEO2 in the river water amounted on average to 7.5 and 8.2 μg L^{-1}, respectively. With increasing distance of the observation well from the bank the levels of both ethoxymers decreased: the mean value for NPEO1 dropped from 1.0 (2.5 m) to

0.1 μg L^{-1} (14 m), while NPEO2 exhibited a slightly better removal during the first 2.5 m (0.4 μg L^{-1}) of the subsoil passage. In the sample taken from the 14 m well NPEO2 was detected at 0.1 μg L^{-1}.

In follow-up work [84] the same research group presented a more detailed description of the fate of persistent NPEO-derived metabolites during infiltration of contaminated river water into groundwater. Analytical determinations of NPEO1 and NPEO2 in water samples were performed by a combined steam distillation/solvent extraction procedure and following quantitative measurements by NP–HPLC–FL. The nonylphenoxy carboxylic acids NPEC1 and NPEC2 were extracted from acidified water with chloroform and the obtained extracts purified on a silica column. After conversion of the analytes into their methyl esters, quantitation was likewise done by NP–HPLC–FL. The detection limits of the methods were 0.01 μg L^{-1} for the ethoxylates and 0.1 μg L^{-1} for carboxylated degradation products. The concentrations found in the river and the adjacent groundwater showed a temporal and spatial variation. Mean levels in the groundwater were considerably lower indicating an efficient removal of nonylphenolic compounds during most of the year. On average the concentrations of NPEO1 and NPEO2 in the river were 7.8 and 8.4 μg L^{-1}, respectively. After reaching the first well at 2.5 m from the bank, the levels were 0.91 μg L^{-1} for NPEO1 and 0.33 μg L^{-1} for NPEO2. During the further underground passage, the values dropped to 0.04 and $<$0.01 μg L^{-1}, respectively. In contrast to the substantial removal of the ethoxylates in the section between the river and the first observation well, the carboxylates NPEC1 and NPEC2 showed a much slower elimination. Upon reaching the first well the concentrations had been reduced on average from 14.7 to 10.9 μg L^{-1} for NPEC1 and from 24.7 to 22.4 μg L^{-1} for NPEC2. In the samples collected from the 14 m well, mean values of 4.5 and 5.1 μg L^{-1} were obtained for the carboxylates. The higher mobility compared to those of NPEO was attributed to a higher water solubility and their resistance to biodegradation under aerobic conditions. As for the occurrence of NPEO in the water from the pumping station situated 130 m from the river, both NPEO1 ($<$0.1–3.4 μg L^{-1}) and NPEO2 ($<$0.1–1.7 μg L^{-1}) were detectable at traces. Examinations of seasonal variations in the groundwater sampled from the 2.5 m well showed a trend of highest levels in winter and lowest in summer. These findings correlated well with the general trend of concentration changes in the river.

In a field study Ahel [85] investigated the infiltration of organic compounds from a heavily polluted river and a wastewater canal into the alluvial aquifer. Sampling wells were situated at distances of between 3 and 120 m from the surface waters. NPEO-derived metabolites were among the most abundant contaminants in the analysed groundwaters. Measurements of NPEO1, NPEO2 and NPEC indicated relatively efficient removal during infiltration reducing the concentrations of 0.4, 0.2 and 5 $\mu g\,L^{-1}$, respectively, found in the river to $<0.1\,\mu g\,L^{-1}$ for NPEO and to $0.05\,\mu g\,L^{-1}$ for NPEC in the well located at 120 m from the bank. On the other field site situated adjacent to the wastewater canal, the elimination efficiencies during a short underground passage (well at 3 m from the river) ranged from 98.3 to 99.8%. Residual concentrations of NPEO1, NPEO2 and NPEC in the groundwater amounted 0.078, 0.01 and 0.005 $\mu g\,L^{-1}$.

The behaviour of SPC during drinking water production was investigated by Eichhorn et al. [86] in two waterworks one of which performed artificial groundwater recharge of surface water previously submitted to flocculation and granular active carbon filtration. The analytical method for the determination of C5- to C12-alkyl homologues comprised isolation of the sulfonates on RP-C18 material at pH 3.0 and subsequent ion-pair HPLC separation with mass spectrometric detection. Whereas the total SPC concentration amounted to 0.25 $\mu g\,L^{-1}$ in the water samples collected before infiltration into the subsoil, a level of 0.12 $\mu g\,L^{-1}$ was determined in the water from the recovery well located at a distance of about 170 m from the infiltration well, which corresponds to a travel time of 4–6 weeks. Although the samples, taken in triplicate over a period of 6 weeks, did not correspond to each other the results indicated that the very polar degradation products of LAS were quite resistant during the subsoil passage.

It was only recently that the use of aqueous film-forming foams (AFFF) containing perfluorinated surfactants was recognised as a source of groundwater pollution (Table 6.7.3).

One of the first groundwater contaminations with perfluorinated surfactants was reported by Levine et al. [87] at a fire-training site, which had served for approximately 11 years for fire fighting exercises using water and AFFF. The site was characterised by a very shallow groundwater table with depths ranging from 0.6 to 3 m below the land surface. The aquifer consisted of clean, fine-grained quartz and clayed sandy soils. The groundwater samples were extracted with diethyl ether employing liquid–liquid extraction and derivatised with diazomethane to yield the methyl esters. Analysis was performed by GC–MS. At a distance of about 45 m downgradient of the fire training pit, the surfactant concentration reached a maximum level of around 11 500 μg L^{-1} indicating surfactant transport in the aquifer.

A more in-depth description of the extent of groundwater pollution with perfluoroalcanoic acids (PFAA) was given by Moody and Field [88] in a study conducted on the site previously investigated by Levine et al. [87]. For chemical analysis the samples were concentrated on SAX disks and the extracts methylated with methyl iodide. GC–MS operated in electron impact mode was employed for quantitative measurements of the methyl esters (LOD: 18 μg L^{-1}), while confirmation of the identity of the target compounds was achieved by electron capture negative ionisation GC–MS. The water samples collected from three wells contained total PFAA concentrations varying between 124 and 298 μg L^{-1} with the highest levels in wells located in the close vicinity of the fire-training pit. The two major components were C6- and C8-PFAA making up 46–52% and 34–40%, respectively; the odd homologue C7-PFAA accounted for the rest. The authors had performed studies on a second field site likewise used for fire-training activities. Monitoring wells were situated within a radius of 120 m from the fire pit, where the groundwater table was between 2 and 3 m below ground surface. The highest PFAA concentration (7090 μg L^{-1}) was measured in the groundwater near the burn pit, whereas in a well located downgradient, a value of 540 μg L^{-1} was determined. On average 89% of the total PFAA was constituted by C8-PFAA. As information on the exact chemical composition of surfactants used in AFFF formulations was not available, the authors discussed as origin of the detected PFAA the presence as active ingredient or by-product in AFFF mixtures, their formation during combustion or through (bio)chemical degradation. Anyway, the PFAA concentrations found at both field sites confirmed the poor biodegradability of perfluorinated surfactants deriving from AFFF, which was attributed to the particular strength of the carbon–fluorine bond and the rigidity of the perfluorocarbon chain.

A possible source of groundwater contamination, which has up to now almost been neglected, is associated with the introduction of surfactants into soils as pesticide additives (Table 6.7.3). Non-ionic surfactants composed of alcohols and fatty acids are most widely recommended as adjuvants to facilitate and enhance the absorbing, emulsifying, dispersing, wetting and penetrating properties of pesticides. Other pesticide adjuvants are silicone-based surfactants, which are gaining popularity owing to their superior spreading ability (see Chapter 2.8).

One case study on the occurrence of AE and alkylamine ethoxylates (AME; Fig. 6.7.1) in soil interstitial water and groundwater, originating from the leaching out of agricultural fields previously treated with pesticide formulations, was presented by Andersen-Krogh et al. [89]. Soil interstitial water taken from 1–2 m below land surface and groundwater collected from wells at 1–2, 2–3 and 3–4 m below the groundwater table were concentrated on polymeric solid-phase material. Liquid chromatographic separation of the C-even alkyl homologues on an RP-C18 column, using distinct mobile phase compositions for AE and AME, preceded the mass spectrometric detection. In all samples analysed the levels of AME were below the LOD ($0.5-6 \ \mu g \ L^{-1}$). However, a series of ethoxymers of C12-AE were determined in a groundwater sample with single concentrations between 60 and 189 ng L^{-1}. In two soil waters the shorter ethoxymers of C12-AE were found at levels in a range from 33 to 73 ng L^{-1}.

In summary, surfactants, which are convectively transported in subsurface soils by considerable volumes of water, exhibit—as could be expected—a much higher mobility than those introduced onto fields through sludge application, whose downward movement is largely controlled by the amount of rainwater. During infiltration of contaminated water both the anionic LAS and the non-ionic NPEO can be eliminated to a certain extent by adsorption onto soil particles or biodegradation, but they appear to be more persistent once they reached the aquifer, in which the prevailing biochemical conditions—low temperature, depleted dissolved oxygen, low density of microbial communities, nutrient limitation—slow down the conversion rates of microbial degradation. With respect to the behaviour of the degradation intermediates belowground, the carboxylated metabolites of NPEO were reported to be more resistant to microbial attack than the parent compounds with one or two ethoxymer groups [82,84]. The primary degradation products of LAS, the SPC, were detected in the plume of infiltrated secondary effluent [72] and in groundwater supplies artificially recharged with processed surface water [86] indicating that these very polar xenobiotics may persist in the environmental compartment. A group of surfactants, which was reported to strongly resist biodegradation in the aquifer, are perfluorinated compounds [87, 88]. The chemical structure of these synthetic components used in fire fighting foams contains a perfluorinated carbon chain, which greatly impedes a biochemical attack.

REFERENCES

1   E. Matthijs and H. de Henau, *Tenside Surf. Det.*, 24 (1987) 193.
2   M.S. Holt, E. Matthijs and J. Waters, *Water Res.*, 23 (1989) 749.
3   M.S. Holt and S.L. Bernstein, *Water Res.*, 26 (1992) 613.
4   R.A. Rapaport and W.S. Eckhoff, *Environ. Toxicol. Chem.*, 9 (1990) 1245.
5   A. Marcomini and W. Giger, *Anal. Chem.*, 59 (1987) 1715.
6   T. Küchler and W. Schnaak, *Chemosphere*, 35 (1997) 153.
7   L. Comellas, J.L. Portillo and M.T. Vaquero, *J. Chromatogr. A*, 657 (1993) 25.
8   J. McEvoy and W. Giger, *Naturwissenschaften*, 72 (1985) 429.
9   J.L. Berna, J. Ferrer, A. Moreno, D. Prats and F. Riuz Bevia, *Tenside Surf. Det.*, 26 (1989) 101.
10   L. Cavalli, A. Gellera and A. Landone, *Environ. Toxicol. Chem.*, 12 (1993) 1777.
11   S.C. Wilson, R. Duarte-Davidson and K.C. Jones, *Sci. Total Environ.*, 185 (1996) 45.
12   J.P. Mieure, J. Waters, M.S. Holt and E. Matthijs, *Chemosphere*, 21 (1990) 251.
13   W. de Wolf and T. Feijtel, *Chemosphere*, 36 (1998) 1319.
14   P. Kloepper-Sams, F. Torfs, T. Feijtel and J. Gooch, *Sci. Total Environ.*, 185 (1996) 171.
15   J. Jensen, *Sci. Total Environ.*, 226 (1999) 93.
16   G. Kuhnt, *Environ. Toxicol. Chem.*, 12 (1993) 1813.
17   S.D. Haigh, *Sci. Total Environ.*, 185 (1996) 161.
18   R.J. Maguire, *Water Qual. Res. J. Canada*, 34 (1999) 37.
19   H. de Henau, E. Matthijs and W.D. Hopping, *Int. J. Environ. Anal. Chem.*, 26 (1986) 279.
20   J. de Ferrer, A. Moreno, M.T. Vaquero and L. Comellas, *Tenside Surf. Det.*, 34 (1997) 278.
21   J. Waters, M.S. Holt and E. Matthijs, *Tenside Surf. Det.*, 26 (1989) 129.
22   L. Carlsen, M.-B. Metzon and J. Kjelsmark, *Sci. Total Environ.*, 290 (2002) 225.
23   A. Marcomini, P.D. Chapel, Th. Lichtensteiger, P.H. Brunner and W. Giger, *J. Environ. Qual.*, 18 (1989) 523.
24   W. Giger, P.H. Brunner, M. Ahel, J. McEvoy, A. Marcomini and C. Schaffner, *Gas, Wasser, Abwasser*, 67 (1987) 111.
25   M. Hawrelak, E. Bennett and C. Metcalfe, *Chemosphere*, 39 (1999) 745.

26 S. Heise and N. Litz, *Tenside Surf. Det.*, 36 (1999) 185.

27 D. Prats, M. Rodríguez, P. Varo, A. Moreno, J. Ferrer and J.L. Berna, *Water Res.*, 33 (1999) 105.

28 A. Moreno, J. Bravo, J. Ferrer and C. Bengoechea, *J. Am. Oil Chem. Soc.*, 70 (1993) 667.

29 N. Litz, W. Doering, M. Thiele and H.-P. Blume, *Ecotoxicol. Environ. Saf.*, 14 (1987) 103.

30 L. Elsgaard, S.O. Petersen and K. Debosz, *Environ. Toxicol. Chem.*, 20 (2001) 1664.

31 R.J. Larson, T.W. Federle, R.J. Shimp and R.M. Ventullo, *Tenside Surf. Det.*, 26 (1989) 116.

32 R.J. Larson, T.M. Rothgelb, R.J. Shimp, T.E. Ward and R.M. Ventullo, *J. Am. Oil Chem. Soc.*, 70 (1993) 645.

33 K. Figge and P. Schöberl, *Tenside Surf. Det.*, 26 (1989) 122.

34 U. Dörfler, R. Haala, M. Matthies and I. Scheunert, *Ecotox. Environ. Saf.*, 34 (1996) 216.

35 U. Branner, M. Mygind and C. Jørgensen, *Environ. Toxicol. Chem.*, 18 (1999) 1772.

36 D.B. Knaebel, T.W. Federle and J.R. Vestal, *Environ. Toxicol. Chem.*, 9 (1990) 981.

37 T.W. Federle, R.M. Ventullo and D.C. White, *Microb. Ecol.*, 20 (1990) 297.

38 D.B. Knaebel, T.W. Federle, D.C. McAvoy and J.R. Vestal, *Appl. Environ. Microbiol.*, 60 (1994) 4500.

39 B. Gejlsbjerg, C. Klinge and T. Madsen, *Environ. Toxicol. Chem.*, 20 (2001) 698.

40 S.T. Giolando, R.A. Rapaport, R.J. Larson, T.W. Federle, M. Stalmans and P. Masscheleyn, *Chemosphere*, 30 (1995) 1067.

41 A. DiCorcia, R. Samperi and A. Marcomini, *Environ. Sci. Technol.*, 20 (1994) 850.

42 P. Eichhorn, M. Petrovic, D. Barceló and T.P. Knepper, *Vom Wasser*, 95 (2000) 245.

43 A. DiCorcia, L. Capuani, F. Casassa, A. Marcomini and R. Samperi, *Environ. Sci. Technol.*, 33 (1999) 4119.

44 W.-H. Ding, J.-H. Lo and S.-H. Tzing, *J. Chromatogr. A*, 818 (1998) 270.

45 E. González-Mazo, M. Honing, D. Barceló and A. Gómez-Parra, *Environ. Sci. Technol.*, 31 (1997) 504.

46 V. León, E. González-Mazo and A. Gómez-Parra, *J. Chromatogr. A*, 889 (2000) 211.

47 R.D. Swisher, *Surfactant Biodegradation*, Marcel Dekker, New York, 1987.

48 P. Eichhorn and T.P. Knepper, *Environ. Toxicol. Chem.*, 21 (2002) 1.

49 B. Thiele, K. Günther and M.J. Schwuger, *Chem. Rev.*, 97 (1997) 3247.

50 European Commission, Council Directive of 21 May 1991 concerning urban wastewater treatment, 91/271/EEC.

51 European Commission, Working document on sludge, 3rd draft, Brussels, 27 April 2000.

52 L. Kravetz, J.P. Salanitro, P.B. Dorn and K.F. Guin, *J. Am. Oil Chem. Soc.*, 68 (1991) 610.

53 K. Yoshimura and F. Masuda, *J. Am. Oil Chem. Soc.*, 59 (1982) 328.

54 P. Gode, W. Guhl and J. Steber, *Fat Sci. Technol.*, 89 (1987) 548.

55 D.C. McAvoy, S.D. Dyer, N.J. Fendinger, W.S. Eckhoff, D.L. Lawrence and W.M. Begley, *Environ. Toxicol. Chem.*, 17 (1998) 1705.

56 J. Steber and H. Berger, In: D.R. Karsa and M.R. Porter (Eds.), *Biodegradability of Surfactants*, Blackie Academic and Professional, 1995, pp. 134–182.

57 A.M. Bruce, J.D. Swanwick and R.A. Ownsworth, *J. Proc. Inst. Sew. Purif.*, 5 (1966) 427.

58 J. Steber and P. Wierich, *Tenside Surf. Det.*, 26 (1989) 406.

59 W. Janicke and G. Hilge, *Tenside Surf. Det.*, 16 (1979) 117.

60 S. Chiron, E. Sauvard and R. Jeannot, *Analusis*, 28 (2000) 535.

61 J. Riu, E. Martínez, D. Barceló, L.I. Ginebreda and A. Tirapu, *Fresenius J. Anal. Chem.*, 371 (2001) 448.

62 M. Petrovic and D. Barceló, *Anal. Chem.*, 72 (2000) 4560.

63 P.L. Ferguson, C.R. Iden and B.J. Brownawell, *Environ. Sci. Technol.*, 35 (2001) 2428.

64 M. Petrovic, A.R. Fernández-Alba, F. Borrull, R.M. Marce, E. González-Mazo and D. Barceló, *Environ. Toxicol. Chem.*, 21 (2002) 37.

65 A. Kreisselmeier and H.-W. Dürbeck, *Fresenius J. Anal. Chem.*, 354 (1996) 921.

66 A. Kreisselmeier and H.-W. Dürbeck, *J. Chromatogr. A*, 775 (1997) 187.

67 W.-H. Ding and J.C.H. Fann, *Anal. Chim. Acta*, 408 (2000) 291.

68 D.Y. Shang, R.W. Macdonald and M.G. Ikonomou, *Environ. Sci. Technol.*, 33 (1999) 1366.

69 S. Valsecchi, S. Polesello and S. Cavalli, *J. Chromatogr. A*, 925 (2001) 297.

70 I. Ferrer and E.T. Furlong, *Anal. Chem.*, 74 (2002) 1275.

71 E.M. Thurman, T. Willoughby, L.B. Barber and K.A. Thorn, *Anal. Chem.*, 59 (1987) 1798.

72 J.A. Field, J.A. Leenheer, K.A. Thorn, L.B. Barber, C. Rostad, D.L. Macalady and S.R. Daniel, *J. Contam. Hydrol.*, 9 (1992) 55.

73 J.A. Field, L.B. Barber, E.M. Thurman, B.L. Moore, D.L. Lawrence and D.A. Peake, *Environ. Sci. Technol.*, 26 (1992) 1140.

74 C.J. Krueger, L.B. Barber, D.W. Metge and J.A. Field, *Environ. Sci. Technol.*, 32 (1998) 1134.

75 U. Zoller, *Water Res.*, 28 (1994) 1625.

76 C. Crescenzi, A. DiCorcia, E. Marchiori, R. Samperi and A. Marcomini, *Water Res.*, 30 (1996) 722.

77 R.A. Rudel, S.J. Melly, P.W. Geno, G. Sun and J.G. Brody, *Environ. Sci. Technol.*, 32 (1998) 861.

78 D.C. McAvoy, C.E. White, B.L. Moore and R.A. Rapaport, *Environ. Toxicol. Chem.*, 13 (1994) 213.

79 C.G. Naylor, B.E. Huntsman, J.G. Solch, C.A. Staples and J.B. Williams, *Proceedings 5th World Surfactant Congress*, Firenze, 2000.

80 U. Zoller, E. Ashash, G. Ayali, S. Shafir and B. Azmon, *Environ. Int.*, 16 (1990) 301.

81 U. Zoller, *J. Environ. Sci. Health*, A27 (1992) 1521.

82 Y. Fujita, W.-H. Ding and M. Reinhard, *Water Environ. Res.*, 68 (1996) 867.

83 C. Schaffner, M. Ahel and W. Giger, *Water Sci. Technol.*, 19 (1987) 1195.

84 M. Ahel, C. Schaffner and W. Giger, *Water Res.*, 30 (1996) 37.

85 M. Ahel, *Bull. Environ. Contam. Toxicol.*, 47 (1991) 586.

86 P. Eichhorn, T.P. Knepper, F. Ventura and A. Diaz, *Water Res.*, 36 (2002) 2179.

87 A.D. Levine, E.L. Libelo, G. Bugna, T. Shelley, H. Mayfield and T.B. Stauffer, *Sci. Total. Environ.*, 208 (1997) 195.

88 C.A. Moody and J.A. Field, *Environ. Sci. Technol.*, 33 (1999) 2800.

89 K. Andersen-Krogh, K.V. Vejrup, B. Bügel-Mogensen and B. Halling-Sorensen, *J. Chromatogr. A*, 957 (2002) 45.

Chapter 7

Toxicity of surfactants

7.1 TOXICITY OF SURFACTANTS FOR AQUATIC LIFE

Julián Blasco, Miriam Hampel and Ignacio Moreno-Garrido

7.1.1 Introduction

The high production volumes and widespread use of surfactants have been an environmental concern since the early 1960s. Surfactants are used in industry and agriculture and are also found in household and personal care products. Subsequently, these substances are released into the environment, directly, or after being processed in wastewater treatment plants (WWTP).

The presence of surfactants and their biodegradation products in different environmental compartments can invoke a negative effect on the biota. The ecotoxicity of surfactants to aquatic life has been summarised in the scientific literature [1–5]. Nevertheless, some information is still lacking in relation to the aquatic toxicity of surfactants, especially knowledge regarding the toxicity of the degradation products, the effect of surfactants on marine species, the ecotoxicity of mixtures of chemical compounds with surfactants, the relationship between toxicity and chemical residue and the effect of surfactant presence in specific environmental compartments (water, particulate matter, pore-water, sediment).

In general, existing information is related to single species and acute toxicity; these types of tests form the basis for the derivation of 'safe concentrations' for a given chemical in the environment, according to EC regulation [6]. The data for surfactants other than anionic

Comprehensive Analytical Chemistry XL
T.P. Knepper, D. Barceló and P. de Voogt (Eds)

and non-ionic, and about mixture species and chronic toxicity, are scarcer.

This chapter presents a summary of the available information regarding the toxicity of surfactants in the aquatic environment and also the new data with special emphasis on the marine environment, the use of microalgae and early life-stages of fish in toxicity assays. In the last few years, one aspect related to the impact of biodegradation products of surfactants in the environment has acquired a significant relevance—the estrogenic effect—and this subject is treated in depth in Chapter 7.3 of this book.

7.1.2 Acute toxicity of surfactants

A broad range of information pertaining to the toxicity of several classes of surfactants including anionic (linear alkylbenzene sulfonates (LAS), alkylether sulfates (AES), alkyl sulfates (AS), non-ionic (alkylphenol ethoxylates (APEO)), cationic (ditallow dimethyl ammonium chloride (DTDMAC)—a group of quaternary ammonium salts of distearyl ammonium chloride (DSDMAC)) and amphoteric surfactants (alkyl-betaines) is available. Several reviews of the scientific literature have been published [3–5,20].

Anionic surfactants
Much more information exists about LAS than about AES, and the available data refers, above all, to green algae, crustacean and fish. On the basis of short-term toxicity data, fish seem to be more susceptible than other taxonomic groups [18], although other authors estimated that fish and *Daphnia* effect concentrations are more or less equal [7]. High EC_{50} values (concentration of the xenobiotic that inhibits a certain parameter of the studied population to 50% when compared with controls not exposed to the xenobiotic) have been reported for AES with short alkyl chain length [18]. The embryos and tadpoles of *Xenopus laevis* exposed to commercial AES demonstrated $LC_{50}72h$ values (lethal concentration for half of the experimental population at 72 h after beginning of the test) of 6.75 mg L^{-1}; thus, at early life-stages, toxicity may overlap with teratogenic effects [8] which may cause death.

The available information about LAS toxicity on freshwater organisms is very extensive. The water flea (*Daphnia magna*) is the most frequently employed species in freshwater acute toxicity tests.

The reported values range between 0.26 mg L^{-1} [18] and 13.9 mg L^{-1} [9]. These ranges are caused by differences in the employed LAS mixtures tested with respect to alkyl chain length, and/or phenyl isomer distribution, as well as test design [18]. Thus LC$_{50}$ values range between 1.22 mg L^{-1} for C$_{14}$ and 13.9 mg L^{-1} for C$_{10}$ [9]. The toxicity values of commercial sulfonic acids are in the same order as their sodium salts [9]. Some data with low EC$_{50}$ concentrations can be found for water fleas, but due to the large database available these can be concluded to be outliers [10].

In Fig. 7.1.1 (top plot), EC$_{50}$ concentrations for a range of different taxonomic groups are plotted. For freshwater species, the majority of EC$_{50}$ values ranged between 1 and 100 mg L^{-1} showing a high interspecies variability. The LC$_{50}$ values for all studied species ranged between 0.02 and 270 mg L^{-1} corresponding to the species *Cirrhina mrigala* [11] and *Asellus sp* (isopods) [2], respectively. The references are very numerous for crustaceans, fish and algae, but scarcer for other species (Fig. 7.1.1). In the majority of acute bioassays with freshwater organisms, compounds with average chain length between C$_{11.6}$ and C$_{11.8}$ were employed. For crustaceans, the LC$_{50}$ ranged between 1 and 10 mg L^{-1}, although some data with values over 100 mg L^{-1} have been reported. Fish showed the smallest variation interval, with EC$_{50}$ data between 1 and 10 mg L^{-1}. The toxicity data for microalgae in fresh and seawater has been addressed in depth in this chapter (Section 7.3). Other approaches, such as the use of gill epithelial cells and hepatocytes to predict acute toxicity in fish have shown that EC$_{50}$ values for cellular tests were higher than the corresponding values for fish and *Daphnia*, indicating that cellular tests are less sensitive than whole organisms. However, a good correlation with acute toxicity on fish has been found and as such cellular tests seem to be a promising alternative [12].

In the case of marine species, the literature is limited with molluscs and algae being the most frequently employed species (Fig. 7.1.2). The number of references for individual homologues and commercial LAS in the scientific literature is similar, and the range of the interspecies variation is also similar to intra variations.

The effect on molluscs shows a wide variability, with reports of significant reductions in the percentage of fertilised oyster eggs developing normally at concentration of LAS as low as 0.025 mg L^{-1} [13]. In *Mytilus edulis*, effects were observed at 0.05 mg L^{-1} and the length attained by larvae was affected by concentrations of 0.1 mg L^{-1} [14]. Nevertheless, the acute toxicity (LC$_{50}$) ranged between 5 and

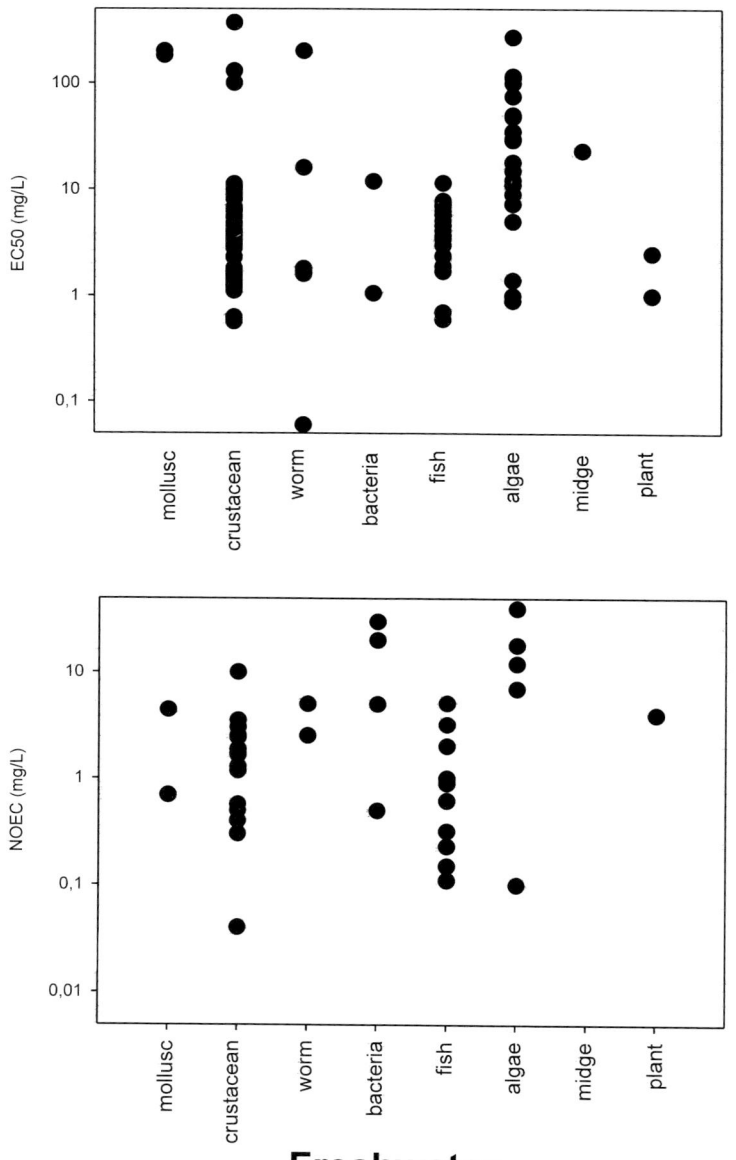

Freshwater

Fig. 7.1.1. Acute toxicity (EC_{50}) and chronic toxicity (NOEC) of LAS for freshwater species.

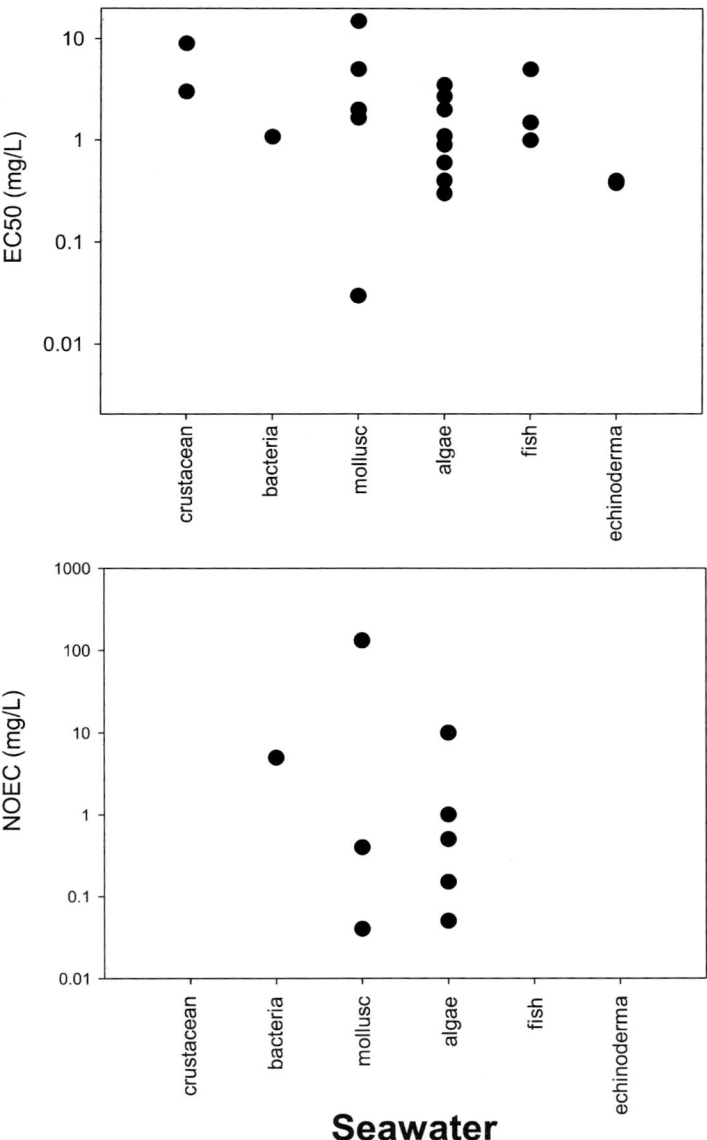

Fig. 7.1.2. Acute toxicity (EC_{50}) and chronic toxicity (NOEC) of LAS for seawater species.

100 mg L^{-1} [22] for adult stages with crustaceans being the most resistant species.

In the sea urchin, *Paracentrotus lividus*, LAS induces a significant decrease of skeletal development at concentrations of 0.30 mg L^{-1}. The effect of LAS on early life-stages in fish are discussed in Section 7.1.4 of this chapter. In general, acute toxicity for fish ranged between 1 and 10 mg L^{-1} for LAS. Selected LC$_{50}$ 96h-values for cod (*Gadus morhua*), flounder (*Pleuronectes flesus*) and plaice (*Pleuronectes platessa*) were 1, 1.5 and 1–5 mg L^{-1}, respectively [15]. Other anionic surfactants such as sodium dodecyl sulfate (SDS) showed a LC$_{50}$ concentration of 6.1 mg L^{-1} for *Sparus aurata* [16]. When the analysed acute endpoint was fertilisation success, the EC$_{50}$ were 2.8 and 2.6 mg L^{-1} for SDS and LAS, respectively.

LC$_{50}$ data for other species were: mysids 0.45–1.4 mg L^{-1}; clams 10.5 mg L^{-1}; pink shrimp 19–154 mg L^{-1} and blue crabs 29.9 mg L^{-1} [15]. Recently, a marine risk assessment of LAS in the North Sea has been carried out [17]; in this study the comparison of LC$_{50}$ for fresh and seawater fish species showed significantly lower values for freshwater organisms, although the comparison with two other taxa (crustacean and algae) showed no significant difference between freshwater and marine species.

Non-ionic surfactants

Several reviews of the available information on the toxicity of APEO have been conducted, especially driven by political initiatives. Assessment of the compounds as Priority Substances has been carried out (i) under the Canadian Environmental Protection Act [5], (ii) in the framework of the Dutch 'Plan Action of Laundry and Cleaning Products' [18] and (iii) in the framework of the project 'Investigating for chemicals in the future'. The latter is a North Sea Directorate project performed by the Rijkswaterstaat Institute for Coastal and Marine Management (RIKZ, NL) aimed at identifying the most important contaminants representing a threat to the North Sea [19]. Although there is lot of literature available, the data are scattered among many phyla and species with significant intra- and interspecies variability, due to the employment of different methods and compounds. The octylphenol ethoxylates (OP$_{11}$EO) varied from very toxic to moderately toxic to algae [19]. Nevertheless, a general pattern in the toxicity can be found; notably an increase of toxicity with decreasing EO chain length. The LC$_{50}$s and EC$_{50}$s reported for NP$_9$EO are much higher than for NP in fish, invertebrates and algae; LC$_{50}$ values ranging from 2500 to 12500 μg L^{-1} have been reported for the higher EO chains in fish [20]. This pattern is related to increasing K$_{ow}$; the extent of branching of the nonyl or octyl group is also known to affect the chemical properties of the compounds, including their solubility, thus affecting their bioavailability and toxicity [5]. Finally, nonylphenol (NP) and octylphenol (OP), the final products of the degradation of nonylphenol ethoxylates (NPEO) and OPEO, are acutely toxic to freshwater fish, invertebrates and algae [5].

In regards to marine organisms the information is scarcer. The studies carried out with twelve marine algae species using branched NPEOs showed total or partial growth inhibition at concentrations above 100 mg L^{-1} [21]. The development stages are in general more sensitive than adults, and the active species more than less active species. In crustaceans, resistance to this type of surfactant decreases immediately after moulting. For fish, the LC$_{50}$ 96h for NP$_{10}$EO in *Gadus morhua* was 6.0 mg L^{-1}, for molluscs ranged between 5 and 12 mg L^{-1} and for crustaceans was > 100 mg L^{-1} [22]. For larvae of *Hya araneus* and *Balanus balanoides* the LC$_{50}$ 96h ranged between 1.5 and 10 mg L^{-1} [22]. The acute toxicity of 4-nonylphenol to early life-stages of four saltwater invertebrate species (*Palaemonetes vulgaris*, *Americamysis bahia*, *Leptocheirus plumulosus* and *Dyspanopeus sayii*) and two fish species (*Menidia beryllina* and *Cyprinodon variegates*) in a 7 day flow through experiment ranged from 17 to > 195 μg L^{-1} [23].

The toxicity (LC$_{50}$ 48h) of the non-ionic surfactant alkylethoxylates (AE) in molluscs [2] ranged between 1.0 and 6.8 mg L^{-1} for *Dugesia* and *Rhabditis*, respectively. The toxicity values for daphnids were the lowest for all invertebrates ranging from 0.29 to 0.70 mg L^{-1} [2]. The LC$_{50}$72h of *X. laevis* embryos for AE surfactant (Lorodac 7 24 from Condea Augusta, Milan Italy) was 4.59 mg L^{-1}, and teratogenic effects were also observed [8]. A review of the short-term toxicity of C$_{12-15}$ EO$_{3-10}$ alcohol ethoxylates on bacteria, algae, diatoms, worms, insects, molluscs, crustaceans, fish and aquatic plants (freshwater as well as marine organisms) showed significant intra- and interspecies variability, especially for algae [11].

Cationic surfactants
The most extensively studied of the quaternary ammonium-based cationic surfactants (QACs) is the hydrogenated tallow dimethyl

ammonium chloride (DHTDMAC), due to its wide use as a fabric softener until few years ago. Several reports on its aquatic toxicity have been published [24,25], and a review of its environmental impact was published by ECETOC [26]. Acute toxicity assays of this surfactant employing the *D. magna* 24 h immobilisation test (IC_{50}) and *P. phosphoreum* 30 min luminescence reduction test (EC_{50}) as endpoints gave values <1 mg L^{-1} [27]. Differences in toxicity of different homologues have not been observed; this was attributed to a lower bioavailability of homologues with longer alkyl chain due to their decreasing solubility [27]. The toxicity of quaternary ammonium chloride (alkyltrimethyl ammonium (TM), dialkyldimethyl ammonium (DM) and alkyldimethylbenzyl ammonium (BM) chloride) has also been evaluated by measuring the production of ^{13}C glycerol in *Dunaliella sp.* [28]; the results showed the toxic potencies followed the order TM $>$ BM $>$ DM, with the EC_{50}-24 h ranging between 0.79 and 0.71 and between 18 and 19 mg L^{-1} for TM and DM, respectively [28]. The concentration at which this surfactant produces observable effects is between 0.1 and 1.0 mg L^{-1} for freshwater and marine algae [29].

Amphoteric surfactants
The amphoteric surfactants are less widely used than the anionic, non-ionic and cationic surfactants, and information about their toxicity is very limited. In assays carried out with *Daphnia magna* (immobilisation) and rainbow trout, the EC_{50} and LC_{50} ranged between 4.8 and 182 mg L^{-1} for *D. magna* and 1.1 and 96.4 mg L^{-1} for trout, with increasing toxicity with increasing chain length observed [30]. The damage on the gills of the common carp *Cyprinus* carpio, exposed for 96 h to lethal concentrations of *N*-alkyl (C_8–C_{18})-*N, N*-dimethyl-3-ammonium-1-propane sulfonate, increases with the number of carbon atoms in the alkyl chain [31]. Amphoteric surfactants have been employed as disinfectants against the ciliate *Colpoda aspera* due to its toxicity [32]. For this type of surfactant no estrogenic activity has been detected [33].

7.1.3   Toxicity to microalgae

Phytoplankton as test organisms in toxicity bioassays with surfactants
Microalgae are the basis of the aquatic trophic chain and at least 30% of the organic carbon planetary primary production is attributed to

unicellular algae [34]. These micro-organisms have been shown to be quite sensitive to different toxicants [35], including surfactants [36]. Damage to cells caused by surfactants seems to occur at membrane level, above all in the mitochondrial electron-transport chain [37]. By affecting membrane permeability, accumulation of xenobiotics may increase and escape of microalgal biopolymers from cells may occur [38]. Due to their small size, microalgae possess a large surface/volume ratio. This leads to a higher exposure to toxicants present in the water than for multicellular organisms.

Recent data demonstrate that among the organisms usually employed microalgae are the group that is most sensitive to synthetic surfactants in ecotoxicologic bioassays [39,40]. Although information from bioassays cannot be readily obtained in any other way, existing data involving microalgae (mainly in marine environments) is unsystematic and difficult to compare, due to the lack of standard procedures and reference organisms [41]. Certainly, toxicity of a selected substance cannot be established solely by the sensitivity of a single species. In fact, the concept 'maximum permissible concentration' (MPC) is always determined according to the responses of several species. This parameter is defined as the concentration of the substance at which at least 95% of the population shows no toxic effects. When the number of species is above 30, good statistical evaluation of the effect of chemicals can be achieved, but a minimum of eight species is necessary for single species testing in order to define the distribution curve of 'no observed effect concentrations' (NOEC) [42]. Nevertheless, comparison of toxicity of substances must take some selected organisms for reference. In marine or freshwater environments the selection of a 'standard' organism should give results comparative with assays using other known organisms or local species. Furthermore, all these data should be complementary.

Most microalgal toxicity tests procedures recommend the use of initial cellular concentrations of 10^4 cells mL^{-1}. This cellular concentration should be selected because it is the minimum cellular concentration that can be measured in haematocytometers (Neubauer chambers). Furthermore, natural cellular concentrations in non-polluted conditions (in marine environments) are often below the concentration mentioned. The importance of cellular density at the beginning of the test has been demonstrated for certain toxicants [43]. The lower the cellular concentration, the higher the sensitivity of the test, at least for certain types of xenobiotics, such as heavy metals.

TABLE 7.1.1

Chemical composition for different synthetic marine waters. Values are expressed as g L^{-1}

Reagent	EPA	SIGMA	ASTM	APHA
NaCl	21.03	23.96	24.53	23.5
$MgSO_4 \cdot 7H_2O$	–	10.346	–	–
$MgCl_2 \cdot 6H_2O$	9.5	6.5	11.103	10.78
$NaSO_4$	3.52	–	4.09	4
$CaCl_2 \cdot 2H_2O$	1.32	0.394	1.536	1.47
NaBr	–	1.029	–	–
KCl	0.61	0.596	0.695	0.7
$NaHCO_1$	0.17	–	0.201	0.2
KBr	0.088	–	0.101	0.1
$SrCl_2$	0.02	0.027	0.042	0.011
NaF	–	0.042	0.003	0.003
H_3BO_3	–	0.006	0.027	0.03
$Na_2B_4O_7 \cdot 10H_2O$	0.034	–	–	–
$Na_2SiO_3 \cdot 9H_2O$	–	–	–	0.02
Na_4 EDTA	–	–	–	0.001

EPA: Environmental Protection Agency; SIGMA: Sigma Chemicals, Plant Culture Cell™, Seawater synthetic basal salt mixture; ASTM: American Society for Testing and Materials; Apha: American Public Health Association.

Accordingly, the use of flow cytometry can improve the design of toxicity bioassays, as the detection limit of this apparatus includes cellular concentrations equal to those of microalgal populations found in natural conditions. Comparison of compositions utilised in some known toxicity tests for microalgaes are shown in Table 7.1.1.

Freshwater species

Toxicity test procedures in freshwater environments are often more standardised than those developed for sea water organisms. Cladocerans like *Daphnia magna* or fishes such as *Pimephales promelas* have received great attention as standard organisms in ecotoxicology [44]. Three freshwater microalgal species have also been selected as the subject of toxicity tests: *Pseudokirchneriella subcapitata* (formerly *Selenastrum capricornutum* and *Raphidocelis subcapitata*), *Scenedesmus subspicatus* or *Chlorella vulgaris*. All the species belong to Chlorophyceae. High resistance of species from this class to different toxicants have been reported [45,46]. In particular for LAS (commercial LAS C_{11}), *Scenedesmus subspicatus* showed a high resistance with an EC_{50} value after 72 h of 240 mg L^{-1} (data supplied by PETRESA). Some other green algae demonstrated significant resistance to surfactants: cells of *Chlamydomonas reindhartii* did not show morphological changes at 20 mg L^{-1} of LAS, and *Plectonema boryanum* showed little inhibition of growth at 30 mg L^{-1} LAS [47]. This is not surprising considering the natural habitats of the two genus: these organisms live in small ponds and stagnant waters [48]. This extreme and unstable biocenosis would invoke strong selection against sensitive organisms. It is possible that inshore environments, which are subject to changes of pH, oxygen level and temperature would select more resistant species than oceanic or even coastal environments. As such other possibilities (such as Cryptophyta or Chrysophyta) should also be tested in order to increase the sensitivity of the assay.

While degradation products of LAS are less toxic to microalgae than the original substances, as well as molecules with shorter carbon chains showing lower toxicity than molecules with larger carbon chains, in the case of NPEO the degradation product NP is known to be more toxic than the ethoxylated forms. Toxicity values of NP found in the literature for freshwater microalgae range from 92 mg L^{-1} (value of NOEC for *Selenastrum capricornutum*) to 2500 mg L^{-1} (inhibiting cell division of *Chlamydomonas segnis*) [49]. In mixtures, NP shows a synergistic toxic effect with oil when tested on freshwater microalgae *Chlorella pyrenoidosa* [50]. Due to the specific toxicity, it is possible that assays in multicellular organisms would yield more valuable results than assays with microalgae. In any case, sensitivity to APEO in microalgal tests seems to be as good as that for other organisms [49], with the advantage of being less expensive and less time-consuming.

Marine species

In contrast to freshwater procedures, no media or even organisms have been the subject of general agreement reached in tests involving marine organisms. Only Standard Methods from the APHA, AWWA and WPCF [51] among the consulted guidelines [44,52–54] mention the option of toxicity tests with marine microalgae; however, it is worth noting here that the three species proposed in these guidelines do not seem to be adequate as candidates for a 'standard' organism: *Dunaliella tertiolecta* belongs to a very resistant genus [55], and furthermore green algae are not of significant ecological relevance in ocean waters. *Thalassiosira pseudonana* and *Skeletonema costatum* are centric diatoms. This group is ecologically the most important in marine environments, and the two described species have shown high sensitivity to toxicants [56–61]. However, both species tend to form chains, and this can present problems if electronic counters are used (i.e. coulter counter or flow cytometry). Significant data reported by Walsh [56] reviewed up to 45 microalgal species used in different marine toxicity tests. This author recommends the use of the diatom *Minutocellus polymorphus*, although caution would be required as this genus is also known to form colonies [62]. The centric diatom *Chaetoceros gracilis* Schütt (Fig. 7.1.3) seems to be a good choice for toxicity testing, as it matches all necessary requirements [63]. The genus is taxonomically complex (up to 180 planktonic marine species) and most of them form colonies, although importantly *C. gracilis* never forms groups. It is easy to culture, does not present specific nutrient requirements, grows rapidly and does not present long lag phases when transferred to fresh media.

Sensitivity of microalgae to xenobiotics is especially appreciable for some marine phytoplanktonic groups. LAS homologue C_{13} at concentrations of 6.25 μg L^{-1} inhibits mobility of the dinoflagellate *Gymnodinium breve* and levels of 26 μg L^{-1} rapidly kills the cells [64]. Higher sensitivity of marine microalgae to surfactants than that showed by freshwater microalgae must be an intrinsic quality of the marine organisms, not related with activity of these substances in one or other

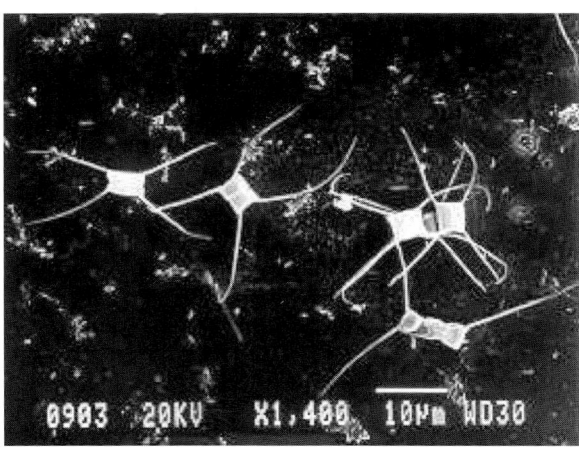

Fig. 7.1.3. Scanning electron microscopy photograph of *chaetoceros gracilis* cells. Scale bras means 10 μm.

media: LAS was demonstrated to be more toxic to copepod species when salinity decreased [65].

The use of synthetic media is preferable to natural filtered and/or sterilised media, as trace substances are expensive to remove from natural waters. In addition to this, the reproducibility of assays is improved if synthetic media are used. Different artificial seawater compositions have been used in toxicity testing. From the media investigated by our group, (see Table 7.1.2) the ASTM 'substitute ocean water' [66] gave the best results for growing microalgae.

Nutrients added to cultures must be carefully taken into account in toxicity tests, especially if chelators and metals are involved. In the case of surfactants, Guillard's f/2 medium invokes a good growth of microalgal populations and does not seem to interfere with the effects of the substances. Use of a light nutrient medium in heavy metal toxicity tests with diatoms, consisting in 5 mg L^{-1} SiO_2^-, 2 mg L^{-1} NO_3^- and 17 μg L^{-1} PO_4^{3-}, added to substitute ocean water [66] resulted in a good growth of cells in 72 h experiments without interference with metals, due to the absence of added chelators [67]. These results are of particular relevance as these nutrient levels are representative of those found in natural conditions [68].

Literature about the effect of LAS on marine microalgae is scarce. EC_{50} values for four marine microalgal species in the presence of the C_{11} and C_{13} LAS homologues have been reported [69], in which growth inhibition and esterase activity were taken as endpoints. Growth inhibition tests demonstrated higher sensitivity than esterase activity tests, with EC_{50} values ranging from 4.4 and 0.4 mg L^{-1} described for the homologues C_{11} and C_{13}, respectively, to 13.4 and 1.2 mg L^{-1}, for *Rhodomonas salina* (Cryptophyceae) and *Tetraselmis suecica* (Prasinophyceae) respectively. EC_{50} values for four marine microalgal species for commercial LAS have been reported [49], ranging between 0.3 mg L^{-1} for *Chaetoceros gracilis* (Bacillariophyceae) and >2 mg L^{-1} for *Pleurochrysis carterae* (Primnesiophyceae). The same authors reported NOEC values for four marine microalgal species for LAS homologues ranging between 0.1 mg L^{-1} for *Coccochloris sp.* and >10 mg L^{-1} for *Pleurochrysis carterae* in the case of the C_{11} homologue, and between 0.05 mg L^{-1} for *Gymnodinium sp.* and >1 mg L^{-1} for *Pleurochrysis carterae* in the case of the C_{13} homologue.

With respect to NP, the lowest EC_{50} values for microalgae were again observed for marine organisms. The EC_{50} for *Thalassiosira pseudonana* to NP is 27 μg L^{-1} which is significantly lower than the EC_{50} values for

TABLE 7.1.2

Some toxicity test conditions for freshwater and marine microalgae

	Species recommended	Init. cell. density (cells mL^{-1})	Media	Time	Endpoint	Temperature	Illumination
Freshwater							
OECD	*Selenastrum capricornutum, Scenedesmus subspicatus, Chlorella vulgaris,* others	10^4 for *S. capricornutum* and *S. subspicatus,* similar biomass for others	OECD[a]	72 h	Growth or growth rate	21–25°C	120 µmol m^{-2} s^{-1}
APHA	*Selenastrum capricornutum, Microcystis aeruginosa, Anabaena flos-aquae, Cyclotella sp. Nitzschia sp. Synedra sp.*	Species-dependent: between 10^3 and 50 × 10^3	APHA[b]	–	Growth rate	24 ± 2	32–65 µmol m^{-2} s^{-1}
Sea water							
APHA	*Dunaliella tertiolecta, Thalassiosira pseudonana Skeletonema costatum*	Species dependent: between 10^2 and 10^3	APHA[b]	–	Growth rate	18 ± 2	32–65 µmol m^{-2} s^{-1}
Our proposal							
Planktonic	*Chaetoceros gracilis*	10^4	ASTM[c] + Nutrients[d]	72 h	Growth	20 ± 1	~75 µmol m^{-2} s^{-1}
Benthic	*Nitzschia closterium*	10^4	ASTM[c] + Nutrients[d]	72 h	Growth	20 ± 1	~75 µmol m^{-2} s^{-1}

[a]Organization for the Economic Cooperation and development.
[b]American Public Health Association.
[c]American Society for Testing and Materials.
[d]5 mg L^{-1} SiO$_2^-$, 2 mg L^{-1} NO$_3^-$ and 17 µg L^{-1} PO$_4^{3-}$.

freshwater chlorophyceae species, i.e. *Scenedesmus subspicatus*, which is 1300 μg L^{-1}. EC$_{50}$ values in a growth inhibition test for *C. gracilis* of 1.0 $\pm$ 0.8 mg L^{-1} for the 10-ethoxylated form of nonylphenol (NPEO$_{10}$), and EC$_{50}$ values around ten times lower for the NP were reported [63]. Data in the literature on the aquatic toxicity of APE is scarce [49] especially for algae [37], but generally toxicity of APEs increases as the length of the EO chain decreases. This effect is also noticeable in experiments involving microalgae [70].

Some green microalgae (*Chlorella*, *Scenedesmus*, *Hydrodyction*) produce natural surfactants and these have been industrially exploited in order to manufacture natural shampoos for children and people suffering allergies. In the natural environment, these surfactants should play an important role in the setting and attachment of algae on substrates [71], but the actual physiological role of these compounds has not yet been conclusively elucidated [36].

7.1.4 Early life-stages. Hatching and larvae development as endpoints of toxicity

A very large number of marine organisms, fish and various invertebrate phyla, inhabiting both pelagic and benthic ecosystems, produce larvae in their life-cycles. Larvae are independent, morphologically different stages developing from fertilised eggs that must undergo a profound change before assuming adult features. They are almost always considerably smaller than the adult stage of the species.

In addition to inter-species variation with respect to susceptibility to chemicals, not all development stages in the life-cycle of an organism are equally sensitive to pollutant exposure. Immature or young neonatal organisms are often more susceptible to chemical agents than the adult form. This is related to various factors, including (i) differences in the degree of development of detoxification mechanisms, (ii) differences in excretion rates of toxic chemicals and metabolite products, and (iii) difference in body size. In some special cases, however, embryos may be less sensitive than adults because, at particular stages, they may be encapsulated by protective or impermeable membranes [72].

Even if the embryonic and larval stage are only transitory growth phases, basic metabolic functions such as nutrition, respiration and excretion still have to be carried out. Excretion of toxic compounds, both exogenous and endogenous, as well as osmoregulation processes, are usually closely related in animals, and are performed by the same organs. In aquatic organisms, these structures are the gills and the kidneys, and although gills function chiefly as respiratory organs, they also play important roles in excretion and osmotic regulation.

The fertilised egg is a single cell, which excretes its toxic products by simple diffusion into the perivitelline space. While successive cell division is occurring, this mechanism is used until the final or transitory excretion organs are formed in the embryo. The process of organ genesis takes a certain length of time and the embryo has to make use of less effective excretion mechanisms until these organs are completely functional. Depending on the species, hatching can occur earlier or later, and consequently newly liberated larva can be at different stages of development [73]. In some species, part of the genesis of the organs can take place during the free-swimming larval stage which can thus cause differences in sensitivity to toxicants in different species.

Living systems have developed mechanisms for metabolising and detoxifying xenobiotics, achieved with different enzymatic processes. Generally, the responsible enzymes are part of the normal biochemical pathways of the organism, but they may be specifically involved in the detoxification of internally generated toxicants. Some of these systems have been previously adapted through exposure and incorporated afterwards into the genome, the result being that the organisms can deal more efficiently with xenobiotics. Low-levels of biotransformation activity of young animals is therefore related to enzymatic deficiencies. Nevertheless, environmental parameters such as temperature and diet have also been shown to have effect on the activity of detoxification enzymes in fish, and increasing stress produced by pollutants can induce or increase detoxification activity.

The body size of an exposed organism has, especially in fish, an important influence on the toxicokinetic effects exerted by a chemical agent. Body size not only affects kinetic rate constants due to important differences in surface area/volume ratios, but also the accessibility of target organs. Before a toxic substance is able to induce its harmful effect, it has to reach the site of action or target organ. The time needed for the pollutant to reach the target organ depends on the size of the organism and the membrane barriers to be permeated. The older an organism, the higher is its level of development and thus the more complex is its body structure. More organs have developed and, as a consequence, more membranes have to be passed.

In fish, the developing embryo and larva are the stages that are most sensitive to many kinds of low-level environmental changes, and even during the early embryonic development at the egg stage, differences in sensitivity to pollutants have been observed. Newly fertilised eggs (e.g. morula and blastula stages) have been shown to be more sensitive to pollutants than after completion of gastrulation [74]. Studies on the effects of surface-active agents on marine organisms showed that the larval stages of fishes, crustaceans and bivalves are more sensitive than the adults [75]. The presence of the anionic surfactant, LAS, in the surrounding environment can affect both the fertilisation process itself, through alteration of the effectiveness of the sperm, as well as the later embryonic and larval development. In exposure experiments with mussels (*Mytilus edulis*), eggs were fertilised and maintained during 10 days in solutions of LAS at concentrations between 0.05 and 10 mg L^{-1}. The fertility rate was reduced from 88% in the control to 61% even at the lowest assay concentration [76]. The reduced surface activity presumably affects the ability of the acrosome of the sperm to penetrate the egg's vitelline membrane. In the cases where fertilisation did occur successfully, the differences to normal development at certain exposure concentrations was observed primarily as a delay in achieving the successive larval stages. Exposure of seabream (*Sparus aurata*) sperm to LAS showed a significant inhibitory effect on fertilisation at concentrations from 0.3 to 6.0 mg L^{-1} after 60 min. These authors reported an EC$_{50}$ value of 2.6 mg L^{-1} [77], but did not specify if the employed LAS was a commercial mixture or a certain homologue. During the development of the rainbow trout *Oncorhynchus mykiss*, the larvae's resistance to the non-ionic surfactant NPEO decreased at a certain stage of development, but rose again later [78]. Other authors [79,80] also reported differences for the reaction of cod (*Gadus morhua*), herring (*Clupea harengus*), and plaice (*Pleuronectes platessa*) embryos when exposed to crude oil extracts.

But even if the relative susceptibility of the early embryonic stages is well documented, the yolk-sac or alevin stage of a fish is known to be 10 to 100 times more sensitive [81]. This fact may be related partly to the protection against external pollutants provided to the organisms by the surrounding eggshell, the chorion. The entry of the sperm into the egg initiates the water hardening of the egg, after which the chorion becomes only slightly permeable to waterborne substances [82]. In experiments involving exposure of dechorinated and chorinated medaka (*Oryzias latipes*) embryos to different concentrations of thiobencarb, the dechorinated embryos showed a slightly higher susceptibility to the pollutant than chorinated embryos at the same development stage ($LC_{50} = 2.5$ vs $1.0 \, \text{mg L}^{-1}$) [83]. Differences in uptake rates of xenobiotics between eggs and larvae can be explained by the lower transport rate of xenobiotics across the chorion compared to the rate of transport across the gill epithelium of the larvae. Although the chorion does not totally prevent the uptake of xenobiotics due to its semipermeable characteristics, their penetration into the embryo tissues is probably lower than in larvae for two reasons: eggs have a smaller surface area exposed to the environment than larvae, and eggs do not actively circulate fluid through or near the chorion while larvae circulate blood through the gills [79]. Furthermore, the level of protection provided by the chorion depends on the specific hatching characteristics of each species, which occurs in different teleosts after variable levels of development have been reached [84]. In some cases, such as in medaka, newly hatched larvae are fully functional with the hatchling being capable of swimming, inflating its swim bladder, and feeding exogenously within 1 day. The maintenance of an apparently intact chorion until organ systems have formed and become functional may afford more protection to the medaka than in other species which may be liberated at an earlier stage of development, such as striped bass (*Morone saxatilis*) [85], fathead minnow (*Pimephales promelas*), and goldfish (*Carassius auratus*). In these species, hatching occurs much earlier in comparison, and at their onset, the newly hatched cannot swim, maintain equilibrium or feed for the first days [86]. This is also the case for the seabream. At the moment of hatching, the larva, equipped with a large yolk-sac containing lipid reserves in its distal part, floats inactively at the water surface with the yolk-sac orientated upwards. Three days after hatching, the larva has consumed the reserves of the yolk-sac and thus requires external feeding thereby necessitating mouth opening [96].

In studies with early life-stages of fish, several authors have observed that the elimination rate of lipophilic xenobiotics, which includes surfactants, is also dependent on the developmental stage of the organism [86–90], with significant differences in the uptake and elimination rates of xenobiotics by eggs and larvae exhibited. The difference in sensitivity between early life-stages and juvenile or adult forms to xenobiotic substances depends on accumulation levels of the compound, which occurs especially in the yolk tissue. Due to the nature of the yolk reserves, the lipid content in early life-stages is higher than in adult forms. For this reason, early life-stages are expected to accumulate higher quantities of lipophilic chemicals, with the yolk acting as a temporary toxicant sink. In combination with the lower metabolic capability, elevated body burdens of chemicals may thus be reached in eggs and larvae with a potential for developing chronic effects in later developmental stages [91] with the yolk absorption process believed to be a major factor in the re-distribution of chemicals causing death in subsequent growth stages [79]. Even if the exposure does not produce any acute effect, such sublethal effects from environmental pollution are believed to reduce the survival potential at some later period in the organism's life-cycle [92].

Knowledge regarding differences in sensitivity of determined development stages in the life-cycle of aquatic organisms has to be taken into consideration particularly when assessing receiving waters which may be spawning grounds for fish and invertebrates, due to the possible effects on the first year generation and inevitable consequences on the whole food chain.

Toxicity tests are necessary tools to evaluate the concentration and duration of exposure of a chemical required to produce certain adverse effects. Molecular processes directly affected by the exposure to the chemical agent are the most liable criterions. Nevertheless, these effects are difficult to detect in aquatic toxicology because the processes are generally not well understood [72]. Alternatively, other end points which fulfil the necessary requirements, namely the need to be unequivocal, relevant, easy to observe, describe and measure, biologically significant and repeatable, are used. These include measures of mortality, which is frequently employed in the early evaluation of the toxicity of a pollutant in acute toxicity tests. This criterion allows comparison of toxicity exerted by chemical agents with very different mechanisms of action.

Although the use of chronic full life-cycle tests for both invertebrates and fish provides more accurate information on potential toxicant effects by exposing the organisms from embryo to the next generation of embryos, the time and costs associated with these tests are extremely high, and much greater than with acute toxicity tests. Early life-stage tests include the continuous exposure of the aquatic organisms as e.g. egg, embryo, larva and fry to various concentrations of a chemical. Different standard procedures have been proposed for this type of assay [93,94]. Although these tests do not provide total life-cycle exposure and lack a complete assessment of reproduction, they do address exposure during the most sensitive life-stages, and have been shown to predict accurately 'maximum acceptable toxicant concentration' (MATC) values defined by the NOEC and the 'lowest observed effect concentration' (LOEC) in fish early life-stage tests. Based on a literature review, 82% of the tests evaluated showed that the LOECs for early life-stages (embryos, sac fry and juveniles) were identical to the LOECs obtained for the complete life-cycle of fish, and both test types can therefore be considered comparable [95].

The marine fish seabream, *Sparus aurata*, economically a very important species in South Spain due to its fast growth and culturable characteristics, is used intensively in local aquaculture. Hatcheries are thus able to provide fertilised eggs of this species over most times of the year, and its whole life-cycle, physiological characteristics and aquacultural needs are well researched and documented. Additionally, this species presents a relatively fast development, completing gastrulation approximately after 24 h after the fertilisation [96] upon which organogenesis is initiated. Approximately 48 h after fertilisation, the embryo emerges from the egg [96] thereby rendering it possible to determine early life-stage toxicity in a few days using this species, which reduces necessary costs and efforts significantly.

Incubation time and hatching itself generally depend on a multitude of factors, such as temperature, photoperiod, oxygen availability, and the presence of hormones, both natural and synthetic, in the medium. These factors may stimulate or inhibit the synthesis by the embryo of the hatching enzyme which is finally responsible for the emergence of the organism from the egg [97].

The exposure of different early development stages of the seabream [98,99] (eggs and larvae) to LAS homologues with alkyl chain lengths between 10 and 14, as well as to commercial LAS with an average chain length of 11.6 carbon atoms, showed that mortality of the organisms is directly related to both surfactant concentration and exposure time. As already described by several authors [100–103], toxicity of LAS homologues increases with alkyl chain length, reflected in decreasing LC_{50} values for the different homologues. Fig. 7.1.4 shows the accumulated mortality percentages for the LAS homologue C_{11} after 24 h with increasing surfactant concentration.

Eggs exposed to effective concentrations producing mortality showed white coloration, which increased with increasing surfactant concentration (Fig. 7.1.5(a), (b) and (c)) and loss of flotability. At very high

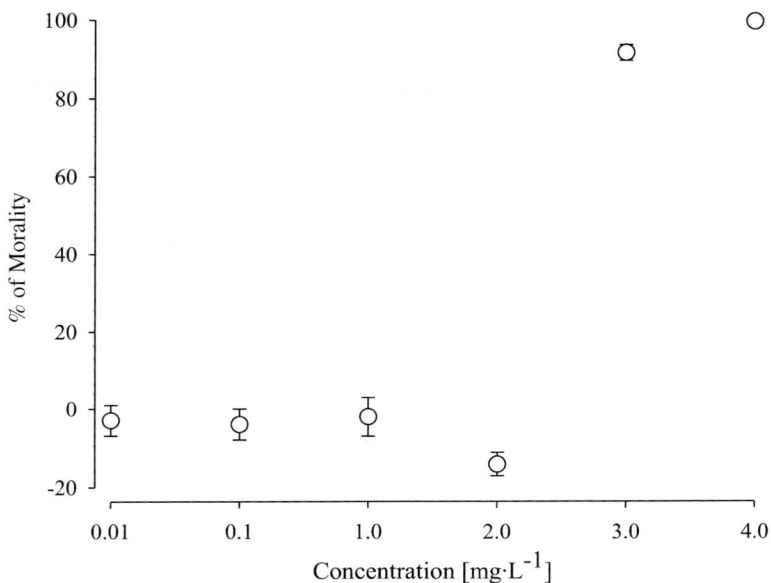

Fig. 7.1.4. Percentages of corrected mortality of the seabream, *Sparus aurata*, versus exposure concentration for LAS homologue C_{11} after 24 h of exposure.

exposure concentrations, the already differentiated embryo appeared as a formless cellular mass (Fig. 7.1.5(d) and (e)). In the case of exposed larvae, mortality is also accompanied with white coloration (Fig. 7.1.6(a) and (b)) and loss of flotability. In comparison, organisms from control assays, both eggs and larvae, are normally transparent and float on the water surface (Fig. 7.1.7(a)–(c)).

LC$_{50}$ values from hatching experiments with eggs after 24 h of exposure as well as from exposure experiments with larvae after 24, 48 and 72 h are shown in Table 7.1.3. The dependency of the toxic effect on exposure concentration and time is demonstrated with decreasing LC$_{50}$ values observed with increasing exposure time for the larvae. The obtained 24 h LC$_{50}$ values range between 0.1 and 3.0 mg L^{-1} for larvae and between 0.1 and 7.7 mg L^{-1} for eggs (Table 7.1.3), with the higher limit corresponding to the shorter, and the lower to the longer chain homologues, confirming that longer homologues are more toxic. LAS toxicity was shown to increase exponentially with an increasing number of carbon atoms in the alkyl chain. An important decrease in the LC$_{50}$ value between the C_{11} and C_{12} homologues is observed, which implies

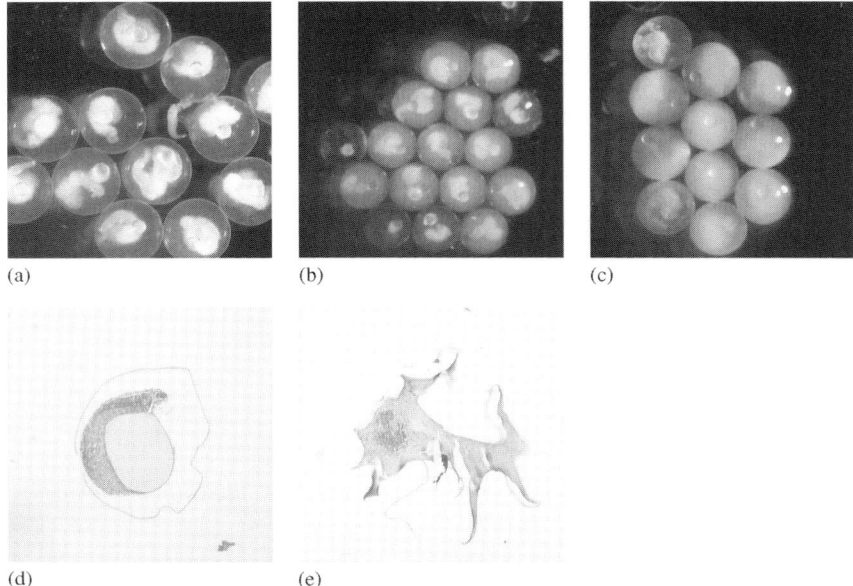

(a)   (b)   (c)

(d)   (e)

Fig. 7.1.5. Seabream eggs after exposure to increasing LAS concentrations; (a)–(c): Eggs after exposure to 0.25, 0.50 and 5.00 mg L^{-1} LAS C$_{13}$, respectively; (d)–(e): Histopathological view of seabream eggs after exposure to 5.00 and 10.0 mg L^{-1} LAS C$_{10}$, respectively.

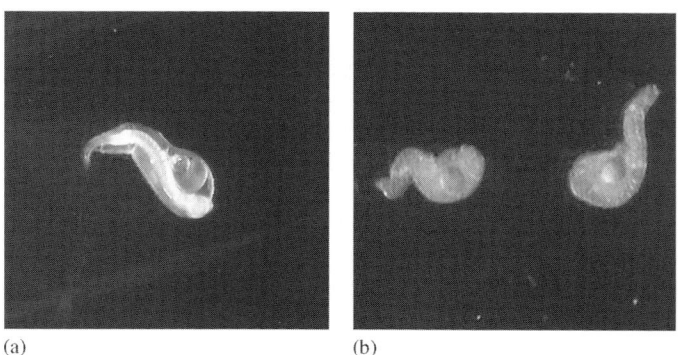

(a)   (b)

Fig. 7.1.6. Seabream larvae exposed to lethal LAS concentrations. (a) Larva exposed to 0.25 mg L^{-1} LAS C$_{14}$, (b) larva exposed to 0.50 mg L^{-1} LAS C$_{14}$.

848

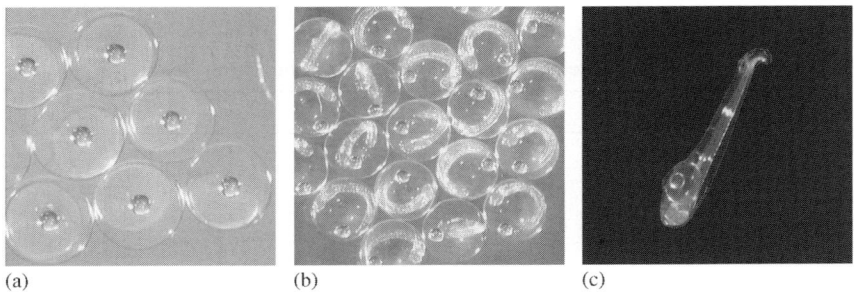

Fig. 7.1.7. Control eggs before (a) and after (b) gastrulation, as well as control larva of the seabream, *Sparus aurata*.

an increase in toxicity exerted on exposed organisms of the homologues with 12, 13 and 14 carbon atoms, respectively, in their alkyl chain. The commercial LAS used is a mixture of homologues with alkyl chain length varying between 10 and 14 carbon atoms, at different proportions (Table 7.1.4), and has an average chain length of 11.6 carbon atoms. The LC_{50} value of 0.7 mg L^{-1} obtained for commercial LAS is more similar to that obtained for LAS C_{12} than for LAS C_{11}, and confirms the exponential behaviour of LAS toxicity with increasing chain length when fitting the LC_{50} values versus number of carbon atoms into an exponential curve (data not shown). Figure 7.1.8 shows the percentages of corrected, accumulated mortality versus exposure time for the LAS C_{14} homologue at three relevant concentrations. Concentrations that are apparently not harmful at 24 h of exposure are able to produce important mortalities when the organisms are exposed for a longer period. However, at very low exposure concentrations, negative mortality percentages can be frequently observed due to the

TABLE 7.1.3

Nominal LC_{50} values obtained in larvae exposure assays with different LAS homologues after 24, 48 and 72 h of exposure

LAS homologue	LC_{50} (24 h)	LC_{50} (48 h)	LC_{50} (72 h)
LAS C_{10}	–	3.7 ± 0.2	2.6 ± 0.3
LAS C_{11}	3.0 ± 0.3	2.9 ± 3.0	2.0 ± 1.2
LAS C_{12}	0.2 ± 0.1	0.2 ± 0.0	0.1 ± 0.1
LAS C_{13}	0.1 ± 0.0	<0.1	<0.1
LAS C_{14}	0.2 ± 0.0	0.2 ± 0.0	0.1 ± 0.0
Commercial LAS	0.7 ± 0.1	0.5 ± 0.0	0.5 ± 0.0

TABLE 7.1.4

Homologue composition of the employed commercial LAS mixture [136]

Homologue distribution	Percentage (%)
<Phenyl C_{10}	1.1
Phenyl C_{10}	10.9
Phenyl C_{11}	35.3
Phenyl C_{12}	30.4
Phenyl C_{13}	21.2
Phenyl C_{14}	1.1
2-Phenyl alkane	15.5

fact that the percentage of corrected mortality is obtained by subtracting the mortality observed in the control assays from the mortality obtained at different exposure concentrations, implying that low surfactant concentrations in the environment can be favourable for survival. Pollutants frequently act as hormetic substances, inducing at low concentrations an enhanced response relative to the responses of control organisms but being toxic at concentrations that exceed the ability of the organism to homeostatically regulate the chemical [104].

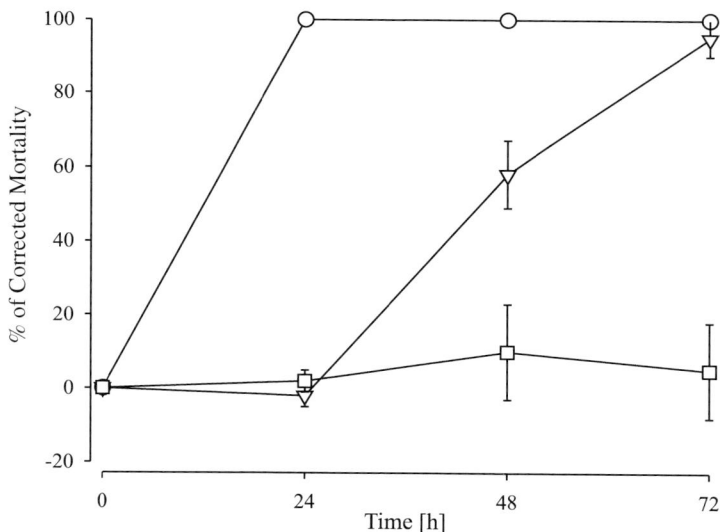

Fig. 7.1.8. Percentages of corrected mortality at three different exposure concentrations for LAS homologue C_{14} versus exposure time (○, 0.2 mg L^{-1}; △, 0.1 mg L^{-1}; □, 0.07 mg L^{-1}).

Calculation of the ratio between LC_{50} (eggs) and LC_{50} (larvae) after 24 h of exposure obtained in hatching assays with eggs and exposure experiments with larvae (Table 7.1.5) allows the comparison of sensitivities of different development stages towards pollution. The closer the ratio is to one, the more similar are both development stages in sensitivity towards the assayed homologue (Table 7.1.5, final column). Values between 3.0 and 1.0 for the different LAS homologues, and 0.7 for commercial LAS, were obtained with the results indicating that larvae are up to 3 times more sensitive to LAS exposure than eggs during the first 24 h of exposure. This difference in sensitivity decreases with increasing numbers of carbon atoms in the alkyl chain of the LAS homologues, reflected in the progressive decrease of this ratio. As such the eggs and larvae exhibit the same sensitivity for C_{13} and C_{14} LAS (ratio = 1.0). The results show an apparent increased resistance for seabream eggs to lower chain length homologues compared to the larvae, but with higher homologues the sensitivities are more similar. This indicates that longer LAS homologues are more capable of penetrating the eggshell than shorter ones, probably due to their higher hydrophobicity. Nevertheless, the ratio of 0.7 obtained for commercial LAS indicates that in this case, eggs were more sensitive to LAS exposure than larvae and furthermore the commercial blend does not show behaviour similar to C_{12} LAS as may have been expected, taking into consideration the average chain length is 11.6 carbon atoms. In any case it should be noted that assays were carried out with two different LAS products with different action indexes, which, although

TABLE 7.1.5

Nominal LC_{50} values and standard deviation (SD) after 24 h of exposure to different LAS homologues and commercial LAS for seabream eggs and larvae, as well as ratio between LC_{50} (eggs) and LC_{50} (larvae)

Compound	LC_{50} (24 h) (mg L^{-1}) $\pm$ SD (eggs)	LC_{50} (24 h) [mg L^{-1}] $\pm$ SD (larvae)	LC_{50}(eggs)/LC_{50}(larvae)
LAS C_{10}	4.4 $\pm$ 1.2	–	–
LAS C_{11}	7.7 $\pm$ 0.8	3.0 $\pm$ 0.3	2.6
LAS C_{12}	0.6 $\pm$ 0.6	0.2 $\pm$ 0.1	3.0
LAS C_{13}	0.1 $\pm$ 0.0	0.1 $\pm$ 0.0	1.0
LAS C_{14}	0.2 $\pm$ 0.0	0.2 $\pm$ 0.0	1.0
Commercial LAS	0.5 $\pm$ 9.5	0.7 $\pm$ 0.1	0.7

taken into consideration when calculating nominal concentrations, could have some influence on the final LC_{50} values obtained. Furthermore, embryos and larvae were obtained from different aquaculture facilities and as such the possibility of different qualities for the eggs or larvae can not be ruled out as a possible source of the observed variations in the LC_{50} values. The influence of egg quality on fry survival is well known and its influence on toxicity tests and obtained results should be submitted to further investigation in order to guarantee exact and secure safety limits.

Ecotoxicological data from single species tests form the basis for the derivation of safe concentrations for a given chemical in the environment. It can be assumed that long-term exposure of test animals will give a more realistic knowledge for this prediction than short-term exposure [35]. Comparisons between observed chronic NOECs with LC_{50} values of tested organisms provide risk assessment factors which can be used in the determination of safe environmental concentrations of a given chemical. Several multiplication factors have been proposed by different environmental organisations to predict environmentally safe concentrations (PNECs) from LC_{50} values obtained in laboratory short-term toxicity tests. Based on the ratio between acute LC_{50} values and chronic NOECs, the German detergent and specialties supplies association TEGEWA proposes a factor of 27.9 [105], whereas the European Center for Ecotoxicology and Toxicology of Chemicals, ECETOC [106,107], is more cautious suggesting a higher factor of 224. Finally, the EU proposed in its Technical Guidance Document in Risk Assessment [108] the highest safety factor at 1000, which has received criticism for being excessively conservative [9].

In the case of the Sancti Petri Channel (San Fernando, South Spain) with a mean environmental LAS concentration of $15\ \mu g\ L^{-1}$, and its different LAS homologue proportions reported by González-Mazo [109] shown in Table 7.1.6, the obtained 24 h LC_{50} values for eggs and larvae are up to two and three orders of magnitude higher, respectively, than the environmental concentration reflected by the ratio between LC_{50} value and environmental concentration of the corresponding homologue. The obtained ratios show that in the case of the longer homologues, some of the safety criterions proposed by the organisations mentioned are exceeded. In the case of the sea-bream eggs, observed environmental LAS homologue concentrations are thus not safe either for the C_{12} or the C_{13} LAS homologue. For the larvae, concentrations of the C_{12} and C_{13} LAS homologues

TABLE 7.1.6

Measured environmental concentrations of the different LAS homologues in the Sancti Petri Channel (San Fernando, Spain), nominal LC_{50} (24 h) values for eggs and larvae, as well as the ratio between environmental concentration and obtained LC_{50} (24 h) values

LAS homologue	Env. concentration ($\mu g \, L^{-1}$)	LC_{50} (eggs) ($\mu g \, L^{-1}$)	LC_{50} (larvae) ($\mu g \, L^{-1}$)	Ratio LC_{50}/Env. concentration (eggs)	Ratio LC_{50}/Env. concentration (larvae)
C_{10}	2	4400	–	2200	–
C_{11}	8	7700	3000	1026	375
C_{12}	5	600	200	134	40
C_{13}	1	100	100	100	100
C_{14}	–	200	200	–	–

cannot be considered as safe for ECETOC, with the ratio between environmental concentration and LC_{50} for C_{12} LAS being 40, which is close to the lowest limit considered safe even using the most lenient criterion, as proposed by TEGEWA. In any case, the selected environment, which receives untreated waste water from a town of about 100 000 inhabitants and has restricted water circulation conditions, can be considered as highly polluted and not representative of average environmental concentrations found in coastal waters where urban and domestic waste waters are more commonly discharged after sewage treatment.

Studies addressing LAS concentrations in marine environments are generally scarce and limited to recent investigations. Most of them have been carried out in different coastal systems, where restricted water circulation conditions and human concentration are often coupled [110–113]. These studies have not always provided information regarding individual homologues, but rather use of total LAS concentrations regardless of selective biodegradation or adsorption mechanisms of the LAS homologues with differing alkyl chain lengths. Identification of admissible threshold concentrations of individual homologues in a given environment is difficult, but without any doubt a useful task providing necessary information for making decisions about possible modifications to product formulations of commercial chemicals which consist of a heterogeneous composition in order to reduce their inherent toxicity.

The EU, in order to protect the environment from the adverse effects of urban and industrial waste water discharges, adopted in 1991

Directive 91/271/EEC concerning the collection, treatment and discharge of urban waste water and certain industrial sectors with respect to the nature of the waste water, number of inhabitants of the area and the characteristics, in terms of sensitivity, of the receiving water body. According to the Commission the choice of the treatment level for a town should take into account the degree of sensitivity of the receiving water bodies situated further downstream from the towns concerned. Taking into account the relatively rapid biodegradation of LAS in both sea and freshwater and the existing legislation and the existence of appropriate WWTP, LAS concentrations in European open coastal waters and fresh waters theoretically should not reach critical values.

Nevertheless, ten years after the adoption of this directive, the vast majority of Member States show important delays and shortcomings in its implementation with regards to deadlines and identification of sensitive areas. Thirty-seven large cities are still discharging untreated wastewater into the environment, and many others are discharging large quantities of effluent after inadequate treatment. The incomplete designation of sensitive areas results in under-assessment of the wastewater treatment for many agglomerations, such as London, Paris, Athens or Dublin. Additionally, in third world countries, where in many cases environmental legislation is not given high priority, industrial and domestic wastewaters are often discharged directly into rivers and coastal waters due to the lack of treatment facilities. Furthermore, certain physical conditions of the receiving water bodies, such as restricted water circulation or closed water bodies, can lead to critical surfactant concentrations being reached which represent a threat to the organisms present with similar sensitivity to LAS as the seabream.

Toxicity assays are carried out with the aim of determining environmentally safe pollutant concentrations in ecosystems which are highly complex assemblages where coexisting species are closely related, with respect to both nutritional and habitat related aspects. Knowledge about the difference in sensitivity of determined development stages in the life-cycle of aquatic organisms has to be taken into consideration especially when assessing receiving waters which may be spawning grounds for fish and invertebrates. This can not only invoke potential adverse effects on the first year generation but also may cause dangerous effects on the whole food chain and the resulting biodiversity of the ecosystem. For this reason, the most susceptible species or development stages within an ecosystem should ideally be selected for toxicity testing in order to guarantee safe and effective environmental assessment.

7.1.5 Chronic toxicity

The long-term and chronic exposure of test animals gives a more realistic picture for forecasting environmental effects than short-term exposure or acute toxicity tests. Within this framework several organisations, such as TEGEWA [7], initiated a program on the long-term effect of surfactants. Also, the determinations of PNECs values (Predicted No Effect Concentrations) and the risk characterisation requires information regarding chronic toxicity. With this objective, the NOEC values are frequently employed. However, the usefulness of the no-observed effect level is questioned by some authors [114], mainly due to measurement aspects, as the quantification of this parameter depends critically on the size and variability of the experiment.

The chronic and sublethal toxicity data is not equal for different classes of surfactants, with chronic toxicity for cationic classes less than for non-ionic and anionic ones. Recently, several works have collected the chronic toxicity data for surfactants with the objective of establishing the PNECs of surfactants in the environment. Thus, for example, in the framework of the project 'Investigating chemicals for the future' the National Institute for Coastal and Marine Management (Netherlands) has studied a selection of alkylphenols (APs) [19]. The Ministry of Health and Environment of Canada has also authorised research initiatives targeting the collection of information on a wide variety of substances, among APs and APEOs [5]. The Dutch 'Plan of Action Laundry and Cleaning Products' also derived the PNEC of these compounds [18]. In all cases, one of the analysed aspects was the chronic toxicity of these compounds. The long-term toxicity data available for different taxonomic groups (blue algae, diatoms, green algae, rotifers, crustaceans, molluscs, fish and worms), showed NOEC values ranging between 0.60 and 1.9 mg L^{-1} with the marine species appearing to be somewhat less sensitive than the freshwater species [18]. Data of chronic toxicity showed that NPEOs are moderately toxic to fish and very slightly toxic to algae and crustaceans. The chronic toxicity of NP is high, thus NOEC values reported for fish are as low as 6 μg L^{-1} and for invertebrates 3.9 μg L^{-1} [115], these data corresponding to chronic studies with early life-stages of rainbow trout and the 28d NOEC of the marine mysid *Mysidopsis bahia*. The chemicals NP and NPEOs have been reported to cause estrogenic responses. However field studies are limited and refer to APs. They are summarised in the review carried out by the RIKZ [19] and other authors [5]. The results in an experiment carried out with 45 taxa showed that cladoceran and copepod taxa were significantly reduced in number in the enclosures exposed to 243 μg L^{-1} NP compared to controls [116] and some more sensitive taxa were reduced in number at NP concentrations of 76 and 23 μg L^{-1}. In experiments carried out with *Lepomis macrochirus* [117], no statistically significant differences were found in mean lengths and weights of fish from the NP-treated enclosures, although some trends in relation to capture success and cumulative mortality could be observed. The relationship between alkyl chain length, number of EO groups and toxicity remains unclear [118], although it has been observed that NP and OP, in general, showed higher NOEC values than EO compounds. Other non-ionic surfactants, such as fatty acid C_{12}–C_{18} di-ethanol amide, showed NOEC values of 0.26 mg L^{-1} for fish (*Oncorhynchus mykiss*), 0.07 mg L^{-1} for *Daphnia magna* and 0.32 mg L^{-1} for algae (*Scenedesmus subspicatus*) [7]. In experiments carried out in outdoor stream mesocosms [119] with the surfactant $C_{14-15}AE_7$ on fish survival, growth and reproduction, no effects on bluegill sunfish growth were observed, even at the highest assayed concentration (330 μg L^{-1}). Nevertheless, the surfactant affected egg production and larval survival at the highest concentrations.

The chronic toxicity of anionic surfactants has been reported in the scientific literature [3,7,17,18], presenting the main volume of information available regarding LAS. The studies carried out with fish, daphnids and algae [7] for anionic surfactants have shown that NOECs were higher for algae than for fish and daphnids. The information about NOEC concentration for the anionic surfactant LAS has been summarised in Figs. 7.1.1 and 7.1.2, where the range for this parameter is plotted for fresh and seawater organisms, respectively. The information for freshwater is more extensive than seawater and the values ranged between 0.04 mg L^{-1} for $C_{13.1}$ LAS in *Mysidopsis bahia* and 80 mg L^{-1} for C_{10} LAS in the algae *Scenedesmus*. *Daphnia magna* has been the most common species, with the more typically reported chronic effect concentration exceeding 0.1 mg L^{-1}; although 0.005 mg L^{-1} has also been reported [120], this data has been considered an outlier [3]. Long-term toxicity data for algae are highly variable, thus NOEC of 0.05 and 0.09 mg L^{-1} were obtained for

Microcystis spp., whereas for *Scenedesmus,* as mentioned above, values reached 80 mg L^{-1}.

The toxic effect of LAS on several planktonic freshwater organisms using small in situ enclosures was studied [121], with the NOEC as low as 0.02 mg L^{-1} for the most sensitive group (protozoans) investigated. This toxicity is notably higher compared with the results derived from laboratory tests. Mesocosm experiments carried out with blue mussel (*Mytilus edulis*) larvae showed a decrease in abundance within two days at concentrations as low as 0.08 mg L^{-1} LAS, due to increases of mortality and settling [122]. For long-term chronic toxicity tests, differences in the sensitivity between fresh and seawater have been described; the mean of NOEC values of marine and freshwater organisms is 0.3 and 2.3 mg L^{-1} LAS [17], respectively. The effect of mixtures of chemicals and environmental modifying factors has been reported. In general, the presence of pesticides and metals on surfactant toxicity is unpredictable, but oil–surfactant mixtures are usually more toxic; the increase of water hardness and temperature increases surfactant toxicity for some compounds, although the outcome depends on the compound and is species-specific [50]. Scale-up microcosms have also been employed to determine the effect on aquatic ecosystems [123], with the NOEC of LAS on population density less than 1.5 mg L^{-1}. The determination of LAS effects on other parameters such as adenosine triphosphate (ATP), nutrients, pH and dissolved oxygen (DO) and the net photosynthesis/respiration ratio were less than 2.5 mg L^{-1}.

The ecotoxicological database for AES on aquatic organisms is small in comparison with LAS; long-term test results are available for green algae, rotifers, crustacean and fish, the NOECs normalized to $C_{12.5}EO_{3.4}$ AES ranged between 0.80 mg L^{-1} for *Brachionus calcyflorus* and 2.4 mg L^{-1} for *Scenedesmus subspicatus* and *Selenastrum capricornutum* [18]. In a chronic test carried out with *Ceriodaphnia dubia* [124], a general trend towards decreased toxicity with increased EO units per alkyl chain length, as well as towards increased toxicity with increased alkyl chain length per EO unit, were observed; similar results were obtained with acute toxicity tests.

In a field study performed with nine surfactants from all groups except amphoteric [125], the laboratory effect concentrations were either lower than or similar to the mean in situ EC_{50} values for cationic and non-ionic surfactants. In contrast, for C_{12} and C_{13} LAS, the laboratory EC_{50} values were higher than the in situ effect concentrations. The short-term photosynthetic response to the same surfactant was variable, and in the field studies, related to seasonal changes. Nevertheless, despite the differences between standard laboratory tests and field responses, the authors pointed out the usefulness of laboratory test to estimate 'safe concentrations'.

The chronic effect data for the amphoteric and cationic surfactants are scarcer than for other types of surfactants. Thus the amphoteric surfactant, fatty acid amidopropyl betaine showed NOECs values below 1 mg L^{-1}, with *Daphnia* and algae being less sensitive than fish [7]. The cationic surfactant DTDMAC was selected as a model surfactant for risk evaluation of cationic surfactants in the Netherlands [126], with the results showing that the algae species (*M. aureginosa*, *S. capricornutum* and *N. seminulum*) are the most sensitive and the investigated bacteria the least sensitive. In the case of algae, the NOEC values reported, derived from algistatic concentration, ranged between 0.21 and 0.71 mg L^{-1}. Low toxicity for the fish species *G. aculeatus* and *P. promelas*, the midge larva *C. riparus*, the crustacean *D. magna* and the water snail *L. stagnalis* were obtained. When the bioassays were carried out with surface water, the determined NOECs were higher, with this fact a consequence of the suspended matter in the water [127].

7.1.6 Biomarker approach

The use of physiological and biochemical responses has been employed as endpoints to determine the effect of pollutants on marine organisms [128]. Thus, mussels exposed to LAS (6 mg L^{-1}) for 15 and 30 days were analysed to determine the impact of this xenobiotic on the digestive gland [129]. The results showed an increased number of vacuoles and a decrease of the antioxidant enzymes superoxide dismutase (SOD), catalase (CAT) and glutathione peroxidase (GPX). The anionic surfactant has been related to the modification of lysosomal activity, thus acid (ACP) and alkaline phosphatase (ALP) activities are inhibited in the liver and gills of the teleost *Channa punctactus* [130]. The fish fingerlings *Cyprinus carpio* exposed to LAS (0.005 mg L^{-1} for 24, 48 and 96 h) showed significant alterations in lactic acid, glycogen, sialic acid and ALP activities, with the decrease in the glycogen level attributed to the impairment of glycogen synthesis processes [131]. In vitro studies of the effect of LAS on ACP activity in *Ruditapes philippinarum* showed a clear inhibition, whereas ALP was not inhibited in the assayed range (0–100 mg L^{-1}).

In assays carried out in vivo (500 µg L^{-1}) no effect was observed for both enzyme activities [132]. LAS is negatively charged and lipophilic and it has been demonstrated to accumulate in lysosomes from rats [133]. In consequence, the subsequent toxic effect on enzymatic activities is associated with its accumulation, although other toxic mechanisms are not excluded. The effect of the anionic surfactant LAS (1.0–2.5 mg L^{-1} for 24–192 h) on the epidermis of catfish *Heteopneustes fossilis* included a marked increase in the number of mucous cells and acidic glycoproteins, the loss of mitochondrial enzyme activities in the superficial cell layers and an increase in glucose and lactate dehydrogenase [134]. This latter fact has been related to the increase in anaerobic oxidation. Pathomorphological changes in the skin of fish exposed to LAS (0.005 mg L^{-1}) was noticed in fish fingerlings *Cirrhina mrigala* [11] where the presence or deposition of mucus on the surface skin indicated interactions between constituents of mucus and LAS. Other anionic surfactants such as SDS provoked changes in the size and number of globet mucous cells in the opercular epidermis of *Rita rita* as a consequence of the profuse secretion by cells [135]. The effect of SDS (at concentrations ranging between 5 and 15 mg L^{-1}) on the intestine of gilthead seabream induced severe alterations in the epithelium, mainly as retractions of the intestinal villi and fusion of their tips, epithelium detachment and hypertrophy of the submucosa [16].

The mitochondrial respiratory parameters have also been employed to determine the toxicity of surfactants, including anionic (LAS), non-ionic (NPEO) and their metabolites, sulfophenyl carboxylates (SPCs), NP and nonylphenoxy carboxylate (NPEC$_1$) [37]. The system employed was the in vitro response of submitochondrial particles from beef heart. The EC$_{50}$ toxicity calculated as the reduction rate of NAD$^+$ ranged from 0.61 mg L^{-1} for a commercial LAS mixture to 18 000 mg L^{-1} for SPCs, and 1.3 mg L^{-1}, 8.2 and 1.8 mg L^{-1} for NPEO$_{10}$, NPEC$_1$ and NP, respectively. These results indicate that from the toxicity perspective, LAS is the compound demanding increased attention, while for NPEO, the parental compound and the metabolites must be quantified.

Although the marine microalgae have great ecological relevance, their use is not frequent in toxicity tests and the employment of biochemical endpoints is less usual. The esterase activity in marine microalgae has been employed as a biomarker of the toxicity to LAS homologues (C$_{11}$ and C$_{13}$) [69] with different marine microalgae (*Rhodomonas salina, Isochrysis aff. galbana, Tetraselmis suecica* and *Nannochloropsis gaditana*). The inhibition of this enzyme activity was increased with chain length and a good correlation was found between growth inhibition and esterase activity.

7.1.7   Future research

Available information regarding surfactant toxicity is generally related to different surfactant groups (anionic and non-ionic) and, above all, to freshwater species, whereas toxicity in marine species is much less frequently described. A high number of investigations deal with the toxicity of commercial surfactants (homologue mixtures), but not much work has been done concerning the relationships between structure and toxicity.

The majority of the existing data is available for species selected for the standardisation of freshwater toxicity tests (*Daphnia magna*, *Scenedesmus subspicatus*, *Pimephales promelas*). However, in marine species, there is no defined criterion regarding species selection, causing additional problems in toxicity research. Moreover, the scientific community is currently debating the suitability of the selection of standardised assays to evaluate environmental toxicity. Nevertheless, the selection of standard species seems to be a convenient and necessary measure in order to perform surfactant toxicity assays in the marine environment and to establish certain security factors whereby enabling the minimisation of their environmental impact, taking into account all limitations and necessary precautions. In this sense, the employment of marine microalgae (*Rhodomonas salina* has shown to be a sensitive species towards LAS exposure), as well as of eggs and larvae of marine fish and molluscs, which represent especially sensitive life-stages in which exposure effects can have implications in the later development, growth and survival offers a valuable proposal for toxicity research.

One aspect to be addressed in order to obtain a realistic vision of the toxicity of these kinds of compounds is their environmental behaviour. Surfactants tend to be adsorbed on particulate matter and thus subsequently to sediment. Consequently, the highest surfactant concentrations are found in sediments, although their distribution is dependent on the partitioning equilibrium between the substrate and interstitial water. This results in two possible routes for uptake (bioaccumulation) and effect. The relative importance of each of these routes depends on the special habits of each benthic organism.

Literature about bioaccumulation of different surfactant groups is limited, specially for benthic organisms. The bioavailability of sediment sorbed surfactants represents one of the key questions for evaluating their toxicity and impact on ecosystems.

Information regarding toxicity of surfactants is in general still somewhat limited, especially that related to surfactants other than Na(LAS) and NPEO. Furthermore, even in the case of these examples, most of the information given is obtained from freshwater species and thus may not be applicable to organisms occupying other habitats, e.g. marine environments. Bioaccumulation data are scarce and practically no information is available about these compounds when adsorbed on sediments, although according to the results, 'bioconcentration factors' (BCFs) do not seem to be very elevated. This situation is even more pronounced for the surfactant types other from the two mentioned above. Further investigation is needed in order to improve the available knowledge regarding the most intensively used of these other surfactants, as well as about their effects when adsorbed on sediments.

7.1.8 Conclusion

The information about the toxicity of surfactants on freshwater organisms is very extensive and it could be considered sufficient to establish environmental risk assessment for these compounds. Nevertheless, the situation for marine organisms is quite different. Although some environmental risk assessments have been carried out in the marine environment, the database for these species should be increased to get more realistic information. The selection of ecological relevant species (microalgae) and sensitive stages (fish early life-stages) can be an adequate way of achieving this objective. Regarding marine sediment, a relevant environmental compartment for surfactants, the database of toxicity assays is very limited, and in many cases performed with ecological species of less relevance. The use of phytobenthos and other benthic, sensitive and representative organisms joined with surfactant concentration data will be necessary to establish adequately the risk of the use of these compounds in the environment.

REFERENCES

1   P. Abel, *J. Fish. Biol.*, 6 (1974) 79.
2   M.A. Lewis and D. Suprenant, *Ecotoxicol. Environ. Saf.*, 7 (1983) 313.
3   M.A. Lewis, *Water Res.*, 25 (1991) 101.
4   C.P. Groshart, P.C. Okerman and W.B.A. Wassenberg, *RIKZ*, (2000) 92.
5   M.R. Servos, *Water Qual. Res. J. Can.*, 34 (1999) 123.
6   Technical Guidance Documentes in Support of The Comisión Directive 93/76 EEC on Risk Assessment for New Notified Substances and The Comisión Regulation (EC) 1488/94 on Risk Assessment for Existing Substances, 1996.
7   N. Scholz, *Tenside Surf. Det.*, 34 (1997) 13.
8   P. Cardellini, *Ecotoxicol. Environ. Saf.*, 48 (2001) 170.
9   C. Verge, and A. Moreno, *Tenside Surf. Det.*, 37 (2000) 172.
10 H. Lal, V. Misra, P.N. Vuswanathan and C.R.K. Murti, *Ecotoxicol. Environ. Saf.*, 7 (1983) 538.
11 V. Misra, G. Chawla, V. Kumar, H. Lal and P.N. Viswanathan, *Ecotoxicol. Environ. Saf.*, 13 (1987) 164.
12 M. Sandbacka, I. Christianson and B. Isomaa, *Toxicol. Vitro*, 14 (2000) 61.
13 A. Calabrese and H.C. Davis, *Proc. Nat. Shellfisheries Assoc.*, 57 (1967) 11.
14 A. Granmo, *Mar. Biol.*, 15 (1972) 356.
15 M. Stalmans, E. Matthijs and N.T. Oude, *Water Sci. Technol.*, 10 (1991) 115.
16 A. Ribelles, M.C. Carrasco, M. Rosety and M. Aldana, *Ecotoxicol. Environ. Saf.*, 32 (1995) 131.
17 A. Temara, G. Carr, S. Webb, D. Versteeg and T. Feijtel, *Mar. Pollut. Bull.*, 42 (2001) 635.
18 E.J. van dePlassche, J.H.M. Bruijn, R.R. Stephenson, S.J. Marshall, T.C. Feijtel and S.E. Belanger, *Environ. Toxicol. Chem.*, 18 (1999) 2653.
19 National Institute for Coastal and Marine Managemente/RIKZ, The Netherlands. Chemical study of alkylphenols, 2000.
20 Environment Canada Health Canada Canadian Environmental Protection Act. Priority substance list assessment report nonylphenol and its ethoxylates. Draft, 2000.
21 R. Ukeles, *J. Phycoly.*, 1 (1965) 102.
22 M. Swedmark, B. Braaten, E. Emanuelsson and A. Granmo, *Mar. Biol.*, 9 (1971) 183.
23 S.M. Lussier, D. Champlin, J. LiVolsi, S. Poucher and J.R. Pruell, *Environ. Toxicol. Chem.*, 19 (2000) 617.
24 M.A. Lewis and V.T. Wee, *Environ. Toxicol. Chem.*, 2 (1983) 105.
25 C.J. Roghair A. Buijze and H.P.N. Schoon, *Chemosphere*, 24 (1992) 1.

26 ECETOC. DHTDMAC. *Aquatic and Terrestrial hazard* (Technical report no 53). European Centre for ecotoxicological and Toxicological Safety Assessment of Chemicals, Brussels, Belgium, 1993.

27 M.T. García, I. Ribosa, T. Guindulain, J. Sánchez-Leal and J. Vives-Rego, *Environ. Pollut.*, 11 (2001) 169.

28 A. Utsunomiya, T. Watanuki, K. Matsuhita, M. Nishina and I. Tomita, *Chemosphere*, 35 (1997) 2479.

29 M.A. Lewis, *Ecotoxicol. Environ. Saf.*, 20 (1990) 123.

30 M. Sandbacka, I. Christianson and B. Isomaa, *Toxicol. Vitro*, 14 (2000) 61.

31 T. Wayusuanwit, and K. Hiroshi, *Fish. Sci.*, 61 (1995) 305.

32 N. Kakiichi, T. Yamamoto, S. Kamata, H. Otsuka and K. Uchida, *Anim. Sci. Technol.*, 64 (1993) 1013.

33 E.J. Routledge and J.P. Sumpter, *Environ. Toxicol. Chem.*, 15 (1996) 241.

34 P.G. Falkowsky, *Primary productivity in the sea*, Plenum press, New York and London, 1980, p. 517.

35 I. Moreno-Garrido, L.M. Lubián and A.M.V.M. Soares, *Bull. Environ. Cont. Toxicol.*, 62 (1999) 776.

36 T.V. Parshikova and Negrutskiy, UDC 574.64:582 (1989) 23.

37 E. Argese, A. Marcomini, P. Miana, C. Bettiol and G. Peri, *Environ. Toxicol. Chem.*, 13 (1994) 737.

38 M.A. Bragadin, G. Perin, S. Raccanelli and S. Manente, *Environ. Toxicol. Chem.*, 15 (1996) 1749.

39 P. Radix, M. Léonard, C. Papantoniou, G. Roman, E. Saouter, S. Gallotti-Schmitt, H. Thiëbaud and P. Vasseur, *Ecotoxicol. Environ. Saf.*, 47 (2000) 186.

40 A. Weyers, B. Sokull-Klüttgen, J. Baraibar-Fentanes and G. Vollmer, *Environ. Toxicol. Chem.*, 19 (2000) 1931.

41 S.Y. Maestrini, M.R. Droop and D.J. Bonin, *Test Algae as Indicators of Sea Water Quality: Prospects*, Algae as Ecological Indicators, Academic Press, 1984.

42 L.N. Britton, *J. Surf. Det.*, 1 (1998) 36.

43 I. Moreno-Garrido, L.M. Lubián and A.M.V.M. Soares, *Ecotoxicol. Environ. Saf.*, 47 (2000) 112.

44 OECD guideline for testing of chemicals: alga, growth inhibition test, 1998.

45 S. Maeda, T. Sakaguchi and Akatsuka (Eds.), *Introduction to Applied Phycology*, Vol. I, SPB Academic Publishing, The Hage, The Netherlands, 1998, p. 109.

46 V.M. Bagnyuk, T.L. Oleynik, N.R. L'vovskaya and L.O. Eynor, *Hydrobiol. J.*, 12 (1976) 35.

47 A.S. Dhaliwal, A. Campione and S. Smaga, *J. Phycol.*, 13 (1977) 18.

48 H. Streble and D. Krauter, *Atlas de los microorganismos de agua dulce*, Ediciones Omega S.A., Barcelona, 1987.

49 M.R. Servos, *Water Qual. Res. J. Can.*, 34 (1999) 123.

50 M.A. Lewis, *Water Res.*, 26 (1992) 1013.

51 APHA-AWWA-WPCF, Clesceri, Greenberg and Trussel (Eds.), *Standard Methods for the Examination of Water and Wastewater*, 17th Edn., 1989.

52 EEC Directive 92/32/EEC (1992/04/30) Seventh amendment of Directive 67/548/EEC. C.3 Microalgae growth inhibition test.

53 EEC 1992 Directive 92/32/EEC (1992/04/30) Seventh amendment of Directive 67/548/EEC. C.3 Microalgae growth inhibition test.

54 EPA Short-term methods for estimation the chronic toxicity of effluents and receiving waters to freshwater organisms. (EPA/600/4-85/014), 1985.

55 I. Moreno-Garrido, L.M. Lubián and A.M.V.M. Soares, *Bull. Environ. Cont. Toxicol.*, 63 (1999) 392.

56 G.E. Walsh and P. Calow (Eds.), *Handbook of Toxicology*, Blackwell Scientific Publications, Oxford, 1993, p. 119.

57 J.R. Reinfelder, R.E. Jablonka and M. Cheney, *Environ. Toxicol. Chem.*, 19 (2000) 448.

58 B.R. Berland, D.J. Bonin, O.J. Guérin-Ancey, V.I. Kapkov and D.P. Arlhac, *Mar. Biol.*, 42 (1977) 17.

59 A. Metaxas and A.G. Lewis, *Aquat. Toxicol.*, 19 (1991) 265.

60 N.R. Rao and G.M. Latheef, *Comp. Physiol. Ecol.*, 14 (1989) 41.

61 R.B. Rivkin, *Mar. Biol.*, 50 (1979) 239.

62 M. Ricard, Atlas du phytoplancton marin, Vol. 2, Editions du CNRS, París, 1987.

63 I. Moreno-Garrido, M. Hampel, L.M. Lubián and J. Blasco, *Fresenius J. Anal. Chem.*, 371 (2001) 474.

64 W.S. Hichcock and D.F. Martin, *Bull. Environ. Contam. Toxicol.*, 18 (1977) 291.

65 M. Bressan, R. Brunetti, S. Casellato, G.C. Fava, P. Giro, M. Marin, P. Negrisolo, L. Tallandini, S. Thomann, L. Tosoni and M. Turchetto, *Tenside Surf. Det.*, 26 (1989) 148.

66 ASTM (American Standard for Testing and Materials), Standard specification for Substitute Ocean Water. Designation D 1141-75, 1975.

67 I. Moreno-Garrido, L.M. Lubián and J. Blasco, SETAC Congress, Brighton, UK, 2001.

68 R. Establier, J. Blasco, L.M. Lubián and A. Gómez-Parra, *Scient. Mar.*, 54 (1990) 203.

69 M. Hampel, I. Moreno-Garrido, C. Sobrino, L.M. Lubián and J. Blasco, *Ecotoxicol. Environ. Saf.*, 48 (2001) 287.

70 R. Ukeles, *J. Phycol.*, 1 (1965) 102.

71 Z. Duvjnak, D.G. Cooper and N. Kosaric, *Biotechnol. Bioeng.*, 1 (1981) 165.

72 G.M. Rand, P.G. Wells and L.S. McCarty, In: G.M. Rand (Ed.), *Fundamentals of Aquatic Toxicology*, Effects, Environmental Fate and Risk Assessment, Taylor & Francis, 1995, p. 3.

73   E. Kamler, *Early Life History of Fish. An Energetics Approach*, Fish and Fisheries Series, Vol. 4, Chapman & Hall, 1992, p. 261.

74   A. Stene and S. Lönning, *Sarsia*, 69 (1984) 199.

75   M. Swedmark, B. Braaten, E. Emanuelsson and A. Granmo, *Mar. Biol.*, 9 (1971) 183.

76   A. Granmo, *Mar. Biol.*, 15 (1972) 356.

77   M. Rosety, F.J. Ordóñez, M. Rosety-Rodriguez, J.M. Rosety, I. Rosety, C. Carrasco and A. Ribelles, *Histol. Histopathol.*, 16 (2001) 839.

78   R. Marchetti, *Ann. Appl. Biol.*, 55 (1965) 425.

79   W.W. Kühnhold, In: M. Ruivo (Ed.), *Marine Pollution and Sea Life*, Fishing News (Books), London, 1972, p. 315.

80   J. Davenport, S. Lönning and L.J. Saethre, *J. Fish. Biol.*, 12 (1979) 31.

81   H. von Westernhagen, H. Rosenthal and K.R. Sperling, *Helgol. Wiss. Meeresunters.*, 26 (1974) 16.

82   J.H.S. Blaxter, In: W.S. Hoar and D.J. Randall (Eds.), *Fish Physiology*, Vol. XIA, Academic Press, San Diego, 1988, p. 1.

83   S.A. Villalobos, J.T. Hamm, S.J. Teh and D.E. Hinton, *Aquat. Toxicol.*, 48 (2000) 309.

84   K. Yamagami, In: W.S. Hoar and D.J. Randall (Eds.), *Fish Physiology*, The Physiology of Developing Fish, Part A: Eggs and Larvae, Vol. XI, Academic Press, San Diego, 1988, p. 447.

85   S.I. Doroshov, *J. Ichthyol.*, 10 (1970) 235.

86   P.D. Guiney and J.J. Lech, *Toxicol. Appl. Pharmacol.*, 53 (1980) 521.

87   S. Galassi, D. Calamari and F. Setti, *Ecotoxicol. Environ. Saf.*, 6 (1982) 439.

88   C. Van Leeuwen, P.S. Griffioen, W.H.A. Vergouw and J.L. Maas-Diepeveen, *Aquat. Toxicol.*, 7 (1985) 59.

89   G. Görge and R. Nagel, *Chemosphere*, 21 (1990) 1125.

90   S. Korn and S. Rice, *Rapp. P.V. Reun. Cons. Int. Explor. Mer.*, 23 (1981) 22.

91   G.I. Petersen, PhD Thesis, Comparative studies of toxic effects and bioaccumulation of lipophilic organic substances in fish early life stages. University of Copenhagen, 1997, Denmark.

92   H. Rosenthal and D.F. Alderice, *J. Fish. Res. Board Can.*, 33 (1976) 2047.

93   OECD Draft Guideline: Fish, acute toxicity test on egg and sac-fry stages. OECD Guideline for Testing of Chemicals. Draft, 1992.

94   EPA: Ecological Effects Test Guidelines, OPPTS 850.1400: Fish Early-Life Stage Toxicity Test, 1996.

95   J.M. McKim, *J. Fish. Res. Board Can.*, 34 (1977) 1148.

96   C. Sarasquete, In: C. Sarasquete and M.L. González-Canales (Eds.), *Histopatología y Biotoxicología en Acuicultura*, 1997, p. 11.

97   J.V. Helvik, PhD Thesis Biology of hatching and control of hatching in eggs of halibut (*Hippoglossus hippoglossus*). University of Bergen, 1991, Norway.

98   M. Hampel, J.B. Ortiz, C. Sarasquete, A. Moreno and J. Blasco, *CESIO, 5th World Surfactant Congress, Firenze*, (2000) 1617.

99   M. Hampel, I. Moreno-Garrido, C. Sobrino, L.M. Lubián and J. Blasco, *Ecotoxicol. Environ. Saf.*, 48 (2001) 287.

100   A.W. Maki, *J. Fish. Res. Bd Can.*, 36 (1979) 11.

101   R.A. Kimerle and R.D. Swisher, *Water Res.*, 11 (1977) 31.

102   K.J. Macek and B.H. Sleight, *Aquatic Toxicity Hazard Eval. ASTM STP*, (1977) 634.

103   W.F. Holman and K.J. Macek, *Trans. Am. Fish. Soc.*, 109 (1980) 122.

104   A.J. Bailer and J.T. Oris, *Environ. Toxicol. Chem.*, 16 (1997) 1554.

105   N. Scholz, *CLER Rev.*, 4 (1998) 62.

106   ECETOC, Aquatic Toxicity Data Evaluation. Technical Report No. 56, Brussel, 1993.

107   ECETOC, The Value of Model Ecosystem Studies in Ecotoxicology. Technical Report, Brussel, 1997.

108   Technical Guidance Documents in Support of the Comisión Directive 93/67 EEC on Risk Assessment for New Notified Substances and the Comisión Regulation (EC) 1488/94 on Risk Assessment for Existing Substances, 1996.

109   E. González-Mazo, J.M. Forja and A. Gómez-Parra, *Environ. Sci. Technol.*, 32 (1998) 1636.

110   H. Takada, R. Ishiwatari and N. Ogura, *Estuar. Coast. Shelf Sci.*, 35 (1992) 257.

111   A. Marcomini, S. Stelluto and B. Pavoni, *J. Environ. Anal. Chem.*, 35 (1989) 207.

112   S. Terzic and M. Ahel, *Bull. Environ. Contam. Toxicol.*, 50 (1993) 241.

113   M. Stalmans, E. Matthijs and N.T. Oude, *North Sea Pollution: Technical Strategies For Improvement*, Amsterdam, September, 1990, pp. 10–14.

114   J.A. Hoekstra and P.H. Ewijk, *Environ. Toxicol. Chem.*, 12 (1993) 187.

115   Canadian Environmental Protection Act Priority substances list, Assessment report nonylphenol and its ethoxylates, 2000, pp. 34–48.

116   S.L. O'Halloran, K. Liber, J.A. Ganga and M.L. Knuth, *Environ. Toxicol. Chem.*, 18 (1999) 376.

117   K. Liber, J.A. Gangl, T. Corry, L.J. Heinis and F.S. Stay, *Environ. Toxicol. Chem.*, 18 (1999) 394.

118   BKH Consulting Engineers, Environmental data review of alcohol ethoxylates (AE). Final Report, The Dutch Soap Association, Delft, The Netherlands, 1994.

119   E.D. Kline, R.A. Figueroa, J.H. Rodgers and B.D. Phillip, *Environ. Toxicol. Chem.*, 15 (1996) 997.

120   H. Lal, V. Misra, P.N. Viswanathan and C.R. Murti, *Ecotoxicol. Environ. Saf.*, 8 (1984) 447.

121 E. Jorgensen and K. Christoffersen, *Environ. Toxicol. Chem.*, 19 (2000) 904.

122 B. Hansen, F.L. Fotel, N.J. Jensen and L. Wittrup, *Mar. Biol.*, 128 (1997) 627.

123 Y. Takamatsu, Y. Inamori, H. Nishimae, T. Ebisuno, R. Sudo and M. Matsumura, *Water Sci. Technol.*, 368 (1997) 207.

124 S.T. Dyer, D.T. Stanton, J.R. Lauth and D.S. Cherry, *Environ. Toxicol. Chem.*, 19 (2000) 608.

125 M.A. Lewis, and B.G. Hamm, *Water Res.*, 20 (1986) 1575.

126 K. Leeuwen, C. Roghair, T. Nijs and J. Greef, *Chemosphere*, 24 (1992) 629.

127 C.A. Pittinger, D.A. Woltering and J.A. Masters, *Environ. Toxicol. Chem.*, 8 (1989) 1023.

128 A. Viarengo, *Mar. Pollut. Bull.*, 16 (1985) 153.

129 L. Da Ros, C. Nasci, G. Campesan, P. Sartorello, G. Stocco and A. Menetto, *Mar. Environ. Res.*, 39 (1995) 321.

130 B.N. Gupta, A.K. Mathur, C. Agarwal and A. Singh, *Bull. Environ. Contam. Toxicol.*, 42 (1989) 375.

131 V. Misra, V. Kumar, S. Dar Pandey and P.N. Viswanathan, *Arch. Environ. Contam. Toxicol.*, 21 (1991) 514.

132 J. Blasco, E. González-Mazo and C. Sarasquete, *Toxicol. Environ. Chem.*, 71 (1999) 447.

133 M. Bragadin, G. Perin, S. Raccanelli and S. Manente, *Environ. Toxicol. Chem.*, 15 (1996) 1749.

134 G. Zaccone, P. Lo Cascio, S. Fasulo and A. Licata, *Histochem. J.*, 7 (1985) 453.

135 D. Roy, *Ecotoxicol. Environ. Saf.*, 15 (1988) 260.

136 Petresa, Internal Report, 2000.

7.2 BIOCONCENTRATION

Mónica Sáez, Pim de Voogt and Eduardo González-Mazo

7.2.1 Introduction

In the 1960s, organic residues (e.g. DDT, PCBs, methyl mercury) began to be detected in several species of shellfish, fish and fish-eating birds [1,2]. Since then, assessment of the bioaccumulation of chemicals has been considered decisive for determining the potential hazard and environmental risk; it is regulated by various official organisations such as the OECD [3], EPA [4,5] and ASTM [6].

Bioconcentration is the process by which an organism accumulates a chemical directly from the surrounding medium, as a result of processes of uptake and elimination. Bioaccumulation is a more general term, which considers the uptake of a chemical directly from the medium (bioconcentration), as well as by other routes (like ingestion of food). When the chemical passes up the food chain and the concentration in tissue increases from one trophic level to the next, biomagnification is said to happen.

Bioconcentration may occur as a result of continuous exposure to a certain compound, even at concentrations sufficiently low for no toxic effect to be observed. This process is very relevant in the case of surfactants, due with their affinity for interfaces, since they tend to associate with biological membranes (i.e. the interfaces between organisms and their medium).

7.2.2 Uptake

Uptake is the process by which chemicals (either dissolved in water or sorbed onto sediment and/or suspended solids) are transferred into and onto an organism. For surfactants, this generally occurs in a series of steps: a rapid initial step controlled by sorption, where the surface phenomenon is especially relevant; then a diffusion step, when the chemical crosses biological barriers, and later steps when it is transported and distributed among the tissues and organs.

Sorption is a physical process in which a chemical binds by electrostatic, covalent or molecular forces to surfaces like biological

T.P. Knepper, D. Barceló and P. de Voogt (Eds)

membranes. This process is an essential parameter for predicting and understanding the environmental behavior of surfactants [7], and is especially important for micro-organisms, as a result of their extremely high surface-to-volume ratio. Equally effective sorption in live and dead microalgae has been observed for DDT [1], and for $C_{11}LAS$ [8], with LAS sorption coefficients in live and dead *Dunaliella salina* of 810 ± 26 and 700 ± 42 L kg^{-1}, respectively. The sorption equilibrium of LAS and SPCs in microalgae was reached in 4 h and the sorption values increased with hydrophobicity [8]. The sorption variation at different concentrations was fitted to a Freundlich isotherm, giving values of $K = 176 \pm 0.02$, and $n = 0.964 \pm 0.02$ for $C_{11}LAS$ on *D. salina* [8], in close agreement with the data reported for marine sediment at similar LAS concentrations ($K = 142 \pm 59$, and $n = 1.38 \pm 0.29$ [9]). Increasing values of K ranging from 28.5 to 323.7 with slopes (n) from 1.001 to 0.8624 have been established for algae pulp of *Gracilaria debilis* exposed to lauryl benzene sulphonate [10]. This high LAS sorption capacity of the algae could be useful for removing LAS in a pre-treatment plant, offering a possible use for the algae pulp waste produced in the agar−agar extraction process [10].

Passive diffusion takes places through semi-permeable membranes if a concentration gradient exists. In general, surfactants are likely to accumulate primarily in gills, due to their enormous area of exposure. The highest values have been observed for anionic surfactants [11−13]; and the lowest for cationic, such as hexadecylpyridinium bromide (HPB), with less sorption capacity [11] and concentration ratios of 165, 130 and 173 L kg^{-1} in clams, fish and tadpoles, respectively [14]. Uptake of anionic (LAS) and non-ionic (NP, OP) surfactants through the gills appears to be very fast, and more important than through skin [13,15], but similar OP values have been observed in fish skin and gills: 259 and 187 L kg^{-1}, respectively [16]. The intestine presents a large area of exposure, but cationic surfactants are not readily absorbed through it [17].

The distribution of a chemical in tissue depends on the binding/partitioning between circulatory blood and tissues, the transfer across biological membranes and the tissue-blood perfusion. After incorporation, contaminants are distributed from blood to high perfusion tissues (e.g. liver, kidney), then to low perfusion ones (skin, muscle) and finally to lipoidal tissues [2], establishing kinetically different compartments (internal organs > liver > head ≈ skin) and different times to equilibrium [18]. Surfactants (LAS, AS, AES, AEO and APEO) have been mainly detected and quantified in gills [12,13,19−23], bile [16,22−26], gall bladder [19,23,24,27,28,30], urine [26,29], liver [16,20,22−26] or hepatopancreas [19,21,30] and kidney [12,22−24]; and to a lesser extent in faeces [16,22], pyloric caecae [16,22], fat [24], stomach [20], intestine [16,22], skin [22,23], brain [12,20,22], spleen [12,22], muscle [20,22−24,26], blood [22,26], heart [12,20,22,24], gonad [12], eye [12] and swim bladder [12]. At the beginning of the uptake, the distribution pattern is gills > hepatopancreas > gall bladder but, as time increases, the concentration in the gall bladder (target tissue) increases by up to two orders of magnitude more than in the whole body [12].

Uptake of surfactants by aquatic organisms occurs rapidly [31,32], with uptake rates (k_1) ranging from 0.4 to 240 L kg^{-1} per day for cationics; from 2 to 500 L kg^{-1} per day for anionics and from 1 to 1660 L kg^{-1} per day for non-ionics. Although wide variations are observed within a group and within the same compound, a general tendency of increase in the uptake with increasing alkyl chain length and decreasing ethoxylate units has been observed [33], leading to the conclusion that uptake is related to hydrophobicity.

7.2.3 Elimination

The elimination rate of a compound (directly or by biotransformation) from an organism determines the extent of the bioconcentration and depends both on the chemical and the organism. Direct elimination includes transport across the skin or respiratory surfaces, secretion in gall bladder bile, and excretion from the kidney in urine. Other processes are moulting (for arthropods), egg deposition (fish, inverte-brates) and transfer to offspring or via lactation (in mammals), which are more specific and not usually contemplated in bioconcentration determination.

Passive elimination of surfactants may occur across the skin and gills by the same partitioning process involved in uptake. This process appears to be more important for non-polar compounds that are not rapidly biotransformed and depends on their affinity for the organism or medium.

Surfactant elimination happens mainly through biotransformation, denoted by the identification of several metabolites. These have been found in liver and to a greater extent in the gall bladder, where they are transported and subsequently discharged with bile. They have also been detected in kidney and faeces, since a significant portion may be reabsorbed in the intestine, returned to the blood, and excreted in the urine [12,16,22–24,26].

The presence of LAS metabolites has been detected in algae [8], in the gall bladder of carp [19,30,34], blue gill [28], fathead minnows [27], in the hepatopancreas of carp [19], and in the urine of catfish [35]. SPCs have been detected and quantified by HPLC–MS in the marine clams *Ruditapes semidecussatus* and *Cerastoderma edule*, and in the crabs *Carcinus maenas* and *Liocarcinus vernalis* [21]. Other authors have also detected possible LAS metabolites in goldfish [12], and in fathead minnows [36], where only C_4SPC has been quantified. In an experiment of $C_{11}LAS$ bioconcentration on *R. semidecussatus* [37], SPCs with alkyl chain lengths ranging from C_4 to C_{11} have been detected and quantified during the depuration phase (see Table 7.2.1). In the first hours SPC with long alkyl chains were detected, the shorter chain homologues remaining at the end of the depuration phase. C_{11} sulphate metabolites, mainly 2-methyl butyric acid-1 sulphate and to a lesser extent the C_7 and C_9 derivatives, have also been presumed to appear in goldfish bile [12]. In the case of alcohol ethoxylates, the appearance of the 3-ethoxy sulphate of glycolic acid or γ-hydroxy butyric acid (from $C_{12}EO_3$ sulphate), the 4-hexethoxy-butyric acid (from $C_{12}EO_6$), and the decaethoxy oligomer (from $C_{12}EO_{10}$) has been suspected [12]. Although NP and OP are already surfactant metabolites, their metabolic by-products, oxidative and glucuronic acid conjugates [16,22–24,26,29,38], and hydroxylates [23,25,26,29] have been detected in fish bile and urine and also observed in snail and weeds [25]. Recently NPEC have been described as NPEO metabolic by-products [39], and several oligomers have been described and quantified in water, from both laboratory (NPE_1C to NPE_7C) and field (NPE_1C to NPE_5C) bioconcentration

TABLE 7.2.1

Average whole body SPC concentrations (ng g^{-1}) in the marine clam *R. semidecussatus* during depuration after 5 days exposition to different $C_{11}LAS$ concentrations under flow-through conditions [21]

Depuration day	C_4SPC	C_5SPC	C_6SPC	C_7SPC	C_8SPC	C_9SPC	$C_{10}SPC$	$C_{11}SPC$
0	0	166	314	3	72	0	15	18
1	0	72	52	1	12	0	0	0
2	14	14	–	–	7	17	0	0

experiments with clams, but they have not been detected in the organisms [21].

Surfactants elimination via bile is faster in fed than in unfed fish [12,40], due to the higher secretion of bile into the digestive tract when fed. The fastest and slowest surfactants excreted were the linear alkylbenzene sulphonates (C_{12}) and the alkyl sulphates (especially the C_{12-15}) [12], respectively. Depuration of NP is quite rapid, 1.8–20 days in fathead minnows [32] and 4 days in the Atlantic salmon [41]. Within the same organism, differences in tissues were observed, with values of 20 h for adipose tissue and 18.5 h for muscle of rainbow trout [24]. The elimination rate of non-ionic surfactants is directly proportional to the ethoxylate unit number and inversely proportional to the alkyl chain length [33]. Average elimination rates (k_2) of $NPEO_{2.8}$ and NP in clams were 1.8 and 1.4 per day, respectively [21], and decreasing k_2 values from 0.19 to 0.001 per day were obtained from butylphenol to dodecylphenol in salmon [42].

7.2.4 Bioconcentration factor determination

The determination of the bioconcentration factor (BCF) can be performed in two different ways: computationally with quantitative structure activity relationship (QSAR) methods, or from experimental measurements [2]. The QSAR methods estimate BCF from the structural or physicochemical properties of the compound, whereas the experimental methods use measured values of uptake and elimination rate constants or concentrations in the steady state.

Computational methods
Estimation is easier and less time-consuming because use is made of empirical relationships between the BCF and physicochemical properties of the compound, such as water solubility (S) [42–48], K_{oc} (solid organic carbon/water partition coefficient) [48], K_{mw} (membrane water partition coefficient), K_{lipw} (liposome water partition coefficient) [49], critical micelle concentration (CMC) [45], steric factors, molecular weight [47,48], and others. The most common regression method is the estimation of BCF from the octanol–water partition coefficient (K_{ow}) [18,42,44–48,50,51].

Log BCF $= a \log K_{ow} + b$

Different values of the parameters a and b have been established for several organisms: for daphnids [44], molluscs [44], and fish [46] log BCF values for LAS of 0.47, 0.49 and 0.70, respectively, were obtained from its log K_{ow} (2.02). With these relationships, the highest BCF values are observed for chemicals with low water solubility and high K_{ow}, but these consider the chemical in its original, not in its ionised form, and do not include its metabolites. Bioconcentration is linearly related to log K_{ow} up to about log $K_{ow} = 6$ or 7, and for superlipophilic compounds (log $K_{ow} > 7$) tends to decrease further with increasing log K_{ow}, being not determined exclusively by that coefficient [45,49], but thermo-dynamically [52]. A more accurate method for estimating BCF also considers the molecular structure [53], dividing compounds into non-ionic, aromatic azo, ionic, and tin and mercury groups. Depending on the K_{ow} value, different equations, even within the same group, have been established. For some groups (like the aromatic azo compounds, the ionics with more than C_{11} alkyl, and the tin and mercury compounds), a correction is applied depending on the presence of specific moiety in their molecular structure. Physicochemical properties of the environment (e.g. pH, pK_a, temperature, salinity) [47], as well as the presence of dissolved organic matter (DOM) [54], should be included in the estimation of the BCF, since these also have influence. Equations have also been developed to estimate the uptake and elimination rate constants [13,44,45,49,55].

Experimental methods
The BCF describes the degree of bioconcentration as the ratio between the chemical concentration in the organism or a specific tissue and the chemical concentration in the medium, in the steady state [1,18,52,55,56] or over a discrete exposure time period [44,45,57]. The steady state is reached when the ratio of concentrations in the organism and in the water no larger vary with time.

The first-order one-compartment model [43,58], which considers the organism as one homogeneous compartment surrounded by a homogeneous medium, provides an acceptable estimation of surfactants bioconcentration, and has been adopted by the OECD and EPA in their guidelines [3–5]. The BCF can be determined as a ratio of the concentrations of the chemical in the organism (C_o) and the medium (C_w) under equilibrium state or as a ratio of the uptake and elimination constants (k_1 and k_2, respectively).

874

Laboratory assays are generally divided into continuous and discontinuous methods [59], with the flow-through bioassays being the most suitable type for testing the bioconcentration of biodegradable compounds, as recommended in the OECD [3], EPA [4,5] and ASTM [6] guidelines, since the main objective is to attain a stable concentration under continuous input conditions. Bioconcentration experiments under flow-through conditions have been performed for LAS [13,21,30,36,59,60], AS [30,62], APEO [15,16,21,53,61] and AEO [21,33,62]. Frequently a pre-test based on an estimated BCF is performed in order to determine the time required to reach the steady state and thus to establish the duration of the definitive test phases. LAS accumulation by clam [21] and by fish [57] reached the steady state at 48 h, as shown in Fig. 7.2.1.

Static experiments have been performed for radiolabeled surfactants [42,63], with microorganisms [8] or contaminated sediment exposure [64], since continuous assays are difficult, expensive (large amount of radiolabeled chemical needed) and restrictive (due to safety regulations) to perform. To avoid surfactant depletion in water, the exposure times are shorter, 15 h [26] or 24 h [12,24], but are valid for determining the BCF when the steady state is reached [8], and the actual exposure concentration (not the nominal) is used.

Intermediate condition bioassays are also found in the literature. Semi-static [2] experiments have been performed for SDS (C_{12}-AS), with renewal of the water every 2 days [65] and intermittent flow-through [32], in which fathead minnows were exposed to NP for 20 days.

Species

Surfactant bioconcentration has been determined in several species, corresponding mainly to aquatic organisms. In the group of freshwater algae and aquatic plants, *Lemna minor* [25], *Cladophora glomerata* [20], *Fontinalis antipyretica* [20] and *Potamogeton crispus* [20] have been tested, whereas for marine microalgae only *Nannochloropsis gaditana* [8] and *D. salina* [8] have been reported. Considerable data exists for invertebrates, but referring to only a few species, mainly *Daphnia sp.* [27], *Mytilus edulis* [63,69], and, to a lesser extent, *Corbicula fluminea* [14], *Ampelisca abdita* [67], *Lymnea stagnalis* [25], and *R. semidecussatus* [21]. The group of freshwater fish is the most extensively studied, with *Pimephales promelas* [13,14,27,33,57,59,62],

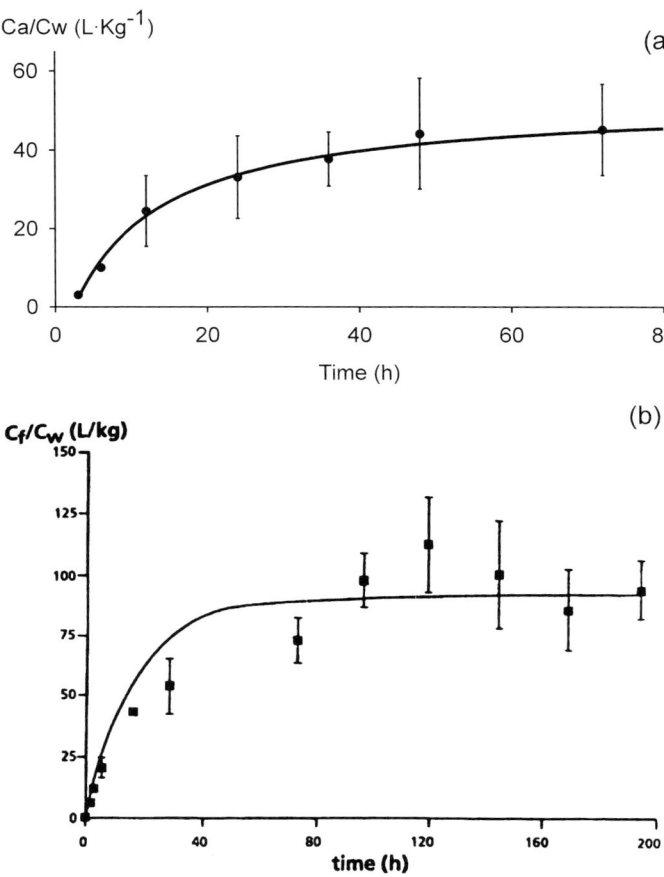

Fig. 7.2.1. LAS uptake kinetic curve for different organisms; (a) C_{11} homologue by the marine clam *R. semidecussatus* (taken from Ref. [21]); (b) C_{12} LAS by the freshwater fish *P. promelas* (taken from Ref. [57]).

Oncorhynchus mykiss [16,22–25,29], *Lepomis macrochirus* [28,31,66, 67], and *Cyprinus carpio* [19,30,34,60] being the species most frequently studied; others include *Salmo salar* [26,41], *Proterorhinus marmoratus* [65], *Carassius auratus* [12] and *Oryzias latipes* [15].

Experiments with various species simultaneously carried out in ponds, streams or mesocosms have been reported for non-ionic, anionic and cationic surfactants. NP bioconcentration has been studied in the shrimp *Crangon crangon*, the mussel *M. edulis* and the fish

Gasterosteus aculeatus [51]; LAS in the clam *C. fluminea*, the amphipod *Hyalella azteca*, snails *Elimia sp.*, and fish *L. macrochirus, P. promelas, Ictalurus punctatus* [68]; and HPB in the clam *C. fluminea*, the fish *P. promelas*, and the tadpole *Rana catesbeiana* [17].

Although the marine environment is the final compartment receiving surfactants, few reports have considered seawater species. To the best of our knowledge, surfactants bioconcentration has been determined in marine algae [8], shrimp [51], mussel [51,66,69] and the stickleback [51]. Although experiments are increasingly being focused on the marine and coastal environments, further work is necessary to measure bioconcentration in these ecosystems.

7.2.5 BCF values

The surfactant bioconcentration data available in the literature show considerable variability, due mainly to the different compounds, species, environmental characteristics and analytical procedures used to determine the BCF. Physicochemical properties of surfactants, such as molecular structure, molecular weight, partitioning coefficients (K_{ow}, K_{oc}), water solubility and sorption rate constants all influence their BCF [47].

The length of the alkyl chain, the number of ethoxylates and the extent of branching are factors that explain the variation found in data for the same compound. In general, bioconcentration increases with increasing alkyl chain length [8,32,33,70], as shown in Table 7.2.2 for the different LAS homologues in several fish species. Bioconcentration values for C_{10} to C_{13} increased from 2–6 to 34–987 for *P. promelas* and from 1 to 372 for *O. mykiss*. For non-ionics, an increase of the BCF is also observed with increasing alkyl chain length [32,33]: from 13 for C_{12}-EO_8, to 32 for the C_{13}, 96 for the C_{14}, and 388 for C_{16}-EO_8 in *P. promelas*. However, the BCF decreases with the number of ethoxylate units: in *P. promelas*, the BCF for C_{14}-EO_x was 237, 96, 16 and less than 5 (where x is 4, 8, 11 and 14 EO units, respectively).

Uptake occurs from the bioavailable fraction, which in almost all cases corresponds to the dissolved fraction. Sorption and binding to suspended solids, sediments, and DOM have a great effect on bioavailability [71,72]; therefore the more hydrophobic surfactants tend to be less bioavailable. Thus, for the same initial concentration, the bioavailable fraction of C_{12}LAS, compared with that of the C_{11}

TABLE 7.2.2

Isomers and homologue variability in experimental BCFs of surfactants in two freshwater fish species, *P. promelas* and *O. mykiss*

Species	Compound	BCF (L kg^{-1})	Reference
P. promelas	HPB	22	[14]
	C_{10}-LAS	2–6	[57]
	C_{11}-LAS	6–32	[57]
	$C_{11.6}$-LAS	269	[57]
	C_{12}-LAS	30–212	[57]
		21–204	[68]
	C_{13}-LAS	34–987	[57]
	C_{12}-EO$_8$	13	[33]
	C_{13}-EO$_4$	233	[33]
	C_{13}-EO$_8$	32	[33]
		31	[62]
	C_{14}-EO$_4$	237	[33]
	C_{14}-EO$_8$	96	[33]
	C_{14}-EO$_{11}$	16	[33]
	C_{14}-EO$_{14}$	<5	[33]
	C_{16}-EO$_8$	388	[33]
	NP	246–434	[61]
O. mykiss	C_{10}-LAS	1	[13]
	C_{11}-LAS	6	[13]
	C_{12}-LAS	6–82	[13]
	C_{13}-LAS	372	[13]
	NP	24–110	[24]

homologue, has been found to be smaller [8]. A reduction of LAS bioavailability in sediments for different benthic invertebrates, such as worms, midge, mussels and copepods [73], has been reported. The surfactant's hydrophobicity will also determine its affinity to lipids, but its structural features will be relevant in determining the kinetics of the linking [2].

Differences are found among the species tested, due mainly to their sensitivity, life stage, feeding and living behavior. Thus deposit-feeding clams may incorporate sediment-associated contaminants but filter-feeding clams may not [2]. Benthic organisms (filtering or bottom-dwelling species) concentrate surfactants to a higher degree, which would be expected considering that sediment concentration can exceed the water concentration by several orders of magnitude [2]. Likewise the highest BCF values of surfactants correspond to bivalves (cf. Table 7.2.3), with values of up to 2700 for NP in *M. edulis* and 26 742 in

TABLE 7.2.3

Interspecies variability in experimental BCFs of two surfactants

Compound	Species	BCF (L kg^{-1})	Reference
C$_{12}$-LAS	P. promelas	30–212	[57]
	P. promelas	21–204	[68]
	O. mykiss	6–82	[13]
	I. punctatus	3–223	[68]
	C. fluminea	3–61	[68]
	R. semidecussatus	372–1118	[21]
	E. sp.	6–71	[68]
	H. azteca	15–146	[68]
	N. gaditana	274	[8]
	D. salina	710	[8]
NP	S. salar	280	[41]
	Oncorhymchus mykiss	24–110	[24]
	C. carpio	0.9–2.2	[80]
	L. macrochirus	2.5–262	[32]
	P. promelas	271–741	[80]
	P. promelas	246–434	[61]
	O. latipes	167	[15]
	G. aculeatus	1250	[51]
	M. edulis	10	[63]
	M. edulis	2700	[51]
	R. semidecussatus	11 838–26 742	[21]
	C. crangon	100	[51]

R. semidecussatus and 1118 for C$_{12}$LAS in *R. semidecussatus,* whereas in *C. fluminea*, bioconcentration of C$_{12}$LAS is about 16 times lower. The algae group also shows high BCF values (up to 710 L kg^{-1} for LAS), followed by fish (up to 232 and 1250 for LAS and NP, respectively), being higher than in amphipods for LAS (up to 146) or in decapods for NP (up to 100) and in gastropods for LAS (71 L kg^{-1}). Significant bioconcentration differences have been observed within the same group, due to differences in species characteristics. The marine microalgae *N. gaditana*, with smaller size (higher surface/area ratio, for biomass normalised) and a cellular wall, which *D. salina* lacks, shows higher LAS and SPC sorption values [8]. Differences between tissues have been observed in the BCF for LAS, as shown in Table 7.2.4, where for carp values up to 1000 have been measured in the gall bladder, whereas in the hepatopancreas and gills, values were no higher than 97 and 13, respectively. For the bluegill, values of 5000 and 104 L kg^{-1} have been observed in the gall bladder and whole body, respectively [28], and for

TABLE 7.2.4

Whole body and tissue (gill and digestive gland) BCF ($L\,kg^{-1}$) for C_{12} LAS

Species	$BCF_{whole\ body}$	BCF_{gills}	BCF_{hp}	$BCF_{gall\ bladder}$	Reference	Comments
C. carpio	16–400	4–13	1.7–97	0.5–1000	[19]	A
L. macrochirus	104		30	9000	[30]	A, B
P. promelas	270–1200			5000	[28]	B
				20 000–70 000	[27]	B
R. semidecussatus	6–987	1.7–684	1.5–505[a]		[13]	
	351–1018	664–2330	1422–1969[b]		[21]	

hp, hepatopancreas; A, ^{35}S-radiolabelled; B, ^{14}C-radiolabelled.
[a]Correspond to liver.
[b]Correspond to the digestive gland.

the fathead minnow, values ranging from 20 000 to 70 000 L kg^{-1} in the gall bladder and from 270 to 1200 L kg^{-1} in its whole body [27] were found. However, these data are overestimates, since they were determined by non-LAS specific analytical methods and with radio-labeled (^{35}S or ^{14}C) compounds. BCF values determined properly in *P. promelas* [13], and *R. semidecussatus* [21] are in general lower, and although differences were found between tissues and the whole body to a lesser extent (about 2–4 times), these are explained by the fact that the gall bladder is the main tissue where biotransformation occurs.

The environmental physicochemical characteristics (temperature, dissolved oxygen content, salinity, pH, hardness, presence, size and composition of suspended material or sediment) influence bioconcentration, but interrelationships between different factors make it difficult to assign the effects. Some organisms show higher sensitivity in winter, but this is due to a reduction in their nutritional state, as reported for grass shrimp to the anionic dodecyl sulfate [72]; others increase their ventilation rate at low oxygen concentration in water and thus have a higher gills uptake [12]. Low pH favors anionic surfactant adsorption because protons act as a bridge between the surfactant and the microalgae surface, both negatively charged [10,74]. Thus an increase of the algae pulp adsorption at equilibrium from 0.5 to 0.7 mg g^{-1} has been reported for a three units pH decrease (from 6 to 3) [10]. In partition between solid and water, the nature and particle size of the solid phase influence the bioavailability of lipophilic chemicals. Tadpole intestine concentrations of HPB were about 7 times higher when exposed to contaminated sediment, compared with clean sediment, and in the presence of bentonite clay, this difference increased by up to 46 times [17]. The presence of artificial humic acids (as an approximation for DOM) significantly decreases the bioconcentration by the fish *P. promelas* of some cationic surfactants, such as dioctadecyl dimethyl ammonium chloride (DODMAC), C_8, C_{14}, C_{16}, C_{18} TMC (trimethyl ammonium chloride), but no effect was detected for the C_{12}TMC. For the anionic C_{12}LAS, an increase of approximately 50% occurred at low concentrations (about 2–6 mg C L^{-1}) but at DOC concentrations of 12–24 mg L^{-1} the effect is inverted with reductions up to -55% [75]. At low concentrations of water-soluble organic carbon (8–20 mg L^{-1}), a reduction of C_{12}LAS uptake by *P. promelas* was also observed but to a lower extent [76].

In the natural environment, organisms are exposed to mixtures of compounds. Therefore toxicological interactions such as competitive inhibition, antagonism or synergism may occur. Interactions between surfactants, and with pesticides or metals, have been reported in the literature. In respect of metals, LAS significantly increases zinc accumulation by the goby (*Proterorhinus marmoratus*), whereas AS does not [65]. LAS also reduces the accumulation of copper and lead by clams (*Ruditapes philippinarum*) [77], and of cadmium by freshwater trout (*Salmo gairdneri*) [78], as does $NPEO_{10}$, but no effect was observed for cadmium in mussels (*Mytilus* sp.) [79]. In the willow shiner (*Gnathopogon caerulescens*), DBS increases the bioconcentration of PCB and the nitrophenylethers: NIP and CNP (4'-nitrophenyl-2,4-dichlorophenyl ether and 4'-nitrophenol-2,4,6-trichlorophenyl ether, respectively) by a factor of about 1.5, but in pre-exposed fishes a reduction of the uptake was observed [74]. Synergies were also found in the bioconcentration of dieldrin in the bluegill sunfish (*L. macrochirus*) [75].

7.2.6 Conclusion

Bioconcentration is a relevant tool in the assessment of environmental risk and fate. The use of different analytical methods (not always specific), the use of mixtures instead of single chemical species, as well as differences in experimental conditions (continuous or static, exposure concentration and time to reach the steady state, if reached) explain the wide spread of existing surfactant BCF data, which therefore cannot always be extrapolated or compared. Only reliable BCF data have been collected in Tables 7.2.2 and 7.2.3, with values ranging from <5 to 400 for AEOs, from 1 to 1100 for LAS and 10 to 27 000 for NP. The wide range of the data is also due to species differences, showing the bivalves group the highest BCF values followed by microalgae, fish and invertebrates (amphipods, decapods and gastropods). In a similar way to the water-sediment partition coefficient, which is normalised to the carbon organic content, the BCF is normalised to the lipid content of the organisms (BCF_{LIP}) [31,33,53,65] and sometimes to both the lipid content and the organic carbon content (BSAF, the biota-sediment accumulation factor [67]). For *R. semidecussatus* [21], *Lepomis machochirus* [31] or *P. promelas* [33,62], normalisation has been performed to the neutral lipid content, but proper normalisation should be performed to the total lipids. Further efforts should be made to improve the experimental and analytical methodology in order to obtain reliable bioconcentration data. Restrictions or limitations of those values, or the methodology employed, should also be added. As for the final fate of surfactants, more research should focus on the marine environment, which has received relatively little attention to date.

REFERENCES

1   A. Spacie and J.L. Hamelink, Bioaccumulation, In: G.M. Rand and E.A. Petrocelli (Eds.), *Fundamentals of Aquatic Toxicology*, 2nd edn, Taylor & Francis, North Palm Beach, FL, 1995, p. 1052.
2   M.G. Barron, Bioaccumulation and bioconcentration in aquatic organisms, In: D.J. Hoffman, B.A. Rattner, G.A. Burton Jr. and J. Cairns Jr. (Eds.), *Handbook of Ecotoxicology*, Lewis Publishers/CRC Press Inc, Boca Raton, FL, 1995, p. 652.
3   OECD, Bioaccumulation fish test. OECD Guidelines for testing of chemicals. Draft guideline 305, OECD, Paris, 1982.
4   EPA, Fish BCF. OPPTS 850.1730. US EPA, Washington DC, 1996.
5   EPA, Oyster BCF. OPPTS 850.1710. US EPA, Washington DC, 1996.
6   ASTM, Standard practice for conducting bioconcentration tests with fishes and saltwater bivalve molluscs. E1022-84. ASTM, Philadelphia, PA, 1985.
7   M.J. Rosen, F. Li, S.W. Morrall and D.J. Versteeg, *Environ. Sci. Technol.*, 35 (2001) 954.
8   M. Sáez, A. Gómez-Parra and E. González-Mazo, *Fres. J. Anal. Chem.*, 371 (2001) 486.
9   J.A. Rubio, E. González-Mazo and A. Gómez-Parra, *Mar. Chem.*, 54 (1996) 171.
10 N.A. Fernandez, E. Chacin, E. Gutierrez, N. Alastre, B. Llamoza and C.F. Forster, *Bioresource Technol.*, 54 (1996) 111.
11 J. Tolls, P. Kloepper-Sams and D.T.H.M. Sijm, *Chemosphere*, 29 (1994) 693.
12 C.S. Newsome, D. Howes, S.J. Marshall and R.A. van Egmond, *Tenside Surfact. Det.*, 32 (1995) 498.
13 J. Tolls, M. Haller, W. Seinen and D.T.H.M. Sijm, *Environ. Sci. Technol.*, 34 (2000) 304.
14 J.P. Knezovich, M.P. Lawton and L.S. Inouye, *Bull. Environ. Contam. Toxicol.*, 42 (1989) 87.
15 T. Tsuda, A. Takino, K. Muraki, K. Harada and M. Kojima, *Water Res.*, 35 (2001) 1786.
16 A.M.R. Ferreira-Leach and E.M. Hill, *Mar. Environ. Res.*, 51 (2001) 75.
17 J.P. Knezovich and L.S. Inouye, *Ecotox. Environ. Saf.*, 26 (1993) 253.
18 M.G. Barron, B.D. Tarr and W.L. Hayton, *J. Fish Biol.*, 31 (1987) 735.

19   M. Kikuchi, M. Wakabayashi, H. Kojima and T. Yoshida, *Ecotox. Environ. Saf.*, 2 (1978) 115.

20   M. Ahel, J. McEvoy and W. Giger, *Environ. Pollut.*, 79 (1993) 243.

21   M. Sáez, Bioconcentración y toxicidad de tensioactivos sintéticos y sus intermedios de degradación sobre organismos marinos. PhD Thesis, University of Cádiz, 2002.

22   N.G. Coldham, S. Sivapathasundaram, M. Dave, L.A. Ashfield, T.G. Pottinger, C. Goodall and M.J. Sauer, *Drug Metab. Dispos.*, 26 (1998) 347.

23   R. Thibaut, L. Debrauwer, D. Rao and J.P. Cravedi, *Mar. Environ. Res.*, 46 (1998) 521.

24   S.K. Lewis and J.J. Lech, *Xenobiotica*, 26 (1996) 813.

25   R. Thibaut, A. Jumel, L. Debrauwer, E. Rathahao, L. Lagadic and J.P. Cravedi, *Analusis*, 28 (2000) 793.

26   A. Arukwe, R. Thibaut, K. Ingebrigsten, T. Celius, A. Goksøyr and J.P. Cravedi, *Aquatic Toxicol.*, 49 (2000) 289.

27   R.M. Comotto, R.A. Kimerle and R.D. Swisher, *Bioconcentration and metabolism of linear alkylbenzene sulfonate by daphnids and fathead minnow*, In: L.L. Marking and R.A. Kimerle (Eds.), *Aquatic Toxicology*, Vol. 667, American Society for Testing and Materials, ASTM STP, 1979, pp. 232–250.

28   R.A. Kimerle, K.J. Macek, B.H. Sleight and M.E. Burrows, *Water Res.*, 15 (1981) 251.

29   R. Thibaut, L. Debrauwer, D. Rao and J.P. Cravedi, *Sci. Total Environ.*, 233 (1999) 193.

30   M. Wakabayashi, M. Kikuchi, H. Kojima and T. Yoshida, *Chemosphere*, 11 (1978) 917.

31   K. Liber, J.A. Gangl, T.D. Corry, L.J. Heinis and F.S. Stay, *Environ. Toxicol. Chem.*, 18 (1999) 394.

32   M.R. Servos, *Water Qual. Res. J. Can.*, 34 (1999) 123.

33   J. Tolls, M. Haller, E. Labee, M. Verweij and D.T.H.M. Sijm, *Environ. Toxicol. Chem.*, 19 (2000) 646.

34   M. Wakabayashi, M. Kikuchi, A. Sato and T. Yoshida, *Bull. Jap. Soc. Fish.*, 47 (1981) 1383.

35   E.J. Schmidt and R.A. Kimerle, New design and use of a fish metabolism chamber, In: D.R. Brandon and K.L. Dickson (Eds.), *Aquatic Toxicology and Hazard Assessment*, Vol. 737., American Society for Testing and Materials, ASTM STP, 1981, p. 436.

36   J. Tolls, M.P. Lehmann and D.T.H.M. Sijm, *Environ. Toxicol. Chem.*, 19 (2000) 2394.

37   M. Sáez, V.M. León, A. Gómez-Parra and E. Gónzález-Mazo, *J. Chromatogr. A*, 889 (2000) 99.

38   A.C. Meldahl, K. Nithipaticom and J.J. Lech, *Xenobiotica*, 26 (1996) 1167.

39   N. Jonkers, T.P. Knepper and P. de Voogt, *Environ. Sci. Technol.*, 35 (2001) 335.

40   P.W.A. Tovell, D. Howes and C.S. Newsome, *Toxicology*, 4 (1975) 17.

41   D.W. McLeese, V. Zitko, D.B. Sergeant, L. Burridge and C.D. Metcalfe, *Chemosphere*, 10 (1981) 723.

42   M. van den Berg, D. van de Meent, W.J.G.M. Peijnenburg, D.T.H.M. Sijm, J. Struij and J.W. Tas, Transport, accumulation and transformation processes, In: C.J. van Leeuwen and J.L.M. Hermens (Eds.), *Risk Assessment of Chemicals: an Introduction*, Kluwer Academic Publishers, Dordrecht, 1995, p. 37.

43   D.W. Hawker and D.W. Connell, *Ecotox. Environ. Saf.*, 11 (1986) 184.

44   J. Tolls and D.T.H.M. Sijm, *Environ. Toxicol. Chem.*, 14 (1995) 1675.

45   D. Mackay, *Environ. Sci. Technol.*, 16 (1982) 274.

46   H. Geyer, P. Sheehan, D. Kotzias, D. Freitag and F. Korte, *Chemosphere*, 11 (1982) 1121.

47   W.J.A. Van den Heuvel, A.D. Forbis, B.A. Halley and C.C. Ku, *Environ. Toxicol. Chem.*, 15 (1996) 2263.

48   D.J.H. Phillips, Bioaccumulation, In: P. Calow (Ed.), *Handbook of Ecotoxicology*, Vol. I, Blackwell Scientific Publications, Oxford, England, 1993, p. 378.

49   M.T. Müller, A.J.B. Zehnder and B.I. Escher, *Environ. Toxicol. Chem.*, 18 (1999) 2191.

50   M.G. Barron, *Environ. Sci. Technol.*, 24 (1990) 1612.

51   R. Ekelund, Å. Bergman, Å. Granmo and M. Berggren, *Environ. Pollut.*, 64 (1990) 107.

52   H.A.J. Govers, H. Loonen and J.R. Parsons, *SAR QSAR Environ. Res.*, 5 (1996) 63.

53   W.M. Meylan, P.H. Howard, R.S. Boethling, D. Aronson, H. Printup and S. Gouchie, *Environ. Toxicol. Chem.*, 18 (1999) 664.

54   M. Haitzer, S. Höss, W. Traunspurger and C. Steinberg, *Aquat. Toxicol.*, 45 (1999) 147.

55   D. Mackay and A.I. Hughes, *Environ. Sci. Technol.*, 18 (1984) 439.

56   H.A. Painter and T.F. Zabel, Review of the environmental safety of LAS, Water Research Centre, Henley Road, Hedmenham CO 1659-M/1/EV 8658, 1988.

57   J. Tolls, M. Haller, I. de Graaf, M.A.T.C. Thijssen and D.T.H.M. Sijm, *Environ. Sci. Technol.*, 31 (1997) 3426.

58   P.D. Abel and V. Axiak, Measuring the toxicity of pollutants to marine organisms, In: E. Horwood (Ed.), *Ecotoxicology and the Marine Environment*, Chichester, 1991, p. 103.

59   G.S. Schuytema, A.V. Nebeker and M.A. Cairns, *Bull. Environ. Contam. Toxicol.*, 56 (1996) 742.

60 M. Wakabayashi, M. Kikuchi, H. Kojima and T. Yoshida, *Ecotox. Environ. Saf.*, 4 (1980) 195.

61 S.A. Snyder, T.L. Keith, S.L. Pierens, E.M. Snyder and J.P. Giesy, *Chemosphere*, 44 (2001) 1697.

62 J. Tolls and D.T.H.M. Sijm, *Environ. Toxicol. Chem.*, 18 (1999) 2689.

63 D.W. McLeese, D.B. Sergeant, C.D. Metcalfe, V. Zitko and L.E. Burridge, *Bull. Environ. Contam. Toxicol.*, 24 (1980) 575.

64 A.A. Fay, B.J. Brownnawell, A.A. Elskus and A.E. McElroy, *Environ. Toxicol. Chem.*, 19 (2000) 1028.

65 S. Topcuoglu and E. Birol, *Pall. Turk. J. Nucl. Sci.*, 9 (1982) 100.

66 W.E. Bishop and A.W. Maki, A critical comparison of bioconcentration test methods, In: J.G. Eaton, P.R. Parrish and A.C. Hendricks (Eds.), *Aquatic Toxicology*, Vol. 707, American Society for Testing and Materials, ASTM STP, 1980, p. 116.

67 A.F. Werner and R.A. Kimerle, *Environ. Toxicol. Chem.*, 1 (1982) 143.

68 D.J. Versteeg, J.M. Rawlings, *Arch. Environ. Contam. Toxicol.*, 44 (2003): 237.

69 C. Wahlberg, L. Renberg and U. Wideqvist, *Chemosphere*, 20 (1990) 179.

70 R.A. Kimerle, *Tenside Surfact. Det.*, 26 (1989) 169.

71 K.J. Buhl and S.J. Hamilton, *Trans. Am. Fish. Soc.*, 129 (2000) 418.

72 M.A. Lewis, *Water Res.*, 26 (1992) 1013.

73 R.J. Larson and D.M. Woltering, Linear alkylbenzene sulfonate, In: G.M. Rand and E.A. Petrocelli (Eds.), *Fundamentals of Aquatic Toxicology*, 2nd edn., Taylor & Francis, North Palm Beach, FL, 1995, p. 859.

74 N.A. Fernandez, E. Chacin, E. Gutierrez, N. Alastre, B. Llamoza and C.F. Forster, *Process Biochem.*, 31 (1996) 333.

75 M. Haitzer, S. Höss, W. Traunspurger and C. Steinberg, *Chemosphere*, 37 (1998) 1335.

76 S.J. Traina, D.C. McAvoy and D.J. Versteeg, *Environ. Sci. Technol.*, 30 (1996) 1300.

77 J. Blasco, E. González-Mazo and C. Sarasquete, *Toxicol. Environ. Chem.*, 71 (1999) 447.

78 P. Pärt, O. Svanberg and E. Bergström, *Ecotoxicol. Environ. Saf.*, 9 (1985) 135.

79 A. Temara, G. Carr, S. Webb, D. Versteeg and T. Feijtel, *Marine Pollut. Bull.*, 42 (2001) 635.

80 C.A. Staples, J. Weeks, J.F. Hall and C.G. Naylor, *Environ. Toxicol. Chem.*, 17 (1998) 2470.

7.3 ESTROGENICITY OF SURFACTANTS

Nicolas Olea, M.F. Fernández, A. Rivas and F. Olea-Serrano

7.3.1 Introduction

In 1991 new insights into the biological activity of some alkylphenols (APs) drew attention to a singular form of toxicity, hormone mimicry. It was discovered that *p*-nonylphenol (NP), used as an antioxidant additive in plastics to retard yellowing, was estrogenic [1]. NP forms part of the chemical structure of an extensive group of APs and alkylphenol ethoxylates (APEOs) that can release these monomers after the environmental degradation of their more complex chemical structures [2]. Soon after the surprising discovery of the estrogenicity of NP, further chemicals with simple chemical structures were found to be estrogen mimics. Bisphenols [3–5], phthalates [6,7], and parabens [8], among others, have now been added to what seems likely to be a long list of estrogenic compounds.

Environmental studies have revealed associations between exposure to chemical contaminants and harmful effects on development and hormonal homeostasis in humans and animals [9]. Chemical substances that are capable of recognising, mimicking, altering, changing, blocking and/or antagonising hormones are now known as endocrine-disrupting chemicals. Endocrine disruptors were defined at the 1996 Weybridge meeting as exogenous substances released into the environment that cause adverse health effects in an intact organism or its progeny and consequent changes in endocrine function (Impact of Endocrine Disrupters on Human Health and Wildlife, 1996). More recently, new evidence presented at the Aronsborg Conference (2001) confirmed international concerns about endocrine disrupting effects on wildlife and human health.

The designation 'potential endocrine disrupter' has been proposed for chemical products with an endocrine-disruption ability that is demonstrated in an in vitro assay but not confirmed in an in vivo animal model. To date, most of the available information on chemical products with endocrine disrupter activity has been generated by in vitro experiments [10]. Various existing tests and bioassays of very different types have been proposed by distinct international bodies to identify hormonal

Comprehensive Analytical Chemistry XL
T.P. Knepper, D. Barceló and P. de Voogt (Eds)

mimics/antagonists. The goal is to assess the risk of exposure to chemical compounds and to use the tests as regulatory instruments for international application and for regulating trade [11]. The development of tests for potential endocrine disruptors has become one of the main priorities of international agencies. In 1997, the European Parliament and Council [12] instructed the European Commission to develop screening tests for endocrine disrupters. The Commission delegated this responsibility [13] to the International Committee of Experts in Endocrine Disruption of the Organisation for Economic Cooperation and Development (OECD). These members discussed, with groups of experts from the US, Japan and other nations, the types of tests and bioassays that should be implemented and how the screening program should be setup. To date, three assays have been approved by the different groups of experts: the uterotrophic test in immature rats; the Hersberger prostate assay; and Protocol 407, extended to assess repeated dose effects.

The uterotrophic assay measures the increase in uterine wet weight and tissue mass after in vivo treatment with the test substance [14]. Over the past 75 years, uterine weight (uterotrophic response), vaginal cornification, age at vaginal opening/puberty and female sexual receptivity have been employed as endpoints to test chemicals for estrogenic activity. The uterotrophic test can be used as one of the set of initial endocrine-screening tests for existing chemicals on which little or no toxicological information is available, in order to establish dose ranges for more extensive endocrine toxicological studies.

Despite the major efforts that have been made over a very short period of time, it has been alleged that the hormonal effects attributable to some chemical compounds suspected of being endocrine disrupters cannot be measured with the toxicological tests currently in use [15]. The following points have been made: (i) a chemical compound may have endocrine disruption-related effects at a lower dose than the currently accepted no observable adverse effect level (NOAEL); (ii) many toxicity tests are not designed to detect effects that occur after exposure during early development; (iii) very few tests evaluate the combined effect of several chemical compounds; and (iv) alterations of endocrine function can have repercussions in any organ and at any moment in the life of an individual, because of the special actions of hormones during development and for the maintenance of homeostasis during critical periods.

Estrogenic effects are the most commonly reported in all of the endocrine systems investigated. Several types of in vitro tests have been used to identify potential estrogenic compounds and are now included in the screening process. Quantitative structure–activity relationship (QSAR) models have also been developed that have the potential for rapidly screening a large number of chemicals. However, the estrogenicity of chemicals cannot be predicted solely from their structure. There have been reports of difficulties in reconciling chemical structure with the wide range of chemical compounds described as having endocrine-disrupting properties. Nevertheless, structure–activity relationship-based studies can often serve as a reasonable basis for further testing [16]. Another classic method to identify possible xenoestrogens is by measuring the relative binding affinity to the estrogen receptor (ER) using competitive ligand-binding techniques.

Cell proliferation is the hallmark of estrogenic action and, over the past decade, bioassays that use hormone-dependent cells in culture have proven a reliable tool for rapidly assessing estrogenicity in a large number of compounds. The E-Screen test (MCF7 cell proliferation assay) is used to analyse chemical products of unknown biology and toxicology in order to determine their biological function. The volume and quality of the data gathered using MCF7 cells as a bioassay warrant the inclusion of this test in any programme for estrogenicity, and it stands out among the different in vitro tests proposed to date. The E-Screen test compares the cell yield after 6 days of culture in the presence (positive control) or absence (negative control) of estradiol and with varied concentrations of a single chemical or a mixture of chemicals [17,18]. The E-Screen test is very sensitive, easy to perform, and can screen many compounds over a wide range of concentrations. Interestingly, most of the information now available on chemicals with endocrine-disrupting activity was produced by the MCF7 in vitro test and later confirmed in whole-animal studies [6].

The objective of the present work was to investigate the toxicity and estrogenicity of selected groups of industrial pollutants discharged into water resources. The ability of surfactants to modify cell kinetics in MCF7 cultures and affect the signal transduction pathway of estrogens was explored. The uterotrophic assay was used as an in vivo estrogenicity test with the same aim.

7.3.2 Materials and standards

Estradiol (E_2), ethynyl estradiol (EE) and synthetic estrogens such as diethyl stilbestrol (DES) were from the Sigma Chemical Company, Poole, Dorset, UK. Stocks were prepared in ethanol at 1 mM concentration and stored in glass vials at $-20°C$.

APEO, NP, NP monocarboxylate (NPE_1C), monoethoxylate ($NPEO_1$), NP monoethoxy carboxylate (NPE_2C) and NP diethoxylate ($NPEO_2$) were provided by Petroquímica Española, San Roque, Spain. Brominated NP derivatives were synthesised by Dr Ventura from Aguas de Barcelona (AGBAR, Barcelona) with the incorporation of a bromine atom to the molecules of NPE_1C, NPE_2C and nonylphenol diethoxy carboxylate (NPE_3C), yielding the brominated derivatives $BrNPE_1C$, $BrNPE_2C$ and $BrNPE_3C$.

Linear alkylbenzene sulfonate (LAS) and the sodium salt Na-LAS (P550), alkyl sulfate (AS) $C_{12}–C_{14}$, alkylether sulfate (AES), alcohol ethoxylate (AE) 1012/60, methylester sulfonate (MES) $C_{16}–C_{18}$, Na-lauryl sulfate, 1-octane sulfate Na, 1-decan sulfonate Na, 1-hexadecan sulfonate Na, Lauric alcohol, 1-octanol (A. Caprílico), 1-nonanol, n-decyl-alcohol, 1-undecanol, 1-tridecanol, miristic alcohol (1-tetradecanol), 1-pentadecanol, cetyl-alcohol (1-hexadecanol), 1-heptadecanol, estearil alcohol (1-octadecanol), nonadecanol, araquidil alcohol (1-ecosanol), heneicosanol, behenil alcohol (1-docosanol), 1-octansulfonate sodium, 1-decan sodium sulfonate, 1-hexadecan sodium sulfonate (Cetyl sodium sulfonate), sodium sulfonate of 2-phenyl butylic acid, sodium sulfonate p-pentyl propyl tetraline and alcohol sulphate $C_{16}–C_{18}$ were provided by Petresa, San Roque, Spain.

Alkyl glucamide (AG) (MW 420) $R–CH_2–CO–N(CH_3)–[CH_2–(CHOH)_4–CH_2OH]$, $R = 10, 12, 14, 16$ and alkyl polyglucoside (APG) (MW 600) were also provided by Petresa, San Roque, Spain.

Both 2,2-bis(4-hydroxyphenyl)propane (BPA, bisphenol-A) and 2,2-bis(4-hydroxyphenyl)heptane (BP-5) were a gift from Dr M. Metzler (Kaiserlautern, Germany). Mono-, di-, tri- and tetrachlorinated BPA were a gift from Dr J.L. Vilchez (Granada, Spain).

7.3.3 Test methods for estrogenicity

The estrogenicity of individual chemicals and chemical mixtures in human, animal and environmental samples was tested using two sets of bioassays recommended by international agencies. These were used to investigate the ability of the chemicals to alter the estrogen transduction signal pathway in MCF7 breast cancer cells in culture. The estrogenicity of the chemicals was also studied with the in vivo uterotrophic test. We present below the methodology employed in these bioassays and their requirements, and briefly discuss the quantitative expression of the results of these tests when applied to surfactants and other environmental chemicals.

The ER transduction signal pathway in MCF7 as a model for estrogenicity

Great attention has deservedly been paid in the last few years to the E-Screen estrogenicity test based on the use of MCF7 breast cancer cells. The selection of MCF7 clones for use in the bioassay has been widely studied [18]. Significant differences in proliferation response have been found among sub-clones, which could not be attributed to culture conditions or cell passage number. The low sensitivity of some clones to estradiol means that rigorous pre-test cell selection and maintenance of a uniform cell stock during the study period are essential to ensure reproducibility of the results.

MCF7 cell line and cell culture conditions. Human breast cancer estrogen-sensitive MCF7 cells originally established by Soule and colleagues were a gift from C. Sonnenschein (Tufts University, Boston); they are known as MCF7-BUS clone and used at passages 70–103 after cloning at the time of study. For routine maintenance, cells were grown in Dulbecco's modification of Eagle medium (DME) supplemented with 5% foetal bovine serum (FBS) at 37°C in an atmosphere of 5% CO_2/95% air under saturating humidity.

MVLN cell line. MVLN cells were a gift from Michael Pons (INSERM, Montpellier, France). MVLN cells were established by transfecting MCF7 cells with pVit-tk-Luc plasmid [19]. In this reporter gene, $331/-87$ of the 5'-flanking region of Xenopus Vitellogenin A2 gene (Vit), alpha-fragment that contains a palindromic ERE, is inserted in front of the herpes simplex virus promoter for thymidine kinase (tk). This regulatory part controls the firefly luciferase structural gene. The MVLN cells were cultured in DME supplemented with 10% FCS,

L-glutamine, penicillin, streptomycin and gentamycin (Gibco, Gent, Belgium) at the usual concentrations.

Cell proliferation in the E-Screen assay. The E-Screen (MCF7 cells) test method was slightly modified from that originally described by Soto et al. [17] and adapted by Villalobos et al. [18]. Briefly, cells were trypsinised and plated in 24-well plates (Limbro, McLean, VI) at initial concentrations of 20 000 cells per well in 5% FBS in DME. Cells were allowed to attach for 24 h; the seeding medium was then removed and replaced by the experimental medium (5–10% charcoal dextran human serum supplemented with phenol red-free DME (CDHuS)). A range of concentrations of the test compounds was added to this medium. The bioassay was stopped after 144 h by removing the medium from the wells, fixing the cells and staining them with sulforhodamine-B (SRB). The staining technique was modified from that described by Skehan et al. [20]. Briefly, the experimental medium was removed and cells were treated with cold 10% trichloroacetic acid (TCA) and incubated at 4°C for 30 min, then washed five times with tap water and left to dry. TCA-fixed cells were stained for 10 min with 0.4% (w/v) SRB dissolved in 1% acetic acid. Wells were rinsed with 1% acetic acid and air-dried. Bound dye was solubilised with 10 mM Tris base (pH 10.6) in a shaker for 20 min. Finally, aliquots were transferred to a 96-well plate and read in a Titertek Multiscan apparatus (Flow, Irvine, CA) at 492 nm. We evaluated the linearity of the SRB assay with cell number before the cell growth experiments.

Surfactants were tested at concentrations ranging from 0.1 mM to 10 nM dissolved in the culture medium, with the concentration of solvent always maintained below 0.1%. Results are expressed as means ± SD. In the proliferation yield experiments, each point represents the mean of three counts from the four culture wells. Mean cell numbers were normalised to the control, equal to 1, to correct for differences in the initial plating density. The proliferative effect (PE) was expressed as the ratio between the highest cell yield obtained with each chemical tested and the hormone-free control. Proliferation experiments were repeated at least three times.

The relative proliferative potency (RPP) is the ratio between the minimum concentration of estradiol needed for maximal cell yield and the minimum dose of the test compound needed to obtain a similar effect. The relative proliferative effect (RPE) describes the ratio between the highest cell yield obtained with the chemical and with estradiol × 100. The RPE indicates whether the chemical of interest is a full (RPE = 100) or partial agonist, depending on whether the proliferative response is similar to or less than that produced with estradiol.

The E-Screen bioassay also enables study of the additive-synergistic effects of chemicals by culturing the cells in the presence of several compounds at suboptimal concentrations. The effects are investigated by testing weak estrogenic surfactants in the presence of estradiol (1 pM). Following the standard protocol of the E-Screen bioassay, the antiestrogenicity and toxicity induced by a chemical can be distinguished. Thus, when cells are cultured in the presence of 10 pM estradiol and a chemical that induces a decrease in cell number when tested alone, the antiestrogenic effect is reversed by the natural estrogen. In contrast, the PE of the estradiol is unable to reverse the effects of toxicity.

Progesterone receptor induction. MCF7 cells seeded in T25 flasks were incubated in 10% CDHuS for 72 h with 0.1–10 nM estradiol and a parallel set of flasks were exposed to 0.1–10 μM of the chemicals. Controls received the vehicle alone. At the end of the experiment, the medium was aspirated and the cells were frozen in liquid N_2. To extract receptor molecules, cells were incubated with 1 mL extraction buffer (0.5 M KCl, 10 mM potassium phosphate, 1.5 mM ethylenediamine tetraacetic acid and 1 mM monothioglycerol, pH 7.4) at 4°C for 30 min. Progesterone receptors (PgRs) were measured in extracted cells by enzyme immunoassay using the Abbott progesterone receptor kit (Abbott Diagnostics, Chicago, IL), following the manufacturer's instructions.

Induction of mRNA for pS2 and secretion of cell-type-specific proteins. pS2 was measured in the culture medium of MCF7 cells with the ELSA-pS2 immunoradiometric assay (CIS Bio International, Gif-sur-Yvette, France). Cells were subcultured in 24-well plates for 144 h in 10% CDHuS. The culture medium was centrifuged at 1200g for 10 min to eliminate floating and detached cells. Results are expressed as ng of secreted protein per million cells. The relative-induced protein potency (RIPP) was calculated as 100 × the ratio between the dose of E_2 and that of the chemical needed to produce maximal expression of cell-type-specific proteins (pS2).

Transfected MCF7 cells MVLN-luciferase assay. MVLN cells were trypsinised and seeded in individual dishes in DME supplemented with 10% FBS. After 4 days of subculture, cells were then incubated for 24 h with the test compounds dissolved. At the end of the incubation, the media were removed and cells washed twice with PBS. Luciferase was measured using the Promega Luciferase Assay System according to manufacturer's instructions (Promega Corporation, USA). The luciferase activity was quantified using a Berthold luminometer (Lumat LB 9507). Data values were expressed as arbitrary luminescence units per mg of protein. Proteins were quantified on cellular homogenate using the Coomassie Protein Assay Reagent (BioRad, USA). The luciferase activity of each experiment was normalised to the steroid-free control cultures in order to correct for differences in the initial seeding density.

In vivo tests: the uterotrophic assay
Immature female Wistar rats (15 days old) were housed individually in wire mesh cages with solid floors. Artificial light was maintained as a 12/12 h light/dark cycle under standard humidity and temperature conditions. Animals were weaned on conventional standardised laboratory diets. Diet and water were available ad libitum. All animals were acclimatised for 24 h before their injection. The study was conducted in accordance with the OECD Good Laboratory Practice Procedures [21].

The uterotrophic activities of chemicals were evaluated in three separate experiments using the previously published test protocol (OECD). The test substances were dissolved in ethanol and dimethyl sulfoxide (DMSO) and administered by subcutaneous injection. The dosing volume was 100 µl. Rats were weighed and divided randomly into seven experimental groups ($n = 6$). Each animal received daily one dose of the test compounds on three consecutive days. After the final injection, each animal was weighed and sacrificed by cervical dislocation, and the uterus was then dissected. The uterus was trimmed free of fat, pierced, blotted to remove excess fluid and then weighed (wet weight). The uterus was then dried to establish the dry weight. Estradiol, DES and ethynyl estradiol were used as positive control agents, and each study included a negative (vehicle) control group. Differences in the weight of wet and dry uteri were assessed for statistical significance using a two-tailed Student's t-test.

For light-microscopic examination, tissues were fixed for 5 h at 4°C in cacodylate-buffered 2.5% glutaraldehyde, postfixed in 1.5% osmium tetroxide in the same buffer, dehydrated in a rising series of ethanol and embedded in Epon 812. Sections were cut on an Ultracut S (Leica) ultramicrotome. Semi-thin sections were stained with toluidine. Microscopic observations were performed with an Olympus BX-40 at the University of Granada Technical Service. This allowed a more meticulous study of the groups through morphometric evaluation, determining the volumetric density of the epithelium of the uterine lumen, and studying the glandular epithelium, connective tissue, muscle tissue, and height of the lumen epithelial cells.

7.3.4 Estrogenicity of environmental chemicals

Natural and synthetic estrogens
The addition of estradiol to CDHuS-supplemented medium increased the number of MCF7 cells in the culture (Fig. 7.3.1). Maximum PE was obtained at concentrations of 10 pM estradiol and higher. Cell yields were 6-fold greater than in control cultures after 6 days (mean $\pm$ SD, 6.10 ± 1.21 in 15 experiments). In the absence of the estrogen (control), cells proliferated minimally. In comparison with the PE of estradiol, DES and ethynyl estradiol showed a full agonist response. Thus, they behave as full estrogen agonists at the range of concentrations assayed.

In order to investigate the response of MCF7 cells in different media, our experimental design also included tests of natural estrogens and pharmaceuticals in medium supplemented with foetal bovine serum and in synthetic medium. Five per cent CDFBS supplemented medium was as effective as CDHuS medium in showing the PE of estradiol. In contrast, when MCF7 cells were maintained in insulin-transferrin supplemented DME (ITDME), estradiol had no effect on cell prolifer-ation (Fig. 7.3.1). This phenomenon may suggest that serum borne factors are necessary to mediate the estrogenic response in the E-Screen bioassay.

Natural estrogens and pharmaceuticals were also tested in the pathway that links hormone exposure of MCF7 to gene expression. MCF7 cells have receptors for estradiol (ER) and progesterone (PgR). The basal concentration of ERs and progesterone receptors (PgR) in MCF7 cells was 183 ± 29 and 15.5 ± 4.3 fmol mg^{-1} of extracted protein, respectively. Treatment with estradiol and the estrogen pharmaceu-

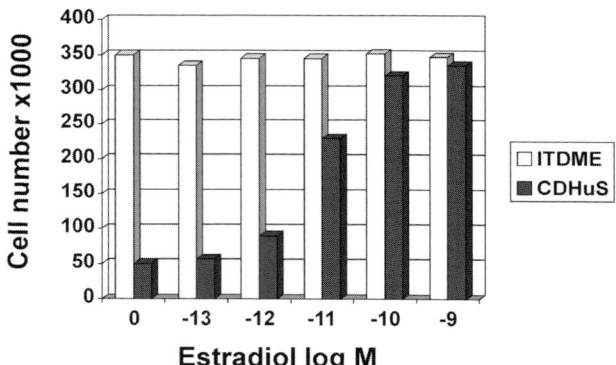

Fig. 7.3.1. Effect of estradiol on MCF7 cell proliferation. MCF7 cells were grown for 6 days in estrogen-depleted medium supplemented with 5% charcoal dextran-treated human serum (5% CDHuS) or with insulin (100 ng ml^{-1}) and transferrine (25 μg ml^{-1}) (ITDME). Estradiol was added to cultures at concentrations ranging from 0.1 pM to 1 nM. The maximal estrogenic response corresponded to 10 pM of estradiol and higher concentrations. Cell yields were six-fold (10 pM, 6.7 ± 1.2) those of control. No proliferative response was observed in cells maintained in ITDME. Each point is the mean of three counts from four culture wells; bars indicate SDs.

ticals increased the concentration of PgR to nearly 10- to 15-fold the basal value (Fig. 7.3.2). This was accompanied by a concomitant decrease in ER levels.

pS2 protein accumulation in the culture medium reflected increases in cell number during the time the experiments lasted. The secretion of pS2 by MCF7 cells was significantly increased by 0.1 nM estradiol and higher concentrations (3.5-fold increase versus controls). For instance, the basal concentration of pS2 (53.6 ± 8.7 ng 10^{-6} cells) increased to 179.4 ± 13.1 ng 10^{-6} cells after treatment with 1 nM estradiol. Exposure of MC7 cells to estradiol also significantly increased the pS2 mRNA expression of the cells (Fig. 7.3.2), and their exposure to pharmaceutical estrogens for 24 h resulted in a similar increase.

Finally, estradiol induced a response in the luciferase activity of MVLN transfected cells. The hormone increased three-fold the basal level of enzymatic activity (Fig. 7.3.2). The luciferase activity of each experiment was normalised to the steroid-free control cultures to correct for differences in the initial seeding density. The activity induced by the antiestrogen RU 58 668 was lower than the response to the control.

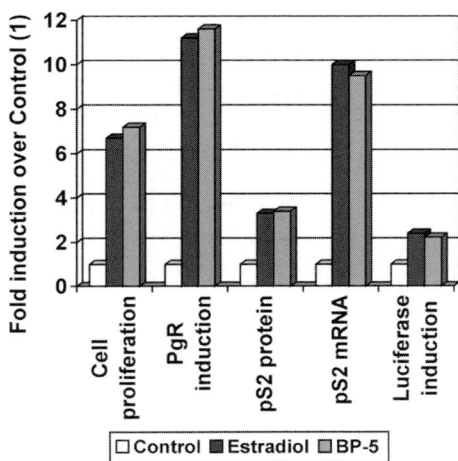

Fig. 7.3.2. Effect of estradiol and BP-5 on MCF7 cell proliferation and gene expression. MCF7 cells were grown in estrogen-depleted medium. Estradiol or BP-5 were added to cultures and cell number (proliferation), protein determination (PgR and pS2), gene expression (pS2 mRNA) and luciferase expression were estimated at the appropriate times, as indicated in the Methods Section.

In all assays, E_2 was used as a positive control and, as expected, induced the highest estrogenic response at the lowest concentrations (Fig. 7.3.2). In the proliferation test, cell yields were six-fold (10 pM, 6.7 ± 1.2) those of the control. The progesterone receptor assay showed an even greater response (9-fold control values), although the concentration of E2 needed for the maximal effect was also higher (10 nM). The response in terms of the pS2 mRNA expression was similar to that shown in the proliferation test. Lower responses (approximately 3-fold control values) were observed in the protein pS2 test and luciferase bioassay. The amplitude of the response to a given dose of E_2 is of interest, because a sufficiently wide difference between treated and control cells enables full and partial E_2 agonists to be differentiated. The time required for the cell treatments and reading process differs between these tests. The pS2 protein induction and cell proliferation assays require a 144-h exposure period, compared with the 24-h period for the luciferase induction or pS2 mRNA assays. The pS2 mRNA assays needs the longest time (4–9 days) for the reading process.

In summary, estradiol induced a specific response in MCF7 cells in all the tests applied. The progesterone receptor assay, E-Screen proliferation test and pS2 mRNA assay showed a wider response range than the other tests, offering a greater ability to differentiate between agonist, partial agonist and antagonist responses.

The rodent uterotrophic assay is an established in vivo method to determine the estrogenicity of a chemical by measuring the increase in wet weight of the uterus after exposure to selected chemicals [22]. The administration of estradiol, DES or ethynyl estradiol at concentrations between 10 and 100 μg kg^{-1} bodyweight had a significant effect on the wet weight of the uterus relative to the control group. Histological examination also showed a significant increase in both epithelial cell height and relative area of epithelium within the uterus.

There is a good correlation between the assays employed in this study, suggesting that any of them would be effective to measure estrogenic responses. However, their advantages and disadvantages must be considered in order to choose the most adequate screening method to detect estrogenic activity. The receptor assay is quite simple to perform, and it allows the identification of chemicals that bind to the ER. However, it cannot discriminate between antiestrogens and estrogens. The E-Screen is one of the most sensitive assays and is recognised as biologically equivalent to the increase of mitotic activity in the rodent endometrium [6]. However, there are drawbacks associated with maintaining the cell line and with the long duration of the test. Reporter gene expression assays in the luciferase bioassay assess the ability of the test compounds to activate the ER. The advantages are the simplicity and rapidity of measurements when compared with the conventional test. However, it has been reported that the antiestrogen hydroxytamoxifen induces a rapid and irreversible inactivation of estrogenic response [19], which may complicate the interpretation of results. The induction of progesterone receptor levels and the increased expression of secreted protein such as pS2 are simple and easy to standardise. Quantification of pS2 mRNA requires only 24 h of exposure to the test chemicals, although measurements of this end-point involve laborious and time-consuming methods such as Northern Blot.

Alkylphenol derivatives
AP and brominated AP derivatives were tested in the E-Screen proliferation bioassay in a concentration range of 100 μM to 0.1 mM.

They were assayed at intervals of a half-order of magnitude. The RPE and RPP were calculated from the estimated PE of each chemical (Table 7.3.1). The PE was significantly different from one for NP and all the compounds tested, except NPEOs, $O_{n>2}$.

It has been suggested that octylphenol (OP) and NP are the most potent AP derivatives and that the length of the ethoxy chain (EO) determines the estrogenicity [2]. In addition, carboxylic derivatives were also estrogenic in some bioassays. NP and OP were shown to be estrogenic compounds by using MCF7 and ZR-75-1 cells [17]. Both chemicals increase cell yield and induce PgR receptors in MCF7 cells. For instance, NP increased PgR levels to a lower extent than estradiol, but this increase in progesterone receptor levels was not associated with ER downregulation. Interestingly, this particular pattern was shown for some catecholestrogens, which increased MCF7 cell proliferation rate and progesterone receptor levels but induced ER processing only during the first 8 h after treatment, after which time ER levels increased to reach baseline values at 24 h.

Soto and co-workers first reported the estrogenicity of NP released from modified polystyrene tubes in 1991 [1]. Soto also proved that

TABLE 7.3.1

Estrogenicity of NPs and brominated alkylphenols in the E-Screen bioassay

Compound	Concentration	RPE (%)	RPP (%)
Estradiol	50 pmol L^{-1}	100	100
NP	1 μmol L^{-1}	75	0.005
BrNPE$_1$C	5 μmol L^{-1}	65	0.0002
Br$_2$NPE$_1$C	5 μmol L^{-1}	48	0.0002
Br$_3$NPE$_1$C	5 μmol L^{-1}	36	0.0002
NPE$_1$C	10 μmol L^{-1}	–	–
NPEO$_1$	10 μmol L^{-1}	25	0.0001
NPE$_2$C	10 μmol L^{-1}	–	–
NPEO$_2$	10 μmol L^{-1}	–	–
NPEO$_{10}$	10 μmol L^{-1}	–	–
BrNPE$_2$C	5 μmol L^{-1}	48	0.0001
BrNPE$_3$C	5 μmol L^{-1}	36	0.0001

RPE was calculated as $100 \times$ (PE-1) of the test compound/(PE-1) of estradiol; PE is the lowest concentration needed for maximal cell yield; RPP is the ratio between estradiol and compound doses needed to produce maximal yield $\times$ 100. All of the compounds that were designated as full or partial agonists (RPE > 15) significantly increased cell yields compared with the controls without hormones ($p < 0.05$). –, no effect observed on proliferation.

phenolic derivatives with para chains from four to twelve carbons were estrogenic in MCF7 cells and in rats. Years before, Dodds and Lawson [23] had tested some AP derivatives in the uterotrophic assay and demonstrated their estrogen mimicry.

APEO have long represented a substantial part of the industrial detergent market. The first report demonstrating the environmental presence of APEO appeared in 1978. Nowadays, they are frequently reported as very common contaminants in surface water, wastewater, water plant effluents and sediments [24,25].

For risk assessment purposes, the relative potency of AP and APEO has been evaluated by several authorities, initially on the basis of aquatic toxicity. In the UK, the Environment Agency used a QSAR approach [26] to derive the potencies listed in Table 7.3.2. Environment Canada used two approaches to characterise risks of NP, NPEO, and NPEC [27]. In the distributional approach, relative toxicities were proposed based on categorising acute and chronic toxicities. These are listed in Table 7.3.2. In addition a conservative approach was used.

Similar to the relative potencies for toxicity, relative potencies for receptor binding, vitellogenin induction and several other ER mediated responses, such as those from the in vitro tests presented in Table 7.3.2

TABLE 7.3.2

Relative potencies of APEO and their carboxylated degradation products proposed by Environment Agency UK and Environment Canada (adapted from Ref. [28])

Substance	Aquatic toxicity		Estrogenicity (relative potency to NP)[a]
	Relative potency to NP [b]	Relative potency to NP [a]	
NP	1	1	1
OP	1	1	4.1
$NPEO_1$	0.77	0.5	0.67
$NPEO_2$	0.59	0.5	0.67
$NPEO_{3-17}$	0.017–0.43 ($NPEO_{15}$–$NPEO_3$)	0.005	0 (0.02)
NPEC	–	0.005	0.63
NPE_2C	–	0.005	0.63
OPEC	–	0.005	0.63
OPE_2C	–	0.005	0.63

[a]From Environment Canada [27].
[b]From Environment Agency UK [26].

and Section 7.3.4.6, have been derived. In the Canadian assessment a distributional approach similar to that for toxicity was used to derive relative estrogenicities. The relative potencies shown in Table 7.3.2 are a weighted mean of several ER mediated responses [27]. Such relative potencies must be used with some care, as they depend on the endpoints considered. Moreover, it has been shown that in vitro potencies can differ substantially from in vivo data, also for APE [28].

The data presented here and results published in scientific journals provide guidance on the structural requirements for the estrogenicity of APs and derivatives: (i) the alkyl chain should contain three or more carbons; (ii) only the p-isomers are estrogenic; (iii) polyalkylation results in a loss of estrogenic activity, as in the case of compounds such as butylhydroxytoluene or Irganox 1640 (Ciba-Geigy), which are not estrogenic; (iv) conjugated aromatic rings, as in the case of naphthols, are not estrogenic, despite their reproduction of the A and B rings of natural estrogens; and (v) the estrogenicity of ethoxylates is less than that of the non-ethoxylated phenols, is dependent on the number of EO units, and is restricted to compounds with one or two groups.

Finally, it is interesting to note that brominated derivatives of NP included in the present study, estrogenic in the E-Screen bioassay, were proposed to be fat-soluble [30]. In fact, as non-ionic surfactants, these molecules have a hydrophilic and a hydrophobic part [31]. The presence of these brominated compounds in fat tissue of humans or animals has yet to be demonstrated. However, if they do bioaccumulate in adipose tissue, as their fat solubility suggests, they may account for the xenoestrogen burden alongside organohalogenated compounds.

Linear alkylbenzene sulfonate

In the E-Screen bioassay, LAS was not effective in promoting cell proliferation (Table 7.3.3). This compound was tested at concentrations of up to 100 μM with no evidence of cellular toxicity. The antiestrogenic effect of this compound was also measured but all samples tested were negative. Because it has been suggested that surfactants of the alkylbenzene sulfonate type are readily degradable and transformed into sulfophenyl carboxylates or SPCs, an important number of SPCs were assayed in the E-Screen test. These SPCs did not induce cell proliferation of MCF7 cells.

Routledge and Sumpter [32] studied the estrogenicity of LAS and SPCs using a biossay based on ERE transfected yeasts. All the

TABLE 7.3.3

Estrogenicity of LAS, SPCs and other compounds in the E-Screen bioassay

Compound	Concentration	RPE (%)	RPP (%)
Estradiol	50 pmol L^{-1}	100	100
LAS	100 μmol L^{-1}	–	–
2C$_2$-SPC	100 μmol L^{-1}	–	–
3C$_3$-SPC	100 μmol L^{-1}	–	–
4C$_4$-SPC	100 μmol L^{-1}	–	–
5C$_5$-SPC	100 μmol L^{-1}	–	–
2C$_3$-SPC	100 μmol L^{-1}	–	–
2C$_4$-SPC	100 μmol L^{-1}	–	–
3C$_4$-SPC	100 μmol L^{-1}	–	–
p-Pentylpropyltetraline sulfonate-Na	10 μmol L^{-1}	–	–
APG	10 μmol L^{-1}	–	–
AG	10 μmol L^{-1}	–	–

RPE was calculated as $100 \times$ (PE-1) of the test compound/(PE-1) of estradiol; PE is the lowest concentration needed for maximal cell yield; RPP is the ratio between estradiol and compound doses needed to produce maximal yield $\times$ 100. All of the compounds that were designated as full or partial agonists (RPE > 15) significantly increased cell yields compared with the controls without hormones ($p < 0.05$). –: No effect observed on proliferation.

compounds tested were negative in the bioassay of estrogenicity. Unfortunately, however, when these chemicals were tested at concentrations higher than 0.01 g L^{-1}, the cytotoxic effect prevailed over any other effect. This cytotoxic effect may mask other specific actions.

LAS was also tested in the uterotrophic bioassay in order to investigate whether the compound has any estrogenic effect in this in vivo model. Rats were treated for three days with LAS concentrations ranging from 250 to 1000 mg kg^{-1} body weight. There was no significant difference in uterine wet weight between the control group and experimental treatment groups. Therefore, these results confirmed the observations in the E-Screen proliferative test, indicating that this chemical is devoid of estrogenic activity both in vivo and in vitro.

Linear alcohols and alkylether sulfates

Linear alcohols were studied using MCF7 cells in the proliferation assay and in the expression of estrogen-dependent genes (Table 7.3.4). The AS C$_{12-14}$, its ethoxylated derivative (ALFONIC 1214) and linear alcohols from nonanol to dodecanol increase cell proliferation in the E-Screen assay. This effect was seen at 10–100 μM concentrations and,

TABLE 7.3.4

Estrogenicity of linear alcohols in the E-Screen bioassay

Compound	Concentration	RPE (%)	RPP (%)
Estradiol	50 pmol L^{-1}	100	100
AE C_{10-12}	10 μmol L^{-1}	–	–
AE C_{12-15} (LYAL 125)	10 μmol L^{-1}	–	–
AE C_{14-15} (LYAL 145)	10 μmol L^{-1}	–	–
AE C_{12-14} (ALFONIC 1214)	10 μmol L^{-1}	32.0	0.0001
AE C_{16-18} (ALFONIC 1618)	10 μmol L^{-1}	–	–
1-Octanol	10 μmol L^{-1}	22.0	0.0001
1-Nonanol	10 μmol L^{-1}	34.5	0.0001
n-Decyl-alcohol	10 μmol L^{-1}	35.0	0.0001
1-Undecanol	10 μmol L^{-1}	25.2	0.0001
1-Dodecanol	10 μmol L^{-1}	23.8	0.0001
1-Tridecanol	10 μmol L^{-1}	–	–
1-Tetradecanol	10 μmol L^{-1}	–	–
1-Pentadecanol	10 μmol L^{-1}	–	–
1-Hexadecanol	10 μmol L^{-1}	–	–
1-Heptadecanol	10 μmol L^{-1}	–	–
1-Octadocanol	10 μmol L^{-1}	–	–
1-Nonadecanol	10 μmol L^{-1}	–	–
1-Ecoxanol	10 μmol L^{-1}	24.8	0.0001
Heneicosanol	10 μmol L^{-1}	26.1	0.0001
1-Docosanol	10 μmol L^{-1}	29.9	0.0001
AS C_{12}	10 μmol L^{-1}	–	–
AS C_{12-14}	10 μmol L^{-1}	31.5	0.0001
AS C_{12-14} dehydrated	10 μmol L^{-1}	29.2	0.0001
AES	10 μmol L^{-1}	–	–
AS C_{16-18}	10 μmol L^{-1}	29.4	0.0001
MES C_{16-18}	10 μmol L^{-1}	–	–
1-Dodecansulfnate-Na	10 μmol L^{-1}	34.1	0.0001
1-Octansulfonate-Na	10 μmol L^{-1}	25.8	0.0001
1-Decansulfonate-Na	10 μmol L^{-1}	28.7	0.0001
1-Hexadecansulfonate-Na	10 μmol L^{-1}	25.6	0.0001

RPE was calculated as 100 × (PE-1) of the test compound/(PE-1) of estradiol; PE is the lowest concentration needed for maximal cell yield; RPP is the ratio between estradiol and compound doses needed to produce maximal yield × 100. All of the compounds that were designated as full or partial agonists (RPE > 15) significantly increased cell yields compared with the controls without hormones ($p < 0.05$); –: No effect observed on proliferation.

despite the low level of proliferation, it was significantly different from the growth of control cultures. When compared to estradiol, the proliferation observed at the highest AS concentrations gives an RPE of 25% and an RPP of 0.0001.

The same chemicals were also effective inductors of other markers of estrogenic action, such as PgR and pS2, in MCF7 cells. They increased

PgR at the maximal dose tested (100 μM), and significantly increased pS2 levels (100 μM). Their affinity to bind to the ER was also measured and was found to be low.

Mixtures of linear alcohols were also tested in the uterotrophic assay (LYAL 125, LYAL 145, ALFONIC 1214, ALFONIC 1618) at concentrations ranging from 250 to 1000 mg kg^{-1} body weight. For uterine wet weight, there were no significant differences between the control group and alcohol sulphate-treated animals. Thus, these experiments did not confirm observations in the proliferation in vitro test. Interestingly, a high mortality rate (50%) was observed in the animals at the highest dose range.

MCF7 cells have the ability to transform some non-estrogenic compounds into estrogenically active derivatives [4]. This is the case with methoxychlor, which is demethylated and therefore activated to bind to the ER. However, in vitro tests have addressed only a part of the metabolic system present in animals. Only a few studies have investigated the hormonal effects of AS and ethoxylated derivatives, and they were designed to investigate the toxicity of these compounds at concentrations at which the estrogenic effect would appear. For instance, Routledge and Sumpter [32] showed the absence of estrogenic effect in a recombinant yeast screen bioassay. However, at a concentration of 0.01 g L^{-1}, the toxic effect of AS destroyed yeast cultures. The same observation was made when this assay was applied to alkylbenzene sulfonates.

Other surfactants
AG and APG were assayed in the E-Screen test for estrogenicity but were ineffective in inducing MCF7 cell proliferation. Table 7.3.3 shows the results of the proliferation test.

Bisphenol-A and chlorinated bisphenols
Bisphenols is a broad term that includes many chemicals with the common chemical structure of two phenolic rings joined together by a bridging carbon. Bisphenol A is a monomer widely used in the manufacture of epoxy and phenolic resins, polycarbonates, polyacrylates and corrosion-resistant unsaturated polyester-styrene resins. It can be found in a diverse range of products, including the interior coatings of food cans and filters, water containers, dental composites and sealants. [4]. BPA and BP-5 were selected for testing by the whole

battery of assays because their estrogenicity was demonstrated many years ago, when Dodds and Lawson [23] were looking for synthetic estrogens devoid of the phenantrenic nucleus. As a consequence, they were considered of use to assess estrogenicity bioassays of proven excellence for the study of environmental estrogens.

The PEs of BPA and BP-5 in a normal experimental medium, 5% CDFBS-supplemented medium and synthetic ITDME medium are presented in Table 7.3.5. The addition of 0.1 μM BPA to 10% CDHuS or 5% CDFBS supplemented medium increases cell proliferation as effectively as estradiol. BPA was also tested in the presence of an antiestrogen, producing inhibition of the PE associated with BPA. The PEs of chlorinated bisphenols are shown in Table 7.3.5. It was significantly greater than one for all the compounds tested. In comparison with the RPE of estradiol, all the positive compounds showed a full to partial agonistic response, producing cell yields that ranged from 85% of estradiol-induced yield for bisphenol-A to 30% for the bisphenol-A derivative tetrachlorine.

BPA was also tested in the uterotrophic test for estrogenicity at doses ranging from 200 to 800 mg kg^{-1} body weight. BPA was effective in increasing the wet uterine weight, confirming previous results.

The estrogenicity of BPA has been well studied and documented for many years. To the best of our knowledge, the first report of the estrogenic properties of BPA was published by Dodds and Lawson [23] in 1936, as mentioned above. Later, Reid and Wilson [33] confirmed

TABLE 7.3.5

Estrogenicity of bisphenols in the E-Screen bioassay

Compound	Concentration	RPE (%)	RPP (%)
Estradiol	50 pmol L^{-1}	100	100
Bisphenol A (BPA)	1 μmol L^{-1}	85	0.005
BP-5	10 nmol L^{-1}	95	0.5
Cl-BPA	10 μmol L^{-1}	71	0.0005
Cl$_2$-BPA	10 μmol L^{-1}	71	0.0005
Cl$_3$-BPA	10 μmol L^{-1}	59	0.0005
Cl$_4$-BPA	10 μmol L^{-1}	28	0.0005

RPE was calculated as $100 \times$ (PE-1) of the test compound/(PE-1) of estradiol; PE is the lowest concentration needed for maximal cell yield; RPP is the ratio between estradiol and compound doses needed to produce maximal yield $\times$ 100. All of the compounds that were designated as full or partial agonists (RPE > 15) significantly increased cell yields compared with the controls without hormones ($p < 0.05$).

the estrogenicity of BPA and some derivatives. More recently, BPA released from polycarbonate flasks during autoclaving [3], from the epoxy resin coating of cans [34] and from dental sealants [4–35] has also been shown to possess an estrogenic effect.

The use of chlorine for disinfecting wastewaters or drinking water has become widespread in this century. Recently, however, the increasing presence and variety of aquatic contaminants has raised the question of the chemical fate of these contaminants when subjected to aqueous chlorination [36]. In fact, the production of organochlorine compounds in chlorinated water, including mutagenic and carcinogenic substances, is well established [37,38]. A number of alternatives to chlorination are used in many parts of the world, but the risks associated with their by-products are even less well established [39].

Perfluorinated bisphenol-A (bisphenol-AF) and other bisphenol derivatives besides BPA have been shown to be estrogenic in MCF7 cells, promoting cell proliferation and increasing the synthesis and secretion of cell-type specific proteins [5]. The estrogenicity of mono- to tri-bromobisphenol-A has been also described [40]. These compounds induce cell proliferation and increase the levels of both progesterone receptors and pS2 protein in MCF7 cells. It was reported that the greater the number of bromo-substitutions, the lower was the potency of these effects. Interestingly, tetrabromobisphenol-A was not effective in the cell proliferation bioassay and showed a strong binding to serum proteins, suggesting that differences in bioavailability contribute to the estrogenic potency. The estrogenicity of mono- to tetra-chlorinated bisphenols decreases as the number of chlorine substituents increases. There is to date little information about the bioavailability of these compounds, although chlorine bisphenols were recently grouped with lipophilic organochlorine pesticides that bioaccumulate in human fat tissues [30]. Because of their estrogenicity, the presence of halogenated (chlorinated and brominated) bisphenols in human tissues warrants further investigation.

Comparison with other in vitro tests
Several data sets from in vitro estrogenicity testing published in the literature can be used to derive relative estrogenicities. For example, Blair et al. have published relative binding affinities to the ER for 188 chemicals [41]. In Table 7.3.6 an overview is given of the relative estrogenic potencies (estradiol equivalent factors, EEF) of different estrogens and xeno-estrogens, observed in the ER-CALUX in vitro assay.

TABLE 7.3.6

Mol-based estradiol equivalency factors of (xeno-)estrogens in the ER-CALUX assay (adapted from Ref. [28])

Code	Compound	ER-CALUX assay	
		EEF [a]	Reference
E2	17β-Estradiol	1	[42]
E2-17α	17α-Estradiol	0.016	[42]
E1	Estrone	0.056	[42]
EE2	17α-Ethynylestradiol	1.2	[42]
BPA	Bisphenol-A	7.8×10^{-6}	[42]
DMP	Dimethylphthalate	1.1×10^{-5}	[42]
DEP	Diethylphthalate	3.2×10^{-8}	[42]
DBP	Di-n-butylphthalate	1.8×10^{-8}	[42]
BBP	Butylbenzylphthalate	1.4×10^{-6}	[42]
DEHP	Di(2-ethylhexyl)phthalate	$<6.0 \times 10^{-7}$	[29]
DOP	Dioctylphthalate	$<6.0 \times 10^{-7}$	[29]
NPE	Nonylphenol ethoxylates	3.8×10^{-6}	[42]
OPE	Octylphenol ethoxylates	$<6.0 \times 10^{-7}$	[29]
NP	4-Nonylphenol	2.3×10^{-5}	[42]
OP	4-t-Octylphenol	1.4×10^{-6}	[42]
BDE47	2,4,2′,4′-Tetrabromodiphenylether	2.0×10^{-7}	[43]
BDE85	2,3,4,2′4′-Pentabromodiphenylether	2.0×10^{-7}	[29]
BDE99	2,4,5,2′4′-Pentabromodiphenylether	2.0×10^{-7}	[29]
BDE100	2,4,6,2′4′-Pentabromodiphenylether	2.0×10^{-5}	[29]
T3-like OH-BDE	1,3,5,3′-Tetrabromo-4′-hydroxydiphenylether	1.0×10^{-4}	[29]
o,p'-DDT	2-(o-Chlorophenyl), 2-(p-chlorophenyl)-1,1,1-trichloroethane	9.1×10^{-6}	[44]
o,p'-DDE	2-(o-Chlorophenyl), 2-(p-chlorophenyl)-1,1-dichloroethene	2.3×10^{-6}	[29]
	Methoxychlor	1.0×10^{-6}	[43]
	Dieldrin	2.4×10^{-7}	[43]
	Endosulfan	1.0×10^{-6}	[43]
	Chlordane	9.6×10^{-7}	[43]
	Genistein	6.0×10^{-5}	[43]

[a]EEF, estrogenic potency relative to estradiol, ratio of EC_{50} (17β-estradiol)/EC_{50} (compound) in the in vitro ER-CALUX test.

7.3.5 Conclusion

The present work aimed to investigate the estrogenicity of industrial pollutants discharged into water resources by studying modifications in the cell kinetics of MCF7 breast cancer cell cultures and effects on the signal transduction pathway of estrogens. The uterotrophic assay, an in vivo estrogenicity test, was also used. Experimental results suggested that the hallmark of estrogen action is the induction of cell proliferation, which is measured by the E-Screen test and uterotrophic bioassay. The effects of estrogens on the induction or downregulation of cell type-specific protein synthesis and on cell hypertrophy are variable, and may be elicited by non-estrogenic agents.

The cell proliferation test is one of the most sensitive assays and is recognised as biologically equivalent to the increase of mitotic activity in the rodent endometrium [6]. However, there are disadvantages associated with maintaining the cell line and with the long duration of the test. Reporter gene expression assays in the luciferase bioassay assess the ability of the test compounds to activate the ER. The advantages are the simplicity and rapidity of measurements when compared with the conventional test. However, it has been reported that antiestrogens induce a rapid and irreversible inactivation of estrogenic response [19], which may complicate the interpretation of results. The induction of progesterone receptor levels and the increased expression of secreted protein such as pS2 are simple methods that are easy to standardise. Quantification of pS2 mRNA requires only 24 h of exposure to the test chemicals, although measurements of this end-point involve laborious and time-consuming methods such as Northern Blot. The cell proliferation-based bioassay is proposed as an initial test for estrogenicity screening.

The ability of surfactants to induce cell proliferation in in vitro and in vivo tests has been demonstrated for APs, some lineal alcohols and bisphenols. In contrast, APEO, LAS, SPCs, AG and APGs are devoid of estrogenic activity. In addition, it has been demonstrated that low chlorination of APs or low bromination of bisphenols does not affect estrogenicity but may affect the biodistribution of the resulting chemicals. It is proposed that surfactants in current use and new chemicals under development should be carefully tested for estrogenic activity.

7.3.6 New directions and perspectives

Existing tests and bioassays of very distinct types have been proposed by different international bodies for the identification of hormonal mimics/antagonists, in an effort to assess the risk of exposure to chemical compounds and establish assays that can serve as regulatory instruments worldwide [11].

The detection of endocrine disrupters is both a simple and complex task. It is not difficult to determine whether a given chemical compound (e.g. any surfactant) is a hormonal mimic, either agonist or antagonist. Indeed, many bioassays used for seventy years to define estrogenic hormones can also be used to identify natural or synthetic chemical compounds with estrogenic activity. Our study has demonstrated the availability of well-established and adequately standardised bioassays to detect hormonal action and classify compounds according to their ability to mimic or antagonise natural hormones. However, the issue of their influence on the pathogeny of endocrine-related diseases is far from simple. The determination of endocrine disruption effects requires testing with longer observation periods and needs to take trans-generational effects into account. Unless considered alongside an extensive battery of toxicological studies, the results of short-term bioassays should be interpreted with caution.

The development of new tests and bioassays is likely to lengthen the list of endocrine disrupters. Recent research on hormonal disruption has not only investigated estrogens and androgens as agonists and antagonists but also considered the development of the individual and the presence of compounds that interfere in other hormonal systems, such as the thyroid system.

ACKNOWLEDGEMENTS

We would like to thank Richard Davies for editorial assistance. This research was supported by grants from the EU Commission (QLRT-1999-01422) and the Spanish Ministry of Health (FIS, 00/543 and 02/1314).

REFERENCES

1 A.M. Soto, H. Justicia, J.W. Wray and C. Sonnenschein, *Environ. Health Perspect.*, 102 (1991) 380.
2 R. White, S. Jobling, S.A. Hoare J.P. Sumpter and M.G. Parker, *Endocrinology*, 135 (1994) 175.

3 A.V. Krishman, P. Stathis, S.F. Permunth, L. Tokes and D. Feldman, *Endocrinology*, 132 (1993) 2279.

4 N. Olea, R. Pulgar, P. Pérez, M.F. Olea-Serrano, A. Rivas, A. Novillo-Fertrell, V. Pedraza, A.M. Soto and C. Sonnenschein, *Environ. Health Perspect.*, 104 (1996) 298.

5 P. Pérez, R. Pulgar, M.F. Olea-Serrano, M. Villalobos, A. Rivas, M. Metzler, V. Pedraza and N. Olea, *Environ. Health Perspect.*, 106 (1998) 167.

6 A.M. Soto, C. Sonnenschein, K.L. Chung, M.F. Fernandez, N. Olea and M.F. Olea Serrano, *Environ. Health Perspect.*, 103 (1995) 113.

7 C.A. Harris, P. Henttu, M.G. Parker and J.P. Sumpter, *Environ. Health Perspect.*, 105 (1997) 802.

8 E.J. Routledge, J. Parker and J. Odum, *Toxicol. Appl. Pharmacol.*, 153 (1998) 12.

9 T. Colborn and C. Clement, In: T. Colborn and C.R Clement (Eds.), *The Wildlife/Human Connection*, Princeton Scientific Publishing, Princeton, NJ, 1992, p. 295.

10 T. Colborn, F.S. vom Saal and A.M. Soto, *Environ. Health Perspect.*, 101 (1993) 378.

11 European Commission communication on a Community strategy for endocrine disrupter. COM 706, Brussels, 1999.

12 Scientific Committee on Toxicity, Ecotoxicity and the Environment (SCTEE). Directorate General for Consumer Policy and Consumer Health Protection, Brussels, 1999.

13 SACO 100EN, European Parliament, Public Health and Consumer Protection Series, Strasburg, 1998.

14 J.S. Evans, R.F. Varney and F.C. Koch, *Endocrinology*, 28 (1941) 747.

15 W.R. Miller and R.M. Sharpe, *Endocrine-related cancer*, 5 (1998) 69.

16 C.L. Waller, D.L. Minor and J.D. McKinney, *Environ. Health Perspect.*, 103 (1995) 702.

17 A.M. Soto, T.M. Lin, H. Justicia, R.M. Silvia and C. Sonnenschein, In: T. Colborn and C.R. Clement (Eds.), *The Wildlife/Human Connection*, Princeton Scientific Publishing, Princeton, NJ, 1992, p. 295.

18 M. Villalobos, N. Olea, J.A. Brotons, M.F. Olea-Serrano, J.M. Ruíz de Almodovar and V. Pedraza, *Environ. Health Perspect.*, 103 (1995) 844.

19 M. Pons, C. Gogne, J.C. Nicolas and M. Methali, *Biotechniques*, 9 (1990) 450.

20 P. Skehan, R. Storeng, D. Scudiero, A. Monks, J. Mcmahon, D. Vistica, J.T. Warren, H. Bokesch, S. Kenney and M.R. Boyd, *J. Natl. Cancer Inst.*, 82 (1990) 1107.

21 OECD, Guideline for testing of chemicals: Inmature rat uterotrophic bioassay screening test, 1998.

22 C.M. Markey, C.L. Michaelson, E.C. Veson, C. Sonnenschein and A.M. Soto, *Environ. Health Perspect.*, 109 (2001) 55.

23 E.C. Dodds and W. Lawson, *Nature*, 13 (1936) 996.

24 F. Ventura, A. Figueras, J. Caixach, I. Espadaler, et al., *Water Res.*, 22 (1988) 1211.

25 J.E. Harries, D.A. Sheahan, S. Jobling, P. Matthiessen, et al., *Environ. Toxicol. Chem.*, 16 (1997) 534.

26 P. Whitehouse and G. Brighty, *Proceedings 3rd SETAC World Congress, Brighton*, 2000, pp. 115, and poster.

27 M.R. Servos, R.J. Maguire, D.T. Bennie, H.-B. Lee, P.M. Cureton, N. Davidson, R. Sutcliffe and D.F.K. Dawn, Supporting document for NP and its ethoxylates, *National Water Research Institute 00-029*, Environment Canada, Burlington, 2000, pp. 1–217.

28 P. de Voogt and B. van Hattum, *Pure Appl. Chem.*, (2003) in press.

29 J. Legler, PhD thesis Wageningen Universiteit, Ch. 3, 2001, p. 50.

30 A. Rivas, M.F. Fernández, I. Cerrillo, J. Ibarluzea, M.F. Olea-Serrano, V. Pedraza and N. Olea, *APMIS*, 109 (2001) 185.

31 T. Czserati, *Environ. Health Perspect.*, 103 (1995) 358.

32 E.J. Routledge and J.P. Sumpter, *J. Biol. Chem.*, 272 (1997) 3280.

33 E.E. Reid and E. Wilson, *J. Am. Chem. Soc.*, 66 (1944) 967.

34 J.A. Brotons, M.F. Olea-Serrano, M. Villalobos, V. Pedraza and N. Olea, *Environ. Health Perspect.*, 103 (1995) 608.

35 R. Pulgar, M.F. Olea-Serrano, A. Novillo-Fertrell, A. Rivas, P. Pazos, V. Pedraza, J.M. Navajas and N. Olea, *Environ. Health Perspect.*, 108 (2000) 21.

36 W. Weinberg, *Anal. Chem.*, 1 (1999) 801.

37 Y. Mori, S. Goto, S. Onodera, S. Naito, S. Takitani and H. Matsushita, *Fresenius J. Anal. Chem.*, 345 (1993) 63.

38 S. Itoh and Y. Matsuoka, *Water Res.*, 30 (1996) 1403.

39 S.D. Richardson, A.D. Thruston, T.V. Caughran, P.H. Chen, T.W. Collette, T.L. Floyd, K.M. Schenck, B.W. Lykins, G.R. Sun and G. Majetich, *Environ. Sci. Technol.*, 33 (1999) 3378.

40 M. Samuelsen, C. Olsen, J.A. Holme, E. Meussen-Elholm, A.A. Bergmann and J.K. Hongslo, *Cell Biol. Toxicol.*, 17 (2001) 139.

41 R.M. Blair, H. Fang, W.S. Branham, B.S. Hass, S.L. Dial, C.L. Moland, W. Tong, L. Shi, R. Perkins and D.M. Sheehan, *Toxicol. Sci.*, 54 (2000) 138.

42 A.J. Murk, J. Legler, M.M.H. van Lipzig, A. Meerman, A.C. Belfroid, A. Spenkelink, G.B.J. Rijs and A.D. Vethaak, *Environ. Toxicol. Chem.*, 21 (2002) 16.

43 I.A.T.M. Meerts, R.J. Letcher, S. Hoving, G. Marsh, A. Bergman, J.G. Lemmen, B. van den Burg and A. Brouwer, *Environ. Health Perspect.*, 109 (2001) 399.

44 J. Legler, C.E. van den Brink, A. Brouwer, A.J. Murk, P.T. van der Saag, A.D. Vethaak and B. van den Burg, *Toxicol. Sci.*, 48 (1999) 55.

7.4 RISK ASSESSMENT OF SURFACTANTS

Pim de Voogt, Peter Eichhorn and Thomas P. Knepper

7.4.1 Introduction

Environmental risk assessment of substances is nowadays based on an evaluation of exposure pathways and concentrations on the one hand and identification and selection of sensitive endpoints on the other. The concept is operationalised by comparing real or estimated (predicted) exposure concentrations (PEC) with calculated no-effect concentrations (NEC or PNEC, predicted NEC). The comparison can be made by calculating the quotient of exposure and no-effect concentration. If the quotient is less than one, then the substance poses no significant risk to the environment. If the quotient is greater than one, the substance may pose a risk, and further action is required, e.g. a more thorough analysis of probability and magnitude of effects will be carried out.

For the derivation of the PNEC several approaches have been proposed. Generally these can be categorised into three distinct assessments: a conservative, a distributional, and a mixture toxicity approach. In conservative approaches, usually the most (realistic) sensitive endpoint such as LC_{50} or the known no observed effect concentration (NOEC) is taken and divided by an uncertainty factor (10–100). The selected uncertainty factor value depends on the type of endpoint and the number of available data, and is applied to account for laboratory to field extrapolations, species differences in sensitivities, and similar uncertainties. In distributional approaches, a series of, or all available, literature data are taken and a selected cut-off value is applied to the distribution of these data. The cut-off value may be, e.g., the concentration value that will protect 95% of the species (tested). In general, again an uncertainty factor (usually of 10) is then applied to take into account species differences. In the mixture toxicity approach, a similar mode of action is assumed for the assessment of the combined (additive) effect of the mixture. All relevant mixture components are scaled relative to the most potent one. This results in relative potencies for each component. The total effect of the mixture is then evaluated by

Comprehensive Analytical Chemistry XL
T.P. Knepper, D. Barceló and P. de Voogt (Eds)

adding the products of concentration and relative potency for each component, and comparing this value to the PNEC.

In addition to the understanding of biodegradability (Chapter 6), toxicological and bioaccumulation data (see Chapters 7.1 to 7.3) are necessary for both surfactants and their recalcitrant metabolites, in order to assess effects of residues to aquatic organisms. In general, the anthropogenic surfactants are relatively highly toxic due to inherent surface-active properties primarily contributing to disruption of biomembranes and denaturation of proteins [1]. The aquatic toxicity of single surfactants was thoroughly studied indicating that the response varies strongly between the used test organisms, such as fish, *Daphnia*, algae, or bacteria [2–5]. Besides the dependence on the exposed biological species, individual constituents of commercial mixtures, e.g. alkyl homologues of linear alkylbenzene sulfonates (LAS), contribute distinctly to the total toxicity observed [6]. The presence of unreacted raw materials from manufacturing or by-products formed during synthesis also has to be taken into account for measuring the toxicity of technical surfactants blends [7].

In this chapter only a brief introduction to the risk assessment thoroughly performed for many surfactants is given.

7.4.2 Risk characterisation of non-ionic surfactants including their metabolites

Alkylphenol ethoxylates and their metabolites
Alkylphenol ethoxylates (APEO) are no longer used in household detergents in the Western world and represent only a minor portion of the whole non-ionic surfactants group. Even if APEO is a group of surfactants of no commercial importance anymore, there is a need for risk assessment since these compounds are still present in the aquatic environment due to the recalcitrant nature of some of their metabolites (see Chapters 6.1, 6.2.1, 6.3.1 and 6.4). More attention should be given in the future to other non-ionic surfactants like alcohol ethoxylates.

No effect concentrations (NEC) for risk characterisation. Several countries and agencies have performed risk assessments for alkylphe-nols (AP) and APEO based on the approaches outlined above. Perhaps the most elaborate has been the assessment carried out by Environment

Canada [8]. In this document all three approaches have been performed after an extensive literature study of effects documented and environmental concentrations measured in the Canadian aquatic environment. Thus, based on the most sensitive endpoint found in the literature (LC_{50} of nonylphenol (NP) for winter flounder: 17 μg L^{-1} [9]), and applying an uncertainty factor of 100, for NP a PNEC of 0.17 μg L^{-1} was derived. Analogously, NEC were derived for nonylphenol ethoxylates (NPEO) and nonylphenol ethoxy carboxylates (NPEC); these are listed in Table 7.4.1.

The risks of NP have been evaluated in a recent risk assessment report from the EC states [10]. According to this document, currently there exists no limit value for NP in the EU. The document provides a PEC from model calculations at the regional scale of 0.6 μg L^{-1}, as well as a PNEC of 0.33 μg L^{-1}. Hence, for NP the risk quotient (PEC/PNEC) amounts to 1.8.

Mixture toxicities. As is obvious from Chapters 2 and 6, APEO and their degradation products occur in the environment as complex mixtures. In risk characterisation studies of AP and APEO, the toxicity of individual constituents of such mixtures, whether assayed in acute and chronic toxicity, or in estrogenicity tests (Chapter 7), is being considered to occur, for each separate endpoint, through the same separate mode of action, and consequently to be additive. Thus, relative potencies can be established for each individual

TABLE 7.4.1

Aquatic toxicity NEC values of NP, NPEO and NPEC proposed for risk characterisation by Environment Canada [8]

Substance	(Hyper)conservative NEC [μg L^{-1}]	Distributional/mixture[a] NEC (μg L^{-1})
NP	0.17	1.0
NPEO$_1$	1.1	2.0
NPEO$_2$	1.1	2.0
NPEO$_{3-17}$	9.0	200
NPE$_1$C	100	200
NPE$_2$C	9.9	200

[a]For all substances except NP, the proposed NEC is based on: first, estimating relative toxicity of substance to NP, and second, applying the distributional approach to the resulting NP toxicities together with those of NP itself.

915

constituent[1] of an oligomer mixture, referring to, e.g., the toxicity of a single[1] compound (usually NP). These potencies are often referred to as toxic equivalent factors (TEF), analogous to the concept which has first been applied to the toxicity of chlorinated aromatic hydrocarbons (e.g. dioxins) that exhibit Ah-receptor mediated toxicity. For the estrogenic activity of APEO, several studies have shown that additivity is indeed observed. In Chapter 7.3 examples are given of several relative estrogenic potencies (estradiol equivalent factors, EEF) of different surfactant oligomer series. Until now there has been no direct evidence in the literature as to whether or not the mode of action for NP, NPEO and NPEC is the same. In fact, higher NPEO oligomers may well have a different mode of action than NP, because the mechanism is likely to be a physical surfactant effect (see Chapter 7.4.1). Despite this, for acute or chronic toxicity endpoints, NP equivalent factors have been proposed for NPEO and NPEC. In one approach [11], NPEO and NP toxicities were collected and combined into a quantitative structure-activity relationship (QSAR) (Fig. 7.4.1). To that end, 26 available acute LC and EC_{50} data were regressed versus the number of EO units (n). The resulting graph is shown in Fig. 7.4.1. The QSAR thus derived is shown in Eq. (7.4.1).

$$\ln EC_{50} = 0.27n + 0.17 \qquad n = 26 \qquad r^2 = 0.933 \qquad (7.4.1)$$

The relative potency can now be derived from this equation. The risk characterisation then proceeds by calculating the total concentration (C_{total}) of the sample (expressed in equivalents of NP) as follows (Eq. (7.4.2)):

$$
\begin{aligned}
C_{total} &= PEC_{NP\text{-}eq.} \\
&= \sum \left\{ [OP] + [NP] + \frac{[NPEO_1]}{1.3} + \frac{[NPEO_2]}{1.7} + \frac{[NPEO_3]}{2.3} + \cdots \right. \\
&\quad \left. + \frac{[NPEO_{15}]}{57.6} \right\}
\end{aligned}
\qquad (7.4.2)
$$

As Eq. (7.4.2) shows, the relative potency of $NPEO_1$, as derived from Eq. (7.4.1), is $(1/1.3) = 0.77$ times that of NP (cf. Table 7.3.2). The toxicity of octylphenol (OP) is considered equivalent to that of NP in

---

[1] The term individual constituents pertains here to separate oligomers e.g. $NPEO_1$, NPE_2C, which in fact themselves are mixtures of several (differently branched alkyl) isomers.

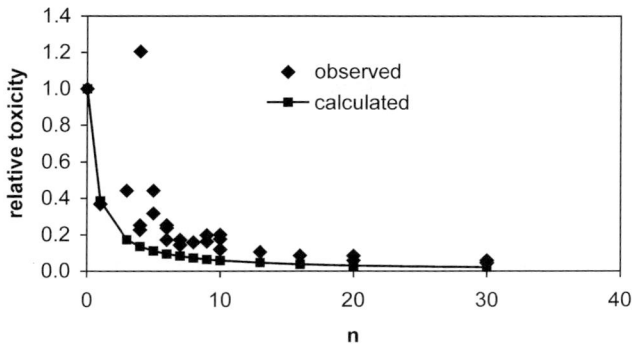

Fig. 7.4.1. QSAR derived from observed toxicities of oligomers of NPEO, used for predicting toxicity and evaluating risks of oligomer mixtures. Diamonds: observed toxicities; line with squares: calculated QSAR; n: number of EO units. For explanation, see text. Adapted from Ref. [11].

this study, and hence its relative potency is equal to one. NPEC were not included in this assessment.

In the Environment Canada distributional approach to characterise risks of NP, NPEO, and NPEC [8], relative toxicities were proposed based on categorising acute and chronic toxicities. A similar approach was used in that study to provide relative estrogenic potencies. Both are listed in Table 7.3.2 (Chapter 7.3).

It can be concluded that, with regard to aquatic toxicities, in the Environment Agency (UK) assessment, the longer chain ethoxylates ($n_{EO} = 4-15$) were attributed higher relative potencies than in the Environment Canada assessment. This may be due to several reasons: the UK assessment considered only acute toxicities, whereas the Canadian considered both acute and chronic ones. Moreover, in the Canadian assessment many more data (i.e. more than 200) were considered than in the UK assessment, and weighing factors were applied related to the confidence in the studies.

Exposure concentrations. Concentrations of NP and NPEO occurring in the aquatic environment in the EU have been summarised in Chapters 6.2.1 and 6.3.1. It can be concluded that in both freshwater and marine surface waters, concentrations of NP still exceed the PNEC of 0.33 μg L^{-1} derived in the EU document cited above [10]. Whereas in several freshwater systems in NW Europe (notably The Netherlands and Germany) in the most recent sampling campaigns observed concentrations of NP tend to be below this PNEC (Chapter 6.2.1), in the Mediterranean countries concentrations well above this level can still be found.

Alkyl glucamide degradation intermediate
A short-term bioluminescence bioassay using the marine bacterium *Vibrio fischeri* was employed to evaluate the acute toxicity of C_4-glucamide acid, which is a biodegradation product of the non-ionic surfactant alkyl glucamide [12]. The IC_{50}, calculated to 603 mg L^{-1}, was fairly high compared to other surfactants or their metabolites such as a technical NPEO mixture (2.7 mg L^{-1}) or NPEC (2.6 mg L^{-1}) [13]. Assessment of the acute toxicity of C_4-glucamide acid with CellSense™ assay, monitoring the effect of toxicants on the metabolic activity of immobilised bacteria *Pseudomonas putida*, showed no toxic potential even in the g L^{-1} range [14]. In view of these outcomes and the expected environmental levels of the transient biodegradation intermediate, which are likely to be one or two orders of magnitude lower than the parent compounds— occurring in the sewage in the mid μg L^{-1} range [15]— it appears to be highly improbable that C_4-glucamide acid might exert toxic effects on aquatic organisms. As far as the ability of the metabolite to interfere with the endocrine system is concerned, estrogenicity testing in the MCF-assay gave negative responses throughout (see Chapter 7.3).

7.4.3 Risk characterisation of linear alkylbenzene sulfonates including their metabolites

In this section a comparison is presented between the PNEC of LAS for the various environmental compartments, based on the calculations in the comprehensive risk assessment study conducted by HERA (Human and Environmental Risk Assessment on ingredients of European household cleaning products) [16], and the concentrations of LAS measured at different locations covering a broad range of distinct site characteristics (hydrological settings, discharge situation, climatic conditions).

Exposure concentrations

Concentration ranges reported for the occurrence of LAS in fresh and marine waters are compiled in Fig. 7.4.2. The levels are frequently found in the low to medium $\mu g\, L^{-1}$ range [17–30]. Values in the high $\mu g\, L^{-1}$ range were determined at locations where sewage treatment was either inadequate [18] or lacking [31]. In comparison to the PNEC of 270 $\mu g\, L^{-1}$, calculated by HERA [16], most environmental concentrations are below this threshold value indicating that LAS residues routinely detected in rivers or coastal waters do not pose a risk to aquatic organisms. With regard to the PNEC of sediment-bound LAS, 8.1 mg kg^{-1} according to HERA [16], the concentrations are generally below this value (Fig. 7.4.3) [17,20–22,32–34], provided efficient removal of LAS during wastewater treatment prior to discharge into the receiving bodies of water. Average concentrations well above the PNEC were reported for sampling sites in the proximity of wastewater treatment plant (WWTP) effluents using trickling filter (TF) technology [20,22]. Elevated LAS levels were also determined in sediment samples in various areas in Spain, where coastal regions were strongly impacted by industrial and urban activities [30,35]. At such hot spots concern is raised about possible adverse effects on benthic organisms. As far as agricultural soils amended with sewage sludge are concerned, the LAS concentrations determined depend strongly on the moment of sampling, i.e. the period of time elapsed after the last sludge application [32,36–38]. The highest levels are observed directly after sludge application yielding maximum levels of between 45 and 150 mg kg^{-1} (Fig. 7.4.4) [36,39–42]. Under aerobic conditions in the soil these values drop rapidly within a few weeks as the anionic surfactant is subject to microbial degradation. Regarding the PNEC of LAS in soils (4.6 mg kg^{-1} [16]), the field concentrations are usually found above this value directly after spreading of the sludge, i.e. as long as biodegradation has not led to a substantial decrease of the LAS level, harmful effects to soil organisms cannot be totally ruled out.

Sulfophenyl carboxylates

The ecotoxicity of sulfophenyl carboxylates (SPC) to aquatic organisms (fish and *Daphnia*) was explored showing that the LC$_{50}$ was increased by a factor of 200–300 by carboxylation of one terminal methyl group of the alkyl chain and by another 10 to 20 times upon shortening of the alkanoate chain from 11 to 7 and to 5 [43]. This was attributed to the fact

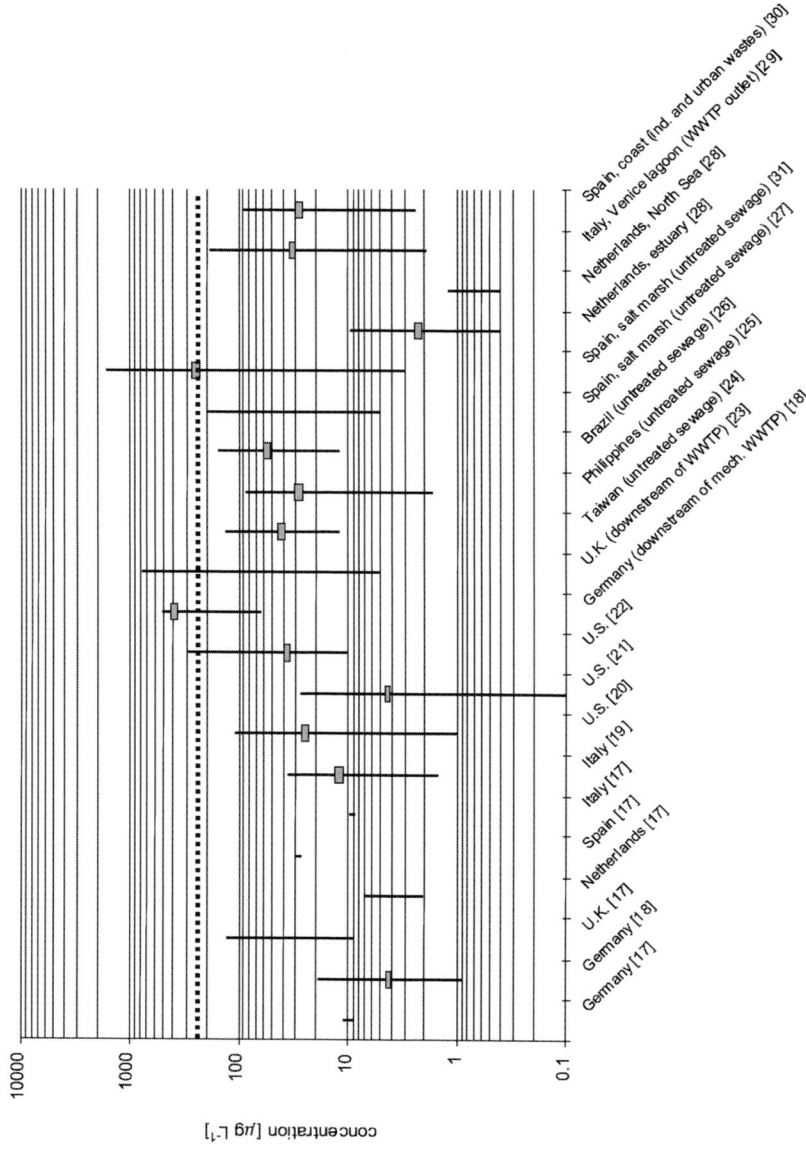

Fig. 7.4.2. Concentration ranges of LAS in surface waters (horizontal bars correspond to mean values). The dotted line at 270 μg L⁻¹ indicates the aquatic PNEC value calculated by HERA Ref. [16].

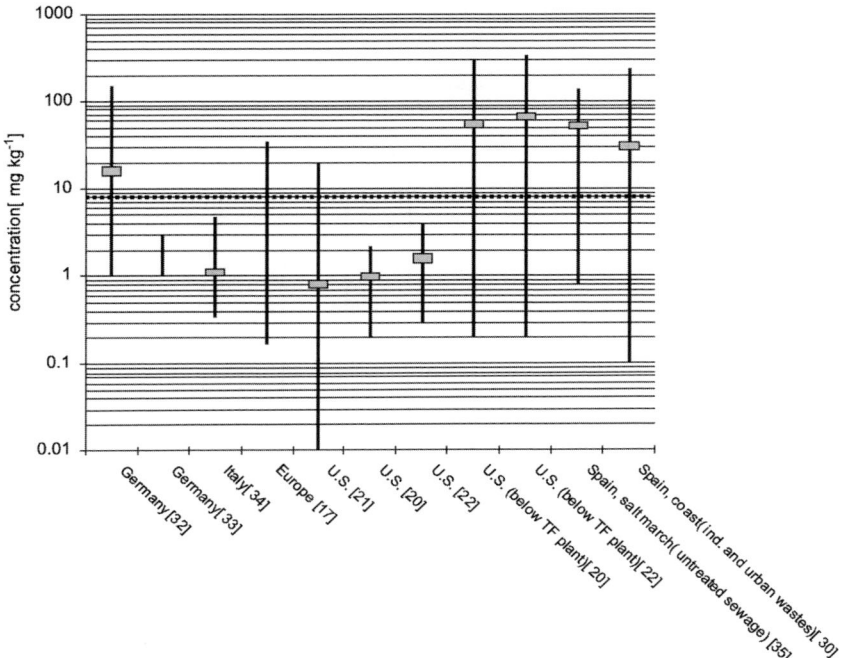

Fig. 7.4.3. Concentration ranges of LAS in sediments (horizontal bars correspond to mean values). The dotted line at 8.1 mg kg^{-1} indicates the terrestrial PNEC value calculated by HERA [16].

that the toxicity of LAS is directly associated with its surface-active properties since oxidation of the alkyl chain leads to the loss of surface activity. This is in line with the findings of Moreno and Ferrer [44] monitoring the toxicity towards *Daphnia magna* during biodegradation of various LAS blends. The metabolites formed during this process were far less toxic than the parent molecule regardless of the mean molecular weight of the type of LAS assayed. The residual material from TF studies of Kölbener et al. [45] showed no detectable impact on the growth of the algae *Selenastrum capricornutum* and the mobility of *D. magna*. As for the ability of SPC to disturb the endocrine system of aquatic organisms or humans, the latter exposed to SPC upon consumption of drinking water containing traces of the carboxylates, no evidence for such effects were found. In the recombinant estrogen yeast screen Routlegde and Sumpter [46] proved that the range of SPC homologues assayed were not

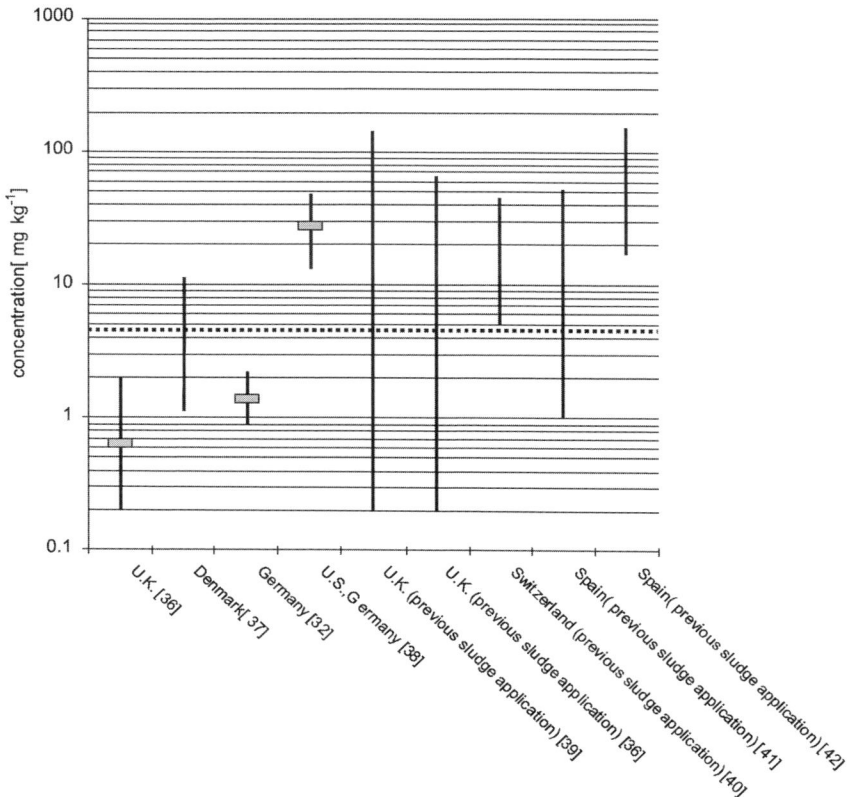

Fig. 7.4.4. Concentration ranges of LAS in sludge-amended soils (horizontal bars correspond to mean values). The dotted line at 4.6 mg kg^{-1} indicates the terrestrial PNEC value calculated by HERA [16].

estrogenic. These findings were corroborated by Navas et al. [47] using the vitellogenin assay as well as the yeast screen. The two SPC homologues C_5 and C_{11} did not induce any positive response in the tests.

7.4.4 Risk assessment of other anionic surfactants

Other widely used anionic surfactants, such as alkyl sulfates and alkylether sulfates (see Chapter 1), have also been thoroughly investigated according to their risk [16]. By means of higher tier

922

exposure and effects data, it could be shown that application of these compounds via household detergents and cleaning products poses no concern in any environmental compartment.

7.4.5 Conclusion

In freshwater and marine surface waters in Europe, concentrations of NP still exceed the PNEC of $0.33\,\mu g\,L^{-1}$ derived in the EU document cited above [10]. Whereas in several freshwater systems in NW Europe (notably The Netherlands and Germany) in the most recent sampling campaigns observed concentrations of NP tend to be below this PNEC (Chapter 6.2.1), in the Mediterranean countries concentrations well above this level can still be found. Elsewhere in the world the situation is between these two, with the highest concentrations of NPEO and NP occurring in developing countries with untreated wastewater discharged directly to surface waters.

As far as LAS is concerned most environmental concentrations are below the threshold value for aquatic organisms, indicating that LAS residues routinely detected in rivers or coastal waters do not pose a risk. Sediments levels of LAS at hot spots raise concern about possible adverse effects on benthic organisms, while in sludge-amended soil harmful effects of LAS on soil organisms cannot be totally ruled out directly after application of sewage sludge. Regarding the risk assessment of SPC, which are several orders of magnitude less toxic than LAS (SPC have higher aquatic LD_{50}), it can be concluded that these polar metabolites pose little or no risk to the aquatic environment since the measured concentrations of SPC in river water are lower than the PNEC values.

REFERENCES

1 R.D. Swisher, *Surfactant Biodegradation*, Surfactant Series No. 11, 2nd edn, Marcel Dekker, New York, 1987.
2 W. Guhl and P. Gode, *Tenside Surfact. Det.*, 26 (1989) 282.
3 M.J. Rosen, F. Li, F.S.W. Morrall and D.J. Versteeg, *Environ. Sci. Technol.*, 35 (2001) 954.
4 N. Scholz, *Tenside Surfact. Det.*, 34 (1997) 229.
5 C.A. Staples, J. Weeks, J.F. Hall and C.G. Naylor, *Environ. Toxicol. Chem.*, 17 (1998) 2470.

6   C. Verge and A. Moreno, *Tenside Surfact. Det.*, 37 (2000) 172.
7   C. Verge, J. Bravo, A. Moreno and J.L. Berna, *Tenside Surfact. Det.*, 36 (1999) 127.
8   M.R. Servos, R.J. Maguire, D.T. Bennie, H.B. Lee, P.M. Cureton, N. Davidson, R. Sutcliffe and D.F.K. Rawn, *Supporting Document for Nonylphenol and its Ethoxylates*, National Water Research Institute 00-029, Environment Canada, Burlington, 2000, pp. 1–217.
9   S. Lussier, D. Champlin, J. LiVolsi, S. Poucer, R. Pruell and G. Thursby, (1996), in [8].
10 JRC (2002), 4-Nonylphenol (branched) and nonylphenol: summary risk assessment report. Special publication I.02.69, Joint Research Centre—EC, Ispra, available at: http://www.europa.eu.int/comm/enterprise/chemicals/markrestr/studies/nonylphenol.pdf.
11 P. Whitehouse and G. Brighty, Proceedings 3rd SETAC World Congress, Brighton, Environment Agency, UK, 2000, p. 115 (and poster).
12 P. Eichhorn and T.P. Knepper, *J. Mass. Spectrom.*, 35 (2000) 468.
13 M. Farré, M.-J. García, L. Tirapu, A. Ginebreda and D. Barceló, *Anal. Chim. Acta*, 427 (2001) 181.
14 P. Eichhorn, PhD Thesis, University Mainz, Germany, 2001.
15 P. Eichhorn, M. Petrovic, D. Barceló and T.P. Knepper, *Vom Wasser*, 95 (2000) 245.
16 HERA, *Human and Environmental Risk Asssessment*, CEFIC, Brussels, 2001, http://www.heraproject.com/ExecutiveSummaryPrint.cfm?ID = 22.
17 J. Waters and T.C.J. Feijtel, *Chemosphere*, 30 (1995) 1939.
18 P. Schöberl, *Tenside Surfact. Det.*, 32 (1995) 25.
19 A. di Corcia, L. Capuani, F. Casassa, A. Marcomini and R. Samperi, *Environ. Sci. Technol.*, 33 (1999) 4119.
20 M.L. Trehy, W.E. Gledhill, J.P. Mieure, J.E. Adamove, A.M. Nielsen, H.O. Perkins and W.S. Eckhoff, *Environ. Toxicol. Chem.*, 15 (1996) 233.
21 C.F. Tabor and L.B. Barber, *Environ. Sci. Technol.*, 30 (1996) 161.
22 D.C. McAvoy, W.S. Eckhoff and R.A. Rapaport, *Environ. Toxicol. Chem.*, 12 (1993) 977.
23 K. Fox, M. Holt, M. Daniel, H. Buckland and I. Guymer, *Sci. Total Environ.*, 251/252 (2000) 265.
24 W.-T. Ding, S.-H. Tzing and J.-H. Lo, *Chemosphere*, 38 (1999) 2597.
25 P. Eichhorn, M.E. Flavier, M.L. Paje and T.P. Knepper, *Sci. Total Environ.*, 269 (2001) 75.
26 P. Eichhorn, S.V. Rodrigues, W. Baumann and T.P. Knepper, *Sci. Total Environ.*, 284 (2002) 123.
27 E. González-Mazo, M. Honing, D. Barceló and A. Gómez-Parra, *Environ. Sci. Technol.*, 31 (1997) 504.
28 E. Matthijs and M. Stalmans, *Tenside Surfact. Det.*, 30 (1993) 29.

29   A. Marcomini, G. Pojana, A. Sfirso and J.-M. Quiroga, *Environ. Toxicol. Chem.*, 19 (2000) 2000.

30   M. Petrovic, A. Fernández-Alba, F. Borrull, R.M. Marce, E. González-Mazo and D. Barceló, *Environ. Toxicol. Chem.*, 21 (2002) 37.

31   E. González-Mazo, J.M. Forja and A. Gómez-Parra, *Environ. Sci. Technol.*, 32 (1998) 1636.

32   E. Matthijs and H. de Henau, *Tenside Surfact. Det.*, 24 (1987) 193.

33   P. Schöberl, R. Spilker and L. Nitschke, *Tenside Surfact. Det.*, 31 (1994) 243.

34   L. Cavalli, G. Cassani, L. Vigano, S. Pravettoni, G. Nucci, M. Lazzarin and A. Zatta, *Tenside Surfact. Det.*, 37 (2000) 282.

35   V.M. León, E. González-Mazo, J.M. Forja Pajares and A. Gómez-Parra, *Environ. Toxicol. Chem.*, 20 (2001) 2171.

36   M.S. Holt, E. Matthijs and J. Waters, *Water Res.*, 23 (1989) 749.

37   L. Carlsen, M.-B. Metzon and J. Kjelsmark, *Sci. Total Environ.*, 290 (2002) 225.

38   R.A. Rapaport and W.S. Eckhoff, *Environ. Toxicol. Chem.*, 9 (1990) 1245.

39   M.S. Holt and S.L. Bernstein, *Water Res.*, 26 (1992) 613.

40   W. Giger, P.H. Brunner, M. Ahel, J. McEvoy, A. Marcomini and C. Schaffner, *Gas Wasser Abwasser*, 67 (1987) 111.

41   J.L. Berna, J. Ferrer, A. Moreno, D. Prats and F. Riuz Bevia, *Tenside Surfact. Det.*, 26 (1989) 101.

42   J. de Ferrer, A. Moreno, M.T. Vaquero and L. Comellas, *Tenside Surfact. Det.*, 34 (1997) 278.

43   R.D. Swisher, W.E. Gledhill, R.A. Kimerle and T.A. Taulli, Proceedings Surfactant Congress, Vol. 7., 1976.

44   A. Moreno and J. Ferrer, *Tenside Surfact. Det.*, 28 (1991) 129.

45   P. Kölbener, U. Baumann, T. Leisinger and A.M. Cook, *Environ. Toxicol. Chem.*, 14 (1995) 561.

46   E.J. Routledge and J.P. Sumpter, *Environ. Toxicol. Chem.*, 15 (1996) 241.

47   J.M. Navas, E. González-Mazo, A. Wenzel, A. Gómez-Parra and H. Segner, *Mar. Poll. Bull.*, 38 (1999) 880.

Chapter 8

Recommendations and future trends

Thomas P. Knepper and Damia Barceló

8.1 MONITORING

Surface water contaminated by municipal and industrial sources and diffuse pollution sources from urban and agricultural areas continue to increase pollution levels in water, soil and sediments. Waste originating from production processes and caused by the use of various products, including surfactants, over their lifetime often has a negative impact on the environment, on ecosystems and on human health. Industrial production processes form together, along with the application of the resulting products and their fate in the environment (from 'cradle to grave'), a very complex system, which is difficult to analyse, to understand and to manage in a sustainable way. Action has to be taken on various levels if sustainability is the target and this action has to be based on a sound scientific knowledge base.

Continuous monitoring of water, soil and sediment is essential in order to identify the most critical/at-risk parts of the overall system and to convert negative trends into positive ones. Measures to remove negative trends can be taken at the level of the production process (e.g. clean technologies), at product level (e.g. clean products), at the end-of-the-pipe (waste and wastewater treatment) and at the consumer level. Furthermore, where negative impacts of past and current activities

Comprehensive Analytical Chemistry XL
T.P. Knepper, D. Barceló and P. de Voogt (Eds)

have already destroyed our resources, remediation action of soil and groundwater has to take place.

For all future activities, and in order to achieve sustainable management of water resources, including related soil and sediment compartments, the strategy to be followed should involve: (i) the precautionary principle, (ii) polluter pays principle and (iii) application of the best available techniques (BATs) and best environmental practices, including where appropriate, clean technologies.

Surfactants (anionics, non-ionics, cationics and amphoterics) are used in different fields such as cosmetics, metal working, mining, agriculture, paper and leather industries and are also employed in large quantities for many applications in households, institutions and industries. A large proportion of this chemical class thus reaches wastewater treatment plants (WWTPs) and, despite being degradable, subsequently reach sewage sludge in high concentrations due to the very large quantities present in wastewaters and incomplete degradation during the sewage treatment process. For example, nonylphenol ethoxylates (NPEOs) are degraded to the mono- and diethoxylates by shortening of the hydrophilic ethoxylate chain during wastewater treatment. These compounds are further degraded in anaerobically stabilised sewage sludge to the fully de-ethoxylated nonylphenol (NP), a more lipophilic and toxic compound which is resistant to further microbial degradation. This compound, along with other alkylphenol ethoxylate (APEO) degradation products, e.g. octylphenol (OP), are considered to be endocrine-disrupting compounds (EDCs) that may potentially alter the normal hormone function and physiological status of animals.

Investigations into the fate of surfactants under different discharge conditions have demonstrated the importance of wastewater treatment in reducing the amounts of residues entering the natural aquatic environment, but also highlight the need for sophisticated and reliable analytical methods in order to obtain monitoring data of both the parent surfactants as well as the quite often more highly concentrated metabolites.

A significant percentage of linear alkylbenzene sulfonates (LASs) is known to be associated with the sludge fraction (15–45% [1]). Since the xenobiotic aryl sulfonates are not degraded during the commonly applied process of anaerobic sludge digestion [2,3], considerable LAS concentrations of $1600-19\,000\,\mathrm{mg\,kg^{-1}}$ have thus been found in anaerobically digested sludge [4–6].

In addition to the monitoring of the parent surfactants, such as NPEO or LAS, studies are required to investigate the possible formation of recalcitrant metabolites (e.g. NP or carboxylated arylsulfonates, SPCs) in sludge-amended soils. This is of particular interest as these very polar intermediates are believed to possess high mobility in soil, resulting in pronounced leachability (confirmed in laboratory experiments by Branner et al. [7]). The persistence of some polar SPC isomers (see Chapter 5.1) is likely to allow their migration into deeper zones, which can thus reach groundwater aquifers, and thereby deteriorate the quality of drinking water sources.

8.2 ANALYTICAL ASPECTS

In view of the inherent resistance of some surfactant metabolite isomers to complete mineralisation, efforts have to be mounted in order to obtain further insight into the reasons behind the persistence of these, such as the SPC and nonylphenol ethoxy carboxylates (NPECs). In order to achieve this, it would thus be indispensable to be able to fully elucidate the chemical structure of individual components, e.g. after isolation from environmental samples. Through the application of, for example, LC–ESI–MS–MS in combination with NMR analyses, this is now possible.

Knowledge regarding the fate of these persistent polar metabolites could be further enhanced by taking a closer look at possibly distinct behaviour of the stereoisomers, such as preferential metabolisation of one diastereomer or enantiomer. Schulz et al. [8] pioneered the first such investigations. The authors were able to isolate an organism which sequentially utilised the two enantiomers of C_4-SPC, which had been separated on a chiral high performance liquid chromatography (HPLC) column. The general importance of stereospecificity on biodegradability and metabolite formation has also been demonstrated by Schowanek et al. [9] in investigations into the degradation of different stereoisomers of the complexing agent ethylene diamine disuccinic acid. Of the three possible isomers examined, only one was rapidly and completely mineralised. Distinct behaviour of enantiomers was likewise reported in the study of Simoni et al. [10] on 3-phenylbutyrate (this is essentially desulfonated C_4-SPC-3). Only the R enantiomer was biodegraded, while the S enantiomer was co-metabolically transformed.

A particularly challenging task for the future, which has so far been almost entirely neglected, will be the analysis of water-soluble synthetic polymers of high molecular weight such as the polycarboxylates in both aquatic and terrestrial compartments. For their determination, selective extraction and pre-concentration methods are required in order to isolate the analytes from environmental matrices and this needs to be complemented by analytical instrumentation allowing accurate quantitative measurements. This is still a strongly limiting factor in matrix-assisted laser desorption time-of-flight (MALDI-TOF) mass spectrometric (MS) analysis, the most commonly applied technique. However, the power of emerging technologies for on-line coupling of MALDI with column separation, such as HPLC, offers a very promising potential approach to this field [11,12].

8.3 LEGISLATIVE REGULATIONS FORCE ACTION

The European Union (EU) environmental policy has put water protection and its sustainable management high on its agenda, reflected by the European Water Framework Directive (WFD), which is now in its implementation phase.

8.3.1 Water framework directive

The WFD has established a framework for the protection of inland surface water, transitional waters and groundwater which aims at achieving good ecological potential and good surface water chemical status within 15 years from the date of commencement of the Directive, i.e. 22 December, 2000 (Directive 2000/60/EC, L327/1, Official Journal of the EU).

One of the key elements of the WFD is the implementation of a 'combined approach' of emission limit values and quality standards (for waters, sediments and biota), and also the phasing out of particularly hazardous substances. Accordingly, a list of 33 priority substances, which represent a significant risk to the aquatic environment at community level, was established. For the first time two surfactant metabolites, namely NP and OP, have been included in this priority list.

Obviously this fact will stimulate the availability of monitoring data for these compounds in the coming years, since untill now neither NP nor OP were included in the routine monitoring programmes of river basin authorities and water treatment companies.

In addition to the WFD the Integrated Pollution Prevention and Control (IPPC) (96/61/EC, OJL 257 October 10, 1996), which is intended to minimise pollution from various point sources, should be considered. The debate about IPPC today focuses on the BAT reference documents, also called BREFs. These BREFS are highly detailed documents outlining correct industrial environmental performance. The BREF documents are being published for the different sectors and just to bring here an example relative to surfactants, we will mention the BREF relative to the tannery sector. In the recent BREF it is clearly indicated that APEOs, still used in large quantities in the south of Europe, especially in Italy, Spain and Portugal, by the tannery industry, should be replaced by alcohol ethoxylates (AEs). This is just one example where surfactants are mentioned in the legislation and that specific action should take place. This directive, which is under the umbrella of WFD, specifies that emission control by the industry is of particular concern and that phasing out of all hazardous substances should take place by 2020 to near-zero levels with intermediate aims of 50% reduction in 2010 and 75% reduction in 2015. This will obviously affect the discharges of surfactants from the different industrial sectors into the aquatic environment.

Another piece of legislative action to be considered is the OSPAR strategy regarding hazardous substances, known as the 'Convention for the protection of the marine environment of North-Atlantic'. Among the OSPAR list of chemicals of priority action, surfactants NPEOs and so-called related substances, such as NP, are included among other persistent organic pollutants (POPs) like PCBs, PAHs and Hg, indicating that these compounds show a risk to the marine environment.

Regarding the specific issue of NP and NPEOs there is still some disparity between the USA and Europe. In Europe, voluntary bans in the industrial sector are in place and in household applications they are already forbidden, with environmental quality standards set at $1 \, \mu g \, L^{-1}$ levels. However, in the USA the situation is quite different with much milder measures being taken such as pollution prevention or source reduction efforts, as opposed to the outright APEO ban in Europe.

8.3.2 Soil and sediment related issues

The WFD is not the only legislative action undertaken by the EU during the last few years. Recently the important needs of protecting soil resources have been recognised, in particular in relation to their quality and functioning. Respective EU policies are under development and are represented in the communication on soil protection 'Towards a thematic strategy for soil protection', which was adopted by the Commission on 16th April 2002 as COM(2002)179. Certainly the objective of this new regulatory document is the sustainable management of soil and water resources. In this document a relevant aspect that is discussed is diffuse soil contamination. Contamination of sewage sludge by a whole range of pollutants, like surfactants and their metabolites, which can thus result in an increase of the soil concentrations of these compounds, deserves special attention. The use of sewage sludge for agriculture has increased during the last few years and this is basically due to the implementation of the *91/271/EEC Directive on urban wastewater treatment*. During the time of implementation of this directive, more than 40 000 WWTPs will be needed in Europe by the year 2005. It has been estimated that the amount of sludge produced in Europe will increase from 6.5 to 11 million tonnes by the year 2005. At present around 40% of the sewage sludge produced in Europe is deposited in landfills, 40% is redirected to land use, below 10% is dumped at sea and the remainder is incinerated. According to the EU, the quantity of re-used sludge represents around 53% of the total sludge produced. In general the EU considers that the re-use of sludge should be encouraged since it represents a long-term solution provided that the quality of the re-used sludge is compatible with public health and environmental protection requirements.

Sludge produced by municipal WWTPs can be considered as one of the most complex matrices to be analysed. However, the chemical composition, and the concentration of pollutants, especially highly lipophilic ones of low biodegradability and transformation products of partly degradable pollutants, should be determined in order to plan sludge disposal in landfills or its re-use in agriculture. The pollution of the terrestrial ecosystem by large amounts of LAS through sludge application, which may result in detrimental effects on soil organisms and flora and solubilisation of hydrophobic sludge contaminants [13] has been demonstrated. This has thus given rise to concern and intensified the research in recent years into the terrestrial fate and the assessment of the risks caused by high levels of anionic surfactant in sludge-amended soils.

Consequently, there is a new Draft Directive on the land use of sewage sludge which requires that organic pollutants be measured in the sludge. In contrast, the present Directive 86/278/EEC only stipulates the measurement of certain metals such as Cr, Cd, Ni, Pb, Hg, Zn and Cu. So, by the implementation of this new directive, different surfactants, including LAS and NPEO and metabolites such as NP, will need to be determined in sewage sludge prior to its use in agriculture. The cut-off level for LAS to be used in agriculture is restricted to 2600 mg kg^{-1} and for NPEOs (with one or two ethoxy groups) and NP, 50 mg kg^{-1} (EU, 2000). Hence, by implementation of this directive, a large percentage of the sludge from anaerobic treatment will no longer be permitted to be applied to fields [14–17].

To mitigate the problem of high LAS concentrations in sludge to be applied on farmland and to fulfil current and projected criteria imposed by EU authorities—keeping in mind that these restrictions are prone to become steadily more severe with the need to dispose of evermore increasing amounts of sewage sludge—the composting of sludge has been proposed as a promising mechanism for efficient reduction of LAS contents. Prats et al. [18] reported on a nearly 100% elimination of LAS after 72-day composting of anaerobically digested sludge. Similar findings were presented by Solbé [19] who demonstrated a quantitative degradation of LAS after composting, and also effective removal of NPEO (78–95%).

The sustainable management of sediments, in addition to water, soil and sludge environmental matrices, in relation to surfactant regulations, is also an important and relevant issue. The US EPA has recently shown concern regarding the levels of surfactants in sediments and has thus released a Draft Contaminated Sediment Science Plan. In this draft, recommendations for the development of analytical methods and evaluations of the toxicity and risk assessment of *Emerging endocrine disrupters* like APEOs and their metabolites in sediment samples are outlined.

8.3.3 EU draft proposal for biodegradation of surfactants

This draft proposal for biodegradation of surfactants is a key issue for the future use of different surfactant chemical groups in detergent formulations. This proposal aims to replace the existing Directive on 'biodegradability of surfactants in detergents' and give recommendations for labelling of detergents and cleaning products—recommendations include the phasing out of several cationic surfactants, such as ditallow dimethylammonium chloride (DTDMAC), DSDMAC and DHTDMA used in fabric softeners, and of NPEOs in cleaning agents. In addition to these, another list of potential candidates for deregulation has been made and includes, among others, alkyl sulfosuccinate, branched AEs, fatty acid alkanolamides, mono- and dialkylamine ethoxylates, dialkyl dimethyl quat, imidazoline derivatives, geuerbet alcohol, alkylated sulphonated diphenyloxide, AEs with more than 20 EO units and alkylated aminoxides. It should be pointed out here that the replacement of these surfactants by the detergent industry should not constitute any important logistic problem in the area of laundry detergents and dishwashing detergents, which constitute a large share of the market.

8.4   FINAL REMARKS AND PERSPECTIVES OF A CONTINUOUSLY CHANGING MARKET

The economic impact of restrictions on surfactants use in business is very relevant. The four main outlets for surfactants are household, cosmetics and toiletries, food processing and industrial and institutional cleaning which already account for half of the total surfactant market. In these markets, a limited number of surfactant types remains dominant, namely LAS, AEs, alcohol sulfates (ASs) and alcohol ether sulfates (AESs). Despite the repetitive declarations of the surfactant industry classifying LAS as intrinsically environmentally safe [20] with sufficient safety margins [21] posing no significant threat in the long term [13], Sweden has adopted a substitution of LAS in detergent products. Due to the comparatively high toxicity and the poor anaerobic biodegradability, this country banned the synthetic compound and replaced it with other surfactants, such as AS or AES, capable of fulfilling the same cleaning function [22]. Denmark has likewise demonstrated a move away from widespread acceptance of LAS use, where the Environmental Protection Agency has placed this anionic surfactant on the 'List of Undesirable Substances' owing to the problem of its removal in the waste cycle [23].

A further example of changes to the detergent market are observed in the Philippines. Increasing public awareness regarding deterioration of the environment and wasting of water resources solely in exchange for cheaper raw materials used for surfactant production has put pressure on manufacturers, which have started to suspend the use of branched alkylbenzene sulfonates (ABSs) [24]. Recently an act has been approved which will prohibit manufacture, distribution, and sale of domestic and industrial detergents composed of hard, petrochemically based surfactants [25]. As a result, new products containing a single coconut-based surface-active ingredient, AS, have been introduced into the market [26].

These few examples presented here illustrate the potential substitution of not fully environmentally sound surfactants as a result of legislative action. In future, we need to be on a constant vigil for adequate and environmentally friendly alternatives, whilst taking into consideration economic, functional and above all ecological viewpoints. For example, the poor anaerobic biodegradation of certain surfactant types, e.g. LAS, should not be considered as single pass/fail criterion for acceptability of use, as the biosphere is predominantly aerobic, and in reality biodegradation may proceed. A rational and holistic approach needs to be adopted, with a sustainable management system being the long- and short-term goals. Only by taking all aspects into account, can we hope to achieve and maintain environmentally sound practises in the use of surfactants and other xenobiotics.

REFERENCES

1   J.L. Berna, J. Ferrer, A. Moreno, D. Prats and F.R. Bevia, *Tenside Surf. Det.*, 26 (1989) 101.
2   R.R. Birch, W.E. Gledhill, R.J. Larson and A.M. Nielsen, *Proceedings of the 3rd CESIO International Surfactant Congress and Exhibition, London*, 26 (1992) 26.
3   H.A. Painter and E.F. Mosey, *Proceeding of the 3rd CESIO International Surfactant Congress and Exhibition, London*, 26 (1992) 34.
4   H. de Henau, E. Matthijs and W.D. Hopping, *Int. J. Environ. Anal. Chem.*, 26 (1996) 2.
5   M.S. Holt and S.L. Bernstein, *Water. Res.*, 26 (1992) 613.
6   J. McEvoy and W. Giger, *Environ. Sci. Technol.*, 20 (1986) 376.

7   U. Branner, M. Mygind and C. Jorgensen, *Environ. Toxicol. Chem.*, 18 (1999) 1772.

8   S. Schulz, W. Dong, U. Groth and A.M. Cook, *Appl. Environ. Microbiol.*, 66 (2000) 1905.

9   D. Schowanek, T.C.J. Feijtel, C.M. Perkins, A.H. Frederick, T.W. Federle and R.J. Larson, *Chemosphere*, 34 (1999) 2375.

10 S. Simoni, S. Klinke, C. Zipper, W. Angst and H.-P. Kohler, *Appl. Environ. Microbiol.*, 62 (1996) 749.

11 A.I. Gusev, *Fresenius J. Anal. Chem.*, 366 (2000) 691.

12 S. Trimpin, P. Eichhorn, H.J. Räder, K. Müllen and T.P. Knepper, *J. Chromatogr. A*, 938 (2001) 67.

13 J. Jensen, *Sci. Total Environ.*, 226 (1999) 93.

14 J.L. Berna, N. Battersby, L. Cavalli, R. Fletcher, A. Guldner, D. Schowanek and J. Steber, 5th World Surfactant Congress, Firenze, Italy, 2000.

15 CEEP, *Scope No. 37–06/2000–Sewage sludge disposal, EU considers tighter sludge spreading rules*, Centre Européen d'Etudes des Polyphosphates, 2000.

16 CESIO, *News Issue 3—New draft directive on sewage sludge*, European Committee of Surfactants and Their Organic Intermediates, May 2000.

17 CLER, In: J.E. Heinze (Ed.), *An update: Science Versus Politics in the Environmental Regulatory Process: March 6, 2000*, Council for LAB/LAS Environmental Research, 2000.

18 D. Prats, M. Rodriguez, M.A. Muela, J.M. Llamas, A. Moreno, J. de Ferrer and J.L. Berna, *Tenside Surf. Det.*, 35 (1999) 294.

19 J. Solbé, Linear alkylbenzene sulphonates (LAS) risk assessment for sludge-amended soils, SPT Workshop, Copenhagen, April 19–20 1999.

20 J. Waters and T.C.J. Feijtel, *Chemosphere*, 30 (1995) 1939.

21 L. Cavalli, G. Cassani, L. Vigano, S. Pravettoni, G. Nucci, M. Lazzarin and A. Zatta, *Tenside Surf. Det.*, 37 (2000) 282.

22 S. Hagenfors, *Good environmental choice—a tool for a better environment*, Swedish Society for Nature Conservation, Chemical Awareness, Issue 3, 1998.

23 DEPA, *"Listen over Unøskede Stoffer"*, The List of Undesirable Chemicals, Danish Environmental Protection Agency, 1998.

24 Senate lauded for environment bill *The Philippine Star*, April 15, 2000.

25 Congress enacts another landmark environment law, *The Philippine Star*, August 26, 2000.

26 Detergents may pollute our sources of potable water, say expert, *The Philippine Star*, May 9, 2000.

Glossary

Surfactants and metabolites

ABS	branched alkylbenzene sulfonate
AE	alcohol ethoxylate
AE/P	alcohol ethoxylate/propoxylate
AES	alkylether sulfate
AG	alkyl glucamide
AOS	α-olefin sulfonate
APEO	alkylphenol ethoxylate
APG	alkyl polyglucoside
AS	alkyl sulfate
BiAS	bismuth active substances
BM	alkyldimethylbenzyl ammonium
$C_6(EO)_3$	triethylene glycol monohexyl ether
CNPEC	carboxylated nonylphenoxy carboxylates
CTAS	cobalt thiocyanate active substances
D	$R_2SiO_{(0.5)2}$
DATS	dialkyl tetralinsulfonate
DCPEG	dicarboxylated PEG
DEEDMAC	ditallow ester of di(hydroxyethyl)dimethyl ammonium chloride
DEQ	Lever's diester quat
DM	dialkyldimethyl ammonium
DTDMAC	ditallow dimethylammonium chloride
EO	ethylene oxide
FADA	fatty acid diethanolamide
LAB	linear alkylbenzene
LAS	linear alkylbenzene sulfonate
M	$R_3SiO_{0.5}$
MBAS	methylene blue active substances
MCPEG	mono carboxylated PEG

M_2D-R	$[Si(CH_3)_3-O]_2-Si(CH_3)-R$
NP	nonylphenol
NPEC	nonylphenoxy carboxylate
NPEO	nonylphenol ethoxylate
OP	octylphenol
PEG	polyethylene glycol
PEMS	polyethermethylsiloxanes
PDMS	polydimethylsiloxanes
PO	propylene oxide
Q	$SiO_{(0.5)4}$
PPG	polypropylene glycol
SAS	secondary alkane sulfonates
SDS	sodium dodecyl sulfat
SPC	sulfophenyl carboxylate
T	$RSiO_{(0.5)3}$
TAM	trialkyl amines
TMC	trimethyl ammonium chloride
TM	alkyltrimethyl ammonium
TMS	$Si(CH_3)_3$
TMS-OH	$(CH_3)_3Si-OH$
$(TMS)_2-O$	$[(CH_3)_3Si]_2-O$
TPS	tetrapropylene benzene sulfonate

Techniques

APCI	atmospheric pressure chemical ionisation
API	atmospheric pressure ionisation
ASE	accelerated solvent extraction
Amu	atomic mass units
BPI	base peak intensity
CE	capillary electrophoresis
CF	continous flow
CGE	capillary gel electrophoresis
CI	chemical ionisation
CID	collision induced dissociation
CV	coefficient of variation
	([Standard deviation/mean] × 100/1)
DAD	diode array detector
EI	electron impact (ionisation)

ESI	electrospray ionisation
FAB	fast atom bombardment
FIA	flow injection analysis
FL	fluorescence
FTICR	fourier transform ion cyclotron resonance
GC	gas chromatography
GCB	graphitised carbon black
GC–MS	gas chromatography/mass spectrometry
HPLC	high performance liquid chromatography
IC	ion chromatography
IR	infra red
IT-MS	ion trap mass spectrometry
LC	liquid chromatography
LC–MS	liquid chromatography/mass spectrometry
LLE	liquid-liquid extraction
LOD	limit of detection
LSD	light scattering (mass) detection
m/z	mass-to-charge ratio
MALDI-ToF	matrix assisted laser desorption/ionization time-of-flight
MEKC	micellar electrokinetic capillary chromatography
MRM	multiple reaction monitoring
MS	mass spectrometry; mass spectrometer
MS–MS	tandem mass spectrometry; tandem mass spectrometer
NCI	negative chemical ionisation
NMR	nuclear magnetic resonance
NP	normal-phase
PB	particle beam
PCI	positive chemical ionisation
Ppb	parts per billion ($\mu g\ L^{-1}$)
Ppm	parts per million ($mg\ L^{-1}$)
RI	refractive index
RIC	reconstructed ion chromatogram
RP	reversed-phase
SAX	strong anion exchange
SCX	strong cation exchange
SD	standard deviation
SDB	styrene-divinylbenzene
SFC	supercritical fluid chromatography
SIM	selected ion monitoring

SRM	selected reaction monitoring
TIC	total ion current
TLC	thin layer chromatography
ToF	time-of-flight
TSI	thermospray ionisation
UV	ultra violet

For combinations of different techniques see the following examples:
        FIA–ESI–MS–MS
        HPLC–UV

Chemicals

ACP	Acid Phosphatase
ALP	Alkaline Phosphatase
ATP	adenosine triphosphate
BP5	2,2-bis(4-hydroxyphenyl) heptane
DES	diethyl stilbestrol
DMSO	dimethyl sulfoxide
E2	17B-estradiol (17β-estradiol)
EE	ethynyl estradiol
EEF	estradiol equivalent factor
HPB	hexadecyl pyridinium bromide
NAD^+	nicotine adenine dinucleotide reduced
PCB	pentachlor benzenes

Others

APHA	American Public Health Association
AWWA	American Water Works Association
BCF	bioconcentration factor
CAT	Catalase
C/C_0	measured concentration/starting concentration
CMC	critical micelle concentration
DO	dissolved oxygen
DOC	dissolved organic carbon
DOM	dissolved organic matter
ECETOC	European Center for Ecotoxicology and Toxicology of Chemicals

EC$_{50}$	effect concentration (50%)
EPA	environmental protection agency
ER	estrogen receptor
GPX	Glutathione Peroxidase
Kow	octanol−water partition coefficient
LC$_{50}$	lethal concentration (50%)
LOEC	lowest observed effect concentration
MATC	maximum acceptable toxicant concentration
MCF7	breast cancer cell line used in E-screen assay
MPC	maximum permissible concentration
NOEC	no observed effect concentration
PgR	progesterone receptor
PNEC	predicted no effect concentration
pS2	expression of cell-type specific proteins
QSAR	quantitative structure activity relationship
S	aqueous solubility
SPM	suspended particulate matter
RIPP	relative induced protein potency
RPE	relative proliferative effect
RPP	relative proliferative potency
SOD	Superoxide Dismutase
TEGEWA	German detergent and specialties supplies association
WPCF	Water Pollution Control Federation
WWTP	wastewater treatment plants

Index

944

ISBN 0-444-50935-6

9 780444 509352